ENCYCLOPAEDIC DICTIONARY
OF PHYSICS

SUPPLEMENTARY VOLUME 4

SUPPLEMENTARY VOLUME 4

ENCYCLOPAEDIC DICTIONARY OF PHYSICS

GENERAL, NUCLEAR, SOLID STATE, MOLECULAR
CHEMICAL, METAL AND VACUUM PHYSICS
ASTRONOMY, GEOPHYSICS, BIOPHYSICS
AND RELATED SUBJECTS

EDITOR-IN-CHIEF

J. THEWLIS

FORMERLY OF HARWELL

Associate Editors

R. C. GLASS   **A. R. MEETHAM**

LONDON   TEDDINGTON

PERGAMON PRESS

OXFORD · NEW YORK · TORONTO
SYDNEY · BRAUNSCHWEIG

Pergamon Press Ltd., Headington Hill Hall, Oxford

Pergamon Press Inc., Maxwell House, Fairview Park, Elmsford, New York 10523

Pergamon of Canada Ltd., 207 Queen's Quay West, Toronto 1

Pergamon Press (Aust.) Pty. Ltd., 19a Boundary Street, Rushcutters Bay, N.S.W. 2011, Australia

Vieweg & Sohn GmbH, Burgplatz 1, Braunschweig

Copyright © 1971 Pergamon Press Ltd.
All Rights Reserved. No part of this publication may be reproduced, stored in a retrieval system, or transmitted, in any form or by any means, electronic, mechanical, photocopying, recording or otherwise, without the prior permission of Pergamon Press Ltd.

First edition 1971

Library of Congress Catalog Card No. 60–7069

Printed in Germany

08 006359 4

ARTICLES CONTAINED IN THIS VOLUME

Acoustic holography	G. A. Massey
Acoustic waveguides	M. Redwood
Acousto-electric effects	E. G. S. Paige
Airborne collision prevention devices	S. Krejcik
Ambient noise in the sea	B. S. McCartney
Astro-archaeology	G. S. Hawkins
Astrodynamics	R. M. L. Baker, Jr.
Auger emission spectroscopy	L. A. Harris
Auroral electrons	R. D. Albert
Biological pressure transensors	C. Compton Collins
Bordoni peak	F. R. N. Nabarro
Bosons, intermediate	W. Lee
Brittleness in ceramics	R. W. Davidge
Cardiac pacemaker, external	D. A. Juett
Ceramics production, physics in	D. S. Dodd
Charge conjugation	C. D. Froggatt
Chart of the nuclides—trilinear	D. J. Rees
Conductivity at high temperatures	S. C. Jain and V. Narayan
Current algebras and their applications	B. Renner
Daylight, power distribution in	A. W. S. Tarrant
Diffusion in non-uniform media	D. H. Kirkwood
Dislocation kinks	F. R. N. Nabarro
Dispersion and fibre strengthening	G. C. Smith
Dynamic electron diffraction, applications of	P. Goodman
Dynamical n-beam theory of electron diffraction	A. F. Moodie
Electric dipole moment of the neutron	C. G. Shull
Electrogasdynamics	L. G. Sanders
Electrohydraulic crushing	K. W. Carley-Macauly
Electrohydraulic forming of metals	K. W. Carley-Macauly
Electron beam machining	J. G. Purchas
Electron diffraction from periodic magnetic fields	M. J. Goringe
Electron metallography, applications of	M. A. P. Dewey
Electroviscous fluid	D. L. Klass
Electroviscous fluids, applications of	D. L. Klass
Elementary particles, nomenclature of	W. W. Bell
Emissivity at high temperatures	S. C. Jain and V. Narayan
Epitaxy, recent work on	D. W. Pashley
Far infra-red lasers	A. Hadni
Far infra-red spectroscopy	A. Hadni
Fast passage in paramagnetic resonance	P. E. Wagner
Ferrohydrodynamics	R. E. Rosensweig
Fluidics	J. L. Nayler

Fourier transform spectroscopy	J. Chamberlain
Fracture of brittle materials	S. F. Pugh
Freeze-coating	R. G. C. Arridge
Geochronology (radiometric dating)	E. I. Hamilton
Glandless reciprocating-jet pump	A. J. Walkden
Glass-ceramics	P. W. McMillan
Glow discharge electron and ion guns for material processing	R. A. Dugdale
Gyroscopic coupler	A. Bloch
Hall effect, applications of	E. Cohen
Hamilton's principle	B. R. Gossick
Helicity	W. W. Bell
High intensity pulsed radiation sources	A. J. Gale
High-lift devices	J. L. Nayler
High polymers, mechanical properties of	M. J. Folkes and I. M. Ward
High resolution electron microscopy	J. W. Sharpe
Holography	H. G. Jerrard
Hydrologic cycle	T. L. Richards
Hydrometeorology	T. L. Richards
Impatt diode microwave generators	M. P. Wasse
Infra-red cameras and infra-red TV	A. Hadni
Infra-red spectroscopy, recent developments in	A. E. Martin
Interferometric spectrometers	A. E. Martin
Interplanetary dust	J. F. James
Ionization and attachment cross-sections of atoms and molecules by electron impact	J. A. Rees
Ion-neutralization spectroscopy	H. D. Hagstrum
Ionosphere, recent observations of	M. J. Rycroft
Ionospheres of Mars and Venus	G. Fjeldbo
Jet flap	J. L. Nayler
Josephson effect	B. W. Petley
Kirchhoff's law (electrical circuits)	A. Bloch
Laser-induced sparks	J. W. Gardner
Lasers, applications of	H. A. Elion
Lasers, tunable	R. G. Smith
Law enforcement science and technology	W. Shaw
Liquid crystals and molecular structure	G. W. Gray
Liquid metals	N. H. March
Luminosity of a spectrometer	J. Chamberlain
Magnetic thin films	M. J. Folkes
Magnetoresistivity	E. Cohen
Melting, modern theories of	B. J. Alder
Metal film preparation by electroless (chemical) reduction	K. A. Holbrook
Metallic colloids in non-metallic crystals	S. C. Jain and G. D. Sootha
Microplasticity	A. Esin
Microwave solid-state oscillators, recent	J. A. Copeland
Mirage and looming, optics of	M. S. Sodha and A. K. Aggarwal
Model ships, photoelectric tracking of	W. Wyslouzil
Momentum transfer cross-section for low energy electrons: determination by swarm techniques	R. W. Crompton
Multiplex principle (of Fellgett)	J. Chamberlain
Neutretto	W. W. Bell
Nuclear quadrupole resonance	T. J. Rowland

Optical information processing	K. R. COLEMAN
Peierls stress	F. R. N. NABARRO
Peltier refrigerators	H. J. GOLDSMID
Permanent magnet technology	D. HADFIELD
Phonon avalanche and bottleneck	P. E. WAGNER
Photoemission and band structure	W. E. SPICER
Physical limnology	R. K. LANE
Piezomagnetic effect	S. BREINER
Pockels effect	H. PURSEY
Polarized proton targets	A. HONIG
Polymer physics	J. W. S. HEARLE
Proteins, structure of	A. MILLER
Proton scattering microscopy	R. S. NELSON
Pulse code modulation	L. K. WHEELER
Pyroelectric detector for infra-red radiation	H. P. BEERMAN
Q-switching	H. A. ELION
Quantum mechanics in biochemistry	F. E. HARRIS and R. REIN
Quantum metrology	B. W. PETLEY
Quarks	S. J. SHARROCK
Radar astronomy	R. B. DYCE
Radiation chemistry, industrial applications of	P. R. HILLS
Radiation quantities, systematics of	W. L. ZIJP
Radioemissions from interstellar hydroxyl radicals	G. DE JAGER
Radioisotopes in food processing	R. M. LONGSTAFF
Radio telescopes, the present position	M. I. LARGE
Refractories, ultra-purification of	S. C. JAIN and V. NARAYAN
Reverberation under water	B. K. GAZEY
Rubber physics	A. SCHALLAMACH
Sailing, physics of	H. M. BARKLA
Satellites in soft X-ray spectra	J. R. CUTHILL
Semiconductors, effects of radiation on	J. H. CRAWFORD, Jr.
Shock pressure in meteorites	R. R. JAEGER
Singing sands	R. A. BAGNOLD
Snoek spectra	I. G. RITCHIE and R. RAWLINGS
Soft X-ray spectroscopy	J. R. CUTHILL
Solar cell	K. H. SPRING
Solar wind, planetary interactions with	E. G. BOWEN
Solid nucleus formation from solution	W. J. DUNNING
Sonic boom, generation and suppression of	H. W. CARLSON
Space power supplies	K. H. SPRING
Speaking machines	K. N. STEVENS
Split-beam microwave beacon	L. G. COX
Sputtering, theories of	R. S. NELSON
Standard Earth, geodetic parameters for	C. A. LUNDQUIST
Streamer chambers	F. VILLA
Submillimetre radiation, applications of	J. CHAMBERLAIN
Submillimetre-wave lasers	J. CHAMBERLAIN
Submillimetre-wave methods	J. CHAMBERLAIN
Temperature-resistant polymers	E. JONES
Textile industry, physics in	J. W. S. HEARLE
Thermionic generation of electricity	K. H. SPRING
Thermoelectric cooling of Infra-red detectors	H. P. BEERMAN

Thermoelectric generation of electricity . . . . K. H. Spring
Thermomagnetic gas torque (the Scott effect) . . . . R. M. Williamson
Thin films, nucleation and growth of . . . . M. J. Folkes
Tungsten halogen lamps . . . . T. Jacobs and G. T'Jampens
Ultrasonic amplification in semiconductors . . . . E. G. S. Paige
Ultrasonic imaging . . . . H. Berger
Ultrasonic spectroscopy . . . . O. R. Gericke
Ultra-violet radiation, some biological effects of . . . . J. Jagger
Vitreous carbons, structure and properties of . . . . . . F. C. Cowlard and J. C. A. Lewis
Walsh functions . . . . N. J. Fine
Weather control . . . . R. C. Sutcliffe
Whole body counting . . . . M. H. Grieveson and D. J. Rees

LIST OF CONTRIBUTORS TO THIS VOLUME

AGGARWAL A. K. (*New Delhi*)
ALBERT R. D. (*Berkeley, Calif.*)
ALDER B. J. (*Berkeley, Calif.*)
ARRIDGE R. G. C. (*Oxford*)

BAGNOLD R. A. (*Edenbridge, Kent*)
BAKER R. M. L. JR. (*El Segundo, Calif.*)
BARKLA H. M. (*St. Andrews, Scotland*)
BEERMAN H. P. (*Stamford, Conn.*)
BELL W. W. (*Aberdeen, Scotland*)
BERGER H. (*Argonne, Illinois*)
BLOCH A. (*Wembley, Middx.*)
BOWEN E. G. (*Sydney, Australia*)
BREINER S. (*Stanford, Calif.*)

CARLEY-MACAULY K. W. (*Harwell*)
CARLSON H. W. (*Virginia*)
CHAMBERLAIN J. (*Teddington*)
COHEN E. (*Manchester*)
COLEMAN K. R. (*Aldermaston*)
COMPTON–COLLINS (*San Francisco, Calif.*)
COPELAND J. A. (*Murray Hill, N.J.*)
COWLARD F. C. (*Towcester, Northants*)
COX L. G. (*Ottawa, Ontario*)
CRAWFORD J. H. (*Oak Ridge, Tennessee*)
CROMPTON R. W. (*Canberra, Australia*)
CUTHILL J. R. (*Washington, D.C.*)

DAVIDGE R. W. (*Harwell*)
DE JAGER (*Macclesfield, Cheshire*)

DEWEY M. A. P. (*Stoke Poges, Bucks.*)
DODD D. S. (*Stoke on Trent, Staffs.*)
DUGDALE R. A. (*Harwell*)
DUNNING W. J. (*Bristol*)
DYCE R. B. (*Stanford, Calif.*)

ELION H. A. (*Cambridge, Mass.*)
ESIN A. (*Ankara, Turkey*)

FINE N. J. (*Pennsylvania*)
FJELDBO G. (*Stamford, Calif.*)
FOLKES M. J. (*Somerset*)
FROGGATT C. D. (*Harwell*)

GALE A. J. (*Lexington, Mass.*)
GARDNER J. W. (*Leicester*)
GAZEY B. K. (*Birmingham*)
GERICKE O. R. (*Watertown, Mass.*)
GOLDSMID H. J. (*Bath*)
GOODMAN P. (*Victoria, Australia*)
GORINGE M. J. (*Oxford*)
GOSSICK B. R. (*Lexington, Kentucky*)
GRAY G. W. (*Hull*)
GRIEVESON M. H. (*Northwood, Middx.*)

HADFIELD D. (*Sheffield*)
HADNI A. (*Nancy, France*)
HAGSTRUM H. D. (*Murray Hill, N.J.*)
HAMILTON E. I. (*Reigate, Surrey*)
HARRIS F. E. (*New York*)
HARRIS L. A. (*Shenectady, N.Y.*)
HAWKINS G. S. (*Cambridge, Mass.*)

HEARLE J. W. S. (*Manchester*)
HILLS P. R. (*Harwell*)
HOLBROOK K. A. (*Hull*)
HONIG A. (*Syracuse, N.Y.*)

JAEGER R. R. (*Ohio*)
JAGGER J. (*Dallas, Texas*)
JAIN S. C. (*Delhi, India*)
JAMES J. F. (*Manchester*)
JERRARD H. G. (*Southampton*)
JONES E. (*Groningen, Holland*)
JUETT D. A. (*Cambridge*)

KIRKWOOD D. H. (*Sheffield*)
KLASS D. L. (*Chicago, Illinois*)
KREJCIK S. (*Montreal, Canada*)

LANE R. K. (*Ontario, Canada*)
LARGE M. I. (*Sydney, Australia*)
LEE W. (*New York*)
LONGSTAFF R. M. (*Wantage, Berks.*)
LUNDQUIST C. A. (*Washington D.C.*)

MARCH N. (*Sheffield*)
MARTIN A. E. (*Newcastle upon Tyne*)
MASSEY G. A. (*Calif.*)
MCCARTNEY B. S. (*Godalming, Surrey*)
MCMILLAN P. W. (*Stafford*)
MILLER A. (*Oxford*)
MOODIE A. F. (*Victoria, Australia*)

NABARRO F. R. N. (*Johannesburg, S. Af.*)
NARAYAN V. (*Delhi, India*)
NAYLER J. L. (*Claygate, Surrey*)
NELSON R. S. (*Harwell*)

PAIGE E. G. S. (*Great Malvern, Worcs.*)
PASHLEY D. W. (*Saffron Walden, Essex*)
PETLEY B. W. (*Teddington*)
PUGH S. F. (*Harwell*)
PURCHAS J. G. (*Harwell*)
PURSEY H. (*Teddington*)

RAWLINGS R. (*Cardiff*)
REDWOOD M. (*London*)
REES D. J. (*Northwood, Middx.*)
REES J. A. (*Liverpool*)
REIN R. (*Amberst, N.Y.*)
RENNER B. (*Pasadena, Calif.*)
RICHARDS T. L. (*Ontario, Canada*)
RITCHIE I. G. (*Cardiff*)
ROSENSWEIG R. E. (*Burlington, Mass.*)

ROWLAND T. J. (*Urbana, Illinois*)
RYCROFT M. J. (*Southampton*)

SANDERS L. G. (*Dorchester on Thames*)
SCHALLAMACH A. (*Herts.*)
SHARPE J. W. (*Glasgow*)
SHARROCK S. J. (*London*)
SHAW W. (*Middletown, N.J.*)
SHULL C. G. (*Cambridge, Mass.*)
SMITH G. C. (*Cambridge*)
SMITH R. G. (*Murray Hill, N.J.*)
SODHA M. S. (*New Delhi, India*)
SOOTHA G. D. (*Delhi, India*)
SPICER W. E. (*Stanford, Calif.*)
SPRING K. H. (*London*)
STEVENS K. N. (*Cambridge, Mass.*)

SUTCLIFFE R. C. (*Reading*)

TARRANT A. W. S. (*Surbiton, Surrey*)
T'JAMPENS G. (*Turnhout, Belgium*)

VILLA F. (*Stanford, Calif.*)

WAGNER P. E. (*Baltimore, Maryland*)
WALKDEN A. J. (*Wembley, Middx.*)
WARD I. M. (*Leeds*)
WASSE M. P. (*Harlow, Essex*)
WHEELER L. K. (*London*)
WILLIAMSON R. M. (*Rochester, Michigan*)
WYSLOUZIL W. (*Ottawa, Canada*)

ZIJP W. L. (*Petten, Holland*)

FOREWORD

In the Foreword to the first of this series of Supplementary Volumes I wrote:

"Of its very nature, a work of the size and scope of the Encyclopaedic Dictionary of Physics can never reach ultimate completion. However, by issuing a continuous series of Supplementary Volumes, we shall strive to keep it as up to date and comprehensive as we can (having regard to the inevitable time lapse between writing and publication), and as free from errors as may be.

"The volumes in this series are intended to form part of a unified whole, and are numbered accordingly. They are designed to deal with new topics in physics and related subjects, new development in topics previously covered and topics which have been left out of earlier volumes for various reasons. They will also contain survey articles covering particularly important fields falling within the scope of the Dictionary.

"The contents of these volumes will be arranged alphabetically, as in the previous volumes. Articles will be reasonably short and will be signed. Cross references to other articles will be incorporated as necessary, and bibliographies will be included as a guide to further study. Each volume will have its own index, prepared on the same generous scale as before; and, in addition, it is intended to issue a cumulative index every five years. Errata and addenda lists will be published, referring to the original Encyclopaedic Dictionary of Physics, and to those Supplementary Volumes which will already have been published.

"In preparing the Supplementary Volumes regard will be had to the changing emphasis in many branches of physics—the invasion of the biological sciences by physics, the possibilities opened out by the increasing use of computers in all branches of science and technology, the ever-increasing scope of theoretical physics, the progress in high energy physics, the emergence of new instrumental techniques etc.; and, at the same time the authors of previously published articles will be given the opportunity of bringing those articles up to date. Naturally there are many articles for which this will not be necessary, and it is certainly not intended that new articles shall be written if there is no need for them. In short, it is our intention to produce a series of volumes in which the high standard already achieved in the Encyclopaedic Dictionary of Physics is fully maintained."

The continuing success of the Supplementary Volumes would seem to indicate that our efforts have not been unavailing. I hope and trust that the present volume, the fourth in the series, will maintain the standard already set. However, in the knowledge that mistakes are inevitable in a work of this nature, I shall be grateful if readers will let me know of any errata or omissions, so that appropriate amendments may be made at a suitable stage.

Once again I should like to express my gratitude to the publishers, and particularly to Mr. Robert Maxwell and Mr. E. J. Buckley for their constant support; to my Associate Editors, Dr. A. R. Meetham, of the National Physical Laboratory, and Mr. R. C. Glass, of the City University, for their help and counsel; to Mr. S. Crimmin, the Assistant Editor at the Pergamon Press, and his successor Mr. M. S. Gale, for their solid work behind the scenes; and to the referees who, as in the past, have helped to ensure the quality of the published articles. Last but not least it is a pleasure to acknowledge the debt I owe to my wife, who has been of invaluable assistance at all stages of the work.

J. Thewlis
Editor-in-Chief

ACOUSTIC HOLOGRAPHY

1. Introduction

It is possible to produce high-resolution, three-dimensional images of objects exposed to acoustic radiation by applying holographic techniques to the detection and display of sound waves. The use of sound instead of light makes available an enormous and continuously variable range of wave-lengths which may be produced by high-powered sources with excellent temporal coherence. Because linear transducers for the conversion of sound energy to electrical energy are readily available, some methods for making acoustic holograms exist without an analogue in conventional optical holography, which must depend on nonlinear electro-optical or photographic detection. However, the absence of a convenient, sensitive medium for recording two-dimensional acoustic fields has led to extensive research in the area of scanning acoustic detectors, sampled arrays, and other devices which correspond to photographic film in optical holography.

Because the subject of acoustic holography is new and rapidly developing, comprehensive survey articles do not exist. This article describes the theory of acoustic holography, with and without an acoustic reference beam, and presents the problems of wave-length scaling, motion of the acoustic medium, and three-dimensional imaging at low resolution. References to a representative sampling of papers in this field are given, but the list is necessarily far from complete.

2. Basic Imaging Theory

2.1. Holography with Offset Acoustic Reference Beam

This method can be developed by direct analogy with the offset reference method of optical holography developed by Leith and Upatnieks (1962, 1963). We assume the scalar theory of diffraction by Huygens' principle applies, and we take the time variation at a single temporal acoustic frequency as understood. Figure 1 shows the basic configuration. Since we are concerned with the details of the reconstruction process for a wide variety of objects which are to be imaged, we can consider only the field present at the hologram recording plane. Using phasor notation, let this field be $U_{\text{sig}}(x, y)$ and let the field from the reference source be $Ae^{-j2\pi f_0 x} = U_{\text{ref}}(x, y)$. Here f_0 is the spatial frequency produced on the recording plane

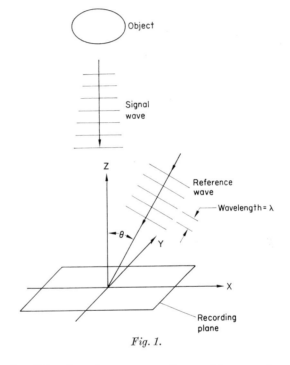

Fig. 1.

by tilting the reference wave off-normal by an angle $\theta = \sin^{-1}(\lambda f_0)$. Now suppose that in the x-y plane there is a recording medium such as a scanning detector which records at each position, with a square-law characteristic, a quantity $I(x, y)$ proportional to the square of the magnitude of the field incident at that point. Then

$$\begin{aligned}I(x, y) &= K_1 \left[U_{\text{ref}}(x, y) + U_{\text{sig}}(x, y) \right] \left[U^*_{\text{ref}}(x, y) + U^*_{\text{sig}}(x, y) \right] \\ &= K_1 \left[U_{\text{ref}}(x, y) U^*_{\text{ref}}(x, y) \right. \\ &\quad + U_{\text{sig}}(x, y) U^*_{\text{sig}}(x, y) \\ &\quad + U_{\text{sig}}(x, y) U^*_{\text{ref}}(x, y) \\ &\quad \left. + U^*_{\text{sig}}(x, y) U_{\text{ref}}(x, y) \right] \\ &= K_1 [|A|^2 + |U_{\text{sig}}(x,y)|^2 + A^* U_{\text{sig}}(x, y)\, e^{j2\pi f_0 x} \\ &\quad + A U^*_{\text{sig}}(x, y)\, e^{-j2\pi f_0 x}].\end{aligned}$$

For reconstruction of the recorded image, the $I(x, y)$ will ordinarily be recorded as a photographic transparency, which can be placed in an optical beam for viewing. If we assume a positive transparency is made, the film recording process can be approximated by the following model:

1. The exposure E of an area of film is proportional to incident intensity and duration of the applied light signal. Therefore define $E = tI$, where t is the exposure time and I is the light intensity.

2. Over a reasonable range of exposure, the intensity transmission τ_i of the positive transparency, as a function of E, is given by:

$$D \equiv \log \frac{1}{\tau_i} = -\gamma \log E + D_0$$

or

$$\tau_i = e^{-D_0} E^\gamma$$

where γ and D_0 are functions of the film type, development process, and exposure level.

The amplitude transmission is therefore

$$\tau_a = \sqrt{\tau_i} = K_2 E^{\gamma/2} = K_3 I^{\gamma/2}$$

where

$$K_2 = e^{-D_0/2} \quad \text{and} \quad K_3 = t^{\gamma/2} e^{-D_0/2}.$$

Now if the hologram information $I(x, y)$ is made equal to I in the exposure relation above, for example, by writing on film with a modulated light beam, the processed transparency can be placed in a beam of collimated light. At normal incidence, with the incident amplitude set equal to unity for convenience, the optical field directly behind the illuminated transparency is simply equal to $\tau_a(x, y) = K_3[I(x, y)]^{\gamma/2}$. Thus if processing is done so that $\gamma = 2$, the reconstructed field $U'(x, y)$ is:

$$U'(x, y) = K_1 K_3 \left[|A|^2 + |U_{\text{sig}}(x, y)|^2 \right.$$
$$+ A^* U_{\text{sig}}(x, y) e^{j 2\pi f_0 x}$$
$$\left. + A U^*_{\text{sig}}(x, y) e^{-j 2\pi f_0 x} \right].$$

Since the reference beam is a plane wave, the quantity $|A|^2$ represents an undeviated transmitted wave normal to the hologram plane. For objects of limited angular extent, or for a field of view much less than 180°, the spatial frequency spectra of $U_{\text{sig}}(x, y)$, $U^*_{\text{sig}}(x, y)$, and $|U_{\text{sig}}(x, y)|^2$ are relatively narrow. Under this assumption, and with the further assumption that f_0 is greater than the spatial bandwidth of the signal, the quantity $|U_{\text{sig}}(x, y)|^2$ is then a group of waves propagating near the normal and the two remaining terms, containing $U_{\text{sig}}(x, y)$ and $U^*_{\text{sig}}(x, y)$, are off-axis groups centered about the angles $\pm \sin^{-1}(\lambda f_0)$ away from the normal. The coefficients A and A^* represent only a constant gain and optical phase shift, and the exponential phase factors imply off-axis propagation. Thus, the original wavefront $U(x, y)$ is reconstructed with only a possible change in signal strength and direction, and an observer viewing the hologram at an angle corresponding to the direction of the term $A^* U_{\text{sig}}(x, y) e^{j 2\pi f_0 x}$ will see a virtual image of the object which produced the original field. The term $A U^*_{\text{sig}}(x, y) e^{-j 2\pi f_0 x}$ has phase variations opposite to the original wave and corresponds to light converging to a real image behind the hologram. This can be understood intuitively by decomposing the object and image into points and considering the phase of the spherical wavefronts corresponding to light from these points. The phase change corresponding to taking the complex conjugate transforms diverging fronts to converging ones; thus the complex conjugate term represents the real image. By offsetting the angle of the reference wave, we have made it possible to separate the light from real and virtual images as well as the direct on-axis component consisting of distortion products. The wavefronts over the aperture of the hologram can be perfectly reproduced; thus high resolution three-dimensional imaging is often possible, depending, of course, on the wavelength and recording geometry.

2.2. Holography without an Acoustic Reference Beam

A fundamental difference between the techniques available for making optical and acoustical holograms is the fact that the acoustic beam can be converted to a useful electrical signal directly with linear transducers, whereas optical detection is always square-law and requires a reference beam to determine phase. As a result, it is possible to avoid the need for an acoustic reference beam by suitable electronic processing of the signal after linear detection.

To illustrate this concept with an example, consider the system diagrammed in Fig. 2. The normal to the

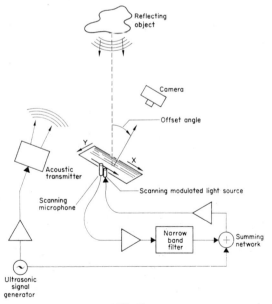

Fig. 2.

hologram plane has been rotated through an angle θ away from the general direction of the object, and the linear transducer is scanned across the tilted plane. The output signal voltage from the transducer will then be

$$U(x, y) = K_4 U_{\text{sig}}(x, y) \, e^{j2\pi f_0 x}$$

where $U_{\text{sig}}(x, y)$ is the acoustic field that would be present without the tilt, and K_4 is the transfer characteristic of the transducer. A voltage of complex amplitude A is also obtained from the signal generator used to produce the original acoustic wave transmitted to the object. The two voltages are added linearly and used to modulate a light source whose output intensity I is proportional to input electrical power, that is:

$$I = K_5 \, |v|^2.$$

Then the time average over a few cycles of the carrier frequency is

$$I(x, y) = K_5 \, [A + K_4 U_{\text{sig}}(x, y) \, e^{j2\pi f_0 x}]$$
$$[A^* + K^*_4 U^*_{\text{sig}}(x, y) \, e^{-j2\pi f_0 x}]$$

$$\bar{I}(x, y) = K_5 \, [|A|^2 + |K_4 U_{\text{sig}}(x, y)|^2$$
$$+ A^* K_4 U_{\text{sig}}(x, y) \, e^{j2\pi f_0 x}$$
$$+ A K^*_4 U^*_{\text{sig}}(x, y) \, e^{-j2\pi f_0 x}]$$

This expression is identical in form to the one derived above using a reference beam. Clearly then, if the average intensity $\bar{I}(x, y)$ is then used to expose film, a hologram can be produced in the same manner as in the reference beam method described earlier.

The tilt could be eliminated if the direct voltage from the oscillator were phase shifted with transducer location so that

$$A \to A e^{-j2\pi f_0 x}.$$

Since it may be more difficult to produce this spatially controlled phase than to tilt the hologram recording plane, the tilted plane technique may be more practical.

3. Wave-length and Hologram Scaling Considerations

The wave-length scaling from acoustical to optical and the hologram size scaling from the original scan area to the film recording affect the size and location of the reconstructed images because of the diffraction angle relations. Obviously if the hologram size scaling is exactly the same as the wave-length scaling, the image size and position are also scaled in the same way, so that all quantities diminish in the ratio optical wave-length/acoustic wave-length. Sometimes this choice of hologram scaling may be inconvenient for viewing, however, because the resulting optical hologram may be extremely small. In general it is desirable to view a scaled hologram that is smaller than the original sound field and larger than that area scaled down by the ratio of the reconstruction and recording wave-lengths.

By considering the hologram as a complex diffraction grating or superposition of Fresnel zone plates, and by applying lowest-order diffraction theory, one can easily show that the distance from the hologram to the reconstructed image of the object varies according to the relation

$$d_2 = \frac{h_2^2 \lambda_1}{h_1^2 \lambda_2} \, d_1$$

where d is the object-(or image-)to-hologram distance, h is the width of the hologram, λ is the wave-length, and the subscript "1" refers to the recording parameter values while "2" refers to values of quantities after scaling in the reconstruction process. From similar considerations it may also be shown that the physical size of the image is given approximately by

$$w_2 = \frac{h_2}{h_1} w_1$$

where w_2 is a dimension across the reconstructed image, and w_1 is the corresponding dimension of the original object.

Because d_2 and w_2 scale differently, various distortions will be present in general. Obviously, for ratios of acoustic wave-length/optical wave-length of the order of 10^3 or so, values of the scaling parameter h_2/h_1 may be chosen to give severe distortion of depth-to-transverse width ratios in the reconstructed image. An additional fact, not considered here in any detail, is the introduction of aberrations into the imaging process. Fortunately, as the resolution of the acoustic hologram is increased by enlarging the w_1/λ_1 ratio, it is usually practical to scale the hologram down by a corresponding greater factor also. This approaches the ideal condition

$$\frac{h_2}{h_1} = \frac{\lambda_2}{\lambda_1}$$

more closely and reduces the distortions and aberrations mentioned above.

A related consideration is the amount of depth illusion present in the reconstructed image. In viewing an ordinary optical hologram, the "three-dimensional" quality of the image is present because the hologram width-to-optical wave-length ratio is sufficiently large to provide good resolution for an observing pupil much smaller than the hologram. Thus the observer may move his point of observation across the hologram aperture and easily resolve substantial differences in the perspective of the image. Putting this in more quantitative terms, the object must lie well within the limit of the near field of the hologram, or

$$d_1 < \frac{w_1^2}{\lambda_1}$$

if any depth illusion is to be obtained. Beyond this limit, the object is imaged essentially as though it were at infinity. Because of coherence problems and the small wave-length, this limit is seldom approached in optical holograms; however, it is easily reached in acoustic work if w_1 is not sufficiently large.

With acoustic holograms it is also possible to realize a condition in which the object is near the recording area but the wave-length is so large that the entire hologram aperture must be covered by the viewing pupil in order to achieve even the minimum required resolution. Depth information remains in this situation in the form of a focusing requirement; in fact, it may be difficult for the observer to determine whether or not the image is correctly focused in the presence of such prominent diffraction effects. A similar situation can exist in the case of high-magnification optical microscopy if the significant object detail becomes comparable to an optical wave-length.

4. Phase Stability Requirements and Effects of Movement of the Acoustic Medium

In any holographic process the relative phases of the acoustic signal and the real or synthetic reference wave must remain constant to within a fraction of a cycle during the recording process. For cases in which all of the waves of interest pass through essentially the same medium, this condition may not be difficult to meet; however, for objects of considerable extent compared to the scale of turbulence in the medium, phase shifts due to motion and thermal inhomogeneities of the medium can be significant. Also, in the case of systems using electronically synthesized reference waves, the medium acts as a delay line many wave-lengths long, and the hologram phase condition may be difficult to achieve in many practical cases. For example, if the total acoustic path difference is $N\lambda$, then a change in acoustic velocity producing an average shift $\Delta\lambda$ will result in a phase shift of $2\pi N \dfrac{\Delta\lambda}{\lambda}$.

If the phase tolerance is set arbitrarily at π radians, we have the requirement

$$\frac{\Delta\lambda}{\lambda} < \frac{1}{2N}.$$

Thus the tolerance on acoustic velocity and oscillator drift is set by the above requirement. Thermal variations in acoustic velocities of water and air under ordinary conditions are of the order of 10^{-3} per deg Centigrade. Thus a 1°C average change along the path would set a path length limit of the order of a few hundred wave-lengths.

In water and air under ordinary conditions the phase disturbances due to motion along the acoustic path can produce significant phase shifts also. The velocity of the medium adds to the sonic velocity to give

$$\Delta\Phi = \frac{2\pi d}{\lambda}\left(\frac{v}{c+v}\right)$$

where the phase shift $\Delta\Phi$ depends on the path length d, the acoustic wave-length λ, and the medium velocity v compared to the acoustic velocity c. As an example, if $d = 10^3\lambda$ and $c = 1450$ m/sec, as in water, a phase shift of π would result from a relative velocity of the medium of 0·7 m/sec. From this it is clear that holography under uncontrolled conditions with large object distances and high resolution (small wave-lengths) may be very difficult. In the case of synthetic reference holography, it may prove necessary to shift the phase of the reference with time to maintain constant phase at one point in the hologram. Such a scheme does not avoid the problems of localized turbulence and temperature variations, however.

5. Conclusion

Thus far, most acoustic holography experiments have been limited to air (Massey, 1967; Metherell et al., 1967) and water (Preston and Kreuzer, 1967; Thurstone, 1967) as the acoustic media, with mechanically scanned linear transducers used to record the sound field on photographic film by means of a modulated optical source. A method of holographic recording which avoids the scanning transducer by reflecting light from a liquid-gas interface (Mueller and Sheridan, 1966) under the influence of acoustic radiation pressure has been demonstrated also. Research for many years in the area of ultrasound image conversion has produced a variety of alternate methods (Kennedy and Muenow, 1967; Berger and Dickens, 1963) which may be useful in acoustic holography. It is hoped that further development of one or several of these approaches will lead to significant applications in the areas of medical diagnosis, underwater imaging, materials inspection, and perhaps to underground mapping of geological features.

See also: Holography.

Bibliography

BERGER H. and DICKENS R. E. (1963) A review of ultrasonic imaging methods with a selected annotated bibliography, Argonne National Laboratory publication No. ANL-6680.

KENNEDY JOHN A. and MUENOW RICHARD (1967) Practical improvements and applications for the ultrasonic image converter, *IEEE Transactions on Sonics and Ultrasonics*, **SU-14**, No. 2, 47–52.

LEITH E. N. and UPATNIEKS J. (1962) Reconstructed wavefronts and communication theory, *Journal of the Optical Society of America*, **52**, No. 10, 1123–1130.

LEITH E. N. and UPATNIEKS J. (1962) Wavefront reconstruction with continuous-tone objects, *Journal of the Optical Society of America*, **53**, No. 12, 1377–1381.

MASSEY G. A. (1967) Acoustic holography in air with an electronic reference, *Proceedings of the IEEE*, **55**, No. 6, 1115–1117.

METHERELL A. F., EL-SUM H. M. A., DREHER J. J. and LARMORE L. (1967) Introduction to acoustic holography, *Journal of the Acoustical Society of America*, **42**, 733.

MUELLER R. K. and SHERIDON N. K. (1966) Sound holograms and optical reconstruction, *Applied Physics Letters*, **9**, No. 9, 328–329.

PRESTON K. JR. and KREUZER J. L. (1967) Ultrasonic imaging using a synthetic holographic technique, *Applied Physics Letters*, **10**, No. 5, 150–152.

THURSTONE F. L. (1967) Three-dimensional imaging by ultrasonic holography, *Digest of the 7th International Conference of Medical and Biological Engineering* (Stockholm, Sweden), p. 313.

G. A. MASSEY

ACOUSTIC WAVEGUIDES. Simple examples of the acoustic waveguide are the airfilled "speaking tube", allowing verbal communication between different parts of a building or ship, and the child's "mechanical telephone" made of a stretched string linking two cans. In the speaking tube a *fluid*, air, carries the acoustic waves, while the relatively rigid solid walls of the tube guide the waves over their route by preventing appreciable radiation of energy into the surrounds. In the mechanical telephone a *solid*, the material of the string, carries the energy, and again guides the waves, there being little radiation into the surrounding air. In both these waveguides there is a large change in both density and velocity (in "acoustic impedance") at the boundary—between air and tube wall, or between string and air—and this has the consequence that any waves in the guide striking the boundary will be reflected into the guide again with only a small loss of energy to the surrounds. As a consequence of this acoustic signals may be guided over appreciable distances. Another example, at much higher frequencies (perhaps 1–10 MHz), is provided by certain types of ultrasonic delay line used in radar and computers. These are used to delay a signal for times of the order of 100 μsec, achievable with relative ease using the travel time of an ultrasonic signal, but difficult to achieve electrically because of the much higher velocity of electromagnetic waves.

In discussing the theory of acoustic waveguides it is usual at first to idealize the boundary conditions, assuming that no energy is transmitted across the boundary. In practice this condition is difficult to achieve completely at lower acoustic frequencies (of 0–100 kHz say) since for no transmission of sound there must be no motion of the tube walls, but the assumption of rigid walls provides a reasonable starting point for first analysis. At ultrasonic frequencies of 1 MHz and above, however, as commonly used in ultrasonic delay lines, it is much simpler to obtain boundary conditions closely approximating to the ideal. A wave guided along a metal wire or strip may be used; ideal boundary conditions would be achieved if the metal were situated in vacuum, but in practice the presence of air has usually a negligible effect. The boundary is here free to move but there is zero stress, in contrast to the fluid waveguide with zero motion but finite pressure. A third possible idealized boundary condition is sometimes found with a liquid in a tube at ultrasonic frequencies; here the boundary of the liquid sometimes behaves as if it were free to move, with zero pressure. The reasons are not fully understood, but in certain instances it is believed to be associated with very short waves propagating in a liquid which does not wet the walls of its container; some early delay lines using mercury in steel containers appeared to behave in this way. There are three cases of ideal boundaries to be considered, then, (i) fluid with boundaries not free to move (zero displacement), (ii) fluid with boundaries completely free to move (zero pressure), (iii) solid with boundaries free to move (zero stress). The fourth possibility, solid with boundaries not free to move, is rarely, if ever, met with, and will not be discussed.

Before treating these waveguides in more detail it is appropriate to point out that guided waves may also occur in systems with boundaries far removed from the ideal forms mentioned. For example, waves may be guided over appreciable distances through the air or through the sea in a layer whose boundaries are formed by gradual changes in density or temperature; ill-defined but nevertheless effective. Other extremely complex waveguide systems may be found in seismology, where guided-wave propagation may be found in structures composed of layers of different solids and liquids.

It is also important to point out that there is little clear distinction between the *waveguide* and the *resonator*. If some degree of reflection of waves occurs at the two ends of a waveguide, the multiple reflections may reinforce to produce marked resonance at certain frequencies. This is, of course, the principle used in all musical instruments, whether they use the simple resonances of waves travelling in air-filled tubes (organs, woodwind, brass) or the more complex resonances of flexural waves travelling on stretched strings, skins, or metal and wood objects of various shapes (violin, keyboard, percussion families). An example at much higher frequencies is the quartz crystal used in filters and frequency control in electronic equipment, making use of the resonances of ultrasonic waves in a quartz plate at frequencies as high as 100 MHz. Many other examples of resonators are found in the mechanical systems studied in vibration analysis, for example the resonant cantilever. Although studies of such resonators frequently commence by analysis of the standing wave system and do not specifically deal in terms of guided travelling waves, it can sometimes be useful to a physical understanding if the problem is thought of in terms of the reflections of guided waves.

We now describe some of the important properties of guided waves in the three idealized systems previously listed.

Fluid Waveguides with Rigid Boundaries

Many of the features of acoustic waveguides are illustrated by this idealized system—approximated to by the air-filled tube. The wave in a fluid is "compressional" or "longitudinal" in nature, transmitted through fluctuations in pressure and in particle velocity (the velocity with which small portions of the fluid move as they are displaced from the position they occupy in the absence of the wave). The pressure at the boundary may take any value, but the particle displacement and particle velocity normal to the boundary must be zero since the boundary cannot move. Analysis shows that many modes of propagation are possible. The most commonly analysed waveguides have circular or rectangular cross-section; for both these curves of the general form illustrated by Fig. 1 are obtained. These display the variation of the phase and group velocities for each mode with the frequency of the sinusoidal wave assumed to be propagating.

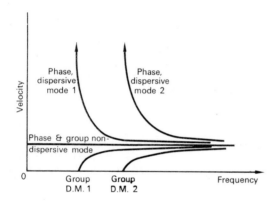

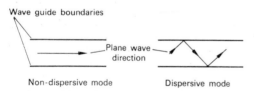

Fig. 1. Waveguide modes in (a) fluid with rigid boundaries, (b) fluid with free boundaries (non dispersive mode absent), (c) shear waves in a solid plate, polarized with motion parallel to plate surfaces, (d) torsional modes in a solid cylinder.

The simplest and most important mode is the "non-dispersive" mode whose phase and group velocities are at all frequencies equal to the velocity of a compressional wave in an unbounded fluid. This mode is made up of a plane wave travelling directly along the waveguide (as indicated by the arrow in Fig. 1); pressure, particle velocity and displacement are all uniform over the cross-section of the waveguide.

In each of the "dispersive" modes (of which the first two are shown, but there are an infinite number), it is possible to regard the mode as a set of plane waves travelling a zig-zag path by successive reflections between the boundaries. The angle of the zig-zag depends on the frequency and on the mode. In the first dispersive mode propagation is only possible if the wave-length is less than about twice the cross-sectional dimension (a more precise statement involves considering the shape of the cross-section). At lower frequencies than this "cut-off" frequency the mode is "evanescent" and waves will not propagate freely, but are attenuated, usually rapidly. At the cut-off frequency the angle of the zig-zag path becomes such that the wave strikes the boundaries at normal incidence and makes no progress; the group velocity (which is the velocity of energy transfer along the waveguide) falls to zero; while the phase velocity (which is the velocity of intersection of the wavefront and boundary) rises to infinity. At high frequencies and short wave-lengths, the angle of the zig-zag path becomes such that the waves travel almost directly along the guide, striking the boundaries at near-grazing incidence; here the group and phase velocities both tend asymptotically to the velocity of waves in an unbounded material. The set of plane waves travelling a zig-zag path combine to produce the cross-sectional variation in pressure, particle velocity and displacement typical of the mode; in this first mode in a system of rectangular geometry, one half-cycle of a sinusoid between boundaries. (Or a Bessel function in a cylindrical system.)

Higher dispersive modes travel a more zig-zag path and show higher cut-off frequencies, corresponding in a system with very wide parallel boundaries to the cross-sectional dimension being an integral number of half-wave-lengths. The variation in pressure, etc., over the cross-section also possesses an integral number of halfcycles of a sinusoid; the higher the mode, the larger this integral number. In other ways their characteristics are similar to those of the first dispersive mode.

The non-dispersive mode of this system is of considerable importance since in it waves of different frequencies travel in the same way, and there is no distortion of a signal of wide frequency spectrum. It is this mode that is used in the speaking tube, since at voice frequencies all other modes are well below cut-off. At much higher frequencies it might be possible to excite also the higher modes, and a signal travelling in any one of these might suffer considerable distortion since its velocity depends on frequency. In addition if several modes were simultaneously excited, even with a single frequency interference between modes would be possible since the velocity in each mode is different.

The extent to which each mode is excited depends on the configuration of the source of waves. Uniform vibration over the whole cross-sectional area should produce only the non-dispersive mode, while another pattern might produce several modes simultaneously; their relative amplitudes can be deduced using Fourier analysis.

Fluid Waveguides with Free Boundaries

If the fluid has free boundaries, with zero pressure, a new feature appears; the waveguide will not sustain a non-dispersive mode. Dispersive modes are again possible, and their general form is much the same as in Fig. 1, but the mode with constant phase and group velocity disappears. Again the dispersive modes may be thought of in terms of reflecting plane waves, and again sinusoidal distributions of pressure, particle velocity, and displacement are found. In such a waveguide, however, unattenuated propagation is not possible at all below a certain minimum frequency—in marked contrast to the fluid with rigid boundaries—and even above this frequency only dispersive propagation is possible.

The characteristics of these modes are somewhat similar to those found in the propagation of electromagnetic waves in hollow electrical waveguides of rectangular or circular cross-section, where again no non-dispersive mode is possible.

Solid Waveguides with Free Boundaries

Reflection of a plane wave at the free boundary of a solid is often considerably more complex than reflection at the boundary of a fluid. The fluid can carry only compressional waves, but the solid will sustain both compressional and shear waves, travelling with different velocities (the shear wave is slower) and with markedly different characteristics. The most important difference is that in the shear wave the particle motion is at right angles ("transverse") to the direction of propagation, and the "polarization" of the motion of the wave with respect to the boundary must therefore be taken into account. Reflection is only simple with a shear wave whose particle motion is parallel to the boundary; this wave is reflected without loss of amplitude, as in the reflection of a compressional wave at a fluid boundary. With any other polarization the shear wave may produce both another (weaker) shear wave and a compressional wave; though for some angles of incidence the latter degenerates into an evanescent disturbance and does not (in theory) withdraw energy from the incident shear wave. In the reflection of a compressional wave both a weaker compressional wave and a shear wave are always produced. Further, in the presence of a free boundary a Rayleigh surface wave may exist; this has a velocity slightly less than that of plane shear waves in an unbounded medium, and amplitude falling exponentially with distance from the surface.

All these factors combine to produce waveguide modes which may be far more complex than those found in fluids, with two important exceptions, which are treated first.

Simple shear and torsional modes

If a shear wave with particle motion parallel to the waveguide boundaries is launched, relatively simple waveguide modes might be expected. There are two practical cases of importance. (i) In the first propagation takes place in a flat strip of width much greater than thickness; here it is often possible to ignore the effects of the small side boundaries and obtain modes in which the particle displacement is indeed parallel to the major boundaries. (ii) In the second propagation of a torsional mode in a cylinder of circular cross-section leads to simple modes; these are analogous to (i) because in torsion the particle motion may again be parallel to the circular boundary. In both cases a set of modes very similar to those of Fig. 1 is obtained, including the important non-dispersive mode. This mode is made use of in ultrasonic delay lines employing (i) shear wave propagation in thin, relatively wide, strips of metal or (ii) torsional waves on wires.

Higher, dispersive, modes again exist, and may be launched from sources of certain geometries. Unlike the air-filled tube, the ultrasonic delay line often operates at frequencies when many modes may be above their cut-off frequencies; the aim then is usually to design a source whose vibration produces only the required mode.

"Longitudinal" and "flexural" modes

If the shear motion is not parallel to the boundary of the waveguide, or if a compressional wave is launched, another set of modes may be excited. The phase-velocity/frequency characteristics of these are shown in Fig. 2; curves of this type would be found in cylinders of circular cross-section and in plates whose width is much greater than thickness, but the situation with rods of rectangular cross-section is more complicated and is not treated here.

Two modes exist at low frequencies and are therefore of prime importance, though there are many additional modes showing cut-off frequencies. One of these low-frequency modes is the well-known "flexural" wave possible in a plate or in a cylinder. Its particle displacement is antisymmetrical with respect to its central plane; its overall "snaking" motion in both plate and cylinder is well known. Its phase velocity rises from zero for very long waves towards the velocity of surface waves at short wavelengths. The other low-frequency mode is given various names: "extensional", "longitudinal", "compressional", or "Young's Modulus", the latter because in the cylinder the low frequency velocity tends to (Young's Modulus $\div$ Density)$^{1/2}$. With wavelengths much greater than the cross-sectional dimensions the whole cross-section moves with almost equal amplitude and the wave propagates with a "concertina" motion, but with short waves a marked exaggeration of motion at the surface is observed and the velocity falls towards the Rayleigh surface wave velocity. The motion characterizing this mode is symmetrical about the central plane of the plate or axis of the cylinder.

In both sets an infinite number of higher modes, exhibiting cut-off frequencies, are possible. It should be particularly noted that all modes, including the

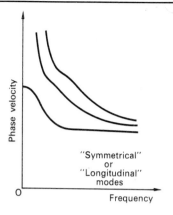

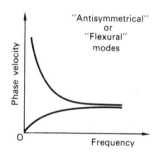

Fig. 2. *"Longitudinal" and "flexural" modes in a solid plate or cylinder.*

low-frequency modes, are dispersive and propagation in any mode may result in appreciable distortion of signals of finite frequency-bandwidth. For this reason these modes find little application in devices making use of ultrasonic travelling waves, though at low frequencies, of course, most plate or bar resonators are carrying waves of one or both these types.

See also: Waveguides, electromagnetic; Radio waves, dispersion of; Radio waves, guided; Elastic waves; Sound vibrations, longitudinal; Sound vibrations, torsional; Sound in pipes; Sound in sea; Sound propagation in the atmosphere; Group velocity; Wave (phase) velocity.

Bibliography

EWING W. M., JARDETSKY W. S. and PRESS F. (1957) *Elastic Waves in Layered Media*, New York: McGraw-Hill.

KOLSKY H. (1953) *Stress Waves in Solids*, Oxford.

MASON W. P. (Ed.) (1964) *Physical Acoustics*, Vol. 1A, New York: Academic Press.

REDWOOD M. (1960) *Mechanical Waveguides*, Oxford: Pergamon.

M. REDWOOD

ACOUSTO-ELECTRIC EFFECTS. Acousto-electric effects are usually restricted to those phenomena which arise from the interaction of free electrons and holes with non-equilibrium acoustic waves, e.g. waves launched from an external source.

An acoustic wave propagating through a solid sets up forces which act on the free electrons. The acoustic wave becomes a combined motion of the lattice atoms (or ions) and the free electrons. As a consequence, the velocity of propagation of the wave is affected by the electrons. So too is the attenuation since a net transfer of energy can occur between wave and electrons because, due to scattering of the electrons, the induced velocity of the electrons is not in quadrature with the periodic forces associated with the wave.

The attenuation of an acoustic wave is accompanied by a transfer of momentum as well as energy. The transfer of momentum is related to a time average force which can be regarded as originating from a time independent effective field, the *acousto-electric field* E_{ae}, within the medium. This field imparts a motion to the electrons in the same direction as the wave is propagating. There is a linear relationship between the attenuation due to free electrons $2k_i$ and E_{ae}

$$2k_i = (nqs) E_{ae}/\phi \qquad (1)$$

where s is the velocity of sound, nq is the charge density of electrons and ϕ is the acoustic energy flux. Under open circuit conditions, E_{ae} will give rise to an *acousto-electric voltage* across the crystal. Under short circuit conditions, an *acousto-electric current* will flow. This relation, called the Weinreich relation, breaks down when both electrons and holes are simultaneously present, when trapping of the mobile carriers occurs and when collision drag effects are important. (The collision drag effect relates to the coherent feedback of energy from electrons to acoustic wave through scattering of the electrons. It is only important when electromagnetic coupling predominates, i.e. in metals.)

Externally applied fields, such as electric, magnetic and stress alter the behavior of electrons and, consequently change the acousto-electric properties.

In an electric field, the attenuation of acoustic waves due to free electrons can become a strong function of field, leading to *ultrasonic amplification* (negative attenuation) when the drift velocity of the electrons exceeds the velocity of sound. Under amplifying conditions the acousto-electric field is reversed in accord with the Weinreich relation. Under amplifying conditions the acoustic flux may build up from the thermal equilibrium level to such a magnitude that the acousto-electric field becomes significant compared with the applied field. Then the current-voltage characteristic becomes non-ohmic due to acousto-electric effects and, under certain conditions, instabilities may appear due to the occurrence of a negative differential conductance.

In the presence of a magnetic field, resonance absorption phenomena appear. They are well defined only if $\omega_c \tau \gg 1$, where ω_c is the cyclotron resonance frequency and τ is the electron relaxation time. They may be subdivided into (i) temporal resonance (cyclotron resonance) given by $\omega = n\omega_c$, where ω is the angular frequency of the acoustic wave and n an integer, (ii) spatial resonance (geometric resonance) which occurs when the diameter of the cyclotron orbit is an integral number of acoustic wave-lengths, and (iii) quantum resonances which only occur in degenerate materials and are associated with the passage of the Landau levels through the Fermi surface. Within the previously stated limitations the Weinreich relation is valid in the presence of a magnetic field, hence these magnetic field induced absorption processes are accompanied by the generation of acousto-electric fields.

Away from resonance, acousto-electric effects appear when an acoustic wave is propagated perpendicular to a magnetic field. The usual acousto-electric force sets the carriers in motion in the propagation direction but the Lorentz force produces a component of motion in a direction mutually perpendicular to the magnetic field and acoustic wave vector. This effect is of particular significance in materials containing similar concentrations of electrons and holes since the normal acousto-electric effects are small because the currents due to the electrons and holes are opposed.

The periodic forces accompanying an acoustic wave which act on the electrons derive from (i) electromagnetic coupling, (ii) deformation potential coupling and (iii) piezo-electric coupling. Electromagnetic coupling is of primary importance in metals. In semiconductors deformation potential coupling usually predominates though, at low frequencies in piezo-electric materials, piezo-electric coupling is most important. The largest acousto-electric effects have been observed in strongly piezo-electric semiconductors, such as cadmium sulphide and zinc oxide, where, at low frequencies, strong coupling, low lattice loss and weak screening by the free carriers all combine to produce favourable conditions.

Bibliography

McFee J. M. *Transmission and Amplification of Acoustic Waves in Piezoelectric Semiconductors.* Vol. 4, part II, *Physical Acoustics*, W. P. Mason (Ed.), Academic Press.

Spector H. N. (1966) Interaction of acoustic waves and conduction electron, *Solid State Phys.* Vol. 19, page 291, F. Seitz and D. Turnbull (Ed.), Academic Press.

Yamada T. (1965) Acousto magneto electric effect in bismuth, *J. Phys. Soc. Japan*, **20**, 1424.

E. G. S. Paige

AIRBORNE COLLISION PREVENTION DEVICES

Introduction

1. The accepted best solution to the problem of potential mid-air collisions is a positive, reliable and fail-safe control of all aircraft, exercised from the ground by appropriate Air Traffic Service unit(s) having a jurisdiction over the airspace utilized. The primary responsibility for maintaining safe separation limits between aircraft in flight—laterally, as well as in the vertical sense—must always rest with the ground Air Traffic Control System, without any reliance on airborne collision prevention devices. Yet, for some two decades already, the aviation industry has been searching for an effective airborne collision prevention device which would ensure that safe separation of aircraft is maintained when, as a result of either human or equipment failures, the Air Traffic Control System does not fulfil its intended role.

2. Research and development efforts in airborne collision prevention have been directed towards two basic categories of airborne device, namely Collision Avoidance Systems and Pilot Warning Instruments. A Collision Avoidance System (CAS) performs collision threat computations and communicates to the pilot avoidance instructions. Its four basic distinct tasks are:

(a) to detect all aircraft which represent a potential danger;
(b) to evaluate continuously the potential collision threat presented by the detected aircraft;
(c) to determine what kind of evasive manoeuvre, if any, is required;
(d) to provide to the pilot(s) an unambiguous command of the appropriate avoiding manoeuvre(s) in time to ensure safe clearance.

The role of Pilot Warning Instruments (PWI) is to assist the pilot in the visual acquisition of another potentially dangerous aircraft; thereafter, the pilot retains the responsibility for determining whether a collision hazard actually exists and, if so, what avoidance manoeuvre is needed. The tasks of PWI are:

(a) to detect the proximity of another aircraft;
(b) to give a broad indication of its relative elevation and relative bearing;
(c) as a desirable but not essential feature, to indicate the rate of closure of the two aircraft.

3. CAS AND PWI airborne devices can be further categorized into non-cooperative or cooperative types. A non-cooperative system is self-contained and aims at offering to equipped aircraft collision protection without requiring other potentially hazardous aircraft to be similarly equipped. A cooperative system requires both protected and intruding aircraft to be correspondingly equipped in order that collision protection functions can be performed; consequently, systems of this kind do not safeguard against collision risks with intruding aircraft not carrying the prerequisite airborne equipment.

4. The development of effective and technically and economically feasible CAS or PWI has proved to be a formidable task. The magnitude and complexity of the problem and the systems requirements were not fully appreciated until after rigorous operational analyses were carried out during 1955–8. These studies clearly demonstrated that for obvious reasons operationally preferable non-cooperative CAS, though theoretically possible, is not attainable because the demands on accuracy of data acquisition are so stringent that they cannot yet be satisfied. Therefore, exploitation of cooperative equipment techniques is essential for an effective CAS at this time and a similar conclusion also appears applicable in the case of PWI.

Collision Avoidance Systems (CAS)

5. Cooperative CAS is based on the concept that all aircraft measure certain collision threat parameters, interchange required information and establish compatible escape manoeuvre commands. The types of CAS system which have been applied experimentally or for other than civil aviation purposes are:

5.1. Relative position-velocity systems. These systems measure and exchange information about range, bearing and both velocity vectors. They have been used primarily in military service operations for aerial rendezvous or close proximity formation flying, in which all aircraft are assumed to be at the same altitude, with a collision threat arising whenever any one aircraft wanders from its predetermined position. Horizontal manoeuvring is utilized to bring the aircraft back exactly to their assigned position.

5.2. Range-altitude systems. Altitude is measured in each aircraft and communicated in order to eliminate all other aircraft not on the same altitude and thus to minimize the computations involved in the prediction of collision hazards. Range measurements are made between all aircraft pairs in the same altitude layers. Range rate is either measured directly or derived from successive measurements of range. In either case, the quantity: measured Range divided by rate of change of range (called "Tau") is used to predict the approximate time remaining before two aircraft reach the point of minimum separation. A minimum value of Tau is chosen so as to provide sufficient time for the execution of a safe escape manoeuvre. In view of the stringent requirements imposed on any CAS technique based on a horizontal turning escape manoeuvre, use is made of vertical manoeuvres to ensure that aircraft manoeuvre directly out of the collision plane.

Several alternative methods of range measurement and data communication can be employed, namely:

(a) Interrogator — transponder technique:
The interrogation signal normally encodes barometric altitude of the transmitting aircraft, and the response contains either the altitude information of the responding aircraft, or altitude difference data. Range is obtained by measuring the time elapsed between interrogation and response. Since two transmissions are required between each pair of aircraft within range, the communication density is rather high and increases nearly as the square of the number of aircraft within range.

(b) Ground bounce technique:
An aircraft periodically transmits a pulse signal encoded with its altitude information. A receiving aircraft measures the time difference between the arrival of the direct and ground reflected signals from the transmitting aircraft, and from this and the altitude data of both aircraft computes the distance from the transmitting aircraft. An experimental system of this kind utilized successive range measurements to smooth the range data and overcome the effects of terrain roughness and slopes. The volume of communications involved increases linearly with the number of aircraft within communications range rather than as its square.

(c) Synchronized time and frequency technique:
Each aircraft transmits at an assigned precise instant of time a message encoding its pressure altitude and other data. Range information is extracted from the times needed by the transmitted radio frequency signal to reach other receiving aircraft within the protected altitude band. Precisely synchronized highly accurate time clocks are used to provide a common time reference and to ensure that each aircraft begins to transmit at exactly the correct instant assigned to it. Rate of change of range is determined by measuring the Doppler shift of the RF carrier frequency. Communications density is directly proportional to the number of aircraft within communication range of each other.

5.3. Projected hazard prediction systems. Hazard prediction technique calls for the measurement and exchange of information about certain flight parameters and future intentions essential to foresee for some preselected time the expected positions and velocity vectors of cooperating aircraft.

CAS for Civil Aviation Application

6. The most promising technique for providing an airborne CAS civil aviation is currently considered to be a Synchronized Time and Frequency Range-Altitude System. An effective system of this kind— EROS I (Eliminate Range Zero System) was developed by McDonnell-Douglas Co. and is now in operational use to protect company-built military aircraft during flight tests. Member Airlines of the Air Transport Association of America (ATA) were sufficiently impressed by the EROS I system to prepare in cooperation with the manufacturers and experts, the

ATA Technical Description and Specification of the CAS system. A programme for simulator and flight testing of the system has also been developed and published, and flight evaluation of developmental airborne CAS is expected to commence during the first quarter of 1969.

7. In view of the fact that the ATA CAS is the result of the cooperative effort of its Member Airlines and the avionics industry, the main features of this system only are briefly described below:

7.1. Frequency and time synchronization: The radio frequency and system time for the CAS is based on the zero magnetic field cesium atomic resonance defined as 9,192,631,770 H_2. The RF carrier frequencies of airborne CAS stations must be maintained to within one part in 10^8 with respect to their assigned frequencies; this is achieved either by using airborne atomic standards or stable airborne crystal standards that are reset and synchronized during communication with ground stations.

The time-synchronization is provided by the master-time complex of ground stations containing highly accurate cesium standards that are checked periodically to ensure synchronization accuracy within ± 0.5 μs of a common time. Ground stations are the source of master time (00 in time hierarchy) and furnish synchronization to airborne stations that are within their communications and synchronization limits. The master time is extended by aircraft to aircraft relays, with each receiving station having to achieve the time accuracy of ± 0.25 μs (three sigma value) with respect to the time of the station that synchronized it. A time hierarchy system of 60 steps (01 to 60) is established to indicate the relative quality of time of aircraft stations, the expected three sigma error at N hierarchy level being equal to $0.25\sqrt{N}$ μs. The time hierarchy number of an airborne station is a function of the number of relays between it and a ground station, or of the time elapsed since its last synchronization by a station of a better hierarchy, the system demoting itself one hierarchy step for each 0.05 μs of accumulated error. The demotion technique is continued only until hierarchy 40 is reached, so as to limit the rundown error to 2 μs and to permit arriving aircraft to join the CAS environment.

The system employs four frequencies—1600, 1605, 1610 and 1615 MHz (F1, F2, F3 and F4); both amplitude and phase modulation are utilized. Each period of 3 sec—called an "epoch"—is divided into 2000 message "slots". To differentiate between synchronizations from a ground master-time source and from aircraft stations, the synchronization cycle extends to 6 sec. The ground station starts the cycle with a ground-to-air epoch start pulse group and answers all requests for synchronization during the remainder of the 3 sec period. During the next epoch, airborne stations transmit their epoch start pulses and thereafter reply to all requests for synchronizations. Epoch start pulses serve to identify the start of ground and airborne cycles and to provide coarse synchronization to within the propagation time of the epoch start pulses from the station which originated them.

Each of the 2000 message slots in a 3 sec epoch is of 1500 μs duration. Each system has its own message slot selected and preset, and transmits once during each 3 sec interval the following elements:

(a) an RF burst of 200 μs giving range and rate of change of range, biphase modulated in the case of full capability systems for airborne propagation of master time and aircraft identity;

(b) a single pulse, time positioned to provide a standard pressure-altitude based on a reference of 29·92 in. Hg;

(c) altitude rate and velocity vectors (for future use).

The 200 μs RF burst acts as a synchronization request. When no bi-phase modulation is present (limited systems), the ground stations give synchronization replies during the ground cycle; when biphase modulated data are included, synchronization response is given only by the addressed station. The replying system or donor make appropriate time adjustments for the propagation delay of the interrogation and for the interrogator's clock error to ensure that the synchronization reply reaches the interrogating station 1419·2 μs after slot time. To provide protection against line of sight interference from an aircraft utilizing the adjacent time slot, frequency shifting to an adjacent frequency is employed after each slot data period. Aircraft stations receive fine synchronization from a station with the best (lowest) time hierarchy designation, and assume that hierarchy number increased by one.

7.2. Threat evaluation technique and separation criteria. The basic threat-sensing parameters are intruder range, intruder rate of change of range, intruder altitude and own altitude. Intruder range is derived from the measured elapsed time between the start of the time slot when the transmission was known to occur and the time of receipt of the leading edge of the 200 μs RF pulse. Intruder rate of change of range is determined by measuring the Doppler frequency shift of the received 200 μs pulse. Intruder altitude is decoded from the measured time interval between the arrival of the leading edge of the 200 μs pulse and its associated altitude pulse. Own altitude is an input from the aircraft altitude encoder or air data system.

During each epoch, the aircraft system evaluates received data for altitude, Tau (range divided by rate of change of range), and minimum range threats. If the system detects a collision threat in the co-altitude band or the presence of another aircraft within the defined adjacent altitude layers, it initiates the alarm and indicates to the pilot the desired evasive manoeuvre. Four manoeuvre commands are used, namely:

(a) "Climb/Descend" command, applied when a co-altitude collision threat is detected. Both aircraft are manoeuvred cooperatively to increase their altitude difference, but safe vertical separation is assured even if only one aircraft obeys the command. When a co-altitude climb/descend or roll-out command is given, the system logic of the aircraft concerned biases the next altitude transmissions so that altitude 200 feet in the direction of the intended escape is encoded.

(b) "Hold Altitude" command, used when three aircraft are involved in co-altitude situations, in which case one aircraft must not manoeuvre in either direction.

(c) "Roll-Out" command to stop the turn(s) and return the aircraft to linear flight when a potential collision situation is sensed by the system.

(d) "Level-Off" command, utilized to stop the climb or descent in ten seconds when the system detects that the protected altitude band will shortly thereafter be penetrated by an aircraft posing Tau or minimum range threat.

The system also provides "Aircraft Above/Aircraft Below" advisory display, using the same Tau zone warning criteria as the Level-Off command and applied when the intruding aircraft is outside of the level-off band of the receiving aircraft, but within its above/below altitude band.

7.3. Versions of airborne CAS. Several versions of airborne CAS are envisaged, namely full capability CAS Air Carrier Systems and cheaper Limited Equipments. For other than air carrier application, use is envisaged of a cheaper, scaled down version operating compatibly with the CAS message format and retaining full hazard logic as well as avoidance manoeuvre command capability; it would receive on all four frequencies but transmit on only one (F1) at a lower power, and would not require bi-phase modulation, use of synchronization hierarchy or propagation of Master Time. A still further simplified minimum cost version would leave out full hazard logic and escape command computations; equipment of this kind would place the escape manoeuvre burden wholly on the air carrier system, and would not provide protection against another aircraft likewise equipped with a minimum system only.

Pilot warning instruments (PWI)

8. The synchronized time-frequency CAS described above is generally recognized as technically sound and workable. However, in view of the complexity and cost, some reservations exist about its efficacy as a widely used airborne collision prevention device. It is believed in many quarters that the CAS is too costly to provide protection for the large numbers of general aviation aircraft currently used or expected in the future. In the United States, NASA Langley Research Center and Electronic Research Center are therefore pursuing two programs to develop and refine the technology for a light-weight, low-cost collision hazard warning system. The techniques under investigation are:

(a) A continuous wave cooperative Doppler radar/transponder system. This concept would provide approximate determination of the range of intruding aircraft from the strength of the retransmitted signal, accurate measurement of the rate of change of range from its Doppler frequency shift, and approximate azimuth and elevation of the intruder would be determined by using a multilobe receiving antenna. Only aircraft carrying both the radar transmitter and receiver would receive warnings of the presence of intruding aircraft; aircraft equipped with only a transponder would be "seen" by fully equipped aircraft but would not receive any alarm, and no protection would be provided against collision of two aircraft carrying transponders only.

(b) An infrared detection system, based on the use of high intensity anticollision Xenon lights and a radiometer receiver in the protected aircraft. The receiver would determine the azimuth and elevation angles of the pulsed infrared emissions from the Xenon lights, and would provide a rudimentary estimate of range by determining their intensity. It would thus be able to trigger an audible alarm when another aircraft approached within a certain distance.

Bibliography

Airborne Collision Avoidance System, April 15 1968, Air Navigation/Traffic Control Division, Air Transport Association of America, ANTC Report No. 117, Revision 5.

BATES M. R., FLETCHER H. K., MICHNIK L. and PRAST J. W. (1968) *History of Time-Frequency Technology.*

BROWN D. A. (1967) NASA seeks low-cost collision warning, *Aviation Week and Space Technology*, Sept. 1967.

JAYCOX R. L. (1967) Collision avoidance system synchronization. Collins Radio Co.

KLASS P. J. (1966) New anti-collision system stirs interest, *Aviation Week and Space Technology*, April 1966.

LAROCHELLE P. J. (1962) Technical feasibility of collision avoidance system. Federal Aviation Administration: U.S. Dept. of Transport.

PERKINSON R. E. (1967) CAS Message Format, McDonnel Douglas Corp.

SHEAR W. G. (1967) Elements of the ATA collision avoidance system, Avionics Division, Bendix Corp.
WHITE F. C. (1968) Airborne Collision Avoidance System. Development, ATA.

S. KREJCIK

AMBIENT NOISE IN THE SEA. Underwater acoustic ambient noise in the sea arises from a number of sources, most of which are dynamic processes dependent on environmental factors. It is easier to measure ambient noise than to measure or isolate the sources, several of which may be contributing simultaneously, and some of which may be interdependent. The noise statistics are not stationary, though over short periods of time they may be considered quasi-stationary; then reasonable agreement is found between the pressure level spectra of some generating processes and the average measured spectrum. The average levels of noise pressure spectra and the departures therefrom have been measured in all seas and oceans, for different depths, weather, seasons and other environmental conditions. A large quantity of data was comprehensively collated by G. Wenz to produce the composite picture (Fig. 1) of prevailing ambient noise and its principal components, considered briefly below. The noise spectrum

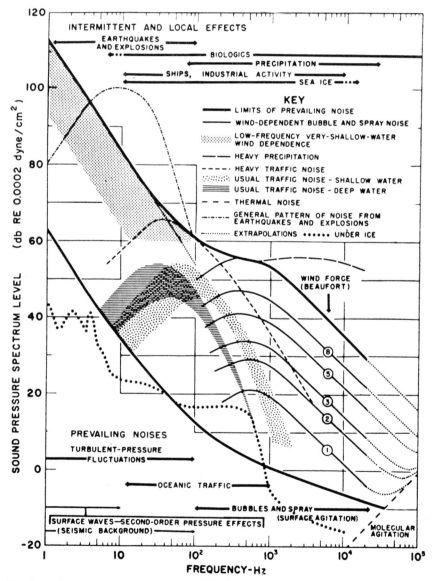

Fig. 1. *Ambient noise sources, spectra and levels. (From Becken, 1964, after Wenz, 1962.)*

level at any frequency is defined as the mean-square pressure in a 1 Hz band centred on the frequency, and is usually expressed in decibels relative to a reference pressure, which may be 0·0002 dynes/cm² or preferably, 1 dyne/cm² (= 1 μbar = 0·1 N/m²). Noise including obvious local or intermittent sources such as passing ships or marine animals is usually excluded from the data, but distant traffic noise has a more continuous nature and is included. Measurements are frequently made in one-third octave bands and referred to the spectrum level in a 1 Hz band by subtraction of $10 \log_{10}$ (bandwidth in Hz). Noise is usually measured with a small omnidirectional hydrophone; however, the noise energy is not always isotropic and so the directional distribution of noise is required for a more complete description of the noise apparent with directional systems.

Thermal Noise

Thermal noise due to molecular agitation of the water is predominant at frequencies above about 100 kHz and can be apparent in otherwise quiet conditions down to 20 kHz. At thermal equilibrium the mean power flow due to thermal noise across any enclosed surface of area A is $4\,kTB$, which is independent of frequency and dimensions, where k is Boltzmann's constant ($1·37 \times 10^{-16}$ ergs/deg), T is the absolute temperature and B the bandwidth in Hz. The mean intensity is thus $4\,kTB/A$, which can be equated to $\langle p^2 \rangle / R_a$, where $\langle p^2 \rangle$ represents the mean square pressure averaged over the surface and R_a is the specific radiation resistance. Thus $\langle p^2 \rangle = 4kTBR_a/A$, which is analogous to the Johnson formula for the thermal noise voltage in an electrical resistance. A hydrophone having a uniform pressure sensitivity thus has a terminal voltage spectrum which is constant with frequency, providing its radiation resistance is constant.

For a spherical volume of radius, a, less than a wavelength, $R_a \cong (2\pi a/\lambda)^2 \varrho c$ and $A = 4\pi a^2$ so that $\langle p^2 \rangle$, which now represents the mean square noise pressure *at a point*, becomes

$$\langle p^2 \rangle = \frac{4\pi kT}{C} \varrho\, f^2,$$

where ϱ (the density of sea water) $\cong 1·02$ g/cc, C (the velocity of sound in sea water) $\cong 1·5 \times 10^5$ cm/sec, and f is the frequency. The thermal noise (acoustic) spectrum level at a point is thus

$$N_T = -115 + 20 \log_{10}(f) \text{ dB relative to 1 dyne/cm}^2$$

in 1 Hz band at normal temperatures and where f is in kHz.

Thermal noise is isotropic and the $+6$ dB/octave characteristic reflects the increased volume density of normal modes, each with one degree of freedom and energy of kT associated with it. If the electro-acoustic efficiency of the measuring hydrophone is less than unity then each loss mechanism will have associated thermal noise, as will the measuring amplifier, thus raising the effective noise pressure level. It is possible, but not customary, in underwater acoustics to refer to an acoustic noise level in terms of an equivalent noise temperature as used for radio aerial work; over the most useful part of the spectrum the noise is not thermal in nature and the equivalent noise temperatures become very great and meaningless in any practical sense.

Wind and Sea-state Dependent Noises

Between 500 Hz and 20 kHz the predominant noise levels are dependent upon wind speed and the associated sea-surface-state. Wind dependence of the noise between 100 Hz and 500 Hz is also noted when residual noises are low. There is a broad maximum between 300 Hz and 1 kHz and a high frequency slope of between -5 and -6 dB/octave. The "Knudsen" curves on Fig. 1, with Beaufort wind force shown are an average, inter alia, of shallow and deep water measurements. For shallow water, taken as 100 fm depth or less and therefore including most of continental shelf waters, the levels are generally 2 to 3 dB higher, and for deep water 2 to 3 dB lower than the levels shown; however, the variability is such that the distributions overlap, and seasonal variations of up to 13 dB can occur at the same location. There is strong evidence to suggest that the sea surface agitation is the origin of this noise by way of bubbles undergoing oscillations in volume, of the generation and collapse of vapour cavities, of spray and spindrift. The broad maxima and -6 dB/octave slope are consistent with a wide range of sizes of resonating bubbles. As the surface agitation is dependent upon the duration, fetch, constancy and direction of the wind in relation to swell, tides, currents and topography in addition to wind speed it is to be expected that large variations from the average levels will occur.

Noise levels due to surface agitation in the band 750–1500 Hz have been shown to possess marked directional distribution, particularly in the vertical plane and at high sea states. Figure 2 indicates the high level received normal to the surface; the waist, which is around horizontal directions, and which may be from 2 to 12 dB lower than the vertical component, is due in part to the reduced radiation by the surface in horizontal directions and in part to the extra distance travelled and thence greater attenuation by absorption suffered by the sound. Noise from the surface layers also reaches the measuring hydrophone via reflection at the sea-bed and the secondary maximum in the distribution at 140° and the minimum at 180° can both be explained by this; the sound reflected forward at glancing angles is generally greater than that reflected back normally, at which incidence transmission of considerable energy into the bed may occur.

Azimuthal distributions measured simultaneously with the above vertical distribution have elliptical shape at high angular elevations when whitecaps are present; the greater levels generally occur along directions parallel to the water wave crests and it is thought that the water wave troughs impede sound transmission from the wavelet sources on the wave crests. Similar distributions would be expected at other frequencies in the band 500 Hz to 20 kHz when sea-noise is predominant, though the detail would depend upon water depth and nature of the sea-bed.

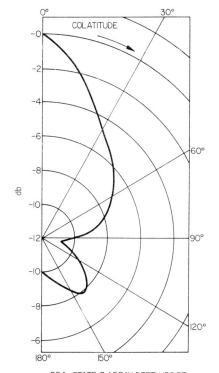

Fig. 2. *A vertical distribution of ambient noise measured at 750–1500 Hz. (After Becken, 1964.)*

Oceanic Traffic Noises

The ambient spectrum between 10 Hz and 1 kHz is very variable and the residual noise at low sea states is believed to be due to distributed oceanic traffic. There is a broad maximum between 20 Hz in deep water and 100 Hz in shallow water and generally a rapid fall off above 100 Hz. It has been calculated that as few as 100 ships within a range of 1000 miles could account for the typical deep water spectrum level at 100 Hz. In shallow water the average range of the sources is probably less so that the higher frequency components are not attenuated to the same extent as in deep water. At frequencies below 50 Hz ships are less effective radiators to long ranges due to the doublet effect with the surface reflected sound; this produces a $+6$ dB/octave slope relative to a simple source. Source spectrum levels vary from ship to ship, traffic is not uniformly distributed and ambient levels are influenced greatly by the acoustic propagation conditions, so that average shipping traffic noise will vary considerably with location.

Turbulent Pressure Fluctuations

At frequencies from below 1 to 10 Hz the observed average ambient pressure spectrum shows the highest levels and the largest slope of -8 to -10 dB/octave, which may continue up to 100 Hz in the absence of traffic noise. Below 10 Hz in very shallow water there is some evidence for wind dependence. It is thought that these levels are predominantly due to the local, nonradiated, but high, pressure level fluctuations associated with turbulent flow in the sea. This is not the self flow-noise experienced by a hydrophone in motion relative to the water mass, but the space and time fluctuations of pressure level around a pressure sensitive hydrophone moving with the mean, turbulent, flow. These pressures are not radiated acoustic waves but, in so far as they set a limit to signal detectability, they are considered inseparable from acoustic noise. The spatial correlation distance of the turbulence noise is comparable with the mean turbulence dimension and not the acoustic wavelength, corresponding to these frequencies. Turbulence is a mechanism by which energy may be transferred from large eddies to higher frequencies. It is reasonable to explain the wind dependence of ambient noise in very shallow water by turbulence of the wind-induced drift current, which is fastest near the surface and negligible in deep water.

Surface Wave Effects

Subsurface pressure fluctuations due to surface waves are attenuated with depth and frequency. Like turbulence pressures they are controlled by inertial forces and do not radiate acoustic waves. Surface wave-height spectra show a maximum below 0·3 Hz and are dependent upon wind force and fetch and consequently first-order pressure effects are limited to very low frequencies and shallow depths.

Second-order pressure fluctuations due to interacting wave trains can occur at twice the frequency of the surface waves and are not attenuated with depth. These may be a further source of ambient noise below 10 Hz.

Seismic Noise

It has been shown that the above second-order pressure fluctuations can contribute to the micro-seismic background noise levels measured on land and similarly other relatively continuous seismic background sources may contribute to low frequency ambient noise in water. This is possible because the acoustic mismatch between the sea and its bed is not too great.

Intermittent Sources of Noise

Earthquakes and volcanic action can generate considerable transient acoustic energy which propagates to great distances especially at frequencies below about 100 Hz.

Passing ships frequently overload noise measuring systems for short periods of time.

Local heavy rain or hail raises noise levels, generating a fairly flat spectrum up to around 10 kHz. Noise levels recorded below cracking sea ice show wind dependence up to 10 kHz, but in the absence of cracking some of the quietest recordings have been made under ice.

Biological Sources

Many localized sources of intermittent noise especially in coastal waters can be attributed to marine animals. Biological sounds have a fascinating variety: they range from the 20 Hz, 1 sec long pulses repeating several times a minute from the common finback whale (*Balaenoptera physalus*) which can be heard over many miles, to the weak 75 μsec long echolocating clicks of the rough-toothed porpoise (*Steno bredanensis*) at rates which can exceed 100 per second and containing energy at 200 kHz.

Many fish on close investigation are found to be sound producers in some situations, though generally the source level of an individual is low so that contributions to background noise are restricted to those close to the hydrophone unless they are vocal in large groups to produce a croaker chorus. The sounds of some fish have a tonal quality believed to be due to resonance of the swim-bladder.

Some marine invertebrates can produce impulsive noise by stridulation of teeth, claws, spines or shell, and snapping shrimps can dominate background noise up to 20 kHz near their beds.

The cetaceans, whales and porpoises, are the most vocal group making distinctive sounds subjectively described, for example, as whistles, grunts, chirps and moans.

Bibliography

BECKEN B. A. (1964) *Sonar*, Chapter 1 in *Advances in Hydroscience*, Vol. 1, Ed. Ven Te Chow, New York: Academic Press.

KNUDSEN V. O., ALFORD R. S. and EMLING J. W. (1948) Underwater ambient noise, *J. Mar. Res.* **7**, 410–429.

TAVOLGA W. N. (1964 and 1967) (Ed.) *Marine Bio-Acoustics*, Vols. 1 and 2, Oxford: Pergamon Press.

WENZ G. M. (1962) Acoustic ambient noise in the ocean: spectra and sources, *J. Acoust. Soc. Am.* **34**, 1936–56.

B. S. MCCARTNEY

ASTRO-ARCHAEOLOGY. Astro-archaeology is the study of ancient structures which appear to have astronomical significance. It is an inter-disciplinary subject where physical science yields information on prehistoric, extinct, non-literature cultures.

Lockyer (1901) showed that the avenue at Stonehenge in southern England pointed toward the direction of midsummer sunrise in the second millennium B.C. and used this fact to determine the date of construction. He also claimed that Babylonian and Egyptian temples were built to point to the rising or setting of specific stars. This early work has been criticized because the dates assumed for the construction of the temples do not agree with radio-carbon and historic dates. The Stonehenge data was confirmed by radio-carbon methods, but the fiduciary mark for the sunrise was probably the heelstone which is set to the east side of the avenue. Marquina (1951) showed that the Plataforma Adosada of the pyramid of the Sun at Teotihuacan near Mexico City was aligned to point to the setting Sun at azimuth 291° east of north. This condition occurs on those two days of the year when the Sun passes overhead at noon at Teotihuacan.

Astro-archaeology made significant advances after Hawkins showed substantial astronomical properties in the total structure at Stonehenge. Post-holes, stone-holes, erect stones and archways were aligned to the directions of the rising and setting of the Sun and Moon at particular dates. These were, for the Sun, the days of extreme declination, the midsummer and midwinter solstices. The extremes of the Moon, which recur with a cycle of 18·61 years, were also marked. It was also suggested that the concentric circles, numbering 29, 30 and 56 holes, were used as a computing device. The circles containing 29 and 30 holes could be used to count lunations. An alternate count of 29 and 30 is a good approximation to the synodic period of the Moon, 29·53 days. If a marker were moved three holes each year around the 56-hole circle, it would keep in phase with the nodal cycle of 18·61 years to three parts in 10^3, thus functioning as an analog computer. The swing of the Moon from one extreme declination to another and the occurrence of eclipse seasons could then be predicted. The computer would require resetting or calibrating every two hundred years or so.

Moon and Sun alignments for over a hundred Megalithic sites in Great Britain have been confirmed by Thom. Evidence has been produced to indicate an awareness of the first perturbation term in the inclination of the lunar orbit which amounts to ±9′. Many of the sites show evidence of the marking of the rising and setting of certain stars. Dow determined that the orientation of Teotihuacan may have been chosen to point to the rising of Sirius and the setting of the Pleiades around A.D. 150. Marshack interprets the carved lines on mammoth tusk and prehistoric bone as day-counts through successive lunations and seasons of the year. Some of these artifacts are 25,000 years old.

At the time of writing, astro-archaeology had thrown some light on the mind of prehistoric man. The physical sciences have contributed to anthropology, prehistory and the humanities. It is to be hoped that further research will yield information of value to physical science, such as astronomical constants, periodicities, etc. In this regard the 56-hole circle at Stonehenge led to a discovery, or perhaps rediscovery, of the basic cycle of eclipse seasons in which eclipses recur at the same lunation of the calendar year in intervals of 9, 9, and 10 years.

Bibliography

HAWKINS GERALD S. (1968) Astro-archaeology, *Vistas in Astronomy*, Vol. 10, pp. 45–88, Oxford: Pergamon Press.

G. S. HAWKINS

ASTRODYNAMICS

1. Definition and Historical Roots

Astrodynamics is the engineering or practical application to astronautics of celestial mechanics and allied fields, such as high-altitude aerodynamics, geophysics, and electromagnetic, optimization, observation, navigation, and propulsion, theory. It is not meant to include conventional aerodynamics or booster propulsion theory.

Astrodynamics is the modern outgrowth of theoretical celestial mechanics as applied to astronomical subjects during the some 300 years since Newton discovered the epoch-making law of gravitation. With his newly invented calculus and fundamental laws of motion, Newton proved that Kepler's empirical laws of planetary motion followed as a necessary consequence of the law of gravitation.

In spite of the differences in approach and application, there are striking parallels between the situations encountered in classical celestial mechanics and the current problems in astrodynamics. The preliminary orbit determination procedures and the numerical integration processes have not greatly changed over the past several decades. On the other hand, modern astrodynamics addresses itself to some problems that are entirely new and it is appropriate to identify them. Non-gravitational forces, especially drag and radiation pressure, often assume a considerable significance. Thus, an understanding of the atmospheric density and its heating due to corpuscular and electromagnetic solar radiation becomes vital. The reflective properties of the Moon and planets must also be more precisely defined in order to reckon radiation pressure. Our guide to the accuracy with which we must model a spacecraft's force environment is directly related to the accuracy with which we can define its location. Clearly, as advanced phased-array radars, lasers, and optical sensors (not available to the celestial mechanics of the past) increase in accuracy, our force models must keep pace. On the other hand new missions, both Earth orbital and interplanetary, set requirements for improved sensor accuracy and coverage so that progress becomes never ending.

To the deterministic celestial mechanics theories of the past must be added new concepts of statistics. Not only must one add an extension of Gauss' statistics of observations, but also one must incorporate an entirely new approach to the probabilistic modelling of missions. Here we find evidence of a new philosophy entering astrodynamics. No longer can we view orbit determination, orbit prediction, etc., as independent disciplines that are only loosely coupled to non-astronomical problems. Today astrodynamics becomes but one important facet of the overall systems approach. Because of this the astrodynamicist must be conversant with other engineering and scientific fields in order to place his work in proper perspective and in order to interface properly his approaches with those of others. Fortunately, he now has a mighty device available to him, the digital computer. The computer must not, however, be viewed as an excuse for inefficient computational algorithms, but rather as a tool to open up new vistas of astrodynamics and to show their origin and interrelationship with the classical celestial mechanics of the past.

2. Orbit Determination

One can identify three major divisions of orbit determination: preprocessing of observational data, preliminary orbit determination, and processing of large quantities of data either in a batch or sequential fashion. The pre-processing division involves data credence evaluation, data editing, and data compression. The credence evaluation process involves determining whether the observation either relates to spacecraft or is spurious. The basis for this decision can, for example, be based upon the electronic environment of a radar-type sensor such as the signal-to-noise ratio of the observation, slewing rate, etc. It can also be based upon estimates of whether or not the observations relate to a gravitational orbit. For example, are the observations compatible with the motion of an object moving nearly in one plane and does the time sequence between observations agree with a constant areal velocity of the radius vector to the spacecraft required by Kepler's second law. The data editor acts upon the credence evaluators results plus other *a priori* or *a posteriori* information on the spacecraft and sensor in order to delete data and/or weight it properly. The data editor may also include a decorrelation processor if data correlation is a problem area. The data compressor's role is ordinarily dependent upon the limits imposed by the communication links to the data-processing computer, the capacity of the data processor, and the data rate of the sensor. All three of these pre-processor functions interface with both the preliminary orbit determination and with the batch/sequential data processor.

2.1. Preliminary orbit determination

We shall illustrate preliminary orbit determination by means of the Laplacian and Gibbsian procedures. In the classical Laplacian one makes use of three angular observations, represented by the unit vectors $\mathbf{L}_1$, $\mathbf{L}_2$, $\mathbf{L}_3$, as reckoned from the observer towards the spacecraft. From this minimum set of data, one computes $\dot{\mathbf{L}}_2$ and $\ddot{\mathbf{L}}_2$ by numerical differentiation. That is,

$$\dot{\mathbf{L}}_2 = \frac{-\tau_{23}^2 \mathbf{L}_1 + (\tau_{23}^2 - \tau_{12}^2)\mathbf{L}_2 + \tau_{12}^2 \mathbf{L}_3}{\tau_{12}\tau_{23}\tau_{13}} \quad (1)$$

and

$$\ddot{\mathbf{L}}_2 = 2\left\{\frac{\tau_{23}\mathbf{L}_1 - \tau_{13}\mathbf{L}_2 + \tau_{12}\mathbf{L}_3}{\tau_{12}\tau_{23}\tau_{13}}\right\} \quad (2)$$

where $\tau_{ij} \triangleq k(t_j - t_i)$, k is a gravitational constant ($k = 0.074,365,74$/min for Earth satellites), and t is measured in minutes and refers to the time of the first, second, or third observation. When large quantities of observational data are available it is more sensible to fit the $\dot{\mathbf{L}}_2$ and $\ddot{\mathbf{L}}_2$ to all of the data by least-squares, i.e. to solve

$$\mathbf{L}_i = \mathbf{L}_2 + \dot{\mathbf{L}}_2(\tau_i) + \ddot{\mathbf{L}}_2 \frac{(\tau_i^2)}{2!} + \dddot{\mathbf{L}}_2 \frac{(\tau_i^2)}{3!} + \cdots \quad (3)$$

$$i = 1, 2, \cdots, n, n \gg 3$$

for $\mathbf{L}_2$, $\dot{\mathbf{L}}_2$, and $\ddot{\mathbf{L}}_2$ (where equation (3) has been truncated at some point). Given the components of position X, Y, Z (in equatorial coordinates relative to some specific equator and equinox) and the acceleration $\ddot{X}$, $\ddot{Y}$, $\ddot{Z}$ of the centre of the Earth relative to the observer, one then computes:

$$D\varrho_2 = A' - \frac{B'}{r_2^3} \quad (4)$$

where the range, ϱ_2, from the observer to the spacecraft and the magnitude of the radius vector, r_2, from the Earth's center to the spacecraft are unknowns. The quantities D, A', and B' are known determinants of the form (subscript, 2, indicating middle date or expansion point, have been deleted):

$$D \triangleq \begin{vmatrix} L_x & \dot{L}_x & \ddot{L}_x \\ L_y & \dot{L}_y & \ddot{L}_y \\ L_z & \dot{L}_z & \ddot{L}_z \end{vmatrix}$$

$$A' \triangleq \begin{vmatrix} L_x & \dot{L}_x & \ddot{X} \\ L_y & \dot{L}_y & \ddot{Y} \\ L_z & \dot{L}_z & \ddot{Z} \end{vmatrix} \text{ and}$$

$$B' \triangleq -\mu \begin{vmatrix} L_x & \dot{L}_x & X \\ L_y & \dot{L}_y & Y \\ L_z & \dot{L}_z & Z \end{vmatrix},$$

where μ is the mass constant (usually taken as unity since the spacecraft's mass is negligible relative to that of the Earth). A word of caution should be mentioned at this point. If the spacecraft is following a nearly great circle path over the observer, then D is nearly zero and A' and B' are small. If D is not significantly different from zero, then one can solve equation (4) directly for r_2. If all determinants are significant, then one solves equation (4) simultaneously with:

$$r_2^2 = \varrho_2^2 - 2\varrho_2(\mathbf{L}_2 \cdot \mathbf{R}_2) + R_2^2, \quad (5)$$

where $\mathbf{R}$ is the vector from the observer to the Earth's centre (components X, Y, Z). Having solved for r_2 and ϱ_2 (for example, by means of a Newton-Raphson root-solving technique) one next forms:

$$D\dot{\varrho}_2 = C' - E'/r_2^3 \quad (6)$$

where $\dot{\varrho}_2$ (range-rate at the second date) is unknown. The determinants are defined as follows:

$$C' \triangleq \frac{1}{2}\begin{vmatrix} L_x & \ddot{X} & \ddot{L}_x \\ L_y & \ddot{Y} & \ddot{L}_y \\ L_z & \ddot{Z} & \ddot{L}_z \end{vmatrix}$$

and

$$E' \triangleq \frac{-\mu}{2}\begin{vmatrix} L_x & X & \ddot{L}_x \\ L_y & Y & \ddot{L}_y \\ L_z & Z & \ddot{L}_z \end{vmatrix}.$$

Again one is cautioned against small determinants. The velocity, $\dot{\mathbf{r}}_2$, at the second date is computed from:

$$\dot{\mathbf{r}}_2 = \dot{\varrho}_2 \mathbf{L}_2 + \varrho_2 \dot{\mathbf{L}}_2 - \dot{\mathbf{R}}_2. \quad (7)$$

Having the position and velocity, $\mathbf{r}_2$ and $\dot{\mathbf{r}}_2$, the orbit is determined and one can predict position and velocity at any other time. It should be recognized that in the case of the Laplacian method (as well as in the case of the Gibbsian method to be discussed next) the equations are valid for unperturbed two-body orbits. Such a restriction is ordinarily not important since during the course of the observations (only a few minutes for Earth satellites as observed from a single sensor) perturbations are not very influential.

The Gibbsian procedure is a purely geometrical one, that is one only assumes a planar, conic-section motion, but no particular force law. It makes use of three radius vectors $\mathbf{r}_1$, $\mathbf{r}_2$, $\mathbf{r}_3$, which, like $\mathbf{L}$ and its derivatives may result from a least-squares reduction of many observed $\mathbf{r}$'s. Using the condition that the radius vectors are coplanar, $\mathbf{r}_2 = c_1 \mathbf{r}_1 + c_3 \mathbf{r}_3$, one computes c_1 and c_2 from:

$$c_1 = |\mathbf{r}_2 \times \mathbf{r}_3|/|\mathbf{r}_1 \times \mathbf{r}_3|$$
$$c_3 = |\mathbf{r}_2 \times \mathbf{r}_1|/|\mathbf{r}_1 \times \mathbf{r}_3|. \quad (8)$$

The parameter or semi-latus rectum is computed from:

$$p = (c_1 r_1 + c_3 r_3 - r_2)/(c_1 + c_3 - 1). \quad (9)$$

The unit vector (normal to the orbit plane), $\mathbf{W}$, can be computed from:

$$\mathbf{W} = \mathbf{r}_1 \times \mathbf{r}_3/|\mathbf{r}_1 \times \mathbf{r}_3|. \quad (10)$$

The choice of $\mathbf{r}_1 \times \mathbf{r}_3$ is not, however, unique: if $\mathbf{r}_1$ and $\mathbf{r}_2$ were more nearly at right angles than $\mathbf{r}_1$ and $\mathbf{r}_3$, then $\mathbf{r}_1 \times \mathbf{r}_2$ would more strongly define $\mathbf{W}$ than would $\mathbf{r}_1 \times \mathbf{r}_3$. Here we find an example that points up the need for *understanding* astrodynamic formulas rather than simply memorizing a set of cookbook equations. The eccentricity, e, multiplied by the unit vector $\mathbf{Q}$ (perpendicular to the perifocus direction in the orbit plane) is a combination that can be evaluated from:

$$e\mathbf{Q} = \{\mathbf{r}_3(p - r_1) - \mathbf{r}_1(p - r_3)\}/|\mathbf{r}_1 \times \mathbf{r}_3|. \quad (11)$$

Thus
$$e^2 = (e\mathbf{Q}) \cdot (e\mathbf{Q}), \quad (1\mathbf{Q}\,1 = 1) \quad (12)$$

and the semi major axis can be obtained from:
$$a = p/(1 - e^2).$$

Having e, one can re-enter equation (12) and solve for $\mathbf{Q}$ and then obtain:
$$\mathbf{P} = \mathbf{Q} \times \mathbf{W}. \quad (13)$$

For elliptical ($a > 0$) orbits:
$$\cos E_1 = \frac{(p - r_i)}{ae} + e$$
$$\sin E_i = \mathbf{Q} \cdot \mathbf{r}_i/a\sqrt{1 - e^2}.$$
$$i = 1, 2, 3 \quad (14)$$

from which either E_1, E_2, or E_3 can be obtained. The mean anomaly of epoch (time t_i) is:
$$M_i = E_i - e \sin E_i. \quad (15)$$

We are now possessed of the more classical orbital elements a, e, M_i, $\mathbf{P}$, $\mathbf{Q}$, $\mathbf{W}$, which, like position and velocity, allow for an orbit prediction.

2.2. Batch data processing

Having obtained a preliminary or first-approximation orbit on the basis of minimum data, one next would like to employ the entire collection of data to refine the orbit. The historically tested procedure is called weighted least-squares differential correction. If we define the vector (or column matrix) $\mathbf{x}$ to be the elements or parameters that specify the orbit (e.g. position and velocity or a, e, M_i, $\mathbf{P}$, $\mathbf{Q}$, $\mathbf{W}$ plus, perhaps, coefficients that specify the force environment such as drag coefficient, gravitational potential coefficients, planetary masses, etc.) and $\Delta\mathbf{x}$ its differential improvement, then:

$$\Delta\mathbf{x} = (A^T W A)^{-1} A^T W \Delta\mathbf{z} \quad (16)$$

where A is a matrix whose elements are the partial derivatives of the observations with respect to the elements or parameters of the orbit, p_j ($j = 1, 2, \cdots, n$), e.g. for a set of range-rate observations $\dot{\varrho}_i$ ($i = 1, 2, \cdots, n$) A exhibits the form:

$$A = \begin{bmatrix} (\partial\dot{\varrho}_1/\partial p_1) & \cdots & (\partial\dot{\varrho}_1/\partial p_j) & \cdots & (\partial\dot{\varrho}_1/\partial p_n) \\ \vdots & & \vdots & & \vdots \\ (\partial\dot{\varrho}_i/\partial p_1) & \cdots & (\partial\dot{\varrho}_i/\partial p_j) & \cdots & (\partial\dot{\varrho}_i/\partial p_n) \\ \vdots & & \vdots & & \vdots \\ (\partial\dot{\varrho}_N/\partial p_1) & \cdots & (\partial\dot{\varrho}_N/\partial p_j) & \cdots & (\partial\dot{\varrho}_N/\partial p_n) \end{bmatrix}, \quad (17)$$

W is the diagonal matrix of the weights, w_i^2, of the observations, e.g.

$$W = \begin{bmatrix} w_1^2 & & & 0 \\ & w_2^2 & & \\ & & w_i^2 & \\ 0 & & & w_N^2 \end{bmatrix}, \quad (18)$$

and $\Delta\mathbf{z}$ is a vector on column matrix representing the observed—minus—computed residuals in the observations, e.g.

$$\Delta z_i = \dot{\varrho}_{0i} - \dot{\varrho}_{ci} \quad (19)$$

where $\dot{\varrho}_{0i}$ is the observed range rate (as obtained, say, by a doppler radar) and $\dot{\varrho}_{ci}$ is the computed range rate (as computed or "represented" from an orbit prediction based upon, say, the original preliminary orbital elements). The subscript, i, indicates that the quantities are all obtained at the ith time.

In order to evaluate the partial derivatives in the A matrix one can employ a numerical approximation, the solution of a higher-order adjoint differential equation, or obtain them analytically. The analytical procedure is best for high-precision work. On the other hand, the numerical approximation offers the clearest illustration of partial derivative generation and, therefore, will be given as follows:

$$\frac{\partial \dot{\varrho}_i}{\partial p_j} \cong \frac{\dot{\varrho}_i(p_1, \ldots, p_i + \Delta p_j, \ldots, p_n) - \dot{\varrho}_i(p_1, \ldots, p_j, \ldots, p_n)}{\Delta p_j} \quad (20)$$

where, for example, if p_j stands for the element, a (semi-major axis) then one might choose $\Delta p_j = \Delta a = 0.000,01$ Earth radii (638 m), and so on for the other n parameters.

The diagonal coefficients of the weighting matrix, w_i, can be:

(a) Based upon *a priori* estimates of the comparative weights of data of different types from different sensors (e.g. a sensor manufacturer may give specifications that show a given range-rate radar has a standard deviation of 0.5 m/sec).

(b) Based upon variation as a function of geometry (e.g. at high angular heights, the azimuth direction approaches indeterminancy and should, therefore, be weighted down).

(c) Based upon a time variation (e.g. earlier data may not be as valuable as newer data).

In principle one could employ the *a priori* estimates of the uncertainties in the orbital elements at some epoch (e.g. initial conditions) and predict forward (along with the nominal or most likely orbit) the uncertainties of the position velocity and of the spacecraft. These uncertainties are often termed an error ellipsoid. One could, in turn, translate these position and velocity errors into errors that one *would* expect to find in the spacecraft observations from some given sensor. We could term these the computed variances (or standard deviation—square root of variance) of

the observational data and symbolize them by, for example, $\sigma\dot{\varrho}_{ci}$:
where

$$\sigma\dot{\varrho}_{ci} = \sum_{j=1}^{n} \frac{\partial \dot{\varrho}_i}{\partial \varrho_j} \sigma p_j, \qquad (21)$$

or in terms of the **x** and **z** notation

$$\sigma \mathbf{z}_{ci} = \sum \frac{\partial \mathbf{x}}{\partial \mathbf{x}_i} \frac{\partial z_i}{\partial x} \sigma \mathbf{x}. \qquad (22)$$

Of course, we have some estimate of what the accuracy of the sensor itself is expected to be and these we will symbolize by $\sigma\dot{\varrho}_{0i}$. Thus the weight would be:

$$w_i = \sigma\dot{\varrho}_{ci}/\sigma\dot{\varrho}_{0i}. \qquad (23)$$

2.3. Sequential data processing

In the preceeding section we described a standard procedure for utilizing a large number of independent observational data to improve a smaller number of orbital elements and constants. Here we accomplished the improvement in one fell swoop by collecting and then processing *all* of the data (a data "batch") by successively cycling through equation (16). One could alternatively process the data sequentially as soon as they are collected. This "continuous" procedure attacks the problem from the statistical-estimation approach, and, although identical to the weighted-least-squares-differential-correction inversion when the observational error exhibits a Gaussian distribution and the problem can be linearized, it is often given the name "filtering", and specifically "Kalman filtering".

A physical feeling for "filtering" can be obtained by viewing a very simplified case in which we are making one measurement of a quantity that can be described by but one variable. As an example we might say that we are observing orbital period, P (our **z** in this case), in order to define semi-major axis, a, under the assumption that only the semi-major axis is uncertain (thus, a is **x**). The process involves the differencing of our observed period less our computed period, thereby yielding $\Delta \mathbf{z}$. The new improvement to a can then be expressed as the best estimate of a, at time t_{k+1}, call it $\hat{a}(t_{k+1})$, and is given by:

$$\hat{a}(t_{k+1}) = a(t_{k+1}) + K(t_{k+1}) \Delta P(t_{k+1}). \qquad (24)$$

In equation (24) the $K(t_{k+1})$ can be viewed as a "weighting factor" that properly adds on the new knowledge of a as contained in the residual ΔP. Let us reformulate equation (23) and employ it as the weighting factor, but employ variances instead of standard deviations, i.e.

$$\hat{a}(t_{k+1}) = a(t_{k+1}) + \left(\frac{\sigma^2 P_c(t_{k+1})}{\sigma^2 P_0(t_{k+1})}\right) \left(\frac{\partial a}{\partial P}\right)_{t_{k+1}} \Delta P(t_{k+1}). \qquad (25)$$

Thus $K(t_{k+1})$ has the form:

$$K(t_{k+1}) = \left\{\frac{\sigma^2 P_c(t_{k+1})}{\sigma^2 P_0(t_{k+1})}\right\} \left\{\frac{\partial a}{\partial P}\right\}_{t_{k+1}}. \qquad (26\text{a})$$

If a changes (e.g. due to perturbations) and we wish to refer it to its value at some epoch value, e.g. a_0, then:

$$K(t_{k+1}) = \left\{\frac{\sigma^2 P_c(t_{k+1})}{\sigma^2 P_0(t_{k+1})}\right\} \left\{\frac{\partial a_0}{\partial P}\right\}_{t_{k+1}} \left\{\frac{\partial a}{\partial a_0}\right\}_{t_{k+1}}. \qquad (26\text{b})$$

In the more general **x** and **z** notation we have:

$$\hat{\mathbf{x}}(t_{k+1}) = \mathbf{x}_c(t_{k+1}) + K(t_{k+1}) \Delta \mathbf{z}_{k+1} \qquad (27)$$

where $\mathbf{x}_c(t_{k+1})$ is the best computed estimate of the state *before* the new observational residual ($\Delta \mathbf{z}_{k+1}$) has been introduced. Thus for the one-dimensional problem (in which the semi-major axis, a, is more generally denoted by the parameter p_1):

$$K(t_{k+1}) = \left[\sigma^2 z_c(t_{k+1}) \left\{\frac{\partial p_1}{\partial z_{k+1}}\right\}^2\right] \left[\frac{\partial z_{k+1}}{\partial p_1}\right] [\sigma^2 z_{k+1}]^{-1}. \qquad (28)$$

In essence the operator $K(t_{k+1})$ represents the filter. The conventional notation for K is:

$$K(t_{k+1}) = Q(t_{k+1}) A^T(t_{k+1}) Y^{-1}(t_{k+1}), \qquad (29)$$

where $Q(t_{k+1})$ is like:

$$\sigma^2 z_c(t_{k+1}) \left\{\frac{\partial p_1}{\partial z_{k+1}}\right\}^2 = \sigma^2 p_1(t_{k+1}),$$

i.e. it is the covariance (variance if only one variable—the one-dimensional problem) matrix of the state. $\partial z_{k+1}/\partial p_1$ is like $A^T(t_{k+1})$, and $(\sigma^2 z_{k+1})^{-1}$ is like $Y^{-1}(t_{k+1})$, i.e. the inverse of the covariance (variance if but one observation) matrix of the observations. The actual form of Y is:

$$Y = AQA^T + \overline{\varepsilon^2}, \qquad (30)$$

where $\overline{\varepsilon^2}$ is the covariance matrix of the new datum (or data).

In the foregoing we have represented the state **x** by only one dimension, p_1; in general it is given by $\mathbf{x}(p_1, p_2, ..., p_j, ..., p_n)$. Note also that in the generation and propagation of a statistical covariance matrix, along with the deterministic orbital state, we have a clear-cut example of the new stochastic-processes approach that differentiates modern astrodynamics from classical celestial mechanics.

3. Orbit Prediction

There are two broad categories of orbit prediction. The first involves numerical integration of the basic differential equations of motion and is termed "special perturbations". The second involves an analytical integration (usually in series form) and is termed "general perturbations". In the case of special perturbations we shall specifically consider the Cowell formulation of the differential equations and the second-sum numerical integration. In the case of general perturbations we shall only present a qualitative discussion of the use of computerized algebra since the method of general perturbations offer such a wide variety.

3.1. Special perturbations

The most straightforward manner in which to formulate the differential equations of motion is termed the "Cowell Method". Basically, one establishes the accelerations in an x, y, z inertial coordinate system and presents them as:

$$\frac{d^2x}{d\tau^2} = \ddot{x} + \dot{x}`$$
$$\frac{d^2y}{d\tau^2} = \ddot{y} + \dot{y}` \quad (31)$$
$$\frac{d^2z}{d\tau^2} = \ddot{z} + \dot{z}`$$

where again τ defines a characteristic time scale (such that accelerations are in units of g's), e.g. $\tau \Delta k_e(t - t_0)$ for Earth (geocentric) orbits ($k_e = 0.074{,}365{,}74$ (min)$^{-1}$ if t is measured in minutes), t_0 is some arbitrarily defined epoch (e.g. time of launch), the "double dot" derivatives indicate two-body accelerations and the "dot-grave" derivatives indicate non-two-body or "perturbative" accelerations. One example of such a perturbation might be atmospheric drag. In this case,

$$\dot{x}` = -D_0^2 \mu^* \alpha^* \gamma \sigma \nu \nu_x \quad (32)$$
$$x \to y, z$$

where

$D_0^2 = C_{D_0} A_0 \varrho_0 V_{c0}^2 / 2 g_0 M_0,$

$\mu^* = M_0/M,$

$\alpha^* = A/A_0,$

$\gamma = C_D/C_{D_0}$ = the drag coefficient variation in the near-free-molecule flow (or "transitional") region and in the supersonic region,

$\sigma = \varrho/\varrho_0,$

C_{D_0} = the reference value of the drag coefficient,

A = the projected frontal area of the spacecraft (i.e. projected against a plane normal to the velocity vector ν),

ϱ_0 = the "sea-level" atmospheric density (1·225 kg/m^3),

ϱ = the atmospheric density at any height (kg/m^3) as affected by solar heating, magnetic storms, solar corona, etc.,

ν = the velocity of the spacecraft with respect to the resistive medium (in characteristic units); note that ν has components ν_x, ν_y, ν_z,

V_{c0} = the characteristic speed in m.k.s. units (for the Earth $V_{c0} = 7905$ m/sec),

g_0 = the acceleration of gravity at unit distance (9·780 m/sec^2),

M = the mass of the spacecraft (in kg), and,

C_D = the variable drag coefficient.

Note that D_0^2 can be interpreted as the theoretical acceleration experienced by a circular satellite at sea level moving at unit speed (7905 m/sec). The subscript 0 indicates an initial value of the quantity in question.

Since equation (32) is explicitly a function of velocity (via ν) and implicitly a function of position (via γ and σ) the integration of equation (31) must proceed in a step-by-step fashion. The two-body portions of equations (31) are given by the classical formulas:

$$\ddot{x} = -\mu x/r^3 \quad (33)$$
$$x \to y, z$$

where μ is the same mass factor introduced earlier ($\mu = m_1 + m_2$), where m_1 is the mass of the most ponderous object, e.g. the Earth, and m_2 is the mass of the spacecraft (as measured in units of the most ponderous object) and usually $= 1$. The magnitude of the radius vector, $\mathbf{r}$, is given by

$$\mathbf{r} = +\sqrt{x^2 + y^2 + z^2}.$$

Given initial values of position and velocity, $x, y, z, \dot{x}, \dot{y}, \dot{z}$ at a number of different times at the beginning of the orbit, one constructs a table of sums and differences following the conventional rules:

$$\sum f_{i+1/2} = \sum f_{i-1/2} + f_i,$$
$$\sum{}^2 f_{i+1/2} = \sum{}^2 f_i + \sum f_{i+1/2},$$
$$\delta f_{i+1/2} = f_{i+1} - f_i,$$
$$\delta^2 f_i = \delta f_{i+1/2} - \delta f_{i-1/2},$$
$$\text{etc., } i = 1, 2, 3, \ldots, \quad (34)$$

where f stands for the $d^2x/d\tau^2$, $d^2y/d\tau^2$, and $d^2z/d\tau^2$ given by equations (31), (32), and (33). If the integration interval is h (typically a fraction of a minute for moderately perturbed Earth-satellite orbits), then at each step one computes x and $\dot{x}$ ($\dot{x} \to \dot{y}, \dot{z}$) from a diagonal of the table of differences constructed by means of relationships (34). The "numerical integration" formulas are as follows:

$$x_i = h^2 \Bigg[\sum{}^2 f_i + \frac{1}{12} f_{i-1} + \frac{1}{12} \delta f_{i-3/2} + \frac{19}{240} \delta^2 f_{i-2}$$
$$+ \frac{3}{40} \delta^3 f_{i-5/2} + \frac{863}{12{,}096} \delta^4 f_{i-3} + \frac{275}{4{,}035} \delta^5 f_{i-7/2}$$
$$+ \frac{33{,}953}{518{,}400} \delta^6 f_{i-4} + \frac{8{,}183}{129{,}600} \delta^7 f_{i-9/2} + O(h^9) \Bigg] \quad (35\,\text{a})$$

and

$$\dot{x}_i = h \left[\sum f_{i-1/2} + \frac{1}{2} f_{i-1} + \frac{5}{12} \delta f_{i-3/2} + \delta^2 f_{i-2} \right.$$
$$+ \frac{251}{720} \delta^3 f_{i-5/2} + \frac{95}{288} \delta^4 f_{i-3} + \frac{19{,}087}{60{,}480} \delta^5 f_{i-7/2}$$
$$+ \frac{36{,}799}{120{,}960} \delta^6 f_{i-4} + \frac{1{,}070{,}017}{3{,}628{,}800} \delta^7 f_{i-9/2}$$
$$\left. + \frac{25{,}713}{89{,}600} \delta^8 f_{i-5} + O(h^9) \right] \quad (35\,\text{b})$$
$$x \to y, z.$$

Possessed with x_i and $\dot{x}_i$ ($x \to y, z$) one can compute f_i and then the $\sum f_{i+1/2}$, $\sum^2 f_{i+1}$, $\delta f_{i-1/2}$, etc. That is one can compute the perturbations equation (32) and the two-body term equation (33), add them up in order to obtain equations (31), which are essentially the f's, and then "build" up a new diagonal. The index, i, then shifts by one and equations (35a) and (35b) will yield the next position and velocity. The numerical integration proceeds step by step in this fashion.

3.2. General perturbations

During the 18th century the Moon had provided a useful laboratory for trying out theories and analytical methods of general perturbations supplementing and completing the pioneering geometrical methods of Newton. The brilliant success of Clairant, d'Alembert, Lagrange, Laplace, and Euler in explaining the large deviations of the Moon from an elliptic orbit due to solar attraction laid a solid foundation for solving the varied problems of minor-planet orbits in the 19th century and the astrodynamic problems of the 20th century.

Any modern-day researcher who has involved himself in the development of a general perturbations theory has no doubt been struck by the tedious algebra and trigonometry associated with the work. Not only does this mountain of analysis tend to discourage the researcher, but often he is faced with the spectre of a small error at the beginning of his work propagating itself through his analysis and vitiating his whole effort. A particularly notorious example of a problem requiring extensive algebraic manipulation is the development of a "theory" of the orbital motion of an artificial satellite. This example is particularly relevant because of the importance of such theories in space exploration, and it is reasonable to propose that development of automated manipulation techniques be encouraged by astrodynamicists, even though applications to other fields of science may be even more extensive.

In order to be a useful tool, automated manipulation should appear to the astrodynamicist as a device by which he need only state the particulars of the problem and of the analysis, which leads to its solution, in order to obtain that solution in the desired form. This device amounts to developing a specific problem-oriented algorithmic language and then to implementing its processor. (The "processor" consists of a set of computer algorithms for reducing source-language problem statements (that is, reducing symbols that the analyst puts into the computer, such as, "add", "multiply", "divide", and so forth) to a form that can be executed by the computer's basic logic.) In computer-program design terms this consists of establishing syntactic elements and grammar, and the agreed upon meaning of these elements (semantics). The syntactic elements and grammatical rules are chosen for the computer to express the field of knowledge of the problem areas for which the program is designed. Thus in the development of a special computer language and in the use of sophisticated input and output devices we find a vista of considerable interest opening up in the general perturbations facet of astrodynamics.

Definitions

Apofocus: the apsis on an elliptic orbit farthest from the principal focus or centre of force.

Argument of perifocus: the angular distance measured in the orbit plane from the line of nodes to line of apsides.

Astrodynamics: the engineering or practical application of celestial mechanics and other allied fields such as high-altitude aerodynamics; geophysics; attitude dynamics; and electromagnetic, optimization, observation, navigation, and propulsion theory to the contemporary problems of space vehicles. Astrodynamics is sometimes meant also to include the study of natural objects such as comets, meteorites, and planets.

Eccentricity: the ratio of the radius vector through a point on a conic to the distance from the point to the directrix.

Ephemeris (pl. ephemerides): a table of calculated coordinates of an object with equidistant dates as arguments.

Epoch: arbitrary instant of time for which the elements of an orbit are valid (e.g. initial, injection, or correction time).

Inclination i: angle between orbit plane and reference plane (e.g. the equator is the reference plane for geocentric and the ecliptic for heliocentric orbits).

k_e^{-1} minute: the characteristic time for geocentric orbits, i.e. the time required by hypothetical satellites to move 1 radian in a circular orbit of radius a_e (equatorial Earth's radius); equal to 13·447,052 min.

Line of nodes: the intersection of a reference plane and the orbit plane.

Longitude of the ascending node: the angular distance from the vernal equinox measured eastward in the fundamental plane (ecliptic or equator) to the point of intersection of the orbit plane where the object crosses from south to north.

Mean anomaly: the angle through which an object would move at the uniform average angular speed n, measured from perifocus.

Orientation angles: the classical orientation elements, i.e. the inclination, longitude of the ascending node, and longitude of perifocus.

Osculating element: the orbital elements of the osculating orbit; usually the two-body elements that such an orbit would have if all non-two-body forces were removed.

Parameter: semi-latus rectum $= a(1 - e^2)$.

Perifocus: the point on an orbit nearest the dynamical centre.

Reference orbit: an orbit, usually, but not exclusively, the best two-body orbit available, on the basis of which the perturbations are computed.

Semimajor axis: the distance from the centre of an ellipse to an apsis; one-half the longest diameter; one of the orbital elements.

Space: the expanse (perhaps limitless) that surrounds the celestial bodies of the Universe.

Time of perifocal passage: the time when a space vehicle travelling upon an orbit passes by the nearer apsis or perifocal point. The date of the latest perifocal passage, T' is $T' = T + NP$ where N is the number of revolutions since the initial time of perifocal passage T and P is the orbital period.

Topocentric: referred to the position of the observer on the Earth, as origin.

Trajectory: a segment of an orbit often differentiated from "orbit" in that a "trajectory" does not necessarily imply a complete circuit.

True anomaly: the angle at the focus between the line of apsides and the radius vector measured from perifocus in the direction of the motion.

Two-body orbit: the motion of a body of negligible mass around a centre of attraction.

Vernal equinox: that point of intersection of the ecliptic and celestial equator where the Sun crosses the equator from south to north in its apparent annual motion along the ecliptic.

Differential correction: a method for finding from the observed-minus-computed residuals or small corrections which, when applied to the elements or constants that define the computed motion, will reduce the deviations from the observed motion to a minimum.

Drag coefficient: The total drag coefficient is defined as total drag force acting on an object divided by one-half of the local atmospheric density, the projected frontal area of the object, and the square of the magnitude of the velocity of the object relative to the resistive medium.

f and g series: a series employed in calculating the coordinates at some arbitrary time in terms of the coordinates at the epoch and their first derivatives.

Kalman filter: a mathematical device utilized in optimally processing observational data in a sequential fashion.

Mean elements: approximately the mean values of the osculating elements over a considerable number of revolutions of an object.

Perturbations: deviations from exact reference motion caused by the gravitational attractions of other bodies or by nongravitational forces.

Representation: the computation of the position of a space vehicle given the orbital elements and the time; it usually is utilized to form computed values of position to be subtracted from actually observed position in order to form residuals, which are employed in differential correction.

Residuals (observed minus computed): small differences between the observed and computed coordinates.

Residuals (observed minus ideal): small differences between precomputed ideal observational data and the actual observed data on, for example, an interplanetary voyage.

Variant orbits: computed orbits in which one of the initial conditions (or parameters) is varied slightly from those of the nominal trajectory—such orbits are utilized to compute numerical partial derivatives or to determine the effects of errors in launch conditions.

Weights: a number assigned to a given observational measurement datum that indicates accuracy or freedom from error relative to a whole set of other data. A low weight indicates a poor observation and a high weight indicates a good observation.

Bibliography

BAKER R. M. L., Jr. and MAKEMSON M. W. (1967) *An Introduction to Astrodynamics*, New York: Academic Press.

BAKER R. M. L., Jr. (1967) *Astrodynamics: Applications and Advanced Topics*, New York: Academic Press.

ROY A. E. (1965) *The Foundations of Astrodynamics*, University of Glasgow Press: Glasgow.

BROUWER D. and CLEMENCE G. M. (1961) *Methods of Celestial Mechanics*, New York: Academic Press.

DUBYAGO A. D. (1961) (Translated from the Russian) *The Determination of Orbits*, New York: Macmillan.

R. M. L. BAKER Jr.

AUGER EMISSION SPECTROSCOPY. It is possible to analyze materials non-destructively by carefully examining their secondary-electron energy distributions for evidence of electrons emitted by the Auger effect. If the primary electron-beam energy is sufficient to ionize inner orbits of the atoms, the energy released when an atom rearranges itself to fill the ionized level is characteristic of the atom. This energy, which may appear as an X-ray photon, may instead go to an outer-orbit electron, ejecting it by the radiationless process known as the Auger effect. The secondary-electron energy distribution thus contains peaks localized at energies which serve to identify the atoms producing them.

Although the Auger effect has been understood for many years, and the closely related form of spectroscopy by X-rays has long been a standard analytical technique, Lander (1953) was the first to suggest the use of Auger electrons for material analysis. But in his measurements the large background of secondary emission on which the small Auger peaks are superimposed made it difficult to observe and locate the peaks. Harris (1968a), by electronically differentiating the energy distribution curve, demonstrated that the peaks can be enhanced so that they are easily observed and characterized.

The form of apparatus used by Harris (1968a) for Auger emission spectroscopy is shown in Fig. 1.

perturbation voltage on the slowly swept deflecting voltage. The corresponding a.c. component in the multiplier output current is detected synchronously by a phase-sensitive (phase-coherent) detector which responds only to signals having the same frequency as the perturbation signal. The a.c. output signal, which has an amplitude proportional to the slope or derivative of the actual energy distribution curve, is rectified and filtered, yielding a direct voltage suitable for recording.

The apparatus shown in Fig. 1 is one of several possible forms the equipment might take. Any of the various other forms of electron spectrometers could replace the 127° electrostatic deflection analyzer. One arrangement of particular interest is that devised by Weber and Peria (1967) and used by Palmberg and Rhodin (1968) as shown in Fig. 2, in which a set of spherically-shaped grids is used to sort the electrons according to energy. These grids are part of the

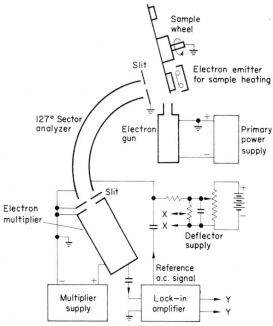

Fig. 1. *Apparatus for secondary electron emission spectroscopy using a deflection type analyzer and means for differentiating the distribution curve.*

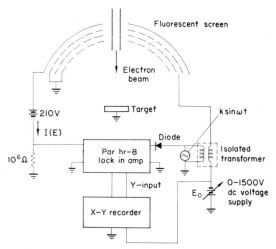

Fig. 2. *Adaptation of LEED apparatus for Auger analysis. The diode generates a second harmonic reference signal for measurement of the second derivative of $I(E)$.*

Specimens to be examined are placed where they can be bombarded by electrons with a few keV energy from the electron gun. A portion of the secondary electron emission from the specimen is received by the energy analyzer, which transmits only those electrons within a determined narrow energy range to an electron multiplier where the current is amplified to easily measured values. The energy range which the analyzer transmits is controlled, with the deflection type analyzer shown, by the deflecting voltage applied to the curved plates. In practice this voltage is slowly varied to provide an energy scan across the entire range to be examined.

The derivative of the distribution curve is obtained electronically by superimposing a small alternating

apparatus for low-energy electron-diffraction (LEED) studies. Electron energies are determined by measuring the current passing the second grid, which is depressed in potential. Since all electrons with energies great enough to overcome this potential barrier are measured, the resulting current as a function of energy is the integral of the energy distribution as measured by a deflection type analyzer. In order to obtain the derivative of the energy distribution current, it is, therefore, necessary to find the second derivative of the transmitted current. This is readily accomplished with a similar perturbation technique to that previously described but in which the phase sensitive detector is made to respond to the second harmonic of the perturbation frequency. Since the ability to identify surface atoms greatly enhances the value of LEED for surface studies, this arrangement is an especially useful one.

The advantage gained by differentiating the distribution curve is illustrated in Fig. 3, which shows the distribution and its derivative as obtained from an oxidized beryllium sample. In the derivative curve

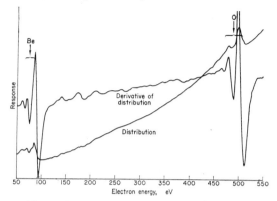

Fig. 3. *Enhancement of Auger spectral features achieved by electronic differentiation of the distribution curve.*

the Auger peaks are much more prominent and reveal structure not easily seen on the distribution curve itself. For this reason we present the Auger spectra routinely in derivative form.

In the case of light elements, having only K and L shell electrons, the Auger lines occur at energies close to those of the corresponding X-rays, differing only by the small portion of its energy which the L-Auger electron loses at the work function barrier in escaping from the material. The heavier elements, having more electron shells, allow a greater variety of transitions, reflected by accordingly more complex spectra. Each element has its own characteristic spectrum or "signature" by which it may be identified. Like X-rays, the emission of Auger electrons involves the deeper lying atomic levels so that the spectra are primarily dependent on the atoms producing them and much less so on the chemical or physical state of those atoms.

Some of the more prominent lines obtained by Auger analysis are illustrated in Fig. 4, where the lines are in their proper energy relation but have arbitrary vertical positions and unrelated amplitudes. Those lines in the top row of the figure are due to transitions of L-shell electrons down to ionized K-shells with the ejection of another L-shell electron. For the elements in the second row the KL–L lines are beyond the observed energy range, but the LM–M lines do conveniently fall in the original range. For even heavier elements the transitions between higher energy shells may be observed, but the chances of overlapping increase. The chart in Fig. 5 shows how the Auger energies depend on atomic number and

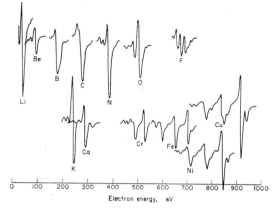

Fig. 4. *Typical differentiated Auger lines of some of the lighter elements. The vertical positions and amplitudes are arbitrary.*

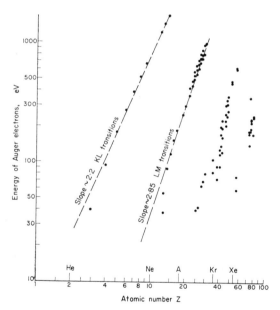

Fig. 5. *Dependence of the energy of the principal Auger lines on atomic number. The energies are those at which the differentiated curve reaches a minimum.*

provides a basis for identification of Auger lines. This chart could not be taken from published values of X-ray energies because for heavier elements the Auger electron energies depart from the X-ray values. One reason for this departure is the increased energy loss of Auger electrons escaping from levels below the valence level; e.g. the KL–L electron from silicon loses about 100 eV (the L shell binding energy) in leaving its parent atom. The second reason is that many Auger lines of the heavier elements do not have counterparts in the X-ray spectrum. Selection rules for the Auger process are less restrictive than those governing the emission of photons.

From the foregoing it would appear that secondary electron emission spectroscopy is simply an alternative to X-ray spectroscopy. In fact, it has features that make it particularly useful in certain circumstances. The most important of these is its sensitivity to the light elements, which are not easily detected by X-ray methods. For elements of low atomic number the X-ray fluorescent yield is very low, and the Auger yield is correspondingly high. Since the rearrangement of an atom ionized in an inner shell involves either the emission of a photon or of an electron, the two processes are complementary. Auger analysis is thus a sensitive indicator for all the light elements except hydrogen, helium and single atoms of lithium, but lithium in the solid state is easily detected.

The other distinctive feature of Auger analysis is its extreme surface sensitivity. Because the Auger electrons must retain most of their original energy to contribute to a spectral line rather than to the background, they cannot afford to be scattered appreciably. Those electrons contributing to an identifiable peak must, therefore, come from a region of the specimen very close to its surface. It has been shown by Palmberg and Rhodin (1968) that the depth of the layer observed by Auger electron spectroscopy is typically only a few atom layers thick. This explains the observation that a thin layer of surface contamination, undetectable by other means, can completely obscure the identity of the surface on which it rests. The method is thus most useful for characterizing surfaces and their constituents, which usually differ appreciably, except for very pure elements, from the constitutions of the bulk portion of the specimens.

Because the radiationless emission of Auger electrons is a relatively high probability process, it is possible to detect very small numbers of atoms on a surface. Weber and Peria have shown that a few percent of a monolayer coverage of alkali atoms (K or Cs) is detectable on a clean silicon surface. This implies a surface density of about 10^{14} atoms/cm^2, and in a typical area of 1 mm^2, detection and identification of about 10^{12} atoms. For portions of single surface layers, where depth effects do not enter into consideration, the peak-to-peak amplitude of the differentiated Auger line is proportional to the quantity of material present on the surface.

The most obvious use of secondary electron emission spectroscopy is the qualitative identification of surface contaminants. In applications to a variety of materials for this purpose it has successfully identified contaminants which could not be labelled by many of the other analytical techniques. Thus residues from organic solvents, from plating baths, and other fabrication processes have been found on a number of conducting and nonconducting surfaces. The method has also been used by Harris (1968b) to detect the surface segregation of minor impurities in metals. For example, sulphur, present in trace amounts in many metals, has a strong tendency to segregate at the surface of the heated metal. When this occurs the apparent sensitivity is greatly enhanced, enabling bulk concentrations of only a few ppm to show up as very prominent features in the Auger spectrum. In a similar fashion the tendency of various elements to dissolve, segregate, or otherwise move through materials may be studied with the help of Auger analysis, making it a valuable tool for metallurgical investigations.

At this writing the technique is a comparatively new one and much remains to be learned about it. In particular it needs to be put on a firm quantitative basis so that amounts of material observed may be measured. For this purpose we need to know not only the efficiencies of excitation of materials by electron bombardment, but also the relative efficiencies of Auger emission for the various transitions in different materials and scattering coefficients of primary and Auger electrons. It is the extreme sensitivity to surface conditions which makes calibration of the method difficult. Without independent knowledge of the surface conditions, by other means as sensitive as Auger analysis, one cannot be certain of what is being measured. The combination of Auger analysis with low-energy electron-diffraction techniques promises to be valuable in this regard, as well as being most fruitful for other forms of surface studies. Even with these uncertainties about the method, secondary electron emission spectroscopy is a useful supplement to the many other analytical methods, particularly for the nondestructive detection of the light elements.

Bibliography

BEARDEN J. A. (1964) *X-ray Wavelengths*. U.S. Atomic Energy Commission, Oak Ridge, Tenn.

BURHOP E. H. S. (1952) *The Auger Effect and Other Radiationless Transitions*, Cambridge, England: University Press.

BUTT D. K. (1961) Auger effect, in *Encyclopaedic Dictionary of Physics* (J. Thewlis Ed.), **1**, 334, Oxford: Pergamon Press.

HARRIS L. A. (1968a) *J. Appl. Phys.* (*U.S.A.*) **39**, 1419.

HARRIS L. A. (1968b) *J. Appl. Phys.* (*U.S.A.*) **39**, 1428.

LANDER J. J. (1953) *Phys. Rev.* **91**, 1382.

LEWIS D. (1962) X-ray fluorescence analysis, in *Encyclopaedic Dictionary of Physics* (J. Thewlis Ed.), **7**, 802, Oxford: Pergamon Press.

NOBLE R. (1962) X-ray spectra, in *Encyclopaedic Dictionary of Physics* (J. Thewlis Ed.), **7**, 835, Oxford: Pergamon Press.

PALEVSKY H. (1961) Beta-ray spectrometers, in *Encyclopaedic Dictionary of Physics* (J. Thewlis Ed.), **1**, 403, Oxford: Pergamon Press.

PALMBERG P. and RHODIN T. N. (1968) *J. Appl. Phys.* (*U.S.A.*) **39**, 2425.

WEBER R. E. and PERIA W. T. (1967) *J. Appl. Phys.* (*U.S.A.*) **38**, 4355.

L. A. HARRIS

AURORAL ELECTRONS

Introduction

The history of auroral observations may be traced back to several hundred years B.C. when descriptions of them appeared in accounts by classical Greek writers. The aurora must have been a rare sight in these regions, because it is ordinarily confined to two, relatively narrow, zones at latitudes 22° from the Earth's magnetic poles. However, during times of intense magnetic storm activity the aurora moves southward and is frequently sighted at Mediterranean latitudes as well as at mid-latitudes of the United States. Sunspot activity and frequency of low-latitude auroral occurrences are assumed to be related because auroral phenomena are more common during the active part of the 11-year solar cycle.

The first scientific investigations of aurora were conducted at the turn of the century by Birkeland, a Norwegian scientist, who used beams of electrons impinging on magnetic spheres to simulate aurorae in the laboratory. Birkeland and Störmer proposed that the aurora originates from rays of energetic electrons that emanate from the Sun and are deflected in the polar atmosphere by the Earth's magnetic field. An exhaustive series of allowed particle trajectories were calculated by Störmer. It was not until 1950, however, that Meinel proved that charged particles descend into the Earth's atmosphere and lose their energy through a series of collisions with atoms and molecules. Part of this energy is converted to excitation energy in atoms and molecules. During the de-excitation processes, light is emitted in the form of visible aurora.

Most of the auroral luminosity observed in the night sky is attributed to the excitation of neutral oxygen atoms, neutral nitrogen molecules, charged nitrogen atoms, and charged nitrogen molecules. Energies of electrons producing aurorae characterized by bright arcs and folded bands were recently measured with rocket-borne instrumentation. The primary electron spectrum was found to be *nearly monoenergetic*, with a sharp peak in the vicinity of 10 keV. This peak suggests that low energy auroral electrons are accelerated by static electric fields, although the origin and the energization mechanism of auroral electrons are unknown. However, significant breakthroughs in our understanding of the auroral process are likely to occur as a result of the discovery of the "monoenergetic" electron spectrum.

Types of Auroral Electron Observations

Many experimental techniques have been employed to investigate the wide variety of observable auroral effects. The early ground-based measurements have now been augmented by rocket, satellite, and balloon measurements.

1. Ground-based measurements

Aurorae were first studied visually, of course, and a large variety of visible auroral forms and structures have been described, photographed, and catalogued in the literature. Auroral forms vary from the sharply defined horizontal arc (which is of the order of 1 km in width and extends hundreds of kilometers in the east–west direction) to the vertically structured curtain, or drapery type, form (which extends over a wide range of altitude). At times the horizontal forms are nearly stationary, but westward movement before local midnight and eastward movement after local midnight usually occur. (Occasionally, pulsating aurorae are observed.) Late in the evening the auroral forms usually break up into a large number of relatively small patches, which move eastward. Auroral motions are usually recorded by all-sky cameras, however, correlated observations using television cinematography and radio and radar soundings are becoming more common.

The primary auroral electron flux may be measured indirectly by studying the vertical patterns of ionization and optical emission produced during its interaction with the atmosphere. The most common visible lines occurring in auroral optical emissions are the 5577 Å wavelength (referred to as the green oxygen line) and the 3914 Å wavelength (which results from de-excitation of the nitrogen ion). Prominent auroral spectral lines are readily detected with ground photometers equipped with optical filters that transmit only the desired wavelength.

The primary electron energy flux occurring during the aurora can be crudely determined if the efficiency of conversion to photons of electron energy deposited in a columnar mass of atmosphere is known. Auroral luminosity is measured with ground-based photometers in Rayleighs (R) (i.e. 10^6 photons/cm^2 sec produced throughout the volume of a 1 cm^2 vertical column of air extending from the top of the atmosphere). An electron energy flux of 3 ergs/cm^2 sec precipitating into the atmosphere will produce roughly 1 kR (10^3 R) of 3914 Å radiation, 1 kR of 5577 Å radiation, and an equilibrium electron density of about 10^5 electrons/cm^3. Thus one can indirectly determine the approximate primary auroral electron flux from ground-photometer measurements.

The visible aurorae are classified according to a luminosity range of 1–1000 kR. A Type I aurora produces a luminosity of 1–10 kR. The brightest intensity (Type IV) is roughly equivalent to the luminosity of the full moon ($\sim$1000 kR).

Indirect information concerning the primary electron spectrum may be obtained by ground-photometer measurements of auroral optical emission. A red, 6300 Å, oxygen line is occasionally observed that appears to originate from about 300 km altitude. The red line requires much less excitation energy than the green line, and it is believed that such high altitude red aurorae are caused by very soft electron spectra.

Ground instrumentation may also provide an approximate vertical profile of auroral ionization produced in the atmosphere. The radioionosonde measures the time from ground transmission of a radio pulse to ground reception of its echo, reflected back from the ionosphere in a vertical direction. The auroral ionization produces absorption of the signal, which is inversely proportional to its frequency squared. A radiosonde is obtained by measuring the strength of the ionospheric echo using different radio frequencies. Another useful ground instrument is the riometer, which detects celestial radio signals and measures their relative intensities at different frequencies. The opacity of the ionosphere to these signals is correlated with changes in atmospheric ionization produced by precipitating charged particles. Radar signals are used to obtain information about irregularities in ionospheric ionization by measuring their reflection at normal incidence to the Earth's magnetic field lines. Data obtained using these various ground-based techniques indicate that the most common homogeneous auroral arc occurs at a height of about 105 km (there is usually a pronounced peak in the intensity of the vertical ionization profile at this altitude). When the vertical ionization profile is broad, and its peak is less pronounced at low altitudes, the curtain, or drapery type, aurora often occurs.

Once the vertical ionization profile has been determined, a crude primary electron spectrum may be obtained. The usual presence of a peak in the vertical ionization density profile at about 100 km altitude is consistent with the range of incident electrons having an energy of about 10 keV.

One of the most commonly used instruments for measuring auroral activity is the ground-based magnetometer. The pattern of ionospheric currents is usually affected during periods of strong auroral activity, and changes in overhead current loops are accompanied by the observation of magnetic field variations on the ground. Magnetic bays, which are positive or negative variations in the Earth's magnetic field, frequently occur in the auroral zone. These variations are usually a few hundred gammas in magnitude (1 gamma $= 10^{-5}$ gauss), although at times of unusually strong activity, changes of the order of a thousand gammas may occur.

2. Rocket measurements

With the advent of the rocket, more sophisticated techniques for investigation of the auroral electron spectra became available. The primary auroral electron spectrum could be measured directly with rocket-borne instruments.

The earliest rocket experiments were performed with primitive instrumentation, by modern standards, which had insufficient energy resolution and sensitivity for measurement of the low energy region of the electron spectrum. (Low energy electrons are known to be the main cause of the common visible aurora.) It was recently discovered (through measurements with greatly improved instrumentation) that many aurorae (particularly the homogeneous arc forms often observed in the evening in the auroral zone) are caused by nearly monoenergetic electrons in the 10 keV region. The electron spectrum appears to harden during and after break-up of the visible aurora, i.e. the monoenergetic spectrum shifts to higher energy and becomes more complex.

The pitch angle distribution of auroral electrons (i.e. the angle between their plane of gyration and the local magnetic line of force) is not yet well known. Knowledge of this distribution would be valuable in studies of auroral electron source and loss mechanisms for the following reasons: If the electron's magnetic moment is assumed to be invariant (which is true if the Earth's magnetic field varies slowly during the gyration period of the electron), then

$$\frac{\sin^2 \alpha}{B} = \frac{1}{B_m}$$

where α is the electron pitch angle, or angle between the particle trajectory and geomagnetic vector B, and B_m is the value of the Earth's magnetic field at the electron mirror point. From this relationship it follows that, when $\alpha = 90°$, the electrons are at their mirror point and are orbiting in a plane normal to the Earth's magnetic field. If the mirror point is beneath the region of visible aurora (about 100 km altitude) the electrons must give up their energy to atmospheric interactions, and, thus, they lose their ability to return.

The pitch angle, α_p, defines a loss cone that is a function of B, or a given location on an auroral zone field line.

$$\sin^2 \alpha_p = \frac{B}{B_{100}}$$

where B_{100} is the value of the auroral zone field line at 100 km altitude. Electrons having a pitch angle less than α_p precipitate into the auroral zone (auroral precipitation) and become permanently lost.

Although pitch angle information for the low energy auroral electrons is sparse, data are available which indicate that electrons give up most of their energy in the auroral production process. It is likely that conjugate auroral events (aurorae that occur simultaneously at opposite ends—i.e. in the northern and in the southern hemisphere—of the same magnetic field lines) are the exception rather than the rule. It appears more likely that the auroral electrons are energized on "open field lines" that extend back to the tail of the magnetosphere and perhaps connect to magnetic field lines emanating from the Sun. This important question remains unanswered for the present [see also Hargreaves (1967)].

The presence of high energy electrons in the auroral zone was first detected by rocket experiments in the late 1950's. The role of the high energy electron fluxes

(>20 keV) is apparently independent of that of the low energy electrons that produce visible aurorae. The high energy auroral zone electron fluxes generally exhibit distinctively characteristic behavior (with relatively fast time variations in intensities, energy spectra, and pitch angle distributions), at the same time that the low energy electron flux appears to be relatively steady and unchanging.

Recent observations of large anomalous return fluxes of auroral zone electrons indicate that local ionospheric processes that accelerate electrons out of the atmosphere are occasionally present.

3. Satellite measurements

Although rocket experiments appear to be the most productive means of obtaining direct measurements of auroral electrons, many interesting and useful data have been obtained with satellite instrumentation. However, the large orbital velocity of the satellite and the narrow width of the auroral forms (~ 1 km) through which it passes result in observation times that are less (<1 sec) than those typical of rocket experiments.

Auroral electron spectra were recently measured with instruments on board a polar-orbiting satellite (~ 400 km altitude). Two zones of daytime auroral precipitation were detected in the northern hemisphere. While a preponderance of low energy electrons (<21 keV) were measured in the higher latitude zone ($\sim 75°$ magnetic latitude), considerably more high energy electrons were found in the lower latitude zone ($\sim 69°$ magnetic latitude). Only one zone of electron precipitation was present on the night side of the Earth (about 65° latitude). On the average, about 80% of the precipitating particles were electrons, although at times large fluxes of protons were observed in the auroral zones.

Satellites orbiting at higher altitudes through the outer radiation belts have carried experiments that have yielded data for low energy electrons. Information on electron pitch angle distribution obtained from these experiments revealed time shifts of the outer belt trapping boundary. It was found that, in general, the high latitude trapping boundary is lower at night ($\sim 69°$) than during the day ($\sim 75°$). Unfortunately, it is still not clear whether the zone of auroral precipitation falls inside, outside, or on the boundary of trapped radiation.

4. Balloon measurements

An advantage of the balloon technique over other high altitude techniques is that under good conditions balloons can stay aloft for 20 hours or more at practically the same location. This facilitates the study of temporal variations of auroral activity. Unfortunately, at balloon altitudes (typically about 30 km) observations of electrons are indirect—i.e. they are obtained through detection of X-rays (electron bremsstrahlung), which diffuse down to the altitude of observation from about 100 km altitude. These X-rays (produced by electrons >30 keV) must therefore penetrate through about 10 g/cm² atmosphere before reaching the detectors. Considerable distortion of the original X-ray spectrum results from photo-electric interactions with the atmosphere. The original angular distribution is blurred (for the most part) by Compton scattering in the atmosphere. Thus, the primary electron energy spectrum and angular distribution cannot be determined with a high degree of accuracy by this procedure.

The balloon experiments have uncovered many new and puzzling auroral phenomena. Auroral X-ray events have been catalogued into many different classifications according to the nature of their periodicities. X-ray activities have been observed to vary from relatively slow changing events to rapid short periodic "microbursts" of about 0.2 μsec duration. The existence of conjugate X-ray activity has been suggested by data obtained from coordinated balloon experiments in the northern and southern hemispheres.

However, auroral X-rays and visible aurorae do not appear to be correlated, and it is likely that the electrons responsible for X-rays do not represent the high energy tail of the auroral electron spectrum. Different precipitation mechanisms must be responsible for the high and low energy portions of the auroral electron spectrum.

5. Comments on the origin and energization of auroral electrons

A correlation between solar and auroral activity has been established. It is therefore tempting to assume that the solar wind is the source of auroral electrons and that the energy required to supply the aurora is provided by the Sun. There is indeed sufficient energy ($\sim 10^{20}$ ergs/sec) in the solar wind at the boundary of the Earth's magnetosphere to replenish the 10^{17} to 10^{18} ergs/sec energy expended by the total auroral process.

A close relationship between loss of outer belt particles and precipitation of auroral zone particles has been observed with correlated satellite and ground-based experiments. Results of these experiments indicate that the auroral zone latitude is nearly contiguous with the latitude of the trapping boundary of outer belt particles. It would appear from this evidence that aurorae are caused by "dumping" of electrons from the large reservoir of outer belt trapped particles. However, the average, world-wide, precipitated auroral zone electron energy flux (4×10^{17} ergs/sec) is several orders of magnitude larger than the flux that would empty the outer belt in a reasonable length of time.

Correlations have been established between precipitating electron fluxes (detected by satellite instruments) and 3914 Å light emitted by a visible aurora directly beneath the satellite (i.e. on the same magnetic field lines). Measurement of the auroral luminosity shows an increase that is clearly correlated in time with the increase of precipitating particles observed by the satellite instruments. This is experimental evidence of a relationship between the loss of particles from the outer belt and the occurrence of aurorae. But the picture is clouded by the observation that the trapped outer belt electron flux increases at the same time. It would appear that an unspecified causative agent increases both the trapped and the precipitating flux of electrons simultaneously.

It is unlikely that trapped radiation belt fluxes provide the energy and/or the source of primary auroral electron fluxes. However, the energy stored in the outer radiation belt may account for the production of higher energy electron fluxes and related auroral X-ray activities.

We may thus eliminate the Van Allen or trapped radiation belt and return to a consideration of the Sun as a possible source of energy and raw material for auroral production. An ample energy supply has been shown to be present in the solar wind. But the process by which low energy solar wind electrons penetrate into the Earth's magnetosphere, and then undergo many orders of magnitude energization to attain the keV energies of auroral electrons, remains to be explained. The presence of a high degree of variability in the auroral electron fluxes ($\sim 10^6$) during a period of relatively constant solar wind activity also requires clarification.

Although our understanding of auroral activity and the nature of auroral electrons has made rapid progress in recent years, the fundamental problems of the origin of primary auroral electrons and the means by which they acquire their energy remain unsolved.

Bibliography

HARGREAVES J. K. (1967) Space physics, in *Encyclopaedic Dictionary of Physics* (J. Thewlis Ed.), Suppl. vol. 2, 230, Oxford: Pergamon Press.

McCORMAC BILLY M. (Ed.) (1967) *Aurora and Airglow*, New York: Reinhold Publishing Corporation.

McCORMAC BILLY M. (Ed.) (1966) *Radiation Trapped in the Earth's Magnetic Field*, New York: Gordon and Breach.

WALT MARTIN (Ed.) (1964) *Auroral Phenomena: Experiments and Theory*, Stanford: Stanford University Press.

R. D. ALBERT

BIOLOGICAL PRESSURE TRANSENSORS. Continuous records of intracranial, intra-ocular and other body pressures can be obtained clinically from human patients and experimentally from unanesthetized, physiologically normal and psychologically responsive laboratory animals by a novel technique of passive telemetry involving microminiature transensors (Fig. 1) (Collins, 1967b). To circumvent some of the

Fig. 1. Photograph of a 5 mm diameter glass pressure transensor with its pickup induction coil. The pressure sensitive diaphragms are 50 μ thick glass membranes.

limitations of manometry and tonometry in the measurement of body pressures, a passive transensor has been developed in order of magnitude smaller than the active, battery operated endoradiosonde (Mackay, 1961) devices previously available. The transensor's ultimate simplicity, consisting only of a pressure-tuned circuit has permitted functioning units as small as 3 μl to be made. These miniature, batteryless, pressure-sensitive radio transensors have been surgically implanted intracranially in human patients, within the eyes, brain and vascular system of small laboratory animals to absorb energy from an active detector outside of the subject at a frequency dependent upon the pressure within the implanted organ. Passive operation allows theoretically unlimited life; implanted transensors have been functioning satisfactorily *in vivo* for over one year. The capsules are biologically inert. Of over one hundred implantations made, none has shown signs of tissue reaction or other visible irritation related to the implant. This atraumatic technique has revealed new phenomena in conscious animals not previously observable. Millisecond response, 0·1 mm Hg pressure resolution, 0·015 μl per mm Hg compliance, and no fluid loss during measurement have permitted continuous, sensitive pressure recordings of both fast transients and slow variations to be made in the eye, brain and artery under physiological conditions.

Over the past century two methods have been employed for measuring biological pressures, manometry and tonometry. The radio pill technique circumvents some of the limitations of the previous methods (i.e. single pressure samples and cannula clotting). With this method nothing pierces or even touches the organ being monitored. Once implanted, the organ need not be touched further. For intra-ocular measurements head and eye movements need not be restricted by force or anesthesia, hence continuous pressure records can be obtained from animals in their alert and psychologically responsive state. This technique has made possible investigations heretofore impracticable.

To monitor pressure, the size of a plastic or glass encapsulated bubble is measured by a novel physical principle. A pair of parallel, coaxial, Archimedean spiral coils constitutes a high Q distributed resonant circuit whose frequency varies sensitively with relative coil spacing (Fig. 2). As pressure acting on the capsule

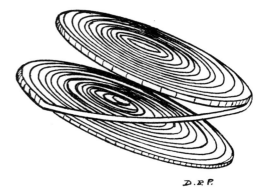

Fig. 2. A pair of flat spiral coils constitutes a distributed resonant circuit. Both stray capacity and mutual inductance are altered, making the resonant frequency vary sensitively with coil spacing.

forces these coils closer together (Fig. 3), both the stray capacity between the spirals and their mutual inductance increase, which lowers the resonant frequency of the configuration. This resonant frequency is sensed through intervening tissue by virtue of its absorption of electromagnetic energy. An external inductively coupled grid dip oscillator has proved to be the simplest method for detecting the passive bubble tonometer at workable distances (Fig. 4). By rapidly sweeping the frequency of the detecting oscillator, the absorption characteristics of the transensor can be displayed on a synchronized oscilloscope (Fig. 5). For automatic transensor frequency tracking, a tunnel diode switch detects the zero crossing of the differentiated absorption curve (Fig. 6). The time to this zero crossing from the beginning of the sweep is demodulated to yield an analog signal representing the transensor resonant frequency. Continuous pressure records are routinely obtained from conscious animals by this method. Calibration is linear (Fig. 7).

The implantable transensor is encased in a short, thinwalled glass tube with mylar or thin glass (Collins, 1967a) diaphragms stretched tightly over its ends. Transensors have been built with diameters of 2 to 6 mm and thicknesses of 1 to 2 mm. The size of a passive transensor is determined by its required working distance from the detector. The absorption signal varies exponentially with distance, and ranges up to 6 spiral coil diameters are possible.

$$V = 3e^{-0.2x/a} \text{ volts}$$

where
V = voltage at receiver,
x = range to transensor,
a = mean radius of transensor.

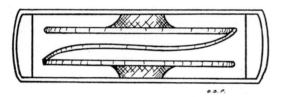

Fig. 3. Cross-section showing the plastic bubble transensor consisting of a small hermetically sealed pillbox with tightly stretched drumhead diaphragms bearing on the spiral coils within. Pressure forces the coils closer together, thus lowering the pill frequency.

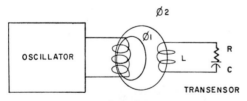

Fig. 4. Principle of the resonant frequency detection of a passive transensor. The common-linkage flux, $\varnothing_2$, from the oscillating grid-dip detector coil may constitute as much as 1 per cent of the total. This electromagnetic link is the basis of passive telemetry.

Fig. 5. The resonant absorption peak of a passive transensor appears as above on the face of the panoramic absorption analyzer oscilloscope.

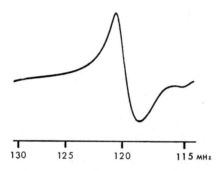

Fig. 6. The peak is more readily located from the zero crossing of the first time derivative below. The inverted frequency scale indicates increasing pressure to the right. The absorption of the transensor shown on this expanded frequency scale indicates a Q of approximately 80.

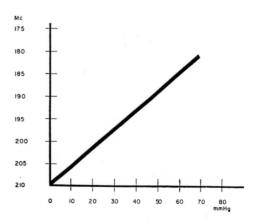

Fig. 7. Calibration curve of a typical plastic pressure transensor at equilibrium immersed in saline. Note 2 per cent linearity covers the physiological range of intra-ocular pressure and extends through zero pressure.

The resonant frequency of these transensors has been chosen in the region of 100 MHz. For a typical 5 mm diameter unit,

$$f_0 = 168 \times 10^6 \, s^{1/2}$$

where f_0 = resonant frequency,
and s = spiral coil separation in mm.

Pressure sensitivities in excess of 1 MHz per mm Hg have been achieved. The mechanical resonance of a typical unit is approximately 1000 cps.

Lord Rayleigh (1930) conducted long-term experiments indicating the required dimensional stability of glass for the purpose of fabricating radio transensors. Glass transensors are sealed in an airtight bomb, fired in a furnace and subsequently annealed to yield an impervious vitreous capsule well tolerated by the eye and other body tissues.

Efforts in this field have been motivated by the increasing incidence of glaucoma leading to blindness and of hydrocephalus and brain damage occurring after severe impact. In glaucoma, vision is usually damaged by the pressure build-up in the eye resulting from an obstruction to aqueous fluid flow out of the eye. A number of methods are available for the management of a glaucoma and hydrocephalus, including drugs and surgery. It is the great hope that better understanding of the mechanisms influencing physiological pressures may lead to additional methods of treatment.

The transensor has been used to study some of the problems relating to glaucoma, hydrocephalus and other diseases with an etiology of pressure change within an organ. Since the initial studies of spinal fluid dynamics in monkeys (Collins, 1966; Collins and Olsen, in preparation; Olsen et al., 1967) and in man (Collins, 1967a; Lindstrom, in preparation) with implanted radio pills, clinical investigation of hydrocephalus has been undertaken with pressure transensors (Atkinson, in press). This self-contained miniature pressure sensor has revealed physiological relationships in conscious animals not previously observable. When recorded over many hours in alert, responsive animals, the salient feature of intra-ocular pressure is its extreme lability. One of the most exciting aspects of this work is a new phenomenon revealed by the radio pill. In the conscious rabbit a slight sensory stimulus can evoke a characteristic intra-ocular pressure rise as great as 10 mm Hg (i.e. rising to 150 per cent of normal pressure). It has been determined that this event is mediated by the sympathetic nervous system during an arousal response, and it appears to be caused by contraction of orbital smooth muscle compressing the globe (Collins, 1967d). Apparently, psychological or nervous excitation can result in transient elevation of both the intra-ocular and general blood pressure. For example, an evoked pressure response can be observed when a male rabbit first sees a female rabbit. This observation may explain the brief pressure plateaus or other transient effects observed in human eyes when patients are distracted during tonography.

Variations of a few millimeters of mercury can be measured in the intra-ocular pressure during eye movements, presumably due to unbalanced efforts of the oculorotary muscles. It has also been found that these striated oculorotary muscles can produce a large intra-ocular pressure change which is proportional to

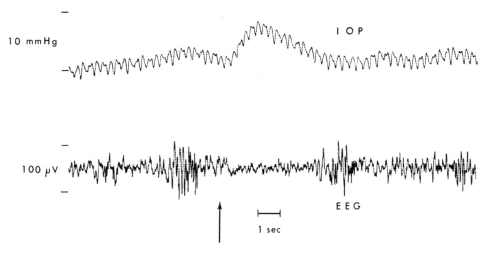

Fig. 8. Sensory evoked pressure response. A typical sensory evoked intra-ocular pressure response recorded with an implanted transensor in a conscious New Zealand rabbit. These characteristic transient variations can be observed following auditory, photic, olfactory, tactile, or thermal stimuli. In some cases it is possible to elicit this response by merely moving a finger at a distance of 1 m from the animal. The evoked pressure response (top record) occurs concomitantly with a general arousal response indicated by a characteristic decrease in amplitude and frequency of the EEG (bottom record). An audible stimulus occurred at the arrow.

the square of the developed muscle tension, apparently by distorting or creasing the eyeball (Collins, 1967c; Collins and Bach-y-Rita, 1967). With the new transensor (bubble tonometer) technique, very large transient intra-ocular pressure increases (exceeding 70 mm Hg, or four times the normal intra-ocular pressure) have occasionally been seen accompanying natural blinks in rabbits. Others have suggested (and it is theoretically possible) that these large pressure fluctuations could be associated with trabecular flushing. Pressure undershoots of some 5 mm Hg have been noted following such transients. The characteristics of the recovery to base line intra-ocular pressure following these undershoots implies the possibility of measuring ocular blood flow and vascular outflow resistance by a technique which might be called "impulse tonography".

Using the radio pill, the quantitative role of blood pressure excursions on the intra-ocular pressure have been determined. The hemodynamic influence on intra-ocular pressure can be expressed in terms of a simple analog model with a straightforward mathematical transfer function (Collins and Loeb, 1967). These findings also bear on the role of acute hypertension to glaucoma, as well as on the psychologically evoked pressure response.

This new technique has demonstrated utility in following sensory and other neurogenically induced pressure phenomena; assessing drug effects in conscious versus anesthetized animals; and monitoring intra-ocular, intracranial and blood pressures under normal physiological conditions. Wherever there is a need to measure pressure close to the body surface, the passive spiral transensor offers a sensitive, wireless method in a small package.

Research continues towards a better biophysical understanding of biological pressure mechanisms. Modeling techniques are being employed in attempts to characterize the relationship of intra-ocular and intracranial pressure to other physiological variables. The roles of nervous, vascular, muscular, and biochemical factors as effectors of physiological pressures are being investigated by use of the "radio pill" transensor.[†]

Definition

Transensor—a passive telemetering transducer (i.e. using no batteries or external wires) for measuring physical quantities (such as pressure, temperature, displacement, force, etc.) at short ranges from an active (powered) detector. Currently, transensors typically consist of a pair of flat spiral coils contained within a small glass or plastic pillbox whose diaphragms produce relative coil displacement when force or pressure is applied. The resulting resonant frequency change is detectable at a distance by means of a panoramic absorption analyzer, and continuous records of the measured variable can thus be made.

Bibliography

ATKINSON J. R., SHURTLEFF D. B. and FOLTZ E. L. Radio telemetry for the measurement of intracranial pressure (in press).

COLLINS C. C. (1966) Microminiature endoradiosonde for intra-ocular implantation, *Proc. Annual Conference on Engineering in Medicine and Biology* **8**, 171.

COLLINS C. C. (1967a) Passive telemetry with glass transensors, *Proc. National Telemetry Conference*, pp. 146–151.

COLLINS C. C. (1967b) Miniature passive pressure transensor for implanting in the eye, *IEEE Trans. Bio-Med. Engineering*, **BME-14**, 74–83.

COLLINS C. C. and LOEB D. R. (1967) Hemodynamic influence on intraocular pressure, *Invest. Ophthal.* **6**, 666 (Abstract).

COLLINS C. C. and BACH-Y-RITA P. (1967) Drug induced muscle co-contracture as a tool for intraocular pressure studies, *Proc. Western Pharmacology Soc.*, **10**, 16–18.

COLLINS C. C., BACH-Y-RITA P. and LOEB D. R. (1967c) Intraocular pressure variation with oculorotary muscle tension, *Amer. J. Physiol.*, **213**, 1039–1043.

COLLINS C. C. (1967d) Evoked pressure responses in the rabbit eye, *Science*, **155**, 106–108.

LINDSTROM P., COLLINS C. C. and LOUGHBOROUGH W. B. Extracranial measurement of intracranial pressure (in preparation).

LORD RAYLEIGH (1930) *Nature* **125**, 311.

MACKAY R. S. (1961) Radio telemetering from within the body, *Science*, **134**, 1196–1202.

OLSEN E. R., COLLINS C. C., LOUGHBOROUGH W. B., RICHARDS V., ADAMS J. E. and PINTO D. W. (1967) Intracranial pressure measurement with a miniature passive implanted pressure transensor, *Amer. J. Surgery*, **113**, 727–729.

C. C. COLLINS

BORDONI PEAK. The Bordoni Peak is a maximum in the internal friction of plastically deformed metals, especially face-centred cubic metals, which is observed when the friction is measured as a function of temperature at a constant frequency. In the frequency range $10^3 - 10^6$ cycles/sec it appears at about one-third of the Debye temperature.

Bibliography

NIBLETT D. H. (1966) *Physical Acoustics*, ed. Mason, W. P., New York: Academic Press. III A 77–121.

F. R. N. NABARRO

BOSONS, INTERMEDIATE. The hypothesis that all presently known weak interactions are actually second-order processes generated through the emission

---

[†] This work was supported by Public Health Service Grants NB 04669, NB 06038 and General Research Support Grant 5 SOL FR-05566.

and absorption of an intermediate boson field has been considered by many physicists since the early days of β-decay theory (Lee and Wu, 1966). The original Fermi theory of β-decay of the weak interaction was made in analogy with the second-order electromagnetic interactions. The role of an intermediate boson (or "W meson") in the weak interactions would, therefore, be that of photons in the electromagnetic interactions, and of pions and other mesons in the strong interactions. One of the main differences between the weak interactions and the electromagnetic interactions lies in their respective range. Unlike the Coulomb field, the interaction in Fermi's β-decay theory is short range. Therefore, the intermediate boson is massive, unlike the photon which has zero mass.

This attractive idea suffered a serious setback in early 1950. As a result of several incorrect experimental measurements, the β-decay coupling was determined to be scalar and tensor. A tensor field cannot be transmitted by either spin 1 or spin 2 field. However, these experiments were redone and, since 1957, β-decay interaction is known to be vector and axial vector. A vector and axial-vector field can be transmitted by a spin 1 boson.

While the intermediate boson is, as yet, undiscovered, its possible existence seems the most promising way of unifying the theory of weak interactions.

1. Basic Properties of W

(a) *Charge.* If the hypothesis of the intermediate boson is valid, the W is coupled to an electron and a neutrino. Since there are two electrons, i.e. negatron and positron, there exist, at least, two charged bosons: W^+ and W^-.

(b) *Spin.* Since the weak interaction is vector and axial-vector, the spin of the W is 1.

(c) *Interaction form and coupling constant.* The interaction between the $W^\pm$ field and the known leptons can be written as

$$\mathscr{L}(W^\pm) = f j_\lambda W_\lambda^* + \text{h.c.}$$

where W_λ is the operator describing the charged W field. The lepton current j_λ is given by

$$j_\lambda(x) = i \sum_{l=e,\mu} \psi_l^+ \gamma_4 \gamma_\lambda (1 + \gamma_5) \psi_{\nu_l}.$$

The coupling constant f can be determined from the β-decay coupling constant g_V

$$\frac{f^2}{m_W^2} = \frac{g_V}{\sqrt{2}}.$$

Since $g_V = 1 \cdot 024 \times 10^{-5}/m^2 p$, we have

$$f^2 = 0 \cdot 124 \, (m_W/m_p) \times 10^{-5}$$

where m_W and m_p are masses of $W^\pm$ and proton, respectively.

(d) *Decays.* The decay rate of $W^\pm \to l^\pm + \nu_l(\bar{\nu}_l)$ can be computed:

Rate $(W^- \to l^- + \bar{\nu}_l) = $ Rate $(W^+ \to l^+ + \nu_l)$
$$= g_V m_W^3 (6\pi/\sqrt{2})^{-1} > 5 \times 10^{18} \text{sec}^{-1}.$$

Thus, the $W^\pm$ meson is not observable in the laboratory due to its short liefetime. Detection of $W^\pm$ depends on observing decay products of W in the laboratory. W decays into the following modes

$$W^\pm \to l^\pm + \nu_l(\bar{\nu}_l)$$
$$W^\pm \to \text{pions}$$
$$W^\pm \to K^\pm + \text{pions}.$$

The relative decay rate to these final states is not known.

(e) *Production.* There are many ways $W^\pm$ can be produced in the laboratory. The following reactions have been seriously considered by experimentalists.

(i) $\nu_\mu + \text{Nucleus}\,(Z, A) \to W^+ + \mu^- + \text{Nuclear fragments}\,(Z, A)$.

(ii) $\pi^\pm + p \to W^\pm p$

(iii) $\bar{p} + p \to W^\pm \pi^\pm$

(iv) $p + p \to W^+ + d$

(v) $p + Z \to W^\pm + Z''$.

In the rest of this article we will discuss the experiments to detect $W^\pm$. To date only reactions (i) and (v) have been investigated.

II. Production of $W^\pm$ by Neutrinos

The search for $W^\pm$ was carried out in high energy neutrino facilities based on an external proton beam both at the CERN and Brookhaven 30-GeV proton accelerators (Bernardini *et al.*, 1965; Burns *et al.*, 1965b). We will describe the Brookhaven experiment. The CERN experiment is very similar to the one performed at Brookhaven. Figure 1 shows the experimental arrangement at Brookhaven. Neutrinos are generated by the decay-in-flight of pions and kaons produced in a 1·5 cm-diameter × 30-cm Be target by 30 BeV/c protons near zero degrees. The 22 m flight path was followed by a 27 m iron shield. All known particles other than neutrinos are stopped in the iron shield. Only the neutrino gets through the shield because it interacts very weakly with matter. The neutrino detector is placed behind the shield. This was an aluminium spark chamber composed of 184 plates, each 6 ft × 6 ft × 1/4 in. and weighing a total of 12 tons. This "production" chamber was followed by a range chamber consisting of 90 plates, each 8 ft × 8 ft × 1 in. thick, with interspersed steel plates designed to measure the range of muons up to about 2 BeV.

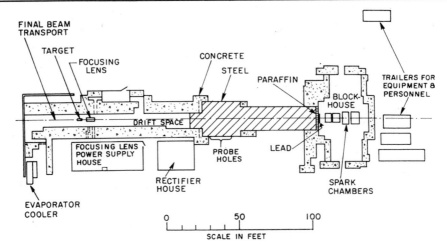

Fig. 1. *General layout of the neutrino experiment at the Brookhaven National Laboratory.*

Experimentalists looked for the reaction

$$\nu_\mu + Z \to \mu^- + W^\pm + Z'$$
$$\hookrightarrow l^\pm + \nu$$

W events would be characterized by a high energy muon and electron in the spark chamber. A muon is identified by its range in the range chamber and an electron is identified by its characteristic shower in the production chamber. Other decay modes of the W were not looked for because of background and calibration problems.

Experimentalists found no candidate for W. If $m_W \approx 2m_p$, they expected to see three events of the type described before. If $m_W < 2m_p$, they expected to see more than three. The result, on the basis of ≥ 90 per cent confidence level, is, then, $m_W > 2m_p$. Similar results were reported by the CERN group. Limitations on the available flux of high energy neutrinos from present accelerators make it difficult to extend this limit to higher values of W mass using this technique. In order to detect W of $m_W > 2m_p$,

it is advantageous to produce and detect W in the interaction of hadrons. This method can hope to observe W's of mass as high as $5m_p$.

III. Production of $W^\pm$ in p-N Collisions

There is no question that, if the W exists, more can be produced in $pN \to W +$ anything than any other reaction. However, because of huge background, it is difficult to detect W in this reaction. However, one can make use of the largeness of the W mass to pick certain kinematical configurations to maximize the signal to background ratios. This is done by looking for muonic W-decay in the large momentum transverse to the beam. Experiments of this kind were carried out at Argonne and Brookhaven National Laboratory (Burns *et al.*, 1965b; Lamb *et al.*, 1965). We will again describe the Brookhaven experiment. The Argonne experiment is very similar to the one at Brookhaven. The experimental arrangement at Brookhaven is shown in Fig. 2. The fast extracted proton beam of the AGS was transported in a vacuum pipe up to the 82-ft steel shield. An 18-in. block of

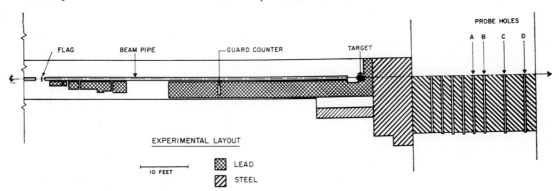

Fig. 2. *General layout of the W production experiment in $p + N$ collisions at the Brookhaven National Laboratory.*

hevimet (90 per cent tungsten) absorbed the bulk of the beam and its secondary particles. Muons were counted by detectors inserted in 3 in. × 3 in. holes located in the steel shield at ranges corresponding to muon momenta greater than 9·6, 11, and 12·5 BeV/c. Muons were counted in each of the three holes and over an angular range from 0° to ~12°. This experiment observed no evidence for W and set an upper limit for W production followed by muonic decay of $\sigma B \leq 10^{-34}$ cm^2, where σ is the W production cross section, and B is the decay ratio $(W \to \mu + \nu)/(W \to$ all modes).

Since there is, at present, no way of calculating the W-production cross section in $p + N$ collisions, this experiment cannot rule out the possible existence of a W. However, this experiment will certainly discourage any further attempts to look for W with existing accelerators.

When the 200 GeV accelerator at Weston, Illinois becomes operational, one of the first things physicists will look for will be the W.

Bibliography

BERNARDINI G. et al. (1965) *Nuovo Cimento* **38**, 608.
BURNS R. et al. (1965a) *Phys. Rev. Letters* **15**, 42.
BURNS R. et al. (1965b) *Phys. Rev. Letters* **15**, 830.
LAMB R. et al. (1965) *Phys. Rev. Letters* **15**, 800.
LEE T. D. and WU C. S. (1966) *Annual Review of Nuclear Science* **15**, 381.

W. LEE

BRITTLENESS IN CERAMICS. Brittleness is the tendency of a material to fracture without appreciable plastic deformation. The important consequence of such deformation is that it provides a mechanism whereby a material can absorb large amounts of energy without fracturing. Metals therefore are generally tough. In ceramics the main energy required to produce fracture is that to create new surface.

Two conditions must be satisfied, however, before brittle behaviour is manifested. Firstly, the fracture stress of the material must be exceeded at some point. This generally results from the application of an external stress which may be continuous (e.g. dead loading) or transitory (e.g. mechanical shock). Secondly, the amount of stored energy at the instant of fracture plus any external work done on the specimen must be large compared with the energy required to produce new fracture face, i.e. the effective surface energy. A quantitative comparison between the behaviour of various brittle materials can thus be made by comparing the energy available for fracture with the effective surface energy. This approach is obviously related to the Griffith theory of fracture which considers the conditions for crack initiation in brittle materials.

Table 1 gives data for three typical but very different ceramic materials. There is a wide range of values of both the fracture strength and Young's modulus. It should be noted, however, that the ratio of strength to modulus, or the critical strain at fracture, is constant at 10^{-3}. Most ceramics show critical strains within the range 0·5 to 1·5 × 10^{-3}. There is also very little variation in the effective surface energy and values lie generally in the range 1 to 5 × 10^4 erg/cm^2. The elastic stored energy in a 1 cm cube of material when raised to the fracture stress is included in the table and this varies greatly between materials. For alumina there is ample energy stored to provide the surface energy required to break the cube in half, i.e. to produce at least 2 cm^2 of fracture face. Alumina will therefore fracture catastrophically when raised to the fracture stress. On the other hand, the energy stored in a 1 cm firebrick cube when raised to the fracture stress is not quite sufficient to break the cube in half. Thus while fracture is still brittle in nature it is not catastrophic.

The brittleness of ceramics at ambient temperatures is due to their inability to undergo significant amounts of plastic deformation. Plastic deformation may occur by a number of mechanisms including slip by dislocation motion, viscous flow or grain boundary sliding. The two latter processes rely on diffusional mechanisms and thus are applicable only at high temperatures. This leaves slip as the remaining alternative. Slip involves the coordinated movement of atoms arranged in a regular atomic array; slip occurs therefore only in crystalline materials. To understand brittleness it is necessary, therefore, to consider the structure of ceramics.

The range of ceramic materials is very broad and so the number of possible structures is large. There are two extreme types of structure: those typified by

Table 1. Mechanical property data for three ceramics

Material	Fracture stress σ_F (dyn/cm^2)	Young's modulus E (dyn/cm^2)	Stored energy in a 1 cm cube $\sigma_F^2/2E$ (erg)	Effective surface energy γ_S (erg/cm^2)
Polycrystalline dense alumina	3×10^9	3×10^{12}	$1·5 \times 10^6$	4×10^4
Whiteware pottery	1×10^9	1×10^{12}	5×10^5	2×10^4
Refractory firebrick	$0·1 \times 10^9$	$0·1 \times 10^{12}$	5×10^4	3×10^4

glasses based on a random silica network, and crystalline structures as typified by pure oxide ceramics such as alumina. A high proportion of ceramics, however, comprise an intimate mixture of both glassy and crystalline phases since fabrication processes often rely on the presence of lower melting point glassy phases to aid densification. The brittleness of ceramics containing glass can obviously be related to their structure but the absence of plastic deformation by slip in crystalline ceramics needs further consideration.

In principle, slip by dislocation motion can occur in any crystalline material. However, slip in ceramic materials is possible only at relatively high stresses because of strong interactions between dislocations and lattice defects such as small impurity precipitates. Nevertheless, small amounts of slip do occur and some single-crystal oxides, e.g. MgO, exhibit appreciable plasticity. But to produce significant amounts of slip in polycrystalline materials the von Mises criterion must be satisfied, i.e. slip must be possible on at least five independent slip systems so that each crystallite can undergo an arbitrary change in shape. The number of slip systems depends on the crystalline lattice structure and in general increases with the degree of symmetry of the structure. The crystalline symmetry of ceramics is such that less than five independent slip systems are available at ambient temperature and so brittle behaviour is expected.

Brittleness in ceramics is therefore a consequence of their structure. Generally, once fracture is initiated there is enough stored energy available to break a body into pieces. These factors must be taken into account when designing ceramic structures.

Bibliography

ASTBURY N. F. (1963) *Advances in Materials Research in the Nato Countries*, p. 369, Oxford: Pergamon Press.
CLARKE F. J. P. and KELLY A. (1963) *Trans. Brit. Ceram. Soc.* **62,** 785.
DAVIDGE R. W. and CLARKE F. J. P. (1966) *Bull. Soc. Franc. Ceram.* **72,** 61.
GROVES G. W. and KELLY A. (1963) *Phil. Mag.* 8, 877.
MCCLINTOCK F. A. (1961) Fracture of solids, in *Encyclopaedic Dictionary of Physics* (J. Thewlis Ed.), **3,** 287, Oxford: Pergamon Press.

R. W. DAVIDGE

C

CARDIAC PACEMAKER, EXTERNAL. A cardiac pacemaker is an electronic device which delivers a sequence of controlled electrical impulses to the heart, causing the muscles to contract regularly, thus artificially producing a steady beat at the normal rate. This method of stimulation is used in the treatment of heart disorders where because of disease the heart rate has become abnormally slowed so that an efficient circulation cannot be maintained. In these cases attacks of fainting known as Stokes-Adams seizures may occur and in severe cases death may result.

There are many similarities between internal and external pacemakers. Electrical impulses can be delivered to the heart by means of wires implanted in the walls of the ventricles. If the unit supplying the impulses is enclosed in the chest then this is a completely internal pacemaker. Alternatively the wires may pass through the chest wall and skin to an external supply unit, although this method has been superseded by inductive coupling across the skin to avoid the risk of infection entering the body along the wires. This is also termed an internal cardiac pacemaker although the majority of the equipment is carried outside the body. Internal pacemakers require surgery for implantation but they are necessary for long term treatment.

There are two ways in which the electrical impulses from an external pacemaker can be applied to the heart. When time and facilities permit catheterization is used. An electrode attached to a wire is inserted via a catheter into a vein in the arm or neck and manoeuvred into the right ventricle of the heart. This electrode is insulated except for the tip and a current of between 5 and 20 mA is passed between this and a large electrode, termed the indifferent electrode, on the surface of the chest. The duration of each current impulse is approximately 2 msec, this being similar to the natural pacing signal, and the rate of impulse supply is adjusted to produce the pulse rate required for the patient, typically 70 pulses/min. A pulse of shorter duration requires a larger current and the converse is also true. The technique of insertion of the catheter, requiring the services of a doctor experienced in cardiac catheterization, involves some risk of infection due to a foreign body in the vein and heart, and the tip of the catheter may touch and irritate the heart causing ectopic beats or fibrillation.

The advantages are that it is simpler and less hazardous than the implantation of an internal pacemaker involving major surgery, it is possible to perform quickly in an emergency, and it can be withdrawn easily when it is no longer required.

When the natural heart action suddenly becomes dangerously inadequate, if the blood pressure falls very low, the veins collapse and it becomes difficult to perform catheterization. However, the same stimulating effect can be achieved by passing a much higher current of between 50 and 150 mA between two electrodes placed on the surface of the chest, along the long axis of the heart. The obvious advantage of this method is its simplicity. Treatment can be commenced very quickly and, if necessary, before the arrival of a doctor. However, this treatment is painful to a conscious patient and is only applicable for short periods of time.

In some cases the heart of a patient undergoing pacemaker treatment may spontaneously return to natural rhythms which conflict with the pacemaker stimulus. Then the heart may lapse into ventricular fibrillation where the muscle fibres contract in a random fashion with the result that the heart merely quivers and is quite unable to maintain the circulation. Death results within a few minutes unless an electric shock of between 50 and 400 joules is administered across the closed chest to depolarize the heart muscles completely. To avoid this risk a method of demand pacing is used. This method detects the occurrence of a natural heart beat and uses it to recommence the timing sequence of the pacemaker. If the natural heart beat is occurring at a normal rate, the pacemaker is prevented from functioning. Whenever the period of the natural heart beat becomes too long, a pacing pulse is automatically supplied.

External pacemakers are available which supply constant voltage, constant current or constant charge pulses to the heart. The constant voltage type has the advantages of simplicity and it is cheap to construct, but the magnitude of stimulus varies with changes in electrode impedance. A range of about 6 to 18 V is needed for the catheter plus indifferent electrode method, and 60 to 180 V for the external electrodes method. For both methods the pulse may be generated using a transistor, thyristor or shockley diode to switch on an adjustable voltage source, or a transformer may be used to step up a low voltage

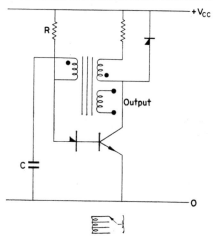

Fig. 1. Constant voltage external pacemaker circuit. Blocking oscillator. The duration of output pulse is determined by the magnetic properties of the transformer. The repetition rate is determined by the time constant RC. The output voltage is adjusted by varying the number of turns in the output winding (with a switch).

pulse to the required level. A blocking oscillator is an example of a suitable circuit for this application (Fig. 1).

The constant current pacemaker (Fig. 2) gives a stimulus that is unaffected by normal changes in electrode impedance but involves greater circuit complexity than the constant voltage method. Negative feedback has to be used to generate the high output impedance without using excessive voltages and this requires a linear system, precluding simple

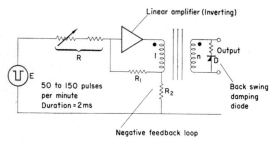

Fig. 2. Constant current external pacemaker circuit. This circuit produces a constant current, constant amplitude positive going pulse of 2 msec duration and with a variable repetition rate of 50 to 150 pulses/min. The output amplitude is adjusted by varying R. If the turns ratio of the transformer is 1 : n and the input pulse amplitude is E, the short-circuit output current is approximately given by

$$I_s \simeq \frac{ER_1}{nRR_2}.$$

saturating or switching circuits such as the blocking oscillator. Again, a transformer may be used to step up a given voltage supply to the value necessary for pacing. With a transformer output the feedback is usually taken from either the primary circuit or a tertiary feedback winding to allow the output winding to remain isolated, which is particularly desirable in a mains powered instrument as it increases the safety.

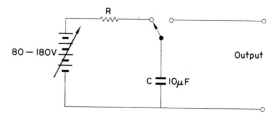

Fig. 3. Constant charge external pacemaker circuit. R limits the initial charging current. The voltage source can be a switched h.t. battery or a controlled d.c. to d.c. inverter.

The constant charge pacemaker (Fig. 3) is less common than the other two and yet it has the advantage of being little affected by changes in electrode impedance and it only requires switching circuitry. The principle of this method is to charge a capacitor to a voltage that can be varied as needed, and to discharge the capacitor directly across the heart of the patient. The amount of charge depends only on the capacitance and the voltage to which the capacitor is charged. The capacitor is selected to give a discharge time of approximately 2 msec for a typical patient. A capacitor of 10 µF charged to between 80 and 180 V is satisfactory for external electrodes. An external electrode pacemaker can be improvised from a single pole changeover switch, a 10 µF capacitor and two 90 V radio batteries. A more satisfactory instrument may consist of a 12 V inverter controlled to produce the 80 to 180 V supply, and a thyristor switching system to charge and discharge the capacitor.

Bibliography

JUETT D. A. (1967) An external cardiac pacemaker, *Bio-Medical Engineering*, April 1967, 168.
KANTROWITZ A. (1964) Treatment of Stokes-Adams syndrome in heart block, *Progress in Cardiovascular Disease* **6**, No. 5, 490.
PELESKA B. and BICIK V. (1966) A review of the status and use of electronics in cardiac stimulation, *Slaboproudy Obzor*, 501.
TAYLOR A. B. (1966) Experience with cardiac pacemaking, *B.M.J.*, Sept. 1966, 543.
TOGAWA T. (1965) Direct induction pacemaker, *Japan Electronic Engineering* **5**, 76.

D. A. JUETT

CERAMICS PRODUCTION, PHYSICS IN

Introduction

The ceramic industry is concerned with the manufacture of solid articles which are, in a large part, composed of inorganic, non-metallic materials. Its present-day scope is much wider than that encompassed by the older definition as an industry manufacturing articles using the action of heat on clay. As well as traditional pottery (china and earthenware), electrical porcelain, refractories, enamels on iron, building bricks and drain pipes, the term ceramics nowadays includes non-metallic magnetic materials, ferroelectrics, oxide refractories, nuclear materials, manufactured single crystals, pyrocerams and a wide variety of other materials.

Clay has been used as a raw material since earliest times and although the pottery industry was one of the first in which factory methods were introduced (Wedgwood, 18th century), it is only since the second world war that significant changes have occurred in the application of physics to the industry. Among the reasons for this are:

(a) The increased mechanization of processes which necessitates more rigid control throughout manufacture.

(b) The increase in research, both co-operative and by individual firms, which has occurred as larger groupings have formed within the industry.

The first stage in ceramic production is the mixing of raw materials in suitable proportions to form a body. In some sections of the brick industry, the clay, as mined, may only need to be crushed and mixed with water before use, but generally the raw materials are ground to give a suitable particle size. In refractory production, relatively coarse particles are incorporated in the mix, the sizes being graded to obtain the maximum packing density. The required amount of water, depending on the shaping method, is then added. In pottery, the materials, ground much finer, are mixed as a suspension in water, known as a slip, which is either used directly, or prepared in a plastic form by removing water in a filter press.

The next process is forming and shaping by:

(a) Slip casting in which the slip, or suspension of powdered materials in water, is poured into a plaster of paris mould, and left for a short time, during which water is absorbed into the mould, thus drying the outer layer of slip. The remainder of the liquid slip is then poured out, leaving a hollow cast, which faithfully reproduces the shape of the inner surface of the plaster mould.

(b) Pressing dry or slightly damp powder in a metal die, at a sufficient pressure for a strong and dense piece to be formed. This method is extensively used for tiles, refractory bricks, spark plug insulators and other products for which large numbers of simple shapes are required. Another advantage is that low water content leads to less shrinkage on drying. Technical and special ceramics are usually pressed dry at higher pressures.

(c) Plastic forming is a method which includes the traditional throwing on a potters wheel, although this is rarely seen nowadays, except in studio potteries. Jiggering is widely used. A lump of plastic clay, with a water content of around 25 per cent, is formed into a flat disk and placed on the surface of a plaster of paris mould. The mould is rotated, and a steel profile tool is pulled down to skim off surplus clay and form the upper surface of the ware. Water is sprayed on during this process to act as a lubricant. Another method suitable for forming bricks, tiles, sewer pipes and other shapes is extrusion of the plastic clay through a suitable orifice, followed by cutting to the required lengths. In the electrical porcelain industry, after extrusion of blanks, the final shapes are produced to accurate dimensions by turning the dried clay on a lathe.

Once formed, ware must be dried in a controlled manner, as removal of water causes considerable shrinkage (up to 25 per cent by volume) and too rapid a rate of drying will cause strains leading to surface cracks. The dried ware, in what is known as the "green hard" state, is then ready for firing. Ceramic wares are fired at temperatures varying from 700°C, for vitreous enamels and some bricks, to over 2000°C for certain materials such as silicon carbide. Firing, once carried out in coal fired intermittent ovens, is now mainly done in tunnel kilns, in which temperature gradients are maintained in a long tunnel lined with refractory bricks, the zone of maximum temperature being at the centre. In the larger kilns, the ware is mounted on trucks with refractory bases, which are slowly pushed through. Recent developments include multipassage kilns in which ware, mounted on refractory slabs typically 1 ft square, is pushed through in times measured in minutes rather than hours. The so-called "top hat kiln" uses the intermittent principle for rapid firing of batches. In this the setting base is stationary, and the heating portion, which makes up the walls and roof, is lowered onto the base for each firing. Firing can be by gas or electricity and by having a number of bases, the heat in the firing mantle can be conserved.

Once they have been fired, certain ceramic products, such as bricks and most refractories, are ready for use. Pottery, however, is glazed by dipping the fired ware in a slip consisting of a suspension of silicate glasses and clay in water. Some water is absorbed by the porous ware leaving a thin layer of solid, which on firing at a temperature some 100°C lower than the first or "bisque" firing vitrifies to give the article a brilliant, impervious layer. For nonporous ware, glazes with a high viscosity have to be used so that the layer of slip will adhere.

Fine china and earthenware can be decorated by patterns and colours, printed on the bisque ware before glazing, but the colours must then withstand the relatively high glost firing temperatures, and the palette is limited. More brilliant decoration is applied on top of the glaze. Finely ground inorganic pigments and low melting glasses, in an oily medium, are fired at a temperature of around 700°C, to allow the glaze to soften sufficiently, so that a strong bond is formed between it and the colour. For expensive items, a dozen or more decoration firings, at successively lower temperatures, may be necessary to develop all the colours, including decoration with gold.

This brief survey covers the standard ceramic production processes. The article on Ceramics in Volume 1 of this Dictionary gives details of some other methods particularly appropriate to the production of Special Ceramics, while refractory manufacturing processes are outlined in the article on Refractories Industry in Volume 6 of this Dictionary.

Application of Physics

The main applications of physics in production come in the tests which are applied during manufacture. These are of two kinds, those which are applied at each stage of manufacture, and those which are specification tests on the finished articles before they are released for service. These latter particularly apply to electrical porcelain, refractory materials and heavy clay products. These tests are in addition to tests and examinations in use in research and development. Test procedures used in production control must generally be simple, quick and reliable, so that there is no delay in guidance to the production department. Ceramics being heterogeneous materials, more than one physical property is often involved in any one physical test, and there tends to be a wide scatter of results due to faults, flaws, etc. Most tests involve setting out a standard practice, using a suitable number of test pieces of agreed shape and size, so that the inevitable scatter gives some estimate of the reproducibility of the results. Physical testing during production will be dealt with in general terms, and then specification tests will be surveyed for each of the major branches of the industry in turn.

Raw Materials

Before being passed for production, each new batch of raw materials must be analysed. Some minor variations in chemical composition may give large variations in the final properties of the ceramic body. In the refractories industry, direct reading spectrographs are used, while pottery analysis is facilitated by spectro-photometric examination of solutions. Physical tests applied to trial bodies in the pottery industry include:

(a) The determination of the thermal expansion as a rod of material is fired to a specific temperature, say 500°C, which is too low for contractions, due to chemical reactions, to mask the expansion.

(b) The determination of the volume of the pores connected to the surface in a piece fired to normal firing temperature.

(c) The determination of the contraction of an ovoid of the material, after the initial drying, and again after firing.

(d) Determination of the modulus of rupture of a bar of specific dimensions, using three point loading in a tensometer, this test being carried out on dried but unfired bars, on fired bars, and on fired and glazed bars.

(e) The modulus of elasticity of a rod of the material is found by a sonic method, involving the determination of the resonant frequency of oscillation, when the bar is excited in one of the possible modes of longitudinal, flexural or torsional vibration.

The raw materials as received must be crushed. The finer particle sizes are commonly produced in ball mills. Extra precautions against contamination of high duty ceramics are required, in the form of rubber lined mills, with grinding balls of approximately the same composition as the material to be ground. A modern development is the fluid energy mill, in which grinding occurs as the result of high speed impacts between particles in an air stream. Noise analysis has been used to control the raw material feed into a grinding mill. A microphone picks up the noise, the resulting electrical signals being fed to a frequency analyser, and as the particles are ground finer, the sound spectrum alters, until a relay is operated to feed in more coarse material.

Grinding must be continued until the optimum particle size range is reached. Sizing of particles greater than $40\,\mu$ is commonly performed by sieve analysis, using a series of sieves, whose openings are in geometric ratio to one another. The total weight passing through any one mesh is plotted against the diameter of the sphere which would just pass through the opening. For smaller particle sizes, the screens tend to clog easily and sedimentation analysis, based on the application of Stokes' law, is used. A standard experimental routine is to take a series of hydrometer readings, at known time intervals, during the settling of the suspension. A newer development is an automatic balance, in which a scale pan is held balanced in the suspension, while the particles, starting with the heaviest, gradually settle on it. The changes in counterbalance weights are recorded, and can be used to determine the particle size distribution.

Particles of iron or iron oxide must be removed from the raw material at this stage, as they would have a very deleterious effect on the colour of the final product. Powerful electromagnets are used to remove these particles, as the slip or powder passes over them.

Control of Slips

The most easily measured characteristics of a slip are its specific gravity, its pH, and its viscosity. The first is measured using suitable hydrometers, or by standard weighing techniques. Standard electrodes are also used to measure pH. The instrument most commonly used for viscosity measurement is the torsion viscometer. A metal cylinder, suspended by a torsion wire, is immersed in the slip, the wire is twisted through 360°, and the cylinder is released. The viscosity is measured by the overswing past the position of rest. Some slips have thixotropic properties, their viscosity falling as they are set in motion. Measurements are then made with the torsion viscometer after a series of settling times, and the variation in over-swing gives a guide to the degree of thixotropy.

Plasticity has proved to be a very difficult property to measure by a method that satisfies the needs of production. Many firms use an empirical relationship between the plasticity of a clay body and the green strength of the dried clay, measured on a modulus of rupture machine.

Firing

Standard temperature control techniques are used, temperatures being measured by chromel/alumel or platinum/platinum-rhodium thermocouples at fixed points along the length of a tunnel kiln. It is also the practice, from time to time, to mount thermocouples on a kiln car and obtain direct readings of the temperatures throughout the length of the kiln. The properties of the fired ware, however, depend both on firing temperature, and on the time for which that temperature is maintained, and so thermoscopes still find a considerable use in the industry. One common type is a ring about 3 in. in diameter, $\frac{1}{4}$ in. thick and with a 1 in. central hole, made from a pottery body of very closely controlled composition. A number of these are placed, with the ware, in positions such that they can be removed through strategic portholes in the kiln; their fired size, measured on a gauge, is an indication of the total effect of the firing on the ware. Refractory cones are also used. These are triangular based pyramids, $1\frac{1}{2}$ in. high, which are mounted with the ware in such a way that, when a certain temperature is reached, they start to fuse and slump over. A range of cones is used, whose composition is varied so that they bend over at a series of increasing temperatures.

Colour

In the tile industry, great importance is attached to maintaining close matching of colours, both during a production run and over a long period of time. Photoelectric tristimulus colour meters are used to assist in detecting a gradual drift from the standard. They also assist when a new batch of colouring agent is made up, since adjustments must be made to the recipe, using fired trials, and it is often very difficult to decide by visual means in what manner the colour of the trial departs from that of the standard.

Tests applied to the finished articles are more conveniently dealt with by reference to each branch of the industry in turn, because of the widely varying nature of the products:

Pottery

In this industry a number of empirical tests are used to simulate conditions likely to be found in actual usage. Among these are:

(a) *Moisture expansion test*. If the body and glaze are not matched they will expand and contract by unequal amounts, thus causing cracking or crazing of the glaze. For a typical test, ware is heated in steam at 50 psi for about 8 hr, allowed to cool, and examined for presence of hair-line cracks. This is repeated for two further cycles.

(b) *Thermal shock tests*. Ware, likely to be used in kitchen ovens, must be capable of withstanding rapid changes of temperature, and a typical test consists in heating ware to a series of temperatures, ranging from 160° to 240°C, and plunging it into cold water. Another such test is to place a plate containing water on a hotplate, allow the water to boil dry, and then transfer the plate to a cold metal surface.

(c) *Abrasion resistance tests*. On-glaze decoration, in particular gold decoration, is liable to be abraded in use. To compare resistance to abrasion, a felt pad, soaked in an abrasive flint slurry, is rubbed backwards and forwards across the gold, using a fixed loading. The number of cycles for, say, 90 per cent of the decoration to be worn away is recorded.

(d) *Chipping resistance test*. A standard pendulum, falling from a recorded height, impacts a point on the rim of the ware. The height from which the pendulum swings is increased, until a chip breaks away.

Refractory Materials

Refractories must meet specifications which are determined by the customers' requirements. The main physical tests applied are measurements of density and porosity, stability of volume, strength and creep at high temperature, resistance to thermal shock, thermal conductivity and thermal expansion. Most of the properties are determined by straightforward physical methods: those peculiar to the industry or of special importance include:

(a) *Determination of refractoriness*. A test piece, made in the shape of a standard test cone, is fired, and its behaviour, as the temperature is raised, is compared with that of a range of standard test cones of the same shape. The number of the cone, whose behaviour most closely approximates to that of the specimen, is the refractoriness.

(b) *Refractoriness under compressive load*. There are two forms of test. In one a cylindrical specimen is heated at a prescribed rate while bearing a load of 2 to 3·5 kgf/cm². The heating is continued, until either the specimen collapses, or a specified deformation occurs. In an alternative test, the loaded specimen is heated to a predetermined temperature, which is maintained, again for a specified time, or until collapse occurs. The dilation is measured with the aid of a micrometer dial gauge to which a refractory extension rod, usually of sintered alumina, is clipped. This rod rests on top of the specimen and corrections must be made for expansion both of it, and the supporting and loading columns.

(c) *Electrical conductivity.* A modern and increasingly important use of refractory materials is in electrical storage heaters where, for safety reasons, the refractory blocks must offer adequate electrical resistance at the operating temperature; since the electrical resistance of refractories falls with increasing temperature, once a leakage current starts to flow through the material, joule heating causes a further decrease in the resistance and the system runs away until the heating element or the refractory melts. Standard methods are used for the measurement of the conductivity.

(d) *Microscopic and X-ray examination.* Thin sections of refractory materials, viewed either by transmitted or reflected light, through the polarizing microscope, can be used to study reactions which occur during initial firing, and which occur as a result of chemical attack during use. The orientation, and segregation, of phases and the formation of glasses, may all be studied. This information is augmented by the use of conventional X-ray powder photography.

Heavy Clay Products

These products, which are extensively used in the building industry, must be capable of lasting in exposed situations for times measured in centuries, and, in large buildings, must be able to support considerable loads. At the same time, they must be produced cheaply in large amounts, and, as a result, economic considerations prevent the use of elaborate physical tests. The major measurements made during production include (a) moisture content, using a simple weighing method, (b) drying shrinkage by direct physical measurement of brick shaped specimens, or alternatively by a mercury displacement method, (c) the measurement of modulus of rupture, and (d) the determination of firing shrinkage which includes shrinkage under load, since bricks are loaded into a kiln stacked in a systematic manner one on top of the other, and the lower bricks have to bear a considerable load.

Many of these products have been the subject of specifications for some time and standard methods of testing have been drawn up by the British Standards Institution. Those with a physical content include:

(a) *Determination of compressive strength of bricks.* A sample of a dozen bricks is immersed in water for three days, and then each is crushed in a machine, which increases the load at a standard rate. The results, as might be expected, show a considerable spread about the mean value. Shaped and hollow bricks are made into parallelipipeds with mortar for this test.

(b) *Determination of water absorption of bricks to give a measure of porosity and bulk density.*

(c) *Transverse strength of roofing tiles.* Normally these do not support heavy loads, but, during building or repair operations, may have momentarily to support the weight of a man. A dozen tiles saturated in water are subjected to three point loading and tested to breaking.

(d) *Pressure tests on salt glazed pipes.* In sewerage systems, these pipes may have to withstand internal pressures during floods. Increasing pressure is applied hydraulically at a specified rate, and the pipes must withstand, without injury or leakage, an ultimate pressure of 20 psi.

(e) *Durability of brickwork.* Frost is an important weathering agent, since, in winter, bricks may become partly saturated with water which then freezes. The resultant expansion may disrupt the goods. Some measure of frost resistance is provided by a measure of the difference between simple water take up, after immersion of a brick for 24 hr, which measures the pores that water can enter easily, and water take up after the brick has been immersed in boiling water for 5 hr, which then includes the pores the water fills with difficulty. Products with a large difference in water absorption values tend to be frost resistant, as the ice, on being formed, expands into the unfilled pores. Direct tests of frost resistance include long-term exposure tests and a simulation test, in which warm sodium sulphate solution is allowed to penetrate the bricks; on cooling, the salt crystallizes in the pores, with similar disruptive effects to ice. Several cycles of heating and cooling are carried out.

(f) *Hardness and resistance to abrasion.* These are properties of importance for flooring tiles and bricks used to line sewers. Hardness can be measured by scratching or diamond indentation. Various types of test are used for abrasion resistance, and these include measurement of loss of weight or volume when a test piece is pressed against a rotating disk, fed with a slurry of sand and water. Alternatively, sand fed into a jet of steam or compressed air is directed against a brick, held a standard distance away.

Electrical Porcelain

The tests applied to insulators are laid down as British Standards. Design tests check the suitability of that particular design of insulator for a given purpose. They include:

(a) An impulse flashover test in which a $1/50$ μsec high-voltage impulse is applied. The voltage at which flash-over occurs is higher in this test than at mains frequency, and the ratio of the two, known as the impulse ratio, measures the ability of the insulator to withstand rapid over-voltage surges.

(b) A dry flashover test, in which the actual voltage of flashover at mains frequency is determined. The ability to withstand a specified (but lower) voltage is also measured.

(c) Wet flashover tests, in which the tests described immediately above are carried out in simulated rain.

Sample tests must be carried out on 0·5 per cent of production from each batch to ensure that specification values are maintained. They include:

(a) A temperature cycle test, in which the insulator is immersed alternatively in baths of water 70°C apart in temperature.

(b) Tensile tests to breaking point on the insulators and any metal attachments.

(c) Over-voltage tests.

(d) Tests to demonstrate that broken pieces are not porous.

Routine tests carried out on all production include (a) an electrical flashover test repeated every few seconds for 5 min and (b) tensile tests.

Bibliography

CLEWS (1955) *Heavy Clays Technology*, Stoke-on-Trent: British Ceramic Research Association.
KINGERY W. (1958) *Ceramic Fabrication Processes*, London: Chapman & Hall.
SINGER and SINGER (1963) *Industrial Ceramics*, London: Chapman & Hall.
VAN VLACK L. H. (1964) *Physical Ceramics for Engineers*, Reading, Massachusetts: Addison Wesley Publishing Co. Inc.
WAYE B. E. (1967) *Introduction to Technical Ceramics*, London: Maclaren & Sons Ltd.

D. S. DODD

CHARGE CONJUGATION. The idea of charge conjugation first arose from the hole interpretation of the negative energy solutions of the Dirac equation. For an electron of charge $-e$ in an external electromagnetic field, described by a vector potential $A_\mu = (\mathbf{A}, i\phi)$, the *Dirac equation* takes the form:

$$[\gamma_\mu(\partial/\partial x_\mu + (ie/\hbar c) A_\mu) + \chi] \psi = 0 \quad (1)$$

where $\chi = mc/\hbar$ is the reciprocal Compton wavelength of the electron. The charge conjugate wave function ψ^c is obtained from the negative energy wave function as follows:

$$\psi^c = C\bar{\psi}^T \quad (2)$$

where $\bar{\psi} = \psi^+ \gamma_4$ is the "adjoint" wave function and C satisfies

$$C^{-1}\gamma_\mu C = -\gamma_\mu^T. \quad (3)$$

In the standard Pauli representation we have, up to a phase, $C = \gamma_4\gamma_2$. By comparison with the transposed adjoint equation

$$[-\gamma_\mu^T (\partial/\partial x_\mu - (ie/\hbar c) A_\mu) + \chi] \bar{\psi}^T = 0, \quad (4)$$

it is easily verified that ψ^c satisfies

$$[\gamma_\mu (\partial/\partial x_\mu - (ie/\hbar c) A_\mu) + \chi] \psi^c = 0 \quad (5)$$

and describes the motion of a particle of the same mass m but positive charge e. This particle, the positron, has been observed experimentally and in the hole theory is interpreted as a vacant negative energy state in an otherwise completely filled sea of negative energy states. It is essential here that the electrons satisfy the Pauli exclusion principle.

The hole theory is a many body theory and is most simply and consistently formulated in terms of a *second-quantized electron field*. This involves changing the one-electron wave function into a set of operators satisfying the equal time anti-commutation relations:

$$\{\psi_\alpha(\mathbf{x}, x_0), \bar{\psi}_\beta (\mathbf{x}', x_0)\} = (\gamma_4)_{\alpha\beta}\, \delta^3(\mathbf{x} - \mathbf{x}'), \quad (6)$$

which are a direct expression of the Pauli principle. In order to remove undesired singularities due to the infinite sea of negative energy electrons, all physical quantities have to be anti-symmetrized in the Dirac field. So the electromagnetic current operator becomes:

$$j_\mu = -iec[\bar{\psi}\gamma_\mu\psi - \psi^T\gamma_\mu^T\bar{\psi}^T] \quad (7)$$

and the interaction Lagrangian for quantum electrodynamics is

$$\mathscr{L}_{\text{int}} = \frac{1}{c} j_\mu A_\mu. \quad (8)$$

Under the charge conjugation operation

$$U_c \psi U_c^{-1} = C\bar{\psi}^T$$
$$U_c A_\mu U_c = -A_\mu \quad (9)$$

the equal time commutation relations of the fields are invariant. Also $j_\mu \to -j_\mu$, as is required when electron $\to$ positron, and $\mathscr{L}_{\text{int}}$ is invariant. Thus U_c can be made unitary and charge conjugation is a symmetry operator for quantum electrodynamics.

It follows that the scattering matrix is invariant under U_c, from which one can derive selection rules and relations between different reactions involving electrons and positrons. Since U_c anti-commutes with the charge operator, the eigenstates of U_c must have zero charge. In particular it follows from equation (9) that the photon has odd "charge conjugation parity". So for an arbitrary n-photon state we conclude

$$U_c |n\gamma\rangle = (-1)^n |n\gamma\rangle. \quad (10)$$

Positronium is a bound state of an electron and positron and for states of definite orbital angular momentum and spin (l, σ) it will be an eigenstate of U_c with the eigenvalue $(-1)^{l+\sigma}$. For the decay into a final state of n photons, charge conjugation invariance requires

$$(-1)^{l+\sigma} = (-1)^n. \quad (11)$$

Thus the singlet $^1S_0$-state is allowed to decay into 2 photons while the 3-photon decay is forbidden, and vice versa for the triplet $^3S_1$-state.

Now it is a consequence of relativistic quantum field theory that to every particle a there corresponds an antiparticle $\bar{a}$, with the same mass and spin but opposite additive quantum numbers such as charge and baryon number. These antiparticles are observed experimentally and charge conjugation has been generalized to mean particle-antiparticle conjugation. Charge conjugation (C) invariance then asserts that to every state of a system there corresponds another state identical in all respects except for the interchange of particles and antiparticles. Thus the reactions

$$a + \bar{b} + \cdots \to \bar{c} + d + \cdots$$

and

$$\bar{a} + b + \cdots \to c + \bar{d} + \cdots$$

would proceed in the same way in all respects. The action of U_c on a single particle state of momentum **P** and helicity (component of spin parallel to **P**) λ is defined by

$$U_c |\alpha, \mathbf{P}, \lambda\rangle = \eta_c |-\alpha, \mathbf{P}, \lambda\rangle \qquad (12)$$

where α denotes all the observables, like charge, which anticommute with U_c and η_c is a phase factor.

The strong and electromagnetic interactions are probably invariant under charge conjugation. In strong interactions the most accurate test of this symmetry comes from annihilation experiments of proton antiproton, $p\bar{p}$, states at rest into pseudoscalar mesons. The reaction is believed to take place from an S-state. It follows that the singlet state has charge conjugation parity $\eta_c = 1$ and decays into an overall S-state, whereas the triplet state has $\eta_c = -1$ and decays into a P-state. Therefore, after angular integration over the final state, there will be no interference from the two possible initial $p\bar{p}$ states and the momentum distribution for the annihilation products should be invariant under interchange of particle and antiparticle.

In the electromagnetic decay of strongly interacting particles, there is a number of results which follow from C invariance. The virtual strong process for the pion $\pi^0 \to p\bar{p}$ gives $\eta_c(\pi^0) = 1$ and, then, charge conjugation invariance allows $\pi^0 \to 2\gamma$, which is the dominant pion decay mode experimentally, but strictly forbids $\pi^0 \to 3\gamma$. The η meson has the same charge conjugation parity as π^0, and any asymmetry between π^+ and π^- in the electromagnetic decay modes $\pi^+\pi^-\pi^0$ and $\pi^+\pi^-\gamma$ would be direct evidence for the violation of charge conjugation symmetry. As already stated, there are at present no data supporting violation of the symmetry in strong or electromagnetic reactions.

In 1956 it was shown from the study of the β-decay of polarized C^{60} that parity and most probably charge conjugation were not conserved in weak interactions. A direct experimental test of C invariance was made the following year in muon decay $\mu^\pm \to e^\pm + \nu + \bar{\nu}$. The respective helicities of the electron and positron were found to be opposite from μ^- and μ^+ decay, in violation of C invariance.

It should be remarked here that when C invariance is violated it is not possible to give an exact definition of the charge conjugation operator U_c. For instance, the free Lagrangian of a neutrino is not C invariant and there is no physical antineutrino state with the helicity required on the right-hand side of equation (12).

The relative phases of the effective vector and axial vector interactions in μ decay and β decay turned out to be relatively real within experimental errors. It therefore seemed that the combined operation CP of charge conjugation and space inversion might be a rigorous symmetry law for all interactions. In 1964 a precise measurement was made of the decay of the long-lived neutral meson K_2^0 which, if the symmetry is valid, is odd under a CP reflection and would not decay into two π-mesons. It was found that about one K_2^0 in every 500 decayed into two π-mesons, conclusively demonstrating the violation of CP invariance in some interaction.

There is still theoretical support for one rigorous discrete symmetry CPT, the combined operation of charge conjugation and space-time reversal, which depends only on general properties of relativistic local field theory. Indeed, the very existence of antiparticles with masses, electromagnetic form factors and lifetimes equal in magnitude to those of the particles is strong evidence in favour of the symmetry. However, these results do not test the full content of CPT invariance, which requires that when one changes particles to antiparticles, but keeps their momenta fixed and helicities reversed, all matrix elements of the interaction should become their respective complex conjugates. The semi-leptonic and strangeness-changing weak interactions have yet to be tested with precision for CPT invariance, and the various decays of $K^\pm$ and the neutral K mesons will provide detailed tests of the symmetry.

Bibliography

GOEBEL C. J. (1961) Mesons and hyperons, decay of, in *Encyclopaedic Dictionary of Physics* (J. Thewlis Ed.), 4, 578, Oxford: Pergamon Press.

LEE T. D. and WU C. S. (1965–66) Weak interactions, *Annual Review of Nuclear Science*, Vols. 15 and 16.

POLKINGHORNE J. C. (1962) Quantum electrodynamics, in *Encyclopaedic Dictionary of Physics* (J. Thewlis Ed.), 5, 735, Oxford: Pergamon Press.

SAKURAI J. J. (1964) *Invariance Principles and Elementary Particles*, Princeton, New Jersey: Princeton University Press.

SKYRME T. H. R. (1961) Dirac equation, in *Encyclopaedic Dictionary of Physics* (J. Thewlis Ed.), 2, 423, Oxford: Pergamon Press.

SOPER J. M. (1961) Beta (β) decay, in *Encyclopaedic Dictionary of Physics* (J. Thewlis Ed.), 1, 394, Oxford: Pergamon Press.

WILLIAMS W. F. (1962) Positronium, in *Encyclopaedic Dictionary of Physics* (J. Thewlis Ed.), 5, 594, Oxford: Pergamon Press.

C. D. FROGGATT

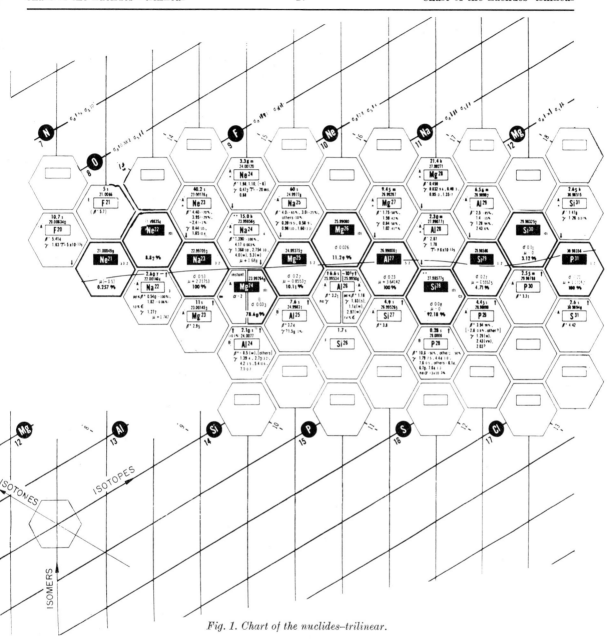

Fig. 1. Chart of the nuclides–trilinear.

CHART OF THE NUCLIDES—TRILINEAR. Charts of the nuclides show many of the important relationships between nuclides, and they enable a considerable amount of information pertaining to nuclides to be concisely presented. Commonly the atomic number (Z) and neutron number (N) are used as abscissa and ordinate in a rectangular coordinate system (Vol. 5, p. 171). Isobars then lie on lines at 45° to these axes. The trilinear form of the chart of the nuclides includes the mass number (A) in such a way that it is given equal weight with N and Z in the representation. A small portion of the chart of nuclides in the trilinear form is shown in Fig. 1 in which the principles upon which the whole chart is constructed are also shown. This should be compared with Vol. 5, p. 171, which illustrates the same portion of the chart in the rectangular Cartesian form.

The solid line which passes almost through the centre of Fig. 1 is the "line of stability", which shows the most stable numbers of neutrons and protons in nuclei.

The trilinear form of the chart of the nuclides can

be considered to be obtained from the rectangular form by adjusting the angle between the N and Z axes to 60°; the A-axis will then be at 60° to each of these and so A will be given equal weight with N and Z. For convenience of printing, the whole chart is rotated through about 30° in the clockwise direction; the effect of this is that the stability line is almost horizontal so the whole chart can be printed on successive pages of a book. Each nuclide is represented by a hexagon which is superimposed on the basic grid as shown in Fig. 1.

Bibliography

STEHN J. R. (1962) Chart of the nuclides, in *Encyclopaedic Dictionary of Physics* (J. Thewlis Ed.), **5**, 170, Oxford: Pergamon Press.

SULLIVAN W. H. (1957) *Trilinear Chart of Nuclides.* Washington, D.C.: U.S. Govt. Printing Office.

D. J. REES

CONDUCTIVITY AT HIGH TEMPERATURES. In many modern branches of technology, refractory materials are used at very high temperatures. A knowledge of thermal conductivity and other physical properties is necessary to determine the behaviour of materials at high temperatures. Some examples of high temperature materials in use are alloys of zirconium and uranium dioxide or particles of this oxide embedded in graphite or carbon for nuclear technology, ablative materials for re-entry vehicles and powders and fibrous materials used as insulating materials at high temperatures. The knowledge of electrical and thermal conductivities of refractory metals is necessary for designing float zone experiments. The thermal and electrical conductivities of metals at high temperatures are also of great academic interest. In particular, the mechanism of scattering due to unfilled shells in transition metals and its influence on the temperature variation of electrical and thermal conductivities is receiving considerable attention. Borides, nitrides and carbides of Ti, Nb, Hf and Zr are probable high temperature materials of the future because of their good mechanical properties at high temperatures.

A great deal of effort is being made to find thermal conductivity standards at high temperatures. Mo, W, Ta and their alloys and many ceramics are being examined as candidates for this purpose. It is necessary to have well characterized samples for this work since density, porosity, impurities, structural defects and environment can influence the thermal conductivity appreciably.

The mechanism of conduction of heat through materials is of two kinds, (1) heat conduction due to phonons and (2) heat conduction due to electrons. For materials which are good dielectrics, the major part of heat conduction is due to phonons. In metals which are electrical conductors, electrons are mainly responsible for the heat conduction. Semiconductors may have either phonon conductivity or electronic conductivity (or both) as major components. Experimentally, the value obtained for the thermal conductivity is the total conductivity, i.e. the total contribution of both electronic and phonon conductivity. If we assume that the electronic thermal conductivity K_e can be obtained from the Lorentz relation

$$\frac{K_e \varrho}{T} = L \qquad (1)$$

(where T is the temperature and L is a constant), and if the electrical resistivity ϱ as a function of temperature is known, the lattice thermal conductivity K_L can be calculated by subtracting the electronic part from the experimental value of total thermal conductivity K.

It has been found that for iron the lattice thermal conductivity determined by the above method is negative. Modifications to the Lorentz function have been suggested to remove this difficulty. However, recent work on cobalt shows that the modified Lorentz ratio is not adequate to interpret data on cobalt.

Considerable attention has been given in recent years to the development of methods for, and improvement of accuracy in measurement of thermal conductivity of materials at high temperatures. The unpublished proceedings of the conferences on thermal conductivity are rich sources of information on the techniques and methods. The factors which determine the choice of a particular method to be used for determining the thermal conductivity in a given case are the order of magnitude of thermal conductivity, the size of the specimen, the accuracy required and the electrical conductivity of the specimen material. Other factors such as ease of fabrication or time available for measurement are also important.

The steady-state methods used at high temperatures can be broadly divided into two classes, electrical methods and non-electrical methods. The electrical methods are those in which the specimen is heated by passing electric current through it. These methods can be used only for materials which have reasonably good electrical conductivity. Though electrical methods were used at low temperatures at the beginning of the century, their use at high temperatures has become popular during the past 10 years or so. Usually the mathematics involved in these methods is somewhat complicated but they are simpler experimentally than the nonelectrical methods. The electrical conductivity and other properties can also be determined simultaneously using these methods. A few years ago some experimental results indicated that the thermal conductivity determined by electrical methods may be in error. It was suggested that this might be due to failure to match the mathematical boundary conditions, or that the assumptions used in simplifying the mathematics may not be valid under practical conditions, or that the approximations involved in linearization of the Boltzmann equation may not be valid when heavy electric currents are passed through the solids. Fortunately more recent work has shown that electrical methods are capable of giving correct and accurate values of thermal conductivity.

In one class of electrical methods, the electric current is passed through a thin filament of the specimen so that the heat flow is essentially one dimensional along the axis. The neck down method, Bode's method and Jain and Krishnan method essentially belong to this class. In an alternative method, used by Angell, Powell and Scofield, the specimen is taken in the form of a large diameter rod, the temperature gradient along the radius being measured near the middle of the rod, assuming that the axial conduction of heat in this region is negligible. These methods have been used at temperatures up to 2500°K.

In the non-electrical methods also, there are two standard techniques, the longitudinal heat flow method and the radial heat flow method. In the longitudinal heat flow method the source and sink of heat are provided at the two ends of a thin rod of the specimen. The radiation loss is suppressed by using guard rings with the temperature matched to that of the specimen. The amount of heat passing through the specimen is measured either by measuring the quantity of heat given by the source or quantity of heat absorbed by the sink. Alternatively, a filament of known thermal conductivity is placed in series with the specimen and the quantity of heat flowing is determined by measuring the temperature gradient across this filament. If the thermal conductivity varies rapidly with temperature, the radial heat flow method is preferable to the longitudinal heat flow method as it uses small temperature gradients. It is also an ideal method for low thermal conductivity materials. An axial heater is placed along the axis of a cylinder of the specimen. The axial heat flow is

Table 1

Temperature °K	Thermal conductivity of metals and alloys, W/cm deg									
	Cobalt	Nickel	Zirconium	Niobium	Molybdenum	Tantalum	Tungsten	Platinum	Platinum + 40% rhodium	Rhenium	Zircaloy*
1000	—	0·718	0·216	0·611	1·12	0·614	1·2	0·73	0·69	—	0·22
1200	0·35	0·761	0·238	0·638	1·06	0·629	1·1	0·73	0·72	—	0·24
1400	0·24	0·804	0·257	0·665	1·01	0·644	1·1	0·75	0·74	—	0·28
1600	—	—	0·272	0·692	0·97	0·659	1·0	0·78	—	0·34	0·35
1800	—	—	0·285	0·719	0·94	0·674	1·0	0·80	—	0·34	0·43
2000	—	—	—	—	0·916	0·690	1·0	—	—	0·33	—
2200	—	—	—	—	0·896	0·705	0·9	—	—	0·33	—
2400	—	—	—	—	0·880	0·720	0·9	—	—	0·32	—
2600	—	—	—	—	0·867	0·735	0·9	—	—	0·32	—
2800	—	—	—	—	0·856	0·750	0·9	—	—	0·32	—
3000	—	—	—	—	—	0·765	0·9	—	—	—	—

* A zirconium alloy containing Sn, Fe and Cr as alloying constituents. The total quantity of the three constituents is very small, ∼1 per cent.

Table 2

Temperature °K	Electrical resistivity of metals and alloys, µohm cm						
	Cobalt	Nickel	Platinum	Tungsten	Molybdenum	Rhenium	Zircaloy*
1200	70·5	50·0	39·4	—	30·0	78·0	115
1400	88·0	56·0	44·5	37·0	36·0	84·0	116
1600	96·0	—	49·5	43·0	42·0	92·0	120
1800	—	—	—	49·0	49·0	96·0	—
2000	—	—	—	56·0	55·0	102·0	—
2200	—	—	—	62·0	62·0	105·0	—
2400	—	—	—	70·0	67·0	108·0	—
2600	—	—	—	77·0	74·0	110·0	—
2800	—	—	—	83·0	—	—	—

* A zirconium alloy containing Sn, Fe and Cr as alloying constituents. The total quantity of the three constituents is very small, ∼1 per cent.

Table 3

Temperature °K	Thermal conductivity of nonmetals, W/cm deg						
	MgO	BeO	Al_2O_3	ThO_2	TiO_2	Quartz fused	Pyroceram*
1000	0·100	0·49	0·080	0·038	0·036	0·028	0·029
1200	0·085	0·35	0·067	0·033	0·034	0·039	0·028
1400	0·075	0·26	0·060	0·029	0·033	—	0·028
1600	0·071	0·21	0·058	0·028	—	—	—
1800	0·072	0·17	0·058	—	—	—	—
2000	0·087	0·16	0·061	—	—	—	—
2200	0·120	0·15	—	—	—	—	—
2400	—	0·15	—	—	—	—	—

* Pyroceram is made from silicon dioxide, aluminium oxide, magnesium oxide and a small amount of titanium dioxide. It is the trade name used by Corning Glass Works.

avoided either by making the specimen cylinder very long and using its central region or by splitting the cylinder into discs and providing insulating materials between the discs. In some cases the heating has been provided at the outside so that inward radial heat flow takes place. Thermal conductivity has been measured at temperatures up to 2700°C by these methods.

If the density d and the specific heat c of the material are known at high temperatures, the thermal conductivity K can be calculated from the diffusivity ($K' = K/dc$) values measured by transient methods. In transient methods, a sinusoidal or pulsed heat input is applied at one end of a long rod or filament of the specimen and diffusivity is measured either by the "velocity" method or the "amplitude decrement" method. The sinusoidal or pulsed heat input is obtained either by electron bombardment or by high intensity light beam, e.g. from a Xenon discharge tube. Considerable improvements in the technique have been made by using a laser beam as the pulsed source of heat in the past three or four years. Switching on or off an electric heater at one end of the rod and measuring the variation of temperature with time at different points along the specimen is another variant of this method.

The measurement of electrical conductivity is much less difficult and can be made much more accurate. All that is necessary is to obtain a reasonable length of the specimen at a uniform temperature and measure the voltage drop across a known length and electric current passing through it. The voltage used is usually alternating and is measured by a high precision a.c. potentiometer. The current is usually measured by using a standard non-inductive resistance connected in series with the specimen and measuring the potential drop across it. The electrical resistivity of metals measured at high temperatures by different workers usually agree with each other within a few per cent, whereas large discrepancies are found in the values of thermal conductivity. These discrepancies are due partly to the differences in the impurity content and history of the samples and partly to the limited accuracy of temperature measurements. The materials that can be used as thermocouples at high temperatures are drastically limited. W : W + Re are the only thermocouples which have been used by many workers. Alternatively an optical pyrometer can be used. However, the calibration of the thermocouples or the optical pyrometer is difficult and a small error in the measurement of temperatures can cause large errors in thermal conductivity values.

The thermal conductivity and electrical resistivity of some selected metals and alloys are given in Tables 1 and 2 respectively.

Cezairliyan has found that the thermal conductivities of 15 metals can be represented by one universal curve if we plot K/K_θ vs. T/θ where K_θ is the thermal conductivity at Debye temperature θ. The curve obtained by him is shown in Fig. 1. Recent work has shown that the thermal conductivity data for iron, cobalt and nickel does not plot on this universal curve. The data for these three metals plot on another single curve below their Curie temperatures. Thermal conductivity of the ferromagnetic metals decreases

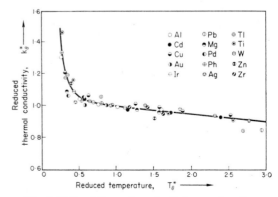

Fig. 1. Reduced thermal conductivity as a function of reduced temperature for fifteen metals at moderate and high temperatures.

Table 4

Temperature °K	Thermal conductivity of gases, mW/cm deg		
	Helium	Nitrogen	Argon
1,500	4·94	0·842	0·561
1,900	5·96	1·080	0·667
2,300	6·93	1·314	0·766
2,700	7·85	1·494	0·864
3,100	8·72	1·691	0·962
3,500	9·58	1·915	1·060
3,900	10·43	—	1·158
4,300	11·27	—	1·256
4,700	12·11	—	1·354
6,000	—	—	1·673
7,000	—	—	1·950
8,000	—	—	2·480
9,000	—	—	3·390
10,000	—	—	4·560

with temperature up to Curie point, shows a minimum at Curie point and increases with temperature at higher temperatures. Considerable further work is necessary before the detailed mechanism of thermal conductivity in these materials is understood.

Below Curie point, the temperature dependence of electrical resistivity of ferromagnetic metals is found to contain a T^3 term. This has been attributed to s-d scattering in the transition metals.

The thermal conductivity values of some selected non-metals and gases are given in Tables 3 and 4 respectively.

Efforts have been made to measure the thermal conductivity of ablative materials: these are used for atmospheric re-entry and for propulsion devices. These measurements are very difficult to relate to the properties of the materials in use because the ablation process is violent, the events taking place during this process are not well defined, the change produced is heterogeneous and the heat conductance takes place by several methods which are difficult to simulate in the laboratory. Efforts have also been made to measure the thermal conductivity of insulating fibrous materials such as Dynaquartz, a heat stabilized silica material; Dynaflex, an aluminium silicate with small chromia additions and sapphire wool. Here again, the transfer of heat takes place by conduction as well as by radiation. Since reproducible values for these materials have not been obtained yet, these are not given in the tables. The electrical resistivities of these materials are extremely high and do not seem to have been measured yet.

Bibliography

ABELES B., CODY G. D. and BEERS D. S. (1960) Apparatus for the measurement of the thermal diffusivity of solids at high temperatures, *J. Appl. Phys*, **31**, 1585.

BACKLUND N. G. (1961) An experimental investigation of the electrical and thermal conductivity of iron and some dilute iron alloys at temperatures above 100° K, *Phys. Chem. Solids*, **20**, 1–16.

Cobalt Monograph (1960) Edited by Centre D'information Du Cobalt, Brussels, Belgium.

HERZFELD CHARLES M. (Ed.) (1962) *Temperature, Its Measurement and Control in Science and Industry*, New York: Reinhold.

JAIN S. C. and GOEL T. C. (1968) Thermal conductivity of metals at high temperatures by Jain and Krishnan method. I. Nickel, *Brit. J. Applied Physics*, **1**, 573–580.

LUCKS C. F. (Program Developer) (1961) *Conference on Thermal Conductivity Methods*, Battelle Memorial Institute, Columbus, Ohio.

MCELROY D. (Chairman) (1963) *The 3rd Thermal Conductivity Conference*, Metals and Ceramics Division, Oak Ridge National Laboratory, Oak Ridge, Tennessee.

MINGES MERRILL L. and DENMAN GARY L. (Co-Chairmen) (1966) *Proceedings of the Sixth Conference on Thermal Conductivity*, Air Force Materials Laboratory, Dayton, Ohio.

PLUNKETT JERRY D. (Chairman) (1965) *Proceedings of the Fifth Conference on Thermal Conductivity*, Department of Metallurgy, College of Engineering, University of Denver, Denver, Colorado.

POWELL R. F. (1966) *Modern Techniques for the Measurement of the Thermal Conductivity of Metals*, Contemporary Physics, London: Taylor and Francis.

RUDKIN R. (1964) *Proceedings of the 4th Conference on Thermal Conductivity*, U.S. Naval Radiological Defence Laboratory, San Francisco, California.

S. C. JAIN and V. NARAYAN

CURRENT ALGEBRAS AND THEIR APPLICATIONS. Current algebra is a branch of contemporary theories of elementary particles, proposed by M. Gell-Mann in 1962. It deals with the currents measured in electromagnetic and weak reactions; these interactions are described in relativistic quantum field theory by Hamiltonians

$$\mathcal{H}^{\text{el}}(x) = e\, j_\mu^{\text{el}}(x)\, A^\mu(x) \qquad (1)$$

$$\mathcal{H}^W(x) = \frac{1}{\sqrt{2}}\, G\, \frac{1}{2}\, [J_\mu^W(x)\, (J^{W\mu}(x))^+ \\ + (J^{W\mu}(x))^+\, J_\mu^W(x)] \qquad (2)$$

where e is the elementary charge; $j_\mu^{\text{el}}(x)$ the electromagnetic-current operator; $A^\mu(x)$ the photon-field operator; G the Fermi coupling constant of weak interactions; $J_\mu^W(x)$ the weak-current operator and $(J_\mu^W(x))^+$ the hermitian adjoint.

The Hamiltonians (1) and (2) are understood to describe electromagnetic and weak interactions as first-order perturbations on the strong interactions, they operate in an idealized space of states, consisting of free leptons and only strongly interacting hadrons. In this framework electromagnetic and weak reactions are described by (1) and (2) as first order transitions between the eigenstates of strong interactions, second order effects like electromagnetic mass shifts are neglected.

In lowest order of e and G, the matrix elements of the currents between lepton states as free particles are easily determined, while for the strongly interacting (hadron) states the matrix elements of the currents cannot yet be computed from basic principles, but have to be taken from experiment. In β-decay one measures for instance:

$$\langle p, e^-, \bar{\nu}_e | \mathscr{H}^W(x) | n \rangle$$
$$= \frac{1}{\sqrt{2}} G \underbrace{\langle e^- \bar{\nu}_e | (J^{W\mu}(x))^+ | 0 \rangle}_{\text{fixed by kinematics}} \underbrace{\langle p | J^W_\mu(x) | n \rangle}_{\text{to be measured}}. \quad (3)$$

Though the matrix elements of the currents are still largely unknown, they can all be measured, at least in principle, through suitable reactions. The theory of current algebra treats the currents as basic observables and proposes specific equal-time commutation relations for their components. At present an algebraic structure of $SU_3 \times SU_3$ appears likely, but only parts of it have so far been tested with some accuracy.

For specifying the algebra, we decompose the currents into hermitian current operators $j^j_\mu(x)$ with different properties in isospin, strangeness and parity; this assignment of quantum numbers (and consequently of selection rules) is based on empirical principles and, apart from the yet unexplained very small CP violation in neutral K-meson decays, it is consistent with the present experimental knowledge. Thus for the electromagnetic current we have the decomposition

$$j^{\text{el}}_\mu(x)^{\text{def}} = \underbrace{j^3_\mu(x)}_{\substack{\text{isovector}\\\text{part}}} + \underbrace{\frac{1}{\sqrt{3}} j^8_\mu(x)}_{\substack{\text{isoscalar}\\\text{part}}} \quad (4)$$

inducing transitions with: $\Delta I = 0, \pm 1 \quad \Delta I = 0$
I: (total) isospin.

The weak interaction current is given by

$$J^W_\mu(x) = \cos\theta [j^1_\mu(x) + ij^2_\mu(x)]$$
$$+ \sin\theta [j^4_\mu(x) + ij^5_\mu(x)] \}^{\text{vector}}_{\text{part}}$$
$$+ \cos\theta [\bar{j}^1_\mu(x) + i\bar{j}^2_\mu(x)] + \sin\theta [\bar{j}^4_\mu(x) + i\bar{j}^5_\mu(x)] \}^{\text{axial}}_{\substack{\text{vector}\\\text{part}}}$$

$$\underbrace{}_{\substack{\text{strangeness}\\\text{conserving part}}} \underbrace{}_{\substack{\text{strangeness}\\\text{changing part}}} \quad (5)$$

inducing transitions with:

$\Delta I = 0, \pm 1 \quad \Delta I = \pm \tfrac{1}{2}$
$\Delta Q = 1 \quad \Delta Q = 1 \quad Q$ electric charge
$\Delta S = 0 \quad \Delta S = 1 \quad S$ strangeness

θ is the Cabibbo angle, whose empirical value of about $15°$ is yet unexplained; the superscripts follow SU_3 notation; the bar denotes axial vectors.

The following equal-time commutators have been proposed by Gell-Mann:

$$\left[\int j^i_0(x)(\mathrm{d}x)^3, j^j_\mu(y)\right] = if^{ijk} j^k_\mu(y) \quad . \quad (6)$$

$$\left[\int j^i_0(x)(\mathrm{d}x)^3, \bar{j}^j_\mu(y)\right] = if^{ijk} \bar{j}^k_\mu(y) \quad (7)$$

$$\left[\int \bar{j}^i_0(x)(\mathrm{d}x)^3, j^j_\mu(y)\right] = if^{ijk} \bar{j}^k_\mu(y) \quad (8)$$

$$\left[\int \bar{j}^i_0(x)(\mathrm{d}x)^3, \bar{j}^j_\mu(y)\right] = if^{ijk} j^k_\mu(y) \quad (9)$$

f^{ijk} are the SU_3 structure constants. These commutators include currents not occurring in (4) and (5), but apart from the axial hypercharge current $\bar{j}^8_\mu(x)$ they are all related by isospin transformations to some observable currents.

Some of these commutators just restate the results of earlier theories. The isospin generators I^i and the hypercharge Y are expressed through electromagnetic and weak currents:

$$Q = \int j^{\text{el}}_0(x)(\mathrm{d}x)^3 = \int j^3_0(x)(\mathrm{d}x)^3$$
$$+ \frac{1}{\sqrt{3}} \int j^8_0(x)(\mathrm{d}x)^3 = I^3 + \frac{Y}{2} \quad (10)$$

$$I^{1,2} = \int j^{1,2}_0(x)(\mathrm{d}x)^3; \quad j^{1,2}_\mu(x) \text{ taken from } J^W_\mu(x). \quad (11)$$

Equation (10) defines I^3 and Y, equation (11) expresses the CVC-theory (conserved-vector-current theory): the strangeness conserving weak vector current and the isovector part of the electromagnetic current belong to the same isotriplet of vector currents. Commutators of I^i and Y with other currents in equation (5) and (6) just restate their behavior under isospin and hypercharge transformations, as given in equation (5). New information is provided in particular through three sets of commutators:

(i) Among the vector operators, the commutators of two strangeness changing currents are not yet specified by established principles. In equation (6) it is proposed that they create an SU_3 algebra. This is in line with considering SU_3 an approximate symmetry group of strong interactions. Cabibbo's parametrization of the current in equation (5) was originally proposed only for an approximate theory in the framework of SU_3 symmetry. In current algebra the assumption is made that both the concept of the Cabibbo angle θ and the commutators (6) are true under the full physical strong interactions and unaffected by SU_3 symmetry breaking. This aspect of current algebra has not yet been convincingly confirmed, but there are indications for its correctness.

(ii) Commutators (7) are to be understood as expressing octet-behavior of the axial current under the SU_3 algebra of vector charges, defined by (6). Commutators (8) are the most regular generalization of (7).

(iii) In equation (9) the hypothesis is made that the commutators of axial currents reproduce the vector currents discussed above. Since there is no symmetry observed with negative parity operators, one needs a nonlinear relation like equation (9) to fix the scale of the axial currents; only then one is able to give a precise meaning to the concept of universal coupling in weak interactions. At least for strangeness conserving currents, this set of commutators has been amply confirmed through sum rules and low energy theorems derived from them.

Wider sets of commutators, which include equations (6)–(9), have been discussed in the literature, but no conclusive statement can yet be made about their validity.

Applications: Two related classes of applications have been developed to test current algebras: sum rules and low energy theorems. A sum rule is derived by taking specific matrix elements of commutators and substituting a complete set of intermediate states:

$$\langle a | [\int j_0^i(x)\,(\mathrm{d}x)^3, j_\mu^j(y)] | b \rangle = i f^{ijk} \langle a | j_\mu^k(y) | b \rangle$$

$$= \sum_n (\langle a | \int j_0^i(x)\,(\mathrm{d}x)^3 | n \rangle \langle n | j_\mu^j(y) | b \rangle$$
$$- \langle a | j_\mu^j(y) | n \rangle \langle n | \int j_0^i(x)\,(\mathrm{d}x)^3 | b \rangle). \quad (12)$$

Though all matrix elements in equation (12) are observable in principle, most sum rules cannot yet be tested, because the experimental data are available only for a few single-particle states $|n\rangle$. A reliable theoretical approximation to these unknown matrix elements exists only when one is dealing with axial charges. First one can express the matrix element of the charge through the current divergence

$$\langle n | \int j_0^i(x)\,(\mathrm{d}x)^3 | \rangle = (2\pi)^3\,\delta^3(\mathbf{p}_a - \mathbf{p}_n)\,\langle n | j_0^i(0) | a \rangle$$
$$= (2\pi)^3\,\delta^3(\mathbf{p}_a - \mathbf{p}_n)\,\frac{\langle n | \partial^\mu j_\mu^i(0) | \rangle}{i(E_n - E_a)} \quad (13)$$

and approximate the matrix element of the divergence by a pseudoscalar-meson pole at momentum transfer $(p_a - p_n)^2$ equal to the mass m_i^2 of the pseudoscalar meson $|M^i\rangle$

$$\langle n | \partial^\mu j_\mu^i | a \rangle \approx \frac{\langle n | T | aM^i \rangle \langle M^i | \partial^\mu j_\mu^i(0) | 0 \rangle}{(p_a - p_n)^2 - m_i^2}. \quad (14)$$

In the pole residue the part $\langle M^i | \partial^\mu j_\mu^i(0) | 0 \rangle$ is known from the weak decay of the meson $|M^i\rangle$, the (strong) transition-matrix element $\langle n | T | aM^i \rangle$ is measured in reactions $M_i + a \to n$. In Feynman diagrams, relation (14) corresponds to

$$(15)$$

This approximation is particularly reliable when using π-meson poles to approximate matrix elements of the strangeness conserving charges. It has been used in the sum rule of Adler and Weisberger, which is so far the most convincing test of current algebra.

A commutator is formed with two axial charges: $\int (j_0^1(x) + i j_0^2(x))\,(\mathrm{d}x)^3$ (as measured in neutron β-decay) and $\int (j_0^1(x) - i j_0^2(x))\,(\mathrm{d}x)^3$ (its hermitian adjoint), and is considered between one-proton states equation (9) gives:

$$\langle p | [\int j_0^1(x) + i j_0^2(x))\,(\mathrm{d}x)^3,\,\int (j_0^1(x) - i j_0^2(x))\,(\mathrm{d}x)^3] | p \rangle$$
$$= \langle p | 2 \int j_0^3(x)\,(\mathrm{d}x)^3 | p \rangle = 2\langle p | I^3 | p \rangle. \quad (16)$$

The right-hand side takes the value 1 (twice the isospin of the proton), on the left-hand side the neutron appears as a possible intermediate state, contributing (in this normalization) the absolute square of the β-decay axial vector coupling constant $g_A = \left(\dfrac{G_A}{G_V}\right) \approx 1\cdot 18$. The contributions of the higher intermediate states $|n\rangle$, notably pion-nucleon continua, can be estimated according to equations (13) and (14) through the pion decay constant $f_\pi \approx 0\cdot 9 m_\pi$ and $\pi^\pm p$ total cross-sections:

$$|g_A|^2 + |f_\pi|^2 \int \frac{\mathrm{d}s}{(s - m_p^2)^2}\,(\sigma_{\pi^-p}^{\mathrm{tot}}(s) - \sigma_{\pi^+p}^{\mathrm{tot}}(s))\,\frac{2 p_\pi^L m_p}{\pi} = 1 \quad (17)$$

where m_p is the proton mass; $p_\pi^L(E_\pi^L)$ pion momentum (energy) in the laboratory frame; $s = m_p^2 + 2 m_p E_\pi^L + m_\pi^2$. Inserting scattering data, $g_A \approx 1\cdot 16$ is predicted from equation (17) within the accuracy expected from the pion pole approximation (equations (13), (14)).

Similar sum rules have been derived for pion photoproduction and for the scattering of K-mesons, but it is not possible to test commutators of vector charges this way, because no scalar mesons are yet established.

As Fubini, Furlan and Rossetti first recognized, current algebra sum rules like equation (17) are formally equivalent to dispersion relations for meson scattering with the commutator specifying the scattering amplitude at the unphysical point of zero meson energy. To derive such low energy theorems, it is not necessary to use dispersion sum rules; one can derive them directly by using the LSZ reduction formalism (Lehman, Symanzik and Zimmerman). In this way it is shown that one can represent a matrix element $\langle a, M^i(p) | \hat{O}(y) | b \rangle$ of a local operator $\hat{O}(y)$ through an interpolating meson field $\varphi^i(x)$ as the Fourier transform of a time-ordered product.

$$\langle a, M^i(p) | \hat{O}(y) | b \rangle$$
$$= i \int (m_i^2 - p^2)\,e^{ipx}\,\langle a | T(\varphi^i(x), \hat{O}(y)) | b \rangle\,(\mathrm{d}x)^4. \quad (18)$$

So far this matrix element is defined only on the meson mass shell ($p^2 = m_i^2$), but with a specific definition of $\varphi^i(x)$, in our case:

$$\varphi^i(x) = \frac{\partial^\mu j_\mu^i(x)}{\langle 0 | \partial^\mu j_\mu^i(0) | M^i(p) \rangle} \quad (19)$$

one may extrapolate (18) down to $p_\mu = 0$, assuming that its dependence on p^2 is negligible. This assumption is practically equivalent to pion-pole dominance (14), and again it is relatively best justified with π-mesons, due to their small mass. By partial integration, the identity is derived:

$$\langle 0 |\partial^\mu j^i_\mu(0)| M^i\rangle \langle aM^i(p) |\hat{O}(y)| b\rangle$$
$$= p^\mu \int (m_i^2 - p^2)\, e^{ipx} \langle a |T(j^i_\mu(x), \hat{O}(y)| b\rangle\, (dx)^4$$
$$- i(m_i^2 - p^2) \int \delta(x_0 - y_0)\, e^{ipx} \langle a |[\bar{j}_0^i(x), \hat{O}(y)]| b\rangle\, (dx)^4 \quad (20)$$

In the limit $p_\mu \to 0$, equation (20) can be exploited by current algebra

$$\langle 0 |\partial^\mu j^i(0)| M^i\rangle \langle aM^i(0) |\hat{O}(y)| b\rangle$$
$$= (-i)\, m_i^2 \langle a |[\int \bar{j}_0^i(x)\, (dx)^3, \hat{O}(y)]| b\rangle$$
$$+ \text{(possible Born terms)}. \quad (21)$$

In some problems contributions remain in equation (21) from Born terms in the first integral of equation (20), which are singular in the limit $p_\mu \to 0$. Apart from these, equation (21) expresses the emission of a soft meson $|M^i(0)\rangle$ in the process described by $\langle a, M^i| \hat{O}(y) |b\rangle$ by the matrix element of the commutator on the right-hand side. To what extent this unphysical limit approximates a physical low energy process is uncertain in most cases. There is little reason to expect an error of less than 10 per cent, sometimes it may even be larger.

The relation of Callan and Treiman will be derived as an example. We choose $|M^i\rangle = |\pi^0\rangle$, $|a\rangle = |0\rangle$; $|b\rangle = |K^-\rangle$, $\hat{O}(y) = (j^4_\mu(y) + i j^5_\mu(y))$, and apply equations (21) and (8):

$$\langle 0 |\partial^\mu j^3_\mu| \pi^0\rangle \cdot \langle \pi^0 |(j^4_\mu(y) + i j^5_\mu(y))| K^-\rangle$$

↑ taken through isospin invariance from the decay $\pi^+ \to \mu^+ + \nu$

↑ measured in the decay $K^- \to \pi^0 + \mu^- + \bar\nu$ and extrapolated to zero pion energy

$$= (-i)\, m_\pi^2 \langle 0 |[\int \bar{j}_0^3(x)\, (dx)^3, (j^4_\mu(y) + i j^5_\mu(y))]| K^-\rangle$$
$$= \frac{-i}{2} m_\pi^2 \langle 0 |(\bar{j}^4_\mu(y) + i \bar{j}^5_\mu(y))| K^-\rangle. \quad (22)$$

↑ measured in the decay $K^- \to \mu^- + \bar\nu$

This low energy theorem has connected parameters of decays ($K \to$ leptons) and ($K \to \pi +$ leptons). An extension has been made to ($K \to 2\pi +$ leptons) by the same technique. Similarly a relation has been established between pion photoproduction and matrix elements of the weak axial vector current between nucleons, known before as the Kroll-Ruderman theorem.

A small modification of the method allows to describe the simultaneous emission or absorption of two pions by current algebras. Results have been obtained mainly on low energy meson scattering: pion–nucleon scattering lengths, pion–pion scattering and vector meson decays. Attempts have also been made to describe the threshold behavior of inelastic reactions. These tests largely confirm the correctness of current algebras, but the extrapolations are quite uncertain in some cases.

Remarkable contributions have been made to the theory of nonleptonic decays by Sugawara and Suzuki. It is an empirical fact that decays satisfying the selection rule $\Delta I = \pm \frac{1}{2}$ are much more frequent than others. The Hamiltonian $\mathscr{H}^W$ in equation (2) does not suggest any mechanism for this effect, and for this reason its correctness had been doubted. The weak current J^W_μ is a sum of a vector and an axial current with equal quantum numbers. As a consequence we have from equations (6)–(9):

$$[\int j_0^i(x)\, (dx)^3, \mathscr{H}^W(y)] = [\int \bar{j}_0^i(x)\, (dx)^3, \mathscr{H}^W(y)]. \quad (23)$$

Inserting (23) into (22) we see that the decay matrix element $\langle a + \pi^i| \mathscr{H}^W(0) |b\rangle$ can be reduced to $\langle a| \times [I^i, \mathscr{H}^W(0)] |b\rangle$ in the limit of zero pion momentum. With this simplification, the emergence of an effective $\Delta I = \frac{1}{2}$ rule, can be demonstrated in the low energy limit. In $K \to 2\pi$ decays, for instance, we have a proportionality

$$\langle \pi^i(p=0), \pi^j(p=0)|, \mathscr{H}^W(0)| K\rangle$$
$$\sim \langle 0 |[I^i[I^j, \mathscr{H}^W(0)]] + [I^j[I^i, \mathscr{H}^W(0)]]| K\rangle. \quad (24)$$

As the K-meson has isospin $\frac{1}{2}$, only the $\Delta I = \pm \frac{1}{2}$ part of $\mathscr{H}(0)^W$ has a nonvanishing low-energy limit. For hyperon s-wave decays a more complicated argument shows that the low-energy limits simulate again a $\Delta I = \pm \frac{1}{2}$ selection rule, for hyperon p-wave decays the current algebra result cannot yet be reliably extrapolated to physical pion momenta, due to unknown Born terms.

More recent applications of current algebra include the derivation of spectral function sum rules (Weinberg) and an estimate of the pion electromagnetic mass difference (Das, Guralnik, Mathur, Low and Young).

Bibliography

ADLER S. L. and DASHEN R. F. *Current Algebra*, Benjamin, Inc.

BERNSTEIN J. *Elementary Particles and Their Currents*, Freeman

GELL-MANN M. and NE'EMAN Y. *The Eightfold Way*, Benjamin, Inc.

B. RENNER

D

DAYLIGHT, POWER DISTRIBUTION IN

Factors Affecting the Spectral Power Distribution

The spectral power distribution of daylight at ground level is affected by several variable factors, e.g. the position of clouds, so that one single curve cannot adequately represent it. A typical SPD† is shown in Fig. 1. Most of the light reaching the Earth

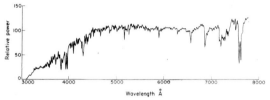

Fig. 1. A typical spectral power distribution of daylight.

originates from the Sun's photosphere, and in view of the complex nuclear and chemical reactions taking place within the central mass of the Sun, it cannot properly be regarded as resembling that from a Planckian radiator. This light passes through the reversing layer of the Sun's atmosphere, where a large number of elements are present in gaseous form. Each element produces its characteristic absorption effects, so that the spectrum of the light that escapes is crossed by many dark lines—the well-known "Fraunhofer lines". It will be seen from Fig. 1 that their effect is quite significant. Some radiation arises in the outer layers of the Sun's atmosphere—the chromosphere—but this is nearly all X-ray and short u.v. radiation which does not penetrate the Earth's upper atmosphere.

On reaching the atmosphere of the Earth, solar radiation suffers absorption and scattering. The tenuous gases of the upper atmosphere absorb heavily in some cases, e.g. oxygen which absorbs most radiation of wavelength shorter than 1900 Å; ozone effectively prevents u.v. radiation shorter than 2950 Å reaching ground level. There is also a great deal of scattering by the actual molecules of the atmosphere, and by minute droplets of water. This is Rayleigh scattering, and in consequence the shorter wavelengths

† The phrase "spectral power distribution" is here abbreviated to SPD.

are scattered to a much greater extent than the longer ones. Some of this light reaches the ground, and is responsible for the blue appearance of the clear sky.

In the lower levels of the atmosphere the light may encounter clouds, which are composed of masses of water droplets whose dimensions are considerably larger than the wavelengths. Enormous numbers of refractions and re-refractions take place, i.e. the cloud acts as a very efficient diffuser, and serves to mix all the light falling on it.

In the lowest levels of the atmosphere pollution occurs in urban areas; there is a great deal of dust present, together with effluent gases from industrial plant. Tarrant has examined this point and reports that no absorption bands, specific to pollution, can be found in an urban atmosphere, and that there is strong evidence that the dust acts as a diffuser. The dust almost certainly absorbs as well, but apparently does so in a roughly neutral manner.

In practice, the SPD of daylight obtained on any surface, or entering an aperture, is chiefly determined by the relative amounts of direct sunlight, Rayleigh scattered light, and cloud scattered light that are received. The most significant factors are the amount and position of cloud in the sky, and the orientation of the receiving surface or aperture. The length of path through the atmosphere (dependent on solar angle) is of less significance as far as the visible range is concerned, though it profoundly affects the infrared region (where some atmospheric gases have strong absorption bands) and the u.v. region around 3000–3300 Å close to the absorption edge of ozone.

Experimental Studies

The spectral distribution has been investigated in several places and the most significant studies of the visible region are listed below (Table 1).

The measurements of Abbot, Fowle, and Aldrich were on direct sunlight only. Gibson reformulated their results to take account of Rayleigh scattering, and Parry Moon's work was a critical review of the knowledge of sunlight existing in 1940. The work of Judd, Macadam and Wysecki was a numerical formulation of the results of Condit and Grum, Henderson and Hodgkiss, and Budde. Budde's results were not published separately.

The combined results put forward by Judd, Macadam, and Wysecki have been adopted by the C.I.E.

Table 1

	Place	Number of observations	Year
Abbot, Fowle and Aldrich	Mt. Wilson, California	20	1923
Gibson	—	—	1940
Parry Moon	—	—	1940
Taylor and Kerr	Cleveland, Ohio	19	1941
Condit and Grum	Rochester, U.S.A.	249	1963–4
Henderson and Hodgkiss	Enfield	274	1964
Budde	Ottawa	99	1964
Judd, Macadam and Wysecki	—	—	1965
Winch, Boshoff, Kok and Du Toit	Pretoria, S.A.	422	1966
Tarrant	Putney	366	1967
	Saffron Walden	25	

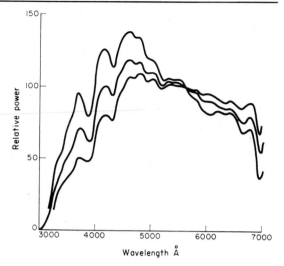

Fig. 2. Typical SPD curves observed in lower resolution than Fig. 1.

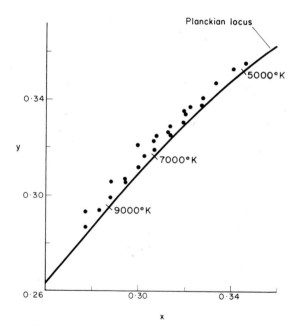

Fig. 3. Mean C.I.E. chromaticities for daylight reported by Henderson, Winch and Tarrant.

for the purpose of specifying standard illuminants to represent daylight. The colour of sky light has also been measured at various places and notable studies have been made by Nayatani and Wysecki at Ottawa, Chamberlin, Lawrence and Belbin at Salisbury (England) and Collins at Teddington.

The introduction of automatic measuring techniques has enabled large numbers of observations to be made in recent years, and with increasing refinement. Winch's measurements were made at 50 Å intervals and Tarrant's at 10 Å intervals. Condit and Henderson and Tarrant made their measurements in places where there was much pollution but Tarrant was able to make some at a country location for comparison purposes. Winch's measurements comprised the only recent study in the Southern hemisphere, and were made at a latitude much nearer the equator (25°S) than the others.

Typical Results

As far as the major part of the visible range from 4000–7000 Å is concerned, all recent studies are in close agreement. The SPD is always found to be one of a family of curves, such as that shown in Fig. 2. These curves very roughly resemble those of Planckian radiators in the visible range, and if the C.I.E. chromaticity is calculated for any observed SPD it lies close to the locus of chromaticities of Planckian radiators in the C.I.E. chromaticity chart (see Fig. 3). It is thus convenient to refer to any particular SPD obtained for daylight by its correlated colour temperature. It is generally held that for a given correlated colour temperature the SPD will be the same in all parts of the world.

For typical results in the range 4000–7000 Å reference may be made to the formulation of Judd et al. (i.e. the figures adopted by the Commission Internationale de l'Eclairage) or any of the recent works mentioned. For wavelengths above 7000 Å, Judd relied on an extrapolation which in some quarters is felt to be unreliable, but both Henderson and Winch have made adequate measurements in this region. For wavelengths below 4000 Å, there is currently some doubt. Winch reported a much higher content of u.v. than the contributors to the Judd formulation, and it was at first presumed to be due to the higher altitude (1400 m) and lower latitude of Winch's observing station; but Tarrant's results confirm Winch's to a striking degree. (At the time of writing further studies of this region were in progress.)

A sample SPD is tabulated below (Table 2); it is based on Tarrant's results and represents a typical SPD whose correlated colour temperature is 6500°K. The power figures represent the mean power in each 100 Å block of the spectrum; Tarrant's measurements were made at 10 Å intervals and each of the power figures quoted here are the mean of 10 such points.

Colour Temperature Range of Daylight

A word of caution is necessary in speaking of daylight SPD's in terms of correlated colour temperatures. Henderson has demonstrated that light from any part of the sky has an SPD roughly resembling that of a Planckian radiator. In certain cases, where light is only taken from a small zone of the sky (e.g. through a north-light window) it is possible to obtain correlated colour temperatures very much higher than those which are obtained when light from the whole sky is used. Consequently in discussing the ranges of colour temperature of daylight observed under various circumstances it is important that the collection conditions are not overlooked.

In general, when light from the whole sky is used the correlated colour temperature will lie in the range 4800°K to 10,000°K. The most frequently occurring value in this country is near 6500°K. Direct sunlight, alone, generally has a colour temperature about 4800°K; when the sky is fully overcast it is near 6500°K; and on a clear day the intense blue sky at the zenith may reach 100,000°K. A histogram indicating the relative frequency of different colour temperatures at Teddington, England, due to Collins, is shown in Fig. 4; the collection conditions were such that light from the entire sky was sampled.

Table 2

Wavelength, Å	Spectral power per 100 Å band (relative to that of 5600 Å, taken as 100)
3000	0
3100	3
3200	8
3300	29
3400	38
3500	48
3600	57
3700	71
3800	68
3900	61
4000	82
4100	98
4200	101
4300	92
4400	105
4500	115
4600	119
4700	116
4800	117
4900	109
5000	110
5100	108
5200	100
5300	103
5400	103
5500	103
5600	100
5700	98
5800	97
5900	90
6000	90
6100	90
6200	88
6300	84
6400	85
6500	82
6600	75
6700	77
6800	79
6900	56
7000	58

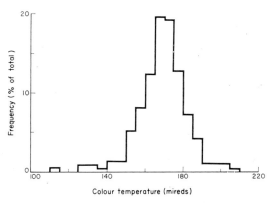

Fig. 4. Histogram showing the frequency of occurrence of different colour temperatures in daylight at Teddington. (After Collins.)

Effects of Local Variables

Attempts have been made in some recent studies to correlate the SPD of daylight with time of year, time of day, degree of cloud cover and wind direction. As far as time of year is concerned it is found that although the probability of a particular SPD occurring varies with time of year, the SPD's obtained for a given colour temperature at widely different times of year are indistinguishable. Tarrant attempted a time-of-day study and the results indicated, but did not prove, that there was a marked effect as the industrial haze built up during the morning—the sky becoming slightly more diffusing—which remained roughly constant during the afternoon. A study of wind direction also indicated a slight but definite increase in diffusion when the wind blew from London over his observing station.

The correlation between cloud cover and SPD is what might be expected. When the sky is heavily clouded there is a great deal of diffusion, so that direct sunlight and sky-scattered light are completely mixed, and the SPD obtained will be near that corresponding to 6500°K. When the sky is clear the SPD may fall anywhere in the range, according to the relative intensities of direct sunlight and sky scattered light that are being collected.

Bibliography

ABBOT, FOWLE and ALDRICH (1923) *Smithsonian Miscellaneous Collection* **74**, No. 7.
CHAMBERLIN, LAWRENCE and BELBIN (1963) *Light and Lighting* **56**, 73.
COLLINS (1965) *Brit. J. Appl. Phys.* **16**, 527.
CONDIT and GRUM (1964) *J. Opt. Soc. Amer.* **54**, 937.
GIBSON (1940) *J. Opt. Soc. Amer.* **30**, 88.
HENDERSON and HODGKISS (1963/1964) *Brit. J. Appl. Phys.* **14**, 125; **15**, 947.
JUDD, MACADAM and WYSECKI (1964) *J. Opt. Soc. Amer.* **54**, 1031.
NAYATANI and WYSECKI (1963) *J. Opt. Soc. Amer.* **53**, 626.
PARRY MOON (1940) *J. Franklin Inst.* **230**, 583.
TARRANT (1967) Ph. D. Thesis, University of Surrey and *Transactions of the Illuminating Engineering Society, London* **33**.
WINCH, BOSHOFF, KOK and DU TOIT (1966) *J. Opt. Soc. Amer.* **56**, 456.

A. W. S. TARRANT

DIFFUSION IN NON-UNIFORM MEDIA. It is convenient to distinguish between diffusion in non-uniform single-phase systems and in multiphase systems. Phase is used here in the wider sense to mean the thermodynamic region of a substance which may be traversed without a discontinuity in any of the properties. We may further distinguish within the first category between diffusing systems where Fick's Law is an adequate description of the process, that is where the diffusion flux reduces to zero as the concentration gradient approaches zero, and systems where other potential gradients exist (a temperature gradient for instance), for which Fick's Law is no longer adequate.

Choice of Reference Frame

In the case of diffusion of a single species, such as a pure gas down its own pressure gradient, the flux may be measured with respect to the container walls without ambiguity and a single diffusion coefficient is sufficient to describe the process. In a binary system, however, we would expect each component to diffuse at a different rate with respect to local inert markers, giving two intrinsic diffusion coefficient, defined by Fick's Law:

$$D_A = -J_A/(dc_A/dx); \quad D_B = -J_B/(dc_B/dx).$$

Here J represents the flux per unit area relative to local markers in the x direction, and c the concentration per unit volume. A binary system may nevertheless be described by a single diffusion coefficient by the choice of a reference frame other than markers embedded in the diffusing medium. This may be achieved by determining the Matano interface in a diffusion system (*see* Fick's Law, Matano-Boltzmann solution of) and defining it as the origin. When no volume changes are involved, this is equivalent to fixing the reference frame to a point outside the diffusion zone. The interdiffusion coefficient D derived in this way is related to the intrinsic diffusion coefficients by the expression:

$$D = X_A D_B + X_B D_A \quad (\textit{see Darken equations})$$

where X_A and X_B are the mole or atom fractions. Even if D_A and D_B are concentration-independent, it is clear that D cannot be, and one may only use this approximation at dilute solution.

Concentration-dependent Diffusion Coefficients

If a concentration gradient exists only in the x direction, we may write Fick's Law as:

$$\partial c/\partial t = \partial(D\partial c/\partial x)/\partial x$$

where c is the concentration of the diffusant, t the time, and D the diffusion coefficient. In many systems, particularly where the concentration differences are not large, D may be assumed constant, and analytical solutions of the above equation for this situation exist for many different boundary conditions. However, in binary metallic systems which form complete solid solutions, order of magnitude changes in D with concentration have been measured. In systems of limited solubility, the concentration effect is even more marked. Again in the case of vapour diffusion in high polymers, the diffusion coefficient is found to be very sensitive to concentration. Presumably this reflects the change of environment of the diffusing species in the medium, and since this must be a continuous function of concentration in a single phase, so also must D be a continuous function of concentration.

Solutions of the diffusion equation in terms of commonly tabulated functions are not available for a variable D. However, for certain relations between concentrations and D of special importance in infinite and semi-infinite media, solutions have been obtained in the form of closed integrals, which have been numerically calculated and tabulated. These include the situation dealt with by Wagner where the diffusion coefficient varies exponentially with concentration (commonly observed in metallic systems), and that of a linear relationship (*see* Crank).

Boltzmann found that the substitution of $\eta = x/2t^{\frac{1}{2}}$ in the diffusion equation, transforms it to an ordinary differential equation:

$$-2\eta \, dc/d\eta = d(D \, dc/d\eta)/d\eta.$$

This substitution may be used where the boundary conditions of the problem can be expressed in terms of η above, and this is possible for infinite and semi-infinite systems with uniform initial concentrations. The integration of the above equation gives the Matano-Boltzmann solution (*see* Fick's Law, Matano-Boltzmann solution of) which allows the concentration dependance of D to be determined from diffusion experiments by graphical integration. On integrating a second time, the following solution is obtained:

$$c = C_0 - \frac{\int_0^\eta (1/D) \exp\left\{-\int_0^{\eta'} (2\eta'/D) \, d\eta'\right\} d\eta'}{\int_0^\infty (1/D) \exp\left\{-\int_0^{\eta'} (2\eta'/D) \, d\eta'\right\} d\eta'}$$

(where $c = C_0$, $x = 0$, $t > 0$; $c = 0$, $x > 0$, $t = 0$) and the integrals can be calculated numerically for any given function of D with concentration.

In finite systems where the Boltzmann transformation cannot be made, it is necessary to solve the original partial differential equation by substituting finite difference approximations for the derivatives (*see* Differential equations, Partial). Where the initial conditions have very steep gradients or discontinuities, one may treat the problem initially as a semi-infinite system by the method above, and introduce the finite difference approximations at the point where this approximation is no longer reasonable.

Diffusion in a Single Phase in the Presence of Other Potential Gradients

In systems where more than one transport process occurs, we might expect an interaction or coupling between the fluxes. This coupling is evident in Thermal Diffusion (Soret Effect), where a heat flux may lead to the unmixing of an initially uniform solution. Similarly, the diffusion flux of one species may be influenced by that of another, or by the current produced by an applied electric field. It is generally assumed that the flux of a species 1 may be written:

$$J_1 = L_{11}X_1 + L_{12}X_2 + L_{13}X_3 \ldots$$

in which the X's represent thermodynamic forces related to the various potential gradients in the system, and the L's are coefficients. This equation allows for the dependence of the flux J_1 on all the other potential gradients in the system (thermal, electric or chemical), as well as on its own potential gradient through X_1. A flux may be written in this form for each transport process, and providing the thermodynamic forces and the frame of reference in the system have been properly chosen, the coupling coefficients are symmetrical, that is: $L_{ij} = L_{ji}$ (Onsager relationships—*see* Thermodynamics, Irreversible). We confine ourselves in what follows to those situations where the forces arise from within the diffusion medium itself, and will not consider the effect of imposed forces such as thermal gradients or electrical fields which produce a net flux of heat and current.

The proper treatment of multicomponent diffusion must take account of the coupling effects between the different diffusing species. However, calculations from simple diffusion models and recent experiments in ternary alloy systems both indicate that the coupling term is probably negligible in most cases, and therefore the only effect of other components on a diffusing atom or molecule is through its own potential gradient. In isothermal diffusion the correct thermodynamic force is the chemical potential gradient, and we may write for the diffusion flux of the ith species:

$$J_i = -L_{ii} \, d\mu_i/dx$$

where μ is the chemical potential. The flux here is measured with respect to inert markers imbedded in the diffusion medium. An alternative treatment of multicomponent diffusion is to extend Fick's Law to make the flux of each component a linear function of the concentration gradients:

$$J_i = -\sum_k D_{ik} \nabla c_k.$$

Clearly this formalism does not require any knowledge of the thermodynamics of the system, but requires a larger number of coefficients to be determined. However, where the thermodynamic properties are known, Onsager has shown that the following relations hold:

$$\sum_j \partial \mu_i/\partial c_i D_{jk} = \sum_j \partial \mu_k/\partial c_j D_{ji}$$

thus reducing the number of independent coefficients.

The process of diffusion itself may develop forces which resist its flow. In electrolytes, for example, the different mobilities of the ionic species set up an electric field producing a drag on the faster moving species. In a crystalline solid, if the different species diffuse at different rates with respect to the lattice, there must be lattice flow for the total number of lattice sites to be conserved. Similar effects occur in non-crystalline solids and liquids. If this flow is restricted in some way, pressure gradients will exist resisting further diffusion. Also pressure gradients will be set up if phase transformations occur in solids in which the products and reactants have different specific volumes. Dislocations are another source of stress field in crystals, and the hydrostatic component will influence diffusion in their vicinity.

Both these problems are treated similarly. The coupling coefficients L_{ij}, $i \neq j$, in the general diffusion equations are again assumed zero, and we may write for an isothermal system, containing ions 1 and 2:

$$J_1 = L_{11}(-d\mu_1/dx - e_1\,d\phi/dx)$$
$$J_2 = L_{22}(-d\mu_2/dx - e_2\,d\phi/dx)$$

where μ and e are the chemical potential and ionic charge respectively, and $d\phi/dx$ is the electric field set up in the solution due to the diffusion process. Assuming dilute solutions and using the condition of electrical neutrality, it is possible to show that these equations may be reduced to:

$$J_1 = -D\,dc_1/dx; \quad J_2 = -D\,dc_2/dx$$

where the single diffusion coefficient

$$D = RTu_1u_2(e_1 - e_2)/(u_1 + u_2)\,e_1e_2$$

and u_1 and u_2 are the mobilities of the ions (their velocities at uniform concentration under unit electric field).

In the case of solid state diffusion in hydrostatic pressure gradient, the diffusion fluxes may be written:

$$J_1 = -L_{11}\,d\mu_1/dx$$
$$J_2 = -L_{22}\,d\mu_2/dx.$$

If B is the mobility of an atom in a unit potential gradient (analogous to u)

$$J_1 = \frac{-B_1c_1}{N}\,d\mu_1/dx; \quad J_2 = \frac{-B_2c_2}{N}\,d\mu_2/dx$$

where the concentration c is in moles/unit volume, N Avogadro's number and the flux is in moles/unit area. The mobility is related to the intrinsic diffusion coefficient through Einstein's equation: $D_i = B_iRT/N_0$. We may also write

$$d\mu_i = (\partial \mu_i/\partial P)\,dP + (\partial \mu_i/\partial n_i)\,dn_i$$

or

$$d\mu_i = \bar{V}_i\,dP + RT\,d\ln a$$

where n is the mole fraction, a the chemical activity and $\bar{V}$ the partial molar volume. For dilute solutions where $d\ln a \approx d\ln c$, we may substitute this equation in the flux equations to obtain:

$$J_1 = -D_1\left\{dc_1/dx + \frac{c_1\bar{V}_1}{RT}\,dP/dx\right\}$$
$$J_2 = -D_2\left\{dc_2/dx + \frac{c_2\bar{V}_2}{RT}\,dP/dx\right\}.$$

Since the diffusion of the two species is necessarily in opposite directions, the effect of the pressure gradient is to assist the diffusion of one, and oppose the other. It should be remembered that the boundary conditions in any actual problem are also affected by pressure.

Diffusion in Multiphase Systems

Diffusion to or from a boundary separating two phases often results in the growth of one and the disappearance of the other, with a consequent movement of the boundary. This occurs during the precipitation of a new phase within a matrix. Phase boundary movement necessarily occurs in binary systems because the interface concentrations are invariant under equilibrium conditions, but in three component systems or higher the boundary may be stationary. In the solutions to the diffusion equations which have been obtained, local thermodynamic equilibrium at phase interfaces is usually postulated since the interface reaction involves only single jumps of atoms or molecules, whereas the diffusion process involves a very large number of jumps requiring about the same activation energy. This is probably very reasonable when the diffusing species are single atoms; where complex molecules are involved requiring the breaking and reforming of chemical bonds to form a new phase at the interface, this assumption may no longer be valid. Further assumptions are that constant diffusion coefficients may be used in each phase, and the heat produced at the interface as a result of transformation or solution may be neglected.

The simplest situation is that involving a stationary planar boundary. An example would be that of a dissolved gas diffusing between two immiscible liquids, where the partial molar volume of the gas in each phase is very small and the gas partitions itself at the interface according to the Nernst partition law. The solution to this problem has been given by Jost for the initial conditions of uniform concentration. For different initial conditions, and systems involving a number of phases—as in composite sheets where each medium or phase has a different diffusion coefficient and each pair of contiguous phases a given partition coefficient—it is necessary to resort to finite difference techniques for solutions. In all cases of diffusion with stationary boundaries, the flux of diffusant must be the same on each side of the boundary:

$$D_A(dc/dx)_A = D_B(dc/dx)_B.$$

The moving boundary problem has been considered by Wagner for a number of cases of practical interest in binary systems (*see* Jost). The solutions, as before, are confined to diffusion in one dimension, and planar interfaces normal to the diffusion direction. Also differences in specific volume between phases is ignored. The cases treated are (a) that where the boundary separates phase A from the phase mixture $(A + B)$; (b) that where the boundary separates single phase regions A and B; (c) that where the single-phase region B develops from the phase mixtures $(A + B)$ and $(B + C)$. An example will help illustrate the method of obtaining a solution. The situation considered is that of a material initially composed only of the phase mixture $(A + B)$, in which the single phase A develops from the surface ($x = 0$) at $t = 0$ by the diffusion into the specimen of one of the species. The surface concentration is maintained at c_0, as shown in Fig. 1. A solution to Fick's Law is assumed for the single phase region of the form:

$$c = c_0 - \beta\,\mathrm{erf}\,(x/2(Dt)^{1/2})$$

where c is the concentration at x, and β a constant. It may be shown by substitution that if the boundary (at $x = \xi$) moves with time according to the relation $\xi = \gamma t^{\frac{1}{2}}$ (where γ is a proportionality constant), then the concentration at the interface is constant. This

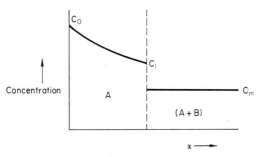

Fig. 1.

allows the second boundary condition to be satisfied. The actual value of β is determined by the equation expressing conservation of mass at the moving interface:

$$(c_1 - c_m)\,\mathrm{d}\xi = -D\,\mathrm{d}t\,(\mathrm{d}c/\mathrm{d}x)_{x=\xi}$$

where c_1 and c_m are given.

Danckwerts (*see* Crank) has derived more general solutions to moving boundary problems, which take account of the difference in specific volume of the two contiguous phases. He also treats the class of problem where the movement of the boundary is not directly proportional to the amount of substance diffusing across it. Such a situation exists, for example, where a substance diffuses into a solution and there reacts with another solute to produce a precipitate (as in the internal oxidation of metals).

So far the discussion has been confined to linear diffusion. Non-equilibrium conditions may be induced in a system by the simple change of temperature, without the need for diffusion from outside. The resulting growth or dissolution of precipitates in the matrix involves diffusion in more than one dimension. In solving the diffusion equations, it is always assumed that the precipitate concentration, and that of the matrix in contact with it, are those of the bulk phases in equilibrium. This cannot be when the particles are of very small dimensions owing to the Gibbs-Thomson effect which tends to lower supersaturation. For the ideal cases of the thickening of plates and the radial growth of cylinders and spheres, the diffusion equations to be solved for the matrix are:

$$\partial c/\partial t = D(\partial^2 c/\partial r^2) + (j-1)(D/r)(\partial c/\partial r)$$

where $j = 1, 2, 3$ correspond to the plate, cylinder and sphere respectively and r is measured from the particle centre. For the given boundary conditions, Zener has shown that the solutions are:

$$c = c_\infty + (c_1 - c_\infty)\frac{\phi_j\{r/(Dt)^{1/2}\}}{\phi_j\{R/(Dt)^{1/2}\}}$$

where
$$\phi_j(x) = \int_x^\infty z^{1-j}\exp(-z^2/4)\,\mathrm{d}z$$

and the compositions are as indicated in Fig. 2. From this one may easily derive the flux at the precipitate

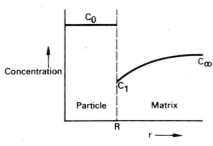

Fig. 2.

interface and determine the growth rate from solute conservation at the moving boundary:

$$\text{Rate} = D(\partial c/\partial r)_{r=R}/(c_0 - c_1)$$

(where the concentrations are expressed in amount per unit volume). A useful approximation to the concentration gradient at the surface of a growing sphere is given by the steady-state solution at each stage, provided the supersaturation of the matrix is not high:

$$(\partial c/\partial r)_{r=R} = (c_\infty - c_1)/R.$$

The diffusion fields around an array of particles in a matrix eventually impinge on one another, and the competition for solute causes a reduction from the above predicted growth rate. The exact treatment of impingement is difficult, and often arbitrary factors are inserted in the growth equations to take account of it. For a detailed account of this, and of the diffusion equations relating to other growth forms such as dendrite needles and Widmanstätten plates, Christian should be consulted.

Bibliography

CHRISTIAN J. W. (1965) *The Theory of Transformations in Metals and Alloys*, Oxford: Pergamon Press.
CRANK J. (1956) *The Mathematics of Diffusion*, Oxford.
DENBIGH K. G. (1951) *The Thermodynamics of the Steady State*, London: Methuen.
JOST W. *Diffusion*, Academic Press.
ONSAGER L. (1945) *Ann. N.Y. Acad. Sci.* **46,** 241.

D. H. KIRKWOOD

DISLOCATION KINKS. If a crystal dislocation lies predominantly along a low-index crystal direction under the influence of the Peierls Stress (q.v.), it advances from one potential trough to another in regions of higher energy density called kinks. A kink is thus a displacement of the line of a dislocation in its own glide plane, while a jog (q.v.) is a displacement of the dislocation line out of its glide plane. The effects of kinks are discussed in the article *Peierls Stress*. The theory is given fully by Rosenfield.

Bibliography

ROSENFIELD A. R. *et al.* (eds.) (1968) *Dislocation Dynamics*, New York: McGraw-Hill.

F. R. N. NABARRO

DISPERSION AND FIBRE STRENGTHENING

Introduction

Resistance to plastic flow and fracture are important in relation to the use of materials. High resistance to flow enables working stresses to be carried elastically, whilst high resistance to fracture ensures that if the design stresses are exceeded at any local point in a structure, failure is difficult and requires the expenditure of considerable energy. Two methods of controlling flow and fracture behaviour are dispersion and fibre strengthening.

To produce plastic deformation in perfect crystals of metals, stresses are required which are an appreciable fraction of the shear modulus. However, more usually dislocations are present in metal lattices and these reduce by several orders of magnitude the resistance of single crystals and polycrystals to plastic flow. Pure polycrystalline metals are therefore easy to deform plastically. However, certain other materials which possess strongly directional bonding, e.g. some oxides, carbides, and nitrides and also pure germanium, silicon and carbon, may be extremely difficult to deform plastically even though they contain dislocations. Such solids are said to be inherently strong to distinguish them from metals such as copper and iron which are only strong in the pure state when free from dislocations.

There are many methods of increasing the resistance of a metal to plastic flow such as work hardening, solid solution hardening and dislocation locking, but two important ways are:
(a) to reduce the mobility of the dislocations by forming barriers in the lattice as in dispersion (or precipitation) strengthening, or
(b) to arrange to carry much of the applied load by high-strength fibres inserted in the metal as in fibre strengthening.

Dispersion Strengthening

Dispersion strengthening has been utilized for many years and a large number of important engineering alloys are strengthened in this way by having particles of a second phase distributed as uniformly as possible through the matrix.

An important requirement is that the barriers to dislocation movement should be as close together as possible, i.e. the interparticle spacing should be small. In this way a dislocation line gliding in the matrix comes in contact with the barriers at many points. Interparticle spacing decreases as the particle size decreases, if the volume fraction is kept constant, and dispersion strengthened metals normally contain second phase particles having sizes in the range from a few tens to a few thousand angstroms. In principle it is possible to produce such a structure synthetically, i.e. by direct incorporation of the second phase into the metal, using for example powder metallurgy techniques, but there are serious technical problems in trying to do this and an alternative method enables certain dispersions to be obtained much more readily. This depends on using a solute metal which has a greater solubility in the solvent metal matrix at high temperature than at low temperature. Slow cooling of an appropriate composition thus results in the precipitation of the solute metal, or more generally some intermetallic compound formed by interaction of the solvent and solute, in the form of rather coarse and hence widely spaced second phase particles. To produce a fine dispersion the alloy is first cooled very rapidly from a temperature at which the two metals form a solid solution. This prevents precipitation taking place during cooling and a supersaturated solid solution is formed. If this is subsequently reheated to a suitable temperature the second phase forms at many nucleation sites, to produce a dispersion of small closely spaced particles. The heat treatment required for particular alloys to develop maximum resistance to deformation depends on the solvent and solute metals; in particular the temperature and time of reheating are very important. The precipitate particles produced by this technique do not necessarily have the equilibrium composition and structure which would result by precipitation on slow cooling. This is because the supersaturated solid solution usually transforms to the equilibrium structure via a series of intermediate precipitates of different composition and structure. Frequently one of these intermediate states produces maximum dispersion strengthening.

Many commercial alloys are dispersion strengthened, e.g. the aluminium–copper alloy which is the basis of duralumin, and in which the precipitate is an intermetallic compound of copper and aluminium ($CuAl_2$) and steels which in the fully heat-treated condition contain dispersions of small carbide particles. In complex alloys several types of particle may be present simultaneously and give enhanced strengthening. Yield strengths, i.e. resistance to initial plastic deformation, can be increased considerably by fine precipitate particles. Thus a dispersion-hardened copper alloy containing about 2·5 per cent of beryllium can develop a yield strength (0·1 per cent plastic deformation) of 50 tsi compared with 3 tsi for pure copper.

The theory of dispersion strengthening has been extensively studied. The particles can interfere with the movement of dislocations in the matrix in different ways depending upon whether the particles are non-deformable or deformable. In the first case they must be inherently very strong so that the dislocations do not penetrate them and are halted at the matrix-particle interface. Extensive plastic deformation can then only occur by the lengths of dislocation between the particles moving forward, or by the dislocations surmounting the obstacles by gliding onto a new slip-plane at an angle to the initial slip-plane. Non-deformable particles thus act as rigid pegs inserted across the slip-plane and they carry an increasing stress as dislocations moving in the matrix try to by-pass them. The initial yield stress in this case depends inversely on the interparticle separation by an expression of the form

$$Y = \alpha \frac{Gb}{l} + Y_m$$

where Y = shear yield stress, b = burgers vector of the dislocation, l = interparticle separation, G = shear modulus, Y_m = shear yield stress of the matrix and α = constant. Values of yield stress which are significant fractions of the shear modulus can thus only be obtained when l is less than a few hundred angstroms. When deformation does take place the stress rises on the particles and on their interface with the matrix, and eventually one or other cracks to initiate fracture. Dispersed particles may not always be able to resist elastically the applied stresses acting on them due to the dislocations at the interface with the matrix. Plastic flow then occurs in the particles and the yield stress is less than that given by the above expression. Its value depends on factors such as (a) whether slip changes the state of order in the lattice of the particle since the work of disordering must be supplied by the applied stress, (b) the amount of interfacial energy between particle and matrix which has to be provided as particle shear increases matrix–particle contact area, (c) the presence of local elastic stresses in particle and matrix due to misfit between them, which interact with the stress fields of the dislocations. The initial yield stresses of alloys known to contain deformable particles have been quantitatively explained in terms of the stresses needed to force dislocations through the particles.

The change of yield stress with temperature for a dispersion hardened material depends on the nature and size of the dispersed particles. It is small for non-deformable particles but greater for small particles which deform as yielding takes place. However, even for non-deformable particles the small temperature dependence is only true if the dispersion remains stable as the temperature is raised. Any increase in particle size or any re-solution of the particles will result in an increase in the average interparticle spacing and a reduction in yield stress. Dispersions produced by precipitation hardening are thus sensitive to the effects of particle coarsening caused by short periods of heating at temperatures above that at which the precipitate was formed initially, and also to the effects of increased solubility of the particles at higher testing temperatures. These problems can be overcome by using as the dispersed phase a compound which has very low solubility in the metal matrix so that it can resist growth and solution at high temperatures. A low particle solubility means, however, that a precipitation hardening technique can no longer be employed to produce the dispersion. Instead a method must be used such as the direct mixing of powders of matrix metal and second phase, followed by compaction and sintering. It is then very difficult to produce the uniformity of the dispersion which is required, combined with small particle sizes and spacings. Numerous techniques have been employed in an attempt to solve this manufacturing problem, and produce metals strengthened by small particles of hard, stable compounds such as oxides and nitrides. Some success has been achieved and the alloys show particularly good properties at high temperatures due to the stability of the dispersions. It is, however, difficult to produce alloys with large volume fractions of dispersed phase so that the strengthening obtained at low temperatures usually compares unfavourably with that achieved by precipitation hardening, which allows much larger volume fractions of precipitate to be obtained.

In general the higher the yield stress produced by dispersion strengthening, the smaller the total plastic elongation to fracture. This can set a limit to the useful strength level of an engineering alloy, since good resistance to crack propagation and fracture are usually equally important to strength. In some dispersion strengthened alloys structural inhomogeneities are responsible for the low ductility. Precipitate particles in grain boundaries are usually coarser than those within the grain interior and a region free from precipitates can occur in a narrow zone adjacent to the boundary. This structure develops as a result of precipitate nucleation and growth taking place at different rates in the grain boundary and its vicinity for reasons associated with the boundary structure. It is thus difficult to achieve a completely uniform distribution of precipitate particles. The larger grain boundary particles can result in easier crack initiation during plastic deformation either through cracking of the particles or their interface with the matrix. The precipitate-free region may also act as a relatively easy path along which a crack can grow by plastic deformation. Because of these effects cracking becomes localized in the grain boundary regions and the energy absorbed to fracture is small. The factors controlling crack initiation and growth in dispersion strengthened alloys are only imperfectly understood compared with the understanding of yielding behaviour and there is room for considerable research.

Fibre Strengthening

Strengthening by dispersed particles depends on limiting the dislocation movement in the matrix by using small closely spaced particles. An alternative approach to prevent plastic deformation of a soft matrix is to incorporate an appreciable volume fraction in fibre form of a strong material which is distributed in such a manner that it carries elastically a large part of the applied load. This is the principle of fibre strengthening. Since the fibres carry the stresses elastically they must have a high resistance to plastic flow, and are usually made from inherently strong substances such as oxides and certain other crystalline materials in which dislocations are very difficult to move. Such materials in the form of fine smooth surfaced filaments, exhibit yield strengths which are an appreciable fraction of the shear modulus. Fibres of glassy materials can also be made which show high strength and these are used in glass-reinforced plastics. In some cases high-strength metal wires made by extensive drawing or other techniques may also be suitable as reinforcement fibres.

In general high-strength fibres only remain strong if they are free from internal defects and surface defects. This is because in highly elastic solids such defects concentrate stress very effectively at their tips so that fracture occurs at low applied stresses. It should be noted that such materials cannot be used in monolithic form, since it would be difficult to make them defect free initially in large volume and even if this could be done any damage occurring on the surfaces due to abrasion or impact during service would cause a catastrophic reduction in strength. High-strength fibres used for reinforcement also suffer loss of strength if their surfaces are damaged and it is therefore necessary to prevent this during service by embedding them in a suitable matrix, such as a metal or a plastic. Another advantage of fine fibres over monolithic sections of strong solids, is that if any defect is present on the surface of one particular fibre this affects only that fibre and does not result in a crack which can immediately propagate over the whole cross-section of the solid.

To obtain the maximum strengthening effect from fibre reinforcement, it is essential that as much as possible of the applied load should be carried by the fibres. This can be achieved by either (a) the matrix having a lower elastic modulus than the fibres so that a given strain applied to the composite specimen produces a higher stress in the fibres, or (b) the matrix deforming a small amount plastically and in this way transferring the load to the fibres. The former is the situation with glass fibres in resin matrices and the latter with high strength non-metal or metal fibres in a metallic matrix.

The elastic modulus of a fibre composite is found to depend by the simple rule of mixtures on the elastic moduli of the components and their volume fractions, provided both components behave elastically. However, when the matrix starts to deform plastically, the elastic modulus contribution from the matrix becomes very small compared with that from the fibres which are still strained elastically, so that the overall modulus then depends mainly on the fibre modulus and its volume fraction.

The maximum stress which a composite can withstand, defined as the maximum load divided by the initial cross-sectional area, depends on the breaking stress and volume fraction of fibres, V_f, and the yield stress of the matrix after it is work hardened to the strain at which the composite breaks. Thus

$$\sigma_c = \sigma_f \cdot V_f + \sigma_m' \cdot (1 - V_f),$$

where σ_c and σ_f are the breaking strengths of the composite and fibre and σ_m' is the yield stress of the matrix at the breaking strain of the composite. This breaking strain will be much less than that which an unreinforced metal undergoes to fracture, so that at composite fracture the stress carried by the matrix will be small compared with its usual ultimate tensile strength. One result of this is that a minimum volume fraction of fibres, V_{crit}, is required in a particular matrix metal, before any increase in strength is obtained relative to the unreinforced metal, i.e.

$$\sigma_f V_{\text{crit}} + \sigma_m'(1 - V_{\text{crit}}) = \sigma_u,$$

which gives

$$V_{\text{crit}} = \frac{\sigma_u - \sigma_m'}{\sigma_f - \sigma_m'}.$$

The critical volume fraction decreases therefore as the fracture strength of the fibres σ_f increases, and as the work-hardening characteristic of the matrix decreases, i.e. $\sigma_u - \sigma_m'$ becomes very small.

The maximum strengthening effect of fibres is obtained when they are continuous throughout the length of stressed material. However, strong discontinuous fibres also give high strength composites provided the aspect ratio, i.e. length to diameter ratio, is sufficiently large. With discontinuous fibres load is transferred to the fibres either by the matrix having a lower modulus than the fibre or by plastic flow of the matrix. The longitudinal stress in a discontinuous fibre is zero at the ends and rises to a maximum value at the centre, whilst the shear stresses in the matrix are a maximum at the fibre ends and zero at the fibre centre. To obtain the maximum strengthening from discontinuous fibres, the stress carried by them at failure should be equal to their inherent fracture strength, i.e. composite failure should occur by fibre fracture rather than flow or fracture of the matrix. An isolated fibre radius r, length l and of strength σ_f embedded in a ductile matrix of shear yield strength τ, will break due to the stress transmitted by flow of the surrounding matrix, when the load at its mid-point is just greater than its breaking load, i.e. $\dfrac{l}{2}\tau 2\pi r = \pi r^2 \sigma_f$.

This gives

$$l = \frac{r \cdot \sigma_f}{\tau}$$

and this length is known as the critical length, l_c. Fibres with an aspect ratio less than that given by

$$\frac{l}{r} = \frac{\sigma_f}{\tau}$$

cannot be loaded to their fracture stress and failure of the composite specimen occurs by general failure of the matrix. For a given diameter of fibre the critical length is a measure of the rapidity with which stress builds up from zero at the fibre ends, and is a function of the shear strength of the matrix. Because of the stress variation along the length, discontinuous fibres strengthen a given matrix less than continuous fibres. However, the difference is small if the discontinuous fibres are much longer than the critical length.

The strength of a fibre composite is greatest when the stress is applied along the axis of the fibres. Transverse strengths are lower due to the fibres carrying negligible stress in this orientation, and composite failure occurs by shear in the matrix or by fibre–matrix interface failure. To obtain a uniform strength in all directions within the plane of a fibre-reinforced plate the fibres must either be arranged randomly in the plane or alternatively thin sheets containing orientated fibres must be stacked together in a random manner. The theoretical strengths of random arrays of fibres have been derived. In two dimensions the strength is about one-third the uniaxial strength, whilst for a random three-dimensional array it falls to about one-sixth.

A particular advantage of fibrous composites is the possibility of achieving high strength combined with good fracture toughness, i.e. an insensitivity to notches so that the growth of cracks involves a substantial adsorption of energy. The toughness arises due to effects such as a crack growing perpendicular to the fibres being deflected into the fibre–matrix interface so that further growth is hindered; alternatively fibres which have a length less than l_c and are thus not stressed to fracture, can be pulled out of the matrix as a crack spreads and if the shear strength of the fibre–matrix interface is appreciable, energy must be absorbed to spread the crack.

Fibre composites can be made in several ways. The most direct and widely used technique is the mechanical incorporation of the fibres by processes such as the casting of liquid metal around them, or alternatively using powder metallurgical methods. Other approaches are the production of fibres *in situ* by controlled solidification of particular alloys to produce fibrous growths of solid, or the extensive unidirectional deformation of two-phase systems, so that one phase is drawn out into fibre form. Mechanical incorporation has many advantages such as flexibility in relation to the composition, aspect ratio and volume fraction of the fibres. There are, however, problems in connection with the handling and incorporation of the fibres in the matrix without fibre damage, and in achieving good alignment without fibre–fibre direct contacts which reduce composite strength.

High-strength fibres are available in various forms and materials. Whiskers of metals and non-metals can be made with near theoretical strengths, but they may be susceptible to damage and difficult to handle because of their small size. Fibres of glassy, metallic, or ceramic materials can be produced by drawing, with reproducible high strengths, and are fairly easily handled. Carbon fibres produced by thermal decomposition of polymer fibres have advantages in terms of a favourable combination of high strength and modulus, coupled with low density.

Fibre strengthening has considerable potential for producing composites which are strong up to temperatures which are a high fraction of the melting point of the matrix. This is because the fibres themselves, if made of a ceramic or similar material, may have significantly higher melting points than the matrix, and thus still retain considerable strength at temperatures where the matrix is readily deformed. However, in such cases it is important that the fibre should not dissolve in, or interact with, the matrix, since this could result in loss of fibre strength.

Bibliography

Composite Materials (1966). See chapters by P. J. E. Forsyth (*Fibre Strengthened Materials*), A. J. Kennedy (*Potential of Composite Materials*), A. Kelly (*Theory of Strengthening of Metals*), and G. C. Smith (*Dispersion Strengthened Materials*), London: Institution of Metallurgists/Iliffe.

Experimental Aspects of Fibre Reinforced Metals (1965). See chapters by A. Kelly and G. J. Davies (*Principles of Fibre Reinforcement of Metals*) and by D. Cratchley (*Experimental Aspects of Fibre-Reinforced Metals*), in *Metallurgical Reviews*, Vol. 10, London: Institute of Metals.

KELLY A. (1966) *Strong Solids*, Oxford: Clarendon Press.

KELLY A. and NICHOLSON R. B. (1963) Precipitation hardening, *Progress in Materials Science* **10**, 149, Oxford: Pergamon Press.

PECKNER D. (Ed.) (1964) Chapters on *Dispersed Phase Strengthening* and *Precipitation Hardening* in *The Strengthening of Metals*, New York: Reinhold.

G. C. SMITH

DYNAMIC ELECTRON DIFFRACTION, APPLICATIONS OF. Present-day applications of dynamic scattering of electrons from solids include studies of imperfections in thin metal films, special aspects of inorganic structure analysis, and the study of diffuse scattering. The electron microscope or alternatively convergent beam diffraction apparatus is used for the collection of single-crystal data for these purposes, and this work has depended largely on recent improvements in these techniques. Another essential problem which has received much attention in recent years is the development of the dynamic theory and methods of calculation.

A solution to the dynamic electron scattering problem in the two-beam approximation was given by Bethe in 1932, but a general solution, the so-called n-beam theory, was not available until the late

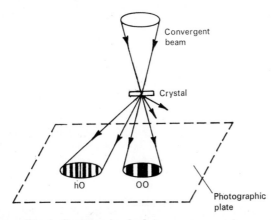

Fig. 1. *Geometry involved in a convergent beam electron diffraction camera using a circular diffraction aperture. The pattern, here drawn for two of the Bragg beams (including the zero-order beam) shows the parallel bands associated with diffraction involving systematic interactions (those between different reflection orders from one set of planes).*

1950's. This development which was important for accurate structure analysis, is dealt with in a separate article. The inaccuracy of the two-beam approximation may be seen to be due to the strong interaction of electrons with matter, so that even reflections having large excitation errors have significance, causing deviations from the extinction length of the two-beam solution of at least several per cent. In some applications, however, where identification of structural features on a semi-quantitative basis is of greatest importance, the two-beam method in which two strong beams are excited experimentally has proved particularly useful, and the simplicity of this solution is such as to encourage its use where possible. The electron microscope study of the faulting and dislocation distribution in metal films has reached its present level of sophistication from work originating principally in Cambridge (Hirsch *et al.*, 1956–60) and has provided a means by which the behaviour of metal under various stresses, at first deduced theoretically from the bulk properties, can be studied in detail. The electron microscope contrast in the unfaulted crystal deviates from uniformity only when the crystal is bent (extinction contours) or has variable thickness (equal thickness fringes). Additional internal boundaries or strain fields associated with stacking faults or dislocations produce contrast by modifying the extinction conditions of the perfect structure. Observation of these defects is dealt with in another article and we are concerned here only to mention it as an important application. Study of defects has led not only to an understanding of their mechanism, but to determination of some properties of the normal structure. The relative strengths of absorption and elastic scattering power for fast electrons, and more recently structural anisotropy, have been measured. With present development many-beam diffraction has offered as yet no clear advantage for this work. The reason appears to be that low or medium resolution has been sufficient for the analyses so far carried out, and indications are that as soon as high resolution microscopy is used to examine dislocation contrast many-beam interactions will need to be considered.

Application of dynamic electron scattering to structure analysis falls naturally into two parts, namely symmetry determination (space-group classification) and refinement of structure potentials. Here the electron diffraction method provides a satisfactory complement to the X-ray method. Both applications depend on the number of beams involved and their interaction strength, so that suitable orientations are those of the principal zone-axes, involving the highest symmetry and the strongest interaction. For inorganic structures these usually involve from 25 to 100 beams of significance. For symmetry determination the convergent beam technique, introduced in its original form by Kossel and Mollenstedt in 1932, is ideal since a display of intensity as a function of angle of incidence of radiation allows certain symmetry elements to be identified by inspection. In X-ray analysis a frequent cause of ambiguity in space-group allocation is inability to distinguish between a centred and a non-centred cell. This distinction is readily made by means of dynamical electron scattering. Centring in a projection results in a diffraction pattern which is centrosymmetric about the zone-axis. In the most complex case when interactions with upper layer reflections as well as those of the zero layer are significant, the pattern of the zero-order beam, otherwise centred independently of the structure, assumes significance. This may arise with 80 kV electrons when the "c" spacing exceeds 10–20 Å. In this case the pattern contains information of the three-dimensional symmetry; however, in most cases the symmetry must be examined from separate projections. Existence of other symmetry elements such as screw axes and glide planes may be determined from the X-ray absences. Parallel to this, dynamical electron absences are generated, provided relatively strict conditions of incidence are maintained. These have so far been only of theoretical interest. The main practical interest at present lies in the ability to distinguish between centred and non-centred structures. However, in cases where single-crystal data is not available from X-ray diffraction a more detailed symmetry analysis may be justified. (See example in Fig. 2; also for illustration of the comparison of the phases of two structure amplitudes by a three-beam method, see Kambe, 1957.)

In the refinement of structure factors, the contrasting behaviour of electron and X-ray scattering is

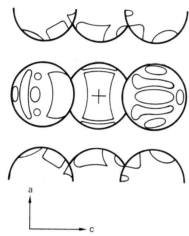

Fig. 2. Zone axis convergent beam pattern, illustrating symmetry determination. With the projected axes shown, the pattern (apart from the central beam pattern which remains centrosymmetric) reveals structural asymmetry in the "c" direction. Position of zone axis is indicated by a cross. (CdS pattern: Goodman and Lehmphul, 1968).

used to advantage. Since the electron method involves a complete dynamical interpretation, those reflections which normally cause most trouble in X-ray analysis, namely the strong inner reflections from inorganic structures, are those determined with the greatest accuracy. On the other hand, for the higher order reflections determined from X-ray intensities relatively free from extinction errors, the electron analysis lacks any great power, since the dynamical coupling is weak. Accurate determination of inner reflections is essential to investigation of the outer electron distribution in the structure. These may be determined to within one per cent or possibly better by a detailed dynamical analysis, as against an error of several per cent in X-ray determinations. In addition there is no barrier in the electron method to examining the scattering power from elements of high atomic number.

For this work two experimental methods have been evolved, distinguished principally by the form in which the data is obtained. The diffraction intensities are either obtained as a function of thickness for a fixed orientation, or as a function of orientation with a constant thickness. However, finally both sets of data require the same type of calculation and general procedure. The method introduced principally by Uyeda and his co-workers depends upon the electron microscope measurement of equal thickness fringes, using a wedge crystal. The contrast from a wedge is sinusoidal according to two-beam theory. In practice the fringes are approximately sinusoidal only when one strong beam is excited or in conditions of high symmetry with several strong beams, but the period is changed from the two-beam solution. The extinction length (fringe spacing) in the bright field image and the dark field images from the low index reflections which are excited may be used to determine the structure factors involved in the pattern, by making n-beam calculations, if the crystal orientation is already known. This latter is usually estimated from a diffraction pattern, but the images from the weak beams (not directly excited) may be used as a check, since they contain fine detail which is sensitive to orientation. Care must be taken to avoid a convergence in the illumination which would destroy the angular resolution. The problems associated with wedge boundaries are avoided by invoking the column approximation (also applied to dislocation contrast in thin metal films), by which it is assumed that columns of the crystal parallel to the incident radiation diffract independently. The dynamical calculation is then made for each thickness as for a parallel-sided crystal. The agreement found so far between calculation and experiment shows that this is an acceptable approximation.

The convergent beam technique has also been adopted for quantitative dynamical work by using the electron microscope objective lens to produce the convergence. In this way a fine cross-over approaching 100 Å is obtained. The fineness of cross-over required is determined by the sample since it is necessary to irradiate a section of plane-parallel sided crystal free from faults. When one strong reflection is excited the original two-beam analysis given by MacGillavry may be employed to estimate the structure factor to within 5% to 15%. The current method may be regarded as an extension of the method to n-beam treatment. When the patterns formed by all the significant beams are examined, certain fringe spacings are found to have strong thickness dependence and relatively little dependence on structure factors, while others are sensitive to both. It is therefore necessary to determine thickness before structure factors can be estimated but the necessary data may frequently be found from detail in a single diffraction pattern.

A trial and error method is used for refinement. The n-beam calculation is made using values of a structure factor deviating by small amounts from the previously accepted value. The results are compared with experiment so that by interpolation and recalculation the correct value may be found. This method is successful when a sufficiently accurate starting value is available. When digital calculation is used, by means of one of the several methods now available, repeated access to a high-speed computer is necessary, and finally perhaps one hundred or so separate calculations need be made. Analogue computation, as introduced recently by A. W. S. Johnson for n-beam calculation, allows the consequences of a change in structure factor to be displayed directly, allowing refinement to be carried to an advanced stage without recourse to digital calculation. For this calculation the n-beam problem is expressed as a set of simultaneous differential equations.

In practice two approaches exist to the problem of n-beam refinement. The first is to calculate for only the systematic interactions (Hoerni, 1956) involving all the higher orders of the main reflection. The validity of this approach depends upon how effectively the non-systematic interactions can be reduced, by suitable choice of orientation. For compounds of light elements, it appears that satisfactory results for the inner reflections may be obtained in this way. For the most accurate determination results should be obtained from orientations of high symmetry, i.e. principal zone-axes, since for these, calculations including all zero-layer reflections may be made. However, since such analyses in general require simultaneous refinement of several structure factors it may be necessary to begin with separate determinations from systematic interactions. The basic accuracy of the n-beam determination of structure factors lies in the number of independent parameters involved. Any number of refinements may be made, by changing the electron wavelength, or in the convergent beam method also by changing the thickness, so that the possibility of obtaining fortuitous agreement, or of hidden errors, may be satisfactorily reduced.

Finally, mention should be made of subjects in which the current or future application of dynamic scattering theory seems most likely to bring new results. Much of the study of inelastic and thermal diffuse scattering to date has been made kinematically, a treatment which owes its success to the relatively low scattering powers involved. These studies have included measurement of angular distributions of scattering, and determination of the vibrational modes of the structure. However, dynamic effects of the diffuse scattering may be observed, by measuring either extinction contrast in the diffuse scattering, or its influence on the Bragg scattering. The former measurement allows a distinction to be made between localized and non-localized scattering sources (the latter occupying an appreciable volume of the crystal), while study of effects on Bragg scattering should allow more detailed analysis of the vibrational states of the lattice.

Another important aspect of dynamic scattering lies in the interpretation of image contrast associated with small detail on a crystal surface. Observation of fine detail such as growth steps may be facilitated by making use of the sensitivity of dynamic intensities, when several beams are involved, to the smallest change in crystal thickness (e.g. Fig. 3). Epitaxially oriented layers of foreign atoms may also be rendered visible by imaging from suitable diffraction patterns, for the same reason. Full development of this branch of high resolution microscopy lies in the future.

See also Dynamical n-beam theory of electron diffraction.

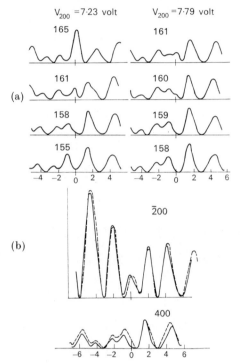

*Fig. 3. Dynamical rocking curves for weak beams from MgO, showing sensitivity of intensities to thickness. (a) Intensity of the 400 reflection with 200 excitation, calculated assuming systematic interactions. Crystal thickness "n" expressed in units of 4·2 Å above each curve; columns represent calculations for two values of V_{200}. (b) Agreement obtained between experiment (broken line) and calculation (for $n = 159$), for $\bar{2}00$ and 400 reflections. (Goodman P. and Lehmpfuhl G. (1967) Acta Cryst. **22**, 14).*

Bibliography

Cowley J. M. (1967) Crystal structure determination by electron diffraction, *Progress in Material Science*, **13**, 6, 269.

Goodman P. and Lehmpfuhl G. (1968) Observation of the breakdown of Friedel's law, *Acta Cryst.* **A24**, 339.

Hirsch P. B. *et al.* (1965) *Electron Microscopy of Thin Crystals*, London: Butterworths.

Howie A. (1966) Dislocations, transmission electron microscopy, in *Encyclopaedic Dictionary of Physics* (J. Thewlis Ed.), Suppl. vol. 1, 61, Oxford: Pergamon Press.

JOHNSON A. W. S. (1968) The analogue computation of dynamic electron diffraction intensities, *Acta Cryst.* **A 24**, 534.

KAMBE K. (1957) Study of simultaneous reflexion in electron diffraction by crystals. *J. Phys. Soc. Japan* **12**, 25.

MACGILLAVRY C. H. (1940). *Physica* **7**, 329.

MOODIE A. F. and WARBLE C. E. (1967) The observation of primary step growth in magnesium oxide. *Phil. Mag.* **16**, 891.

WHELAN, H. J. (1961) Dynamical theory of electron diffraction, in *Encyclopaedic Dictionary of Physics* (J. Thewlis Ed.), **2**, 540, Oxford: Pergamon Press.

P. GOODMAN

DYNAMICAL N-BEAM THEORY OF ELECTRON DIFFRACTION. When a beam of electrons, with an energy in the range extending from several keV to a few MeV is incident on a perfect crystal, many orders of diffraction are observed to be excited simultaneously and further, to show strong coupling effects. The reasons for this arise essentially from the combination of high scattering probability with short wavelength: the former leading to high order interactions of appreciable weight, which are enhanced by the favourable geometry due to the latter.

In general, therefore, neither the kinematical, nor the two-beam approximation of X-ray diffraction will be applicable. Inspection of atomic scattering factors shows, however, that over the range of wavelengths considered, back scattering can be omitted with negligible error for the practically important case of transmission, provided that glancing incidence is excluded.

The Multislice Model

The crystal, in the form of an infinite parallel sided plate, is imagined to be divided into slices parallel to the bounding surfaces and of arbitrary thickness. There is no requirement that the slices should relate to specific entities within the crystal, for instance, unit cells or layers of atoms; they are simply a device to clarify a limiting process. The potential within each slice is now projected on an arbitrary internal plane, so that the model for the crystal is a set of parallel gratings separated by the thickness of the slices.

A recurrence relation is obtained for the wavefunction on the basis of the phase-change resulting from propagation between gratings, and through gratings. Here superposition is invoked. Because back scattering is neglected the recurrence relation is two-point, that is, the wavefunction at the nth grating is completely determined by that at the $(n-1)$th grating.

In order to exploit the periodicity of the crystal the relation is transformed to momentum space, the terms collected, and the resulting wavefunction evaluated under the limiting conditions that the slice thickness approaches zero as the number of slices approaches infinity in such a way that the product of the two approaches the thickness of the crystal. The solution for the continuous crystal is thus recovered without any approximation other than the neglect of back scattering; provided that the limiting processes can be justified and the model shown to be consistent.

Quantitative Considerations

The origin is chosen on the entrance face, with the z axis normal to the surface.

The incident wave is represented by a wavefunction having a value $\psi_1(x, y)$ at the point (x, y) of the entrance face of the crystal. If q_{n-1} represents the transmission function of the $(n-1)$th slice, then the effect of the material in this slice on the wavefunction will be given by $\psi_{n-1} q_{n-1}$, where ψ_{n-1} is the wavefunction representing the wave incident on the $(n-1)$th slice. The phase change between slices can be described by a propagation function, p_{n-1}, so that applying the super-position principle, the recurrence relation between slices, in the absence of back scattering, is given by

$$\psi_n = \psi_{n-1} q_{n-1} * p_{n-1}, \qquad (1)$$

where the symbol * represents convolution, that is

$$f_1(x) * f_2(x) = \int_{-\infty}^{\infty} f_1(u) f_2(x-u) \, du.$$

If the thickness of the crystal is H, and the thickness of a slice is Δz, then the number of slices is $H/\Delta z$. Error of the order $(\Delta z)^2$ in each slice will not accumulate beyond the order $(\Delta z)^2$. $H/\Delta z = H\Delta z$, which vanishes in the limit, so that no error results if the functions p and q maintain this approximation. This function p may then be chosen in the advantageous form $\exp\{k(x^2 + y^2)/2\Delta z\}$, where $k = 2\pi/\lambda$, and λ is the wavelength of the electron corrected for relativistic effects. This function has properties under the operations of convolution and Fourier transformation which greatly simplify manipulation.

A direct argument shows that q_{n-1} can be represented by

$$\exp\left\{i\sigma \int_{z_{n-1}}^{z_{n-1}+\Delta z} V(x, y, z_{n-1}) \, dz\right\},$$

where

$$\sigma = \frac{\pi}{W\lambda} \frac{2}{1 + (1-\beta^2)^{1/2}},$$

$$V(x, y, z) = \sum_h \sum_k \sum_l E(h, k, l) \exp\left\{2\pi i \left(\frac{hx}{a} + \frac{ky}{b} + \frac{lz}{c}\right)\right\}$$

is the potential within the crystal with the unit cell sides a, b, c; W is the energy of the incident electron, v is its velocity, and $\beta = v/c$. On the above criterion this function can be further approximated as, $1 + i\sigma V(x, y, z_{n-1})\Delta z$.

Adapting an argument due to Feynman the relation (1) can be shown to limit to the Schrödinger equation without back scattering.

Repeated application of the recurrence relation leads to an expression,

$$\psi(x, y) = q_N(x, y)$$
$$\times [q_{N-1}(x, y) \underset{N-1}{[} \cdots \underset{N-1}{]} * p_{N-1}(x, y)] \underset{N}{*} p_N(x, y), \quad (2)$$

where $\psi(x, y)$ is the wavefunction of the electron at a point (x, y) of the exit surface of a crystal of thickness H. This expression is exact in the limit $\Delta z \to 0$, $N \to \infty$, $(N-1)\Delta z \to H$.

Transforming equation (2) to reciprocal space, the Fraunhofer diffraction pattern is given by

$$\Psi(u, v)$$
$$= [Q_N(u, v) \underset{N}{*} [Q_{N-1}(u, v) \underset{N-1}{*} [\underset{N-1}{\cdots}] P_{N-1}(u, v)] P_N(u, v)], \quad (4)$$

where $\Psi(u, v)$ is the amplitude of an emerging wave for which the wave vector $2\pi/\lambda$ has components $2\pi u$ and $2\pi v$ parallel to the exit-face of the crystal, and capital letters indicate Fourier transforms, that is,

$$Q_N(u, v) = \mathscr{F} q_N(x, y)$$
$$= \int_{-\infty}^{\infty} q_N(x, y) \exp \{2\pi i(ux + vy)\} \, dx \, dy.$$

Again, equation (4) is to be evaluated under the limits given by (3).

For a perfect crystal the Q_N are sets of delta functions, so that the convolutions reduce to summations, and terms can be readily arranged in increasing order of interaction. On proceeding to the limit, the evaluation of standard integrals yields the solution.

The mathematical justification for this procedure is provided by the theory of Volterra's product integral.

Analytical Solution

The solution is in the form of a series,

$$\Psi(h, k) = \sum_{n=0}^{\infty} \Psi_n(h, k), \quad (5)$$

where $\Psi(h, k)$ is the amplitude of a wave for which the wave vector $2\pi\lambda$ has components $2\pi h/a$ and $2\pi k/b$ parallel to the exit surface of the crystal, n is the order of interaction, that is, pictorially, the number of times the beam has been scattered, and $\Psi_n(h, k)$ is the contribution from the aggregates of nth order processes. This solution is therefore in the form of a Born series.

Explicitly,

$$\Psi_n(h, k) = (i\sigma)^n \sum_l \sum_{h_1} \sum_{k_1} \sum_{l_1} \cdots \sum_{h_{n-1}} \sum_{k_{n-1}} \sum_{l_{n-1}}$$
$$\times \frac{E(h_1, k_1, l_1) \cdots E\left(h - \sum_j h_j, k - \sum_j k_j, l - \sum_j l_j\right)}{\cos \theta_1 \cdots \cos \theta_h}$$
$$\times \frac{\exp\{-2\pi i H \zeta\}}{(2\pi i)^n}$$

$$\times \left[\sum_{m=1}^{n-1} \frac{\exp\{2\pi i H \zeta_m\} - 1}{\zeta_m(\zeta_m - \zeta_1) \cdots (\zeta_m - \zeta_{m-1})(\zeta_m - \zeta_{m+1}) \cdots (\zeta_m - \zeta)} \right.$$
$$\left. + \frac{\exp\{2\pi i H \zeta\} - 1}{\zeta(\zeta - \zeta_1)(\zeta - \zeta_2) \cdots (\zeta - \zeta_{n-1})]} \right]. \quad (6)$$

Here the ζ_m are the n-beam excitation errors, distances in the reciprocal lattice space of the crystal. Specifically, ζ_n is the distance, measured in the l direction in reciprocal lattice space, of the Ewald sphere from the reciprocal lattice point with coordinates $\sum_{j=1}^{n} h_j, \sum_{j=1}^{n} k_j, \sum_{j=1}^{n} l_j$ corresponding to a beam which has been diffracted n times, the indices of the individual reflecting planes being $h_1 k_1 l_1$, $h_2 k_2 l_2$, $\cdots$, $h_n k_n l_n$.

ζ is the excitation error of the plane hk. The θ_n are the direction cosines of the beam relative to the surface normal, in practice, factors which depart little from unity.

Thus, the contribution from a single nth order process is of the form $E_p Z_p$, where E_p depends only on the structure of the crystal, and Z_p depends only on the excitation errors, that is, only on the geometry of the process. The quantities E_p thus play the part of generalized structure amplitudes, while the Z_p play the part of generalized shape functions.

Equation (5) can be rewritten in symbolic form as,

$$\Psi(h) = \sum_{n=1}^{\infty} \mathbf{E}_n(h), Z_n(h),$$

where $\mathbf{E}_n(h)$ is an operator involving all the summations of (6) and multiplications by all the structure amplitudes $E(h, k, l)$.

Equation (5) can then be written,

$$\Psi(h) = \sum_{n=1}^{\infty} \sum_{r=0}^{\infty} \mathbf{E}_n(h) \frac{(2\pi i H)^{n+r}}{(n+r)!} h_r(\zeta, \zeta_1, \cdots, \zeta_{n-1}), \quad (7)$$

where $h_r(\zeta, \zeta_1, \cdots, \zeta_{n-1})$ is the complete homogeneous symmetric polynomial of order r in n variables. Various forms and approximations can be classified in terms of this equation.

(a) Born series

The first summation is made over r, giving a series in ascending values of n. Thus, the curvature of the Ewald sphere is first taken fully into account for each order of interaction, and then the orders are summed.

The first term of this series,

$$\Psi_{\text{kin}}(h, k) = \sum_l i\sigma E(h, k, l) \exp\{-\pi i H \zeta\} \left[\frac{\sin \pi H \zeta}{\pi \zeta} \right],$$

is the kinematical or single scattering approximation, correct in the limit $V \to 0$.

(b) Phase grating series

The first summation is made over n, giving a series in ascending values of r. All orders of interaction are first taken fully into account for each power of the excitation errors, and then the powers summed. This amounts to successive approximation to the Ewald sphere.

The first term of this series is,

$$\Psi_{\mathrm{tpg}}(h, k) = \sum_l \Big[\delta(h, k, l) + (2\pi i \sigma H)\, E(h, k, l)$$

$$+ \frac{(2\pi i \sigma H)^2}{2!} \sum_{h_1} \sum_{k_1} \sum_{l_1} E(h_1, k_1, l_1)$$

$$\times E(h - h_1, k - k_1, l - l_1) + \cdots \Big]$$

which is the (h, k) coefficient of the transform

$$= \mathscr{F} \left[1 + (2\pi i \sigma)\, \overline{V} + \frac{(2\pi i \sigma)^2}{2!}\, \overline{V}^2 + \cdots \right]$$

$$= \mathscr{F} \exp\{2\pi i \sigma \overline{V}\}, \qquad (8)$$

where $\delta(h, k, l) = 1$ if $h = k = l = 0$
$\qquad\qquad\qquad = 0$ otherwise,

and $\qquad \overline{V} = \int_0^H V(x, y, z)\, dz.$

This is the thin phase grating approximation which includes all orders of interaction, but is zero order in the excitation errors, that is, the Ewald sphere is approximated by a plane. This approximation is correct at zero wavelength and is therefore a high voltage limit (HVL). To establish this, a typical term of equation (7) is evaluated as $\lim \lambda \to 0$, giving

$$\lim_{\lambda \to 0} \eta^r \left(\frac{2\pi m_0}{h^2}\right)^n \left(\lambda^{2 + \frac{2r}{n}} + \frac{h^2 \lambda^{\frac{2r}{n}}}{m_0^2 c^2} \right)^{\frac{n}{2}},$$

where η is a constant.

This has the value

$$\begin{cases} 0, & r > 0, \\ \left(\dfrac{2\pi m_0}{h^2}\right)^n \left(\dfrac{h}{m_0 c}\right)^n = \left(\dfrac{2\pi m_0 \lambda_c}{h^2}\right)^n = \sigma_c^n, & r = 0, \end{cases}$$

where λ_c is the Compton wavelength.
Hence

$$\lim_{\lambda \to 0} \Psi(h, k) = \mathscr{F} \exp\{2\pi i \sigma_c \overline{V}\} = \Psi_{\mathrm{HVL}}(h, k).$$

Physically this result derives from the fact that wavelength limits to zero with increasing voltage whereas interaction limits to a finite value.

(c) *Thickness series*

If the initial summation is over terms for which $n + r$ is constant an expansion in powers of H is obtained.

(d) *Two-beam approximation*

If it is assumed that all ζ_r but one, ζ_h, are so large that terms containing them can be neglected there results on putting

$$S = 2\pi \zeta_h, \quad t^2 = \sigma^2 E(h) E(\bar{h}),$$

$$\Psi_{2b}(h) = it \{ E(h)/E(\bar{h}) \}^{1/2} \Bigg[\sum_{n=0}^{\infty} \frac{(Hst^2)^n}{n!} \sum_{m=0}^{\infty} C_m^{n+2m}(it)^{2m}$$

$$- \exp\{-iSH\} \sum_{n=0}^{\infty} \frac{(-Hst^2)^n}{n!} \sum_{m=0}^{\infty} C_m^{n+2m}(it)^{2m} \Bigg]$$

$$= i\sigma E(h) \exp\{-2\pi i \zeta_h H\}$$

$$\times \frac{\sin H \{\pi^2 \zeta_h^2 + \sigma^2 E(h)\, E(\bar{h})\}^{1/2}}{\{\pi^2 \zeta_h^2 + \sigma^2 E(h)\, E(\bar{h})\}^{1/2}},$$

the standard two beam result.

Multiple Scattering Diagrams

A single term, $E_p Z_p$ in the general solution can be represented by a multiple scattering diagram (Fig. 1), a linear graph in reciprocal space terminating on the scattering vector. The directed segments of the graph are the scattering vectors of the constituent processes,

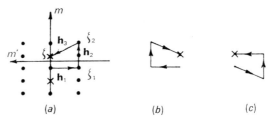

Fig. 1. *Multiple scattering diagrams for third order processes associated with the kinematically forbidden reflection marked X. (a) Typical third order diagram. (b), (c), symmetry related diagrams associated with (a) in defining the loci of zeroes listed in Table 1.*

the sequence of the segments being that of the interactions. The form of the solution is such, that with a conventional construction for the Ewald sphere the vertices of the graph have the excitation errors appropriate to the term. Since a diagram is defined as a graph with labelled vertices, it represents a specific sequence of interactions at a particular angle of incidence. These graphs are useful in the classification of interactions and the analysis of symmetry.

Symmetry

The symmetry of electron diffraction patterns depends on a parameter, the angle of incidence, that is, the relevant group is that of the crystal and electron, a Lie group. Thus the symmetry of the diffraction pattern over a sufficient angular range uniquely determines the symmetry of the crystal. In dynamical n-beam diffraction, the symmetry of the entire crystal is effective, that is, including boundaries and any defects within the irradiated volume. Only if the crystal is perfect is the symmetry that of the unit cell

Typical derivations are sketched for a few symmetry elements, exhaustive treatments are available in the various references.

(a) Reciprocity

This symmetry first derived by von Laue for the particular case of the two-beam approximation, in fact, holds for the general solution, and is present in all diffraction patterns.

Reciprocity refers to the interchange and reversal of the directions of incidence and diffraction (Fig. 2),

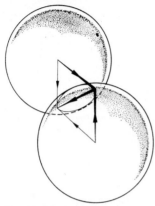

Fig. 2. The reciprocity configuration, the centres of the Ewald spheres are related by reflection across the diffraction vector.

the observed intensities being equal for both configurations. In fact, every individual scattering process in an arbitrary orientation is equal in amplitude and phase to the process with reversed interactions in the reciprocity configuration. The proof follows from the invariance of the Z function under the transformation, $\zeta_m \to \zeta_m - \zeta, \zeta \to -\zeta$.

(b) Inversion

Friedel's law states that the absence of a centre of symmetry cannot be detected by kinematical scattering.

This ambiguity is removed in dynamical n-beam diffraction, even in the high voltage limit; a result which is obtained immediately on separating the potential distribution into symmetric and antisymmetric parts and substituting into the series expansion.

(c) Screw axes and glide planes

These are the symmetry elements which, under kinematical scattering conditions lead to the general space group forbidden reflections.

For n-beam scattering the resulting symmetries can be determined by means of multiple scattering diagrams (Fig. 1). Typically the series is summed in pairs of terms corresponding to diagrams related by an appropriate symmetry. At certain angles of incidence all pairs cancel for the reflections that would be forbidden kinematically. Specific loci of zeros therefore exist in the angular space of those reflections. A summary of results is given in Table 1.

Table 1. *Conditions under which kinematically space-group forbidden reflections remain at zero intensity when dynamic interactions are included.*

A   Incident beam in the plane defined by the reciprocal lattice vector of the forbidden reflection and the zone axis to which the reflection belongs.
B   Bragg condition satisfied by forbidden reflection.

Axis of incident beam relative to symmetry element	Conditions for absence	
	Only zero-layer interactions	Including higher layer interactions
Perpendicular to screw axis	A or B	B
Parallel to glide plane	A or B	A
Perpendicular to glide plane	always (trivial)	A and B

(d) Rotor axes and planes of reflection

Multiple scattering diagrams can be used in the analysis of any combination of these elements. The most striking departures from kinematical symmetries occur where upper layer lines are important, so that, for instance, scattering from a cubic crystal in [111] orientation will, in general, show trigonal rather than hexagonal symmetry.

(e) Zero order of diffraction

Every multiple scattering diagram for the zero order of diffraction forms a closed loop, so that the rocking curve for this order differs in symmetry from all others. For instance, in the absence of higher layer line interactions this rocking curve is centrosymmetric whether the crystal is or not.

Relationship Between Formulations

The elastic scattering of electrons by a perfect crystal, while a one body problem, exhibits many of the characteristics of N body problems, and, in fact, may be regarded as a model made exactly soluble by the relationships connecting the excitation errors, considered as coordinates for thus purpose. Techniques effective in N body problems have, accordingly, proved useful and the relationships between most are known from this field.

At the expense of considerable over-simplification a brief classification is possible in terms of Tournarie's semi-reciprocal formulation. In this method a Fourier transform of the Schrödinger equation is made over the x and y coordinates to give, on neglecting back scattering the set of first order equations,

$$\frac{d}{dz}|\Psi(z)\rangle = i\left[\mathbf{K} - \frac{1}{2}\mathbf{K}^{-1}\mathbf{V}(z)\right]|\Psi(z)\rangle = i\mathbf{A}(z)|\Psi(z)\rangle, \tag{9}$$

where $\mathbf{K}$ and $\mathbf{V}$ are matrices defined by,

$$K_{hh'} = \delta_{hh'} K_h, \quad V_{hh'} = 2k\sigma E(h-h'),$$

K_h is the z component of the wave vector of the beam h in vacuum, and h is used as an abbreviation for the indices (h, k). Tournarie's methods permit the inclusion of back scattering, but this degree of generality is not required here.

If upper layer lines are neglected so that $\mathbf{V}$ and hence $\mathbf{A}$ is independent of z, the solution is

$$|\Psi(z)\rangle = \exp[iz\mathbf{A}]|a_0\rangle, \qquad (10)$$

where $|a_0\rangle$ defines the boundary conditions. This, essentially, is Sturkey's solution in his matrix formulation, although derived by a different technique. Sturkey has developed the method into a tool of wide application.

When $\mathbf{A}$ depends on z the solution of (9) is

$$|\Psi(z)\rangle = \Omega_0^H |a_0\rangle,$$

where

$$\Omega_0^H(\mathbf{A}) = \mathbf{E} + \int_0^H i\mathbf{A}(z)\,dz + \int_0^H i\mathbf{A}(z_2)\,dz_2 \int_0^{z_2} i\mathbf{A}(z_1)\,dz_1 + \cdots \qquad (11)$$

is the Peano matriser, $\mathbf{E}$ being the unit matrix.

The matrix formulation of Niehrs, in effect, establishes this form and its transformations, although, again by entirely different methods from those indicated here.

The matriser may be expressed in terms of Volterra's multiplicative integral.

$$\Omega_0^H(\mathbf{A}) = \int_0^H (\mathbf{E} + i\mathbf{A}\,dz)$$

$$= \lim_{\Delta z_k \to 0} [\mathbf{E} + i\mathbf{A}(z_n)\,\Delta z_n] \cdots [\mathbf{E} + i\mathbf{A}(z_1)\,\Delta z_1]. \qquad (12)$$

Equation (12) summarizes the multi-slice formulation outlined in the present article.

The matriser development can be carried through by the method of Green's functions. In this way Fujiwara obtained n-beam solutions for both the Schrödinger and Dirac equations. He was thus able to establish that at least in the voltage range considered here, magnetic interaction is negligible for non-ferromagnetic crystals, but that relativistic corrections for both wavelength and mass are important.

The prototype of all dynamical formulations for electron diffraction is due to Bethe. This is a dispersion method established by an expansion in Bloch functions. Fujimoto showed that the matrix solution (10) could be derived from the dispersion equations and that the Born series could be recovered as one of the expansions of the matrix.

Takagi has developed methods for generalizing the matrix formulations to describe scattering from distorted crystals. This is important in many applications, and has particular relevance to problems in which "column" approximations are invoked, but lies outside the scope of the present discussion.

For many problems a combination of two or more of these techniques has proved advantageous.

Absorption

Phenomenological absorption coefficients can readily be included in all of the formulations. Since they are in general orientation dependent very considerable caution is required for all but rough approximations. Detailed discussion of this point, however, lies outside the scope of this article.

Numerical Methods

The convergence of the Born series (6) is so slow that it is entirely unsuitable for numerical work. So far most computations have derived from the dispersion equation, matrix, or multi-slice formulations. All are sufficiently rapid, and can be made sufficiently accurate to constitute practical procedures for digital computation.

In practice stringent requirements in speed must be met since intensity distributions vary appreciably over angular ranges of 20 sec of arc, and with steps of one unit cell in thickness. Further, particularly about axes of high symmetry rather large (> 70) numbers of beams are required for accurate estimates.

See also: Dynamic electron diffraction, applications of.

Bibliography

BACON G. E. (1961) Atomic scattering factor, in *Encyclopaedic Dictionary of Physics* (J. Thewlis Ed.), **1**, 326, Oxford: Pergamon Press.
COWLEY J. M. and MOODIE A. F. (1957) *Acta Cryst.* **10**, 609.
COWLEY J. M. and MOODIE A. F. (1962) *J. Phys. Soc. Japan*, **17**, Suppl. B-II, 86.
FEYNMAN R. P. (1958) *Selected Papers on Quantum Electrodynamics*, edited by J. Schwinger, New York: Dover.
FUJIMOTO F. (1959) *J. Phys. Soc. Japan* **14**, 1513.
FUJIWARA K. (1961) *J. Phys. Soc. Japan* **16**, 2226.
GANTMACHER F. R. (1959) *Application of the Theory of Matrices* (English trans.), New York: Interscience.
GJONNES J. and MOODIE A. F. (1965) *Acta Cryst.* **19**, 65.
GOODSTEIN R. L. (1961) Lie groups in differential equations, in *Encyclopaedic Dictionary of Physics* (J. Thewlis Ed.), **4**, 273, Oxford: Pergamon Press.
NIEHRS H. (1959) *Z. Phys.* **156**, 446.
STURKEY L. (1962) *Proc. Phys. Soc.* **80**, 321.
TOURNARIE M. (1962) *J. Phys. Soc. Japan* **17**, Suppl. B-II, 98.
TURNER G. L'E. (1962) Reciprocal lattice, in *Encyclopaedic Dictionary of Physics* (J. Thewlis Ed.), **6**, 212, Oxford: Pergamon Press.
VOLTERRA V. and HOSTINSKY B. (1938) *Opérations Infinitésimales Linéaires*, Paris: Gauthier-Villars.
WHELAN M. J. (1961) Dynamical theory of electron diffraction, in *Encyclopaedic Dictionary of Physics* (J. Thewlis Ed.), **2**, 540, Oxford: Pergamon Press.

A. F. MOODIE

E

ELECTRIC DIPOLE MOMENT OF THE NEUTRON.
Until a few years ago, it had long been considered that nuclei and elementary particles could not possess intrinsic electric dipole moments, EDM (as distinct from induced or polarization moments caused by external field influence) because this would violate parity conservation. This conclusion was questioned in 1950 by Purcell and Ramsey (1950) who pointed out that such arguments were based on theoretical assumptions not completely tested by experiment. Moreover, Purcell and Ramsey proposed an elegant and very sensitive experiment to be carried out with neutrons which would search for the presence of a finite EDM. This experiment was executed successfully by Smith, Purcell and Ramsey about 1952 and published in 1957 with the finding that the neutron EDM, if it exists, must be smaller than 5×10^{-20} electron charge-cm.

Just prior to this publication date, the celebrated theoretical and experimental advances showing that parity was not always conserved in some interactions had occured and this had raised new interest in the existence of a neutron EDM. New theoretical arguments were soon advanced by Landau (1957) showing that even with parity violation it would be necessary to have time-reversal-invariance fail if the neutron were to possess an EDM. There were few advocates of this dual violation scheme until the very significant experimental finding of Christensen, Cronin, Fitch and Turley (1964) on the K^0-particle decay which was to be interpreted as arising from simultaneous failure of both charge conjugation and parity conservation. This implied, again by theoretical reasoning, that time-reversal-invariance might indeed be in violation and hence the question of the existence of a particle EDM was again an uncertain one.

This state of affairs has aroused a flurry of interest among both theorists and experimentalists with the former estimating the possible magnitude of the neutron EDM based upon various interaction models and the latter designing and executing new experiments of improved sensitivity. A summary of the theoretical results was given by Boulware in 1965 with more recent findings appearing periodically.

Among the new experimental results are those reported by Miller, Dress, Baird and Ramsey (1967) and by Shull and Nathans (1967) representing experiments which are completely different in concept. The former is a much improved version of the original Smith, Purcell and Ramsey (1957) experiment in which is sought a perturbation on the Larmor precession frequency of a neutron in a magnetic field caused by the simultaneous presence of an electric field. The electric field through its interaction with an EDM will change this frequency and a frequency shift was searched for when the electric field direction was reversed relative to the neutron spin axis. The second experiment of Shull and Nathans was a scattering experiment in which the additional scattering amplitude term arising from interaction of the EDM with the atomic Coulombic field was sought. Polarized neutrons were Bragg reflected by atoms in a crystal and the imaginary EDM amplitude term was searched for through its coherence with other imaginary amplitude terms in the nuclear scattering. This was to be sensed in very sensitive fashion by reversing the neutron polarization direction. Since in both experiments the sought-for effect is very small, it was necessary to insure that other extraneous effects could not be introducing artificial results and much effort was expended in studying and establishing the contaminant contributions.

Both experiments were negative in the sense of finding a real effect larger than experimental error but they agreed with each other in setting a new upper limit of the neutron EDM value. The resonance beam experiment yielded a dipole length of $(-2 \pm 3) \times 10^{-22}$ cm and the scattering experiment gave the value $(+2 \cdot 4 \pm 3 \cdot 9) \times 10^{-22}$ cm with either one representing an improvement of two orders of magnitude over the original resonance beam experiment. Additionally a second resonance beam experiment has been performed by Cohen, Lipworth, Silsbee, Nathans and Ramsey with unpublished results which confirms the above upper limit values. The experiments are continuing and further refinements are to be expected.

Bibliography
Boulware D. G. (1965) *Nuovo Cimento* **40**, 1041.
Christensen, Cronin, Fitch and Turley (1964) *Phys. Rev. Letters* **13**, 138.
Landau L. (1957) *Nucl. Phys.* **3**, 127.
Miller, Dress, Baird and Ramsey (1967) *Phys. Rev. Letters* **19**, 381.

PURCELL E. M. and RAMSEY N. F. (1950) *Phys. Rev.* **78**, 807.
SHULL C. G. and NATHANS R. (1967) *Phys. Rev. Letters* **19**, 384.
SMITH, PURCELL and RAMSEY (1957) *Phys. Rev.* **108**, 120–122.

C. G. SHULL

ELECTROGASDYNAMICS.

1. Introduction

Electrogasdynamic (EGD) generators have been proposed for the direct conversion of heat to electricity or simply as devices for the production of high voltage. Although the feasibility has been established the hardware stage remains one of exploratory research.

The basic principle can be understood by reference to Fig. 1. Ions are produced within a moving gas stream, at the entrance to a duct of insulating material, and collected at the exit. This results in a voltage difference between entrance and exit, and hence an electric field along the duct which opposes further movement of charge. Collisions between ions and neutral molecules of the moving gas cause transport of ions against the field and towards the collector. Thus, work is done on the ions by the gas stream, part of the gas enthalpy thereby being converted to electrical energy which is expended in the external load. The reduction in enthalpy is manifested as a decrease in temperature or pressure of the gas (or both), depending on the flow conditions.

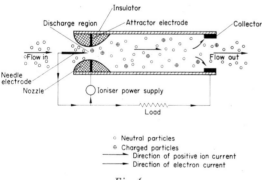

Fig. 1.

Several methods are available for seeding the gas with ions. The one shown in Fig. 1 is a corona discharge which is present between the needle electrode and the attractor electrode. Since the electric current in the duct is due to the passage of unipolar ions, space charge effects are present, and the generator is characterized by a low current output at high voltage, and has a high internal impedance. This is similar to the output of a Van de Graaff high voltage generator, which the EGD generator resembles. However, the latter can carry more charge in its gas volume than can a Van de Graaff machine on the surface of its belt. Moreover, gas speeds can be much greater than belt speeds.

The power supplied by the ionizer is $(1 - \eta_I) J_I V_I$ while the output power from a single-stage duct is $\eta J_I V_L$. Thus, for the output power appreciably to exceed the input power it is required that $V_L \gg V_I(1 - \eta_I)/\eta_I$. This can readily be achieved in practice. There are, however, other possible sources of electric power loss which reduce the output power. The positive ions must have a resultant movement towards the collector and therefore their drift velocity towards the entrance, kE, due to the electric field, must be less than their stream velocity (the gas velocity) towards the exit. In addition, the electric field should not be allowed to change direction anywhere within the duct and drive ions towards the collector. This effect dissipates electric energy as Joule heating within the gas. Both these requirements are included in the statement that the ion slip, S, has the limits

$$-1 \leq S \leq 0$$

where $S = kE/v$.

Although an EGD machine can employ several types of thermodynamic cycle the most usual is one or other of the variety based on the Brayton cycle. This cycle is also employed in gas turbines and MHD devices (cf. Swift-Hook, Magnetohydrodynamic generation of electricity, *Encyclopaedic Dictionary of Physics*,

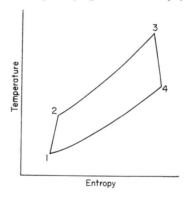

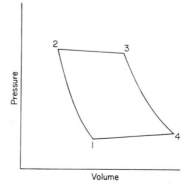

Fig. 2.

Supplementary volume 1, p. 162). The basic form is illustrated in Fig. 2, the steps of the cycle being

1—2, compression of working fluid;
2—3, addition of heat, through heat exchanger or by combustion;
3—4, interaction within the EGD duct and consequent reduction of enthalpy;
4—1, rejection of heat, either to cooler or direct exhaustion of working fluid to atmosphere.

A simplified conceptual layout for an open-cycle plant is illustrated in Fig. 3. Here the fuel mixture would be pulverized coal or fuel oil along with com-

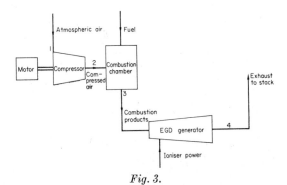

Fig. 3.

pressed air in the region of 10–30 atm. The scheme for a closed-cycle plant is illustrated in Fig. 4. A gas-cooled reactor is shown as heat source but the intermediate source could be a heat exchanger receiving

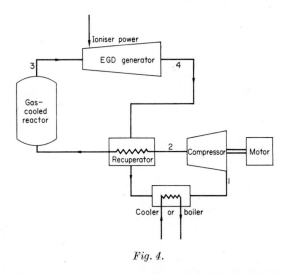

Fig. 4.

heat from some other type of primary supply. The heat dump (cooler), which completes the thermodynamic cycle, could be replaced by a boiler to provide steam for a conventional steam power plant.

2. Working Fluid

Two important requirements of the working fluid are that it should have a high electric breakdown strength and that the mobility of the charge carriers should approach zero. Unipolar charge carriers are employed, but these can be of either sign. Thus, positive ions produced in a discharge, or negative ions formed by electron attachment are possible choices. However, they have high mobility values and hence high slip, a disadvantage which may be overcome by employing particulate matter as the charged seed. The particles may be in the solid (dust) or liquid (aerosol) phase. Charging is achieved either by ion attachment or by condensing liquid onto ions.

The mobility of the ions, in ionised air at atmospheric pressure, is about $2 \cdot 1 \times 10^{-4}$ m²/V sec. In fact, under similar conditions, any molecular ion, with mass greater than 20, has a mobility within about 30 per cent of this value. As an approximate guide, macroscopic particles have mobilities which are about two orders of magnitude lower. This leads to greatly improved viscous coupling between gas and charge, negligible slip, and hence negligible ohmic losses.

If the diameter of the particles is much greater than a molecular mean free path in the gas then the mobility is independent of the pressure for a given temperature. This would be expected for a coal-burning plant, the fly ash in the combustion products having an average diameter of the order of 10 μ. When the diameter is about a mean free path or less, mobility decreases as pressure increases. For example, particles of diameter 10^{-2} μ have a mobility of about 10^{-6} m²/V sec in air at atmospheric pressure, but this decreases to 10^{-7} m²/V sec at 30 atm pressure.

Air, as a carrier gas, has a breakdown strength of about 3×10^6 V/m at NTP. The breakdown field is directly proportional to pressure over a wide range but can be increased by adding an electronegative gas such as sulphur hexafluoride.

3. Operating Characteristics

For determining the general performance characteristics of an EGD generator one-dimensional analysis, which neglects radial variations, provides a useful guide. Further refinements then deal with radial velocity profiles, wall effects, space-charge spreading, and similar more detailed phenomena.

In the one-dimensional case conservation of mass and charge, assuming all the charge in the flow is removed by the collector, implies constant values for dm/dt and J where

$$dm/dt = \varrho v A,$$

and

$$J = jA.$$

Equating the rate of change of momentum to the

electrostatic and pressure forces leads to
$$\varrho v\, dv/dx = qE - dp/dx,$$
while the conservation of energy equates the electrical output power per unit volume of duct to the rate of decrease of total enthalpy density,
$$\varrho v\, d(c_p T + \tfrac{1}{2} v^2)/dx = Ej.$$
To these relations can be added the equation of state
$$p = \varrho \mathbf{R} T,$$
the expression for the net current density
$$j = q(v + kE),$$
and Gauss's law
$$dE/dx = q/\varepsilon\varepsilon_0.$$

These seven relations have eight unknowns; there is therefore freedom to specify one unknown.

In such an approximate approach it is assumed that the fluid is inviscid and that no heat is lost by conduction through the duct walls. Remembering that E is negative, the equation for the net current density illustrates that the current is severely limited by large values of mobility. Field distortion by space-charge also defines operating limits. The effect can lead to reversal of the field, a condition to be avoided since such a region would absorb electrical power through Joule heating.

The general characteristics of the flow can be appreciated through the procedure adopted by Khan, of Curtiss-Wright Corporation, who solved the equations above numerically for constant duct area and introduced the following non-dimensional parameters, all of which refer to conditions at inlet,

$D_i = kE_i/v_i$, ion drift velocity/flow velocity;

$K_i = q_i L/\varepsilon_0 E_i$, electric charge density/electric flux density;

$N_i = q_i E_i L/\varrho_i v_i^2$, a measure of electric body force/inertia force.

All three parameters are negative under conditions of power generation.

For molecular ions which have high mobility $-D_i$ lies between 0·5 and unity. For charged colloidal particles $-D_i$ has much smaller values, say 10^{-1}–10^{-3}. For very small values of $-D_i$ the electric field approaches zero at the output end. As $-D_i$ approaches values around unity the electric field approaches constancy along the duct. Molecular ions can thus be expected to yield higher output voltages than colloidal ions. The upper limit for $-K_i$ is around unity, a value corresponding to colloidal ions, higher values resulting in electric field reversal.

The electric field distribution is only weakly dependent on N_i. However, N_i is a measure of how well the charge interaction slows down the flow. For this to be significant $-N_i$ should be greater than about 0·1.

The output voltage as a function of current density has the general form shown in Fig. 5. For increasing values of mobility the "knee" occurs at increasing values of current. The knee indicates a marked increase of internal impedance as current increases. At a given mobility and flow velocity, a current increase is accompanied by a charge density increase.

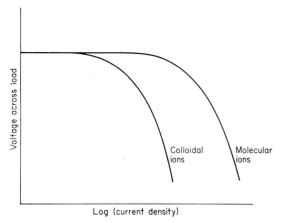

Fig. 5.

This reduces the average field, the reduction becoming very marked when conditions result in a reversal of field direction spreading from the output end, and the voltage output is correspondingly reduced. The current is determined by the boundary conditions at the inlet charge source. For a given load the intersection of the load line with the voltage/current characteristic yields the operating point. This point may be determined by the operator's choice, maximum output power for example.

The generator conversion efficiency can be defined as
$$\eta_g = JV_L/(h_i\, dm/dt),$$
neglecting the ionizer power. It is the fraction of the total enthalpy input that is converted to useful electrical power. In general the conversion efficiency is greater for colloidal ions than molecular ions and could be as much as 10 per cent for a duct under optimum conditions at 20–30 atm pressure. For a high value of η_g several ducts must be connected in series.

The generator isentropic efficiency is given by
$$\eta_T = \Delta H/\Delta H',$$
the ratio being a comparison of the actual and isentropic enthalpy changes for the same pressure change (cf. Fig. 2). The isentropic efficiency is small for molecular ions, due to their high mobility which results in excessive Joule heating. For colloidal ions, having negligible Joule losses, isentropic efficiencies in the range 70–90 per cent might be expected, depending on the wall losses and viscous effects.

The effects of pressure are important. For a working fluid at 1 atm pressure reasonable output parameters would have the orders of magnitude—axial electric field 1 MeV/m; current density 1 A/m^2; power density 1 MW/m^3. As the working pressure is increased the breakdown field strength increases, the charge density can be increased, and the mobility decreases. Pressurizing to 20 or 30 atm could have a very marked effect, increasing the field and current density by an order of magnitude or more and, correspondingly, the output power density by at least two orders of magnitude.

In general the overall thermal efficiency (heat to electricity) of an EGD plant increases with increase of inlet temperature and decreases as inlet to outlet pressure ratio increases. However, the output power increases with increase of pressure ratio. For instance, an open cycle device with 1400°C inlet temperature and a pressure ratio of 20 could have a thermal efficiency approaching 50 per cent. This assumes reasonable losses, but also that all operational problems have been solved.

4. Charged-particle Source

In the corona-discharge type of ion source, the high electric field near a needle electrode, or array of such electrodes, ionizes the nearby gas. If the attractor electrode is negative with respect to the needle, positive ions tend to move towards the attractor but a fraction of them are swept downstream by the moving gas. Fluid-dynamic energy is consumed in removing ions usefully, only those ions which reach the attractor constitute a current drain on the corona power supply and therefore require electric power input. Thus, the ionizer power input approaches zero as η_i approaches unity. This ideal is emphasized by the fact that the fluid-dynamic energy required to remove one ion may be in the order of 10^{-4} of the output energy due to one ion finally reaching the collector.

The forward kinetic energy of a neutral molecule in the duct is around 10^{-2} eV. For an output potential of 1 MV it follows that the flow energies of about 10^8 neutral molecules are required to transport a singly-charged particle to the collector. The number of neutrals will therefore exceed the number of charged particles by a factor of about this order.

A positive potential at the attractor electrode removes electrons from the vicinity of the needle and these form heavy ions by attachment. Thus, if the gas is seeded with particulate matter, the particles will become negatively charged. An alternative method of forming heavy ions is by providing an expansion nozzle at the duct entrance, the working fluid being a mixture of gas and vapour or simply a vapour. According to the sign of the attractor potential, positive or negative ions will appear in the fluid stream and serve as condensation nuclei as vapour condenses on expansion. Negative point coronas (positive attractor), which seed the gas with negative ions, are expected to yield greater current over a wider pressure range than positive point coronas. A mixture of air and water vapour is often used in experimental rigs.

A corona discharge is currently the most favoured method of providing a sufficient density of unipolar charged particles in the gas stream. The maximum injection rate is limited by the electric breakdown field in the gas and by space charge effects, while its detailed behaviour is governed by the properties of the moving fluid and the geometry of the discharge region. Experiments have included several types of needle geometry, including a flattened tip ("nailhead") configuration examined by Decaire and Wifall. This end disc induces a fan of supersonic flow, with corresponding pressure reductions, and thus favourable local conditions. These have resulted in molecular ion currents increasing linearly with total pressure up to at least 80 atm. In the case of charged droplets the linear variation extends to less than half this value, but this is not necessarily the achievable optimum.

Amongst other types of source under consideration are thermionic emitters and external injectors. The former would simply produce an electron space charge in the vicinity of a heated electrode. The high mobility of electrons implies little viscous coupling with the gas stream, but electrons which became attached to molecules or particles would be swept away by the flow. Electrons or positive ions could be injected from an external discharge by arranging for its anode or cathode to be a gridded structure in the duct wall. Since the source discharge would be screened by the grid, the rate of injection would have the desirable characteristic of being independent of the electric field in the duct and hence independent of output voltage.

The required rate of charge injection increases as duct area increases. Large ducts would therefore probably be provided with multiple injectors.

The design of charge collectors requires further detailed study to evolve an optimum design. Experimental evidence suggests that direct contact with the charge is not necessary in order to collect it, some processes taking place which are the inverse of those at the injector.

5. Practical Problems

The major development problem for EGD generators is the small percentage of the enthalpy flow of the working fluid which is converted into electrical power in a single duct. A high conversion efficiency demands a high density of charged particles having strong viscous coupling with the bulk gas. These requirements themselves impose further restraints on operating parameters.

Conversion efficiencies of perhaps 1 per cent per metre suggest that a practical system would require in the order of 100 stages in series while, to increase output power, each stage might have several ducts in parallel. One possible approach would employ ducts a few centimetres in diameter, having many intermediate electrodes disposed along the length in series connection. A multiplicity of ducts does have the advantage of redundancy, the loss of one or two having small effect on the total output. Moreover, if breakdown is due to internal build-up of a conducting film the fault is, in some cases, self-correcting. In a high-voltage system a large current will flow through a conducting layer which may thus be burned away. This is not necessarily so, however, when organic materials are present since a carbon deposit can form a permanent tracking site.

The performance of an EGD generator improves markedly as the pressure is increased. A fossil-fueled open-cycle device, with exhaust therefore near atmospheric pressure, does not appear promising. Economics would be improved by coupling the generator exhaust, while still at relatively high pressure, to a gas turbine. In this case it would be necessary to include a high performance precipitator between the systems in order to free the turbine feed of particulate matter.

A closed-cycle configuration would have greater flexibility in the choice of pressure range but a major problem is then the provision of a high temperature heat exchanger between system and heat source. In addition, the advantage of there being no requirement for cooling water is lost.

The formation of conducting deposits on the walls of the duct is one part of the more general electric breakdown problem. Insulating duct walls must be provided which remain effective insulators and retain their integrity at high temperature and high voltage. The walls must also be tailored to minimize erosion effects and pressure losses due to friction. External breakdown problems could be alleviated by surrounding the ducts with an electronegative gas.

The ideal requirement for negligible slip and small loss of charged particles demands further development of charge sources and channel geometry. The concept of slender-channel geometry, that is with length very much greater than width, could improve the conversion efficiency but it conflicts with the aim of restricting radial space charge spreading. If the radial spread extends to the duct walls before the collector is reached, charge loss is enhanced. In addition, if charged particles are present in the stagnant fluid boundary layer at the wall, they will dissipate energy by drifting upstream in the axial electric field.

6. *Applications*

In the role of an electric power source an EGD device suffers the disadvantage of an output at high voltage and low current. During the evolution of electric power plant a combination of economics and technical feasibility led to plant with output voltage well below that of EGD capability. However, similar reasons have led to transmission voltages of several hundred thousand volts, with a continual upward progression. Since d.c. transmission is also favourable, for long lines, an EGD large-scale electric power supply may well have output parameters which become more advantageous with time. Expensive transforming and rectifying equipment, at the station, could be eliminated.

The adoption of the EGD concept in large power stations can only be expected in those cases where considerable advantages are evident. This is an inevitable consequence of the competition being with power plant already in an advanced stage of development. If open-cycle EGD generators could completely replace steam plant at a given site then a further advantage is that such a site would no longer be dependent on large supplies of cooling water.

A particular advantage of an EGD generator with a nuclear reactor heat source is the reasonable pressure matching. A high-temperature reactor, cooled with helium at 45 atm pressure and 1000 °C outlet temperature, could be associated with a developed EGD generator operating with a pressure ratio of 4 and yielding around 40% thermal efficiency.

If an a.c. output is required this could be achieved by employing an a.c. ionizer. An EGD device has practically no reactive impedance (the small amount present being capacitive). This reduces the probability of instabilities and results in a very rapid response to the ionizer, though the periodic time of an a.c. ionizer should not be less than the charged particle transit time in the duct.

Scaling to large sizes for power station application decreases the relative effect of wall losses. It is to be expected that radial space-charge fields would have to be kept below 10^8 V/m, but this figure would only be approached in very large ducts.

Smaller power applications can be envisaged, the most likely being in space. Here an EGD generator could provide the voltage to impart high exhaust velocity to the charged particles in a colloidal electrostatic thruster. Also, for on-board rocket power, the heat source is so large that the small EGD conversion efficiency per stage would not be a drawback.

Since the working fluid is seeded with ions, an EGD interaction is present whether or not the bulk gas is at elevated temperature. Thus, an EGD device could be considered as a high-voltage machine, and the gas could be set in motion by means other than a local heat source—for example, an electrically-driven compressor. Provided adequate currents can be achieved, such a device could be the high-voltage supply for charged-particle acceleration or irradiation processes.

7. *Status*

The concept of EGD generators for power production appears to be feasible but the economic justification has yet to be convincingly demonstrated. Experimental results at low pressure (≈ 1 atm) are satisfactorily interpreted by theory but more experimental and theoretical investigations are required at high pressures (20–30 atm).

Experiments by Gourdine and Malcolm have convincingly demonstrated the reduction of mobility between the case for dry air and that for air containing charged water droplets. The condensation of water vapour was produced by expanding the working fluid in a supersonic nozzle at the duct entrance. Some of the data of Gourdine and Malcolm, which apply to the slender channel experiments with water saturated air, are listed in Table 1. These are illustrative of the range in which there is good correlation between theory and experiment.

Table 1. Slender channel experiments of Gourdine and Malcolm

Channel length	3 cm
Channel average diameter	2 mm
Mach number	1·5–2
Ion mobility	10^{-5}–10^{-6} m²/V sec
Corona input	0·2 W at 4 kV
Axial duct field	4 MV/m
Output parameters	120 kV max.
	2×10^5 A
	2 W max.
	20 MW/m³ of duct
Kinetic to electric conversion efficiency	>1%
Isentropic efficiency	>2%

On the basis of past experiments it has been suggested that little further development would be required to build a generator with output in the order of 100 µA at 1 MV. Experimental programmes in the U.S. aim to test full-scale channel segments up to 30 atm pressure with coal-burning plant. The upward trend in power output from research equipment has been further advanced by experiments with high-pressure air in a closed loop, at the U.S.A.F. Aerospace Research Laboratory, which have yielded 0·5 kW output.

Symbols

A cross-sectional area of duct,
c_p specific heat at constant pressure,
E electric field intensity in duct,
ΔH actual enthalpy change through duct,
$\Delta H'$ enthalpy change in duct for isentropic process,
h enthalpy per unit mass,
i suffix referring to duct inlet,
J electric current in duct,
J_I discharge current,
j electric current density in duct,
k mobility,
L length of interaction region,
m mass of working fluid,
p pressure,
q electric charge density,
R gas constant of unit mass,
S ion slip,
T absolute temperature,
v gas velocity,
V_I ion source voltage,
V_L load voltage,
x distance along duct from inlet,
ε_0 vacuum permittivity,
ε relative permittivity,
η_g conversion efficiency,
η_I fraction of discharge current removed by gas stream,
η_T isentropic efficiency,
ϱ density of working fluid.

Bibliography

Advances in Energy Conversion Engineering (1967), Intersociety Energy Conversion Engineering Conference, Aug. 13–17 1967 (Am. Soc. Mech. Eng., New York), session VIII.

BRANDMAIER H. E. and DIMMOCK T. H. (1967) Factors influencing electro-fluid dynamic power generation, *J. Spacecraft & Rockets* **4**, No. 8, 961.

BRANDMAIER H. E. and KAHN B. (1966) Recent advances in electro-fluid dynamic power generation, *Inst. of Electrical & Electronics Engineers International Convention Record*, p. 28, New York.

KHAN B. and GOURDINE M. C. (1964) Electrogasdynamic power generation, *J. Am. Inst. of Aeronautics & Astronautics* **2**, 1423.

KHAN M. P. (1964) High voltage electrogasdynamic generators. Am. Soc. of Mech. Eng., Winter Annual Meeting, New York, Nov. 29–Dec. 4, 1964, paper 64-WA/ENER-11.

L. G. SANDERS

ELECTROHYDRAULIC CRUSHING

Definition

Electrohydraulic crushing comprises the use of repeated underwater sparks to cause the breakage of adjacent brittle materials. Fracture is caused by the shock wave from the spark which is virtually the equivalent of the detonation of a small explosive charge. For this to be the case, the discharge must take place primarily in the water, and hence the process is preferably used on non- or poorly conducting materials such as ceramics or natural rocks. Thus it is distinguished from spark erosion where the material being treated is part of the electrical discharge circuit, and also the product particles are much smaller.

State of the Art

The technique appears to have originated in the U.S.S.R., where the crushing of chalk, coal, optical glasses and other materials has been reported. The construction of a prototype for preparation of road aggregate was also reported from the U.S.A. However, in neither case do details of operation and performance seem to be available, and information on energy requirements is known mainly from experimental investigations in Great Britain. It is believed that the process may establish itself for fine crushing (down to 500 μ or below) in some special cases where its relatively higher costs are justified by particular advantages such as ease of application and control of contamination; but that for coarse crushing applications (down to 1 cm, say) it may prove to be competitive with conventional crushers. The technique has also been applied experimentally to the removal of sand cores from castings.

Spark Circuits

The discharge circuit has three main requirements:
(a) The voltage gradients and electrode configuration must be adequate to cause electrical breakdown of the water gap.
(b) The energy released must be sufficient to cause effective fracture of the largest particles present.
(c) The energy release in the spark should take place in the shortest possible time in order to provide the highest and most effective shock pressures. A typical discharge circuit is shown in Fig. 1

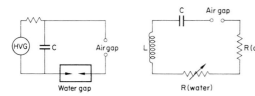

(a) Basic circuit   (b) Equivalent discharge circuit
Fig. 1.

The capacitor C is charged from a suitable high-voltage generator, and the air gap is adjusted to breakdown when the desired sparking voltage is reached on the capacitor plates. On breakdown of the air gap the capacitor voltage is thrown instantaneously across the electrode points of the underwater cell causing breakdown of the water gap. This circuit is the same as that used for electrohydraulic forming, though for the latter larger energy pulses are generally used, of about 10,000 joules; for crushing brittle materials the spark energy required experimentally is indicated in Table 1. Because of the mechanism of crushing (see below) the use of sparks of too low an energy is ineffective, but there is also evidence that the energetic efficiency of the crushing process decreases as the spark size is increased beyond the minimum required for effective crushing of the particular feed size.

Exploding wires have been used to initiate the underwater spark channel, especially in experiments on crushing larger pieces of rock, but this technique is probably useful only in the laboratory because it is unlikely to allow the rapid repetition rate required

Table 1. Experimental crushing conditions

Particle size cm		Capacitance μF	Voltage kV	Spark energy J	Spark gap length cm
Feed	Product				
20	<2	1·5	40	1200	6
1	<0.02	0·02	70	25	1·5

for a production unit. If a wire is not used, there is found to be a delay time of some microseconds between the arrival of the voltage pulse on the underwater electrodes and the breakdown of the gap. During this delay time some loss of charge (hence of voltage) occurs, chiefly by ionic conduction between the electrodes, and this sets a limit to the length of delay time permissible, and hence a lower limit to the voltage gradient across the gap which controls this delay. Working values for two conditions are indicated in Table 1. Lower voltages are acceptable for the larger capacitance values because a given loss of charge results in a smaller voltage reduction as the capacitance rises.

When the spark occurs the resistance of the water gap drops readily; but for the maximum rate of energy release it is desirable that the current pulse should correspond to that for a nearly damped discharge; i.e. that

$$R_m \cong \sqrt{\frac{L}{C}}, \text{ in practice,}$$

where R_m is the effective mean value of the water gap resistance during discharge. R_m can be directly varied by altering the spark gap length, and it has been shown experimentally that for fixed values of L (the inductance of the discharge circuit), C and V, the above approximate condition gives the most effective spark for both crushing and metal deformation. In practice, some compromise may be necessary between this criterion and the voltage gradient required for spark initiation.

In addition a higher shock pressure is obtained if the period of the discharge $\sqrt{(LC)}$ is smaller, and this requires that L, the inductance, should be as small as possible.

For lower energy discharge circuits ($C = 0.02$ μF), inductances of about 0·2 μH have been achieved, giving a discharge time in the damped condition ($\sim 3 \sqrt{(LC)}$) of 0·2 μsec.

A theoretical prediction of peak shock pressure in the spark channel gives

$$P = 0.82 \, V/\sqrt{(lL)} \quad \text{(kilobars)}$$

with V in kV, L in μH and the gap length l in cm. This indicates that capacitance, which affects both the energy of the spark and the time for the discharge, has no nett effect on the pressure generated. However, as it alters the width of both the shock pulse and (presumably) of the spark channel, the capacitance will influence the pressures experienced at fixed points away from the channel.

Mechanism of Crushing

The rapid release of energy in the shock channel creates a pressure discontinuity which travels outwards as a shock pulse. A pressure–distance plot through the pulse would show a vertical front and behind this the compressive stress falls off gradually to give an approximately triangular form. The shock front as a whole also expands on a roughly spherical or cylindrical front about the source of the explosion. Any account of the mechanism of solid fracture must explain how the compressive stress pulse produces tensile or shear stresses in the solid, since it is these stresses which cause failure in brittle solids. Methods which have been suggested are:

(a) Formation of transverse tensile forces at the head of flaws or bubbles in the compressive zone leading to "Griffith-crack" failure. This is probably responsible for the crush zone observed adjacent to explosions. The tensile forces here are proportional to the compressive load and depend on the flaw structure.

(b) The circumferential or "hoop" strains produced by movement of the material in an expanding (spherical or cylindrical) shock front. For this effect the explosion must be within the solid, or close to it relative to the solid dimensions. The magnitude of this effect depends more on the energy of the shock wave than on its initial pressure, and also on the elastic (or plastic) properties of the solid.

(c) The reflection of compressive waves as tensile waves when the former are passing from the solid into water or other fluids. This arises because of the different impedances of liquid and solid to the shock wave, and because at any instant during transmission the pressure must balance across the boundary of solid and fluid. With a plane shock wave this effect gives fracture near the back face of the solid (i.e. the face remote from shock wave entry); this is called scabbing or Hopkinson fracture and is thought to be the primary cause of fracture in crushing small particles. With an expanding shock front, reflections may also occur at the sides of the solid to produce tensile waves which reinforce one another in the solid. The presence of a gas in place of a liquid on the faces of the solid remote from the spark has been shown to increase the extent of fracture, because of increased tensile reflection.

(d) A less important cause of failure is from the possible tensile component of shear waves which will be generated when a shock front meets a solid surface at an oblique angle.

A further shock pulse is produced by rebound during the final collapse of the discharge bubble; this produces the well-known cavitation effect which can also cause local fracture. Although underwater discharges have in fact been used to generate bubbles for cavitation studies, it is considered that for discharges which approximate to the damped condition mentioned above, the cavitation shock will be appreciably less than the initial expansion shock.

Crushed Product

Efficient crusher operation requires prompt removal of the fine product from the crushing zone: here again an element of compromise is involved because it is also required to surround the spark gap with as dense as possible a bed of uncrushed solid, which is bound to impede the removal of the product. This factor is likely to affect the apparent energetic efficiency of the process.

Experimental determinations of energy consumption indicate values of around 50 kWh/tonne in crushing gravel from 1·5 cm to below 350 μ; and 1–2 kWh/tonne for quartzite rock from 20 cm to below 2 cm. (The energy required depends of course on the size reduction ratio as well as the absolute size of product.) These values relate to crushing lumps *en masse* under water: lower values have been achieved where the spark is specially positioned close to a single sample, and where the sample projects into the air. Other features of the crushed product are:

(i) The relatively compact shape of particles compared to the flakey product produced from a jaw crusher for instance.

(ii) The reported incidence of incipient cracks.

(iii) Somewhat narrower size distribution than for conventional crushing.

(iv) Relative lack of undesirable contamination: most impurities are introduced from electrode wear and the electrode composition can be selected within certain limits of durability.

These have led to suggestions for application of the process for preparation of special concrete aggregates and (at the fine crushing end) for analytical crushing uses or preparation of powders free from undesirable impurities such as iron. The fact that crushing takes place in a confined zone gives the process advantages in crushing radioactive materials, and should also give advantages in an integrated crushing operation for reducing conventional materials over a large size range.

See also: Electrohydraulic forming of metals.

Bibliography

BERGSTROM B. H. (1961) The electrohydraulic crusher, *Eng. and Mining Journal*, p. 134.

CARLEY-MACAULY K. W., HITCHON J. W. and MAROUDAS N. G. (1966) Energy consumption in electrohydraulic crushing, *Trans. Inst. Chem. Eng.* **44,** T 395–T 404.

Ephsteyn Ye. A., Arsh E. I. and Vitort G. K. (1960) *New Methods of Crushing Rocks*, Moscow: Goztoptekhizdat.
Fruengel F. (1965) *High Speed Pulse Technology*, New York: Academic Press.
Gardner G. H. D. et al. (1964) The electrohydraulic spark cleaning of castings, *Brit. Cast Iron Research Assn. Journal* **12** (5), 573.
Linke G. (1968) Electrohydraulic comminution—the state of the art, *Chem. Ing. Technik* **40** (3), 117–120.
Maroudas N. G. (1967) Electrohydraulic comminution, *Brit. Chem. Eng.* **12** (4), 558–562.
Maroudas N. G., Johnston H. A. and Yigit E. (1967) The mechanism of electrohydraulic crushing, *Dechema-Monographien, Verlag Chemie* **57**, 551–582.
Shutler N. D. and Mesler R. B. (1965) A photographic study of the dynamics and damage capabilities of bubbles collapsing near solid boundaries, *Trans. A.S.M.E. J. Basic Eng.* **87**, 511–517.
Yigit E. (1967) Ph. D. Thesis, Electrohydraulic comminution studies, Univ. London.
Yutkin L. (1955) *The Electrohydraulic Effect*, Moscow: Mashgiz.

K. W. Carley-Macauly

ELECTROHYDRAULIC FORMING OF METALS. Electrohydraulic forming is an example of explosive working in which the explosion is generated by an underwater electric spark. (Other methods of forming by an electric discharge have also been proposed, particularly discharge through inductive coils around the part to be formed, i.e. electromagnetic forming.) The electrical circuits used are similar to those for electrohydraulic crushing but to avoid a discharge through the working material the spark is separated from the metal blank by some thickness of water, i.e. the operation is a "stand-off" one. The intense and narrow shock pulses desirable in crushing are not so advantageous for forming and may indeed cause spalling of the metal.

The optimum electrical parameters for forming operations have not been completely explored, but there is evidence that for a given energy of discharge, the greatest deformation results when the discharge is concentrated in a single pulse approximately at critical damping. (A series of sparks gives less deformation than a single spark of the same total energy.) For single sparks an empirical relationship has been found for the depth d of a simple bulge formed in a plate:

$$d = K \frac{D}{tR} W^{0.5}$$

where D is the diameter of the bulge (defined by a hole in a support plate),
t is the thickness of plate,
R is the standoff, and
W is the energy of discharge.

The constant K will vary with both gap width and voltage. It is probable that a relationship of this type would hold over a considerably wider range than the limited one for which results are available.

In forming operations, the required repetition rate of sparks is relatively slow, and hence wires may be used for spark initiation. This application is described in the article on exploding wires in Supplementary volume 2.

Spark discharges should be competitive with forming by conventional explosives for relatively large throughputs of small components where the forming energy is moderate, say up to 40,000 joules. For much larger discharge energies the high capital cost of the condenser bank is likely to rule it out. One American firm have reported a development using sparks of 170,000 joules to form 42-inch blanks. However, at the time of writing application has been reported mainly to forming small components and items such as ashtrays and coffee pots; also to piercing holes, forming ribs and bulges on tubes, and to expanding tubes into tube plates. The characteristic, as with explosive forming, is the precision and ease with which these operations can be carried out.

See also: Electrohydraulic crushing.

Bibliography

Brown R. (1967) Shaping metal with underwater sparks, *New Scientist* **36** (575), 671–673.
Chace W. G. (1967) Exploding wires, uses of, in *Encyclopaedic Dictionary of Physics* (J. Thewlis Ed.), Suppl. vol. 2, 85, Oxford: Pergamon Press.
Hodgson G. (1967) *Design Factors in Electrohydraulic Forming*, National Engineering Laboratory, Glasgow, Scotland. NEL Report No. 280.
Kirk J. W. (1962) Impulse forming by electrical methods, *Sheet Metal Industries* **39** (424), 533–540.
Pearson J. (1966) Explosive working of metals, in *Encyclopaedic Dictionary of Physics* (J. Thewlis Ed.), Suppl. vol. 1, 98, Oxford: Pergamon Press.
Sandefjord (Oslo) and Lillehammer (1964) *High Energy Rate Working of Metals*, NATO Advanced Study Institute, 2 vols.

K. W. Carley-Macauly

ELECTRON BEAM MACHINING. Today electron beam welding and melting are well-known processes. Many hundreds of electron beam welding machines are being used throughout the world and most of the organizations producing reactive and refractory metals use the electron beam melting process. In contrast the machining of materials with electron beams is restricted to a very few applications, despite the fact that in the late 1930's electron beams had been used for drilling the apertures for electron microscopes, and in 1948 Steigerwald of Carl Zeiss developed the modern electron beam milling machine capable of drilling holes, cutting slots and profiles in a number of materials including ceramics.

The equipment used is similar to that used for electron beam welding and comprises an electron optical column mounted on a vacuum chamber in which the workpiece is mounted. The chamber contains equipment for precision mechanical movement of the workpiece. A schematic drawing of the electron

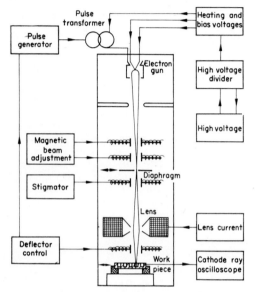

Fig. 1. *Block diagram of electron beam generating column and electrical supplies.*

beam column together with the supplies is shown in Fig. 1. The electron gun is located in the top section of the column and consists of a filament, control electrode and anode. Below the gun are coils for electromagnetic beam adjustment, a mechanical diaphragm and a stigmator, further down are the magnetic lens and the beam deflector. An accelerating potential of 150 keV is applied between the heated filament and the grounded anode and with the specially designed gun produces a collimated beam of electrons which leaves the anode and passes down the column; the beam intensity is controlled by adjusting the bias voltage. The beam is aligned by the adjustment coils and shaped by the diaphragm, the normally oval section of the beam is corrected to a circular section by the stigmator.

Final focusing of the beam onto the workpiece is done by the magnetic lens. Its focal length is adjustable and the deflector coils can move the beam across the workpiece on both the x and y axes.

The maximum total power of a typical machine is 100 W but as the diameter of the beam spot impinging on the workpiece is less than 0·001 in. the beam power density is in the range 10^9 W/in^2. Because of the very high beam power density the surface layers of almost all materials will be sublimated when the beam impinges on them, this enables metal to be removed without heating the surrounding area unduly. In this manner drilling and milling operations can be performed with a high degree of accuracy. To improve the efficiency of the process the beam is pulsed at frequencies of 10,000–50,000 cps with a pulse duration of 2–70 μsec. The intermittent action is obtained by means of a pulse generator mounted in the top of the optical column, this position is essential as the capacity of any appreciable length of connecting cable could affect the pulsing signals.

In order to cut round or shaped holes of a cross-section greater than that of the beam, or for treatment of a surface area, the beam and the workpiece are

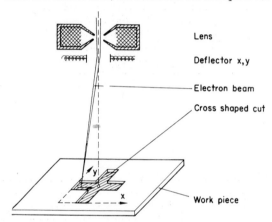

Fig. 2. *Cutting of a shape with a deflected electron beam.*

moved with respect to each other. For large shapes the workpiece is moved under a fixed beam, for smaller shapes the beam is swept across the stationary workpiece, in some cases both motions are used simultaneously. Figure 2 shows how the beam is deflected to cut a cross-shaped hole. The shape followed by the beam is determined by control of the deflector coils current, the current can be programme-controlled to produce complex shapes or a large number of one shape. In addition to the perforation of thin foil deep holes can be drilled up to 0·3 in. depth.

An electron-beam machining equipment provides a means of accurately cutting holes in metal for nozzles, dies, and spinnerets for synthetic fibres. The optimum performance range is:

round holes 0·002 in.–0·008 in. diameter
depth of cut 0·020 in.–0·040 in.
accuracy of shape ±0·0002 in.

The operating range extends on either side of the optimum range with lowered efficiency. The smallest holes which can be drilled are a few microns in diameter. The speed of cutting is very high, e.g. 0·004 in. diameter holes with 26,000 holes to the square inch can be cut in 0·008 in. stainless steel at the rate of 10 μsec per hole.

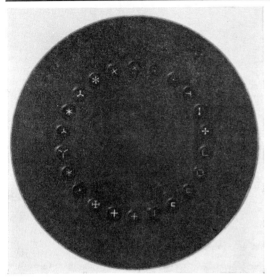

Fig. 3. Die steel with apertures cut for spinnerets. Recess 0·020 in. diameter. Magnification × 25.

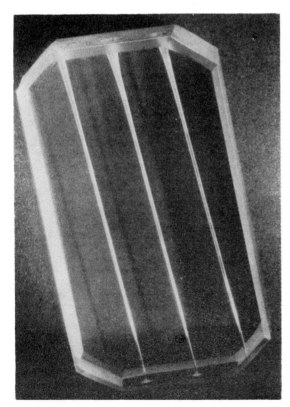

Fig. 4. Electron beam drilling in fused quartz. Drilling time 3 sec. Mean diameter 0·1 mm or 0·004 in. Length of hole 15 mm or 0·6 in.

Figure 3 shows examples of the intricately shaped apertures that have been cut in die steel for spinnerets.

The application of the process is not, however, limited to metallic or electrically conducting materials. An important application is the cutting of insulators and refractories as well as vitreous and crystalline material. In most cases preheating is necessary to reduce thermal stress. This can be done by the use of a small furnace or a separate electron gun. Bearing jewels such as sapphire can be readily drilled. Figure 4 shows an example of fused quartz with 0·004 in. holes drilled through it.

A further application is in the production of small electronic components where it can be used for dicing wafers, cutting of contours or fine lines on the surface of semiconductor materials, also films deposited on substrates for micro-miniature components can be shaped and circuit patterns cut.

Bibliography

BAKISH R. (1960) *Introduction to Electron Beam Technology*, New York: Wiley.
GREEN R. E. (1962) Applications of electron beams, *Machining* **100**, Jan. 1962.
OPITZ W. (1960) Electron beam machining laboratory developments, *Proceedings of the Second Symposium on Electron Beam Technology*, Cambridge, Mass.: The Alloyed Corporation.
STEIGERWALD K. H. (1961) Electron beam milling, *Proceedings of the Third Symposium on Electron Beam Technology*, Cambridge, Mass.: The Alloyed Corporation.

J. G. PURCHAS

ELECTRON DIFFRACTION FROM PERIODIC MAGNETIC FIELDS. When an electron beam crosses a region of magnetic field, it experiences changes of phase, which depend on the flux enclosed by different parts of the beam. These changes of phase are responsible for the contrast effects by which magnetic domain structure can be directly observed in the electron microscope (see, for example, Hirsch, Howie, Nicholson, Pashley and Whelan, 1965). To explain these contrast effects, it is convenient to regard the electron beam as having suffered a change of direction caused by the "Lorentz force" rather than a change of phase. This geometrical approach is useful when the electron beam can be regarded as incoherent. However, a more rigorous approach is needed for coherent plane waves. Such an approach has been used by several authors (see, for example, Cohen, 1967, and Wohlleben, 1967) to discuss the effect of increasing coherence on electron microscopic images of magnetic specimens. In the present article, we shall discuss the effect of non-uniform magnetic fields on the diffraction patterns produced by coherent electron beams. In particular, we shall deal with periodic magnetic fields whose effect on the electron beam is similar to that of a phase grating on light. In practice periodic fields may be found, for example, in ferromagnetic materials with uniaxial crystal anisotropy, in antiferromagnetics, and in the mixed state of type II superconductors. In this article we develop the theory in a very simple form and apply the results to the materials mentioned above. Extension to more general situations is simply a matter of greater numerical or algebraic complexity.

General One-dimensional Phase Grating

We consider first the magnetic fields which are experienced by an electron beam passing through a thin magnetized specimen in, for example, an electron microscope. This situation is depicted schematically in Fig. 1, where it can be seen that, for simplicity,

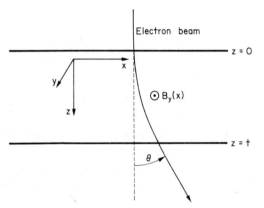

Fig. 1. Schematic representation of a thin foil containing a magnetic field.

only the y-component of the magnetic induction **B** is considered. Further, the magnetic field is assumed to be confined to the specimen, i.e. to a region of thickness t in the direction of the incident electron beam (z-direction). The phase difference $\phi(x)$ between the beam crossing the specimen at $x = 0$ and at x is then given by

$$\phi(x) = -\int_0^x \frac{2\pi e t}{ch} B_y(x') \, dx' = \frac{-2\pi e}{ch} \times \text{flux linked} \quad (1)$$

where $-e$ is the electronic charge, c is the velocity of light and h is Planck's constant. The angular distribution of intensity in the Fraunhofer diffraction pattern after the beam has crossed the region of magnetic field is given by

$$I(\theta) \propto |\int \exp\{i[\phi(x) - 2\pi s x]\} \, dx|^2 \quad (2)$$

where $s = \sin\theta/\lambda$, λ is the electron wavelength and θ is the angle relative to the beam direction (see Fig. 1). The integration is carried out over the coherently illuminated region and thus, if $\phi(x)$ is known for all x, $I(\theta)$ can be evaluated for all θ.

If the magnetic field B_y is taken to be a periodic function of x (period a) then $\phi(x)$ is also a periodic function of x, apart from a term which varies linearly with x (arising from the mean value of B_y over a period). It then follows that the diffracted intensity is given by

$$I(s) \propto I_1(s) \times I_2(s) \quad (3)$$

where

$$I_1(s) = \frac{\sin^2\{N[\phi(a) - 2\pi s a]/2\}}{N^2 \sin^2\{[\phi(a) - 2\pi s a]/2\}}, \quad (4)$$

$$I_2(s) = \frac{1}{a^2} \left| \int_0^a \exp\{i[\phi(x) - 2\pi s x]\} \, dx \right|^2, \quad (5)$$

and N is the number of repeat units (slits) of the coherently illuminated grating. The normalizing factors N^{-2} and a^{-2} are included in order to provide convenient ordinate scales for I_1 and I_2.

$I_1(s)$ corresponds to the interference pattern produced by a diffraction grating with N infinitely narrow slits of spacing a, where each successive slit introduces an additional phase change $\phi(a)$ compared with the previous slit. This function has principal maxima at

$$s = \frac{\phi(a)}{2\pi a} \pm \frac{n}{a} \quad (n = 0, 1, 2 \ldots). \quad (6)$$

$I_2(s)$ corresponds to the diffraction pattern from a single slit of width a with the appropriate phase variation across it. Thus the intensities of the maxima of I_1 are modulated in the usual way. The integral in equation (5) can be evaluated if $\phi(x)$ is known in the region $0 \leq x \leq a$.

An approximation which is physically reasonable and enables equation (5) to be integrated explicitly is that of two linear phase functions, i.e.

$$\left.\begin{array}{l}\phi(x) = \alpha_1 x, \quad 0 \leq x \leq b_1 \\ = \alpha_1 b_1 + \alpha_2(x - b_1), \, b_1 < x \leq a\end{array}\right\} \quad (7)$$

as shown schematically in Fig. 2. The integral in equation (5) can then be evaluated analytically.

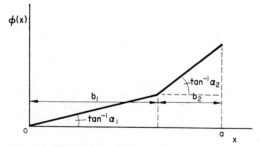

Fig. 2. Illustration of the two-component linear phase approximation.

$I_2(s)$ then becomes

$$I_2(s) = A_1^2 + A_2^2 + 2 A_1 A_2 \cos\psi \quad (8$$

where
$$A_i = \sin(\pi s_i b_i)/\pi s_i a \quad (i = 1, 2),$$
$$\psi = \pi(s_1 b_1 + s_2 b_2),$$
$$b_2 = a - b_1,$$
and
$$s_i = s - \alpha_i/2\pi \quad (i = 1, 2).$$

Figure 3(a) is a typical plot of A_1 and A_2 against s while Fig. 3(b) shows the resultant envelope of $I_2(s)$.

Important general properties of equation (8) are (i) A_1 and A_2 [Fig. 3(a)] have principal maxima at $s_1 = 0$ and $s_2 = 0$ respectively, i.e. at the values of s corresponding to the angle of "Lorentz deflection". Provided that these maxima are sufficiently separated $I_2(s)$ [Fig. 3(b)] shows maxima at these same angles. (ii) The widths of the principal maxima (to first zero) of A_1 and A_2 ($\Delta s = 2/b_1$ and $2/b_2$ respectively) are equal to r_1 and r_2 times the separation of the principal maxima of the interference term $I_1(s)$ ($\Delta s = 1/a$), where $r_1 = 2(1 + b_2/b_1)$ and $r_2 = 2 \times (1 + b_1/b_2)$. (iii) At the principal maxima of the interference term the cross term $2A_1A_2 \cos \psi$ is equal to $\pm 2A_1 A_2$, being positive for values of s between the principal maxima of $I_2(s)$ and negative outside.

The implications of these general principles are illustrated in the practical applications cited below.

Experimental Applications

(i) *Magnetic domains in thin cobalt foils.* In certain orientations cobalt specimens may be found with long, narrow domains separated by 180° Bloch-walls. Then in equations (7) and (8)

$$\left. \begin{array}{l} b_1 = b_2 = a/2 \\ \alpha_1 = -\alpha_2 = -2\pi e t B_s \sin \gamma / h \end{array} \right\} \quad (9)$$

where b_1 is the domain width, t is the foil thickness, B_s is the saturation magnetization induction, and γ is the angle between the magnetization direction in the domains and the foil normal. Thus the separation of the principal maxima of A_1 and A_2 [equation (8)] is $\Delta s = 4\pi e t B_s \sin \gamma / ch$ and the resultant intensity distribution is symmetrical about the origin. A typical distribution is shown in Fig. 4, where it can be seen that the widths of the "Lorentz peaks" are such that only two to four interference spots are visible under each.

Cobalt foils of this type have been studied by Goringe and Jakubovics (1967) and Wade (1967) using low angle electron diffraction techniques in conventional electron microscopes (see, for example, Ferrier and Murray 1966) and these predictions verified.

(ii) *Antiferromagnetic materials.* The phase variation produced by antiferromagnetic materials in certain orientations should be of the same form as that given by equation (9), where a ($= 2b_1$) is now the antiferromagnetic period. However, in contrast to the ferromagnetic case discussed above, no effects have as yet been observed in the electron microscope. The explanation of this is that although a is typically of the same order of magnitude as (or less than) the value for ferromagnetic domains, a is typically two or three orders of magnitude smaller. Thus, in contrast to the ferromagnetic case, the separation of the "Lorentz peaks" is smaller than the interference spot spacing, as can be seen in the schematic plot of Fig. 5. As none of the interference spots except zero-order lies between the principal maxima of A_1 and A_2, the discussion above shows that the terms in equation (8) tend to cancel out. As their magnitudes are nearly equal the resulting intensities are very small, i.e. antiferromagnetic "superlattice" diffraction spots are not expected to be visible, in agreement with experiment.

(iii) *Possible application to type II superconductors.* In the mixed state of a type II superconductor there is a periodic variation of magnetic field, i.e. a periodic

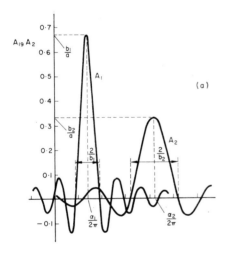

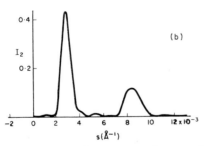

Fig. 3. Illustration of some general properties of $I_2(s)$ derived from equation (8) and the phase function of Fig. 2, with the following values: $b_1 = 1000$ Å, $b_2 = 500$ Å, $a = b_1 + b_2$, $\alpha_1 = 1.76 \times 10^{-2}$ Å^{-1} and $\alpha_2 = 3\alpha_1$. (a) Variation of the amplitude terms A_1 and A_2 with s. (b) Variation of I_2 with s.

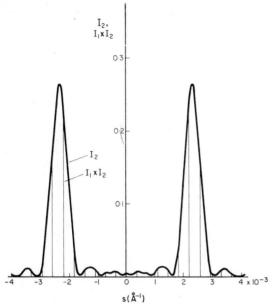

Fig. 4. Graph showing the variation of I_2 and $I_1 \times I_2$ with s, for a uniaxial ferromagnetic thin foil, with the following values: $a = 2720$ Å, $\alpha_1 = -\alpha_2 = 1\cdot 48 \times 10^{-2}$ Å.

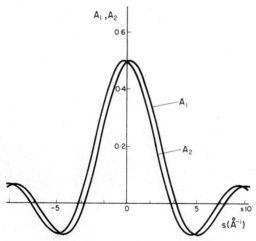

Fig. 5. Schematic plot of the amplitude terms A_1 and A_2 [equation (8)] for an antiferromagnetic specimen. The positions of the principal maxima of I_1 [equation (4)] are indicated on the abscissa.

distribution of Abrikosov "flux lines" (see, for example, de Gennes, 1966). Assuming that such a distribution of flux lines can exist in thin foils in the electron microscope the above theory may be applied to predict the diffraction effects expected in an idealized situation where the phase change of π (one flux quantum $hc/2e$ per line) is assumed to occur linearly over a distance of the order of the penetration depth and to remain unchanged between flux lines. Thus in equations (7) and (8) $\alpha_1 = 0$ and $\alpha_2 = \pi/b_2$, where $b_2 =$ penetration depth and $a =$ separation of flux line centres.

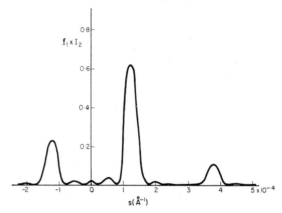

Fig. 6. Computed intensity distribution in the diffraction pattern which might be obtained from a type II superconductor containing a periodic array of flux lines ($b_1 = 3000$ Å, $b_2 = 1000$ Å, $\alpha_1 = 0$, $\alpha_2 = \pi \times 10^{-3}$ Å^{-1}, $N = 5$).

Figure 6 shows a plot of the diffracted intensity distribution expected under very favourable conditions of flux lattice spacing and angular resolution. It can be seen that under these conditions some diffraction effects should be visible. However, the phase variation is certainly not linear across the flux line, nor is it zero anywhere unless the flux lines are very widely spaced. Both these considerations would tend to decrease the visibility of the effect.

Conclusions

So far we have discussed applications of equations (3) to (5) in the particularly simple case of a grating of two linear phase components. Obviously many situations exist where the phase variation is far from linear for which the intensity distribution can be computed for particular models and compared with experiment. Conversely iterative procedures may be applied to experimentally measured intensities to calculate $\phi(x)$ from equation (5), although little work has yet been done in this direction. In all cases the angular distribution of the diffraction envelope is only sampled at the grating interference maxima.

See also: Electron diffraction.

Bibliography

COHEN M. S. (1967) *Journal of Applied Physics* **38**, 4966.

FERRIER R. P. and MURRAY R. T. (1966) *Journal Royal Microscopical Society* **85**, 323.

DE GENNES P. G. (1966) *Superconductivity of Metals and Alloys*, New York: Benjamin.
GORINGE M. J. and JAKUBOVICS J. P. (1967) *Phil. Mag.* **15**, 393.
HIRSCH P. B., HOWIE A., NICHOLSON R. B., PASHLEY D. W. and WHELAN M. J. (1965) *Electron Microscopy of Thin Crystals*, London: Butterworths.
WADE R. H. (1967) *Physica Status Solidi* **19**, 847.
WOHLLEBEN D. (1967) *Journal of Applied Physics* **38**, 3341.

M. J. GORINGE

ELECTRON METALLOGRAPHY, APPLICATIONS OF.

The application of electron microscopy to the study of the microstructure of metals and alloys is termed electron metallography. As the metallurgical information which can be obtained depends to a large extent on the procedure used for making the electron micrograph, it is necessary to divide the discussion according to the different types of electron microscopic technique.

Transmission Microscopy

As electrons are readily scattered by metals, the materials and components used industrially cannot in general be examined directly in the transmission microscope, so that special preparation techniques are required. In the *replica technique* the surface topography of the metal is reproduced in a thin film of a substance which is not decomposed by the action of the electron beam, commonly carbon. Generally the technique involves polishing and etching a section through the metal, as for optical metallography, when the structure so developed can be reproduced either using a two stage plastic/carbon or single stage direct carbon replica. In the former case the plastic most often used is cellulose triacetate, sheets 25–250 μ thick of which are softened in acetone, laid on to the prepared metal surface and as the acetone evaporates the plastic hardens and contracts into the surface features. It is then carefully removed from the metal and a layer of carbon, 200–300 Å thick, is evaporated in vacuo on to the replica face, after which the plastic is dissolved in acetone leaving a secondary carbon replica. The more direct technique is to evaporate the carbon layer directly on to the prepared metal surface. Then because the carbon layer is somewhat porous it may be removed by re-etching the surface to dissolve the metal away from the carbon film. A particularly useful variation of this technique is the carbon extraction replica, which is used for the examination of multi-phase alloys. The etching of the specimen, both prior to and after deposition of the carbon film, is controlled so that one phase, usually the matrix, is dissolved at a much faster rate than the other phase(s). In this way the second phase precipitate particles present in most alloys, except those of aluminium, can be extracted from the matrix on to the carbon replica film and because these particles are attached to the carbon, their distribution in the replica is normally identical to that in the original bulk metal sample. Prior to the examination in the microscope the pieces of carbon replica are collected on to fine mesh support grids, usually made of copper. The resolution attainable in carbon replicas is typically 20 Å and they may be usefully examined at all magnifications up to about 40,000 times.

The second type of specimen preparation procedure, the *thin foil technique* involves reducing the thickness of the metal until it can be penetrated by the electron beam of the microscope. The limiting thickness depends on the atomic number of the metal and the electron accelerating voltage at which the microscope is operated. Thus for conventional instruments operating at 100 kV the thickness decreases from about 2000 Å for aluminium to about 300 Å for uranium, whereas using a high voltage microscope at 1000 kV an aluminium specimen of up to about 9 μ can be penetrated. The usual method for reducing the thickness of the specimen is by electropolishing and a widely used procedure requires that the starting specimen be in the form of a disc up to 3 mm diameter and 0·5 mm thick, which can be readily prepared from the bulk material by processes such as grinding or machining. These discs are then electropolished until a small hole(s) is formed near the centre, when it is usually found that the metal adjacent to the edge of the hole is sufficiently thin to transmit electrons. Using these disc techniques thin foils can easily be produced from selected areas or from a required distance with respect to a reference surface. The resolution attainable in a thin foil is usually limited to ~20 Å by chromatic effects due to the energy loss of the electrons incident on the metal film. High resolution can only be attained in thin foils of ferromagnetic materials such as ferritic steels, by ensuring that the bulk of specimen is as small as possible with the perforation occurring in the cent e of the sample, to minimize the effects of the foil on the magnetic field of the objective lens. Modification of the structure of a metal may occur during the preparation of thin foils or during subsequent examination in the microscope. Thus defect structures may be altered because of the relaxation of long range stresses during thinning or due to mechanical damage of the thin foil by careless handling. Also metastable alloys may undergo a phase transformation when in thin foil form, either because the surface acts as a preferential nucleating site for nucleation and growth transformations or because in materials which undergo a phase transformation by shear, the large stresses produced by the transformation may be relaxed at the surface of the foil.

Selected area electron diffraction analysis is extensively used for the identification of second phase particles in carbon extraction replicas and thin foils and individual particles down to ~1000 Å diameter can be analysed. Diffraction patterns from extraction replicas are normally calibrated (to determine the camera constant) by evaporating a substance whose interplanar spacings are known accurately, such as gold or thallium chloride, on to the carbon film. In the case of second phase particles in thin foils the diffraction patterns are usually calibrated using the matrix phase. Spot patterns are obtained from single crystal particles and the patterns are effectively an enlarged image of a plane of the reciprocal lattice of the crystal. If a large number of small particles are contributing to the diffraction pattern, a ring pattern is obtained. In either case the interplanar spacings can be calculated from the equation:

$$d = \frac{L\lambda}{r},$$

where L is the camera length, λ the wavelength of the electrons, and r is half the distance separating corresponding spots in the single crystal spot pattern, or the radius of the diffraction ring in a polycrystalline ring pattern. $L\lambda$ is usually obtained from the calibration pattern. By calculating the d values for 3 or 4 sets of diffraction spots or rings and determining the symmetry of the diffraction pattern, it is possible to index the spots or rings in terms of their Miller indices and therefore to determine the crystal structures and parameters of the diffracting phase. The best accuracy possible with selected area electron diffraction is about 0·1 per cent but is more commonly in the range 0·5–1·0 per cent.

However, to identify unambiguously the structure of a single particle it is often necessary to tilt the diffracting crystal into two or three prominent diffracting zones. Also, as the contrast in images of thin foils arises mainly from Bragg diffraction of the electrons, the contrast from fine second phase particles and defect structures depends critically upon the orientation of the foil with respect to the electron beam. Therefore modern instruments incorporate special stages which allow the specimen to be tilted and rotated. It is also useful to have a facility for cooling the specimen, to reduce the amount of contamination which can occur if one area of the foil is examined for a prolonged period. Stages designed to enable the specimen to be heated or strained are of only limited value in electron metallography, because due to the high surface area : volume ratio, the rate and morphology of structural changes which occur during heating or straining of thin foils are usually different from those occurring in bulk material.

Replicas and thin foils have been extensively used to study the relationship between the structure of metals and alloys and their bulk mechanical and physical properties. Wherever possible both specimen preparation techniques are used to obtain the maximum amount of information from a single sample. Replicas reveal the morphology and distribution of second phase particles over areas as large as 3 mm diameter, whereas thin foil areas are usually much smaller and are used to study the crystallography of the matrix/second phase particle relationships as well as the type and distribution of the atom packing defects present in industrial metals and alloys. Probably the most extensive use of electron metallography has been in studies of the phase transformations which take place during the heat treatment of precipitation-hardening alloys and to follow the structural changes occurring when steels are quenched and tempered. These studies have greatly increased the metallurgist's understanding of the structural requirements for alloys to be strong and tough at different service temperatures, and this knowledge is assisting in the development of new and improved alloys, particularly for applications where resistance to brittle fracture, or creep, or fatigue damage is important.

Transmission microscopy has been applied to the investigation of processes which essentially attack only the surface of a metal, including oxidation, chemical and electro-chemical attack and mechanical wear. It is important to establish the structural sites at which these various forms of attack are initiated, and the way in which the attack propagates. The various forms of replica have been extensively used but more recently the thin foil technique has been adapted to this type of investigation. The technique consists of preparing the foil by a conventional electropolishing process, followed by examination in the microscope. The thin foil is then subjected to oxidation, corrosion or stress corrosion attack for a short period, followed by re-examination in the microscope and in this way the early stages of attack can be followed in fine detail. It is sometimes possible to study the early stages of continuous oxide film formation by stripping this film from the metal substrate, followed by direct examination in the microscope.

The fracture surfaces of failed metallic components can be examined using replica transmission microscopy to provide information on the mode of fracture. This can be used to provide a diagnosis of the cause of premature service failure so that steps can be taken to ensure that such a failure does not occur again. Thus the relevant features of the fracture topography of failure occurring due to cleavage, ductile, fatigue and intergranular cracking have been characterized. Some forms of cracking occur due to local liquation in the structure, particularly during welding, and examination of these fractures using the extraction replica technique, combined with electron diffraction and X-ray micro-analysis, can allow the elements responsible for the liquation to be identified.

Reflection Microscopy

Compared to transmission microscopy this technique has the advantage that no special form of specimen preparation is required, and the surface of bulk samples can be examined directly, but a major disadvantage is the inferior resolution of about 300 Å. It is also possible to obtain diffraction patterns in reflection instruments, and the advantage of these compared to X-ray diffraction is that the penetration of the electrons into the surface is much lower, so that thinner surface films can be analysed without interference from the substrate. The major application of reflection microscopy is in problems associated with the lubrication, wear, corrosion and oxidation of metallic surfaces, particularly for examining attacked regions or deposits which have formed on a coarser scale than can be readily examined by transmission microscopy. The technique has been a useful intermediary between optical and transmission microscopy, but is likely to be largely superseded by scanning microscopy.

Scanning Microscopy

This is a rapidly developing technique which has been extensively used in recent years. A beam of electrons with energies in the range 1 to 30 kV is focused to a spot ∼100 Å diameter and made to scan the surface to be examined in a rectangular raster. Primary electrons are reflected and secondary electrons emitted from the surface and these are focused with an electro-static electrode on to a biased scintillator. The light produced is transmitted via a perspex light pipe to a photomultiplier and the signal generated is used to modulate the brightness of an oscilloscope spot which transverses a raster in exact synchronism with the electron beam at the specimen surface. The image observed on the oscilloscope screen is similar to the optical image in that the signal at the collector depends on the angle between the primary electron beam and the normal to the surface. The magnification of the projected image is easily adjusted, because it is simply the ratio of the rasters of the electron beam and oscilloscope. Since the angle of divergence between electron paths in the probe is small, very small apertures may be used, so that the depth of focus is very large, at least 300 times that of an optical microscope. The resolution depends upon the specimen being examined, but is typically between 200 and 500 Å. No special specimen preparation is required for metallic samples, but for non-conducting materials which tend to charge-up it is necessary to deposit a thin conducting film by evaporation in vacuo.

The main uses of scanning microscopy in metallurgy are for the examination of fracture surfaces and for studying surfaces which have been subjected to chemical or electro-chemical attack or mechanical damage. Its particular advantage over replica transmission microscopy for fracture surface studies is the ability to examine much more uneven fracture profiles and also the ease of carrying out comprehensive surveys of comparatively large areas of fracture, which may contain more than one type of fracture. The technique can also be used to obtain much useful information from oxidized fracture surfaces, such as those resulting from high temperature creep fracture, which cannot be readily examined by replica techniques because the oxide films are partially extracted and often opaque to the electron beam, so that their presence tends to obscure other more relevant details.

In studies of corroded surfaces the most striking advantage of the scanning microscope arises from the easy appreciation of the three-dimensional nature of the oxide growths which can be achieved by using a rotating stage. This allows the morphology of the oxide deposits to be readily established and this can often be particularly important in deciding the circumstances of its formation.

Bibliography

BRAMMAR I. S. and DEWEY M. A. P. (1966) *Specimen Preparation for Electron Metallography*, Oxford: Blackwell Scientific Publications.
HIRSCH P. B., HOWIE A., NICHOLSON R. B., PASHLEY D. W. and WHELAN M. J. (1965) *Electron Microscopy of Thin Crystals*, London: Butterworths.
KAY D. H. (Ed.) (1965) *Techniques for Electron Microscopy*, Oxford: Blackwell Scientific Publications.
MARTON L. (1961) Electron microscopy, in *Encyclopaedic Dictionary of Physics* (J. Thewlis Ed.), **2**, 799, Oxford: Pergamon Press.
NUTTING J. (1963) *The Electron Microscope and Its Application to the Study of Metals*, Iron and Steel Institute Special Report No. 80, 154.
THOMAS G. (1962) *Transmission Electron Microscopy of Metals*, New York: Wiley.
THOMAS G. and WASHBURN J. (Eds.) (1963) *Electron Microscopy and Strength of Crystals*, New York: Interscience.

M. A. P. DEWEY

ELECTROVISCOUS FLUID. A fluid which undergoes an essentially instantaneous reversible change in apparent viscosity when subjected to an externally applied electric field is called an electroviscous fluid. The fluids which exhibit substantial electroviscosity, sometimes to the point of solidification, are generally dispersions of particles in a low-dielectric-constant liquid that contains surface-active materials. Electroviscosity is dependent on many parameters such as composition of the dispersion, characteristics of the applied field, temperature and shear rate. The phenomenon appears to be caused by field-induced interfacial polarization of the double layers surrounding the particles or particle aggregates dispersed in the medium. The resulting electrostatic interactions of the distorted double layers in the medium of low dielectric constant require the dissipation of additional energy on application of a shear stress normal to the direction of the field-induced distortion. This increased energy dissipation manifests itself as a change in apparent viscosity.

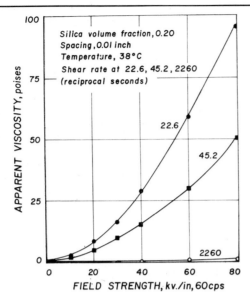

Fig. 1. The apparent viscosity of a typical electroviscous dispersion varies directly with the strength of the applied electric field and inversely with the shear rate.

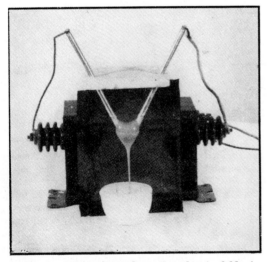

Fig. 2. Effect of an alternating electric field of about 13,000 volts applied across a globule of electroviscous fluid.

Bibliography

ANDRADE E. N. DA C. and HART J. (1954) *Proc. Royal Soc. (London)* Ser. A **225**, 463.
BJÖRNSTÅHL Y. and SNELLMAN K. O. (1939) *Kolloid-Z.* **86**, 223.
KLASS D. L. and MARTINEK T. W. (1967) *J. Appl. Phys.* **38**, 67, 75.
WINSLOW W. M. (1949) *J. Appl. Phys.* **20**, 1137.

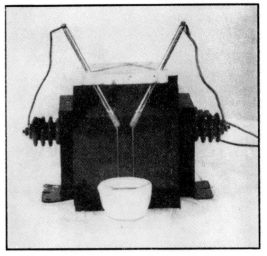

Fig. 3. When the field (see Fig. 2) is removed, the fluid reverts to its normal free-flowing liquid state.

D. L. KLASS

ELECTROVISCOUS FLUIDS, APPLICATIONS OF.

The unusual characteristic of electroviscous fluids, namely, the essentially instantaneous and reversible change in apparent viscosity under the influence of an externally applied electric field, suggests that these fluids offer practical means for the transport and control of mechanical power in couplings and hydraulic devices.

Consider, for example, how the simplest form of coupling could function. Two concentric electrically conducting cups in a hypothetical clutch are separated by a relatively thin layer of electroviscous fluid. One cup is the driving member of the coupling and is rotated at constant speed; the other cup is the driven member. Without an externally applied electric field, the clutch operates under slip conditions and the torque transmitted depends upon the viscous drag of the fluid between the cups. When an electric field is applied across the coupling, the fluid undergoes an instantaneous increase in apparent viscosity, the magnitude of which is a function of the properties of the fluid, the characteristics of the coupling and the applied field, and the environmental conditions. The immediate result is the transmission of torque. The coupling can be operated under non-slip or partial slippage conditions and torque transmission is controlled at will by changing the applied field. Termination of the applied field causes the fluid to revert to its normal liquid state and torque transmission ceases.

Electroviscous fluids that exhibit substantial changes in apparent viscosity on application of an external electric field also exhibit another unusual characteristic. If the driving member of the hypothetical clutch is rotated without an electric field impressed across the cups, the electrical resistance of the fluid decreases markedly from that of the fluid under zero shear conditions. A potential difference and current flow occur between the cups. Small d.c. power outputs that depend upon the external resistive load and the shear rate, or rate of rotation of the driving clutch member when the distance separating the cups is constant, are transmitted through the external circuit. Mechanical power is thus converted to electric power. The coupling behaves like a small electric cell when the electroviscous fluid is subjected to a mechanical stress, and its electrical properties are temporarily modified. The theoretical implications of this electrokinetic effect on the relationships of the rheological and electrical properties of fluids are significant. But from the point of view of electroviscous fluid applications, a second major parameter is introduced that offers additional variability and flexibility in the design of electroviscous fluid devices. Devices are suggested that use both electroviscous and electrokinetic effects separately or together.

Other unusual properties of electroviscous fluids include rapid response times to the applied field of the order of microseconds, the ability to control the character of the mechanical output by responding to electrical wave-shaping techniques, and low electric power consumption. These properties complement the inherent properties of liquid-state systems such as the case of filling the void space in an irregularly shaped container.

These deceptively simple descriptions of electroviscous fluid couplings and properties have numerous ramifications that can lead to more sophisticated applications and devices, some of which are now under commercial development.

The field-induced rheological properties of electroviscous fluids appear to make them suitable for many types of motion control, detection, and monitoring applications; torque, tension, and speed control; overload protection; rapid cycling and synchronization; and damping. A large number of devices such as clutches, brakes, transmissions, valves, shock absorbers, chucks, servomechanisms, shock shakers, viscometers, mixers, particle separators, loudspeakers, pumps, sonic generators, and high-speed relays have already been proposed. Other applications in which the amount of electroviscosity induced in the fluid is used to measure field characteristics have been suggested.

The electrokinetic properties of electroviscous fluids have also resulted in what might be called a family of liquid-state electronic devices such as pressure transducers, electrostrictive transducers, audio frequency doublers, seismometers, phonograph pickups, fluid-flow controllers, pulsating d.c. generators, electrokinetic weighing scales, and accelerometers. Combination electroviscous-electrokinetic fluidic amplifiers have been proposed.

The ability to control mechanical outputs by electrical means and to modulate and stop the flow of fluid in "valves" with no moving parts presents advantages not only in the design, cost, and precision control of devices, but also in new functionality. As an illustration, electrically controlled hydraulic vibrators utilizing electroviscous fluids have been developed that operate at frequencies and forces beyond the capability of ordinary mechanical and hydraulic vibrators. Phase shifts between input control voltages and output acceleration are low enough in these devices to permit several vibrators to be used in series at high frequencies, with all units kept in phase by automatic control circuits. It is possible to feed back a signal from the vibrating piston to indicate its instantaneous position, direction, and speed so that during a single cycle, corrections or changes can be made electrically if desired. High acceleration outputs and programmed vibration can also be provided over a wide frequency range with low distortion. Good sinusoidal, triangular, and square waves of acceleration have been produced

Development of practical applications for electroviscous fluids requires a systems approach. That is, the device should be designed for the intended application as well as around the properties and limitations of the electroviscous fluid to be used in the device. For example, clutching applications in which the electroviscous fluid is subjected to constant shear stresses at higher-than-ambient temperatures might require low-viscosity, low-volatility fluids with good thermal stability and life characteristics on prolonged operation. On the other hand, chucking applications, in which a film of electro-

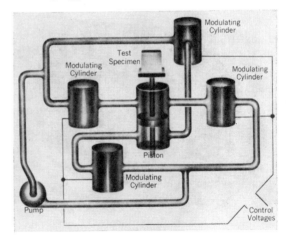

Fig. 1. This schematic diagram of Stanford Research Institute's vibration and shock exciter system shows the flow of electroviscous fluid through the four modulating cylinders, or "valves", the piston, and the generalized circuits by which the control voltages are applied. Variation of the potential applied to the cylinders in the proper sequence causes pressure drops across the cylinders and vibration of the piston. The modulating cylinders are compactly nested within each other in the actual operating units.

viscous fluid is used to temporarily hold an object for machining on a chucking surface, might require fluids that exhibit extremely high electroviscosities on short-term usage after which the fluid is discarded.

Commercialization of the first generation of electroviscous fluid devices is expected in the near future. The Rucker Company, Control Systems Division, of Oakland, California recently began a development and marketing program for an electroviscous fluid vibrator designed by The Boeing Company. This device can be used for heavy-duty testing of full-size space booster and missile components. Rucker's vibrator uses direct electrical control alternately to block and release the flow of electroviscous fluid in valving chambers surrounding the output shaft. The output shaft, the only moving part in the vibrator, moves first in one direction and then the other as the flow of electroviscous fluid is electrically modulated. A conventional hydraulic pump circulates the electroviscous fluid through the system.

The Rucker Company is also developing an electroviscous fluid heart pump valve for use in an implantable energy conversion system designed by Hamilton Standard, a Division of United Aircraft Corporation. The electroviscous fluid valve is to be used in conjunction with a high frequency flexing diaphragm element (bimorph). Two valves are used to direct the flow of fluid from the bimorph into and away from two pressure sacs which alternately massage the heart muscle. The valves open and close at the rate of the bimorph vibration, 500 to 1000 Hz, and operate 180° out of phase from one another. Relative phase between the valves alternates with every heart beat.

Indeed, it appears that the application of electroviscous fluid technology can have a significant impact on many different industries and markets.

Bibliography

EIGE J. J. and FRASER E. C. (1962) *Vibration and Shock Exciter Using Electric-Field Modulation of Hydraulic Power*, U.S. Air Force Systems Command Technical Documentary Report No. ASD-TDR-62-536.

KLASS D. L. (1967) *New Scientist* **33,** No. 538, 664.

KLASS D. L. (1968) Electroviscous fluid, in *Encyclopaedic Dictionary of Physics* (J. Thewlis Ed.), Suppl. vol. 3, Oxford: Pergamon Press.

STRANDRUD H. T. (1966) *Hydraulics and Pneumatics* **19,** No. 9, 139.

D. L. KLASS

Fig. 2. *This photograph illustrates how electroviscous fluids can be used for chucking applications. Fourteen pounds of lead-shot in the glass beaker are supported by a thin film of electroviscous fluid between the white metal disc and the adjacent electrically energized chucking surface. When the electric field is turned off, the fluid will immediately revert to its liquid state, and the beaker will fall.*

ELEMENTARY PARTICLES, NOMENCLATURE OF. Given below are brief definitions of various terms occurring frequently in elementary particle physics. For ease of reference they are given in alphabetical order.

Antiparticle. All elementary particles have associated with them an antiparticle which is related to the corresponding particle by changing the sign of as many quantum numbers as possible (baryon number, lepton number, charge, strangeness, hypercharge and isotopic spin component I_3). Thus the cascade particle Ξ^- has charge -1, baryon number $+1$, hypercharge -1, strangeness -2 and third component of isotopic spin $-\frac{1}{2}$, whereas the corresponding antiparticle $\overline{\Xi}^-$ has charge $+1$, baryon number -1, hypercharge $+1$, strangeness $+2$ and third component of isotopic spin $+\frac{1}{2}$. Some particles, having all appropriate quantum numbers equal to zero, are their own antiparticles; these are the photon, the π^0 meson and the η^0 meson.

Baryon. This name was originally invented to describe the heavier elementary particles, but now it would be more correct to describe the baryons as fermions with mass greater than or equal to that of the nucleons, since some of the boson resonances have mass greater than the nucleon mass. With each baryon it is possible to associate an additive quantum number B called the baryon number. B takes the value $+1$ for baryons, -1 for antibaryons (and 0 for all other particles). All processes known to date conserve total baryon number.

Boson. Particles of integral spin are called bosons.

Fermion. Particles of half odd integral spin ($\frac{1}{2}$, $\frac{3}{2}$, $\frac{5}{2}$, ...) are called fermions. They obey the Pauli exclusion principle: no two fermions may occupy identical quantum states.

Hadron. This is the generic name given to particles which interact via the strong interaction, namely the baryons and the mesons.

Hypercharge. The hypercharge Y of a particle is related to the difference between its charge Q and the third component of its isotopic spin I_3: $Q = I_3 + \frac{1}{2}Y$. It has the same value for all members of an isotopic multiplet.

Hyperon. This is the name given to the baryons with non-zero strangeness.

Isotopic spin. It is observed that certain sets of particles have very similar properties apart from those associated with their different charges (e.g. n, p and π^-, π^0, π^+). It is useful then to think of the set as consisting of different charge states of a single particle. For example, the nucleon can exist in two charge states (n, p) and the pion can exist in three (π^-, π^0, π^+). The mathematical description of the consequences of this idea then involves the introduction of a quantity analogous to ordinary spin—the so-called isotopic spin, isobaric spin or isospin. Just as a particle with total spin S can exist in $2S + 1$ spin states one component of the spin (say S_3) taking on values differing by integral steps from $-S$ to $+S$, so a particle with total isospin I can exist in $2I + 1$ charge states with I_3 taking on values differing by integer steps from $-I$ to $+I$. I_3 is related to the charge Q of the particle by $Q = I_3 + \frac{1}{2}Y$, Y being a quantum number known as the hypercharge. (Y has a constant value for any particular isotopic multiplet.) Thus in the case of the nucleons we have $I = \frac{1}{2}$, $I_3(n) = -\frac{1}{2}$, $I_3(p) = +\frac{1}{2}$ and $Y = +1$. For the pions we have $I = 1$, $I_3(\pi^-) = -1$, $I_3(\pi^0) = 0$, $I_3(\pi^+) = +1$ and $Y = 0$.

Lepton. This is the name given to members of the low mass group of fermions, viz. the neutrino, electron and muon; these are particles which do not participate in the strong interactions. An additive quantum number called lepton number may be introduced. For a lepton it takes the value $+1$, for an antilepton -1 (and for all other particles 0). Then all processes known to date conserve total lepton number.

Meson. This name was originally used to describe particles intermediate in mass between the electron and the nucleons, but is now used to describe all bosons with the sole exception of the photon.

Nucleon. The particles which constitute the nucleus are called nucleons. Thus a nucleon is either a proton or a neutron.

Strangeness. The strangeness of a particle is a quantum number S related to its hypercharge Y and baryon number B by $S = Y - B$.

W. W. BELL

EMISSIVITY AT HIGH TEMPERATURES. In recent years the usage of materials at high temperatures has become very important. The importance of radiative transfer of heat from solids increases with temperature. A spacecraft travelling through interplanetary space attains some mean temperature which depends upon the mean rate of loss and gain of thermal energy which itself will depend upon the radiative properties of the surface of the spacecraft. For measurements in photometry and pyrometry, a standard source of radiation with a well-defined area of uniform brightness and known radiative properties is needed. A knowledge of radiative power is also necessary in the design of incandescent lamps, thermionic valves, etc. The efficiency of solar cells depend upon the radiative power of the surface of the solar cells. For maintaining a stable floating liquid zone in zone refining experiments, knowledge of emissivity of the surface of the specimen is important.

The quality and quantity of the thermal radiations emitted by a blackbody depend only upon the temperature of the body. The radiations emitted by actual substances depend upon the temperature of the substances and upon their emittance. The spectral radiant energy $J_\lambda d\lambda$ emitted by the blackbody of unit area in one second at temperature T in the wavelength region λ and $\lambda + d\lambda$ is given by Planck's law

$$J_\lambda d\lambda = c_1 \lambda^{-5} \left(e^{\frac{c_2}{\lambda T}} - 1 \right)^{-1} d\lambda, \qquad (1)$$

where C_1 and C_2 are known constants.

In the visible and infrared wavelength regions eq. (1) reduces to Wien's law,

$$J_\lambda d\lambda = c_1 \lambda^{-5} e^{-\frac{c_2}{\lambda T}} d\lambda. \qquad (2)$$

The total energy W emitted by unit area of the blackbody per second over the whole spectral range is given by Stefan–Boltzmann law

$$W = \sigma(T^4 - T_0^4) \qquad (3)$$

where σ is the Stefan's constant and T_0 is the ambient temperature.

In practice, the radiations emitted by almost all bodies differ appreciably from those given by eqs. (1) and (3). Emittance is the term used to signify this difference and is expressed as the ratio of the rate of emission of radiant energy from the surface to that from a blackbody of equal area and at the same temperature. Spectral emittance ε_λ is thus defined as the ratio of the energy emitted by the surface in the spectral range λ and $\lambda + d\lambda$ to that which would be emitted by a blackbody of the same area at the same temperature and in the same spectral range. Emissivity is the emittance for the optically smooth surface of a body with a thickness sufficient to be opaque. The total emissivity is defined as the ratio of the total energy emitted by the surface to the total energy emitted by the blackbody of the same area and at the same temperature.

For opaque bodies the relation between emissivity ε and reflectivity r is

$$\varepsilon + r = 1. \qquad (4)$$

Since emissivity is directly related to the reflectivity, the theory of emissivity can be derived from that of optical constants of materials. Hagen and Rubens obtained an expression for the spectral emissivity of metallic surfaces in terms of electrical resistivity and wavelength of light. Their relation, known as Hagen–Rubens relation, is

$$\varepsilon_\lambda = 0\cdot 365 \left(\frac{\varrho}{\lambda}\right)^{\frac{1}{2}}. \qquad (5)$$

Experimentally this relation is obeyed by most metals at wavelengths greater than 4 microns and at moderate and high temperatures. At low temperatures the electron mean free path becomes longer than the radiation skin depth. The effective conductivity of the surface layer becomes less than the bulk conductivity and the reflectivity measured experimentally becomes lower than that calculated by the Hagen-Rubens formula. The theory of anomalous skin effect gives results in reasonable agreement with experiments.

At short wavelengths when the radiation frequency is more than the collision frequency of the electrons, Hagen–Rubens formula again breaks down. Recent work has shown that the observed optical constants of any metal can be fitted by Drude's formula (see p. 529 of vol. 2 of this encyclopaedia) over a wide range of wavelengths, if, in addition to bound electron resonance term, at least two free electron terms are included. The concept of more than one class of free electrons is not inconsistent with the modern band theory.

The measurements of spectral emissivity of refractory metals show that at lower wavelengths the temperature coefficient of emissivity is negative. As the wavelength increases, the numerical value of the temperature coefficient decreases and at a certain wavelength, known as X point, the temperature coefficient becomes zero. At longer wavelengths the temperature coefficient becomes positive. A good explanation for this behaviour can be given in terms of more than one class of free electrons in a metal.

Some theoretical work has also been done on the optical constant of polar and nonpolar non-metallic crystals, with some success for crystals such as diamond and germanium.

Attempts have been made to derive theoretical expressions for total emissivity. Both normal and hemispherical emissivities increase with temperature in qualitative agreement with the theory. The theoretical expressions have many limitations. It is difficult to take into account the finite relaxation time of electrons. Further complications arise due to the contribution by bound electrons at shorter wavelengths. The emissivity depends upon the effective value of the resistivity within the penetration depth of the electromagnetic waves. At low temperatures the anomalous skin effect mentioned above makes the theory more complicated.

The theory of radiative heat transfer from solid surfaces is based on the assumption that the surfaces are optically smooth, free from contamination, chemical impurities and mechanical strains. Considerable attention has been given to the effect of surface roughness on the emittance. If the surfaces are slightly rough the theoretical expression based on diffraction theory indicates correctly the trend of change of emittance with roughness. A good deal of work remains to be done for rough and particularly for very rough surfaces.

The reflectivity of monochromatic light can be easily measured for polished surfaces at low temperatures and the spectral emissivity can be calculated from eq. (4). The reflectivity cannot be measured at high temperatures because the light emitted by substances at high temperatures interferes with the reflected light. At these temperatures, the surface brightness temperature T_B is usually measured with the help of a precision optical pyrometer using a filter of wavelength of interest and the correct thermodynamic temperature T is measured either with the help of suitable thermocouples or with the help of an optical pyrometer if a blackbody can be constructed and maintained at the same temperature. ε_λ is given by

$$\ln \varepsilon_\lambda = -\frac{c_2}{\lambda}\left(\frac{1}{T_B} - \frac{1}{T}\right). \qquad (6)$$

Alternatively, the radiations from the experimental body and blackbody can be compared with a suitable detector and their ratio gives the spectral emissivity ε_λ.

The most convenient method of determining the total emissivity is by heating the filament of the substance by passing electric current, measuring the voltage drop V across a known length of uniform temperature and current I through the filament. The total emissivity is calculated from the equation

$$VI = s\varepsilon\sigma(T^4 - T_0^4) \qquad (7)$$

where S is the surface area of the filament. Knowing V, I, T, T_0, S and σ, ε can be calculated.

A recent method for the determination of total emissivity is to determine the specific heat of the specimen. If a refractory metal is heated in vacuum to a certain temperature and then suddenly allowed to cool and if the temperature of the surroundings is much lower (which is the case at high temperatures), the experimental relation between time θ and temper-

Table 1

Substance	Temperature °K	Total emissivity ε	Wavelength λ_x at X point	λ_x	Spectral emittance* at λ_x and other wavelengths (μ)						
					0.5	0.65	0.7	0.9	1.0	3.0	5.0
Iron	1200	—	1.5	2.27	—	—	—	—	0.31	0.21	—
	1500	—			—	—	—	—	0.30	0.22	—
Cobalt	1200	0.21	1.5	0.25	—	0.35	—	—	0.27	0.18	—
	1500	0.22			—	0.33	—	—	0.26	0.19	—
Nickel	1200	0.16	1.5	0.26	—	0.45	—	—	0.30	—	—
	1500	0.22			—	0.40	—	—	0.29	—	—
Molybdenum	1600	0.27	1.35	0.22	0.45	0.43	0.41	0.38	0.33	0.11	—
	2800	—			0.37	0.35	0.34	0.28	0.25	0.18	—
Tantalum	1700	0.18	0.67	0.41	0.56	0.47	0.41	0.33	0.29	0.15	—
	2800	0.30			0.47	0.42	0.40	0.35	0.32	0.24	—
Tungsten	1600	0.28	1.3	0.33	0.47	0.45	0.44	0.41	0.39	—	—
	2800	0.36			0.45	0.43	0.42	0.38	0.37	—	—
Rhenium	1537	0.26	0.9	0.36	0.43	0.42	0.41	0.36	0.35	—	—
	2772	0.38			0.41	0.40	0.39	0.36	0.35	—	—

Table 2

Substance	Temperature °K	Total emissivity ε	Wavelength λ_x at X point	λ_x	Spectral emittance* at λ_x and other wavelengths (μ)						
					0.5	0.65	0.7	0.9	1.0	3.0	5.0
Zirconium diboride	1604	0.41	0.55	0.85	0.92	0.75	0.70	0.60	0.56	0.40	0.38
	2480	0.64			0.88	0.83	0.82	0.75	0.71	0.62	0.60
Niobium diboride	1594	0.30	1.0	0.48	0.75	0.62	0.62	0.53	0.48	0.28	0.26
	2415	0.41			0.67	0.64	0.60	0.51	0.48	0.34	0.34
Titanium diboride	1648	0.49	1.15	0.85	0.96	0.94	0.94	0.89	0.86	0.45	—
	2022	0.67			0.89	0.89	0.88	0.87	0.86	0.60	—
Zirconium carbide	2100	0.43	2.25	0.42	0.68	0.67	0.66	0.62	0.60	0.34	—
	2670	—			0.56	0.54	0.53	0.51	0.49	0.40	—
Tantalum carbide	1830	0.22	0.65	0.47	0.60	0.47	0.41	0.34	0.31	0.20	—
	2880	0.32			0.53	0.47	0.44	0.38	0.37	0.28	—
Zirconium nitride	1895	0.41	0.85	0.70	0.93	0.91	0.87	0.60	0.54	—	—
	2287	0.70			0.84	0.78	0.76	0.73	0.72	—	—
Tantalum nitride	1648	0.72	—	—	0.85	0.84	0.82	0.80	0.80	0.71	0.68
	2070	0.73			0.83	0.83	0.82	0.81	0.79	0.73	0.73

* Since at high temperatures the surfaces are not usually optically smooth the term emittance rather than emissivity has been used. The effect of surface roughness is much less pronounced on the total energy radiated over the whole spectral range.

ature T is found to be

$$e + b\theta = \frac{1}{T^3}, \quad (8)$$

where e and b are constants and the relation

$$\frac{c_p}{\varepsilon\{1 + 2\alpha(\Delta T) + 2\beta(\Delta T)^2\}} = \frac{3\sigma A_0}{bm} \quad (9)$$

between specific heat C_p and ε can be established. Here α and β are the first and second coefficients of linear thermal expansion, ΔT is $T - T_0$, σ is the Stefan's constant, A_0 is the area of sample at room temperature and m is the mass of the sample. If C_p is known as a function of temperature, ε can be determined accurately from the experimentally observed ratio C_p/ε. This method has been used by many authors for determination of emissivity for many refractory metals.

In the absolute method of measurement of emissivity, the temperature and the radiating area of the specimen must be known and the radiative power must be measured accurately. The measured energy is then compared with that expected from a blackbody of equal area and at the same temperature and ε is determined.

Thermal radiation properties of solids have been measured for a long time. However, attention to high temperature properties has been given only recently. The data for seven refractory metals and seven refractory compounds (nitrides, borides and carbides) are given in Tables 1 and 2 respectively.

Bibliography

BLAU H. and FISHER H. (1962) (Eds.) *Radiative Transfer from Solid Materials*, Macmillan, N.Y.

CROWLEY D. (1959) *Emittance and Reflectance in the Infrared*, University of Michigan, Willow Run Laboratories, IRIA Report 2389-15-S.

JACOB L. (1961) Emissivity, in *Encyclopaedic Dictionary of Physics* (J. Thewlis Ed.), **2**, 849, Oxford: Pergamon Press.

Proceedings of the Symposium on Thermal Radiation of Solids (1964) National Aeronautics and Space Administration, Research and Technology Division U.S.A.F., San Francisco, California. (Unpublished, can be obtained from Clearing House for Federal Scientific and Technical Information, Department of Commerce).

RICHMOND J. C. (1963) *Measurement of Thermal Radiation Properties of Solids*, National Aeronautics and Space Administration: Washington, D.C.

Temperature, its Measurement and Control in Science and Industry (1962), American Institute of Physics, Reinhold Publishing Corporation, N.Y.

WARD L. (1956) The variation with temperature of the spectral emissivities of iron, nickel and cobalt, *Proc. Phys. Soc. Sec. B*, **69**, 339.

S. C. JAIN and V. NARAYAN

EPITAXY, RECENT WORK ON

Introduction

The original article on epitaxy in this Dictionary (Wilman, 1961) defined the term according to the original definition of Royer. This should still be regarded as the strict definition, despite the increasing use of the term epitaxy to describe the growth of surface layers of silicon on silicon single crystals. Although no general agreement has been reached, it is probably better to use the term autoepitaxy to describe the oriented growth of a substance on itself.

The more recent of the review articles referred to in the original article were written just before the electron microscope techniques were first applied to the study of epitaxy. Since 1955 a considerable amount of new information relevant to our understanding of epitaxy has been obtained, particularly as a result of electron microscopy. The need for more detailed information on the nucleation and growth of surface layers was very apparent in 1955, in order that the mode of growth of epitaxial deposits could be better understood.

Although most of the early observations by optical microscopy were made on crystal overgrowths formed from supersaturated solutions, the more recent work has been concentrated on studies of relatively thin surface layers formed by various deposition techniques (e.g. electrodeposition and sputtering), but particularly the vacuum evaporation technique. The appropriate vapour is formed by heating the solid phase in a vacuum, and the vapour is allowed to condense on the single-crystal substrate. The transmission electron microscope is used for studying the layers in one of three ways: (1) a thin shadowed replica of the deposit is made, so that the surface geometry of the deposit is revealed; (2) the deposit is detached from its substrate, so that it can be examined by direct transmission; (3) a thin film substrate is used so that both the substrate and deposit can be simultaneously examined.

The Mode of Formation and Growth of Deposits

Although many of the electron microscope observations have concerned deposition of metals, studies have also been made of the formation of deposits of simple chemical compounds (e.g. metal oxides, sulphides and halides). In the vast majority of cases, the first observable stage of deposition consists of the formation of discrete separated nuclei on the substrate surface, as previously deduced from electron diffraction observations. The nuclei are commonly much less than 100 Å in diameter when first detected, and they form when the average thickness of the deposit (i.e. total volume of deposit/area of substrate) is very low, often less than one atomic or molecular layer. A saturation density (i.e. number/unit area) of nuclei forms at an early stage, and subsequent deposition leads to the growth of these nuclei until they become sufficiently large to join together. Eventually the joining results in the formation of a continuous hole-free deposit layer.

The size and shape of the nuclei or islands, and their density, varies appreciably from one substrate-deposit combination to another. Also, for a given substrate-deposit combination there is a major influence of the growth parameters such as substrate temperature and deposition rate. The detailed manner in which the islands grow together to form a continuous film is also strongly dependent upon these parameters.

Surface diffusion processes are prominent during the deposition. The initial formation of nuclei results from the migration of deposit atoms over the substrate surface, and their subsequent aggregation. When two islands join together, or coalesce, shape changes can occur in the compound island because of transport of atoms over the surface of the islands. These shape changes can occur extremely rapidly, immediately the islands touch, and they then give the impression of a liquid-like coalescence, although the islands are normally solid and crystalline throughout the process.

Lattice Parameters of Epitaxial Deposits

When the lattice networks at the interface between the substrate and deposit do not match exactly (i.e. there is a lattice misfit) some lattice strain is expected. In special cases where the misfit, as deduced from bulk lattice parameters, is quite small, there are reasons to believe that the strain will be such as to remove the misfit so that a perfect match of the two lattices occurs at the interface. This phenomenon is known as basal-plane pseudomorphism.

Many studies have been made in an attempt to reveal the occurrence of basal-plane pseudomorphism. In the vast majority of cases, basal-plane pseudomorphism does not occur, although lattice strain at the interface does frequently lower the value of the misfit. In a few special cases, notably with very thin layers of metals deposited on metals, the absence of any misfit at the interface has been revealed by the absence of misfit dislocations, which can normally be rendered visible in the electron microscope. As the deposit thickness is increased, misfit dislocations do appear at the interface, and so basal-plane pseudomorphism can occur in certain circumstances, but only if the lattice strains required are no more than one or two per cent.

With most substrate-deposit combinations, the deposit has its normal bulk crystal structure. However, several clear examples of the occurrence of abnormal structures have been observed, and these are dependent upon the substrate employed. Both the crystal structure and the orientation of the substrate can be important, so that a deposit grown on two different crystal faces of the same substrate will not necessarily have the same structure. The matching of symmetry at the interface seems to be more important than matching in size between the two crystal lattice networks. The most common examples involve:
(a) normally f.c.c. metals depositing in h.c.p. form;
(b) normally b.c.c. metals depositing in f.c.c. form;
(c) compounds normally having the CsCl structure depositing with the NaCl structure.

Deposition Conditions for Obtaining Epitaxy

It has long been recognized that various growth parameters can have a major influence on the occurrence of epitaxy. The temperature of the substrate during deposition is often very important, and this fact led to the concept of a so-called "epitaxial temperature". According to this concept, good epitaxy will not occur unless the substrate is above the critical epitaxial temperature. In recent years, much evidence has shown that the epitaxial temperature is by no means a unique temperature, for a given substrate-deposit combination. The rate of deposition can have a major influence on the observed epitaxial temperature, as can the general condition of cleanness of the system. This has been revealed in a striking and surprising way by studies on film growth in different degrees of high vacuum.

Systematic studies of this kind have been carried out, in particular, with f.c.c. metals deposited on (100) cleavage surfaces of rocksalt. In order to eliminate the possible influence of the exposure of the rocksalt surface to the atmosphere, fresh cleavage surfaces are prepared in the vacuum immediately prior to the deposition of the metal. With some metals (e.g. gold and silver) the epitaxy is improved (i.e. there is a lower epitaxial temperature) as the residual vacuum is improved from 10^{-4} to 10^{-7} torr. However, further improvement to 10^{-9} or 10^{-10} torr results in poor epitaxy at all substrate temperatures. The poor epitaxy consists of a mixture of nine orientations, the normal parallel orientation and eight orientations with (111) metal planes parallel to the rocksalt cleavage. Exposure of a rocksalt cleavage surface to the atmosphere before the deposition is carried out at 10^{-9} torr gives good epitaxy.

Although there is no detailed understanding of these results, two different effects seem to be important. First, contamination can somehow favour good epitaxy in the initial nuclei of the deposit. There is some evidence that this might be important in some cases. Second, there is clear evidence that recrystallization during post-nucleation growth of the film can have a controlling influence. The occurrence of recrystallization during growth (for gold or silver on rocksalt) is demonstrated by the considerable improvement of the epitaxy as growth proceeds. The mechanism of recrystallization is revealed directly when deposits are grown inside the electron microscope, so that structural changes occurring during growth are observed as they take place. The recrystallization occurs by the migration of grain boundaries, and if conditions favour the growth of grains with the appropriate epitaxial orientation, the epitaxy is improved. Thus the recent experiments have shown that epitaxy is controlled not only by the nucleation stage of growth, but also by the post-nucleation growth processes. There is still need for much more experimental evidence to show which stage is more critical when particular growth conditions are employed for a particular deposit. This applies especially in relation to the understanding of the influence of impurities or contamination (e.g. vacuum environment).

Some remarkable effects have been observed when the surface of the substrate is bombarded with electrons during the condensation of a deposit. Very much improved epitaxy can be obtained, and the continuity of the deposit occurs at an earlier stage (i.e. lower thickness). Also, the general crystalline perfection of the layers is much improved. Most of the evidence is based upon the use of alkali halide substrates. One of the important effects of the electron irradiation is to cause a marked increase in the initial density of the nuclei, which influences the post-nucleation processes because joining together of the nuclei or islands occurs when they are smaller. The reason for the increased nucleation density is not clearly understood, but seems most likely to be associated with radiation damage to the alkali halide surface.

Recently, some experiments have shown that epitaxy can be produced when the single crystal surface is first covered by an amorphous layer (e.g. carbon) of several hundred angstroms in thickness. These results, so far few in number, suggest that relatively long-range forces can give rise to epitaxy, but it is probable that such effects only occur under special experimental conditions.

The Internal Structure of Epitaxial Deposits

The transmission electron microscope has allowed very detailed studies to be made of the structure of deposited layers up to one or two thousand angstroms in thickness. The grain structure, if any, is clearly revealed and selected area electron diffraction allows the misorientation between neighbouring grains to be studied. Even when there is no grain structure, and hence the epitaxy is very perfect, there is almost always a very high density of lattice imperfection present.

The lattice imperfections are of the same types as those found in bulk crystals, i.e. dislocation lines, stacking faults, twins, etc. The numbers of defects of a particular kind vary considerably from one system to another, or possibly for a given system when the growth conditions change. Almost always, however, the total imperfection content is very high and equivalent to a fairly heavily worked crystal. The only known exception is for the case of autoepitaxy, especially the growth of silicon on silicon. In this special case, for sufficiently clean growth conditions, perfect deposits with no lattice imperfections can be grown.

Electron microscope studies of the mechanisms by which the lattice imperfections are introduced into the deposit during growth suggest that the imperfections arise largely as a result of the lattice misfit between substrate and overgrowth. The imperfections appear when neighbouring nuclei or islands grow together. These mechanisms do not operate when the misfit is zero, and hence this is why no imperfections are introduced when silicon is grown on silicon. The imperfections which do arise in this latter case are always associated with contaminating conditions, usually contamination existing on the substrate surface before deposition is commenced.

Applications

Much of the stimulus for studies on epitaxy results from the desirability to grow single-crystal layers for special applications. Single-crystal metal films are required for fundamental scientific studies (e.g. of oxidation or alloying), and also because electrical or magnetic properties are of potential use for electronic devices of various kinds. Single-crystal semiconductor films are of most interest, from the applications point of view, because of the possibility of selective doping of the layers for the control of the properties of devices. For this reason, very extensive studies have been made of the epitaxy of silicon on silicon, although, as already indicated, this would be better termed autoepitaxy.

Bibliography

FRANCOMBE M. H. and SATO H. (Eds.) (1964) *Single Crystal Films*, Oxford: Pergamon Press.
MATTHEWS J. W. (1967) *Physics of Thin Films* **4,** 137.
PASHLEY D. W. (1965) The nucleation, growth, structure and epitaxy of thin surface films, *Advances in Physics* **14**, 327.
WILMAN H. (1961) Epitaxy, in *Encyclopaedic Dictionary of Physics* (J. Thewlis Ed.), **3,** 1, Oxford: Pergamon Press.

D. W. PASHLEY

F

FAR INFRA-RED LASERS. Maser oscillations have now been excited at a large number of far infra-red wavelengths between 15 and 774 μ (Fig. 1).

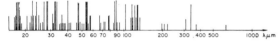

Fig. 1. Far infra-red laser lines uncovered up to 1968 (adapted from Garrett, 1966).

I. Rare Gas Lasers

The first studies concerned rare gases. At the First Quantum Electronic Symposium (1959), Javan showed (1959, 1960) the possibility of obtaining a state of a negative temperature between two of the excited states of the atoms by electron impact. At the Second Symposium (1961) the coherent emission of a simple discharge tube containing 1 Torr He and 0·1 Torr Ne at 1·15 μ (Javan, 1961) was described as originating from the $2s_2 \to 2p_4$ transition (Fig. 2), the neon atoms being brought to the $2s_2$ level by collisions of the second kind with metastable helium atoms. It is worth noting (Javan, 1961) that in the early 1930's a similar inverted population had been observed in Na–Hg gas mixtures under discharge conditions where Hg (6^2P_0) metastable atoms excite Na atoms to their $7S$ state, with an inverted population ratio as high as 4 : 1. So with the tools available to the experimentalist around 1930 and their understanding of the processes involved in a discharge, the inverted population and its implications could have been uncovered long ago. Using windows transparent to the far infra-red a large number of laser lines have been observed between 1·15 and 133 μ for neon, two of which are shown in Fig. 2 (Rigden and Mcfarlane, 1963). The rare gas infra-red masers operate continuously, the power obtainable from them falling rapidly as $h\nu$ becomes comparable with kT, and 133 μ appears close to the practical limit (Garrett, 1966).

II. Molecular Lasers

II. 1. H_2O

A few simple molecules have been used recently in far infra-red molecular lasers. They have been chosen from the most polar ones. The first molecule found that was able to lase in the far infra-red was the water molecule H_2O (Gebbie, Stone, Findlay and Robb, 1964; Mathias, Crocker and Wills, 1965). Figure 3 gives the scheme of the experiment. The laser emission

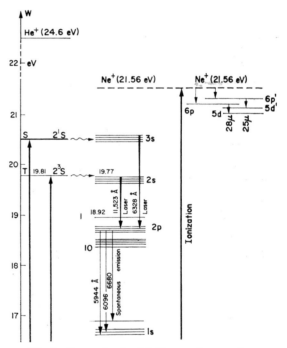

Fig. 2. Energy levels of helium and neon, from Rigden and McFarlane (Quantum Electronics Symposium, 1963, pp. 500 and 575).

is analysed with a Michelson interferometer. Gebbie was investigating laser emission with diatomic polar molecules like HCl and DCl and was intrigued always to get the same lines. In fact they even occurred without any gas and the explanation was the residual water vapour. It is now possible with water vapour to obtain nine lines between 23 and 200 μ with peak powers as high as 1 kW and a duration of a few microseconds. Figure 4 shows a speculative mechanism for the emission.

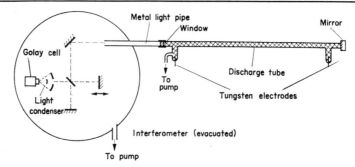

Fig. 3. Far infra-red laser and Michelson interferometer.

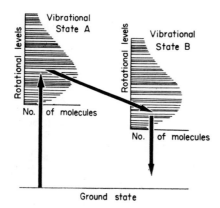

Fig. 4. *Speculative mechanism showing it is possible for transitions to occur between a particular rotational component of one vibrational state and a different rotational component of a second vibrational state, even when there is no over-all inversion between the two vibrational states in the system of water-vapour molecules (Garrett, 1965).*

II. 2. HCN

Several years ago, maser oscillations from HCN were observed at 88·6 kMc/s (Marcuse, 1961). More recently HCN, CH_3CN, $(CH_3)_2NH$ and $(C_2H_5)_2NH$ have also given laser lines located at 126–135, 284–310, 335–337, 373–538 and 774 μ (Mathias et al., 1965; Steffen et al., 1967). They were first ascribed to CN but most of them are due to excited HCN molecules (Lide et al., 1967). Nearly all H_2O and HCN laser lines are now obtained continuously. Precise frequency measurements of these lines by harmonic frequency mixing with V-band klystrons have been made by Javan (1967).

The accuracy is high enough to expect a better determination of the velocity of light $c = \lambda \cdot \nu$. It also allows one to check the various possible mechanisms of the emission. Lide and Maki (1967) have shown that most of the lines are emitted by the HCN molecule. Figure 5 gives the scheme of the mechanism of the transitions around 337 μ.

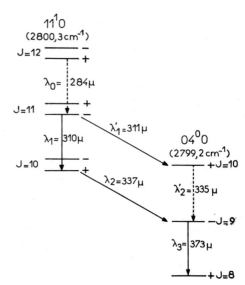

Fig. 5. *Vibration–rotation energy levels of hydrocyanic acid in the vibrational states. 11^10 and 04^00 with $\gamma_1 = 2062\ cm^{-1}$, $\gamma_2 = 712 \cdot 1\ cm^{-1}$ and $\gamma_3 = 3312\ cm^{-1}$.*

II. 3. Resonator modes

The frequency spectrum of the laser radiations can be studied using the laser resonator itself as a tunable

Table 1. Frequency of HCN laser lines

Approximate wavelength (μ)	373	337	335	311	310
Measured frequency (Gc/s)	804·7509	890·7607	894·4142	964·3134	967·9658

Fabry–Perot interferometer (Steffen et al., 1967). Usually it is difficult to decide if modes of one or more laser emissions with similar wavelengths are present. An additional selection with a grating is useful and the interferograms corresponding to the different wavelengths can be separated if several detectors are used simultaneously (Hadni, 1968), (Fig. 6). In the case of HCN only the strongest line

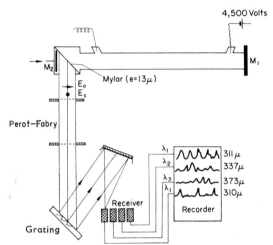

Fig. 6. Experimental arrangement for recording the interferograms of four CH_3CH_2CN lines at 337, 310, 311 and 373 microns as a function of the laser length. The mirrors M_1 and M_2 are parallel in all except the uppermost graph where they are slightly inclined and the non-axial modes much clearer.

($\lambda_2 = 337\ \mu$) gives an interferogram with the expected period $\lambda_{2/2}$ (Fig. 7). For $\lambda_1 = 310\ \mu$, and $\lambda_3 = 373\ \mu$ the interferograms are distorted with a beat frequency equal to $1/\lambda_1 - 1/\lambda_2$, and $1/\lambda_2 - 1/\lambda_3$. This arises from the λ_2 emission needed to depopulate the $J = 10^+(11^10)$ level, and to populate the $J = 9^-(04^00)$ level. On the other hand, it is clear from Fig. 5 that there is competition between

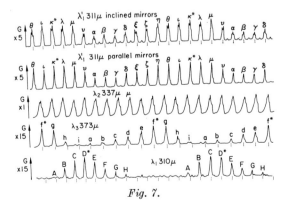

Fig. 7.

$\lambda'_1 = 311\ \mu$ and $\lambda_2 = 337\ \mu$, and on Fig. 7 the position of the maxima for $\lambda' = 311\ \mu$ corresponds to minima for $\lambda_2 = 337\ \mu$.

In conclusion there are generally several conditions to be observed for the length L of the laser in order to get a specified emission.

II. 4. Single Wavelength Operations

The above conditions are useful for making the laser resonator frequency selective. It is also possible to employ a diffraction grating to make the resonator frequency selective. In the case of water vapour Jeffers (1967) has recently obtained single wavelength operation in twenty five emission lines in the range 23 to 57 μ by using successively three diffraction gratings. This technique when correctly used leads to new emission lines, enhanced output on weak lines, and identical pulse-to-pulse emission intensities. These results show that there are strong interactions among the transitions responsible for laser action. They are quantitatively consistent with the assumption that laser action is due to vibration-rotations of the H_2O molecule.

III. Solid State far Infra-red Lasers

Solid state masers are known both for microwaves and visible light. Though there is a lack of host crystals transparent in the far infra-red at room temperature, it has been shown that nearly all crystals become transparent at liquid helium temperature (Hadni et al., 1962). On the other hand a lot of electronic transitions have now been found in this part of the spectrum, especially in rare earth ions. The main problem seems to be to find long-life electronic levels. The use of electronic levels located in a band forbidden to elastic vibrations seems a promising possibility (Nelson et al., 1967). Others can be considered from the fact that phonons interact with electronic states via electric fields (Nelson et al., 1967) and do not directly perturb electron spins. Hence the possibility that magnetic dipole electronic transitions would be allowed but phonon transitions could be weak. This might be the case if Kramers doublets could be split with far infra-red separations, either by external magnetic fields or internal exchange fields (Nelson et al., 1967).

Such a possibility could also exist in a cubic crystal with inversion site symmetry. Electric dipole transition are forbidden, but magnetic dipole (T_{1g}) electronic transition may be allowed. However, even if there are many phonons of the correct energy to connect electronic levels in the crystal, strong selection rules arising from the highly symmetrical situation in the cubic crystal prevent such a coupling (Nelson et al., 1967).

Electroluminescent semiconductor lasers tunable over a large spectral interval are known in the near and middle infra-red. Their applications might be extended to the far infra-red.

IV. Applications of Far Infra-red Masers to Spectroscopy

In spectroscopy, far infra-red masers will be useful for the investigation of weak, narrow spectral lines, such as molecular rotation lines and certain absorption lines in solids. None of the existing far infra-red laser oscillators is tunable, so that we cannot expect to use them to replace present-day incoherent sources in existing infra-red spectrometers. A far infra-red laser, however, has unique possibilities for the investigation of rotation, rotation inversion lines of gases, electronic transitions in solids, local phonon modes. Figure 8 gives the scheme of a laser spectrometer for examining a Zeeman tuned line

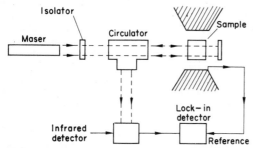

Fig. 8. *Far infra-red laser spectrometer (Garrett, 1966).*

(Garrett, 1966), and good results have recently been obtained with germanium (Bradley et al., 1968).

In addition to the linear optical properties of materials in the far infra-red, masers should make possible the investigation of certain nonlinear effects, which ought to appear at much lower field strengths than are required for the observation of nonlinear effects in the visible (Garrett, 1966). Various possibilities exist for nonlinear optic experiments in the infra-red, for instance those associated with lattice anharmonicity (Garrett, 1966).

Bibliography

BRADLEY C. C., BUTTON K. J., LAX B. and RUBIN L. G. (1968) Quantum effects in cyclotron resonance using a CW cyanide laser, *International Quantum Electronics Conference.*

GARRETT C. G. B. (1965) Far infrared masers, *Science and Technology* **39**, 39.

GARRETT C. G. (1966) Far infrared masers and their applications to spectroscopy, *Physics of Quantum Electronics*, edited by Kelley, p. 557, Lax, Tannewald: McGraw Hill.

GEBBIE H. A., STONE N. W. B., FINDLAY F. D. and ROBB J. A. (1964) Interferometric observations on far infrared stimulated emission sources, *Nature* **202**, 169.

HADNI A., CLAUDEL J., GERBAUX X., STRIMER P. (1962) Spectres d'absorption dans l'Infrarouge lointain, à la température de l'hélium liquide: Iodure de Cesium, Quartz, Germanium et Nitrate de Neodyme, *Comptes Rendus* **255**, 1595.

HADNI A. (1964) Far Infrared electronic transitions in ions and pairs of ions, *Phys. Rev. A* **758**, 136.

HADNI A., THOMAS R., WEBER J. (1967) Lasers pour l'Infrarouge lointain, *J. Chim. Phys.* **1**, 71.

HADNI A., CHARLEMAGNE D., THOMAS R. (1968) Polarisation, compétition entre modes et exaltation de certaines raies dans l'émission d'un laser à propionitrile vers 337 microns, *Comptes Rendus* **266**, 1230.

HOCKER L. O. and JAVAN A. (1967) Absolute frequency measurements on new CW HCN submillimeter laser lines, *Phys. Letters* **25A**, 7, 489.

JAVAN A. (1959) Possibility of production of negative temperature in gas discharge, *Phys. Rev. Letters* **3**, 87.

JAVAN A. (1960) Possibility of obtaining negative temperature in atoms by electron impact, *Proceedings of Quantum Electronics Symposium*, p. 564, New York: Columbia University Press.

JAVAN A. (1961) Optical maser oscillations in gaseous discharge, *Proceedings of Quantum Electronics Symposium*, p. 18, New York: Columbia University Press.

JEFFERS W. Q. (1967) Single wavelength operation of a pulsed water-vapor laser, *Appl. Phys. Letters* **11**, No. 6, 178.

LIDE D. R. and MAKI A. G. (1967) *Appl. Phys. Letters* **11**, 62.

MARCUSE D. (1961a) Stimulated emission from HCN gas maser observed at 88·6 Mc, *J. Appl. Phys.* **32**, 743.

MARCUSE D. (1961b) *Proc. I.R.E.*, Nov. 1961, 1706.

MATHIAS L. E. S., CROCKER A. and WILLS M. S. (1965) Laser oscillations at submillimeter wavelengths from pulsed gas discharges in compounds of hydrogen, carbon and nitrogen, *Electronics Letters*.

NELSON E. D., WONG J. Y. and SCHAWLOW A. L. (1967) Far infrared spectra of $Al_2O_3 : Cr^{3+}$ and $Al_2O_3 : Ti^{3+}$, *Phys. Rev.* **156**, 298.

RIGDEN and McFARLANE (1963), *Proceedings of Quantum Electronics Symposium*, pp. 500 and 575.

STEFFEN H., MOSER J. F. and KNEUBUHL F. K. (1967) Resonator modes and splitting of the 0·337 mm emission of the CN laser, *J. Appl. Phys.* **38**, 3440.

YOSHINAGA H. et al. (1967) Far infra-red lasers, *Science of Light*, **16**, 50.

A. HADNI

FAR INFRA-RED SPECTROSCOPY. Far infra-red spectroscopy developed immediately after Hertz's oscillator had emitted wavelengths of about 10 cm. Several scientists like Lebedev, Glayolewa Arkadiewa, Nichols and Tear tried to reduce the size of Hertz's oscillator in order to obtain shorter wavelengths. They claimed success down to wavelengths as short as 0·3 mm. This was important because other physicists, Rubens for instance, were able to reach these wavelengths with an optical source, the high-pressure mercury arc. Thus, a junction was made between optics and electromagnetism in the submillimetre wave spectrum region. However, experiments using miniature Hertzian oscillators are not convenient or easy, and many of these early experiments (for example, those of Nichols and Tear) have not been successfully repeated. It is thanks to the mercury arc that far infra-red spectroscopy has developed, though very slowly, with about 15 papers every year up to 1960 including some developments such as pure rotation spectra, eigenfrequencies of crystals, the tunnel effect in the ammonia molecule and the calculation of the pure rotational spectra of some asymmetric top molecules such as H_2O in good agreement with the observed lines. Since 1960, there has been greater activity in the instrumentation research and application of far infra-red spectroscopy.

I. Instrumentation in the Far Infra-red

I.1. Detectors

The metal bolometer has been replaced by the pneumatic detector with a noise equivalent power 10 times lower (10^{-10} W (c/s)$^{-1/2}$). Invented by Golay, this is a fragile instrument, but nevertheless the most sensitive detector at room temperature. The noise equivalent power corresponds to a collodion membrane displacement less than 1 Å. Pyroelectricity has recently been used to make very fast and very robust detectors, most of them working at room temperature. The time constant can be as small as 1 μsec and the NEP for $\lambda = 10$ μm is of the same order as that of a gold-doped germanium working at liquid nitrogen temperature (Hadni et al., 1969). Cooled detectors, indium antimonide and gallium-doped germanium have recently been made. They are an order of magnitude more sensitive than the Golay detector. Their noise equivalent powers are about 10^{-12} W (c/s)$^{-1/2}$ at 2°K, and 10^{-14} W (c/s)$^{-1/2}$ at 0·5°K is widely thought to be possible for the germanium bolometer. These performances should mainly arise because of the diminution of heat capacity at low temperature. At 0·5°K, the germanium detector would have the same performance in the whole infra-red range as a photomultiplier in the visible. The gain would be 10^4 over the Golay detector. Heterodyne detection is now possible in the far infra-red where coherent sources have been available for a few years (see below). If P_L be the power of the local oscillator, the gain in sensitivity (Leiba, 1968)

$$\frac{(NEP)_{Direct}}{(NEP)_{Heterod.}} = \frac{P_L}{(NEP)_{Direct}}.$$

With $P_L = 1$ W, and $(NEP)_{Direct} = 10^{-5}$ W for a 100 kc/s band width, the gain (10^5) is considerable. However, the device is only useful if the detector band width is large enough to allow detection of frequencies widely different from those of the local oscillator.

I.2. Sources

From Planck's law, a black body at 3000°K would be a tolerable source of radiations in the far infra-red. Over a spectral range of 1 cm^{-1}, at 1000 μm, it could give 10^{-8} W to a Golay detector (the optical span of the beam is 4×10^{-2} sd cm^2) which can detect 10^{-10} W.

This is not checked by experiment and we must admit that the globar source (silicium carbide) differs widely from the black body. The mercury arc is still the most widely used source of far infra-red. It works at frequencies lower than the plasma frequency, and however it does not appear as a good black body, probably because of the molten quartz envelope. Recently lasers have been developed in the far infra-red. They are molecular lasers mostly using rare gases, or small polar molecules like water H_2O, or hydrocyanic acid HCN. About 100 wavelengths are now available from 20 to 774 μm. The peak power for pulsed ones may be higher than 1 kW (for the 28 μm water vapour line), and for continuously operating ones it is higher than 0·1 W and may reach 1 W for the 337 μm wavelength.

I.3. Dispersion of radiation

Grating spectroscopy and Fourier transform spectroscopy are the most used. The usefulness of Fourier transform spectroscopy results mainly from the Felgett advantage (see the articles on Fourier Spectroscopy and Michelson interferometer). If we consider, for instance, the spectral range extending from 10 to 40 cm^{-1} with a limit of resolution of 0·2 cm^{-1}, we have 150 spectral elements resolved and the Felgett advantage in signal to noise ratio is $\frac{1}{2}\sqrt{150} = 6\cdot1$.

High resolution grating spectroscopy cannot therefore compete with Fourier transform spectroscopy. However, a limit of resolution of 1 cm^{-1} is sufficient for most studies of liquids and solids. Then the Felgett advantage goes down to 2·8, which is still worth while but of the same order of magnitude as the spread of sensitivities of the various detectors commonly used in the far infra-red. In fact as far as the system is still detector noise limited in the near infra-red it is clear that the Felgett advantage is much higher in the near than in the far infra-red where the number of spectral elements considered in one run of the instrument is much smaller. Both Michelson and lamellar gratings interferometers are used in the far infra-red (see Interferometer). The advantage of the lamellar grating instrument is that the reflection coefficient is 100% for both beams if $\nu < \nu_c$, where $\nu_c = \frac{1}{2}b$, b being the groove distance, and hence has an efficiency of 100%. Unfortunately it needs a mechanical accuracy which increases with ν_c and its applications are practically limited to $\lambda > 100$ μm.

Grating spectrometers are still used in the far infra-red. The far infra-red is generally located in the first order spectrum and higher orders have to be eliminated. This is a difficult problem since the luminosity of most sources of radiations is much larger in the near than in the far infra-red. Several solutions are possible, one of them is to use echelette gratings in the null order as reflection filters. Figure 1 represents the optical path of one of our vacuum grating spectrometers for the study of both absorption and reflection of solid samples at low temperatures. Mirrors M_2, M_3, M_5 and M_6 are reflection filters.

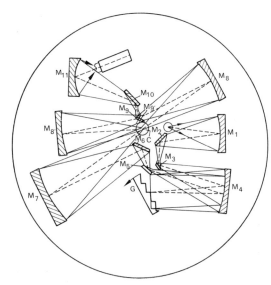

Fig. 1. Far infra-red spectrometer for the study of both absorption and reflection of solid samples at low temperature (from Décamps and Herbeuval, in Hadni, 1967).

II. Research and Application

We shall first consider application of the far infrared to molecular spectroscopy and then more specific applications.

II.1. Molecular spectroscopy in the far infra-red

II.1.1. Molecular vibrations of low frequency. We know that a molecule containing n atoms has $3n - 6$ normal vibrations (see the article on Spectra-molecular) generally rectilinear, where all the atoms vibrate in phase, with the same frequency. Initially it was thought that this whole spectrum was obtained in the near infra-red, the one which was accessible, but gradually interpretations had to be revised and often the fundamental vibrations were assigned to greater wavelengths. This is, among others, the case with the CH vibrations localized first around 1·5 microns by Ellis, where, in fact, the first harmonic is located.

Bending and stretching vibrations of heavy atoms are often found in the far infra-red. The infra-red spectrum of calomel Hg_2Cl_2, or stannous iodide SnI_4 is located entirely in the far infra-red (in so far as we are not concerned with harmonics of combination bands). Figure 2 gives for instance the whole infrared spectrum of stannous iodide. Puckering vibrations and torsions are also occasionally active in the far infra-red.

In the case of torsions the anharmonicity is generally so large that hot bands have weaker frequencies than the fundamental band, and at room temperature a series of lines is observed.

II.1.2. Pure rotations. Rotations are active in the far infra-red if the molecule has a dipole moment. Figure 3 gives the pure rotation spectrum of PH_3 at ordinary temperature. It consists of three series of equidistant lines corresponding to molecules in the fundamental vibrational state (the most intense series) and in the excited states $v_2 = 1$ and $v_2 = 2$.

At high pressures, the rotation spectrum of molecules with no permanent electric dipole moment may be observed as a broad envelope (Fig. 4). In the liquid

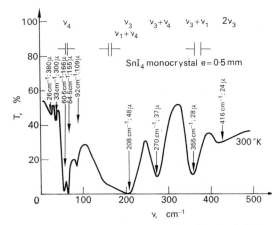

Fig. 2. Absorption spectrum of solid tin iodide from 20 to 500 microns (from Décamps and Herbeuval, in Hadni, 1967).

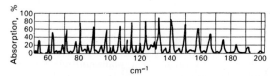

Fig. 3. Pure rotation spectrum of PH_3 (from Oetjen et al., in Hadni, 1967).

state a broad absorption may also be observed. With liquid nitrogen, for instance, a broad absorption band with a maximum near 70 cm^{-1} has been described.

II.1.3. External vibrations in molecular crystals. The far infra-red absorption spectra of several organic single crystals, naphthalene, biphenyl, anthracene, durene, phenanthrene, benzophenone, and acenaphtene, etc., have been run from 50 to 700 microns using polarized radiations. A comparison with their solution spectra in CCl_4 or C_6H_6 allows an interpretation of several frequencies as being either external vibrations, molecular librations, or translations within the unit cell. Figure 5 gives, for instance, the absorption spectrum of a naphthalene plate cut parallel to the (001)

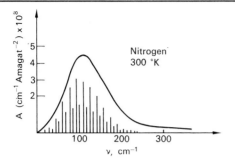

Fig. 4. Induced translational-rotational band of nitrogen. The vertical bars show the location of an unbroadened induced rotational spectrum (from Gush and Bosomworth, 1964).

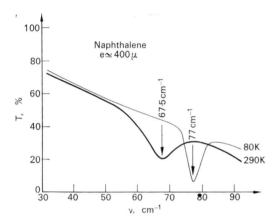

Fig. 5. Absorption spectrum of a naphthalene plate at 80K and 290K (from Gerbaux, Morlot and Wyncke, in Hadni, 1967).

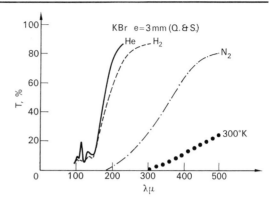

Fig. 6. Supertransmission of potassium bromide single crystal at liquid helium temperature (Hadni et al., 1965).

the absorption of crystals in the very far infrared. It appears as an addition process involving two phonons (Genzel et al., 1960), there is a collision (Fig. 7) between a quantum of infra-red radiation (phonon), and a phonon of thermal agitation of the lattice. In this collision, the photon disappears and gives a more energetic phonon, but with the same amount of angular momentum. In other words, there is creation

Fig. 7. A far infra-red photon can disappear in the same time as a phonon of the lattice to give another phonon with greater energy.

plane. The band at 77 cm^{-1} is very sensitive to temperature. It occurs from the translational frequency parallel to the monoclinic axis.

II.1.4. Far infra-red absorption of liquids. All liquids exhibit a broad absorption band in the far infra-red. It is most intense for most polar liquids. It is, however, only partially explained by Debye relaxation processes.

II.2. Absorption of crystals and glasses in the far infrared

All alkali halides show one intense absorption band in the far infra-red. The considerable absorption around 100 μm is followed by a continuous absorption up to millimetre waves. This additional absorption of crystals in the very far infra-red always disappears when they are cooled. This is, for example, the case with potassium bromide (Hadni et al., 1965) (Fig. 6). On the other hand, we have an idea of the reason for

of a mechanical wave of frequency $v_0 = v_a + v$, carried out by the same wavelength as the one of frequency v_a from which it originated. When we cool the crystal, the number of phonons in the crystal decreases and the absorption disappears. This is a fundamental process in the far infra-red.

The addition of impurities gives a weak additional absorption which would be completely hidden at room temperature by the two-phonon processes. Electronic transitions in an ion (Fig. 8) or a pair of ions can thus be observed. The mechanical agitation of the impurity itself may also be observed, and will reveal the waves of thermal agitation in the lattice. In both cases we are concerned with one-phonon processes, allowed by isolated impurities. Their vibrations are localized in space and there is no need to introduce either mechanical wavelength or any law of conservation for the wave vectors, which are no longer defined. Impurities or defects make active all the lattice frequencies which would otherwise not appear in a homopolar crystal. In the case of silicon, the neutron irradiation makes all the elastic vibrations of the lattice appear (Fig. 9). Analogous effects have recently been found in potassium bromide doped with different impurities (Fig. 6).

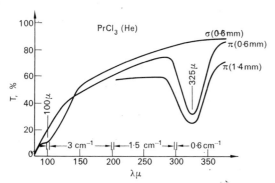

Fig. 8. *Transmission of two $PrCl_3$ plates (0·6 mm and 1·4 mm) cut parallel to the axis, with polarized radiations: σ spectrum and π spectrum (Hadni, 1964).*

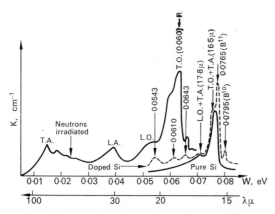

Fig. 9. *One phonon absorption in silicon irradiated with 9.3×10^{18} fast neutron cm^{-2}; local modes introduced by B^{10}, B^{11} and P and combination bands in pure silicon (from Angress, in Hadni, 1967).*

The absorption of glasses in the far infra-red is more important than that of crystals and differs in that it does not depend on temperature. It is again a question of one phonon processes where the far infra-red quantum disappears and gives a phonon of thermal agitation with the same frequency. These processes generally forbidden in a pure crystal, are allowed here by the disorder inherent to the vitreous state.

II.3. *Magnetic resonance*

Paramagnetic crystals acquire new properties when they are cooled below a certain temperature T_N (Néel temperature). Some order is then established in the crystal. If all the magnetic moments align themselves in the same direction, the crystal becomes ferromagnetic. If they align themselves in antiparallel pairs, there is antiferromagnetism if the resultant is null, and ferrimagnetism if it is not. This is the case, for example, when the two ions of the pair are different and have different magnetic moments. In this ordered state of the crystal, each atomic magnetic moment μ is oriented by a field H_E called the exchange field and has an energy μH_E. At the Néel temperature, the available thermal energy is of the same order as μH_E and if $kT_N = \mu H_E$, this energy may be able to re-establish the disordered state of the elementary magnetic moments and hence the classical paramagnetism. One might then expect that a photon of far infra-red radiation with energy $h\nu = kT_N$ would be able to reorient an elementary magnetic moment. In fact, an electromagnetic wave does not act selec-

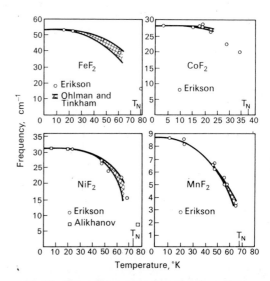

Fig. 10. *The solid curves show the observed temperature dependence of antiferromagnetic resonance in four iron-group fluorides. The cross-hatched area represents the limits of uncertainty (90% confidence) in the resonance frequency (Richards, 1963).*

tively on one elementary magnetic moment (Richards, 1963) but excites all the magnetic spins together. Without going into the details, let us say that there will be infra-red absorption at a frequency which depends on T and H. Figure 10 shows, for example, how the frequency of absorption of four crystals varies with T in the absence of magnetic field.

II.4. *Applications to superconductivity*

The transmission T_S of thin superconductive films is controlled mainly by the high reflectivity and so

$T_S \simeq 0$. Let T_N be the transmission in the normal state. Figure 11 gives T_S/T_N in terms of the frequency for a film of lead. As one might expect, T_S/T_N is zero at low frequencies. It tends towards 1 for $\nu > kT_c/h$, T_c being the highest temperature at which superconductivity is observed. These experiments have been explained by the work of Bardeen, Cooper and Schriefer and show that superconductivity vanished for high frequencies.

II.5. Applications to ferroelectricity

It is now admitted that the key to ferroelectricity is generally a "soft" vibration mode. Its frequency should be proportional to $\sqrt{T - T_c}$ and would vanish for $T = T_c$, T_c being the Curie point. Figure 12 gives the plots of the real and imaginary parts ε' and ε'' of the dielectric constant of $SrTiO_3$ at 300° and 93 °K, as inferred from far infra-red reflectivity. Note how the peak in ε'' shifts to lower frequencies when the temperature is lowered.

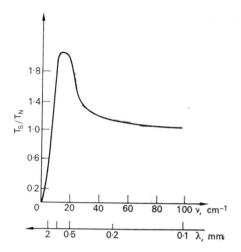

Fig. 11. T_S/T_N in terms of the frequency for a lead film (Glover and Tinkham, 1956).

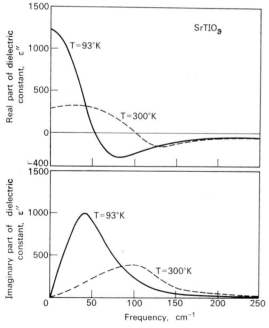

Fig. 12. Plots of the real and imaginary parts ε' and ε'' of a dielectric constant of $SrTiO_3$ at 300° and 93°K as inferred from far infra-red reflectivity. Note how the peak in ε'' shifts to lower frequencies when the temperature is lowered (from Tinkham, in Hadni, 1967).

Bibliography

GENZEL L., BILZ H. and HAPP H. (1960) *Z. Phys.* **160**, 535.
GLOVER R. E. and TINKHAM M. (1956) Transmission of superconducting films at millimeter-microwave infra-red frequencies, *Phys. Rev.* **104**, 844.
GUSH H. P. and BOSOMWORTH D. R. (1964) The absorption spectrum of compressed glasses in the far infra-red. Conference at the I.C.O. Meeting, Tokyo.
HADNI A. (1964) Far infra-red electronic transitions in ions and pairs of ions, *Phys. Rev.* **136**, A758.
HADNI A. (1967) *Essentials of Modern Physics Applied to the Study of the Infra-red*, Oxford: Pergamon Press.
HADNI A., MORLOT G., GERBAUX X., CHANAL D., BREHAT F. and STRIMER P. (1965) Absorption induite dans l'infrarouge lointain par les impuretés et les défauts d'un solide, *Comptes Rendus* **260**, 4973.
HADNI A., THOMAS R. and PERRIN J. (1969) Sur la sensibilité d'un récepteur pyroélectrique d'infrarouge, modulé entre 10 et 300 000 cps. *Comptes Rendus* **269**, 325.
LEIBA E. (1969) Hétérodynage optique avec un détecteur pyroélectrique, *Comptes Rendus* **268**, 31.
MARTIN D. H. (1967) *Spectroscopic Techniques*, Amsterdam: North Holland Co.
RICHARDS P. L. (1963) Far infra-red magnetic resonance in CoF_2, NiF_2, $KNiF_3$ and YbIG, *J. Appl. Phys.* **34**, 1237.
YAMADA Y., MITSUISHI A. and YOSHINAGA H. (1962) Transmission filters in the far infra-red, *J. Opt. Soc. Amer.* **52**, 17.

A. HADNI

FAST PASSAGE IN PARAMAGNETIC RESONANCE. In a spin resonance experiment, a paramagnetic sample is exposed simultaneously to a large steady magnetic field **H** and an oscillatory magnetic field **h** of much smaller magnitude (see *Paramagnetic resonance phenomena*). Fast passage, sometimes called adiabatic fast passage, is the name of a process which occurs when **H** is changed in a prescribed way and **h** satisfies certain conditions.

The field **H** splits the energy levels of the individual magnetic ions or nuclei. In the simplest case there are only two levels, which we label plus and minus. Their separation in energy, $\hbar\omega_0$, obeys the law $\omega_0 = \gamma H$, where γ is a quantity called the magnetomechanical ratio (see *Nuclear magnetic resonance*). The levels are broadened by a number of mechanisms, leading to a spread in ω_0 which we denote by $\Delta\omega_0$. Often, the only important broadening agencies are magnetic dipole and exchange interactions between spins (see *Spin–spin interaction*).

If the angular frequency ω of the field $\mathbf{h}$ lies close to ω_0, that is, if $|\omega - \omega_0| \lesssim \Delta\omega_0$, and if the polarization of $\mathbf{h}$ is correct, the oscillatory field can induce transitions between levels, thereby disturbing the thermal populations n_{+0} and n_{-0}. This disturbance is counterbalanced by other transitions of a thermalizing nature, which themselves are induced by lattice vibrations in solids (or collisions in gases and liquids). The conventional technique of measurement involves a steady sweep of H through the value ω/γ while $\mathbf{h}$ is held at such a low magnitude that the thermal populations are not seriously disturbed by the electromagnetically induced transitions. The quantity measured is the small oscillatory magnetization induced by $\mathbf{h}$.

Fast passage can be thought of as a variant of this procedure. One applies a rather large $\mathbf{h}$ and, at the same time, quickly sweeps the magnitude of $\mathbf{H}$ from a value well away from $(\omega \pm \Delta\omega_0)/\gamma$, through the resonant value ω/γ, and onward to a final value far on the "opposite side". The direction of the sweep is immaterial. The magnitude of $\mathbf{h}$ must be large enough to satisfy certain requirements, as discussed below, and the entire procedure must be completed in an interval t which is deliberately made too short to allow thermal forces to act; hence the word "fast". The sweep, however, must not be made too fast, a matter which is also explained below.

The end result of the sweeping procedure used for conventional paramagnetic resonance is, of course, that the populations are undisturbed. After a fast passage, however, the populations of the levels are actually interchanged, or "inverted". If, initially, $n_+ = n_{+0}$ and $n_- = n_{-0}$, then, after a successful fast passage, $n_+ = n_{-0}$ and $n_- = n_{+0}$. The spin temperature has become negative (see *Spin temperature*). The ability to produce negative spin temperatures is in fact one of the most important features of the fast passage technique.

Although the technique itself was known from the earliest days of spin resonance, it was not correctly understood until 1955, when Redfield provided the proper description in a classic paper. He argued that the spin–spin interactions mentioned above are difficult to treat analytically when the Hamiltonian contains an explicit dependence on time. The most accurate treatment requires as a starting point that the Hamiltonian be subjected to a transformation which removes as much of the time dependence as is possible.

It is well known that the field $\mathbf{h}$ should be polarized perpendicularly to $\mathbf{H}$, in order to interact strongly with the spins, and that only one of the circularly polarized components contained in $\mathbf{h}$ is active; to an excellent approximation, the oppositely rotating circular component can be disregarded. Thus, for all practical purposes, the entire Hamiltonian contains only a steady component due to $\mathbf{H}$, one component due to $\mathbf{h}$, which rotates in a plane perpendicular to $\mathbf{H}$, and the spin–spin interactions. The part of the Hamiltonian which describes the forces of thermalization is omitted altogether, because of the short time scale chosen for fast passage.

It is easy to see physically that if the Hamiltonian is now referred to a new coordinate system, which rotates about the $\mathbf{H}$ axis in synchronism with $\mathbf{h}$, the rotating field will appear to be fixed, and the parts of the Hamiltonian which depend on *fields* will become stationary. The quantum mechanical operation corresponding to a rotating coordinate transformation is well understood. When this procedure is carried out, some of the spin–spin terms remain time independent and some acquire a new variation with time. Redfield's central assumption is that the time varying terms can be neglected and the spin system can be assumed to exist in a canonical distribution of states with respect to the stationary part of the transformed Hamiltonian. We shall denote this operator by $\mathscr{H}$. Stated differently, the spin system is said to be in statistical equilibrium in the rotating reference frame.

Fast passage can be thought of as a steady change in $\mathscr{H}$ (caused by the sweep of $\mathbf{H}$) which leads the spin system through a sequence of equilibrium states. The process must be slow enough that the spins remain in equilibrium at all times. This kind of change of a parameter is often called adiabatic but should more accurately be called isentropic. The various measurable quantities which characterize the spin system (magnetization, energy) are calculated by equating the work done on the spins to their gain in internal energy. All computations are performed in the rotating frame, and the results are then transformed back into the laboratory coordinate system.

The results are most conveniently expressed in terms of the average magnetization of the sample, $\mathbf{M}$. It turns out that $\mathbf{M}$ has two components, one parallel to $\mathbf{H}$, which we call M_H, and the other fixed in the rotating coordinate system and parallel to the direction of the rotating field, which we label M_h. We take the initial condition, where $|\omega - \omega_0| \gg \Delta\omega_0$, to be $M_H = M_0$ and $M_h = 0$, as would be the case if the passage were started from a condition of thermal equilibrium in the laboratory frame. The magnetization for an arbitrary point in the passage is found to be,

$$M_H = \pm M_0(\omega_0 - \omega)/[(\omega - \omega_0)^2 + (\gamma h/2)^2 + \Delta\omega_0^2]^{\frac{1}{2}}$$

$$M_h = \pm M_0(\gamma h/2)/[(\omega - \omega_0)^2 + (\gamma h/2)^2 + \Delta\omega_0^2]^{\frac{1}{2}}$$

The sign of **M** is adjusted so that $M_H = +M_0$ for the given initial condition. For example, if the passage started from $\omega_0 > \omega + \Delta\omega_0$, the plus sign would be used.

The point is that a complete passage from $\omega_0 \ll \omega - \Delta\omega_0$ to $\omega_0 \gg \omega + \Delta\omega_0$, or the reverse, always reverses **M** from its initial direction. In the language of populations, one writes $M_H = \mu(n_- - n_+)$, where μ is the moment of an individual ion or nucleus. A reversal of **M** implies that $n_+ \to n_-$ and $n_- \to n_+$. It should be pointed out that the notion of populations is useful before the passage and afterwards, but that these parameters are poorly defined, in a quantum mechanical sense, during the process itself. Thus it is impossible to trace the details of the time evolution of the n's.

The requirements on the time of passage t can be stated formally, in the following way. The rate of passage must be fast enough to permit the neglect of thermalizing transitions but slow enough to allow the spin system to come into internal equilibrium (not to be mistaken for thermal equilibrium) at each point in the passage. If τ_1 is the characteristic time for thermal transitions and W is the transition probability which governs the rate of approach to equilibrium in the rotating frame, we require $\tau_1 \gg t \gg 1/W$. The analogous requirement for ordinary spin resonance is $W_{em}\tau_1 \ll 1$, where W_{em} is the electromagnetic transition probability.

The expression for W has been given by Franz and Slichter. It is $W = W_{em}[(\omega - \omega_0)^2 + (\gamma h/2)^2 + \Delta\omega_0^2]/[(\gamma h/2)^2 + \Delta\omega_0^2]$. Since W_{em} itself contains a shape function, in the sense that it decreases with increasing $|\omega - \omega_0|$, the rotating frame transition probability W can be thought of as a product of two shape functions which have the opposite behavior with increasing $|\omega - \omega_0|$.

The condition on t is really a condition on **h** as well, because **h** enters into the limit $t \gg 1/W$. In practice, it is relatively easy to satisfy the requirements on both parameters. This was not thought to be the case before the appearance of Redfield's theory. An earlier theory, based on the original Bloch model of spin resonance, gave the condition $\tau_1, \Delta\omega_0^{-1} \gg t \gg 1/\gamma h$ which implied $\gamma h \gg \Delta\omega_0$. These requirements are much more severe.

An enormous body of experimental evidence verifying the Redfield model has accumulated since 1955. For a guide to this literature, the reader is referred to the references listed in the paper by Franz and Slichter.

Bibliography

FRANZ J. R. and SLICHTER C. P. (1966) *Phys. Rev.* **148**, 287.
REDFIELD A. G. (1955) *Phys. Rev.* **98**, 1787.

P. E. WAGNER

FERROHYDRODYNAMICS

1. Introduction

Ferrohydrodynamics refers to the fluid mechanical study of strongly magnetizable fluid media in the presence of applied magnetic fields. Compared to magnetohydrodynamics, no current flow is required in the medium to generate the forces which arise due to the magnetic property of matter.

Fluid media composed of magnetic particles of subdomain size colloidally dispersed in a liquid carrier have been synthesized recently in the laboratory. The particles in the fluid are prevented from clustering to each other by a monomolecular layer of surfactant which envelops each particle. The layer also confers compatability with the carrier solvent and thermal agitation prevents settling. The particle concentration is enormous, of the order of 10^{18} particles/cm^3, with a typical lineal dimension being 100 Å. As a result the mixture may nearly always be treated as a fluid continuum. These fluids are to be distinguished from earlier magnetic clutch slurries which contained micron size particles and had the property of solidifying in a magnetic field.

Theory admits the possibility of true solutions possessing ferromagnetism or even a liquid or gaseous medium in which every molecule participates in the ferromagnetic phenomenon. Although such substances are presently unknown, the theoretical considerations given in the following would remain equally valid.

2. Elements of the Basic Theory

2.1. Description of magnetic stress

The fluid mechanics of polarizable fluid media differs from that of ordinary fluids in that stresses of magnetic origin make an appearance and exert a combination of body and surface force. While theoretical expressions are known for the forces acting between isolated sources of electromagnetic field, in considering the nature of stress within a material medium there exists no universal law. However, it has been found that satisfactory stress relationships may be derived from the principle of energy conservation. The stress relationships are then found to depend on detailed characteristics of the material such as the dependence of magnetization on state variables. For a quite arbitrary dependence of magnetization on applied field under the condition that the local magnetization vector is colinear with the local applied field in any volume element, a suitable stress tensor σ has elements given by

$$\sigma_{ij} = -\left\{\mu_0 \int_0^H \left(\frac{\partial Mv}{\partial v}\right)_{H,T} dH + \frac{\mu_0}{2} H^2\right\}\delta_{ij} + H_i B_j \quad (1)$$

where in cartesian coordinates j is the component of the vectorial force per unit area (traction) on an infinitesimal surface whose normal is oriented in the i direction. δ_{ij}, the Kronecker delta, is unity when the subscripts are the same but otherwise vanishes. B_j and H_i are components of the magnetic field of induction **B** and the magnetic field intensity **H** respectively. In the m.k.s. system the units of **B** are webers/meter² and **H** is given in ampere-turns/meter. The conversion constant μ_0 has the value $1/4\pi \times 10^{-7}$ henries/meter.

The magnetization **M** describes the polarization of the fluid medium. According to the defining equation

$$\mathbf{B} = \mu_0(\mathbf{H} + \mathbf{M})$$

For a ferromagnetic fluid the magnitude of **M** denoted by M has the properties:

$$M = f(H)$$
$$f(0) = 0 \quad\quad\quad (2\,\text{a, b, c})$$
$$f(\infty) = M_s$$

where M_s is the saturation magnetization. As a consequence of colinearity,

$$\mathbf{B} = \mu \mathbf{H}$$
$$\mathbf{M} = \chi \mathbf{H}$$

with μ the permeability and χ the susceptibility representing scalar quantities generally dependent on H. At a given point $H_i B_j = \mu H_i H_j = H_j B_i$ hence σ_{ij} is symmetrical so the fluid medium is free of torque.

Applied to a linear medium wherein μ is independent of H the stress tensor for the nonlinear medium reduces as it must to the Korteweg–Helmholtz expression,

$$\sigma_{ij} = -\left(-\frac{H^2}{2}\varrho\frac{\partial\mu}{\partial\varrho} + \frac{\mu}{2}H^2\right)\delta_{ij} + \mu H_i H_j$$

where ϱ the mass density is the reciprocal of v. In free space where $\mu = \mu_0$ and $\varrho = 0$ this further reduces to the classical Maxwell stress tensor

$$\sigma_{ij} = -\frac{\mu_0}{2}H^2\delta_{ij} + \mu_0 H_i H_j$$

2.2. Equation of motion

The stress tensor is to be used in a statement of Newton's law of motion. Any magnetic tensor or expression of body force is ambiguous unless accompanied by its equation of motion which in this instance is

$$\varrho\frac{D\mathbf{q}}{Dt} = \mathbf{f}_p + \mathbf{f}_g + \mathbf{f}_v + \mathbf{f}_m \quad\quad (3)$$

where **q** is the vector velocity of a fluid element and $D/Dt = \delta/\delta t + \mathbf{q} \cdot \text{grad}$ is the substantial derivative familiar in fluid mechanics. The right side is the sum of the body forces acting upon a unit volume. The familiar terms are:

$\mathbf{f}_p = $ pressure gradient $= -\text{grad }p(\varrho, T)$
$\mathbf{f}_g = $ gravity force $= -\text{grad }\psi$, with $\psi = \varrho g h$
$\mathbf{f}_v = $ viscous force $= \eta\,\nabla^2 \mathbf{q}$

where $p(\varrho, T)$ is the thermodynamic pressure when the applied field is isothermally removed; alternate forms including an adiabatic definition have also been given.

The magnetic body force has components $f_i = \delta\sigma_{ij}/\delta x_j$. For dielectric media so that **H** is curl free there is obtained from (1)

$$\mathbf{f}_m = -\text{grad}\left(\mu_0\int_0^H\left(\frac{\partial Mv}{\partial v}\right)_{H,T}dH\right) + \mu_0 M\,\text{grad }H \quad (4)$$

Substituting the force expressions into (3) gives for the equation of motion,

$$\varrho\frac{\partial\mathbf{q}}{\partial t} + \varrho(\mathbf{q}\cdot\text{grad})\,\mathbf{q} = -\text{grad }p^* + \mu_0 M\,\text{grad }H$$
$$-\text{grad }\varrho gh + \eta\nabla^2\mathbf{q}$$

where p^* is defined as

$$p^* = p(\varrho, T) + \mu_0\int_0^H\left(\frac{\partial Mv}{\partial v}\right)dH$$

2.3. Generalized Bernoulli theorem

For inviscid flows the equation of motion may be rewritten with the aid of vector identities as

$$\varrho\frac{\partial\mathbf{q}}{\partial t} - \varrho\mathbf{q}\times\boldsymbol{\omega} = -\text{grad}\left(p^* + \varrho\frac{q^2}{2} + \varrho gh - \mu_0\int M\,dH\right)$$
$$-\text{grad }T\int\mu_0\left(\frac{\partial M}{\partial T}\right)_{H,v}dH$$

where $\boldsymbol{\omega} = \text{curl }\mathbf{q}$ is the vorticity. For irrotational flow $\boldsymbol{\omega} = 0$ there will exist a velocity potential ϕ such that $\mathbf{q} = -\text{grad }\phi$. Then if $\text{grad }T = 0$ or $\delta M/\delta T = 0$ for the incompressible media there is obtained as the integral of the equation of motion the generalized Bernoulli equation which follows:

$$\varrho\frac{\partial\phi}{\partial t} + p^* + \varrho\frac{q^2}{2} + \varrho gh - \mu_0\overline{M}H = g(t) \quad (5)$$

$\overline{M}$ denotes the field average magnetization defined by

$$\overline{M} = \frac{1}{H}\int_0^H M\,dH \quad\quad (6)$$

Asymptotic values of $\overline{M}$ may be found from (2) and (6)

$$\overline{M} = \begin{cases} \frac{1}{2}\chi_i H & H \ll M_s/\chi_i \\ M_s & H \gg M_s/\chi_i \end{cases} \quad (7\,\text{a, b})$$

χ_i is the initial susceptibility, $(\delta M/\delta H)_0$. A series of magnetic fluids in which magnetite (Fe_3O_4) is the magnetic constituent have been produced for which the parameters are in the range $0 < \overline{M} < 8 \times 10^4$ AT/m (1000 gauss) and $0 < \chi_i < 4$. The nature of the magnetization vs. applied field is shown in Fig. 1.

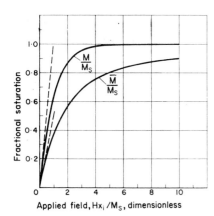

Fig. 1. *Typical magnetization curves for magnetic liquid display superparamagnetic behaviour. Hysteresis is absent in static fields.*

For time invariant flow $\delta\phi/\delta t = 0$ and $g(t) = $ const. so the generalized Bernoulli relationship reduces to

$$p^* + \varrho \frac{q^2}{2} + \varrho g h - \mu_0 \overline{M} H = \text{const} \tag{8}$$

In the absence of an applied field $p^* = p$ and the term proportional to $\overline{M}$ disappears. With one or another term absent the remaining terms provide several important examples from ordinary fluid mechanics. With $h = $ const., the remaining relationship between pressure and velocity describes the operation of the venturi meter, pitot tubes, and the pressure at the edge of a boundary layer. In hydrostatics with $q = 0$ the pressure term combined with the gravity term describes, for example, the pressure distribution in a tank of liquid while the gravity term combined with the term containing speed yields an expression for the efflux rate of material from a hole in the tank, etc. In the similar manner, combination of the "fluid magnetic pressure" $p_m = \mu_0 \overline{M} H$ with each of the remaining terms produces additional classes of fluid phenomena. Some examples are examined below for which it is necessary to develop a boundary condition relationship.

2.4. Boundary conditions

From (1) the traction on a surface element with unit normal $\mathbf{n}$ is

$$\sigma \cdot \mathbf{n} = -\left(\mu_0 \int \frac{\partial M v}{\partial v} dH + \mu_0 \frac{H^2}{2}\right) \mathbf{n} + \mathbf{HB} \cdot \mathbf{n}$$

The difference of this magnetic stress across an interface between media is a force oriented along the normal which may be expressed as follows:

$$[\sigma \cdot \mathbf{n}] = -[\sigma_{nn}] \mathbf{n} = -\mathbf{n}\left[\mu_0 \int_0^H \frac{\partial M v}{\partial v} dH + \frac{\mu_0}{2} M_n^2\right] \tag{9}$$

[ ] denotes difference of the quantity across the interface and subscript n denotes the normal direction. It is noted that the argument of the bracketed quantity vanishes in a non-magnetic medium. In deriving this relationship use is made of the magnetic field boundary conditions $[B_n] = 0$ and $[H_t] = 0$, where t denotes the tangential direction. When the adjacent media are both fluids the stress difference from (9) may be balanced by pressures giving the following result when one medium is non-magnetic.

$$p^* = p_0 - \frac{\mu_0}{2} M_n^2 \tag{10}$$

p_0 is pressure in the non-magnetic fluid medium and p^* was defined previously. While it is familiar to require continuity of pressure across a fluid boundary when considering ordinary fluids in the absence of surface tension effect, this is no longer the case with fluids possessing magnetization. Here it is seen that the magnetic stress at an interface produces a traction of $(\mu_0/2) M_n^2$.

3. Illustrations of Ferrohydrodynamic Problems

3.1. Free surface response to a line current

Consider in a gravity field a pool of magnetizable fluid surrounding a steady line current I producing therefore an azimuthal field with magnitude

$$H = \frac{I}{2\pi r} \tag{11}$$

where r is radial distance (see Fig. 2). At the free surface s from (9) since $[M_n^2] = 0$ there is obtained $p_s^* = p_0$. With $q = 0$, the constant of (5) is evaluated

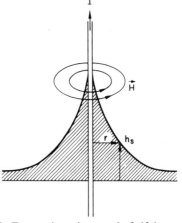

Fig. 2. *Free surface of magnetic fluid in response to the magnetic field of a line current source.*

from $H(\infty) = 0$ giving after minor rearrangement,

$$\Delta h = h_s(r) - h_s(\infty) = \frac{\mu_0}{\varrho g} \overline{M} H. \qquad (12)$$

Then from (7a) and (11) for small applied fields

$$\Delta h = \frac{\mu_0}{\varrho g} \frac{\chi_i I^2}{8\pi^2 r^2}$$

while from (7b) and (11) for saturated fluids,

$$\Delta h = \frac{\mu_0}{\varrho g} \frac{M_s I}{2\pi r}$$

the latter predicting a hyperbolic cross-section. Test has confirmed the general nature of this response.

3.2. Jet flow

Consider a steady two dimensional magnetic fluid jet of uniform speed q_1 that enters and passes through a region of uniform magnetic field. It is desired to determine an expression for the speed at a cross-section of the jet within the magnetic field region. Boundary condition (10) is applicable so in the approach flow $p_1^* = p_0$ and in the magnetic field region $p_2^* = p_0 - (\mu_0/2)\, M_n^2$. Then from (5) with $H_1 = 0$ and $\delta/\delta t = 0$

$$q_2^2 - q_1^2 = \mu_0 \left(\overline{M} H + \frac{1}{2} M_n^2\right) \qquad (13)$$

Since the right-hand side of (13) is inherently positive the speed q_2 in the field region is greater than the speed of the approach stream so the cross-section area must be reduced proportionally in the field. The stream is predicted to widen and slow to the original speed upon leaving the magnetic field region. The extent to which a jet retains its uniformity under these conditions remains to be determined.

3.3. Surface stability

A perpendicular magnetic field has a destabilizing influence on a flat interface between a magnetizable and a non-magnetic fluid. In the photograph of Fig. 3 illustrating the onset of transition the interface takes a new form in which the elevation has a regular hexagonal pattern. The glints indicate local regions where the surface was flat enough for direct reflection of the light. The pattern, remarkable for its high stability, may be stirred with a rod as the protuberances which develop remain in the liquid state. At higher field the surface has the appearance of a porcupine.

The magnetostatic field in this phenomenon may be represented for small deflections of the surface by:

$$B = B_0 + b$$
$$M = M_0 + m$$

where B_0, M_0 are the initial uniform values and b, m are perturbation quantities with orientation in the z direction. The induction field at the interface is

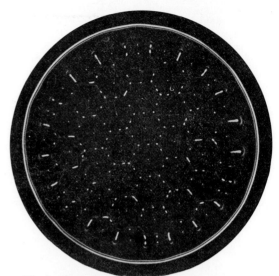

Fig. 3. Photograph of magnetic fluid surface when perturbed by a uniform normally applied magnetic field. The glints indicate local flats of crests, troughs and saddle points.

given by

$$b = \frac{k z_0 \mu_0 M_0}{1 + 1/r}$$

which is equivalent to a distribution of magnetic sources with strength proportional to the height of perturbations. k is a constant, z_0 is height of the perturbations, $r = (\hat{\mu}\mu/\mu_0^2)^{1/2}$ is a geometric-mean relative permeability with $\mu = B_0/H_0$ and $\hat{\mu} = \delta B/\delta H$ at $H = H_0$. The difference of stress at the interface may be supported by surface tension T leading to the condition

$$[\sigma_{nn}] = \left(\frac{\partial^2 z_0}{\partial x^2} + \frac{\partial^2 z_0}{\partial y^2}\right) T = k^2 z_0 T$$

Elimination of $[\sigma_{nn}]$ with equation (9), then p^* with equation (5) and recognizing that to good approximation $M^2 \cong M_0^2 + 2M_0 m$ where $[m] = -[h]$ gives upon use of the expression for b the following condition to be satisfied in order that z_0 not equal zero.

$$M^{*2} = \frac{1}{k^*} + k^* = F(k^*) \qquad (14)$$

The starred variables are dimensionless groups defined as follows:

$$M^{*2} = \text{magnetic group} = \frac{\mu_0 M_0^2}{(g \Delta \varrho T)^{1/2}(1 + 1/r)}$$

$$k^* = \text{wave number group} = \frac{k}{g \Delta \varrho T}$$

The locus of values of M^{*2} vs. k^* defines a boundary separating the regions of stability and instability.

The critical value of magnetization occurs at the calculus minimum of $F(k^*)$, a criterion that corresponds to $k^*_{\text{crit}} = 1$ and for which

$$M^{*2}_{\text{crit}} = 2 \tag{15}$$

This is the principal result of the theory of this surface instability. As a numerical example, with $\Delta \varrho = 1200 \text{ kg/m}^3$, $T = 0.025 \text{ N/m}$, $r = 2$ and $g = 9.8 \text{ m/s}^2$ the critical magnetization is $M_c = 6400 \text{ AT/m}$ (80 gauss). If the fluid's saturation magnetization were less than this value it could not display the instability regardless of the magnitude of the applied field. Likewise, a magnetizable fluid with a great enough saturation value will not undergo the transition until the applied field is large enough to induce the critical magnetization level in the medium given by (15).

A uniform tangential field has been shown theoretically and experimentally to suppress the formation of surface waves whose nodes are normal to the applied field direction.

3.4. Energy conversion

A direct conversion of thermal energy to energy of fluid motion can be based on the change of saturation magnetization with temperature which is pronounced for most ferromagnetic materials near the Curie temperature. The conversion process may be analyzed with reference to flow in a tilted uniform cross-section tube in which cold fluid enters at section 1, isothermal entry into a magnetic field is complete at section 2, heat addition at constant magnetic field is accomplished between sections 2 and 3, and isothermal flow of heated fluid out of the field is completed at section 4. From (5) with $\delta/\delta t = 0$ and $q = \text{const.}$ there is obtained for sections 1 to 2 and 3 to 4 respectively,

$$p_1 = p_2^* + \varrho g(h_2 - h_1) - \mu_0 \overline{M}(T_2) H$$

$$p_4 = p_3^* + \varrho g(h_3 - h_4) - \mu_0 \overline{M}(T_3) H$$

where generally $\overline{M}(T_3) < \overline{M}(T_2)$. In the heated zone from 3 to 4 with grad $H = 0$ the direct utilization of the equation of motion gives the relationship

$$p_3^* - p_2^* = \varrho g(h_2 - h_3)$$

and hence by eliminating the starred variables the overall pressure increase is found to be

$$(p_4 - p_1) = \mu_0 H \Delta M - \varrho g(h_4 - h_1) \tag{16}$$

Analysis of the process conducted as a closed cycle with provision for regenerative heat transfer reveals that conversion efficiency can approach the Carnot limit set by the second law of thermodynamics.

3.5. Seals

Ferromagnetic liquid has been applied to the sealing of gaps to prevent flow or leakage of gas or liquid. An advantage of this seal is the ability to accommodate motion of a moving surface, e.g. a rotating shaft.

Considering parallel or concentric walls bridged by a slug of fluid of finite length held in the gap by a uniform magnetic field which is impressed tangential to the interface at one end, identify interface adjacent sections, 1 in the non-magnetic medium out of the field, 2 in the medium at the interface free of magnetic field, 3 in the medium at the interface where the field is present, and 4 in the field in the non-magnetic medium. From (9) with $[\sigma_{nn}] = 0$ and $[M_n] = 0$ the boundary conditions at the interface are $p_2^* = p_1$ and $p_3^* = p_4$ while from the generalized Bernoulli equation, (5), $p_2^* = p_3^* - \mu_0 \overline{M} H$. Thus the pressure difference $(p_4 - p_1) = \Delta p$ at mechanical equilibrium maintained by the liquid slug is given by

$$\Delta p = \mu_0 \overline{M} H \tag{17}$$

Using permanent magnets pressure differences up to 1 atm have been demonstrated in a single stage. If the applied field is normal to the interface the pressure difference is increased by an amount due to the magnetic surface stress difference with the result given by

$$\Delta p = \mu_0 \overline{M} H + \tfrac{1}{2} \mu_0 M_n^2 \tag{18}$$

Typical values of pressure differential computed from this relationship are plotted in Fig. 4.

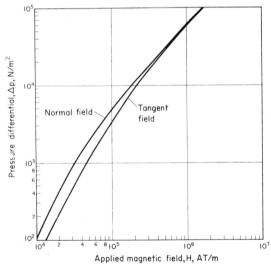

Fig. 4. Pressure capacity of magnetic fluid seals as function of applied field magnitude and direction. $M_s = 5 \times 10^4 \text{ AT/m}$, $X_i = 1$.

3.6. Levitation

Magnetizable fluid acted upon by a magnetic field possessing a gradient of field magnitude experiences a body force that creates a distribution of apparent specific weight throughout the medium. The magnetic force can oppose gravity creating a low or zero g condition or can add to gravity and produce what is effectively a high-density liquid medium. A non-magnetic object whose density is greater than the ordinary density of the medium may be buoyantly floated by introduction of the magnetic field. Perhaps a more striking phenomenon is the levitation of a non-magnetic object at an interior point of the fluid by a magnetic field containing a local minimum of field magnitude (Fig. 5). Any displacement of the levitated object is accompanied by positive restoring forces equivalent to a three-dimensional spring yet the support is non-mechanical. The suspension system constitutes a novel bearing requiring no input of energy. Such a bearing is perfectly free of mechanical friction to the degree that the magnetic fluid medium retains the property of a true fluid that any small shear stress results in a perceptible rate of strain. Then any small torque applied to the supported body will produce a rotation.

A corollary phenomenon to the levitation of a non-magnetic body is the self-levitation of a body containing a source of magnetic field when immersed in the magnetic fluid medium. For instance a ceramic magnet of density four times greater than the density of the fluid medium was demonstrated to produce self-levitation, and seek an equilibrium position removed from the fluid boundaries (see Fig. 6). A description of the phenomenon can be made in analogy to image forces in dielectric media.

Both phenomena are illustrations of the general tendency for a magnetic source and a non-magnetic object to experience a mutual repulsion when separated by a ferromagnetic fluid medium.

Earnshaw's theorem propounded in 1839 denies the possibility of stable levitation of isolated collections of charges (or poles) by static fields. In 1938 Braunbeck elaborated on the theorem and deduced that levitation is impossible in fixed magnetic fields except for diamagnetic materials or superconductors. The ferrohydrodynamic levitation phenomena offer additional means for circumvention of the theorem. Application has been made to the study of inertial sensors of acceleration.

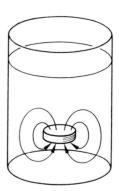

Fig. 6. Stable levitation of a permanent magnet in magnetic fluid under the influence of its own field.

To develop a general formulation for the buoyancy and levitational forces consider an arbitrarily shaped body with surface S immersed in the magnetic fluid medium. The force $\mathbf{F}$ experienced by the body is given as the sum of pressure force and magnetic stress difference integrated over the surface.

$$F = \oiint \{[\sigma \cdot \mathbf{n}] - \mathbf{n}p\} \, dS \qquad (19)$$

From (9), the definition of p^*, and the generalized Bernoulli theorem of (5) the integrand for statics becomes (const. $-\varrho g h + \mu_0 \overline{M} H + (\mu_0/2)(M_n^2)$. By the divergence theorem $\oiint (\text{const.} - \varrho g h) \mathbf{n} \, dS = \iiint_v \mathrm{grad}\,(\text{const.} - \varrho g h) \, dV$ so for a constant density fluid medium the force is,

$$\mathbf{F} = -\varrho \mathbf{g} V - \oiint \mu_0 \left(\overline{M} H + \frac{M_n^2}{2} \right) \mathbf{n} \, dS \qquad (20)$$

The first term represents Archimedes principle that buoyant lift is equal and opposite to the weight of the displaced fluid. The surface integral expresses the additional force due to magnetic pressures. With gravity as the only external force on a body of density ϱ' a volume force $\iiint_v \varrho' \mathbf{g} \, dV$ must be introduced to the right side of (20). Then with $\mathbf{F} = 0$ there is obtained as the condition for neutral buoyancy

$$(\varrho' - \varrho) \mathbf{g} V = \oiint \mu_0 \left(MH + \frac{M_n^2}{2} \right) \mathbf{n} \, dS$$

The integral term depends on geometric shape and size as well as the magnetic field variables. Present magnetic fluids acted upon by permanent magnet sources are able to levitate, against the force of gravity, any non-magnetic element in the periodic table.

The condition for stable levitation at an interior point of the fluid space may now be simply developed. In the absence of gravity it is required in addition to

$$\oiint \mu_0 (\overline{M}H + \frac{M_n^2}{2}) \mathbf{n}\, dS = 0$$

at the equilibrium point that a positive restoring force accompany any small displacement from that point. Since $(\overline{M}H + M_n^2/2)$ increases asymptotically with H it follows that a magnetic field to function as a levitating field must possess a local minimum of field magnitude, i.e. for any displacement

$$\delta H > 0 \qquad (21)$$

Equal bar magnets with like poles opposed produces such a field and in this manner it is possible to levitate a non-magnetic body. A permanent magnet, however, will tend to levitate itself as remarked above.

The magnetic force on a non-magnetic immersed object can be given explicit expression in the case of a uniform gradient of H and with $H \gg M_s$ where M_s is the saturation magnetization. Under this condition

$$\frac{M_n^2/2}{\overline{M}H} \ll 1$$

so with uniform grad p_m the magnetic force from (20) is

$$\mathbf{F}_m = -\oiint p_m \mathbf{n}\, dS = -V \,\text{grad}\, p_m$$
$$= -V\mu_0 M \,\text{grad}\, H.$$

Due to the minus sign this force is equal and opposite to the magnetic body force on an equivalent volume of fluid. This demonstrates for a special case that an immersed non-magnetic object experiences repulsion from a source of magnetic field.

4. Torque Driven Flow

In this discussion the assumption has been made throughout that the field vectors are colinear, e.g. $\mathbf{M} \times \mathbf{H} = 0$. Recently it was demonstrated that in rotating magnetic fields the induced magnetization lags the applied field leading to generation of torque per unit volume $\mathbf{T} = \mu_0 \mathbf{M} \times \mathbf{H}$. The torque generation produces an organized flow of the magnetic fluid medium.

Bibliography

COWLEY M. D. and ROSENSWEIG R. E. (1967) The interfacial stability of a ferromagnetic fluid, *J. Fluid Mech.* **30**, part 4, 671–688.

LANDAU L. D. and LIFSCHITZ E. M. (1960) *Electrodynamics of Continuous Media* (Transl. J. B. SYKES and J. S. BELL), Oxford: Pergamon Press.

NEURINGER J. L. and ROSENSWEIG R. E. (1964) Ferrohydrodynamics, *Physics of Fluids* **7**, 1927.

PENFIELD P., Jr. and HAUS H. A. (1967) *Electrodynamics of Moving Media*, Research Monograph No. 40, The M.I.T. Press, Cambridge, Massachusetts, Table C. 2, p. 255.

ROSENSWEIG R. E. (1966) Buoyancy and stable levitation of a magnetic body immersed in a magnetizable fluid, *Nature* **210**, No. 5036, 613–614.

ROSENSWEIG R. E. (1966) *Magnetic Fluids*, International Science and Technology, No. 55.

WOODSON H. H. and MELCHER J. R. (1968) *Electromechanical Dynamics*, New York: John Wiley and Sons.

ZELAZO R. (1967) *Interfacial Ferrohydrodynamics*, S.M. Thesis, M.I.T.

R. E. ROSENSWEIG

FLUIDICS. Fluidics is a term used to define a technology involving the use of fluids, gaseous or liquid, in motion for control purposes such as switching, sensing and amplification. The fluid energy is manipulated in a manner similar to that used for electricity in electronic devices with the advantage that there are no moving parts. The control components are also highly reliable under extreme environmental conditions and are impervious to any form of radiation, magnetic or nuclear, which explains why nuclear and aerospace engineers are exhibiting a widespread interest in their use, an interest which has spread to mechanical and other forms of engineering, commercial and industrial, on account of the low economic cost plus the reliability of fluidic components.

Like many a modern development the ideas behind fluidics are by no means new. Professor L. Prandtl, the Gottingen aerodynamicist, showed in 1904 that the flow separation of a fluid from a boundary of a wide-angled diffuser was achievable by the application of suction: in fluidic terminology it is called a "NOT" function. Nicola Tesla patented in 1920 a device whereby a low resistance to fluid flow in one direction could be accompanied by a higher resistance in the reverse direction: it was called a valvular conduit by him and in fluid terminology is defined as a "DIODE". Henri Coanda described in 1932 how a moving gas stream attached itself to a suitably contoured wall in its immediate vicinity on account of the pressure differential thus created: the phenomenon is known today as the "Coanda effect" and has a distinct relevance to modern fluidics. This characteristic of wall attachment was, however, originally reported by T. Young in 1800. Kline and Moore drew attention in 1938 to the stable states of fluid flow in a wide-angled diffuser enabling the flow to be separated from one or other wall or to stay between the two walls, being stable in all three states. The above four fundamental concepts were published without any of the originators apparently recognizing their possible applications to the design of fluid amplifiers and to the inherent capabilities of control and of signal gain, as now demonstrated in the development of fluidic logic.

The Harry Diamond Laboratories (then the Diamond Ordnance Fuze Laboratories) in the U.S.A. released information on pure fluid amplification in 1960 and this release, followed by a Fluid Amplification Symposium at these laboratories in October 1962, stimulated investigations into the subject so that by 1963 over 100 companies were actively engaged in America and more than twenty million U.S. dollars had been invested. This interest in the subject has since greatly increased. Thus research and development work started and has continued on a large scale in America, and to a less extent in Russia and in the United Kingdom, where symposia have been held on this subject. Whereas the stimulant has come from the simplicity of construction, reliability, absence of moving parts, freedom from fire and explosion hazard, and considerations of hygiene in industrial controls, there are certain characteristics which may limit the usefulness of fluidic devices including (1) the need for continuous fluid flow in the "standby" condition, (2) the slower response time compared with analogous electronic devices and (3) the poor pressure recovery. It must be appreciated that a fluid circuit, as opposed to the fluidic device which is one element only, may contain moving parts such as actuators, diaphragms, valves, etc.

A number of terms have been adopted in describing fluidic logic devices as follows, the term "signal" being adopted from its electronic analogue:

An "IF" element has a signal present at its output, if, and only if, a signal is present at its sole input.

An "OR" element has a signal present at its output if a signal is present at one or more of its multiple inputs.

A "NOR" element has a signal present at its output only when there is no signal present at any of its inputs.

A "NOT" element has an output signal only if a signal is not present at its sole input.

An "AND" element has an output signal only if a signal is present at both one and the other of its two inputs, the two signals combining vectorially to form a common output.

An "ALL" or "MULTIPLE AND" is a corresponding device to an "AND" in a device with more than two inputs.

In a discussion of fluid amplifiers the following terms are often used:

Pressure gain

$$G_P = (p_{LO} - p_{RO})/(p_{LC} - p_{RC})$$

Flow gain

$$G_Q = (Q_{LO} - Q_{RO})/(Q_{LC} - Q_{RC}) \quad (1)$$

Power gain

$$G_{PQ} = (p_{LO}Q_{LO} - p_{RO}Q_{RO})/(p_{LN}Q_{LC} - p_{RC}Q_{RC})$$

where G = gain, p = pressure, Q = volume flow, c = control, o = output, L = left and R = right.

An AMPLIFIER is an active device in which the output signal is modulated by a control signal of lower energy level.

FAN OUT CAPABILITY is the ability of a device to provide sufficient output for a number of succeeding similar devices connected in parallel.

DIGITAL ELEMENTS are devices suitable for operation in a binary digital mode.

PROPORTIONAL ELEMENTS are devices in which the signal level of the output is a continuous function of the input signal.

Fluidic elements can be classified on a physical basis as active and passive, somewhat analogous to electronic circuit practice, the former having a pressure or flow source and the latter no such force, with subdivisions describing the basis of operation by turbulence effect, etc. The choice of working fluid is usually air, water or an hydraulic fluid, some devices being operable successfully by more than one fluid.

Active Elements

Active elements are operated by various effects due to momentum, turbulence, a vortex, a wall or an edge tone. Passive elements include fluid diodes, some "AND" elements, laminar restrictors and filters.

Vented momentum amplifier. The modulation of a power stream or flow is effected by the momentum of a control stream or streams as shown in Fig. 1

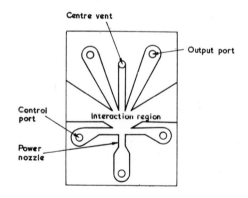

Fig. 1. Vented momentum amplifier (proportional).

for a two-dimensional vented momentum amplifier or bistable fluid amplifier. The jet from a power nozzle proceeds across and out by the centre vent. If a jet is introduced by the left-hand control port, the power jet will be deflected in the interaction region and enter the right-hand receiver to the right-hand output port. Alternatively, a jet from the right-hand control port will deflect the power jet into the left-hand output port. The fluidic element of this type can be constructed by profiling the passages required in a flat sheet of material sandwiched between two other sheets with the necessary tube connections. This type of amplifier will operate high impedance loads since the excess flow can be spilt by vents from the interaction region without affecting output pressure recovery. Its typical characteristics are given in Fig. 2, with various output impedances.

In a similar device the interaction region is closed instead of vented and becomes fundamentally a flow rather than a pressure amplifier.

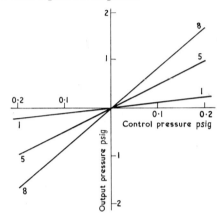

Fig. 2. *Typical characteristics of two-dimensional vented momentum amplifier.*

Proportional fluid amplifier. A diagrammatic beam-deflection fluid amplifier is shown in Fig. 3. The fluid enters at the power chamber and the control jets on each side with suffixes R and L. High-energy fluid flows out of the power chamber as a jet and low energy fluid out of the control chambers, mixing in the interaction region indicated as side chambers which

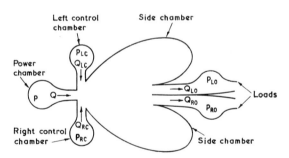

Fig. 3. *Proportional fluid amplifier.*

may or may not be vented. The control jets of fluid apply pressure forces, momentum-flux forces or a combination of both, depending on their position and size. Pressure forces predominate when the control nozzle is close to the side of the power jet. The resulting jet flow is usually collected in two outputs which may be adjacent as shown or separated by a vent. When the control jets on right and left are equal, the power jet is not deflected and each output collects the same amount of fluid. A small change of control energy on one or other side deflects the power jet so that one output collects more fluid than the other. With proper design the changes in output pressure and flow is greater than the changes in control pressure and flow. The gains are as defined in the equations (1) above. In the proportional fluid amplifier there are no side walls between the interaction region and the output aperture so as to prevent the power stream from attaching to the walls and making the amplifier bistable. In the open-type amplifier units, fluid not collected by the outputs is vented; in closed units, all the fluid supplied to the power and control jets must pass through the outlet apertures and, to accomplish this, connections are made to equalize the pressure across the power jet. Open type units are usually pneumatic; closed units operate with either liquids or gases as the working fluid.

Bistable amplifier. To obtain an alternative right- and left-hand flow, a dividing portion, called a splitter, has a different effect according to its position. Splitters have pointed, blunt or cusped ends. For short splitter distances the flow divides evenly and has little bistability. As the splitter is moved downstream, the flow becomes bistable due to the increased influence of the side walls and the greater distance available for the stream to adjust to a differential pressure. The gain increases very rapidly, being a maximum and the stability a minimum just prior to the stream's diversion wholly to one output without a control signal. The tip of the splitter is then just upstream of the region of attachment of the flow to the boundary wall. For a much longer splitter distance instability sets in and the flow oscillates.

The diagrams in Fig. 4 illustrate how a control diverts the stream from the right to the left outlet and returns to the right outlet when a stopper is placed over the left outlet.

Bleeds. Bleeds are sometimes introduced downstream of the two controls and upstream of the splitter. The

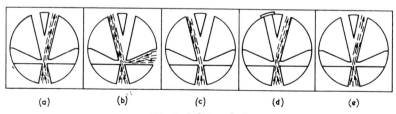

Fig. 4. *Splitter effect.*

bleeds remove the back pressure when there is a load on the outlet and loading has very little effect on gain. The bleeds are vents which connect the internal cavities to a reference pressure, the maintenance of which is an important function in an amplifier. They accomplish this by regulating the amount of fluid removed as a function of the load impedance. When this impedance is high, static pressure builds up at the entrances to the output apertures and the fluid that is prevented from leaving through the outputs is then channelled through the bleeds. The parallel combination of load and bleed impedance is small enough to remove all the fluid entering without raising the internal pressure, that is, the bleed decouples the input of the amplifier from the output load.

Bleeds also shape the pulses of fluid that pass through the load and in bistable fluid amplifiers, the position of the bleeds may affect the switching characteristics. In proportional amplifiers, a bleed connection between two output apertures will remedy a loss of gain when fluid prevented from leaving one output aperture spills over into the other aperture.

Bleeds are usually at approximately right angles to the flow. With a high velocity of the stream and a low static pressure in the region of the bleed, there will be little flow into or out of a bleed, but as the velocity decreases and the pressure increases there will be flow out of the bleed. Consequently, bleeds are placed as far as is practicable where the velocity is high consistent with being downstream of the attachment region. In general, a bleed must provide an area greater than the area of the stream for diverting the full flow of the stream without affecting the interaction region. On the other hand, an excessive area will result in energy loss at low loads.

Impact modulator. In the above the control jets acted at right angles to the main power jet. In a Direct Impact Modulator two opposing circular coaxial supply jets impinge so that the flow turns normally outward in a thin radial flat sheet. A control signal can be used to prevent the dissipation of the power jet and to drive the radial impact sheet from just outside into a cone-shaped jet just inside an output recapture edge. This makes the device into a high gain proportional pressure amplifier; using several devices in series (cascaded) has resulted in a pressure gain of up to 8000 to 1 while operating with an input signal of 0·0001 atm (0·0125 psig).

Induction amplifier. An Induction Amplifier has control input ports which introduce the control flow tangential to a wall boundary. Its action decreases the circulation on the side where it is introduced so that the main flow is pulled to the control flow side where it remains after the control flow is removed. This contrasts with the momentum-exchange or pressure-controlled amplifiers which increase the circulation and push the main jet to the opposite wall.

This device can be arranged so that it acts bistably with the control ducts open to ambient pressure.

Turbulence amplifier. A Turbulence Amplifier, one of the simplest fluid devices, uses the principle of converting a power stream from laminar to turbulent flow by a control jet with a marked reduction of the pressure recovered at the output (Fig. 5). If the pipe is sufficiently long the laminar stream can be projected

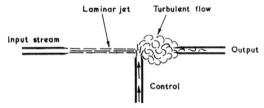

Fig. 5. Turbulence amplifier.

a distance up to 100 diameters before becoming turbulent. With a nozzle 0·1 in diameter, output pressures over 10 cm of water are possible. Since the output signal exists only in the absence of control it is a logic "NOR", suitable for an on–off device. Pressure gains up to 1000 can be achieved in a single stage. The output of a single turbulence amplifier will provide adequate signal outputs for ten or more others and thus have a high fan-out capability. Because of the low power consumption more than 10 units require only about one horsepower to operate and because of the ease of impedance matching are well suited to the construction of logic circuits. The general characteristics of the device are shown in Fig. 6.

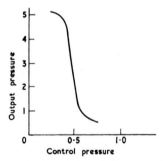

Fig. 6. Characteristic turbulence amplifier.

Vortex modulator or valve. A vortex modulator has a power stream entering via a circumferential wall and flowing radially inwards to an axial output duct to give a uniform radial stream of low pressure loss when the element is in the "ON" condition (Fig. 7). The control signal, applied via a tangential port, imparts a circulatory motion to the fluid in the annulus and results in a combined vortex flow. Thus the power stream mass flow is modulated and the power consumption is reduced in the "OFF" condition, a feature which compares favourably with the momentum type of proportional amplifier where the output power is modulated but the input power remains constant. The working fluid is preferably a liquid and the device is being applied to gas turbine engine fuel control systems.

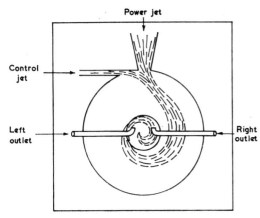

Fig. 7. Vortex sensor.

An angular rate sensor works on a similar vortex effect, with the supply entering through a porous wall, flowing radially inwards and leaving via an axial output duct. By rotating the whole element, a rapidly rotating flow results in the central core and output duct, and the resulting low static pressure in the centre forms the output signal from the device.

Vortex oscillator. A simple type of vortex oscillator consists of two vortex chambers with the fluid flow driven in opposite directions by tangential inputs and connected by a pipe, the second chamber having a central exhaust. When the helical flow passes through the central core of the second chamber it will at one moment rotate in one direction and then reverse to rotate in the other direction. As the core flows out of the exhaust it will do so in pulses which are periodic and produce an oscillator tone. This counter-vortex oscillator has been demonstrated to have a frequency stability of 15 cps over a pressure range from $1/10$ to 4 atm. In a single-chambered vortex whistle the frequency is a function of the inlet pressure. The above are examples of many types of vortex devices that are under development in several countries.

Edgetone amplifier. A jet will oscillate about a sharp edge and has been used to reduce greatly the switching time in a bistable amplifier such as Fig. 8 where a gas jet impinging on a sharp edge causes the shedding of eddies alternately on each side of the obstruction. The pressure fluctuations are heard as acoustic waves and a jet can be found to oscillate between the splitter and the right cusp at a frequency of about 10,000 cps. A control signal can change the jet over to oscillate from the right to the left cusp and vice versa. The switching time is roughly equal to that of the oscillation period which gives a very high speed of response, and there is a high gain. "FAN IN" and "FAN OUT" capability are obtainable.

Wall effect. Wall effect devices are two-dimensional. The boundaries of the interaction region contain and direct the flow with a switching or amplifying function by virtue of an aerodynamic effect. A control flow can be used to entrain fluid from a power stream and thus

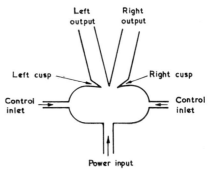

Fig. 8. Edgetone amplifier.

deflect it to the same side as the control flow. Alternately, many fluidic devices are based on a controlled separation effect in which a fluid flow from a passive supply port ensures separation from an outer wall. The direction of the active supply stream is altered and the combined effect sweeps the power stream across the output ducts to result in a device which is a very effective flow amplifier, called a Controlled Separation (Elbow) Amplifier.

Wall interaction fluid amplifiers can easily be made to produce a flow gain of 10 and a pressure recovery of 25 per cent. They can switch four other like units at the same power jet pressure to give, what is called, a fan-out of 4. Fan-out gains in excess of 16 with excellent stability have been achieved. Mass flow gains around 220 and power gains around 37 have been reported.

Of the same type of fluidic control is the "Aerofoil Amplifier" shown in Fig. 9, the control blowing off the power stream into the receiver. With the output duct situated to catch the attached stream, the device

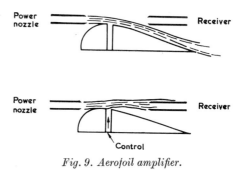

Fig. 9. Aerofoil amplifier.

is a logic "NOT" and if the detached stream is the one detected, the device is a logic "AND". An unconfined Elbow Amplifier can reach a maximum frequency of about 1000 cps.

Passive Elements

Passive elements include many devices such as those briefly described below.

Fluid Diodes are passive elements with a common characteristic, namely, that of offering a far greater resistance to fluid flow or pressure signal in one direction than in a second. A *Jet Diode* has two tubes arranged at an angle so that the forward flow enters the output tube but a reverse flow is vented. A *Turbulence Diode* or amplifier has a laminar flow in one tube, a laminar free jet between two tubes facing each other and trips to cause turbulence in the second tube. A *Wall Attachment Diode* has a forward flow with attachment to side walls in the reverse flow direction. In a *Vortex Diode* the flow resistance from the centre port to a tangential port is less than that in the case of reverse flow. A *Counterflow Diode* is essentially a Tesla's diode as described earlier. A *Linear flow Resistance* or *Laminar Restrictor* has a flow through a tube which is widened out at one part to include a large number of small flow passages. *Filters* range from a gauze packed chamber to more elaborate devices designed to improve signal-to-noise ratio or to modify the frequency response characteristics of fluid circuits.

Applications

Fluid Amplifiers are at about the same stage of development as transistors were some fifteen years earlier. Their response is slower than that of an electronic device but much faster than that of a pneumatic or hydraulic device. Operations are possible in temperatures ranging from near cryogenic to white heat. The basic amplifier is nothing more than a block of metal or hard plastic with passages for the flow of fluid. Applications of fluid elements will mainly be at the intermediate speeds between the better known devices already in use. They include typically missile guidance control, respirators, ramjet intake control by movement of internal plug, hydrofoil control, gas turbine engine control, aircraft propulsion systems, multivibrators, machine tool operations and computers. An exceptional development has been that of puff-blow control developed for the disabled at Stoke Mandeville Hospital, Bucks., to operate typewriters and make phone calls. In explanation, it has been found that a patient's mouth is an efficient pneumatic pulse source giving up to ten pressure-suction cycles per second at 7 to 14 cm mercury.

Noise. Fluidic device amplifiers give rise to "Noise" which is always associated with any form of turbulence. This is not dealt with here. It has been studied by workers at the Harry Diamond Laboratories.

Bibliography

CATCHPOLE B. G. *A Survey and Classification of Fluidic Elements*, Australian Dept. of Supply Mech. Eng. Note ARL/ME 278.

GOTTRON R. N. and WEINGER S. D. (1965) *Parameters Affecting the Noise in no Moving Parts Fluid Devices*, Harry Diamond Labs., TR-1283, April 1965.

KINSKNER J. M. (Ed.) (1966) *Fluid Amplifiers*, McGraw-Hill.

OSBORN M. J. (1967) *Fluid Amplifier—State-of-the-Art Report*, Vol. II A, Bibliography, N.A.S.A. CR-734, May 1967.

J. L. NAYLER

FOURIER TRANSFORM SPECTROSCOPY. This is the name given to the means of finding a spectrum by recording a two-beam interference pattern as a function of path-difference within the interferometer and then performing a numerical Fourier transform on this function. The spectrum can be constructed from the results of the Fourier transform. The technique offers the greatest advantage over conventional methods in the infrared spectral region and particularly in the far infrared, submillimetre and millimetre-wave region. This is because the mechanical and optical requirements are less stringent at long wavelengths and because the high radiation grasp due to the *luminosity* advantage of Jacquinot and the greater signal-to-noise ratio due to the *multiplex principle* of Fellgett offset the low luminance of long-wave sources. An additional consideration making short-wave applications difficult is the need to measure the path-difference between the beams of the interferometer at distances equal to at least half the shortest wavelength present.

Suppose the Michelson interferometer shown in Fig. 1 is irradiated with monochromatic radiation of luminance $B(\sigma_0)$ at wave-number σ_0. Then it is easy to show that the signal reaching the detector is

$$I_0(x) = \tfrac{1}{2}B(\sigma_0) + \tfrac{1}{2}B(\sigma_0) \cos 2\pi\sigma_0 x$$

where x is the path-difference $PP_1 - PP_2$ between the beams. Cosine fringes are therefore observed. If, now, the source is quasi-monochromatic and of luminous density $B(\sigma)$ in the interval σ to $\sigma + d\sigma$ the intensity is

$$dI(x) = \tfrac{1}{2}B(\sigma)\,d\sigma + \tfrac{1}{2}B(\sigma) \cos 2\pi\sigma x\,d\sigma.$$

By now allowing the source to be polychromatic and to occupy a broad spectral range we simply sum the contributions from all the elements $d\sigma$ in the band to give

$$I(x) = \tfrac{1}{2}\int_0^\infty B(\sigma)\,d\sigma + \tfrac{1}{2}\int_0^\infty B(\sigma) \cos 2\pi\sigma x\,d\sigma.$$

The intensity $I(x)$ in the interference pattern is, therefore, composed of two terms one of which is a constant

$$I = \tfrac{1}{2} \int_0^\infty B(\sigma) \, d\sigma,$$

a measure of the total intensity emitted, while the other represents the variation of the observed intensity with path-difference. This variable term

$$I(x) - I = \tfrac{1}{2} \int_0^\infty B(\sigma) \cos 2\pi\sigma x \, d\sigma$$

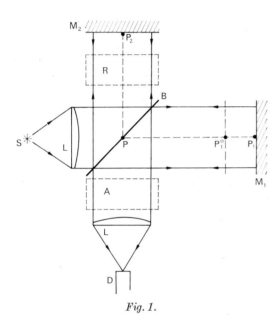

Fig. 1.

is called the interferogram $F(x)$; hence

$$F(x) = \tfrac{1}{2} \int_0^\infty B(\sigma) \cos 2\pi\sigma x \, d\sigma$$

$$= \tfrac{1}{4} \int_{-\infty}^\infty B(\sigma) \cos 2\pi\sigma x \, d\sigma.$$

By Fourier's inversion theorem we find the spectral distribution

$$B(\sigma) = 4 \int_{-\infty}^\infty F(x) \cos 2\pi\sigma x \, dx$$

is given as the cosine Fourier transform of the interferogram $F(x)$.

In practice the interferogram cannot be recorded from $-\infty$ to $+\infty$ but only from $-D$ to $+D$ (say); neither can $F(x)$ be recorded and stored as an analogue function prior to the numerical evaluation of the integral. In fact, the evaluation is usually done digitally and the interferogram is, consequently, sampled at discrete intervals of path-difference β and the detector signal at each path-difference $x = r\beta$ is coded in suitable form for the computer. Because of these practical circumstances the computed spectrum is given by

$$B_c(\sigma) = 4 \int_{-D}^{+D} F(x) \, w\!\left(\frac{x}{D}\right) \, \text{III}\!\left(\frac{x}{\beta}\right) \cos 2\pi\sigma x \, dx$$

where

$$\text{III}\!\left(\frac{x}{\beta}\right) = \sum_{r=-\infty}^{\infty} \delta(x - r\beta)$$

is the infinite sampling comb and $w\!\left(\frac{x}{D}\right)$ is a weighting or apodizing function necessary to offset the effects of truncating the integral at $x = \pm D = \pm N\beta$. In the absence of the weighting function each computed monochromatic feature has a total half-width $1/2D$ and large side-lobes, alternate ones being negative. A suitable weighting function virtually eliminates these side lobes and gives a smooth profile of total half-width $1/D$ which is consequently the resolution R in the spectrum. The resolving power is $\mathscr{R} = \sigma D$. The smooth profile is called the apparatus function or the spectral window and is equivalent to the slit function in a conventional spectrometer. A typical weighting function is

$$w\!\left(\frac{x}{D}\right) = \left\{1 - \left(\frac{x}{D}\right)^2\right\}^2$$

which leads to the apparatus function

$$A(\sigma D) = \frac{J_{5/2}(2\pi\overline{\sigma - \sigma_0}D)}{(2\pi\overline{\sigma - \sigma_0}D)^{5/2}}$$

where $J_{5/2}(2\pi\overline{\sigma - \sigma_0}D)$ is a Bessel function.

In a well-adjusted interferometer the interferogram is symmetrical about $x = 0$ and need only be recorded for $x > 0$. $N+1$ samples of the interferogram are taken from $x = 0$ to $x = D = N\beta$. The magnitude of β is determined by the highest wave-number (smallest wavelength) present and must be no greater than $\beta = \tfrac{1}{2}\sigma_{\max}$. N is determined by the resolution $R = 1/N\beta$ required in the spectrum. With this choice of β there is no overlapping of spectra as in a grating spectrometer. On supplying the interferogram ordinates $F(r\beta)$ to a suitably programmed computer the spectrum

$$B^c(\sigma) = 2\beta \sum_{r=0}^{N} F(r\beta) \, w\!\left(\frac{r}{N}\right) \cos 2\pi\sigma r\beta$$

may be evaluated. If a specimen of transmission factor $\tau(\sigma)$ is placed in front of the source at A the interferogram is modified to $F_a(x)$ because of the modified spectral distribution

$$B_a(\sigma) = 4 \int_{-\infty}^{+\infty} F_a(x) \cos 2\pi\sigma x \, dx.$$

It is clear that the transmission spectrum $\tau(\sigma) = B_a(\sigma)/B(\sigma)$ may be calculated, to an excellent approximation, from $\tau^c(\sigma) = B_a^c(\sigma)/B^c(\sigma)$, where

$$B_a^c(\sigma) = 2\beta \sum_{r=0}^{N} F_a(r\beta)\, w\left(\frac{r}{N}\right) \cos 2\pi\sigma r\beta,$$

in a manner analogous to the methods of conventional spectroscopy.

If the specimen has an optical thickness $2n(\sigma)d$ where $n(\sigma)$ is its refractive index and d its thickness, and if it is placed in the static arm of the interferometer at R it introduces a path difference

$$\Delta(\sigma) = 2\{n(\sigma) - 1\}d$$

which is different for each wave number. The bright fringe formerly at zero path difference is displaced to

$$\bar{x} = \bar{\Delta} = 2(\bar{n} - 1)d$$

where $\bar{n}$ is some mean refractive index, and the interferogram, measured from $x = \bar{x}$ with the variable $y (= x - \bar{x})$, is

$$G_\psi(y) = \frac{1}{4} \int_{-\infty}^{+\infty} B_\psi(\sigma) \cos\{2\pi\sigma y - \Psi(\sigma)\}\, d\sigma$$

where $\Psi(\sigma) = 2\pi\sigma\{\Delta(\sigma) - \bar{x}\}$ is the phase difference due to the specimen. Fourier transformation shows that, in this case, the cosine transform

$$B_\psi(\sigma) \cos \Psi(\sigma) = P_\psi(\sigma) = 4 \int_{-\infty}^{+\infty} G_\psi(y) \cos 2\pi\sigma y\, dy$$

and the sine transform

$$-B_\psi(\sigma) \sin \Psi(\sigma) = Q_\psi(\sigma) = 4 \int_{-\infty}^{+\infty} G_\psi(y) \sin 2\pi\sigma y\, dy$$

can be combined to give the intensity

$$B_\psi(\sigma) = \sqrt{\{P_\psi(\sigma)\}^2 + \{Q_\psi(\sigma)\}^2}$$

and the phase

$$\Psi(\sigma) = -\arctan \frac{Q_\psi(\sigma)}{P_\psi(\sigma)}.$$

It is easily shown that for a given specimen $B_a(\sigma) = B_\psi(\sigma)$. In practice, therefore, we can find the transmission and the refraction spectrum from

$$B_\psi^c(\sigma)/B^c(\sigma) = \frac{\sqrt{\{P_\psi^c(\sigma)\}^2 + \{Q_\psi^c(\sigma)\}^2}}{B^c(\sigma)}$$

and

$$n^c(\sigma) = 1 + \frac{\bar{x}}{2d} - \frac{1}{4\pi\sigma d}\left\{\arctan \frac{Q_\psi^c(\sigma)}{P_\psi^c(\sigma)} + m\pi\right\}$$

where m is an integer necessary when the true value of the inverse tangent strays from the principal value and

$$P_\psi^c(\sigma) = 4\beta \sum_{-N}^{N} G_\psi(r\beta)\, w\left(\frac{r}{N}\right) \cos 2\pi\sigma r\beta$$

$$Q_\psi^c(\sigma) = 4\beta \sum_{-N}^{N} G_\psi(r\beta)\, w\left(\frac{r}{N}\right) \sin 2\pi\sigma r\beta.$$

Michelson interferometers are commonly used for Fourier transform spectroscopy with optical components made of materials appropriate to the spectral region under study. Other devices, such as lamellar grating spectrometers are outlined in some reviews and detailed in some original papers.

Bibliography

GEBBIE H. A. and TWISS R. Q. (1966) *Rept. Progr. Physics* **29**, 729.
VANASSE G. and SAKAI H. (1967) *Advances in Optics*, edit. E. Wolf, **6**, 261.

J. CHAMBERLAIN

FRACTURE OF BRITTLE MATERIALS

1. Brittle Fracture

In engineering usage, brittle fracture is defined as fracture not preceded by plastic yielding across the whole section of the component under stress, but there is no reference to the atomic mechanisms whereby the crack propagates. When a "brittle" crack propagates the crack tip is completely contained in an elastic stress field. With some empirical adjustment originally suggested by Orowan, the particular part of elastic continuum theory developed by Inglis and Griffiths can be used to calculate the strength of a structure containing a crack. The rapidly growing subject of fracture mechanics treats this problem and supplements design procedures based on a plastic yield criterion of failure.

Brittle cracks in the sense defined above can be propagated by cleavage of individual grains or by cleavage-type processes; there are, however, other mechanisms including the formation of microvoids by plastic flow in a small volume near the crack tip. Conversely in a material which ultimately fails by cleavage the fracture may not be brittle in the engineering sense. For example, if an MgO single crystal is pulled it is often possible for slip to occur across the whole specimen before a cleavage crack is initiated. Or in an unnotched mild steel specimen considerable plastic flow throughout the whole specimen may occur before a cleavage fracture is initiated. Thus an essential requirement for brittle fracture in the engineering sense is the presence of a notch or flaw or other strain-concentrating feature which either creates the local stress required to initiate cleavage fracture or confines plastic flow to a small region near the crack tip.

A. A. Griffith first realized that an infinitely sharp crack in a linear elastic continuum which gave an infinite local stress concentration under load would not propagate unless the rate at which energy was lost from the elastic stress field was always greater than the rate of increase of surface energy of the faces of the crack. He showed that for propagation a crack must be greater than a critical size which was smaller the higher the applied stress and the smaller the surface energy. For a crack one atom diameter in length the Griffith equation indicates that fracture occurs at a stress of one tenth of Young's Modulus. If flaw free material does not yield plastically under such a stress it will not fracture. Freshly drawn glass fibres and flame polished alumina single crystals have approached such strengths. For propagation of a slit shaped crack in an infinite linear elastic medium, Griffith derived the equations $\sigma^2 c \geqq 2E\gamma/\pi$ for plane stress (1) and $\sigma^2 c \geqq 2E\gamma/\pi(1-\nu^2)$ for plane strain (2), where E is Young's Modulus, γ is the surface energy, that is the free energy per unit area of crack surface, σ is the stress at infinity applied normal to the plane of the crack, ν is Poissons ratio and $2c$ is the crack width. The stress required to propagate a crack decreases as the crack width increases, so that propagation is unstable. During crack propagation in conventional structural metals an amount of energy a few orders of magnitude greater than the surface energy is absorbed by nonconservative processes such as plastic deformation in the highly stressed region of the crack tip. When a tensile stress is applied to a structure containing a pre-existing crack there is plastic flow and a redistribution of the stress field at the crack tip and a change in the shape of the crack front so that the stress required for propagation may first rise to a critical maximum value and then fall. For a crack of this type Orowan suggested substitution of a quantity which takes into account the extra source of dissipated energy in place of the surface energy term in the Griffith equations. The strain energy release rate with crack extension per unit width of crack front is represented in fracture mechanics by the symbol G. The energy required to create unit area of crack is for equilibrium propagation also equal to G. The subscript I represents the opening mode of fracture and the subscript c represents the critical maximum value reached when the stress for propagation reaches its maximum value. The parameter K is the stress intensity factor of the elastic stress field near the crack tip, again used with the subscripts I and c. K is related to G thus: $G_c = K_c^2/E$ for plane stress (3) and $G_c = K_c^2(1-\nu^2)/E$ for plane strain (4). Substituting in the Griffiths modified equations with G_c in place of 2γ gives the expression $\sigma = K_{Ic}(\pi c)^{-\frac{1}{2}}$ (5). The term K_{Ic} is thus directly proportional to the stress required to propagate an opening mode of crack of a given size to the point of instability. The quantities G_{Ic} and K_{Ic} are used as fracture toughness parameters in design and safety analysis to indicate the resistance of a given material to the opening mode of crack propagation in the presence of penetrations or of flaws. Failure by plastic shear is predicted by conventional design methods but the opening mode of fracture occurs at stresses well below those for general yielding in thick sections or in very high strength materials containing flaws or fatigue cracks. There is considerable evidence that in practice for a given structural material the modified Griffith equation is not obeyed and there is uncertainty over the value of c to be used in the case of a crack with a large plastic zone at its edges. These features have required elaboration of equation (5).

2. The Cleavage Crack

Cleavage occurs in most crystalline solids at or below room temperature, but is significantly absent in f.c.c. metals and in some b.c.c. and h.c.p. metals. The cleavage crack propagates by separation of two adjacent planes of atoms. In a real material Hooke's law breaks down when the displacement between atoms is about 10 per cent. The stress at the crack tip does not rise above a value of about $E/5$ (Fig. 1.) Elliott and Barenblatt used models which limit the stresses near the crack tip, and found the same conditions for crack stability as those derived from linear elastic continuum models. The surface free energy is the work done in separating the crack surfaces against attractive interatomic forces and is given by the area under the curve in Fig. 1. The energy should be recoverable. Partial reversibility of fracture is observed in mica. There are many examples of perfect elastic fracture where the work of fracture does not depart significantly from the increase in surface energy.

The nature of the interatomic forces is different in metals, ionic solids and covalent solids. The true surface energies of materials of these three types at the same homologous temperature are not, however, very different. If W is the energy of vaporization per unit volume and a is the lattice parameter, then the surface energy per unit area is about $Wa/6$. For alumina the heat of vaporization is about 4500 cal g^{-1}; hence the surface energy is about 10^3 erg cm^{-2}. The process of elastic fracture involves the breaking of bonds between atoms. For refractory solids the bond strength between pairs of atoms is about 1 eV, which again corresponds to a surface free energy of about 10^3 erg cm^{-2}. Cleavage of some single crystals occurs with dissipation of energy of this order, but usually the work of fracture of polycrystalline material is at least an order of magnitude larger than the surface energy (Table 1).

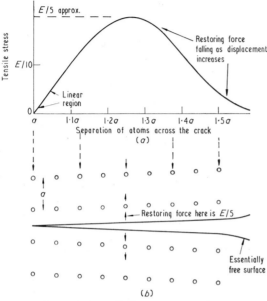

Fig. 1. (a) A force–displacement curve for atoms at the tip of a crack in a metal, showing the linear elastic region, the maximum stress of about $E/5$ and the descending region for strains above 25%. The area under the curve indicates a surface energy of about 10^3 erg cm^{-2} if $E = 2 \times 10^{12}$ dyne cm^{-2} and $a = 2 \times 10^{-8}$ cm. (b) A normal section through a crack tip, showing atomic positions corresponding to the curve in (a).

Table 1

Material	Grain diameter (μm)	γ_F (erg cm^{-2})
Firebrick		3 ×10^4
Polycrystalline aluminas		2–5 ×10^4
Single-crystal alumina		1.2 ×10^4
Polycrystalline magnesia		
Density { 3.48 g cm^{-3}	10	1.6 ×10^4
3.51 g cm^{-3}	50	1.95 ×10^4
3.56 g cm^{-3}	100	3.5 ×10^4
theoretical density	7	4.2 ×10^3
theoretical density	13	8.9 ×10^3
theoretical density	23	1.6 ×10^4
theoretical density	38	1.7 ×10^4
theoretical density	130	1.4 ×10^4
theoretical density	150	7.9 ×10^3

Cracks can be propagated by the same atomic mechanism along surfaces of low cohesive strength such as certain grain boundaries, twin interfaces, or interfaces between different phases in polycrystalline solids.

3. Cleavage Crack Stopping Processes which Raise the Work of Fracture

Examination of a typical fracture surface indicates that cleavage usually takes place on a number of parallel crystal planes with steps between, running approximately in the direction of crack propagation. Steps are thought to begin when the crack intersects screw dislocations. The steps usually form a river pattern, since there is a tendency for steps to run together and join up. Formation of the steps increases the work of fracture since there is extra surface energy and energy is dissipated by plastic flow occurring during the formation of the step. In MgO steps require very little energy when they form by cleavage on the other two planes of the form {100} forming a castellated edge. In zinc large steps form by bending of the bridge forming two sets of mechanical twins, cleavage within the twins then forms a typical serrated edge. A more effective barrier to cleavage crack propagation is provided by a grain boundary that has a twist component. Plastic processes absorbing high energy may occur at the boundary, and the cleavage crack in the next grain may have many steps (Fig. 2).

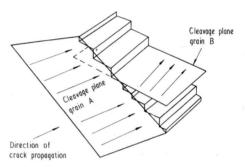

Fig. 2. Intersection of cleavage planes in neighbouring grains at a point. Steps formed on crack surface.

The absence of grain boundaries is partly responsible for ease of crack propagation in glasses. In fine-grain material the crack may be halted after traversing a single grain. The stress at the ends of the crack is then proportional to $d^{\frac{1}{2}}$, where d is half the grain diameter. In zinc and steel under certain conditions the cleavage fracture strength is a function of $d^{-\frac{1}{2}}$ and the stress at the crack tip when it traverses the first grain boundary is approximately constant.

The expected increase in toughness with decreasing grain size in brittle materials is not always found. The opposite relationship can be found in non-porous flaw-free specimens of polycrystalline magnesia and group VIa metals chromium, molybdenum and tungsten. These materials all have the property of readily transmitting cleavage type cracks along the grain boundaries. The cohesive strength of the grain boundary is very dependent on structure and purity, and is lower than that of cleavage in some grades of material.

The presence of second phase particles finely distributed throughout the matrix can have an important effect in making crack propagation more difficult. The extra energy is required either to fracture the particles or to pull out second phase particles or fibres, or energy is absorbed by the creation of extra river lines due to local displacement of the crack front. A common feature seen on the fracture surface of a zinc single crystal is a line normal to the local direction of crack propagation where a large number of river lines appear to originate. This feature arises when the crack front temporarily stops due to relaxation of the applied stress. Plastic deformation occurs at the crack tip in such a way as to introduce screw dislocations ahead of the crack. When the crack intersects this local high concentration of screw dislocations a large number of river lines are initiated. Also the crack tip is blunted and the new crack may form at different levels at the root of the blunted crack. This process again introduces steps on the surface of the crack. The lines run together and are usually annihilated because there are equal numbers of steps of both signs.

Extra work is absorbed by the requirement to shear, tear or fracture material which is to form the steps. Because of the extra energy requirement the crack travels more slowly thus allowing more plastic deformation at the crack tip. The individual processes absorbing energy are thus not separable, but are cooperative, so that a sharp brittle to ductile transition is possible.

4. Crack Interactions

A. A. Griffith first showed that a crack propagating under a uniaxial tensile stress could be stopped by a barrier crack lying on a plane parallel to the crack front and parallel to the tensile stress. The applied stress does not produce stress concentrations at the tips of the barrier crack (Fig. 3a). Barriers of this kind interfere with crack propagation in two-phase structures if the constituents are poorly bonded. If there is a preponderance of cracks parallel to the applied tensile stress, as is often the case with fibrous material, the material will be very resistant to fracture (Figs. 3b and 3c). In polycrystalline aggregates of a material whose individual grains have a single well-marked cleavage plane, fracture in tension can be impeded by the presence of grains with cleavage planes lying parallel to the direction of the applied stress (Fig. 3d). The effectiveness is increased if the grain size is small. Thus coarse-grained zinc is brittle, while fine-grained material is tough. In the case of graphite the crystals are distorted such that as a cleavage crack runs along a basal plane it becomes diverted into a non-propagating direction (Fig. 3e). Graphite has a high work of fracture, probably because of this phenomenon and the occurrence of multiple cracking. If a crack runs between parallel crack nuclei the stress at the original crack tip is relaxed and new cracks begin to propagate, leaving bridges to be broken (Fig. 3f).

When material from the environment penetrates a crack and screens the attractive forces between the partly separated ions at the tip, stresses at the crack

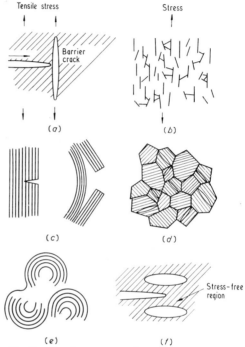

Fig. 3. Barriers to crack propagation. Diagrammatic sections: (a) barrier crack parallel to tensile stress, (b) multiple internal cracks, (c) notch-tough material, (d) cracks stopped at grain boundaries, (e) curved cleavage planes, (f) stress-free region.

tip are reduced. The work of fracture of mica in air is lower than that in vacuo, and the fracture of glass is greatly assisted by the presence of water.

Some dielectrics develop electrical charges of different sign on opposite surfaces of cracks. The long-range attraction increases the work of fracture. Again the effectiveness of this contribution depends on the environment, which may either cause leakage of the charges or reduction of the interaction between them.

Plastic deformation may cause blunting of a crack tip and relaxation of the stress field near the crack tip. Whether the crack continues to propagate then depends partly on the rate of work hardening in the plastic zone.

5. Crack Nucleation

Besides the presence of a pre-existing crack there are other ways in which fracture may be initiated. For example, a void lying on a grain boundary has at equilibrium a lenticular shape with a sharp edge lying along the boundary and so could act as a crack nucleus. This may explain why polycrystalline ceramics are more fragile when porous and when they are coarse grained. Any form of heterogeneous plastic flow is also likely to give rise to intense local stresses. A coarse slip band meeting an obstacle such as a grain boundary creates a high local tensile stress and nucleates a crack unless the stress is dissipated by further plastic flow. This process has been observed in a transparent MgO bicrystal. Over a range of conditions mild steel undergoes cleavage fracture at the stress at which yield occurs, so providing circumstantial evidence that slip nucleates fracture. A crack can also be nucleated when two edge dislocations interact. In b.c.c. iron $\frac{a}{2}[\bar{1}\bar{1}1] + \frac{a}{2}[111] \to a[001]$, i.e. two slip dislocations can coalesce to form an edge dislocation lying in the cleavage plane which can crack open under the applied tensile stress (Fig. 4).

At a free surface of a solid there can be no component of tensile stress normal to the surface. The

6. Fracture by Microvoid Formation

Ductile fracture propagates in a manner entirely different from that of cleavage. In the region of triaxial tension just ahead of the crack-tip voids are formed by plastic flow (Fig. 5), usually nucleated by

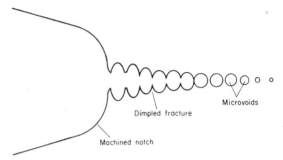

Fig. 5. Diagram of a section through a crack tip propagating by microvoid formation. The voids are typically 1–10 µm diameter and are irregular in shape.

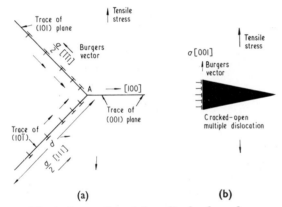

Fig. 4. Intersection of two slip bands and coalescence of edge dislocations to form a crack in the (001) plane in body-centred cubic metals, illustrating the reaction $5\frac{a}{2}[111] + 5\frac{a}{2}[\bar{1}\bar{1}1] \to 5a[001]$. (a) Section showing 5 partial dislocations on each of two slip planes moving towards the point A. The Burgers vectors do not lie in the plane of the diagram. (b) Dislocations coalesced to form a crack.

hard particles which disturb the pattern of plastic flow. The presence of particles may also assist void formation by providing a surface of low cohesive strength which separates in tension, or by undergoing fracture with the halves of the fractured particles moving apart by plastic flow of the matrix. The individual voids then join up by necking down of the matrix material between the voids. In the presence of a notch or flaw plastic deformation is limited to a thin layer along the path of the crack, and in the engineering sense the fracture is brittle.

Dimpled fracture in plane strain or plane stress occurs at low temperatures in f.c.c. metals and alloys, in most h.c.p. and some b.c.c. metals and alloys. While dimpled fracture occurs at temperatures above the so-called brittle–ductile transition temperature for each metal showing cleavage at low temperatures. Even when cleavage occurs dimpled fracture is common at the grain boundaries.

The work of fracture for ductile dimple formation depends largely on the rheological properties of the matrix, and is likely to be high when both the yield stress and the rate of work hardening are high. Under plane stress conditions there is a large contribution to the work of fracture from macroscopic plastic shear in the matrix which does not contribute directly to the propagation of the crack.

7. Grain Boundary Fracture

At elevated temperatures cracks in polycrystalline solids tend to run along the grain boundaries. The cracks propagate slowly—at a rate controlled by some thermally activated process. Typical examples of such behaviour are shown by UO_2 above 800°C and by austenitic stainless steels at 650–950°C. The

maximum tensile stress of the surface is therefore, limited to a value related to the plastic yield stress, and may not be high enough to initiate fracture. Within the solid particularly at a position below a notch the lateral contraction that accompanies plastic elongation is prevented and tensile stresses normal to the applied tensile stress develop. The maximum stress can therefore be higher before yielding occurs and may be high enough for fracture to occur.

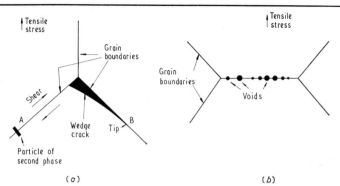

Fig. 6. Grain boundary fracture: (a) wedge crack growing from a triple point, (b) cavitation.

incidence of grain boundary fracture is extremely sensitive to trace impurities and to heat treatment. Fracture is associated with grain boundary shear and is of two main types, wedge-type cracking and cavitation (Fig. 6). Wedge-type fracture occurs at the lower end of the critical temperature range and at higher stresses, while cavitation takes over at the higher temperatures where creep stresses are lower.

The wedge crack develops typically from the junction of three grain boundaries. Sliding along one boundary throws a tensile stress across the second boundary, causing a wedge-shaped crack to propagate along the latter boundary. Whether the wedge crack grows in the same way as a cleavage crack or by surface diffusion is not known. A treatment formally similar to the Zener slip band pile-up model for cleavage fracture can be applied, but the rate of growth is controlled by the rate of thermally activated shear of the boundary at A and perhaps by surface diffusion at the crack tip at B (Fig. 6). Crack growth can be reduced by pegging the boundary at A with hard precipitate particles or reducing the grain size so that the length of unsupported boundary is reduced. Crack growth can be affected by the presence of material at the boundary which alters the grain boundary energy, or which alters the surface energy of the crack surfaces so that the work of fracture is reduced. Traces of lead promote grain boundary fracture of copper and traces of lead, bismuth, thallium and lithium in amounts of 10 ppm or by considerably smaller amounts of helium affect high temperature fracture of Nimonics. The impurities are assumed to be largely segregated in the grain boundaries forming a monatomic layer on the boundary, or on precipitates in the boundary, or forming non-wetted precipitates which nucleate wedge-cracks or voids. The presence of an inert gas under pressure in a cavity may also initiate fracture.

In the case of cavitation failure, work done by the applied stress is converted to surface energy by causing the growth of near spherical cavities. For equilibrium $\sigma \, dV = \gamma \, dS$ and hence $\sigma = 2\gamma/r$ where σ is the equilibrium stress, r the cavity radius, V the volume and S the surface. If $r = 10$ Å and the surface energy $\gamma = 10^3$ erg cm^{-2} then $\sigma = 2 \times 10^{10}$ dyne cm^{-2}, while at a nucleus of radius 10^{-4} cm the equilibrium stress is only 2×10^7 dyne cm^{-2} (300 lb in^{-2}).

In the temperature range where surface diffusion occurs at a rate sufficient to allow the shape of a cavity to change under stress, but where volume diffusion is negligible, the Griffith criterion will show whether a particular applied stress would cause the cavity to change in shape. The criterion for flattening when a uniaxial tensile stress is applied is the same as that for propagation of the penny-shaped crack. The propagation mechanism, however, is quite different, and hence the rate of propagation is also different.

8. Fatigue Fracture

Under alternating stresses a sharp crack can form in a local region such as a slip band where alternating plastic flow is occurring. The crack then propagates a short distance during each cycle by mechanisms involving ductile or cleavage processes or both caused by alternate plastic stretching and compression in a small volume near the crack tip. Eventually as the section bearing the load decreases the crack may propagate as a fast tensile crack by the microvoid mechanism. This is a common cause of failure in service, but the early stages are usually initiated by the presence of a defect within the material such as a quenching crack or stress corrosion crack or a hard inclusion or by a scratch on the surface.

9. Stress Corrosion Cracking

Certain solids when stressed and immersed in certain fluids undergo brittle fracture at low stresses, and in some instances normally ductile materials such as f.c.c. metals and alloys fracture with very little general plastic flow. These effects are observed with fluids which dissolve or chemically attack the solid very slightly and perhaps selectively. In some f.c.c. alloys which are normally very ductile failure can occur by cracking along grain boundaries or along slip bands on {111} planes when they are immersed in warm aqueous salt solutions. Phenomena which are thought to play a part in stress corrosion cracking and which are related to features important in normal cleavage fracture are listed below.

(i) When the medium attacks the grain boundaries fairly severely in the absence of stress, notches are formed which act as stress concentrators.
(ii) Coarse slip in a metal may cause cracking of a brittle corrosion film on its surface.
(iii) The effective surface energy of a solid may be decreased by an adsorbed layer.
(iv) Preferential solution of material at the root of a notch may occur because it is either highly stressed or is undergoing plastic strain. The crack propagates by physical or chemical solution of the material at its tip.
(v) The root of the crack may be preferentially attacked because the chemical state of the fluid is different there. Pitting attack at certain points is associated with a protective film over the rest of the specimen. The specific action of certain ions in promoting cracking needs special explanation. Stress or strain at the root of the notch can increase the rate of attack by increasing the potential difference between the anodic and cathodic regions.
(vi) In the case of ceramics which normally fracture in a brittle manner the nature of the environment may either make fracture easier or more difficult. If the fluid attacks or dissolves the solid fairly rapidly and uniformly it may eliminate sharp microcracks from the surface layers and raise the fracture stress.

See also: Brittleness in ceramics.

Bibliography

GREENOUGH G. B. (1961) Fracture, normal stress law for, in *Encyclopaedic Dictionary of Physics* (J. Thewlis Ed.), **3**, 286, Oxford: Pergamon Press.
McCLINTOCK F. A. (1961) Fracture of solids, in *Encyclopaedic Dictionary of Physics* (J. Thewlis Ed.), **3**, 287, Oxford: Pergamon Press.
PUGH S. F. (1967) Fracture of brittle materials, *Brit. J. Appl. Phys.* **18**, 129–162.
TETELMAN A. F. and McEVILY A. J. (1967), *Fracture of Structural Materials*, London: Wiley.

S. F. PUGH

FREEZE-COATING. Freeze-coating is a method of applying a coating of one material on to another in a continuous process involving the removal of latent heat from the coating material by using the substrate as a heat sink. It is therefore related to *continuous casting* but differs from it in the use of the substrate rather than a cooled mould as the heat sink and by the fact that the thickness of the coating is a function of the physical properties of the materials involved and of the time of passage of the substrate through the bath of molten coating material.

It is applicable to any pair of materials in which the material used for coating has a reasonably sharp melting region, and the substrate can provide a heat sink sufficient to produce freezing of the coated layer.

Although the process belongs to technological antiquity there appears to be no use of it other than for candle-making until quite recently. The art of making candles by dipping is well known but there is no record of the thickness produced being other than empirically determined. This may be because the problem of heat transfer involved (Stefan's problem (Carslaw and Jaeger, 1959)) is only capable of analytical solution under very restricted conditions and numerical solutions have had to await the modern computer.

Two familiar processes which involve dipping a cold substrate into a bath of molten metal are tinplating, dating back to the 16th century, and galvanizing. In both of these, however, the thickness of the final coat is controlled by rollers which wipe off excess metal, the aim being to produce the thinnest coating compatible with the purpose to which it is to be put. In addition, interaction between coating and substrate is usually desirable in these processes and relatively long times of dwell in the coating bath are normal.

Two recent applications of freeze-coating have been discovered in the last five years. These are (a) dip-forming, a process for continuously producing a strand of copper rod at high speeds (70 m/min) developed by the General Electric Company of the United States, and (b) the freeze-coating of fine filaments of silica or stainless steel, developed by Rolls-Royce Limited, Derby, as part of an investigation of fibre reinforced metals. The large difference in the diameters of the substrate rods (9·5 mm for the copper and 50 μ for the silica) in the two processes illustrates the universal validity of the principles involved.

Theory of Freeze-coating

To a first approximation the thickness of the frozen layer is proportional to the square root of the time. This follows from the form of the equation of heat conduction in one dimension

$$\frac{\partial v}{\partial t} = \varkappa \frac{\partial^2 v}{\partial x^2}$$

(where v is the temperature at point x and time t and $\varkappa$ the diffusivity) and is, in fact, an exact relation for the solution (Neumann's solution) of the problem when the substrate surface is maintained at a constant temperature. The initial growth rate of the coating in any practical case should therefore follow the square root law to a first approximation and this appears to be so for the published curves for both the G.E. and the R.R. processes. It is also true for the initial growth rate of the shell of frozen metal in a continuously cast ingot where cooling is from the outside.

In the general case, the equation to be solved, for cylindrical symmetry, is

$$\frac{\partial v}{\partial t} = \varkappa \left(\frac{\partial^2 v}{\partial r^2} + \frac{1}{r} \frac{\partial v}{\partial r} \right)$$

where v is now the temperature at radius r and time t and $\varkappa$ the diffusivity, with the boundary conditions

$$K_1 \left(\frac{\partial v_1}{\partial r} \right) = K_2 \left(\frac{\partial v_2}{\partial r} \right)$$

(at the substrate interface)

and

$$K_2 \left(\frac{\partial v_2}{\partial r} \right) - K_3 \left(\frac{\partial v_3}{\partial r} \right) = L\varrho \frac{dR}{dt}$$

(at the solid–liquid interface)

In these equations K_1, K_2, K_3 are the thermal conductivities for the substrate, the frozen coating and the liquid coating respectively and L, ϱ the latent heat and density respectively of the coating material.

The radius of the solid–liquid boundary is denoted R.

There is no analytical solution known to this equation under these boundary conditions, the few exact solutions involving rather unrealistic boundary conditions. Murray and Landis (1959) review the methods and solutions and propose a method for numerical integration. Horvay (1965) uses an analogue computer method while the author (1967) has used a simple finite difference method with a digital computer.

All numerical solutions approximate to the physical result which is that a cold bar or rod (fibre) entering a molten metal bath freezes on to its surface a thin layer of the metal. If it is left in the bath for a sufficiently long period of time thermal equilibrium will be attained and the metal will melt off again. The same argument applies of course to any frozen-on material such as wax or ice, but to date only metal-coating has been studied.

General Features of the Thickness–Time Curves (Fig. 1)

1. The initial rate of rise is proportional to the diffusivity of the substrate rod or fibre. For example, it is 40 times greater for steel than it is for a poorly conducting glass such as silica. The factor $\varkappa t/a^2$, where a is the radius of the substrate rod, must be assumed to be near unity for maximum coating thickness. Hence the time of immersion required for maximum coating varies as the square of the radius of the substrate rod.
2. The maximum coating thickness is proportional to the temperature differential between fibre and bath.
3. The melt-off rate, which is linear soon after the maximum has been passed (because the entire fibre cross section is now at bath temperature), is proportional to the degree of superheat.

The G.E. Dip Forming Process

In this process, originally developed for copper, a 9·5 mm diameter rod enters the lower end of a crucible containing the molten copper and is removed together with the frozen-on layer from the upper end.

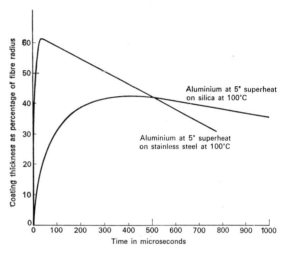

Fig. 1. *Calculated thickness–time curves for two different substrates.*

The process operates at speeds of 70 m/min or more and the 9·5 mm rod becomes coated to a total thickness of 16 mms in a continuous manner.

The process is directly coupled to a preceding melting operation and to subsequent hot rolling and coiling so that, if desired, the 16 mm rod may be rolled down to 9·5 mm for recycling into the crucible.

Yields of up to 5 tons/hour are claimed for the process which is particularly applicable to small, low investment units.

G.E. point out that the concept applies also to rods, bars and slabs; for copper, brass, aluminium and steel and for cladding a steel core with metals of lower melting point such as copper, brass or aluminium.

With the relatively large diameters of rod used in the G.E. process the main technological difficulties are the design of and material for the orifice at the lower end of the crucible (since metal must not be allowed to flow out), the cleanliness of the input rod and the melt (in particular as regards gaseous impurities) and the avoidance of gas being carried in along the rod as it enters the crucible—this is done by using a vacuum chamber below the crucible.

Comparison of the G.E. process with other continuous casting processes is made in a symposium on continuous casting held in 1965 (University of Birmingham Symposium, 1965).

The Freeze-coating of Fine Filaments

In the Rolls-Royce process for coating fine filaments with metal the fibre, of diameter 50–120 μ, is passed downwards through the molten metal pool at 100 m/min. A requirement in the first development of the process was that the silica glass fibres used should not be damaged by contact with any surface before they received the metal coating. This led to the use of a twin bore slotted tube to feed metal to both sides of the fibre, the main pool being supported below this tube as a meniscus. Surface tension is sufficient to prevent the metal from flowing freely and no die is therefore necessary as is the case with the larger diameter rods in the G.E. process. An inert gas blanket was used to prevent "blobbing" (large lumps of metal adhering to the fibre at irregular intervals).

Very accurate control of temperature proved to be necessary in order to control the coating quality since too high a temperature at the slot led to patchy coating and poor fibre strength while too low a temperature gave freezing and fibre breakage.

Coatings of about 50 per cent by area (of the composite fibre) were obtained and the process, while primarily used for making aluminium coated silica fibre for compaction into a fibre reinforced aluminium composite, was also used for coating stainless steel wire with aluminium.

Provided that the fibres can be obtained in suitable lengths freeze-coating is a useful preliminary to the making of fibre reinforced metal composites. This is because the surface is protected, controlled chemical reaction at the interface may be obtained and, above all, a uniform coating leads to a uniform dispersion of fibres when the composite is made.

Conclusions

Recent work, both theoretical and technological, has established freeze-coating (or dip-forming) as a basic industrial process capable of considerable refinement and a reasonable measure of accuracy. It remains to be seen whether it is of economic value in the fields where it could be applied such as wire-making, cladding or the continuous casting of metals.

Bibliography

ARRIDGE R. G. C. and HEYWOOD D. (1967) The freeze-coating of filaments, *Brit. J. Appl. Phys.* **18**, 447.

CARSLAW H. S. and JAEGER J. C. (1959) *Conduction of Heat in Solids* (2nd ed.) Oxford: O.U.P.

HORVAY G. (1965) The Dip-forming process, *Trans. A.S.M.E.* Series C (*J. Heat Transfer*) **87**, 1–16.

MURRAY W. D. and LANDIS F. (1959) Numerical and machine solutions of transient heat conduction problems involving melting or freezing, *Trans. A.S.M.E.* Series C (*J. Heat Transfer*) **81**, 106.

UNIVERSITY OF BIRMINGHAM SYMPOSIUM (1965) *Continuous Casting of Ferrous and Non-ferrous Materials* (published by Light Metals and Metal Industry).

R. G. C. ARRIDGE

GEOCHRONOLOGY (RADIOMETRIC DATING).
Prior to the discovery of radioactivity, no reliable method existed by which either the true ages of rocks and minerals of the Earth's crust or the age of the Earth itself could be determined. Attempts to determine the ages of rocks were often restricted to estimates of the time required for a certain thickness of sediments to be deposited in a marine environment. Further information concerning relative age was obtained by a study of the fossil record: the process of organic evolution makes it possible to correlate fossiliferous rocks at widely separated parts of the Earth's crust.

The discovery of radioactivity marked the beginning of a new era in geochronology and for the first time the age of a geological event (e.g. the time in the past when an igneous rock crystallized from a molten magma, when sediments were deposited, or when both were affected by later thermal or dynamic events—metamorphic rocks) could be determined by utilizing the known rate of decay of a radioactive nuclide. By determining the concentration of both the parent radionuclide and the daughter decay product, together with a knowledge of the decay constant, the time required for the daughter product to accumulate in a mineral can be calculated. The principal natural radioactive nuclides used in geochronology together with relevant geochemical data, are given in Table 1.

In isotopic geochronology it is convenient to measure time backwards from the present. Consequently, the basic equation of geochronology is

$$\frac{D}{P} = e^{\lambda t} - 1 \qquad (1)$$

where D is the present number of atoms of daughter nuclide in a sample of rock or mineral formed since time t,
P is the present number of atoms of parent nuclide in a sample of rock or mineral,
λ is the decay constant of parent nuclide,
t is the time measured from the present.

Hence we have:

$$t = \frac{1}{\lambda} \log_e \left(1 + \frac{D}{P}\right) \qquad (2)$$

where t is the apparent age of the rock or mineral.

Provided that there has been no net migration of parent or daughter into or out of the mineral since it was formed, t is the true age. For a few favourable minerals it is relatively simple to undertake quantitative assay for both parent and daughter nuclides such as in the instance of measuring the total amount of radiogenic lead in a pure uranium mineral (e.g. uraninite). In most instances, the radiogenic isotopes are "contaminated" by the presence of an isotope of the "common element", and a correction for this component is necessary: e.g. radiogenic ^{40}A is inevitably contaminated with a small amount of atmospheric argon which contains 99·6 per cent of ^{40}A.

Decay Constants

In many minerals it is possible to measure the concentration of both parent and daughter nuclides accurately, but the remaining crucial value for the decay constant is not always known with sufficient accuracy. The half-lives for ^{238}U and ^{235}U are accepted as being reliable although improvements are still possible. The half-life of ^{232}Th is known with less certainty. At present, the half-lives of ^{238}U, ^{235}U and ^{232}Th are probably accurate to about 2 per cent. There is still appreciable uncertainty for the best values for the decay rates for ^{40}K, ^{87}Rb and ^{187}Re. Potassium-40 presents special difficulties by having two decay constants, one for electron capture (λe) and the other for β decay ($\lambda \beta$). In the potassium/argon method the calculated age of a rock is most sensitive to errors in λe which is the more difficult to measure. By measuring the appropriate daughter : parent ratios in minerals of known age, i.e. co-genetic with accurately dated uranium or thorium bearing minerals, slightly greater estimates of the half-lives have been obtained than those from nuclear counting techniques. The half-life for rubidium-87 obtained by the geological approach compared with the best values from counting methods differs by 6 per cent.

In calculating the age of a rock or mineral from the measured daughter : parent ratio and the appropriate decay constant, a number of assumptions must be made. The calculated age will only be equal to the true age if:

(a) The rock or mineral was formed in an interval of time short compared with its age.

Table 1. Naturally occurring radionuclides used in geochronology

Parent radioactive nuclide	Isotopic composition parent "common" element (%)	Daughter nuclide	Isotopic composition daughter "common" element (%)	Half-life (years)	Materials used for dating enriched in radioactive parent nuclide	Average crustal concentration
^{40}K	^{39}K 93.08 ^{40}K 0.0119 ^{41}K 6.91	^{40}A	^{36}A 0.337 ^{38}A 0.063 ^{40}A 99.600	λ_e 1.19×10^{10}	Muscovite, Feldspar, Hornblende, Biotite, Pyroxene, Nepheline, Leucite, Glauconite,	K : 2.1%
		^{40}Ca	^{40}Ca 93.08 ^{42}Ca 0.64 ^{43}Ca 0.145 ^{44}Ca 2.06 ^{46}Ca 0.0033 ^{48}Ca 0.185	λ_β — 1.47×10^9	Sylvite, Whole rock	
^{87}Rb	^{85}Rb 72.15 ^{87}Rb 27.85	^{87}Sr	^{84}Sr 0.56 ^{86}Sr 9.86 ^{87}Sr 7.02 ^{88}Sr 82.56	λ_β — 4.7 to 5.0 $\times 10^{10}$	Biotite, Muscovite, K-Feldspar, Glauconite, Whole rock	Rb : 90 ppm Sr : 375 ppm
^{232}Th	^{232}Th 100	^{208}Pb	^{204}Pb 1.55 ^{206}Pb 22.51 ^{207}Pb 22.60 ^{208}Pb 53.34	1.39×10^{10}	Thorianite, Monazite, Zircon	Th : 10 ppm
^{235}U ^{238}U	^{234}U 0.0058 ^{235}U 0.71 ^{238}U 99.285	^{207}Pb ^{206}Pb		7.13×10^8 4.51×10^9	Uraninite, Monazite Zircon	U : 3 ppm Pb : 13 ppm
^{187}Re	^{185}Re 37.07 ^{187}Re 62.93	^{187}Os	^{184}Os 0.018 ^{186}Os 1.59 ^{187}Os 1.64 ^{188}Os 13.3 ^{189}Os 16.1 ^{190}Os 26.4 ^{192}Os 41.0	4.3×10^{10}	Molybdenites, Copper ores	Re Os : <0.001 ppm

Notes: Radiogenic daughter underlined. λ_e refers to decay by electron capture. λ_β refers to β decay.

(N.B. The following examples of natural radioactivity have not been used for radiometric dating

Decay	Half-life (years)
^{115}In (95.77%) $\xrightarrow{\beta^-}$ ^{115}Sn	$\sim 6 \times 10^{14}$
^{130}Te (34.49%) $\xrightarrow{\beta^-}$ ^{130}Xe	$\sim 10^{21}$
^{138}La (0.089%) $\begin{cases} \xrightarrow{e} {}^{138}\text{Ba} \\ \xrightarrow{\beta^-} {}^{138}\text{Ce} \end{cases}$	$\sim 7 \times 10^{10}$
^{176}Lu (2.6%) $\begin{cases} \xrightarrow{e} {}^{176}\text{Yb} \\ \xrightarrow{\beta^-} {}^{176}\text{Hf} \end{cases}$	$\sim 2.4 \times 10^{10}$
^{147}Sm (15.1%) $\xrightarrow{\alpha}$ ^{143}Nd	$\sim 1.25 \times 10^{16}$
^{209}Bi (100%) $\xrightarrow{\alpha}$ ^{205}Tl	$\sim 2.7 \times 10^{17}$

(b) There has been no gain or loss of either parent or daughter nuclide in a rock or mineral other than that by natural radioactive processes.
(c) Appropriate corrections are made to allow for any daughter product in the rock or mineral at the time of its formation.
(d) The decay constant is known with sufficient accuracy.

Method of Measurement

Apart from carbon-14, together with various cosmogenic nuclides, the determination of parent and daughter nuclides are generally made by mass spectrometry. Routine isotopic analyses using the technique of isotope dilution analysis can be made on a few micrograms of lead, about 0·1–1 μg of strontium, while much smaller quantities of uranium, rubidium, potassium or argon can be measured with ease.

Radiometric Dating Methods

In radiometric dating, the term "radiogenic" may be defined as the difference between the atomic abundance at the present day and the initial atomic abundance of a daughter nuclide, e.g.

$$\left\{\begin{array}{l}\text{Atomic} \\ \text{abundance of} \\ \text{daughter, today}\end{array}\right\} - \left\{\begin{array}{l}\text{Atomic} \\ \text{abundance of} \\ \text{daughter, initial} \\ \text{(e.g. age of earth,} \\ \text{rock, mineral,} \\ \text{or an event)}\end{array}\right\} = D_{radiogenic}$$

$$D_{radiogenic} = P \frac{\lambda_D}{\lambda} (e^{\lambda t} - 1)$$

where P is the atomic abundance of the parent today,
λ_D is the decay constant for the decay Parent → Daughter
λ is the total decay constant

Single radiometric ages are of limited value and in terms of investigating geological problems it is essential to date several rocks or minerals by more than one method. The U-Pb and Rb-Sr methods are particularly susceptible to errors arising from failure to account for the incorporation of the daughter elements in a rock or mineral at the time of its formation. Both lead and strontium have non-radiogenic isotopes which can be used to make the appropriate corrections. Strontium-87 in a rock is derived from two sources: that present as a constituent isotope of common strontium $^{87}Sr_{initial}$ and radiogenic $^{87}Sr^*$ formed by the decay of ^{87}Rb.
Therefore,

$$^{87}Sr_{total} = {}^{87}Sr_{initial} + {}^{87}Sr^*_{radiogenic}$$
$$= {}^{87}Sr_{initial} + {}^{87}Rb\,(e^{\lambda t} - 1). \quad (3)$$

Using non-radiogenic ^{86}Sr as a monitor for common strontium

$$(^{87}Sr/^{86}Sr)_{today} = (^{87}Sr/^{86}Sr)_{initial} + (^{87}Rb/^{86}Sr)_{today}\,(e^{\lambda t} - 1) \quad (4)$$

Therefore

$$t = \frac{1}{\lambda} \log_e \left[1 + \frac{(^{87}Sr/^{86}Sr)_{today} - (^{87}Sr/^{86}Sr)_{initial}}{(^{87}Rb/^{86}Sr)_{today}}\right] \quad (5)$$

A similar age equation can be derived for the system U → Pb, and Th → Pb using non-radiogenic ^{204}Pb, e.g.

$$t = \frac{1}{\lambda} \log_e \left[1 + \frac{(^{206}Pb/^{204}Pb)_{today} - (^{206}Pb/^{204}Pb)_{initial}}{(^{238}U/^{204}Pb)_{today}}\right] \quad (6)$$

In the above equations it is not necessary to know how much strontium or lead was incorporated into the rock or mineral when it was formed, but only the isotopic composition of the common strontium or lead. In many instances it is not always possible to obtain a rubidium or uranium free mineral to determine the initial isotopic composition of the common Sr or Pb, but provided the isotopic composition has a unique value at $t = 0$ it is possible, by selecting minerals having different Rb/Sr or U/Pb ratios, to solve arithmetically or graphically for age and initial isotope ratios. In the graphical solution (equation 4) when the abundance ratios for two or more pairs of nuclides are plotted for a number of samples of the same age the plot is referred to as an *isochron*. Conformity of the graph, in practice a straight line,

$$(y = mx + c)$$

where $y = (^{87}Sr/^{86}Sr)_{today}$
$m = $ (slope) $= e^{\lambda t} - 1$
$x = (^{87}Rb/^{86}Sr)_{today}$
$c = $ (intercept on y axis) $= (^{87}Sr/^{86}Sr)_{initial}$

to a predicted model for isotope development serves to test both the validity of the model and presumed contemporaneity of the samples.

A least squares analysis for several samples having different Rb : Sr ratios provides an age from the slope, and the initial $^{87}Sr/^{86}Sr$ ratio at the time the rock or mineral was formed is obtained from the intercept of the x-axis as illustrated in Fig. 1.

Potassium/Argon

Of the three isotopes of potassium, ^{39}K, ^{40}K, ^{41}K, only ^{40}K is radioactive. Approximately 11 per cent of all transmutations will yield ^{40}A, the remainder give rise to calcium-40. Except for a few calcium free minerals, such as sylvite (KCl), the $^{40}K : {}^{40}Ca$ ratio cannot be used for dating as the abundance of ^{40}Ca constitutes 93·08 per cent of all calcium isotopes, making the accurate measurement of any radiogenic component very difficult.

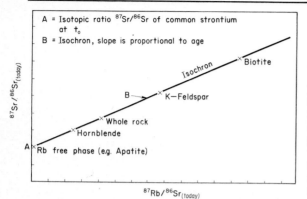

Fig. 1. Isochron plot for a; granite (a typical granite consists of 20–50 per cent quartz, 15–50 per cent feldspar, 1–15 per cent biotite, hornblende, 1 per cent apatite).

The age equation for potassium/argon dating is

$$t = \frac{1}{\lambda_e + \lambda_\beta} \ln\left[1 + \frac{\lambda_e + \lambda_\beta}{\lambda_e} \frac{^{40}A}{^{40}K}\right] \quad (7)$$

Potassium is often determined by flame photometry with an estimated precision of 0·5 per cent and an accuracy of ±1·5 per cent and argon by mass spectrometry with a precision of ±2 per cent and an accuracy of ±1 per cent. The overall accuracy of the method is ±10 per cent for samples that are not contaminated by atmospheric argon.

The K/A ages are based on the assumption that any ^{40}A derived from the decay of ^{40}K is not lost from the mineral since the time of the event being studied. This also assumes that there is no inherent ^{40}A in the system. In a few minerals such as chlorite, beryl, cordierite, and tourmaline, extraneous argon is incorporated as a result of crystal structure. However, these minerals are not commonly used for dating. Argon is a noble gas and its retention in minerals will be dependent upon crystal structure and degree of chemical or physical alteration. The retention of argon and diffusion rates of argon are characteristic for individual mineral species.

The K/A method can be used to date the time at which a magma solidified such that diffusion of argon is insignificant, or subsequent periods of re-heating (metamorphic events). As different minerals retain argon under the influence of various external stimulii to varying degrees the choice of a mineral is important to date a specific event.

The K/A method can be applied to the whole range of geological time, but Pre-Cambrian dates generally indicate the effects of post crystallization secondary heating processes. The method is particularly valuable for Mesozoic, Cenozoic volcanic rocks, and those of Pleistocene age for which ages of about 6000 years have been obtained, providing an overlap to the range of dating possible by the ^{14}C method.

Rubidium/Strontium

No common minerals contain rubidium as a major element and this trace element occurs in potassium rich minerals such as micas and feldspars. Ages are calculated using equation 5 while the isochron approach (see equation 4) permits dating of rocks having a wide range of Rb/Sr ratios and also provides valuable information concerning the initial isotopic composition of strontium which has an important bearing on the genesis of magma types. The Rb/Sr method is also very useful for dating metamorphic events; during such events strontium tends to re-distribute itself among the mineral components of a rock, and by using the radiogenic strontium as a tracer the nature of the re-distribution may be elucidated. In many cases both the time of a later metamorphism and the time of original intrusion can be determined. Radiogenic strontium in a potassium mineral occupies the site of parent rubidium atoms in the mineral latice and is in an unfavourable position from which it is easily released during metamorphism.

Uranium/Thorium methods

The minerals most amenable to U–, Th–Pb dating are uraninite, monazite and zircon. The three parent isotopes yield ages (see equation 6) derived from the following in decays relations:

$$^{206}Pb = {}^{238}U(e^{\lambda t} - 1) \quad \lambda = 1·54 \times 10^{-10} \text{ years}^{-1}$$

$$^{207}Pb = {}^{235}U(e^{\lambda' t} - 1) \quad \lambda' = 9·72 \times 10^{-10} \text{ years}^{-1}$$

$$^{208}Pb = {}^{232}Th(e^{\lambda'' t} - 1) \quad \lambda'' = 0·499 \times 10^{-10} \text{ years}^{-1}$$

Parent and daughter nuclides are determined by mass spectrometry or radioactivation techniques. Isotopic compositions are determined by mass spectrometry using lead-204 to correct for non-radiogenic lead. As the isotopic composition of common lead varies considerably with both time and place, it is difficult to make accurate corrections if it constitutes more than about 25 per cent of the lead in a particular mineral sample.

If all intermediate members of the decay families are in equilibrium with their parent, ages can be calculated by measuring intermediate members of the same decay series, e.g.

$$^{210}Pb/^{206}Pb, \quad ^{231}Pa/^{230}Th, \quad ^{234}U/^{238}U.$$

The use of helium,

e.g.
$$^{238}U \rightarrow {}^{206}Pb + 8\ {}^{4}He + \text{Energy}$$

$$^{235}U \rightarrow {}^{207}Pb + 7\ {}^{4}He + \text{Energy}$$

$$^{232}Th \rightarrow {}^{208}Pb + 6\ {}^{4}He + \text{Energy}$$

as a measure of the decay of uranium and thorium is

only possible for a few minerals that retain this inert gas to an appreciable extent, e.g. magnetite. Measurements of U/Pb and Th/Pb ages show that there are relatively few for which agreement is obtained; in Pre-Cambrian rocks the following sequence is observed

$$t\left(\frac{^{207}Pb}{^{206}Pb}\right) > t\left(\frac{^{207}Pb}{^{235}U}\right) > t\left(\frac{^{206}Pb}{^{238}U}\right). \quad (8)$$

This discordancy can be explained by consideration of continuous diffusion of lead from the host mineral, diffusion as a result of re-heating, the loss of an intermediate decay member, or the selective loss of uranium, thorium and lead by leaching. Th/Pb ages often differ appreciably from U/Pb and Pb/Pb ages particularly when the thorium content is low and are generally regarded as being inferior.

Nuclear fission track method

In this recently developed method use is made of the fact that ^{238}U decays by spontaneous fission ($\lambda = 6.9 \times 10^{-17}$ years). As a result of fission the heavy fragments give rise to very small tracks of radiation damage in a mineral which can be enlarged by simple selective etching techniques. By determining the number of naturally formed spontaneous fission tracks in an area of a mineral and then determining the uranium concentration of the same area, the relation of daughter to parent is established and an age can be calculated. Fission track ages are susceptible to periods of re-heating and the loss of fission tracks by annealing is a feature of a particular mineral.

Common lead method

This method is used to date lead ores such as galenas (PbS) and common lead extracted from minerals enriched in lead but depleted in uranium and thorium. Any lead ore can be considered as a mixture of primeval lead of fixed isotopic composition and radiogenic lead produced by the radioactive decay of uranium or thorium. In undertaking common lead age determinations it is assumed that very early in the history of the earth the isotopic composition of lead was everywhere the same. When the earth became rigid at time t_0 differing U:Pb and Th:Pb ratios were established. At a time t_m (mineralization) the lead may concentrate to form a lead ore body, the isotopic composition of which has remained constant until the present as, during the geochemical process, lead was efficiently separated from uranium and thorium.

If we assume that since the earth was formed U–Th–Pb have remained within closed chemical systems until a time t_m when lead ores were formed, the number of ^{206}Pb atoms present at time t_m will be

$$N(^{206}Pb)_{t_m} = N(^{206}Pb)_{t_0} + N(^{238}U)_{t_0} - N(^{238}U)_{t_m}. \quad (9)$$

Dividing both sides of the equation by the non-radiogenic lead isotope at mass-204 we have

$$\frac{N(^{206}Pb)_{t_m}}{N(^{204}Pb)} = \frac{N(^{206}Pb)_{t_0}}{N(^{204}Pb)}$$
$$+ \frac{N_0(^{238}U)}{N(^{204}Pb)} e^{\lambda t_0} - \frac{N_0(^{238}U)}{N(^{204}Pb)} e^{\lambda t_m} \quad (10)$$

N_0 = number of atoms at present time,
N = number of atoms at any time, i.e. $N = N_0 e^{\lambda t}$.

In terms of the symbols shown in Table 2, equation (10) may be written:

$$x = a_0 + \mu(e^{\lambda t_0} - e^{\lambda t_m})$$
$$y = b_0 + \frac{\mu}{137.8}(e^{\lambda' t_0} - e^{\lambda' t_m}) \quad (11)$$
$$z = c_0 + W(e^{\lambda'' t_0} - e^{\lambda'' t_m})$$

Table 2. *Symbols used in galena dating*

Isotope ratios	At $t = 0$ (formation of earth)	At present time
$\frac{^{206}Pb}{^{204}Pb}$	$a_0 = 9.56$	x
$\frac{^{207}Pb}{^{206}Pb}$	$b_0 = 10.42$	y
$\frac{^{208}Pb}{^{204}Pb}$	$c_0 = 30.0$	z
$\frac{^{238}U}{^{204}Pb}$	$2.0124\,\mu$	μ
$\frac{^{238}U}{^{235}U}$	3.33	137.8
$\frac{^{232}Th}{^{204}Pb}$	$1.2548\,W$	W

The values for a_0, b_0, c_0 and the age of the Earth are obtained from measurements of the meteorite-Earth system. If at time t_m a lead mineral such as galena is formed, the values for x, y, z will be unique for this event. A plot of y vs. x or z vs. x will define a series of growth curves illustrating how the isotope ratios change with time (Fig. 2). By eliminating $^{238}U/^{204}Pb$, μ from equation (11)

$$\frac{y - b_0}{x - a_0} = \phi = \frac{(e^{\lambda' t_0} - e^{\lambda' t_m})}{137.8\,(e^{\lambda t_0} - e^{\lambda t_m})} \quad (12)$$

and a plot of y vs. x for equation (12) defines a straight line (isochron)

$$y = b_0 + \phi(x - a_0). \quad (13)$$

The isochron passes through the primeval abundances a_0, b_0 and has a slope determined by t_0, t_m. The use of a model (Holmes–Houtermans) obtained from growth rates from an isotopic homogeneous system requires modifications in instances when isotope

Table 3. Geologic time-scale

Era	Period	Duration, million years	Evolutionary time scale	Million years
CENOZOIC	QUATERNARY	~ 1	Age of mammals	0
	TERTIARY	~ 60	Mammals begin to dominate the earth	50
MESOZOIC	CRETACEOUS	~ 70	Primitive mammals Dinosaurs extinct	100
	JURASSIC	~ 45	Peak for dinosaurs First mammals	150
	TRIASSIC	~ 45	Vertebrates replace invertebrates First dinosaurs	200
PALEOZOIC	PERMIAN	~ 50	Vertebrates develop Modern insects appear	250
	CARBONIFEROUS	~ 65	First reptiles	300
	DEVONIAN	~ 60	First terrestrial land trees Abundance and variety of marine vertebrates	350 400
	SILURIAN	~ 20	Rudimentary plants. Few primitive animals	
	ORDOVICIAN	~ 75	First vertebrates Marine invertebrates and arthropods	450 500
	CAMBRIAN	~ 100	Marine arthropods Molluscs Sponges	550 600
PRE-CAMBRIAN			First fossil algae, fungi, sulphur bacteria ↗↘ First biogenic limestones → Oceans formed → Formation of earth →	1,000 2,000 3,000 4,000 5,000

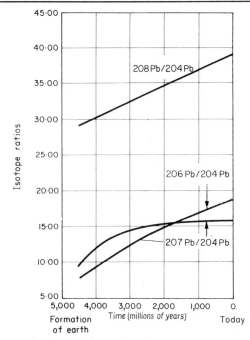

Fig. 2. Curves illustrating relationships between the isotopic composition of lead ores and geologic age.

variations occur and for lead ores that have passed through several periods of complete or partial mixing with other U/Pb, Th/Pb systems. Lead isotopes, like strontium isotopes, may be used to trace the geochemical evolution of continents and represent the integrated geochemical history of the elements lead, uranium and thorium up to the time of the final mineralization.

Rhenium/Osmium method

Rhenium is a rare element and at present its use for dating is restricted to minerals enriched in rhenium such as molybdenites and some copper ores; osmium is restricted to basic rocks and occurs with platinum enriched minerals. The estimated half-life of ^{187}Re is about $4.3 \pm 0.5 \times 10^{10}$ years and when this value has been determined with sufficient accuracy together with an improvement in the method of isotope assay for osmium, this method will be very useful and supplement the U/Th-Pb and Rb/Sr methods.

Carbon-14

Carbon-14 is being continually produced by cosmic rays in the upper atmosphere by the reaction $^{14}N(np) \rightarrow {}^{14}C \rightarrow {}^{14}N$ and becomes incorporated into carbon dioxide of the earth's atmosphere. Carbon dioxide is then available to enter living biological matter which continues until death when the exchange process ceases. The time that has elapsed since death can be determined from the difference between the sample ^{14}C content and that measured for identical living matter today. The half-life for carbon-14 is $5,568 \pm 30$ years and the oldest samples that can be adequately dated are about 70,000 years old. The accuracy of the carbon-14 method is dependent upon the constancy of ^{14}C content of the atmosphere in relation to the total carbon dioxide content of the atmosphere. Recent studies have clearly shown that fluctuations in the ^{14}C content of the atmosphere during historic times have occurred, while fluctuations in prehistoric times have not been established.

Time Scale

One of the purposes of radiometric dating is to establish a quantitative time scale in order to place precise boundaries between the geological periods. The Post-Cambrian time scale, given in Table 3 is now well documented and can be related to the evolution of organic life. A Pre-Cambrian time scale, covering the period 600–3500 million years similar to that developed for Post-Cambrian rocks, does not exist. At present it is only possible to record profound events manifested on a world-wide scale.

Bibliography

AHRENS L. H. (1965) *Distribution of the Elements in our Planet*, New York: McGraw-Hill.

FAUL H. (1966) *Ages of Rocks, Planets and Stars*, New York: McGraw-Hill.

FLEISCHER R. L., PRICE P. B. and WALKER R. M. (1967) *Dictionary of Geophysics* (S. K. Runcorn Ed.), **1**, 529–532, Oxford: Pergamon Press.

GEISS J. and GOLDBERG E. D. (Eds.) (1963) *Earth Science and Meteorites*, Amsterdam: North Holland.

HAMILTON E. I. (1965) *Applied Geochronology*, London: Academic Press.

HAMILTON E. I. and FARQUHAR R. M. (1968) *Radiometric Dating for Geologists*, New York: Interscience.

HOLMES H. (1945) The construction of a geological time scale, *Trans. Geol. Soc. Glasgow* **21**.

HURLEY P. M. (1959) *How Old is the Earth?*, London: Heinemann.

POOLE J. H. J. (1961) Age determination by radioactivity, in *Encyclopaedic Dictionary of Physics* (J. Thewlis Ed.), **1**, 95, Oxford: Pergamon Press.

RANKAMA K. (1954) *Isotope Geology*, Oxford: Pergamon Press.

RANKAMA K. (1963) *Progress in Isotope Geology*, New York: Interscience.

RUSSELL R. D. and FARQUHAR R. M. (1960) *Lead Isotopes in Geology*, New York: Interscience.

SCHAEFFER O. A. and ZÄHRINGER J. (1966) *Potassium Argon Dating*, Berlin: Springer-Verlag.

STARIK I. E. (1961) *Yadernaya Geokhronologiya*, Akad. Nauk S.S.S.R., Moscow.

TILTON G. R. and HART S. R. (1963) Geochronology, *Science* **140**, No. 3565, 357–366.

The Phanerozoic Time Scale (1964) Symposium, *Quart. Journ. Geol. Soc., London* **120**.

E. I. HAMILTON

GLANDLESS RECIPROCATING-JET PUMP

1. Introduction

A pump designed to operate with corrosive liquids at extremely high temperatures will usually be constructed from refractory oxides, and hence, if possible, must not rely on solid moving parts such as valves, pistons and rotors. A simple method of achieving this is to make use of some non-reciprocal hydrodynamic effect.

A conventional jet pump is valveless and uses a fluid jet separately pumped into a suitably shaped main pipe, which carries the fluid to be pumped. The main body of fluid is propelled along this pipe by a momentum transfer and mixing process. The injected fluid must, however, be suitable for pumping by conventional means, be compatible with the pumped fluid, and be acceptable as a diluent. At high temperatures these conditions will usually eliminate all fluid combinations apart from a gas or vapour to pump a liquid, and vice versa. Unfortunately there is then a large mismatch in density which causes considerable energy loss during momentum mixing.

2. Principles of Operation

The technique used in the reciprocating-jet pump is to extract from the main pipe an appropriate amount of the fluid to be pumped, in readiness for re-injection. During this extraction or "suck" stroke the fluid mass removed at a sink or jet nozzle contains very little momentum prior to entering the nozzle. During the injection or "blow" stroke, however, the jet nozzle acts as a source of momentum directed as desired. This is then coupled to the surrounding fluid in the main pipe, with an unavoidable but not excessive energy loss. Over a complete "suck–blow" cycle a net forward thrust is created, which may be used to generate a pressure rise or a velocity increase in the main body of fluid, depending on the shape chosen for the main pipe. A similar principle has been studied for use in ship propulsion.

The non-reciprocal action of an alternating jet depends upon the fact that the flow pattern for a steady suck stroke is fundamentally different from that for a steady blow stroke. This is shown in Fig. 1 and has been observed in a two-dimensional water tank using polystyrene beads as markers. The left-hand jet is sucking and the right-hand one blowing. During the suck stroke irrotational potential flow occurs with a convergent "scalar" flow pattern in which curl $\mathbf{v} = 0$, where $\mathbf{v}$ is the fluid velocity. At the start of the blow stroke a divergent flow pattern occurs momentarily, but because of viscous forces in the boundary layers the liquid cannot sustain the large velocity gradients around the edges of the jet nozzle, and a rotational motion is initiated. The transient potential flow pattern therefore breaks down as inertial forces take over control. Ring vortices are generated around the periphery of the jet nozzle and are shed downstream, so that in the steady state there is a gradual erosion of the jet associated with a widening of its influence as ejected fluid enters a turbulent mixing region. A steady thrust is the ultimate result. The flow pattern of the blow stroke might be regarded as a "vector" pattern in which curl $\mathbf{v} \neq 0$.

The problem of pumping a steady stream of fluid has now been transformed to that of cycling fluid through a nozzle. In a severe environment the latter problem is much more simple to solve, for example by use of a vertically-reciprocating gas–liquid interface stabilized by gravity. This forms a piston in a chamber coupled to the jet nozzle by means of a suitable diffuser. In the case of a pump for liquids, the gas pressure above the liquid surface may be cyclically varied by means of a reciprocating mechanical piston or bellows or by a valve-controlled supply of compressed gas, so as to blow and suck alternately through the jet orifice.

3. Some Possible Designs

The simplest form of pump comprises a pipe with injector tubes or jet nozzles arranged at various points along the main pipe of the pump, and directed in the flow direction (Fig. 1). The pipe and jet nozzles are filled with liquid. Suitably phased alternating pressure pulses are applied to the jet nozzles by one of the methods just described. During the blow stroke, the pressure in the nozzle is greater than in the pipe so that liquid flows out to form a jet with "vector" flow along the pipe. During the suck stroke, the pressure is reversed and an equal amount of liquid flows back into the nozzle to produce a convergent or "scalar" flow pattern.

In the first practical design of reciprocating-jet pump (Fig. 2) the jet tubes are fitted to the pipe like

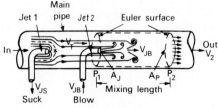

Fig. 1. Simple pump showing "suck" and "blow" patterns.

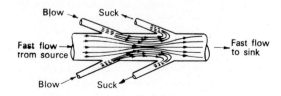

Fig. 2. "Tree" pump for high velocities.

the branches of a tree, with no extension within the pipe. This design has the advantage of simple construction, low resistance to flow in the pipe, and least surface area open to corrosion by the liquid being pumped. The design will be most useful when the requirement is for a high flow rate across a small pressure gradient and when efficiency is of little importance.

The second practical design is the "bulb-venturi" pump, shown in Fig. 3, which includes the air/

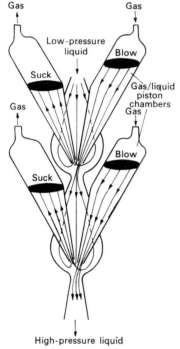

Fig. 3. *Two stage twin-jet bulb-venturi pump.*

liquid piston chambers. The pipe is shaped so as to provide a venturi and diffuser mixing chamber. The main purpose of the former is to accelerate the main liquid stream to a velocity nearly matching that of the jet so that less energy is lost during momentum mixing; the diffuser converts the velocity head back to pressure during the mixing. Each stage incorporates two jets cycling in anti-phase, so as to produce a more even thrust and continuity of mass. A larger number of suitably phased jets per stage would produce smoother flow, but space is limited close to the venturi. The jet nozzles have a rounded profile and a 10° diffuser angle so as to reduce losses and circulation during the suck stroke. Streak photographs of two-dimensional flow patterns in a shallow tank have shown that, with the correct geometry, there is negligible interference between the blowing and sucking jets. This design will be most useful in a system requiring low flow rates across large pressure gradients.

A design, which in section might be similar to Fig. 2 rotated clockwise through 90°, consists basically of a series of similar inverted cones which are concentric with a vertical axis. Each conical cavity thus formed is sealed at the top, with small gas pipe connections, and has a shaped orifice at the bottom directed into the cavity beneath. Initially the top half of each cavity is filled with air and the lower half with liquid which is continuous throughout all cavities. During the first half of a pump cycle, a high air pressure is applied to alternate cavities and a low pressure to the intermediate cavities. Liquid is thus forced into the low-pressure cavities. Owing to the shaping of the orifice and to the creation and ejection of momentum downwards from the high-pressure cavity, liquid is forced preferentially into the cavity below. The second half of the pump cycle involves the reversal of the whole operation in a given cavity, although for the pump as a whole there is a net transfer of liquid downwards against the pressure gradient. Alternative phasing of the "travelling wave" type could be advantageous. This design may be suitable for viscous liquids.

4. Theory

A simple theory, based on separate consideration of quasi-static blowing and sucking, has been found to give reasonable agreement with the results, provided boundary layer corrections are applied. The pump head h_P may be expressed:

$$h_P = nk\bar{v}_{JB}^2 \alpha(1 - \alpha/2 - 2v_P/\bar{v}_{JB})/g \qquad (1)$$

where n is the number of jets,

k is the ratio of blow period to the period of one cycle,

$\bar{v}_{JB}^2$ is the mean square jet velocity during T_B,

α is the effective area ratio of jet nozzle to main pipe (Fig. 1),

v_P is the liquid velocity in the main pipe,

$\bar{v}_{JB}$ is the mean jet velocity,

g is acceleration due to gravity.

At high frequencies, or short wavelengths as defined by the ratio of pipe velocity to frequency, the steady state treatment is likely to be less accurate. In order to allow for transients during switch-over, a more sophisticated approach might be required which involves consideration of ring vortex streets.

Equation (1) relates pump performance to the jet velocity either as measured, or as programmed by hydraulic coupling. However, it does not consider the way in which the jet velocity was achieved, for example by pneumatic coupling to a mechanical piston with a sinusoidal volume displacement. This has required a more extensive treatment since the drive pressure is related to the results it achieves, so the differential equations are non-linear and must be solved with the help of an analogue computer. If, however, the pressure variation as a function of time is known, either from a recorder or as a result of programming a valve controlled compressed air supply, it is possible to estimate pump performance.

A simple mathematical model may be used to describe the behaviour of two piston chambers attached to nozzle/diffusers, which feed liquid alternately from one to the other via the pump body. Assuming gravitational and inertial effects within the piston chambers to be small, but taking into account typical nozzle/diffuser losses, the mean square of the jet blow velocity available from a twin-jet combination at the rth stage may be expressed:

$$\bar{v}_{JBr}^2 = 2C_J \overline{\Delta p_r}/\varrho \qquad (2)$$

where C_J is the nozzle/diffuser loss coefficient ($= 0.67$ for example),

$\overline{\Delta p_r}$ is the mean of the modulus of the difference in pressure between the two rth piston chambers during an average cycle,

ϱ is the liquid density.

Substituting (2) in (1) for each stage, with $k = \frac{1}{2}$, we have the pump stagnation head ($v_P = 0$) for n twin-jet stages given by:

$$h_{P0} = 2C_J \alpha(1 - \alpha/2) \left[\sum_1^n \overline{\Delta p_r}\right]/\varrho g \qquad (3)$$

Experience has shown that the suck contribution to the pressure difference $\overline{\Delta p_r}$ sometimes has little effect, especially when cavitation occurs. Under these circumstances the use of the mean blow pressure alone in equation (3) gives satisfactory agreement with experiment.

5. Experimental results

A small hot single-stage twin-jet model, similar in principle to Fig. 3, but using piston chambers curved through 90°, has been constructed from silica glass with a jet diameter of 0.7 cm venturi bore of 1 cm. This has pumped molten potassium sulphate at 1500°K at flow rates up to 0.18 l/sec across a head of 13.5 cm. This corresponds to a stagnation head of 51 cm (6.8 cm Hg). The pump was pneumatically coupled via a pair of water-cooled pipes to two mechanical pistons of 60 cm³ displacement, which were sinusoidally cycled in anti-phase at frequencies in the range 2 to 5 Hz.

A somewhat larger pump for molten aluminium has been constructed from graphite with a design similar to Fig. 2, by A. E. Pengelly and S. Pilkington of British Aluminium Company Limited Research Laboratories. There were four jets of 2 cm diameter fitted to a main pipe of 7.5 cm bore. The drive was by means of compressed air actuated mechanical pistons, which were coupled pneumatically to the piston chambers containing molten aluminium at 1000°K. The pump achieved a stagnation head of ca. 25 cm (5 cm Hg) and should provide a flow rate of 10 l/sec across a 5 cm head.

Water experiments have been carried out using a two stage model closely similar to Fig. 3, with jet and venturi dimensions identical to the silica pump. This pump was driven by four servo-controlled spool-valves switching air pressures of up to 1.3 atm (100 cm Hg). At this drive pressure, a stagnation head of 50 cm Hg (8 cm for the first stage) was achieved at a frequency of 0.86 Hz, without the need for a sub-atmospheric suck stroke. This is because a small static pressure head at the first stage is sufficient to provide a reasonable jet suck velocity when assisted by the dynamic pressure rise in the nozzle/diffuser. A much greater head is produced by the second stage (42 cm), since the suck stroke is helped by the pressure developed by the first stage. The suck stroke of downstream stages is more effective due both to the increased pressure, and to the consequent alleviation of any limitation in suck velocity as a result of cavitation.

Considerable experimental work has been carried out using a pump of similar jet and venturi geometry, but with long straight sided 1.6 cm bore piston chambers driven by the pneumatically coupled mechanical drive described earlier. A stagnation head of 32.5 Hg has been achieved at 2.5 Hz for two stages each with a mean blow pressure of 0.45 atm (34.5 cm Hg).

6. Efficiency

There are basically three efficiencies which require consideration with the reciprocating-jet pump. Firstly there is the hydrodynamic or pumping efficiency, the efficiency of conversion from liquid energy in the air/liquid piston chamber to work done on the pumped liquid. Secondly there is the pneumatic efficiency, the efficiency of conversion of pneumatic energy to liquid energy in the piston chamber, and finally there is the efficiency of gas compression in creating the pneumatic energy.

The main drawback to conventional jet-pumps is that the pumping or hydrodynamic efficiency is not more than 38 per cent mainly as a result of mixing losses. There are, however, two important differences when the reciprocating-jet pump is compared with this. Firstly there is a suck stroke, which consumes power while providing no output, but secondly there is much less restriction on the allowed values of the ratio of jet to venturi area. This ratio could approach unity with a reciprocating-jet, which could result in a considerable increase in efficiency to near that for a diffuser (say 70 per cent).

If a mechanical piston, bellows or alternative closed system is used, the pneumatic efficiency could be near 100 per cent but unfortunately the efficiency with which the pneumatic energy is created is likely to be low. Alternatively compressed gas drive may be used which has been efficiently produced. However, it is likely that some pneumatic energy will be discarded to the exhaust on the suck stroke.

In a hot system the liquid enthalpy could perhaps provide some power by expanding injected cold gas or generating steam, but these possibilities remain to be investigated. It is difficult to generalize, but the overall efficiency is unlikely to be greater than 40 per cent.

7. Applications

There are many applications where efficiency is less important than some other consideration such as the absence of solid moving parts. The reciprocating-jet pump should be considered for application to the pumping of the following materials: molten metals, slags, glasses and salts, where high temperatures and corrosion are a problem; blood, where contamination and damage must be avoided; solid suspensions, where blockage and erosion can occur; and liquid gases such as helium at low temperatures.

Bibliography

DICKMANN (1950) Ship propulsion with nonsteady propulsion organs, *Schiff und Hafen* 2, No. 10, 252–265.

MUELLER (1964) Water jet pump, *Journal of the Hydraulics Division of the Proceedings of the American Society of Civil Engineers* 90, No. HY 3, 83–113.

WALKDEN (1967) Reciprocating-jet pump, *Nature* 213, (5073), 318.

WALKDEN and EVELEIGH (1969) An experimental and theoretical study of a reciprocating-jet pump using pneumo-hydraulic drive (submitted to *Trans. Inst. Chem. Eng.*).

WALKDEN and KELL (1967) Reciprocating-jet pump for hot corrosive fluids, *G.E.C. Jnl. of Science and Technology* 34 (1), 9–16.

A. J. WALKDEN

GLASS-CERAMICS. Glass-ceramics are polycrystalline solids made by the controlled crystallization of glasses. This is achieved by subjecting suitable glasses to a carefully regulated heat-treatment process which brings about the nucleation and growth of crystal phases within the glass. A residual glass phase is present between the crystals, but usually this constitutes the minor part of the material (5 to 30 per cent by volume).

Glass-ceramics are distinguished from glasses, which are of course amorphous, by the presence of major amounts of crystals. The distinction between glass-ceramics and traditional ceramics is that for the former materials the crystalline phases arise exclusively by crystal growth from a homogeneous glass phase, whereas in traditional ceramics most of the crystalline material is introduced when the ceramics are prepared.

The object in the production of a glass-ceramic is to bring about crystallization of an initially homogeneous glass and it is relevant therefore to consider briefly the factors which result in the formation of glasses during the cooling of certain melts rather than in crystallization. The failure of a melt to crystallize when it is supercooled below the liquidus temperature (the temperature at which crystals and the liquid phase can co-exist in equilibrium) can be attributed to one of two causes or possibly to a combination of both. These are first that the rate of nucleation in the melt is too low and second that the rate of crystal growth is too low. Practical experience suggests that for many glasses the first of these is probably more important and that insufficient nuclei are formed in the temperature range below the liquidus temperature where fairly rapid crystal growth is possible. Since the production of a successful glass-ceramic (i.e. one with high mechanical strength) demands the achievement of a very fine-grained microstructure, it is necessary to achieve the highest possible nucleation rate to ensure that a large number of very small crystals are formed rather than a relatively small number of coarse crystals. Except perhaps for glasses of very simple compositions (e.g. binary glasses) the rates of homogeneous nucleation which can be achieved are too low and therefore the inclusion in the glass of a nucleating agent to bring about heterogeneous nucleation is necessary.

The optimum method for making a glass-ceramic is to develop the nucleation catalyst particles in the glass at a temperature below that at which major crystalline phases can grow at a significant rate. Growth of these phases can then be achieved in a reheating process which permits proper control of the grain size of the final glass-ceramic. A satisfactory nucleating agent will therefore be soluble in the molten glass and will remain in solution during cooling of the glass through the temperature range where significant crystal growth can occur. The development of the catalyst particles will occur either during the further cooling of the glass or more usually during reheating of the glass after cooling. The rate of homogeneous nucleation of the catalyst itself must be high which requires that the free energy of activation for nucleation should be low. This is favoured by a high degree of supersaturation of the catalyst on cooling and by a low interfacial energy between the dissolved and precipitated form.

In the preceding discussion, the assumption has been made that the first stage in the glass-ceramics process is the precipitation of submicroscopic crystalline particles which act as nucleation sites for the subsequent development of major crystal phases. For certain nucleating agents, notably metallic agents which will be described later, this is almost certainly true but for other nucleating agents, principally oxides, a somewhat different route towards crystallization of the glasses is followed. This second method of promoting crystallization depends upon the occurrence of immiscibility in certain glasses. Many glass compositions exist which in their molten states are homogeneous single-phase liquids but which on cooling separate into two liquid or glass phases. Other compositions can be cooled from the molten state as single phase glasses, but during subsequent reheating, glass-in-glass phase separation occurs leading to the formation of two distinct glass phases. One of these takes the form of droplets dispersed in a matrix of the other phase. With binary lithia-silica glasses, for example, heat treatment leads to the production of silica-rich droplets in a lithia-rich matrix for compositions in the high silica region of the system.

Thus for many glasses, separation into two glass phases occurs prior to crystallization and this exerts a marked influence on the subsequent course of crystallization. For example, homogeneous nucleation in one of the phases may be more probable than in the original glass because the structure of this phase is less disordered than that of the original glass and also because its chemical composition may approach that of a crystalline phase more closely than did that of the original glass. If homogeneous nucleation and crystallization of the dispersed phase takes place, the crystals so formed could heterogeneously nucleate crystallization of the matrix phase. This sequence of events would result in glass-ceramics having a fine-grained structure. Some of the oxide nucleating agents are thought in fact to promote separation of the glass into two microphases for compositions where this would occur only with great difficulty or not at all and subsequently these microphases crystallize to yield the desired glass-ceramic microstructure.

Examples of heterogeneous nucleating agents include the metals copper, silver and gold. It has long been known that these metals are capable of existing in glasses as dispersions of colloidal particles and this has formed the basis of certain coloured glasses (red in the case of copper and gold; yellow in the case of silver). It has recently been found that these metallic dispersions will catalyse the crystallization of a range of glass compositions. In the case of copper, the formation of the colloidal suspension involves reduction of the cuprous ions initially present in the glass to copper atoms followed by aggregation of these during heat-treatment of the glass. The reduction process is accomplished by reactions of the type:

$$2\ Cu^+ + Sb^{3+} \rightarrow 2\ Cu + Sb^{5+}$$

Similar processes are involved in the formation of the colloidal silver or gold dispersions.

It has been found that the chemical reduction process can be catalysed by exposing the glass to ultraviolet radiation before the heat-treatment process. Reactions of the following type occur:

$$Cu^+ + Ce^{3+} + h\nu \rightarrow Ce^{4+} + Cu$$

After irradiation, metal atoms are therefore present and suitable heat-treatment causes aggregation of these to give colloidal particles of a sufficient size (about 80 Å) to act as heterogeneous nucleation sites for crystallization of the glass. The use of the irradiation treatment permits selective crystallization of the glass since during a suitable heat-treatment process only those parts of the glass which are exposed to the ultra-violet irradiation are crystallized, the unexposed regions remaining as clear transparent glass.

As mentioned earlier, certain oxide nucleating agents function by inducing separation of glass into two glassy phases. A good example of this type of agent is titanium dioxide. This oxide is soluble in the molten glass and remains in solution when the glass is cooled. When the glass is subjected to a reheating process, however, a titania-rich phase is precipitated in the form of fine droplets. Further heat-treatment appears to cause homogeneous nucleation and crystallization of the dispersed phase and subsequently to cause heterogeneous nucleation and growth of major crystalline phases from the glass. Other oxide nucleating agents such as phosphorus pentoxide are thought to function in a similar fashion.

Though the inclusion of a nucleating agent in the glass represents the key to the glass-ceramic process it is also of great importance to select glasses having the correct characteristics for conversion into fine-grained glass-ceramics.

The overriding consideration is that the glass shall be capable of being crystallized without the use of prohibitively long heat-treatments. However, there are other important considerations such as the need for the glass to be melted economically and to be capable of being shaped by available techniques. Also, of course, the glass composition has to be chosen so that the types and proportions of crystals developed will confer the desired physical properties upon the final glass-ceramic.

It has been found that many different types of glasses can be converted into microcrystalline glass-ceramics. Space does not permit the full range of compositions to be mentioned but Table 1 lists some important composition types and gives the major crystal phases present in the glass-ceramics.

At this point a brief summary of the practical steps involved in the preparation of a glass-ceramic will be useful. The process begins with the melting of a mixture of the appropriate raw materials together with the nucleating agent to give a homogeneous, bubble-free melt. Articles are prepared from this and are shaped by conventional glass-working methods such as casting, pressing, blowing, drawing or rolling. The articles are annealed and cooled and at this stage are in the form of a transparent glass. The glass articles are next subjected to a controlled heat-treatment schedule. In this the temperature is raised at 3° to 5°C per minute to a first holding temperature which is maintained usually for a period of $\frac{1}{2}$ to 3 hr depending on the glass composition. At this temperature which usually lies within a range of temperatures corresponding with glass viscosities of 10^{11} to 10^{12} poises the phase separation and nucleation processes are initiated. The temperature is then raised further at a rate of about 5°C per minute to a second holding temperature which is maintained for a few hours.

Table 1. *Composition types and crystal phases present in glass-ceramics*

Major constituents	Major crystal phases developed
Li_2O–Al_2O_3–SiO_2 (Al_2O_3 less than 10 wt.%)	Lithium disilicate ($Li_2O \cdot 2\,SiO_2$) quartz and/or cristobalite (SiO_2)
Li_2O–Al_2O_3–SiO_2 (Al_2O_3 10 to 25 wt.%)	β-spodumene ($Li_2O \cdot Al_2O_3 \cdot 4\,SiO_2$) β-eucryptite ($Li_2O \cdot Al_2O_3 \cdot 2\,SiO_2$) and solid solutions of these with silica
Li_2O–ZnO–SiO_2 (ZnO less than 10 wt.%)	Lithium disilicate quartz and/or cristobalite
Li_2O–ZnO–SiO_2 (ZnO 10 to 50 wt.%)	Lithium zinc silicate crystals ($Li_2O \cdot ZnO \cdot SiO_2$ and $4\,Li_2O \cdot 10\,ZnO \cdot 7\,SiO_2$) Willemite ($2\,ZnO \cdot SiO_2$)
MgO–Al_2O_3–SiO_2	Cordierite ($2\,MgO \cdot 2\,Al_2O_3 \cdot 5\,SiO_2$) Cristobalite Clino enstatite ($MgO \cdot SiO_2$) Forsterite ($2\,MgO \cdot SiO_2$)
Na_2O–BaO–Al_2O_3–SiO_2	Nepheline ($Na_2O \cdot Al_2O_3 \cdot 2\,SiO_2$) Celsian ($BaO \cdot Al_2O_3 \cdot 2\,SiO_2$)
BaO–TiO_2–SiO_2–Al_2O_3	Barium titanate ($BaTiO_3$)

During this part of the process, growth of major crystal phases occurs and the glass is converted into a microcrystalline glass-ceramic. The upper holding temperature is chosen so as to ensure maximum crystallinity without encountering deformation of the material. For this reason this temperature is usually about 50°C below the liquidus temperature of the predominant crystal phase in the glass-ceramic. Figure 1 illustrates a typical heat-treatment schedule.

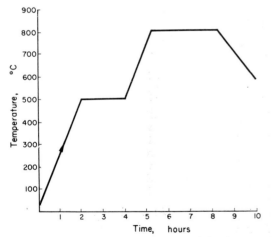

Fig. 1. *Typical heat-treatment schedule for a Li_2O–ZnO–SiO_2 glass-ceramic.*

A number of changes occur in the properties of the material during the heat-treatment process. In many but not all cases the material becomes opaque to visible light owing to scattering of light at grain boundaries. A small change in specific gravity often occurs because the crystal phases have different densities from that of the parent glass. However, the volume change is usually small and corresponds generally with a linear dimensional change of 1 per cent or less. This is of considerable practical value in enabling fairly close dimensional tolerances to be achieved. Other changes include increase in electrical resistivity and reduction of dielectric losses in many cases. Very often the coefficient of thermal expansion is markedly altered as a result of the crystallization process and also the glass-ceramic has a much higher deformation temperature than the parent glass.

Many of the advantageous properties of glass-ceramics derive from their dense, fine-grained microstructure. The average crystal size in a glass-ceramic is in the region of 1 micron or less whereas in conventional ceramics the particle sizes are frequently 20 to 50 microns. Glass-ceramics have a smooth glass-like surface and can be polished to a mirror-like finish. Most glass-ceramics are opaque to visible light owing to scattering of light at phase boundaries but for a range of compositions containing beta-eucryptite as a major crystal phase, high transparency in the visible region of the spectrum can be achieved. The transparency arises because the crystals are very small (2000 Å or less) and because their refractive index is close to that of the residual glass phase. Glass-ceramics which are opaque to visible radiation can be transparent to infra-red radiation as shown in Fig. 2. In such cases, transparency is attained when the wavelength of the infra-red radiation exceeds the crystallite size in the glass-ceramic.

Glass-ceramics have zero porosity and are impermeable to liquids and gases. Their specific gravities are generally similar to those of ordinary glasses and ceramics being in the range 2·4 to 2·6 but compositions containing oxides of the heavy metals such as lead can have specific gravities as high as 5·8.

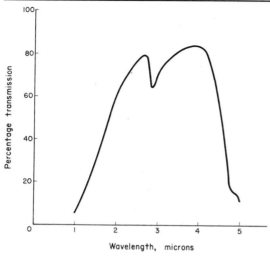

Fig. 2. Infra-red transmission curve of a Li_2O–ZnO–SiO_2 glass-ceramic (specimen thickness 1 mm).

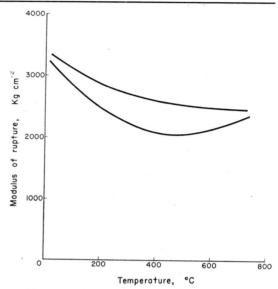

Fig. 3. Modulus of rupture versus temperature for two glass-ceramics.

Perhaps the most outstanding property of glass-ceramics compared with glasses and many conventional ceramics is their high mechanical strength. This is usually measured as modulus of rupture in a 3- or 4-point bending test owing to the difficulties of accurately measuring the true tensile strengths of brittle (i.e. non-ductile) materials. The moduli of rupture of glass-ceramics range from about 1000 to 4000 kg cm^{-2}. Glass in its "normal" condition, that is glass for which no special steps have been taken to prevent surface damage during handling, has a strength of about 600 to 700 kg cm^{-2} and the strengths of ceramics range from about 700 to 4000 kg cm^{-2}.

There are several possible reasons for the high mechanical strengths of glass-ceramics. One of these is that the materials have a high abrasion resistance and therefore they are less susceptible than glass to surface damage which leads to the production of stress-raising flaws. The complete absence of porosity is also a factor contributing to the high strength. Fracture propagation is also likely to be more difficult in a glass-ceramic than in a glass because of the effects of grains boundaries in diverting or slowing down the crack travelling through the material. An important factor is the very fine grain size of glass-ceramics since this may limit the size of stress-raising microcracks which can be present.

As shown in Fig. 3, the moduli of rupture of glass-ceramics are reduced as the temperature increases. However, even at high temperatures (in the region of 600–700°C) the strengths are very much higher than those of ordinary glasses at room temperature.

Other important mechanical properties include hardness and wear resistance. The few tests which have been reported in the literature suggest that glass-ceramics are fairly hard materials having diamond pyramid hardnesses of 630 to 700 kg mm^{-2}. Wear resistance tests have shown that some glass-ceramics are better in this respect than 18-8 stainless steel by a factor of about ten.

The elastic moduli of glass-ceramics are in the range about 8 to 14 $\times$ 10^5 kg cm^{-2} and thus they are considerably stiffer than normal glasses for which the moduli are in the range 6 to 7·5 $\times$ 10^5 kg cm^{-2}.

In general glass-ceramics are good electrical insulators and have volume resistivities in the region of 10^{14} to 10^{16} ohm-cm at 20°C. The small conductivity exhibited by glass-ceramics is due to transport of ions, mainly alkali metal ions, through the structure and the compositions having high resistivities are those which are substantially alkali-free. The resistivity falls with increasing temperature in the typical manner for ionic conductors.

Dielectric loss in glass-ceramics arises chiefly by ion migration and again the alkali metal ions are the ones principally involved. Thus alkali-free glass-ceramics can exhibit very low loss tangents. For example, a titania-nucleated glass-ceramic of the MgO–Al_2O_3–SiO_2 type having cordierite and cristobalite present as the major crystal phases has a loss tangent of 0·0002 at a frequency of 10,000 MHz at 20°C and the loss tangent remains below 0·001 for temperatures up to 500°C.

The permittivities of most glass-ceramics are between 5 and 7. The effect of increase of temperature on permittivity is marked at low frequencies but at high frequency (10^{10} Hz) the effect is very small (see Fig. 4). This is a useful characteristic which has led to the adoption of glass-ceramics for applications where constancy of permittivity despite temperature variations is a necessity. Such an application is in

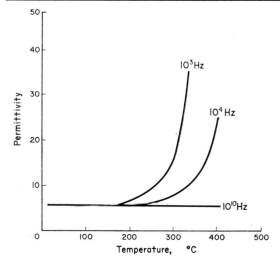

Fig. 4. *Dependence of permittivity upon temperature for a Li_2O–ZnO–SiO_2 glass-ceramic.*

radomes for rocket missiles since variations of the permittivity of the radome material, arising from friction heating of the nose cone as it passes through the atmosphere, would cause aberration of the radar signal.

Special glass-ceramics having permittivities as high as 2000 have been developed for the manufacture of capacitors for electronics. In materials of this type a ferroelectric crystal such as barium titanate is present as a major phase.

A valuable property of glass-ceramics is their high dielectric breakdown strength. Values measured in a standard test (German Standard DIN 40685) range from 28 to 50 kV mm^{-1}. These may be compared with a value of 25 kV mm^{-1} for a typical high-voltage electrical porcelain and values of 14 to 23 kV mm^{-1} for various alumina ceramics.

A notable characteristic of glass-ceramics is the ease with which the thermal expansion coefficient can be varied over a very wide range. At one extreme materials having negative expansion coefficients of -40×10^{-7} per °C are available. These contain β-eucryptite as a major crystalline phase. At the other extreme materials having high positive thermal expansion coefficients (150 to 160×10^{-7} per °C) similar to those of metals such as copper and steel are possible. Glass-ceramics of this type contain crystal phases such as lithium disilicate, lithium zinc silicate, quartz and cristobalite. By variation of the chemical composition of the glass and hence of the crystallographic constitution of the glass-ceramic, materials having thermal expansion coefficients between the two extremes are possible. Materials having zero coefficient of expansion for example are readily produced. In these the major crystalline phase is usually a β-eucryptite-silica solid solution. Figure 5 shows thermal expansion curves of glass-ceramics

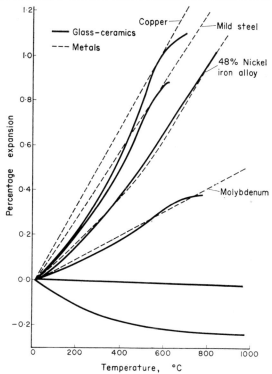

Fig. 5. *Linear thermal expansion curves of glass–ceramics and various metals.*

together with those of various metals for comparison purposes.

The thermal conductivities of glass-ceramics are generally in the range 0·005 to 0·01 cal sec^{-1} cm^{-1} °C^{-1} and they are therefore higher than those of normal glasses (0·0035 to 0·004 cal sec^{-1} cm^{-1} °C^{-1}) but appreciably lower than those of high alumina ceramics (0·05 to 0·07 cal sec^{-1} cm^{-1} °C^{-1}). The specific heats of glass-ceramics are similar to those of glasses and are generally in the range 0·22 to 0·28 cal g^{-1} °C^{-1}.

The important factors which determine the thermal shock resistance of a glass-ceramic are the linear thermal expansion coefficient, the modulus of elasticity and the mechanical strength. As would be expected, the thermal shock resistances of materials having near zero thermal expansion coefficients are very good.

Acknowledgement

The author thanks Dr. E. Eastwood, Director of Research, The English Electric Company Limited for permission to publish this article.

Bibliography

EMRICH B. R. (1964) Technology of new devitrified ceramics—a literature survey. Air Force Materials Laboratory Report ML-TDR-64-203, Washington D.C., Office of Technical Services, U.S. Department of Commerce.

McMillan P. W. (1964) *Glass-ceramics*, London: Academic Press.

Porai-Koshits E. A. (1964) *The Structure of Glass*, Vol. 3, *Catalysed Crystallisation of Glass* (translated from the Russian), New York: Consultants Bureau.

Stookey S. D. and Maurer R. (1962) Catalysed crystallisation of glass in theory and practice, *Progress in Ceramic Science*, pp. 77–107, Oxford: Pergamon Press.

P. W. McMillan

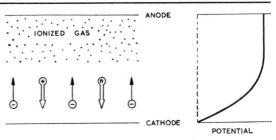

Fig. 1. Simplified schematic diagram of the glow discharge.

GLOW DISCHARGE ELECTRON AND ION GUNS FOR MATERIAL PROCESSING. Glow discharge electron and ion beam techniques originated towards the end of the 19th century. Thus Sir William Crookes (1879, 1891) demonstrated both intense heating by focused cathode rays and sputtering effects at the cathode. Early applications of beam-producing devices were empirical and practical difficulties due to pressure regulation and the glow-to-arc transition were experienced. The discovery and development of thermionic, high-vacuum electron guns during the early part of the present century diverted attention from glow discharge electron beam generators. Ion beam generators, exploiting the glow discharge, persisted into the 1930's in applications for mass spectrography and early nuclear physics experiments (e.g. those of Rutherford and Oliphant, 1933). They have now been superseded in these spheres by sources giving better energy resolution, current and choice of ion.

Recent research by the author and his colleagues at Harwell, however, indicates that glow discharge beam techniques have considerable scope for application in the field of material processing. Beams of electrons or ions may be conveniently generated in glow discharge apparatus operating in the range 1 to 30 kV (or higher voltage) and at pressures of the order of 10^{-1} mm Hg. This high-voltage, low-pressure glow discharge is associated with the left-hand branch of the Paschen gas breakdown curve (Klyarfel'd *et al.*, 1966). Apparatus may be simple and cheap and vacuum systems and techniques may be crude and inexpensive. The beams can be applied to both electrically conducting and electrically insulating materials, i.e. metals, ceramics, glasses and organic materials.

An over-simplified, but adequate, schematic diagram illustrating the main features of the discharge is given in Fig. 1. Ionized gas is found on the anode side of the discharge space. Owing to its high electrical conductivity there is little change in potential in this region. Almost all the applied potential appears across the cathode fall, i.e. the region between the ionized gas and the cathode. The electric field in the cathode fall accelerates electrons towards the anode and positive ions towards the cathode. Some of the ions undergo charge exchange with atoms of the gas present and become fast neutrals. A positive feedback mechanism, in which each electron liberated from the cathode ultimately leads to the production there of another electron at a later time, sustains the discharge.

At low voltages, say 1 to 2 kV, the width of the cathode fall region is small compared to the separation of anode and cathode. The electric field distribution in the cathode fall and consequently the fast electron trajectories are therefore largely governed by the geometry of the cathode. The geometry of the anode and boundary conditions at the sides of the discharge are relatively unimportant. The fast electrons coast through the ionized gas. A concave cathode, whose radius of curvature is larger than the width of the cathode fall, can be used under these conditions to focus the fast electrons at the centre of curvature within the ionized gas. A target can be placed at this point without too much effect on the discharge and may be strongly heated by the concentrated electron bombardment. It does not matter whether the target is isolated electrically, or even an insulating material, since the conductivity of the ionized gas will maintain its surface at near anode potential. Two or more cathodes may be operated independently to control the distribution of fast electrons falling upon the target. Typically 30 to 50 per cent of the applied power input can be liberated as heat in the target under these conditions.

As the glow discharge voltage is increased the cathode fall width increases at the expense of the region of ionized gas near the anode. Now the electric field distribution becomes sensitive to the anode geometry and boundary conditions at the sides become important. The fast electrons tend to be mono-energetic and become penetrating, owing to a falling cross-section for interaction with the gas. They can be taken out of the discharge region completely by allowing them to pass through a perforated anode. Outside the discharge they can be manipulated magnetically. At high voltages (20 to 30 kV) 80 per cent of the input power can be brought out as fast electrons in this way under favourable conditions (aluminium cathode or other metal having a high coefficient γ of secondary electron emission under ion bombardment).

Ions and fast neutrals can also be brought out of

the discharge through a perforated cathode. The majority can be brought out through a cathode of high transparency since the discharge can be sustained by the small proportion which hit the cathode. A low coefficient γ is desirable for high yield. Extraction of these particles is not quite so straightforward as for the fast electrons, however. An instability arises because ionized gas which forms outside the perforated cathode supplies extra electrons which may diffuse back into the cathode fall region to take part in the discharge. Thus positive feedback leading to large amplitude oscillations occurs. A stable state can be restored, however, by biasing the external plasma positively to cut off the electron back diffusion. The magnitude of the bias required depends on the dimension of the space charge sheath needed to pinch off the electrons. This in turn is related to the dimensions of the cathode perforations, gas pressure, etc. By keeping perforation dimensions small in relation to cathode fall dimension, quite moderate biasing of about 100 V is adequate.

Owing to relatively large cross-sections for interaction with the gas, ions and neutrals passing through the cathode have a wide spread in energy and are not suitable for precise external manipulation with magnetic fields. However, their trajectories are sensitive to discharge geometry which may be designed accordingly.

Discharge current depends mainly on pressure, although other factors such as the nature of the gas, electrode material, etc., play a part. Owing to the fact that the ionization and excitation processes in the gas, which sustain the discharge, depend on the electric field, a given discharge geometry can be scaled to operate with the same current voltage characteristic provided the product of pressure and linear dimension is kept constant. A limitation on scaling is set by field effects at electrode surfaces or by thermal dissipation effects. Subject to these limitations, this principle can be employed to allow several guns (both electron and ion guns) to be operated independently and simultaneously in the same vacuum system.

An undesirable event, which tends to occur more readily at high current, is the glow-to-arc transition (von Engel, 1955). According to an investigation made by Maskrey and Dugdale (1966), volatile contamination of the cathode surface plays an important part in the transition. Thus contamination such as adsorbed water or organic material should be avoided. A simple solution, effective under many operating conditions, is to maintain cathode temperature at 200 to 300°C. Thus successful beam-producing devices may be said to exploit the warm cathode glow discharge rather than the cold cathode discharge (a term often used in referring to the glow discharge). Cathode temperature can in fact be much higher and is only limited by the onset of excessive evaporation or thermionic emission. An arc cannot usually be sustained at currents of less than a few amps. Simple current-limiting techniques can therefore be incorporated to throttle the arc should one occur.

Three pieces of apparatus will now be briefly described which illustrate the application of glow discharge beam techniques to material processing. Figure 2 shows an apparatus working at lower vol-

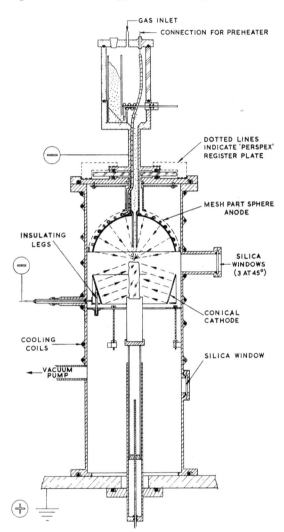

Fig. 2. Crystal growing apparatus for sapphire and other refractory materials.

tages, i.e. 1 to 4 kV, in which the target, a growing sapphire crystal, is immersed in the ionized gas at near anode potential. The Verneuil process of feeding powder, or in this case granulated alumina, into the pool of molten alumina at the tip of the crystal is here combined with heating by electron bombardment. A second cathode anneals the growing crystal. Each cathode operates at a current of about one amp.

and is controlled independently. All other metal parts of the apparatus are at anode potential.

Figure 3 shows an electron gun for welding and thermal milling applications. The gun is designed to

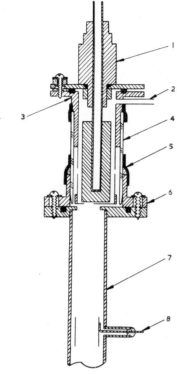

Fig. 3. Hollow anode electron gun: (1) Wilson seal carrying off-set aluminium cathode assembly, (2) gas inlet, (3) grid assembly, (4) glass tube insulator with resin shoulder, (5) rubber seal, (6) anode alignment flange, (7) hollow anode, (8) auxiliary anode control electrode.

work at voltages up to about 30 kV and at currents up to about 75 mA. A hollow anode glow discharge (Dugdale, 1966) generates a beam of electrons travelling down the axis of the anode. A diaphragm or grid near the cathode serves as the middle component of an immersion lens. Its position and potential control the trajectories of the electrons. The electron beam passes through an auxiliary anode whose potential in relation to the main anode (about 100 V) controls the discharge current over a range of about 5 to 1. The current can be regulated against minor pressure fluctuations with such an electrode. The penetrating, approximately mono-energetic beam can be subsequently passed through magnetic lenses and deflection coils for manipulation outside the discharge. It can be applied to an insulating target as readily as to a metal or conducting target owing to the presence of ionized gas which forms near the target. The ionized gas electrically connects the target surface to the environment. Heat fluxes approaching 10^6 W/cm^2 in a focussed beam have been achieved with this type of gun.

Figure 4 shows a broad beam ion gun of focus fixed by the geometry and approximately independent of operating conditions. This gun is intended to sputter insulating materials for deposition as thin films in micro-electronic applications. The silica tube insulation is adequate for operation to 20 kV and current is limited only by electrode temperature. The environment must be positively biassed in relation to the

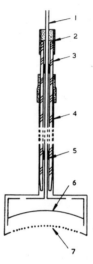

Fig. 4. Broad beam ion gun: (1) gas inlet, (2) rubber seal, (3) silica insulator, (4) lead through case, (5) space, (6) anode, (7) cathode mesh.

cathode to maintain stability. An essentially similar geometry may be used to generate a fixed focus electron beam (reversed electrode potentials).

To sum up: glow discharge beam techniques are understood sufficiently well that beams of any shape can be designed. Power densities up to 10^6 W/cm^2 are possible at present and higher power density may be achieved with further development. The associated vacuum system may be crude and cheap since operating pressures of the order of $1/_{10}$ mm Hg may be designed (with higher pressure at the target if differential pumping is employed). Pump-down time is short and fast access to the target is possible. The beams can be applied equally well to electrically conducting and insulating materials. They are suitable for many applications in material processing involving heat treatment, melting, evaporation, radiation chemistry and sputtering.

Summary

The high-voltage, low-pressure glow discharge can be made to operate as an efficient generator of electron or ion beams. Many geometrical arrangements of anode and cathode are possible which produce thin or broad beams or sheets of particles, converging or diverging, over a wide range of current and voltage. Modulation and regulation can be achieved with auxiliary electrodes. Operation at gas pressure of *ca.* 100 microns can be designed. Extracted electron beams tend to be mono-energetic and penetrating as the voltage is increased and can be manipulated magnetically outside the discharge. The beams may be applied to insulating materials without electrical charging difficulties, owing to the presence of ionized gas. The technique is illustrated with a brief description of apparatus for crystal growing (sapphire), welding and thermal milling of metals and ceramics and sputtering of insulators.

The writer wishes to acknowledge the contribution of his colleagues Mr. J. T. Maskrey, Mr. S. D. Ford, Mr. P. R. Harmer and Mr. J. A. Desport to these techniques.

Bibliography

CROOKES W. (1879) *Phil. Trans. Roy. Soc.* **170**, 135.
CROOKES W. (1891) *Proc. Roy. Soc.* **50**, 88.
DUGDALE R. A. (1966) *Proc. 7th Int. Conf. on Phenomena in Ionized Gases*, Vol. 3, p. 334 (Gradevinska Knjiga, Beograd).
VON ENGEL A. (1955) *Ionized Gases*, Oxford: Clarendon Press.
KLYARFEL'D B. N. *et al.* (1966) *Sov. Phys.—Tech. Phys.* **11**, 520.
MASKREY J. T. and DUGDALE R. A. (1966) *Brit. J. App. Phys.* **17**, 1025.
OLIPHANT M. L. E. and RUTHERFORD E. (1933) *Proc. Roy. Soc.* A **141**, 259.
WEHNER G. (1961) Discharge glow, in *Encyclopaedic Dictionary of Physics* (J. Thewlis Ed.), **2**, 437, Oxford: Pergamon Press.

R. A. DUGDALE

GYROSCOPIC COUPLER (GYRATOR). Gyroscopic coupling terms are those terms in the differential equations of a dynamic system which are linear in the velocities and which obey a relation of anti-reciprocity ($\alpha_{ik} = -\alpha_{kl}$) in place of the more usual reciprocity relation ($\alpha_{ik} = \alpha_{kl}$). They occur in systems containing actual gyroscopes or, more generally, in cyclic systems or in systems with time-varying constraints. Cyclic systems are systems where "every particle on leaving its position is replaced after an infinitesimally short time interval by another particle of the same kind and velocity" (Boltzmann, 1891). Examples for such systems are gyroscopes, endless chains, stationary fluid motions and above all electric currents (Maxwell, 1873). For the mathematical treatment of both kinds of dynamic systems the reader is referred to Thomson and Tait.

An ideal gyroscopic coupler (Bloch, 1944) is an idealized two-port device which performs "physically" what the gyroscopic coupling terms carry out mathematically, i.e. they convert a "velocity" into a "force" either way.

As an example let us consider a gyrostat (flywheel in gymbals at right angles to each other) schematically represented in Fig. 1 where the arrows indicate the

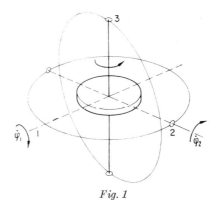

Fig. 1

direction of rotation of the flywheel and the direction in which angular velocities $\dot{\varphi}_1$ and $\dot{\varphi}_2$ of axes 1 and 2 of the gymbals and of the torques applied thereto are counted positive. Coupling between these axes is due to the moment of momentum of the flywheel $H_0 = I_0 \Omega_0$. If I_1 and I_2 denote the moments of inertia of the device about axes 1 and 2 (measured with the flywheel at rest) we obtain the following differential equations for the response to the applied torques M_1 and M_2

$$I_1 \ddot{\varphi}_1 + \dot{\varphi}_2 H_0 = M_1 \qquad (1)$$

$$-\dot{\varphi}_1 H_0 + I_2 \ddot{\varphi}_2 = M_2 \qquad (2)$$

Evidently from the point of view of studying the nature of the coupling between axes 1 and 2 the values of I_1 and I_2 are accidental and we will therefore assume that both are zero. This in turn implies $I_0 = 0$. However, by letting $\Omega_0 \to \infty$ as we let $I_0 \to 0$ we are allowed to assume a finite value of H_0.

The idealized coupler is thus described by the equations

$$+\dot{\varphi}_2 H_0 = M_1 \qquad (3)$$

$$-\dot{\varphi}_1 H_0 = M_2 \qquad (4)$$

To bring out its characteristic behaviour we will from now on assume that all the variables in (3) and (4) are sinusoidal functions of time of frequency ω so that we can use complex notation and write phasor equations in place of differential equations. We will further assume that M_2 is not an impressed torque but arises from the fact that axis 2 is loaded with a mechanical impedance Z_2, i.e. that we have $M_2 = -z_2 \dot{\varphi}_2$. It follows then that

$$\dot{\varphi}_2 = -\frac{M_2}{z_2} \qquad (5)$$

$$M_1 = \dot{\varphi}_2 H_0 = +\dot{\varphi}_1 \frac{H_0^2}{z_2} = +\dot{\varphi}_1 z_1 \qquad (6)$$

i.e. the "generator" M_1 sees on axis 1 an input impedance

$$z_1 = \frac{H_0^2}{z_2} \qquad (7)$$

Equation (7) leads to a number of most intriguing results.

If axis 2 is loaded for instance with a flywheel ($z_2 = j\omega I_2$), then

$$z_1 = \frac{H_0^2}{j\omega I_2} = \frac{c_1}{j\omega} \qquad (8)$$

i.e. axis 1 appears to be loaded with a spring of constant $c_1 = H_0^2/I_2$. Conversely, a spring load on axis 2 $\left(z_2 = \dfrac{c_2}{j\omega}\right)$ appears as a flywheel load on axis 1 for we have

$$z_1 = \frac{H_0^2 j\omega}{c_2} = j\omega I_1 \qquad (9)$$

i.e. $I_1 = \dfrac{H_0^2}{c_2}$. We will call the system elements on axis 1 derived in the foregoing manner the images of the original elements on axis 2.

From eq. (7) we can also derive rules concerning the images of a combination of elements. If axis 2 is loaded with two elements in such a way that their admittances add up, i.e. if we have

$$y_2 = y_2' + y_2'' \qquad (10)$$

it follows that

$$z_1 = H_0^2 y_2 = H_0^2 y_2' + H_0^2 y_2'' = z_1' + z_1'' \qquad (11)$$

i.e. the images have to be combined in such a way that their impedances add.†

Conversely, if on axis 2 we have two elements combined so that their impedances add up, we have to combine their images so that their admittances add, for we have

$$y_1 = \frac{z_2}{H_0^2} = \frac{z_2'}{H_0^2} + \frac{z_2''}{H_0^2} = y_1' + y_1'' \qquad (12)$$

The kind of image which we thus obtain as a result of eq. (7) is well known in electrical network theory: it is the dual of the original system. The origin of this dualization process is of course the interchange of the parameters "force" and "velocity" prescribed by eqs. (3) and (4) (corresponding to the interchange of "voltage" and "current" in the electrical dualization).

After we have replaced all the elements on axis 2 by their images on axis 1 we can study the interaction

---

† Electrically we would say that if the original elements are connected in parallel their images are connected in series. The author has pointed out elsewhere (Bloch, 1944; 1945) that in mechanical systems the terms "series" and "parallel" connections are best avoided and replaced by the terms "co-resistive" (if the forces add up) and "co-yielding" (if the displacements add up).

of these equivalent images with the elements originally on axis 1 without having to think about the gyroscopic coupler. This procedure is quite analogous to the transformation used in the analysis of mechanical systems containing a gear train where one refers all the elements to one of the axes, or to the analysis of systems containing transformers where one refers all components to one side of the transformers. Having obtained the solution of the simplified system there remains of course the task—not a very difficult one—of interpreting it in terms of the original system.

As an example let us consider the dynamical system shown in Fig. 2a which is typical for many systems

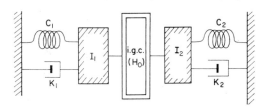

Fig. 2a

containing gyroscopes. It illustrates for instance the gyroscopic vibration damper suggested by R. N. Arnold (1941) for the damping of oscillations which afflicted the turntables of large milling machines. This damper consists of a horizontal gyroscope capable of tilting around a horizontal axis. The tilting moment due to the rotation and oscillation of the turntable is balanced by a spring and damped by a dash pot. The same Figure also illustrates the gyroscopic pendulum with two axes of freedom such as discussed by Parks and Maunder (1961).

It will be noted that seen from the gyroscopic coupler the mass I_2, spring c_2 and dashpot K_2 are co-resistively combined. On transfer to axis 1 we obtain therefore the co-yielding combination of their images shown in Fig. 2b. It is obvious from these

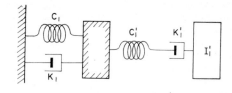

Fig. 2b

considerations that the gyroscopic damper can be replaced in principle by a non-gyroscopic one (though possibly of inconvenient dimensions) and that a certain invariant property of the gyroscopic pendulum (Parks and Maunder, 1961) is in fact also possessed by non-gyroscopic oscillatory systems (Bloch, 1961).

The concepts so far developed can also be traced in

the analytical treatment of such systems. We have then the following equations

$$z_{11}(p)\,\dot{\varphi}_1 + z_{12}\dot{\varphi}_2 = M_1(t) \tag{13}$$

$$z_{21}\dot{\varphi}_1 + z_{22}(p)\,\dot{\varphi}_2 = M_2(t) \tag{14}$$

where $z_{11}(p)$ and $z_{22}(p)$ represent in operational form the impedances connected to axes 1 and 2 and where $-z_{21} = +z_{12} = H_0$. We have also added for the sake of generality on the right-hand side the driving torques $M_1(t)$ and $M_2(t)$.

Substitution from (14) into (13) yields

$$\dot{\varphi}_1\left(z_{11} + \frac{H_0^2}{z_{22}}\right) = M_1(t) - \frac{H_0 M_2(t)}{z_{22}} \tag{15}$$

Here we recognize in the second term inside the bracket the impedance transferred according to eq. (7). We also note that $M_1(t)$ is "augmented" by the transferred torque $-\dfrac{H_0 M_2(t)}{z_{22}}$ and we have to show that this transfer has taken place in accordance with the duality prescription of eqs. (3) ff. For the sake of brevity we shall use the language of electric network theory. Seen from the terminals of the gyroscopic coupler we have a series combination of impedance z_{22} and of a voltage generator $M_2(t)$. The dual counterpart of these elements is according to eqs. (7) and (4) a parallel combination of an admittance $y'_{11} = \dfrac{z_{22}}{H_0^2}$ and of a current generator $-\dfrac{M_2(t)}{H_0}$. However, according to Norton's theorem this parallel combination is equivalent to a series combination of an impedance $z'_{11} = \dfrac{H_0^2}{z_{22}}$ and of a voltage generator $M'_1(t) = -\dfrac{M_2(t)}{H_0}z'_{11}$
$= -\dfrac{M_2(t) H_0}{z_{22}}$.

Other Types of Gyroscopic Couplers

We stated right at the beginning that gyroscopic coupling terms are not limited to systems containing actual gyroscopes. The idealization of these more general systems leads of course to other types of gyroscopic couplers. On the other hand, once we have a gyroscopic coupler such as the one based on Fig. 1 we can use it to synthesize a system obeying any given differential equation with antireciprocal coupling, even if this equation no longer relates to a problem in torques and angular velocities. Thus it requires no great mechanical ingenuity to change the device of Fig. 1 into a device that converts a translational velocity into a force P and vice versa. Obviously, we can also add to it some electro-mechanical transducer and adapt it to some electrical problem.

We shall return to this latter variation in a moment but will first discuss the problem of coupling due to Coriolis forces (Bloch, 1944). Let us suppose that we have a roundabout rotating with angular velocity Ω and on this a mass point (mass m) oscillating in a straight frictionless tube according to the time law $v_1 = v_{10} \sin \omega t$. On account of Coriolis law this tube will then experience a sideways force

$$F_2 = 2m\Omega v_1 \tag{16}$$

If the tube itself can move sideways it can be used to drive some other device, say, one with an impedance z_2. It will therefore move with a sideways velocity

$$v_2 = \frac{2m\Omega v_1}{z_2} \tag{17}$$

provided we include in z_2 the impedance $j\omega m$ offered by the mass point itself. However, this velocity in turn gives rise itself to a Coriolis force at right angles to it, i.e. in the direction of v_1, of amount

$$F_1 = \frac{(2m\Omega)^2}{z_2} v_1 \tag{18}$$

This means that in order to produce v_1 we have to overcome not only the original impedance of the mass point $j\omega m$ but also the reflected impedance

$$z_1 = \frac{(2\Omega m)^2}{z_2} \tag{19}$$

An electro-dynamic transducer of the type used in loudspeakers consists of a current carrying coil in a magnetic field. If B denotes the flux density and l the length of the wire a current i_{1d} produces a driving force

$$f_{2d} = Bl i_{1d} = b i_{1d} \tag{20}$$

Conversely if the coil has a velocity v_2 it will produce across its terminals a voltage

$$e_{1d} = Bl v_{2d} = b v_{2d} \tag{21}$$

(We assume the resistance and self-inductance of the coil to be negligibly small.) If the coil is used to drive a mechanical impedance z_m we have

$$v_{2d} = \frac{f_{2d}}{z_m} = \frac{b i_{1d}}{z_m} \tag{22}$$

and hence

$$e_{1d} = \frac{b^2 i_{1d}}{z_m} \tag{23}$$

i.e. to enforce the flow of current i_{1d} into the coil we have to overcome an input impedance

$$z_{1eld} = \frac{b^2}{z_m} \tag{24}$$

If we apply this transformation rule to specific examples (e.g. a spring $z_m = \dfrac{c_m}{j\omega}$ transforms into $z_{1eld} = \dfrac{b^2 j\omega}{c_m}$, i.e. into an inductance $L = \dfrac{b^2}{c_m}$) we shall see immediately that the equivalent images produced by this transducer at its electrical input terminals are identical with the analogues formed according to the "inverse" analogy (Bloch, 1944; 1945). This is of course to be expected because the eqs. (20) and (21) are the mapping rules of this analogy.

Let us now compare the case of the ideal electrostatic type of transducer. Here we have a driving force

$$f_{2s} = e_{1s} a \qquad (25)$$

and an input current

$$i_{1s} = v_{2s} a \qquad (26)$$

If we have

$$v_{2s} = \frac{f_{2s}}{z_m} \qquad (27)$$

we have

$$z_{1els} = \frac{e_{1s}}{i_{1s}} = \frac{z_m}{a^2} \qquad (28)$$

and it is easily seen that the electrical images produced by this type of transducer are those of the direct analogy (Bloch, 1944; 1945)—as eqs. (25) and (26) are the mapping equations of this type of analogy. We note that the electrical images of these two types of analogy are duals of each other, as indeed is evident from the following relations derivable from (20), (21) and (25), (26)

$$i_{1d} b = e_{1s} a \qquad (29)$$

$$\frac{e_{2d}}{b} = i_{1s} a \qquad (30)$$

$$z_{1els} a^2 = \frac{b^2}{z_{1eld}} = y_{1eld} b^2 \qquad (31)$$

It follows that we can replace the electro-dynamic transducer by an electrostatic one provided we interpose between z_m and the electrostatic transducer a gyroscopic coupler which carries out the dualization. However, we can just as well carry out this dualization on the electrical side of the electrostatic transducer, if we have a device which interchanges current and voltages according to eqs. (29) and (30), i.e. a device which is the electrical image of the gyroscopic coupler preceding the electrostatic transducer. The "existence" of such a device was first explicitly stipulated as a mere mathematical concept by Jefferson (1944). Later on Jefferson (1945) drew attention to the fact that such a device could be realized by joining in tandem the mechanical "terminals" of an electrostatic and an electrodynamic transducer. Tellegen (1948) again stipulated the existence of this electrical version of the gyroscopic coupler and introduced the catching name "gyrator" for it.

Nonreciprocal Systems

If the action of a gyroscopic coupler which obeys a relation of antireciprocity $\alpha_{ik} = -\alpha_{ki}$ is added to the action of an "ordinary" coupler which obeys a relation of reciprocity $\beta_{ik} = \beta_{ki}$ we obtain a resulting coefficient of coupling which belongs to neither class and which we will call non-reciprocal. We have

$$\gamma_{ik} = \alpha_{ik} + \beta_{ik} \qquad (32)$$

$$\gamma_{ki} = \alpha_{ki} + \beta_{ki} = -\alpha_{ik} + \beta_{ik} \qquad (33)$$

Conversely, if we have a system of differential equations with non-reciprocal coupling coefficients we can synthesize an appropriate dynamic system if we split these coefficients into a symmetrical and anti-symmetrical part according to the following equations:

$$\alpha_{ik} = \frac{\gamma_{ik} - \gamma_{ki}}{2} \qquad (34)$$

$$\beta_{ik} = \frac{\gamma_{ik} + \gamma_{ki}}{2} \qquad (35)$$

Provided the α_{ik} come out as real constants they can be realized by gyroscopic couplers. The β_{ik} represent couplings due to mutual resistance, mass or elasticity if they are of the form

$$\beta_{ik} + p m_{ik} + \frac{1}{p} c_{ik} \quad \left(\text{with } p \equiv \frac{d}{dt}\right)$$

or

$$\gamma_{ik} + j\omega m_{ik} + c_{ik}/j\omega.$$

Transfer of this statement to electrical systems is obvious.

An example of a system with simultaneous coupling by a gyroscopic coupler and a mutual resistance was given by Bloch (1944) and is illustrated in Fig. 3. Its electrical counterpart was published by Mcmillan (1946)

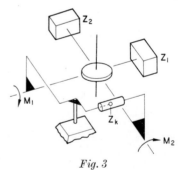

Fig. 3

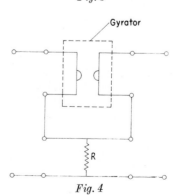

Fig. 4

and is illustrated in Fig. 4 where the device inside the dotted rectangle represents a gyrator and R is an ohmic resistance.

Bibliography

ARNOLD R. N. (1947) The tuned and damped gyrostatic vibration absorber, *Proc. Inst. Mech. Eng.* **157**, 1–19.

BLOCH A. (1944) A new approach to the dynamics of systems with gyroscopic couplings terms, *Phil. Mag.* Ser. 7, **35**, 315.

BLOCH A. (1945) Electromechanical analogies and their use for the analysis of mechanical and electromechanical systems, *J. Inst. Electrical Engineers* **92** (I), 157–169.

BLOCH A. (1961) Letter, *Nature* **191**, 1086–1087.

BOLTZMANN L. (1891) *Vorlesungen über Maxwells Theorie der Elektrizität und des Lichtes*, Vol. I, p. 5, Leipzig.

JEFFERSON H. (1944) Notes on electromechanical equivalence, *Wireless Engineer* **21**, 563,

JEFFERSON H. (1945) Letter, *Phil. Mag.* Ser. 7, **36**.

MCMILLAN, E. M. (1946) *J. Acoust. Soc. of America* **18**, 347.

MAXWELL J. C. (1873) *A Treatise on Electricity and Magnetism*, 3rd ed. Vol. II, Article 572, Oxford.

PARKS R. and MAUNDER L. (1961) Letter, *Nature* **190**, 710.

TELLEGEN B. D. H. (1948) The gyrator, a new electrical circuit element, *Phil. Res. Report* **3**, 81–101.

THOMSON SIR W. and TAIT P. G. *A Treatise on Natural Philosophy*, Vol. I, Article 319, Examples F and G. New edition.

A. BLOCH

H

HALL EFFECT, APPLICATIONS OF. The ability of the Hall effect to form a product of two input signals lends itself to a vast range of applications and presents the electrical engineers with a new circuit element which simplifies complex electronics circuitry for many applications.

Modern requirements for materials of which Hall effect devices are made, are as follows:

1. A high Hall coefficient and carrier mobility.
2. A resistance of the Hall plate such that a high control current can be tolerated without excessive power dissipation.
3. The plate must be as thin as possible.
4. An appropriate temperature coefficient. (The Hall constant and resistivity should be largely independent of temperature.)

These requirements stem from the fundamental equations

$$V_{HO} = \frac{R_H IB}{t} = B(\mu_n \varrho_n^{\frac{1}{2}}) \cdot \left(\frac{h}{t}\right)^{\frac{1}{2}} \cdot \omega$$

where the symbols have their usual significance (Harris, 1961). For technological applications in which amplification of the Hall output is required, transistor circuitry is now employed almost exclusively; this implies a fairly low source impedance, and the ability of the Hall plate to furnish power is an additional advantage. Indeed with Hall plates of germanium with no prior amplification of the output, it is essential to use moving coil meters of microampere range. On the other hand, Hall plates of indium arsenide allow, in consequence of their high electron mobility, the use of milliampere range instruments hence more rugged movements. The majority of present-day devices utilize Hall plates made of either indium arsenide or gallium arsenide, the use of indium antimonide being precluded as this compound has poor temperature and galvanomagnetic characteristics.

Applications

The temperature invariance of the extrinsic Hall coefficient of indium arsenide and gallium arsenide and the high electron mobility (20,000 cm^2 V sec and 7500 cm^2 V sec respectively) make these materials particularly suitable for Hall effect devices. The particular properties of Hall effect devices which have aroused much attention are:

1. Response to instantaneous magnetic induction.
2. Provision of vector multiplication.
3. Linear amplification.
4. Provision of non-reciprocal properties.

In considering the application of Hall devices it is convenient to group them according to the structural arrangement as follows:

1. Hall effect probes.
2. Probes with associated magnets.
3. Probes with flux concentrators.
4. Multiplier units.

Each structural arrangement has several well known and undoubtedly many unexplored applications in which various input–output relationships are of interest.

In the case of the Hall probe unit alone, there are three broad applicational areas worthy of special note, they are:

1. Measurement of magnetic fields (gaussmeters).
2. Sensors for magnetic field control.
3. Sensors for microwave power.

Hall effect probes in combination with associated magnets used to generate magnetic fields necessary for function operations, yield:

1. Gyrators.
2. Isolators.
3. Circulators.
4. Displacement transducers.
5. Function generators.
6. Resolvers.

Another area of application of Hall probes is their use with a flux concentrator which integrates or enhances the magnetic induction in various arrangements. In some cases flux concentrator may be used as a method of controlling the path of flux lines available. Primary interest in flux concentrations have been in the following:

1. Ammeters.
2. Magnetic compasses.
3. Magnetometers.
4. Magnetic tape read-out heads.

The applicational effort to-date has been associated with the magnetic circuit arrangement in which the Hall effect device and an electromagnet are used to secure the advantages of two current inputs to yield output functions. Several of the interesting applications of such an arrangement are as follows:

1. Analogue multipliers.
2. Squaring units.
3. Wattmeters.
4. Phase detectors.
5. Low level current transducers.
6. Brushless d.c. motors.
7. Modulators and demodulators.
8. Frequency deviation meters.
9. Digital to analogue converters.
10. Frequency spectrum analysers.
11. Ternary logic devices.

Some Examples

One of the outstanding characteristics of the Hall effect multiplier is its ability to form products of two signals over a large range of amplitudes without the generation of spurious signals. Hence, unlike all other electronic multipliers which utilize the non-linear transfer characteristic of the device, the Hall effect multiplier is a true product-forming device. On the other hand, the practical Hall effect multiplier suffers from a number of limitations, such as:

(a) A relatively small output signal which requires amplification.
(b) The need to provide one of the input signals in the form of a magnetic field results in an inefficient use of power, in particular when a wide range of frequencies of such signal is desired.
(c) An induced potential at the output of the device due to the rate of change of the magnetic field.
(d) A potential difference at the output terminals due to their misalignment, or due to the non-uniformity of the specimen.

An obvious application of the Hall effect multiplier is for use in analogue computers: if the frequency response of the applications is limited, one input signal can be utilized in the form of a coil wound on a ferromagnetic core to provide the magnetic field for the Hall plate, the other signal is providing a current supply to the plate.

As the Hall voltage will be linearly related to the currents in the coil and plate, it is essential that these should be supplied from high impedance sources to obviate the effect of coil reactance and the magnetoresistance effect in the plate (see *Magnetoresistivity*); in this respect a plate of gallium arsenide would be superior as its mobility is lower than that of indium arsenide. Although Hall effect multipliers can be made with the use of air-cored coils where operation at high frequencies is desired, the output of such multipliers is small owing to the low value of flux density obtained with the normal values of coil current. Normally to concentrate the flux, a ferrite pot core assembly is used, containing a small gap; the high resistivity ($>1\,\Omega\text{m}$) of the ferrite enabling the Hall plate to be mounted directly on the magnetic material without the need for insulation, which would increase the gap and reduce the flux. An additional advantage in comparison with the use of silicon steel, is the greater initial permeability and lower hysteresis losses; however, a disadvantage is the low saturation flux density ($B_{\max} \cong 0\cdot 34$ T at 20°C). The multiplier shown in Fig. 1 has an output of 100 mV with a 1 per cent accuracy at maximum ratings. A later design is shown in Fig. 2.

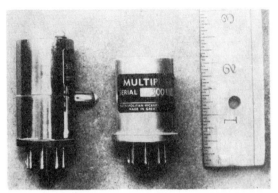

Fig. 1. Left: Hall effect radio-frequency modulator. Right: Hall effect multiplier (1960).

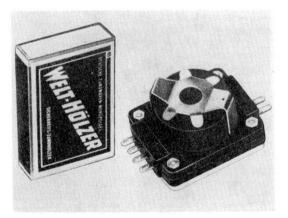

Fig. 2. Modern Hall effect multiplier (1964).

High-frequency Applications

When a Hall effect field is established in a semiconductor a brief time elapses before the current carriers redistribute themselves. Welker has discussed theoretically the frequency dependence of the Hall effect and Harris (1961) has shown experimentally that substantial Hall voltages can be produced in germanium at frequencies up to several thousand megaherz. The frequency limitation in a practical multiplier is shown, not in the Hall effect itself, but in the inevitable stray capacitances and inductances of the phase and its associated wiring. Of especial interest is the fact that since the Hall voltage is a product of specimen current and applied magnetic field then these are the basic requirements of a modulator. This is of great importance in the field of telecommunications, where a need has long been experienced for a distortionless pure-product modulator with absence of harmonic sidebands and carriers. Let I_c and I_m represent the currents through the semiconductor plate and coil respectively, thus:

$$I_c = I_1 \sin \omega_1 t$$

and

$$I_m = I_2 \sin \omega_2 t$$

where I_c and I_m represent the currents through the semiconductor plate and coil respectively.

The output expression then becomes

$$V_h = \frac{K_1 K_2 I_1 I_2}{2} (\cos(\omega_1 - \omega_2) t - \cos(\omega_1 + \omega_2) t)$$

where K_1 and K_2 are constants and the output contains the sum and difference signals but no component of either input signal alone; double sideband suppressed carrier signal is then obtained.

Figure 1 shows a high-frequency version of the multiplier. In practice well over 50 db carrier suppression has been measured.

It is interesting to note the various measures taken to ensure very low carrier break-through; one such measure devised by the writer is shown in Figs. 3a and 3b, which depicts an equivalent circuit representation of a Hall plate showing the effects of distributed resistive coupling as well as injection by mutual inductance between input and output circuits. Since the injected r.f. voltage is in phase quadrature with respect to the exciting plate current, it is possible to compensate for it by means of a capacitor connected as shown in dotted lines. In this method, the capacitor (a back-biased silicon $p-n$ junction) is connected between the input and one Hall electrode so that the current injected into the secondary circuit establishes a voltage whose phase relationship to the induced voltage ($j\omega M I_c$) is such as to compensate the latter. As the Hall effect represents a perfect modulator, the output consists ideally of two sidebands with complete suppression of carrier and modulation frequencies and with a 100 per cent modulation index. For telecommunication purposes it is highly desirable that all the modulating power be confined solely to the two sidebands, none being wasted in transmission of carrier frequencies. However, in a practical Hall modulator the presence of residual voltages and harmonic and modulation products arising from the magnetic circuit may lead to modulating envelope distortion, thus causing the transmitted signal to occupy a wider channel than is necessary. By connecting the Hall output to the vertical plates

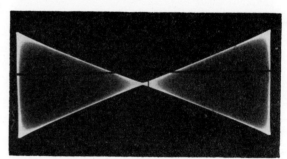

Fig. 4. Modulation characteristic of R. F. Hall modulator.

of an oscilloscope (the amplifier being assumed distortion free) and the modulation to the horizontal deflection plates, a twin trapezoidal ("bow-tie") pattern is obtained. The exceptional linearity of the Hall modulation process and the absence of residual and parasitic voltages in the centre is clearly shown in Fig. 4.

Single Sideband Generator

The considerable advantages of single sideband transmission are well known and the Hall effect lends itself admirably to the formation of a single sideband. A method used to achieve this is shown in Fig. 5 in which a high frequency ferrite assembly contains two gaps at right angles to each other, each Hall plate being excited by a coil carrying carrier frequency currents in quadrature with each other. Modulating

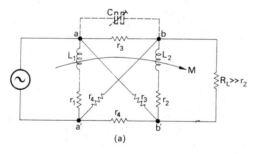

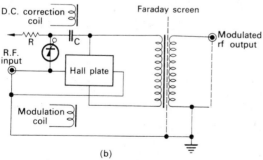

Fig. 3. (a) Equivalent circuit of Hall plate showing effects of distributed resistive coupling with compensating capacitor between input and on Hall electrode. (b) Circuit representation of improved Hall modulator.

currents are fed to the Hall plates again in quadrature such that a rotating magnetic vector exists.

Let I_{c1} and I_{c2} be the carrier currents of angular frequency ω_c flowing through coils 1 and 2 and I_{m1} and I_{m2} be the modulation currents of angular

Fig. 5. Top left: R. F. Hall modulator (1960). Top right: R. F. Hall modulator (1962). Bottom centre: Single sideband generator (after Saraga). For clarity the two Hall plates have been removed from gaps in ferrite.

frequency ω_m flowing through Hall plates 1 and 2. Then the respective magnetic fluxes may be represented by $B_1 = B_{01} \sin \omega_c t$ and $B_2 = B_{02} \cos \omega_c t$.

Hence the Hall voltage Vh_1 from plate 1 is represented by

$$Vh_1 = k \sin \omega_c t \sin \omega_m t$$

and

$$Vh_2 = k \cos \omega_c t \cos \omega_m t$$

where k is a constant.

If the Hall output voltages are added we obtain

$$Vh_1 + Vh_2 = k \cos(\omega_c - \omega_m) t$$

in other words the lower sideband is immediately achieved. Similarly if the Hall output connections are arranged for opposing voltages

$$Vh_1 - Vh_2 = k \cos(\omega_c + \omega_m) t$$

i.e. the upper sideband can be obtained.

The simplicity and elegance of this method is apparent.

Improved Forms of Hall Effect Probes

There is often a requirement in the technological applications of Hall effect devices for a probe to furnish large voltage outputs. For a given flux density and given semiconductors, there are two parameters which can be varied in order to obtain larger outputs, these are:

1. The thickness t occurring in the expression

$$Vh = \frac{R_h}{t} I.B.$$

2. The control current through the plate.

Requirement (1) may be met by grinding and lapping a bulk specimen but semiconductor wafers which are uniformly as thin as 0·001 in. are difficult to achieve. However, it is now possible to grow evaporated films of the compound semiconductors with thicknesses of the order of microns and possessing useable mobilities of say 10,000 cm²/V sec for indium arsenide. These thin films are at the moment polycrystalline in nature—some examples are shown in Fig. 6.

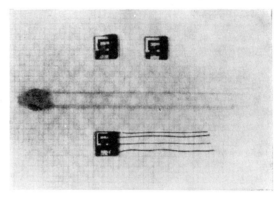

Fig. 6. Some examples of evaporated n-InAs Hall plates on alumina or ferrite substrates.

A third method of attack is the use of epitaxial growth on an insulating substrate which is crystallographically isomorphous with the epitaxial growth. For example, epitaxial gallium arsenide of a few microns thickness can be grown on a substrate of semi insulating gallium arsenide and moreover possesses a single crystal morphology. Requirement (2) involving higher control currents results in a higher power dissipation in the Hall plate (it should be noted that for a given current a reduction of plate thickness as above would also lead to enhanced power dissipation), and such dissipation would result in excessive temperature rises, with subsequent degradation of electrical properties. It is therefore essential that the semiconductor wafer be attached to a substrate whose properties are matched to those of the semiconductor.

Modern substrates covering the majority of applications are the high thermal conductivity ceramics such as alumina and beryllia (BeO) and in special cases diamond (possessing five times the thermal conductivity of copper). Figure 7 shows an indium arsenide Hall plate of thickness 0·0015 in. on a beryllia substrate. The leads are of evaporated silver and attached to a flexible printed circuit. It is interesting to compare the sensitivities of an indium arsenide and a gallium arsenide Hall plate with the same length and width but with an epitaxial gallium arsenide thickness of 3μ, 0·5 mV/mA kg and 78·5 mV/mA kg respectively. For 1 W dissipation the Hall voltage of the gallium arsenide device would be 16 V in a field of 1 Tesla.

Fig. 7. n-InAs Hall plates mounted on beryllia wafer.

Watt Transducers

The utilization of the Hall effect to measure power is not a new concept. The development of highly reliable practical Hall effect wattmeters is, however, relatively new. Figure 8 shows a wattmetric element utilizing a hall plate on a beryllia substrate. The line current $I \cos(\omega t + \delta)$ flows through a coil on the

Fig. 8. Top centre: Hall effect wattmetric element. Bottom: Two views of assembled wattmeter.

magnetic circuit and sets up a magnetic field in the ferrite core which is proportional to this current. A potential transformer supplies a control current to the Hall element which is proportional to the line voltage. Since the output of the device is, ideally, proportional to the two currents, the output voltage is

$$Vh = KE \cos \omega t I \cos(\omega t + \theta)$$

or

$$Vh = KE\, I \cos \theta + KE\, I \cos(2\omega t + \theta)$$

The first term is a d.c. term and is proportional to true power, the second double frequency term is proportional to volt-amps. In practice, there are some error terms which may be introduced by phase shifts caused by the core hysteresis and the errors due to the potential transformer. If δ_1 and δ_2 are the phase shifts caused, respectively, by the core and transformer, then

$$Vh = KE\, I \cos(\theta + \delta_2 - \delta_1)$$
$$+ KE\, I \cos(2\omega t + \theta + \delta_2 + \delta_1)$$

For the d.c. term to be a true indication of power it is necessary that $\delta_2 = \delta_1$. The wider the power factor range over which the transducer must operate the more critical the relationship between the two phase errors become. However, since a great many applications of such a power transducer is at a fixed power line frequency external phase correction can be inserted in the control current circuit.

There are numerous applications where Hall effect power transducers offer advantages over some of the other power measuring techniques. Since the transducers are all solid state devices, they are extremely rugged and the type of transducer shown in Fig. 8 can furnish several milliamps capable of driving meters, recorders or serving as the input to control circuitry. With regard to accuracy the wattmeter conforms to B.S. 89 (1954) for the limits laid down for an induction type wattmeter, and, furthermore, large overloads will not damage the instrument.

The Hall effect has thus far furnished a great variety of devices possesisng at least one important advantage over most semiconducting devices, i.e. Hall effect devices require the use of majority current carriers only and therefore involve no junctions. There are no areas of concentrated electric field, so that the surfaces need not be elaborately protected. Figure 9 illustrates a variety of Hall probes commercially available. In practice there are particularly intriguing applicational aspects and certain limitations imposed by the compromises in design and technological limitations related to the degree of experience in the field. It should be stressed that although the Hall effect has been extant for many years serious applications have been attempted only during the last 5–10 years.

Bibliography

HARRIS, D. J. (1961) Hall effect, in *Encyclopaedic Dictionary of Physics* (J. Thewlis Ed.), **3**, 560, Oxford: Pergamon Press.

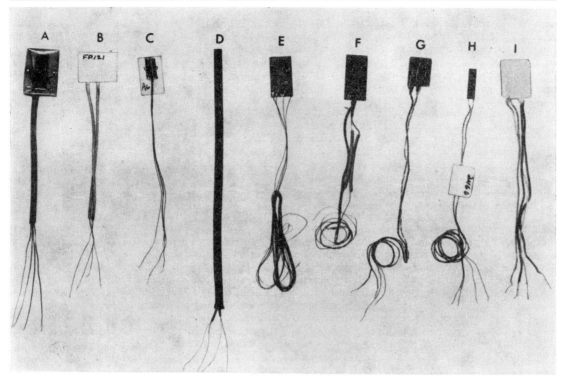

Fig. 9. Present-day Hall probes.

HILSUM C. and ROSE-INNES A. C. (1964) *Semiconducting II–V Compounds*, Oxford: Pergamon Press.
MADELUNG O. (1964) *Physics of the II–V Compounds*, London: John Wiley.
PUTLEY E. H. (1960) *The Hall Effect and Related Phenomena*, London: Butterworths.
Conference on Hall effect and galvanomagnetic devices, *Solid State Electronics* (1966) **9**, No. 5, Oxford: Pergamon Press.
Hall effect issue, *Solid State Electronics* (1964) **7**, No. 5, Oxford: Pergamon Press.

E. COHEN

HAMILTON'S PRINCIPLE. Hamilton's principle is a famous landmark of classical mechanics. A key part in this principle is played by the Lagrangian integrand function $\mathscr{L}(y_1, y_2, ..., y_n, \dot{y}_1, \dot{y}_2, ..., \dot{y}_n, t)$ which may be associated with a physical system whose behaviour can be characterized by the coordinates $y_i, i = 1, 2, ..., n$, their time derivatives $\dot{y}_i$ and the time t. In brief, Hamilton's principle consists in applying the calculus of variations to determine the stationary value of the integral,

$$\mathscr{P} = \int_{t_1}^{t_2} \mathscr{L}(y_i, \dot{y}_i, t) \, dt, \qquad (1)$$

termed the *principal function*. The stationary value of the principal function establishes the path $y_i(t)$ traversed by the system between two fixed points $y_i(t_1)$ and $y_i(t_2)$. A necessary and sufficient condition for the principal function to be stationary is that the following n differential equations be satisfied.

$$\frac{d}{dt}\frac{\partial \mathscr{L}}{\partial \dot{y}_i} - \frac{\partial \mathscr{L}}{\partial y_i} = 0, i = 1, 2, ..., n. \qquad (2)$$

In (2) we have the *equations of motion of the system*.

With external forces J_i applied to the system, the component J_i being associated with the coordinate y_i, then (2) is modified as follows:

$$\frac{d}{dt}\frac{\partial \mathscr{L}}{\partial \dot{y}_i} - \frac{\partial \mathscr{L}}{\partial y_i} = J_i. \qquad (3)$$

The Lagrangian of an isolated conservative system may be written,

$$\mathscr{L} = T - U, \qquad (4)$$

with T the kinetic energy and U the potential energy. By using some common parameter τ between all the coordinates including the time, it can be shown that both the Lagrangian and the principal function are invariant under rheonomic transformations of coordinates. It is in order to hold reservations about the nature of the stationary value of the principal function that one is concerned with whether it establishes a maximum, saddle point, or a minimum, stable behaviour being associated only with the latter. The stability of physical systems has been treated by Routh (1877 and 1913).

The coordinates $y_i(t)$ do not necessarily refer to points in space, and neither do the time derivatives $\dot{y}_i(t)$ necessarily correspond to velocities in the usual sense. In fact, the coordinates y_i may represent (a) linear displacement (mechanics, acoustics, hydraulics), (b) angular rotation (mechanics), (c) electric charge (electric networks), etc. Nevertheless, it goes without saying that the terms in the Lagrangian depending on linear displacement are not exactly analogous to terms depending on angular rotation nor electric charge, i.e. a Lagrangian for linear motion does not in general serve for angular rotation by merely substituting angular rotations for linear displacements. But, by paying due regard to the distinctions between different types of coordinates, Lagrangian functions may be written for a wide variety of physical systems. Furthermore, by giving special consideration to the scope of problems, the selection and handling of coordinates, it has been possible to extend Hamilton's principle into modern physics, e.g. into general relativity and quantum electrodynamics.

Since it is fairly clear that care must be exercised in the formulation of the principal function, and in the choice and handling of coordinates, it may not seem surprising that a non-existence proof has been given for a general variational treatment of a steady state nonconservative system (Gage et al., 1966). In this variational treatment, a single space coordinate and generalized thermodynamic potentials play the roles of the time and space coordinates respectively in the usual formulation of the principal function. It has been shown that the *Onsager relations* cannot be obtained from such a general variational treatment.

Nevertheless, Hamilton's principle can be applied to many nonconservative systems, e.g. a nonconservative linear system may be treated by separating the Lagrangian into two parts, one for the dissipative and the other for the conservative properties of the system. The Lagrangian term for energy dissipated within the system is summarized by the functions F and Π, e.g.

$$\mathscr{L}'' = F + \Pi \qquad (5)$$

in which†

$$F = \dot{y}_i \frac{R_{ij}}{2} \dot{y}_j \qquad (6)$$

is Rayleigh's dissipation function (half the power consumed by relaxation processes) and

$$\Pi = \ddot{y}_i \frac{P_{ij}}{2} \ddot{y}_j \qquad (7)$$

† The summation convention of repeated subscripts is used throughout.

is one-half the power lost by radiation. The matrix elements R_{ij} and P_{ij} are real constants and symmetric. The Lagrangian term for the conservative properties of the system has the same functional form as (4) and is designated

$$\mathscr{L}' = T - U. \qquad (8)$$

It can be shown that any nonconservative linear system has the following equations of motion:

$$\left\{ \frac{\partial \mathscr{L}''}{\partial \dot{y}_j} - \frac{d}{dt} \frac{\partial \mathscr{L}''}{\partial \ddot{y}_j} \right\} - \left\{ \frac{\partial \mathscr{L}'}{\partial y_j} - \frac{d}{dt} \frac{\partial \mathscr{L}'}{\partial \dot{y}_j} \right\} = J_j. \qquad (9)$$

In order to reduce (9) to a more specific statement, we take U as the quadratic function of the coordinates $y_s(U_{sj}/2) y_j$ and T as the sum of a quadratic term $\dot{y}_s(M_{sj}/2) \dot{y}_j$ and a gyroscopic term $\dot{y}_s(G_{sj}/2) y_j$; then we have

$$-P_{js} \dddot{y}_s + M_{js} \ddot{y}_s + B_{js} \dot{y}_s + U_{js} y_s = J_j \qquad (10)$$

with $B_{js} = R_{js} + G_{js}$, and with the matrices M_{js} and U_{js} symmetric and G_{js} antisymmetric, and with the elements of M_{js}, G_{js}, and U_{js} real.

We shall treat the external forces, the coordinates and their time derivatives as Fourier series, taking the duration of the principal function as the fundamental period. Thus, the frequencies of the Fourier series may be written

$$\omega_k = \frac{2\pi k}{t_2 - t_1}, \qquad k = 0, \pm 1, \pm 2, \pm 3, \ldots \qquad (11)$$

We write the forces, the coordinates and their time derivatives as the complex functions,

$$J_j(t) = \sum_k J(k) \, e^{i\omega_k t}$$

$$y_j(t) = \sum_k y_{jk}(t) = \sum_k y_j(k) \, e^{i\omega_k t}$$

$$\dot{y}_j(t) = \sum_k \dot{y}_{jk}(t) = \sum_k \dot{y}_j(k) \, e^{i\omega_k t}$$

$$\ddot{y}_j(t) = \sum_k \ddot{y}_{jk}(t) = \sum_k \ddot{y}(k) \, e^{i\omega_k t}$$

etc.

Now we write the Lagrangian of the system,

$$\mathscr{L} = \sum_k [\mathscr{L}_k'' - i\omega_k \mathscr{L}_k'], \qquad k = 0, \pm 1, \pm 2, \ldots \qquad (12)$$

in which the nonconservative and conservative terms of the Lagrangian have been decomposed into the following spectral components:

$$\mathscr{L}_k'' = \underbrace{[\mathrm{Re}\, \dot{y}_{sk}(t)] \frac{R_{sj}}{2} [\mathrm{Re}\, \dot{y}_{jk}(t)]}_{F_k}$$

$$+ \underbrace{[\mathrm{Re}\, \ddot{y}_{sk}(t)] \frac{P_{sj}}{2} [\mathrm{Re}\, \ddot{y}_{jk}(t)]}_{\Pi_k}$$

$$\mathscr{L}_k{}'$$
$$= \underbrace{[\operatorname{Re} \dot{y}_{sk}(t)] \frac{M_{sJ}}{2} [\operatorname{Re} \dot{y}_{jk}(t)] + [\operatorname{Re} \dot{y}_{sk}(t)] \frac{G_{sJ}}{2} [\operatorname{Re} y_{jk}(t)]}_{T_k}$$
$$- \underbrace{[\operatorname{Re} y_{sk}(t)] \frac{U_{sJ}}{2} [\operatorname{Re} y_{jk}(t)]}_{U_k.}$$

It has been shown that the following relation holds:

$$\frac{2}{t_2 - t_1} \sum_k \int_{t_1}^{t_2} [\mathscr{L}_k{}'' - i\omega_k \mathscr{L}_k{}'] \, dt$$
$$= \frac{1}{t_2 - t_1} \sum_k \int_{t_1}^{t_2} \frac{J_j^* y_j}{2} \, dt. \quad (13)$$

The real part of the expression on the right side of (13) gives the power irreversibly supplied to the system, and that equals the power dissipated by the system as given by the real term on the left side. The imaginary part of the expression on the left refers to the rate at which energy alternates between kinetic energy and potential energy, and any excess of one over the other is exchanged with external forces as indicated on the right. The factor 2 on the left enters because $\mathscr{L}''$ was originally specified as half the rate of energy dissipation, and because the kinetic and potential energies in $\mathscr{L}'$ vary with frequency components $2\omega_k$ (while the coordinates themselves have frequency components ω_k). We recognize, from the expression on the right, that (13) refers to the *complex power* of the system, and therefore we may write

$$\mathscr{P}^* = \frac{2}{t_2 - t_1} \int_{t_1}^{t_2} \mathscr{L} \, dt = \frac{1}{t_2 - t_1} \sum_j \int_{t_1}^{t_2} \frac{J_j^* \dot{y}_j}{2} \, dt. \quad (14)$$

Thus, with nonconservative linear systems, the complex power $\mathscr{P}^*$ serves as the principal function. Therefore, the stationary value of the principal function for a stable operating path of the system makes the energy dissipation a minimum by making the real part of $\mathscr{P}^*$ a minimum, and it makes the difference between kinetic and potential energy a minimum by making the imaginary part of $\mathscr{P}^*$ a minimum.

Bibliography

A collection of papers in memory of Sir William Rowan Hamilton (1945), *Scripta Mathematica*, New York: Yeshiva College.

GAGE D. H., SCHIFFER M., KLINE S. J. and REYNOLDS W. C. (1966) The nonexistence of a general thermokinetic variational principle, in *Non-equilibrium Thermodynamics Variational Techniques and Stability* (R. J. DONNELLY et al., Eds.), Chicago: University of Chicago Press.

GOSSICK B. R. (1967) *Hamilton's Principle and Physical Systems*, New York: Academic Press.

ROUTH E. J. (1877) *The Adams Prize Essay* and (1913) *Dynamics of a System of Rigid Bodies*, 8th ed., London: Macmillan.

B. R. GOSSICK

HELICITY. In elementary particle physics, the helicity of a particle is defined to be the component of its spin along the direction of its motion. Thus, if spin and momentum are represented respectively by the operators Σ and $\mathbf{p}$ then the helicity operator is $H = \Sigma \cdot \mathbf{p}/|\mathbf{p}|$.

The helicity is most useful in the description of weak interactions. For spin $\frac{1}{2}$ particles, with spin represented by the Pauli spin matrices σ, the helicity is often defined by $\sigma \cdot \mathbf{p}/|\mathbf{p}|$ (differing from H above by a factor 2) the eigenvalues of which are ± 1. According to the two-component theory of the neutrino, the neutrino occurs in nature in only one helicity state, whereas the anti-neutrino occurs in the opposite helicity state. This has been tested experimentally, and found to hold true to a high degree of accuracy. It turns out that the neutrino has negative helicity (or is left-handed) while the anti-neutrino has positive helicity (or is right-handed).

W. W. BELL

HIGH INTENSITY PULSED RADIATION SOURCES

Introduction

Over the decade 1940–50, flash X-ray equipments were developed mainly for the examination of explosive ordnance. X-ray voltages in the region of 100 kV (Breidenbach, 1949; Criscuolo and O'Connor, 1953) were used and flash durations progressively reduced from 5 μsec to less than one microsecond. The field emission tube WL 389 (Westinghouse) typifies the development of this period. Outputs on the order of 10 millionroentgens at one metre from target were obtained.

The demand for extremely short intense bursts of radiation for testing has spurred development since about 1960 in the supervoltage region (1–10 MV) with consequent emphasis on tube, energy store, and switch design.

The Field Emission Tube

A flash X-ray system consists essentially of the energy store, the field emission tube, and the switch system to deliver the stored energy to the tube (Fig. 1).

Energy store and tube may be regarded as having impedance Z_s and Z_t. Switch system impedance is considered negligible in most systems.

The field emission diode consists in its simplest form of an anode, usually planar, and a "point" cathode. Field emission diodes were first developed for relatively small currents (about 10^{-4} A)—for example, in field emission microscopes. Fowler and Nordheim in 1928 developed a wave mechanical theory for field emission applicable to small current densities. Under these circumstances the current is an exponential function of the field or applied voltage,

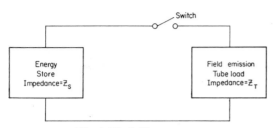

Fig. 1. Flash X-ray system.

and is critically dependent on the work functions of the cathode and its microscopic geometry. Extremely clean and outgassed points of approximately one micrometre radius are required for stable operation and, therefore, for a stable impedance. Good and Muller (1956) and W. P. Dyke (1964) and his co-workers have investigated this subject most thoroughly.

At the supervoltages and very large currents (about 10^4 A) required for the production of intense radiation pulses, another factor dominates. This is the space charge due to the beam itself. For example, a one centimetre slab of a 30,000 A beam contains one microcoulomb of charge—creating at its periphery a negative potential of about one million volts. The net result of this space charge effect is that the design of such field emission diodes is, as J. C. Martin and his co-workers have shown, no longer critically dependent on the microscopic geometry of the cathode or its surface condition. The factor governing tube impedance is its macroscopic geometry, and even this is relatively non-critical.

Tube designs have been made over the range 600 keV and 5000 A to 10 MeV and 0·5 MA. At the lower end of the range sealed tubes are used (Field Emission Corporation) but, for the most part, limitations on tube life, as well as fabrication criteria, demand pumped demountable arrangements.

A major feature of supervoltage tube design again due to the J. C. Martin group concerns the tube envelope. They have shown that a preferred angle of the dielectric/vacuum interface relative to the tube axis minimizes flashover. A theoretical justification of this structure has been given by Watson and Shannon (1966).

One typical tube structure (Ion Physics Corporation) used in conjunction with a coaxial gas capacitor and shown in Fig. 2 consists of insulators (polymethylmethacrylate or glass) and metal planes. The cathode post, mounted from the cathode termination, is provided with a replaceable tip that is the cathode proper. The anode, for X-radiation, is a tantalum disc of suitable thickness. Other materials of high atomic number (e.g. tungsten, or even uranium) may also be used, and the anode is replaceable by a 0·005 cm titanium window for electron beam use.

In another tube design (Physics International Corporation) a single (epoxy)insulator is used that not only conforms with the J. C. Martin criteria, but also serves to give an approximately correct Brewster angle between the liquid dielectric of the stripline and the vacuum insulation of the tube (Fig. 3).

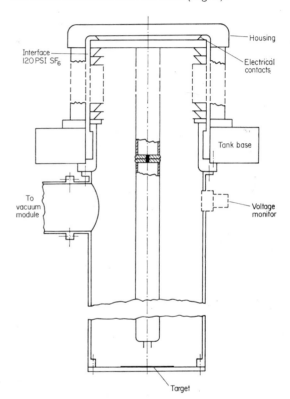

Fig. 2. Typical flash-tube structure.

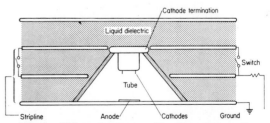

Fig. 3. Tube structure with Brewster angle envelope (Physics International Corporation).

Pulsed Discharge Supplies

Several forms of fast discharge energy stores are used with field emission tubes. At relatively low voltages a pulse transformer and pulse forming network (Fig. 4) is the simplest arrangement. Only

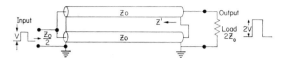

Fig. 4. The stacked line transformer.

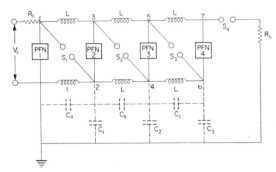

Fig. 5. Schematic circuit for the Marx–Surge generator, indicating energy storage elements, switches, isolation charging impedance, lead and "stray" capacitances.

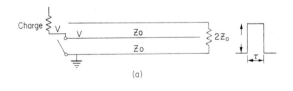

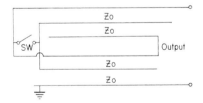

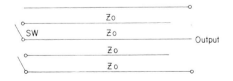

Fig. 6. Methods of stacking the Blumlein circuit.

one switch is needed to transfer energy from store to tube. The Marx circuit (Fig. 5) is used for tube voltages up to about 2 MV. Multiple spark gaps in series are required to pulse the tube. Figure 6 illustrates the Blumlein circuit developed by J. C. Martin and his co-workers at AWRE. Extensions of this principle have been described by Fitch and Howell (1964). The principle of operation of the Blumlein circuit is illustrated in Fig. 7. The central conductor or plate

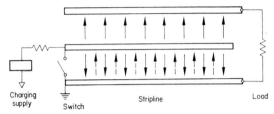

Fig. 7. Blumlein circuit-operating principle.

is elevated to potential V and establishes an electric field in the direction of the solid arrows to the outer conductors, which are grounded—the upper through a high impedance. When the outer ends of the inner and lower conductors are shorted, an equal and opposite field is progressively established (at the velocity of light in the dielectric) between them. A potential $2V$ is thereby established across the load for a discharge time determined by the dimensions (length) of the system. Blumlein or multiple stripline arrangements require pulse transformers or Marx generators to energize them so that sequential switching of the slow and fast energy stores is required.

The coaxial gas capacitor is illustrated in Fig. 8. Energy is stored in the high pressure gas dielectric between the inner conductor, the long terminal of an electrostatic belt generator, and the pressure vessel or outer conductor. The coaxial capacitor is also a transmission line whose discharge time t (into a matching load) is given by

$$t = \frac{2L}{c},$$

where L is the length of the terminal and c the velocity of light.

The impedance of the transmission line is given by

$$Z_s = 60 \ln \frac{r_2}{r_1},$$

where r_2 is the inner radius of the pressure vessel, and r_1 is the radius of the inner conductor (terminal). When switched into a matching load ($Z_T = Z_s$) the voltage occurring across this load equals $V_{s/2}$, where V_s is the storage (terminal) potential. However, the problem is one of maximizing X-radiation output. Since X-radiation varies as $V^{3.4}$ in the voltage range of interest, some mismatch between the tube and the coaxial transmission line is desirable. The derivation

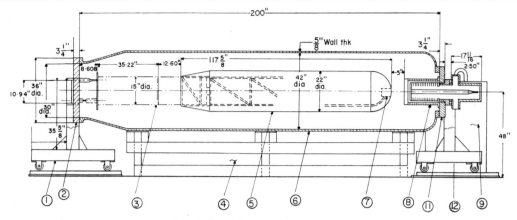

Fig. 8. Coaxial gas capacitor system.

of optimum tube impedance shows that it should be about 3·4 times the line impedance.

The coaxial gas capacitor has merit in that the dielectric is self-healing, only one switch is required, namely, between the high voltage conductor and the tube, and the in-line arrangement and absence of inductive components minimize extraneous electrical noise. Also substantially all the stored energy is delivered to the tube, avoiding potentially disruptive energy losses. A major disadvantage exists in the lower energy stored per unit volume compared with liquid (or solid) dielectrics.

The potential demand for still higher rates of radiation energy delivery has encouraged the study of pulsed liquid dielectrics. The most interesting is water with its high dielectric constant (about 80) and ability to withstand high electric fields for short times ($< 3 \times 10^{-6}$ sec). Systems in which a water dielectric Blumlein circuit is charged within 1 or 2 μsec and discharged in a few tens of nanoseconds into a suitable X-ray tube, therefore, hold good promise. The fire hazard and practical difficulties of handling and decontaminating large volumes of expensive oil dielectric are avoided.

Switching

The technique for switching the stored energy into the field emission tube generally consists of triggered gas insulated spark gaps. Arrangements are illustrated for various stores in Figs. 3, 4 and 5.

Illustrative of various techniques are three methods employed with the coaxial gas capacitor store. The first method, used when the experiment can be triggered by the radiation pulse, or where non-dynamic (electrical or mechanical) systems are being investigated has the merit of extreme simplicity. It consists simply of energizing a solenoid to cause a metal plunger to protrude through the terminal shell toward the tube. As the plunger moves toward the tube, a breakdown condition is reached at which the terminal energy is transferred to the tube.

The second method, used where the radiation pulse must be critically synchronized to an experimental event is an adaptation of the trigatron. At the apex of the terminal is a small disc or "pin" insulated from the remainder of the terminal. When a voltage is applied to the "trigger pin" from a pulse transformer within the terminal, an electrical discharge occurs to the terminal proper and then propagates to the field emission tube. The pulse transformer is activated by a photocell (and associated amplifier) that observes (through a polymethylmethacrylate light "pipe") a fast light source (spark gap) triggered at ground. The "jitter" of this electronic system is on the order of a few nanoseconds and in general is significantly less than the radiation pulse duration.

The third method uses a laser beam to ionize the gas gap between the high voltage terminal and the tube.

Use of the Electron Beam

The electron beam may be used by substituting an electron window—a titanium foil, 0·005 cm thick—for the anode.

With a good vacuum (10^{-3} torr) outside the window, the electron-beam cross-section rapidly expands owing to space charge effects. As the pressure is raised to about 2×10^{-1} torr (for air) the Bennett (1955) pinch is exhibited (Graybill and Nablo, 1966). In brief, the incident high current beam ionizes the gas molecules, radially repels the low energy electrons, leaving the slower moving ions to neutralize the beam charge. The self-magnetic field due to the beam current constrains the beam, exhibiting the "pinch" effect. The effect of pressure on beam energy density is shown in Fig. 9. Figure 10 shows the beam cross section at various distances from the window.

This gas focusing technique can be used to convey the beam a considerable distance in a drift field. If the beam approaches a conducting surface at a large angle of incidence, the combination of self- and image-magnetic fields causes the beam to be reflected from the conductor.

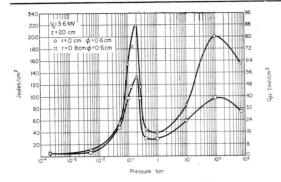

Fig. 9. The pressure dependence of the beam energy flux monitored calorimetrically 20 cm from the entrance window.

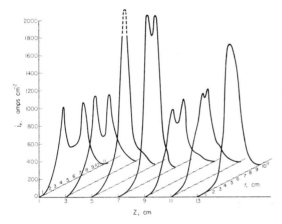

Fig. 10. Drifting electron beam in vertical profile.

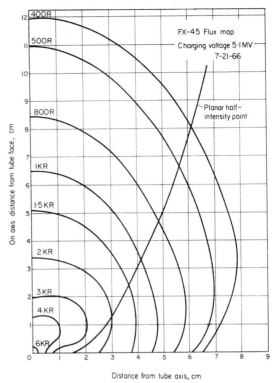

Fig. 11. Distance from tube axis (cm) (Ion Physics Corporation).

Diagnostics and Dosimetry

The high powers (10^{11}–10^{12} W) and short time durations (10^{-9}–10^{-7} sec) associated with flash beam equipments and their applications requires sophisticated measurement techniques.

1. Tube potential can be displayed on a fast (1 nsec or less rise-time) oscilloscope, using a capacitive potential divider across the last tube section. Calibration can be accomplished by a pulse generator delivering a few tens of volts.

2. Beam current may be determined and displayed by substituting a "Faraday cup" for the anode. However, some uncertainty arises due to secondary electrons. A more accurate method is to use a pickup loop around the "pinched" electron beam.

3. X-radiation intensity is displayed by a photodiode and fast oscilloscope. This technique is used to determine the X-radiation pulse length and waveform. Total X-radiation dose per pulse is determined by thermoluminescent dosimeters. The radiation contours of Fig. 11 have been determined this way. X-ray film dosimeters may also be used since the reciprocity law extends down to at least the pulse lengths and radiation intensities of interests. Development of the films makes the method somewhat tedious. X-ray calorimeters have been used as monitors. Ionization dosimeters saturate at intensities much less than the output of flash radiation equipments.

4. The pinched electron beam current density profiles (Fig. 9) are determined by the bleaching of blue cellophane (Graybill and Nablo, 1966) film arranged as shown in Fig. 12.

5. Electron beam energies have been determined by a magnetic spectrometer, Fig. 13. The spectrum (Fig. 14) obtained by this means suggests that lower energy components exist due to reflections on the coaxial line due to mismatch with the tube.

Applications

The major purpose for the development of super-voltage flash radiation machines has been to examine electronic systems and components exposed to intense X-radiation.

The X-rays may also be used for flash radiography of explosive motions in heavy sections, and other fast-moving events.

The X-ray intensity is sufficient at the tube face

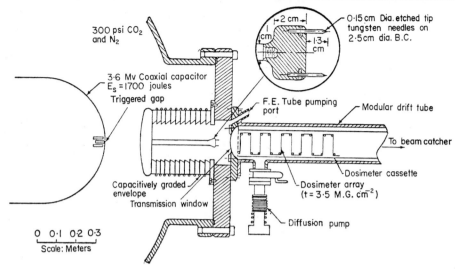

Fig. 12. Field emission tube/beam diagnostics geometry.

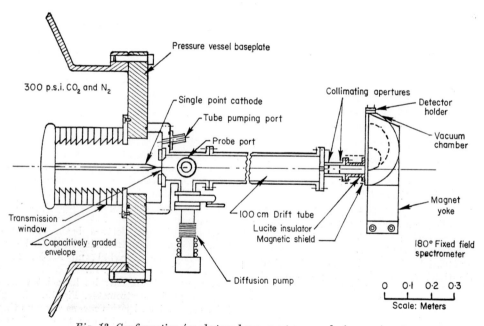

Fig. 13. Configuration for electron beam spectrum analysis experiment.

for much radiolysis research. However, both radiographic and radiolysis applications gain from using the pinched electron beam bombarding a target of high atomic number—the former from the smaller spot size and the latter from the higher X-ray intensity.

The electron beam may be directly used in radiolysis research, particularly because it can be constrained to approximate a pencil beam over several tens of centimeters. This pinched electron beam can also be used for materials investigations, and silicon (Oswald, Eisen and Conrad, 1966) fracture has been examined under such conditions. Extremely high energy deposition rates are obtained. Figure 15 shows an example in which the pinched electron beam produces explosive evaporation from the front of a tantalum plate and at the same time creates a shock wave to spall material from the rear face.

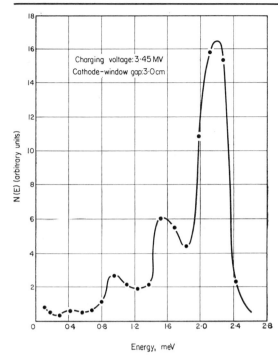

Fig. 14. Electron spectrum $V_0 = 3.5$ MV.

such accelerators is a pulsed field emission diode. Energy is switched into the X-ray tube from either a pulse transformer, a Marx circuit, a Blumlein circuit, or a multimegavolt coaxial gas capacitor. These machines are used for the study of the behaviour of materials and assemblies under intense short-time radiation, for the high-speed radiography of mechanical systems, and for the investigation of high dose-rate effects in chemistry and biology (pulse radiolysis).

Large space-charge forces within the electron beam limits magnetic focusing to small machines. In large machines constraint is obtained by passing the electron beam into a low pressure (c. 0·2 torr) gas. The gas is ionized, neutralizes the beam space charge and permits the "Bennett pinch" (1955) effect, due to the self-magnetic field of the intense beam ($>10,000$ A). In addition to its ability to ionize and heat gaseous plasma, the pinched electron beam can deliver energy to a material at an extremely high density (about 10^{12} W cm^{-2}). These high densities create and permit the study of shock waves.

Monitoring instruments include thermoluminescent dosimeters (TLD) and calorimeters for the X-ray dose, and photodiodes to determine waveform. Novel methods are required to "map" the pinched electron beam. One system uses radiation sensitive film (blue cellophane) with isodensitometric readout of the resultant bleaching.

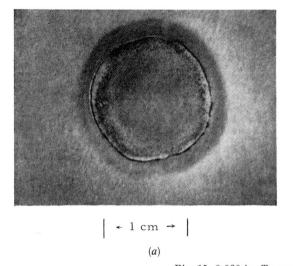
(a)

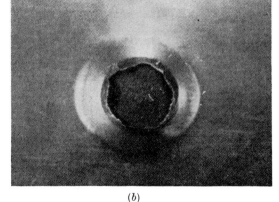
(b)

Fig. 15. 0·030 in. Ta exposed to 450 J/cm^2 beam.

Summary

Within the last decade "flash" supervoltage electron accelerators have been actively developed to produce intense bursts of X-radiation (1 to >1000 R at 1 m distance from the target) in very short time intervals ($<10^{-7}$ sec). The common component of

Bibliography

BENNET W. A. (1955) Magnetically self-focusing streams, *Phys. Rev.* **98**, 890; (1934) **45**, 1584.

BREIDENBACH H. I. Jr. (1949) Fractional microsecond X-ray pulse generator for studying high explosive phenomena, *RSI* **20**, 899.

CRISCUOLO E. L. and O'CONNOR D. T. (1953) The development of a fine focus flash X-ray tube, *RSI* **24**, 944.

DYKE W. P. (1964) Advances in field emission, *Scientific American* **210**, 108.

FITCH R. A. and HOWELL V. T. S. (1964) Novel principles of transient high voltage generation, *Proc. I.E.E.(U.K.)* III, No. 4, 849.

GOOD R. H. Jr. and MÜLLER B. W. (1956) *Handbuch der Physik* **21**, 176, Springer-Verlag, Berlin, Germany.

GRAYBILL S. E. and NABLO S. V. (1966) Techniques for the study of self-focusing electron streams, *Proc. Eighth Annual Symposium, Electron and Laser Beam Technology*, Ann Arbor, Michigan.

MARTIN J. C. AWRE, Aldermaston, U.K. Private communications.

OSWALD R. B., EISEN H. A. and CONRAD E. E. (1966) *Proc. I.E.E.E. Radiation Effects Conference*, Palo Alto, California.

WATSON A. and SHANNON J. (1966) Ion Physics Corporation, Pulsed flashover of insulators in vacuum, *Proc. Int. Symposium on Insulation of High Voltages in Vacuum*.

A. J. GALE

HIGH-LIFT DEVICES. Lift in aerodynamic terms is the force opposed to gravity which can be used to carry a body in the air. Lifting forces can be generated in a number of ways, statically in balloons and airships and dynamically by aeroplane wings, by jet lift and by the rotating arms of rotorcraft. High-lift devices are here assumed to apply to dynamic reactions excluding rotorcraft.

The lift of an aerofoil or wing section is the air reaction in a direction normal to that of the airstream. Per unit span it has the value $k\varrho v$, where k is the circulation round the aerofoil, ϱ the density of the medium, and v the velocity of the main stream, a result known as the Joukowski formula. The choice of the circulation for a given aerofoil is usually determined from this formula and a general account of the circulation theory is given in volume 4 of the *Dictionary of Physics*. The object of a high-lift device is to increase the maximum value of this circulation. Any increase above that given by the formula is called supercirculation.

These devices are many and include flaps, slots, slotted flaps, blown flaps, jet flaps, suction, blowing and rotating surfaces. A brief description of the aerodynamics of slots is given in volume 6 and the increase of the maximum lift of a wing obtainable by their use is mentioned without specifically referring to a slot as a high-lift device. A separate article on jet flaps is included in the present volume.

Flaps

The simplest device for increasing lift is a plain flap (Fig. 1) involving the bending downwards of the rear portion of the aerofoil, as demonstrated by wind tunnel experiments as early as 1914. Lowering the flap increases the lift at all incidences with usually no appreciable change in the stalling angle at which the lift decreases (Fig. 2). Typically wind tunnel experi-

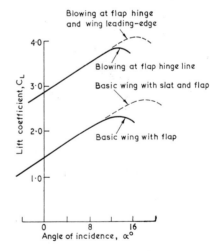

Fig. 2. *Generalized effect of flap slat, suction and blowing upon lift of basic wing.*

ments on the aerofoil R.A.F. 34 at a Reynolds number of 4×10 increased the maximum value of C_L, the lift coefficient, from 1·3 to 2·2. What happens is that the airflow breaks away from the sharp rear edge of the flap, a region of low pressure is formed behind it and an increase of suction is induced over the whole of the upper surface of the aerofoil. Another simple device invented independently by Sir Frederick

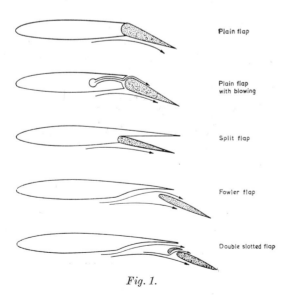

Fig. 1.

Handley Page and Dr. Lachman is to make a slot through the aerofoil or wing from the high pressure under surface to a region of high suction on the upper surface somewhat forward of the maximum camber. The flow of air through the slot delays the stall. With the slot open behind the leading edge slat as in Fig. 3, bottom sketch, the lift below that at which

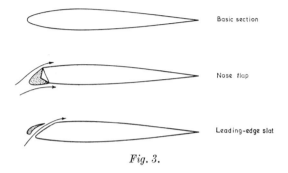

Fig. 3.

stalling begins is slightly lower than for the normal section. But as the angle of incidence of the wing increases the lift of the slotted wing continues to increase to a much larger angle. As a typical result the aerofoil R.A.F. 31 has a maximum lift of 1·15 at 12° incidence and when slotted this is increased to 1·82 at 23° incidence. Handley Page in his lecture to the Royal Aeronautical Society in 1921 quotes a case in which a very high lift section known as R.A.F. 19, divided into eight sections by seven slots equally spaced along the chord length, gave a maximum lift of 3·4 at 43° angle of incidence compared with the normal coefficient of 1·64 at 14°. In one experiment with only 6 slots open C_L reached the figure of 3·92 at a 45° stalling angle. Considered generally, the effect of a single slot is to continue the linear part of the lift incidence curve of the plain aerofoil to a much higher incidence with the same slope. With slotted flaps and a leading edge Handley Page slot it is comparatively easy to design a modern wing section to give a maximum lift of about 3·0. Analogous experiments were made earlier on rider planes by Dr. Thurston at East London College in 1914, and were reported at the above lecture.

The general development of flaps as high-lift devices has been wrapped up with that of slots, suction and blowing. Turning to Fig. 1, the plain flap which was first developed became a split flap so as not to alter the contour of the upper surface. The Fowler flap followed with the flap portion of the wing moved back to provide a larger total area for the wing and to open a slot. This was succeeded by the double and multiple slotted flap, a development of the multi-slotted wing of Handley Page, referred to above. In the meantime the leading edge slat was developed partly to gain lift and also for use as an aid to an aileron control. In 1928 the Air Ministry ruled that all British service aeroplanes should be fitted with wing-tip slots which, as auto-slots, automatically opened when the wing incidence reached a certain value, to increase the lift of the wing tip and thus prevent premature wing stalling with loss of control and a risk of involuntary spinning. Nose flaps (Fig. 3) were also tried with a view of increasing lift but were not widely adopted when first proposed. The adoption of many of these devices in the late 1960's is illustrated in Fig. 4, showing the de Havilland Trident with a nose flap, a double-slotted rear flap and an upward moving flap or spoiler, the last a device to decrease lift for control purposes. In the same Fig. 4 the Boeing 727

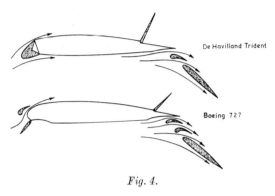

Fig. 4.

has still more devices with a triple-slotted rear flap, a spoiler, a leading edge slot and a nose flap which can be retracted into the under surface of the wing.

There are other types of flap in addition to those described above. A blown flap, for example, is a flap over which air is discharged in order to prevent boundary layer separation and thus to improve the lift of the flap. It is somewhat similar to the jet flap mentioned above. The use of suction and blowing is dealt with below.

Suction

The increase of lift on a surface over which fluid (air) is flowing is bound up with the problem of preventing the flow from separating from the surface. As early as 1904 Prandtl showed that separation from the surface of a circular cylinder could be prevented by sucking the surface fluid into a slot. Over the next 40 years or so this idea was followed up by several investigations in which suction was applied through slots, or porous surfaces, at various positions on conventional aerofoils. Slot suction near the leading edge was found to increase maximum lift and like suction through a porous nose, postponed separation of the flow from the surface to higher angles of incidence. The position of the slot has to be chosen with care and it appears that, on the whole, narrower slots with a width of one hundredth of the chord and less are more economical in the total quantity of fluid to be removed. Typical results are shown in Fig. 5 of the variation of maximum lift with momentum coefficient for an N.P.L. 434 aerofoil with slot widths varying from 0·02 to 0·05 in., the chord length being 18 in. The position of the slot near the leading edge

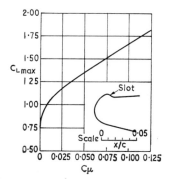

Fig. 5. *Maximum lift for an NPL 434 aerofoil*

is indicated on the figure. The momentum coefficient C_μ is defined for incompressible flow by

$$C_\mu = 2(c/h)\,(C_\varphi)^2$$

where h is the width of the slot and C_φ is the total quantity of fluid per unit span divided by the product of the stream velocity and the chord c. Figure 5 shows that $C_{L\max}$ has been increased from about 0·76 to 1·80, with C_μ increasing from zero to 0·125.

The suggestion that suction should be used to increase lift by application through a slot at the hinge of a flap came later than for the leading edge slot. Schrenk made tests in 1953 and H. B. Freeman in 1946 followed by extensive German studies on an aerofoil with a range of thickness/chord ratios between 0·15 and 0·18. The results of both sucking and blowing at trailing edge flaps and with suction through nose slots were summarized by John Williams in 1949, the suction quantities being of the order of 0·01 to 0·03. Of greater interest were later tests by Cook *et al.* on a 33° swept-back wing when with 70° flap deflexion the lift increment reached 90 per cent of its theoretical value when C_φ was equal to 0·0009.

Blowing

Separation over a trailing edge flap can be mitigated by blowing and some results are given in the above paper by John Williams. In this method the slot is arranged so that air is discharged through it at high speed over the top surface of the flap (Fig. 1). With increase in the rate of discharge, the lift coefficient rises rapidly at first until the flow over the deflected flap has become fully attached. After separation has been suppressed, the lift continues to rise less rapidly, due mainly to the additional circulation induced by the jet. From a practical standpoint, blowing as well as suction at the hinge of a flap gives rise to a nosedown pitching moment which may need trimming by the tailplane. The space that would be needed to accommodate the ducting within the wing of an aeroplane for suction or blowing is, however, large, but can be reduced somewhat by decreasing the slot width and increasing the power for suction or blowing. Attinello proposed that the high-pressure air from the engine compressor should be discharged through the slot without prior expansion so that the slot ran choked and this proposal stimulated interest in blown flaps that have been mentioned above. Experiments were made during the Second World War in Germany on the Dornier 24 flying boat and the Arado 232 troop transport followed by later developments in America. The engineering difficulties are considerable although the discharge rates needed to achieve valuable lift increments are modest and within the engine bleed that designers are prepared to allow. Blowing over simple trailing edge flaps has been applied to a few production aircraft including the Blackburn Buccaneer (Fig. 6). A possible high-lift scheme of the future is shown in Fig. 7, indicating

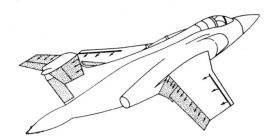

Fig. 6. *Blackburn Buccaneer.*

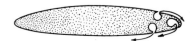

Fig. 7. *A possible high-lift scheme.*

that jets near the trailing edge blow backwards on the top surface and forwards on the under surface to increase the circulation round the wing: jets of cold air are being tried in this scheme.

Turning to the jet flap, this consists of a high-velocity jet in the form of a thin full-span sheet issuing from the trailing edge at an angle to the direction of the undisturbed stream. It induces a lift over the whole wing independently of wing incidence. Sometimes the angle of discharge is controlled by means of a small trailing edge flap (shrouded jet flap) through what is known as the Coanda effect and this method must be distinguished from the blown flap decribed above, because the discharge rates in the latter device are usually little more than is needed to prevent boundary-layer separation.

Magnus Effect

Rotating surfaces have been known as a means of producing high lift for some centuries. Newton (1671) was aware of this phenomenon and much later Magnus (1853) and Rayleigh (1877) made systematic and theoretical investigations. The downward force on a top-spun ball is well known to golfers and to tennis players and the modern game of tennis owes its speed almost entirely to this downward force. Flettner succeeded in 1924 in demonstrating the application of a vertical rotating cylinder as a replacement for a sail on a boat in which he crossed the North Sea. Several persons have experimented later with this phenomenon in the laboratory, notably Prandtl in 1925 and 1931, who predicted that the lift produced by a rotating cylinder cannot exceed a certain upper limit, but M. B. Glauert's investigation in 1957 suggested that the lift would increase almost indefinitely with the rotational speed as has since been demonstrated.

Wind tunnel experiments on rotating cylinders show how the lift effect increases with rotational speed. Figure 8 gives the results of Reid and Betz, the former on a cylinder with aspect ratio 13·3 and with Reynolds

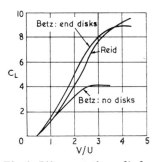

Fig. 8. Lift on rotating cylinder.

number varying from $3·9 \times 10$ to $1·16 \times 10$, and the latter for an aspect ratio of 4·7 and a Reynolds number of $5·2 \times 10$; where U is the wind speed and V the peripheral speed of the cylinder. The diagram shows that there is no lift for small rotational speeds with the ratio V/U less than 0·5. Betz obtained two-dimensional flow over the cylinder by fitting end discs 1·7 times the cylinder diameter, thereby raising the lift coefficient from 4 to 9. Reid with the much larger aspect ratio of his longer cylinder obtained results as good as Betz without end discs. The drag coefficients of the cylinders increased greatly at the higher V/U ratios, but were less at the small ratios with a minimum at about V/U equal to unity.

Experiments in Holland on the effect of incorporating a rotating cylinder in the front of an aeroplane wing gave a performance measured by the lift/drag ratio not as good as that of a conventional wing although the maximum lift coefficient was larger. Favre in Germany has also made experiments on an aerofoil in which the upper surface contained an endless band of silk driven so as to move with the air. The stalling angle in this latter experiment was delayed and a maximum lift coefficient of 3·7 was reached. Neither of these ideas have been further pursued.

Rotating Flap

Under certain conditions, an aerofoil set at a small angle of incidence to the wind will autorotate in a wind tunnel. The earliest known paper on the aerodynamic properties of autorotating bodies was by J. C. Maxwell in 1853. The action is similar to that of an oblong visiting card dropped by the hand and allowed to descend freely rotating about its longer axis, that is turning over and over, at a steep angle to the ground. Knowledge on this subject has been summarized by L. F. Crabtree in his paper on the "rotating flap" as an attractive high lift device; alternatively, if the flap is rotating in the opposite direction it could be an efficient air-brake system with the possibility of controlling changes of trim. This work has been followed by a theoretical discussion by S. Neumark on mounting a lamina near and below the trailing edge of a mainplane in which he quotes experimental data obtained by Küchemann at low Reynolds numbers with a $C_{L_{max}}$ up to 3·8, in general agreement with theory. No attempt has been made to apply this idea in a full-scale aeroplane and it seems that the promise of jet flaps is so much greater that the rotating flap idea is unlikely to be taken up.

Thwaites Flap

Large lift coefficients independent of rotation and incidence can be gained by a device known as a Thwaites flap fitted to a body with a rounded trailing edge. A thin flap is placed perpendicular to the surface to fix the position of the rear dividing streamline and thus maintain an unseparated boundary layer. The first experiments were made on a circular porous cylinder with a flap extending a distance of about one-fifth of the radius. For an angular deflection β of the flap, the inviscid value of the lift coefficient is given by $C_L = 4\pi \sin \beta$. Pankhurst and Thwaites measured a lift coefficient of nearly 9 at $\beta = 60°$ and higher values would have been expected with greater suction. The observed pressure distributions over the centre section of the cylinder for $\beta = 0°$ and 60° and $C_\varphi \sqrt{R}$ of 21·9 and 33·1 respectively are shown in Fig. 9. With the flap present the flow was steady, but without it was unsteady. It is of interest that a wire of diameter less than one hundredth that of the cylinder, placed on the surface, was adequate for stabilization of the flow with suction. A flap has even been tried on the lower surface of a circular fuselage when a projection 10 per cent of the diameter gave a lift coefficient of 1·4 for a lift/drag ratio of 2·5.

Other wind tunnel experiments have been made by the N.A.S.A. on a lifting circular cylinder 6 in. diameter with a fineness ratio of 8 and a number of slots. The lift coefficient could be varied from -6 to $+24$ with three slots 45° apart, the latter being with a momentum coefficient of 6 and a Reynolds number of 4 × 10. With more than three slots the lift coefficient was independent of the slot position, but its maximum value was relatively low. The results of British experiments on tangential blowing through slots on a circular cylinder have been reported to the effect that lift coefficients have been measured up to twenty times the momentum coefficient.

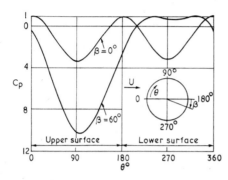

Fig. 9. Pressure distributions on a porous circular cylinder with a Thwaites flap.

VTOL Aircraft

High-lift devices as such are generally considered as applicable to the conventional aeroplane, but mention should be included here of the classes of aircraft known as VTOL and STOL, that is, vertical and short take-off and landing. Both types need a high lift to take off with no run or with a short one respectively. VTOL ability demands the generation of lift without forward motion, thus eliminating the aeroplane wing which is the most efficient lifting system. A mechanical device is needed to produce a jet and the most efficient jet is provided by the helicopter rotor with next the propeller, ducted fan, turbojet and last that new class of vehicle called a hovercraft or ground cushion vehicle, the weight of which is sustained by a cushion of air. It can be said that increasing jet velocity results in progressively reduced efficiency as a lifting medium (see Fig. 10).

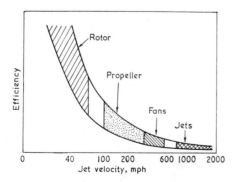

Fig. 10. Propulsive efficiency.

Illustrated here are two sketches out of some 16 possible configurations of the VTOL family showing the deflected jet, which is a high-lift device in a general sense and has become a practical proposition (Fig. 11); also direct lift either by vertically sited jet engines in the fuselage (Fig. 12) or similarly by lift fans in the under surface of an aeroplane's wings. Such fans are

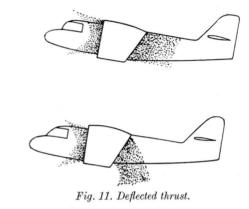

Fig. 11. Deflected thrust.

Fig. 12. Vertical jet.

driven by peripheral turbine blades to which gas is diverted from a turbojet. This last has not progressed further than a moderately successful experimental aircraft. A like remark applies equally to other experimental types dependent on the rotation of the propeller axis from vertical to horizontal and/or the engines driving the propellers, or of the whole aircraft.

Supersonic Flight

The above paragraphs relate mainly to subsonic speeds. In addition, the maximum lifting ability of a supersonic aeroplane can be improved by placing the wing entirely above the local part of the body and thus obtain some lift therefrom. This extra lifting pressure is contained within the shock waves trailing behind the nose of the aircraft beneath the leading edges of the wings, and to contain the pressure field the wing tips are hinged downwards in flight as in the B-70 aeroplane. This extra pressure supports some 30 per cent of the aircraft's weight in cruising flight so that the wing tips have thus become high-lift devices.

Bibliography

Aero. Res. Coun. C.P. 480, 1951.
Aero. Res. Coun. R. &. M. 110 (iv and v).
Aero. Res. Coun. R. & M. 2787.
GOLDSTEIN, *Mod. Devel. Fluid Dynamics*, Vol. II, p. 223, Oxford, 1938.
LACHMANN G. V. (1962) Slot (Aerodynamics), in *Encyclopaedic Dictionary of Physics* (J. Thewlis Ed.), 5, 509, Oxford: Pergamon Press.
N.A.S.A. Res. Memo. A53E06.
N.A.S.A. T.N. D-170, 1959.
N.A.S.A. T.N. D-244.
N.P.L. Aero Note 175.
Proc. Roy. Soc. A **242**, 108.
R. Aero. Soc. J. Jan. 1960.
R. Aero. Soc. J. June 1921, pp. 263–89.
SHAWE D. J. (1961) Circulation, in *Encyclopaedic Dictionary of Physics* (J. Thewlis Ed.), 1, 686, Oxford: Pergamon Press.
THWAITES, *Incompressible Aerodynamics*, Oxford, 1960.
THWAITES B. (1961) Kutta-Joukowski hypothesis, in *Encyclopaedic Dictionary of Physics* (J. Thewlis Ed.), 4, 189, Oxford: Pergamon Press.

J. L. NAYLER

HIGH POLYMERS, MECHANICAL PROPERTIES OF

1. Introduction

Polymers include a very wide range of materials, both solids and liquids, and both synthetic and natural, whose chemical structures often consist of interpenetrating molecular chains with large molecular weights in the range from about 10^3 to values greater than 10^6. This general definition includes synthetic plastics, rubbers and textiles and many natural occurring materials such as proteins, cellulose and gums.

Plastics are now common place in domestic life but their application to more technical problems has been limited. One factor which limits their performance in practical applications is that the mechanical properties of polymers depend critically on not only temperature but time as well. This means that the interrelationship between the observed mechanical properties and the molecular structure of the particular polymer must be studied more comprehensively than in the case of more conventional materials.

This article will be mainly concerned with the mechanical properties of solid high polymers and their relationship to structure. Polymers can be crystallizing or non-crystallizing according to their chemical structure. Non-crystallizing polymers are often referred to as "amorphous" to indicate their lack of crystalline structure. Crystallizing polymers are manufactured by processes which usually produce partially crystalline material together with an "amorphous" content. The mechanical properties will vary greatly depending on the crystallographic texture and the degree of molecular orientation. Contrary to expectation, the semi-crystalline polymers are not necessarily stronger than the amorphous polymers, for example, at room temperature polymethylmethacrylate (perspex) compares favourably in ultimate tensile strength with say polyethylene although the former is brittle whereas the latter is ductile.

The mechanical properties of polymers are very temperature and frequency dependent, i.e. they may be described as viscoelastic materials. At low temperatures a polymer will be glass-like with a Young's modulus of 10^{10}–10^{11} dynes cm^{-2} and will fracture at low strains. At high temperatures the polymer will be rubber-like with a modulus of 10^7–10^8 dynes cm^{-2} and will give a large extension for a given stress without plastic deformation occurring. At even higher temperatures, uncross-linked thermoplastics exhibit viscous flow.

In the intermediate temperature range, commonly called the glass-transition range, the viscoelastic behaviour is most evident and the polymer will exhibit properties which are intermediate between those of a viscous liquid and an elastic solid. At low values of strain, the viscoelastic behaviour is linear and this can be represented on an exact basis theoretically. At larger values of strain, non-linear behaviour is observed. This region of strain is of great importance technologically if polymers are to replace conventional materials in highly stressed components.

The mechanical properties of a polymer are not usually isotropic and not only can anisotropy be present in the unstressed state but anisotropy can be induced, for example, by drawing an otherwise isotropic specimen. Various methods exist for studying the details of the molecular orientation so produced and in fact anisotropy can often be used to enhance specific properties of a polymer during a fabrication process.

Rubber-like state. The physical basis of rubber-like behaviour is relatively well understood, at least qualitatively, and originates from the mobility of the individual monomer units of the polymer chain. The chains of monomer units respond to a Brownian movement of their individual atoms which is thermally activated. The production of cross-links between neighbouring polymer chains results in the polymer increasing its stiffness and decreasing its ability to flow under a given stress. This forms the basis of the vulcanization process in rubber, in which sulphur or chemical compounds are used to promote cross-linking to produce articles of practical utility. Progressive increase in the number of cross-links eventually suppresses all rubber-like behaviour and results in a product like ebonite.

The quantitative calculation of the stress–strain curves for a rubber network originated with Meyer *et al.* (1932)—see also Treloar (1958). The most elementary calculation is based on a model in which the rubber consists of molecular chains interlinked to form a network, but the number of cross-links is supposed small so that the Brownian motion is not severely affected. The polymer chains are regarded as flexibly jointed links and the configuration of the chains is thus random with the proviso that any random configuration cannot exceed the total chain length. The problem then involves the calculation of the configurational entropy of the whole network of chains as a function of the strain. The result of the calculation is given by

$$F = NkT \left\{ \lambda - \frac{1}{\lambda^2} \right\}$$

where F is the force per unit area of the unstrained cross-section, k is Boltzmann's constant, T is the absolute temperature, N the number of active chains per unit volume (an active chain is defined as that portion of a polymer molecule joining two cross-links) and λ is the extension ratio.

The curve predicted from this theory for vulcanized rubber is compared with a typical experimental curve in Fig. 1. It is immediately seen that there are two regions of deviation from the theory given above.

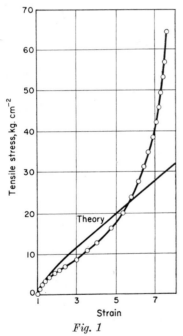

Fig. 1

At moderate strains, the stresses fall below the theoretical values and at very large strains the experimental stresses greatly exceed the predicted values. The latter discrepancy can be attributed partly to the failure of the assumption that the molecular chains have random configurations when the rubber is stretched and partly to strain induced crystallization—see Treloar (1958).

The disagreement between the theoretical and experimental stress–strain curves at moderate strains may arise from the inadequacies of the molecular theory of rubber elasticity. Formal extensions of the theory of finite elasticity have been invoked to explain these discrepancies—see Green and Adkins (1960). A classic series of experiments conducted by Rivlin and Saunders (1951) on deformations in thin rubber sheets enabled an empirical stress–strain curve to be obtained.

The simple molecular theory also makes two assumptions which are not strictly justified. First it was assumed that all the molecular configurations have identical internal energies and secondly the thermodynamic formulae derived are only applicable to measurements at constant volume, whereas most experimental results are obtained at constant pressure.

Glassy state. If a polymer is rapidly cooled, many properties such as the specific volume or volume coefficient of expansion undergo abrupt changes at some temperature known as the glass-transition temperature. Below this temperature, the polymer is essentially glassy and its volume coefficient of expansion decreases less rapidly with temperature than in the rubber state. The mechanical properties also change abruptly—in tension the polymer will often fracture at about 2 per cent extension and the capacity of the material to undergo large reversible deformations is lost.

This change in mechanical properties is associated with the severe reduction in molecular mobility which progressively inhibits the rubberlike behaviour outlined above. However, this inhibition of molecular movement does not itself change abruptly at the glass-transition temperature but is lost over a temperature interval of some tens of degrees.

The position of the glass-transition temperature is critical in determining the stiffness of the polymer. If the glass-transition temperature is lower than room temperature the polymer is soft and rubber-like at room temperature, whereas if it is higher than room temperature the polymer is hard and brittle. The physical properties of the polymer are often changed artificially by mixing with *Plasticizers* which effectively lower the glass-transition temperature so that the material can be made soft at room temperature where otherwise it may have been glassy.

Although in the rubber state the elastic properties of a material are appreciably changed by increase in the cross-linking of molecular chains, the effect is negligible in the glassy state. In a similar way, increasing the molecular weight has hardly any effect on the modulus in the glassy state, whereas at low molecular weights in the rubber state the polymer flows as a normal viscous liquid and increasing the molecular weight ensures the rubber-like behaviour of the polymer.

Viscoelasticity. Polymers in general are rarely entirely rubber-like or entirely glassy but show a combination of these characteristics. The stress–strain relationships for such a viscoelastic material will depend not only on temperature and strain, but also on strain rate and higher derivatives of the strain with respect to time.

At small values of strain (less than 1 per cent) the

stress and strain are related by a linear differential equation and the behaviour is known as linear viscoelasticity. At higher values of strain, however, non-linear behaviour occurs and the mathematical formulation of the stress–strain relationship is usually rather empirical.

Linear viscoelasticity. The stress and strain of a linear viscoelastic material can be related by a linear differential equation

$$a_0\sigma + a_1\frac{d\sigma}{dt} + a_2\frac{d^2\sigma}{dt^2} + \cdots$$
$$= b_0\varepsilon + b_1\frac{d\varepsilon}{dt} + b_2\frac{d^2\varepsilon}{dt^2} + \cdots$$

where σ is the stress, ε is the strain and the a's and b's are constants. Often it is sufficient to take only one or two terms in this equation and in this case the viscoelastic response can be represented by models composed of mechanical "circuit elements"—perfectly elastic Hookean springs E and viscous dashpots η which obey the Newtonian law of viscosity. An approximate representation to the linear viscoelastic behaviour of a polymer in the glass-transition range is afforded by the model shown in Fig. 2 and is known as the "standard linear solid". If a stress is applied in the directions of the arrow in Fig. 2 and then suddenly removed the resulting strain will relax exponentially with a time constant known as the strain–relaxation time. To represent viscous flow another dashpot could be added in series with the standard linear solid illustrated in Fig. 2.

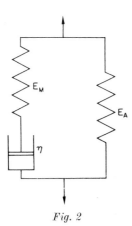

Fig. 2

To describe the behaviour of polymers more accurately it is necessary to construct more impressive models composed of combinations of the types of models depicted in Fig. 2. This is mathematically equivalent to defining a relaxation time spectrum or distribution. This distribution can be obtained experimentally by measurement of the dynamic elastic constants. Such measurements have been carried out, in particular on polyisobutylene by Marvin (1954).

An alternative way of relating stress and strain in a linear viscoelastic material is to use the Boltzmann superposition principle. This regards a complex stress system applied to a polymer as equivalent to the sum total of the individual effects due to a series of infinitesimal loading steps. This is a very general mathematical treatment but can be developed to describe non-linear behaviour as well.

The mechanical properties of viscoelastic materials depend on both temperature and time (or frequency). A complete stress–strain–frequency–temperature relationship would require a formidable amount of experimentation but simplification is possible by use of the concept of time-temperature equivalence. It has been found that for many polymers the effect of changing the temperature on the mechanical properties is equivalent to changing the frequency of measurement. Hence by carrying out measurements over a restricted temperature range at a series of values of frequency, the behaviour of the polymer at one value of frequency but over a wider temperature range can be estimated.

Rouse (1953) and Zimm (1956) are largely responsible for the development of the molecular theories of viscoelastic behaviour. They originally sought to explain the behaviour of dilute solutions of polymers but their theories can be extended to explain the behaviour of polymers in the glass-transition region. They effectively extend the molecular theory of rubber elasticity by assuming the polymer chains are not only affected by Brownian diffusional movement but the joints between successive monomer units are subject to the viscous forces of the solvent molecules.

Non-linear viscoelasticity. The possibility of finite strains leads to deviation from the ideal linear viscoelasticity. This is important technologically because creep over long periods of time is often a severe limitation to the use of a polymer in a particular application.

Attempts are being made to extend the theory of linear viscoelasticity to cover the non-linear situation but the description of non-linear behaviour is chiefly made by empirical expressions derived from experimental creep curves. This form of representation is already well established for metals. For plastics, the expression:

$$\varepsilon = A\sinh B\sigma + Ct^n \sinh D\sigma$$

where A, B, C, D and n are constants for a particular material, has been found to fit data on polyethylene and polyvinyl chloride—see Findley and Khosla (1955) while simpler relations, for example $\varepsilon = K\sigma^n t^m$, have been used to represent the behaviour of other polymers over limited ranges of time and stress.

Although other attempts have been made to explain non-linear viscoelasticity by introducing non-Newtonian dashpots into the mechanical models described earlier, the most satisfactory approach has been to attempt a generalization of the Boltzmann superposition principle. One possible approach is to assume that the strain at a given time depends not only on the value of the stress at the same time but also on contributions arising from infinitesimal loading steps at all previous times. Rivlin (1964) has called materials of this type *hereditary materials*.

It should be noted, however, that this theory is so general that it effectively allows for an infinite number of curve-fitting constants and so could be regarded as unsatisfactory an approach as the empirical curve-fitting technique.

Anisotropic effects in polymers. High polymers quenched under stress free conditions from the molten state would be expected to be isotropic, that is, the molecules would be randomly coiled in amorphous polymers and in addition the crystalline and amorphous regions would be randomly distributed in semi-crystalline polymers. In single crystals of polymers, grown from dilute solution, anisotropy can be present simply due to the formation of a definite lattice structure. However, anisotropy can also be induced in the initially isotropic polymers by processes such as drawing uniaxially or by rolling. If a material in the rubber-like state is cooled below its glass-transition temperature while the molecular chains are partially oriented due to drawing, then the orientation becomes "frozen-in" and an anisotropic polymer results. Such orientation can have a large effect on mechanical properties. Examples of fabrication methods which give rise to anisotropy are the extrusion process used in the production of synthetic fibres and films and injection moulding.

Most of the studies of mechanical anisotropy have been restricted to drawn fibres and uniaxially drawn plastic films. A relatively successful model for the interpretation of the variation of elastic modulus with draw ratio is due to Ward (1962). Here the partially oriented film or fibre is regarded as an aggregate of anisotropic units of structure and the drawn material shows mechanical isotropy in a plane perpendicular to the draw axis. The theory is successful in predicting the overall pattern of mechanical anisotropy in many polymers including low density polyethylene, where there is an initial reduction in the room temperature Young's modulus with draw ratio.

Not only do mechanical measurements give information on the degree of molecular orientation but measurements of the birefringence, infrared absorption and anisotropy of the second moment of wide-line nuclear magnetic resonance curves all enable a statistical distribution of molecular orientations to be obtained.

Yield and fracture of polymers. The brittle fracture of polymers in the glassy state has been extensively studied. It has been shown that the Griffith theory of fracture can be applied to glassy polymers such as polymethylmethacrylate and polystyrene. As in metals the derived surface energies are much higher than those predicted by assuming that the energy required to form the new surface originates in the breaking of chemical bonds only. The discrepancy is attributed to plastic work expended in the alignment of polymer chains ahead of the crack. This explains the observation of interference bands on the fracture surface, which have been associated with a *crazing* process where orientation proceeds without macroscopic reduction in cross-section. Crazing appears to be an important feature in high impact polymers; the high impact energies are attributed to extensive crazing.

The tensile strength of rubbers is not explicable in terms of the Griffith theory. Molecular theories have therefore been developed which treat the failure as a critical stress phenomenon. It is accepted that the strength of the rubber is reduced from its theoretical strength in a perfect sample by the presence of flaws. The strength can then be related to the failure of a cross-linked network, and related to the degree of cross-linking and the primary molecular weight. A subsidiary problem of great technological importance is the influence of reinforcing fillers such as carbon black and silica, on the tensile strength. It is believed that these increase the tensile strength by allowing the applied load to be shared amongst a group of chains.

The tearing of rubber is explicable in terms of a modified Griffith energy balance criterion, in which the tearing energy includes the energy dissipated irreversibly in viscoelastic and flow processes (Rivlin and Thomas, 1953). The tearing energy is related to the work done in rupture, but there is no direct correlation with the tensile strength because rubbers of identical tearing energies can possess very different stress–strain curves.

Strain rate and temperature greatly affect the tensile properties of polymers. For elastomers and amorphous polymers it has been shown that the principle of time–temperature equivalence still holds as proposed for linear viscoelastic behaviour.

Brittle fracture and the fracture of a rubber are the two extreme types of failure observed in polymers. In general there will exist an intermediate temperature range where ductile behaviour is observed. There exists a brittle–ductile transition, as in metals often this is at low temperatures as in polythene, nylon and polyethylene terephthalate contrasting with polymethylmethacrylate and polystyrene. The temperature of this transition is generally increased by notching the specimen as in metals. Thus a notch impact test is a severe test of behaviour.

In the ductile range a yield stress can be defined conventionally as the maximum observed load divided by the cross-sectional area. There is little definite knowledge concerning the variation of yield stress with test conditions. It does appear, however, that in isotropic polymers the yield process can be affected by hydrostatic pressure. This suggests a yield criterion similar to that proposed by Coulomb for soils where the shear yield stress is expressed as a constant plus a friction term proportional to the pressure on the shear plane.

In a rather narrow temperature range the remarkable phenomenon of cold drawing is often observed. Following the yield drop accompanied by the formation of a neck, further increase in applied strain causes the neck to propagate along the specimen. Thus the specimen thins from its initial cross-sectional area to a final cross-sectional area until the whole specimen reaches a given degree of plastic extension. This is called the natural draw ratio. At very low speeds of extension the drawing process can be related to physical variables such as the extension of a molecular network. At the high speeds obtained in synthetic fibre processing much heat is generated in the drawing process, and the stable situation is determined also by the energy balance associated with heat generation and conduction.

Oriented sheets of polymer (produced by uniaxial drawing) subjected to tensile loading, show a phenomenon similar to the necking of isotropic polymers: the formation of deformation bands. The plastic strain is concentrated into a narrow region (~ 0.1 cm in width) and is predominantly a simple shear strain, with the shear direction approximately parallel to the initial draw direction. This behaviour can be discussed satisfactorily in terms of conventional plasticity theory and a modified von Mises yield criterion (Brown, Duckett and Ward, 1968).

Bibliography

BROWN N., DUCKETT R. A. and WARD I. M. (1968) *Phil. Mag.* **18**, 153, 483.
GREEN A. E. and ADKINS J. E. (1960) *Large Elastic Deformations*, Oxford: Clarendon Press.
MARVIN R. S. (1954) *Proc. Second Int. Congr. of Rheology*, p. 156, London: Butterworths.
MEYER K. H., VON SUSICH G. and VALKO E. (1932) *Kolloidzeitschrift* **59**, 208.
NIELSEN L. (1962) *Mechanical Properties of Polymers*, Reinhold.
RIVLIN R. S. (1964) Office of Naval Research, Washington Report NR–064–450.
RIVLIN R. S. and SAUNDERS R. W. (1951) *Phil. Trans. Roy. Soc.* **A 243**, 251.
RIVLIN R. S. and THOMAS A. G. (1953) *J. Polym. Sci.* **10**, 291.
ROUSE P. E. (1953) *J. Chem. Phys.* **21**, 1272.
TRELOAR L. R. G. (1958) *The Physics of Rubber Elasticity*, Oxford: Clarendon Press.
WARD I. M. (1962) *Proc. Phys. Soc.* **80**, 1176.
ZIMM B. H. (1956) *J. Chem. Phys.* **24**, 269.

M. J. FOLKES and I. M. WARD

HIGH RESOLUTION ELECTRON MICROSCOPY. The resolving power of the first generation of electron microscopes was limited to about 30 Å, mainly because of the aberrations of the lenses. The four principal lens defects, spherical aberration (δS), chromatic aberration (δC), diffraction error (δD) and astigmatism (δA) combine to give a total error

$$\delta = \sqrt{(\delta S^2 + \delta C^2 + \delta D^2 + \delta A^2)}$$

which represents the radius of the disc of confusion corresponding to these individual errors.

While it was possible to reduce the first three aberrations quite effectively in the earlier electron microscopes, it was not until 1947 that Hillier and Ramberg achieved a major breakthrough in the correction of astigmatism, which arises due to a lack of rotational symmetry in the lens field. This was achieved by introducing eight iron screws into the region between the pole-pieces of the objective lens, so affecting the field distribution in the gap. The screws were adjusted to correct the pole-piece astigmatism and for the first time electron micrographs showing detail down to 12 Å were obtained. This marked the beginning of high resolution electron microscopy.

A modern high resolution electron microscope (Fig. 1) has a minimum of five electron lenses, each of which requires to be of variable focal length and at the same time held in accurate alignment with one another. By the use of transistorized stabilizing circuits, water-cooled lenses and other refinements it is possible to reduce fluctuations in lens currents and accelerating voltage to less than three parts in a million over a period of a minute, with the result that δC becomes almost negligible. The objective lens astigmatism corrector (the "stigmator") can be controlled electrically and adjusted while observing a specimen at high magnification, so reducing δA to less than 1 Å. The remaining aberrations δS and δD are related to the semi-angular aperture angle (α) the former being proportional to α^3 while the latter is inversely proportional to α. The best theoretical performance of the instrument is found by calculating the optimum value of α and the corresponding minimum value of δ. These are given by

$$\alpha_{\text{opt}} = \frac{1 \cdot 45 \times 10^{-2}}{[C_S f V^{\frac{1}{2}}]^{\frac{1}{4}}} \text{ rad}$$

and

$$\delta_{\min} = \frac{600 \, (C_S f)^{\frac{1}{4}}}{V^{\frac{3}{8}}} \text{ Å}$$

For a typical high resolution objective lens (Philips EM 300) the spherical aberration constant C_S is 1·6 mm and the focal length is 1·6 mm, so that at an accelerating potential of 100 kV, $\delta_{\min}$ is 3 Å. The stability of this instrument is such that under good operating conditions a limit of 2·3 Å can in fact be achieved and useful magnifications of 500,000× are possible.

In most high resolution work an anti-contamination device is necessary in order to prevent build-up of carbonaceous layers on the specimen arising from hydrocarbon vapours from the vacuum pumping

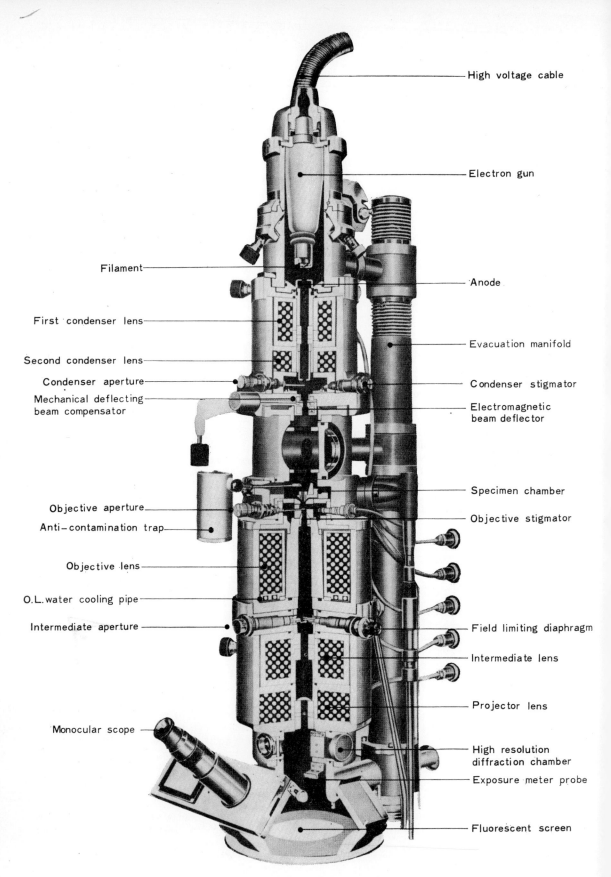

Fig. 1. Electron microscope column.

system. These vapours are removed in the vicinity of the specimen by a cooling device comprising two copper blades, one above and one below the specimen, maintained at liquid nitrogen temperature. In a few instruments a special design of specimen chamber allows ultra-high vacuum conditions to be obtained in the immediate region around the specimen. This is of particular importance if the specimen is to be kept under observation for a lengthy period as, for example, in crystal growth and nucleation studies.

One of the most striking applications of high resolution electron microscopy has been in the direct observation of crystal lattice periodicities. This was first achieved by Menter in 1956 when he photographed the individual planes of atoms in platinum phthalocyanine crystals with a spacing of 12 Å. More recently Komoda has succeeded in resolving the (110) planes in MoO_3 with a spacing of 3·8 Å and the (111) and (200) planes in a gold crystal with spacings of 2·35 Å and 2·04 Å respectively. The imaging of these lattice planes depends on "diffraction contrast" and the essential condition that has to be fulfilled is that the first order diffracted beam from the set of crystal planes be allowed to pass through the objective aperture in addition to the transmitted incident beam. For normal illumination this condition is $2\theta < \alpha$ where θ is the Bragg angle corresponding to the lattice planes. With the small objective aperture ($\alpha \sim 5 \times 10^{-3}$ rad) used in high resolution work it is only possible to ensure this condition for small lattice spacings if the illumination is tilted so that the incident and diffracted beams make equal angles with the axis of the objective lens, as shown in Fig. 2. The condition now becomes $\theta < \alpha$ and using this method it should theoretically be possible to resolve down to 1 Å. During the past year photographs of the (220) planes in gold at spacing 1·44 Å and in copper at 1·27 Å have in fact been obtained. Tilting the illumination also has the effect of reducing chromatic aberration arising from fluctuations in gun potential, as the path lengths are now identical for both transmitted and diffracted beams.

The high resolution electron microscope can also be used to reveal imperfections in the crystal lattice and the path of a dislocation line itself can be seen, enabling a detailed analysis to be made of the strain fields in single crystals. This has, for example, already resulted in a greatly improved understanding of the role played by the secondary slip systems during deformation.

A method of by-passing the resolution limit in the case of very fine crystal lattice spacings is to make use of moiré patterns formed by two overlapping crystals. In the first method a parallel moiré pattern is formed by growing a single crystal of, say, palladium epitaxially on a gold film. The moiré spacing is given by $\dfrac{d_1 d_2}{d_1 - d_2}$ where d_1, d_2 are the spacings of the lattice planes in the two crystals. For example, the

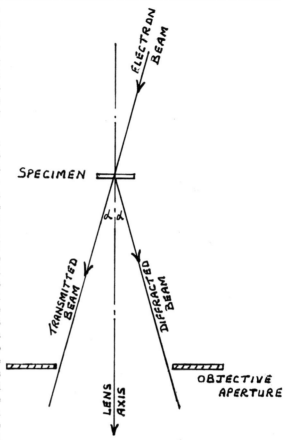

Fig. 2. Tilted illumination method for lattice imaging.

(4$\bar{2}\bar{2}$) spacing in palladium is 0·798 Å which is beyond the limit of resolution of the electron microscope; but when combined with a parallel gold film this yields a moiré spacing of 17 Å. In the second method a rotation moiré pattern is formed by superimposing two identical films with a small relative rotation ε. This gives rise to a pattern with a spacing d/ε, where d is the spacing of the lattice planes.

More important, however, is the fact that dislocations can be observed and studied from moiré patterns, the method showing clearly the errors in continuity of the crystal planes associated with a disturbance in the lattice. For example, an edge dislocation in one of the crystals gives rise to an extra terminating halfline in the moiré pattern and in this way it is possible to extend the study of lattice imperfections to metal crystals where the spacings would otherwise be too small for direct observation. Figure 3 shows a dislocation in gold revealed by a parallel moiré pattern formed by a (111) gold layer deposited on MoS_2 (Pashley, 1965).

The ultimate goal of high resolution electron

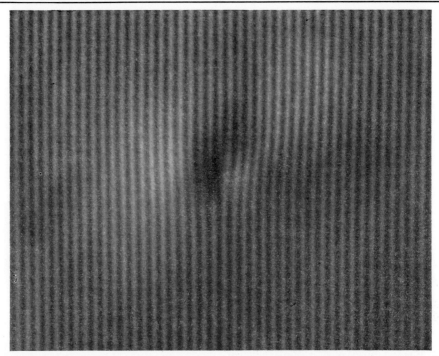

Fig. 3. Moiré pattern showing dislocation in gold film, magnification ×1,600,000 (Pashley, 1965).

microscopy is to reveal the atomic or molecular structure of a solid object, but while modern instruments are capable of achieving a point to point resolution of better than 3 Å there is an additional factor to be considered, that of adequate contrast in the image. There are two main mechanisms for producing image contrast at high resolution, diffraction contrast and phase contrast. The former is the dominant factor in the study of crystal lattices while the latter assumes importance for object distances less than 10 Å in films less than 100 Å in thickness and for which the scattering amplitudes are weak. In this case electrons scattered coherently through small angles are used and the waves must pass through the objective aperture to recombine with the transmitted wave at the image plane; the image point intensity distribution is then a phase map in which points of equal intensity are points of equal phase amplitude. The interpretation of phase detail requires a knowledge of the defocus conditions at the objective lens and an added vernier fine focus control is fitted giving steps down to about 10 Å. The amount of defocus for maximum phase contrast between points of spacing in the object is given by

$$\delta f = \frac{a^2}{2\lambda}$$

A "through-focal" series of micrographs must be taken and the results require skilful interpretation. Spurious phase contrast effects such as those arising from the structure of the supporting film and from artefacts must be disregarded and very frequently the most "acceptable" image may not be the one which shows the actual structure under investigation. A further complication in this work arises from charging up of the specimen, which interferes with the astigmatic correction of the objective lens, frequent adjustments being necessary. Heidenreich has shown, however, that a point to point resolution of 2 Å can be achieved by this method, and he has obtained phase micrographs of thin hexagonal graphite crystals showing cell structure down to 5 Å. One of the main fields of application is in the study of synthetic polymers where the arrangement of polymer chains can be examined directly. There is clearly a problem of possible damage to such fragile specimens and this is partly overcome by using a "cold finger" to keep the specimen at low temperature and also by maintaining the specimen chamber at ultra-high vacuum. The "contrast to damage" ratio improves with increase in gun voltage, reaching an optimum at about 200 kV.

The high resolution electron microscope has opened up many new lines of research in solid state physics, metallurgy and biology and nowadays a very high proportion of commercial instruments can achieve a resolving power of 5 Å or less. Perhaps the extremely thorough investigations of the distribution of dislocations in metals which have undergone plastic deformation provides one of the most impressive examples of the use of the instrument. For this work very thin foils of the metal have to be prepared, usually by thinning down from the bulk material, using electropolishing techniques. The relationship between the internal distribution of dislocations and slip lines can then be investigated in great detail. The effects of fatigue have been studied in a number of metals and in the case of copper it has been shown that the dislocations are arranged in extremely dense patches of highly irregular networks occurring in bands parallel to the {111} planes. The presence of many dislocation loops and the irregular nature of many of the dislocations suggests that a large number of point defects are created during fatigue.

Studies of damage by α-particles, fission fragments or neutron irradiation have been made on a number of materials. As a result of collisions in the lattice, clusters of vacancies and interstitials are formed at the early stages of irradiation. These are revealed as dislocation loops; black spots or regions of strain where they condense in sufficiently large numbers. In this way the instrument has greatly assisted research into radiation damage in copper, steel, graphite and other reactor materials. A technique for the determination of the vacancy or interstitial nature of point defect agglomerates in quenched or irradiated specimens has recently been developed and applied to a variety of problems including the study of particle tracks in crystals.

Some of the most striking advances have been in the field of biology; but while many specimens of physical and chemical interest can be prepared by simply spreading a dispersion of minute particles on a supporting film of carbon, biological specimens do not in general allow such easy preparation. In the cases where such methods are available, a negative staining procedure is generally used, the details of the particles under examination standing out against a more densely stained background. In this way the fine structure of virus particles and bacterial flagella can be examined. The structure of single macromolecules such as nucleic acid and proteins can also be seen with detail down to about 5 Å in size.

More commonly in biology it is necessary to examine pieces of tissue which must be preserved by fixation, embedded and cut into sections thin enough to allow the electron microscope to resolve the details of the structure of the cells. Even with modern techniques of ultramicrotomy, it is doubtful whether sections thinner than 500 Å can be consistently produced. Owing to the great depth of focus of the instrument, the entire thickness of such a section is in focus, giving superimposition effects in the micrograph, an effect which can be well illustrated if a stereo-pair of micrographs is taken showing detail of the picture in depth.

This superimposition effect is not the only problem in high resolution electron microscopy of biological thin sections. The tissue to be studied must be fixed in a chemical preservative, dehydrated by alcohol, infiltrated by other solvents and embedded in a plastic monomer, which is subsequently polymerized by heat. The sections obtained from this block of tissue must be stained to increase the contrast, using heavy metal salts, such as lead, which scatter electrons to produce the electron image. The most common fixatives often contain osmium which also binds to tissue components and helps to enhance contrast. Thus the detail recorded in the final micrograph is the sum of a number of systematic artefacts which are indispensable if any high resolution results are to be obtained. The question of interpretation raised is not so much technical as theoretical, involving the biological significance of electron images. When a membrane, for example, is seen in a micrograph as a trilaminar structure, consisting of two dense laminae 25 Å thick, separated by a pale interspace of about 25 Å, the living state of that membrane is far from being demonstrated. The image is certainly a consistent artefact but it must be interpreted with caution, since it represents a pattern of osmium and lead deposition in tissues altered by preparative procedures.

Further technical difficulty is caused by the evaporation of the embedding material in the electron beam. This causes dirt to be deposited on apertures and pole pieces, thus increasing the problem of correcting astigmatism, and also causes contamination on the section being examined, so decreasing the quality of the image and the possible resolution. These two effects can be minimized by using the anti-contamination device, which prevents evaporation from the embedding material. Thus it is seen that biological electron microscopy is attended by difficulties arising from preparative techniques which limit resolution as well as by the standard problems of electron image interpretation which are common to all high resolution electron microscopy. Despite this the electron microscope is at present being used in the study of membrane structure, cell components, crystalline biological material of various types, DNA strands, micro-organisms, in particular, viruses and a whole range of macromolecular substances. In many cases, however, the problems of specimen preparation and the difficulty of assessing the biological significance of high resolution micrographs have set limits to work in this field.

Bibliography

HEIDENREICH R. D. (1967) *Siemen's Review* **34**, 4.
HOWIE A. (1965) Dislocations, transmission electron microscopy of, in *Encyclopaedic Dictionary of Physics* (J. Thewlis Ed.), Suppl. vol. 1, 61, Oxford: Pergamon Press.
KAY D. (Ed.) (1965) *Techniques for Electron Microscopy*, Oxford: Blackwell.
KOMODA T. (1964) *Optik* **21**, 93.
MARTON L. (1961) Electron microscope, in *Encyclopaedic Dictionary of Physics* (J. Thewlis Ed.), **2**, 795, Oxford: Pergamon Press.

PASHLEY D. W. (1965) *Reports on Progress in Physics* **28**, 291.

TONER P. G. and CARR K. E. (1968) *Cell Structure—an Introduction to Biological Electron Microscopy*, Edinburgh: E. and S. Livingstone.

J. W. SHARPE

HOLOGRAPHY. Holography is the technique of recording an image of an object using the entire content of the light reflected, diffracted or transmitted by that object. The name is derived from the Greek words *holos* meaning whole, and *grapho* to write. The method, which is a two-step lensless imaging or photographical process, was proposed by Gabor in 1948 but its development was delayed until the coming of the laser. By the entire content of the light is meant the amplitude, wavelength and phase. Photographic methods of recording images record the intensity distribution of the light originating from the object. In conventional photography the light from a three-dimensional scene is focussed on the film emulsion by lenses and/or mirrors and only amplitude and wavelength are involved to form a two-dimensional image. In holography the phase content of the light is also used. This is accomplished by separating a beam of electro-magnetic radiation into a coherent object-illuminating beam and a coherent phase-related reference beam and allowing the light transmitted by, reflected by or diffracted from the object to interfere with the reference beam so that the phase information is manifest as an intensity variation and a photograph of the resultant pattern without focussing carries information about both amplitude and phase. The pattern which is a record of the light waves rather than of the object is called a hologram and by illuminating it with the original coherent reference beam a diffracted wave identical in amplitude and phase distribution with the original wave from the object is produced so that an image of the object can be obtained which can either be photographed or viewed directly by looking through the hologram. The image is three-dimensional. Any diffusely reflecting surface illuminated by coherent light exhibits speckle, i.e. random brightness variations resulting from the interference of the diffusely scattered waves so the reconstructed image shows this. It is prominent for direct observation where the pupil of the eye defines the resolution, but is insignificant if a recording medium having a resolution limit larger than the speckle grain is used.

The light source used to record holograms must have a coherent length (velocity of light divided by spectral width) at least equal to the maximum difference between the reference beam path length and the corresponding signal path length. A typical value for a helium-neon laser giving 50 mW at 6328 Å is about 20 cm. Pulse ruby lasers give about 1 cm. The recording medium for the hologram must have very high resolution of the order of 1000 lines per mm and shrinkage must be at a minimum. In addition to photographic emulsions alkali halide crystals coloured with F centres and photochromic glasses have been used.

It is not essential for the reconstructing beam to be identical with the original reference beam but it should preferably bear some simple relationship to it. However, the hologram may also be viewed in moderately coherent light because the small entrance pupil of the human eye much reduces the information collected from it at any time.

The hologram itself looks quite different from the object and appears as an unintelligible smudged photographic plate. Often concentric circle patterns are present but these are caused by dust and are not part of the hologram. It is a combination of fringes and diffraction patterns, the details of which usually become apparent upon enlargement. An example is shown in Fig. 1. Holograms may be classified as planar (thin) or volume (thick). In the planar class all the information from the three-dimensional object is stored in a planar layer: the spacing between the fringes is large compared to the thickness of the recording medium and they result when the angles between the reference and object waves are small. The hologram behaves as a planar diffraction grating and the images are reconstructed by transmitting light through them. Volume holograms are formed with the spacing between the fringes small compared with the thickness of the recording medium and the information is stored in layers throughout the entire volume of the emulsion which may be up to 20 microns in thickness. They result when the angles between the reference and object waves are large and behave as three-dimensional gratings. Volume holograms are sometimes known as Bragg-effect holograms and when they are recordings of the interference between object and reference waves travelling in nearly opposite directions so that the interference surfaces act as partially reflecting periodic surfaces they are often termed Lippmann-Bragg holograms. From these multi-coloured images may be reconstructed with white light illumination because the sharp diffraction conditions select only those wavelengths and beam directions which give rise to diffracted beams corresponding to the coloured image.

The efficiency of reconstruction of a hologram image is not high because it is an amplitude grating formed by the silver grains of the photographic emulsion. Since the image is reconstructed only as a result of the diffraction of light by the hologram pattern and not because of the density of the photographic emulsion, the silver image in the latter may be removed by bleaching. The refractive index and thickness differences in the gelatin between the originally exposed and unexposed parts will then produce the diffraction and give clearer images. Only the phase of the light is now used for reconstruction: the amplitude grating is thus converted into a phase grating and such a hologram is called a phase hologram. It should be noted that because in a hologram every small part of

(a)

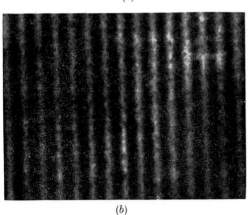

(b)

Fig. 1. Typical hologram made with an angle between the signal and reference beams of 7° appears to contain little except a fairly uniform tonal density. Microscopic inspection reveals the carrier frequency interference at 200 lines per millimetre. (By courtesy of Science Journal, London.)

the photographic material contains all elements of the image as seen from the angle of view by that part the hologram may be cut into small pieces and it will still be possible to reconstruct the entire image from each piece. The only limiting factor is the resolving power of the photographic material. Holograms always produce positives. Thus if a hologram normally obtained is contact printed the image reconstructed from the print is the same. This is so because the information on the holograms is recorded in the form of a modulated spatial carrier and printing only reverses the polarity of the carrier but does not effect the signal data carried.

In general four types of hologram may be produced: (1) Fresnel, (2) Fraunhofer, (3) side band, and (4) Fourier transform. A Fresnel hologram is one in which the interference occurs between the light which has suffered Fresnel or near-field diffraction by transmission through the object and a co-axial coherent background reference beam. The first hologram produced by Gabor was of this type.

In a Fraunhofer hologram, the Fraunhofer or far-field diffracted light is combined with the reference beam. Side-band holograms occur when interference arises between the diffracted light, which may be either Fresnel or Fraunhofer, arising from transmission through or reflexion from the object and a non-coaxial coherent background—i.e. an off axis reference beam. They are sometimes called offset reference or Leith–Upatnieks holograms. Fourier transform holograms result from the interference between a Fourier transform of the planar object amplitude distribution and a coherent background from an off axis point reference in the plane of the object. The Fourier transform hologram differs from the Fraunhofer in that the former gives a diffracted wave consisting of a Fourier transform of the object, while the latter has a diffracted wave consisting of this same Fourier transform multiplied by a quadratic phase factor. The resolution is better than for Fraunhofer holograms. In Figs. 2–5 the formation of the holo-

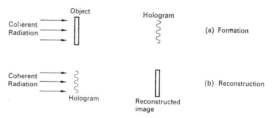

Fig. 2. Formation and reconstruction of Fresnel hologram.

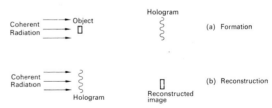

Fig. 3. Formation and reconstruction of Fraunhofer hologram.

grams and the reconstruction are shown schematically.

The mathematical analysis of the holographic process may be considered in two parts—the hologram formation process and the image reconstruction procedure.

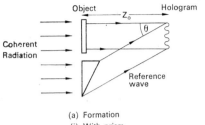

(a) Formation
(i) With prism

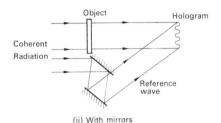

(ii) With mirrors

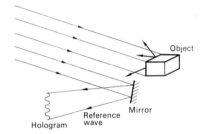

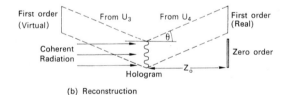

(b) Reconstruction

Fig. 4. Formation and reconstruction of side-band hologram.

1. Mathematical analysis of hologram formation

Let the coherent monochromatic wavefront amplitude of the signal to be recorded be given by $A_s(x, y)$ where

$$A_s(x,y) = a_s(x,y) \exp[-i\phi(x,y)] \quad (1)$$

and the reference wavefront amplitude by $A_\gamma(x, y)$ where

$$A_\gamma(x,y) = a_\gamma(x,y) \exp[-i\psi(x,y)] \quad (2)$$

If $A(x, y)$ is the total amplitude, then dropping the co-ordinates (x, y) for convenience of writing

$$\begin{aligned} A &= A_s + A_\gamma = a_s e^{-i\phi} + a_\gamma e^{-i\psi} \\ &= e^{-i\phi}[a_s + a_\gamma e^{i(\phi-\psi)}] \end{aligned} \quad (3)$$

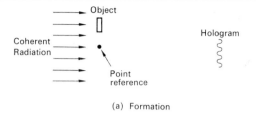

(a) Formation

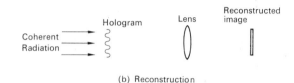

(b) Reconstruction

Fig. 5. Formation and reconstruction of a Fourier transform hologram.

Now the recording medium is sensitive to intensity which, if denoted by $I(x, y)$, is given by

$$\begin{aligned} I = AA^* &= (A_s + A_\gamma)(A_s^* + A_\gamma^*) \\ &= A_s A_s^* + A_\gamma A_\gamma^* + A_s A_\gamma^* + A_s^* A_\gamma \end{aligned} \quad (4)$$

or alternatively,

$$\begin{aligned} I &= (a_s e^{-i\phi} + a_\gamma e^{-i\psi})(a_s e^{-i\phi} + a_\gamma e^{-i\psi}) \quad (5) \\ &= a_s^2 + a_\gamma^2 + a_s a_\gamma [e^{i(\psi-\phi)} + e^{i(\psi-\phi)}] \\ &= a_s^2 + a_\gamma^2 + 2 a_s a_\gamma \cos(\psi - \phi) \end{aligned} \quad (6)$$

In equation (6) the first two terms depend only on the intensities of the two waves, but the third involves their relative phases and so it is seen that the phase ϕ of the signal wavefront has not been lost as it would have been if the reference beam was not present.

The recording medium is assumed to provide a linear mapping of intensity incident during exposure into amplitude transmittance t_f after development whence if it be assumed that a_γ^2 is uniform across the recording surface then

$$\begin{aligned} t_f &= \beta t\, A_\gamma A_\gamma^* + \beta t\, (A_s A_s^* + A_s A_\gamma^* + A_s^* A_\gamma) \\ &= t_b + \beta t\, (A_s A_s^* + A_s A_\gamma^* + A_s^* A_\gamma) \end{aligned} \quad (7)$$

in which β is the slope of the time exposure curve and t is the exposure time. t_b represents a uniform bias transmittance established by the reference beam.

2. Mathematical analysis of image reconstruction

Let the developed transparency be illuminated by a coherent reference beam $B(x, y)$ so that the light transmitted is given by

$$\begin{aligned} Bt_f &= Bt_b + B\beta t A_s A_s^* + B\beta t A_s A_\gamma^* + B\beta t A_\gamma A_s^* \\ &= U_1 + U_2 + U_3 + U_4 \end{aligned} \quad (8)$$

In this U_3 is the reconstructed wave and if $B = A_\gamma$, then

$$U_3 = \beta t\, |A_\gamma|^2 A_s \quad (9)$$

which is, except for a multiplicative constant, an except duplication of the original wave front A_s: it generates a virtual image of the original object.

Similarly if $B = A_\gamma^*$, i.e. the conjugate of the reference wave, then

$$U_4 = \beta t \, |A_\gamma|^2 \, A_s^* \qquad (10)$$

which is proportional to the conjugate of the original wavefront: it generates a real image which corresponds to an actual focussing of light in space. In both cases then there are three additional field components which may be regarded as extraneous interference and it is desirable to suppress or to separate these from the real or virtual image.

Fresnel and Fraunhofer holograms

In the Fresnel hologram the reference wave is transmitted through the object and it is, in general, strong compared to the scattered wave. From equation (8) the transmitted wave is represented by U_1, where $U_1 = \beta t A_\gamma^* |A_\gamma|^2$ and U_2 given by $\beta t A_\gamma |A_s|^2$ is, since $A_\gamma \gg A_s$, negligible in comparison, so that there are two images accompanied by a coherent background. Thus the reconstructed image is formed but is partly degraded by the presence of another image just of focus. If the reference beam is not much stronger than the signal beam, U_2 becomes significant and so it is possible, for example, to obtain an image of an object consisting of opaque letters on a transparent background but not one of transparent letters on an opaque background.

The analysis for Fraunhofer holograms is essentially the same but the reconstructed image is less troubled because the second image is located at infinity.

Side-band holograms

In these the reference beam is introduced at an offset angle (see Fig. 3) so that ψ in equation (2) is $2\pi\alpha y$ in which α the spatial frequency of the reference wave is given by $\alpha = \sin \theta/\lambda$, whence by equation (6)

$$I = a_s^2 + a_\gamma^2 + 2a_s a_\gamma \cos(2\pi\alpha y - \phi)$$

The cosine term shows that the real part of the wavefront to be recorded on the hologram has been modulated from the temporal carrier wave obtained from the object onto a spatial carrier wave of frequency α on the hologram produced by the reference beam. In the reconstruction process the interaction of the coherent light beam with the hologram causes the modulation present in the spatial carrier contained in the hologram to be transferred to the light beam. Then with reference to equation (8), the term U_1 represents a plane wave travelling down the transparency axis, U_2 is spatially varying,

$$U_3 = \beta t \, BA_s A_\gamma^* e^{2\pi\alpha y} \text{ and } U_4 = \beta t B A_\gamma A_s^* e^{-2\pi\alpha y}.$$

The reconstructed wave U_3 generates a virtual image at a distance z_0 from the hologram and the term $2\pi\alpha y$ indicates that it is deflected off the axis at an angle θ. The term U_4 refers to the real image. If the bandwidth of the signal wave is small compared with α, the energy in the wave U_2 remains sufficiently close to the transparency axis to be spatially separated from the image (either U_3 or U_4) of interest. Thus the sideband hologram differs from the Fresnel and Fraunhofer type in that the two images are angularly separated from each other (Fig. 3) and there is no coherent background. The zero-order term is of no interest because it carries no phase modulation.

Holograms may be formed and images reconstructed with spherical waves and reconstruction waves of a wavelength different from the reference wave. With such systems magnification results. An analysis is given by De Velis and Reynolds.

The earliest application of holograms was in the area of microscopy and a two step optical microscope has been developed which possesses greater depth of field than a conventional microscope. Other applications include removal of aberrations, three-dimensional display, information storage and coding character and pattern recognition, small particle size measurement, creating beams of light with special distribution of intensity and direction, holographic interferometry with special reference to engineering and vibration and strain analysis, flow visualization. It is also possible to extend the wavelengths used from the optical into the ultrasonic region, so that acoustic holograms may be recorded at the wavelength of sound and be reconstructed at optical wavelengths. This is ultrasonic holography. Also possible is the construction of computer-generated "synthesized" holograms capable of displaying three-dimensional images of objects that never existed in reality.

Bbliography

CHAMBERS R. P., COURTNEY–PRATT J. S., STEVENS B. A. and LATTA J. N. (1966) Bibliography on holograms, *Journal S.M.P.T.E.* **75** (4), 373–435; (8), 759–809.

DE VELIS J. B. and REYNOLDS G. O. (1967) *Theory and Applications of Holography*, Addison-Wesley Publ. Co.

ENNOS A. E. (1967) *Contemporary Physics* **8**, 153–170.

GABOR D. (1948) *Nature*, **161**, 777–778.

GABOR D. (1949) *Proc. Roy. Soc.* **A197**, 454–487.

LEITH E. N. and UPATRIEKS (1964) *J. Opt. Soc. Amer.* **54**, 1295–1301.

STROKE G. W. (1966) *An Introduction to Coherent Optics and Holography*, Academic Press: London.

H. G. JERRARD

HYDROLOGIC CYCLE. The hydrologic cycle is the circulation of water from the oceans, through the atmosphere back to the oceans, or to the land and thence to the oceans again by over-land and subterranean routes.

This cycle may be visualized as beginning with evaporation from the oceans. The resulting water vapour is transported to all parts of the earth by the general circulation of the atmosphere. At times, under proper conditions, the water vapour condenses to form clouds from which precipitation may result. The rain or snow which falls on the land portions of the earth is further distributed in several ways. A large portion is temporarily retained in the soil and is eventually returned to the atmosphere through evaporation or evapotranspiration. Another part finds its way over the earth's surface to streams and rivers while still another portion penetrates into the ground to be at least temporarily stored as ground water.

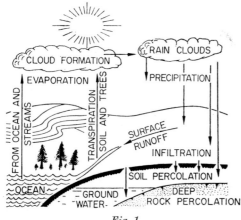

Fig. 1.

Both surface water and ground water move to lower elevations under the influence of gravity and may eventually reach the ocean. However, a large part of both surface and ground water returns to the atmosphere by evaporation before reaching the ocean.

The science of meteorology deals with the movement of water in the atmosphere both in vapour and liquid form and the science of hydrology is concerned with the distribution and occurrence of liquid water on and under the earth's surface.

See also: Hydrometeorology.

T. L. RICHARDS

HYDROMETEOROLOGY. Although the term "hydrometeorology" has become generally accepted throughout the world during the past several decades its scope is still subject to a variety of interpretations. These range from the very limited view that restricts the term to the study of storms required for the design of hydrologic structures to the equally broad usage prevelant in a number of countries where the whole of both hydrology and meteorology are incorporated in a single Hydrometeorological Service.

The World Meteorological Organization, in its *Guide to Hydrometeorological Practices*, defines hydrometeorology as being "concerned with the study of the atmospheric and land phases of the hydrologic cycle with emphasis on the inter-relationships involved". By this quasi-official definition hydrometeorology includes such primary elements of the hydrologic cycle as precipitation, snow cover characteristics, water levels of lakes and streams, streamflow and storage, evaporation and evapotranspiration, soil moisture and that portion of ground water that interacts with surface water. Other elements considered to fall at least partially within the scope of the subject include river and lake ice, frost in the ground, water temperatures, chemical quality, sediment discharge, radiation, air temperature, humidity and wind. The last four elements are included because of their importance in studies of snowmelt and evaporation.

A widely accepted and useful definition of the term describes hydrometeorology as "the application of meteorology to hydrological problems". This increasingly popular textbook interpretation provides a functional division of the subject based on the ways meteorological data and theory may be used in the solution of hydrological problems. Projecting this concept, the subject matter falls into the following broad functional categories:

(i) Analyses of meteorological data to solve hydrological problems.
(ii) Extension of the time series of hydrological data through the use of meteorological data.
(iii) Application of meteorological data and theory to hydrological problems associated with:
  (a) engineering design and water management,
  (b) river forecasting, (c) lakes and large reservoirs.
(iv) Man's influence on the hydrologic cycle.

(i) *Analyses of Meteorological Data to Solve Hydrological Problems*

In many instances analyses of meteorological data are required for the solution of hydrological problems. For example, frequency analyses of rainfall measured at a single location (*point rainfall*) are often needed for the design of relatively small hydraulic works such as storm sewers or drainage systems. These statistical analyses are based on data recorded during heavy rainfalls. Extreme value frequency analyses methods (such as Gumbel) are applied to observed rainfall intensity data for a number of specified durations. Results are generally in the form of a family of curves (Fig. 1) defining return periods of 2, 5, 10 and 25 years plotted on graphs showing rainfall intensity and duration.

More frequently, estimates are often required of *areal precipitation* over a river basin. Such estimates are based on observations taken at a number of gauges in the basin. Methods for estimating areal precipitation include the finding of the simple arithmetic average, the objective *Thiessen polygon* technique and the subjective *isohyetal analysis* method. The easiest to apply is the arithmetic average of the observed values which is satisfactory only if there are a large number of stations which give a representative sample of the rainfall patterns. The Thiessen polygon method, described in most hydrology textbooks, has advantages for basins with a relatively few, unevenly distributed, precipitation stations. In addition, the method does not require a professional analyst and after the initial calculation of weighting factors may be carried on by simple machine processes.

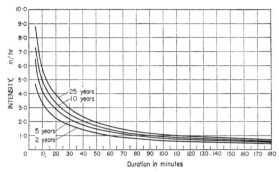

Fig. 1. Frequency analysis of point rainfall at Toronto (Malton), Ontario.

The isohyetal method of drawing lines of equal precipitation amounts is potentially the most accurate if employed by an experienced meteorologist who is able to take into account both the observed data and a physical knowledge of the natural distribution of precipitation.

For some hydrological purposes it is necessary to know not only the depth–area relationship of the total storm rainfall, but also the maximum distribution in time of this rainfall on an areal basis. Such *depth–area–duration analyses* usually entail the preparation of isohyetal maps, *mass curves of rainfall* and *maximum depth–area curves* developed for the major historical storms in the area. A detailed description of the technique is given in a U.S. Weather Bureau Manual (see Bibliography).

(ii) *Extension of Time Series of Hydrological Data through the Use of Meteorological Data*

The hydrologic regime of a stream must frequently be predicted when little or no streamflow data are available. It is fortunate that in many cases a relatively long record of precipitation data may be available for the basin, or its vicinity, and techniques have been developed to use these data to extend or synthesize the streamflow.

The simplest practical method for relating annual or seasonal runoff and rainfall is by a linear regression equation. If there is a substantial storage carried over from one year to another a multiple regression equation is more consistent. There are other non-linear relationships which have been found useful for specific watersheds and for certain climatic regions. Rather complicated empirical relationships have also been brought forward to predict streamflow volumes from melting snow. These generally include the water equivalent of the snow pack, precipitation, temperatures and, quite frequently, indices of soil moisture.

Sophisticated methods have also been developed to estimate flood runoff and peak discharge of a storm based on climatic data and streamflow records of short duration. The development of computer capabilities for handling large numbers of calculations and great quantities of data have recently made feasible the estimation of daily flows from meteorological and hydrological data.

(iii) *Application of Meteorological Data and Theory to Hydrological Problems*

(a) *Engineering Design and Water Management*

It is generally conceded that one of the most important functions of hydrometeorology is the provision of criteria based on meteorological data and theory for the design and operation of many types of water control structures such as dams and storage reservoirs, storm sewers, dykes, bridges and irrigation systems.

For *reservoir design* the most common problem is the design of water supply systems to meet a variety of requirements such as municipal water supply, flood control and the needs of power development, navigation and recreation. Reservoir design requires a thorough knowledge of at least two basic meteorological parameters—water supply provided through precipitation and water loss through evaporation.

The design and operation of a dam spillway is often an extension of the same problem and in each case there is a requirement for the *design flood* of the river. If a sufficient length of record of streamflow is available an extreme value frequency analysis will produce reliable estimates of flood magnitudes for return periods up to and a little beyond the period of observed or synthesized flows on the river. If, however, there are not adequate streamflow data to statistically or even empirically define the design flood, meteorological data and frequently meteorological theory are often employed. This hydrometeorological approach involves physical analyses of major storms in the region over the complete period of record. If geography so dictates, additional analyses of snowmelt must also be made.

In reservoir and spillway design two sets of safety criteria are usually considered. The maximum safety factor must be employed where the exceedance of spillway capacity could result in failure of the structure and where such failure would be disastrous in terms of life and property damage. Under these conditions the dam is designed to pass a physical upper limit of flood flows on the river called the *probable maximum flood*. The probable maximum flood results from the *probable maximum storm* on the basin or from a combination of the probable maximum rainfall and the maximum snowmelt which can occur at the same time.

In many cases, although a relatively high degree of safety is still required, the situation does not warrant the maximum safety factor associated with the probable maximum flood. In these cases the structure is designed to accommodate the flood resulting from the largest recorded storm over or near the basin. This is referred to as the *project storm*. At other times, when safety factors are even less critical, economic considerations dictate the use of a somewhat smaller design storm.

Determination of both the probable maximum storm and the project storm is based on the concept of *storm transposition*. This technique increases the storm experience of a basin by considering not only the storms which occur over the basin, but also those which release the heaviest rainfall on adjacent areas which are meteorologically "similar". This approach presents two difficult problems. The first is to precisely define the region of meteorological similarity—a subjective step to be undertaken only by an experienced synoptic meteorologist familiar with the area, while the second concerns the permissible change and orientation of the storm rainfall pattern. In this latter case it has been suggested that the orientation should not be adjusted by more than 20°.

The determination of the probable maximum storm involves an additional step—*storm maximization*. The objective of storm maximization is to determine by how much the precipitation from a particular storm could have been increased by physically possible increases in the meteorological factors producing the storm. This approach is based on the concept that the precipitation production of a storm is due to two factors, (i) the water vapour content of the rain-producing air mass (usually indicated by the dew points in the warm sector of the storm system) and (ii) the efficiency with which the storm mechanism converts this water vapour into precipitation. On the theory that at least one of the storms of record for the region has occurred at near maximum efficiency the actual procedure concerns itself only with the maximum possible water vapour content of the rain-producing air mass. In effect, the storm is maximized by calculating the total amount of precipitation the storm could have produced under conditions of the highest dew points ever recorded in the region for the same time of year. The probable maximum precipitation for a region is determined by drawing enveloping curves of maximized storm rainfall for various areas and durations as shown in Fig. 2.

In some parts of the world the largest floods are caused by *snowmelt* or more likely by a combination of rainfall and snowmelt. In assessing the greatest contribution to flood flows that can arise from snowmelt, there are two main factors to be determined—maximum snowpack accumulation and critical melting rates.

There are several complimentary methods for estimating maximum snow accumulation. The *partial season method* assumes that a combination of greatest observed snowfalls for discrete intervals (i.e. 4 days, a week, a fortnight or a month) during the period of record can be combined, regardless of the year of occurrence, to produce a synthetic year of very high

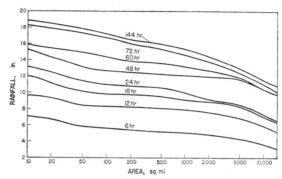

Fig. 2. Probable maximum precipitation (Saint John River Basin, N.B.).

snowfall. The selection of the length of the discrete interval should depend on the frequency of occurrence of extra-tropical storms producing snow in the area under consideration and the intervals between such storms.

A second approach, the *snow storm maximization method*, involves a maximization of moisture content for each of the storms which occurred over the basin during the several seasons of greatest recorded snow accumulation. This technique is similar to that used in maximizing storm rainfall. The two approaches tend to support each other and it is recommended that both be used in any investigation of maximum snow accumulation.

The *critical snowmelt rate* is that snowmelt rate which, when combined with the probable maximum storm for the snowmelt period, produces the probable maximum flood. The critical snowmelt rate is derived by relating snowmelt to the pertinent meteorological parameters and then ascertaining those values of the parameters which will produce the critical rate. One of the most important of these parameters is the air temperature. It is generally assumed that relatively long temperature records of stations within the basin will have yielded temperature sequences near the maximum possible—i.e. the critical temperature sequence. This approach is deemed valid because there are physical upper limits imposed on warm air mass temperatures by the source region of the air mass and by the trajectory of the air at that time of the year.

Critical snowmelt rates may be calculated by the degree index method or by energy balance melt equations. The degree-day method correlates an index of the daily maximum temperature with the daily snowmelt, which is usually determined from streamflow records. The energy balance melt equations compute snowmelt by calculating the energy flow into the pack. In this case such parameters as conduction from the ground, conduction and advection from the air, radiation and heat from the raindrops are considered. Various basin parameters such as slope of the land and type of forest cover are derived empirically by fitting the computed snowmelt to the streamflow. This method requires maximization of mean air temperatures in addition to consideration of such other meteorological parameters as insolation, dew point and wind.

In many parts of the world hydrometeorological problems are associated with too little, rather than too much water. When considering design capacity of an irrigation system both gross water storage and demand are often critical when viewed in economic terms. An estimate of maximum water needs requires a thorough study of climatological records of precipitation and evaporation. The use of a computer now makes practical the assessment of not only a simple daily water budget of soil moisture but also allows the consideration of such complex conditions as the withdrawal of soil moisture from several layers of soil by potential evaporation.

(b) *River Forecasting*

Forecasts of river flow volumes and stage of water elevations are necessary in order to make the most efficient use of the river and to minimize damage due to floods. On rivers with control structures reliable river forecasts permit the efficient operation of the dams—often built for many and varied purposes. On rivers without control structures the river forecast is a basis for any necessary flood warning.

For rivers with small drainage areas there are two basic steps in the preparation of river forecasts and each includes meteorological data and forecasts. The first step is to predict the volume of runoff by means of rainfall-runoff correlations and the second is to forecast the distribution in time of the runoff volume. Rainfall-runoff correlations are derived from records of storms and resulting river flows in the basins. Required data include depth and rate of storm rainfall, an index of pre-storm moisture conditions, seasonal factors and the duration of the storm. Snowmelt must also be considered if geographical location and the season so dictate.

Time distribution of runoff is predicted on the basis of the *unit hydrograph* of the stream. The unit hydrograph is a chronological representation of stage or discharge of the river as related to the specified unit of "excess" precipitation (i.e. 1 in. in 6 hr). The hydrograph (Fig. 3) incorporates the integrated effects of all the basin constants such as drainage area, shape, pattern and capacities of the channels, land slopes, etc.

With the introduction of high speed computers the trend in river forecasting methods is towards the direct use of more meteorological and hydrological parameters in mathematical models of the river.

Forecasts for large rivers generally make use of the dependable and consistent relationships between factors involved in streamflow routing. On larger rivers the time between the end of rainfall or snowmelt and the peak of the resulting hydrograph at a downstream location is frequently measured in days. However, the basic approach is still one of rainfall-runoff correlation,

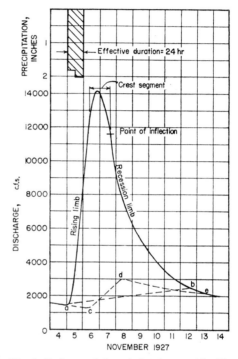

Fig. 3. *Hydrograph from isolated storm* (*St. Mary River at Stillwater, N.S.*).

streamflow routing and the development of the unit hydrograph. As in the case of smaller streams an accurate meteorological forecast is essential to a good river forecast.

The presence of various forms of *ice* is a problem not only in river forecasting but also for the operation of hydraulic structures. *Frazil ice* forms as clusters of disc-needle particles in swiftly running streams and plays a spectacular role in clogging water intakes. *Anchor-ice* forms on the bottom of relatively fast flowing streams reducing the flow under certain meteorological conditions by as much as 25 per cent. *Sheet ice* is the most common variety and is to be found on more slowly moving streams as well as on lakes and reservoirs. Techniques for forecasting the occurrence and extent of each type of ice are under development and references may be found in the literature and a few textbooks.

(c) *Lakes and Large Reservoirs*

There are a number of hydrometeorological problems associated specifically with lakes and large reservoirs, namely: wind effects, such as waves, set-up and seiche, currents and littoral drift; ice formation and dissipation; and the role of precipitation and evaporation in what can be large variations of water levels.

Wind Effects. For many reasons, varying from ship's design to shore erosion, there are requirements for wave analyses for most large bodies of water. Frequently there are not enough observed wave data on which to base such analyses and a significant wave analysis must be synthesized from empirical relationships involving the speed, duration and direction of wind, depth of water, and length of over-water fetch. Because the over-water winds can be quite different from the simultaneous over-land winds, techniques have been developed to estimate the over-water wind on the basis of air–water temperature differences and the length of the fetch.

An important application of this wave analysis technique can be employed when estimates for freeboard allowances are required for proposed dams or dykes. In such cases the wind record of the closest representative meteorological station is modified to make it representative of the wind regime over the reservoir. The over-water wind speeds for critical directions are analysed by extreme value frequency techniques and analyses of synthesized waves are then developed from the wind analyses in terms of return periods for maximum significant waves affecting the dam for each month of the year. Because the width of the reservoir is a factor in wave formation a calculated "effective fetch" should be employed in reservoir wave synthesis.

Wind set-up is another example of the influence of the wind on the surface of a large reservoir or lake. The wind exerts a horizontal stress on the water raising the level at one end of the water body and lowering it at the other. This effect is especially significant if the lake or reservoir tends to be shallow and elongated and it can be large enough to seriously affect human safety, the design and operation of dams and the flow of outlet rivers. The set-up lasts as long as the wind stress is applied. When the stress is withdrawn the water body oscillates back and forth with a regular period as it regains its equilibrium. The oscillation of the water mass is known as a *seiche*. The magnitude of the set-up and seiche can be estimated from empirical relationships involving wind speed, the depth and physical configuration of the water body and the length of the over-water fetch. The frequency of occurrence of large set-ups at any given location may also be estimated by extreme value frequency analysis techniques based on the wind record in terms of return periods for each month of the year.

Currents in a lake or reservoir are mainly wind-induced, and empirical relationships linking wind speed and direction to water motion are available. However, such wind-induced currents are also closely interlinked with set-up and seiche effects because these phenomena also generate water motion.

Littoral drift is the transport, primarily by waves and shoreline currents, of beach material along the shores of lakes and reservoirs. Since these are primarily wind-induced effects most studies of littoral drift must also begin with analyses of wind speed, duration and direction.

Ice. Ice on lakes and large reservoirs is frequently divided into two main categories: sheet ice which, except for cracks, is a smooth homogeneous cover, and *agglomeritic ice* which is produced by the jamming together of separate ice and snow masses or by the freeze-thaw cycle acting on the ice. The physical factors associated with ice formation are linked with the winter energy balance of the water body and are difficult to evaluate. They involve a large number of meteorological factors including temperature, humidity, wind, cloud cover, radiation, evaporation, condensation, and even the stability of the atmosphere. Many empirical equations have been developed relating ice formation or ice thickness to some, or all, of these meteorological parameters. Recent studies of ice on large lakes have established a relatively simple relationship between ice cover and two meteorological factors, namely, the severity of the winter and the temperature regime of the previous summer. In addition, however, each lake has its own ice cover characteristics dependent upon its physical features and geographic location.

The Role of Precipitation and Evaporation in Lake Level Variations. One of the more complex problems associated with lakes and large reservoirs is that of large variations in water levels and the role of the meteorological factors that influence them. Although some of the changes in the levels of large lakes are no doubt due to such man-made alterations as diversions and channel deepenings, and to such physical factors as geological tilting and the scouring of river outlets, it is recognized that these effects are comparatively small and it is generally conceded that the major variations in levels are due to gains from precipitation and losses by evaporation. However, in the case of reservoirs and smaller lakes man-made controls may very well mask the effects of both the physical and meteorological factors.

Since precipitation is the source of the earth's fresh water supply, it is only natural to seek a relationship between the levels of water bodies and the precipitation falling on the basins. Recent computer-oriented studies have indicated that 70 per cent of the variance of the levels of the Laurentian Great Lakes is associated with basin precipitation. Areal estimates of precipitation over the land areas of a watershed can be readily calculated using methods described in section (i). However, a major problem arises when dealing with large lakes since there are indications that annual precipitation measurements over the water are five to ten per cent less than the corresponding measurements over land. This deficiency has a seasonal variation and is greatest in the summer. Unfortunately the matter of uniform exposure of the gauges is so critical that the problem is still largely unresolved.

The estimation of evaporation losses from a large body of water is an even more difficult task. While standard evaporation pan data may be representative of losses from very small water bodies, other approaches are necessary to estimate monthly and annual losses from larger lakes because of the heat storage factor. Three basic methods are in general use, each requires its own set of hydrologic or meteorological data and each has its own merits and shortcomings.

The *water budget method* is an accounting of the gains and losses of water in the water body. A simplified form of the equation is:

Inflow + Precipitation = Outflow + Evaporation + Change in water level.

This method is very useful when dealing with small reservoirs if there is no ground water exchange but serious shortcomings develop in large lakes when inflow and outflow figures are very large compared to those for evaporation. In such cases it has been calculated that a 1 per cent error in flow measurement can introduce errors of up to 25 per cent in evaporation estimates.

The *energy balance method* is an accounting of energy gains and losses in the lake with the evaporation energy as the unknown. Although scientifically sound, the method is less frequently used because of the scarcity of year-round surveys of water temperature to depth which are required for the evaluation of the heat storage term.

The *mass transfer method* is used most frequently because it requires observations of only three factors: wind speed, the humidity of the air and surface temperature of the water body. However, before the method can be used with confidence the *evaporation coefficient* in the quasi-empirical equation should be calculated for the specific water body by comparison with the results from other techniques.

(iv) *Man's Influence on the Hydrologic Cycle*

Man frequently modifies various phases of the hydrologic cycle either consciously or unconsciously. Changes in streamflow on a basin may be brought about by a change in vegetation. Draining a swamp for agricultural purposes may not change the total annual flow of a river, but it usually does change the seasonal flow regime. There are also a number of examples of increasing runoff volumes by removing stands of trees and thus eliminating high rates of evapotranspiration. However, although the denuding of a forest drainage area may increase total flow, the additional water may come at an unwanted time such as during flood periods. This is a most important area of hydrometeorology which still requires much research.

There are other areas where man is consciously attempting to influence the hydrologic cycle such as through induced precipitation by cloud seeding, the induced melting of snow and ice covers by dark dusting techniques and evaporation control by introducing a monomolecular film. However, although productive results have been reported there is still some uncertainty as to the degree of effectiveness and the cost–benefit economics of the various methods.

It has also been suggested that man is inadvertently affecting the hydrologic cycle through urbanization and may even be responsible for bringing about climatic trends as a result of the prolonged and widespread burning of fossil fuels. Although much research is required to substantiate these theories and to delineate the degree of actual change the evidence is noteworthy and deserves the increasing attention that it is receiving.

See also: Hydrologic cycle.

Bibliography

BRUCE and CLARK (1966) *Introduction to Hydrometeorology*, Oxford: Pergamon Press.

CHOW (1964) *Handbook of Applied Hydrology*, New York: McGraw-Hill.

LINSLEY and FRANZINI (1964) *Water Resources Engineering*, New York: McGraw-Hill.

Manual for Depth-Area-Duration Analysis of Storm Precipitation (1946) U.S. Weather Bureau Co-op. Studies, Tech. Paper No. 1.

WORLD METEOROLOGICAL ORGANIZATION (1965) *Guide to Hydrometeorological Practices*, No. 168, TP. 82, Geneva.

T. L. RICHARDS

IMPATT DIODE MICROWAVE GENERATORS.

The term IMPATT (IMPact Avalanche Transit Time) is applied to a variety of pn, or more complex, semiconductor diodes which, when reverse biased, generate power at frequencies above 100 MHz. These diodes are a comparatively recent development and follow a prediction made by Read in 1957 of a method by which a negative resistance at microwave frequencies could be produced across the terminals of the device. This occurs by a two-stage process in which electrons and holes are first generated by impact ionization in the high fields of the pn junction and then drift out of the diode through the depletion layer.

The details of the process are most easily understood by reference to the $n^+ pip^+$ diode of Fig. 1 which was originally proposed by Read and subsequently tested experimentally. This particular structure and the exact equivalent $p^+ nin^+$ have the advantage that the two stages, generation and drift of carriers, occur almost independently of one another.

Fig. 1. Structure of the Read diode.

When a reverse bias is applied to the diode a depletion layer will form at the pn junction and with increasing voltage this will expand to encompass the whole of the p and i regions. Applying Poisson's equation div $D = \varrho$, we obtain the variation of field through the diode which is shown in Fig. 2. It will be seen that a narrow region of high field is formed around the junction.

In this region electrons and holes leaking across the device will be accelerated to very high speeds, so high indeed that on collision with an atom of the lattice they can give up sufficient energy to enable an electron hole pair to be created. This is the process which is known as impact ionization. The new carriers are then able themselves to pick up energy from the field, to make collisions and to produce further electron hole pairs. Thus we have the possibility, provided that the field is high enough, of creating an avalanche of electrons and holes around the pn junction.

Consider now what will happen if the device is to operate as a microwave generator in a resonant circuit. A time-varying voltage will appear across the diode, the field around the junction will likewise increase and decrease with time and the avalanche will grow and decay. The maximum *rate* of carrier generation in the avalanche will occur while the field is around maximum. This is because the rate is a very strong (approximately E^6) function of the field, which gives energy to the ionizing electrons and holes, but is only simply proportional to the number of these in the avalanche. By a similar argument the rate will drop to zero when the r.f. voltage swings the total field in the device below that critical for the sustaining of the avalanche. The number of carriers in the avalanche region will then drop to zero as they are swept out by the field. The process is illustrated in Fig. 3. It can readily be seen that the number of carriers in the avalanche will lag the field or the applied voltage by a phase of about $\pi/2$.

Of the carriers which are generated in the avalanche region the electrons will be almost instantaneously removed from the diode by way of the n^+ contact leaving the holes to drift through the intrinsic region to the p^+ contact. Since the field in the intrinsic region is sufficiently high that the holes drift under saturated velocity conditions, and hence do not diffuse, the pattern of space charge will preserve its original form as it drifts from the junction to the p^+ contact. During the whole of the time τ that the holes are drifting they contribute to the conductive current flow through the diode and thus the generation of carriers at time t gives rise to a conduction current which flows from time t to $t + \tau$. Since the holes are produced periodically this current will also be periodic, but with a

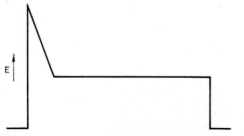

Fig. 2. Electric field profile through Read diode.

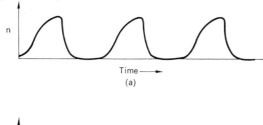

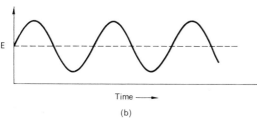

Fig. 3. Number of carriers (a) generated by applied field, (b) in the avalanche region of Read diode.

time lag of $\tau/2$ behind the avalanche. By a suitable choice of the dimensions of the diode we can arrange for $\tau/2$ time lag to be equivalent to $\pi/2$ phase lag at the circuit resonant frequency. Imposing this condition on the diode we find that the conduction current lags behind the phase of the voltage giving rise to it by a total of around π radians. In practical terms this means that, instead of power being absorbed resistively from the circuit by the diode, power from the device bias supply will be pumped into the resonant circuit by way of the conduction current. In addition to the conduction current a capacitive current will also flow in the circuit. This serves to charge and discharge the depletion layer of the diode but need not concern us here since it is in quadrature with the varying voltage and will not contribute to the power transfer between diode and a correctly matched circuit.

Thus we have seen that Read's $n^+ pip^+$ diode in a resonant circuit acts as a self-sustaining oscillator at a frequency determined by the transit time for holes crossing the depletion layer. For clarity the foregoing argument has been kept as simple as possible. Using a more detailed analytical approach Read was able to predict the following performance figures.

500 MHz	25 W–5 kW	pulsed 30% efficiency
5 GHz	50 W	pulsed 30% efficiency
50 GHz	2 W	pulsed 30% efficiency

In the subsequent practical verification of the operating principle of the $n^+ pip^+$ dipole 19 mW CW at 5 GHz with only 1·4% efficiency was achieved. Any complete theory for the IMPATT diode must take the following complications, some of which were neglected by Read, into account.

(1) The avalanche usually occurs in an extended rather than a localized region so that drift of both electrons and holes becomes important.

(2) The velocities of electrons and holes are field dependent and different as are their powers of ionization.

(3) The field patterns in the diode are modified by the space change produced in the avalanche and are therefore time dependent. In consequence, detailed solutions are in general possible only with the aid of a digital computer. Such calculations have now been performed for the Read structure and show that efficiencies should indeed approach the 30% predicted by Read. This is particularly so if the structure is allowed to degenerate somewhat towards that of the more simple pn diode.

In 1965 added impetus was given to the study of avalanche microwave generators when it was discovered that many of the pn diodes already in production as varactors would in fact deliver small amounts of microwave power when reversed biased into avalanche in a resonant circuit. The theoretical existence of a microwave negative resistance in such structures has since been demonstrated by Misawa and it is to the avalanching pn diode that most workers have now turned their attention. This is not only because of the simpler technology but also because calculations promise greater power handling capability, and hence power output, from the pn diode as opposed to the $n^+ pip^+$ structure.

While silicon, because of the advanced state of its technology, has so far been the most commonly used material for the construction of IMPATT diodes both germanium and gallium arsenide have operated well. Diodes made from these latter materials in fact give better noise performance than the equivalent silicon diodes. For example, when operated as amplifiers, phase locked to an external source, typical noise figures for silicon IMPATT diodes are 36–45 dB, while for germanium and gallium arsenide figures in the range of 27–35 dB are obtained. Operated as free running oscillators the f.m. noise is far greater than the a.m. noise. It is the rather high level of this noise, originating as it does in the inherently noisy avalanche process, which may prove to be a limiting factor in many of the potential applications of the IMPATT diode. There are, however, encouraging signs that the noise can be reduced by the use of high Q circuits. For instance an increase in circuit Q from 100 to 1000 is reported to decrease f.m. noise by 20 dB.

The power output from IMPATT diodes is proportional to the cross-sectional area employed and for this reason it is possible to produce in very similar manner devices which will operate CW with low power output and devices which will give high peak powers at low duty cycle. In the laboratory the efficiency of operation has ranged up to 15 per cent but the CW silicon IMPATT diodes at present in production rarely offer efficiencies of more than 1 or 2 per cent. The power input conditions for IMPATT diodes vary greatly according to their construction. The applied voltage is that capable of producing avalanche breakdown, usually between 10 and 100 depending on frequency and current densities of 100–1000 A cm^{-2} are typical. The removal of heat from the diode is one of the chief factors which at present limit power output and efficiency and it is likely that as constructional techniques improve these will both rise.

As in the Read structure the frequency of operation of the *pn* IMPATT diode depends primarily on the dimensions of the depletion layer, although factors such as the applied current density also play a part. The following achievements illustrate both the present state of development of these diodes and the wide frequency range which they cover.

Power	Frequency	Efficiency	Material
CW pn IMPATT diodes (July 1968)			
—	6 GHz	15%	Ge
250 mW	8–10 GHz	6%	Ge
1 W	10.2 GHz	9.6%	Si
2.7 W	13.8 GHz	10.2%	Si
50 mW	14.2 GHz	9%	GaAs
800 mW	35 GHz	6%	Si
130 mW	80 GHz	2%	Si
37 mW	106 GHz	—	Si
Pulsed pn IMPATT diodes (July 1968)			
50 W	3–4 GHz	6%	Si
1.3 W	6.5 GHz	3.6%	GaAs
2.1 W	16.52 GHz	5.62%	GaAs

Although the performance cited suggests that the IMPATT diode will find uses in a wide range of microwave equipment, the recently announced "anomalous mode" promises to offer true solid state competition to the established microwave valves such as the magnetron. In this mode silicon *pn* IMPATT diodes which normally give 50 W pulsed at 3–4 GHz with 6 per cent efficiency can be operated in a lower frequency resonant circuit to yield 100 W at around 1 GHz with about 60 per cent efficiency. The reasons for this dramatic increase in efficiency are not yet fully understood but it appears that the phenomenon is limited to certain diodes with higher than normal avalanche voltage.

Similar behaviour has also been found in both germanium and gallium arsenide and in this case the behaviour has been reproduced in computer studies. These have shown that to give such high efficiencies the diode must be operated in a circuit capable of supporting oscillation at both the normal IMPATT transit time frequency and at a lower frequency which is a sub harmonic of this. The high-frequency circuit is lightly loaded and power is extracted at the lower frequency. The result is that oscillations build up in the circuit which lead to the diode being driven into a very large avalanche once per low-frequency cycle. The generated carriers flood the diode and, because of the large space charge created, the field in the diode is greatly reduced. This has two effects; the carriers are not swept quickly out of the diode but remain until they are slowly removed by diffusion and the residual electric field and secondly, while the carriers are in the diode, it appears as a closed switch which passes a large current with a low voltage drop. Eventually the carriers are removed and the diode reverts back to its low current, high voltage state. During the low-frequency cycle time the high-frequency circuit being lightly loaded can sustain an oscillation which apparently not only serves to initiate the low-frequency resonance but which prepares the diode to receive the full low-frequency voltage by removing any carriers remaining from the previous cycle. Since in this mode of operation the diode experiences large current and voltage excursions the efficiency of conversion of d.c. to r.f. can be very high. For example, germanium diodes have been reported to give 5 W CW at 500 MHz with 43 per cent efficiency, which has increased to 60 per cent when pulsed.

It remains to be seen how repeatable and useful the high efficiency "anomalous mode" will prove to be but there is already no doubt that the conventional IMPATT diode will make a considerable impact on the whole field of microwaves.

Bibliography

MISAWA T. (1966) Negative resistance in *pn* junctions under avalanche breakdown conditions, Parts I and II. *IEEE Transactions on Electron Devices* **ED-13**, No. 1 (this volume contains a number of other papers dealing with theoretical and experimental work on IMPATT diodes); *IEEE Transactions on Electron Devices* **ED-14**, No. 9 (this volume also contains a collection of theoretical and experimental papers on IMPATT diodes).

PRAGER H. J., CHANG K. K. N. and WEISBROD S. (1967) High power, high efficiency silicon avalanche diodes at ultra high frequencies. *Proc. IEEE* **55**, 586.

READ W. T. JR. (1958) A proposed high frequency negative resistance diode, *Bell System Technical Journal*.

M. P. WASSE

INFRA-RED CAMERAS AND INFRA-RED TV

I. Introduction

I.1. Black-body emission

We are used to seeing things by the light they reflect or scatter. Such an object has to be illuminated to be seen. The source of light may be a sunbeam, or the light diffused by the sky, or an artificial source. There are, of course, also those sources which are directly seen by the eye, for instance a black body at a temperature higher than 800K. The reason for this need of illumination lies in the very narrow spectral range of the eye, from 0.4 to 0.8 μm. In this wavelength interval the electromagnetic emission of any body at a temperature lower than 800K is negligible, and cannot be detected by the eye. Hence the necessity of illumination and its big technical and industrial applications.

I.2. Vision in the dark

However, detectors of invisible radiations are known and have permitted, among other things, the development of infra-red spectroscopy. The first used were thermopiles and bolometers which are really a kind of very sensitive thermometer. More recently, photoconductive cells using pure semiconductors have been made for the near infra-red. Their range has been extended to 40 μm with doped materials. They have to be cooled to liquid nitrogen, hydrogen or helium temperatures. In some cases they are more sensitive and show a much faster response than thermal detectors. Besides their application to spectroscopy where a beam of large cross-section but narrow spectral range is focused on the infra-red detector, they can be used to replace the eye for seeing things at room temperature in "the dark", that is without any source of light. In this case all wavelengths emitted by a small area element are received on a mirror and focused on the detector. Here it is the case of a large spectral range, and a very narrow beam to get a narrow field of view. This is "thermal vision" and to look at different elements of a chosen area, the mirror has to be rotated, or a large number of detectors has to be displayed in a convenient array. So we shall distinguish two types of infra-red camera either using scanning systems and one detector, or an infra-red "retina" made of a number of small detectors. In both cases we shall get "thermal photographs" and if the number of images available per second could exceed a few unities, it would really constitute "thermal or infra-red television".

II. Scanning Systems

They are practically the only ones used at this time to look at wavelengths longer than 3·5 μ. They can use either thermal detectors (mainly thermistors), or quantum detectors.

II.1. Thermal detectors

Probably the first scanning system to be described was the Barnes infra-red camera, commercially available for several years and using one thermistor bolometer. The camera's optical system contains a target scanning mirror (Fig. 1), which views the target in small increments, corresponding to a field of view of 1×1 mrad. These increments measure $0·12 \times 0·12$ in at a distance of 10 ft from the camera. The plane mirror is oscillated horizontally to deflect the optical beam of the radiometer in a horizontal pattern over the plane of the target while slowly tilting in the vertical direction. As the mirror returns quickly to its initial position, the electronic picture is blanked out. Thus the camera produces a horizontal raster similar to that seen on a television receiver. The image scanning section contains a glow modulator tube, a collimating lens, an image-scanning mirror, and a camera back. Visible light is eliminated by a germanium filter which screens out wavelengths shorter than 1·8 μm. A mirror chopper alternately shows the target radiation and a black-body temperature reference. This produces an a.c. signal proportional to the difference between the unknown radiation and the known one. With the Barnes camera it is possible to detect temperature variations of 0·1K in a target made of 200×300 elements which are mapped in 13 min. This is a long time, owing to the long time constant of thermistor detectors. The 0·1K sensitivity concerning the target temperature corresponds roughly to 10^{-4}K for the bolometer temperature variation and could be improved in case of necessity. Very recently the thermistor bolometer has been replaced by the l triglycine sulphate pyroelectric detector, and the time needed for mapping has been reduced down to 30 sec.

II.2. Quantum detectors

Here the time constant may be several order of magnitude lower, and the detectivity an order higher

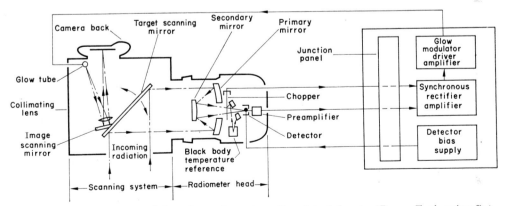

Fig. 1. Scheme of the Barnes infra-red scanning using a thermistor bolometer (Barnes Engineering Co.).

The scanning time can be reduced to a fraction of a second and is now limited mainly by the mechanical system. Figure 2 gives the scheme of the Aga camera where a high speed is obtained by the use of a rotating prism. Here the frequency of 400 cps must be achieved. We can compare the performance of the Aga camera No. 652 using an InSb photoconductor working at 77K, and the CSF infra-red camera No. 815 using mercury doped germanium working at 20K.

	Aga No. 652	CSF No. 815
Minimum temperature variation detected	0·5K	0·1K
Number of elements scanned	100 × 100	400 × 400
Number of images per second	16	1/4

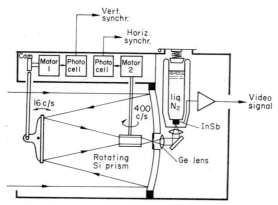

Fig. 2. Scheme of the Aga high-speed scanning infra-red camera using an InSb detector (A.G.A.).

There is a considerable increase of speed over the thermistor camera. However, low temperature has to be employed. This may be a problem for field operations in spite of the successful development of such devices as "minicoolers" which are very small refrigerators able to cool a small detector down to 10K for instance.

III. "Retina" Systems

These avoid all those mechanical difficulties which are especially troublesome for high scanning speed as we have seen. They also bring the advantage of integrating the signal on each element during the whole time needed for scanning the entire image. This ought to give a comparable gain of sensitivity to the one obtained when Zworykin's iconoscope replaced Farnsworth's image dissector in 1934, which indeed marks the beginning of television in the visible spectral range. Several systems have been described either with an electron beam analysis of the retina, or with a direct visual display of the infra-red image.

III.1. Direct visual display

We shall consider the evaporograph, the edgegraph, the thermoradiograph, and the possibilites of thermochroism.

III.1.1. Evaporography. Figure 3 gives the scheme of an evaporograph. The method consists in preparing a thin film, solid or liquid, and forming a thermal image on its surface. We know that the speed of evaporation is a highly temperature-sensitive function. The thickness of the film will therefore diminish at the points which are most illuminated and the thermal image will be translated into interference colours. The method is an old one. It was used by Czerny to record absorption spectra in the region of 10 μm. All military authorities have shown interest in it for other purposes. It has been taken up by Baird Atomic (Fig. 3). Ten seconds are needed to photo-

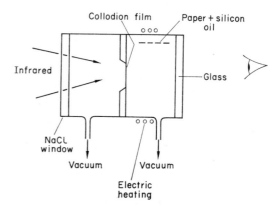

Fig. 3. Scheme of an evaporograph. There is a deposit of silicon oil on the collodion film which is the cooler part of the silicon oil gas chamber. The thickness is non-uniform and gives a pattern of the infra-red image formed on the film (Baird Atomic).

graph an object at 37°C against a background at 18°C. The smallest temperature difference which can be detected is 0·5°C. This is the same as with the Aga camera but the speed is much lower. A rough calculation gives the speed of evaporation of a film in terms of the temperature:

$$\phi_0 \simeq A T^{-1/2} e^{-L_0/RT},$$

A being a constant, T the absolute temperature, L_0 the latent heat of evaporation, $R = 8\cdot 32 \times 10^7$ ergs and ϕ_0 representing the speed of evaporation per cm², evaluated as the number of molecule-particles. It is seen that

$$\frac{d\phi_0}{\phi_0} = \frac{1}{2}\frac{dT}{T} + \frac{L_0\,dT}{RT^2} \simeq \frac{L_0\,dT}{RT^2}$$

Hence the relative variations of the speed of evaporation in terms of the temperature are the greater, the greater L_0 and the smaller T. Furthermore, the noise will probably decrease at the same time as T. Low-temperature evaporography therefore appears to be one of the most promising methods.

III.1.2. Edgegraphy. We know that the absorption coefficient of a semiconductor in the intrinsic cut-off is a very sensitive function of the temperature. It arises from the fact that the potential barrier E_g separating the valence and the conduction bands is a very sensitive function of the temperature: an increasing function for Se and PbS, a decreasing one for InSb.

This property is used to give thermal images visually. Selenium is generally chosen, its main advantage being that it permit work in the yellow, where the eye is most sensitive. We get $dK/dT = 190 \text{ cm}^{-1} \text{ deg}^{-1}$ for $\lambda = 0.589$ μm. Moreover, the technology of selenium films is easy even for thicknesses of about a micrometre. Finally the thermal conductivity is poor. The absorbing layer is a chromium metallic film thin enough to remain transparent to yellow. The infra-red radiations should arrive on the side of the dielectric, the incidence should remain weak and the retina should be in a vacuum (Fig. 4). The fact that visible

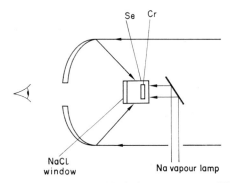

Fig. 4. Scheme of Harding's image converter. The energy gap of selenium corresponds to yellow radiations and is very temperature dependent (Harding et al., 1958).

images are obtained directly is particularly advantageous. Objects can be seen at 15°C above the ambient temperature in a time of about 0.5 sec, but there is a lack of contrast.

III.1.3. Thermoradiography. Fluorescence can be temperature sensitive and used as a thermoelectric property. Urbach uses low yield "poisoned" phosphors, that have to be strongly excited with ultra-violet. He gives as example a phosphor with 49 per cent ZnS, 49 per cent Cds, 15 per cent NaCl, 0.4 per cent Ag and 0.1 per cent Ni, excited with $\lambda = 0.365$ μm. The quantum yield is then very temperature-sensitive. Visible fluorescence is used and

observation is direct. Objects are detected at 10°C above the ambient temperature (Fig. 5). Account is taken of the heterogeneity of the sensitive surface by taking successively two exposures of the excited fluorescent surface, with and without thermal image.

III.1.4. Thermochromism. A few compounds display visible absorption bands which are highly temperature-sensitive. "Thermochromism" is often found in a remarkable way for liquid crystals, for instance several derivatives of cholesterol. It has been used

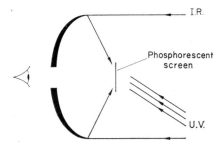

Fig. 5. Thermoradiography from Urbach. The yield of phosphorescence must be very temperature-dependent (Urbach, 1949).

recently by Westinghouse to produce a thermal photographic device. This is reminiscent of the evaporograph. Although it is less sensitive, at least at the moment, it is useful for studying the infra-red distribution in a laser beam.

III.1.5. Silver halide photographic emulsions. It is known that the speed of development for an illuminated emulsion may be very temperature-sensitive. This could find applications to reveal a temperature distribution.

III.2. Electron beam analysis

Practically all devices developed include a photo-conductive retina and are not sensitive to wavelengths longer than 4.5 μm. Other retinae have been considered for longer wavelengths.

III.2.1. Photoconductive retinae. A photoconductive retina is made of a thin film and on to it is projected an infra-red image. If the range of radiations used lies within the sensitive band of the semiconductor, the thermal image is transformed into a relief of conductivities which has to be analysed by an electron beam. The main difficulty seems to be the production of a high resistivity semi-conductor. Morton (1959) prepares a PbS retina by evaporation of PbO followed by some doping with sulphur. In this way, the resistivity increases from 2 Ω to 10^{10} Ω-cm. As the limit of sensitivity for PbS is at about 2 μm where room temperature emission is negligible, illumination with near infra-red is necessary. A lens can then pair the retina and the scene illuminated, and a relief of conductivities is obtained on the retina, which exactly reproduces the infra-red image (Fig. 6). The retina is then swept by a beam of slow electrons. Each element of the retina catches some electrons, and its potential decreases until it reaches nearly that of the cathode. The extra electrons are then repulsed and collected by an electronmultiplier, coaxial with the cathode.

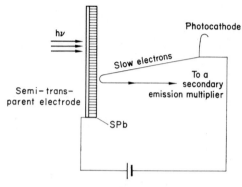

Fig. 6. PbS image convertor from Morton (Morton, 1959).

When the spot leaves the element, the surface discharges slowly (dark current of the photoconductor), then is charged positively. If the element considered is illuminated the operation is more rapid, and the potential reached when the spot comes back to the element is higher. The spot therefore takes up more electrons, and the multiplier, receiving less, will give a weaker video signal. This signal modulates the spot intensity of a second cathode ray oscillograph whose scanning is synchronized with the first one. A negative image is thus obtained. Here again, the advantage of the device is that the signal is integrated during a whole image cycle. Recent improvements have been made by RCA. A "vidicon" sensitive to wavelengths up to 4·5 μm has been described under the name of "Iricon". The retina is made of PbTe and has to be cooled down to 80K. With 100×100 spatial elements, and a scanning speed of 30 images per second, the limit of sensitivity for temperature variations is 1 °C. The extension to still longer wavelengths (8–14 μm) has been considered by Westinghouse with a retina made of As_2Se_3–Sb_2Se_3 ("Thermicon").

Organic semiconductors give also promising possibilities. Triphenylphosphene complexes, for instance $[(C_6H_5)_3P = 0]_2CuCl_2$, show both the high resistivity and the temperature dependence needed for photoconductive retina. From 25 to 50 °C the resistivity decreases from 10^{13} to 10^9 ohm cm. The complex keeps its properties with a still higher resistance when incorporated in various synthetic resins. Thin films are thus available to make retinae for vidicon tubes. The mechanism for the thermal conductivity is not known but it appears to be independent of the wavelength, and could be used up to the far infra-red. In some cases electron beam analysis can be replaced by a direct conversion of the relief of conductivities into a visible image thanks to the remarkable properties of electroluminescent screens. Figure 7 gives a scheme for such a device. Here again the thermal image is formed on the photoconductor. The points illuminated become less resistant and the electric field in the

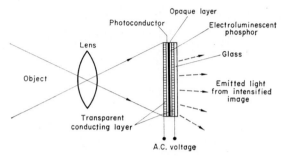

Fig. 7. Amplifier and image convertor with a double layer, from Kazan (Kazan, 1959).

elements facing the electroluminescent screen increases. Visible luminescence is observed, corresponding point by point with the infra-red image.

III.2.2. Pyroelectric retinae. A thin plate of a pyroelectric crystal cut perpendicular to the pyroelectric axis can be used as a retina. A thermal image focused on such a retina gives both a charge distribution and a dielectric susceptibility distribution below the Curie point. Only the last one will subsist above the Curie point. These distributions have to be analysed with an electron beam, and this is a new problem which has been considered recently (Le Carvennec, 1968). Figure 8 gives the scheme of a possible vidicon tube.

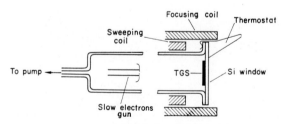

Fig. 8. Scheme of a vidicon tube for thermal TV using a ferroelectric retina.

One difficulty arises from the considerable direct current resistivity of most pyroelectric materials. The charges deposited on the crystal cannot be eliminated by conductivity. Fortunately the secondary emission coefficient for some ferroelectric materials like triglycine sulphate can be made higher than unity (Fig. 9) and gives a possibility for discharging the retina after an electron beam analysis. Some experiments and a theoretical examination give a sensitivity to flux variations less than 10^{-6} W cm^{-2}, with 200×200 spatial elements, a high scanning speed, and no spectral limitations. This means the possibility of detecting 0·3°C on images every second.

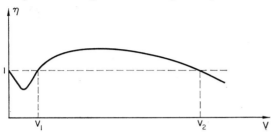

Fig. 9. *Variation of the secondary emission coefficient η vs. the voltage V of impeding electron. For $V_1 < V < V_2$, η is larger than unity, V_1 and V_2 are the cross-over voltages (Le Carvennec and Charles, 1968).*

III.2.3. Photo-emissive retinae. Photo-emissive retinae are not sensitive beyond the red. But some of them, for instance a Sb–Cs cathode, have a very temperature dependent photo-emissivity coefficient when they are used with radiations corresponding to their spectral limit of sensitivity (red radiation). This property is particularly suited for the production of a thermal image converter. The image is projected on to a retina formed of three thin layers (Fig. 10).

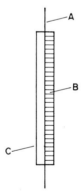

Fig. 10. *Scheme of the retina used in Garbuny's image convertor for thermal radiation. A, support membrane made of SiO. B, SbCs photocathode whose emission is very thermosensitive. C, membrane absorbing infra-red and transmitting visible radiations (Garbuny, 1961).*

A lens followed by a red filter pairs the screen of a cathode ray oscilloscope with the retina. A luminous red spot of constant illumination can then sweep the retina point by point. Each element of the retina then gives an electronic signal which is a function of its temperature. This video signal is transmitted and will give a visible image on a television screen, a transposition of the thermal image. The retina of Fig. 10 is surrounded on all sides apart from the solid angle needed to receive the image, by cold screens at 150K. Thirty images per second are obtained with a definition of 130,000, and a signal-to-noise ratio $S/B = 5$ for objects at 35°C above zoom temperature. The performance is obviously due to the fact that variations of flux on each element are integrated during the whole time of sweeping, that is about 1/30 sec.

IV. Applications

Most of the research concerning infra-red cameras and infra-red TV has been sponsored for military application. The first ones at the end of World War II consisted of an infra-red camera sensitive in the very near infra-red, and used with an infra-red projector. Now the extension to much longer wavelengths has made possible "passive" devices which do not need any lighting. They see the radiation emitted by objects to be detected. Meanwhile a lot of peaceful applications have been developed. Processing plants employ furnaces, heat exchangers, tanks and flow lines. Abnormal thermal patterns of the surface of a furnace can arise from wear (hot points), or soot (cold points). Heating or cooling systems can be checked for uniformity. Internal wear or malfunction in a machine generally causes variations in the thermal patterns appearing on the surface. Aircraft and missile testing are easier with thermal photography either in static testing, or dynamic in wind tunnels. In electrical engineering incidents may be prevented if hot points are located. In electronics, experimental printed circuit boards for instance can be rapidly temperature-tested. Medical and biological investigations by means of infra-red cameras are beginning to yield results. Some correlations have been found between discontinuities in skin temperature and underlying growths. Unexplained temperature differences exist over supposed uniform areas of plant life. In oceanography the use of infra-red cameras to get a rapid survey of temperature distributions seems very promising.

In conclusion, infra-red cameras and infra-red TV now allow a new vision of the world, completely different from the one the eye sees. When its sensitivity, speed of response, definition and spectral range have been improved, a considerable amount of new information will be available, and the applications can only be guessed at the present time.

Bibliography

A. G. A. (Lidingö 1–Sweden). Aga Thermovision Model 652. Documentation.
BABITS V. A. (1963) Infra-red TV tube, *Electronics* (13 Sept.).
BARNES ENGINEERING Co. Barnes infra-red camera, *Bulletin* 12–600.
GARBUNY M. (1961) *J. Opt. Soc. Amer.* **51**, 261.
HADNI A. (1963) Possibilités actuelles de détection du rayonnement infrarouge, *J. Phys.* **24**, 694.

HARDING W., HILSUM C. and NORTHROP D. (1958) *Nature* **181**, 691.

KAZAN B. (1959) A feedback light-amplifier panel for picture storage. *Proc. I.R.E.* **47**, 12.

LE CARVENNEC F. and CHARLES D. (1968) Recherche d'un dispositif nouveau de Télévision Thermique. Rapport Final C.S.F.

MORTON G. A. (1959) *Proc. I.R.E.* **47**, 1607.

URBACH F. (1949) *J. Opt. Soc. Amer.* **39**, 1011.

A. HADNI

INFRA-RED SPECTROSCOPY, RECENT DEVELOPMENTS IN. Since the first wide-range double-beam grating spectrometer was introduced in 1955, steady progress has been made in the further development of these instruments to meet the needs of several different types of user.

Low-cost grating spectrometers of modest resolution are coming into wide use for routine analysis and cost has been kept down by substituting low-pass interference filters for the prism unit previously used for order selection. Two or more gratings are usually employed to cover the desired spectral range and ingenious mechanisms make the change from one grating to another without producing any discontinuity in the spectrum. Use of moderately priced instruments of fairly high resolution (1 cm^{-1}) is also growing rapidly and for advanced research work expensive spectrometers of high resolution (~ 0.25 cm^{-1}, 1 to 25 μm) or somewhat lower resolution (~ 0.5 cm^{-1}, 2.5 to 50 μm) are available. Even higher resolution (~ 0.1 cm^{-1}), over restricted ranges, has been attained by individual workers with specially built equipment.

Two novel instruments (Beckman) make use of a rotary interference filter wheel, developed by Optical Coating Laboratory, Inc., the selected wavelength varying as the wheel rotates. This filter has three circular segments to cover the total range 2.5 to 14.5 μm, and the thicknesses of the layers composing each section are tapered to give at any point along a track a wavelength linearly related to the angular position of the wheel. Such instruments have a low resolving power (56 to 77) and are capable of scanning a spectrum quickly (shortest time 5 sec with 1 sec between scans) as well as slowly (e.g. 7.5 min): a useful spectrum can be obtained without using a beam condenser from as little as 2 μg of sample. The fast-scan instrument is intended for use with a gas chromatograph, whereby a component eluting from the column can be identified "on the fly" from its spectrum. Even greater speed is required, for example, in studying fast reactions and with special equipment low resolution spectra, over a limited range, have been recorded in as short a time as 1/150 sec. A highly refined version of the variable filter is the tunable Fabry-Perot etalon which enables a narrow spectral region to be scanned at high resolution.

Conventional spectrometers operate without difficulty to 50 μm and the range can be extended to 1000 μm provided that steps are taken to make the best possible use of the small amount of energy obtainable from available sources and to eliminate interference from stray radiation of the shorter wavelengths. Between 50 and 100 μm a Globar is probably the best source but beyond 100 μm a high-pressure mercury lamp is more effective since the silica envelope becomes increasingly transparent and energy radiated from the plasma gives a useful addition to that coming from the hot silica. Furthermore, wavelengths shorter than 20 μm which comprise much of the stray radiation contribute a smaller proportion of the total energy than is the case for "black body" radiators. Larger slits are needed to compensate for low energy and these in turn demand detectors with an increased receiving area; in consequence the Golay cell with a receiving element of about 3 mm diameter gives a larger signal than a thermocouple beyond 50 μm and, in addition, the difficulty of "blackening" the receiving element at long wavelengths is overcome since the Golay employs an extremely thin film of unblackened aluminium which absorbs radiation from the visible region to the far infra-red.

To reduce stray radiation to an acceptable level some or all of the following expedients are employed.

(1) Grating reflectors which give an angular separation between diffracted shorter wavelengths and specularly reflected longer wavelengths.

(2) Scatter plates, i.e. roughened plane metal surfaces, which scatter shorter wavelengths while acting as a mirror for longer ones.

(3) Fine wire gauze which reflects without loss wavelengths longer than twice the wire spacing while transmitting or scattering shorter wavelengths.

(4) Choppers with blades of crystalline material which chop long wavelength radiation while transmitting (and therefore not chopping) shorter wavelengths. For example, a chopper with KBr blades becomes effective beyond 30 μm.

(5) Reststrahlen filters, i.e. crystalline reflectors which reflect preferentially over limited spectral regions:

NaCl	45– 60 μm
KCl	55– 70
KBr	70– 85
KI	85–100
CsBr	100–140
CsI	130–170
KRS-5	120–220

(6) Low-pass filters which reject high frequencies. For example, a crystal quartz window fitted to the Golay detector eliminates most of the incident radiation between 4 and 40 μm and black polyethylene sheet is useful for rejecting short wavelengths transmitted by quartz. A low-pass

interference filter cutting on at 58 μm has recently been produced by Grubb Parsons.

Rigid polyethylene sheet makes an inexpensive and convenient window material for the far infra-red and polystyrene is also useful beyond 100 μm. Many new window materials for general spectroscopic use are now available and some details of the more important are given in Table 1.

besides its output at 6328 Å can also provide radiation at 1·152 and 3·391 μm, and with lower intensity at five other wavelengths between 1 and 3·4 μm. Laser radiation can be very nearly monochromatic and the 6328 Å linewidth (single mode) is only about 3×10^{-7} cm^{-1}; this radiation can be varied in frequency over a small range, of the order of 1 Gc/s, and amplitude modulation is possible. Other gas lasers of particular

Table 1. Some Newer Optical Materials

Name	Formula	Longwave limit, μm (approx.)	Mean R.I. n, μm	Remarks	
IRTRAN 1	MgF$_2$	9·2	1·35, 4	I	h
IRTRAN 2	ZnS	14·7	2·22, 8	I	h
IRTRAN 3	CaF$_2$	11·5	1·40, 5	S	h
IRTRAN 4	ZnSe	21·8	2·42, 11	I	h
IRTRAN 5	MgO	9·4	1·64, 5	I	h
IRTRAN 6	CdTe	30	2·67, 10	I	u 300
Silicon	Si *	>300	3·42, 10	I	h
Germanium	Ge †	>300	4·00, 16	I	h
KRS-5	TlBr/Tl I	42	2·34, 20	S	m.p. 414
	As$_2$S$_3$ (glass)	11	2·40, 6	I	s 210
	BaF$_2$	12·5	1·42, 10	S	h
	CsBr	41	1·64, 20	VS	h
	Cs I	55	1·72, 25	VS	h
	AgCl	23	1·97, 12	I	m.p. 458

IRTRAN, polycrystalline material manufactured by the Eastman Kodak Company.
I, insoluble
S, slightly soluble } in water.
VS, very soluble
u 300, usable at 300°C.
s 210, softens at 210°C.
h, m.p. above 600°C.

* Transmission depends on purity; appreciable absorption 10—20 μm.
† Transmission depends on purity; appreciable absorption 15—150 μm.

As an alternative to the spectrometer, particularly for wavelengths longer than 20 μm, the Michelson type of interferometric spectrometer has attracted a great deal of interest in the last ten years or so. Another interferometer which is of particular use at the lowest frequencies, down to 2 cm^{-1} (5000 μm), is the lamellar variety (RIIC—Research and Industrial Instruments Company). Richards has given an interesting comparison between these two interferometers and the echelette grating spectrometer in the far infra-red.

Monochromatic radiation sources. Well over 100 "spot" frequencies extending from the visible region to 1000 μm or more have been reported. Most of these are obtained by laser action in gases at low pressure and in some cases a considerable amount of power, continuous or pulsed, is available. A notable example of a c.w. source is the ubiquitous He/Ne laser which

interest are the CO$_2$/N$_2$/He variety which can generate up to 300 W c.w. at 10·6 μm (and from this the harmonic at 5·3 μm) and the CN laser, developed mainly at the NPL, in which a pulsed high-voltage discharge through the vapour of a nitrile at low pressure in a resonant cavity produces a wavelength of 337 μm with a peak power of 10 W. Much of the pioneer work on stimulated emission at long wavelengths was carried out at SERL and successful results were obtained there with H$_2$O, D$_2$O, NH$_3$, N$_2$O and compounds of hydrogen, carbon and nitrogen.

Solid lasers also provide infra-red wavelengths, but these are usually not much longer than 1 μm. For example, a pulse power of 50 MW at 1·06 μm is obtainable from a Q-switched neodymium-doped glass laser. More modestly, gallium arsenide at room temperature produces radiation from 0·84 to 0·90 μm, depending on circumstances, with a pulse power of 10 W or c.w. power of 125 mW; when cooled to 77° K, a c.w. power of 1 W has been obtained with an efficiency of light production of more than 50 per cent. Most work has been carried out on GaAs, but other semiconductors have been used, e.g. Ga(As-P), (Ga-In)As, InAs, InP and InSb, and wavelengths as long as 3·1 μm have been observed at low temperatures. Conditions are not so critical as with a gas laser and simple cleavage of a crystal often provides a satisfactory resonator. Radiation is much less monochromatic than for gas lasers, due to multi-mode operation, and a linewidth of 100 Å is not unusual, although individual modes have a width of much less than 0·008 Å. Amplitude modulation up to 200 Mc/s is possible and frequency may be varied somewhat by application of a magnetic field or by applying stress to the crystal, e.g. by ultrasonic means.

Complete overlap of optical and microwave frequencies has now been achieved since commercial microwave equipment operates at wavelengths down to 1·35 mm (e.g. the French carcinotron generates 1 W at 2 mm wavelength) and harmonic generation from microwave frequencies gives appreciable power down to 500 μm (NPL); resolution of about 2×10^{-3} cm^{-1} is claimed. An exciting prospect for the infra-red spectroscopist is a variable frequency source to cover the 1 to 25 μm region and some progress in this direction has been made; spectra in the near infra-red have already been produced by this means.

Detectors. While a great many new detectors have appeared in recent years little impact has been made on the design of infra-red spectrometers. One reason is that these detectors operate at low temperatures (in some cases down to 1·5 °K) and the cost and difficulty of continuous refrigeration tend to restrict their use. Another consideration is that these detectors have a very fast response, compared with the usual thermal types, and suitable beam chopping and ratio recording systems to take advantage of their speed have yet to be developed. When these are available it will be possible to record a 1 to 25 μm spectrum at 1 cm^{-1} resolution in 1 sec or less. Spectra produced at this speed can be recorded with ultra-violet print-out techniques which require no chemical processing, or by electrostatic means (Varian).

Accessories and techniques. The value of infra-red spectroscopy to the chemist increases year by year, due in no small measure to the exertions of accessory manufacturers. Beam condensers have been available for some time, but the latest models are considerably simplified by making use of two replicated off-axis ellipsoids (see Fig. 1). With a 6:1 reduction at the primary focus a spectrum of 1 μg of sample can be obtained, while with an 8 mm path length microcell (RIIC) as little as 0·15 ml of gas may be examined.

The KBr disc technique has extended considerably. It is now possible to obtain a small die in which KBr powder is compressed between the polished faces of two opposed stainless steel bolts; no special press is required and the die with the disc inside is placed in the sample position in the spectrometer. As an alternative to circular discs RIIC provides dies for making rectangular plates, the smallest being 5×1 mm with a

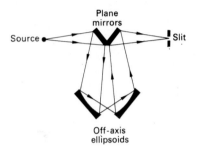

Fig. 1. Reflecting microscope using two off-axis ellipsoids.

sample content of 100 μg; yet another die produces a disc with a recess 0·1 mm deep, which with a plane KBr disc forms a microcell for liquids. Wedged plates of KBr, 13×13 mm, can also be produced and these are used in two ways: firstly, to obtain a variable path length by moving the wedge (with sample) across the sample beam, or secondly, by incorporating one or more major components of a sample in a wedge and placing this in the reference beam in the appropriate position, absorption bands due to these components in the sample can be cancelled out and a difference spectrum obtained for the minor components.

ATR (or internal reflection) techniques, introduced by Fahrenfort in 1961, have expanded considerably. When radiation strikes, at an angle of incidence θ, an interface between a medium of high refractive index n_1 and another of low refractive index n_2, reflection occurs when $\sin \theta > n_2/n_1$, but if the second medium absorbs radiation the reflected ray is weakened, due to some penetration into the absorbing region. The depth of penetration is of the order of one wavelength and naturally increases as the angle of incidence is reduced. In early forms of ATR attachment a single reflection was employed and high sensitivity obtained by allowing θ to approach the critical value. However, it was found that spectra produced in this way suffered considerable distortion from the substantial changes of refractive index which frequently occur in the vicinity of an absorption maximum (anomalous dispersion). This difficulty is overcome and sensitivity improved in more recent ATR units by increasing the angle of incidence and using multiple reflections; a common number is 9, but this may be increased to 25 or more in some cases. Figure 2(a) shows the ray path in such an ATR reflector (which may be made of KRS-5, refractive index ~2·4); the sample may be in contact with one or both surfaces and, for example, might be obtained by washing a spot on a paper chromatogram (say 20 μg) from the paper onto the plate and allowing the solvent to evaporate. In Fig. 2(b) a variant of the multiple reflection technique is shown in which two reflector plates are placed together with a film (~0·002 mm) of

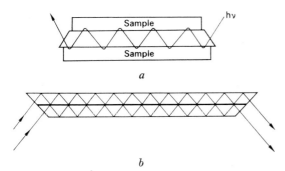

Fig. 2 (a) *Multiple internal reflection effect.*
(b) *Capillary internal reflection cell.*

liquid sample between them; the quantity required is about 0·4 μl.

Spectroscopic examination of GC fractions emerging from the column of a gas chromatograph has led to the development of special sampling devices. In one (RIIC) such a fraction is fed through a heated line via a self-sealing septum into an extruded silver chloride cell (transparent to 23 μm). Liquid collects in a wider part of the cell and is centrifuged into the narrower part used for absorption measurements; solvent can also be introduced if the material is solid at room temperature or the quantity very small; minimum cell volume is 0·2 μl. An alternative technique makes use of the capillary cell [Techmation, Fig. 2(b)] which is hinged to allow exhaust gas from the chromatograph to sweep over the cooled reflector plates. Thermoelectric cooling is employed (−20° C) and the cell with condensed fraction (0·4 μl) removed for spectroscopic examination. Yet a further use for ATR is in a pyrolysis unit (Techmation) which consists of an evacuated gas cell (horizontal) with a nichrome filament at the bottom of the unit and a port at the top through which sample is placed on the filament. The port is fitted with an ATR plate, thermoelectrically cooled, on which some of the products of pyrolysis condense. Operating conditions (firing temperature, up to 1000° C, and duration of heating) can be varied to suit the nature of the sample, and the products, both gaseous and condensed, may be examined spectroscopically without opening the pyrolysis chamber.

Temperature and pressure ranges over which spectra are recorded tend steadily to increase and a unit is now available (RIIC) for the accurate temperature control of solid and liquid samples from −190° to 250 °C; a "diamond squeezer" from the same source allows liquids and solids to be studied at pressures in excess of 100 kbar.

Storage of spectral data in a computer and the use of sophisticated methods for comparing spectra and identifying a compound from its spectrum are making progress, but, unfortunately, the bulk of available spectra (100,000 or so) are of prism quality and need to be re-run with modern grating instruments. Norman Jones and his colleagues have used a computer to improve spectra by eliminating known instrumental errors; they have also shown how a spectrum may be reproduced with a comparatively small number of parameters, four for each separate band and three more to allow for baseline shape. Signal-to-noise ratio in difficult conditions can be improved by multiple scanning followed by averaging, the improvement being proportional to the square root of the number of scans; it is claimed that a recognizable spectrum of 10^{-8} g of phenacetin has been obtained in this way. It seems probable that spectroscopists will find many uses for computers in the future and a programming language for them has already been proposed—SPECTRAN.

Applications. Ever-increasing resolution is being sought to cope with the minute line widths found in rotational spectra, etc. (~0·01 cm^{-1}) and the whole subject of band shapes and intensities in solids, liquids and gases (including pressure broadening) is being actively pursued. Other important topics include solute–solvent interactions (sometimes with unusual solvents such as liquid HF, SO_2 and NH_3); hydrogen bonding in special molecules or conditions of particular interest, e.g. the vapour phase; pressure-induced, electric field-induced and defect-induced absorption, e.g. in solid argon with xenon as impurity; absorption in compounds of the inert gases, e.g. xenon dichloride; internal rotation and restricted rotation in a variety of molecules and circumstances; rotation of molecules in a viscous medium; computer aided normal coordinate analysis.

Sheppard and others have studied the spectra of a number of different molecules adsorbed on a variety of surfaces, e.g. metals and porous glass, while other workers have obtained spectra over a wide range of temperatures and pressures: Price *et al.*, for example, have covered the range 4° to 4000 °K and at the lowest temperature observed an absorption band with a half-width of ~0·5 cm^{-1} for a polyatomic ion isolated in an alkali halide lattice. Free radicals and transient species, often trapped in a matrix (e.g. trichloromethyl in solid argon), have also been investigated. Vibrational spectra of molecular ions in the gas phase are being profitably studied by the technique of photoelectron spectroscopy. The value of infra-red spectra to the inorganic chemist has rapidly increased as the technique for dealing with aqueous solutions has been developed by Goulden and others; it is claimed that most polyatomic inorganic ions can be unequivocally identified. Vibrational frequencies and intensities have been studied in polymers, both theoretically and experimentally, and such properties as hardness and elasticity can be related to molecular configurations.

Several thousand steroids have now been examined and the data provide a unique opportunity for studying group frequencies in a wide variety of molecular structures. Spectra of bacteria, viruses, yeasts, fungi and algae have been recorded and a useful review has been given by Norris. Atmospheric pollution has been investigated in the U.S. with a multi-reflection gas cell (path-lengths up to nearly 3 miles are available with such cells) and light has been thrown on the chemical reactions which occur in a city atmosphere.

Availability of long-wave instruments has led to much activity in this region; studies include far infra-red dispersion, pure rotational spectra, pressure-induced absorption (in N_2 at 100 atm, for example), lattice vibrations and absorption in ionic crystals and in organo-metallic and related complexes; vibrational frequencies associated with metal–ligand and metal–halogen bonds. Long-wave measurements provide information on the liquid state, on semiconductors (impurity-induced and deformity-induced absorption), ferroelectricity, superconductivity and plasma diagnostics. Torsional modes in long-chain molecules (e.g. ketones) and barriers to internal rotation have been studied.

Military applications such as the fusing and homing of missiles, surveillance and thermal mapping have accelerated the development of detectors and detecting systems. As a by-product thermal pictures of areas of the human body are now obtained on a routine basis and breast growths and other abnormalities can be detected at an early stage through a small elevation of temperature (as little as 0·5 °C) above surrounding tissue.

Studies of planetary atmospheric absorption indicate the presence of CO_2 on Mars and Venus, CH_4 on Jupiter, Saturn, Neptune and Uranus, and NH_3 on Jupiter. Rank has tried to estimate the gas pressure on the latter by broadening the bands of CH_4 and NH_3 with hydrogen in a laboratory experiment. Radiometric measurements of the surface temperatures (or cloud temperatures) of the planets have been made, and S. D. Smith and colleagues have constructed a radiometer to operate from a satellite and measure the Earth's atmospheric temperature in six layers from the ground to 40 km.

Industrial applications are legion and include the following: control of manufacture of epitaxial planar devices by the measurement of thickness of a deposited silicon layer from the spacing of interference bands; infra-red gas analysers (up to three gases with one instrument); analysers for plant control; radiometers for measuring the temperature of moving or otherwise inaccessible surfaces, e.g. plastic films during manufacture, rapidly checking temperature distribution in a microcircuit under operating conditions, and fault detection in overhead power cables. Lack of knowledge of emissivity when measuring surface temperature is largely overcome by comparing the amounts of radiation for two wavelength regions which are fairly close together. Sophisticated analysers for process control are typified by that developed by Howarth and Stanier (Brit. pat. appn. 37623/64), in which the desired measurement is obtained by ratioing two signals from a single detector to give $(A - B/A + B)$, where B represents energy transmitted by absorbing sample in one spectral region and A the energy transmitted in an adjacent region where absorption is slight. This result is, to a large extent, independent of instrumental changes and satisfactory operation can be obtained even when 90 per cent of the radiation is obscured, due for example to fouling of the windows of the sample cell.

Another novel instrument is the milk analyser, described originally by Goulden (NIRD) and further developed by Grubb Parsons. Energy transmitted by a thin layer of homogenized milk is compared with that transmitted by an equal thickness of water at 5·72 μm (fat), 6·5 μm (protein) and 9·6 μm (lactose); percentage concentrations are indicated in succession (well under 1 min for three components) on a digital voltmeter and may be punched out automatically on paper tape, if desired, for subsequent processing by a computer.

See also: Interferometric spectrometers.

Bibliography

BRÜGEL W. (1962) *An Introduction to Infra-red Spectroscopy*, London: Methuen.
CARTER J. B. R. (1961) Interference filter, in *Encyclopaedic Dictionary of Physics* (J. Thewlis Ed.), **3**, 872.
DAVIES M. (Ed.) (1963) *Infra-red Spectroscopy and Molecular Structure*, Amsterdam: Elsevier.
HENDERSON E. (1961) Filters, spectroscopic, in *Encyclopaedic Dictionary of Physics* (J. Thewlis Ed.), **3**, 143.
HOUGHTON J. and SMITH S. D. (1966) *Infra-red Physics*, Oxford: Clarendon Press.
MARTIN A. E. (1966) *Infra-red Instrumentation and Techniques*, Amsterdam: Elsevier.
POTTS W. J. (1963) *Chemical Infrared Spectroscopy*, Vol. 1, New York: Wiley.
RICHARDS P. L. (1964) *J. Opt. Soc. Am.* **54**, 1474.
SZYMANSKI H. (1967) *Progress in Infrared Spectroscopy*, Vol. 3, New York: Plenum.
THOMPSON H. W. (Ed.) (1961) *Advances in Spectroscopy*, Vol. 2, New York: Interscience.

A. E. MARTIN

INTERFEROMETRIC SPECTROMETERS. Interferometers such as the Fabry-Perot, Michelson and Jamin have been in existence for periods varying from 70 to more than 100 years, and their uses include refractive index measurement, accurate wavelength determination and investigation of the hyperfine structure of spectral lines. The idea of using an interferometer to replace a spectrometer *completely* is, however, a relatively new one which attracted some attention in 1950 when Fellgett pointed out that a significant signal-to-noise advantage should accrue from the use of interferometric techniques. Because of practical difficulties little progress was made until 1956, when Gebbie and Vanasse published the first long-wave spectrum obtained with a Michelson interferometer. Application to the far infra-red region was understandable since at that time spectrometers for use at long wavelengths were not very satisfactory, owing to the low energy available from black-body sources and the large amount of stray radiation invariably present. Conversion of an interferogram into a spectrum by means of a Fourier transformation is a matter of some difficulty and the most accurate result is obtained by using a digital computer. In the last ten years or so interferometric spectroscopy has attracted a great deal of interest, and effective instruments are now available for obtaining interferograms of gases, liquids and solids, in a wide variety of experimental conditions and over a wavelength range of 20 to 5000 μm (500 to 2 cm^{-1}). Extension to shorter wavelengths has also been achieved but mechanical and other difficulties increase, while grating spectrometers already provide an excellent performance. However, when the available radiation is of low intensity, as in astronomical applications, the improved signal-to-noise of an interferometer makes its use worthwhile and P. and J. Connes have obtained impressive planetary spectra around 1·6 μm.

Michelson interferometric spectrometer. A typical optical system (Grubb Parsons) is shown in Fig. 1 in which mixed radiation is chopped and, after travers-

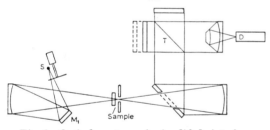

Fig. 1. *Optical system of simplified interferometric spectrometer.*

ing any sample present, passes through a circular aperture, and having been collimated is directed at an angle of 45° onto a beam splitter T which may conveniently be a taut film of polyethylene terephthalate (Melinex or Mylar). Part of the radiation is transmitted and following reflection from a fixed plane mirror is partially reflected from T into a condensing system which forms a reduced image of the entrance aperture on the detector D (e.g. a Golay cell). The remaining part of the radiation from the collimating mirror is reflected from T onto a plane mirror which can be moved along a line perpendicular to its surface, and after passing through T is condensed onto the detector as before. Since radiation received by the detector has followed alternative paths, there is in general, for a particular wavelength λ, a phase displacement $2\pi x/\lambda$, where the path difference x and λ are both measured in centimetres. Replacing $1/\lambda$ by a frequency k, the phase difference becomes $2\pi kx$; when the path difference is zero all frequencies are exactly in phase and the amplified and rectified output from the detector is a maximum. As the path difference is steadily increased from zero, detector output falls to a minimum and thereafter passes through many subsidiary maxima and minima, with a tendency for the amplitude of the fluctuations to diminish. The resulting curve, $I(x)$, is known as an interferogram (Fig. 2) and the

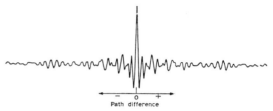

Fig. 2. *Interferogram from Michelson interferometer.*

spectrum, $G(k)$, is obtained by performing the process of Fourier transformation. Mathematically,

$$G(k) = \int_{-\infty}^{\infty} I(x)\, e^{2\pi i k x}\, dx$$

but provided that the interferogram is symmetrical about the central maximum, the simpler cosine transformation can be employed:

$$G(k) = 2\int_{0}^{\infty} I(x) \cos 2\pi k x\, dx;$$

the factor of 2 will be ignored from now on.

Modern instruments vary in construction and one recent form follows the design of Gebbie, in which the heart of the interferometer (comprising beam splitter and fixed and movable plane mirrors) is enclosed in a relatively small cubical box, while other component parts such as radiation source (mercury lamp), detector, and a variety of sampling facilities can be attached to make up the complete instrument. This modular construction leads to flexibility and reduction of size; moreover, since it is necessary to evacuate the interferometer to avoid absorption by water vapour in normal air, the pumping operation can be speeded up. The moving mirror is commonly displaced by means of a micrometer screw driven by a stepping motor. Alternatively, uniform increments of displacement may be obtained by using moiré fringes from a pair of transmission gratings (one fixed and the other attached to the moving mirror), or interference fringes (e.g. Fabry–Perot) from a nearly monochromatic light source.

Mathematical details. For radiation of a given frequency, the resultant displacement for two superimposed rays of unit amplitude but differing in phase by θ can be written as

$$e = \cos \alpha + \cos (\alpha + \theta),$$

where any value for α is equally likely. Thus

$$e = 2\cos(\alpha + \theta/2) \cos \theta/2$$

and since the energy, E, associated with the resultant ray is proportional to e^2, E is proportional to $\cos^2 \theta/2$, i.e. to $1 + \cos \theta$, when $\cos^2 (\alpha + \theta/2)$ is averaged over all values. A single frequency k_1 therefore gives an infinite cosine wave plus a constant term which we shall ignore for the moment, and if we calculate the cosine transform for a path difference x_1 rather than infinity, and assume an amplitude a, we obtain

$$G(k) = a \int_0^{x_1} \cos 2\pi k_1 x \cos 2\pi k x \, dx$$

$$= a/2 \int_0^{x_1} [\cos 2\pi(k_1 + k) x + \cos 2\pi(k_1 - k) x] \, dx$$

$$= \frac{a}{2} \left[\frac{\sin 2\pi(k_1 + k) x_1}{2\pi(k_1 + k)} + \frac{\sin 2\pi(k_1 - k) x_1}{2\pi(k_1 - k)} \right]$$

This expression becomes of interest when $k \approx k_1$ and is then very nearly equal to

$$\frac{a x_1}{2} \frac{\sin 2\pi(k_1 - k) x_1}{2\pi(k_1 - k) x_1}$$

which is of the form $\sin X/X$ [see Fig. 3(a)] and reaches its maximum, $a x_1/2$, when $k = k_1$. It will be seen that a crude representation of the monochromatic frequency k_1 is given and that as x_1 is increased the func-

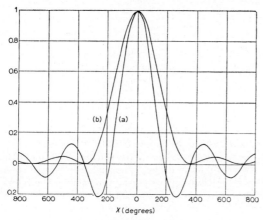

Fig. 3. (a) Function $\dfrac{\sin X}{X}$

(b) Function $\left(\dfrac{\sin X/2}{X/2} \right)^2$.

tion more and more closely approaches a single frequency. Furthermore, the maximum value $a x_1/2$ gives a measure of a, the amplitude associated with frequency k_1. Now an actual spectrum is the sum of a very large number of narrow spectral intervals with different intensities, and each will give rise to its own train of cosine waves; the interferogram is therefore the sum of this whole series of waves. If the frequency of the multiplying cosine function $\cos 2\pi k x$ is varied over the complete spectral range, each frequency will pick out, as it were, the same frequency in the interferogram, and the area of the resulting curve taken to a suitable value of x is proportional to the intensity of this particular frequency in the spectrum.

Thus an infinite cosine wave transforms to a single frequency and, conversely, a single frequency transforms to an infinite cosine wave. This reciprocal relationship may be applied quite generally to any spectrum, however complicated, and the corresponding interferogram. A single frequency cannot actually occur and a more realistic situation is provided by the Gaussian frequency distribution shown in Fig. 4. The

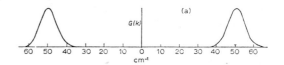

Fig. 4. (a) Gaussian frequency distribution; (b) half of corresponding interferogram; $k = 50$ cm^{-1} and half-width $= 10$ cm^{-1}.

corresponding interferogram is a cosine wave with amplitude diminishing according to a Gaussian function. A wide Gaussian frequency distribution gives rise to a quickly diminishing interferogram while a narrow Gaussian leads to one which falls away slowly. It can readily be shown from a consideration of these Gaussian functions that the smallest resolvable frequency interval, Δk, is effectively given by $1/x_1$.

In practice, the interferogram is sampled at uniform intervals of path difference, Δx, and values of $I(x)$ are recorded on magnetic or punched tape, or placed directly in a computer store. Information about $I(x)$ between sampling points is lacking so that the interferogram becomes equivalent to

$$I'(x) = I(x) \times \text{|||||} \cdots \times \text{▭}$$

where the comb-like symbol represents the sampling process and the rectangle stands for a truncating function corresponding to abrupt termination of the interferogram at path difference x_1. From the convolution theorem it is known that the Fourier transform of $I'(x)$, F. T. $I'(x)$, is equal to

F. T. $I(x)$ ✱ F. T. |||||

✱ F. T. ▭

where ✷ means in convolution with. It follows that

F. T. $I'(x) = G(k)$ ✷ ⊥⊥⊥⊥⊥ ✷ $\sin X/X$,

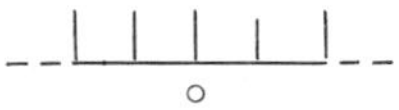

since the comb-like function transforms into a similar comb with the spacing Δx replaced by a frequency interval $1/\Delta x$ (see below), and the transform of the rectangle is $\sin X/X$. This result agrees with that obtained above for a single frequency, except that the spectrum $G(k)$ repeats at intervals of $1/\Delta x$ (a phenomenon known as aliasing), and, from a practical point of view, the Fourier transform of $I(x)$ can be replaced by the transform of $I'(x)$, with the consequence that

$$\text{F. T. } I(x) = \tfrac{1}{2}I(0) + \sum_{n=1}^{n_1} I(x) \cos 2\pi k n\, \Delta x,$$

where $n_1 \Delta x = x_1$ and $I(0)$ is the value of the central maximum; half of $I(0)$ is attributable to each side of the interferogram. It is obvious that such a computation requires the calculation of a large number of cosines but if Δx and Δk are suitably chosen, the same numerical values will appear again and again. Sophisticated methods of computation have been evolved to reduce the time required, one in particular being due to Cooley and Tukey. Another method which may be adopted when an on-line computer is available, is to compute $G(k)$ for all the different frequencies as data are fed in. The summations are carried out in parallel for all of these values of k and only the up-dated totals are stored. This procedure has the advantage that a spectrum can be obtained from the stored totals, with a recorder or graph-plotter, as soon as the interferogram is completed, or even at an earlier stage without stopping the interferometer, so that a decision can be made regarding the length of interferogram (and resolution) necessary for a particular quality of spectrum.

Since each frequency produces a train of cosine waves in the interferogram, it is possible to determine $G(k)$ from the stored values of $I(x)$ by cycling the store and using a wave analyser to measure the relative amplitude associated with each frequency. This technique is employed by Research & Industrial Instruments Company (RIIC) in equipment available for use with their interferometer.

To obtain a transmittance spectrum of a sample it is of course necessary to compute $G_1(k)$ without sample and $G_2(k)$ with it, and then calculate the ratio $G_2(k)/G_1(k)$. This aspect is dealt with in some detail later.

Reverting to the Fourier transform of an infinite comb, this follows automatically when the function is expanded in the form of a series by Fourier's theorem:

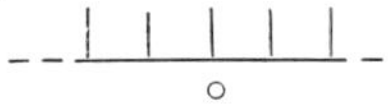

$$\text{--} \bot\bot\bot\bot\bot \text{--} = \cos 2\pi x/\Delta x + \cos 4\pi x/\Delta x + \cos 6\pi x/\Delta x + \cdots.$$

Each cosine function transforms to a single frequency, k, $2k$, $3k$, etc., and the result is a frequency comb with spacing $1/\Delta x$. A spectrum is equally valid for $-k$ as for k, so that successive spectra overlap if $2k_m > 1/\Delta x$, where k_m is the maximum frequency present (Fig. 5). If the number of sampling points is doubled (assuming that sufficient points were taken originally), the sum will of course be doubled also; therefore if the additional sampling points only are used the original sum will be obtained (see Fig. 6). It follows that the

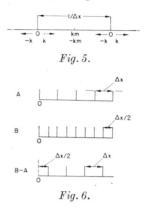

Fig. 5.

Fig. 6.

central maximum need not coincide with a sampling point but may be placed midway between two such points. Again, expanding the comb function, we obtain

$$\text{--} \bot\bot\bot\bot \text{--} = -\cos 2\pi x/\Delta x + \cos 4\pi x/\Delta x - \cos 6\pi x/\Delta x + \cdots$$

and if this is added to the previous expansion, the result is

$$2[\cos 4\pi x/\Delta x + \cos 8\pi x/\Delta x + \cos 12\pi x/\Delta x + \cdots]$$

which is a comb similar to the original one but with half the spacing. Both sets of sampling points with Δx spacing give rise to aliasing, but with the second set alternate spectra are reversed in sign. To be able to use large sampling intervals and thereby reduce the amount of data to be handled and in consequence the computing time, it is good practice to filter off all unwanted higher frequencies and efficient filters for this purpose are being developed. Filtering also leads to improved accuracy, particularly at extremely low frequencies where energy is unavoidably low and information of interest may otherwise be swamped by signals (with noise) from unwanted higher frequencies.

An intensity error which can be quite serious arises if the maximum of an interferogram does not coincide exactly with a sampling point or occur halfway between two sampling points. No difficulty arises if the two sides of the interferogram are taken and a complex transformation applied, but of course both the scanning and computing times are then increased. Mathematical techniques are available for dealing with a one-sided interferogram with a positional error, while the author and colleagues have pointed out that the error can be made negligible by sampling both sides of the interferogram and adding together pairs of ordinates, one from each side and equidistant (apart from sampling error) from the central maximum. A normal cosine transform is then computed from the data so obtained and it is easy to show that the result is $G(k) \cos 2\pi k\delta$, where δ is the positional error. If $\cos 2\pi k\delta$ is significantly different from unity, δ can be estimated from the interferogram and a correction applied.

Other errors may be caused by the "constant" term which accompanies the alternating signal of the interferogram since, clearly, the interfering radiations can never lead to a negative value. If this term really has a fixed value no harm is done provided that x_1 is always equal to an integral number of half wavelengths; it follows then that $2k$ must be an integral number of times Δk, the smallest resolvable interval. If the output signal is subject to drift (assumed to be linear with time) or the interferogram is apodized according to a triangular function (see below), it is necessary for x_1 to be equal to an integral number of wavelengths (Fig. 7) and this requires that k should be an integral

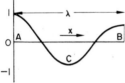

Fig. 7. Ordinate $= (1 - constant \times x) \cos 2\pi kx$; $-area\ C = area\ A + area\ B$.

number of times Δk. Alternatively, of course, a constant amount can be subtracted from all values of $I(x)$ before carrying out the cosine transformation, with or without apodization.

The beam splitter strongly polarizes radiation reflected from it and also has another important characteristic.

Owing to multiple reflections within the film (Fig. 8) reflectance can vary from zero to a maximum value dependent on refractive index and angle of incidence.

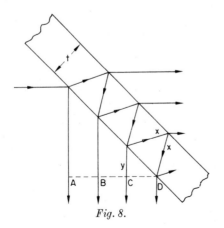

Fig. 8.

It can be shown, in absence of absorption, that when rays B, C, D, etc., are all in phase their resultant is exactly equal and opposite in phase to ray A, which suffers phase reversal on reflection. Therefore when the film thickness t is small compared with λ, the beam splitter fails to reflect and the signal from the interferometer becomes zero. To overcome this difficulty the film thickness must be increased as the frequency falls. Similarly, the film loses its reflectivity when the optical path difference between adjacent rays is an integral number of wavelengths. The condition is $2\mu x - y = n\lambda = (2\mu - 1/\mu m)t/\cos r$ when the angle of incidence is $45°$; since $\mu \sin r = 1/\sqrt{2}$, $n\lambda = t(4\mu^2 - 2)^{\frac{1}{2}}$. A $6\cdot 25\mu m$ Melinex film has a minimum ($n = 1$) at 500 cm^{-1} while maximum efficiency is obtained around 250 cm^{-1}. Several beam splitters of different thickness are used to cover the range ~ 2 to 500 cm^{-1} as efficiently as possible. A beam splitter of metallic mesh (mesh constant 50μ and wire thickness $7\cdot 5\mu$) has been employed by Genzel and co-workers in the region 25 to 100 cm^{-1}, but its reflectivity is of course strongly wavelength dependent and efficiency varies considerably over this range.

Ratio recording. Unfortunately, the Michelson interferometer has no simple analogue corresponding to the double-beam spectrometer and the accuracy of transmittance measurements, $G_2(k)/G_1(k)$, tends to be low unless great care is taken to maintain constant output from the radiation source, together with constant sensitivity of the detector and gain of the associated amplifying system.

As already explained, an actual interferogram consists of an alternating part $I(x)$ plus a "constant" term which varies if instrumental parameters change, so that the interferogram may be represented as $I(x)H(x) + aH(x)$, where a is a true constant and $H(x)$ includes effects due to variations in the measuring conditions. The author has shown, in principle at least, that the effects of such instrumental changes can be eliminated by interleaving three sets of measurements: from source, source plus sample, and source plus standard. On carrying out the Fourier transformation, we obtain:

$$G_1(k) + a_1 F(k) = P$$
$$G_2(k) + a_2 F(k) = Q$$
$$G_3(k) + a_3 F(k) = R$$

while $G_3(k)/G_1(k) = S$, corresponding to the known spectrum of the standard. $F(k)$ is the transform of $H(x)$ while the transform of $I(x)H(x)$ is very close to $G(k)$. $F(k)$ can now be eliminated and it turns out that the true transmittance of the sample, $G_2(k)/G_1(k)$, is

$$\frac{\dfrac{a_2}{a_3}(PS - R) + Q\left(1 - \dfrac{Sa_1}{a_3}\right)}{P - \dfrac{Ra_1}{a_3}};$$

a_1/a_3 and a_2/a_3 are readily obtainable from the interferograms at such a distance from the maxima that fluctuations have practically ceased.

A more practical approach is due to Haswell (Brit. Pat. 1,102,061, 1968), the idea being to add a third beam, derived from the source but by-passing the interferometer. The detector alternately receives the output from the interferometer and direct radiation from the third beam, so that the a.c. signal obtained is proportional to the difference between the two. After being amplified the signal is fed to a servo-motor which varies the position of an attenuator in the third beam so that the signal is maintained close to zero. By continuously recording the attenuator position a new interferogram is obtained which is corrected to a large extent for variations in source, detector and amplifier, and also the unwanted constant term can be eliminated.

Comparison with grating. Theoretically, the difference between a Michelson interferometer and an echelette grating is not great; in practice the difficulty of ruling a grating is exchanged for the difficulty of sampling an interferogram at equal increments of

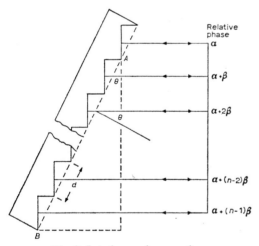

Fig. 9. Interference from grating.

path difference. Assuming monochromatic radiation, the amplitude of the disturbance at the detector resulting from the combined contributions from n steps of a grating (Fig. 9) is

$$A = a[\cos \alpha + \cos (\alpha + \beta) + \cos (\alpha + 2\beta) + \cdots$$
$$+ \cos \{\alpha + (n-1)\beta\}]$$
$$= a \cos [\alpha + (n-1)\beta/2] \sin \frac{n\beta}{2} / \sin \beta/2,$$

where α is a random phase angle and β the change of phase coresponding to a single step. Summing the n terms and remembering that energy E is proportional to A^2 and that all values of α are equally probable,

we have

$$E \propto \sin^2 (n\beta/2)/\sin^2 \beta/2 \propto n^2 \sin^2 (n\Delta\beta/2)/(n\Delta\beta/2)^2,$$

very nearly, when β is close to 2π, 4π, etc., corresponding to the first and higher orders. This is the scanning function for a grating spectrometer with infinitesimal slits, i.e. the instrumental profile for a monochromatic input. The corresponding function for the interferometer is $\sin n\Delta\beta/n\Delta\beta$ but this can easily be changed to the grating function (Fig. 3) by suitably apodizing the interferogram, i.e. multiplying each ordinate value by the function $(1 - n/n_1)$, where $n\Delta x = x$ and $n_1\Delta x = x_1$. In effect, the echelette grating is a self-apodizing interferometer, for

$$\tfrac{1}{2} \sin^2 n\beta/2 / \sin^2 \beta/2$$
$$= n/2 + (n-1) \cos \beta + (n-2) \cos 2\beta + \cdots$$
$$+ \cos (n-1) \beta$$

and it can be seen that the grating behaves as a combination of interferometers, with one contribution from the first and last steps, two from the first and last but one and from the second and last steps, and so on with $n-1$ contributions from adjacent steps. It is sometimes useful to be able to apodize an interferogram in order to reduce the fluctuations to zero at path difference x_1, even though some resolution is sacrificed (Fig. 3). By this means spurious oscillations in the spectrum, associated with the $\sin X/X$ function when an interferogram is abruptly terminated, are avoided.

Refractive index measurement. The Michelson is well suited to the determination of refractive index in the far infra-red region since parts of the beam paths are well separated and it is a simple matter to provide a focus in one beam for the accommodation of small samples. If a sample, uniform thickness t, is placed in one beam, an interferogram can be obtained in the usual way but it will be found that the maximum has shifted, relative to the interferogram without sample. Also, the maximum is less sharp than before and the interferogram is no longer symmetrical, both effects being due to the variation with wavelength of the refractive index of the sample. The problem is to measure accurately the change of optical path length caused by one or two traversals of radiation through the sample, $t(\mu - 1)$ or $2t(\mu - 1)$, and Gebbie and Bell and co-workers have been able to determine this quantity with marked success. Of course, the measurement is made easier if radiation is filtered to provide a narrow band of frequencies or a "spot" frequency source is used.

Lamellar grating interferometer. This two-beam instrument has been discussed in detail by Strong and Vanasse. The grating has a rectangular groove form and consists of two sets of interleaved plates (~1 cm thick) which move relatively to one another so as to vary the groove depth. Radiation falls on the grating perpendicularly, half being reflected from the front and half from the back surfaces, so that there is a path difference between the two halves of the beam of twice the groove depth. The path difference is continuously variable from, say, 0 to 10 cm, and Fourier transformation is employed to obtain a spectrum from the interferogram; mathematical details are similar to those given for the Michelson. This instrument avoids difficulties associated with beam splitters and is highly efficient at the lowest frequencies. However, at higher frequencies the interleaved members should be made thinner and, apart from the nuisance of having to change from one stack to another, it becomes increasingly difficult to maintain the necessary mechanical accuracy. Richards gives an interesting comparison between the echelette grating and the Michelson and lamellar interferometers at long wavelengths.

Unlike the Michelson, the lamellar interferometer may also be regarded as a grating spectrometer and it can indeed be used as such if the angle of incidence on the grating face is variable. If we consider a lamellar grating with zero depth of groove and radiation at normal incidence [Fig. 10(a)], the instantaneous

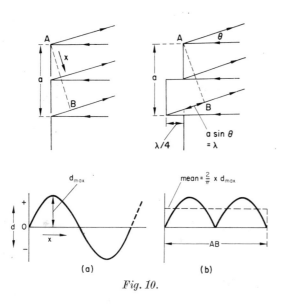

Fig. 10.

displacement, d, across a wavefront AB will be represented by the sine wave as shown and the average displacement over many waves will be zero. Therefore the reflected energy will be concentrated in the zeroth order, as of course it must be since in this instance the grating behaves as a plane reflector. If the depth of groove is an integral number of times $\lambda/2$, the situation is unchanged since the displacements across any diffracted wavefront will be represented by the sine wave as before; in consequence, the second and higher even orders are absent. The first order of diffraction

is obtained when the back surfaces produce a phase reversal [Fig. 10(b)], i.e. a groove depth of $\lambda/4$ (very nearly, since θ is small). The disturbance across the wavefront is as shown and the mean displacement is $2/\pi$ times the maximum attributable to the incident radiation. A similar line of argument leads to the conclusion that the mean displacement for the third order of diffraction is $2/3\pi$ times that of the incident radiation, in agreement with the summation

$$2\left(\frac{2}{\pi}\right)^2\left(1+\frac{1}{3^2}+\frac{1}{5^2}+\cdots\right)=1.$$

It will be noticed that when the zeroth order is strong the odd orders are weak, and *vice versa*. For most efficient operation the detector should receive only the zeroth order and for this to be so, $f\theta \geq w$, where f is the focal length of the condenser which forms an image of the entrance aperture on the detector, diameter w (a factor of 2 is included to allow also for aperture size). But for the first order, $a\theta = \lambda$, nearly, and $wa \leq f\lambda$; it follows that $a \propto \lambda$, i.e. narrower slats are required as the wavelength is reduced.

The instrument is marketed by RIIC.

Fabry–Perot interferometric spectrometer. This instrument consists essentially of two parallel plates of high reflectivity and low absorption, with means for varying the spacing between them. With parallel monochromatic radiation passing through the interferometer at normal incidence, interference of the multiply reflected beams produces an emergent beam which exhibits successive sharp intensity maxima as the separation of the plates, t, is steadily increased. For wavelength λ, the change of spacing, Δt, corresponding to one complete cycle—maximum to maximum—is given by $\lambda/2\mu$, where μ is the refractive index of the medium between the plates. The condition for obtaining a maximum is $2\mu t = n\lambda$, where n, the order of the maximum, is an integer. By differentiation, resolving power, $\lambda/\delta\lambda$, is equal to $-n/\delta n$, $\delta\lambda$ being the smallest resolvable spectral interval and δn the smallest change of order that can be recognized.

The spectral range over which a Fabry–Perot can be used without overlap of orders, $\Delta\lambda$, is inversely proportional to the plate separation, since if the $(n+1)$ order of λ coincides with the nth order of $\lambda + \Delta\lambda$, $\lambda/\Delta\lambda = n = 2\mu t/\lambda$ and $\Delta\lambda = \lambda^2/2\mu t$. To extend the useful range, an auxiliary filter or even a grating monochromator may be used to overcome problems of overlapping. Chabbal *et al.* have employed two or more Fabry–Perot interferometers in series, the purpose being to eliminate all but one of the pass bands of the highest resolution instrument.

A spectrum may be scanned by using straightforward means to vary the separation of the Fabry–Perot plates, but mechanical imperfections are liable to introduce errors and methods in which a rigid Fabry–Perot unit is maintained are preferred. In one design the spacer between the plates consists of a piezoelectric ceramic tube to which an electric field is applied radially; a useful change of length is given by 200 V. Alternatively, the spacing of the plates can be varied by passing current through coils around three Invar magnetostrictive legs; the plates may be made accurately parallel by passing small currents through additional coils. Beer, Bovey and Ring have achieved interesting results in the region of 2μm by pressurizing the gas between the interferometer plates (up to 40 atm of N_2) so as to alter the refractive index; a spectral interval of $0·02$ cm^{-1} was achieved.

At wavelengths to 10μ or so, reflecting stacks composed of alternate $\lambda/4$ layers of high and low refractive index materials are very effective, while at long wavelengths (100 to 600μ) Renk and Genzel have used a pair of metallic meshes to form a Fabry–Perot.

Miscellaneous. Both the Michelson and lamellar grating interferometers can be operated in a periodic manner, with a to and fro variation of path difference at constant velocity in each direction (Strong and Genzel and co-workers). A tuned amplifier or phase-sensitive detecting system may be used to pick out in turn each frequency present.

A novel version of the Michelson has the fixed and moving mirrors replaced by similar gratings set at the same angle to the incident radiation (P. Connes). A small periodic variation of path difference serves to modulate the output signal and by synchronized rotation of the gratings a direct spectrum may be obtained. High luminosity of the interferometer is combined with the high resolving power of the grating ($\sim 0·1$ cm^{-1}); however, the instrument requires very high optical and mechanical accuracy.

Several workers have tried to increase the luminosity of a spectrometer by increasing the number of entrance apertures and using some form of coding to disentangle superposed images of different wavelengths (Golay, Girard, Davies *et al.*). Ring and Selby have constructed a so-called mock interferometer (original concept due to Mertz) in which a grating spectrometer designed to have entrance and exit slits in the same focal plane has both slits replaced by a rotatable grid of parallel strips of equal width, alternately transparent and opaque. In operation this grid is rotated and a monochromatic image, formed by the grating and associated optics, is superimposed on the grid (image and grid are always parallel) and rotates about a centre displaced from the centre of rotation of the grid itself, the displacement depending on wavelength. The effect is that each wavelength is modulated at a different frequency and a mock interferogram, similar to that produced by a Michelson, is obtained. A Fourier transformation is necessary to obtain a spectrum but a few hundred cycles in the interferogram suffice to give good resolution. Since luminosity is high, the instrument may be used in series with a Fabry–Perot to provide even higher resolution without troublesome overlap of orders.

See also: Infra-red spectroscopy, recent developments in.

Bibliography
CARTER J. B. R. (1961) Interference filter, in *Encyclopaedic Dictionary of Physics* (J. Thewlis Ed.), **3**, 872.
CONNES J. (1961) *Rev. Opt.* **40**, 45, 116, 171 and 231.
GEBBIE H. A. (1961) *Advances in Quantum Electronics*, Part II, 155, J. R. Singer (Ed.), Columbia: University Press.
HOUGHTON J. and SMITH S. D. (1966) *Infra-red Physics*, Oxford: Clarendon Press.
JENNISON R. C. (1961) *Fourier Transforms*, Oxford: Pergamon Press.
MARTIN A. E. (1966) *Infra-red Instrumentation and Techniques*, Amsterdam: Elsevier.
RICHARDS P. L. (1964) *J. Opt. Soc. Am.* **54**, 1474.

A. E. MARTIN

INTERPLANETARY DUST. There is very little direct evidence of the existence of an interplanetary dust cloud and most of our knowledge of it has been gained by deductions made from astronomical observations. In recent years these observations have been supplemented by experiments made with artificial satellites and space probes and none of the information gained from them has been at variance with the results of Earth-bound observations.

The major piece of evidence for the existence of a dust cloud is the Zodiacal light. This phenomenon was first ascribed to scattering of sunlight by interplanetary dust by J. D. Cassini who wrote of his observations in the *Journal des Scavans* of May, 1683. Subsequent observations have amply confirmed this view. Measurements of the polarization of the Zodiacal light led to the conjecture that a significant fraction of it might be due to scattering by free electrons in space, but this was refuted by the observations of the spectrum by Blackwell and Ingham. Their evidence was later supported by direct measurements of the electron density in deep space. This density is now known to be far too low to make a significant contribution to the Zodiacal light or to its polarization.

The dust cloud appears to have the form of a lenticular disc or very oblate spheroid, with the number of particles per unit volume diminishing with increasing distance from the Sun and from the plane of symmetry. The minor axis of the spheroid appears to be parallel to the vector which represents the total angular momentum of the Solar System, although the measurements of this axis are not yet sufficiently accurate for a definite assertion to be made.

The Zodiacal light then seems to be due entirely to the scattering of sunlight by small dust particles in orbit around the Sun, with a possible contribution made by particles in orbit around the Earth. Fluorescence caused by the impact of the solar wind on the dust is about ten orders of magnitude smaller, since the solar photon flux at one astronomical unit (A.U.) from the Sun is about 4×10^{17}/cm^2 sec, and the proton flux at the same distance is about 7×10^{7}/cm^2 sec in the normal "quiet" solar wind condition. Fluorescence caused by ultra-violet light or X-rays from solar flares may account for the small variations in brightness that are reported from time to time, but present information is too scanty for any firm conclusion to be reached.

Nature of the Dust

Occasional samples of dust recovered from space vehicles may be taken as typical of the dust cloud about the Earth. Their appearance under the electron microscope is very convoluted, so that their scattering areas may well be much greater than that of a sphere of the same material. The samples examined so far have had dimensions of the order of a few microns and it is not known if this structure is typical of the larger and smaller particles that may be present.

The constitution of the particles is inferred by the analyses that have been made of meteoric material. They may be metallic, consisting largely of iron and nickel; or of the same siliceous material that is found in the "chondritic" meteorites. They may even be the chondrules themselves. So long as chemical and petrological examinations are not possible, theories of the constitution of the dust must depend on the theories of its origin. These will be discussed presently. The likely structure of the particles, however, can be inferred from the results of microphone experiments and of foil puncturing experiments that have been flown in satellites, and a material density lower than 1 has been suggested by the failure of one foil experiment to show any punctures, despite the registration of particles by their sounds in microphones. The convoluted structure observed later seems to support this idea.

Forces acting on the Dust Particles, and their Orbits under these Forces

The Newtonian theory of gravitation gives a complete account (apart from trivial relativistic corrections) of the motions of the larger bodies in the Solar System. Other significant forces operate when the bodies are small and the orbits that are followed by the dust particles may be quite different from those of planets, asteroids, or comets. The forces which affect their motion are (a) gravitation, (b) radiation pressure and (c) the solar wind. The solar wind is insignificant unless the particle size is as small as 10^{-8} cm, but then it is dominant, and single molecules may be ejected from the Solar System. This effect results in the tails of comets pointing away from the Sun. The solar wind is not important for particles in the micron to millimetre size range, and it is radiation pressure that must be studied in detail when the motion of the dust is considered. If the radiant energy flux per unit area at a distance R from the Sun is E/R^2, the momentum that passes through this unit area per second is E/cR^2. If this flux is intercepted by a particle and is absorbed or scattered isotropically, and if the particle has a projected area A, there will be a force $\dfrac{AE}{cR^2}$ acting on it. If the scattering is not isotropic, and if diffraction must be allowed for, the force will be $\dfrac{Q(a)\,AE}{cR^2}$ where $Q(a)$ is an efficiency factor, which depends, obviously, on the size of the particle. The quantity E/c is generally taken as $1\cdot035 \times 10^{22}$ dynes when R is measured in centimetres.

It is convenient to classify particles by their "radius" a even though they may not be spherical. The radius is defined in terms of the mass by $m = \dfrac{4}{3}\pi\varrho a^3$ and the effective value of the density may be taken to be the same as that found in foil penetrating experiments. The projected area is then defined as πa^2, and the forces acting can be described as functions of the particle "radius".

Three regions need to be considered separately. These are:

(1) $a \gg \lambda$, the Tyndall or geometrical scattering region.

(2) $a \simeq \lambda$, the Mie scattering region.

(3) $a \ll \lambda$, the Rayleigh scattering region.

Region 1. Although, according to Babinet's theorem, diffraction accounts for as much scattering as geometrical interception of the light, the scattering angles are very small and there is no significant momentum transfer. Thus the scattering cross-sectional area is πa^2, and the efficiency, $Q(a)$, which depends on the nature of the surface, is 1 for a perfectly reflecting sphere, and is equal to 4/3 for a white lambertian reflector. As the radius a decreases the diffraction scattering angles become larger and $Q(a)$ may rise to be about 2.

Region 3. For these small sizes the amount of scattering depends not on the shape of the particle but only on its volume and its polarizability. In the case of a particle made of an insulating material the polarizibility is given by the Lorentz–Lorenz formula, and this, together with the formula for the radiation rate from an oscillating dipole, gives an exact expression for the scattering cross-section, and for the force:

$$F = \frac{1}{3}\left(\frac{2\pi}{\lambda}\right)^4 8\pi \frac{E}{cR^2} a^6 \left(\frac{\mu^2-1}{\mu^2+2}\right)^2$$

where μ is the refractive index, and λ can be taken as 5×10^{-5} cm. The inverse fourth power of the wavelength is apparent, but more important here is the dependence on the sixth power of the particle radius.

If the particle is made of a conducting material, a similar sum can be done, but the *absorption* cross-section is now much greater than the scattering cross-section, and is equally effective in transferring momentum to the particle. The force resulting is:

$$F = \frac{8\pi^2}{\lambda} \frac{E}{cR^2} a^3 \left\{\frac{6\alpha}{(\varepsilon+2)^2+4\alpha^2}\right\}$$

where ε is the dielectric constant of the material (at the frequency of the light waves) and $\alpha = \sigma\lambda/c$ where σ is the electrical conductivity. The size dependence here is on a^3.

Region 2. The intermediate case where a is comparable to λ cannot be treated by elementary means. The dipoles excited in different parts of the particle may no longer be considered to be in phase, and the general theory must take account of the shape of the particle and of the direction in which the light is scattered. Mie gave the general solution to this rather difficult problem, but only specific cases such as homogeneous spheres, ellipsoids and cylinders yield "easily", and there are resonances to be found in the scattering diagrams. When the particles are of unknown and random shapes there is little profit in trying to study them, but the general conclusion can be applied: that the scattering and extinction that takes place does not differ markedly from that which happens in cases (1) and (3).

Figure 1 shows, on logarithmic scales, the way the various forces that act on the interplanetary dust

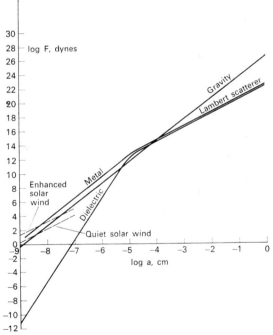

Fig. 1. The forces acting on particles in circular orbits. F is the force, in dynes, when $R = 1$ cm; a is the particle radius, in centimetres.

depend on the particle radius a. The curves correspond to

$\log F_{\text{grav}} = 26\cdot745 + \log \varrho + 3 \log a$

$\log F_{\text{rad press}} = 22\cdot512 + 2 \log a$ for a reflecting sphere, $a \gg \lambda$

$\log F_{\text{rad press}} = 22\cdot712 + 2 \log a$ for a white reflector, $a \gg \lambda$

$\log F_{\text{rad press}} = 43\cdot037 + 6 \log a$ for a dielectric, $\mu = 1\cdot4$, $a \ll \lambda$

$\log F_{\text{rad press}} = 27\cdot620 + 3 \log a$ for iron, $a \ll \lambda$

$\log F_{\text{solar wind}} = 18\cdot37 + 3 \log a$

The interesting region is of course that where the radiation pressure force exceeds the gravitational attraction. It may be supposed that all the material in this size range is missing from the dust cloud.

The Poynting–Robertson Effect

Because of the relativistic aberration of light, the outward force on the particles will not be precisely away from the Sun but will have a tangential component Fv/c, where v is the orbital velocity of the particle. This "drag" results in the particle descending in a slow spiral into the Sun, and there is an accompanying Doppler shift in the wavelength of the scattered light.

The lifetime of a particle in an initially circular orbit was given by Robertson, but a more general study by Whipple included the more realistic cases of particles that have been dislodged from comets and are in highly elliptical orbits.

The lifetime of a particle in an initially circular orbit is:

$$t = 7 \times 10^6 R^2 \varrho a \text{ years}$$

where a is the particle radius, in centimetres, ϱ its density and R its initial distance from the Sun in astronomical units.

Numerical integration is needed in the cases of initially elliptical orbits. A typical result quoted by Whipple may be of interest since it allows one to estimate the importance of the effect: with a perihelion distance of 1 A.U. and a semi major axis of 3 A.U., $t = 2\cdot9 \times 10^7 a$ years, a being again the particle radius.

The implication is that in the time intervening between the condensation of the inner planets and the present time (some $4\cdot5 \times 10^9$ years), all particles of diameter less than about 3 cm must have been removed, even from highly elliptical orbits.

A Model of the Dust Cloud

It is not possible, from the observations, to deduce an equation for the variation of the particle number with distance from the Sun or with particle radius. The custom is to assume that the number-density is a function of particle size, of the form:

$$n(a) \, da = c a^{-p} \, da,$$

where $n(a) \, da$ is the number of particles/cc with radii between a and $a + da$.

The variation of number density with distance is similarly expressed as:

$$n(r) = D r^{-\alpha},$$

where

$$D = c \int_{a_1}^{\infty} a^{-p} \, da$$

and is the total number of particles/cc at 1 A.U. from the Sun. There is no justification, except convenience, in assuming that the number density is separable in this way into a function of a and a function of r.

The surface brightness and its variation with elongation can be computed from this model and the constants C, D and α adjusted to fit the observations. Ingham has done this for various values of p and an example of his results is:

with
$$p = 4, \quad a_1 = 10^{-4} \text{ cm}, \quad \alpha = 1$$
then
$$C = 6 \cdot 2 \times 10^{-26}, \quad N = 300/\text{km}^3.$$

The values to be given to p and α are deduced from considerations of meteor data (where $p = 3$ for larger meteor sizes and 4 for smaller ones) and from measurements of the polarization of the Zodiacal light.

The differential scattering cross-sections may be included when calculating the surface brightness that is to be expected from a particular model. Rayleigh scattering, for example, shows a differential scattering function of the form

$$I(\theta) = \tfrac{1}{2}(1 + \cos^2 \theta)$$

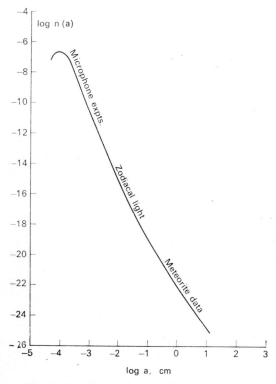

Fig. 2. *The size-number distribution at 1 A.U.*

where θ is the angle through which the light is scattered. Reflecting spheres show isotropic scattering, and other assumptions can give more or less complicated scattering functions. None of them has any significant effect on the calculated brightness distribution of the Zodiacal light, and they are normally ignored when making computations.

While the data for the larger particle sizes is derived mainly from the studies of meteors and micrometeors, the direct evidence for the distribution of the smaller particle sizes comes from the microphone and foil penetration studies. So far only the microphone studies have been extended into interplanetary (as opposed to near circumterrestrial) space. The data gathered is necessarily confined to the smaller particle sizes, since the impact rate for the larger sizes is too small for a statistically significant sample to have been collected. A composite model of the size distribution at 1 A.U. is shown in Fig. 2, but it must be stressed that the numbers are only very approximate. Data is gathered from different sources for different parts of the graph, as indicated on the figure.

The Origin of the Dust

Two main theories contend. Neither seems to be completely satisfactory. They are:

(1) The Primaeval Dust-Cloud Theory.
(2) The Cometary Debris Theory.

The first of these theories is a corollary of the "accretion" theory of the origin of the Solar System first proposed by Kant, and restated with attendant mathematical details by Laplace, and more recently by Hoyle, McCrea and others. The planets are supposed to have condensed by gravitational accretion and molecular adhesion from a dust cloud either at the same time as or after the formation of the Sun. The Zodiacal light represents scattering by the dust left over when this process was virtually complete. (It is still in fact going on, as the capture of meteors by the Earth shows.) The age of the Solar System is generally thought to be in the region of $4 \cdot 5 \times 10^9$ years, so that the Poynting–Robertson effect presents a serious objection, as Whipple has pointed out. In this time all particles of diameters less than about 3 cm should have been removed from the inner part of the Solar System.

The cometary debris theory assumes that the Poynting–Robertson effect acts as a broom, continually sweeping dust into the Sun, and that the dust is replaced sporadically by the disruption of comets as they pass perihelion. Order-of-magnitude calculations indicate that sufficient material could be contributed from such a source. The objection in this case is that the total energy of a particle of mass m_1 when it is part of the comet is:

$$E = \frac{-G_{m_1 m_\odot}}{2A},$$

where G is the gravitational constant, $m_\odot$ the mass of the Sun and A the semi-major axis: and that this is very nearly equal to zero. When the fragment parts from the comet it experiences the radiation pressure and suddenly acquires, in effect, a small extra positive potential energy. This is sufficient to make its subsequent orbit hyperbolic, and the particle is removed from the Solar System before the Poynting–Robertson effect has a chance to operate. The orbit will be hyperbolic provided that the particle radius a is less than $1 \cdot 2 \times 10^{-4} \dfrac{A}{\varrho R}$, where R is the distance of release.

The replacement rate required for dust is of the order of several tons per second. Other possible sources for material are:

(1) Abrasion of larger particles by collision.
(2) Condensation of gas.
(3) Accretion by capture from interstellar space.

The least unlikely of these is (1) and this can be divided into two subdivisions:

(a) *Meteorite and asteroidal impact with the Moon or a planet.* Large new craters should appear on the Moon roughly once every 10^5 years. It would require each to excavate about 10^{19}g into interplanetary space, a volume of about 1000 km³, which seems unreasonably large.

(b) *Collisional fragmentation of asteroid-sized bodies.* This seems to be the most likely source of Zodiacal dust. Piotrowski has calculated that dust production from this process may well be in the range 20–600 tons/sec. An examination of the spectrum of the Zodiacal light should yield some evidence to support one or other of these theories. Cometary debris from long period comets should be distributed equally in direct and retrograde orbits, whereas debris from short period comets and dust resulting from asteroidal collision must be expected to be entirely in direct orbit. An examination of the Zodiacal light spectrum for doppler shifts of the Fraunhofer lines, and a search for possible doubling of them, may indicate how much of the dust, if any, is in retrograde orbit.

The Dust-free Zone

The equilibrium temperature of an object in the Solar System rises as the object approaches more closely to the Sun. This temperature does not depend on the material or the emissivity of the body, but only on its distance from the sun. The point must arrive where this equilibrium temperature is greater than the melting or sublimation temperature of the material, and the space around the Sun must therefore be free of dust. The radius of this zone is about 4·5 solar radii for a dust melting point of 2000 °K. The still solid dust close to the edge of this zone is possibly dense enough and hot enough to emit a measurable amount of radiation on its own account, apart from that which it scatters. At the 1966 solar eclipse, Peterson, using an infra-red detector, found a discontinuity in the curve of intensity versus elongation at the appropriate point, and this may well be ascribed to the edge of the dust-free zone.

Bibliography

BLACKWELL D. E. (1962) Zodiacal light, in *Encyclopaedic Dictionary of Physics* (J. Thewlis Ed.), **7**, 863.
BLACKWELL D. E. and INGHAM M. F. (1961) Observations of the zodiacal light from a very high altitude station, *Monthly Notices of the Royal Astronomical Society*, Vol. 122.
VAN DE HULST H. C. (1957) *The Scattering of Light by Small Particles*, Chapman & Hall, London.
WHIPPLE F. L. (1955) A comet model. III. The Zodiacal light, *Astrophysical Journal* **121**, 751.
The Zodiacal Light and the Interplanetary Medium. Proceedings of a conference held at the University of Hawaii, Jan./Feb. 1967, NASA SP-150.

J. F. JAMES

IONIZATION AND ATTACHMENT CROSS-SECTIONS OF ATOMS AND MOLECULES BY ELECTRON IMPACT. For the proper understanding of many problems in widely differing areas of plasma physics, e.g. aeronomy, astrophysics and thermonuclear research, it would be of great assistance to have available accurate values of ionization and attachment cross-sections. Over the years a large body of experimental data has been accumulated and there have been a number of discussions of the experimental techniques and results (see for example, Craggs and Massey, 1959; Fite, 1962; and McDaniel, 1964). In the present article the discussion will be restricted to some aspects of cross-section measurements for ionization of atoms, molecules and ions by electron impact and for the attachment of electrons to atoms and molecules.

In spite of the numerous investigations of these cross-sections it is still exceptional to find data, even for a particular total ionization cross-section, which are reliable to better than 10 to 15 per cent and even in the case of the ionization of inert gases uncertainties of 25 to 30 per cent still occur. Kieffer and Dunn (1966) concluded from an excellent critical evaluation of the available data for absolute ionization cross-sections that such uncertainties are a good indication of the range of systematic errors afflicting the measurements.

Experimental Methods

(a) *Total ionization cross-sections of stable atoms and molecules*

For many purposes it is sufficient to know the total cross-section for the production of all positive ions, irrespective of charge, i.e. it is not necessary to know whether the positive ions formed are singly, doubly or more highly ionized. The vast majority of the available data for total ionization cross-sections has been obtained by the method of Tate and Smith (1932). The method can be understood by reference to Fig. 1. A beam of electrons from the filament F passes through the slits $S_1 \cdots S_5$, which serve to collimate the beam and define its energy, into the

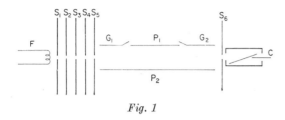

Fig. 1

collision region between S_5 and S_6. An axial magnetic field ensures a well-defined beam between S_5 and S_6. The total cross-section is deduced from $\sigma_T = (I_+/I_e)\left(\dfrac{1}{LN}\right)$ where I_e is the electron current to the collector C, I_+ is the positive ion current collected by plate P_1, L, the axial length of P_1, is the length of the collision path, and N is the gas number density. The positive ions are collected on P_1 by maintaining a potential difference of a few volts between P_1 and P_2. The guard electrodes G_1 and G_2, which are at the same potential as P_1, help to define L. The end-plates S_5 and S_6 are at a potential which is mid-way between those of P_1 and P_2 so that the energy of the primary electron beam is unchanged between S_5 and S_6. The gas pressures are chosen to ensure single-collision conditions, i.e. the probability of a primary electron undergoing two ionizing collisions in the length L is negligibly small.

Although the experimental method is apparently straightforward, it is by no means easy to eliminate all systematic errors. For example, it is difficult to ensure that the measured values of I_+ and I_e do not include spurious currents from a number of possible sources. Kieffer and Dunn (loc. cit.) have enumerated the necessary conditions for obtaining accurate cross-sections. Perhaps one of the major sources of error in the past has been the measurement of the gas number density N by means of a McLeod gauge. Where McLeod gauges have been used without adequate precautions to eliminate errors due to mercury vapour pumping effects the resultant cross-sections should be treated with caution.

(b) Partial ionization cross-sections

Although measurements of total ionization cross-sections provide valuable data, it is often important to have available data for so-called partial ionization cross-sections, i.e. the cross-sections for ionization from the ground state X to the ionized states X^+, X^{2+}, X^{3+} and so on (e.g. Bleakney, 1932). Recent work has included that of Adamczyk *et al.* (1966). The cross-sections obtained in such investigations involve measurements of the charge-to-mass ratio of the various ionic species and are usually relative values, e.g. Adamczyk *et al.*'s relative cross-sections for ionization of Ne to Ne^+, Ne^{2+} ... were normalized by reference to the total cross-sections determined in the same laboratory.

(c) Ionization cross-sections for unstable atoms and molecules and for ions

For studies of the ionization of a large number of chemically unstable atomic species, e.g. atomic oxygen, nitrogen, hydrogen, and of ions such as H^+ and He^+ the most suitable methods are undoubtedly crossed-beam techniques. Since the early work of Boyd and Green (1958) and of Fite and Brackmann (1958) there have been many investigations. The basic features of the method may be understood by reference to Fig. 2. A beam of electrons intersects at right angles, in a vacuum chamber, a well-collimated beam of the target species (either neutral or ionic) and a fraction of the ions produced is collected—often at right angles to both the electron and target beam directions—and measured. If the target number density, electron current and ion current are all

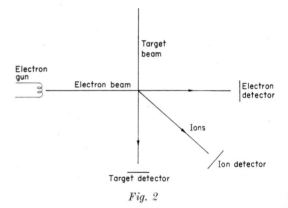

Fig. 2

accurately measured and if the geometry of the interaction region is carefully determined the ionization cross-section can be calculated. In practice, it is often difficult to ensure that the ions collected are those produced by the process of interest and have not originated in collisions between the residual gas in the vacuum chamber and either the electron or target beams. It is usual, therefore, to divide either or both the latter into a sequence of pulses and to use suitable detecting equipment to measure the ion current. Good examples of techniques of this type were described by Dolder *et al.* (1963). For ionization of neutral targets it is difficult to measure the target number density and the data have often been normalized with the aid of cross-sections for more stable species which have been studied by, for example, the Tate and Smith technique.

(d) Measurements of cross-sections near ionization threshold and of fine structure

Electron beams having a broad spread of energies—such as the Maxwellian distribution obtained when using a hot filament source—are not adequate for examining the exact form of cross-sections near ionization threshold nor for looking for possible fine-structure of particular cross-sections resulting from the production of excited ions or from auto-ionization. Various methods of obtaining electron beams of very low energy spread have been introduced. One of the most widely used methods is the so-called retarding potential difference (R.P.D.) method introduced by Fox et al. (1955) (see A. Weber, 1961). More recently, electrostatic energy selectors have been used, the $127°$ cylindrical analyser and the hemispherical analyser being the most popular. A survey of designs for sources of mono-energetic electrons has been given by J. A. Simpson (1967). A recent application is described by Brion and Thomas (1968). Attempts have also been made (e.g. Winters et al., 1966) to remove by various mathematical techniques called "deconvolution" methods the effects of energy spread from experimental data.

For some time it has generally been believed that the cross-section for producing singly-charged atomic ions increases linearly with the excess energy above threshold of the ionizing electron, as was predicted by Geltman (1956) and others. However, recent data, such as that of Brion and Thomas (loc. cit.) for the production of He^+ suggests that the number of ions produced increases as the 1·127th power of the excess energy as predicted by Wannier (1953). The present position (Rudge, 1968) suggests more experimental data are required.

Intensive studies have been made in recent years of the fine structure of ionization cross-sections for electron impact and of possible correlations with the corresponding photo-ionization data. The early results and the techniques employed have been described by McDowell (1963) and by Kieffer and Dunn (loc. cit.). The experimental investigations have made use of energy selectors (e.g. Marmet and Kerwin, 1960; McGowan and Finemann, 1965) with either molecular beams or volume gas samples. Analytical techniques such as obtaining the first and second derivatives of the ionization efficiency curve have also been used (e.g. Morrison, 1964). Three possible mechanisms leading to structure in ionization efficiency curves are (a) the formation of the ions in excited states, (b) auto-ionization, i.e. the formation of an ion through excitation of the neutral molecule to a short-lived level above the ground-state energy of the ions, (c) Auger ionization.

(e) Attachment cross-sections

Much of the above discussion applies equally to the measurement of attachment cross-sections, in particular to the determination of the cross-sections for dissociative attachment which is one of the major attachment processes under ordinary laboratory conditions, particularly at low gas pressures (a typical reaction is $e + CO_2 \rightarrow CO + O^-$). Many of the experimental techniques have been the same, e.g. the Tate and Smith method has been applied by Asundi et al. (1963) and others. Poor energy resolution due to electron energy spread has also to be overcome as, for example, in the applications of the R.P.D. method by Schulz (1962).

The problem of establishing an absolute energy-scale for cross-section measurements has not yet been mentioned here. For both ionization and attachment measurements the standardization of the relative energy scale has often been done by examining near threshold the ionization cross-section for a gas whose ionization potential has been well-established spectroscopically. The method is more difficult than it appears. In a number of investigations (e.g. Frost and McDowell, 1958) the resonance capture of electrons to form SF_6^- in SF_6 has been used both to establish the electron energy-scale for a particular apparatus and to establish the energy spread of the electron beam. The method relies on the observation by Hickam and Fox (1956) and others that the resonance capture cross-section is centred at 0 eV and has a half-width of approximately 0·1 eV. Schulz (1960) has pointed out the precautions that need to be taken when using this method. An alternative method of determining electron-capture cross-sections has been introduced by Christophorou et al. (1965) who use a combination of electron-swarm and electron-beam techniques.

Bibliography

ADAMCZYK B., BOERBOOM A. J. H., SCHRAM B. L. and KISTEMAKER J. (1966) *J. Chem. Phys.* **44**, 4640.

ASUNDI R. K., CRAGGS J. D., and KUREPA M. V. (1963) *Proc. Phys. Soc.* **82**, 967.

BLEAKNEY W. (1932) *Phys. Rev.* **40**, 496.

BOYD R. L. F. and GREEN G. W. (1958) *Proc. Phys. Soc.* **71**, 351.

BRION C. E. and THOMAS G. E. (1968) *J. Mass Spec. and Ion Physics*, **1**, 25.

CHRISTOPHOROU L. G., COMPTON R. N., HURST G. S. and REINHARDT P. W. (1965) *J. Chem. Phys.* **43**, 4273.

CRAGGS, J. D. and MASSEY H. S. W. (1959) *Handbuch der Physik*, **37**, 314.

DOLDER K. T., HARRISON M. F. A. and THONEMANN P. C. (1963) *Proc. Roy. Soc.* **A274**, 546.

FITE W. L. (1962) In *Atomic and Molecular Processes* (Ed. Bates D. R.) New York: Academic Press.

FITE W. L. and BRACKMANN R. T. (1958) *Phys. Rev.* **112**, 1141.

FOX R. E., HICKAM W. M., GROVE D. J. and KJELDASS T. (1955) *Rev. Sci. Inst.* **26**, 1101.

FROST D. C. and MCDOWELL C. A. (1958) *J. Chem. Phys.* **29**, 964.

GELTMAN S. (1956) *Phys. Rev.* **102**, 171.

HICKAM W. M. and FOX R. E. (1956) *J. Chem. Phys.* **25**, 642.
KIEFFER L. J. and DUNN G. H. (1966) *Rev. Mod. Phys.* **38**, 1.
MARMET P. and KERWIN L. (1960) *Can. J. Phys.* **38**, 787 and 972.
MCDANIEL E. W. (1964) *Collision Phenomena in Ionized Gases*, New York: Wiley.
MCDOWELL C. A. (1963) *Mass Spectrometry*, New York: McGraw-Hill.
MCGOWAN J. W. and FINEMAN M. A. (1965) *Phys. Rev. Letters*, **15**, 179.
MORRISON, J. D. (1964). *J. Chem. Phys.* **40**, 2488.
RUDGE M. R. H. (1968) *Rev. Mod. Phys.* **40**, 564.
SCHULZ G. J. (1960) *J. Appl. Phys.* **31**, 1134.
SCHULZ G. J. (1962) *Phys. Rev.* **128**, 178.
SIMPSON J. AROL (1967) *Methods of Experimental Physics*, vol. 4A, section 1.1.7, Academic Press.
TATE J. T. and SMITH P. T. (1932) *Phys. Rev.* **39**, 270.
WANNIER G. H. (1953) *Phys. Rev.* **90**, 817.
WEBER A. (1961) Ionization potential, in *Encyclopaedic Dictionary of Physics* (J. Thewlis Ed.), **4**, 60, Oxford: Pergamon Press.
WINTERS R. E., COLLINS J. H. and COURCHENE W. L. (1966) *J. Chem. Phys.* **45**, 1931.

J. A. REES

ION-NEUTRALIZATION SPECTROSCOPY. Ion-neutralization spectroscopy (INS) is a method of deriving information concerning the electronic states in solids and at solid surfaces. Like any spectroscopy of an atomic system it is based on an electronic transition process. Its results will thus depend on the relative probability of involvement of electrons as a function of their energy in the allowed band in the solid. In the INS method a function of band energy is derived called the transition density which depends both on density of allowed electronic states and transition probability of the transition process.

The basic electronic transition process upon which INS is based is the two-electron, Auger-type process which occurs when a slow ion of large neutralization energy (ionization energy of the parent atom) is neutralized at a solid surface. This process is illustrated in Fig. 1. In the presence of the incoming ion outside the surface two electrons in the filled valence band of the solid interact, exchanging energy and momentum. One electron, called the neutralizing or "down" electron, tunnels through the barrier into the ion well and drops to the vacant atomic ground level which lies at an energy $E'_i(s)$ below the vacuum level. The energy released in this transition (1 or 1' in Fig. 1) is taken up by the second interacting electron, termed the excited or "up" electron. This exciting transition (2 or 2' in Fig. 1) carries the second electron to the energy E above the vacuum level. The pairs of transitions (1, 2) and (1', 2') bear an exchange relation to one another.

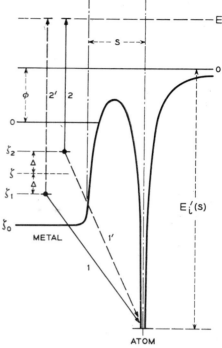

Fig. 1. Electron energy diagram showing electron transitions of the two-electron, ion-neutralization process and defining energy scales and energetic quantities.

Equating the energies of transition of the up and down electrons we obtain:

$$E = E'_i(s_t) - (\zeta_1 + \Phi) - (\zeta_2 + \Phi)$$
$$= E'_i(s_t) - 2\zeta - 2\Phi. \quad (1)$$

$E'_i(s)$ is the energy required to ionize the parent atom at a distance s from the surface and is evaluated at the distance, s_t, at which the electronic transitions occur. E and ζ are the energies defined by Fig. 1 and Φ is the work function of the solid taken to be a metal in this discussion. Since the initial electrons at ζ_1 and ζ_2 may lie anywhere in the filled band we expect a band of excited electrons covering a range of energies on the E scale. Some of these excited electrons may, if properly directed, surmount the surface barrier and leave the solid. Outside, their kinetic energy distribution may be measured.

Examples of a number of electron kinetic energy distributions taken for He$^+$ ions of incident kinetic energy $K = 5$ eV are shown in Fig. 2. Distributions for atomically clean surfaces of Ge(111), Ni(100), Cu(100) are shown as well as a Ni(100) surface on which a centred (2 × 2) structure of oxygen atoms has been formed. It is clear from these distributions that the ion-neutralization process is sensitive to the nature of the solid and to the presence of foreign atoms on its surface. INS extracts spectroscopic information from these kinetic energy distributions.

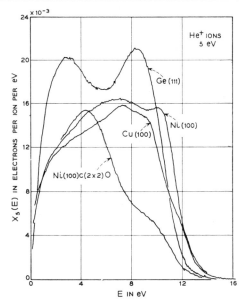

Fig. 2. Representative kinetic energy distributions for electrons ejected by 5 eV He⁺ ions incident on several surfaces illustrating sensitivity to the nature of the solid and the state of the surface.

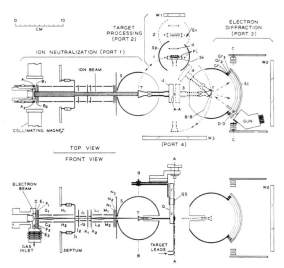

Fig. 3. Schematic diagram of experimental apparatus used in ion-neutralization spectroscopy. Details are explained in the text.

The kinetic energy distributions $X_K(E)$, like those in Fig. 2, must be obtained under a demanding set of experimental conditions. A schematic diagram of the apparatus used is depicted in Fig. 3. The apparatus is enclosed in a metal vacuum envelope in the form of a three-dimensional cross of tubing diameter about 16·5 cm. There are four horizontal flanged ports, one port on top, and another on the bottom. The top port carries a target-turning mechanism which (1) can present the target face to any one of the horizontal ports by rotation about the axis A–A and (2) can turn the target about the axis B–B through its face. The bottom port connects to the pumps (sputter-ion and diffusion) and a gas-inlet system.

Port 1 carries the ion-neutralization apparatus. In it ions formed by electron impact in an electron beam are focused in two lens systems (G, H and L, M) onto the face of the target, T. Ejected electrons are collected at S and the kinetic energy distribution determined as the slope of the retarding potential curve of electrons collected at S.

Port 2 carries the target-processing apparatus. Here the target may be enclosed in a retractable sphere Sp. When inside this sphere the target may be sputtered by ions from an arc discharge run between the filaments Fi and the ring R in a pressure of neon or argon of 10^{-2} to 10^{-3} Torr.

At port 3 the target may be examined by low-energy electron diffraction (LEED). This apparatus is essentially of the display-type in current use except that here we look past the gun at the pattern on the phosphor-coated glass screen Sc, rather than past the target and through the grids as in the conventional system. The gun has been placed to one side giving excellent screen visibility.

The basic experimental requirements of INS are as follows:

(1) The incident ions must be slow in order to reduce the inherent energy broadening of the neutralization process. Usually two distributions at incident ion kinetic energies, K, of 5 and 10 eV are obtained for extrapolation to a distribution having relatively small energy broadening.

(2) The apparatus must have sufficiently high resolving power in the energy distribution measurement. Since a retarding potential curve of electron current to electrode S is differentiated to produce the electron kinetic energy distribution, resolving power depends on the relative sizes of the target T and the electron-collector S. S is a sphere 80 mm in diameter (d_s) and an average dimension of the target is 10 mm (d_t). The degradation in resolution inherent in the measurement amounts to convoluting the distribution by an instrumental broadening function whose width at half-maximum is of the order of $\frac{1}{2}(d_t/d_s)^2 E \sim 0{\cdot}0075\,E$. At $E = 10$ eV this corresponds to an instrumental broadening of less than 0·1 eV.

(3) The data should have as little noise, particularly low frequency noise, as possible because digital deconvolution is required. Both analogue and digital data are taken. A number of runs are stored and averaged in a multichannel scaler to obtain data of the requisite quality. Digital

data smoothing with some consequent degradation of energy resolving power is also required.

(4) The surface conditions at the target must be strictly controlled. For study of bulk properties the surface must be atomically clean to within a few per cent of a monolayer of any foreign adsorbed species. Background pressure in the apparatus below 1×10^{-10} Torr is required and achieved. Flash filament accumulation measurements indicate that the target can be sputtered clean, annealed, and the LEED pattern observed and the kinetic energy distribution measurement made before 2 per cent of a monolayer of foreign gas has accumulated on the surface. Known foreign atoms can be adsorbed on the surface to study their effect on the local band structure at the surface.

The end product of the experiment is a kinetic energy distribution like those of Fig. 2 obtained under known and controlled conditions. It is necessary now to discuss the factors upon which the form of such a distribution depends. We introduce the functions needed to discuss the ion-neutralization process. These are plotted in Fig. 4 on an electron energy diagram like that of Fig. 1. Curves for He^+, Ne^+, and Ar^+ ions are shown. We consider first the curves for He^+ and simplify the discussion at this point by assuming that all transition-probability factors are constant, independent of band energy, ξ, and independent of the symmetry character of the band electrons. We also neglect, for the moment, the effects of exchange-matrix-element cancellation and final-state interactions, and assume equal transition probability factors for the up and down electrons. Under these conditions the band function $U(\xi)$ represents the state density in the valence band.

The probability of occurrence of a process involving electrons at $\zeta_1 = \zeta + \varDelta$ and $\zeta_2 = \zeta - \varDelta$ will be proportional to the product function $U(\zeta + \varDelta)U(\zeta - \varDelta)$. We wish to determine the relation of $U(\zeta)$ to $F(E)$, the internal distribution of electrons excited in the neutralization process. This relation is readily derived as follows. Excited electrons lying in the range dE at energy E (end of arrow 2 in Fig. 1) may be obtained from any neutralization process for which the initial states of the two participating electrons are symmetrically disposed on either side of the energy ζ which lies half-way between E and the ground state of the atom at $-E'_i(s_t)$. Values of E and ζ meeting this requirement satisfy equation (1). The total probability of producing an excited electron in dE at E is then the integral of the probability $U(\zeta + \varDelta)U(\zeta - \varDelta)$ over all possible $\varDelta$. This defines a pair probability function

$$F(\zeta) = \int_0^\zeta U(\zeta + \varDelta) \, U(\zeta - \varDelta) \, \mathrm{d}\varDelta . \qquad (2)$$

$F(E)$ is obtained from $F(\zeta)$ by change of variable using equation (1) and by normalizing the area of $F(E)$ above the Fermi level to one electron per incident ion.

$F(\zeta)$ is the self-convolution, convolution square, or fold of $U(\zeta)$. If transition probability effects are included, $U(\zeta)$ becomes the transition density for a single electron and $F(\zeta)$ the transition density for all pairs of electrons producing excited electrons at $E = E'_i = -2(\zeta + \varPhi)$. $F(\zeta)$ extends to a maximum ζ which is half the maximum ζ to which $U(\zeta)$ is known (Fig. 4). $F(E)$ could be modulated by final-state density variation but the INS method has means of checking for such modulation. We also recognize that, a priori, the two U factors in equation (2) need not be the same function making the integral the convolution product of different functions $V(\zeta)$ and $W(\zeta)$. It can be shown, however, that the transition probabilities for up and down electrons are sufficiently close to the same magnitude as to make $V \sim W \sim U$ and thus justify the use of the self-convolution integral.

The external kinetic energy distribution, $X(E)$, of electrons observed outside the solid is related to the

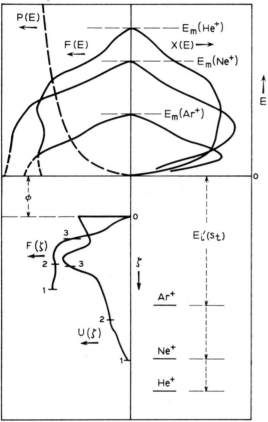

Fig. 4. Electron energy level diagram like Fig. 1 on which is indicated the functional dependences involved in the ion-neutralization process.

internal distribution $F(E)$ through the probability of escape over the surface barrier, $P(E)$, Thus

$$X(E) = F(E) P(E). \qquad (3)$$

Derivation of $X(E)$ from $U(\zeta)$, as done here, is called the forward model.

INS attempts to extract information concerning $U(\zeta)$ from measured $X_K(E)$. The success of INS thus depends upon the feasibility of reserving the forward model just discussed. The steps of the INS procedure are as follows:

(1) $X_K(E)$ distributions are obtained at two different ion energies K, usually $K_1 = 5$ eV and $K_2 = 10$ eV. These distributions are measured under the stringent experimental conditions outlined above.

(2) $X_{K_1}(E)$ and $X_{K_2}(E)$ are extrapolated to a "debroadened" distribution $X_0(E)$ having relatively small energy broadening.

$$X_0(E) = X_{K_1}(E) + [R + R'(E_m - E)/\Delta E] \times \\ \times [X_{K_1}(E) - X_{K_2}(E)]. \qquad (4)$$

R and R' are parameters depending on K_1 and K_2, E_m is the maximum electron kinetic energy before broadening and ΔE is the energy interval between data points. R may be calculated if energy broadening depends on incident ion velocity as has been shown to be the case. R' takes account of final state broadening and increases the "strength" of the extrapolation linearly with increasing band energy, ζ. Typical values of R and R' for $K_1 = 5$ eV and $K_2 = 10$ eV are 3·11 and 0·11, respectively. $X_0(E)$ functions are shown for He$^+$, Ne$^+$, and Ar$^+$ ions in Fig. 4.

(3) $X_0(E)$ is divided by a parametric $P(E)$ function to obtain $F(E)$. The parameters of $P(E)$ are chosen so that the pieces of $F(\zeta)$ obtained for the three ions are essentially coincident. He$^+$, Ne$^+$, and Ar$^+$ distributions yield data to the points 1, 2, and 3, respectively, as indicated on the $F(\zeta)$ and $U(\zeta)$ functions of Fig. 4. Data in the range $0 < E < 2$ eV are not used since both $P(E)$ and $X_0(E)$ are rapidly rising and less well known in this range. We note that structure in $F(\zeta)$ and $U(\zeta)$ which is found in curves obtained for the three ions cannot possibly be final state structure whose position is constant on the E scale since the $X(E)$ curves for the three ions lie at different positions on the E scale.

(4) $F(\zeta)$ is now unfolded or deconvolved to produce $U(\zeta)$, the transition density function. This deconvolution is performed sequentially on the digital data sequence F_n, $n = 1, m$ ($F_0 = 0$) to produce the digital unfold sequence U_{2n-1}, $n = 1, m$. The unfold equations are:

$$U_1 = (F_1/\Delta\zeta)^{\frac{1}{2}}$$

$$U_3 = (1/2U_1)(F_2/\Delta\zeta)$$

$$U_{2n-1} = (1/2U_1)\left[F_n/\Delta\zeta - \sum_{p=1, n-2} U_{2n-2p-1} U_{2p+1}\right], \\ n > 2. \qquad (5)$$

In the course of the unfolding it is necessary to test for the true origin of the F_n sequence by manipulation of the data in its initial portions. This avoids the introduction of spurious features owing to an incorrect origin position. Detailed discussion of this and other important characteristics of the unfolding procedure are beyond the scope of this article and are discussed elsewhere. It is found that the unfold procedure works well for the class of functions for which $F'(0) \neq 0$ corresponding to $U(0) > 0$.

Results of applying the INS procedure to data for He$^+$ ions on the atomically clean (110) face of copper is shown in Fig. 5. $U(\zeta)$ results for atomically clean Ni(100) and for this surface with a $c(2 \times 2)$ structure of oxygen on it are shown in Fig. 6. Although INS is relatively new, work has been published for three faces each of Cu and Ni and for oxygen adsorbed in ordered arrays on nickel, copper and tungsten single-crystal faces. Work is in progress on clean semiconductors and refractory metals and on extending the studies of ordered monolayers to sulphur and selenium. Sufficient is known about the method so its characteristics can be specified and comparisons made with other spectroscopies of the solid state, namely the soft-X-ray spectroscopy (SXS) and the photoelectric-emission spectroscopy (PES).

1. *Density of initial states*. Although there are strong and characteristic transition probability factors,

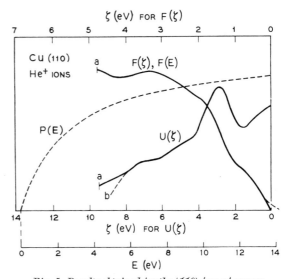

Fig. 5. *Results obtained for the (110) face of copper using He$^+$ ions. The fold function $F(\zeta)$, the unfold function or transition density $U(\zeta)$, and the electron escape probability $P(E)$ are shown.*

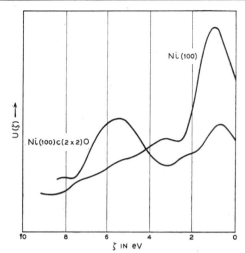

Fig. 6. Transition density or unfold functions for atomically clean Ni(100) and for this surface with an adsorbed, centred c(2 × 2) structure involving oxygen.

when the crystal is atomically clean the bulk density of initial states plays a large role in determining the structure of the $U(\zeta)$ function. d-band electrons are clearly visible in the $U(\zeta)$ for copper (Fig. 5) and nickel (Fig. 6), but their intensity is reduced relative to the s-p band. Unlike PES, the INS results for copper and nickel appear to corroborate the rigid band model. Similarly the degenerate p band in semiconductors can be seen. It is also probable that with the INS method it will be possible to detect the increased differential density of states in certain crystal directions caused by critical points in the band structure of both metals and semiconductors. In certain cases INS results corroborate bulk state density information from SXS and PES. Transition-probability factors show interesting differences, however. INS has not revealed the many-body resonance found in nickel by PES. INS should prove to be a fruitful method of studying the band structure of solid materials. It is worthwhile to have several methods differing as they must in transition probability factors but involving the same state density factors.

2. *Density of final states.* Effects of modulation by the density of final states will be revealed in a failure of agreement among the results for different ions. A change of ion in INS is the analog of changing photon energy in PES. There is no evidence in the results of INS to date of final-state-density modulation. This is most likely because the final states involved lie so high in energy as to be free-electron-like and not to involve strong critical points.

3. *Transition probability factors for the clean surface.* There are three important factors. These are:
   (a) A factor decreasing with increasing depth in the band (ζ) which arises because of reduced wave function tunnelling into the atomic well.
   (b) A factor depending on wave function symmetry which decreases as the character proceeds from s to p to d, etc.
   (c) A factor depending on **k** vector direction which favors states whose **k** vector is normal to the surface used in the experiment. Results are consistent with approximately equal transition probability factors for up and down electrons and no effects have as yet been observed which are attributable to interactions in the final state of the ion-neutralization process.

4. *Additional transition probability factor for a surface containing foreign atoms.* Foreign atoms incorporated in a crystallographically ordered surface monolayer or in an amorphous layer can alter the dependence of transition probability on band energy. The foreign atom produces a virtual bound state at the surface which is a region on the energy scale in which wave functions from the solid tunnel more easily through the monolayer than they do elsewhere. This region is seen in Fig. 6 for an oxygen monolayer on nickel. Thus INS also offers the attractive possibility of studying (1) the electronic band structure of extremely thin layers of one material or another and (2) the nature of the band structure transition from one solid to another.

Bibliography

HAGSTRUM H. D. (1953) *Review of Scientific Instruments* **24**, 1122.
HAGSTRUM H. D. (1966) *Physical Review* **150**, 495.
HAGSTRUM H. D. and BECKER G. E. (1967) *Physical Review* **159**, 572.
HAGSTRUM H. D., PRETZER D. D. and TAKEISHI Y. (1965) *Review of Scientific Instruments* **36**, 1183.

H. D. HAGSTRUM

IONOSPHERE, RECENT OBSERVATIONS OF. It is intended that this article should be read in conjunction with other articles published in this Dictionary, referenced below under the heading Bibliography —Background.

Introduction. The *ionosphere* is formed, in the main, by photoionization of the tenuous upper atmosphere by solar ultra-violet and X-radiations.

Electron densities in the D, E and F regions of the *bottomside ionosphere* have been investigated by various ground-based radio techniques, including studies of the *waveguide* modes of propagation of low frequency (LF) radio signals, the *reflection, absorption, forward scatter* and *Faraday rotation* of HF signals, and the *incoherent backscatter* of VHF radar pulses. Direct measurements of ionospheric parameters have been made by *rocket*-borne probes.

Limited information on the ionosphere, above the F region peak, namely the electron content along the propagation path weighted by the geomagnetic field, has been obtained from ground-based studies of the

Faraday rotation of beacon satellite signals. *Topside* electron and ion densities and temperatures, plus the ionic composition, have been found as functions of altitude, by examination of the spectrum of the incoherent backscattered radar signal. Electron densities, at several earth radii in the equatorial plane, have been deduced from *whistlers*, dispersed VLF signals from lightning discharges. Instrumentation carried aboard space craft has recently made possible many observations of this otherwise inaccessible region.

Bottomside ionosphere. C region. This term is being used increasingly to refer to the lowest part of the ionosphere, at altitudes between about 50 and 70 km. Electrons are produced by the collision of *galactic cosmic rays* with atmospheric molecules, the electron density being typically 10^2 cm^{-3}. During a polar cap absorption (*PCA*) event caused by a flux of protons $\sim$20 MeV from the sun, electron densities $\sim 10^3$ cm^{-3} are observed. The formation of negative ions, whose density is more than an order of magnitude greater than the electron density, and their effects on radio wave propagation, are being studied.

D region. Rocket experiments confirm that the normal D region at times of minimum solar activity is explained by the photoionization of the trace constituent, neutral nitric oxide, by solar Lyman-α radiation at 1216 Å. During solar activity X-rays may make an important contribution. Deeks (1966) presents a series of mid-latitude electron density profiles at different local times and seasons, at sunspot minimum and maximum, and during an eclipse, derived by a theoretical "full-wave" technique from observations of the reflection of VLF and LF waves from a land-based transmitter.

The excess D region absorption of HF waves reflected by the mid-latitude ionosphere in winter, the *winter anomaly*, may be associated with changes in stratospheric or mesospheric meteorology, that is temperature or chemical composition changes. However, Maehlum (1967) believes that precipitated energetic electrons can be important. Abnormally large electron densities, giving rise to enhanced radio wave absorption, are produced in auroral regions at times of an enhanced flux of precipitating $\sim$100 keV electrons.

E region. A rocket-borne ions mass spectrometer, launched from a mid-latitude site, shows the predominance of nitric oxide ions and molecular oxygen ions. Any molecular nitrogen ions formed are rapidly removed by the fast dissociative recombination. A small proportion of metallic ions (for example magnesium), possibly of meteoric origin, is observed. Changes in density and temperature of neutral molecules can cause important changes in the electron-neutral collision frequency.

Direct measurements confirm that there is a correlation between the occurrences of mid-latitude *sporadic E* and horizontal wind shear, as observed by observation of rocket released vapour trails, thus tending to support the wind shear generation theory.

Rocket-borne magnetometers show that the eastwards current flowing in the daytime *equatorial electrojet* reaches a maximum density of up to 10 amp km^{-2} at an altitude of approximately 107 km.

F region. Some electron density observations can be explained by assuming the ionized gases to be in a state of local as distinct from diffusive equilibrium, the electrons being lost at 300 km altitude by the two-stage process $O^+ + N_2 \to NO^+ + N$; $NO^+ + e \to N + O$. However, *diffusion* of the plasma takes place in the geomagnetic field under the action of electrodynamic forces. This can explain the Appleton, or geomagnetic, anomaly, that is the minimum value of electron density at a particular altitude ($\sim$300 km) at zero geomagnetic latitude, with two maxima at that altitude at geomagnetic latitudes of $\sim 15°$ North and South. The importance of movements due to the coupling of neutral winds to the ionospheric plasma is stressed by Kohl and King (1967). High latitude phenomena occurring at certain Universal Times may be understood in this manner, as may changes of altitude of the peak of the $F2$ layer.

Processes by which ionization in the F region is maintained during the night are not fully understood. Charged particles precipitating from the Van Allen radiation belts can cause extra ionization; however, the neutral atmosphere is also heated, which results in a dramatic increase of the electron recombination rate, and the electron density is thus reduced.

Rocket measurements show that marked differences exist during daylight between electron, ion and neutral *temperatures* ($T_e \sim 2T_i$, $T_i \sim T_n$). Observations made by an ion energy analyzer aboard a polar orbiting satellite at altitudes of between 150 and 300 km reveal ion temperatures of 400 to 1000°K at mid-latitudes by day, and irregular fluctuations around 2000°K in the auroral zones. Harris, Sharp and Knudsen (1967) also report the daytime dominance of atomic oxygen ions over molecular ions above 200 km altitude. The total ion concentration (which equals the electron concentration) reaches a minimum value at night near 60° geomagnetic latitude, $L \sim 4$, known as the *trough*, which is a demarcation region between temperate and polar ionospheres.

Topside ionosphere. To date, topside parameters can be investigated directly by rocket and satellite experiments of one of the following *three* types: sampling of the local plasma, propagation and reflection of HF radio signals transmitted by the spacecraft, and reception of natural VLF phenomena.

Of the *first* type, electrostatic Langmuir probe measurements of the local *electron density* (N_e) *and temperature* (T_e) are reported by Brace, Reddy and Mayr (1967). Aboard the *Explorer 22* satellite in a near-polar orbit at an altitude of 1000 $\pm$ 100 km, observations are taken at latitude increments of

Table 1

Geo-magnetic latitude	N_e, cm^{-3}		T_e, °K	
	Local noon	Local midnight	Local noon	Local midnight
0°	2.8×10^4	1.5×10^4	1900	850
35°	1.5×10^4	2.7×10^4	2300	1000
55°	1.1×10^4	0.7×10^4	2400	1800

about 10°. Typical results, for the vernal equinox of 1965, are given in Table 1. Similar data are available from the *Tiros 7* satellite, whose orbit plane is inclined at an angle of 58° to the geographic equatorial plane, at an altitude of 640 ± 15 km. A comparison of observations made at times when the planetary index of magnetic activity, K_p, is greater than 3o with those taken at quiet times reveals marked enhancements of N_e at local geomagnetic latitudes between 20° and 60°, and slight reductions of T_e. At latitudes above 70°, T_e is increased during a magnetic disturbance.

Knudsen and Sharp (1966) report retarding potential analyzer measurements of *ion temperatures*. From a rocket launched during the local night, T_i at mid-latitudes is found to be 1200 (± 200) °K at apogee (1014 km). Between 700 and 1000 km altitude, T_i increases linearly with altitude, at a rate of 0·8°K/km.

Positive ions in the topside ionosphere, namely H$^+$, He$^+$ and O$^+$, are observed at night by the rocket-borne magnetic mass spectrometer of Hoffman (1967). The lower ionosphere consists primarily of O$^+$, with the transition to H$^+$ as the predominant ion at an altitude of 430 km. He$^+$ never becomes the predominant species, its maximum density being 5×10^3 cm^{-3}, 15 per cent of the total ion density, at 450 km altitude.

Turning to the second type of experiment, the Canadian satellites *Alouette I* and *Alouette II*, together with the U.S. satellite *Explorer 20*, have contributed much information on the topside ionosphere. As the frequency of a swept-frequency transmitter in a plasma is increased, propagation first in the Z mode becomes possible. At a slightly higher frequency, a resonance at the local plasma frequency is observed and propagation in the ordinary mode commences. At still higher frequencies resonances appear at harmonics of the electron gyrofrequency and the

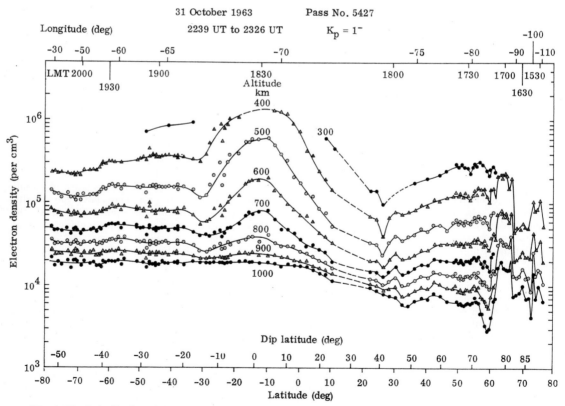

Fig. 1. *The latitudinal variation over the Americas of topside ionospheric electron density at 100 km altitude increments.*

extraordinary mode propagates. Energy in this mode propagates vertically down into the ionosphere, and an echo is received. A display of the time delay between transmission and reception as a function of frequency is known as an *ionogram*. Inversion of this relation, using a digital computer, enables an electron density profile below the satellite to be calculated. *Alouette I* travels in its near polar orbit at an altitude of 1025 ± 25 km, and a topside ionogram is produced every 18 sec at latitude intervals of about 1°. Figure 1 (Thomas, Rycroft, Colin and Chan, 1966) shows the latitudinal variation over the Americas of topside ionospheric electron density at 100 km altitude increments. The curves are symmetrical about the dip equator. At a dip latitude of 74°, the trough is seen to mark the division between the smoothly varying mid-latitude ionosphere and the large N_e fluctuations characteristic of the auroral and polar ionospheres. Diurnal variations of electron density are less well known than latitudinal ones, since much scatter is introduced during the necessary three months of accumulation of data as the satellite orbit precesses.

The manner in which electron density changes with altitude, h, are conveniently depicted, as in Fig. 2, in terms of variations of the vertical plasma scale height, $H_v = -N_e/(\partial N_e/\partial h)$. At magnetic latitudes, λ, above $\sim 25°$, a plasma in *diffusive equilibrium* is, to a good approximation, distributed according to the relation $H_v = k(T_e + T_i)/\overline{m}_i g$, $\overline{m}_i$ being the mean ionic mass. Near the geomagnetic equator, where the dip angle is small, the interpretation of H_v is more complicated since the plasma is constrained to diffuse along the geomagnetic field lines. A field-aligned plasma scale height, proportional to the ratio of the electron and ion temperatures to the mean ionic mass, is defined to combine the vertical scale height with a latitudinal scale length, the latter weighted by $\cot \lambda$. Using a model latitudinal variation of electron and ion temperatures, which increases monotonically with latitude up to the auroral zone, $\overline{m}_i$ is found to increase with increasing latitude, except, sometimes, within a few degrees of the magnetic equator.

Similar topside soundings are now available from *Alouette II*, in a polar orbit at altitudes of between 500 and 3000 km. Hagg (1967) has observed electron densities as low as 10 cm^{-3} at apogee at geomagnetic latitudes near 60° in the vicinity of the trough.

Explorer 20 carried a complementary experiment, namely a sounder with six fixed frequencies between 1·5 and 7·2 MHz which are pulsed in sequence, the entire cycle time being only 0·1 sec. Scattering from

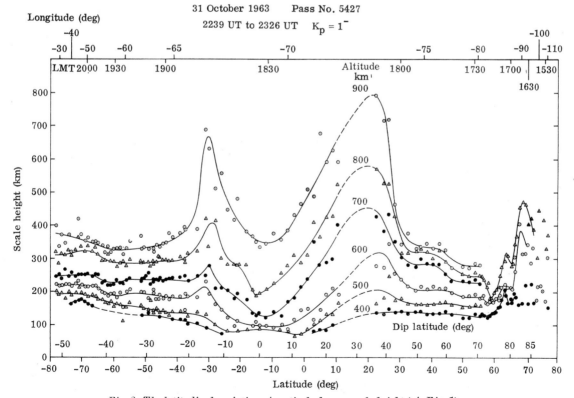

Fig. 2. The latitudinal variation of vertical plasma scale height (cf. Fig. 1).

irregularities of scale ~100 m is observed in high latitudes. Ducting by field-aligned tubes of enhanced or reduced electron density (relative to the ambient plasma) of scale ~ few kilometers is observed between geomagnetically conjugate regions at lower latitudes. Predominantly field-aligned *spread F* irregularities are also observed by the Alouette topside sounders.

As for the *third* type of experiment, spectrograms of both short fractional-hop and long fractional-hop *whistlers* are received aboard *Alouette, Injun* and *OGO* satellites, and can be compared with the integral-number-of-hops whistlers observed on the ground. Belrose and Barrington (1965) report a band of VLF noise having a sharp low frequency cut off ~5 kHz, which tends to decrease as the *Alouette I* satellite moves to higher latitudes. Brice and Smith (1965) postulate that the cut off of this noise band, which is triggered by both short and long fractional-hop whistlers and is thus generated in the vicinity of the satellite, defines the limit for propagation transverse to the geomagnetic field. This occurs at the lower hybrid resonance frequency, which is proportional to the square root of the sum of the fractional abundances of the constituent positive ions weighted by the inverse of their atomic masses. Thus, at 1000 km altitude, heavier ions are more prevalent at higher latitudes; oxygen ions become ~95 per cent predominant in the daytime auroral zone. Combining this result with *Alouette I* electron density profiles, interpreted on a diffusive equilibrium theory, the mean of electron and ion temperatures in the auroral zone is deduced to be ~3000°K.

A new type of whistler is sometimes observed, after a short fractional-hop whistler, as a tone which initially shows a rapid rise of frequency and asymptotically approaches the local proton gyrofrequency. Known as a *proton cyclotron whistler*, this arises from coupling of the whistler mode signal into the ion cyclotron mode (having the opposite sense of polarization to that of the initial whistler). The percentage abundance of hydrogen ions can be deduced, and is found to increase with increasing altitude and decreasing latitude. Typical values at 1000 km altitude are 75 per cent at a local geomagnetic latitude of 52° during the winter night and 25 per cent at 32° during the summer day. From the abrupt termination of the proton cyclotron whistler signal, attributed by Gurnett and Brice (1966) to cyclotron damping, the proton temperature can be estimated to within ± 20 per cent, and typically falls within the range 600 and 1100°K. The simultaneous occurrence of a proton and a helium cyclotron whistler is reported by Barrington, Belrose and Mather (1966). From measurements of the frequency-time characteristics of this signal, the percentage composition of the ternary ion topside ionospheric plasma is found without recourse to the assumption of diffusive equilibrium. Ion cyclotron whistlers provide a powerful technique for studying the high ionosphere, their only limitation arising out of the random occurrence of lightning discharges which produce them.

Ground based techniques. Faraday rotation. Liszka (1966) reports differential Faraday observations of the multi-frequency beacon satellite S-66, at ~1000 km altitude, taken at Kiruna, Sweden. Variations in total electron content up to the satellite, both diurnally and latitudinally (from 50° to 75°N latitude), are presented. By night the trough is evident. Measurements are now being made using a beacon satellite in a geostationary orbit, at an altitude of ~5·5 earth radii.

Incoherent backscatter. The power backscattered, by the Thomson scattering mechanism, by electrons in the ionosphere depends on their number density, a factor depending on the ratio of electron to ion temperatures and a factor depending on the ratio of the Debye shielding distance to the transmitted wavelength. The spectrum of the backscattered signal is determined by kinetic motions of the ions, and thus gives information on the ion mass and the ion temperature. From spectra recorded with different time delays, corresponding to scattering from different altitudes, it is possible to deduce the electron density, electron and ion temperatures, and ionic composition as functions of height. The backscatter technique is thus seen to be a most worthwhile, if rather expensive, research tool. Such observations are being made at the six observatories listed in Table 2.

Table 2

Station	Geomagnetic latitude, °N
Jicamarca, Peru	0°
Arecibo, Puerto Rico	30°
Saint-Santin de Maurs, France	48°
Millstone Hill, M.I.T., U.S.A.	53°
Malvern, U.K.	54°
Prince Albert, Canada	62°

At the geomagnetic equator, N_e can be inferred at altitudes up to ~5000 km. By day, T_e reaches a maximum value of 1500 to 2000°K at 240 km, above which it decreases, becoming equal to $T_i \simeq T_n$ ~ 900°K above ~300 km. Above ~450 km, T_e and T_i gradually increase together. At sunrise T_e ~ 3000°K, $\gtrsim 3T_i$; at night $T_e \simeq T_i \sim 600$°K. The concentration of O^+ ions falls to 50 per cent at ~600 km altitude at night, or $\gtrsim 800$ km by day. The concentration of He^+ ions is, at all times and altitudes, $\lesssim 10$ per cent.

Carlson (1966) interprets a pre-sunrise T_e increase at altitudes of 300 to 500 km, observed at Arecibo during winter, as showing local heating by photoelectrons produced in the already sunlit magnetic conjugate region. T_e/T_i is ~ 3 during the day, and $\sim 1\cdot 2$ at night. The altitude dependence of ionic composition is not dissimilar to that deduced at the equator, although He$^+$ ions may constitute up to ~ 25 per cent of the positive ions in the plasma.

Observations over Massachusetts at altitudes of up to 750 km are reported by Evans (1966). By day T_e/T_i is $\gtrsim 1\cdot 6$, but is larger at sunrise; the value of this ratio falls to $\gtrsim 1\cdot 2$ at night. Evidence is presented of ionospheric heating, especially at night, during magnetically disturbed conditions.

Such measurements, complementing as they do *in situ* observations, provide much information on the ionosphere.

All these ionospheric variations can be broadly interpreted as indicating the expansion of the entire ionosphere and neutral atmosphere above ~ 120 km altitude in response to the greater daytime flux of solar ultra-violet and X-radiations. This causes the mean ionic mass at a certain altitude to increase, and the lighter ions and atoms to float higher and higher. Other sources of ionization and heating, of corpuscular origin, are important at high magnetic latitudes.

Whistlers. Observation of whistlers that have been guided along several paths through the magnetosphere enables estimates of equatorial plane electron densities to be made at several radial distances ($\sim$few earth radii). At times a whistler component that has travelled to a large radial distance arrives earlier than expected, indicating a much reduced electron density at large distances. Carpenter (1963) calls the sudden decrease of electron density, by an order of magnitude within a fraction of an earth radius, the *knee*, or *plasmapause*. By studying variations of the

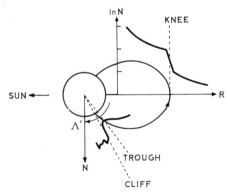

Fig. 3. Diagram illustrating typical night-time variations of electron density in the outermost ionosphere.

positions of the plasmapause and the topside ionospheric trough with changing magnetic activity and local time, Rycroft and Thomas (1967) show that both tend to lie on the same line of force of the geomagnetic field, a result illustrated in Fig. 3.

Bibliography—Background

BEYNON W. J. G. (1962) Ionosphere, in *Encyclopaedic Dictionary of Physics* (J. Thewlis Ed.), **4**, 70, Oxford: Pergamon Press.

ELLISON M. A. (1962) Solar-terrestrial relationships, in *Encyclopaedic Dictionary of Physics* (J. Thewlis Ed.), **6**, 537, Oxford: Pergamon Press.

ELLISON M. A. (1962) Sudden ionospheric disturbance, in *Encyclopaedic Dictionary of Physics* (J. Thewlis Ed.), **7**, 83, Oxford: Pergamon Press.

HARGREAVES J. K. (1962) Radio waves, propagation of, in the ionosphere, in *Encyclopaedic Dictionary of Physics* (J. Thewlis Ed.), **6**, 156, Oxford: Pergamon Press.

HARGREAVES J. K. (1967) Physics of the upper atmosphere, in *Encyclopaedic Dictionary of Physics* (J. Thewlis Ed.), Suppl. vol. **2**, 236, Oxford: Pergamon Press.

HARGREAVES J. K. (1967) Space physics, in *Encyclopaedic Dictionary of Physics* (J. Thewlis Ed.), Suppl. vol. **2**, 342, Oxford: Pergamon Press.

HARGREAVES J. K. (1967) Space research, physics of, in *Encyclopaedic Dictionary of Physics* (J. Thewlis Ed.), Suppl. vol. **2**, 348, Oxford: Pergamon Press.

KERTZ W. (1962) Geomagnetic field, transient variations of, in *Encyclopaedic Dictionary of Physics* (J. Thewlis Ed.), **3**, 445, Oxford: Pergamon Press.

SMITH E. K. (1966) Sporadic-E ionization, in *Encyclopaedic Dictionary of Physics* (J. Thewlis Ed.), Suppl. vol. **1**, 329, Oxford: Pergamon Press.

Bibliography

FRIHAGEN J. (Ed.) (1966) *Electron Density Profiles in Ionosphere and Exosphere*, Amsterdam: North-Holland Publishing Company.

GORDON W. E. (1967) F region and magnetosphere, backscatter results, *Reviews of Geophysics* **5**, 191, Washington, D.C.: American Geophysical Union.

SCHMERLING E. R. (1966) Advances in ionospheric physics in the rocket and satellite era, *Reviews of Geophysics* **4**, 329, Washington, D.C.: American Geophysical Union.

M. J. RYCROFT

IONOSPHERES OF MARS AND VENUS. The first conclusive experimental evidence that Mars has an ionosphere was obtained on July 15, 1965 as the U.S. spacecraft *Mariner 4* flew behind that planet. The experiment was conducted by utilizing the 2·2 GHz telemetry link between the spacecraft and the earth to probe the ionosphere and neutral atmosphere of the planet. By observing the variations in the carrier frequency as the radio link swept tangentially through the atmosphere, one determined how the refractivity of this medium changed with altitude. (The refractivity N is equal to $(n - 1)\,10^6$, where n is the refractive index.)

Figure 1 shows the refractivity profiles obtained from the analysis of the *Mariner 4* occultation data. The day side profile was derived from the immersion

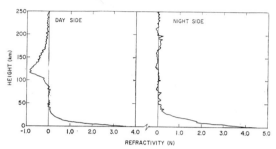

Fig. 1. Atmospheric refractivity profiles for Mars.

measurements made over Electris near 50°S, 177°E at 13^{00} hours local time in late winter. The solar zenith angle was 67° at the immersion point.

From the magnetoionic theory, one can show that, for the conditions of this experiment, the refractivity N of free electrons is given by:

$$N = -40.3 \times 10^6 \times \frac{n_e}{f^2}$$

where f is the radio frequency (hertz) and n_e is the electron number density (m^{-3}). The negative portion of the refractivity profile corresponds to the ionosphere.

The day side ionization reached a peak density of 10^5 el/cm^3 near 120 km altitude. The plasma scale height on the topside of the ionosphere was of the order of 30 km. A minor layer was detected near 95 km.

No ionization was detected during emersion on the night side of Mars as illustrated by Fig. 1. This measurement was made near 60°N, 36°W at 23^{40} hours local time in late summer. The solar zenith angle was 104° at the emersion point.

By analogy with the formation of ionization layers in the terrestrial atmosphere one can interpret the day side ionization peak observed on Mars with three types of models. These three classes of ionospheric models may be designated as F_2, F_1 and E.

The F_2 models are based on the assumption that the observed ionization peak is a so-called β-Chapman layer produced by a rapid upward decrease of the ion-recombination loss coefficients together with downward diffusion of plasma. In the F_2 models, the electron density peak is above the region where most of the electron production and recombination occurs.

The F_1 and E models are based on the assumption that the observed ionization profile is a so-called α-Chapman layer in which the peak coincides with the electron production peak caused by solar extreme ultraviolet and X-rays, respectively.

The simplest of these ionospheric models assumes that CO_2, which is known to be the principal constituent in the lower neutral atmosphere, also predominates in the upper atmosphere. The ionization peak is an F_1 layer with principal ion CO_2^+. No F_2 peak develops because the electron ion recombination rate coefficient is independent of altitude in this model. With CO_2^+ being the principal ion on the top side, the plasma temperature would have to be about 300°K in order to explain the scale height of this region.

The F_1 model mentioned above is attractive because it requires few assumptions in order to explain the observations, however, more independent evidence is necessary before one can rule out the other alternatives. Mass spectrometric identification of major ions and neutral constituents would be particularly useful in this regard.

The ionosphere of Venus has also been explored with the radio occultation technique. The measurements were made on October 19, 1967 when the U.S. spacecraft *Mariner 5* was occulted by Venus. The resulting electron density profiles are shown in Fig. 2.

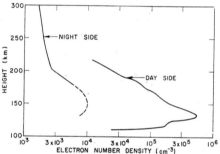

Fig. 2. Ionospheric electron number density profiles for Venus. The height scale assumes a planetary radius of 6050 km.

The day side profile was determined from the frequency variations on the telemetry carrier. The measurements were made near 30°S, 10°W. This region had been in direct sunlight for about 24 terrestrial days and the solar zenith angle was 33° at the time of the measurement. The night side profile was obtained from the dual radio frequency experiment. In this last experiment the ionosphere was probed with two harmonically related frequencies, 50 and 425 MHz, which were transmitted from the Earth and received in the *Mariner 5* spacecraft. Dispersive doppler measurements of the change in the amount of plasma along the Earth spacecraft propagation path were made by comparing the 2/17 subharmonic of the higher frequency with the lower frequency in the spacecraft. The night side ionosphere was probed near 35°N, 170°E. The solar zenith angle was 142° and this area had been out of direct sunlight for about 24 terrestrial days.

Important additional information about the upper atmosphere of Venus was obtained from the ultraviolet experiment conducted with *Mariner 5* and the composition measurements made with the instrument capsule released from the U.S.S.R. spacecraft *Venera 4*. The ultra-violet experiment was designed to look for Lyman-alpha airglow from hydrogen and ultra-violet airglow from atomic oxygen. The preliminary results indicate the presence of both atomic and molecular hydrogen but no atomic oxygen was detected.

The U.S.S.R. capsule, which entered the atmosphere of Venus on October 18, 1967, also detected atomic hydrogen in the upper atmosphere and, in addition, sampled the composition of the lower atmosphere as it parachuted down towards the surface. The preliminary results of the gas analyzer experiments indicate that the lower atmosphere consists of at least 90 to 95 per cent CO_2. The absence of atomic oxygen in the upper atmosphere shows that CO_2 remains largely undissociated by solar ultraviolet. These findings suggest that the ionization peaks observed on Venus are analogous to the terrestrial F_1 or E region rather than to the F_2 peak.

The dual radiofrequency measurements indicate that the daytime ionosphere terminates in an abrupt plasma pause near 500 km altitude. On the night side, the ionization appears to have extended to an altitude of 4000 km or more. This difference between the day and night side is believed to be caused by the interaction between the upper atmosphere and the solar wind.

Bibliography

BARTH C. A., PEARCE J. B., KELLY K. K., WALLACE L. and FASTIE W. G. (1967) *Science* **158,** 1675–8.

DAVYDOV V. D. (1968) *Science and Technology* **73,** 49–54.

KLIORE A., CAIN D. L., LEVY G. S., ESHLEMAN V. R., FJELDBO G., and DRAKE F. D. (1965) *Science* **149,** 1243–8.

KLIORE A., LEVY G. S., CAIN D. L., FJELDBO G. and RASOOL S. I. (1967) *Science* **158,** 1683–8.

Mariner Stanford Group (1967) *Science* **158,** 1678–83.

G. FJELDBO

THE JET FLAP. The jet flap has been described by the main originator of the device, I. M. Davidson (1956), as "depending on nothing more than the aerodynamic resynthesis of the lifting and propulsive means, the entire propulsive jet being ejected in the form of a thin full span sheet from the trailing edge of the wing" (of an aeroplane). More simply it has been defined as a thin sheet of air or gas discharged at high speed close to the trailing edge of a wing so as to induce lift over the whole wing independently of wing incidence; if the angle of discharge is controlled by means of a small trailing-edge flap, it is called a shrouded jet flap. To this definition was added the comment that the quantity of air used with a jet flap is large compared with blowing for boundary layer control.

The underlying physical principle had been employed earlier in experiments on boundary layer control by blowing over the upper surface of a trailing edge flap. Several papers were written on this subject and references are given in Davidson's paper (1956) referred to above, but the emphasis had been in all cases on the use of a jet for boundary control. After some crude exploratory experiments, wind tunnel tests were made at the National Gas Turbine Establishment on an 8-in chord aerofoil of elliptical section which had three alternative trailing edges with exits for the jet at angles of 31·4°, 58·1° and 90·0° to the chord. Davidson describes the behaviour of the jet flap as in Fig. 1. The results of the earliest historic experiments are given typically in Fig. 2, where C_{L0} is the lift coefficient at zero incidence and C_J is the jet coefficient, being the total jet reaction J divided by the product of the dynamic pressure and the wing area. The forces on the model were calculated from the pressures recorded at static pressure holes on the surface since the introduction of air into the model to create the jet would have invalidated any measurements made by wind tunnel balances.

It is significant that a jet-flapped wing will give a lift that can be increased almost indefinitely and the aerodynamic properties will improve smoothly with C_J until there is a full-chord separation bubble over the top of the wing. The reason for the improvement is the influence of the jet in its mixing entrainment of the main flow which has been demonstrated by photographing the streamlines made by smoke flowing over the wing. As the incidence of the wing is in-

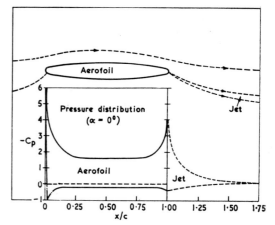

Fig. 1. Basic principle of jet flap operation.

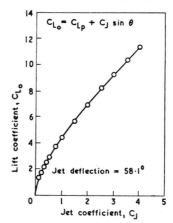

Fig. 2. Measured lift at zero incidence.

creased, the separated main stream is ultimately thrown clear of the sphere of influence of the jet so that the bubble bursts to form an open wake and the wing consequently stalls. At any given value of the lift coefficient, C_L, of the wing the danger of an incidence stall is reduced by increasing the strength of the jet C_J; on the contrary, a jet at a larger angle will make a wing more prone to stall than a jet at a smaller angle, the incidence of the wing remaining unaltered (see Fig. 3).

The jet flap has applications for STOL and VTOL aircraft on account of the extra lift for take-off and landing. This is, however, reduced by the presence of an adjacent landing surface, though the loss of lift is very much less than that obtainable by any other lifting system, the flow pattern being like Fig. 4. Typically, with a ground clearance of a height equal to the chord of the wing about 20 per cent is lost and

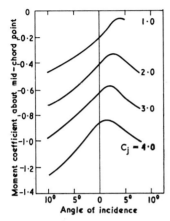

Fig. 3. Stalling stability of 31·4° model.

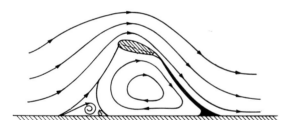

Fig. 4. Flow field of jet flap near the ground.

at a height equal to the half chord the loss may reach 60 per cent for the large value of C_J of 4 and a jet angle of 58°: the percentage losses are much less for smaller values of C_J. For large values of C_J, coupled with ground effect, a large suction is induced on the under surface of the wing towards the rear especially for small values of the ratio of height over chord, thus contributing to the reduction of lift.

The jet stream will normally give rise to extra drag and the amount will depend upon whether the jet or wake velocity is greater or smaller than that of the main stream, or, since the densities may be different, such as when the jet stream is produced by the exhaust from the engines, on the product of density and velocity. In general, in cruising flight this product will approximate closely to that for the main stream of air past the aeroplane and in that case the jet drag will be zero: it could be positive or negative in value.

The applications of jet flaps will be in practice to wings of finite aspect ratio. The first wind tunnel experiments were made by Hagedorn and Ruden (1938) on a model of 144 mm chord and 4·28 aspect ratio. This work was in connection with boundary layer control. The results gave an increasing lift coefficient with increase of C_J up to 1·78 over a range of angles of incidence of the aerofoil. Moreover, the active principle was correctly explained, namely, that blowing gave rise to a pressure difference at the trailing edge which altered the pressure difference on the aerofoil and thereby increased the lift. An aeroplane's wings generate the lift by throwing down the main stream as a downwash. In the jet-flapped wing the jet has an entrainment action and throws down a much larger mass flow at a smaller angle than when the jet is absent. Thus the induced drag of a jet-flapped wing may be less than that of a conventional wing with the same conditions of loading, but this would not be the case under cruising conditions. The early experiments of Hagedorn and Ruden were not followed up.

In practice a jet flap on a wing in its simplest form would affect the longitudinal control of an aeroplane and a small hinged flap or jet shroud movable by the pilot has been introduced. This is a small flap with a curved upper surface having a typical length of 3 to 10 per cent of the wing chord, placed at or just under the exit of the jet. The jet will follow the upper surface of the shroud by what is called the "Coanda" effect, and as the shroud is rotated the jet also will be rotated. If suitably located a jet can be rotated by the shroud through an angle of 90°. The effect of the shroud acting as a flap and of the jet are additive vectorially.

The production of the jet needs engine power and one of many suggestions that have been made for a practical aeroplane is to embody a number of small engines in the wings. Whatever system is adopted for the power units to supply the air for the jet, there will necessarily be a considerable complication in engineering construction as well as addition to the total weight of the aircraft. This is a matter for design development. In addition, the problem of possible engine failure has to be met, which would prove difficult as regards stability and control with a jet only. In the case of the shrouded jet, satisfactory control can probably be maintained solely from the operation of the shroud as a flap. Similar remarks apply in the case of landing when gusts are present. The same maximum lift is clearly not available if one or more engines providing the jet do fail, but this would have to be countered by increasing the forward speed to give adequate lift for landing and retaining longitudinal control by use of the shroud as well as the elevator. Some idea of the restoring moment with a jet flap can be gauged from Fig. 3, which relates to wind tunnel tests on an aerofoil as given in Davidson's paper (1956).

The introduction of jet flaps will involve an integration of the work of engine and airframe designers and not least is the need for a large amount of ducting from a circular shape at the engine to a relatively narrow slot that would use up available volume space that is in some cases already restricted by the demand made for fuel capacity.

As regards noise, with the propulsive jet more spread out when used as a jet flap, it is anticipated that there will be a reduction of as much as 10 to 20 dB. This is important since fluctuating pressures must be expected in engines of the future amounting to $\frac{1}{2}$ psi which, assuming a frequency of 300 cycles, would mean a million reversals per hour, thus introducing important considerations of fatigue. The noise would also be pushed up the frequency spectrum.

The general position of knowledge on jet flaps has been summarized in an unclassified R.A.E. paper (Williams *et al.*, 1961) and abbreviated as follows: "The aerodynamic principles are well established as a result of wind tunnel and theoretical investigations. Useful estimates can be made of the lift and other forces and of the stability of wings of moderate aspect ratio for full-span jet deflection. The existing theories are based on simple thin aerofoil and lifting-line treatments assuming inviscid flow. Thus, semi-empirical approaches have to be used to account for the effect of viscosity. There is need for further experimental and theoretical research on many aspects including the reduction of thrust associated with skin friction, turning and mixing of the jet, and on the determination of oscillatory derivatives.

"With the development of turbojet engines of high bypass ratio and of light-weight units as gas generators, mixed jet-flap/round-jet configurations have become attractive for short take-off and landing, the jet flaps being used primarily for lift and the conventional round nozzles for thrust. Again, jet-flaps might usefully be incorporated to lower the transition speed of new types of VTOL aircraft, or to simplify the lifting rotor and improve helicopter performance at all speeds."

Most of the experimental work reported in the above was conducted for a jet angle of approximately 50° although some tests at angles as low as 20° and as high as 80°, over a range of incidence of the aerofoil and for C_J values up to 3·96. The aspect ratios of the aerofoils were 6 and 9. Later experiments in Canada (Korbacher) have favoured a jet angle of 55° for two-dimensional jet-flapped wings. This optimum angle was hardly affected over a range of aspect ratios from 6 to 20 (Korbacher). Model experiments on a Delta wing in Australia gave an optimum jet angle of 60°.

A theory of a two-dimensional thin jet-flapped wing was published by D. A. Spence (1956) and was extended by E. C. Maskell and D. A. Spence two years later (1958) to the case of a jet flap in three dimensions. Spence used the methods of thin aerofoil theory in incompressible flow and showed that the lift coefficient C_L of a thin two-dimensional wing at zero incidence, with a narrow high-velocity jet of momentum-flux coefficient C_J issuing from its trailing edge at a small downward inclination θ is given by

$$C_L = 2(\pi C_J)^{\frac{1}{2}} \cdot \theta,$$

and the loading on the chord line ($0 < x < 1$) by

$$\Delta C_p = 2[C_J/\pi \varkappa (1 - \varkappa)]^{\frac{1}{2}} \cdot \theta$$

for small values of C_J, except near the trailing edge, x being the length as a percentage of the chord. These formulae agree well with known measurements.

For larger values of C_J, the following interpolation formulae were found:

$$\delta C_L/\partial \theta = 2(\pi C_J)^{\frac{1}{2}} (1 + 0 \cdot 151 C_J^{\frac{1}{2}} + 0 \cdot 139 C_J)^{\frac{1}{2}}$$

$$\delta C_L/\partial \alpha = 2\pi (1 + 0 \cdot 151 C_J^{\frac{1}{2}} + 0 \cdot 219 C_J).$$

In the extension to three dimensions the theory was developed for a thin wing of finite aspect ratio with a deflected jet sheet of zero thickness emerging with a small angular deflection at its trailing edge and with the restriction that transverse momentum-transport was excluded. The downwash field was assumed to arise from horseshoe vortices proportional in strength to the local lift distribution with the ability of the sheet thus formed to sustain a pressure difference on account of the longitudinal flux of the jet momentum. The loading and downwash distribution were assumed to depend only on the ratio x, where x measures distance from the leading edge divided by the chord length and $c(y)$ is the local chord. This requires both c and the jet-momentum flux per unit span to be elliptically distributed, and the deflection θ and incidence α to be constant over the span.

$$C_{Di} = C_L^2/(\pi A + 2 C_J)$$

where C_{Di} is the coefficient of induced drag, defined as the difference between the thrust and the (constant) flux of momentum in the jet, and A is the aspect ratio. The ratio of the lift $C_L^{(3)}$ in the three-dimensional theory to that for two dimensions $c_L^{(2)}$ was found to be

$$C_L^{(3)}/C_L^{(2)} = \{A + (2/\pi) \, C_J\}/\{A + (2/\pi) \, (\partial C_L^{(2)}/\partial \alpha) - 2(1 + \sigma)\}$$

where α is the incidence of the wing and $\sigma = 1 - \alpha_i/(\frac{1}{2}\alpha_{i\infty})$; α_i, $\alpha_{i\infty}$ are the induced downwash angles at the wing and at infinity downstream.

The results when compared with experimental measurements on an 8:1 elliptic cylinder of rectangular planform with aspect ratios of 2·75, 6·8 and infinity gave close agreement for C_L and the differences between the total and induced drag coefficients were virtually independent of aspect ratio.

In addition to the above theory, an exact solution has been obtained for the unsteady linearized flow past a thin two-dimensional wing of chord c at zero incidence in an incompressible stream of density ϱ and undisturbed velocity v with a thin jet of momentum-flux $2\varrho v^2 c \mu$ emerging from its trailing edge at an oscillating deflection angle $\theta \exp(i\omega vt/c)$. Computations of jet shape and lift force for a range of values of $v = \mu\omega$ are presented and the solutions for periodic plunging and pitching motions of the wing are derived from that for deflexion (R.A.E. Report Aero. 2690, 1964). The effect of the proximity of the ground has not been studied theoretically, but it has been shown experimentally to be similar to that indicated in Fig. 4.

Bibliography

DAVIDSON I. M. (1956) *R. Ae. S. Jrnl.*, Jan. 1956, 25–41.

HAGEDORN and RUDEN (1938) Inst. Tech. Hochschule A 64, Hannover: Lilienthal G. B.

KORBACHER G. K. *Toronto UTIAS* 90.

KORBACHER G. K. *Toronto UTIAS* 97.

MASKELL E. C. and SPENCE D. A. (1958) *Aero. Quart.* **9**, 395.

NAYLER J. L. (1961) Jet flaps, in *Encyclopaedic Dictionary of Physics* (J. Thewlis Ed.), **4**, 133, Oxford: Pergamon Press.

R.A.E. report Aero. 2690, 1964.

SPENCE D. A. (1956) *Proc. Roy. Soc.* **A 238**, 46–68.

WILLIAMS J., BUTLER S. F. J. and WOOD M. N. (1961) *R.A.E. report* Aero. 2646 (A.R.C. 22, 823), Jan. 1961.

J. L. NAYLER

THE JOSEPHSON EFFECT

1. Introduction

The Josephson effect was predicted in 1962 by Brian D. Josephson while working at the Mond Laboratories of the University of Cambridge. It is associated with superconductivity in metals or superfluidity in liquid helium, is intimately linked with the phenomenon of magnetic flux quantization, and its fascination stems from its being a macroscopic manifestation of quantum effects.

In a normal conductor the conduction of electrons may be understood by treating them as a gas whose particles obey the Fermi-Dirac statistics, the quantum statistics for spin $\frac{1}{2}\hbar$ particles, and in which, as a result of the Pauli exclusion principle, only one electron can occupy each quantum state. At absolute zero temperature and in zero electric field, all energy levels below a certain value E_F (the Fermi energy) are occupied and levels immediately above E_F are empty. As the temperature is increased from zero the separation between occupied and unoccupied levels becomes less sharp, existing over a region of roughly kT about E_F (k is Boltzmann's constant, and T the absolute temperature). In a superconductor, however, there is a forbidden energy zone of width Δ on either side of E_F in the spectrum of allowed energy values. At absolute zero, Δ has a maximum value $\Delta(0)$ and decreases with increasing temperature until it becomes zero at the transition temperature—the temperature at which superconductivity ceases.

According to the Bardeen, Cooper and Schrieffer theory of superconductivity, in addition to the normal Coulomb repulsive force between the electrons, which has a range of about 10^{-5} mm, there is a longer range attractive force between the electrons, of range about 10^{-3} mm, which usually occurs via the lattice vibrations (the phonon interaction) and is between pairs of electrons having equal and opposite spins and electromagnetic momenta. (This should not, however, be taken to imply that their total mechanical momentum or the velocity of their common centre of gravity is also zero.) For superconductivity to occur, the attractive electron–phonon interaction must exceed the Coulomb repulsion. The energy of such a pair is slightly lower than it was before the interaction occurred and 2Δ may be thought of as being the pair binding energy.

The coupled electrons, which are known as Cooper pairs, are separated by up to a thousand times the metal crystal lattice spacing and so their associated wave or probability function overlaps those of many other pairs. A direct consequence of this is that the Pauli exclusion principle can only be satisfied if the motions of the electron pairs are correlated, in particular if the sum of the two electron momenta is the same for all pairs, and hence the wave functions associated with all pairs must have both the same frequency and phase. This is the long range ordering which was envisaged in London's original theory of superconductivity and which accounts for the stability of the superconducting state.

When two superconductors are separated by some distance their wave functions need have no particular frequency or phase relationship to each other, but Josephson considered what would happen if the separation was decreased to about 10^{-6} mm (the metals being separated by a medium which is usually an oxide layer), in which case the wave functions of the two metals could overlap. He predicted that under certain conditions the frequencies and phases of the two sets of wave functions would be related and that there would be tunnelling of electron pairs across the gap (tunnelling refers to the quantum mechanical process which allows electrons to penetrate potential barriers). A further prediction was that there would be interactions between this tunnelling current and magnetic and electromagnetic fields in which the junction was situated. Analogous effects were also expected in superfluid helium. These effects are discussed briefly below but they have excited so much theoretical and experimental work in the last six years that the reader should refer to the review references which are given at the end. Sections 2 to 6 are concerned with superconductivity and section 7 with superfluidity.

2. The Basic Josephson Equations

The first equation leads to an expression which relates the current density $j(\mathbf{r}, t)$ in the insulating region and $\phi(\mathbf{r}, t)$ the relative phase difference of the Cooper pairs which are opposite one another at position $\mathbf{r}$ and time t as:

$$j(\mathbf{r}, t) = j_1 \sin \phi(\mathbf{r}, t) \qquad (1)$$

The constant j_1 is a measure of the tunnelling probability between the two superconductors, which decreases exponentially with increasing oxide layer thickness and increases as the temperature is reduced. According to Ambegaokar and Baratoff, j_1 has a maximum value of $\pi\Delta(0)/2eR$ at absolute zero (e being the electronic charge and R the resistance per unit area of the junction above the transition temperature). This relation only applies when the two superconductors are the same metal, otherwise the value must be evaluated numerically.

The second and third Josephson equations determine the spatial and temporal dependence of ϕ and are:

$$\dot{\phi}(\mathbf{r}, t) = (2e/\hbar)\, V(\mathbf{r}, t) \qquad (2)$$

and $\operatorname{grad}\phi(\mathbf{r}, t) = (2ed/\hbar)\,(\mathbf{B} \wedge \mathbf{n}) \qquad (3)$

where $V(\mathbf{r}, t)$ is the total voltage across the junction at $\mathbf{r}$ and time t, and (see Fig. 1) $d = l + 2\lambda$, l being the thickness of the oxide and λ the depth to which the magnetic flux $\mathbf{B}$ penetrates the superconductor, $\hbar = h/2\pi$, h being Planck's constant, and $\mathbf{n}$ is the unit vector in the positive x-direction.

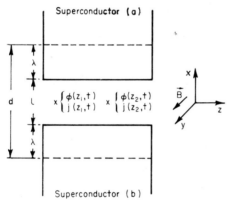

Fig. 1. *A tunnel junction, illustrating the symbols used in the text.*

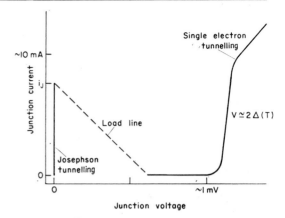

Fig. 2. *Illustrating how the Josephson current passes through the junction at zero volts. At a current i_J the junction switches across the load line to follow the normal single electron tunnelling characteristic.*

3. The Zero-voltage Current Step

The simplest effect occurs when the magnetic flux applied to the junction is zero. It then becomes possible for the Cooper pairs to cross the insulating oxide layer without a potential being developed across it until a maximum current i_J is reached. The junction then suddenly switches along the load line (which is determined by the source impedance and the resistance of the leads into the cryostat), to the normal single electron tunnelling curve, that is electrons can no longer tunnel through the junction as Cooper pairs. The voltage–current characteristic of such a junction is therefore similar to that shown in Fig. 2, the values shown being typical of those obtained experimentally.

4. The Effect of a Magnetic Field

The zero-voltage current step is of course difficult to distinguish experimentally from the behaviour that would be observed with single electron tunnelling if for example there was a whisker of metal bridging the otherwise insulating gap. However, if a magnetic flux is applied to the junction the behaviour is markedly different, for the maximum supercurrent $i_{\max}$ which can be passed is related to the total magnetic flux linking the junction by:

$$i_{\max} = i_J \left| \frac{\sin(\pi\Phi/\Phi_0)}{\pi\Phi/\Phi_0} \right| \qquad (4)$$

where Φ_0 is the quantum of magnetic flux, which in superconductivity is associated with paired electrons and is equal to $h/2e (= 2.06 \times 10^{-15}$ webers). A consequence of equation (4) is that whenever the flux through the junction is an integral multiple of the flux quantum, the maximum supercurrent that can be passed through it falls to zero.

For the junction shown in Fig. 1, $\Phi = (2\lambda + l)\,LB$ where L is the length of the junction in the z-direction which is normal to B. For typical junctions $L \cong 0.1$ mm, and $\lambda \cong 5 \times 10^{-8}$ m, so that flux densities of only a few in 10^{-4} Tesla are required to reduce the supercurrent through the junction to zero. This is a factor of more than a hundred times smaller than that required to reduce the supercurrent in a superconducting metal to zero, moreover this behaviour is repetitive with the maximum value of $i_{\max}$ decreasing with increasing flux. The envelope of $i_{\max}$ with increasing flux predicted from equation (4) is shown in Fig. 3. The values shown are for typical junctions, although equation (4) is not obeyed exactly in practice since the current through the junction produces a magnetic field which slightly distorts the characteristic.

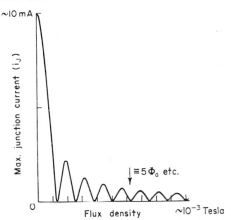

Fig. 3. The envelope of the maximum d.c. Josephson current as a function of the applied magnetic flux density.

It was the observation of the above behaviour in a tunnel junction by Rowell and Anderson in 1963, within a year of its prediction, that clinched the discovery of the Josephson effect.

5. The a.c. Effect

Equations (1) to (5) predict that the pair phase ϕ may vary with time as well as spatially. The variation of ϕ with time is seen from equation (2) to be associated with the appearance of a fluctuating supercurrent, for integrating (2) for a constant potential $V = V_0$ across the barrier yields $\phi = \phi_0 + (2eV_0/\hbar)t$ and hence $j = j_1 \sin(\phi_0 + (2eV_0/\hbar)t)$. The observation of this fluctuating current is less straightforward than the d.c. effect and it was discovered in 1963 by Shapiro using a method envisaged by Josephson.

If in the presence of a static voltage V_0 across the junction a microwave voltage $v_0 \sin(\omega t + \theta)$ is induced across it, then the relative pair phase is given by $\phi = \omega_0 t + (\omega_0 v_0/\omega V_0) \sin\{(\omega t + \theta) + \phi_0\}$, where $\omega_0 = 2eV_0/\hbar$. The Josephson supercurrent is obtained by substituting for ϕ in equation (2). The resulting expression represents frequency modulation of the frequency ω_0 with sidebands $\omega_0 \pm n\omega$, where $n = 1, 2 \ldots$. When $n\omega = \omega_0$, a d.c. component of the current density is present:

$$j_{d.c.} = j_1(-1)^n J_n(nv_0/V_0) \sin(\phi_0 - n\theta) \quad (5)$$

where J_n is the nth order Bessel function. The beat frequency between the wave functions in the two superconductors thus synchronizes to a harmonic of the applied frequency. When it changes from synchronizing with one harmonic to another, the supercurrent between the two superconductors switches from one stable configuration to another. The effects at the junction are highly non-linear and synchronizing with a sub-harmonic of the applied frequency has also been observed.

The current–voltage characteristic for such a junction in the presence of radiation—usually in the microwave region, resembles a non-uniform staircase with the voltage increasing in constant steps of amplitude $\hbar\omega/2e$, and the current in steps given by equation (5). Since $2\Delta \lesssim 2$ mV, the maximum voltage across the junction which allows the passage of a supercurrent cannot be much greater than this and hence there is a maximum frequency for which steps can be observed, although of course the current–voltage characteristic is still affected by higher frequencies. The current amplitude of a particular voltage step depends, as may be seen from equation (5), on the microwave power level and so it may be varied from zero to a maximum by varying the incident power. The characteristic is also affected by an applied magnetic flux and this may also be used to tune the steps to a maximum.

Langenberg, Parker and Taylor made careful measurements of the steps induced by microwave radiation in the region of 9 GHz and although the voltage was small ($2e/h = 483 \cdot 6$ MHz/μV), by measuring the separation of a number of steps and reversing the current through the junction they were able to obtain a measurement of this constant with an estimated standard deviation of < 6 ppm. This precision was considerably better than had been achieved previously using methods from other branches of physics. They also verified that the value obtained did not depend on the nature of the superconductors, the temperature, the current, the microwave power or the frequency. The method appears to be capable of being extended to achieve an even greater precision

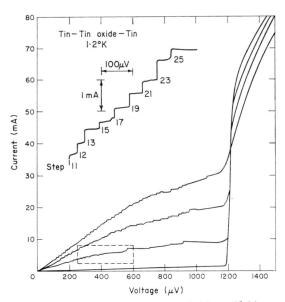

Fig. 4. The I–V curves for a tin/tin oxide/tin tunnel junction displaying microwave induced, constant-voltage, current steps. The frequency of the incident radiation is in the X band, ~ 10 GHz (after Taylor et al.).

than that quoted above and it may be possible to use it in standards laboratories to maintain the volt. Work is already in progress or contemplated at the National Physical Laboratory, the National Bureau of Standards, the National Research Council and others. One of the characteristics observed by Langenberg et al. is illustrated in Fig. 4 and the staircase is seen to steepen as the microwave power is increased, the variations in the step structure may also be seen with the current steps having a variable amplitude and the voltage jumps a constant amplitude.

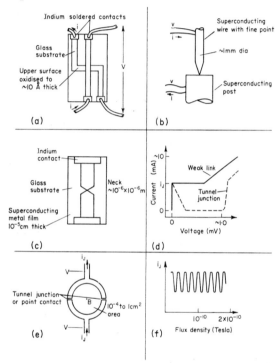

Fig. 5. Types of junction and typical characteristics: (a) tunnel junction, (b) point contact weak link, (c) thin film weak link, (d) their d.c. characteristics, (e) a double junction, and (f) the envelope of the maximum d.c. current with flux density for a double junction $\sim 10^{-4}\,m^2$ area.

Some of the above group, together with Scalapino, observed the a.c. effect by detecting the emitted radiation, although the emitted power was only in the region of 10^{-11} W. The observation of this radiation required tuning the junction to its naturally resonant frequency by varying both the applied magnetic flux and the voltage across the junction.

6. Types of Junction

Josephson effects are observed not only with tunnel junctions but also with other junctions in which there is a weak link between the two superconductors. Examples of these are shown in Fig. 5 and are notably the evaporated film type having a narrow constriction as used by Dayem, or the point contacts used by Mercereau who also formed double junctions. Since the double junctions enclose a larger area than a single junction they are affected to a considerably greater extent by magnetic fields and hence may be used as a very sensitive fluxmeter. As such they have been used by Mercereau and others in a series of elegant interference and flux quantization experiments. The interference effects observed with a double junction are in many ways similar to the diffraction of electrons by a double slit, especially if with the latter the pattern is viewed at a particular location and scanned by varying a magnetic field close to the slits.

Goldman, Kreisman and Scalapino demonstrated that persistent currents could flow around a ring which contained two Josephson junctions. They, and also Smith with a larger ring, observed persistent currents for periods of the order of an hour with 1, 2 and 3 quanta of magnetic flux through the ring. Clarke made a junction quite simply by forming a drop of solder on an oxidized niobium wire. Such junctions have been used to show all of the Josephson effects, and Clarke and others have used them for flux-meter and sensitive galvanometer applications.

There are a number of applications of the Josephson effect to practical devices, such as harmonic generation, microwave oscillators, computer memory elements, fluxmeter or galvanometer applications. So far, however, they are still in the laboratory stage of development. The most promising application to date has been the measurement of $h/2e$ with the possible future use of the method for voltage standardization.

7. The Effect in Liquid Helium

The Josephson effect is also associated with the anomalous flow properties of liquid helium II, known as superfluidity, but it is more difficult to realize it experimentally. However, the helium analogue of the superconducting point contact weak link has been used to demonstrate the existence of the effect. Richards and Anderson used a small orifice to obtain the a.c. analogue and their experiments were later repeated by Khorana and Chandrasekhar with increased stability. In the latter experiment a volume of helium II was contained in a beaker having a thin nickel sheet, 4×10^{-2} mm thick as its base, with a small orifice 8×10^{-3} mm diameter in it. The beaker was placed inside another vessel containing liquid helium which was cooled to below the λ-point. The only contact between the two liquids was via the orifice. The liquid flowed to equilibrium between the respective levels in the usual way, taking 15 min to change by about 5 mm. When the liquid was subjected to phonon vibrations from a transducer which was excited at 99·722 kHz, the level of the helium could be maintained constant for periods of as long as 45 min and excited to a different level by changing the transducer power. The spacing between the steps was as expected, being integral multiples of $z_0 = h\nu/mg = 1\cdot01$ mm, m being the mass of the helium atom, g the acceleration due to gravity, h the Planck constant and ν the transducer frequency. As with the electrical effect steps were sometimes observed for $(n_1/n_2) z_0$ where n_1 and n_2 were integers, $n_2 > 1$.

It seems likely that further experiments will be designed to show the other analogues of the electrical effect and that the Josephson effect will continue to make important contributions to our understanding of the physics of superconductivity and superfluidity as well as leading to useful technological applications.

Bibliography

ANDERSON P. W. (1967) *Progr. in Low Temp. Phys.* **5**, 1–43 (North Holland, Amsterdam); also (1966) *Rev. Mod. Phys.* **38**, 298.

COHEN E. R. and DUMOND J. W. M. (1967) Fundamental constants of atomic physics, in *Encyclopaedic Dictionary of Physics* (J. Thewlis Ed.), Suppl. vol. **2**, 97, Oxford: Pergamon Press.

ENDERBY J. (1962) Superconductivity, in *Encyclopaedic Dictionary of Physics* (J. Thewlis Ed.), **7**, 104, Oxford: Pergamon Press.

JOSEPHSON B. D. (1962) *Phys. Letters* **1**, 251; also (1965) *Adv. in Phys*.

KHORANA B. M. and CHANDRASEKHAR B. S. (1967) *Phys. Rev. Letters* **18**, 230.

LANGENBERG D. N., SCALAPINO D. J. and TAYLOR B. N. (1966) *Proc. I.E.E.E.* **54**, 560.

MACKINNON L. (1962) Superfluid, in *Encyclopaedic Dictionary of Physics* (J. Thewlis Ed.), **7**, 112, Oxford: Pergamon Press.

PARKER W. H., TAYLOR B. N. and LANGENBERG D. N. (1967) *Phys. Rev. Letters* **18**, 287; **3**, 89.

SQUIRES G. L. (1962) Phonon, in *Encyclopaedic Dictionary of Physics* (J. Thewlis Ed.), **5**, 366, Oxford: Pergamon Press.

TAYLOR B. N., PARKER W. H., LANGENBERG D. N. and DENENSTEIN (1967) *Metrologia* **3**, 89.

B. W. PETLEY

KIRCHHOFF'S LAWS (ELECTRICAL CIRCUITS).
Kirchhoff's Laws refer to the voltages and currents in an electric network and form the basis of electric network theory. According to this theory each network consists of branches connected to each other at nodes (each branch conveniently so defined that it contains only one component). Each component is characterized by an equation of performance, say,

$$e_k = f_k(i_k, i_l, i_m, ..., t) \quad (1)$$

which defines the voltage e_k across it in terms of the current i_k through it and in terms of the currents i_l, i_m... through other branches (e.g. in the case of mutual inductances). An equation of performance may, of course, also have the form

$$i_k = g_k(e_k, e_l, e_m, ..., t) \quad (2)$$

These components may be linear or non-linear, with or without hysteresis and may show a behaviour independent of or dependent on time; further these may be active or passive [for ideal voltage generators $e_k = f_k(t)$, for ideal current generators $i_k = g_k(t)$].

Kirchhoff's Laws are then as follows:

(a) "Current Law." The sum of all currents flowing into a node of the network must be zero:

$$\Sigma i = 0. \quad (3)$$

(b) "Voltage Law." When traversing the network along a closed loop the sum of all voltages encountered must be zero:

$$\Sigma e = 0 \quad (4)$$

(the voltages in question are the voltages which, according to eq. (1) exist across the branches of the closed loop).

(a) follows from the fact that a node by itself is devoid of capacity and cannot therefore store or supply electric charge.

(b) follows from the fact that after such a traverse we must have returned to a point of the same potential. The case of time-varying magnetic flux, linking meshes in specified lumped components, i.e. mutual inductances is included in this formulation. If, however, such linkage is specified only as regards magnitude it is impossible to define unique potentials along the meshes concerned and the zero on the right-hand side of eq. (4) has to be replaced by $-d\phi/dt$, the rate of flux change through the loop.

These laws were originally formulated (1845) for d.c. only, and for networks consisting of linear resistances. It is, however, easily seen that they are valid for the instantaneous values of currents of all wave shapes and for non-linear or time-varying components. Furthermore, these laws are also valid if the summations are applied to the vector symbols (phasors) used in a.c. theory for the description of sinusoidal currents and voltages (as otherwise they could not be valid for every instant). As a consequence they are also valid independently for the in-phase and the quadrature components of such voltage and current waves with respect to any arbitrary common reference phase.

Non-sinusoidal currents or voltages can always be resolved into their sinusoidal components; since such a component of one frequency cannot annihilate a component of a different frequency it follows that Kirchhoff's Laws must be valid for each individual frequency. More generally: if O denotes a linear operator, i.e. an operator for which holds $O(a) + O(b) \equiv O(a + b)$, then

$$\Sigma O[i(t)] = O[\Sigma i(t)] = O[0] = 0$$

i.e. Kirchhoff's Laws quite generally are valid for all those mathematical entities which result from the application of such an operator to e or i. Examples for such operators are:

(1) The differential operator $D = \dfrac{\mathrm{d}}{\mathrm{d}t}$. Thus Kirchhoff's Laws are valid for the first or higher time derivatives of $e(t)$ and $i(t)$.

(2) Integral operators of the form $\int_a^b \mathrm{d}t$. Thus the instantaneous values of e and i in eqs. (3) and (4) can be replaced by their arithmetic mean values (but not by their r.m.s. values).

(3) Integral transforms like Laplace or Fourier Transforms. Thus if

$$\int_0^\infty i_k(t)\, e^{-pt}\, \mathrm{d}t = L_k(p),$$

eq. (3) can be rewritten as $\sum L_k(p) = 0$.

Kirchhoff has shown that these laws yield in a network with b branches exactly b equations which together with b linear performance equations (of type 1 or 2) define completely the b currents and the b voltages of the network. In particular if the number of nodes and independent meshes in the network are n and m there will be

(i) exactly $n - 1$ independent equations of type (3) and

(ii) $m = b - (n - 1)$ independent equations of type (4).

To prove (i) we write down the n equations of type (3); each current appears here twice, with opposite signs. The sum total of all these equations is therefore identically zero, showing that at least one of these equations was a consequence of the others. This argument cannot be applied to any group of these equations containing less than the complete number of nodes, for in every such group there is at least one node with one or more currents whose second appearance is in the excluded group of nodes. The equations for these "unbalanced" nodes cannot be obtained by combining the other members of the incomplete group.

The following proof of (ii) is written so as to serve as an introduction to other basic network concepts: we remove from the network a sufficient number of branches so as to leave the minimum number b_t just sufficient to link all the n nodes. Such a minimum configuration (which usually can be chosen in several ways) is called a "tree" and can obviously contain no meshes. The b_c branches that were removed are called "chords". Each chord together with the tree (without any other chords) defines a unique one-chord mesh (as the two nodes joined by the chord must have been linked once—and once only—by the tree).

Building up such a tree from its individual branches it can be seen that for n nodes there must always be $b_t = n - 1$ branches in the tree and therefore the number m of meshes associated with the b_c chords (the so-called "chord loops") must be

$$m = b_c = b - b_t = b - (n - 1).$$

As each of these chord loops contains a chord branch and therefore a chord voltage which does not occur in any other chord loop, none of the equations of type (4) based on these chord loops can be obtained by a combination of the others, i.e. these equations form a set of independent constraints. However, in a complex network it is usually possible to trace loops additional to these chord loops and one may ask if the associated equations of type (4) are independent of those already derived.

It can be shown that this is not the case. For brevity we introduce the self-explanatory terms "tree path" and "chord path" to classify the alternative ways in which we can go from a point, say E, to another point, say F, on a loop. Once the tree of a network has been defined, there can be for such a given pair of points only one "tree path" between these points. The existence of a second path would violate the definition of a tree. There may be, however, several "chord paths" to choose from and we may speak of chord path "a", "b" and so on according to the designation of the chord involved. Now, additional loops of the kind mentioned in the previous paragraph can only be traced if it is possible to discard, say, the tree path between points E and F of a chord loop and replace it by a chord path, say "b" so that we have now a loop consisting of two chord paths. But this kind of loop can be traced by successively tracing chord path "a", tree path EF, tree path FE and chord path "b". This means that we have traced successively the complete loops "a" and "b" and the voltage sum around the new loop can therefore be obtained by adding the equations of type (4) for the loops "a" and "b", the partial sum associated with the tree path EF cancelling in the result. Thus the new loop does not yield an independent constraint. Additional loops consisting of more than two chord loops can be dealt with by a repetition of this type of argument.

In toto there exist then $b_t + b_c = (n - 1) + b - (n - 1) = b$ equations based on the topological structure of the network which together with the b equations of performance [of type (1) or (2)] are just sufficient for the determination of the b voltages and b currents in the b branches. There are two well-known and systematic ways of solving these two b equations. In the first method all branch currents are expressed as sums or differences of b_c independent mesh currents. A mesh current is a current flowing in a closed loop or mesh superposed on any other currents that may be using the same branches. This concept as well as the method of solution arises naturally if we first consider only the tree of the network. No current can flow in the tree by itself. Currents can flow only after we have added the chords one by one and we can visualize in each chord a current generator controlling the (mesh) current in the associated chord loop. The total current distribution of the network can evidently be conceived as arising from the superposition of these b_c independent current systems.

Now, as any mesh current flowing in a closed loop must necessarily leave each node which it enters, all equations of type (3) are identically fulfilled by our system of b_c auxiliary unknowns and we need therefore solve only the remaining b_c voltage equations of type (4). For this purpose we express all voltages in terms of the unknown (mesh) currents by the aid of equations (1). Solution of the voltage equations yields then the magnitude of the mesh currents from which the branch currents are easily found.

In a simple network it is usually possible to determine by inspection an independent set of meshes required for this type of solution. In complicated cases it is advisable to proceed systematically, e.g. by selecting a mesh, demolishing it by taking away one branch only and continuing this process until only a tree is left. The branches removed must have been the chords of this tree.

The nodal method of solution is the dual counterpart of the method just outlined. Here we ascribe to each node of the network a potential. All branch voltages can then be expressed in terms of differences of these potentials. As one of the nodes can be used as reference point for these potentials this amounts to the introduction of $n-1$ new unknowns in place of the original b branch voltages. Now, when we add all the potential steps around a closed loop we arrive necessarily at the starting potential, i.e. the sum of all these potential steps is identically zero and thus the new unknowns fulfil identically all equations of type (4). Only the $n-1$ equations of type (3) remain. If we express in these all currents in terms of branch voltages, i.e. potential differences, we find from their solution the $n-1$ potentials.

It depends on the nature of the network which of these two methods leads to a smaller system of simultaneous equations. Attention might be drawn to a imitation of the nodal method. Potential differences must by definition be independent of the path of measurements. This means that in cases where time-varying fluxes are linked with meshes of the network (instead of individual components) it will not be possible to allocate potentials to the nodes of such meshes. As an example let us consider a transformer core with an encircling loop consisting of node A, a resistor of 3 Ohms leading to node B and another resistor of 5 Ohms leading back to node A. If the "turn voltage" of the transformer is, say, 800 mV we have in this loop a circulating current of 100 mA and along these two resistors voltage drops of 300 mV and 500 mV respectively. If the transformer has negligible magnetic leakage the space around the core is free of magnetic fields and thus if we place a voltmeter along the 3 Ohms resistor and connect it to nodes A and B it will indicate 300 mV independent of the positioning of its leads (provided we do not encircle the core). Another and identical instrument placed alongside the 5 Ohm resistor and connected to the same pair of nodes will indicate 500 mV, thus demonstrating that in this case it is meaningless to speak of the potential of B relative to A.

Bibliography

GUILLEMIN E. A. (1953) *Introductory Circuit Theory*, New York: Wiley.

HELMHOLTZ H. (1851) *Poggendorff's Annalen der Physik und Chemie*, **88**, 505; *Coll. Papers*, **1**, 429.

KIRCHHOFF G. R. (1845) *Poggendorff's Annalen der Physik und Chemie*, **64**, 513.

KIRCHHOFF G. R. (1847) Ibid., **72**, 487.

LEY, LUTZ and REHBERG (1959) *Linear Circuit Analysis*, New York: McGraw-Hill.

MAXWELL J. C. (1892) *Treatise* (3rd Edn.), Vol. 1, Art. 280, Oxford.

A. BLOCH

L

LASER-INDUCED SPARKS

Introduction

The production of sparks is one of the more spectacular "parlour tricks" in the laser's repertoire. Since it was first demonstrated in air, in 1962, electrical breakdown by pulsed, focused laser beams has been produced in a wide variety of gases and some solid dielectrics (Raizer, 1965). The main interest, to date, of this phenomenon has been its relevance to fundamental studies of the interaction between intense radiation and matter; more recently, however, the possibility of certain practical applications (e.g. to switchgear) has begun to emerge.

An experimental arrangement for studying laser sparks is illustrated schematically in Fig. 1. Typically

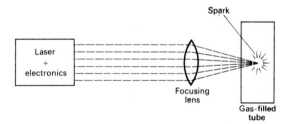

Fig. 1. Experimental arrangement for producing a laser spark.

the Q-switched ruby laser delivers around 100 MW for 20 nsec into a focal volume which may be anything from 10^{-6} down to 10^{-9} cm^3. Actually the energy concentration in the gas, although high, is probably not quite so extreme as suggested by these figures. A likely sequence of events would be that gas in the focal volume becomes completely ionized during the first nanosecond or so and the resultant, rapidly expanding, plasma absorbs the laser beam to a considerable degree.

Several characteristic features of the laser-generated air spark are apparent to the naked eye: it is an extremely intense, bluewhite source of light radiating more or less uniformly in all directions. However, if viewed along the axis, looking back towards the laser, the spark appears quite red, indicating appreciable forward scatter of the laser light.

More detail may be obtained from high-speed photographs: as shown in Fig. 2 the spark is divided into a number of distinct regions. At the centre is an intensely luminous core in the form of a narrow cylinder coaxial with the focusing lens. Surrounding this core is a region of lesser intensity, normally divided into striations or lobes. The shape and number of these lobes depend on the available laser power

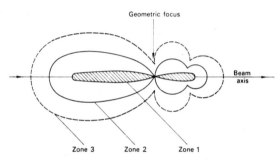

Fig. 2. Schematic representation of the appearance of the laser air spark, showing the three zones described in the text.

and the optical arrangement, but for any given experimental set-up they are essentially reproducible from spark to spark. Finally, surrounding the second region is a structureless cloud consisting probably of excited atoms rather than ions and electrons.

The main quantitative properties of a typical laser spark are summarized in Table 1 below.

Table 1. Properties of Typical Laser Spark

Energy	~1 joule
Initial volume	$10^{-6} - 10^{-9}$ cm^3
Initial temperature	~10^5 °K
Initial expansion speed	$10^6 - 10^7$ cm/sec
Total number of ions	~10^{13}

Theory and Fundamental Work

The term "spark" has been applied to the laser-induced breakdown of gases because the phenomenon is similar, in its more obvious features, to that occurring in a conventional spark-gap: ionization and intense light are produced to the accompaniment of a sharp sound generated by an expanding blast wave. Nevertheless, the above-described, detailed structure of the laser spark is quite different from that of a conventional one, and this probably arises from the fundamentally different process by which the laser spark develops.

Conventional spark discharges in air and other gases are initiated by the acceleration, in a strong electric field, of the random electrons originally present, up to energies at which they can ionize neutral atoms by collision. The electrons resulting from such ionization are accelerated in their turn, reinforcing the original ones so that a Townsend cascade (ionizing avalanche) develops. In air the process is usually facilitated by the presence of a few impurity atoms having low ionization potentials.

The equilibrium density of free electrons in air under normal conditions is of the order of 1000 per cm^3 and their production rate is around 10 per cm^3 per second, so that there is no shortage of free electrons to initiate the ionizing cascade in the volume of air between the electrodes of a typical spark gap. On the other hand, as we have seen, laser sparks can be initiated in volumes as small as 10^{-9} cm^3 by pulses lasting a few nanoseconds. The chance that a free electron occurs naturally in the focal volume at the time of the pulse is therefore negligible. (The presence of a random electron in a volume of 10^{-9} cm^3 would imply a mean charge density comparable with that observed in tenuous glow discharges!) Consequently an alternative mechanism must be invoked for the triggering of laser sparks.

Given an initial priming of one or more free electrons, these could be rapidly accelerated by the intense, albeit transient, electromagnetic field of the laser radiation and, while the pulse lasts, an ionizing avalanche may then develop. Even so, diffusion and recombination losses would still prevent the ionization level from reaching breakdown point unless the focal volume contained at least around 10^{18} atoms; and in any case the origin of the first free electron still has to be explained.

Evidently we must seek some mechanism akin to the photo-electric effect by which light interacts directly with an atom to eject one or more of its orbital electrons. Gases in which laser sparks have been observed do not exhibit a photo-electric effect with ordinary (visible or ultra-violet) light because their electrons are too tightly bound to be ejected by the quantum energy available in the radiation. An extreme example is helium whose ionization energy (24·58 eV) is over 13 times the energy of a photon of ruby laser light (1·78 eV). Other gases in which laser sparks have been produced include neon (ionization energy 21·6 eV), argon (ionization energy 15·8 eV) and caesium vapour (ionization energy 3·9 eV).

If an atom is already in an excited state it may—depending on its energy level scheme—proceed to higher excited states and, ultimately, ionization by successive absorption of single photons. But in none of the above examples could even the first excited state be reached from the ground state by the absorption of a single ruby laser photon; and the energy level scheme of helium in particular is such that no intermediate excitation is possible, so ionization must occur direct from the ground state by simultaneous absorption of at least 14 photons.

"Simultaneous" in this context means within a characteristic time interval τ, equal to ν^{-1}, where ν is the radiation frequency. This result is a consequence of the uncertainty principle, according to which the interaction time between an electron and a photon is uncertain by an amount τ. For ruby laser light τ has the numerical value $\sim 10^{-15}$ sec. Photons whose arrival times at a particular atom are separated by less than this interval arrive simultaneously as far as that atom is concerned, and it absorbs them concurrently rather than consecutively. This is known as multi-photon absorption. (Multi-photon emission is also possible in special circumstances: thus an atomic transition between a higher and a lower energy state might be accompanied by the emission of several photons each carrying part of the transition energy instead of, as is more usual, a single photon carrying all the energy.)

For typical atomic cross-sections the arrival of several photons within the time interval τ implies a light flux vastly in excess of what could be achieved before the advent of the laser—which explains why this type of breakdown had not hitherto been observed. For example, for a helium atom to become ionized by absorbing 14 photons within 10^{-15} sec the instantaneous photon flux at the atom must be nearly 10^{32} photons cm^{-2} sec^{-1}. Actually, because of large fluctuations in the radiation intensity, an instantaneous flux of this value might correspond to an average flux, over the whole pulse duration, of only about 10^{31} photons cm^{-2} sec^{-1}; but even this is many orders of magnitude above the highest intensity obtainable with conventional optical sources.

Fluctuations in the laser beam intensity are both spatial and temporal: this means that the local "instantaneous" flux (i.e. taken over time τ and the volume occupied by a single atom) can be very much greater or less than the space-time average over the whole focal region, obtained from experiment. Any theory of multi-photon ionization must obviously take account of these statistical fluctuations which may carry the instantaneous photon flux at a single atom across the necessary threshold value. Because of the coherence of laser radiation the fluctuations in photon arrivals at an atom are not wholly random, but are significantly correlated; and a realistic statistical model should allow for these correlations. Thus a simple Poisson distribution law, which describes only random events, cannot predict the photo fluctuations with any accuracy and a more sophisticated, two-parameter, distribution is needed to describe the correlations. It has been found that the Polya (negative binomial) distribution meets the case (Gardner, 1966 and 1967).

Besides giving information about the laser radiation itself, the study of laser sparks can sometimes help to elucidate atomic energy level schemes. As already indicated, in most of the observed cases of laser-induced breakdown the atom is known to be excited directly from the ground state to the ionized state. Occasionally, however, it can proceed via an intermediate (resonance) level. Such a situation obtains in the case of caesium which has a resonance level at 3·56 eV (twice the ruby photon energy) above the ground state. Since a caesium atom needs to absorb at least three ruby photons to carry it across the ionization threshold of 3·9 eV, this can occur either directly, or via the resonance level with a single photon absorption following the initial absorption of two photons (Fig. 3). The relative importance of these two competing processes can be assessed from studies of laser-induced breakdown in caesium vapour at various pressures.

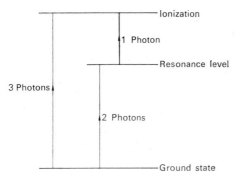

Fig. 3. Alternative routes to ionization for a caesium atom.

Applications

For experimental research into breakdown phenomena the laser spark offers a unique and valuable means of producing a well-metered quantity of ionization at a point in a gas, without the complication of leads or electrodes which may distort the local electric field. For example, it is already known that, with non-uniform field geometries, pre-breakdown corona pulses can appreciably influence the eventual breakdown of the gap; much more might be learnt if similar bursts of electrons could be produced at selected points in the gas, rather than just at the electrode surfaces.

Arising logically from these considerations comes the suggestion that laser sparks might be exploited to initiate breakdown in triggered spark-gaps such as are used in high-speed switchgear where moving metal contacts, because of their inertia, would be too slow-acting. An important advantage of using an optical device such as a laser for this purpose would be that the triggering electronics could be completely isolated from the switch. Other optical triggering devices, involving photocells and conventional light sources, would also achieve this end, but where the operating voltage must be held accurately to a predetermined level the laser would appear to have the advantage. In many cases the choice must doubtless be an economic one: a pulsed laser with its associated electronics could cost upwards of £2000 and the crystal, costing say £500, may well need replacing after 100 or so shots. Nevertheless, as part of a seldom used protective system, a 100-MW laser could be arranged to trigger as many as 20 switches, thereby replacing 20 pulse transformers and ancillary equipment costing much more than the laser installation.

An exciting potential application of laser sparks, currently being explored, is to the production of controlled thermonuclear reactions by heating a gas such as deuterium up to temperatures around 10^8°K. To date most attempts at achieving such reactions have relied on discharging the stored electrical energy of a charged condenser bank into a small volume of gas in a very short time. In terms of the specific power dissipation per unit volume this method is several orders of magnitude inferior to employing a focused laser beam, the bulk of whose energy is absorbed within a volume of, say, 10^{-6} cm^3 around the focus. Unfortunately the very smallness of this volume means that conduction losses are severe and there is little hope of reaching thermonuclear temperatures in the plasma resulting from a single laser spark. A more promising line of attack is to exploit the collision of shock waves from two or more contiguous, simultaneous sparks, and techniques are being developed for this. By employing a weakly focused laser beam of sufficiently high power and short pulse duration (~15 joules for 18 nsec) it is possible to produce over 100 closely grouped, virtually simultaneous, sparks whose shock waves generate cumulative high-temperature effects (Basov et al., 1967).

Less spectacular, but probably more feasible at the present time, is the application of laser sparks to the measurement of gas velocity. When a spark is created in a moving gas stream its centre moves with the local gas velocity. However, the luminosity has sufficient duration and intensity for the spark velocity to be recorded by standard streak camera techniques, and this method has been employed to measure the velocities of both ionized and unionized gas flows. Since velocities can be resolved over distances as short as 5 mm it should also be possible to study boundary layer and turbulence effects by this technique. The method is inherently more suited to the investigation of supersonic flows as the spark, or rather its blast wave, expands with a velocity equal to or just greater than the local sound velocity. Even so, for some of the more demanding, slightly sub-sonic applications, e.g. observing flow conditions in the duct of a magnetohydrodynamic (MHD) generator, laser sparks may provide the only feasible approach. Accuracy could still be preserved in such a case, provided that the progress of the blast wave were recorded (e.g. by Schlieren or interferometric techniques) rather than that of the somewhat diffuse spark. For the MHD application this refinement would offer the quite considerable advantage of supplying a temperature measurement (from the blast wave velocity) as well as the flow velocity.

None of the above applications fully exploits what is perhaps the most obvious property of the laser spark, namely its intense brilliance. Although only a small fraction of the incident laser light is converted to visible and ultra-violet emission from the plasma, this emission is extremely intense, especially in the vicinity of the sharp line spectra, and should itself prove useful as a light source. Its spectral radiance exceeds that of the flashlamp which pumps the laser. The air spark in particular should prove especially useful in experiments where it is desired to mix visible or ultra-violet light with the Q-switched laser radiation, since synchronization is automatically secured.

Future Work

While laser sparks will no doubt continue to be used to investigate breakdown phenomena in gases (including organic and metal vapours) there will probably be a growing emphasis on the production and study of sparks in condensed media. Already it has been found that laser-induced damage in glasses shows many analogies with laser-induced breakdown in gases: for example ionization by multi-photon absorption appears to be a primary triggering mechanism in both cases. Clearly the method will also lend itself to the study of breakdown mechanisms in other (solid and liquid) dielectrics.

Laser-spark excitation of homogeneous powdered materials is another developing technique. The laser microprobe, as this application is called, provides *in situ* spectrochemical analysis of microsamples, such as inclusions and surface deposits (Whitehead *et al.*, 1968).

At the fundamental level, further study of laser sparks should extend and consolidate our information about the nature of laser radiation itself, particularly in regard to the photon correlations and the extent to which spatial coherence is preserved during the entire pulse time.

Acknowledgements

Figures 1 and 3 are reproduced by courtesy of the Editor of *The New Scientist*. The present article is published by kind permission of the English Electric Company Limited.

Bibliography

BASOV N. G. *et al.* (1967) *Soviet Physics-Doklady* **12,** 248.
DER AGOBIAN R. (1969) Laser cascades, in *Encyclopaedic Dictionary of Physics* (J. Thewlis Ed.), Suppl. vol. 3, 183, Oxford: Pergamon Press.
GARDNER J. W. (1966) *Int. J. Electronics* **21,** 479.
GARDNER J. W. (1967) *Ibid.* **22,** 123.
MAGYAR G. (1968) *Nature* **218,** 16.
RAIZER YU. P. (1965) *Usp. Fiz. Nauk.* **87,** 29.
TROUP G. J. (1967) Laser, in *Encyclopaedic Dictionary of Physics* (J. Thewlis Ed.), Suppl. vol. 2, 141, Oxford: Pergamon Press.
WHITEHEAD A. B. and HEADY H. H. (1968) *Applied Spectroscopy* **22,** 7.

J. W. GARDNER

LASERS, APPLICATIONS OF. The main properties of lasers that make them desirable, in principle, for applications are that laser radiation can be collimated and focused to a degree limited only by diffraction, it can be made relatively monochromatic and can therefore be modulated and it is spatially coherent so that different parts of the wavefront can produce interference over relatively large distances. The techniques of Q-switching and pulsed-transmission-made operation can produce high-power pulses of extremely short duration.

Since literally hundreds of applications of lasers have been applied in the laboratory, and many have evolved and are evolving as practical systems, this description of laser applications lists at least *five basic ways* of applying laser radiation and cites illustrative examples. Some apparent future applications of major consequence are also outlined. The fuller meaning of each of the subtitles below is detailed in the text and the particular virtue lasers have in this respect is then given.

1. *Energy concentration.* To project a high concentration of energy into a small area to weld, melt, machine, perforate; to induce specific chemical, biological, medical or photochemical action; to illuminate; to provide a cogent military weapon.

2. *Spectral performance.* To determine spectra of materials, including absorption and transmission as a function of wavelength, or of time; to use the wavelength and interference properties for colour photography, micrography and holography. (Of particular importance in such spectral performance as in holography are the spatial and temporal coherences. Also as compared with incoherent thermal light there is the different probability distribution of the field and higher order coherence function.)

3. *Distance, direction and rates of change.* To determine distance, velocity and direction for ranging and surveying, gravity measurements, laser radar and underwater laser ranging; to apply lasers to the science of metrology for precision measurements of large structures; lens studies and lens production; for navigation, inertial guidance, ring lasers; to make velocity measurements.

4. *Communications.* To transmit meaningful signals for communication such as sound, picture or data between different locations.

5. *Computer or data processing.* To actually do computer or data processing; to perform linear and non-

linear operations and make possible such devices including harmonic generation and parametric effects in solids; to display information in large or data-type displays for producing meaningful signals.

6. *Some apparent future applications of major consequence.*

1. Energy Concentration

Although higher energy rates will be reported in the future present Q-switched lasers in the U.S.S.R. and France produce 20–50 GW pulses of about 5 nsec duration using a high-quality neodymium glass. Powers in the hundreds of gigawatts and picosecond pulse duration range are also being reported. One purpose of such high power lasers would be to study gas heating of focused radiation to produce quasi-thermonuclear reactions in microscopic volumes. Detectable numbers of neutrons can be produced in such a manner. Continuous wave gas or chemical lasers can readily operate in the range of several kilowatts and considerably higher.

Typical applications of high power densities include machining, welding, soldering, drilling, interaction with materials (e.g. electric propulsion), dynamic balancing, chemical and photochemical applications including photopolymerization, biomedical applications (e.g. ophthalmology, micro-irradiation, cancer and tumour research, experimental dentistry, dermatology) and weapons (anti-personnel, anti-vehicle and anti-missile).

For a continuous laser beam of radius a and power W impinging upon the surface of a material thick in comparison with the laser beam radius and its lateral extent equally large, the equilibrium temperature rise at the centre of the circular area is approximately $T = \dfrac{W}{\pi \varkappa a}$, where $\varkappa$ is the thermal conductivity. For a pulsed beam, the transient conditions for short time durations t, of the pulse onto the surface of a material of specific heat c and density ϱ, the surface temperature for a thick material may be approximated as $T = \dfrac{2W}{\pi a^2} \times \left(\dfrac{t}{\pi \varrho c \varkappa}\right)^{\frac{1}{2}}$. The following advantages may be cited for the pulsed laser for hole punching:

(a) No physical contact between material and punching tool eliminating drill breakage and wear. Sometimes simpler jigging and fixturing of article to be drilled.
(b) Possibility of having article moving while drilling process is taking place due to short duration of laser drilling process.
(c) Minimized thermal damage to adjacent regions due to short time duration limiting thermal conduction.
(d) Common use of the locating optical system to focus the laser radiation as well.
(e) Since the laser beam may be focused to a small spot, very small diameter holes may be drilled.

For pulsed laser welding the following advantages may be cited:
(a) No physical contact eliminates need for certain maintenance, such as correction of electrode wear. Article transport and jigging or fixturing may be simple.
(b) Welding may take place in air, vacuum or a controlled atmosphere. The beam may even enter through a transparent window.
(c) Common use of the locating optical system to focus the laser radiation for welding.
(d) Since the laser beam may be focused to a small spot, very small welds can be made.
(e) Welds may be in regions not readily accessible as compared with conventional welding equipment.

Lasers are supplementing other techniques in surgical operations on the eye and in the repairing of detached retinas. In the laser photocoagulation technique, scar tissue can be caused to form at several points along a detached retina to establish once again close contact with the choroid so that nutrients may be obtained therefrom. The laser beam is directed through the eye to the retina. Pulsed glass or ruby lasers have been used for this purpose.

Cancer research is being studied using laser irradiation. In certain cases large malignant tumours have apparently disappeared entirely. The more usual case is the apparent regression of small tumours. Many processes take place during laser irradiation including chemical and heating reactions that may impede tumour growth. Considerable research is in progress. Some exceptionally fine work has been done in certain early experiments by Dr. Paul E. McGuff and associates at the Tufts New England Medical Center Hospital on malignant and nonmalignant tumours, and by Dr. Leon Goldman and associates at the University of Cincinnati (see Bibliography). These are examples only as the relative merits of the considerable research now extant belong to the judgement of medical science.

2. Spectral Performance

The field of holography depends on optical wavelength interference patterns. Such patterns are readily produced or manipulated by lasers. In laser holography rapid events in three dimensions have been recorded which have then been examined later in a more leisurely manner. For example, individual particles in a mist have been studied by this technique. Holograms have been used as three-dimensional computer outputs, training devices for three-dimensional instruction, mask production for exposing photoresists in the manufacture of microcircuits, and strain measurement of machine parts. The argon laser has been particularly effective due to its high power coherent output as compared with the helium–neon laser. (There has also been a fair amount of work in trying to use the ruby laser for holography.)

Some people claim that the first holograms readily defined as such were made by Gabriel Lippmann in 1891, and together with his later work, formed the basis for his getting the Nobel Prize in 1908. These were holograms of mirror surfaces. (It is also a process used in some of the current three-dimensional colour holograms.) Since 1948 Dennis Gabor has made substantial contributions to holography.

Holograms may also be made in ordinary light using a Lippmann concept described in 1908 using a "fly's eye lens". Holograms using a fly's eye lens technique have been made by Dr. Robert V. Pole of I.B.M. as reported in 1967. The fly's eye lens has also been used to make Fourier transform holograms as suggested by W. L. Bragg in 1939 and used to produce recognizable images of a crystal structure (diopside) in this way (Bragg, 1939). Although not widely used, it started to receive acceptance particularly after the work of Huggins' Fourier transform holograms in 1944. Bragg's hologram device was exhibited at an X-ray conference in Cambridge, England, where C. W. Bunn coined the name "fly's eye". Dr. Dennis Gabor, who also coined the name "hologram", points out that his work was inspired by the work of W. L. Bragg. Gabor's work was refined to an extent by E. Leith and J. Upatnieks to postulate a recording emulsion as a two-dimensional model. Apparently holograms recorded as thick emulsions and regarded as three-dimensional media, and the first attempts to record three-dimensional media holograms (neglecting Lippmann's and Bragg's work) were by Yu. N. Denisyuk of the U.S.S.R.

Fatigue, strain and vibration may also be measured using laser holography. For example, several laboratories have measured small deformations of three-dimensional objects of arbitrary shape. This can be achieved by first recording a hologram of the unstrained object and then viewing the object through the hologram reconstruction. Interference fringes are produced by light reflected from the object and the reconstructed hologram image. The interference fringes contour any differences. The deformation can be measured by locating two points on the object and counting the number of fringes in a plane lateral to the line of sight to each point. The deformation can then be trigonometrically computed.

If it is desired to preserve the partially polarized light from an object, a vectorial hologram can be recorded. Such a hologram can reconstruct by using light of two independent states of polarization to illuminate the object and deriving two reference beams which have different inclinations for each of the polarization states. Each polarized reference beam can interfere only with light of its polarization from the object beam. By illuminating the hologram with the two polarized reference beams, the reconstructed image will contain all the object detail as well as the polarized light from the object.

The principle of wavefront reconstruction can be applied to the formation of visual pictures of acoustical waves as shown by A. F. Metherell, H. M. A. El-Sum, R. K. Mueller, H. A. Elion and others, and further described in the proceedings of the First International Symposium on Acoustical Holography. Such systems can potentially achieve better results than conventional acoustic lens systems (particularly effective in regions of turbulence and turbidity). Such a system can investigate sound fields and even render the sound field in three-dimensional colour. (Thus three holograms of a sound field at three sonic frequencies can reconstruct the sound field in colour by using a light wavelength proportional to the acoustical wavelength originally used.)

One should carefully differentiate between holography and spatial filtering. When an object is coherently illuminated as with laser light, every region in the exit pupil has a region in the corresponding two-dimensional frequency spectrum. Blocking or selectively passing light in this region eliminates or selectively passes the corresponding frequency component from the image. This is referred to as spatial filtering. By using a spatial filtering technique in reverse, one can generate a complete image from a uniform wavefront. This may be done by introducing a hologram into the exit pupil. The hologram converts the wavefront into one that corresponds to the desired image.

Other non-holographic vibration and strain measuring laser systems include the modulation of laser beams by microwave subcarrier amplitude modulation (and measurement of phase shift on the demodulated beam), phase shift measurement by coherent optical detection, measurement of the angular position or interference in the reflected light, and interference mapping to obtain gradients to a normal displacement (similar to holography).

Among the conventional applications of lasers is high-speed photography. Micrography is enhanced by lasers since the illumination permits the highest magnifications at very short exposures, and elimination of chromatic defects due to the monochromatic nature of the light.

There are at least two kinds of Raman scattering known to date. The classical or spontaneous Raman scattering was discovered in 1928. The stimulated Raman scattering, discovered in 1962, has light scattered in the forward direction and has frequencies that relate to the classical effect. The intense monochromatic light of lasers permit Raman spectroscopy to be carried out for solids, liquids and gases. The advent of the Raman spectroscopy of solids occurred in 1967 and is due primarily to the power densities available with the laser.

3. *Distance, Direction and Rates of Change*

In ranging, surveying and tracking, the laser is coming into wide use. The intensity and directionality of the radiation make it particularly suitable for ranging and long-distance measurement. For distance measurement, the laser light is transmitted and then received in one of three basic systems:

(1) The light pulse is timed over the distance traversed and the distance computed.

(2) The laser beam is modulated with a high-frequency radio signal and the phase-modulated transmitted wave is compared with the modulated received wave and the distance computed.

(3) Use is made of the relatively high coherence of a laser and an interference pattern is produced making possible distance measurements in terms of a number of wavelengths.

Continuous wave laser altimeters have even been used for high resolution ocean measurements. Thus, ocean wave formation studies are possible as well as measuring and forecasting sea spectra for offshore petroleum operations, and sea-state determinations in connection with rocket and space vehicle recovery. Many laser ranging systems are in operation, both land, airborne and spaceborne. Underwater ranging is under investigation, particularly in the blue-green region of the light spectrum. For airborne surveillance, laser line-scanning devices are reported. Other applications include laser geodesy.

Lasers may be used to introduce several types of sound vibrations in materials. For example, the laser light may be used to induce coherent lattice vibrations in crystals (called stimulated Brillouin scattering). In stimulated Brillouin scattering, the acoustic vibration may be of the order of 3×10^4 MHz and the lattice vibration amplified as it travels through the crystal with the incident laser light beam. A scattered light beam is frequency shifted, such that the optical frequency is equal to the sum of the acoustic frequency plus the shifted frequency. High acoustic intensities are possible and may be made to occur in liquids as well. Lasers may also be used to generate sonic or ultrasonic waves in a material for identification or location of flaws or fatigue.

In metrology, interferometric laser systems are in general use for linear measurement standards. Lasers are also in use for optical tooling, particularly in the aircraft industry. They may also be used for digital angular readout using laser interferometry.

One of the principal uses of the ring laser is for inertial sensing. In the ring laser the radiation travels around a closed path formed by three or more reflectors. The frequency of oscillation of this optical oscillator is affected by the rotation of the ring. The angular velocity can be measured over a dynamic range exceeding 100,000 to 1. The output of the laser is a pulse train generated externally. This pulse train has a repetition frequency that is directly proportional to the angular rate input. The ring laser is unresponsive to linear velocity or acceleration, yet used in a special way, it can measure fluid flow rates. This makes it quite unlike other inertial sensors.

If we have an inertial frame of reference, then a body in this space, or reference line fixed in the body, rotates if this reference line changes pointing direction. In celestial navigation, the fixed stars are an inertial frame of reference, and define changes in direction or rotation. Instruments for sensing inertial rotation in a closed environment include the mechanical gyroscope and now the laser gyroscope (ring laser).

In the ring laser, light is circulated in opposite directions around the same closed path and the waves undergo nonreciprocal phase shifts when the path is rotated (or when a dielectric moves along the path). By using a laser, optical frequency differences are produced by the phase shifts between the two waves. By heterodyning the two frequencies at a common photodetector, a heat frequency is obtained that is proportional to the rate of rotation (or linear velocity of the dielectric fluid).

Applications of the ring laser include its use on radar tracking pedestals for azimuth and elevation. It may also be used for altitude reference systems for target designation and fire control in aircraft. By using three rings in a rigid orthogonal configuration, the angular rates of a body may be monitored continuously and altitude data relative to a pre-selected coordinate system may be obtained for navigation. By optically contacting a fluid, flow may be measured over a great dynamic range with high resolution. This would be of particular use for corrosive liquids and gases. On the other hand, if the velocity of a gas is known, then the density may be determined.

Laser seismometers of a wide variety are under development. One type uses a laser interferometer. Variations in the phase of two beams indicate earth movements to less than a millionth of a centimetre. Another type uses an optical corner reflector array located a great distance (many miles) from the laser transmitter. The optical path difference is accurately measured and used to determine earth strain in fault areas.

Another method of measuring velocity of an object is a straightforward use of the Doppler shift method. This may also be applied to measuring localized velocities in gases and liquids. By measuring three noncoplanar components of the velocity vector simultaneously, the magnitude and direction of the velocity vector can be measured in a given flow field. The velocity is measured by detecting the Doppler shift in the monochromatic light and detected by optical heterodyne techniques. Measurements indicate gas flow velocity ranges from 1 cm/sec to over 200 m/sec. Liquid flow velocities have been made that include 1 to 80 cm/sec.

4. Communications

The communications industry is expected to be a principal beneficiary of the coherence and wide-bandwidth characteristics of laser beams. Optical frequencies may exceed 10^{14} Hz with the result that the available information bandwidth is large as compared with microwave systems which may have carrier frequencies of the order of 10^{10} Hz. The result could be the transmission of many more messages over a single carrier network. The realization of this depends upon further improvement of modulation, amplification, detection, and guiding of light. Closed circuit laser transmission of television signals has also been demonstrated.

Like the microwave system, the laser has many limitations that must be overcome for mass communications. Direct, open use of unshielded laser beams means that the atmosphere could have a severe effect for earth ground-to-ground communications. In outer space, without this type of long distance atmospheric consideration, the use of lasers for inter-space communications is highly attractive. Communication over short distances, if shielded, are of course particularly promising. Guided optical conductors of various types are under development. One such means of guided conductors is the fibre optics bundle. (Single fibres may be a kilometre in length.) Fibre amplifiers of extremely low noise also improve the possibilities of commercialization on a large scale.

5. Computer or Data Processing

The capability of electro-optically controlling the output of lasers are critical to two other potentially large applications: data processing and large-display presentation. Semiconductor lasers would be particularly appropriate for use in computer circuits because of their small size. Optical data processing can provide extremely high switching speeds and access rates, and three-dimensional optical information storage can permit a high data storage capability. Development of a wide variety of optical memory surfaces is under way and some include photoconductive or photochromic materials. There are a wide variety of such surfaces and an early example will be cited as manganese bismuth which is a ferromagnetic material. For example, it has been estimated that based upon present measurements, such a thin film memory could reach packing densities of at least 10^6 bits/cm^2 and a readout rate of 10^8 bits/sec. The basic process is simple and involves altering the state of magnetization of MnBi by raising its temperature above the Curie temperature. This can be done by using a focused switchable laser beam and applying a magnetic field of suitable orientation during the subsequent cooling below the Curie point. Real-time laser data processing (or nearly real-time processing), already accomplished in the laboratory, opens the way to real-time filtering, computing and presentation of data.

Optical digital information is readily available as light or no light, and black and transparent areas. Such information may be used for symbolic information as graphic sketches, line drawings, circuit diagrams, patent illustrations, engineering tracings, etc. In pattern recognition or photographs, one encounters a gray scale and less sharply defined but recognizable forms.

Some fundamental needs for data processing may be found in reducing the dull, and therefore often error producing, human labour of scanning files, interpreting photographs, scanning indices, abstracts, records, etc. Symbolic data may readily be transformed to digital data and back to symbolic data. Optics may also provide high computer input and output speeds as well as generating complex types of outputs. Optics is integrally tied in with magnetics and electronics both fundamentally (e.g. non-linear devices) and practically as merely another portion of electromagnetic spectrum, and meeting the requirements of high-speed recordings, mass storage, mass search and processing. One method of overcoming the volume per bit obstacle is the use of the fibre or the semiconductor laser devices, another is to use three-dimensional optical storage materials. The Fourier or other transforms permit ready conversion of optical information from digital to analogue form or vice versa for optical computing

As indicated above, temporary and permanent optical storage is available and rapidly improving for optical data processing. Ease of packing and high density of information on an area or volume basis make optics excellent for permanent storage. Temporary storage is a developing field. The high intensity of lasers relieves the high photosensitivity requirement, with high resolution, for many applications. Laser beam switching is available and improving for deflecting a laser beam about precisely and rapidly.

In problems of input and output, optical processing offers an improvement to the problem of injecting information into data processing. Both types, namely, identification of symbols belonging to defined set and recognition, and another type which requires pattern recognition can be now handled optically. Lasers are attractive for both the digital and analogue fields, and the outputs of optical computers may also be analogue or digital.

6. Some Apparent Future Applications of Major Consequence

It is not the purpose of this article to speculate far into the future. There are a few applications whose natural evolution is currently in progress and apparent progress is a predictable continuity.

In information processing, the optical computer family is a foregone conclusion for the 1970's with laboratory experiments already in progress in the late 1960's. Automation can thus readily be applied to acquiring, processing, storing, retrieving and using information rather than merely applying automation to machines. Aggregate information (as opposed to serial) can be processed rather than be merely a serial process of time varied information in a circuit. One can process and develop circuit propagation of complex, spatially unique currents using coherent radiation. The transmission of mosaic information or information in parallel can be accomplished by electromagnetic waves.

Holography, both with and without lasers, will develop rapidly for three-dimensional real-time displays. Radar systems can have holographic displays or displays improved or selective by spatial filtering. Computer generated holograms may be applied to solving engineering problems in three dimensions. Increased information capacities will result from optical processing of optical images. Holographic microscopes have already been produced and are commercially available. Since coherent X-rays generated through the use of lasers have recently been produced, one can readily anticipate the display of the three-dimensional structure of proteins and other complex molecules. This should have a profound effect on drug synthesis.

The rapid advance of high continuous power lasers will permit economic use for certain applications in chemical processing, military use, special furnaces, drilling, cutting, welding and gas or atmosphere processing.

Tunable lasers are under development and some have been produced. This permits advanced spectroscopy work, particularly in Raman spectroscopy of solids, liquids and gases. A generation of devices using non-linear optics has resulted from high-power density laser work. This includes harmonic generation and parametric effects in solids.

Bibliography

ARTHUR D. LITTLE INC. (1968) *Industry Comment on Laser Applications and Markets*, Cambridge: Massachusetts.
ELION H. A. (1967) *Laser Systems and Applications*, New York: Pergamon Press.
ELION H. A. (1970) *Lasers, Computers and Data Retrieval*, New York: Pergamon Press.
FISHLOCK D. (Ed.) (1967) *A Guide to the Laser*, New York: Amer. Elsevier Publishing Co., Inc.
GOLDMAN L. (1967) *Biomedical Aspects of the Laser*, Berlin: Springer-Verlag.
GUNSTON W. T. (1966) Practical uses of lasers, *Science Journal*, June 1966, pp. 32–43.
The Institute of Electrical and Electronics Engineers, Inc. (1968) Inter-National Quantum Electronics Conference, *Digest of Technical Papers*, also *IEEE Journal of Quantum Electronics*, Vol. **QE-4**, No. 5 (May 1968).
McGUFF P. E. (1966) *Surgical Applications of Laser*, Springfield, Illinois: Charles C. Thomas, Publisher.
METHERELL A. F. et al. (Ed.) (1969) *Proceedings of the First International Symposium on Acoustical Holography*, New York: Plenum Press.
SALTONSTALL R., Jr. et al. (1965) *The Commercial Development and Application of Laser Technology, A Report*, New York: Hobbs, Dorman & Co., Inc.
TROUP G. J. (1967) Laser, in *Encyclopaedic Dictionary of Physics* (J. Thewlis Ed.), Suppl. vol. 2, 141, Oxford: Pergamon Press. HERBERT A. ELION

LASERS, TUNABLE. In general, laser action, which is due to stimulated emission between levels of atomic or molecular systems, is limited to a number of discrete frequencies or bands of frequencies as a result of the quantized nature of these levels. Through the use of different laser media the number of such discrete frequencies is quite large, covering the region from the ultra-violet to the far infra-red. Even though the number of such laser lines is large there are many regions of the spectrum where there is either no laser emission or such emission, if it exists, is extremely weak. A number of means for achieving tunability of laser emission have been considered. These techniques may be broadly classified into two general categories: nonlinear optical effects and the direct tuning of the emission wavelength of a laser. In the former, the use of optical nonlinearities in materials allows the mixing of optical frequencies, thereby producing new frequencies. Two such processes, namely stimulated Raman scattering (SRS) and optical parametric effects will be considered. Direct tuning of laser emission involves either the shifting of the energy levels of the laser material by temperature or other external means thereby shifting its emission wavelength or the use of a laser medium with a broad emission band with selection of the portion of that band where laser action occurs. An example of the latter, the dye-laser, will be considered.

Stimulated Raman Scattering

When a light beam of frequency v_i is incident upon a medium which may be either a solid, liquid or gas it is observed that light is scattered from the medium differing in frequency from the incident light. It is further found that the difference between the incident frequency and the scattered frequency is characteristic of the medium and corresponds to the difference between some pair of energy levels in the material. This effect is known as Raman scattering after C. V. Raman who first observed it. Under normal illumination, using either a classical source such as a mercury arc or a low power laser the total amount of Raman scattered light is small, perhaps 10^{-6} times the incident power. It is found, however, that as the intensity of the illuminating light is increased a point is reached where the fraction of the incident light scattered increases drastically. This intensity, termed threshold, separates the region where spontaneous emission is dominant and where stimulated emission is dominant. The latter is referred to as stimulated Raman scattering whereas the former is usually referred to simply as Raman scattering. It is in the regime where stimulated emission predominates that high scattering efficiencies are observed.

A typical experimental setup for observing stimulated Raman scattering is shown in Fig. 1. The exact nature of the laser is undefined except that it have sufficient power to achieve threshold for SRS. The Raman active medium is shown surrounded by a pair of mirrors used to provide the optical feedback. Once the incident power exceeds threshold the scattered light bouncing back and forth between the mirrors grows in intensity and is transmitted through one of the mirrors which is assumed to be partially transmitting. In some Raman active media, notably in liquids, the incident laser beam is observed to focus itself due to a nonlinear dependence of the refractive index on the incident laser intensity. In these materials the gain provided by the "trapped" beam is so high

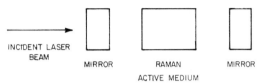

Fig. 1. *Simplified schematic for the stimulated Raman oscillator.*

that mirrors need not be employed. The basic principles are, however, the same. The output beam has the characteristics of a laser, being highly collimated and quasi-monochromatic. The frequency produced in this manner is shifted from the incident frequency by ν_r which is characteristic of the medium. It is in fact observed that a number of frequencies are produced which are related to the incident frequency, ν_i, by

$$\nu_s = \nu_i \pm n\nu_r \quad (1)$$

when n is an integer. Those frequencies higher than the incident frequency are termed anti-Stokes lines and those lower in frequency are termed Stokes lines. Frequencies for which $n \neq 1$ can be thought of as arising from an iteration of the above process. Thus starting with a single frequency ν_i, a series of new frequencies can be generated. Since ν_r is characteristic of the Raman medium used and it has been found that a large number of media are Raman active, each with one or more frequencies, ν_r, the total number of frequencies which can be produced is extremely large. It is further possible to cascade a number of such Raman oscillators using different materials to further increase the multiplicity of available wavelengths. In addition, by changing the basic frequency of the incident laser or by using the second harmonic frequency of such a laser, generated by optical second harmonic generation, a further diversity of wavelength is achieved. Using the neodymium laser ($\lambda = 1.06\,\mu$), the ruby laser ($\lambda = 0.6943\,\mu$) and their second harmonics in conjunction with a number of Raman active materials it is possible to cover a wavelength spread from approximately 8.9 microns to 0.3 micron with over 2500 lines with no more than two iterations.

Thus the stimulated Raman effect can be used to select a number of wavelengths. The tuning is not, however, continuous as it is necessary to change the materials for each new wavelength selection. Although in principle continuous operation is possible, in practice the power levels required to achieve threshold normally limit SRS to pulsed operation with high power lasers.

Optical Parametric Oscillator

The optical parametric oscillator is similar to the parametric amplifiers used at microwave frequencies. The principles of operation are the same, the only differences being the frequency involved and the fact that at optical frequencies the gain provided by the parametric interaction is used along with feedback to provide an oscillator rather than an amplifier. The basic process involves the interaction of three electromagnetic waves termed the pump, signal, and idler at frequencies ν_p, ν_s, and ν_i respectively. These frequencies satisfy the relation

$$\nu_p = \nu_s + \nu_i \quad (2)$$

The parametric or variable parameter interaction involves an interchange of energy between these waves with energy flowing from the high frequency pump wave into the two lower frequency signal and idler waves, thereby amplifying them. For such an exchange of energy to take place between waves of different frequencies it is necessary that the medium in which the interaction takes place have a nonlinear polarizability of the form

$$P_i = d_{ijk}E_jE_k \quad (3)$$

where d_{ijk} is a tensor, characteristic of the medium, and P_i, E_j and E_k are the components of the nonlinear polarization and electric fields respectively. Provided the medium has such a property the three waves can interact and gain can be produced at any pair of frequencies ν_s and ν_i satisfying eq. (2). Because of dispersion, in most crystals the waves will be able to interact only over an extremely small distance limiting the amount of gain which can be achieved. In some crystals which also are birefringent it is possible to find directions of propagation through the crystal where waves of particular frequencies can stay in synchronism throughout the full crystal length provided one or two of the waves propagates as an extraordinary ray. The condition where synchronism is achieved is termed "phase matching" and can be expressed as the condition,

$$\mathbf{k}_p = \mathbf{k}_s + \mathbf{k}_i \quad (4)$$

where $\mathbf{k}_p$, $\mathbf{k}_s$, and $\mathbf{k}_i$ are the propagation constants in the medium. Conditions (2) and (4) are seen to be the equivalent of conservation of energy and momentum. Since $\mathbf{k}_s$ and $\mathbf{k}_i$ will be functions of ν_s and ν_i as well as the indices of refraction, (i.e. $k = 2\pi\nu n/c$, where n is the index of refraction and c is the speed of light) and further the indices are functions of frequency, simultaneous satisfaction of eqs. (2) and (4) determines the pair of frequencies ν_s and ν_i which will achieve the maximum gain and which will oscillate when incorporated in the proper structure, described below.

Tunability of such a parametric process implies the ability to vary the frequencies ν_s and ν_i for a given value of ν_p which will in general be some strong laser line. The means for achieving tunability comes from the ability to vary the indices of refraction seen by the waves propagating in the crystal medium. If we consider only collinear interactions so that eq. (3) may be written in scalar form and drop common multiplicative factors, eq. (3) becomes:

$$\nu_p n_p(\nu_p, \theta, T, E) = \nu_s n_s(\nu_s, \theta, T, E) \\ + \nu_i n_i(\nu_i, \theta, T, E) \quad (5)$$

where we now write the implicit dependence of the indices of refraction on frequency, the common angle of propagation in the crystal relative to some crystalline axis, temperature and an applied electric field. In general any means for varying the indices could be included in such an expression. If through the variation of any one or a combination of such independent quantities the indices are unequally varied, then for a given ν_p, ν_s and ν_i can be changed, subject to eq. (2), thereby achieving tunability.

The schematic diagram of the optical parametric oscillator is identical to that of the stimulated Raman oscillator with the exception that the Raman active medium is replaced by the phase-matchable nonlinear medium described above. The mirrors M are made to be reflecting over the range of frequencies covered by both ν_s and ν_i. These mirrors provide the feedback necessary for the oscillations to take place. The pump wave is introduced into the cavity through one of the mirrors which must be transparent to the frequency ν_p. Oscillation takes place when the gain provided by the parametric interaction is sufficient to overcome the losses in the resonator. Once oscillation is achieved, energy at both the signal and idler can be coupled out of the mirrors which in practice have finite transmission at ν_s and ν_i. As described above, tunability is achieved by varying one of the quantities affecting the indices of refraction such as the temperature of the crystal.

There are several important features of the optical parametric oscillator. The most important is that the amount of pump power required to achieve threshold under typical conditions is less than 1 W and for an extremely low-loss system can be as low as a few milliwatts. Hence an optical parametric oscillator can be pumped by a continuous laser source providing for continuous operation. The powers required for observing stimulated Raman effect and dye laser action (to be described below) are typically several orders of magnitude higher. One further aspect of the parametric oscillator is important; namely the pump frequency ν_p must be highly monochromatic requiring a laser as a source for the pump. Because of the modest powers required, many lasers can be used as pumps.

The degree of tunability and hence the spectral regions covered by optical parametric oscillators depends upon the frequency of the pump, ν_p, and the nonlinear materials used. By the proper choice of pumping wavelengths and materials, it is possible to construct tunable sources which can cover the visible and infra-red regions of the spectrum.

Dye-lasers

The dye-laser is different in principle from the above two methods of achieving a tunable source since it does not rely on optical nonlinearities and is characteristic of the second broad class of tunable lasers. In its operation it is similar to a crystalline solid state, gaseous, or semiconductor laser in that it makes use of stimulated emission between an inverted transition in an organic dye. The dye is excited via optical excitation either using another laser or by means of a high intensity flash-lamp. Provided a sufficient inverted population is achieved so that the resulting gain exceeds the cavity losses, oscillation can occur. The cavity structure for the dye laser is essentially identical to that described in the section on lasers, consisting of the gain medium along with a pair of mirrors used to provide the feedback.

Because the fluorescence linewidth of the dye is broad and also because the peak wavelength is influenced by the solvent, a given dye is capable of oscillating over a broad spectral interval. The exact magnitude of the bandwidth is, of course, dependent on the dye used. The wavelength at which oscillation occurs can be influenced in several ways. In some dyes where there is considerable self-absorption due to an overlap of the absorption and fluorescent bands the wavelength region where oscillation occurs is found to be concentration dependent; tunability can thus be achieved by varying the concentration of the dye. The stimulated emission spectrum from dye-lasers is generally broad (a few tens to several hundred angstroms) but the use of frequency selective elements within the cavity has been shown to produce considerable spectral narrowing. One example of this is the use of a high resolution grating as one of the mirrors in which oscillations with a bandwidth of 0·6 Å have been achieved. Tunability is easily obtained by rotation of the grating. Wide-range tunability can be achieved by using a combination of dyes, solvents and frequency selecting elements along with an excitation source capable of exciting the dye molecule.

The pumping densities required to achieve threshold are comparable to those required for stimulated Raman scattering. Because of the possibility of the population of triplet states and the subsequent excited state absorption it is usually found necessary to provide the excitation in a time short compared to that required for the singlet–triplet transition. This further requires that the excitation be faster than the order of 1 μsec thereby limiting operation to the pulsed regime for dye lasers.

The three examples given above have been chosen because they have been the most thoroughly studied and because they permit tunability over large wavelength ranges. There are, however, a number of other methods by which tunability can be achieved. For example, in a number of lasers the wavelength of emission is temperature dependent with the result that variation of the temperature of the laser medium gives rise to a variable output frequency. The application of magnetic or electric fields or externally applied stress can give similar results. Tunability is also possible in the class of phonon-terminated lasers because of the potentially broad gain profile. Another method is to translate the frequency of a laser's emission via the doppler effect. This has been achieved in practice by passing a laser beam of short time duration through an electro-optic crystal to which has been applied a rapidly varying electric field. Through the electro-optic effect the index of refraction and hence the optical path length of the crystal is varied by the applied field. The resulting shift in the frequency of the laser pulse is proportional to the rate of change of the index of refraction.

Bibliography

BLOEMBERGEN N. (1965) *Nonlinear Optics*, New York: Benjamin.
BLOEMBERGEN N. and GRIVET P. (Eds.) (1964) *Quantum Electronics III*, Vols. I and II, Paris: Dunod.
BUTCHER PAUL N. (1965) *Nonlinear Optical Phenomena*, Columbus: Ohio State Engineering Station Bulletin 200.
GLASS R. C. (1962) Parametric amplifier, in *Encyclopaedic Dictionary of Physics* (J. Thewlis Ed.), **5**, 302, Oxford: Pergamon Press.
MENZIES A. C. (1962) Raman effect, in *Encyclopaedic Dictionary of Physics* (J. Thewlis Ed.), **6**, 177, Oxford: Pergamon Press.
SOROKIN P. P., LANKARD J. R., HAMMOND E. C. and MORUZZI M. L. (1967) *IBM Journal of Research and Development*, **11**, 130.

R. G. SMITH

LAW ENFORCEMENT SCIENCE AND TECHNOLOGY

1. Introduction

It is strange to note that while science and technology has been employed by the law enforcement profession for many years, nowhere can one find a reasonably comprehensive definition of the subject. Yet it is not for the want of published works on some of the more well-known areas of law enforcement science and technology such as complex communications systems, criminalistics and electronic surveillance devices, but rather because of the very nature of police work itself. Thus, to provide a comprehensive reference to what has now become an accepted technology three major areas must be covered: one, a brief history; two, the application of technology to law enforcement; and three, the role that science and technology is about to play in reshaping the field of law enforcement and criminal justice. What follows is based primarily on U.S.A. practice, although much, if not all, is applicable elsewhere.

2. Brief History

The entrance of the scientist and technologist into the field of law enforcement cannot accurately be ascertained. Enough is known, however, to say that police officials sought the aid of the scientist long before that of the technologist. During the early part of the 20th century a small band of scientists formulated a series of analytical techniques that became the basis for forensic science laboratory methods. Although the techniques employed, namely chemical identification, high powered optical studies and spectrograph analysis were standard scientific procedures it was the application of the results that brought forth the scientific discipline of criminalistics. The science of criminology differs from all other sciences inasmuch as it deals not with the likely or usual but with the unlikely and unusual. The exact identification of an item of evidence is but the first step in the forensic process, it holds little value if its association with a crime, a victim or suspect can neither be established nor ruled out. The growth of forensic science has been rather slow and the reasons will be explained shortly.

As for the technologist, his contributions to law enforcement have been in most part by the way of his development of devices for society in general. This becomes obvious when one realizes that with few exceptions the mechanical and electronic equipment employed by police organizations are merely standard items (sometimes a little modified) available to the general public or the armed services.

There have been, however, a few notable exceptions in this area. Technology has developed specifically for the law enforcement field such items as radar speed control devices and breath analyzers for determining the alcohol content in the blood.

It is most important to realize why science and technology have not been efficiently employed by the law enforcement profession and why this situation is rapidly changing. The two major restraining factors are limited budgets and lack of appreciation for technology by law enforcement officials.

However, the technologically oriented society we live in, coupled with a tremendous increase in urban population, has not only out-moded traditional police methods but has presented the criminal element with many new areas of exploitation. Overpopulation of urban centers, high speed transport, and almost instantaneous dissemination of news has provided the criminal with ease of concealment, rapid transit to and from the scene of a crime and the means to learn quickly if the crime he has committed has been detected. Even more complex are the problems of combating organized crime and the control of large-scale civil disorders.

Thus understaffed and technically weak, law enforcement has not been able to prevent a rapid increase in crime rates, check the constant growth of organized crime or effectively cope with civil disorders. This situation has become so bad in recent years that the private citizen is now demanding high level government action.

3. Application of Science and Technology to Law Enforcement

The responsibilities of the police have changed little in the past 50 years. These responsibilities include the maintenance of public order, the detection and suppression of crime, the apprehension of the criminal, and, in a different rôle, the control of vehicular traffic.

Structurally police organizations are normally sub-divided into several specialized divisions:

Patrol activities   Communications
Traffic control   Forensic Science Laboratory
Detective bureau   Administration and records
    Riot control

It should be noted that police organizations vary greatly from country to country, and the size of each of the seven divisions listed will vary and in some instances certain divisions may not exist at all.

Patrol Activities

Before the advent of the telephone and radio, police patrols were, for all practical purposes, police departments within themselves. Isolated from higher command they could neither be directed by nor call for assistance from headquarters or other patrols. The telephone provided the first limited form of communication between the patrol officer and his command. Police call-boxes were installed at strategic points within each patrol area, and the call-box, although it was the first communications aid available to the patrolman, is still actively employed even though it has now been augmented by two-way radio.

Early radio networks were strictly one-way arrangements; headquarters being equipped with a means to transmit voice transmissions while vehicles could only receive. The need for two-way communications was obvious but law enforcement had to wait patiently for technology to provide it.

Modern police two-way radio networks range from a single headquarters station and a few mobile radios to vast highly complex communications systems. Large cities and state police agencies employ several independent networks to reduce the congestion that would result with a multitude of mobile radios communicating with one headquarters station. Range limitation between mobiles and headquarters is another reason for employing more than one headquarters station. The number of network variations and the types of equipment employed are many and to describe them would be beyond the scope of this survey.

In recent years the major advances in the communication field of patrol work have been in the area of headquarters' efficiency. Speed of dispatch, coordination of mobile units, and response to requests for assistance have demanded specialized message-handling procedures and equipment. The large city communications centers are equipped with illuminated status boards that permit the communicator to tell at a glance what sectors of the city have active patrols, how many patrols are in or out of service, and which patrols are actively engaged on a specific assignment. The complaint desk has become an integrated part of the communications system permitting request for action to be quickly supplied to the radio operator responsible for the sector from which the call was received. Switching arrangements have been incorporated that permit mobiles in one sector to communicate directly with patrols in another sector via a headquarters relay. Several large city communications complexes have provided the dispatcher with direct access to computer-stored criminal files thereby permitting him to answer quickly those requests from mobile units which relate to stolen vehicles or wanted persons. Experimental programs are currently being conducted to assess the practicability of teletype communications between headquarters and mobile units. These could provide the patrol officer with a printed record of lengthy messages such as stolen car lists that he would normally have to transcribe; could eliminate the need to stop his patrol activities during these periods; and would allow messages to be sent to the mobile unit even though the patrolman might be out of the car. Additional security would also be given to the communication.

In addition to communications equipment the patrol division employs a few other technical aids. One instrument which is finding wide acceptance is the "Breathanalyzer" or "Breathalyzer". This device which is used to detect the amount of alcohol in the blood provides the patrol officer with a scientifically proven method of determining a driver's sobriety. Its employment is gaining popularity because the results of a breath test have become accepted evidence in the court of law.

Limited use has been made of night vision devices such as near infrared viewers for scanning inaccessible areas such as fenced-in industrial establishments.

Research and development studies have been conducted that recommend the design of vehicles specifically for police work, however, the complexity and cost are inhibiting actual construction.

To complete the patrol division's picture it should be noted that conventional aircraft and helicopters are employed by police departments for highway and harbor surveillance work.

Traffic Control

Except for special situations, the direction and control of vehicular traffic is no longer handled by the individual. Electromechanically-timed traffic lights are currently in widespread use; however, urban areas have been forced to centralized control of traffic lights to permit the changing of timing cycles to favor morning and evening traffic patterns.

Increased traffic densities have degraded all of the conventional traffic control arrangements and the current approach is to employ fully automated traffic control systems which are made up of three major component sub-systems: vehicle sensors, a special purpose computer, and a communications network to connect the sensors and traffic lights with the computer. The traffic controller is specially designed to assimilate the variations of traffic density and with the aid of an internally stored program regulate traffic flow accordingly. The flexibility of computer control allows a set of subroutines to be used when unusual traffic problems develop. Depending upon the size and complexity of the computer used and the skill of the systems programer, additional functions can be obtained, such as a self-learning process which in essence accumulates a historical record of day-by-day busy hours, analyzes this information and up-dates the program accordingly. As a by-product the computer can supply statistical data. At present the number of fully operational computerized traffic control systems is few; however, as more and more cities approach vehicular traffic saturation the automated traffic control system will become an absolute necessity.

In general, the traffic division is also responsible for the operation of speed control equipment although in smaller departments, this function may be shared by the patrol section.

Two basic instruments are employed to measure the speed of a moving vehicle: a radar transmitter-receiver combination that uses the Doppler principle; and the pneumatic-tube-activated speed-timer.† These two devices are unique in the sense that they were developed specifically for police use and have no civilian counterparts.

Detective Bureau

The detective bureau provides the manpower for the investigation of crimes that have been or are in the process of being committed. This division is usually sub-divided by crime types, e.g. homicide, arson, robbery and confidence rackets. The technology employed by this group is usually obtained from other divisions. Vehicles assigned to the detective division are usually equipped with two-way radios which operate on the local police frequency. Hand-carried radios, preset to special frequencies for surveillance work are usually obtained from the

---

† The Pneumatic Tube Activated Speed Timer is basically an elapsed time indicator employing two air filled rubber tubes (accurately spaced along the highway) coupled to a chronograph type stopwatch by means of pressure sensitive switches. The speed timer provides a fairly accurate measure of a vehicle's average velocity by the simple means of having the weight of the vehicle start and stop the chronograph as it passes over the rubber tubes. The chronograph dial is calibrated directly in miles per hour.

communications division which also provides any of the special wire-tapping or eavesdropping instruments needed by the detective division. In smaller departments electronic devices may be under the jurisdiction of the crime laboratory.

The modern well-equipped police department possesses a large variety of devices designed to aid the practicing detective to accumulate, from a distance, with a minimum possibility of being detected, both visual and audible information. These devices include telephone-tapping equipment, miniature transmitters which can be planted in hidden locations, ultrasensitive audio amplifier microphone combinations for listening through walls or at great distances, self-contained voice-operated tape recorders, miniaturized television pickup cameras for visual surveillance from difficult vantage points and telephoto still-and-movie-camera equipment. It is of interest to note that none of these devices was originally developed for law enforcement but rather for private detective agencies.

The detective division, because of the very nature of its work, is the principle user of the crime laboratory, the facilities of which will be described shortly.

The newest tool available to the criminal investigator is the computerized criminal data file. Centrally located computerized data files provide the investigator with the ability to ascertain quickly if a suspect has a criminal record or if he is wanted for a crime outstanding. Lists of stolen vehicles or guns may also be obtained from this file.

Computerized criminal data files like automated traffic control systems are limited in number but are rapidly becoming a necessity.

Communications

The primary responsibilities of the communications division are to operate and maintain the various communications systems in use by their department. These systems include local two-way radio networks, interdepartmental radio networks, and teletype networks. In the strictest sense, the communications division does not actively engage in any of the basic police activities *per se* but rather provides a service. It is, however, responsible for the operational standards and message-handling procedures. Depending on the size of the department, communications technicians may be required to operate the various types of electronic surveillance equipment for the detective division. It is interesting to note that to this date there is no communications equipment unique to law enforcement; all the existing network configurations, be they radio, telephone or teletype, have industrial or military counterparts.

Forensic Science Laboratory

The role of the forensic laboratory has been, and is still to a large degree, misunderstood by both the layman and technologist. Defined in the simplest of terms, forensic science is the scientific examination of evidence, the findings of which are used to assist the police investigator in:

1. Linking the crime scene or the victim to the criminal.
2. Establishing an element of the crime.
3. Corroborating or disproving an alibi.
4. Exonerating the innocent.
5. Providing expert testimony in court.

The scientific equipment employed by the crime laboratory includes basic chemistry for isolating and identifying materials considered as evidence related to a particular crime; optical comparators for ballistics evaluation; compound lens systems for visual evaluation of macro and microscopic sized substances; various types of illumination such as ultraviolet and infrared to make visible or identifiable unusual markings. These basic techniques are further enhanced by spectrographic, X-ray diffraction and neutron activation analysis.

Currently several research organizations are attempting to devise automated fingerprint identification systems. As with the problems of vehicular saturation and the gross inefficiency of manual record-keeping, the manual process of fingerprint identification must, in the not too distant future, become automated.

Administration and Records

In addition to standard responsibilities of budget control, purchasing, payroll accounting and personnel files, the records and administration section must maintain, to the highest possible degree of accuracy, those criminal files directly related to their own department. While large police departments maintain electronic data processing equipment for payroll accounting and personnel records, only the largest have installed special purpose computers for the storage and retrieval of criminal information. In the United States the Federal Bureau of Investigation maintains and operates a National Crime Information Center which currently permits 15 state police agencies, several large cities and the Royal Canadian Mounted Police to contribute and retrieve criminal information. Future plans of the NCIC call for the development and installation of terminals in all 50 states.

Riot Control

Much effort has been devoted to developing concepts and equipment that will disperse or contain, restrain or deter rioting groups. What has emerged is the "minimum force concept" and a family of nonlethal devices to supplement it.

For many years riot control equipment in many countries consisted of tear gas, batons and small arms; protection equipment for the police officer being limited to gas masks, helmets and hand-carried shields.

Technology is currently developing a new family of tear gases that will both limit physical coordination and inhibit aggressive spirits. To reduce the mobility of a mob and make concerted action extremely difficult, two chemical schemes have been devised. One scheme employs an oily compound which, when sprayed on sidewalks and roadways, makes walking hazardous and running almost impossible. Surfaces coated with this compound become so slippery that a firm stance cannot be taken from which to hurl a projectile effectively. The second scheme is really nothing more than an oversized soapsuds generator and projector. This device can completely engulf limited areas to a height of eight to ten feet with a foam of soap which does not affect breathing but makes visual contact between rioters most difficult, not to mention the discomfort associated with taking a bath with one's clothes on.

Although these two schemes have been applied to actual riots their real effectiveness is yet to be established.

In line with the program of developing nonlethal riot control equipment, police batons which incorporate high voltage prods have been produced. Experimental work is currently being conducted with high intensity sound and light projectors in the hope that these devices can, by adding elements of confusion, reduce, if not completely disorganize, a militant mob.

4. The New Rôle of Science and Technology as it Relates to Law Enforcement and Criminal Justice

Crime has become a major source of concern in the major countries of the world. In the United States the government enacted the Law Enforcement Assistance Act of 1965 which empowered the Attorney General to provide funds for educational training and research and development in criminal justice, and in 1967 the first National Symposium on Law Enforcement Science and Technology was held.

At this time it is most difficult to identify those research proposals which will produce tangible results. There are urgent problems confronting law enforcement, the solutions to which merely require the development of specialized equipment hardware and operational programs to make them work. A prime example is the need for a comprehensive network of computerized criminal data files with remote terminals in all levels of law enforcement, the skeletonized model of which is already operational. Communications equipment and systems designers are faced with the challenging problem of providing every active police officer with instantaneous two-way communications that will provide far better efficiency and portability than exists today.

Still other problem areas of law enforcement such as nonlethal weaponry, foolproof automated fingerprint identification, infallible lie detection and voice identification that does not require interpretation will require intensive research and one element that cannot be predicted, that being inventive genius.

The most basic problem, that of understanding the criminal mind, might never be solved. However, concentrated efforts in the field of psychology could very possibly provide the means for detecting criminal tendencies and the therapy to cure it.

Definitions

Criminalistics. That profession and scientific discipline directed to the recognition, identification, individualization, and evaluation of physical evidence by application of the natural sciences in law-science matters. California Association of Criminalists definition adopted at the 21st semiannual seminar at Ventura, California, May 1963.

Polygraph. The polygraph, also improperly referred to as a "lie detector", is a device that measures and records on graph paper the physiological variations of a human being when placed under emotional stress. The commercial polygraph, designed for interrogation work, records the pulse rate, relative blood pressure, the rate and depth of breathing and often the resistance of the skin to the conduction of electricity. Although widely employed by large business firms and law enforcement agencies as an aid in interrogations, the instrument itself merely records physiological changes, any detecting of lies is done by the examiner. Until such time as a scientific technique is devised that will, with complete accuracy and without human interpretation, indicate a truthful statement from a lie the polygraph must be considered a physiological measuring device only.

Bibliography

Task Force Report: *Science and Technology* (1967), Superintendent of Documents, U.S. Government Printing Office, Washington, D.C.

Proceeding of the First National Symposium on Law Enforcement Science and Technology, vol. 1 (1967), Academic Press.

(*Special note.* These volumes provide the only comprehensive bibliography of law enforcement and criminal justice science and technology.)

W. SHAW

LIQUID CRYSTALS AND MOLECULAR STRUCTURE.

An organic compound is normally characterized by a well-defined melting point at which it passes from the three-dimensional, ordered arrangement of the crystal to the disordered isotropic liquid. Certain compounds do not behave in this simple manner, and on heating, the crystals pass through one or more intermediate states before the isotropic liquid is produced. These opaque intermediate states are fluid in varying degree and exhibit many optical properties of crystals. Since their properties are intermediate between those of the crystal and the isotropic liquid, they are termed liquid crystals or mesomorphic states. The molecular organization and general properties of the three main types of liquid crystal (smectic, nematic, and cholesteric) have been discussed in a previous volume. The relationships between molecular structure and the liquid crystalline properties of compounds are now summarized.

For liquid crystalline properties to occur, a compound must undergo the overall process of melting in steps. The three-dimensional order of the crystal must break down in stages, and this is most likely to occur if the molecules are geometrically anisotropic and give rise to anisotropic intermolecular attractions. For example, the intermolecular attractions between the sides and the ends of elongated molecules should be different, and if these attractions weaken selectively, then liquid crystalline states with smaller degrees of molecular order than the crystal may be produced on heating. Conversely, a crystal comprised of approximately spherical molecules is unlikely to yield liquid crystals. Indeed, all liquid crystalline substances consist of highly geometrically anisotropic molecules, i.e. rod-shaped molecules with a high length : breadth ratio. Furthermore, a change in molecular architecture causing a bifurcation of the linear molecules will destroy or reduce the liquid crystalline properties. Examples of this are given later.

Consider what may happen if we heat a crystal comprised of elongated molecules arranged parallel to one another with their ends in line, i.e. in layers, and with the major molecular axes approximately perpendicular to the layer interfaces. If the terminal intermolecular attractions are weak, the layers may move apart and slide over one another yielding the smectic mesophase. The layers may remain plane, or become rolled up as in the focal-conic texture. Further heating may either disrupt the layer arrangement giving the isotropic liquid or simply weaken the lateral attractions sufficiently for the molecules to slide out of the layers yielding the nematic state with a parallel arrangement of molecules whose ends are not in any definite order. Similarly, a layer or non-layer arrangement of parallel molecules in the crystal may give a nematic melt on heating; it is impossible for a non-layer crystal lattice or a nematic mesophase to produce a smectic mesophase on heating. The situation is summarized below.

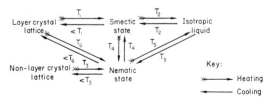

Notes: (1) Crystallization of liquid crystalline states is usually accompanied by supercooling, but within experimental error, transitions T_2, T_4, and T_5 are precisely reversible.

(2) Selective loosening of intermolecular attractions occurs at the various transitions.

T_1: terminal attractions weaken giving layer flow;
T_2: lateral and residual terminal attractions loosen giving disorder;
T_3: lateral and terminal attractions weaken giving the nematic state;
T_4: lateral attractions weaken giving interpenetration of layers;
T_5: residual lateral and terminal attractions loosen giving disorder;
T_6: lateral and terminal attractions weaken giving the nematic state.

(3) Transitions are readily detected and the temperatures evaluated by heating a sample of the substance in a melting point capillary.

Molecular Structure in Relation to Smectic and Nematic Liquid Crystals

Liquid crystals may be produced if there is a suitable imbalance of the attractions between geometrically anisotropic molecules. It is not sufficient simply to have elongated molecules, as shown by the fact that neither the n-alkanes nor the open chain homologues of acetic acid exhibit liquid crystalline properties. Such molecules are either too flexible or the intermolecular attractions are not of the right magnitude to maintain an ordered arrangement after melting. The introduction of multiple bonds and aromatic rings raises molecular polarizability, increasing intermolecular attractions; such functions also reduce the flexibility of the molecule. The majority of liquid crystalline compounds are therefore aromatic, and many contain additional double or triple bonds. The introduction of dipolar groupings also increases intermolecular attractions, but only if the length : breadth ratio is not reduced too much by the grouping. By the same reasoning, the majority of liquid crystalline, aromatic compounds have the substituents *para* to each other, preserving the linearity of the molecule. *Ortho*- and *meta*-disubstituted benzenes are not liquid crystalline. Some examples for consideration are given at foot of page.

The melting point of a compound is an important factor in determining whether a compound is liquid crystalline. The temperature of melting is difficult to predict, and if this is much higher than anticipated, no mesomorphic properties may occur with a potentially mesomorphic system. Therefore, groups which give rise to polymeric intermolecular hydrogen bonding, e.g. $-OH$ and $-NH_2$, are unlikely to occur in a liquid crystalline compound. However, hydrogen bonding between carboxyl groups yields linear dimers and raises the mesomorphic potential.

In general, a compound comprised of linear molecules containing readily polarizable aromatic rings and multiple bonds, and dipolar substituents which do not broaden the molecule is potentially liquid crystalline. Many liquid crystalline compounds contain alkyl or alkoxy groups, but these are always

Structure	Property
Me.CO.O–⟨⟩–CH₂–⟨⟩–O.CO.Me	Not mesomorphic; non-linear molecule
Me.CO.O–⟨⟩–CH=CH–⟨⟩–O.CO.Me	Mesomorphic; essentially linear molecule; polarizable double bond
Me.CO.O–⟨⟩–C≡C–⟨⟩–O.CO.Me	Mesomorphic; linear molecule; polarizable triple bond
AlkylO–⟨⟩–N=N(→O)–⟨⟩–OAlkyl	Polar azoxy grouping; essentially linear molecule
AlkylO–⟨⟩–⟨⟩–C(=O)OMe	Mesomorphic; the shorter benzene analogues are not mesomorphic
AlkylO–⟨⟩–C(O⋯H–O)(O–H⋯O)C–⟨⟩–OAlkyl	Mesomorphic; linear dimer; alkyl branching destroys or reduces the liquid crystalline properties
AlkylO–⟨⟩(Br)–C(=O)(O–H)	Not mesomorphic; linear dimer, but the bromo-substituent broadens the molecule; biphenyl analogues are mesomorphic
O₂N–⟨⟩–⟨⟩–⟨⟩–OMe	Mesomorphic; linear with polar substituents; benzene and biphenyl analogues are not mesomorphic

attached terminally to a more rigid, polarizable part of the molecule.

In the smectic state, the molecules lie parallel to one another in layers. Clearly, the lateral intermolecular attractions must be adequate to retain the layer arrangement once the terminal attractions have weakened. Dipole moments operating across the long molecular axis will assist in this respect, and a terminal ester grouping (—COOAlkyl) appears to be particularly effective, conferring purely smectic properties on many systems. Similar considerations apply to the aliphatic salts of long chain carboxylic acids, e.g. thallous stearate, which exhibit smectic properties although the acids are not liquid crystalline.

The smectic arrangement is adversely affected by broadening the molecule, since the lateral attractions which are essential for preservation of the order may be reduced. Therefore, the introduction of substituents with dipoles which operate across the long axes of the molecules will frequently not have the desired effect of enhancing smectic properties if the breadth of the system is increased.

Lengthening the alkyl chain always enhances smectic properties and disfavours nematic properties. The rôle of other terminal substituents is discussed later.

Nematic liquid crystals are most likely to arise if the molecule has the basic requirements already defined for liquid crystalline behaviour, but lacks the features favouring smectic properties, i.e. if it has no strong dipoles acting across the long molecular axis and no very long alkyl chains occupying terminal positions in the molecule. Methoxy substituents and other short alkoxy groups occupying terminal positions strongly favour nematic properties.

Whether a compound is or is not liquid crystalline and whether it is smectic rather than nematic depends critically upon the strengths of the lateral and terminal intermolecular attractions. When molecular structure is varied widely, many molecular parameters influencing intermolecular attractions are altered, and it is not easy to predict the consequences upon liquid crystalline behaviour. Consequently, only broad generalizations about the relationship between structure and liquid crystalline properties can be made.

However, if the greater part of the molecular structure is kept unaltered, and minor structural changes are made, it is possible to relate more precisely the structural cause to the effect upon the liquid crystalline properties. Minor structural changes may not destroy or create liquid crystalline properties in a compound, and it is necessary to use the temperature of a given transition as an index of whether a liquid crystalline state is rendered more or less stable. Thus, increase in a smectic–nematic transition temperature reflects an increase in the thermal stability of the molecular arrangement in smectic layers, and this may be related to the change in structure occasioning the change in transition temperature.

Studies of this kind have given information on the following topics.

(1) The effect of increasing molecular breadth

By introducing lateral substituents into a molecule which is linear, (a) molecular breadth is increased, tending to decrease mesomorphic thermal stability, and (b) molecular polarity and polarizability are increased, tending to increase mesomorphic thermal stability. Of the two opposing effects, (a) always predominates. The 4′-n-alkoxy biphenyl-4-carboxylic acids provide a typical example.

$$\text{RO}-\underset{}{\overset{3'}{\bigcirc}}-\bigcirc-\text{CO}_2\text{H}$$

Substituents, X, in the 3′-position reduce the smectic and nematic transition temperatures. The nematic-isotropic transition temperatures fall in proportion to the breadth increasing effect of X, showing that physical separation of the long molecular axes is an important factor in the nematic melt. The smectic transitions are less simply related to molecular breadth. The effect of a large substituent such as —NO$_2$ is counteracted to some extent by the ring-NO$_2$ dipole, and the smectic thermal stability is reduced less than expected. Thus, a combination of breadth and polar effects is important when considering the smectic state.

Only when a lateral substituent fills some pocket in the side of a molecule and does not exert its full breadth increasing effect does lateral substitution enhance liquid crystalline properties, e.g. the smectic mesophases of 5-iodo-6-n-alkoxy-2-naphthoic acids are more stable and the nematic mesophases less stable than those of the parent acids. This again illustrates the greater influence of the dipole of the lateral substituent upon smectic thermal stability.

If lateral substitution imposes a steric effect which further increases intermolecular separation, much greater decreases in mesomorphic thermal stability result. Consider the liquid crystalline Schiff's bases:

$$\text{RO}-\bigcirc-\text{CH=N}-\underset{}{\overset{2}{\bigcirc}}-\underset{}{\overset{2'}{\bigcirc}}-\text{N=CH}-\bigcirc-\text{OR}$$

The biphenyl nucleus is planar, but a 2- or 2′-substituent twists the system about the inter-ring bond and increases the intermolecular separation. Such sterically affected biphenyl derivatives have much reduced liquid crystalline transition temperatures, smectic thermal stabilities being the more markedly affected.

(2) The effect of ascending an homologous series

Increase in the length of the carbon chain of a terminal alkyl or alkoxy group gives regular changes in liquid crystalline transition temperatures, and when these are plotted against the number of carbon atoms in the alkyl chain, one of a limited number of types of plot is obtained. The commonest case is one in which the smectic–nematic transition temperatures, which do not alternate, lie on a steeply rising curve; this levels off gradually and merges with the two curves, one for odd and one for even carbon chain members, which may be drawn through the alternating nematic–isotropic transition temperatures (see Fig. 1). Thus, the early members of a series are usually

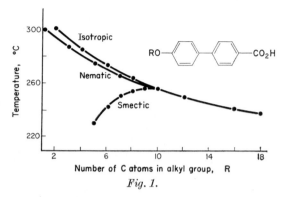

Fig. 1.

nematic, the later members are smectic, and intermediate members exhibit both smectic and nematic properties. An increase in alkyl chain length therefore increases the stability of the smectic state relative to the nematic state. With lengthening alkyl chain, small increases in smectic–nematic transition temperature are observed if the lateral intermolecular attractions are high, but if these are low, e.g. in sterically affected systems, the smectic–nematic temperature curve rises steeply. Therefore, the contribution of the alkyl chain to lateral attractions in the smectic state must be quite small and should be considered relative to the lateral attractions arising from the rest of the molecule. Lengthening the alkyl chain normally reduces the nematic thermal stability if this is already quite high, i.e. if the intermolecular attractions are reasonably high. If however the system is sterically affected and the nematic–isotropic temperatures are low, the nematic–isotropic curve rises. A number of opposing factors is probably involved.

Since nematic–isotropic temperatures alternate, whilst smectic–nematic temperatures do not, this suggests that loosening of terminal intermolecular attractions occurs at nematic–isotropic transitions.

(3) The effect of altering the nature of the terminal substituent

The effects of alkyl chain length and of ester groups upon smectic properties have already been mentioned.

If the terminal hydrogen of a liquid crystalline system is replaced by another substituent, the nematic thermal stability is normally enhanced. Polarizable groups like phenyl, and polar groups like —CN, —OMe, and —NO_2 are most effective; substituents like fluoro- and s-propyl are least effective. These substituents will lengthen the molecule and must increase the overall attractions making randomization of the nematic molecular order more difficult.

Most terminal substituents also increase smectic thermal stability, except —NO_2 and —OMe. The reasons are not obvious, but possibly the dipole moments of these two groups give repulsions which reduce the lateral intermolecular attractions.

Cholesteric Liquid Crystals

Cholesteric liquid crystals are formed on heating a number of esters of cholesterol and other sterols. The only compounds which do not contain the sterol skeleton and give cholesteric liquid crystals are Schiff's bases formed from the amyl ester of p-amino-cinnamic acid and a p-substituted benzaldehyde; the amyl group is 2-methylbutyl, containing an asymmetric centre, and a well-known example is shown below.

NC—⟨⟩—CH=N—⟨⟩—CH=CH·CO·OC_5H_{11}

The Grandjean plane texture of the cholesteric mesophase consists of sheets in which the molecules lie parallel as in the nematic melt. In the next sheet above, the long molecular axes are rotated through a small angle and this is perpetuated through a stack of sheets giving a screw axis through the preparation. This explains the high optical activity of the plane texture, but it is not clear what facet of molecular structure causes the molecules to lie on top of one another with a progressive rotation of their long axes. Replacement of the cyano group in the above Schiff's base by —Cl, —NO_2, or —OMe retains cholesteric properties, but an alkyl group, a dimethylamino group, or an alkoxy group other than methoxy gives smectic behaviour. The shorter chain aliphatic esters of cholesterol are purely cholesteric, longer chain members are purely smectic, and intermediate members exhibit both mesophases; a combination of cholesteric and nematic properties is never observed. Therefore quite subtle factors determine the occurrence of cholesteric properties in a compound.

Bibliography

BROWN G. H., DIENES G. J. and LABES M. M. (1966) Proceedings of the International Symposium on Liquid Crystals, Kent State University, August 1965, published in *Molecular Crystals* **1** and **2**. See particularly Influence of molecular structure on liquid crystalline properties, **1**, 333, New York: Gordon and Breach. Proceedings of the Second International Conference on Liquid Crystals, Kent State University, August 1968, published in *Molecular Crystals and Liquid Crystals* **7** and **8**, New York: Gordon and Breach.

BROWN G. H. and SHAW W. G. (1957) *Chemical Reviews* **57**, 1097.

FERGASON J. L. (1964) *Scientific American* **211**, No. 2, 77.

FRIEDEL G. (1922) *Annales de Physique* **18**, 273.

GRAY G. W. (1962) *Molecular Structure and the Properties of Liquid Crystals*, New York: Academic press.

GRAY G. W. *et al.* (1965) Mesomorphism and chemical constitution, Part XIII, *Journal of the Chemical Society* 3076 (and preceding parts).

Liquid crystals and anisotropic melts (1933) *Transactions of the Faraday Society* **29**, 881–1085.

WOOSTER W. A. (1961) Liquid crystals, in *Encyclopaedic Dictionary of Physics* (J. Thewlis Ed.), **4**, 332, Oxford: Pergamon Press.

G. W. GRAY

LIQUID METALS. Recent progress in this field has come about for two reasons:

(i) The technological importance of liquid metals as coolants in nuclear reactors.

(ii) The interest in the fundamental problem of the nature of electron states and of the electron–ion interaction, in a system with only local order.

We shall focus attention mainly on the points raised in (ii). These will, of course, eventually clarify the basic physics, and hence lead to a much better understanding of the materials most favourable under (i). Three aspects of the problem are already singled out in (ii). The local order, or the liquid structure, is first discussed, with reference to its determination by X-ray and neutron scattering. The essential features of the pair interaction between ions are then summarized, this interaction incorporating the conduction electron shielding of the ions which must occur in a conducting fluid. The way in which the electron–ion interaction, when combined with the structure, can be used to understand the electrical properties of liquid metals is then considered.

1. Liquid Structure

Debye (1915) first demonstrated that in X-ray diffraction in liquids one had to consider two atoms whose scattered rays interfere with one another. As a result, an interference pattern is obtained which depends crucially on the relative separation of the two atoms or the radial distribution function $g(r)$, defined precisely below. While, in principle, $g(r)$ could be calculated from statistical mechanics, if the interionic forces were given, the major progress in understanding the structure of liquids has come from analysis of the X-ray or neutron data to yield $g(r)$.

(a) *Intensity of X-ray scattering*

We shall consider an assembly of N identical atoms irradiated by monochromatic radiation of wavelength λ. If f is the appropriate atomic scattering factor then the angular distribution of the coherent scattered radiation is given by

$$I_{\text{coh}}(K) = f^2 \left\langle \left| \sum_{i=1}^{N} \exp(i\mathbf{K} \cdot \mathbf{r}_i) \right|^2 \right\rangle \quad (1.1)$$

where $K = |\mathbf{K}| = (4\pi \sin \theta)/\lambda$, $\mathbf{r}_i$ is the position of the ith atom, θ is half the scattering angle and the brackets denote a time average.

(b) *Radial distribution function and structure factor*

From this discussion of the intensity of X-ray scattering, it is natural to define a liquid structure factor $S(K)$ by the relation

$$S(K) = \frac{I_{\text{coh}}(K)}{Nf^2} = \frac{1}{N} \left\langle \left| \sum_{i=1}^{N} \exp(i\mathbf{K} \cdot \mathbf{r}_i) \right|^2 \right\rangle. \quad (1.2)$$

The absence of long-range order in a liquid means that the concept of a unit cell is no longer useful and following Zernike and Prins (1927), the radial distribution function $g(r)$ is defined by setting $\varrho_0 g(r) 4\pi r^2 \, dr$ equal to the number of atoms in a spherical shell of radius r and thickness dr, ϱ_0 being $\dfrac{N}{V}$, the number of atoms per unit volume. Then from (1.2) we find

$$S(K) = 1 + \varrho_0 \int_V g(r) e^{i\mathbf{K}\cdot\mathbf{r}} \, d\mathbf{r}. \quad (1.3)$$

Equations (1.2) and (1.3) would be the ones to use if we wished to *predict* the X-ray intensity from a calculated radial distribution function.

However, a more fruitful approach to the problem has been to invert the Fourier transform in (1.3), when we obtain

$$g(r) - 1 = \frac{1}{2\pi^2 \varrho_0 r} \int_0^\infty [S(K) - 1] K \sin Kr \, dK. \quad (1.4)$$

In principle, then, the radial distribution function $g(r)$ can be obtained from the observed structure factor $S(K)$. Schematic forms of $S(K)$ and $g(r)$ in a typical liquid metal are shown in Figs. 1 and 2 respec-

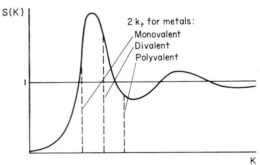

Fig. 1. Schematic form of structure factor $S(K)$. Fermi sphere diameter $2k_f$ is indicated for monovalent, divalent and polyvalent metals.

tively. The peaks in $g(r)$ which flatten out as r increases reflect the short-range order in the liquid.

2. Interionic Forces

One of the objectives of current research on liquid metals is to use the measured structure factor $S(K)$ or the radial distribution function $g(r)$ to obtain the pair potential $\emptyset(r_{12})$ representing the interactions between ions 1 and 2 at separation r_{12}. Actually, although there is an exact relation in classical statistical mechanics relating $g(r)$ and $\emptyset(r)$, the three-atom distribution function is also needed and this function is not yet accessible to experiment.

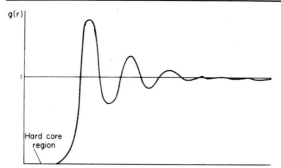

Fig. 2. Schematic form of radial distribution function $g(r)$, showing short-range order in liquid metal, as well as hard core region in which $g(r)$ is very small.

Nevertheless, by relating the three-atom distribution function approximately to the two-atom correlation function $g(r)$, it is possible to gain information on $\emptyset(r)$ from the measured structure (Johnson and March, 1963; Johnson, Hutchinson and March, 1964). Though our knowledge of $\emptyset(r)$ is as yet fairly primitive, we summarize the available information below.

(i) The structure is not too badly represented, in its gross features, by the assumption of hard sphere atoms. Certainly, in liquid metals, there is a strong repulsive potential as we try to push atoms close together, though nevertheless a substantially softer potential in liquid sodium than in an insulator like liquid argon.

(ii) The potential $\emptyset(r)$ is substantially deeper in liquid metals than in liquid argon. Thus, at the minimum of the potential between argon atoms, $\emptyset(r) \sim 0.01$ electron volt (eV). In metals, this value is often 10 times as large, typically 0.05 to 0.2 eV it seems.

(iii) The van der Waals forces characteristic of the attraction at large distances between argon atoms appear to play no role in liquid metals. This is because we are really dealing with the interaction between ions, but shielded by the conduction electrons in the liquid metal. Theory predicts that because of the scattering of the electron waves off the ions, the electrostatic potential as we go away from an ion shows a characteristic diffraction pattern, and oscillates with a wavelength π/k_f, where k_f is the Fermi wave number. This is half the de Broglie wavelength for an electron at the Fermi surface. The oscillations in the electrostatic potential around an ion lead to a "shocking" violation of classical electrostatics, and tell us that if we place another ion at any appropriate distance from the first ion we can get attraction, rather than repulsion.

However, solely from analysis of the structure data, it can be shown that there is a repulsive hump in the potential $\emptyset(r)$ in liquid metals, beyond the principal minimum. The height and extent of this region, plus, of course, the depth and shape of the first minimum and the rate of rise of the potential in the repulsive regime at short distances are the most important quantities we would like to know in elucidating further the ion–ion interaction in liquid metals, though the longer range oscillations predicted by electron theory are of considerable theoretical interest.

3. Conduction in Liquid Metals

Although it has been known for over one hundred years that the ability of metals to conduct electricity is relatively unaffected by the change from the solid to the liquid state, it is only quite recently, particularly through the work of Ziman (1961), that we have come to a fairly fundamental understanding of why this should be so.

To see the origin of the difficulty more clearly, we remind the reader that a piece of pure solid metal, say sodium, will have a quite negligible electrical resistance as we approach the absolute zero, for the electron waves can move freely through a perfectly ordered crystal lattice. However, as we increase the temperature, the ions vibrate about their mean positions and this thermal disorder scatters the electrons and leads to electrical resistance. Obviously the disorder increases with increasing temperature and in this way we can understand the temperature dependence of the resistivity of pure metals.

But now, looked at solely from this point of view, the experimental results which show that the electrical resistivity is relatively unaffected by melting are hard to understand. For surely, the evident disorder which is associated with melting ought to lead to a very substantial increase in electron scattering, and hence make the liquid metal a rather poor electrical conductor.

3.1. Weak scattering (pseudo)-potentials

The ideas of Ziman (1961) were put forward as a result of solid-state physicists gaining more insight into the surprising success of an almost free electron treatment of the electron wave functions and energy levels in many solid metals. The reason for surprise was that the lattice potential created by the ions appears at first sight to be very strong and to alter the character of the electron states greatly from the predictions of a nearly free electron theory.

However, Phillips and Kleinman (1959) focussed attention again on the use of a repulsive potential which simulated the effects of the Pauli exclusion principle, and which prevents conduction electrons falling down into core states. This idea is not new, being used by Hellmann and many other workers previously, but Phillips and Kleinman emphasized the way in which this repulsive potential, when added to the true ionic potential, could lead to quite a weak scattering potential, a pseudopotential.

As we saw from our discussion of X-ray scattering, formulation is in **K** or momentum space, rather than **r** space, and the shape of the pseudopotential, U say, as a function of K, is by now well established in its general features, and is shown schematically in Fig. 3.

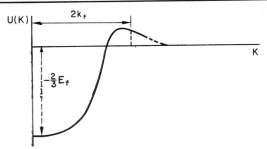

Fig. 3. Schematic form of pseudopotential $U(K)$, required in evaluating electrical resistivity and thermoelectric power.

The $K = 0$ value is closely connected with the shielding arguments referred to earlier, and is two thirds of the Fermi energy E_f in magnitude.

3.2. Scattering from assembly of weak pseudo-potentials

Ziman saw that the concept of the pseudo-potential ought to allow the application of conventional diffraction theory (as in Bragg scattering) to calculate the electron scattering and hence the electrical resistance of liquid metals. In fact such an approach had been advocated earlier by Krishnan and Bhatia (1945), but the concept of the pseudopotential was lacking at that time and it was not possible for them to press their theory through in a quantitative manner.

It now follows readily that we must discuss the scattering of electrons off an array of weak pseudo-potentials, arranged, of course, in a manner compatible with the measured structure factor $S(K)$ discussed in section 1 : here K represents the momentum transfer. Ziman's result for the electrical resistivity of a liquid metal may now be expressed in the form

$$\varrho_{\text{liquid}} = \frac{3\pi}{\hbar e^2 v_f^2 \varrho_0} \int_0^1 S(K) |U(K)|^2 4\left(\frac{K}{2k_f}\right)^3 d\left(\frac{K}{2k_f}\right) \quad (3.1)$$

where v_f is the Fermi velocity. Without doing detailed calculations it is now possible to see that many of the gross features of the resistivity of liquid metals follow from this account. Thus (cf. Enderby, 1968) take as an example the well-established fact that polyvalent liquid metals like lead and tin have a higher resistance than monovalent liquid metals such as sodium or silver. This can be explained by noting where $2k_f$ lies on the structure factor $S(K)$. Figure 1 shows that for lead the integral determining the resistivity includes the main peak in $S(K)$, whereas for silver it does not.

This prompts the speculation that when $2k_f$ falls near the top of the main peak, one might expect anomalous behaviour, and this is borne out by the fact that the resistivity of liquid zinc actually decreases as the temperature increases, reflecting the circumstance that at higher temperatures the principal peak in $S(K)$ is lowered and broadened.

More quantitatively, when we insert $U(K)$'s like that of Fig. 3, together with the measured $S(K)$'s, into the formula (3.1), we come out with resistivities of a few microhm-cm, which is the normal metallic range and the good conducting properties of liquid metals no longer seem a puzzle. It is also possible to explain the variation of resistivity among the monovalent metals and we come out with a natural enough explanation then that, for example, sodium is a much better conductor than liquid caesium.

In addition to the resistivity, the theory can be used to calculate the thermoelectric power Q. Q is usefully defined in terms of the dimensionless quantity ξ, given by

$$\xi = Q \bigg/ \left(\frac{\pi^2 k_B^2 T}{3 e E_f}\right) \quad (3.2)$$

where k_B is Boltzmann's constant, T is the absolute temperature and E_f is the Fermi energy. Using the Ziman theory, an approximate result for ξ may be written as

$$\xi = 3 - \frac{2 S(2k_f) |U(2k_f)|^2}{\langle SU^2 \rangle} \quad (3.3)$$

where $\langle SU^2 \rangle$ is an abbreviation for the integral in (3.1). The most detailed calculations available to date have been made by Sundström (1965): and the agreement with experiment turns out to be reasonably satisfactory.

4. Other Aspects of Liquid Metals

Though we have spoken here only about pure metals, the Ziman theory has been applied to alloys with considerable success (Faber and Ziman, 1965), and further work is clearly going to be forthcoming here.

Two other aspects of liquid metals are going to be of great importance in furthering our basic knowledge. The first is the nature of electron states in liquid metals. Here the pioneering work of Edwards (1962) on the theory has led to a good deal of progress, and to the use of a variety of techniques to study this problem (e.g. positron annihilation, photoelectric emission, soft X-ray emission, etc.). The second is the problem of the dynamics of atoms in liquids. This leads to a time-dependent generalization of the structure factor $S(K)$ which is again accessible to experiment through inelastic neutron scattering. This also leads to some understanding of transport properties, particularly self-diffusion and viscosity (cf. Brown and March, 1968).

Bibliography

BROWN R. C. and MARCH N. H. (1968) *Physics and Chemistry of Liquids*, Vol. **1**, 141.
DEBYE P. (1915) *Ann. d. Physik* **46**, 809.
EDWARDS S. F. (1962) *Proc. Roy. Soc.* **A 267**, 518.
ENDERBY J. E. (1968) *New Scientist* **37**, 313.
FABER T. E. and ZIMAN J. M. (1965) *Phil. Mag.* **11**, 153.

Krishnan K. S. and Bhatia A. B. (1945) *Nature* **156**, 503.
Johnson M. D. and March N. H. (1963) *Phys. Letters* **3**, 313.
Johnson M. D., Hutchinson P. and March N. H. (1964) *Proc. Roy. Soc.* A **282**, 283.
Phillips J. C. and Kleinman L. (1959) *Phys. Rev.* **116**, 287.
Sundström L. J. (1965) *Phil. Mag.* **11**, 657.
Zernike F. and Prins J. A. (1927) *Z. Phys.* **41**, 184.
Ziman J. M. (1961) *Phil. Mag.* **6**, 1013.

General references (not referred to in text)

Cole G. H. A. (1967) *An Introduction to the Statistical Theory of Classical Simple Dense Fluids*, Oxford: Pergamon Press.
Egelstaff P. A. (1967) *An Introduction to the Liquid State*, New York: Interscience.
March N. H. (1968) *Liquid Metals*, Oxford: Pergamon Press.

N. H. March

LUMINOSITY OF A SPECTROMETER.

The luminosity $\mathscr{L}(\sigma)$ of a spectrometer for radiation of wavenumber σ is defined (Jacquinot, 1960) as the ratio

$$\mathscr{L}(\sigma) = \frac{I(\sigma)}{B(\sigma)}$$

of the luminous flux $I(\sigma)$ detected to that $B(\sigma)$ leaving the source. If the transmission factor of the spectrometer is $\tau_0(\sigma)$ the detected intensity is

$$I(\sigma) = A\Omega\tau_0(\sigma)\, B(\sigma)$$

and the luminosity is

$$\mathscr{L}(\sigma) = A\Omega\tau_0(\sigma)$$

where A is the cross-sectional area of the output beam and Ω is the solid angle subtended by the exit aperture. If we introduce a quality factor

$$Q(\sigma) = \frac{E\mathscr{R}(\sigma)}{S}$$

where $E = A\Omega$ is the étendue or throughput of the system, $\mathscr{R}(\sigma)$ is the resolving power and S is the area of some characteristic component in the optical system then

$$\mathscr{L}(\sigma) = \frac{SQ(\sigma)\,\tau_0(\sigma)}{\mathscr{R}(\sigma)}.$$

The luminosities of different spectral systems can be compared if $SQ(\sigma)$ is evaluated for each. Three of the most common spectrometers are the two dispersive types, having as their dispersive elements the prism and the reflection grating, and the Michelson interferometer used in Fourier spectrometry.

Jacquinot (1954, 1958) showed that for the prism

$$S_p Q_p(\sigma) = S_p \beta \sigma\, \frac{dn(\sigma)}{d\sigma}$$

where β is the angular height of the slits and S_p is the area of the *base* of the prism. The angular height of a slit is typically of the order of 0·02 and the term $\sigma\dfrac{dn(\sigma)}{d\sigma}$ is rarely greater than 0·15 (Jacquinot, 1954) hence

$$\mathscr{L}_p(\sigma) = \frac{0.003 S_p \tau_0(\sigma)}{\mathscr{R}(\sigma)}.$$

For a grating spectrometer

$$S_g Q_g(\sigma) = S_g \beta \quad \text{approximately,}$$

where S_g is the area of the grating and the assumption of a Littrow mounting with equal angles of incidence and reflection near 30° is made; the corresponding luminosity is approximately

$$\mathscr{L}_g(\sigma) = \frac{0.02 S_g \tau_0(\sigma)}{\mathscr{R}(\sigma)}.$$

For a Michelson interferometer or Fabry-Perot etalon

$$S_i Q_i(\sigma) = 2\pi S_i,$$

where S_i is the area of the emergent beam ($= A$), and the luminosity is approximately

$$\mathscr{L}_i(\sigma) = \frac{6 S_i \tau_0(\sigma)}{\mathscr{R}(\sigma)}.$$

The luminosities are compared by making $S_p = S_g = S_i$ and assuming equal transmission factors and resolving powers. The ratio of the luminosities

$$\mathscr{L}_p(\sigma) : \mathscr{L}_g(\sigma) : \mathscr{L}_i(\sigma) = 1 : 7 : 2000$$

is then equal to that of the quality factors. It is clear that the radiation grasp of an interferometer is, chiefly on account of its cylindrical symmetry about the principal optical axis, superior to that of a conventional grating spectrometer by more than two orders of magnitude, while that of a prism spectrometer is nearly an order of magnitude worse than that of a grating instrument.

The enormous superiority of the radiation grasp of an interferometric spectrometer over that of any conventional instrument employing slits is known as the luminosity advantage.

Bibliography

Jacquinot P. (1954) *J. Opt. Soc. Amer.* **44**, 761.
Jacquinot P. (1958) *J. Phys. Rad.* **19**, 223.
Jacquinot P. (1960) *Rep. Progr. Phys.* **23**, 267.

John Chamberlain

MAGNETIC THIN FILMS.

Introduction

In recent years there has been a continued interest in thin films of magnetic alloys and in particular those films prepared from a binary alloy system, typically nickel and iron.

Magnetic films have retained their interest due mainly to two reasons. The first is that in general thin films have a particularly simple magnetic configuration as compared with bulk magnetic materials and this has given the research worker a useful tool with which to study fundamental ideas of magnetism. The second reason is the growing importance of magnetic thin films in the commercial field, for application as very high-speed computer memories. The potential offered by the thin film device compared with the conventional method of storing binary information—that is, using the core or toroid of magnetic ferrite—will undoubtedly continue to provide a stimulus to all those interested in expanding the field of computer memories.

In general, magnetic films have a thickness of rather less than 1000 Å and their linear dimensions are usually between 0·5 mm and 1·0 cm either circular or rectangular geometrically. The most common method of preparation used is to heat a magnetic alloy of the desired composition in a crucible and in a vacuum of about 10^{-5} torr. Once the alloy has melted it will evaporate and is allowed to condense onto a suitable substrate. If the substrate is polycrystalline or amorphous then the deposited film will also be polycrystalline, that is, the film will consist of a random distribution of grains and within each grain the magnetization vectors will be approximately parallel. The film will therefore be in a demagnetized state. If, however, the substrate is a single crystal, then providing the deposition conditions are favourable the film will also be a single crystal. This is not a general rule because the substrate will never be a complete single crystal but will contain a multitude of imperfections and dislocations and so therefore will the deposited film. Also epitaxial growth of a single-crystal film will only be successful providing the lattice constants of the substrate material and film are similar (percentage difference in lattice constants less than 10 per cent).

It is not surprising, therefore, that the magnetic properties of polycrystalline and single-crystal films will be rather different and it is without doubt that up to the present time considerable attention has been paid to polycrystalline magnetic films, whereas single-crystal films have taken somewhat low priority. One common feature of both types of film is that they show magnetic anisotropy and the various types that exist are now discussed.

Magnetocrystalline Anisotropy

It is logical to begin with magnetocrystalline anisotropy as this is a phenomenon related not only to thin films but to bulk media as well. When a material is magnetically anisotropic we mean that if a magnetic field is applied, the induced magnetic moment/unit volume or magnetization in the material depends on the direction of the applied field in relation to the crystallographic axes or some other set of reference axes. If the material is single crystal then an anisotropy arises from the coupling of the magnetic spins and lattice upon which the crystal structure is built. Thus it requires more energy for magnetization in some directions than in others. This is seen clearly if one measures the magnetization of, for example, iron as a function of applied magnetic field in various crystallographic directions as shown in Fig. 1. The $\langle 100 \rangle$ is known as the easy direction of magnetization since it is the direction along which the material is most easily magnetized.

In a polycrystalline film no magnetocrystalline anisotropy will exist overall because although it will exist in each individual grain, the grains will be randomly oriented and so the net effect will average out to zero.

The $\langle 111 \rangle$ in iron is referred to as the hard direction

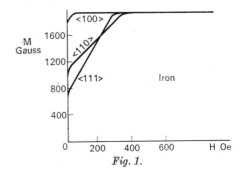

Fig. 1.

as it is the direction along which it is most difficult to magnetize the material. The difference in energy involved in magnetizing the material in the hard and easy directions is known as the anisotropy energy. For a material with cubic symmetry the anisotropy energy E_k is given by

$$E_k = K_1(\alpha_1^2\alpha_2^2 + \alpha_2^2\alpha_3^2 + \alpha_3^2\alpha_1^2) + K_2\alpha_1^2\alpha_2^2\alpha_3^2 + \cdots$$

where the α's are the direction cosines of the magnetization with respect to the crystal axes and K_1 and K_2 are known as the anisotropy constants of the material. The values of K_1 and K_2 depend on a number of factors—the composition of the material, the temperature and stress. For iron at room temperature $K_1 = 4.8 \times 10^5$ ergs cm^{-3} and $K_2 = 1.5 \times 10^5$ ergs cm^{-3}.

For a uniaxial crystal, that is, one having but a single easy direction, E_k can be simplified to give

$$E_{ku} = K_{u_1}\sin^2\varphi + K_{u_2}\sin^4\varphi + \cdots$$

and φ is now the angle the magnetization makes with the single easy axis. Cobalt is an example of such a material since it has hexagonal symmetry about the c axis and this therefore will be the easy direction.

The actual physical origin of magnetocrystalline anisotropy has been the subject of much speculation for the last thirty years. The most outstanding contribution is due to Van Vleck (1937) who laid the groundwork for a quantum theory of anisotropy. Normally in a magnetic crystal the orbital magnetic moment of a particular ion is zero or "quenched" due to its interaction with the remaining crystalline field and magnetic effects would be due entirely to electron spin. Electron spin on its own has no "memory" of a particular direction in the crystal and so cannot alone account for anisotropy. Van Vleck argues that providing the quenching is incomplete some coupling between the spin and orbital wave functions will result and since the orbital wave functions are locked in place by the crystal lattice the electron spin will have a preferred direction. Some success has been achieved using this theory and indeed the values of K_1 are of the correct order of magnitude as compared with the experimental results of Boyd (1960) who has measured K_1 and K_2 on single-crystal films of iron, nickel, nickel–iron and nickel–cobalt when the stresses in the films were relieved by removal of the film from its parent substrate.

Uniaxial Anisotropy

So far the discussion could well have applied to bulk materials as well as magnetic thin films. However, it is the subject of uniaxial anisotropy that really gives thin films their true importance. It has been shown that in polycrystalline films the contribution of the magnetocrystalline anisotropy averages to zero. Nevertheless, it has been known since 1950 that such films can have a uniaxial anisotropy particularly if the film is deposited in a uniform magnetic field. In addition single-crystal films can have a uniaxial anisotropy and the easy axis of the uniaxial component need not coincide with the easy direction of magnetocrystalline anisotropy. It is apparent, therefore, that some other mechanism is needed to explain the uniaxial component.

A film with only uniaxial anisotropy is a particularly simple configuration—one easy direction parallel to the applied field during deposition and one hard direction perpendicular to the easy direction and lying in the plane of the film. The anisotropy energy for this system is given by: $E_k = K_u'\sin^2\varphi$ and now K_u is the uniaxial anisotropy constant and φ is the angle between the magnetization vector and easy axis. It is customary to define an anisotropy field $H_K = \dfrac{2K_u}{M_S}$ and M_S is the saturation magnetization. Physically it is the field one needs to apply in the hard direction, after the film has been deposited, to rotate the magnetization vectors through 90°. This is in fact the basis of the method used to measure H_K.

Little progress has been made in recent years in understanding the origin of the field-induced anisotropy (or M-induced anisotropy as it is often referred to). Any stress in the film could contribute to the M-induced anisotropy providing the stress is uniaxial and that the composition of the film is such that the magnetic vectors are easily aligned by it. As will be seen in the next section it is quite possible that this latter condition is not satisfied and thus one is forced to look for an alternative explanation of M-induced anisotropy.

Néel (1953) and Taniguchi (1954) proposed a theory of directional ordering and to date their work is the most outstanding contribution in this field. The basis of their theory is that when a film is deposited, some anisotropic grouping of the respective atoms takes place due to the presence of the magnetic field. Thus if one considers a binary alloy system consisting of atoms A and B then because the atomic coupling of A and B atoms together will be different from A and A or B and B atoms and if anisotropic grouping has taken place then the film will show an induced anisotropy. The result of their calculation is shown in Fig. 2 where the temperatures on the three curves are the substrate temperatures during preparation of the film.

A further contribution to the M-induced anisotropy can also arise from the presence of non-magnetic impurities such as oxygen in the film. In particular, in the nickel–iron system it is believed that regions of nickel oxide are formed which is an antiferromagnetic material, having a high coercive field. This effectively acts as an aligning field for the remainder of the film and thus produces a uniaxial anisotropy.

Magnetostrictive Anisotropy

The magnetization within a film can also be a function of mechanical stress. This property is described by means of a magnetostriction coefficient λ. Positive magnetostriction is defined in such a way that application of a uniaxial stress causes the magnetization to increase and vice versa.

Magnetostriction can have a number of effects on the magnetic properties of the films. If the stress is uniaxial then an anisotropy will be developed parallel to the stress providing λ is isotropic. If λ is anisotropic, that is, the tensor describing λ has more than one nonzero coefficient, then the easy axis produced will not be parallel to the stress.

The magnetostriction coefficient λ is a function of the composition of the film. In most alloy systems a composition exists where $\lambda = 0$ and a ternary diagram is shown in Fig. 3 for the Ni–Fe–Co system. The line corresponding to $\lambda = 0$ is shown. This data was obtained from bulk material but it has been shown using ferromagnetic resonance that within experimental error the results do not differ from thin film samples.

Physically it is useful to think of magnetostriction as arising from the dependence of the magnetocrystalline anisotropy energy on the state of strain of the lattice: thus it may be energetically favourable for the crystal to deform slightly from its normal shape if doing so will lower the anisotropy energy by more than the elastic strain energy is raised.

Using an energy equation for a stressed single crystal, Blades (1959) has obtained an equation for the magnetic energy of a polycrystalline film which has a planar isotropic plus a uniaxial stress configuration. He obtained $E = \dfrac{3}{2} \lambda \sigma \sin^2 \varphi$ and σ is the stress and λ and φ have their usual meanings. Comparing this equation with the equation for the uniaxial anisotropy energy, enables a stress anisotropy constant K_S to be identified given by $K_S = \dfrac{3}{2} \lambda \sigma$.

Providing one knows the magnitude of the stress within the film and the dependence of λ on the composition then K_S versus composition can be calculated. This, together with any other anisotropies in the film, provides the basis of evaluating the variation of the total anisotropy with composition (and the temperature if necessary).

A good deal of quantitative information is still needed, however, about the actual stress systems present in the films and indeed the origin of the stresses themselves together with values of σ for the most common alloy systems.

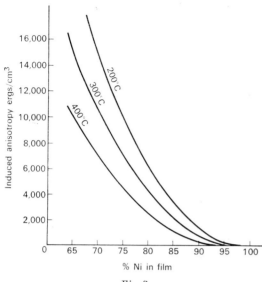

Fig. 2.

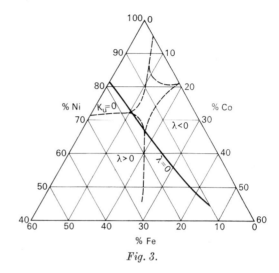

Fig. 3.

Anisotropy Dispersion

The composition of a magnetic thin film will vary on a small scale from place to place within the film. This implies that even though the magnetostriction as measured using macroscopic techniques may be zero the local value may be non-zero. As mentioned in the last section there are sure to be isotropic stresses built into the film as a result of the method of growth of the film during preparation. Thus the easy direction of magnetization will vary from place to place in the film, but the mean value of all the local easy directions will be parallel to the applied field during deposition. This property is known as easy axis dispersion.

It is also possible for the local anisotropy field H_K to vary from place to place in the film and in this case we refer to the property as magnitude dispersion.

The distribution of easy directions or anisotropy fields, as the case may be, can be measured by a variety of experimental techniques and for a general review the reader is referred to Prutton (1962). The distribution is usually Gaussian, but only for films whose linear dimensions are greater than about 5 mm. The reason for this is that when the linear dimensions of the film become small, the demagnetization field arising from the free poles on the ends of the film is large enough to cause a further increase of the angular and magnitude dispersions over and above the intrinsic value assigned to the growth process. In the case of magnitude dispersion the effect of shape of the film is to skew the distribution to high H_K values.

The intrinsic dispersions are a function of the grain size of the deposited material, and because of this it is possible to identify two types of angular dispersion. If the grain size is less than 100 Å then the exchange coupling between adjacent grains will be large and will prevent all but the smallest variation in magnetization direction. Little anisotropy dispersion is therefore observed and any fluctuations that take place in the magnetization direction will be long wavelength variations and are referred to as ripple structure. However, when regions larger than 100–1000 Å are considered, the short range exchange fields become small and are replaced by long range magnetostatic fields which produce high dispersion.

It is difficult to generalize as to orders of magnitude of the angular and magnitude dispersions because the values obtained are critically dependent on preparation technique and the associated variables. However, angular dispersions can vary from about 1° to 10° and the magnitude dispersion can vary from 2 to 10 Oe for the most commonly prepared films of nickel-iron.

Coercive Field

Theoretically speaking, the coercive field H_C of ferromagnetic materials is not a well understood parameter, and consequently the understanding in ferromagnetic films is not as good as that of the anisotropy or the dispersion. Once again the majority of work in this field has been concentrated on polycrystalline films exhibiting a uniaxial anisotropy. In this context the definition of coercive field is simple—being the easy direction field needed to completely reverse the magnetization in a film from one remanent state parallel to the easy axis to its equivalent state in the opposite sense to its original magnetization direction.

H_C is dependent on quite a number of parameters—the thickness of film, the size of the film, the temperature of the substrate during deposition, the rate of evaporation of material and the nature of the substrate.

Néel has suggested that the coercive field should depend on thickness t in the following manner: $H_C \propto t^{-4/3}$. This was derived on the basis of the variation of domain wall energy density in the film. Experimental results indicate that the exponent is <1.25. It is not surprising that such a discrepancy should arise because the mobility of domain walls will be affected by impurity atoms within the film which act as locking centres for domain walls, and the number of impurities present will be a critical function of the deposition parameters.

In general the coercive field will increase with increasing angular dispersion. This is due to the effect of grain size of the material because both coercive field and angular dispersion increase with increasing substrate temperature during deposition. Higher substrate temperatures increase the mobility of the atoms as they are deposited and thus larger grains are formed.

The rôle of the nature of the substrate surface is still not completely understood but variations in the topography of the surface on a scale of the grain size of the film certainly have profound effects on H_C. As the surface roughness increases in this region then H_C tends to increase. Anisotropic surfaces have little effect on H_C if the variations involved are large compared with the grain size.

Applications

A good deal of interest in magnetic thin films stems from the fact that providing the thickness of the film is less than 1000 Å the film will remain as a single domain, once it is placed in a remanent state parallel to its easy direction. This is the situation in a polycrystalline film exhibiting uniaxial anisotropy. There will be two possible directions of stable magnetization —parallel to the easy direction and pointing → or ←. This constitutes a binary computer storage element and this particular aspect of magnetic thin films is at present being exploited as a possible contender to the conventional ferrite core. Magnetic films have a great advantage over cores because the magnetization can be reversed by a rotation process rather than the process of domain wall movement as occurs in cores. It turns out, that the magnetization in a thin film can be reversed in as little as 1 nsec (1 nsec = 10^{-9} sec) compared with 100 nsec for the fastest core available.

Ferrite thin Films

Magnetic ferrites and garnets, have, up to the present time, been of great interest in the bulk rather than thin film form, as exemplified by their extensive application to the production of cores or toroids for use as conventional bistable computer storage devices.

A small but significant interest is now being shown in ferrite thin films mainly from an academic point of view. Ferrite thin films are often prepared by high temperature oxidation of the appropriate metal alloy. The metal alloy is usually sputtered onto a suitable substrate and is then oxidized, for example, Francombe and Flanders (1964) report the production of films of the ferrite $CuFe_2O_4$ from the alloy $CuFe_2$ and Brook and Kingery (1967) report the production of the corresponding nickel ferrite.

Depending on the preparation conditions either a polycrystalline ferrite film is formed or if the original alloy was grown epitaxially on a single crystal substrate then the corresponding ferrite is found by X-ray analysis to comprise well-crystallized spinel-type phases.

The saturation magnetization of ferrite films is smaller than that for permalloy thin films. Often it is of the order of 300–400 Oe compared to 800 Oe for Ni–Fe films. The coercive field is much larger, approximately 300–500 Oe, whereas for permalloy it is usually 1–10 Oe.

Bibliography

BLADES J. D. (1959) *J. Appl. Phys.* **30**, 260 S.
BOYD E. L. (1960) *I.B.M. J. Res. Dev.* **4**, 116.
BROOK R. J. and KINGERY W. D. (1967) *J. Appl. Phys.* **38**, 3589.
FRANCOMBE M. H. and FLANDERS P. J. (1964) *Proc. Int. Conf. on Magnetism* (Nottingham), 803.
NÉEL L. (1953) *Comp. Rend.* **237**, 1468 and 1613.
PRUTTON M. (1962) *Ferromagnetic Thin Films*, London: Butterworth.
TANIGUCHI S. (1954) *Sci. Repts. Res. Inst. Tohuku Univ.* **A6**, 330.
VAN VLECK J. H. (1937) *Phys. Rev.* **52**, 1178.

M. J. FOLKES

MAGNETORESISTIVITY. The magnetoresistivity effect is the increase in the resistance of a conductor when a magnetic field is applied. A physical explanation for this effect can easily be obtained if it is assumed that the mean free path of electrons is independent of the applied electric and magnetic fields. The justification for this assumption is the fact that the perturbing effect of the applied fields upon the electronic distribution will under normal circumstances be small compared with the restoring forces. When a magnetic field is applied, the trajectories of electrons between collisions will acquire a curvature. This means that the component of motion in the direction of the applied electric field will be reduced by the presence of the magnetic field and therefore the resistance will be higher. Calculation shows that for an isotropic system in which the electrons obey classical Boltzmann statistics (as they usually do in semiconductors) this effect should be observed when a magnetic field is applied at right angles to the direction of the electric field (as in the configuration used in measuring the Hall effect, but no effect obtained if the magnetic field is applied in the same direction as the electric field. If the direction of the magnetic field is the same as that of the velocity of the electron (consideration of the Lorentz force on an electron) the force produced by the magnetic field will vanish. When this conclusion is checked by experiment it is not always confirmed.

In germanium and silicon the longitudinal effect is found to be comparable with the transverse effect. This means that energy bands occupied by the conduction electrons exhibit a degree of anisotropy, although they must still exhibit cubic symmetry. It is found that the magnitude of the effect depends upon the crystalline orientation of the specimen. Indeed it must for if it did not, we should have an isotropic medium and therefore no longitudinal effect.

Taking the matter further, for small magnetic fields, isotropic bands and mixed conductivity, the change in resistivity is given by:

$$\frac{\varrho(B) - \varrho(0)}{\varrho(B)} = \frac{\varDelta \varrho}{\varrho(B)} = \left[A_1 \frac{\mu_n^3 n + \mu_p^3 p}{\mu_n n + \mu_p p} - A_2 \left(\frac{\mu_n^2 n - \mu_p^2 p}{\mu_n n + \mu_p p} \right)^2 \right] B^2,$$

where $\varrho(B)$, $\varrho(0)$ are resistivities in fields B and 0 respectively, and for scattering mechanisms with a relaxation time, the factors A_1 and A_2 are constants. It is seen that $\varDelta \varrho / \varrho(B)$ is proportional to the square of the magnetic field perpendicular to the direction of current flow. For pure extrinsic conduction

$$\frac{\varDelta \varrho}{\varrho(B)} = (A_1 - A_2)(\mu B)^2.$$

With increasing magnetic field $\varDelta \varrho / \varrho(B)$ saturates, according to the classical theory.

In the above expression the change of resistance of a semiconductor by a magnetic field depends on two terms. The first describes the deflection of the charge carriers by the Lorentz force of the electric and magnetic fields; the second describes the opposite deflection by the Hall field. The two contributions almost compensate as long as there is only one kind of charge. The resistivity is changed because the Lorentz force is different for carriers of different velocities, while the Hall field can only compensate the average deflection. This fact means that the current paths of individual carriers have a component perpendicular to the sample axis. Only the sum of all these components is zero.

This picture applies only in the case of an infinitely thin rod, where the total transverse component of current is really zero. For an infinitely wide sample current compensation is not required and the resistivity change is given by the first term of the equation alone. This means that the magnetoresistance depends wholly on the geometry of the sample. The magnetoresistance increases with decreasing ratio of sample length to sample width (u), while the Hall voltage sinks and disappears for $u = 0$.

For certain cases the expressions can be given explicitly ($R(B)$ is the sample resistance in the field B):

$$R(u, B) = R(\infty, B) F(u, \tan \theta)$$

$$F(u, \tan \theta) = 1 \text{ for } u = \infty$$
$$= 1 + \tan^2 \theta \text{ for } u = 0$$
$$= \sqrt{(1 + \tan^2 \theta)} \text{ for } u = 1$$

where θ is the Hall angle ($= \tan^{-1} \mu B$).

No explicit expressions can be given for other values of u. The geometry function F is plotted vs. Hall angle in Fig. 1 for various u values. The corresponding expression for the Hall voltage is

$$Vh(u) = Vh(\infty)\, G(u, \tan \theta)$$

and the geometry function G is shown in Fig. 2. The geometry dependence of the galvanomagnetic effects for mixed conductivity has not yet been investigated theoretically. The limit $u = 0$ (infinitely wide sample) can be realized experimentally by using a circular disc (Fig. 3) with one electrode in the centre and the other along the periphery (Corbino disc). The electric field has then only a radial component, and the current paths are logarithmic spirals which make an angle equal to the Hall angle with the radial direction. According to equation (2) the influence of the sample shape is larger, the larger the Hall angle is, i.e. the larger the product of mobility and magnetic field. It is therefore of particular importance for semiconductors of high mobility such as n-InSb and n-InAs.

In practical applications of the magnetoresistive effect indium antimonide is almost exclusively used. In n-InSb, deviations from the approximations used in simple galvanomagnetic theory already appear at very small magnetic fields. The range of applicability of the approximations can easily be estimated. In the theory of the galvanomagnetic effects the magnetic field always appears as the product μB. An expansion in increasing powers of B therefore means an expansion based on the parameter μB. The termination of this expansion after the first non-vanishing term is justified only if $\mu B \ll 1$. In this condition μB must be chosen dimensionless so that, when μ is given in cm^2/V sec, B is given in V sec/cm^2 or 10^8 gauss. In pure n-type material $\mu_n = 77{,}000$ cm^2/V sec at room temperature and $> 600{,}000$ cm^2/V sec at 77°K. The magnetic field dependence of the galvanomagnetic

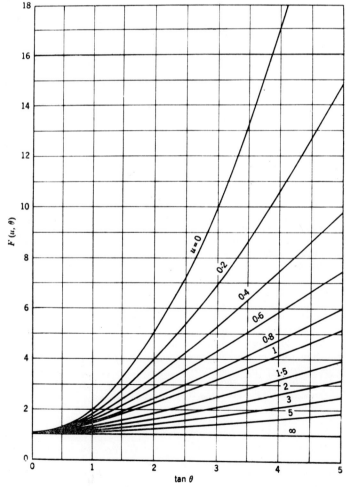

Fig. 1. Geometry factor $F(u, \theta)$ as a function of the tangent of the Hall angle for different length-to-width ratios.

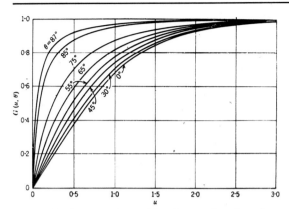

Fig. 2. Geometry factor $G(u, \theta)$ as a function of the length-to-width ratio for different Hall angles.

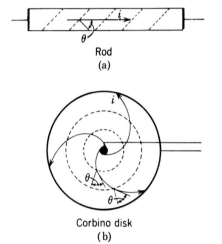

Fig. 3. Equipotential lines and field lines for a long thin rod and for the Corbino disk in a magnetic field.

effects will therefore already be substantial at $B = 1300$ gauss and 160 gauss respectively.

We have already pointed out the dependence of the galvanomagnetic effects on the geometric shape of the sample. This relation is more important for the larger values of the Hall angle. Figure 4 shows this influence clearly. The relative resistance R_B/R_0 is shown for four differently shaped InSb samples of equal purity. This figure shows the importance of the geometry effect in semiconductors of high mobility. Even for long thin rods the resistance change is due mainly to geometry effects at the rod ends. It is possible to create an artificial "geometric" increase in the resistance by two methods.

1. The "rasterplatte" whereby parallel conducting strips are deposited transversely to a long strip of the semiconductor (Fig. 5). Essentially the strips short circuit the Hall voltage and hence to increase the magnetoresistive effect.

2. The introduction of highly-conducting oriented inclusions into the lattice. To demonstrate this, Weiss and Wilhelm produced InSb with needle-shaped NiSb inclusions by directional freezing of the InSb–NiSb eutectic. This material exhibits very large magnetoresistance under suitable orientation and devices made from this material are known as field plates.

Applications of the Magnetoresistive Effect

As early as 1895 it was known that magnetic fields could be detected and measured by bismuth spirals

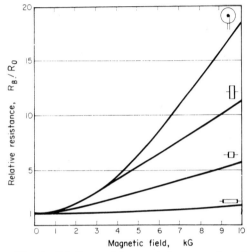

Fig. 4. Relative resistance of four samples of InSb of equal purity but different shape.

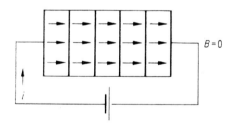

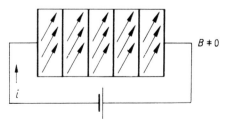

Fig. 5. The pattern of current paths in a "rasterplatte" in schematic form. Top: without a magnetic field; bottom: in a magnetic field perpendicular to the plane of the diagram.

(Fig. 6) and indeed up to the present time the technology has produced high impedance, thin film, flexible bismuth magnetoresistive elements known as "mistors". However, its linearity is poor and its room temperature characteristics are likewise so. It will,

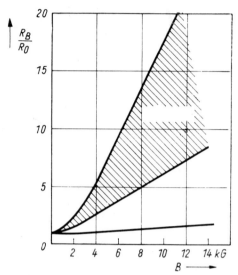

Fig. 6. Comparison of resistance change of bismuth probes (bottom curve) with field plates (shaded area).

however, find increasing use at liquid helium temperatures where large magnetic fields may be encountered in superconducting solenoids.

The advent of indium antimonide with its exceptional mobility has provided the impetus for a large number of applications. As it stands the material presents two disadvantages.

1. As the electrical conductivity is high, the resistance of such a magnetoresistive element would be unduly low, less than 1 ohm, which would make matching difficult.
2. The temperature coefficient of the resistive change is poor.

The use of field plates appears to solve these difficulties as resistance values vary from 10 ohms to 500 ohms, which in a field of 10 kg could change as much as 15 times. A photograph of two typical field plates is shown in Fig. 7, where their small size facilitates inclusion in electronic circuits. A meander-shaped ribbon of semiconductor material about 20 μm thick is deposited on an insulating substrate about 0·5 mm thick and is terminated in a pair of leads. The nickel antimonide metallic inclusions are too small to be seen. The characteristic curves of the materials used in field plates are shown in Fig. 8 plotted at 22°C. The resistance R_B in a magnetic field referred to the resistance R_0 with field, is a quadratic function of the magnetic flux B in fields of less than 3 kg. In larger fields the relation to B gradually becomes linear.

Fig. 7. Two field plates (magneto resistors).

With suitable materials this linearity holds good for fields above 100 kg. In Fig. 8 the curves for the various materials are identified by letters. They differ in the relationship of resistance to magnetic induction as well as to temperature. For suitable designed components the typical dependence of R_B/R_0 on the magnetic induction applies to frequencies well into the gigahertz range.

Magnetoresistors can be used as probes to measure and control magnetic fields at temperatures down to −60°C. Certain probes can be used as low as 4·2°K (liquid helium).

One of the primary applications of magnetoresistors is as contactless and infinitely variable resistors. The control function is performed either with a permanent magnet whose position with respect to the semiconductor is changed or by varying the current in an electromagnet with the semiconductor placed in the airgap. Figure 9a exemplifies the former case, where the right hand side shows a field plate with leads attached. The rectangle on the left represents the magnetic flux of a permanent magnetic acting on the field plate; its effective surface area is determined by the shape of the polecaps of the magnet. If the magnetic flux is displaced from left to right then the resis-

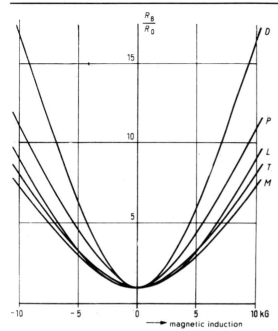

Fig. 8. Resistance relation R_B/R_0 as a function of the magnetic induction B for various semiconductive materials.

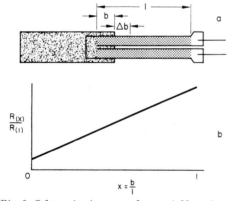

Fig. 9. Schematic of a contactless variable resistor.

tance increases as a larger portion of the semiconductor is covered by the flux. The end point of the increase is reached when the flux has reached the terminals, i.e. $b = l$. If the semiconductor is homogeneous, and if the two legs of the field plate are constant in width and thickness, then for a displacement Δb, an additional portion of the field-dependent resistance proportional to Δb is introduced into the magnetic field. Furthermore, if the magnetic induction within the driving flux is homogeneous, so that the already covered portions do not change their resistance, then the resistance increases linearly with b. Figure 9b shows the relative dependence of the resistance on the displacement b, referred to the length l of the field plate for $R_B/R_0 = 9$.

The Hall effect has been used for measuring electric power. The magnetoresistance effect in semiconductors can also be used as a wattmeter with much higher gain and simpler construction than the corresponding Hall effect. In general, the resistance of a semiconductor element shows an increase at first proportional to the square of and then linear with the applied magnetic field. The principle is that the time average of the voltage drop across a magnetoresistor is always proportional to the net product of a current passed through it and a magnetic field or a magnetizing current passed through a magnetizing coil when the semiconductor is properly biased magnetically in the linear region of the magnetoresistance characteristics. In Fig. 10

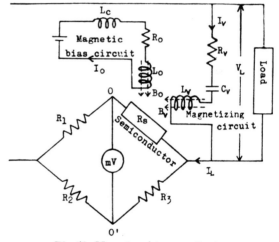

Fig. 10. Magnetoresistance wattmeter.

a choke L_C and a resistor R_0 in the bias circuit are to prevent an alternating current induced by the magnetizing current I_v. Capacitor C_v and resistor R_v are chosen to give a magnetic flux B_v in phase with the load voltage V_L and a very high impedance in the circuit compared with the load. Under the proper condition of the phase adjustment at a.c. operation, the resistance of the field plate can be expressed by

$$R_s = R_{s0} + A V_0 \sin \omega t$$

where R_{s0} is the resistance of the semiconductor under the d.c. magnetic bias only, $V_0 \sin \omega t$ is the load voltage and A a proportionality constant. Then the voltage drop across the semiconductor is

$$v = (R_{s0} + A V_0 \sin \omega t) k \times I_0 \sin (\omega t + \phi)$$

where $\cos \phi$ is the power factor of the load and k is a proportionality constant. Its time average

$$\bar{v} = A k \frac{V_0 I_0}{2} \cos \phi.$$

Hence the active power can be measured by the time-average voltage across the semiconductor element. The accuracy is of the order of 2 per cent. The above application requires biasing on to the linear part of the characteristic.

We may also use the initial quadratic dependence of R_B/R_0 on field to construct a three-dimensional multiplier. For small magnetic fields the resistance R may be expressed as $R = R_0 + mB^2$ where R_0 is the initial resistance at zero magnetic field and m is a constant. If a magnetoresistance unit consists of a field plate accommodated in the air gap of a magnetic core provided with two exciting windings and another identical unit is available then a bridge circuit may be arranged similar to a push–pull amplifier as shown in Fig. 11a, where the external resistance is much larger than that of the specimens, with one of the specimens subjected to the influence of the sum of two variable magnetic fields $B_1 + B_2$ while the other subjected to their difference, $B_1 - B_2$, the resistance of the specimens R_1 and R_2 then becomes:

and
$$R_1 = R_0 + m(B_1 + B_2)^2$$
$$R_2 = R_0 + m(B_1 - B_2)^2.$$

Under the conditions that the magnetic cores are well below saturation and the reluctance of the core is negligibly small in comparison with that in the air gap, the inductions B_1 and B_2 may be considered as proportional to the excitation currents I_1 and I_2. By passing currents of equal amplitude I_3 through the semiconductor specimens, the voltage difference V between the terminals of specimens R_1 and R_2 will be obtained as:

$$V = I_3(R_1 - R_2) = 4mB_1B_2I_3$$
or
$$V = KI_1I_2I_3$$

where K is a constant which includes the factor $4\,m$ and also takes care of the proportionality between the induction and the exciting current. The result is a three-dimensional electronic multiplier whose input currents represent three parameters and whose output voltage is proportional to product. In Fig. 11b, the dynamic characteristic curve of the multiplier is illustrated. This kind of multiplier finds application as an analogue computing element carrying out a large number of mathematical operations, as a polyphase wattmeter capable of showing the total power (real, reactive or apparent) in an electrical network with a single indication, as amplifiers and oscillators and as mixers, modulators and demodulators. It should be noted in comparing similar devices based on the Hall effect that magnetoresistance-electronic devices have the advantages of higher sensitivity, better efficiency and greater ease of adjustment. However, to obtain higher efficiencies, the magnetic field must be sufficiently large and the magnetic core should be composed of material other than ferrite, so that larger flux densities can be applied to the field plates without approaching the saturation of the core.

We can therefore see from the foregoing that although the galvanomagnetic effects are old solid state phenomena they have been increasingly applied to the systematic study of semiconductors, and—from the point of view of clarifying and enhancing our knowledge of the structure of solids—the magnetoresistive effect has become one of the most powerful and elegant tools for studying the electronic properties of such materials. Concomitant with this enrichment of our understanding has been the rapid progress made in utilizing the effect in electronic circuitry, where its peculiar properties have facilitated new applications to technology.

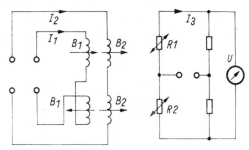

Fig. 11a. Schematic of a three-dimensional multiplier with field plates R 1 and R 2 (after Sun). The semiconductor elements of rectangular form are shown with two resistances in a bridge circuit.

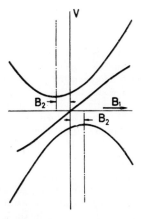

Fig. 11b. Fundamental circuit arrangement and characteristic curves of a three-dimensional multiplier utilizing magnetoresistance of semiconductor. The parabolic curves beyond the horizontal axis represent the characteristic curves of individual magnetoresistance units, while the straight line through the origin is the characteristic curve of the multiplier.

Bibliography

Conference on Hall effect and galvanomagnetic devices, *Solid State Electronics* (1966) **9,** No. 5, Oxford: Pergamon Press.

Hall effect issue, *Solid State Electronics* (1964) **7**, No. 5, Oxford: Pergamon Press.

LEE E. W. Magnetoresistance, in *Encyclopaedic Dictionary of Physics*, **4**, 476.

MADELUNG O. (1964) *Physics of the III–V Compounds*, London: John Wiley.

WEISS H. (1965), Field plates—magnetically controllable resistors, *ETZ*, Bd. 17.

E. COHEN

MELTING, MODERN THEORIES OF. In spite of many ingenious attempts, there is still no satisfactory theory of melting. Such a theory would require the evaluation of a partition function which yielded an equation of state with fluid and solid branches connected by a flat tie line representing the coexistence region. Although there is little doubt that the exact partition function would contain the accurate description of the melting process, the mathematical difficulties connected with the phase transition singularities in the partition function represent a formidable barrier. In order to bypass these difficulties, it is natural to turn to computer simulation of the melting process. This method, in which the detailed motion of a few hundred particles is followed, has led in recent years to a great deal of insight not only into the essential characteristics of melting, but into the mechanism as well.

The most important result of the computer studies is that attractive forces are not required for melting, in contrast to their crucial role in the gas–liquid phase transition. Thus it has been shown that a system of several hundred hard spheres at sufficiently high density will be stable as a solid. Although this may seem at first surprising, the result was essentially predicted long ago by the empirical Lindemann law. This law states that to a good approximation all substances melt when their volumes are larger by a constant fraction than their close-packed volume. Hence, the detailed nature of the attractive forces, whether they are of the metallic, the ionic, or the van der Waals dispersion type, is unimportant. This last observation was also long ago incorporated into another theory by van der Waals who, in his treatment of liquids, made the observation that the attractive forces could be considered as forming a uniform potential sea in which the hard spheres are immersed. Thus, the attractive forces merely determine the cohesion between the spheres and so determine the absolute temperature of melting, but the density of melting is determined by the geometric arrangement of the spheres.

Although all atoms experience a strong repulsive force at small separations resulting from the Pauli exclusion principle for electrons, these forces are weaker than those derived from the hard sphere model. The question therefore might be asked whether melting can also be observed for less strong but still purely repulsive forces. As an extreme case purely repulsive Coulomb forces between particles have been studied on the computer in order to simulate the exceedingly high pressure situation encountered in some astrophysical problems. The Coulomb forces, representing the most weakly repulsive forces imaginable in a real system, arise between the nuclei of atoms once the pressure and temperature are so high that the atoms are completely ionized and the electrons can be considered to form a uniform background. The computer study of this system has shown indeed that at a density for which the parameter $Z^2e^2/kT\sigma$ exceeds 125, a solid would be stable. Ze is the charge on the nucleus, kT is the Boltzmann constant times the temperature and σ is the internuclear distance. Thus, although the protons in the center of the sun are not sufficiently dense to form a solid, the interior of some white dwarfs might be. Concerning the question of melting, however, the above results confirm not only the geometric aspect of melting dominated by the repulsive forces, but also the absence of a solid–fluid critical point for any real substances.

Such a critical point was unlikely anyway in view of the powerful argument that its existence would allow a continuous transition from an ordered (solid) to a disordered (fluid) system. Although it is possible to change continuously the degree of order, as in the gas–liquid transition, it seems improbable that a continuous qualitative change of order (from long range order to none) can take place. The experimental investigation of the melting line to extreme conditions to verify the absence of a critical point has been significantly advanced in recent years by the development of the shock wave method into a quantitative tool for investigating simultaneous high temperature and high pressure conditions. Pressures up to 10 million atmospheres and 50,000°K can be generated, although the melting transitions are hard to detect. No doubt these methods will be refined in the coming years, so that, for example, the melting temperature of iron under conditions as it exists in the interior of the earth can be determined. This would be of enormous help in determining the temperature distribution in the core of the earth. Although solid iron has been shocked to more extreme conditions than in the earth's core and the experiment no doubt entered the fluid region, no special effort has been made to detect melting. Similar experiments on argon have already demonstrated melting under conditions more extreme in terms of the melting temperature reached, relative to the normal value, than any static experiments. These argon experiments confirm the validity of Lindemann's law even under very high pressure conditions. In fact, they show its greater accuracy there because at high pressures the attractive forces play an even lesser role.

Although the computer studies deal satisfactorily with both the liquid and solid phases, they do not give an accurate picture of the coexistence region. This is characteristic of any calculation which evaluates the partition function under a constraint. This constraint results generally in a van der Waals-like loop in the coexistence region, which means that the solid and liquid phases persist beyond their thermodynamic stability limits. In the computer calculations the relatively minor limitation of dealing with only a few hundred degrees of freedom instead of a very large number, leads to a very small van der Waals-like loop for two dimensional hard disks, but in three dimensions it prevents the coexistence of the solid and fluid phase of spheres for the largest system studied so far. In one dimension it can be proven that the transition is not possible even for infinite systems. Very small systems of about 10 particles fail to show the transition at all in either two or three dimensions. Larger two and three dimensional systems of about 100 particles show that only rarely does the system jump back and forth between the solid and fluid phase. Thus the two phases cannot simultaneously coexist. Still larger systems of about 1000 particles in two dimensions yield coexisting phases, and the resulting van der Waals-like loop can be considered as being due to the effects of interfacial tension which stabilizes the predominant phase. Still larger systems of spheres should also show coexisting phases.

The further deduction from the above results is that melting corresponds to the sudden opening of a new region of phase space as the density decreases. The "bottleneck", by which the system has access to the new volume in phase space, corresponds to the relatively rare density fluctuation which can break up the crystal. This density fluctuation becomes sufficiently probable at the hard sphere and disk volumes, relative to their close-packed volumes, as suggested by Lindemann. The nature of this density fluctuation is clear. It must break up the long range order of the crystal and cause the disappearance of the low frequency shear waves in the fluid; that is, the crystal becomes unstable to a long wavelength transverse shear mode. A simple model for disks illustrates these features of melting and leads to nearly quantitative agreement with the computer experiments, although as might be expected, it also shows a van der Waals-like loop. In this model alternate rows of particles are allowed to move relative to each other; and at the melting discontinuity the rows can suddenly slide past each other.

The existence of such sliding motion has been confirmed by computer experiments in a narrow density range of about 1 per cent previous to melting. For disks it can be observed, for example, that a row of particles will slide occasionally one lattice spacing. For spheres, whole planes of particles will slide while the system remains in the solid phase. This phenomenon is observed in the absence of any vacancies in the crystal, showing that they are not essential to the melting process. Of course, in real crystals vacancies will exist, and their number will increase with temperature up to the melting point, but they have no essential role in the shear mode instability.

This shear mode instability may also have been recently observed experimentally in a measurement of the constant volume heat capacity of the body-centered-cubic phase of helium-4 just prior to melting. It has been found that beginning at $0.01°K$ previous to melting the heat capacity of helium rises very rapidly and then jumps discontinuously to its value in the fluid phase. Previous attempts at detection have failed because melting usually appears in thermodynamic measurements as a sharp transition resulting from its cooperative character. This is because thermodynamic functions are sensitive to the high frequency motion of the particles, and the appearance of a highly cooperative low frequency sliding motion affects the thermodynamic properties only minutely. In helium, however, the melting temperature, and therefore the heat capacity, is so low that the high frequency modes are not yet fully excited, and the relative contribution of the rather sudden appearance of a new excitation is more significant. Furthermore, it should be particularly prominent in the body-centered-cubic phase because the transverse {110} mode is well known to be of low frequency. Of course, any other crystal structure should show such an instability upon melting, but it would probably be less prominent than in helium. It should be added that from an experimental point of view the helium system also presents a particularly favorable situation.

Progress towards a more satisfactory model for melting will depend largely an further detailed experiments including those involving scattering and sound wave attenuation. In scattering experiments one would look for the appearance of very anisotropic motion of the particles just previous to melting. This anisotropic motion corresponds to the onset of the cooperative motion of particles leading to sliding in a particular crystal direction. With sound waves one might also attempt the very difficult experiment of feeding energy into the mode which leads to melting and thus shift the melting temperature to slightly lower values.

From the theoretical point of view an attempt to look for such an instability in the crystal is hampered because of the difficulty of dealing with anharmonic forces exactly. Anharmonic forces are, of course, essential to melting, and the hard sphere system represents the anharmonic extreme. Exact calculations on the hard sphere system would in fact yield very realistic results, as pointed out previously and as, for example, is shown by the accurate comparison of the entropy of fusion to experiment; but such calculations do not seem possible at present. What does seem possible at present is the calculation of instabilities of the more unrealistic model of a harmonic crystal with small anharmonic forces which can be treated by perturbation. The difficulty with instability calculation is that it is not possible to identify the stable phase that the instability leads to. Thus, the instability could lead to another crystal structure and it would not be possible to distinguish it from melting.

Nevertheless, the difference in the ability to support a long wavelength transverse mode is the only general way to distinguish between a solid and a fluid. The frequently used alternative of the degree of order does not distinguish between a solid and a fluid in two dimensions. In that case the solid does not have long range order, because in an infinite system a particle can still make infinite excursions from its lattice site. The solid behaves like a gel but nevertheless has anisotropic properties.

Finally, it should be mentioned that the theoretical search for the analogous instability in the fluid phase via the integral equation approach has proved rather elusive. The one integral equation which has shown a singularity for hard spheres seems to have this feature because of a mathematical approximation unrelated to freezing. In general, the integral equations extend the fluid phase into the metastable glass region. It is interesting that the computer experiments show this feature as well. If the freezing region is traversed (i.e. the density is increased) too rapidly, the bottleneck in phase space is choked off and the rare fluctuation leading to crystallization does not take place. The particles then remain jammed into a disordered glass-like structure.

Bibliography

BORN M. and HUANG K. (1954) *Dynamical Theory of Crystal Lattices*, Oxford: Clarendon Press.
Physics of Simple Liquids, Ed. H. N. V. TEMPERLEY, J. S. ROWLINSON, and G. S. RUSHBROOKE (1968) North-Holland Publishing Co.: Amsterdam.
Solids Under Pressure, Ed. W. PAUL and D. M. WARSCHAUER (1963) McGraw-Hill: New York.
ALDER B. J., GARDNER W. R., HOFFER J. K., PHILLIPS N. E. and YOUNG D. A. (1968) *Phys. Rev. Letters* **21**, 732.

B. J. ALDER

METAL FILM PREPARATION BY ELECTROLESS (CHEMICAL) REDUCTION. The method of electroless metal film deposition differs from electroplating processes in that it takes place in the absence of an electric current. The electrons necessary to reduce metal ions to the metal are not provided from an external source via a cathode, but by a chemical reducing agent which is itself oxidized in the process. The overall reaction can therefore be represented by the sum of two processes

$$M^{n+} + n \text{ electrons} \rightarrow M \qquad (1)$$

$$R_1 - n \text{ electrons} \rightarrow R_2 \qquad (2)$$

where M is the metal and R_1 and R_2 are respectively the reduced and oxidized forms of the reducing agent.

The energies associated with processes (1) and (2) are normally expressed in terms of the standard electrode potential (E_1^0) of the M/M^{n+} system and the standard oxidation potential (E_2^0) of the reducing agent. For process (2) as written above, the associated potential is $-E_2^0$ and hence the overall equation is

$$M^{n+} + R_1 \rightarrow M + R_2 \qquad E^0 = E_1^0 - E_2^0.$$

Some values of E_1^0 and E_2^0 for typical electroless plating systems are given in Table 1. It will be noticed that for all of these systems E^0 is positive. This will occur always provided that E_2^0 has a sufficiently large negative value. Since the standard free energy

Table 1

System	Conditions	E_0^1 (volts)	E_2^0 (volts)
Nickel/hypophosphite	Acid	−0·25	−0·50
Nickel/hypophosphite	Alkaline	−0·25	−1·57
Cobalt/hypophosphite	Acid	−0·28	−0·50
Cobalt/hypophosphite	Alkaline	−0·28	−1·57
Iron/hypophosphite	Alkaline	−0·44	−1·57
Iron/borohydride	Alkaline	−0·44	−1·24
Nickel/borohydride	Alkaline	−0·25	−1·24
Palladium/hydrazine	Alkaline	+0·99	−1·16
Copper/formaldehyde	Alkaline	+0·34	−1·07

change (ΔG^0) is related to the potential by the equation

$$\Delta G^0 = -nFE^0$$

where n is the number of electrons involved and F is the Faraday, it follows that all of these processes are thermodynamically favourable under conditions of standard concentration.

The values of E_1^0 and E_2^0 are readily obtainable from reference books, and so the thermodynamic feasibility of any overall reduction process is easily calculated. This is, however, no real criterion of whether the reaction will occur at a measurable rate under any particular set of conditions. The reduction of cobalt ions by hypophosphite ions is, for example, theoretically possible under acid conditions but in practice has only been observed under alkaline conditions.

Since the major part of the experimental work on metal plating by chemical reduction up to the present has involved the nickel/hypophosphite system, the development of the method will be described with reference to that system. The basic principles, insofar as they are understood, will also be discussed mainly in terms of the reactions involved in the nickel/hypophosphite system. Finally the extension to systems involving other metals and other reducing agents will be considered.

The Development of Electroless Plating

The discovery of electroless plating and the origin of the term is due to A. Brenner and G. Ridell of the National Bureau of Standards in the United States. These workers found increased electroplating efficiency of nickel from solutions containing hypophosphite ($H_2PO_2^-$) ions. It became obvious that a secondary plating process was occurring (in the absence of an electric current) involving nickel ions and hypophosphite ions.

It had been known for some time that certain metal salts, notably those of the transition metal series, cause the bulk deposition of precipitates from solutions containing hypophosphite ions. These precipitates were reported to contain phosphorus in varying amounts up to 15 per cent by weight, depending upon the reaction conditions. Oxygen was also suggested to be present.

In the case of a plating reaction, the deposition occurs preferentially at the surface rather than in the bulk of the solution. For nickel and hypophosphite ions, two major overall processes occur

$$H_2PO_2^- + Ni^{++} + H_2O \xrightarrow{\text{Surface}} H_2PO_3^- + 2H^+ + Ni \quad (3)$$

and

$$H_2PO_2^- + H_2O \xrightarrow{\text{Surface}} H_2PO_3^- + H_2 \quad (4)$$

Reaction (4) represents a simultaneous heterogeneous oxidation of the reducing agent, hypophosphite ion, which reduces the amount available to be used in the plating reaction (3).

Further reactions possibly occur which account for the inclusion of phosphorus in the deposited metal.

From the work of Brenner and Riddell, the factors necessary to produce stable plating solutions became established. It was shown that a primary requirement is for the surface which is to be plated to be a catalytic surface for the reduction of metal ions to metal (equation 3). Many metals were found to be capable of being plated directly. Gold, cobalt, nickel and palladium are genuine catalysts, iron and aluminium probably function by first becoming coated with a thin film of nickel by electrochemical displacement. Non-metals such as glass and plastics can be plated provided that they are first "activated" by dipping into stannous chloride solution followed by washing well with water and then dipping into palladium chloride solution. This treatment is thought to produce a thin film of palladium on the non-metallic surface, which then acts as a catalyst for the plating process. The process, once started, is autocatalytic, since the nickel deposited is itself a catalyst for the reduction process. Relatively thick films can thus be built up at the surface, although the adhesion depends upon a number of factors including the composition of the plating solution.

It was found that plating can be carried out under acid or alkaline conditions. The optimum temperature is about 90°C. A typical acid bath (pH 4–6) used by Brenner and Riddell has the composition

Nickel chloride $NiCl_2 \cdot 6H_2O$	30 g/l
Sodium hypophosphite $NaH_2PO_2 \cdot H_2O$	10 g/l
Sodium glycollate CH_2OHCO_2Na	50 g/l

and is reported to give a plating rate of 15 μ/hr. Faster rates are obtained from alkaline baths, although the loss of ammonia by evaporation at elevated temperatures causes some problems with those baths containing this reagent.

A bath of improved pH stability for the plating of nickel/cobalt/phosphorus alloy uses glycine and a boric acid buffer with the following molar concentrations

$Ni^{++} = 2.05 \times 10^{-3}$ [Co^{++}] $= 1.5 \times 10^{-3}$

Sodium citrate $= 3.08 \times 10^{-1}$, [H_3BO_3] $= 3.3 \times 10^{-1}$

$H_2PO_2^- = 7.3 \times 10^{-2}$, [Glycine] $= 2.66 \times 10^{-1}$

The pH is adjusted to 9·00 using solid sodium hydroxide. It will be noticed that both acid and alkaline baths contain added organic anions such as glycollate and citrate. These fulfil several functions and there is not yet a complete understanding of their role. In acid solution it is clear that some of these additives function by buffering the solution against the decrease in pH brought about by the protons produced by the reduction (equation 3). In alkaline solution, however, their prime role is undoubtedly to complex the nickel ion and prevent precipitation of the hydroxide. The complicated shapes of curves of plating rate against organic additive concentration make it clear that additives probably act in more than one way in the plating process.

Factors affecting the Rate of Plating

A full understanding of the mechanism of electroless plating will only come from considerations of the basic chemistry of the process, namely a study of the products and rates of the reactions taking place at the catalyst surface. Once the surface is suitably "activated" and covered with a monolayer of the deposited metal, it is assumed that the substrate has no further effect upon the kinetics of the process. This seems a reasonable assumption although some physical properties of the deposits have been found to vary with the nature and pre-treatment of the substrate. Plating rates are usually quoted as measurements of thickness of deposit on a given area per unit time and are measured, for example, in Angstrom units or millimetres per minute. On glass substrates it is convenient to weigh the specimen before and after plating or after removing the film by dissolution in acid. Alternatively, the composition of the film can be determined by a variety of spectroscopic or spectrophotometric methods and the amounts of nickel, phosphorus or other components can be measured directly as a function of time. In all cases, it is usual to observe a linear relationship between the amount of plating and the time. This is, in general, a consequence of the low ratio (area plated/volume of solution) which results in a negligible decrease in the concentration of reagents due to the occurrence of reaction.

Among the variables which have been found to affect the rate of electroless plating are concentrations of reagents, the concentrations of additives such as organic anions used as buffers, chelating agents, or "stabilizers", the pH, temperature, and the external pressure. Since the plating rate is in general affected by all of these variables and since some variables exert independent effects upon each other, it is not easy to isolate the effects of a particular variable upon the reaction rate. The following general conclusions, however, appear at present to be established.

(a) The concentration of reagents

The plating rate in nickel/hypophosphite systems is first-order in hypophosphite concentration under both acid and alkaline conditions, provided that the ratio area plated/volume of solution is small, and that the solution is well buffered. (The rate of plating is said to be first-order in the concentration of a particular reactant X, if when all other reactant concentrations are kept constant, the initial rate of plating is proportional to the first power of the concentration of X.) The plating rate in nickel/hypophosphite systems is frequently observed to be independent of the nickel ion concentration under both acid and alkaline conditions. It is, however, possible to show that at very low nickel concentrations, particularly in the absence of chelating agents, the rate is approximately first-order in nickel ion concentration.

(b) The concentrations of additives

The majority of electroless plating baths incorporate organic additives such as the anions of acetic, citric, glycollic, lactic or malonic acids. These undoubtedly play a complicated role in the process which is not fully understood. Experimentally it is observed that plots of plating rate versus additive concentration show maxima, the position of which depends upon the additive. Such maxima are observed under both acid and alkaline conditions for the nickel/hypophosphite system.

(c) The pH

Since the overall plating process produces hydrogen ions, it is found, as would be expected, that an acceleration is produced by a rise in pH. Commercial users are limited in the range of pH which can be used, by the degree of control of pH attainable and by the low solubility of nickel phosphite at high pH values. It is well established that the plating rate is an exponential function of pH and that the phosphorus content of the deposit decreases with increase in pH.

The minimum pH below which deposition of a nickel-phosphorus alloy does not occur is just less than 3·0. The phosphorus content at this pH value is about 15 per cent by weight, falling to 2–3 per cent by weight at pH values around 11.

(d) The temperature

For solutions containing relatively low concentrations of complexing and buffering additives, the rate of plating shows a dependence on temperature which fits the Arrhenius equation. From a plot of logarithm of plating rate of nickel/phosphorus versus reciprocal of absolute temperature, an activation energy of 17·7 kcal/mole has been derived for a particular solution composition.

This value cannot be regarded to be well established at present and should not be taken to be a typical value for other electroless plating systems.

(e) The external pressure

A little work has been done to examine the effect of increased pressure (of either hydrogen or nitrogen) above the solution upon the rate of plating. The experiments showed that either an acceleration or a retardation can result from the application of increased pressure, depending upon the composition of the solution. The effects were similar using either hydrogen or nitrogen and are not thought to be due to any chemical interaction.

The Composition and Structure of the Deposit

Many of the factors which influence the rate of plating also affect the composition of the deposited film. Of these the most influential is the pH which can cause a marked change in the phosphorus content. The mechanism of phosphorus inclusion at present is uncertain, but one suggestion which would appear to fit the pH effect has been that the hypophosphite ion is directly reduced to elemental phosphorus by atomic hydrogen

$$H + H_2PO_2^- \xrightarrow{\text{Catalyst}} P + H_2O + OH^-. \quad (5)$$

An attempt has been made to determine the origin of the phosphorus in the deposit by using plating solutions containing hypophosphite ions labelled with radioactive ^{32}P. The results were interpreted as indicating that phosphite ions (products of the reaction) are the source of phosphorus in the deposit. It is not certain whether hypophosphite ions are a direct source of phosphorus or whether they must first be converted to phosphite ions. Unfortunately the conclusions of these experiments are based on rather limited experimental data and more work is undoubtedly needed to confirm the results.

The nickel–phosphorus alloy when freshly prepared is shown by X-ray methods to be an amorphous solid. Microphotographs of cross-sectional samples of the deposit show a multilayer structure and there is some evidence that this is paralleled by a periodic variation in phosphorus content. The deposit when freshly prepared is quite hard, having an average Knoop hardness number of 500 compared with values of 120 to 450 for electroplated nickel. Unlike electroplated nickel, however, the hardness increases on heating. It has been shown by X-ray methods that a transition occurs on heating electroless nickel/phosphorus, from an amorphous solid to a solid having some degree of crystal structure. Heat treatment above 400°C produces a mixture of the compound Ni_3P and a solid solution of phosphorus in β-nickel.

The Mechanism of the Plating Reaction

From the experimental observations of the influence of various factors upon the rate of plating, it is possible to formulate mechanisms for the process.

The Gutzeit mechanism (1959) involves the catalytic removal of both hydrogen atoms from a hypophosphite molecule as the initial rate-controlling step followed by

$$H_2PO_2^- \xrightarrow{\text{Catalyst}} 2H + PO_2^- \quad \text{(slow)} \quad (6)$$

followed by

$$PO_2^- + H_2O \xrightarrow{\text{Catalyst}} H_2PO_3^- \quad \text{(fast)} \quad (7)$$

$$Ni^{++} + 2H \xrightarrow{\text{Catalyst}} Ni^0 + 2H^+ \quad \text{(fast)} \quad (8)$$

$$2H \xrightarrow{\text{Catalyst}} H_2 \quad \text{(fast)} \quad (9)$$

Under certain conditions, this mechanism predicts the observed first-order dependence of the rate of plating upon hypophosphite concentration and its independence of nickel concentration.

It was suggested that the increased rate of plating observed in the presence of some chelating agents could be explained by the formation of a chelating agent–hypophosphite complex. Such a complex could weaken the phosphorus–hydrogen bonds and increase the rate of step (6).

A reaction which is always present in electroless plating systems involving hypophosphite ions is the catalytic oxidation of hypophosphite (reaction 4). This reaction occurs independently at nickel and nickel–phosphorus surfaces in the absence of nickel ions in the solution.

A thorough comparison of the effects of added organic anions on the catalytic oxidation of hypophosphite and on the plating system has shown that both reactions are influenced in a similar way. This suggests a common initial step, possibly adsorption of hypophosphite at the catalyst surface. Chelating agents at high concentrations may influence this process by competitive adsorption whereas at low concentrations they may accelerate the rate of plating by acting mainly as buffers.

Alternative mechanisms suggested have been due to Ishibashi (1960) and Lukes (1964). The Ishibashi mechanism proposes that electrons liberated at the catalyst surface are the effective reducing agents. This can be written as

$$H_2PO_2^- + H_2O \xrightarrow{\text{Catalyst}} H_2PO_3^- + 2H^+ + 2e^- \quad (10)$$

$$2H^+ + 2e^- \xrightarrow{\text{Catalyst}} H_2 \quad (11)$$

$$Ni^{2+} + 2e^- \xrightarrow{\text{Catalyst}} Ni \quad (12)$$

Lukes has taken up an idea first proposed by Hersch of a mechanism based on the hydride ion as the reducing agent, viz.

In acid solution

$$H_2PO_2^- + H_2O \rightarrow HPO_3^{--} + 2H^+ + H^- \quad (13)$$

while in alkaline solution

$$H_2PO_2^- + 2OH^- \rightarrow HPO_3^{--} + H_2O + H^- \quad (14)$$

and then

$$Ni^{2+} + 2H^- \rightarrow Ni + H_2 \quad (15)$$

Hydrogen is produced by the reactions

$$H^+ + H^- \rightarrow H_2 \quad \text{(acid solution)} \quad (16)$$

or

$$H^- + H_2O \rightarrow OH^- + H_2 \quad \text{(alkaline solution)} \quad (17)$$

While it is difficult to prove or disprove the Ishibashi mechanism, the Lukes mechanism is supported by some experimental work by Gorbunova and co-workers using deuterated hypophosphite.

By incorporating deuterium either into the reducing agent ($D_2PO_2^-$) or into the solvent (D_2O) it is possible to provide new evidence for the mechanism by then examining the deuterium content of the liberated hydrogen.

The extent of deuteration of the gas was found to depend upon whether nickel ions are present or not. This fits the Lukes mechanism better than previous mechanisms in which nickel deposition and hydrogen evolution are regarded as independent processes.

Other Electroless Plating Systems

Numerous formulations for plating baths exist in the patent literature concerning a variety of metal ions, complexing agents and reducing agents. Of these, many are not genuine electroless processes in the sense that an autocatalytic deposition of metal occurs at the surface rather than in the bulk of the solution. Some baths are essentially immersion plating baths functioning by electrochemical replacement. The main electroless plating systems established at present are indicated in Table 2.

Table 2
Reducing agent

Metal	Hypo-phosphite	Boro-hydride	Hydrazine	Form-aldehyde	Alkyl-amine boranes
Fe	×	×			×
Co	×	×	×		×
Ni	×	×	×		×
Cu				×	
Au			×		
Cr	×				
Pd				×	

The copper–formaldehyde system

Second in commercial importance to electroless nickel is electroless copper. The favoured reducing agent is formaldehyde, which functions as a reducing agent in highly alkaline solutions (pH 10–12) according to the equation

$$HCHO + 3OH^- \rightarrow HCO_2^- + 2H_2O + 2e^-$$
$$(E_2^0 = -1.07\,V) \quad (18)$$

Electroless copper deposits autocatalytically according to the equation

$$Cu^{++} + 2\,HCHO + 4\,OH^- \xrightarrow{Catalyst} Cu + H_2$$
$$+ 2\,H_2O + 2\,HCO_2^- \quad (19)$$

The reaction is initiated on a surface of electroless nickel, of palladium or silver. Modern electroless copper plating baths contain copper salts, formaldehyde, alkali to give a pH 10–12, and a complexing agent such as a tartrate or salicylate. The formaldehyde is present as a reducing agent but in practice more formaldehyde is used than is indicated by equation (19). An early report suggested that the catalysed decomposition of formaldehyde occurs, viz.

$$HCHO + OH^- \xrightarrow[Catalyst]{Cu} HCO_2^- + H_2 \quad (20)$$

This, however, has not been confirmed, and it is believed that the main side-reaction is the Cannizarro reaction

$$2\,HCHO + OH^- \rightarrow HCO_2^- + CH_3OH \quad (21)$$

A bulk deposition of copper with evolution of hydrogen sometimes occurs spontaneously, possibly initiated by the precipitation of cuprous oxide

$$2\,Cu^{++} + HCHO + 5\,OH^- \rightarrow Cu_2O$$
$$+ HCO_2^- + 3\,H_2O \quad (22)$$

followed by

$$Cu_2O + H_2O \rightleftharpoons Cu + Cu^{++} + 2\,OH^- \quad (23)$$

Relatively few mechanistic studies have been made of the deposition of electroless copper. Lukes has suggested a mechanism involving reduction by hydride ions at the surface.

Hydrazine as a reducing agent

Hydrazine has found some application as an alternative reducing agent to hypophosphite in the production of electroless nickel. The overall reaction in alkaline aqueous solution is represented by

$$2\,Ni^{++} + N_2H_4 + 4\,OH^- \xrightarrow[Catalyst]{Ni} 2\,Ni + N_2 + 4\,H_2O \quad (24)$$

It is possible by the use of hydrazine to deposit nickel of high purity (97–99·2 per cent Ni compared with 90–92 per cent Ni from hypophosphite baths).

The use of hydrazine for the deposition of palladium films has also been reported. A typical electroless palladium solution contains palladium ions complexed as the ammine $[Pd(NH_3)_4]^{++}$, hydrazine as the hydrate or sulphate and the disodium salt of ethylenediamine tetracetic acid, which functions as a stabilizer. Temperatures in the range 30–70°C are possible, although spontaneous decomposition occurs at higher temperatures. Palladium can be directly plated onto many metals and also on certain non-metals such as carbon, glass and ceramics after suitable pre-treatment. The purity of electroless palladium is quoted as at least 99·4 per cent with a Knoop hardness number in the range 150–350.

Borohydride and Amino-boranes as reducing agents

The use of borohydrides and amino-boranes as reducing agents is relatively new. The action of borohydride ion as a reducing agent in aqueous solution (pH 12–14) is represented by

$$BH_4^- + 8\,OH^- = H_2BO_3^- + 5\,H_2O + 8e^-$$
$$(E_2^0 = -1\cdot24\ V) \quad (25)$$

The use of high pH is dictated by the acid hydrolysis of borohydride which occurs according to the equation

$$BH_4^- + 3\,H_2O \xrightarrow{Catalyst} H_2BO_3^- + 4\,H_2 \quad (26)$$

One advantage of borohydride plating systems is the low operating temperature, 25–40°C.

The product of electroless plating using borohydride as reducing agent is usually an alloy containing 90–97 per cent of the metal with the remainder mainly boron. The structure resembles the nickel–phosphorus alloy from the hypophosphite solution in that it is initially amorphous. On heating to 450–475°C metal boride is shown to be formed.

Films of nickel, cobalt and iron containing boron have been prepared from borohydride solutions. As in most electroless plating baths, complexing agents are present such as ammonia, various amines or hydroxy-carboxylic acid salts.

Related reducing agents are the amino-boranes which are compounds with the general formula XBH_3 where X is NH_3 or an amine RNH_2, $RR'NH$, or $RR'R''N$.

In general the most suitable reducing agents for use in plating systems are found to be those involving secondary or tertiary amines such as N-dimethyl-amino borane and N-trimethyl amino borane. An advantage of these reducing agents over borohydride is the lower pH required which is usually between 2 and 6·5. The preparation of nickel–boron, cobalt–boron, iron–boron alloys and ternary alloys such as nickel–iron–boron has been described in patents.

The Uses of Electroless Metal Films

Electroless metal films offer many advantages over electrodeposited films particularly where a uniform hard metal coating is required. The ability to plate on non-metallic surfaces such as glass and plastics, thereby making them conductors of electricity, has led to many applications in the electronics industry. Electroless nickel is manufactured on a large scale in the commercial "Kanigen" process which is the property of the General American Transportation Corporation. Some properties of Kanigen nickel alloy coating are given in Table 3.

Typical uses of electroless nickel are for coating transporters for corrosive liquids such as caustic soda and for protective coating of steel in many items of equipment such as valves, pumps, compressors and heat-exchangers.

Table 3

Some properties of "Kanigen" nickel coating

Composition		
	Nickel	90—92%
	Phosphorus	8—10%
	Carbon	0·04%
	Oxygen	0·002%
	Nitrogen	0·005%
	Hydrogen	0·002%
Melting point		890°C
Specific gravity		7·85
Expansion coefficient (per °C)		$1·3 \times 10^{-5}$
Electrical resistivity micro ohm cm		~60
Thermal conductivity (cal/cm sec °C)		0·0105—0·0135

The discovery of interesting magnetic properties of nickel–cobalt–phosphorus alloys stimulated much research into their method of production. Nickel–cobalt–phosphorus alloys have been found to possess the low coercivities and high saturation magnetization necessary to make storage elements in high-speed computers. Some experimental computers have been constructed using such materials in place of the conventional ferrite stores.

Cobalt–phosphorus alloys on the other hand have been shown to possess quite different magnetic properties which make them more suitable for magnetic recording, e.g. on discs or tapes.

Electroless copper has found many applications in the electronics industry, for example, in the manufacture of printed circuits, in the R.F. shielding of coaxial cables and in the coating of plastic waveguides.

Other metal films for which a considerable market exists if only suitable electroless processes could be developed include chromium (for decorative purposes) and some of the less common metals of the platinum group such as rhodium (for its hardness and wear characteristics).

Bibliography

BRENNER A. and RIDDELL G. (1946) *J. Res. Nat. Bur. Stds.* **37**, 31.
BRENNER A. and RIDDELL G. (1947) Ibid. **39**, 385.
GORBUNOVA K. M. and NIKIFOROVA A. A. (1963) *Physicochemical Principles of Nickel Plating*, Israel Program for Scientific Translations, Jerusalem.
GUTZEIT G. (1959) in *Symposium on Electroless Nickel Plating*, American Society for Testing Materials.
LEVY D. J. (1963) *50th Ann. Tech. Proc. Amer. Electroplaters*, p. 29.
SAUBESTRE E. (1962) *Metal Finishing* **60** (6), 67.

K. A. HOLBROOK

METALLIC COLLOIDS IN NON-METALLIC CRYSTALS

Formation of colloidal particles

In 1905 Siedentopf observed that additively coloured NaCl crystals containing atomically dispersed F centres turned blue if annealed at a temperature near 400°C and showed Tyndall scattering. The blue colour and Tyndall scattering were attributed to colloidal particles of sodium formed in NaCl crystals by aggregation of F centres during the annealing process. Subsequent work has shown that colloidal particles of alkali metal are formed in additively coloured KCl, KBr and KI crystals also if annealed in dark at 300°, 250° and 160 °C, respectively. In LiF crystals, the colloidal particles of lithium are formed when the crystals are irradiated with high energy neutrons (dose $\gtrsim 10^{18}$ nvt). Metallic colloids of silver, lead and cadmium in alkali halide crystals have also been produced recently. The silver colloids are produced by heating the silver doped NaCl crystals in potassium vapours, lead colloids are produced by X-irradiation of the lead doped crystals and cadmium colloids are produced by heating the additively coloured alkali halide crystals doped with cadmium at 520°–600°C.

The metallic colloidal particles have also been produced in many other crystals and in glass, e.g. Li colloids in LiH crystals and Pb colloids in lead halide crystals by irradiation with ultra-violet light, Ag colloids in silver halides by irradiation with X-rays and with ultra-violet or visible light, alkali metal colloids in alkali azides by irradiation with X-rays or ultra-violet light and heating to ~300°C. Colloidal particles of gold (or silver) in glass are produced by heating the gold (or silver) chlorides with glass at 1400°C for several hours, cooling the melt rapidly, irradiating the cooled sample with ultra-violet light to develop nucleation sites for the particles and then heating it to 400°–630°C.

Optical studies of colloidal particles

The metallic colloidal particles when present in non-metallic crystals (or glass) show new optical absorption bands, photoconductivity, excess electronic conductivity at high temperatures and ESR and NMR absorption. The presence of metallic colloids has also been confirmed by X-ray and microthermal analysis as well as by electron microscopic observations.

The dependence of photoelectric emission on the wavelength of the exciting light from a clean sodium metal surface is shown by curve *a* and that of photoconductive current in a natural rock salt crystal containing sodium colloids by curve *b* in Fig. 1. At the long wavelength limit the curve *b* is displaced by 0·5 eV with respect to curve *a* in the direction of lower energy. This difference in the energy is the lowering of the work function of sodium metal in contact with the NaCl crystal and hence it is equal to the electron affinity χ of the salt crystal.

The optical absorption and scattering of light by metallic colloidal particles of the order of the wavelength of light and embedded in a dielectric medium

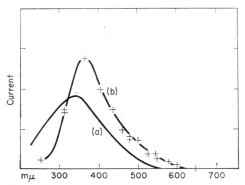

Fig. 1. (a) Photoelectric emission from a clean sodium surface, (b) Photoconductive current in rock-salt containing sodium colloids (see Mott and Gurney, p. 73).

have been treated by Mie from the classical viewpoint. The theoretical results show that the peak position due to these colloids should move towards longer wavelengths as the particle size increases. The calculated and observed peak positions of the band due to small metallic colloidal particles (size much smaller than the wavelength of light) in different non-metallic media are given in Table 1. Typical optical absorption of gold particles about 12 nm in diameter in glass is shown in Fig. 2.

The alkali metal colloids in additively coloured alkali halide crystals dissolve and F centres are formed at higher temperatures. In additively coloured KCl above 300°C, the F centres and the potassium colloids are in heterogeneous equilibrium, i.e. the colloids behave as the condensed phase and the F centres as the vapour phase. At any temperature between 300° and 450°C, the concentration of F centres is a unique function of temperature and does not depend upon the concentration of the excess alkali metal in the crystal.

ESR and NMR studies

The metallic colloidal particles when present in non-metallic crystals give conduction electron spin resonance (CESR) and nuclear magnetic resonance (NMR) absorption characteristic of the metal. A typical CESR line of lithium colloids superimposed on an F centre ESR line in neutron irradiated LiF crystal is shown in Fig. 3. The characteristic features of the observed CESR spectra of alkali metal colloids are narrow half-width, a g-value close to the free

Table 1. Calculated and observed absorption peak positions due to small metallic colloids in non-metallic crystals and in glass

Crystal	Metal of which colloids are formed	Calculated peak position λ (nm)	Observed peak position λ (nm)
KCl	K	730	730
	Cd	—	275
KBr	K	760	790
	Cd	—	275
KI	K	800	840
NaCl	Na	512	865*
	Ag	—	450
	Cd	—	275
LiH	Li	—	650
AgBr	Ag	Shortwave infra-red	—
NaN$_3$	Na	—	550
Glass	Au	525	527
	Ag	396	406

* The observed peak position due to sodium colloids in NaCl is at a wavelength much longer than the calculated value for small colloids indicating that the size of these colloids in NaCl is large.

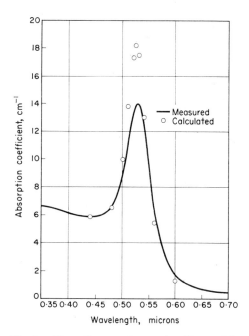

Fig. 2. Absorption spectrum of gold particles about 12 nm in diameter in glass (R. H. Doremus, J. Chem. Phys. **40**, 2389, 1964).

electron g-value and Lorentzian line shape consistent with the theoretical predictions for small metal particles. The CESR absorption spectra due to conduction electrons in lithium colloids in LiH, silver colloids in silver halides, alkali colloids in alkali azides, lead colloids in lead halides and cadmium colloids in alkali halides have also been reported. The observed half-width and g-values and the line shape for several cases are collected in Table 2.

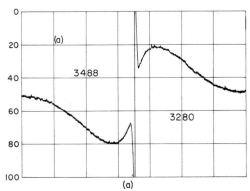

Fig. 3. *ESR spectra of Li metal in neutron irradiated LiF (5×10^{18} nvt). The sharp CESR line superimposed upon the broad line due to F centres is Lorentzian in shape and have a half-width of 1·5 gauss (Kirn, Kaplan and Bray, Phys. Rev.* **117**, *740, 1960).*

Table 2. *The half-widths, line shape and the g-factors for the CESR absorption in metallic colloids*

Crystal	Metal of which the colloids are present	Half-width (peak to peak) gauss	Line shape	g-factor
KCl	K	2·5	Lorentzian	1·9998
	Cd	150	Lorentzian	1·999
KBr	K	6	Lorentzian	1·9997
	Cd	150	Lorentzian	1·98
NaCl	Na	10	—	2·0
LiF	Li	1·5	Lorentzian	2·0023
LiH	Li	0·3	—	2·002
KN_3	K	105	Lorentzian	1·9982
NaN_3	Na	8·8	Lorentzian	2·0017
AgBr	Ag	7	—	2
$PbCl_2$	Pb	23	—	2·0006

The total concentration of alkali metal colloids formed in additively coloured alkali halide crystals is sensitive to the presence of divalent cation impurities and/or exposure of the crystal to light. In Fig. 4 the area under the CESR line due to potassium colloids in additively coloured KCl crystals containing different divalent cation impurity concentrations and heated in dark or in light is plotted as a function of the area under the corresponding optical absorption band. The straight line passing through the points for highly pure crystals not exposed to light does not pass through the origin. This is due to the fact that the

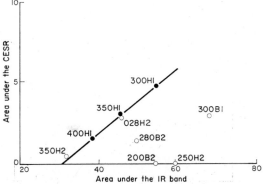

Fig. 4. *Plot of the area under the CESR absorption line due to potassium colloidal centres as a function of the area under the corresponding infra-red (IR) bands in KCl crystals quenched from different temperatures. The letters H and B indicate the highly pure (background divalent cation impurity $C_i \sim 1$ ppm) and "pure" ($C_i \sim 10$ ppm) crystals. The number placed before and after each letter indicates, respectively, the temperature in °C from which the crystal was quenched and the batch number from which the crystal was taken.* ● *heating and quenching done in dark;* ○ *heating and quenching done in light (Jain and Sootha, Phys. Rev.* **171**, *1075, 1968).*

optical absorption under the "coloidal band" is not due to pure potassium colloids but a part of it is due to Scott R' centres. The Scott R' centres are formed by the interaction of divalent cation impurities present in the crystal with the F centres and give optical absorption in the same wavelength region as the colloidal particles. The other points in the figure show that that part of the optical absorption not related with the metallic colloids increases rapidly as the concentration of background divalent cation impurity increases and/or the coloured crystals are exposed to visible light.

Electrical effects associated with colloidal particles

Colloidal particles of alkali metal in KCl, KBr, KI and LiF crystals give rise to large electronic conductivity. The electrons are emitted by thermionic emission from the colloidal particles into the conduction band of the salt crystal. The typical plots of σ_c/σ_n (σ_c is the conductivity of the crystal containing colloids and σ_n that of an identical uncoloured crystal) as a function of increasing temperature for four KCl crystals containing $3·8 \times 10^{17}$, $4·2 \times 10^{17}$, $7·2 \times 10^{17}$

and 1.0×10^{18} cm^{-3} excess metal concentration are shown in Fig. 5. Similar enhanced conductivity is observed above 250°C in KBr, 160°C in KI and above 100°C in neutron irradiated LiF crystals. The electronic conductivity σ_e is given by

$$\sigma_e = CT^{3/2}F(T)\,e^{-\frac{\phi-\chi}{KT}} \quad (1)$$

where ϕ is the thermionic work function of potassium particles for emission of electrons into vacuum, χ is the electron affinity of the host crystal, $F(T)$ is dependent on the concentration of colloids and therefore on the temperature and C is a constant. $F(T)$ is taken to be proportional to $\Delta(T)$, the area under the CESR line. The logarithmic plot of $\sigma_e/\Delta(T)T^{3/2}$ as a function of $1/T$ in the temperature range 300°–450°C is found to be a straight line. The value of $(\phi - \chi)$ can be obtained from the slope of the line. From these measurements the values of χ (0·23 eV for KCl, 0·36 eV for KBr, 1·2 eV for KI and 1·18 eV for LiF) have been determined.

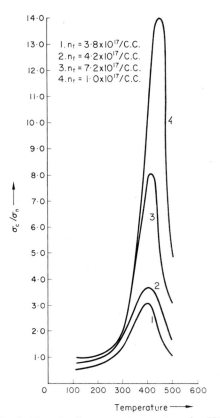

Fig. 5. Plot of σ_c/σ_n versus temperature for KCl crystals in dark, n_f is the concentration of F centres at room temperature in the quenched crystal (Jain and Sootha, J. Phys. Chem. Solids 26, 267, 1965).

The observed shift in the peak position of curves 1 to 4 in Fig. 5 towards the higher temperature with the increase in the initial concentration of colloids is a consequence of heterogeneous equilibrium between the F centres and the potassium colloids. The peak position T_g and the concentration n_K of excess metal colloids are related by

$$n_K = \text{constant.}\ e^{-\Delta Hc/kT}g \quad (2)$$

where ΔHc is the energy required to form one F centre from a colloidal particle. The plot of log n_K versus $1/T_g$ gives $\Delta Hc = 0.8$ eV for KCl.

Metallic colloids in alkali halide crystals have also been detected by several other methods. Electron diffraction lines of metallic sodium in NaCl coloured blue by 40–50 keV electron bombardment have been obtained. The lithium metal colloids in LiF crystals have been detected by X-ray diffraction and also by differential thermal analysis. The Overhauser effect and the day-shift (a shift of electron spin resonance resulting from the field of the nuclei) have also been measured in lithium colloids in LiH. A small concentration of alkali or silver colloids in alkali halide crystals reduces appreciably the thermal conductivity of the crystal at low temperature.

Bibliography

ARENDS J. and VERWEY J. F. (1967) *Phys. Stat. Sol.* **23**, 137.
DOREMUS R. H. (1964) *J. Chem. Phys.* **40**, 2389; (1965) **42**, 414.
JAIN S. C. and SOOTHA G. D. (1965) *J. Phys. Chem. Solids* **26**, 267; (1968) *Phys. Rev.* **171**, 1075.
MOTT N. F. and GURNEY R. W. (1940) *Electronic Processes in Ionic Crystals*, Oxford University Press.
SCHULMAN J. H. and COMPTON W. D. (1962) *Color Centers in Solids*, Pergamon Press.
SEIDEL H. and WOLF H. C. (1965) *Phys. Stat. Sol.* **11**, 3.
SEITZ F. (1954) *Rev. Mod. Phys.* **26**, 7.
SMITH M. J. A. and INGRAM D. J. E. (1962) *Proc. Phys. Soc.* **80**, 139.
WALKER C. T. (1963) *Phys. Rev.* **132**, 1963.
WORLOCK J. M. (1966) *Phys. Rev.* **147**, 636.

S. C. JAIN and G. D. SOOTHA

MICROPLASTICITY. The theories of *Elasticity* and *Plasticity* describe the mechanics of deformation of most solids. Both theories, as applied to the engineering materials, are based on experimental studies of the relations between stress and strain in a polycrystalline aggregate under simple loading conditions. Thus they are of a phenomenological nature on the macroscopic scale and owe, as yet, little to a knowledge of the microstructure of a metal.

The classical formulations of the mechanical behaviour of engineering materials are based on the analysis of infinitesimally small elements of a continuous medium which is generally accepted to be homogeneous and isotropic. However, a real continuous medium, contrary to these assumptions, exhibits a certain amount of inhomogeneity and anisotropy. Therefore, an infinitesimal element of a real material varies considerably from that of a theoretical medium. Consequently, the true mechanical behaviour of a material deviates from the observed macroscopic behaviour by an amount dictated by the inhomogeneity and anisotropy exhibited by the microstructure of the material. Such deviations have a growing importance in view of the failures encountered in engineering, such as the metal fatigue which initiates at microscopic level. It has become essential to understand the limitations so imposed on the classical theories due to the local variations of the mechanical properties.

The mechanics of deformation has been investigated in great detail by physicists, metallurgists and engineers at different levels (depending upon the size of the structural elements selected) as indicated below:

1. Submicroscopic or atomic elements of the order of magnitude less than 10^{-7} cm.
2. Microscopic elements of the order of magnitude between approximately 10^{-7} and 10^{-3} cm.
3. Macroscopic elements which conform with the classical assumptions.

Stress–strain properties of materials which are commonly used deal with the macroscopic structural behaviour, and are based on experiments using specimens of appreciable size. However, the following three major points emerge from the available literature:

(i) The infinitesimal elements of a real medium do not necessarily exhibit the macroscopic properties and the stress–strain behaviour of the material.
(ii) The stresses and the strains of a mathematical element, nominally under uniform stress, are not uniform on a microscopic scale owing to the microstructural properties.
(iii) The yield point or proportional limit is an insufficient definition as an indication of the onset of plasticity, because a metal can undergo local, isolated plastic flow at stresses below this point.

The significance of the last point in connection with the fatigue failures has been pointed out by many investigators.

This plastic flow which is of the same order of magnitude as the elastic strains is termed *Microplasticity*. Since no conventional technique exists for measuring the microplastic strains, and since for most metals the readily observable plastic deformation takes place rather abruptly, its significance has, then, been overlooked in the analysis of stress–strain curves.

The limitations of the classical approaches to the mechanics of deformation can be improved by introducing the concept of "macroscopic" and "microscopic" elements. (Terms "domain", "elastoplastic element" have also been used to convey the same meaning as the latter.) This approach would be intermediate between the crystalline level which is too complex to yield a quantitative understanding and the macroscopic level which is known to have important limitations in certain applications.

The *macroscopic element*, which may be very small in comparison to the body under load, conforms with the classical assumptions and repeats the nominal mechanical properties of the material. The term nominal is used to emphasize the fact that stress–strain or mechanical properties of materials are approximations. Engineering materials are generally polycrystalline or amorphous solids with properties represented by the average of the individual characteristics of the component parts which may properly be called *microscopic elements*. All mechanical properties of the microscopic elements are microstructure sensitive and random. The random nature of the microscopic properties makes the structure of polycrystals considerably more complex than the structure of single crystals. It is for this reason that predictions of macroscopic behaviour of polycrystals from single-crystal behaviour have been generally unsuccessful. The size of a microscopic element is bounded in the upper limit by the condition that the material constants and the components of the stress and strain tensors of the element must be single valued. The lower limit is the dimensions at which it still retains the identity of the medium.

The random properties of the microscopic elements in a polycrystalline solid tend to produce a material on a macroscopic scale which is homogeneous and isotropic. In the mathematical sense, the behaviour of a macroscopic element is the integrated behaviour of microscopic elements and represents a statistical average of the latter.

$$\sigma_{lm} = \bar{\xi}^k_{lm}$$
$$e_{lm} = \bar{\varepsilon}^k_{lm} \quad (l, m = x, y, z) \quad (1)$$
$$C_{ij} = \bar{c}^k_{ij} \quad (i, j = 1, 2, ..., 6)$$

In the above equations, σ, e and C represent the components of the stress and strain tensors and the elastic constants of a macroscopic element in the order. ξ, ε and c are the equivalents of the same terms at microscopic level. The superscript, k, is used to indicate that all microscopic quantities are single valued but random, and the statistical average is denoted by the bar sign.

Consider a macro-homogeneous and isotropic body under uniaxial load. The random potential elastic energy, u_k, of a microscopic element is

$$u_k = \tfrac{1}{2}\xi_k \times \varepsilon_k \quad (2)$$

Since the elastic strain energy of a macroscopic element is an average of the microscopic energies,

then
$$U = \bar{u}_k$$
$$= \tfrac{1}{2}\overline{\xi_k \times \varepsilon_k} \tag{3}$$

The right-hand side of the equation (3) is a statistical identity which can be written as

$$\overline{\xi_k \times \varepsilon_k} = \bar{\xi}_k \times \bar{\varepsilon}_k + \lambda^2_{\varepsilon\xi} \tag{4}$$

where $\lambda^2_{\xi\varepsilon}$ is termed the covariance of ξ and ε. Using equations (1) and (4), and substituting in equation (3),

$$U = \tfrac{1}{2}[\sigma \times e + \lambda^2_{\xi\varepsilon}] \tag{5}$$

Equation (5) shows that the true macroscopic elastic energy would be different from the theoretical one by an amount $\lambda^2_{\xi\varepsilon}$, i.e. by how much micro-inhomogeneity of the stresses and strains is exhibited by the material.

The work done by the applied load is transformed into the potential elastic energies of the microscopic volumes which, together with the existing residual microstresses and strains, form a random field of internal stresses and strains; the equilibrium and compatability requirements are thus satisfied. As the load is increased, the local variations of elastic energy will exceed that required for plastic deformation in some elements and these elements will yield. The extent of plastic flow of any one element is limited by the surrounding elastic matrix. If the matrix is strong enough, upon the release of load, the plastic deformation is restored to its origin—the polycrystalline material exhibiting *Anelasticity*.

The dependence of the deformation of a microscopic element on the neighbouring elements makes it extremely difficult to describe the aggregate behaviour starting with the behaviour of a single element. The influence of the random orientations of the grains and the interaction at the boundaries makes the stress–strain response of a microscopic element, as a component of the aggregate, considerably different from the single crystal behaviour. Figure 1 of the silicon–iron alloy below, showing the *dislocation* etch pits (signs of plastic flow at microscopic level), illustrates the localized nature of the microplasticity and the influence of the grain boundaries in the continuity of the strain field. This also indicates that the concept of a microscopic element cannot be applied to a grain if strain gradients exist across a grain as in the case shown, which is indicated by the fluctuating density of the etch pits.

The foregoing facts compel all mathematical approaches to use statistical functions for a realistic representation of the stress and strain of a macroscopic element in terms of microscopic variables. From the experimental evidence, which suggests that the micro inhomogeneity of stresses and strains must be considered together, there would be a great loss of generality if the dispersion of stresses and strains are not approximated by two statistical functions where means obey the nominal stress–strain diagram of the material.

The elasto-microplastic situation created by the yield of some elements imposes upon the elastic dispersion of strains an additional inhomogeneous state of strain. Upon further loading, the lower yield elements undertake larger strains and some of the stronger elements also undergo plastic deformation. As the elements are not free to deform independently,

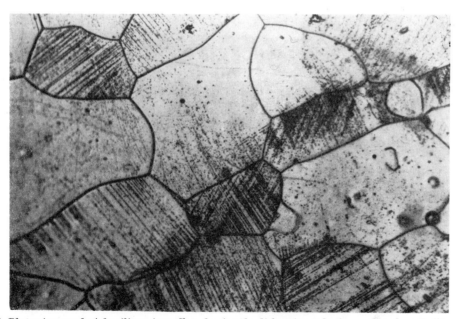

Fig. 1. Photomicrograph of the silicon–iron alloy showing the dislocation etch pits present above the "true elastic limit" and establishing that the metal is microplastic. Magnification $\times 270$.

then, the plastic elements cannot fully deform to take up the stresses imposed on them and, as a result, transmit the load to the stronger elements becoming partially redundant. This creates an additional state of micro inhomogeneity of stresses. When the deformations of some elements are sufficiently large, the elastic matrix cannot restore the plastic flow upon the release of load and a local, permanent set forms. The stress level at which this phenomena occurs is termed as the *True elastic limit* of the material. Although elastic behaviour dominates the apparent deformation of the macro-volume, the true mechanical behaviour consists of the elastic, anelastic and plastic contributions of the micro-elements.

The number and the contributions of the anelastic and plastic elements increase with further loading and the nominal elastic limit defines the threshold value from whereon the observable behaviour is dominated by the plastic elements. At loads above this point, the number of the elastic elements diminishes and when all of the microscopic elements of the macroscopic section have yielded, fully developed plastic flow is observed.

Bibliography

COTTRELL A. H. (1956) *Dislocations and Plastic Flow in Crystals,* Oxford: University Press.

FREUDENTHAL A. M. (1950) *The Inelastic Behaviour of Engineering Materials and Structures,* Wiley.

Materials Science Research (1963) Vol. 1, New York: Plenum Press.

SCHMIT E. and BOAS W. (1950) *Plasticity of Crystals,* London: Hughes and Co. Ltd.

VOLKOV S. D. (1962) *Statistical Strength Theory,* New York: Gordon and Breach, Science Publishers.

A. ESIN

MICROWAVE SOLID-STATE OSCILLATORS, RECENT

Introduction

The transistor is rapidly replacing conventional control-grid electron tubes (triodes, pentodes, etc.) in many applications; however, due to inherent physical limitations the transistor is unable to compete with the specialized electron tubes developed for use at microwave and millimetre-wave frequencies (klystrons, magnetrons, travelling-wave tubes, etc.). For use at these high frequencies, several types of solid-state devices have been developed within the last several years for use as sources of power. Presently the primary solid-state oscillators are the IMPATT diode, the Gunn oscillator, and the LSA oscillator.

There are many reasons for wanting solid-state devices for the microwave frequency range which is becoming rapidly more important for long-range high-capacity communications systems. The primary reason is the long lifetime and high reliability associated with solid-state devices. Lifetimes on the order of 10^9 hr are expected. This advantage is quite apparent when one considers that microwave electron tubes generally have lifetimes of several hundred to several thousand hours. These solid-state devices will also be much more economical since the associated circuitry is much simpler. In many applications it will be possible to operate directly from the dc voltages commonly available in aircraft, automobiles, satellites, and telephone installations.

The IMPATT Diode

The IMPATT diode (IMPact Avalanche Transit-Time) has several possible structures which are shown in Fig. 1. The first structure (PNIN) was proposed

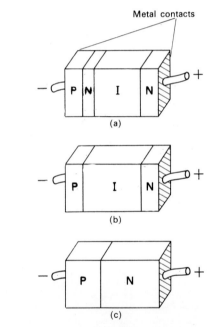

Fig. 1. Three types of IMPATT diodes made by alternating layers of p-type (P), n-type (N), or intrinsic (I) semiconducting material, usually silicon or germanium.

by W. T. Read of Bell Telephone Laboratories in 1958. This structure was relatively easy to treat theoretically but quite difficult to fabricate. The first operation of an IMPATT diode was achieved in 1964 at B.T.L. by R. L. Johnston, B. C. DeLoach and B. G. Cohen using a simpler *p-n* junction diode. Figure 2 shows the maximum powers obtained by various research organizations as a function of frequency. IMPATT oscillators which produce 100 mW of continuous power at 10 GHz (10^{10} c/s) are presently commercially available.

All IMPATT devices operate on the basis of avalanche current multiplication which is the current gain that occurs when a high reverse bias voltage is applied to a p-n junction. As voltage is applied to the junction, the resulting electric field accelerates the carriers (electrons and holes) moving across the junction. If

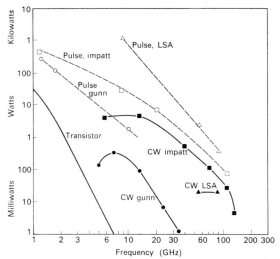

Fig. 2. *A plot of power achieved by four types of semiconductor devices over the microwave frequency range in both pulse and continuous operation.*

the voltage is high enough (above the so-called avalanche threshold), the velocities of the carriers may increase until they attain enough energy to collide with and free other electrons in the crystal structure. Each such collision produces a hole–electron pair. The carriers thus created can in turn produce other carriers, and the process snowballs to produce an avalanche of current that is theoretically unlimited.

If the voltage applied to a p–n junction is held constant, the electric field across the junction is limited to the strength at which avalanche breakdown occurs. If the applied field is varied rapidly above and below the breakdown value, however, a current delay, or phase shift, results. This occurs because the avalanche takes time to build up as the voltage rises, and time to subside after the voltage has passed its maximum. With relatively small signals and high frequencies, this phase shift can approach 90 degrees.

Additional phase shift can be obtained with transit time delay—the time it takes carriers to travel across a region of the diode. At microwave frequencies this time is long enough so that the carriers crossing a field region can reach the opposite side out of phase with the applied voltage.

If circuit components are properly matched to each other, and if the physical dimensions of the diode are just right, the phase delay from both avalanche multiplication and transit time delay can be between 90 and 270 degrees—the amount needed to sustain oscillations. If capacitance and inductance in the circuit are resonant in the frequency range where the diode exhibits negative resistance, the combination of the resonant circuit and the negative resistance instability produce oscillations at the resonant frequency.

Gunn and LSA Oscillators

The Gunn and LSA oscillators operate on the basis of the bulk negative differential resistivity that appears in certain semiconductors such as GaAs, InP, CdTe, and ZnSe. This effect was predicted in 1961 by B. K. Ridley and T. B. Watkins at Mullard Research Laboratories and independently by C. Hilsum at Services Electronics Research Laboratory. However, it was a few years before semiconductor technology advanced to the point where the effect could be observed.

In 1964, while working on electrical noise emitted by semiconductors, J. B. Gunn of I.B.M. Research Laboratories observed that microwave noise powers of the order of a watt were emitted from GaAs and InP when they were subjected to pulsed electric fields of several thousand volts per centimetre. When short samples were used (less than 0·2 mm), the noise changed to coherent oscillation with a period close to the time needed for carriers to drift from one contact to the other (transit-time oscillation).

Gunn then developed an elaborate capacitive probe that plotted the electric field distribution within samples while they oscillated. These measurements showed that as the voltage was increased past a threshold, a high field domain formed near the cathode that reduced the electric field in the rest of the diode and caused the current to drop to about two-thirds of the maximum value. The high field domain would then drift with the carrier stream across the sample and disappear at the anode contact as shown in Fig. 3. As the old domain disappeared at the anode,

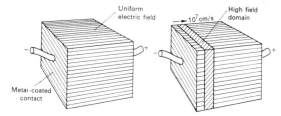

Fig. 3. *Gunn and LSA diodes can be made by contacting a uniform piece of n-type GaAs. The high field domain formation which causes Gunn-effect oscillation is suppressed when the diode is used as an LSA oscillator.*

the electric field behind it increased (to keep the voltage, $\int E\,dx$, constant) until the threshold field was reached and the current had increased back to the threshold value. At this time a new domain would form at the cathode, the current would drop, and the cycle would begin anew.

A current waveform produced by a Gunn oscillation is shown in Fig. 4. The flat valley occurs as the domain drifts across the sample. The upward spikes begin as

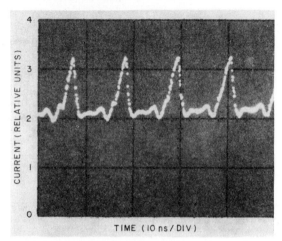

Fig. 4. Current waveform due to Gunn-effect oscillations. Current rises each time a domain drifts into anode and drops again as a new domain forms at cathode.

a domain reaches the anode, and a new domain forms at the cathode.

Peak powers obtained from experimental Gunn devices are shown in Fig. 2. Gunn oscillators which produce 10 mW of continuous power are presently commercially available.

In 1966, J. A. Copeland of Bell Telephone Laboratories predicted that by the proper design of a bulk n-GaAs diode and corresponding resonant circuit, it would be possible to prevent domains and other types of space charge from building up within the diode. This mode of operation, called LSA for limited space-charge accumulation, was soon verified experimentally at B.T.L.

For LSA operation of a bulk n-GaAs diode, the voltage must swing below the threshold voltage each cycle long enough to dissipate space charge. Also, the part of each cycle during which the voltage is above the threshold must be too short for space charge to build up and form a domain. Since the speed of space-charge dissipation and growth is proportional to the doping, the ratio of the doping to frequency must be within 2×10^4 to 2×10^5 cm^3/s. The circuit must be lightly loaded to achieve the necessary ac voltage swing. For high efficiency, the doping should be uniform within 10 per cent.

Devices such as transistors and IMPATT (avalanche) diodes, as well as Gunn-effect oscillators, must be thin enough that carriers can move through the active region during one cycle or less. This means that the thickness and voltage must be decreased to raise the frequency of operation. To maintain a reasonable impedance, the current also must be decreased. The result of these considerations is that the maximum power for a given transit-time or subtransit-time device falls off faster than $1/(\text{frequency})^2$, as can be seen in Fig. 2. The LSA oscillator is the first practical solid-state oscillator that is free of this limitation, since in principle it can be made much thicker than the distance a carrier drifts during one cycle.

Preliminary experiments with epitaxial LSA oscillators have produced 0.7 W and 9 per cent efficiency at 50 GHz on a pulse basis and 0.02 W with 2 per cent efficiency at 88 GHz on a continuous basis. Experiments with bulk-grown diodes have yielded 1000-W pulses at 10 GHz and have led to predictions that 250 kW pulses from a single block of n-GaAs are possible. LSA oscillators which produce 100 W pulses at 10 GHz are presently commercially available.

Conclusion

The transistor has not only replaced many conventional control-grid electron tubes but has made entirely new electronic systems possible. Even more so, these solid-state microwave oscillators will greatly broaden the range of applications that can economically utilize microwaves.

Bibliography

COPELAND J. A. (1967) Bulk negative-resistance semiconductor devices, *IEEE Spectrum* **4,** 71–77.

DELOACH B. C. (1967) Recent advances in solid-state microwave generators, in *Advances in Microwaves* (Leo Young Ed.), Vol. 2, New York: Academic Press.

HILSUM C. (1967) Miniaturizing radar, *Science Journal*.

KRÖMER H. (1968) Negative conductance in semiconductors, *IEEE Spectrum* **5,** 47–56.

SMITH K. D. (1966) Generating power at Gigahertz with avalanche-transit time diodes, *Electronics*.

Special issues on semiconducting bulk-effect and transit-time devices, *IEEE Transactions on Electron Devices*, **ED-13,** No. 1, Jan. 1966 and **ED-14,** No. 9, Sept. 1967.

J. A. COPELAND

MIRAGE AND LOOMING, OPTICS OF

1. Introduction

The refraction of light through air in which the refractive index changes with height (due to the variation of temperature) gives rise to the interesting optical phenomena of mirage and looming. The phenomenon of mirage, i.e. observation of inverted images of objects, occurs on land when the temperature decreases (i.e. the refractive index increases) with increasing height; in looming, which occurs on the sea, one can see an elevated erect image of an object because of the increase of temperature with height. A graphical description of these phenomena has been given by Humphreys (1946) and Gordon (1959). The present discussion will be confined to the formation of images by refraction through air in which the refractive index varies with height in a simple manner. The treatment may be extended to the case of more complicated variations of refractive index with height which give rise to more picturesque manifestations of these phenomena. The present analysis is primarily based on the work of Sodha, Aggarwal and Kaw (1967).

2. Idealized Variation of Refractive Index with Height

The refractive index μ of air for visible light may be expressed as (Fleagle and Businger, 1963)

$$\mu - 1 = a\varrho,$$

where a is a constant and ϱ is the density of air. Combining the above equation with the relations

$$-\frac{\partial p}{\partial y} = \varrho g \quad \text{(hydrostatic equation)}$$

and

$$p = \varrho RT \quad \text{(equation of state)}$$

one obtains

$$-\frac{\partial \mu}{\partial y} = \left\{\frac{g}{R} + \frac{\partial T}{\partial y}\right\} \frac{(\mu - 1)}{T},$$

where the symbols have their usual meanings, y being the height.

From this equation the dependence of μ on y can be obtained for a given variation of T with y. However, for further analysis and to simplify the approach we may assume that

$$\mu = \mu_0 \exp(\alpha y) \tag{1}$$

Thus a positive value of α (increasing refractive index with height) leads to the observation of a mirage, while a negative value of α leads to the observation of looming.

3. Ray Tracing in a Stratified Medium

A Cartesian coordinate system is chosen with the origin on the surface of the ground, the x-axis being along the earth's surface (assumed to be plane) and the y-axis pointing vertically upwards. Let the coordinates of an object S be (x_0, y_0) and that of an observer E be (x_e, y_e) as shown in Figs. 1 and 2. Let $P(x_m, y_m)$ be the point where the direction of the ray (making an angle i_1 with the vertical at the object) is horizontal; this point will also be an extreme position (either the maximum or the minimum) in the curved ray path along which this light beam travels.

From Snell's Law

$$\mu = \mu_1 \frac{\sin i_1}{\sin i} \tag{2}$$

where i is the angle the ray (its tangent) makes with the vertical at any point (x, y) and μ is the refractive index at that point.

Combining equations (1) and (2) gives

$$\frac{\sin i}{\sin i_1} = \exp\{\alpha(y_0 - y)\}. \tag{3}$$

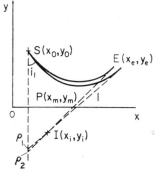

Fig. 1. Path of the rays in mirage.

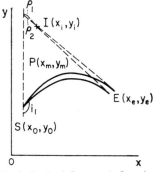

Fig. 2. Path of the rays in looming.

From Figs. 1 and 2

$$dy/dx = \pm\cot i. \tag{4}$$

Using equations (3) and (4), integrating, and evaluating the constant of integration by imposing the condition that

$$y = y_0 \quad \text{at} \quad x = x_0$$

(which means that the ray should pass through the point $S(x_0, y_0)$ the equation obtained for the curved ray path is

$$\exp\{\alpha(y_0 - y)\} = \sin\{\pm\alpha(x - x_0)$$
$$+ \sin^{-1}(\sin i_1)\}(\sin i_1)^{-1} \tag{5}$$

choosing new dimensionless coordinates (X, Y) instead of the usual (x, y) such that

$$X = \alpha x \quad \text{and} \quad Y = \alpha y,$$

(the signs of X and Y are the same as those of α, being positive and negative respectively for mirage and looming), converts (5) to

$$\exp(Y_0 - Y) = \sin\{\pm(X - X_0) + \sin^{-1}(\sin i_1)\}(\sin i_1)^{-1}. \quad (6)$$

The plus and minus signs within the parentheses correspond to the regions on either side of the point $P(x_m, y_m)$. Referring to Figs. 1 and 2 it follows that the physically significant values of $\sin^{-1}(\sin i_1)$ are i_1 and $\pi - i_1$, respectively for positive and negative values of α.

The coordinates of the point P are found by imposing the condition for an extreme value:

$$\left(\frac{dY}{dX}\right)_{X=X_m, Y=Y_m} = 0.$$

From (6) we then obtain

$$X_m = X_0 + \tfrac{1}{2}\pi - \sin^{-1}(\sin i_1) \quad (7\text{a})$$
$$Y_m = Y_0 + \ln(\sin i_1). \quad (7\text{b})$$

4. Position and Form of the Image

Consider two rays with slightly different angles of incidence. Tangents drawn to these curves at the point of observation give the image by the point of intersection of two tangents for the limiting case, when the difference in the angles of the incidence of the two rays is vanishingly small. Thus the coordinates of the image I are given by

$$X_i = X_e - \tfrac{1}{2}[\tan i_1 \cot\{X_0 - X_e + \sin^{-1}(\sin i_1)\} - 1]$$
$$\times [\sin 2\{X_0 - X_e + \sin^{-1}(\sin i_1)\}] \quad (8\text{a})$$

and

$$Y_i = Y_e - [\tan i_1 \cot\{X_0 - X_e + \sin^{-1}(\sin i_1)\} - 1]$$
$$\times [1 - \exp\{2(Y_0 - Y_e)\}\sin^2 i_1]. \quad (8\text{b})$$

To express the coordinates of the image in terms of the coordinates (X_e, Y_e) of the observer i_1 is eliminated from (8) by using the relation

$$\exp(Y_0 - Y_e) = [\sin\{\pm(X_e - X_0) + \sin^{-1}(\sin i_1)\}](\sin i_1)^{-1} \quad (9)$$

which means that the curved ray path passes through the point of observation.

From the variation of Y_i with Y_0 it follows that, for a vertical object, the image is inverted in the case of mirage (positive α) and erect in the case of looming (negative α); as borne out by experience of these phenomena. It is further to be noted from equation (8) that the coordinates of the image are very different from what would be expected on the basis of the usual explanation of these phenomena. It may be mentioned that the position of the image is a function of the position of the observer.

Using equations (8) and (9) the form of the image for an object of general shape can be investigated. Consider an object which may be represented in the coordinate system by the equation

$$Y_0 = \phi(X_0). \quad (10)$$

One can, in principle, eliminate the three quantities i_1, X_0 and Y_0 to obtain an equation between X_i and Y_i involving the coordinates of the observer only obtaining an equation which gives the form of the image.

5. Maxima and Minima in the Range of Observation and Height of the Object

The present analysis predicts some limits on the range of observation, for points outside which the phenomena are not observed; there are similar limits on the height of the object.

5.1. Maximum range

From Figs. 1 and 2 it is seen that the range of observation is larger for the rays starting with a smaller inclination to the vertical (at the object). The least inclination for a mirage corresponds to the least value of i_1, whereas the condition for looming corresponds to a maximum value of i_1. For a mirage, equation (7b) sets a lower limit to the angle of incidence with which a ray can start from the object in order that it may be seen because the least value of Y_m is zero. This gives

$$\sin i_{1\min} = \exp(-Y_0). \quad (11)$$

This case corresponds to the situation wherein the ray becomes horizontal exactly at the earth's surface. For values of $i_1 < i_{1\min}$ the minimum in the curved ray path would occur at a virtual point apparently under the earth: such rays will therefore not be bent unwards at all. Thus a maximum range of observation exists beyond which a mirage cannot be seen. On the other hand, for looming Y_m is the Y-coordinate for a maximum in the curved ray path and, since there is no upper limit for Y_m, i_1 can take any value up to π; thus there is no maximum limit to the range of observation of looming. For any height Y_e of the observer the maximum range $R_{\max}$ of the observation of a mirage is given by

$$\exp(Y_0 - Y_e) = \frac{\sin\{R_{\max} + \sin^{-1}(\sin i_{1\min})\}}{\sin i_{1\min}}.$$

This gives

$$R_{\max} = \pi - \sin^{-1}\{\exp(-Y_e)\} - \sin^{-1}\{\exp(-Y_0)\}. \quad (12\text{a})$$

There will be an absolute maximum in the range of observation of a mirage, which corresponds to the case for which $Y_e = \infty$, therefore

$$R_{\text{abs max}} = \pi - \sin^{-1}\{\exp(-Y_0)\}. \quad (12\text{b})$$

The magnitude of $R_{\text{abs max}}$ is completely determined by the height Y_0 of the object. A simple physical argument can be given for the existence of this maximum observation range for a mirage. In this case when the rays are travelling upwards they are moving from an optically rarer to an optically denser medium, and therefore they are gradually bent towards the vertical.

Finally, at some distance from the object, the rays travel vertically and clearly for observation points beyond this range a mirage cannot be seen by an observer at any height.

The existence of an absolute maximum in the observation range arises in this theory because it has been assumed that the refractive index continues to increase with height at an exponential rate *ad infinitum*. In a real atmosphere, the exponential gradient is limited to a certain altitude above which the gradient of refractive index becomes normal. Thus this theoretical result is not of much practical value.

5.2. Minimum range

The minimum in the observation range will correspond to the extreme value position in the curved ray path (i.e. the minimum for mirage and maximum for looming) because for points before this extreme the reflected image of the object cannot be seen. Thus, since for a given height of the observer the angle i_1 of the ray incident at the object is automatically fixed, it is given by the relation, in accordance with equation (7b),

$$Y_e = Y_0 + \ln(\sin i_1).$$

Correspondingly, the X-coordinate for this extreme position, which defines the minimum range of observation, is given by

$$R_{\min} = \frac{1}{\pi} - i_1 = \frac{1}{2}\pi - \sin^{-1}\{\exp(Y_e - Y_0)\}. \quad (13)$$

Thus, for a mirage

$$R_{\min} = \frac{1}{\pi} - \sin^{-1}\{\exp(|Y_e| - |Y_0|)\}$$

and for looming

$$R_{\min} = \sin^{-1}\{\exp(|Y_e| - |Y_0|)\} - \tfrac{1}{2}\pi.$$

$R_{\min}$ should always be positive. Thus when $Y_e \geq Y_0$, then $R_{\min}$ is either zero or does not exist. This means that there is no lower limit for the range of observation for observers having a height equal to, or greater than, that of the object.

5.3. Minimum and maximum height of objects

The concept of a maximum height for the object is closely related to that of the minimum range of observation; thus a maximum height of an object will exist for a mirage as well as for looming. A maximum height of the object $Y_{0\max}$, means that for a particular position of the observer totally reflected images of points above that height cannot be seen. If the position of the observer is fixed and the height of the object is increased, then, for a certain height, the minimum of curved ray path for mirage and the maximum for looming will fall on the point of observation. This object height corresponds to $Y_{0\max}$. For points above this height the minimum of the curved ray path (for mirage) will be displaced either vertically upwards or towards the right-hand side, while the maximum of the curved ray path (for looming) will be shifted either towards the left or vertically downwards. The actual situations for $R_{\min}$ and $Y_{0\max}$ are identical; therefore if $R_{\min}$ is replaced in equation (13) by R simply, then Y_0 will be equal to $Y_{0\max}$.

A similar argument shows that for mirage the maximum range of observation and the minimum height of the object have a similar reciprocal relation, and therefore the latter can be determined from (12a) by replacing $R_{\max}$ by R. A minimum height of the object does not exist for the case of looming.

6. Discussion

To obtain an appreciation of the orders of magnitude of the various ranges and heights and their variation with the relevant parameters, some numerical results are presented graphically for $R_{\max}$ and $R_{\min}$ against Y_0 (for mirage) for different values of Y_e (Figs. 3 and 4) respectively. The minimum range of observation has a larger magnitude for higher objects.

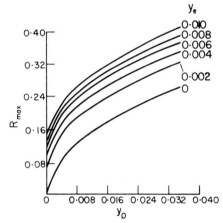

Fig. 3. *Variation of $R_{\max}$ with Y_0 for various values of Y_e.*

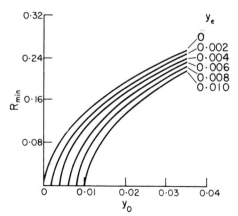

Fig. 4. *Variation of $R_{\min}$ with Y_0 for various values of Y_e.*

By increasing the height of the observer the minimum range of observation is reduced. The corresponding variation of $R_{\min}$ with the height of the object and observer for looming can be obtained from that for mirage by replacing Y_0 by $-|Y_0|$, $R_{\min}$ by $-|R_{\min}|$ and Y_e by $-|Y_e|$, i.e. the curves will lie in the third quadrant. For mirages the maximum range of observation has a similar variation with Y_0; however, the variation with Y_e is reversed, i.e. for higher observers the maximum range of observation increases. The graphs facilitate determination of the maximum and minimum heights of the object for a particular position of the observer by the method mentioned in section 5.

Suppose an observer of height 4 m is situated 150 m from a vertical object. The maximum and minimum heights of the object beyond which points are not visible to the observer are determined thus. Let the rate of increase of the refractive index with height be such that $\alpha = 0.001$, which corresponds to a variation of 10 per cent in the refractive index for a height of about 100 m, a reasonable value in view of the experiments of Fleagle (1950, 1956, 1959). Then the coordinates of the observer in the dimensionless coordinate system are

$$X_e = 0.150, \qquad Y_e = 0.004.$$

Figures 3 and 4 indicate that for $X_e = R_{\max}$ and $R_{\min}$ respectively the corresponding heights of the object are 2 m and 15 m. The former gives the minimum height of the object and the latter the maximum height.

The present treatment has been mainly confined to the phenomenon of image formation by rays in a vertical plane. Such a treatment, although not completely adequate, leads to a number of correct and interesting conclusions. It is worth recollecting that the image is astigmatic in nature. By symmetry all the rays making an angle of incidence i_1 at the object and travelling in different vertical planes appear to diverge from a common point on the same vertical line as the object (when viewed at the same height). Thus if the rays entering the eye of the observer are limited by a horizontal slit, the image will be located on the same vertical line as the object. If the rays are limited by a vertical slit the image will be at a different point, not along the vertical line through the object.

Raman and Pancharatham (1959) and Raman (1959) have, however, pointed out that the change of the path of light from a horizontal direction to a direction travelling upwards cannot be explained on the basis of geometrical optics; a description of this phenomenon on the basis of wave theory has been given by these authors, who also point out the remarkable similarity that exists between the situations in the vicinity of the limiting horizontal stratification and that in the neighbourhood of a caustic surface formed by the reflection or refraction of light. The gradual veering around of the rays (which according to purely geometrical optics would continue to move parallel to the horizontal stratification) can then be explained in terms of the propagation of a cusped wavefront along this limiting horizontal layer. However, for most purposes use of geometrical optics suffices because the laws of total reflection are obeyed. The help of wave theory is sought only to overcome the difficulty of explaining the bending of light in the limiting horizontal stratification. Once the bending of the rays in this layer is understood the problem of image formation may be satisfactorily treated by geometrical optics.

Bibliography

FLEAGLE R. G. (1950) *Bull. Amer. Met. Soc.* **31**, 51–55.
FLEAGLE R. G. (1956) *J. Met.* **13**, 160–5.
FLEAGLE R. G. (1959) *Geophys. Res. Papers* **2**, 128.
GORDON J. H. (1959) *Smithsonian Ann. Rep.*, pp. 327–45.
HUMPHREYS W. J. (1946) *Physics of the Air*, pp. 469–75, New York: Dover.
RAMAN C. V. (1959) *Curr. Sci.* **28**, 309–13.
RAMAN C. V. and PANCHARATNAM S. (1959) *Proc. Ind. Acad. Sci.* **A 49**, 251–61.
SODHA M. S., AGGARWAL A. K. and KAW P. K. (1967) *Brit. J. Appl. Phys.* **18**, 503–11.

M. S. SODHA and A. K. AGGARWAL

MODEL SHIPS, PHOTOELECTRIC TRACKING OF

Introduction

At the Ship Laboratory of NRC's Division of Mechanical Engineering model ships are used to gain valuable insight into the parameters that influence the performance characteristics of full-size vessels.

To study the behaviour of model ships, one of the essential data required is a permanent record of the course and speed of the model.

Since the distances involved are less than 500 ft and trials are conducted only under favourable weather conditions, an optical tracking system determining the model's position by triangulation was considered the simplest and most promising approach.

Description

The principle of operation of the system is quite straightforward and can best be explained by reference to Fig. 1.

A light source comprising a scintillating Xenon lamp is mounted on the model ship. Two light-seeking tracking telescopes are installed on shore to track the light source automatically as the model ship moves about the basin. To generate the directional data necessary to determine the model's position each telescope is coupled to a three-phase synchro transmitter whose output is fed via a cable to a corresponding control transformer in the display unit.

To prevent the model from coming too close to the

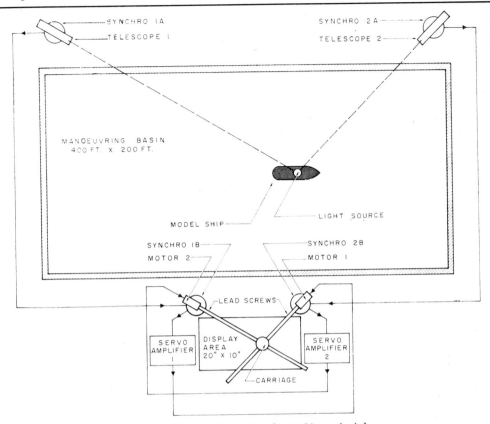

Fig. 1. Diagram illustrating the tracking principle.

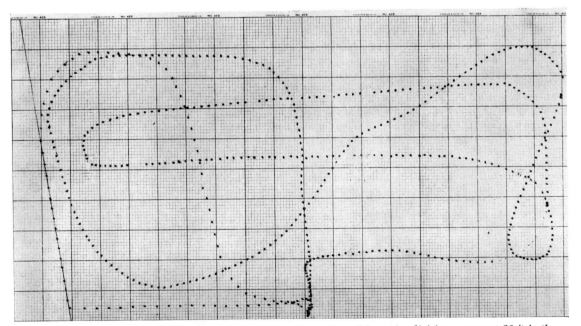

Fig. 2. Typical record produced by photoelectric tracking system. The major divisions represent 20 ft in the manoeuvring basin; the model's position is indicated at 1 sec intervals.

line joining the tracking telescopes, where its position would be indeterminate, the telescopes are positioned some distance back from the edge of the manoeuvring basin.

The display unit incorporates the automatic plotting board which is in effect a scale model of the manoeuvring basin. The display area measures 20 in. by 10 in. so that 1 in. on the display corresponds to 20 ft in the basin. Two arms, extending over the display area, pivot about points corresponding to the position of the tracking telescopes. When the direction of each arm is made to correspond to that of the corresponding telescope by means of a servo system utilizing the directional information transmitted from the telescopes, the point of intersection of the arms will indicate the model's position.

To produce a permanent record of the model's manoeuvres, paper is stretched over the display area and a metal probe extends downwards from the carriage supported at the point of intersection of the arms. An electric discharge through the paper at regular time intervals (say 1 sec apart) not only marks the model's position but also makes it possible to determine its speed. A typical record produced by the tracking system is shown in Fig. 2.

Performance

The equipment has been used successfully since the summer of 1965. The overall dynamic accuracy of the system was measured by marking the indicated position of the model as it crossed and recrossed lines of sight established between stationary points on shore. It was found that the model's position as indicated on the display consistently fell within ± 2 ft of the established lines.

Bibliography

WYSLOUZIL W. (1965) *Photoelectric tracking of model ships*, NRC Report ERB-719.

W. WYSLOUZIL

MOMENTUM TRANSFER CROSS-SECTION FOR LOW-ENERGY ELECTRONS: DETERMINATION BY SWARM TECHNIQUES. When low-energy electrons collide with neutral atoms or molecules, several types of interaction are possible. In a molecular gas, provided the electron energy is insufficient to produce electronic excitation or ionization of the target molecules, the majority of the collisions will be those in which elastic scattering occurs, while a small fraction of the collisions are inelastic leading to rotational and vibrational excitation of the molecules. In electron-atom collisions, all the collisions are, of course, elastic in this energy range.

The elastic scattering cross-section, or more precisely, the momentum transfer cross-section, plays a major role in determining the transport properties of the electrons in the gas. For example, if the electrons drift through the gas as a result of an applied electric field E, the drift velocity W and diffusion coefficient D are given by the relations:

$$W = -\frac{Ee}{3N}\left(\frac{2}{m}\right)^{\frac{1}{2}} \int_0^\infty \frac{\varepsilon}{q_m(\varepsilon)} \frac{df(\varepsilon)}{d\varepsilon} d\varepsilon \quad (1)$$

$$D = \frac{1}{3N}\left(\frac{2}{m}\right)^{\frac{1}{2}} \int_0^\infty \frac{\varepsilon f(\varepsilon) d\varepsilon}{q_m(\varepsilon)}, \quad (2)$$

where $q_m(\varepsilon)$ is the energy dependent momentum transfer cross-section,
N is the gas number density,
e, m and ε are the electronic charge, mass and energy respectively, and
$f(\varepsilon)$ is the energy distribution function for which the normalizing relation is:

$$\int_0^\infty \varepsilon^{\frac{1}{2}} f(\varepsilon) d\varepsilon = 1.$$

There have been two contrasting approaches to the problem of measuring elastic collision cross-sections, one set of experiments determining the total elastic scattering cross-section q_s, and the other the momentum transfer cross-section q_m. In some instances the distribution of angular scattering has been determined in addition to the total elastic scattering cross-section and a direct comparison can then be made between the results of the two types of experiment by using the relation:

$$q_m = q_s(1 - \overline{\cos \theta}).$$

In this equation $\overline{\cos \theta}$ is the mean of the cosine of the angle through which the electron is scattered as measured in the CM frame.

The two techniques were developed almost concurrently. C. Ramsauer on the one hand initiated a series of experiments in which q_s and the angular distribution were measured, while J. S. Townsend deduced values of q_m from diffusion measurements; the success of these physicists in demonstrating independently the relatively strong energy dependence of the cross-section in low energy electron-atom collisions is recognized in the name "Ramsauer–Townsend effect" given to this phenomenon.

The two methods are quite different in principle. In the single collision experiments of Ramsauer and others, a monoenergetic beam of electrons traverses the collision chamber with the target gas at a sufficiently low pressure (10^{-2} to 10^{-3} torr) that the number of multiple collisions is negligibly small. The total collision cross-section can then be determined from the attenuation of the beam. In more elaborate experiments the angular distribution of the elastically scattered electrons is also determined. The principle advantage of these methods is their directness, which arises from the fact that the cross-section may be measured for a set of well defined energies. On the other hand, the low energy limit has so far been relatively high (~ 0.2 eV) and there are difficulties in determining the absolute magnitudes of the cross-sections as a result of uncertainties in the effective number density along the collision path.

In the methods of deducing cross-sections from swarm experiments, as devised initially by Townsend, ease of interpretation of the experimental data is exchanged for the ability to perform well-controlled experiments with electrons having energies of 0.01 eV or less. In these experiments an ensemble or swarm of electrons drifts and diffuses in an electric field through a gas at pressures from several torr to an atmosphere or more. Owing to the very large number of collisions within the collision chamber, typically of the order of 10^6, an energy distribution is established for the ensemble for which the mean value and form are determined by the electric field strength and the nature, pressure and temperature of the gas. The use of low field strengths (~ 1 V cm^{-1}) and high gas number densities results in an energy distribution close to that of the gas molecules; the lowest energy collisions are therefore studied by lowering the gas temperature. Alternatively low gas number densities associated with much higher field strengths result in "electron temperatures" greatly exceeding the gas temperature. It is thus possible to vary the mean energy of the ensemble conveniently from about 0.01 to 10 eV. Collisions in this energy range can therefore be studied without the necessity of using extremely low accelerating potentials as would be the case for the single collision experiments. A further advantage of these methods is that the gas number density, upon which the absolute magnitude of the measured cross-section depends, can be measured accurately because of the high gas pressures employed.

The method of determining the momentum transfer cross-section from swarm experiments is based on the measurement of one or more transport coefficients (for example, the drift velocity W, the ratio of diffusion coefficient to mobility D/μ, etc.) over a range of mean energies, followed by the determination, using an iterative process, of an energy dependent cross-section that is consistent with these data. In practice neither the mean energy $\bar{\varepsilon}$ of the swarm nor the form of the energy distribution need be determined explicitly. As described above, the energy distribution is varied by varying E/N. The experimentally determined transport coefficients are therefore tabulated as functions of this ratio, rather than as functions of $\bar{\varepsilon}$, and the iterative process entails a comparison of the experimental and calculated curves of these coefficients plotted as functions of E/N.

The procedure adopted depends on whether or not inelastic collisions play a significant part in determining the energy distribution of the swarm.

(a) *Elastic Scattering Only*

When elastic scattering is the only energy exchange process, the cross-section can be determined from measurements of one transport coefficient only. Such is the case for the monatomic gases when the mean energy of the swarm is such that a negligible fraction of the collisions produces electronic excitation. The coefficient that has usually been used is the drift velocity and the iterative process therefore rests on the application of equation (1). In this instance, the energy distribution function $f(\varepsilon)$ is given by:

$$f(\varepsilon) = A \exp\left[-\int_0^\varepsilon \frac{M}{6m}\left(\frac{E\,e}{Nq_m}\right)\frac{1}{\varepsilon} + kT\right]^{-1} d\varepsilon, \quad (3)$$

where M is the atomic mass and k is Boltzmann's constant. Specification of $q_m(\varepsilon)$ therefore completely determines the function for a given set of experimental conditions, that is, for a given gas at a given temperature and pressure and applied electric field strength. It follows from equation (1) that W is also determined. The iteration, therefore, consists of:

(1) choosing a set of values of $q_m(\varepsilon)$ and calculating $f(\varepsilon)$, and hence W, for a suitable set of values of E/N,
(2) comparing calculated and experimental values of W as a function of E/N,
(3) adjusting the values of $q_m(\varepsilon)$ until adequate agreement is obtained.

In some cases there may be advantages in using a transport coefficient other than W. It can be shown that, for constant cross-section and for $\bar{\varepsilon} \gg 3/2\,kT$, the coefficient D/μ is twice as sensitive as W to variation in q_m. Measured values of D/μ rather than W should therefore be used at higher energies for those gases in which q_m is not strongly energy dependent, provided comparable accuracy can be achieved in the measurements.

(b) *Elastic and Inelastic Scattering*

It is no longer possible to determine the momentum transfer cross-section from measurements of a single transport coefficient when inelastic collisions are significant in determining the energy distribution. Since $f(\varepsilon)$ is no longer solely dependent on $q_m(\varepsilon)$, the method described above cannot lead to unique values of $q_m(\varepsilon)$; however, uniqueness can be achieved provided all the other (inelastic) cross-sections are determined concurrently with $q_m(\varepsilon)$. To do so requires the analysis of data for a sufficient number of independent transport coefficients. In practice a more elaborate iterative procedure is adopted which has as its starting point a numerical solution of the Boltzmann equation to calculate $f(\varepsilon)$; the transport coefficients can then be calculated as before. Some degree of separation between the effects of the various cross-sections can be achieved by choosing suitable combinations of the transport coefficients when making the comparisons between calculated and experimental data (Frost and Phelps, 1962). In some cases, especially in the region of higher energies, the energy dependence of one or more of the inelastic cross-sections may be taken from independent measurements. Adjustments are then made to the assumed sets of values of $q_m(\varepsilon)$ and the remaining unknown inelastic cross-sections to obtain the best overall fit with the experimental data.

The methods that have been described for deriving cross-sections from swarm data were not possible before high speed computing techniques were developed. Nevertheless, much useful information was derived from the experimental data, it being possible to obtain reasonably close estimates of collision cross-sections in an energy range inaccessible to other techniques. The early methods of deducing cross-sections made use of more tractable forms of equations (1) and (2) that follow from use of simplifying assumptions. Most of the earlier results were expressed in terms of a "*mean free path*" which is in fact an effective "momentum transfer free path". The momentum transfer free path $l_m(\varepsilon)$ is related to $q_m(\varepsilon)$ through the equation $l_m(\varepsilon) = 1/Nq_m(\varepsilon)$.

If it is reasonable to assume that $q_m(\varepsilon)$ does not vary much over the significant portion of the distribution function, equations (1) and (2) reduce to:

$$W = \frac{2}{3} \frac{Ee}{m} l_m \overline{c^{-1}} \qquad (4)$$

$$D = \frac{1}{3} l_m \bar{c}, \qquad (5)$$

where c is the electron speed, l_m has been assumed constant and the bar signifies an average taken over the velocity distribution function. These equations can be solved to give:

$$l_m = \frac{3}{N} (m/2e)^{\frac{1}{2}} [\overline{cc^{-1}}]^{\frac{1}{2}} W(D/\mu)^{\frac{1}{2}}/(E/N),$$

and

$$\bar{c} = (2e/m)^{\frac{1}{2}} [\overline{cc^{-1}}]^{\frac{1}{2}} (D/\mu)^{\frac{1}{2}}.$$

The dimensionless factor $[\overline{cc^{-1}}]$ can be calculated on the assumption of a particular form of the energy distribution. It is therefore possible to calculate l_m and $\bar{c}$ from data for W and D/μ measured at a set of values of E/N and to plot l_m as a function of $\bar{c}$. Data derived in this way were those first used to compare cross-sections derived from swarm experiments with those measured by single collision techniques.

Although there is a large body of data available which has been derived using the approximate treatment described in the previous paragraph (see Brown, 1959; Healey and Reed, 1941; McDaniel, 1964), the more precise work resulting from the application of computer techniques has appeared only comparatively recently. Most of this work can be found only in the original papers, some of which are listed in the bibliography.

Bibliography

ALLIS W. P. (1956) in *Handbuch der Physik*, Vol. 21, p. 383, Berlin: Springer-Verlag.
BROWN S. C. (1959) In *Basic Data of Plasma Physics*, New York: The Technology Press and John Wiley and Sons.
CROMPTON R. W., ELFORD M. T. and JORY R. L. (1967) *Aust. J. Phys.* **20,** 369.
FROST L. S. and PHELPS A. V. (1962) *Phys. Rev.* **127,** 1621.
FROST L. S. and PHELPS A. V. (1964) *Phys. Rev.* **136,** A 1538.
HEALEY R. H. and REED J. W. (1941) in *The Behaviour of Slow Electrons in Gases*, Sydney: Amalgamated Wireless Ltd.
HUXLEY L. G. H. and CROMPTON R. W. (1962) in *Atomic and Molecular Processes* (D. R. Bates Ed.), p. 335, New York: Academic Press.
LOEB L. B. (1955) *Basic Processes of Gaseous Electronics*, Univ. California Press.
MCDANIEL E. W. (1964) *Collision Phenomena in Ionized Gases*, New York: Wiley.

R. W. CROMPTON

MULTIPLEX PRINCIPLE (OF FELLGETT). Let a spectrum extend from wave-number σ_1 to σ_2 and let the resolution in the spectrum be R. We can define the number M of independent elements in the spectrum as $M = \dfrac{\sigma_2 - \sigma_1}{R}$. When this spectrum is observed with a spectrometric system the observation may be made in one of two ways: the elements may be observed sequentially or simultaneously. A sequential

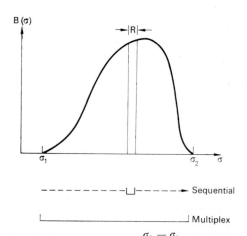

Fig. 1. A spectrum of $M = \dfrac{\sigma_2 - \sigma_1}{R}$ independent elements, each of width R. A sequential spectrometer scans from σ_1 to σ_2 with a window of width R in a total time T; a multiplex spectrometer observes the whole range σ_1 to σ_2 simultaneously for a time T.

spectrometer scans the spectrum element by element from σ_1 to σ_2 as exemplified in infra-red grating spectrometers; a simultaneous observation of all the elements occurs in a (photographic) spectrograph or in an interferometric spectrometer where all the radiation is observed by the detecting element for the total duration of the experiment (see *Fourier transform spectroscopy*).

If the spectrometric system is limited by detector noise the magnitude of the detected luminous signal is proportional to the observation time t while the noise observed is proportional to $\sqrt{t}$. Suppose the total duration T of the observation in the sequential spectrometer and the simultaneous (or multiplex) spectrometer is the same, then in the former case each element is observed for a time T/M while in the latter each is observed for a time T. The noise associated with each of these elements is proportional to $\sqrt{(T/M)}$ and $\sqrt{T}$, respectively, and the signal-to-noise ratio in each spectrum is therefore $S_s = \sqrt{(T/M)}$ in the sequential case and $S_m = \sqrt{T}$ in the multiplex case. The ratio $s_m/s_s = \sqrt{M}$ indicates that for a given resolution R in a given spectrum of M elements observed for a given total time T the signal-to-noise ratio in a multiplex record of the spectrum is better than that in a sequential record by the factor $\sqrt{M}$. This advantage was first explicitly proposed by Fellgett in 1951 and is often known as the multiplex principle or Fellgett's advantage (Fellgett, 1951, 1958, 1967).

It is important to note that Fellgett's advantage is lost if photon noise dominates detector noise. In this case the noise is proportional to the square root of the total incident flux which is, to a first approximation, proportional to the total number of spectral elements if we assume them all to be of similar intensity. The noise signal in the multiplex case thus becomes proportional to $\sqrt{(MT)}$ and the signal-to-noise ratio is now $S'_m = \sqrt{(T/M)}$ thus making S'_m/S_s unity and so exactly cancelling the multiplex advantage (Jacquinot, 1960).

Bibliography

FELLGETT P. B. (1951) Thesis, University of Cambridge.
FELLGETT P. B. (1958) *J. Phys. Radium* **19,** 187.
FELLGETT P. B. (1967) *J. Phys.* **28,** supp. C-2, 165.
JACQUINOT P. (1960) *Rep. Prog. Phys.* **23,** 267.

JOHN CHAMBERLAIN

NEUTRETTO. There are two kinds of neutrino, one associated with the electron and the other associated with the muon. The one associated with the muon is often called the "neutretto".

The existence of two kinds of neutrino was first postulated in 1960 to account for the non-occurrence of the decay $\mu \to e + \gamma$. This process should have been observed, since in the known decay $\mu \to e + \nu + \bar{\nu}$ the neutrino and anti-neutrino should have been able to annihilate one another to produce a photon. If, however, the two neutrinos in muon decay are different, one ν_e associated with the electron, and the other ν_μ associated with the muon, then the decay would have to be of the form $\mu^- \to e^- + \bar{\nu}_e + \nu_\mu$, $\mu^+ \to e^+ + \nu_e + \bar{\nu}_\mu$ and the two neutrinos could not annihilate.

An experimental test of this hypothesis is provided by using a high energy neutrino beam produced by the decay of pions in flight ($\pi^+ \to \mu^+ + \nu_\mu$ or $\pi^- \to \mu^- + \bar{\nu}_\mu$) and then looking for events of the type:

$$\nu_\mu + n \to p + \mu^- \qquad \bar{\nu}_\mu + p \to n + \mu^+$$
$$\nu_\mu + n \to p + e^- \qquad \bar{\nu}_\mu + p \to n + e^+.$$

If ν_e and ν_μ are identical particles then both processes should occur, whereas if the particles are different the processes involving electron production will not appear. Such experiments have been performed at Brookhaven (1962) and CERN (1963). The results show that the probability of ν_e and ν_μ being identical particles is less than 1 per cent.

According to theory both neutrinos should be uncharged, massless and have the magnitude of their helicity equal to unity. The experimental results to date (1967) are as follows.

	ν_e	ν_μ
Charge	$<10^{-19}e$	$<5 \times 10^{-6}e$
Mass	< 0.2 keV	<3 MeV
Magnitude of helicity	> 0.95	>0.80

W. W. BELL

NUCLEAR QUADRUPOLE RESONANCE. The phenomenon of nuclear quadrupole resonance (NQR), like that of nuclear magnetic resonance (NMR), is the result of the selective absorption of energy by a nuclear spin system in response to applied radiation. In the case of the nuclear quadrupole resonance a non-spherically symmetric distribution of nuclear charge interacts with a non-uniform electric field at the nucleus to produce a torque on the nucleus. The energy states of the system depend upon the nuclear spin I, the nuclear electric quadrupole moment Q (which measures the asphericity of the nuclear charge distribution) and the electric field gradient; each state corresponds to a particular orientation of the nucleus.

The electric field gradient is a tensor quantity whose elements are normally referred to its principal axis system. By convention the maximum component $(\partial E_z/\partial z) = eq$, and the symbol q is called the field gradient. Under the condition of axial symmetry, $(\partial E_x/\partial x) = (\partial E_y/\partial y)$, the energy levels are given by

$$E_m = A[3m^2 - I(I + 1)]$$

where $A = e^2qQ/4I(2I - 1)$. The energy difference between adjacent m levels is

$$E_{m+1} - E_m = 3A(2|m| + 1).$$

A nucleus of spin $\frac{3}{2}$, such as ^{35}Cl or ^{37}Cl, may assume four possible orientations represented schematically in Fig. 1. The energy of the $m = +\frac{3}{2}$ state is the same

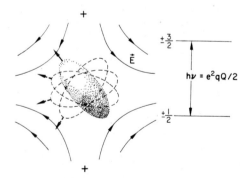

Fig. 1. A nucleus of $I = 3/2$ and $Q > 0$ in an axially symmetric electric field.

as that of the $m = -\frac{3}{2}$ state because of symmetry, likewise the $m = +\frac{1}{2}$ and $-\frac{1}{2}$ states. A single absorption, characteristic of the $m = |\frac{3}{2}| \to |\frac{1}{2}|$ transition is thus observed. The quantum number m may in general take values $-I, -I+1, ..., I$, and transitions of the type $\Delta m = \pm 1$ are observed. In a gradient of axial symmetry there are thus $I - \frac{1}{2}$ distinct transition frequencies when I is an odd multiple of $\frac{1}{2}$, and I frequencies when the spin is integral. Any nucleus which has a spin $I > 1$ may have an electric quadrupole moment; nuclei for which $I = 0$ or $\frac{1}{2}$ do not exhibit quadrupole effects.

The field gradients encountered in solids are often not axially symmetric. An asymmetry parameter $\eta = [(\partial E_x/\partial x) - (\partial E_y/\partial y)]/(\partial E_z/\partial z)$ is used to describe the departure from axial symmetry and to specify completely the gradient in its principal axis system. Energy levels and transition frequencies are then expressible in terms of the field gradient q, and the asymmetry parameter. In solids three additional constants are required to specify the orientation of the principal axes x, y, z, with respect to the crystal axes x', y', z'.

The experimental methods used to observe quadrupole resonance are essentially identical with those used for NMR with the notable difference that no appreciable static magnetic field is applied. Nuclear quadrupole resonance in zero or near zero magnetic field is sometimes referred to as *pure* quadrupole resonance or the "low field" case to distinguish it from the "high field" case in which the interaction energy of the nuclear magnetic moment with an external magnetic field is large compared with the electric quadrupole interaction. The transitions are produced by the action of an oscillating magnetic field on the nuclear magnetic dipole moment in both cases since it is not feasible experimentally to generate very large, time-dependent electric field gradients, and since nuclei have no electric dipole moment with which an homogeneous electric field might interact. Typical transition frquencies lie in the radio frequency region from about one MHz for ^{23}Na in inorganic crystals to over 600 MHz for ^{127}I in some organic compounds. Above approximately 100 MHz ultra-high frequency techniques, utilizing resonant transmission lines or cavities, replace the lumped circuits generally used below that frequency.

Although it is possible to observe transitions in molecules in the gaseous state, pure quadrupole resonance presently finds its principal application in the investigation of solids. In covalently bonded molecules large field gradients arise from the directional character of the bonding electron orbitals. In ionic crystals the gradient at a nucleus arises primarily from ions external to that site, while in metals it is produced by the anisotropic non-s part of the conduction electron distribution. The quantities of direct experimental interest are the transition frequencies, from which the quadrupole coupling constant e^2qQ is deduced. These quantities have been determined for many organic and inorganic compounds, and metals. Studies of the temperature and pressure dependence of the field gradient provide additional structural information about solids. Calculations of electric field gradients in solids are usually not sufficiently reliable to allow this technique to be used for the accurate determination of Q.

Bibliography

ABRAGAM A. (1961) *The Principles of Nuclear Magnetism*, Oxford: Clarendon Press.

DAS T. P. and HAHN E. L. (1958) *Nuclear Quadrupole Resonance Spectroscopy*, New York: Academic Press.

SAHA A. K. and DAS T. P. (1957) *Theory and Applications of Nuclear Induction*, Calcutta: Saha Institute of Nuclear Physics.

SLICHTER C. P. (1963) *Principles of Magnetic Resonance*, New York: Harper and Row.

T. J. ROWLAND

O

OPTICAL INFORMATION PROCESSING. Optical information processing is a term which can be taken to cover the function of any system which processes data in a parallel rather than serial fashion, and which uses light at some stage as the carrier of the data. This definition avoids distinguishing between those systems in which the raw data is in the form of light and those in which a transformation takes place later into light. In addition it excludes systems which are fundamentally serial but in which certain functional components use modulation of a light beam to carry one or perhaps a few channels of information. The exclusion of such systems is arguable and in fact this exclusion has not been carried out in some of the many symposia on the subject held in the past few years. The expression "opto-electronics" however is becoming more accepted as a description of the investigation and use of devices which can replace electronic devices or interconnections in, for example, present computer designs.

The examples given here do not go beyond the simplest of parallel systems, i.e. those in which the raw data can be represented as a surface distribution of either one variable or two connected variables. An obvious example is a photographic transparency. Using the methods of Abbe, Duffieux and later workers, such a distribution on a plane can be considered either as a function of photographic density or of spatial frequency. In other words a photographic transparency can be considered either as a group of small areas each with a transmission number, or as a group of grating-like patterns covering all frequencies. Some of these frequencies contain only noise, for example the frequencies near the inverse of the grain size.

Multiplication and Integration

The simplest operation carried out by optical processing is that of multiplication. Consider two different photographic transparencies placed close together and evenly illuminated by a collimated beam of intensity I_0. If the transparencies have constant transmission over their surfaces T_1 and T_2, the emerging intensity I is given by:

$$I = I_0 T_1 T_2.$$

If the transmission factors, as is usual, vary over the surfaces of the transparencies the relationship is then the parallel multiplication given by:

$$I(x, y) = I_0 T_1(x, y) T_2(x, y).$$

If then the transmitted energy is measured and is denoted by G we have:

$$G = \iint_A I_0 T_1(x, y) T_2(x, y) \, dx \, dy$$

where the integration area A is the common area of the two transparencies that is illuminated. In this way the process of two-dimensional integration is carried out for the functions represented by $T_1(x_1, y_1)$ and $T_2(x_2, y_2)$.

Further extensions of these processes can be obtained by noting that the transparencies can be translated or rotated with respect to each other and furthermore, that these arguments are not restricted to two transparencies. It is clear that optical systems can be used to copy one upon another, and also, in theory, any number of transparencies, i.e. functions, could be handled.

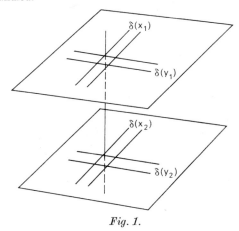

Fig. 1.

The latter point leads to a formal problem in that unless phase is used in the processing there is no positive and negative. In the examples above any function with negative values has to be biased so that it can be represented on film.

Processing with Coherent Light

The use of incoherent light poses certain difficulties in phase which formally can be overcome, as in interferometry, by splitting the beam and recombining with a phase difference. In a loose way this can be called the use of "negative" light.

The realization of the laser has led to a great expansion in the use of coherent light as a carrier in processing systems, aided by the existence of the early pioneer work of Abbe, Zernicke, Gabor, Marechal and O'Neill. A number of examples is given here of recent developments in the use of coherent light.

The difference between coherent and incoherent optical information processing is a fundamental one from the physicist's point of view. In the first case he starts with a "pure" carrier, the coherent beam, which contains information concerning only the source and its apertures. This beam can then be operated upon by transparencies, thus impressing the raw data in the form of amplitude or phase distribution or both according to whether the transparency contains information expressed in amplitude or phase. In the second case, however, the implication is that the light contains information both from the source and the raw data, combined in a way that is not controllable by the observer. In addition, there is the likelihood that the processing must be in real-time.

Examples of Systems

The diffractometer is an instrument which has been used for many years to take advantage of the Fourier transform relationship between the distribution of amplitude and phase at the pupil of a lens and the distribution at the focal plane. In principle the instrument consists of a point source of nearly monochromatic light, a collimator and a telescope objective. Between the collimator and the telescope there is a space into which objects such as transparencies, screens, grids or phase plates can be put. Using the notation of Born and Wolf the disturbance at the focal plane of the telescope is then given by:

$$U(pf, qf) = C_1 \iint_A F(x, y)\, e^{-ik[px + qy]} dx\, dy$$

where F is the transmission function of the object, C_1 is a constant and the integration is taken over the area covered by the object. The quantities p and q represent the spatial frequencies in the directions of x and y respectively and f is the focal length of the telescope objective.

The relationship above is, with the exception of a phase factor, a two-dimensional Fourier transform. This can be confirmed by considering the situation where a sinusoidal grating is the object. In this case two spot images are formed equidistant from the telescope objective axis, the distance between them being given by the product of the focal length and the tangent of the angle of diffraction of the grating for the given wavelength and spatial frequency. Thus the greater the spatial frequency, the further is the light spot from the axis. It should be noted that the position of the object does not affect the position of the transform.

Further processing operations can be carried out by adding to this simple system. For example a lens may be placed at the focus of the telescope objective (the Fourier plane) to reconstruct the object. With such an arrangement a simple annular mask can be applied in the Fourier plane to carry out the operation of spatial filtering. A discussion of the extension to systems like these is given in a paper by Cutrona et al., and a particularly good review of spatial filters and filtering is given by Vander Lugt.

Cutrona in a later publication discusses several systems in which cylindrical lenses are used in order to separate the variables so that integration of one only may be carried out. One of his optical systems is shown schematically in Fig. 2. The y direction is in the plane of the diagram and the x direction is perpendicular to that plane. For cross-correlation between two functions $f(x, y)$ and $g(x, y)$, a transparency corresponding to $f(x, y)$ is placed at plane P_1

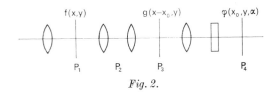

Fig. 2.

and is imaged at plane P_3 where the $g(x, y)$ transparency is mounted in a holder which can traverse in the x-direction. The astigmatic lens system between P_3 and P_4 which has the property of showing spatial frequency only in the x-direction, forms a light distribution in plane P_4 given by:

$$\phi(x_0, y, \alpha) = \int f(x, y)\, g(x - x_0, y)\, e^{i\alpha x}\, dx$$

where $\alpha = \dfrac{2\pi}{\lambda} \sin u$ (representing the x dimension of plane P_4), and x_0 is the displacement at plane P_3 of the $g(x, y)$ transparency.

The cross-correlation function of f and g is given by:

$$\phi_{fg}(x_0) = \int f(x)\, g(x - x_0)\, dx$$

and this function is represented by the distribution of energy at a slit at P_4 in the y direction.

In the paper by Cutrona et al. the authors give a generalized system similar to the one described above in which the operation can be described by the expression:

$$I(\omega_x, X_0, y) = \int_{a(y)}^{b(y)} f(x - x_0, y)\, g(x, y)\, e^{-i\omega_x x}\, dx.$$

They further show that several well-known mathematical operations can be carried out within this operator. The operations are Fourier and Laplace transformation, cross- and auto-correlation and convolution.

Applications of systems as mentioned above have been found in vibration analysis, contrast improvement and noise rejection in photographs, and also in many cases where correlation techniques are used.

Recent Systems Working with Non-coherent Light

A system developed by Hawkins and Munsey is certain to be of importance in the future so it should be mentioned even though it is not strictly optical. As was pointed out earlier, the processing is much more satisfactory if both positive and negative values are allowed. In the Hawkins and Munsey system the optical input is transformed into an electron distribution by the photocathode of an image tube. The image tube contains an intermediate store in the form of a secondary emission target which has the property of storing the input picture as negative charge or as positive charge according to the accelerating voltage.

The tube is used in a particularly interesting way, to carry out a convolution with what the authors term an aperture function. Firstly, the picture is stored in positive charge then the whole picture is displaced by one picture point and stored in negative charge but at one-eighth the level. By repeating the negative charge operation with the other seven possible shift positions within one picture point of the first all density gradients in the picture are enhanced. It can also be seen immediately that such a convolution removes all the low spatial frequencies. More complex "shifts and prints" can be used to represent other aperture characteristics.

The development of partially coherent systems for data-processing is at a very early stage, but interesting suggestions have been made on the use of the degree of coherence as a variable. The realization into modulation of intensity would be carried out in an interferometer where fringes would appear wherever temporal coherence existed. An example of this process is Michelson's stellar interferometer.

Papers and other works mentioned in the text are contained in the following list.

Bibliography

BORN M. and WOLF E. (1964) *Principles of Optics*, Pergamon Press, London and New York.

CUTRONA L. J., LEITH E. N., PALERMO C. J. and PORCELLO L. J. (1960) *I.R.E. Trans. Inf. Theory*, **6**, 386.

HAWKINS J. K. and MUNSEY C. J. (1967) *J.O.S.A.*, **57**, 914.

POLLACK D. K. *et al.* (Eds.) (1963) *Optical Processing of Information*, Cleaver-Hume Press, London, Spartan Books Inc., Baltimore.

TIPPETT J. T. *et al.* (Eds.) (1965) *Optical and Electro-Optical Information Processing*, M.I.T. Press.

VANDER LUGT A. (1968) *Optica Acta*, **15**, 1.

K. R. COLEMAN

PEIERLS STRESS

PEIERLS STRESS. In one of the three pioneering papers which introduced the idea of an edge dislocation in a crystal, Polanyi (1934) considered such a dislocation to be a vernier in which $n + 1$ atoms above the glide plane faced n atoms below. Polanyi estimated (wrongly) that the shear stress under which a dislocation would move would be $1/n$ of the shear strength of a perfect lattice, but he gave no way of estimating n. This problem was given an approximate solution by Peierls (1940), and the resolved shear stress required to move a dislocation through a perfect lattice is known as the Peierls Stress or Peierls Force.

Following a suggestion by Orowan, Peierls considered a simple square lattice of parameter b, assumed to be elastically isotropic with shear modulus μ and Poisson's ratio ν. The glide plane $y = 0$ lies midway between two planes of atoms (Fig. 1). The atoms

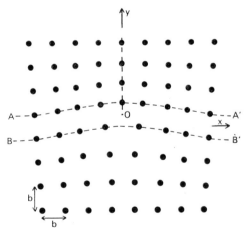

Fig. 1. The Peierls model of an edge dislocation in a cryostat. The glide plane lies between the rows of atoms AA' and BB'.

immediately above the line of the dislocation are under compression, while those immediately below are in tension. They are held in these strained positions by the interaction between the atoms immediately above the glide plane, at $y = \tfrac{1}{2}b$, and those immediately below, at $y = -\tfrac{1}{2}b$. We start with two half crystals which are unstressed, but placed so that the atoms at $y = \tfrac{1}{2}b$ are offset by $\tfrac{1}{2}b$ in the x direction from those at $y = -\tfrac{1}{2}b$. We then turn on the forces of attraction between the atoms which face each other across the glide plane. The displacements of the atoms immediately above the glide plane are taken to be $u(x)$, $v(x)$, and those immediately below the glide plane to be $-u(x)$, $v(x)$. The reference position is chosen so that $u(\infty) = -\tfrac{1}{4}b$, $u(-\infty) = +\tfrac{1}{4}b$, while in the core of the dislocation the region of maximum misfit is represented by $u(0) = 0$.

The shearing stress p_{xy} between the atoms immediately above the glide plane and those immediately below is taken to be a sinusoidal function of u, having period $\tfrac{1}{2}b$, so that the stress is unchanged when the relative displacement $2u$ increases by b. In order that the stress should agree with that given by Hooke's law when the relative displacement from the equilibrium position is small ($u \approx \pm\tfrac{1}{2}b$), we must have

$$p_{xy} = -(\mu/2\pi) \sin (4\pi u/b). \tag{1}$$

Equating this stress to the surface stress on the upper half of the crystal, regarded as an elastic half space, we obtain Peierls's integral equation

$$\int_{-\infty}^{\infty} \frac{\mathrm{d}u(x')}{\mathrm{d}x'} \frac{\mathrm{d}x'}{x - x'} = \frac{1-\nu}{2} \sin \frac{4\pi u}{b}. \tag{2}$$

The simplest solution satisfying the prescribed boundary conditions is

$$u(x) = -(b/2\pi) \tan^{-1}(x/\zeta), \tag{3}$$

where

$$\zeta = b/2(1-\nu). \tag{4}$$

The width ζ of the dislocation is therefore only a few b, and the approximations used are not really self-consistent.

The calculation has been improved in many ways to allow more realistically for the actual laws of force between atoms, and to consider screw dislocations, dislocations in finite crystals, etc. Nabarro (1967) gives a detailed account of these extensions.

We now consider the dislocation gliding through the crystal. If it starts in the symmetrical configuration of Fig. 1, it passes through unsymmetrical positions such as that of Fig. 2(a) to the different symmetrical configuration of Fig. 2(b), and back through unsymmetrical configurations to one equivalent to Fig. 1, but with the x coordinate of the core of the dislocation increased by b. The symmetrical

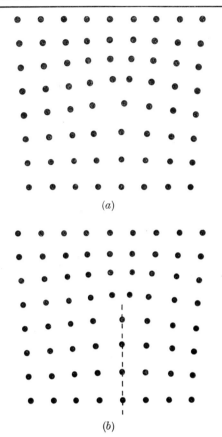

Fig. 2. (a) *An unsymmetrical configuration of the edge dislocation as it glides through the lattice.* (b) *The symmetrical configuration which is reached when the dislocation has moved $\tfrac{1}{2}b$ from the configuration of Fig. 1.*

configurations are in equilibrium (stable or unstable) under zero applied shear stress. Other configurations are in equilibrium under an applied shear stress P. The Peierls stress is the largest stress which has to be applied to enable the dislocation to pass through all the configurations which separate two positions of stable equilibrium a distance b apart. Peierls calculated P by calculating the energy of the dislocation $U(\alpha)$ as a function of the displacement αb of the core of the dislocation from its position in Fig. 1. The force acting on unit length of dislocation is given by $Pb = -\partial U/b\,\partial \alpha$. Peierls determined $U(\alpha)$ by finding the potential energy of each atom on the glide plane corresponding to the force (1), and summing the energy over all the atoms. He also made the self-consistent assumption that the surfaces of the elastic half bodies lying above and below $y = \pm\tfrac{1}{2}b$ were stressed at the positions of the ion cores. Nabarro (1947) corrected a numerical error in Peierls's calculation, but omitted the contribution of the elastic

strain energy. The dominant term in $U(\alpha)$ is

$$U(\alpha) = [b^2\mu/2\pi(1-\nu)] \cos 4\pi\alpha \exp(-4\pi\zeta/b). \quad (5)$$

In this model, the potential $U(\alpha)$ and the Peierls stress decrease exponentially with ζ, so that the Peierls stress depends very sensitively on the nature of the interatomic forces. In face-centred cubic metals the dislocations are wide, and the Peierls stress is very small in comparison with μ; in the covalent diamond structure we expect the dislocations to be very narrow, and, although the mathematical approximations have broken down, we predict a large Peierls stress.

Equation (5) shows some surprising features. $U(\alpha)$ is proportional to $\cos 4\pi\alpha$ and not to $\cos 2\pi\alpha$ as was expected, and the symmetrical configurations of Figs. 1 and 2(b) represent maxima of the energy. Almost any refinement of the calculation introduces terms of the form $\cos 2\pi\alpha \exp(-2\pi\zeta/b)$. Nabarro (1967) has reviewed these refinements, and lists the more elaborate calculations which have been made to treat narrow dislocations. The influence of zero-point energy on the Peierls stress may be important (Suzuki, 1968).

The Relation between the Peierls Stress and the Flow Stress

We have tacitly considered the dislocation to lie along a low-index crystallographic direction in a trough of the potential $U(\alpha)$. If it does so, it can advance into the next trough at zero temperature only if an external stress equal to the Peierls stress is applied. At finite temperatures it can advance under smaller stresses by the nucleation of a pair of kinks (q.v.) of opposite sign. This process may control the rate of plastic flow in crystals such as germanium, in a manner which has been worked out in detail by Haasen and his colleagues (Haasen, 1968; Alexander and Haasen, 1968). If an alternating stress of small amplitude is applied, the alternate creation and annihilation of pairs of kinks gives rise to the Bordoni Peak (q.v.) in internal friction.

If the dislocation line does not lie along a low-index direction, but is anchored at nodes, it necessarily contains a minimum number of kinks, which are

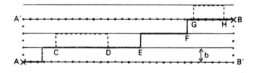

Fig. 3. A dislocation anchored by nodes at A and B, and crossing from one Peierls trough to another by kinks at points such as E and F. The Bordoni peak is associated with the nucleation of pairs of kinks at points such as C and D. The dislocation can move outside the rectangle $AA'BB'$ only by nucleating kink pairs such as GH.

known as geometrical kinks. These geometrical kinks allow the dislocation to sweep over a limited area of the glide plane (Fig. 3) without overcoming Peierls barriers. The dislocation can move outside this area only by nucleating pairs of kinks against the Peierls stress.

The Modified Peierls Stress

It is probably desirable to let the term Peierls stress describe any variation of the energy of a dislocation which has the periodicity of the lattice. However, in some structures, dislocations lying along particular directions are able to dissociate in a way which produces a relatively large reduction in energy, and the corresponding stress is sometimes called the Modified Peierls Stress. For example, a screw dislocation lying along a $\langle 111 \rangle$ direction in a body-centred cubic lattice can dissociate on two or three $\{11\bar{2}\}$ planes, on two or three $\{101\}$ planes, or on one $\{11\bar{2}\}$ plane and one $\{10\bar{1}\}$ plane (Foxall, Duesbery and Hirsch, 1967; Duesbery and Hirsch, 1968). The dislocation can only move on its glide plane if it contracts so that it is no longer dissociated, or until all of its partial dislocations lie in the glide plane. Models of this kind can explain many of the curious plastic properties of body-centred cubic metals, such as the rapid increase in flow stress as the temperature is reduced, the anisotropy of the flow stress in push-pull experiments and the failure of the Schmid Law of resolved shear stress (see, for example, Duesbery and Hirsch, 1968; Dorn and Rajnak, 1964). However, they do not allow for the strong interaction between dislocations and interstitial impurities, and most of the increase in strength on cooling to moderately low temperatures must be attributed to this interaction (Fleischer, 1967).

The Dynamical Peierls Stress

The Peierls stress has been calculated as the stress required to move a dislocation quasi-statically through the lattice. A moving dislocation may use its kinetic energy to surmount the maxima of $U(\alpha)$, and may therefore propagate through the lattice under a lower stress (for references, see Nabarro, 1967). Hart showed that when the speed v of the dislocation approached the speed of sound c in the medium, the necessary stress was proportional to v^{-2}. Theory suggests that very complicated effects will occur when $v \ll c$.

Bibliography

ALEXANDER H. and HAASEN P. (1968) *Solid State Physics*, New York: Academic Press, **22**, 27–158.
DORN J. E. and RAJNAK S. (1964) *Trans. Met. Soc. AIME* **230**, 1052–64.
DUESBERY M. S. and HIRSCH P. B. (1968) in *Dislocation Dynamics*, New York: McGraw-Hill.
FLEISCHER R. L. (1967) *Acta Met.* **15**, 1513–19.
FOXALL R. A., DUESBERY M. S. and HIRSCH P. B. (1967) *Canad. J. Phys.* **45**, 607–29.
HAASEN P. (1968) in *Dislocation Dynamics*, New York: McGraw-Hill.
NABARRO F. R. N. (1947) *Proc. Phys. Soc.* **59**, 256–72.
NABARRO F. R. N. (1967) *Theory of Crystal Dislocations*, Oxford: Clarendon Press.
PEIERLS R. E. (1940) *Proc. Phys. Soc.* **52**, 34–37.
POLANYI M. (1934) *Z. Phys.* **89**, 660–4.
SUZUKI H. (1968) in *Dislocation Dynamics*, New York: McGraw-Hill.

F. R. N. NABARRO

PELTIER REFRIGERATORS. The reversible heating and cooling effects, which occur when an electric current crosses the junction between two different conductors, were first observed by Jean Peltier, a French watchmaker, in 1834. It was obvious that the Peltier cooling effect could, in principle, be utilized in refrigeration, but this did not seem to be a very practicable idea, since it was difficult enough even to demonstrate the effect for most thermocouples. A theoretical analysis by Altenkirch in 1911 showed why the observed lowering of temperature was always rather small; Altenkirch also indicated the properties of thermocouple materials that would be needed if a worthwhile cooling effect were to be obtained.

The theory of Altenkirch shows that the suitability of a thermocouple for thermoelectric refrigeration can be expressed in terms of a figure of merit Z defined as

$$Z = \frac{(\alpha_p - \alpha_n)^2}{KR} \quad (1)$$

where α_p and α_n are the values of the Seebeck coefficient (thermoelectric power) for the positive and negative branches, K is the thermal conductance of the two branches in parallel and R the electrical resistance of the two branches in series. This figure of merit expresses quantitatively the obvious requirements that the reversible thermoelectric effects should be large and the irreversible effects of heat conduction and Joule heating should be small. The Seebeck coefficient is related to the Peltier coefficient π through the Kelvin relationship.

$$\pi = \alpha T. \quad (2)$$

It should be noted that the figure of merit can be optimized for a given couple by selecting the dimensions of the branches so that KR has its minimum value, $[(\varkappa_p/\sigma_p)^{\frac{1}{2}} + (\varkappa_n/\sigma_n)^{\frac{1}{2}}]^2$, where $\varkappa$ is thermal conductivity and σ is electrical conductivity.

The coefficient of performance for refrigeration, defined as the ratio of cooling capacity to electrical power consumption is given by

$$\eta = \frac{T_c}{T_H - T_c} \cdot \frac{M - T_H/T_c}{M + 1} \quad (3)$$

where T_H is the sink temperature, T_c the source temperature and M is equal to $\{1 + Z(T_H + T_c)/2\}^{\frac{1}{2}}$. Here the current flow is assumed to be that for maximum coefficient of performance. Since the figure of merit Z appears multiplied by the average temperature T, one often comes across the term "dimensionless figure of merit", ZT, in the literature. It can be shown readily from eq. (3) that the maximum temperature depression that can be achieved by a single-stage thermoelectric refrigerator is

$$\Delta T_{max} = \tfrac{1}{2} Z T_c^2. \tag{4}$$

Substantial depression of the temperature can be obtained only if ZT has a value approaching unity (or greater).

Thermoelectric Materials

In searching for thermoelectric materials it is useful to employ a figure of merit for a single material defined by

$$z = \frac{\alpha^2 \sigma}{\varkappa}. \tag{5}$$

In most practical situations the figure of merit Z for the optimised couple is close to the average of the values of z for the two branch materials.

Most metals have Seebeck coefficients of a few $\mu V\ deg^{-1}$, while according to the Wiedemann-Franz law the ratio $\sigma/\varkappa$ is always equal to about $4 \times 10^7/T\ deg\ V^{-2}$. Thus for metals and metal alloys zT is typically of the order of 10^{-5} to 10^{-4}, that is much less than unity. zT remains much less than unity even for such thermocouple alloys as chromel and constantan.

Reasonable figures of merit can be reached only using semiconductors, since these substances can have Seebeck coefficients of hundreds of $\mu V\ deg^{-1}$ or even a few $mV\ deg^{-1}$. The sign of the Seebeck coefficient is positive or negative depending on whether the semiconductor is p-type or n-type. It is not immediately obvious that semiconductors will be better than metals since the ratio $\sigma/\varkappa$ is usually much lower than that given by the Widemann-Franz law. This is because the electronic thermal conductivity $\varkappa_e$ (which does more-or-less obey the Wiedemann-Franz law) is low, and usually much less than the lattice thermal conductivity $\varkappa_L$. The total thermal conductivity is the sum of $\varkappa_e$ and $\varkappa_L$. It turns out that, in spite of an inferior ratio $\sigma/\varkappa$, some semiconductors are better thermoelectric materials than are any of the metals.

Semiconductor theory allows us to express the Seebeck coefficient, electrical conductivity and thermal conductivity in terms of the Fermi energy and a number of basic materials parameters. This theory predicts that (a) the Fermi level should lie very close to the edge of the energy band that contains the charge carriers (i.e. the Seebeck coefficient should be about 200 to 300 $\mu V\ deg^{-1}$) and (b) the semiconductor should be chosen to have the highest value of the product $(\mu/\varkappa_L)\,(m^*/m)^{3/2}$, where μ is the carrier mobility and m^*/m the ratio of the (density-of-states) effective mass to the free electron mass. It is also important that the conduction should be due to one type of carrier only. If electrons and holes are present simultaneously their thermoelectric effects act in opposition, while they give an increased thermal conductivity due to a dipolar diffusion mechanism.

The best materials are usually formed from elements of high atomic weight, primarily because such semiconductors have low values of the lattice thermal conductivity. One such material is the compound bismuth telluride, Bi_2Te_3, for which $\varkappa_L$ is only about $0.015\ W\ cm^{-1}\ deg^{-1}$ at room temperature. Semiconductors of high carrier mobility must have low inertial effective masses but it is still possible for the density-of-states effective masses to be reasonably large if the energy bands are of the so-called many-valley type. Bismuth telluride has conduction and valence bands each with 6 valleys, giving values for $\mu_n(m_n^*/m)^{3/2}$ and $\mu_p(m_p^*/m)^{3/2}$ that compare well with any other materials.

Actually, the best materials for thermoelectric refrigeration are alloys of Bi_2Te_3 with Sb_2Te_3 and Bi_2Se_3 rather than the pure compound itself. The formation of a solid solution generally lowers the lattice thermal conductivity quite appreciably but can have little or no effect on the electron or hole mobility. Thus, typical refrigerating thermocouples consist of positive elements made from $Bi_{0.5}Sb_{1.5}Te_3$ doped so as to have a Seebeck coefficient of about $+200\ \mu V\ deg^{-1}$, and negative elements made from $Bi_2Te_{2.7}Se_{0.3}$ doped to give a Seebeck coefficient of $-200\ \mu V\ deg^{-1}$. Such a couple, made from single crystal materials aligned so that the current flows along the basal planes, has a value of Z of just over $3 \times 10^{-3}\ deg^{-1}$ (i.e. $ZT \approx 1$ at 300°K). Commercial refrigerating modules usually incorporate sintered materials with randomly oriented grains; these have figures of merit that are perhaps two-thirds of the single crystal values. The reduction in figure of merit for non-aligned material is due to the fact that the ratio $\sigma/\varkappa$ is smaller for current flow perpendicular to the basal planes than for current flow along the basal planes, particularly for n-type crystals.

At one stage of development it was thought that lead telluride, $PbTe$, and its alloys might be better than Bi_2Te_3 and its alloys but this is now known not to be so, though the $PbTe$ alloys are used in some thermoelectric generators. The performance of the Bi_2Te_3 alloys falls off rather rapidly as the temperature is lowered although they still provide the best positive materials down to liquid nitrogen temperature. The best negative thermoelectric material below about $-100°C$ is an alloy of bismuth with antimony, which is a semiconductor with a very small energy gap. When such an alloy is placed in a magnetic field its figure of merit rises and z values of about $10^{-2}\ deg^{-1}$ have been reported at liquid nitrogen temperature by Wolfe and Smith.

Applications

Thermoelectric refrigerators are capable of giving a maximum lowering of temperature of about 80 deg, with a heat sink at room temperature, using a single stage. They can provide a temperature differential of about 40 deg with a coefficient of performance of about 50 per cent. In principle, one can obtain any required cooling capacity (provided that $\Delta T < \Delta T_{max}$) using a single thermocouple of the appropriate dimensions passing a suitably large current. However, in practice a refrigerating unit consists of a module of several couples acting thermally in parallel but connected in electrical series. Typically, with a current of 10 A one obtains a cooling capacity of about $\frac{1}{4}$ W per couple for a temperature differential of about 40 deg. The individual elements are rectangular blocks having dimensions of a few mm in each direction; the connections are made by metal bridges soldered to the ends of the elements. Care must be taken to insulate these bridges from one another electrically during the operation of the device.

Thermoelectric refrigerators are not so efficient as conventional compressor-type machines though they do not compare too unfavourably with absorption refrigerators. They seem particularly attractive for applications where the required cooling power is less than about 10 W. Thus thermoelectric cooling has proved very useful in the control of temperature of electronic equipment and scientific instruments of all kinds. It has the advantages of no moving parts, no harmful gases, long life without deterioration and small size. It is possible that thermoelectric materials with improved properties will become available over the coming years and if this happens we can expect many more fields of application to become worthwhile.

Bibliography

GOLDSMID H. J. (1964) *Thermoelectric Refrigeration*, New York: Plenum Press.
IOFFE A. F. (1957) *Semiconductor Thermoelements and Thermoelectric Cooling*, London: Infosearch.

H. J. GOLDSMID

PERMANENT MAGNET TECHNOLOGY

Introduction

At the beginning of the century, permanent magnet materials were martensitic high-carbon or tungsten steels and magnets were made by hot-working the cast ingot into bar form, these being subsequently cut and bent or machined to size. Following the development of the cobalt steels some 20 years later, cast magnets of non-uniform cross-section were made in addition to those produced from rolled and forged bars. With the introduction of the brittle carbon-free Ni–Al–Co–Cu–Fe dispersion-hardening alloys Alni and Alnico in the middle 1930's, only castings or sinterings could be produced in such materials, as their mechanical properties were not suitable for the earlier hot-worked methods of fabrication. The application of field orientation to the same alloys, i.e. the alloys in column 1 (e–n), Table 1, was successfully investigated in 1940, and they were produced commercially shortly afterwards resulting in the anisotropic Alcomax series of alloys. Production of these alloys in columnar crystal form began in 1950.

In addition to substantial improvements in energy product as shown in Fig. 1a, these developments in cast materials were accompanied in every case by an

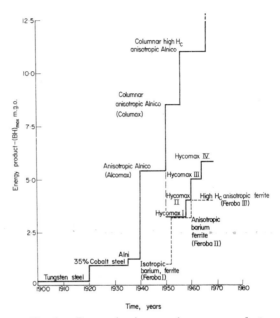

Fig. 1a. Progressive increase in energy product.

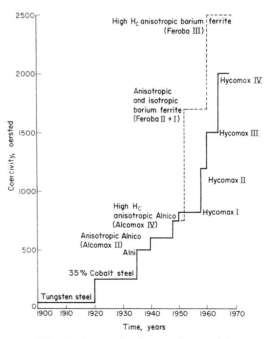

Fig. 1b. Progressive increase in coercivity.

Table 1

1	2	3	4	5	6	7	8	9	10	11
	Magnetic properties							General physical properties		
Material	Remanence B_r gauss	Energy product $(BH)_{max}$ mega-gauss-oersteds	Coercivity H_c oersteds	Recoil permeability μ_r	Optimum working point B_d gauss	Optimum working point H_d oersteds	Max. recoil energy mega-gauss-oersteds	Density g/cm³	How produced	Machinability
(a) Columax	13,500	7.5	740	1.8	11,700	640	2.05	7.35		
(b) Alcomax II S.C.	13,700	5.9	600	2.1	11,600	510	1.61	7.35		
(c) Alcomax III S.C.	13,200	6.1	700	2.8	10,700	570	1.75	7.35		
(d) Alcomax IV S.C.	13,200	5.2	780	3.8	9000	580	1.65	7.35	Castings hard and brittle	Grinding only
(e) Alcomax II	13,000	5.4	580	2.6	11,000	490	1.53	7.35		
(f) Alcomax III	12,600	5.5	650	3.1	10,200	530	1.58	7.35		
(g) Alcomax IV	11,500	4.5	750	4.4	8350	540	1.48	7.35		
(h) Alnico (High Remanence)	8000	1.7	500	7.0	5200	327	0.656	7.3		Generally grinding only, but can be drilled with great difficulty if essential
(i) Alnico (Normal)	7250	1.7	560	6.0	4700	362	0.671	7.3		
(j) Alnico (High Coercivity)	6500	1.7	620	5.2	4250	400	0.688	7.3		
(k) Hynico II	6000	1.8	900	4.2	3450	520	0.90	7.3	Very brittle	Grinding only
(l) Alni (High Remanence)	6200	1.25	480	5.5	4030	310	0.477	6.9		
(m) Alni (Normal)	5600	1.25	580	5.0	3480	359	0.528	6.9	Castings hard and brittle	Grinding only
(n) Alni (High Coercivity)	5000	1.25	680	4.0	3040	411	0.548	6.9		
(o) 35% Cobalt Steel	9000	0.95	250	12.0	5930	160	0.332	8.2	Rolled or cast	Difficult to tap
(p) 3% Cobalt Steel	7200	0.35	130	18.5	4200	83	0.124	7.7		Suitable for all operations when annealed
(q) Hycomax I	9000	3.2	825	4.0	5930	540	1.13	7.25	Castings hard and brittle	Grinding only
(r) Hycomax II	8500	4.0	1200	2.8	5300	750	1.51	7.3		
(s) Columnar Hycomax II	10,500	7.0	1250	2.0	7500	950	2.4	7.3		
(t) Hycomax III	8800	5.0	1500	2.4	5000	1000	1.9	7.3	Very brittle	Grinding only
(u) Columnar Hycomax III	10,400	8.0	1500							
(v) Hycomax IV	7800	5.8	1900	2.0	4500	1290	2.4	7.3		
(w) Columnar Hycomax IV	9000	8.0	1900							
(x) Feroba I	2200	1.0	1700	1.2	1110	900	Up to 0.8	4.8	Pressed and fired	Hard and brittle
(y) Feroba II	3900	3.2	1700	1.1	1400	1330	1.4	5.0		Grinding only
(z) Feroba III	3400	2.7	2500	1.05	6290	1600	1.8	4.7		

increase in coercivity as shown in Fig. 1b, which indicates the accelerating rate of this improvement; ceramic ferrite materials are also shown on both figures for comparison. Increasing interest in coercivity is due to greater emphasis being placed on the magnetic stability of permanent magnets which improves with increase in coercivity. The next impressive advance in high coercivity materials from 1950 onwards came with the commercial introduction of the ceramic ferrite permanent magnet materials, but the energy product was well below that of Alcomax (Hadfield, 1962a).

Properties of Alloy Magnets

In the development of alloy permanent magnet materials the demand for higher field strength or alternatively smaller volume of material has required an increase in energy product so that the emphasis has been usually on the improvement of this parameter. However, the coercivity of the Alnico alloys which was of the order of 500 oersted was gradually increased concurrently, and in the anisotropic Alcomax series eventually achieved 750 oersted in Alcomax IV and Columax alloys. All these properties, together with those of the earlier martensitic steels, are listed in Table 1. Earlier attempts to increase coercivity succeeded usually at the expense of the energy product (see Table 1, column 3 (f) and (g)). With Alcomax IV the energy product was reduced to 4·5 m.g.o. compared with Alcomax III having 5·5 m.g.o. The more recent high coercivity anisotropic alloys named Hycomax, Ticonal 1500 and 2000, and Sermalloy Al were developed with the specific objective of achieving the highest possible coercivity whilst retaining reasonable values for energy product and remanence. Three of these alloys, Hycomax I, II and III, have coercivities of 825, 1200 and 1500 oersted respectively, with remanence between 8500 and 9000 gauss and energy products of 3·2, 4·0 and 5·0 m.g.o. respectively. The latest in this series Hycomax IV has coercivity increased up to ~2000 oersted whilst still retaining an energy product of up to ~6 m.g.o.

Like Columax, the alloys Hycomax II, III and IV also may be produced in columnar crystal form by additions of sulphur or selenium in the case of Hycomax III (Harrison, 1965) and tellurium in Hycomax IV (Hadfield and Harrison, 1967) and in such form have similar coercivities to the equiaxed materials but with energy products of from ~7 m.g.o. up to ~10 m.g.o. depending on the alloy. Such coercivities are technically comparable to those of the ferrite materials Feroba II and III, whilst the energy products are well above in the case of the columnar versions.

The magnetic properties of these new alloys in both equiaxed and columnar form are included in Table 1, column 1 (s), (u), (w). Hycomax I and II require only conventional solution and precipitation hardening treatments (Hadfield, 1962b), but in order to develop the optimum properties in the cases of Hycomax III and IV, both in the equiaxed and columnar states, it is necessary to carry out an isothermal heat treatment following the normal solution treatment. The isothermal heat treatment involves a rapid blast cooling, and then quenching in a salt bath at ~800 °C and holding at this temperature for about 10 min with the simultaneous application of a magnetic field having a minimum field strength of 3000 oersted. The precipitation hardening follows in the same manner as for the other Hycomax alloys.

It will be shown in the next section how coercivity development has proceeded in parallel between alloy and ceramic magnet materials the latter being in the lead (see Fig. 1), whilst the energy products of alloy magnets, particularly in columnar form, have easily surpassed those of the ferrites (see Table 1, x, y, z, and P.M.A. Technical Bulletins Nos. 1, 4, 6 and 9).

Properties of Feroba Magnets

Hard magnetic ferrites may contain barium, strontium, or lead and have unique characteristics which have established non-metallic magnets in specific fields of utilization where their mechanical properties and susceptibility to temperature are acceptable and their advantages of lightness and inexpensiveness are fully exploited. Barium ferrite permanent magnets, produced commercially since 1955 and marketed under the name Feroba, are in increasing demand, sometimes replacing existing cast or sintered Alnico or Alcomax magnets, but also in new applications where permanent magnets have not been utilized before. In the former case the substitution of Feroba is usually for cheapness and it inevitably necessitates a redesign of the magnetic circuit. In the latter case, however, it is the unusual magnetic characteristics of the material, whether in the isotropic or anisotropic form, which enable a permanent magnet to be incorporated in certain applications at all, to replace electromagnets.

Nevertheless, where minimum size is all-important irrespective of cost, cast anisotropic magnets usually have an over-riding claim.

For any permanent magnet material, the effect of a demagnetizing field on the retained magnetization can be recorded and used for design purposes in at least three different ways, each one serving conveniently for a different kind of application. Starting in each case from the remanence value after previous saturation, these three relationships are:

(i) The curve of flux density, B, against demagnetizing field, H. This is the most widely used curve, and is the normal design basis for static permanent magnet circuits with relatively small airgaps and calculable permeances.

(ii) The curve of intrinsic magnetization, B_i or $4\pi J$ against demagnetizing field, H. This is the directly applicable relationship when magnetic moment is required, or for the calculation of the field produced by a magnet at a relatively large distance from a pole face, or for the design of permanent magnet motors and generators subject to heavy armature reaction, or for the estimation of repulsion forces.

(In the c.g.s. systems of units, $B_i + H = 4\pi J + H$, remembering that in the demagnetizing quadrant B and H are of opposite sign; J is the moment per unit volume or the pole strength per unit area of pole face.)

(iii) The curve of recoil or closed circuit flux density, B_c, against H, where B_c is the flux density remaining after the application of a demagnetizing field, H, and then reducing the field to zero. This is the useful relationship for the design of recoil or dynamic magnet systems such as holding devices.

(The value of maximum recoil energy (see Hadfield, 1961) is given as $1/4(B_0 H_0 + \mu_r H_0^2)$ on recoil from open-circuit values B_0, H_0, to $1/2 H_0$; it will be seen that $B_c = B_0 + \mu H_0$, therefore maximum recoil energy can also be expressed as $1/4 B_c, H_0$).

Little attention has previously been given to (ii) and (iii) above because for the cast alloys the differences in the curves, and also in the desirable workingpoints, are relatively small. But for ferrite magnets these differences are very marked. Figure 2 shows the three different relationships for Feroba I and III, and also for cast Alcomax III and Hycomax III, the latter being the high-coercivity alloy which most nearly approaches the ferrites in magnetic properties. These curves make it clear that if either high magnetic moment or high recoil energy is required from a magnet in Feroba I or III then the desirable working point is much lower down the normal demagnetization curve than the $(BH)_{max}$ point.

The completely different proportions of magnet section and length for ferrite as compared to alloy magnets is indicated by the very different working values of B and H in Fig. 2, but is perhaps better shown, for static working, by the plots of $B.H$ product against unit permeance in Fig. 3. The alloys

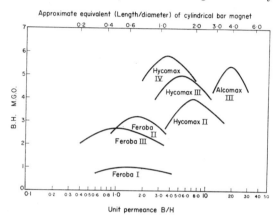

Fig. 3. Static energy product $(B.H)$ v. working point (B/H).

have the higher energy, but very useful energy values can be obtained from the ferrites provided that the working permeance is correctly adjusted, and there are many cases in which the very short length required for ferrites is technically and economically advantageous. As a rough indication of the change in section and length with material, Fig. 3 also gives the approximate length/diameter ratio of a simple cylindrical magnet to have the required permeance.

Figure 4 gives similar plots of recoil energy against open-circuit working point, and now the real field of

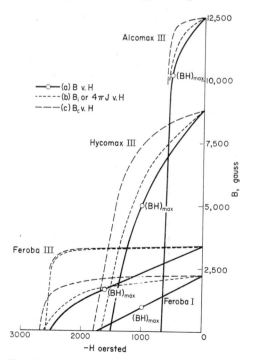

Fig. 2. Demagnetization curves of (a) normal, (b) intrinsic and (c) recoil flux density.

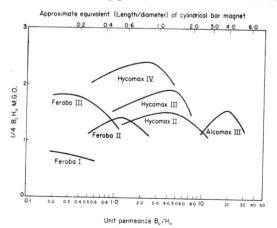

Fig. 4. Recoil energy $(1/4 B_c H_0)$ v. open circuit working point $(B_0 H_0)$.

advantage of the higher coercivity magnets is evident both with high energy values, and with rather less critical design requirements as indicated by the width of the curves. Feroba III is outstanding, provided that the working point is reasonably correct, while for Feroba I, it is technically impossible to have too short a magnet for recoil working. Again, length/diameter ratios for cylinders give a rough guide to proportions.

Economic Comparison of Ceramic and Alloy

Because of the much lower raw material costs of barium carbonate, strontium carbonate and iron oxide against pure nickel, cobalt, copper, aluminium, iron, etc., it would appear that the ceramic magnets should be somewhat cheaper than their cast magnet counterparts for the same application. However, this is not necessarily so, because the lower $(BH)_{max}$ of the ferrites necessitates a larger volume of material, and more particularly because of the high cost of processing. Also, because of the extremely brittle nature of the ceramic barium ferrite (it may be regarded as "black porcelain"), there is sometimes a higher scrap factor due to breakages, chips, cracks, etc., with difficult shapes, compared with Alcomax magnets and this must obviously add to the production costs. Tooling also is an expensive item, particularly for anisotropic Feroba II and III. In the process for optimum magnetic properties, a slurry containing about 40 per cent of water is fed to the presses. During the forming operation in the die cavity in the presence of a strong magnetic field to provide alignment, the water must be filtered off. Thus, tooling becomes very complicated when it has to incorporate drainage channels, holes and filter pads in both magnetic and non-magnetic steels.

The material is also extremely abrasive and tool parts have to be replaced rather frequently. The price of Feroba magnets must cover these high tooling charges.

Conversely, the provision of patterns for castings and the setting up of a moulding plate are relatively inexpensive.

Large loudspeaker magnet assemblies used in high fidelity public address and stereophonic systems have almost entirely changed over from Alcomax rings or slugs to Feroba II rings, because of compactness rather than cost. In some cases, the change from a cast magnet has reduced the back to front depth of the speaker assemblies by as much as two inches. Conversely, the large increase in outside diameter of the permanent magnet assembly is not a disadvantage, as it is still less than the cone or the chassis diameter of the speaker. Although barium ferrite has a lower density than the cast alloys (5 g/cm³ as against 7·3 g/cm³), the increase in volume required because of the lower $(BH)_{max}$ usually results in the Feroba magnet having about the same weight as the cast magnet it replaces. Because of the unusual magnetic characteristics, the internal surface of the Feroba II ring can be brought very close to the mild steel centre pole of these assemblies, and, in fact, the radial gap need only be a few thousands of an inch more than the working gap of the speaker magnet. In comparison with cast ring assemblies the efficiency is therefore improved considerably, and typical leakage factors for a medium gap field strength magnet of this type are 3·5 for Alcomax and 2·2 for Feroba.

A comparison of characteristics of ceramic and alloy permanent magnet materials is given in Table 2.

Table 2. Comparison of characteristics of ceramic and alloy permanent magnets

Feature	Advantageous material	
Mechanical strength	Alloy	Alnico, Alcomax
Corrosion	Ceramic	Feroba I, II, III
Electrical resistivity	Ceramic	Feroba I, II, III
Energy product	Alloy	Alcomax, Columax
Coercivity	Ceramic	Feroba I, II, III
Remanence	Alloy	
Useful recoil energy	Ceramic	Feroba I, III
Temperature coefficient	Alloy	Alnico, Alcomax, Columax
Stability (against magnetic fields)	Ceramic	Feroba I, II, III
Stability (against temperature)	Alloy	Alnico, Alcomax, Columax
Low density	Ceramic	Feroba I, II, III
Raw material cost	Ceramic	Feroba I, II, III
Dimensional control	Alloy especially sintered	Alnico, Alcomax

Interchangeability of Alloy and Ceramic Designs

A series of loudspeaker magnets of Alcomax castings with their Feroba II replacements is illustrated in Fig. 5. Fixing holes to facilitate assembly are provided in the Feroba II rings or segments.

Many types of cast magnets in Alnico and Alcomax alloys cannot possibly be replaced directly by a Feroba magnet. A selection of such magnets is shown in Fig. 6. Intricate shapes are virtually impossible in ceramic magnets because of the limitations of the powder technique, and also the difficulty in providing other than linear particle alignment for anisotropic ceramics. Feroba magnets are therefore confined to simple rectilinear or cylindrical shapes, whereas castings can be much more complicated, particularly in the isotropic Alnico alloys.

A serious disadvantage of barium ferrite is the very high negative temperature of coefficient of magnetization. The performance of Feroba magnets is affected by increase or decrease of temperature which mitigates against its use in certain applications. Since the negative temperature coefficient of magnetization for

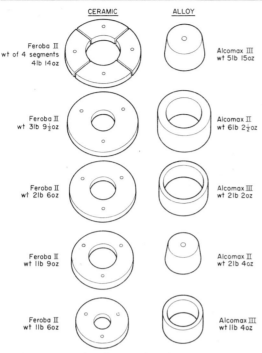

Fig. 5. *Equivalent designs of loudspeaker magnets.*

Feroba (0·19 per cent/°C) is approximately ten times that of the Ni–Al–Co–Fe type alloys (0·010–0·022 per cent/°C), it is undesirable and extremely difficult to use Feroba in measuring instruments. Thus disc-damping magnets in energy meters, and permanent magnet systems for moving coil instruments, etc., are not suited to Feroba magnets. It must be borne in mind that these magnetization changes are only partly reversible; to ensure that extremes of ambient temperature produce no permanent loss of magnetic strength it is necessary to design with increased volumes, dependent upon the upper and lower temperature limits. These remarks also apply to metallic magnets, but to a much lesser extent. In either case, the best stabilizing treatment against irreversible losses is to thermally cycle the magnet a few times between the upper and lower temperature limits.

Where an alloy magnet is essential but the high coercivity of Feroba is also desirable, cast magnets in Hycomax II, III and IV alloy can be utilized.

The material preferred for general utilization in a number of applications is given in Table 3.

Unique Design Examples

The availability of ceramic magnets with different magnetic characteristics and certain economic advantages has opened up new fields of applications

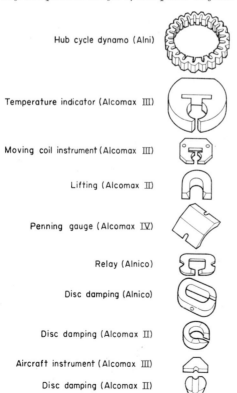

Fig. 6. *Cast alloy magnets.*

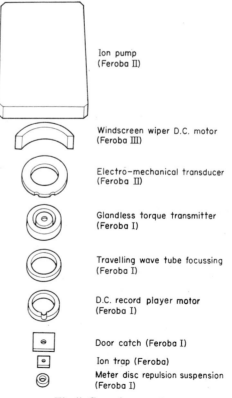

Fig. 7. *Ceramic magnets.*

Table 3. Preferred material for general utilization

Magnet application	Material	Magnet application	Material
Loudspeakers ($H_g < 10000$ Oe.)	Alcomax II	Relays	Alcomax III
Loudspeakers ($H_g > 10000$ Oe.)	Feroba II	Signal devices	Alcomax III
Instrument	Alcomax III	Microwave (TWT)	Feroba I and III
Meter damping	Alcomax II		Hycomax II
Meter suspension	Feroba I	Microwave (magnetron)	Alcomax III
Toys	Feroba I	Attraction devices	
	Alcomax II	(advertising and novelties)	Feroba I and III
D.C. motors (FH)	Feroba II–III	Workholding	Alcomax III
D.C. motors (large)	Alcomax IV	Repulsion devices	Feroba I and III
Alternators (small)	Feroba I, III	Flexible material	Bonded Feroba
Alternators (large)	Alcomax IV	Telephones (ringers)	Feroba I
Separators	Feroba III	(receivers)	Alcomax
	Alcomax		

where magnets have not been used hitherto. A selection of some typical Feroba magnets of this category is shown in Fig. 7. These designs exploit the high resistance to demagnetization of Feroba, particularly for magnets used in door catches and certain repulsion devices (P.M.A. Technical Bulletins Nos. 1, 4, 6, 9), such as small supporting rings for magnetic suspensions in energy meters. The high useful recoil energy of high coercivity Feroba III has enabled the copper wire-wound stator field of d.c. motors to be replaced by ceramic rings or segments. Co-axial magnets for motion transmission through an air gap are also possible in Feroba, with a multiplicity of magnetized poles on the internal surface of the outer magnet and the periphery of the inner magnet.

With regard to the high coercivity permanent magnet alloys, increased coercivity has promoted their use for centre-core magnets in compact high sensitivity moving coil instruments. Another application in which they have proved to be extremely efficient is in units for travelling wave tubes as assembled magnets and pole pieces. For very specialized instruments of a variety of types where space and weight are at a premium, as in space vehicles and capsules, the columnar versions of these alloys are used, particularly since cost is not important.

This latter demand has encouraged and stimulated practical work to overcome the difficult problem of producing columnar structures in alloys containing high percentages of titanium (Hadfield and Harrison, 1967). It is in this region where the cobalt content is approaching 50 per cent with titanium of the order of 8 per cent, that the highest energy products are likely to be obtained, and laboratory work has already indicated energy products approaching 13 m.g.o. for columnar crystal Hycomax IV.

Technical developments in both types of material is continuing at a high pace, as with each improvement in coercivity there immediately follows efforts to increase the fullness factor of the demagnetization curve, thereby obtaining higher values of energy product. There is little doubt that alloy magnets will always lead ceramic magnets in terms of energy product, if not in coercivity. It has already been proved that the highest energy product for ceramic materials will be of the order of 5 m.g.o., whereas it has been postulated that for alloy magnets this product could exceed 20 m.g.o. and already an experimental alloy of Samarium and Cobalt has reached over 18 m.g.o. (see Buschow (1968)).

The proprietary names in this article are used exclusively by the Permanent Magnet Association, Sheffield, England. Similar alloys and permanent magnet materials are manufactured throughout the World in a variety of trade names too numerous to annotate.

Bibliography

Buschow K. H. J., Luiten W., Naastepad P. A. and Westendorp F. F. (1968) *Philips Technical Review*, 29, No. 11, 336.

Hadfield D. (1961) Magnets, recoil curves of, in *Encyclopaedic Dictionary of Physics* (J. Thewlis Ed.), 4, 488, Oxford: Pergamon Press.

Hadfield D. (Ed.), (1962a) *Permanent Magnets and Magnetism*, London: Iliffe.

Hadfield D. (1962b) Magnetic materials, heat treatment of, in *Encyclopaedic Dictionary of Physics* (J. Thewlis Ed.), 4, 418, Oxford: Pergamon Press.

Hadfield D. and Harrison J. (1967) B. P. 1,085,934.

Harrison J. (1965) B. P. 987,636 and B. P. 999,523.

Permanent Magnet Association (1966–8), *Technical Bulletins* Nos. 1, 4, 6, 9. Sheffield: P.M.A.

D. Hadfield

PHONON AVALANCHE AND BOTTLENECK. The terms *avalanche* and *bottleneck* refer to two related phenomena which may occur in paramagnetic crystals at low temperatures. Under certain conditions, it is possible for acoustical phonons to interact more strongly with magnetic ions than with one another or with the thermal bath which surrounds the crystal. When this situation exists, both the bottleneck and the avalanche can take place. They come about in the following way.

Consider an array of magnetic ions which have only two energy levels, separated by a value $\hbar\omega$ (typically 10^{-4} eV) and labelled plus and minus. We imagine that the populations n_+ and n_- of these levels are initially altered from the values n_{+0} and n_{-0} characteristic of thermal equilibrium at the ambient temperature. If the temperature is low enough, equilibrium can be restored only by the process of direct spin-lattice relaxation, in which the ions absorb and emit resonant acoustical phonons. By "resonant" is meant phonons which occupy lattice modes sufficiently close in frequency to ω that the ion–phonon transition probability is large. A resonant phonon which has just been emitted by an excited ion can interact with another ion, or can be destroyed (by collision with a second phonon, an imperfection, or the surface of the specimen). In the former case, the target ion may be either plus or minus. If, however, the initial condition (just after the populations were perturbed) was $n_- > n_+$, the phonon encounters predominantly minus ions and, at least on the average, is reabsorbed. A typical phonon will in fact be absorbed and emitted many times during its outward migration, providing only that there are enough ions and the ion–phonon cross-section is large enough to compete successfully with the various other mechanisms of phonon destruction. The net effect of interaction with the ions is to impede the escape of non-equilibrium phonons; likewise, the equilibration of the populations is also inhibited. The situation is said to constitute a *phonon bottleneck*. It is closely analogous to a process in which electromagnetic radiation is trapped in some gases (see *Resonance radiation, imprisonment of*).

The temporal development is quite different if the initial condition is $n_+ > n_-$. Here, a given phonon interacts with a net excess of *plus* ions. Instead of absorption, the phonon induces the emission of more phonons. With an increase in the number of phonons, the rate of decay of excess ionic energy is also increased, generating still more phonons, and the situation quickly becomes catastrophic. This acceleration of the population decay rate and concomitant enhancement of the density of resonant phonons is what is termed the phonon avalanche.

The bottleneck and avalanche can be described in thermodynamic language. We assign an instantaneous ionic temperature T_s to the populations (see *Spin temperature*), in the sense that

$$\frac{n_+ - n_-}{n_+ + n_-} = \coth(-\hbar\omega/2kT_s),$$

and we similarly define a temperature T_p for the resonant phonons by means of

$$p = [\exp(\hbar\omega/kT_p) - 1]^{-1},$$

where p is the average excitation number of the resonant modes. The avalanche is viewed as a flow of heat from an ionic reservoir, initially at $T_s < 0$, into a reservoir of resonant phonons which, at first, have a temperature equal to that of the thermal environment. Because phonons can never exist at a negative temperature, the heat flow must continue until the two systems reach a common temperature which necessarily is positive. In most cases, the ionic heat capacity exceeds that of the resonant phonons by many orders of magnitude; consequently, the final temperature is extremely high. In favourable cases, values as high as 10^{4}°K are expected.

At this point, the avalanche is over. The two systems, however, are much hotter than the surrounding heat bath (typically at $2°$K); the subsequent cooling represents the bottlenecked decay.

It should be noted that in the limiting case of *no* interaction between resonant phonons and bath, the avalanche would assume its fastest possible rate, but the bottlenecked decay would continue forever. Coupling of the resonant phonons to the thermal environment competes directly with the avalanche and thereby mitigates the fast decay.

It is perhaps worthwhile to point out that there is no particular requirement that the ionic levels be split by a magnetic field. In principle, the nature of the splitting is irrelevant; in practice, however, the two effects have been studied only with magnetic levels.

Theoretical picture. A proper theory for the avalanche and bottleneck involves the solution of coupled, non-linear partial differential equations and, as a consequence, has not been constructed. A simplified model, however, has been developed and has provided a useful qualitative explanation for experimental observations. The model retains the non-linearity (which is essential to an explanation of the avalanche) but leads to total differential equations that can be solved. Rate equations are written for n_+, n_-, and p, under two simplifying assumptions. First, it is assumed that the destruction of resonant phonons by processes not involving ions can be represented by a linear damping term in the rate equation for p. Second, the populations and p are assumed to be uniform throughout the sample. This assumption is undoubtedly wrong, but is nevertheless essential; otherwise the rate equations are completely intractable.

The equations are

$$d(n_+ - n_-)/dt = -2A[(p+1)n_+ - pn_-]$$
and
$$dp/dt = -(2\beta)^{-1} d(n_+ - n_-)/dt - (p - p_0)\tau_p^{-1},$$

(1)

where A is the rate of transitions induced by one phonon, for one ion, β is the number of lattice states which can interact with the ion, τ_p^{-1} is the damping rate for phonons, and the subscript zero denotes thermal equilibrium. The various factors of two occur because one transition ($n_+ \to n_+ - 1$, $n_- \to n_- + 1$,

$p \to p+1$) causes a change of two units in the difference $(n_+ - n_-)$; otherwise the equations are self-explanatory. For convenience, the equations may be rewritten in the following form:

$$du/dt = -(u+1)\tau_1^{-1} - uy\tau_1^{-1}$$
and
$$dy/dt = -y\tau_p^{-1} - S\,du/dt, \quad (2)$$

where $y = (p - p_0)/(p_0 + 1/2)$, $u = (n_+ - n_-)/(n_{-0} - n_{+0})$, $S = (2\beta)^{-1}(n_{-0} - n_{+0})/(p_0 + 1/2)$, and $\tau_1^{-1} = 2A(p_0 + 1/2)$. These four parameters represent, respectively, normalized measures of phonon excitation and population difference, the ratio of energy stored by the ions to energy stored in the resonant lattice modes at thermal equilibrium, and the direct ion-lattice relaxation rate, also at thermal equilibrium.

There is no straightforward solution to (2); however, general solutions can be obtained easily with a digital computer. Some insight can also be gained by an examination of special cases. Notice first that initially, before the phonons have been heated, the logarithmic decay rate of u is τ_1^{-1}. Later, at the point $u = 0$, the rate again becomes τ_1^{-1}. If $u > 0$ initially, these points refer to the instants just before and after an avalanche; therefore the avalanche begins and ends with the same rate. What happens in between is illustrated in Fig. 1, which shows several solutions to (2), as obtained from a computer for representative experimental conditions. (On the time scale depicted in the figure, which is about three orders of magnitude faster than τ_1, the starting and ending slopes are too small to see.)

It is also instructive to examine (2) in the limit $\tau_p^{-1} \to 0$, for $1 \geq u > 0$. The rate equations can easily be integrated to give

$$u(t) = u(0)/\{1 + \exp[(t - \tau_{1/2})/\tau_f]\},$$
$$y(t) = Su(0)/\{1 + \exp[-(t - \tau_{1/2})/\tau_f]\} \quad (3)$$

where
$$\tau_f^{-1} = Su(0)\,\tau_1^{-1} = A[n_+(0) - n_-(0)]/\beta$$

and
$$\tau_{1/2} = \tau_f \ln\{Su^2(0)/[u(0) + 1]\} \gtrsim \tau_f.$$

Equations (3) assume $S \gg 1$, which is nearly always the case. The solutions introduce a characteristic time τ_f, which can be interpreted as the mean time taken for a phonon to stimulate the emission of another phonon in the presence of $[n_+(0) - n_-(0)]$ excited ions. Equations (3) can be shown to be valid so long as the ratio $\tau_f/\tau_p \gg 1$. In practice, both time constants are usually of the same order, and the decay curves are somewhat blunter than (3) would predict.

After the avalanche is complete and the two systems have achieved a common temperature, dy/dt becomes small enough to be neglected. Thus, from (2),

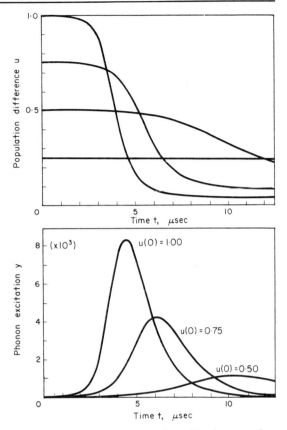

Fig. 1. Solutions to equations (2) for several values of $u(0)$ and for $S = 25{,}000$, $\tau_p = 10^{-6}$ sec, $\tau_1 = 10^{-2}$ sec.

$y \cong -S\tau_p\,du/dt$ and u can be found by direct integration:

$$\{[u(t) + 1]/[u(t_0) + 1]\}\exp\{-[u(t) - u(t_0)]\sigma/(1 + \sigma)\}$$
$$= \exp\{-(t - t_0)/(\sigma + 1)\tau_1\}. \quad (4)$$

Here, we have introduced a parameter commonly called the bottleneck factor, $\sigma = S\tau_p\tau_1^{-1}$. The subscript zero on t_0 refers to the initial instant of time. A typical solution is depicted in Fig. 2. To obtain curves of this kind, it is of course not necessary first to generate an avalanche; the only requirement is that the ions and phonons have the same (elevated) temperature at $t = t_0$.

Eventually, as $u \to -1$ (its value at thermal equilibrium in the surrounding heat bath), the decay in (4) becomes exponential, with a characteristic time $(\sigma + 1)\tau_1$. The factor σ completely characterizes the bottleneck. If $\sigma \gg 1$, this asymptotic decay will be prolonged in comparison with the case $\sigma \ll 1$, which corresponds to the absence of a bottleneck.

Experimental situation. Nearly all of the experimental work to date has dealt only with the behaviour of the populations n_+ and n_-, and not with the phonon excitation p. The parameter invariably measured is the magnitude of the spin resonance absorption signal, which is directly proportional to $(n_- - n_+)$. Standard microwave techniques are used to follow the absorption in time (see *Paramagnetic resonance phenomena*) after the populations have been disturbed by a transient microwave perturbation.

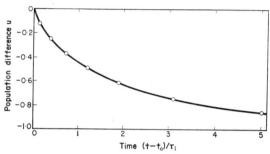

Fig. 2. Solid line is a solution to equation (4) using $\sigma = 2.65$, $\tau_1 = 1.64 \times 10^{-3}$ sec; circles are experimental points for copper dipyrromethene. Taken from Standley, K. J. and Wright, J. K. (1964) Proc. Phys. Soc. 83, 361.

To produce the condition $n_+ > n_-$ prerequisite to an avalanche, the spin-resonance technique termed fast passage (see *Fast passage in paramagnetic resonance*) has been employed with some success. A typical set of decay curves generated in this fashion are presented in Fig. 3. Data of this kind have shown that the avalanche depends qualitatively on various experimental parameters more or less as expected. Increasing $u(0)$ and decreasing the ambient temperature both should increase S; and increasing the size of the specimen should lengthen τ_p, if the primary mechanism of phonon destruction occurs at the surface of the sample. All are found to sharpen the avalanche. The curves of Fig. 3 resemble those of Fig. 1, but the data cannot be fitted in quantitative detail by the theory. This finding is most certainly a consequence of the naïveté of the model discussed above.

The bottlenecked decay has been studied more exhaustively, primarily because experimental measurements are greatly simplified with the much longer time scale. In addition, it is much easier to produce an initial condition which does not require $n_+ > n_-$. To observe the decay, one applies an intense microwave field to the sample for a time long enough to bring about the condition $n_+ = n_-$ (see *Paramagnetic materials, saturation effects in*), and, after removing the pulse, measures $(n_- - n_+)$ as a function of time. A representative decay curve is shown in Fig. 2, together with a theoretical curve based on (4). Clearly, theory and experiment agree closely in the case of the bottlenecked decay. Values of σ can be recovered from data like Fig. 2; it is found that σ increases with increasing sample size, increasing ionic concentration, and decreasing ambient temperature, all of which are expected to enhance the bottleneck. One factor which has *not* been studied carefully is the surface finish. Phonons might be expected to be less effectively destroyed at smooth boundaries than at surfaces which are rough on the wavelength scale of the phonon.

The direct observation of avalanche and bottleneck phonons has proved to be extremely difficult. One worker, N. S. Shiren, was able to generate an avalanche in one end of a crystal of iron-doped MgO and detect the avalanche phonons after they had migrated to the other end, where the iron ions served as a so-called phonon–photon quantum counter. The details of the time evolution of $y(t)$ were not measured, because the response time of the quantum counter was slow enough to distort the phonon signal; nevertheless, the existence of hot phonons was clearly proved.

Bottlenecked phonons have been observed by Brya, Geschwind, and Devlin (1968). These authors used a method called Brillouin scattering, in which light from a laser is inelastically scattered by phonons. One of the spin-resonance absorption transitions of nickel in nickel-doped MgO was continuously excited by a microwave field. The transition decayed by the emission of resonant phonons which, for this material, were severely bottlenecked. Thus, in steady state, one phonon was created for every microwave electromagnetic photon absorbed. The intensity of the Brillouin-scattered light, which is proportional to the phonon excitation p, was measured, and phonon temperatures T_p as high as 60°K were found. Since the ambient temperature was 2°K, the resonant

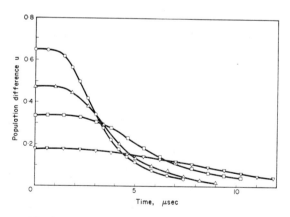

Fig. 3. Experimental behaviour of $u(t)$ during the avalanche in dilute $Ce_2Mg_3(NO_3)_{12} \cdot 24H_2O$ at an ambient temperature of 1.39°K. For this condition, $\tau_1 = 7.75 \times 10^{-3}$ sec. The decay profiles cannot be fitted accurately by equations (2) but have the same qualitative behaviour. (Compare to Fig. 1.)

phonons were heated by a factor of 30. The experiments of Shiren and Brya are especially significant because they represent the only cases in which, respectively, avalanche and bottlenecked phonons were detected in ways that were independent of the back reaction of the hot phonons on to the emitting spin systems.

Unfortunately, there is no one bibliographical source to which the reader can be referred for a summary of the literature; however, a fairly comprehensive picture can be obtained by examining the papers listed below.

Bibliography

BRYA W. J., GESCHWIND S. and DEVLIN G. E. (1968) *Phys. Rev. Letters* **21**, 1800.
BRYA W. J. and WAGNER P. E. (1967) *Phys. Rev.* **157**, 400.
FAUGHNAN B. W. and STRANDBERG M. W. P. (1961) *J. Phys. Chem. Sol.* **19**, 155.
SCOTT P. L. and JEFFRIES C. D. (1962) *Phys. Rev.* **127**, 32.
SHIREN N. S. (1966) *Phys. Rev. Letters* **17**, 958.

P. E. WAGNER

PHOTOEMISSION AND BAND STRUCTURE. The theme of this article is the relationship between the band structure of solids and photoemission. For many years, this relationship was not clearly appreciated due to the fact that photoemission was looked upon as a surface rather than a volume effect. In the last few decades, it has become apparent from a number of careful experiments that in most cases photoemission is a volume rather than a surface effect (Sommer and Spicer, 1965).

Since photoexcitation is a volume process, there are several important consequences of volume photoexcitation. First, since the excitation is between quantum states characteristic of the volume, it can be used to study these quantum states and thus the band structure of the solid. Conversely, the occurrence of structure due to the volume band structure in photoemission gives added evidence that the photoemission event is a volume effect. Another important consequence is the fact that the electrons excited in the volume of the crystal must pass through a finite amount of crystal before escaping; thus, they can suffer scattering events with lattice vibrations (phonons) and valence electrons (electron–electron scattering). Any possible scattering effects must be taken into account before band structure information can be obtained from photoemission data. The photoemission measurement which contributes most to the investigation of band structure is that of the energy distribution of the electrons excited by monoenergetic radiation. Curves giving this information will be called EDCs in this article. Typical curves are presented in Figs. 1 and 2. It is through the detailed study of structure in the EDCs and their movement with $h\nu$ that one obtains quite detailed information on band structure from photoemission measurements.

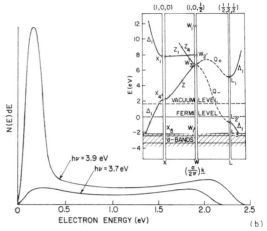

Fig. 1. EDCs for $h\nu = 3.7$ eV and $h\nu = 3.9$ eV. The low-energy peak for $h\nu = 3.9$ eV is due to excitation from the d band. The insert gives E vs. k diagram for Cu according to the calculations of Segall. In band theory, the quantum numbers E and k define a state; thus, the curves give the allowed quantum states along the indicated crystal directions. The narrow bands located 2 to 6 eV below the Fermi level are derived principally from the atomic d states.

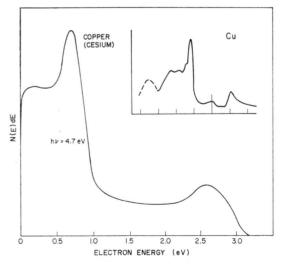

Fig. 2. An EDC for $h\nu = 4.7$ eV which shows excitation from deeper in the d band as well as the direct transition to final states near 2·6 eV which occurs between the states near the L symmetry point. The insert shows the optical density of states determined from photoemission and optical studies obtained by Krolikowski (1967).

It is impossible to survey the large number of materials that have been studied to date using photoemission; therefore, we will concentrate our attention on one material, Cu, to illustrate the method. Conventionally, the band structure of a solid is presented in terms of Bloch wavefunctions of the form

$$\psi(\bar{r}, t) = u_{\bar{k}, E}(\bar{r})\, e^{i[\bar{k}\cdot\bar{r} - (E/\hbar t)]} \quad (1)$$

In this case each quantum state is defined by two quantum numbers: E the energy, and k the propagation vector. If the initial and final states of an optical excitation are well approximated by single Bloch wave functions and there are no other interactions, e.g. lattice or electron–electron interactions, the only transitions which are allowed are those in which the $\bar{k}$ of the initial and final states are practically identical. Such transitions are called direct (or vertical) transitions since the final state lies directly above the initial state on an E vs. $\bar{k}$ diagram. For the purpose of illustration, part of the band structure of Cu as calculated by Segal (1962) is presented as an insert in Fig. 1. The bands derived principally from the atomic d states lie 2·1 eV or more below the Fermi level E_f. They are quite dense. For the sake of simplicity, they are represented by the cross-hatched area in the insert. The bands lying above the d bands are derived to a much larger extent from the atomic s and p states. Because these atomic states are much less tightly bound than the d states, the bands are much wider and "free electron" like.

If we concentrate on the s- and p-derived bands lying above the d band, we see that the lowest energy direct transition is that from the L_2 level near the Fermi surface to the L_1 level about 5 eV above it. Note the way in which the initial and final state energies must change with increasing photon energy for this direct transition. Since the energy of the final state may be measured directly in the photoemission, one might expect to see characteristic structure in the EDCs corresponding to these direct transitions. In particular, a strong peak would appear in the EDCs abruptly at the threshold for the direct transition at a final state energy corresponding to the L_1 point. For higher photon energies, the peak moves to higher energies but at a rate less than the increase in $h\nu$ since the initial state energy must decrease with increasing $h\nu$. One obtains experimentally (Bergland and Spicer, 1964) the results suggested by this analysis. If transitions from the d states were also well represented by transitions which were simply between single Bloch initial and final states, one would expect to see similar behavior for those transitions; however, as we will also see, this has not been found to be the case experimentally. Rather the experimental data for excitation from the d states (as well as a significant fraction of the excitation from the s- and p-derived states) can be explained in terms of optical transitions, which require only conservation of energy without any apparent dependence on conservation of k (Berglund and Spicer, 1964; Spicer, 1967). Such transitions depend principally on a product of the optical density of states at E and $(E - h\nu)$. The term "nondirect" has been coined for such transitions (Spicer, 1967).

In Figs. 1 and 2 energy distributions from Cu with a monolayer of Cs are presented (Berglund and Spicer, 1964). One curve of Fig. 1 is for $h\nu = 3{\cdot}7$ eV so that no photoemission is possible from the d states; rather, the excitation must be from the high-lying states. As can be seen from the insert in Fig. 1, no direct transitions are possible from these states; thus, the transitions must be nondirect. The small peak that occurs a few tenths of an electron-volt above the vacuum level is due to a peak in the final density of states just above the vacuum level, whereas the small peak which appears 0·35 eV below the maximum energy is due to a peak in the initial density of states just below the Fermi level. For $h\nu = 3{\cdot}9$ eV we see the effect of excitation from the d band to states above the vacuum level; the large, low-energy peak for $h\nu = 3{\cdot}9$ eV is due to this excitation. For increasing values of $h\nu$ the number of d electrons escaping increases until the complete d spectrum is exposed. As $h\nu$ is increased, one sees the abrupt appearance of a strong peak near the maximum energy at $h\nu = 4{\cdot}4$ eV. This peak can be seen near the maximum energy in Fig. 2. This is due to the direct transition near L predicted previously. For $h\nu > 4{\cdot}4$ eV, the peak persists but drops farther from the maximum energy as $h\nu$ increases just as was expected due to the fact that the band slopes downward from L_2 (see the insert in Fig. 1). From the final state energy of the direct peak at $h\nu = 4{\cdot}4$ eV, we can locate the L_1 point approximately 4·1 eV above the Fermi level in good agreement with band calculations. Thus, in Cu one sees the expected direct transition between the s- and p-derived bands but nondirect transitions from the d states. The latter result is thought to be a consequence of the more localized nature of a hole in the narrow d bands (Berglund and Spicer, 1964; Spicer, 1967).

The distinction between direct and nondirect transitions is even more obvious if the EDCs are plotted versus the energy of the initial state from which the electron is excited (Berglund and Spicer, 1964; Spicer, 1967). This is due to the fact that the initial state energy will change with $h\nu$ for a direct transition, whereas for nondirect transitions the initial state will not change with $h\nu$. Where nondirect transitions predominate, an optical density of states can be obtained from photoemission and optical measurements. The optical density of states has been obtained for many materials. As an insert in Fig. 2, the optical density of states obtained by W. F. Krolikowski is presented (Krolikowski, 1967).

To summarize, photoemission provides a powerful tool for investigating band structure because the absolute energy of the initial and final states involved in optical transitions can be determined through measurement of the distribution in energy of the photoemitted electrons. Where conservation of k

vector (see equation 1) provides an important optical selection rule, important features of the E vs $\bar{k}$ diagrams (see the insert in Fig. 1) can be determined. Where conservation of $\bar{k}$ does not provide an important optical selection rule, the optical density of states can be obtained (see insert in Fig. 2).

Bibliography

BERGLUND C. N. and SPICER W. E. (1964) *Phys. Rev.* **136**, A 1030 and A 1044.
KROLIKOWSKI W. (1967) Ph. D. Dissertation, Stanford University.
SEGALL B. (1962) *Phys. Rev.* **125**, 109.
SOMMER A. H. and SPICER W. E. (1965) in *Photoelectric Materials and Devices* (S. Larsch Ed.). D. van Nostrand Co., Inc., Princeton, N. J.
SPICER W. E. (1958) *Phys. Rev.* **112**, 114.
SPICER W. E. (1958) *RCA Rev.* **19**, 555.
SPICER W. E. (1960) *J. Appl. Phys.* **31**, 2077.
SPICER W. E. (1967) *Phys. Rev.* **154**, 385.

W. E. SPICER

PHYSICAL LIMNOLOGY. This is one of the disciplines of limnology, the science of lake studies. It involves the application of the principles of physics to the examination of the properties and processes within lakes and at the surface and bottom interfaces of lakes and their surroundings. Pure and applied research programs in this discipline involve techniques and approaches well known to physicists, oceanologists, hydrologists, meteorologists, hydraulic and civil engineers, geographers and cryologists and, in general, include both the physical nature of lake water and phenomena which occur in lakes as a consequence of physical processes. Examples of particular research areas are: physical properties of lake water, fluid dynamics, waves, density structure, air–lake interactions, ice, lake levels, and electromagnetic radiation exchange processes.

As in other natural sciences, important interrelationships exist between the interests and activities of workers in each of the limnological disciplines. Thus, most physical limnologists are interested and frequently competent in fields such as chemical limnology, geological limnology, and biological limnology. Indeed, a great deal of interaction exists between physical, chemical, biological, and geological lake processes.

The properties of pure water are well known. The changes in these properties due to the effects of the impurities found in naturally occurring waters are of great interest. Density, specific heat, electrical and thermal conductivity, viscosity, thermal expansion, transparency, latent heat (of fusion and vaporization), surface tension, elasticity, and the colligative properties of lake and ocean waters have been studied and continue to attract the attention of researchers. In many cases, changes in measurable properties are used to determine the concentrations of impurities causing the changes.

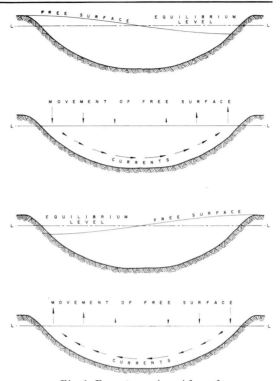

Fig. 1. *Four stages of a seiche cycle.*

The temporal and spatial changes in the energy content of a lake determine the characteristics of each of the physical processes which occur within it. Most of the basic aspects of the hydrodynamics and thermodynamics of lakes closely resemble those of the oceans, and many of the approaches adopted by limnologists and oceanologists in their examinations of them are similar (and also similar to those of meteorology).

The many types of motion found in lakes have not in general received the same degree of attention that has been applied to the oceans, although the principles are essentially identical. Because of the great variety of physical dimensions of lakes, the approaches to lake dynamics problems vary considerably. In large lakes, such as the North American Great Lakes, the consideration of many dynamic forces is desirable because individual forces rarely consistently dominate in all parts of these lakes but may be locally or periodically dominant. However, because of the complexities which arise in the solution of the nonlinear equations of motion, it has been necessary to examine time dependent and steady state situations independently and in simplified forms. For a discussion of the analogous problems involved in oceanology, see Fofonoff (1962).

Large-lake circulation studies have generally involved either preparations of statistics of measured current velocities or the derivation of solutions to the equations of motion to fit simplified dynamic and boundary conditions. The best research technique involves the measurement of water motion to test models of lake dynamics. The driving forces which are of most significance are the wind stress at the lake surface and surface heating. Usually of lesser importance are atmospheric pressure gradients and the effect of river inflow and outflow. Forces which evolve in response to driving forces are: inertial and time dependent accelerations, the effect of the earth's rotation (Coriolis force), pressure gradients due either to nonuniform changes in lake level or horizontal density gradients, turbulent friction, and molecular friction. In smaller lakes, the Coriolis force is less important, while the effects of boundary conditions and bottom stress are of considerable importance.

Various categories of motion, as they pertain to appropriate physical dimensions and environmental conditions have been studied (for general treatments, see Proudman, 1953; and Hutchinson, 1967). Some examples are: geostrophic motion (balance between a horizontal pressure gradient and the Coriolis force); inertial motion (balance between Coriolis force and centrifugal force); Ekman motion (balance between surface wind stress and internal friction); and gradient motion (balance between a horizontal pressure gradient and internal friction).

A good deal of attention has also been paid to the response of lakes to external forces which cause tilting of the lake surface. The periodic decay of such dynamic situations, following the relaxation of the driving forces, results in a seiche in which characteristic oscillations and their harmonics occur. These are also described by Proudman and by Hutchinson, who have each given important examples.

While the density of any fluid is of basic physical importance, the role of the density of lake water is particularly significant. Changes in lake water density are principally dependent upon temperature changes and are therefore closely related to surface heat exchange processes. The input of heat at the surface of lakes warmer than the temperature of maximum density (4°C) results in thermal expansion near the surface and a stabilizing effect. The combination of seasonal heating and wind-mixing results in a concentration of heat in the upper part of the lake (epilimnion), a pronounced vertical thermal gradient (thermocline), and a cooler lower layer (hypolimnion) in lakes where the depth exceeds the depth of the lower limit of the thermocline. Although the depth of the epilimnion normally increases through the subsequent cooling season, the loss of heat at the lake surface results in a degradation or eventual destruction of the thermocline.

When surface cooling at temperatures above 4°C (temperature of maximum density) or surface heating at temperatures below 4°C occur, vertical instability results and mixing takes place until stability is re-established. The consequences of this phenomenon are many: freezing normally begins at a lake's surface, chemical and biological processes which may be vertically confined by stability restrictions during summer will not be so hampered during winter and may be thoroughly mixed throughout the lake, and the dynamic adjustments which take place during periods of strong mixing affect both vertical and horizontal circulation features.

The occurrence of two well-defined layers in some lakes presents dynamic complications, and separate

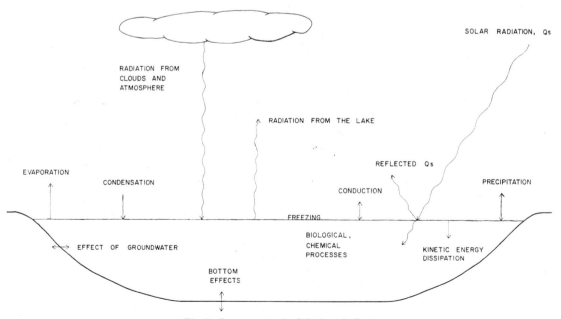

Fig. 2. Components of a lake heat budget.

circulation patterns which are coupled at the interface may occur. Further, perturbations of the interface result in periodic motions termed "internal waves". Internal waves may in some cases dominate the transmission of mechanical energy within a lake (or any other body of water which possesses an internal density gradient) and therefore constitute an item of major research interest.

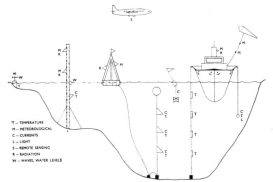

Fig. 3. Examples of types of physical limnology field measurements.

The interactions between lakes and the overlying atmosphere are of considerable importance to both fluid media. Heat, mass, and momentum exchanges with the atmosphere are of significance to all lakes, especially where the atmospheric climatology is highly variable. The relative effect of such exchanges on the atmosphere is naturally also dependent on the magnitude of the surface area of lakes involved.

A high percentage of the mechanical energy transferred to lakes from wind action is transmitted by surface wave movement to the shorelines where much of it is dissipated. Wave formation, transmission, and decay processes, which profoundly affect vertical mixing, sediment transport, the behaviour of lake vessels, and the integrity of shoreline structures have received much attention, primarily in the oceans and large lakes. Wave fore- and hindcasting techniques, based upon wind strength and duration, and the length and depth of the lake or ocean fetch region, have been developed basically for engineering problems related to the effects of wave action. However, the highly complex physics of wave formation at the air–lake interface continues to receive the attention of oceanologists and limnologists.

Measurements of important air–lake exchange phenomena, including precipitation, evaporation, condensation, conduction, and momentum, are extremely difficult. Except for precipitation, which is usually estimated from adjacent land measurements, these processes are examined by two quite different approaches. A great deal of effort is being made to try to understand the complex "small-scale" interaction processes which take place in the turbulent boundary layers at the air–lake (or air–sea) interface. This research, usually at fixed towers, continues to face problems of both theory and instrumentation. The "large-scale" approach, normally adopted in long-term exchange studies, relies upon empirically prepared formulae based upon a more simplified theory. These formulae require data of the type obtained in routine weather observations, occasionally from ships on large lakes but more frequently at nearby land locations. Because of the sophistication involved in small-scale measurement procedures, it is unlikely that the use of routinely gathered meteorological data for this purpose will become obsolete. However, a more complete understanding of the basic exchange processes may result in significant improvements in the empirical approaches now employed and in new criteria for obtaining the necessary data.

Evaporation from lakes may also be estimated by water budget and energy budget techniques. The former requires accurate data of inflow and outflow volumes, and precipitation. In occasional cases, the net inflow of groundwater may be also significant.

In energy budget studies, limnologists are again usually faced with a lack of directly measured values. Lake temperature measurements are used to determine changes in heat content. However, the energy contributions due to net solar radiation, net long wave "terrestrial" radiation, and the conduction of sensible heat, are generally estimated with empirical formulae (see, for example, U.S. Geological Survey, 1954), although solarimeter and radiometer data may be available for local purposes. When necessary, heat advection values are determined from inflow and outflow measurements.

Electromagnetic radiation is of considerable interest to limnologists for other than heat exchange reasons. Light, which penetrates the lake surface mostly in the visible portion of the spectrum, is essential for biological photosynthesis. Thus, the spectral characteristics and the physical behaviour (scattering, absorption) of light in lakes constitute important areas for research. In addition, limnologists are becoming increasingly aware of the usefulness of radiant energy as a remote sensing medium. Although photographic images have often been used in limited ways in lake studies, the availability of new films and multispectral techniques, and the developments of infrared and ultraviolet scanning devices, present new and exciting remote sensing possibilities. While these techniques are limited to sensing the surface or near-surface region, they provide detailed information on the location (and to some extent the nature) of surface thermal features, algae blooms, turbidity plumes, and various types of pollutants.

Studies involving prediction of lake ice have been somewhat oriented to the formation and dissipation of ice which is hazardous to navigation on large lakes. Ice appears in a variety of forms, depending upon environmental conditions and the stages in its development. These forms (e.g. tabular, columnar, granular, abrasional, accretional, frazil, anchor) and

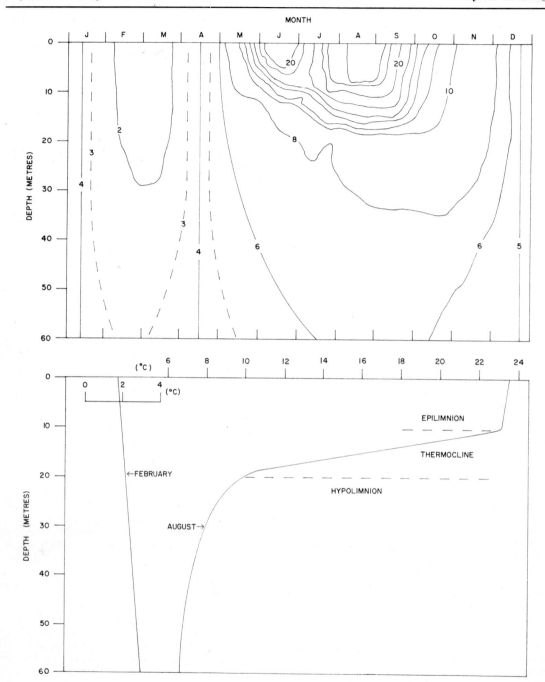

Fig. 4. Examples of temperature variations in a mid-latitude lake and of temperature–depth curves.

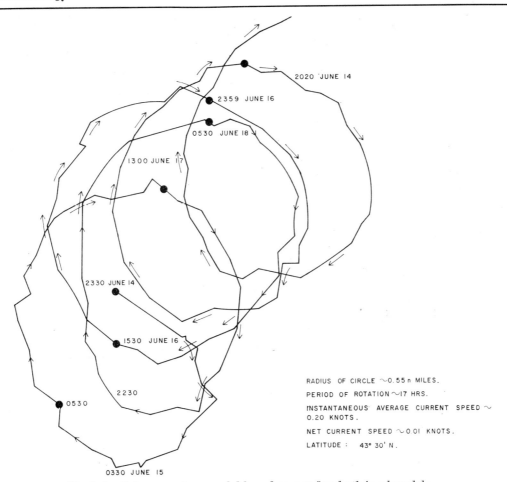

Fig. 5. Inertial movement as revealed by a drogue at 5 m depth in a large lake.

the physics of their occurrence, have been examined and classified.

Research in physical limnology has until recently been stimulated largely by pure scientific interest in this important part of the study of lakes. However, social problems related to pollution, fisheries, navigation, shore erosion, wave damage, flooding, and the recreational usage of lakes have prompted particular emphasis on those aspects pertinent to each situation. Thus, there has been a recent great stimulus to applied research in physical limnology. The use of large ships designed particularly for research, instrumented moorings and towers, and airborne sensors is now common in the research of large lakes. Vehicles such as hovercraft and submersibles will undoubtedly be part of lake research fleets for large lakes in the near future.

Bibliography

FOFONOFF N. P. (1962) Dynamics of ocean currents, in *The Sea* (M. N. Hill Ed.), New York: Interscience.

GREAT LAKES RESEARCH DIVISION, Univ. of Michigan Publ. 1–15, Michigan: Ann Arbor.

HUTCHINSON G. E. (1957) *A Treatise on Limnology*, Vol. 1, London: Wiley.

Limnology and Oceanography. Lawrence, Kansas: Am. Soc. of Limn. and Ocean.

Limnology in North America (1963) Madison, Wisconsin: Univ. of Wisc. Press.

Proceedings Int. Assoc. of Theor. and Applied Limnology, Stuttgart: E. Shweizerbart'sche Verlagsbuchhandlung.

PROUDMAN J. (1953) *Dynamical Oceanography*, London: Methuen.

U.S. ARMY CORPS OF ENGINEERS PROJECT 2030 (1954) Snow, Ice, and Permafrost Establishment, Wilmette, Ill. (Rep. 5, pt. 1).

U.S. GEOLOGICAL SURVEY (1954) Water loss investigations: Lake Hefner studies, Tech. Rep. (Prof. Paper 269).

R. K. LANE

PIEZOMAGNETIC EFFECT. The magnetic properties of ferromagnetic materials are stress dependent. The piezomagnetic effect is an inverse magnetostrictive phenomenon and refers to the variation of magnetic susceptibility and remanent (permanent) magnetization as a function of the applied stress. Piezomagnetism has been investigated experimentally in ferromagnetic specimens and ferrimagnetic rocks and minerals and has also been observed to occur naturally in seismically active areas by virtue of tectonic stress variations within the ferromagnetic-mineral-bearing rocks. It is within this context that the term is most often used. The naturally occurring piezomagnetic effect is also described as the seismomagnetic, magneto-seismic or magneto-elastic effect.

The piezomagnetic effect in rocks, in particular, is a consequence of the stress-dependent magnetic properties of the spinel minerals, especially magnetite, which the rock contains. In the absence of stress, magnetite exhibits cubic symmetry in its magnetocrystalline parameters. Application of a compressive stress creates magnetocrystalline anisotropy so that the easy directions of magnetization, along which the magnetic domains are oriented, are deflected away from the axis of compression. This phenomenon has been investigated in magnetite and magnetite-bearing rocks by Kapitsa, Grabovsky, and Parkhomenko whose results agree, in general, with the theory. From thermodynamic consideration of the energies associated with stress and with the magnetization of the domains, the stress sensitivity, s, can be expressed as

$$\frac{3\lambda_s k_m}{I_s^2}$$

where λ_s is the magnetostriction constant at saturation, I_s is the saturation magnetization, and k_m is the susceptibility. For magnetite the range of s should be

$$-1 \cdot 1 \times 10^{-4} < s < -7 \cdot 5 \times 10^{-4} \text{ cm}^2/\text{kg}$$

for susceptibility measured in the direction of the applied stress.

The corresponding stress sensitivity for the remanent magnetization involves a more complicated phenomenon and has been investigated, albeit inconclusively, in connection with paleomagnetic studies. Quantitatively, many investigators have simply assigned the stress dependence of the remanent magnetization as being approximately equal to the effects of susceptibility so that the net change in both remanent and induced magnetization can be estimated, then, by doubling this value of s. Certain irreversible effects have been observed in connection with both remanent and induced magnetizations which appear to be the result of a non-equilibrium state of elastic stress. After a few stress cycles most irreversible effects cease to appear and in natural environments undergoing numerous stress cycles, the irreversible effects are not usually considered to be important. For investigations of the magnetic effect of stress on a given sample of rock, the stress sensitivity, s, is expressed as the proportionality constant relating a change of stress to the change in magnetic susceptibility, i.e., the slope of the curve

$$\Delta k = s k_0 \Delta \sigma$$

where Δk and k_0 are the change in susceptibility and initial susceptibility of the rock and $\Delta \sigma$ is the change in the applied stress in the direction of the measured susceptibility. Typical behaviour of several rock samples is shown in Fig. 1 where the susceptibility parallel to the stress decreases by approximately 1 per cent for 100 kg/cm² and increases by one half this amount in a direction normal to the applied stress.

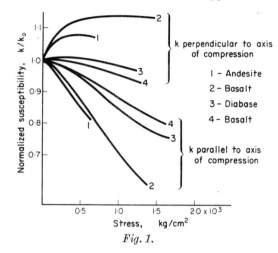

Fig. 1.

The naturally occurring piezomagnetic perturbation, ΔF, one would observe with a magnetometer on the surface of the earth in the vicinity of stressed rock, can be expressed as the integrated effects from an infinitesimal volume dV, or

$$\Delta F = \frac{s k_0 \Delta \sigma F \, dV}{r^3} f(\sigma, \phi)$$

where ΔF is the local change in field intensity, F is the ambient earth's field intensity, r is the distance between the magnetometer and centre of the volume, and $f(\theta, \phi)$ is a function of the geometry of a magnetic dipole field perturbation. This expression is normally integrated over the volume under stress assigning the appropriate components of stress and values of the anisotropic stress sensitivities.

Numerous reports have appeared in the literature through the past 100 years describing the occurrence of magnetic events at the time of large earthquakes. Observations of magnetic field changes which are termed piezomagnetic effects, however, are those local magnetic field variations which are a direct result of changes in sub-surface stress but are specifically *not* those magnetic effects which arise from any of the following: thermal effects of the ferromagnetic properties of minerals; apparent events due to mechanical vibrations of the moving parts of a magnetometer; changes in an associated telluric (terrestrial electric) current caused by stress-dependent resistivity; changes attributed to the normal secular variation of the magnetic field. Observations, as such, of the piezomagnetic effect under natural conditions have been reported especially by Kato in 1950, Breiner in 1964 and 1967, and Rikitake in 1966.

The amplitude of these piezomagnetic events as observed in seismically active areas range from values less than 10^{-5} oersted to larger than 10^{-4} oersted. The signatures thus far observed appear either as step changes or quasi-sinusoidal variations with periods of tens of minutes to an hour or more. In Japan these events appear to be associated indirectly with periods of high seismicity. In California the piezomagnetic events have been observed repeatedly to precede by tens of hours the aseismic displacement of a major transcurrent fault system.

Bibliography

BOZORTH R. M. (1951) *Ferromagnetism*, New York: Van Nostrand.
BREINER S. (1964) *Nature* **202**, 4934, 790.
BREINER S. and KOVACH R. L. (1967) *Science* **158**, 116.
CHEVALLIER R. Magnetic properties of minerals, in *Encyclopaedic Dictionary of Physics* (J. Thewlis Ed.), **4**, 430, Oxford: Pergamon Press.
GRABOWSKY M. A. and PARKHOMENKO Y. I. (1953) *Bull. (Izv.) Acad. Sci. USSR, Geophys. Ser.* **5**, 405.
KAPITSA S. P. (1955) *Bull. (Izv.) Acad. Sci. USSR, Geophys. Ser.* **6**, 493.
KATO Y. (1950) *Science, Rep. Tohoku Univ. Fifth Ser.* **2**, 2, 149.
KERN J. W. (1961) *J. Geophys. Res.* **66**, 3807.
LEE E. W. Magnetization, effect of pressure on, in *Encyclopaedic Dictionary of Physics* (J. Thewlis Ed.), **4**, 457, Oxford: Pergamon Press.
RIKITAKE T., YAMAZAKI Y., HAGIWARA Y., KAWADA K., SAWADA M., SASAI Y., WATANABE T., MOMOSE K., YOSHINO T., OTANI K., OZAWA K. and SANZAI Y. (1966) *Bull. Earthquake Res. Inst.* **44**, 363.
STACEY F. D. (1962) *Phil. Mag.* **7**, 551.

S. BREINER

POCKELS EFFECT. Alteration of the refractive properties of an optical medium by the application of a strong electric field may occur through a number of different effects, the best known being the so-called "linear electro-optic effect" and the Kerr effect. The latter is described by a fourth-rank tensor, and is the dominant electro-optic effect in liquids and in centro-symmetric solids where the "linear electro-optic effect", being described by a third-rank tensor, vanishes on account of symmetry. The high frequency, or constant strain component of the linear electro-optic effect is due to electronic processes within the material, and it is this term which is called "Pockels effect". It is not the same as the d.c. electro-optic effect which, as described later, has an additional term due to piezo-electric deformation of the crystal. The practical interest of this and other electro-optic effects lies mainly in the possibility of using them to produce light modulation, either of phase or intensity, by using an electroded liquid cell or crystal plate between polarizers in such a way as to produce electrically variable phase changes between two polarized interfering light beams.

If a non-centro-symmetric crystal is transmitting a light beam, the application of an electric field (assumed static for the moment) will cause an additional polarization $\mathbf{P}^\omega$ at the optical frequency ω, which may be described in terms of a third-rank polar tensor X_{ijk} by the formula

$$P_i^\omega = \varepsilon_0 X_{ijk} E_j^\omega E_k^0,$$

where P_i^ω is the component of $\mathbf{P}^\omega$ in the i-direction, E_j^ω is the component of the incident optical electric field in the j-direction, E_k^0 is the component of applied field in the k-direction and ε_0 is the free-space permittivity (S.I. units).

Symmetry considerations show that suffixes relating to terms at the same frequency are interchangeable, so that $X_{ijk} = X_{jik}$, and the tensor may therefore be written in condensed two-suffix notation X_{ij}, the first suffix running from 1 to 6 ($1 \equiv 11$, $4 \equiv 23$, etc.).

The field-induced polarization contributes a perturbation component to the dielectric tensor which may be denoted by $\Delta\varepsilon$, and calculated from the electro-optic tensor by means of the formula

$$(\Delta\varepsilon)_{ij} = \varepsilon_0 X_{ijk} E_k^0.$$

The effect of the perturbation may be to change either the magnitudes of the principal dielectric constants, or the directions of the principal axes, or both, and it may also destroy some of the symmetry elements of the crystal. For example, an electric field along the optic axis (z-axis) of a crystal of potassium dihydrogen phosphate (symmetry class $\bar{4}2$ m) will change it from uniaxial to biaxial, so that it will now show birefringence for light along the z-axis. The phase delay Γ between the fast and slow waves in travelling a distance l is given by

$$\Gamma = \frac{2\pi V X_{123}}{\lambda n_0},$$

where Γ is the delay in radians, V is equal to El and is the potential difference along the optical path, λ is the optical wavelength and n_0 the ordinary refractive index. For a given applied voltage the retardation is independent of optical path length (but only if the field is in the same direction as the optical path), and the voltage required to produce a phase delay of π^c is then referred to as the half-wave retardation voltage for the material. Evidently

$$V_{1/2} = \lambda n_0 / 2 X_{123}.$$

It is usual to specify the electro-optic coefficients r_{ijk} in terms of the indicial ellipsoid, which is equivalent to specifying them in terms of the reciprocal dielectric matrix. It may be shown that

$$r_{ijk} = \varepsilon_0^2 X_{ijk} / \varepsilon_{ii} \varepsilon_{jj},$$

the reference axes having been chosen so that the unperturbed optical dielectric matrix is diagonalized. The expression for the half-wave retardation voltage then becomes
$$V_{\frac{1}{2}} = \lambda/2n_0^3 r_{63}.$$

In the case of potassium dihydrogen phosphate at the red wavelength of a helium-neon laser (6328 Å) the numerical value of $V_{1/2}$ is about 8·6 kV, but much lower values have been quoted for some recently peveloped materials.

The linear electro-optic effect may be considered as having two components. The first arises from the mechanical strain caused by the piezo-electric properties of the material, and the second is a direct effect which remains when the crystal is clamped so that no deformation occurs. The direct effect is most readily observed by applying a very high frequency field, above about 100 MHz, and observing the resultant modulation of the optical signal. The proportion of the total electro-optic effect which is due to the direct effect varies from one material to another, and in the case of ammonium dihydrogen phosphate, for example, is about 65 per cent, while for the potassium isomorph the direct contribution is about 90 per cent of the total. The existence of a direct effect was first inferred by F. Pockels (1894), who calculated the magnitude of the electro-optic coefficient in quartz from values of the piezo-electric and photo-elastic constants, and showed that the value so obtained was considerably lower than that measured with a static electric field. If we denote the direct, or Pockels component of p_{ijk} by r''_{ijk}, and the component due to piezo-electric strain by r'_{ijk}, then in terms of the photo-elastic tensor p_{ijlm} and the piezo-electric tensor d_{klm} (where l and m are interchangeable suffixes and the piezo-electric strain is given by $S_{lm} = d_{klm}E_k$) we have

$$r'_{ijk} = p_{ijlm} d_{klm},$$

and the Pockels coefficient is therefore

$$r''_{ijk} = r_{ijk} - p_{ijlm} d_{klm}.$$

The terms "Pockels effect" and "linear electro-optic effect" are not always correctly distinguished, and to avoid confusion it is preferable to refer to the clamped and unclamped electro-optic coefficients. The term "linear electro-optic effect" is somewhat misleading, since it is in fact as much a second-order effect as are the related phenomena of optical rectification and second-harmonic generation.

Theoretical expressions for the Pockels coefficients have been derived from quantum mechanical perturbation theory. If ψ_g, ψ_n and $\psi_{n'}$ are wave functions for the ground state and two excited states of the system, ω_{ng} and $\omega_{n'g}$ the frequencies corresponding to the excited state energies and the coordinates x, y and z are represented by x_i, x_j and x_k we find that

$$X_{ijk} = \frac{(eN)^3}{\varepsilon_0 \hbar^2} \sum_{n,n'} [\{\langle x_i \rangle_{nn'} \langle x_j \rangle_{ng} + \langle x_i \rangle_{ng} \langle x_j \rangle_{nn'}\} \times \langle x_k \rangle_{n'g} A_{nn'} + \langle x_i \rangle_{ng} \langle x_j \rangle_{n'g} \langle x_k \rangle_{nn'} B_{nn'}],$$

where
$$A_{nn'} = \frac{1}{\omega_{n'g}} \left(\frac{1}{\omega_{ng} + \omega} + \frac{1}{\omega_{ng} - \omega} \right),$$
$$B_{nn'} = \frac{1}{(\omega_{n'g} + \omega)(\omega_{ng} + \omega)} + \frac{1}{(\omega_{n'g} - \omega)(\omega_{ng} - \omega)},$$
$$\langle x_i \rangle_{nn'} \equiv \langle \psi_n | x_i | \psi_{n'} \rangle \text{ etc.}$$

N is the number of electrons participating per unit volume.

The symmetry in i and j of the first term of the expression for X is self-evident; that of the second term may be inferred from the interchangeability of n and n' in the double summation.

The relevant wave-functions for practical systems are seldom known with sufficient accuracy to enable numerical calculations of the Pockels coefficients to be made, but the form of the expressions does suggest that a large increase in non-linear susceptibility may occur near an absorbing band. This may have little practical importance because of the high optical attenuation in such a region.

A semi-empirical rule due to R.C. Miller indicates that the second-order susceptibilities may be proportional to the product of the linear susceptibilities for the corresponding frequencies, so that

$$X_{ijk} = \alpha x_{ii}^\omega x_{jj}^\omega x_{kk}^0,$$

where α varies by less than an order of magnitude from material to material. (It is assumed, of course, that the coefficient in question is not forbidden by symmetry considerations.) In practice, "Miller's rule" is not as well obeyed in the case of the Pockels coefficients as is the corresponding rule for the coefficients relating to optical second harmonic generation. Nevertheless it seems likely that large Pockels coefficients are concomitant with large dielectric constants at both d.c. and optical frequencies.

The requirement that the crystal be non-centrosymmetric follows from the fact that a centre of symmetry implies zero values for all elements of an odd-order tensor. Furthermore, the cubic class with point-group symmetry $0 = 432$ has no third-rank tensor although it is not centro-symmetric. Hence, the Pockels effect occurs in only twenty of the thirty-two point group symmetry classes. In practical applications such as optical modulators it is usually desirable to avoid biaxial crystals, and this eliminates those in the triclinic, monoclinic and orthorhombic groups, leaving fifteen classes from which suitable materials may be chosen. Further restrictions are that the crystal should be relatively large in order to give a long interaction path between the light beam and the low-frequency modulating field, have good optical quality, chemical stability and mechanical strength, and also be capable of taking a high surface polish. One of the few naturally occurring crystals which satisfies these criteria is quartz, which falls in the trigonal class 32, but has, unfortunately, rather small electro-optic coefficients. Its application is also complicated by "optical activity", which rotates the plane of polarization of light travelling along the *c*-axis by some 20° per mm. A much better natural crystal is the cubic form of zinc sulphide ("sphalerite") which is not optically active and is isotropic to normal first-order optical effects so that careful alignment is not necessary. However, good natural crystals of adequate size are rare, and synthetic crystals of high optical quality are at present too costly for general applications. "Hexamine" (hexamethylinetetramine) also crystallizes in the cubic $\bar{4}3$ m class, and large crystals of excellent quality are relatively easy to grow, either from solution or from the vapour phase. This material suffers from the disadvantages of mechanical fragility and chemical instability (it is markedly hygroscopic), but may nevertheless have important applications when complete encapsulation is possible.

The ideal crystal for optical modulation remains to be found; most commercially available modulators use either KPD or ADP, which are uniaxial and therefore require accurate alignment of the light beam along the optic axis, but these materials will probably be superseded by ferro-electric crystals as suitable ones become more readily available. A particularly promising crystal is the cubic potassium tantalum niobate (KTN), although this is in fact a centrosymmetric crystal which does not have an intrinsic "Pockels effect". However, terms analogous to the Pockels coefficients can be induced by applying a strong d.c. biasing field, thereby destroying the centre of symmetry, and the crystal then behaves as if it had tetragonal class 4 mm symmetry, with a half-wave voltage of about 50 V for a biasing field of the order of 10^6 V/m.

Bibliography

CADY W. G. (1946) *Piezo-electricity*, Vols. 1 and 2, Dover.

FRANKEN P. A. and WARD J. F. (1963) *Rev. Mod. Phys.* **35**, 23.

GAY P. (1961) Crystal optics, in *Encyclopaedic Dictionary of Physics* (J. Thewlis Ed.), **2**, 219, Oxford: Pergamon Press.

KAMINOW I. P. and TURNER E. H. (1966) *Appl. Optics*, **5**, 1612.

MINCK R. W., TERHUNE R. W. and WANG C. C. (1966) *Appl. Optics*, **5**, 1595.

POCKELS F. (1894) *Abband. der Gesellschaft der Wissen, zu Göttinen*, **39**, 1.

WOOSTER W. A. (1962) Piezoelectricity, in *Encyclopaedic Dictionary of Physics* (J. Thewlis Ed.), **5**, 503, Oxford: Pergamon Press.

H. PURSEY

POLARIZED PROTON TARGETS. Together with mass and electric charge, intrinsic angular momentum (spin) and magnetic moment are fundamental properties of nuclear particles. Electromagnetic interactions among particles, including the dependence on magnetic moment, are well understood. The nuclear interactions and their dependence on spin, however, are far from perfectly known. Thus, scattering experiments in which the scattered particles (beam) and scattering particles (target) are in selected spin states are desirable in order to elucidate the nature of the nuclear interactions. When the selected spin states are such that the net projection of spin along a given direction differs from zero, the particles are said to be spin polarized. The highest degree of spin polarization attainable corresponds to maximum spin projection in a given direction.

Beams of energetic particles with appreciable polarization may be obtained by scattering an accelerated beam from a suitable preliminary target, or by employing low energy deflection techniques to select the spin state of the beam particles prior to acceleration. Targets with a substantial degree of nuclear polarization have been achieved with a variety of nuclei and host materials, but their development is still in an early stage and thus far only one type has been generally employed in nuclear scattering experiments.

In general, a useful target should be in the solid or liquid state, in order to ensure a high density of scattering centers. It is also desirable for the polarized target nuclei to be protons, deuterons or He^3 nuclei, since more complex nuclei compound the difficulties in the analysis of the scattering phenomena. We will confine our discussion here to proton targets.

The maximum degree of polarization attainable by equilibrium or steady state methods for a nuclear spin system depends on the ratio H/T, where H is the magnetic field intensity and T is the Kelvin temperature. Thus, for high nuclear polarization, large values of magnetic field intensity and low temperatures are required. The conceptually simplest procedure for obtaining polarized protons is that of bringing the proton spin system to thermal equilibrium in a magnetic field. Applying the exact expression for the equilibrium polarization of a system of spin $\frac{1}{2}$ particles, such as protons or free electrons, namely $P_{eq} = \tanh \frac{\mu H}{kT}$, where μ is the magnetic moment of the particle and k is the Boltzmann constant, one calculates that for 90 per cent proton polarization, H/T must equal 1.4×10^7 Oe/°K, requiring, for example, a temperature of 0·007°K in a magnetic field of 100 kOe. This method has not yet been utilized for practical targets, since the means of attaining high magnetic fields with *superconducting magnets* and of obtaining low temperatures conveniently and steadily in the millidegree range with He^3–He^4 *dilution refrigerators* have only recently been developed. In addition, spin-lattice relaxation mechanisms which are necessary to achieve the equilibrium (maximum) polarization condition have not been sufficiently studied in the millidegree temperature region. Although the magnetic field intensity and low temperature requirements are technologically severe for equilibrium polarization of nuclei, they are much less so for electrons, since the latter possess a magnetic moment about 1000 times larger than that of nuclei. For 90 per cent polarization of approximately free electrons, an H/T ratio of only $2 \cdot 1 \times 10^4$ Oe/°K is necessary, which is satisfied by a magnetic field and temperature combination of 25 kOe and $1 \cdot 2$°K. These conditions are readily obtainable with an iron core magnet and liquid helium refrigerant. Since dynamic nuclear polarization methods have been developed which permit nuclei under certain circumstances to attain a steady state polarization very close to that of the electrons with which they weakly interact, polarized nuclear targets have been successfully constructed on this basis. Most dynamic polarization methods require the presence of electronic paramagnetic centers and microwave radiation to effect transitions in these centers, as well as restrictive conditions on electronic resonance line width, electronic spin-lattice relaxation times and the relaxation times of the nuclei to be polarized. However, other dynamic polarization methods have been employed to achieve polarized nuclear targets. These include optical pumping for He^3 nuclei and change of orientation of magnetically anisotropic crystals with respect to the applied magnetic field which yields nuclear spin refrigeration and appreciable proton spin polarization.

The first experiment using a polarized proton target was reported from Saclay, France, in 1962. The particular dynamic polarization technique called "solid-effect", involving microwave resonance of electronic paramagnetic centers which are weakly coupled to the nuclei via the dipolar interaction, was used to polarize the protons residing in the water of hydration of a small single crystal of lanthanum magnesium nitrate, frequently abbreviated by LMN. The crystal, whose chemical formula is $La_2Mg_3(NO_3)_{12} \times 24\ H_2O$, was prepared with a fraction of a percent cerium impurity to provide the electronic paramagnetic centers. Shortly thereafter, a relatively large ($\sim$15 cm^3) polarized target, suitable for high energy physics experiments, was constructed at Berkeley, California, using LMN with neodynium impurity in place of the cerium. Subsequently, many laboratories throughout the world have employed similar targets, and polarizations in excess of 70 per cent have been obtained.

In the LMN targets, the weight fraction of "free" protons to bound nucleons (protons and neutrons) is only about 3 per cent. This means that in experiments where scattering from bound and scattering from free nucleons are not easily distinguishable, the effective polarization is much lower than the free proton polarization. In such cases, increased effective polarization would be possible with targets that are richer in free protons. Some of the proton containing materials which have been considered as potentially superior dynamically polarized proton targets are hydrogen fluoride [HF], frozen toluene [$C_6H_5(CH_3)$], alcohol–water mixtures, lithium hydride [LiH], frozen ammonia [NH_3], polyethylene [$(CH_2)_n$], frozen methane [CH_4], hydrogen deuteride [HD], and hydrogen [H_2], in ascending order of free proton weight fraction. In these materials, the electronic paramagnetic centers are introduced either by addition of free radicals or by radiation damage centers. Partial success has been reported with HD ($\sim$1·5 per cent polarization), polyethylene ($\sim$6 per cent polarization), frozen toluene ($\sim$30 per cent polarization) and alcohol–water mixtures ($\sim$35 per cent) but as of the present, none of these is superior to the LMN, although this situation may soon change. In magnetically anisotropic ytterbium doped yttrium ethyl sulfate, the 5 per cent free proton weight fraction has been polarized to 35 per cent by the nuclear spin refrigeration method, and it is almost competitive with LMN as a proton target.

Some of the above materials are also under consideration as candidates for polarized proton targets without dynamic nuclear polarization, since the technology of millidegree temperatures and superconducting magnets has come of age. A few of the problems inherent in the direct equilibrium nuclear polarization targets are the effectiveness of the relaxation mechanisms at very low temperatures, the size and accessibility of the polarized proton sample (enclosed in a fairly elaborate refrigerator) to the high energy particle beam, and beam induced heating effects, which are more serious at the very low temperatures. A method utilizing a relaxation-mechanism-switch with an HD target has been proposed which overcomes most of the above difficulties. High polarizations would be achieved at millidegree temperatures under conditions of an effective spin-lattice relaxation mechanism, whereas the nuclear scattering experiment would be performed with the target at the higher temperatures of ordinary liquid helium after the efficient relaxation mechanism was switched off, leaving the initial high spin polarization metastably frozen-in. Such a switching concept may be applicable to other materials also.

Once polarized proton targets have been produced, the degree of polarization must be accurately measured. Some of the techniques used for this purpose include measurement of the nuclear magnetic resonance signal intensity, direct measurement of the external magnetic field produced by the polarized protons, measurement of the shift in the electronic resonance frequency due to the local magnetic field produced at the location of the electrons by the polarized nuclei, and calibration with a known polarization dependent nuclear scattering process. Other problems of special concern involve degeneration of the polarized targets in time as the incident high energy beam produces radiation damage and modifies the relaxation properties and possibly the obtainable polarization.

Production and use of polarized proton targets involve cryogenics, magnetic fields, microwave and radio-frequency resonance, all trademarks of the very low energy physicist. It is noteworthy that these targets provide a unique tool for investigation of the highest energy processes within our control.

Bibliography

BARKER W. A. Dynamic nuclear polarization, in *Encyclopaedic Dictionary of Physics* (J. Thewlis Ed.), Suppl. vol. 2, 64, Oxford: Pergamon Press.

BARKER W. A. Nuclear polarization, in *Encyclopaedic Dictionary of Physics* (J. Thewlis Ed.), Suppl. vol. 2, 206, Oxford: Pergamon Press.

SHAPIRO GILBERT (1965) Polarized Targets, *Progress in Nuclear Techniques and Instrumentation*, 1, 173–220, North Holland: Amsterdam.

Polarized targets and ion sources, *Proc. Int. Conf. on Polarized Targets and Ion Sources*, Saclay, France, Dec. 5–9, 1966. Edited by La Direction de la Physique, Centre d'Etudes Nucleaires de Saclay, Boite Postale 2, 91 Gif-sur-Yvette, France.

A. HONIG

POLYMER PHYSICS.

The Physical Nature of Polymers

By current convention, polymers are defined as substances whose molecules contain large numbers of atoms—in the simplest view, the number of atoms may be indefinitely large. In thermodynamic terms, molecular weight and molecular length become extensive properties. In addition to the obvious chemical implications, this feature of polymers brings about a fundamental physical difference by introducing a new level of structure. In physics, we are used to dealing with the following levels:

fundamental particles
↓
atoms
↓
molecules
↓
equilibrium phases
↓
macroscopic structures.

But in polymers, we introduce the additional level:

atoms
↓
basic units of polymer chain or network
↓
whole polymer molecule.

As a result, the physical situation is different. Individual polymer molecules can take up many conformations; it is rational to discuss the statistical mechanics of a single molecule and to define thermodynamic parameters, like entropy, for a single molecule; single molecules can exist in different phases. The mere introduction of this additional level of structure would change the physics of the material. But the situation is further complicated because the levels of structure do not remain distinct. If we follow some paths in a polymer material over lengths of 1000 Å or more, the atoms will be linked by covalent bonds; in other directions we shall find intermolecular forces; and in the course of either of these paths, we may pass through separate phases and structural features. The intra-molecular, the inter-molecular, and the macroscopic all merge together. Theoretical polymer physics is concerned with finding ways of analysing these situations and their formation.

Experimental polymer physics deals with the observation of the properties of polymeric materials and with the investigation of polymer structures.

Although, in a sense, polymer physics can be said to have begun with such studies as the experimental work on thermal and mechanical effects in rubber by Gough (1805) and Joule (1859) and the thermodynamic analysis by Kelvin (1857), it did not really exist until chemists had established and accepted the molecular character of polymers in the 1920's. Previously it had been thought that these materials were mere colloidal associations of typical small molecules.

Early physical studies then demonstrated by means of X-ray diffraction the crystallization of polymer chains; established a reasonable model of the fine structure of partially crystalline fibres and bulk polymers; and for amorphous polymers, developed basic theories of rubber elasticity and of the transition from the rubbery to the glassy state. The subject developed steadily during the 1930's and 1940's, and then more rapidly in the 1950's. Journals devoted specifically to polymer physics have appeared in the 1960's.

The Physical Classification of Polymers

The first physical distinction between different classes of polymers is concerned with the basic topology of the molecules. The molecule may be capable of representation as a three-dimensional, two-dimensional or a one-dimensional system, in contrast to the zero-dimensional form of small molecules; and the arrangement of the atoms may be regular or irregular.

Diamond is a regular three-dimensional polymer: extremely hard, rigid, and infusible, and with special optical properties. The study of such materials can also be regarded as a branch of crystal physics, and has developed independently of the rest of polymer science.

Thermoset plastics (such as phenol-formaldehyde or polyester resins) are irregular three-dimensional networks. They are formed from multi-functional components which react together to give a network structure on the action of heat or a catalyst. These materials are hard, insoluble, and infusible. Once a structure has been formed it cannot be re-formed: any major change disrupts the molecule. These materials are intractable both to experimental study and to theoretical analysis, and, apart from some empirical measurement of properties, their physics has not been much studied.

Graphite is a two-dimensional polymer which occurs in a regular crystalline form; and, in some forms of carbon, irregularly distorted graphitic sheets may occur. But, in general, two-dimensional polymers are rare. They will be formed only when a multi-functional component reacts in such a way that its branches remain (in a topological sense) in single planes. They demonstrate characteristic features of slip, cleavage, and the rolling-up of planes.

On the other hand, one-dimensional polymers are common. They are formed whenever a bifunctional component reacts to form a chain. The study of linear polymers has been the main activity in polymer physics, and the rest of this article will discuss their behaviour. Some systems in which there is a small amount of chain branching or cross-linking can be included as modified forms of linear polymers.

The principal forms in which solid linear polymers can exist are rubbery, with a disordered assembly of flexible chains; glassy, with a disordered assembly which is rigid either because the chains are stiff or because they are firmly attracted to one another; and crystalline in which, at least partially, the molecules are in a regular crystal lattice.

The States of a Single Polymer Molecule

In order to bring out some of the main features, it is worth considering the behaviour of a single linear molecule. This could be regarded as existing in free space, but having just enough interaction with its surroundings to take up a defined temperature.

We can see three main types of freedom which may be possible in the system. Firstly, bond lengths may change, but as these depend on the interaction between two neighbouring atoms, only small elastic deformations are likely—large deformation would rupture the chain. Secondly the angles between bonds may alter: this deformation is also difficult, though not quite as difficult as change of bond lengths, since it depends on the relative positions of three chain atoms. Thirdly, a bond may rotate in a cone around the chain direction, without change of length or angle: since we have to bring in the relative positions of four chain atoms to define this rotation angle it is obvious that the interactions will be weaker and that this will be the softest mode of chain deformation.

If the energy barriers to rotation are low (or the temperature, determining the energy of vibration, is high) then the chain will have a high degree of flexibility. Consequently a free chain will take up a random conformation of maximum entropy. Figure 1 shows a calculated distribution of lengths between chain ends. This is based on random walk theory and is a Gaussian distribution. Figure 2 shows a model of a single molecule.

When a chain is extended, it has a more ordered conformation, its entropy is reduced, and hence work must be done in order to raise the free energy. Because of the ease of rotation, the modulus of the chain will be very low. By considering the statistical thermodynamics of the degree of alignment of individual links, we get an equation, analogous to that found by Langevin for the alignment of dipoles:

$$f = (kT/a)\,\mathscr{L}^{-1}(r/R)$$
$$= (kT/a)\,[3(r/R) + (9/5)(r/R)^3 + \cdots]$$

where F is the time-average force needed to hold the chain at a length r,

R is extended length of chain,
a is length of an effective free link,
T is absolute temperature,
k is Boltzmann's constant,

and $\mathscr{L}^{-1}$ is the inverse of the Langevin function,

$$\mathscr{L}(x) = (\coth x - x^{-1}).$$

We note that the origin of the chain tension is entirely entropic, and hence the behaviour of an ideal

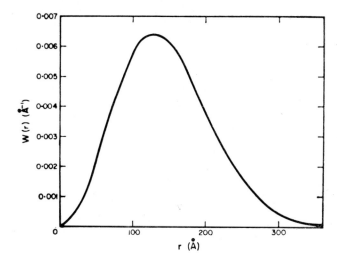

Fig. 1. Distribution function for end-to-end distances for a polyethylene chain 2500 Å long. $W(r)\,dr$ is the probability of finding a chain in the length interval between r and $(r + dr)$.

rubbery chain has many analogies with the behaviour of an ideal gas. Corrections must be made to allow for the differences in energy of different conformations. An alternative approach to the whole problem is given by Russian workers who consider the transitions between specific preferred isomeric forms.

When the barriers to rotation are high the chain will be rigid and will normally have an irregular conformation. Any given chain segment will have a single form, which can suffer elastic deformation with a

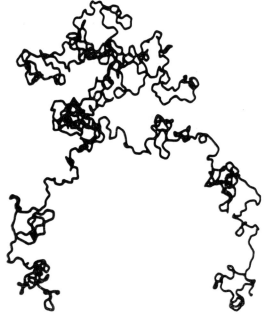

Fig. 2. Model of a single chain (after Treloar).

relatively high modulus, or, under larger forces, will deform plastically to take up a new conformation. At a certain temperature, when the energy of vibrations is high enough, the chain will change from the glassy to the rubbery state; the change would show a fairly high degree of co-operation and so be relatively sharp with the character of a second order transition. The temperature at which the charge occurs will be influenced by the time-scale of the experiment: if the rate of jumping barriers is too slow, the chain will appear rigid.

If a certain conformation has an energy minimum at a much lower level than the minima corresponding to other conformations, the chain will take up a specific regular form at low temperatures, and will become a rigid rod-like molecule.

So far the "intermolecular" forces existing between chain elements remotely separated along the chain (and which might equally be in different molecules) have not been mentioned. These intermolecular forces will cause the chain to take up a condensed form with a volume determined by the usual balance of attractive and repulsive forces. We may thus get forms which can be regarded as rubbery liquids or condensed glassy chains.

If the chain is sufficiently regular in form and by virtue of shape or chemical constitution can pack to give strong specifically localized, attractive forces, neighbouring elements will tend to pack together in a crystal lattice. A single molecule can only crystallize by folding back and forth, and it is possible to calculate the equilibrium fold length, at which the lowering of energy due to satisfying attractive forces is just balanced by the added energy where the chains are distorted in the fold.

In addition to this rather arbitrary folding, attractive forces can also be satisfied by the chain taking up specific forms. This is particularly likely where there are rather strong forces due to hydrogen bonds. For example, a helix may be stabilized by hydrogen bonds between the turns. Some biological molecules, such as haemoglobin, are so constituted that they take up specific globular forms.

Polymers in Solution and Melt

Isolated chains in space cannot be studied, but we can get chains dispersed in solution. Theoretical studies concentrate on the range and probability of different conformations and on the interactions with the solvent and between polymer molecules. Predictions of the size and shape of the conformations can be checked by experimental studies using methods such as sedimentation in an ultracentrifuge, viscosity, and light scattering.

In the molten state, the conformations which can be taken up are restricted. Knowledge of the actual structure of a melt is limited, as with all liquid systems. It is to be expected that viscous flow of molten polymers or concentrated solutions will show anomalous effects, since forces can be transmitted along single chains over appreciable distances.

Rubber Elasticity

An ideal rubber may be defined as an assembly of flexible chains with very weak interactions between chains, except for a low concentration of cross-links needed to prevent the chains just sliding past one another as in a liquid. To some extent chain entanglements function as temporary cross-links, giving rise to elastic behaviour for fast changes but viscous behaviour for slow changes.

Because of the ease with which the chains can extend by altering their conformation, changes of shape are easy; but, as in any condensed system, volume changes will be difficult. In real rubbers (for instance, lightly vulcanized natural rubbers), this behaviour is observed: the bulk modulus, with a value of over 100 g/mm^2, is almost 10^4 times greater than the shear or Young's moduli. By taking up new conformations, the chains can also extend by large amounts without rupture: breaking extensions in the range 500 to 1000 per cent are observed in rubbers.

The simple theory of rubber elasticity is based on a determination of the change of free energy on affine deformation of a cross-linked network of chains, each of which follows the equation for a single chain quoted above. This theory gives a good prediction both of the absolute level and the non-linearity of form of stress–strain curves (Fig. 3). Later developments have

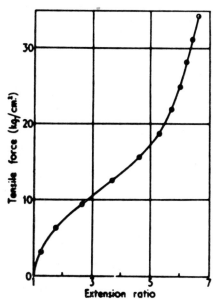

Fig. 3. *Typical rubber stress–strain curve.*

paid attention to the significance of small volume changes and of the contribution of internal energy changes.

Thermodynamically, we have, for an ideal rubber in which there are no significant internal energy changes:

$$\text{work done in extension} = F\,dL$$
$$= \text{change in free energy} = -T\,dS$$

Using a relation analogous to one of Mawell's relations, we find

$$F = -T(\partial S/\partial L)_T = T(\partial F/\partial T)_L$$

The latter differential coefficient, representing the change in tension with temperature at constant length, can be measured experimentally. Slight deviations from the above relation are evidence that small energy changes are involved in the deformation of real rubbers.

Other thermodynamic effects can also be studied. Because the entropy decreases on extension, it follows that heat will be evolved on extension and that stretched rubber will contract on heating. At very low strains, the effect is reversed (at the thermoelastic inversion point) because of the ordinary volume expansion with rise in temperature.

Polymers in the Glassy State

In contrast to rubbers, glassy plastics like polystyrene, equally devoid of any regularity of structure, have a much higher modulus. A typical value would be 200 g/mm², and, as with other rigid solids, all the moduli are of the same order of magnitude. This rigidity of structure is due either to the inherent stiffness of the chains or to the presence of strong attractive forces which inhibit the movement of chains past one another.

Under high loads, glassy plastics will show, at strains of a few per cent, either brittle fracture, if excessive chain breakage occurs first, or yield and the start of cold drawing, if energies barriers between chain conformations are surmounted. The detailed behaviour will depend both on the molecular and morphological character and the stress distribution in the system.

The Transition Region and Viscoelasticity

The transition region between rubbery and glassy plastics occurs when the time rate of jumps between chain conformations becomes comparable with experimental times. At longer times (or higher temperatures) the jumps are effectively instantaneous and the chains take up equilibrium conformations; at shorter times (or lower temperatures) the jumps are very rare, and only small elastic deformations of the chain conformations need to be considered.

The transition region is characterized by a change in the real part of the modulus, a peak in the loss modulus, and by marked creep and stress relaxation behaviour. A fairly extensive empirical theory of viscoelasticity has been developed in order to rationalize the various observed time-dependent features of mechanical behaviour. In addition to the effects of rate at a given temperature, it is also possible to bring in the effects of change of temperature. A rise in temperature has the same effect as a reduction of frequency. For instance, a particular transition will occur at a higher frequency at a higher temperature, so that a displacement to one side of the ridge on an energy loss diagram can be made by increasing temperature or lowering frequency; and vice versa for displacement to the other side. Often the effects of time and temperature can be interchanged by the use of time–temperature superposition, which involves shifting results obtained under different conditions on to a common curve against an axis representing a combination of time and temperature. This is most notably done by the use of the W–L–F (Williams, Landel and Ferry) equation, which, in one form, is given as:

$$\log\left(\frac{\eta_T}{\eta_{T_g}}\right) = -\frac{C_1(T - T_g)}{C_2 + T - T_g}$$

where η_T is the coefficient of the viscous part of the deformation at any temperature T up to 100° above the glass transition temperature, and C_1 and C_2 are constants which were thought to have the universal values of 17·44 and 51·6 but, in fact, vary somewhat in different materials.

In addition to the mechanical changes at the glass transition temperature there are also changes in specific heat, coefficient of expansion, and some other quantities. The change has the character of a second-order transition, without latent heat or volume change, since it involves no change in structure, but only a change in the response of the structure. However, the transition is not always sharp, though whether this is due to its fundamental character or to the effects of local variations in structure is not clear.

The various regions of visco-elastic behaviour in a simple polymer system are illustrated in Fig. 4. Some polymers show multiple transitions: nylon, for example, shows two low temperature transitions associated with chain flexibility and a transition above room temperature due to disassociation of intermolecular hydrogen bonds.

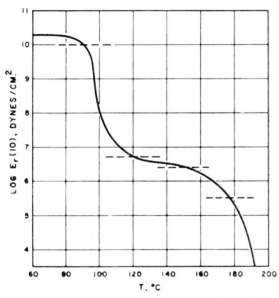

Fig. 4. Transitions in polystyrene as shown by changes in modulus. The relaxation modulus $E_r(10)$ is the ratio of stress to strain, taken 10 sec after the application of a fixed strain (after Tobolsky).

Crystallization

Another major area of polymer physics is the study of crystalline polymers. It used to be thought that polymer chains could never sort themselves out so as to give perfect crystals, and, for a long time, the only evidence of crystallinity came from X-ray diffraction studies of fibres and bulk polymers. A typical view of the fine structure was the fringed micelle model illustrated in Fig. 5.

Then, in 1955, single crystals of polyethylene were prepared from dilute solution. Since then there has been an enormous amount of work on polymer crystallization, covering both morphological and kinetic investigations. Single crystals are sometimes flat sheets, but often they are hollow pyramids with steps consisting of layers spiralling round the pyramid. The layer thickness is typically about 100 Å, and since the chain alignment is perpendicular to the layers, this can only be explained by chain folding. Theoretically this is reasonable, since a folded form is not only the equilibrium minimum energy form for a single chain in appropriate circumstances, but also for a single chain crystallizing on a substrate. Except with short chains, it is impossible to get to the true equilibrium state with all chains extended, because of the difficulty of bringing all the chains together in the extended form; once the chains start to fold up, they get locked in an effectively stable crystal.

Crystallization in bulk shows a characteristic spherulitic structure. From a central nucleus, a sheaf-like form develops by branching; and then, as branching continues, growth becomes equal in all directions. It has been shown that branching, which brings this about, occurs when crystals grow to a size greater than a critical length D/G, where G is a measure of crystal growth rate and D of diffusion rate. Growth tends to denude the neighbouring region of the crystallizing material, and if this is not evened out by diffusion over the whole face of the crystal, another growing point forms in a remote position of higher local concentration. We may note that the fringed micelle structure may be regarded as a limiting form of spherulitic crystallization when the density of nuclei is very high.

Crystallization in bulk polymer systems can be regarded as occurring in two stages. First there is a relatively rapid growth of crystals through the material, collecting chains locally together into crystal lattices. The difficulty of sorting out the chains, and the fact that the same chain may be caught in more than one place means that crystallization cannot be complete. Some further secondary crystallization occurs more slowly, but at the end an appreciable degree of disorder remains.

In spherulitic crystallization, it is common to find evidence of crystalline lamellae stacked on top of one another. In other systems, particularly natural fibres, fibrillar crystals are found. In these real systems, crystallization occurs either rapidly under commercial conditions or in a specific biological environment in natural systems. Very complicated structures may result. These are all metastable forms, occupying particular energy minima among the thousands which can be regarded as existing in a multi-dimensional landscape: it is easy to change from one metastable state to another by force or annealing. The difference from other substances, is that metastable states do not occur only at the level of macroscopic structure, they exist also at the molecular level.

Because of this complexity there have been many views of the fine structure of bulk crystalline polymers

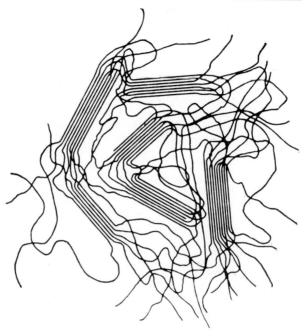

Fig. 5. Fringed micelle structure.

and fibres. These include continuous forms such as an amorphous structure with some correlation between neighbouring chains, or a paracrystalline structure; structures containing crystal defects; and structures composed of a mixture of crystalline and non-crystalline regions. It is possible, however, to view all these models as useful idealizations, and to regard the range of actual structures as covering a continuous region of varying average degrees of order, of varying degrees of localization of order and disorder, and of varying ratios of length to width in ordered regions—to name but three of the thousands of parameters needed to specify the structure completely.

In an isotropic form, and above their second-order transition temperature, partially crystalline polymers display a lower modulus than glassy polymers and are more extensible and tougher. They can be drawn out to give oriented forms, stabilized by crystallization, which display high birefringence, high strength in the direction of orientation, good elasticity, toughness, and breaking extensions of the order of 20 per cent. The properties are very sensitive to the exact degree of orientation.

The understanding of deformation mechanisms is complicated because of the many modes of deformation. These include elastic (spontaneously recoverable) modes such as bond length and angle changes in crystal lattices, glassy chain segments and highly cross-linked networks; phase transitions such as the $\alpha \to \beta$ (helix $\to$ extended chain) change in protein crystals; and elasticity of rubbery chain segments. There are also plastic modes associated with recrystallization, chain unfolding and refolding, and movement of crystal defects; and with molecular effects such as change of glassy chain conformation, slip at chain ends, chain breakage, and cross-link breakage. The art in trying to understand any particular aspect of the subject is to choose a reasonable model of structure and to isolate the most important modes of deformation. Progress is being made.

Other Aspects of Polymer Physics

In this article it has only been possible to indicate some of the main features of polymer physics. There are many other branches. The study of structure makes use of many specialized techniques: microscopy in all its forms; X-ray diffraction; nuclear magnetic resonance; and the response to various electromagnetic or mechanical vibrations.

Other properties of importance are density, which is a measure of molecular packing; the absorption and transport of small molecules taken up by the polymer; the whole range of mechanical properties; dielectric constant, electrical conductivity, static electrification; thermal properties; optical properties; friction; the destructive or constructive effects of irradiation and so on.

The understanding of all these phenomena is helped when the fundamental nature of polymer systems is appreciated: the merging of the molecular and macroscopic; the common occurrence of metastable states, with the possibility of a rate dependent shift from one to another; the importance of effects within chain segments and along the length of molecules.

The number of parameters needed to describe the state of a polymer system strictly is reflected in an attempt to apply the phase rule, $P = C - F + 2$, where P is the number of phases in equilibrium, C is the number of components and F is the number of degrees of freedom. In a typical polymer system it would be common to find degrees of polymerization ranging from a low value up to several thousand. Since each molecular length is a different component, C would be several thousand! Fortunately polymer physicists have found ways of simplifying the theoretical treatment of the problem.

Definitions

Affine deformation: a deformation in which the change of shape of any small element of the structure is similar to the change of shape of the whole structure. (page 334)

Real modulus: the ratio of the in-phase component of stress to an alternating strain. (page 334)

Loss modulus: the ratio of the out-of-phase component of stress to an alternating strain. (page 334)

Degree of polymerization: the number of repeating units making up a whole polymer molecule. (page 337)

Bibliography

BIRSHTEIN T. M. and PTITSYN O. B. (1966) *Conformation of Macromolecules*, New York: Interscience.

BUECHE F. (1962) *Physical Properties of Polymers*, New York: Interscience.

FERRY J. D. (1961) *Viscoelastic Properties of Polymers*, New York: Wiley.

LINDENMEYER P. H. (Ed.) (1968) *Supramolecular Structure of Fibres*, New York: Interscience.

MORTON W. E. and HEARLE J. W. S. (1962) *Physical Properties of Textile Fibres*, London: Butterworths.

NIELSEN L. E. (1962) *Mechanical Properties of Polymers*, New York: Reinhold.

RITCHIE P. D. (Ed.) (1965) *Physics of Plastics*, London: Iliffe.

TOBOLSKY A. V. (1960) *Properties and Structure of Polymers*, New York: Wiley.

TRELOAR L. R. G. (1958) *The Physics of Rubber Elasticity*, Oxford: O.U.P.

J. W. S. HEARLE

PROTEINS, STRUCTURE OF. Proteins form a large part of all living organisms from mycoplasmas and bacteria through plants to mammals and man, and their function varies from providing a scaffolding or support for the organism to comprising the intricate metabolic machinery characteristic of life. The significance of the *structure* of proteins is that it is this feature of the molecules which determines and limits their biological *function*. In this article we shall describe the present state of knowledge about the molecular structure of proteins and discuss what generalizations seem to be possible concerning the factors which determine and stabilize protein structure.

In common with many biological molecules the proteins are polymers, their characteristic monomers being amino-acids. An amino-acid has the general formula

$$H_2N-\underset{\underset{R}{|}}{\overset{\overset{H}{|}}{C}}-COOH$$

in which different amino acids vary only in their R-group and polymerization starts when the $-COOH$ group of one amino acid links with the $-NH_2$ of another forming the *peptide bond* and expelling water

$$-\underset{\underset{H}{|}}{N}-\underset{\underset{R}{|}}{\overset{\overset{H}{|}}{C}}-\underset{\|}{\overset{}{C}}-\underset{\underset{H}{|}}{N}-\underset{\underset{R}{|}}{\overset{\overset{H}{|}}{C}}-\underset{\|}{\overset{}{C}}-$$
$$+H_2O$$

Proteins are composed of long chains of amino acids linked by peptide bonds and the R-groups or side-chains branch out at regular intervals from the main chain. When the number of monomers in the chain is not very large (less than about 50) the molecule is referred to as a polypeptide and only the larger molecules are strictly termed proteins although their main chains are frequently called polypeptide chains; this account of structure will apply to both proteins and polypeptides even when the latter have been artificially synthesized, and indeed a lot of structural information has been obtained from the simple, regular synthetic polypeptides. In nature there are 20 different types of sidechain ranging in molecular weight from 1 to about 130; more significantly they may be classified into non-polar and polar side-chains and the latter subdivided into acidic, basic and hydrogen bond forming groups.

The term structure may be applied to proteins in at least four different ways which are generally understood as follows: The *primary structure* refers to the order or sequence of amino acids along the main chain.

The next two types of structure involve a specification of the three-dimensional disposition of the whole molecule at an atomic level. This is most completely recorded by a list of (x, y, z) coordinates of all the atoms with respect to some selected set of axes. When the arrangement of the main chain is determined by hydrogen bonds this is referred to as *secondary structure* and when by interaction between side-chains (by means of ionic bonds, salt links, covalent bonds or hydrophobic bonds) it is termed *tertiary structure*. This division into secondary and tertiary structure was made a number of years ago and since then accumulating information has made these particular criteria less significant. In lysozyme, for example, hydrogen bonds do play a part in determining tertiary structure; there is a case for regarding the secondary–tertiary division as unhelpful and placing them together in one category, namely, that which describes the three-dimensional disposition of the atoms in the whole protein molecule. The secondary–tertiary division may still be useful, however, if secondary structures are taken as those which have regular relative orientations of neighbouring amino-acid units (i.e. a common value of (ϕ, ψ)—see later) while tertiary structures are those with little or no regularity in inter-amino-acid orientation though tertiary structures may contain short sections of secondary structure.

Quaternary structure is produced when complete polypeptide chains themselves aggregate into still higher levels of organization as in viruses, muscles and cell spindles.

Two other terms remain to be defined. When one molecular structure is connected to another without breaking any covalent bonds the structures are said to have different *conformations*; when it is necessary to break a covalent bond in the transition the structures are said to have different *configurations*.

Proteins may be classified on the basis of function into two broad groups—the fibrous or structure proteins and the globular proteins. The fibrous proteins are those which form a supporting skeleton for the organism or its parts and are exemplified by the silks, collagen of tendon and keratin of hair, wool and feather, while the globular proteins comprise the molecular machinery of the cell such as the enzymes, haemoglobin, etc. In some proteins these two functions are combined as in myosin of muscle which acts in both enzymatic and structural capacities and the division must be regarded as didactic rather than absolute. However, this classification of proteins on the basis of function is also reflected in the molecular structure. In terms of primary structure the fibrous proteins are often formed by the repetition of a small group of amino-acids (two or three) all along the main chain. In some silks, for example, it is -(ala-gly)$_n$ and in collagen -(gly-pro-hypro)$_n$; synthetic polypeptides with these sequences give X-ray diffraction patterns closely similar to the patterns obtained with native silk or collagen thus showing that the structures are similar. Fibrous protein molecules are usually long and rod-like. In the globular proteins no type of regularity has been observed in the order of occurrence of amino acids along the main chain. The main chain usually folds up in what appears to be an arbitrary fashion (though since one particular arrangement is generally energetically favourable, molecules of the same protein normally have identical structures) and the overall shape of the molecule is roughly spherical or ellipsoidal.

One other important structural feature about the fibrous proteins is that the spatial relationship between any two neighbouring amino acids in the main chain is the same all along the chain; in globular proteins this relationship is generally variable except for occasional short regular sections. The spatial relationship between two neighbouring amino acids is illustrated in Fig. 1. Since the peptide group (—CO—NH—) is known to be planar $w = 0$ so only two angles ϕ and ψ are required to define the relationship. In fibrous structures one pair of values of ϕ and ψ is normally sufficient to describe the whole structure

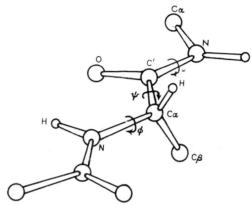

Fig. 1. Angles defining the spatial relationships of two neighbouring amino acids in a polypeptide chain.

(apart from the side-chains) but in globular proteins each pair of amino acids has its individual (ϕ, ψ) value.

Because of the regularity in fibrous protein structure many of the standard configurations were first recognized in them so we shall now consider the various groups into which fibrous protein structures fall and follow this by examining how these express themselves in the structure of globular proteins.

1. The Fibrous Proteins

The three main classes of fibrous protein—having the α-structure, the β-structure and the collagen structure respectively—were first recognized by their characteristic X-ray diffraction patterns and so when describing these structures we shall always relate molecular geometry to the corresponding X-ray reflections. Some terms, used in discussing X-ray patterns from proteins are defined at the end of this article.

A. α-structures

(a) *α-polypeptides.* The standard configuration known as the α-helix was determined by building molecular models which were consistent not only with the X-ray patterns but also with as much known stereochemical information as possible (e.g. that the peptide group is planar) as possible. The α-helix may be fully described in terms of Fig. 1 as having $\phi = 133°$ and $\psi = 122°$. More helpfully, it is a helix having 18 amino-acid residues in 5 turns of the helix; the axially projected separation of the amino-acid residues is 1·5 Å, thus giving a helix pitch of 5·4 Å. The true repeat of the structure is, of course, after 5 turns, i.e. 27 Å. The diagnostic X-ray reflections are at 1·5 Å on the meridian, 5·4 Å off meridional and occasionally a few lower-angle reflections indexing on 27 Å.

A number of homo-polypeptides adopt the α-helix structure such as poly-L-alamine, poly-γ-methyl-L-glutamate, poly-L-tyrosine, etc., and it has also been found in mixed sequential polypeptides. Studies of synthetic sequential polypeptides have indicated that bulky side-chains may cause the α-helix to break down by stereochemical overcrowding and the distorting interactions are frequently not between adjacent side-chains but between side-chains separated by three or four residues on the main chain which the α-helical geometry (3·6 residues per turn) would bring into close proximity. There are inter-chain hydrogen bonds in the α-helix which run parallel to the helix axis and link one —C—O group in the main chain with an —N—H group three residues further along the chain.

The α-polypeptides may be converted to β-polypeptides by stretching and to random coils by solvents which disrupt the α-helix (see discussion of stability of structures).

(b) *α-proteins*. These are a little more complicated than the α-polypeptides in that they are formed by two α-helices coiling round each other to form a coiled-coil of pitch around 200 Å. This super-coiling may be regarded as a tertiary structure and is probably due to attractive interaction between non-polar side-chains. The 1·5 Å X-ray reflection from the α-helix now occurs at 1·48 Å due to tilting the molecule in coiling, and for similar reasons the 5·4 Å offmeridional reflection becomes meridional at 5·1 Å. Equatorial and near-equatorial reflections at a distance corresponding to 10 Å from the meridian are generated by the super-coil.

The α-protein structure occurs in α-keratin, epidermin, muscle and fibrinogen (the *k-e-m-f* group) and it is possible to discern some relation between the molecular structure and the function required by the organ which it comprises. The fact that the chemical bonds which are parallel to the molecular (and fibre) axis are the weak hydrogen bonds, means that the structures have some degree of flexibility as is required for muscles—in the α-keratin structure the molecules cluster into rings of nine (quaternary structure) called filaments of about 70 Å diameter with some 100 Å between filament centres and the space between the filaments is filled with a protein of random coil structure which makes the whole assembly waterproof and gives the flexible fibrils added toughness.

α-proteins may be converted to β-proteins by stretching, normally in some extreme conditions such as steam.

B. *β-structures*

(a) *β-polypeptides*. In contrast to the α-helix which is rod-like, the β-polypeptides have a sheet structure. The polypeptide chains instead of coiling are in zig-zag extended form with $\phi = 38°$ and $\psi = 325°$. The —C=O and —NH groups of the main chain form intra-chain hydrogen bonds with the —NH and —C=O groups of neighbouring chains thus forming a pleated sheet of chains (secondary structure) held together by hydrogen bonds at right angles to the main chain direction. The side chains stick out above and below the sheet at right angles to it so that the structure may be completed in a third dimension by an appropriate packing of the sheets (tertiary structure) determined by the side chains. Neighbouring chains in a sheet run both parallel and anti-parallel to one another giving a random arrangement of chain sense within a sheet. The X-ray reflections characteristic of β-polypeptides lie on a series of layer-lines indexing on 7·00 Å with the first meridional reflection on the second layer-line. This means that the axially projected distance between amino-acid residues is 3·5 Å but, since the main chain has a glide-plane of symmetry, the true repeat is 7·0 Å.

The inter-chain spacing in a sheet is 4·7 Å and since it is determined by hydrogen bonding, is fairly constant—the inter-sheet distance, however, depends on the size of the amino-acid side-chains and can vary between 3·5 and 15·0 Å. Reflections corresponding to these two spacings appear on the equator of X-ray patterns.

The β-pattern is given by quite a number of homo-polypeptides such as poly-*L*-alanine, polyglycine I and poly-S-carbobenzoxy-*L*-cysteine.

(b) *β-proteins*. There are a few different types of β-protein structure.

(i) *Silks*. Two of the silks have been studied in detail— *Bombyx mori* and Tussah silk. They are both made up of anti-parallel pleated sheets but differ slightly in their mode of packing of the sheets. In Tussah all the sheets are 5·3 Å apart but in *Bombyx mori* the amino-acid sequence is (ala-gly) for a considerable fraction of the main chain and this means that on one side of the chain there are mainly gly residues whereas on the other side ala predominates. The sheets then pack with their gly sides together (giving a sheet separation of 3·5 Å) and their alanine sides together (giving a sheet separation of 5·7 Å), making a total repeating unit in the inter-sheet direction of 9·2 Å.

(ii) *Cross β*. This structure is found in the egg-stalk silk of a lace wing fly and the X-ray diagram gives reflections of similar spacing to standard β-structures but the main chain repeat reflection occurs on the equator and the inter-chain hydrogen bond repeat is on the meridian. This suggests that the main chain runs perpendicular to the fibre axis and that they are hydrogen-bonded together in the fibre axis direction. Other evidence indicates that the main chain is in short straight sections which double back on themselves after about every eight residues.

(iii) *β-keratin*. This structure is given when α-proteins are stretched. The detailed molecular arrangement is still a matter of discussion though it is fairly well established that it contains β-pleated sheets within which the chain direction is random.

(iv) *Feather keratin*. α-keratin occurs mainly in mammals while feather keratin is the keratin of reptiles and bird forming beak, claw, horn, hoof, skin (of lizards and crocodiles) and feather. There are still many uncertainties about its detailed molecular structure. In the electron microscope filaments of 30 Å diameter are visible embedded in a matrix of apparently amorphous protein. Two main structures have been suggested for the filaments; one involves crystallites of β-pleated sheet forming a helix of pitch 95 Å with 4 residues per turn and the other is a three-strand coiled-coil (cf. collagen) of pitch 71 Å. Both of these models account for the available X-ray data and further work is clearly required.

The β-pleated sheet with the weakest bonds along the fibre axis being covalent (cf. hydrogen bonds in the α-structures), forms a very inextensible framework and this property is exploited in the insect world and elsewhere in the silks. When such molecular structures are supported in a waterproof matrix as in feather keratin, it provides an ideal component for epidermal tissue.

C. The collagen structure

The characteristic X-ray pattern from collagen has a reflection on the meridian at 2·85 Å, two layer-lines at 9·5 and 4·0 Å, a diffuse equatorial and near-equatorial group of reflections around 4·5 Å and a moisture sensitive equatorial reflection at 11·0 Å. There is, in addition, a series of low-angle meridional reflections which index on approximately 640 Å though this value may also be slightly moisture sensitive. Studies on primary structure indicate that probably every third residue is gly and sequences of the type (gly-pro-hypro) and (gly-pro-R) are common. A synthetic sequential polypeptide of sequence (gly-pro-hypro) gives a similar high-angle X-ray pattern to collagen.

The low-angle X-ray pattern along with electron microscope evidence showed that the so-called tropocollagen molecule was 2600 Å long and about 15 Å in diameter, the 640 Å reflection being generated because neighbouring molecules are staggered by this amount in the quaternary structure, but the details of this longitudinal packing and also the manner of transverse packing have not yet been unequivocally determined.

The high-angle X-ray pattern is accounted for by a triple-strand coiled-coil where the uncoiled minor helices are left-handed, about 2 Å in diameter and have a three-fold screw axis with a 2·91 Å rise per residue. To form the major helix three of these minor helices may be regarded as grouped together 4·5 Å apart or 2·6 Å from a central axis and then the bundle given a right-handed twist to make one complete turn after 30 residues. The projected rise per residue in the major helix then becomes 2·86 Å which generates a meridional X-ray reflection of this spacing and the pitch of the minor helix is 9·5 Å with a new value of $3\frac{1}{3}$ residues per turn. For every three residues in the chain there are one or two hydrogen bonds between the minor helices.

As in the α-proteins this coiled-coil structure with hydrogen bonds approximatey parallel to the molecular axis results in tendon fibres of suitable rigidity and flexibility.

2. The Globular Proteins

Within the last decade complete structure determination, i.e. location of the positions of all the atoms in the molecule, has been done for some six globular proteins, viz. myoglobin, haemoglobin, lysozyme, ribonuclease, chymotrypsin and carboxypeptidase. These studies are known with very much greater precision than the structure of the fibrous proteins because the method of X-ray diffraction becomes more powerful when the specimen examined is more ordered and these globular proteins form the most perfect order possible, namely, crystals. Working out the complete structure of a protein is a major project in X-ray crystallography and involves many man-hours and generally a considerable degree of teamwork. Before such a project is feasible the protein in question must satisfy conditions. First, of course, it must crystallize and the crystals must be of sufficient size (minimum about 0.25×0.1 mm) and regularity; secondly, because of a technical requirement in the X-ray method it is essential that there be at least two (and qreferably more) isomorphous derivatives of the protein which also give satisfactory crystals.

On the basis of the six known molecules it appears possible to make some general statements about protein structure and these are summarized below, but they must be regarded with some caution in view of the vast number of proteins about which there is still no structural knowledge.

(a) The main chain of the molecule follows an irregular course with relatively few straight parts which are mostly either short sections of α-helix or β-pleated sheet. This again emphasizes these standard conformations which are adopted unless prevented by unfavourable stereochemistry or by the demands of other structural or functional criteria (see (b)). The largest amount of α-helix occurs in myoglobin with no β-sheet while carboxypeptidase is largely composed of β-sheet with less α-helix.

(b) The side-chains in the interior of the molecule are almost invariably non-polar while those on the surface may either be polar or non-polar though the former greatly predominates. Also, it appears likely that the non-polar surface patches may be the areas where neighbouring molecules come into contact when quaternary structures are formed. A study of myoglobins from different species indicates that out of more than 140 amino acids only 9 remain invariant; yet even this variability of primary sequence is limited by this principle of non-polar residues only in the interior of the molecule.

(c) As mentioned above there is a degree of variability of primary structures of proteins which have unimpaired function. This degree of variability ranges from that in myoglobin to that in cytochrome c where about 40 per cent of residues have so far been found invariant. Since molecules are selected in nature on the basis of function, the constancy of function is not surprising, but it is significant that it should be preserved throughout a wide range of primary structure.

The tertiary structures are also very constant through a range of different crystal types, crystallizing solvents, etc., and yet the function, of, say, the haem group of haemoglobin is extremely sensitive to small changes in structure. From this it follows that the tertiary structure is thermodynamically stable and function depends on structure. In fact it is possible that a large fraction of the globular protein molecule performs what is essentially a structural task in maintaining a molecular geometry which is favourable for the required biological activity.

(d) Some information may be obtained about the structural mechanism for enzyme action. The active site of lysozyme lies in a cleft in the molecule, this cleft being lined with non-polar residues and one or two strategically placed polar groups which take part in the enzyme action. Studies on crystals of enzyme plus inhibitor show that the inhibitor is held in place by hydrophobic and hydrogen bonds but that covalent bonds are not involved and that slight (1–2 Å) movement of some of the structure occurs on binding. Model building studies reveal that the substrate may similarly be held and hydrolysis of the polysaccharide chain could take place by charge transfer from the polar groups in the cleft.

Stabilization of Protein Structures

Information from known protein structures shows that the important forces involved in the maintenance of structures are provided by the cooperative action of many weak bonds—van der Waals attractions are of first importance followed by hydrogen bonds. Covalent bonds play only a small part and the role of ionic bonds appears to be minimal.

The prime importance of van der Waals interactions has been demonstrated as follows. On a set of orthogonal axes with coordinates ϕ and ψ (Fig. 1) Waals the energy of various relative dispositions of adjacent side chains may be represented by contour lines. If the energy of each (ϕ, ψ) is evaluated for poly-L-alamine using only the formulae for van der Waals interactions then deepest energy minima are found to occur at the (ϕ, ψ) values which correspond to the right-handed α-helix and an antiparallel extended chain. There is also a shallower minimum corresponding to the left-handed α-helix. Calculations on bulky side-chains like poly-L-valine show that the α-helix minimum is removed. In aqueous solvent the hydrogen bonding does not give great stabilization because in the random coil state as many hydrogen bonds can be made between protein and solvent as within the protein in the α-helix, but hydrogen bonding does have some effect because solvents which will not form hydrogen bonds are found to increase the stability of the α-helix and solvents which form hydrogen bonds avidly give a random coil structure. It is likely then that the observation of the large number of van der Waals contacts in the interior of protein molecules is because of their part in stabilizing the structure. It is likely also that this type of interaction between the side chains of α-helices leads to the coiled-coiling in the α-proteins and also, incidentally, the same interaction is the significant stabilizing force in the double-strand structures of the nucleic acids.

In thermodynamic terms the driving force of the molecular folding which tucks as many non-polar groups as possible into the interior of the molecule, is thought to be entropic. Exposed non-polar groups have an ordering effect on adjacent water molecules and this ordering will be minimized and so the entropy maximized by removing as many as possible of the non-polar residues to the interior of the molecule.

Nearly all our information about protein structure comes from X-ray diffraction studies on crystals and fibres so the question frequently arises about the relation between these structures and the structure of the molecules in solution particularly in physiological conditions. Although it is not possible to rule out absolutely molecular rearrangements between the two states, there is some evidence which suggests that in many cases the change will be small. First of all the conservatism of tertiary structure, desgite varieties of crystallizing solvent, crystal form, etc., already mentioned, is in favour of no change as is the evidence from constancy of function throughout a range of primary structure. More direct evidence is provided by the fact that proteins give "solid state chemistry", that is, ribonuclease and chymotrypsin possess enzymic activity in the crystal state and crystals of haemoglobin will react with oxygen and show a change in molecular structure similar to that which is observed if the oxygen is added before crystallization. Finally, optical rotatory dispersion studies give an estimate of the fraction of amino acids which are in the helical form in solution and such estimates have been found to tally well with the fraction observed in structure determined by X-ray diffraction.

Some Terms used in discussing X-ray Fibre Patterns of Proteins

The *meridian* is the line through the origin of the X-ray pattern and parallel to the fibre axis. The first truly meridional reflection gives the repeat of the structure in axial projection.

The *equator* is the line through the origin of the X-ray pattern at right angles to the meridian.

Layer-lines are parallel to the equator but not through the origin. The first layer-line spacing gives the true repeat of the structure along the axis.

Row-lines are parallel to the meridian but not through the origin.

Sampling on the equator or a layer-line means that there are sharp intensity maxima on these lines (due to a regular side-by-side interaction between the molecules) as distinct from a smooth intensity streak (which implies no inter-molecular regularity.) If the equator only is sampled then the molecules are arranged with some regularity among their positions in axial projection but they are randomly arranged in the longitudinal (axial) direction. If there is some regularity in the longitudinal direction as well as the transverse direction then the layer-lines will be sampled as well as the equator.

The diffraction pattern from a helical structure (very common in biological molecules) gives a series of reflections on the arms of a cross centred on the origin and from this the radius, pitch, etc., of the helix may often be derived. Because of the usual random rotational positions of the molecules in a fibre, the X-ray fibre pattern is normally that from a cylindrically averaged structure.

Bibliography

MacArthur I. (1961) Fibre structure, in *Encyclopaedic Dictionary of Physics* (J. Thewlis Ed.), **3**, 116, Oxford: Pergamon Press.

Mahler and Cordes (1966) *Biological Chemistry*, New York: Harper & Row.

Neurath H. (Ed.) (1953) *The Proteins* (in 4 volumes), New York: Academic Press.

Wilson H. R. (1966) *Diffraction of X-rays by Protein, Nucleic Acids and Viruses*, London: Arnold.

A. Miller

PROTON SCATTERING MICROSCOPY. Beams of protons, electrons and X-rays all became available to the experimental physicist in the early years of the 20th century. However, from then until the present day, only electrons and X-rays have been used to explore the crystalline nature of solids. On the other hand it was not until the 1960's that the regular nature of the crystal lattice was shown to influence the trajectories of protons and charged particles.

Suppose a beam of protons, collimated to form a spot about 0·5 mm diameter, is directed at some convenient angle onto the surface layers of a single crystal target. If the proton's energy is chosen to be of the order of 20 keV, a convenient energy both from the experimental and economical view points, a substantial fraction will be scattered from the crystal by collisions within the first 100 atomic layers. These back scattered protons can subsequently be detected by the light they produce in a fluorescent screen placed nearby, as shown in Fig. 1. If the target has no regular crystal structure, the scattered protons will emerge in a completely random manner and the eye will see a uniform glow on the fluorescent screen. However, if the target is a single crystal the trajectories of scattered protons can be influenced by the regular nature of its structure—for instance, Fig. 2 illustrates two possible trajectories taken by protons scattered from a crystalline lattice. Proton A suffers a large angle collision at atom X and is scattered out of the crystal without significant further deflection. On the other hand, proton B, which is scattered by atom Y into the direction of a close-packed atomic row (YY^1) has its path blocked by atom Z (in the row) and is consequently deflected. This situation will be true for all protons which are scattered into the direction of a close-packed row, with the result that it is virtually

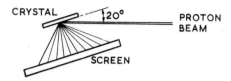

Fig. 1. Proton scattering chamber, showing the relative positions of the ion beam, target and fluorescent screen.

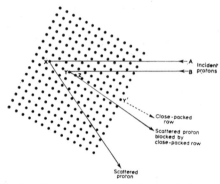

Fig. 2. Protons scattered from single-crystal targets have their trajectories blocked by densely packed atomic rows such as YY^1, so that regular patterns are produced on the fluorescent screen.

impossible for a proton to emerge from the crystal in one of these directions. This phenomenon is known as "blocking" and occurs not only in the case of rows but also for densely populated planes—which are, after all, only a succession of rows with different interatomic distances. The net effect will be to produce a regular proton deficient blocking pattern on the fluorescent screen and this will therefore contain information on the crystalline structure of the crystal.

To illustrate this phenomenon, Fig. 3(a) shows a direct photograph of the pattern produced by scattering 20 keV protons from a gold single crystal oriented to have its {111} planes parallel to its surface. The simplicity of the pattern is evident from the fact that the most significant crystal axes and planes appear directly on the screen at their correct angular separation; compare this with the standard radial projection of the {111} face of the f.c.c. lattice shown in Fig. 3(b).

The theoretical description of blocking has advanced mainly on account of the work of Lindhard (1965) and in this article we shall briefly describe the basic concepts. A beam of protons incident on a solid will, in general, be scattered via a series of statistically independent screened Coulomb interactions. Some colli-

Fig. 3(a). *Proton blocking pattern from the {111} face of a gold single crystal, the proton energy was 20 keV.*

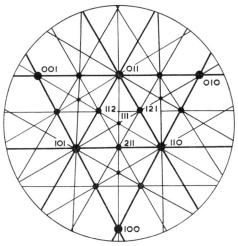

Fig. 3(b). *Radial projection of the {111} face of the f.c.c. lattice.*

sions will result in large angle scattering events with complete reversal of momentum. These protons will suffer further small angle scattering on their way to the crystal surface, and in a non-crystalline solid will emerge in a random manner completely devoid of correlation. However, in a crystalline solid small angle scattering events can be correlated as a direct consequence of the regular atomic array. Consider a row of atoms each separated by a regular distance d, Fig. 4. Suppose an energetic proton suffers a large angle collision with atom A and is elastically scattered towards its nearest neighbour B close to the line of centres. In the case of protons greater than 10 keV,

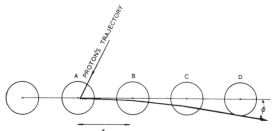

Fig. 4. *Proton blocking by a close-packed row of atoms.*

the impact parameter† for a collision of this nature is generally less than 10^{-9} cm. and the scattered proton appears to come right from the centre of the atom. Then providing the proton passes sufficiently close to the nucleus of atom B it will be deflected farther away from the row. Further small angle scattering will occur at atoms C and D and so on until the proton's trajectory eventually emerges from the influence of the row at some small angle ϕ.

In order to provide an analytical treatment of blocking, it is a useful approximation to smear out the interatomic potential from the individual atoms into a continuum so that the proton sees regular rows and sheets of high potential. Using this approximation and basing the interaction on the familiar screened Coulomb potential $V(r)$ suggested by Thomas-Fermi, i.e.

$$V(r) = \frac{Z_1 Z_2 e^2}{r} \Phi(r/a)$$

where $\Phi(r/a)$ is the Fermi function with screening length $a = a_0\ 0 \cdot 8853\ (Z_1^{2/3} + Z_2^{2/3})^{-1/2}$, Z_1 and Z_2 are the atomic numbers of the projectile and target atoms respectively, and e is the electronic charge; Lindhard has shown that the critical blocking angle subtended by a row of equi-spaced atoms is given by

$$\psi = \left(\frac{2Z_1 Z_2 e^2}{E\,d}\right)^{1/2} \quad \text{if} \quad \psi < a/d$$

where E is the particle energy and d is the interatomic spacing. On the other hand, at large Z_2 or when the proton's energy is low the characteristic blocking angle smoothly changes to

$$\psi = \left\{\left(\frac{a}{d}\right)\left(\frac{2Z_1 Z_2\, e^2}{E\,d}\right)^{1/2}\right\}^{1/2} \quad \text{for} \quad \psi > a/d$$

Returning now to the use of proton scattering microscopy as a tool for the study of crystalline solids, it is clear that such a technique can be a useful complement to electron and X-ray diffraction. In some instances, where the conventional techniques are slower and more difficult to apply, the proton scattering technique can be used to advantage. For example, the fact that crystal structures can be observed dynamically and identified immediately suggests a

---

† The "impact parameter" is defined as the shortest distance between the extension of the original trajectory of the scattered proton and the struck atom.

quick and simple method of orienting single crystals. The crystal to be examined is fixed to a goniometer, which can be operated remotely from outside the vacuum system. A proton beam of the desired energy and intensity is fired at the crystal so that a blocking pattern is easily visible on the fluorescent screen. Then, in a matter of seconds, the crystal orientation can be adjusted so as to produce the correct pattern on the screen relative to a known mark. In this way crystals can be dynamically oriented to at least within one degree without difficulty by a relatively inexperienced operator. Figure 5 illustrates the blocking pattern from a crystal of tungsten which has been oriented in this fashion so as to bring the {110} planes parallel to the screen. A further advantage is that there is no radiation hazard as there is with X-ray diffraction, and the apparatus can be operated safely in the open laboratory.

Perhaps the main application of proton scattering microscopy is to the study of surface layers of only a few tens to hundreds of atom layers thick. Films of this thickness are not readily studied by conventional techniques; X-rays usually examine depths to the order of millimetres, and conventional electron diffraction introduces the disadvantage that the

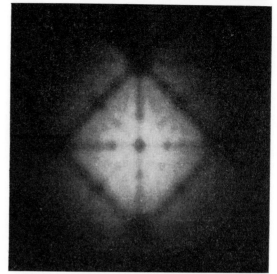

Fig. 6. A 20 keV proton blocking pattern from a 1000 Å single-crystal gold film evaporated epitaxially on to NaCl.

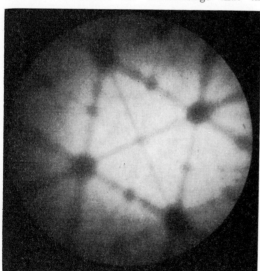

Fig. 5. Proton blocking pattern from a tungsten crystal dynamically oriented so that its {110} planes lie parallel to the screen, the proton energy was 20 keV.

sample must often be an isolated film removed from its underlying substrate. In the present case, however, the protons are scattered from less than 100 atom layers below the surface and do not necessitate removal from the substrate. Thin crystalline metallic films grown by evaporation on to a single-crystal substrate such as rock-salt, can therefore be studied very successfully even while the evaporation takes place *in situ* in a high vacuum. Figure 6 shows the proton blocking pattern from a gold film which has been grown by evaporation on to the cleavage face of rock-salt. The gold is still in intimate contact with the salt and its crystalline perfection is readily apparent.

Bibliography

LINDHARD J. (1965) *Mat. Fys. Medd. Dom. Vid. Selsk* **34** (14).
NELSON R. S. (1966) Brit. Pat. Appl. No. 56081/66.
NELSON R. S. (1967) *Phil. Mag.* **15,** 845.

R. S. NELSON

PULSE CODE MODULATION.

Introduction

The transmission of information may generally be divided into two classes: analogue or coded. Conventionally, telephony is an analogue transmission system, the variations in pressure caused by sound propagated through air are converted by a microphone into analogous currents, which may be transmitted by various media and methods to a receiver, to be reconverted into variations of air pressure. The signal may undergo a number of transformations in its passage, e.g. a high-frequency carrier wave may be modulated by the speech currents, and the carrier may be sinusoidal or a pulse train, but one parameter of the modulated wave (e.g. amplitude, phase or frequency) has a continuous relationship with the original variations of sound pressure or the microphone current. Telegraphy is generally a coded system of communications. An aplhabet of discrete symbols may be signalled by permutations of pulses, e.g. an alphabet of 32 symbols may be transmitted by a five-unit binary code. At the transmitter, the selection of a symbol causes pulses of one of two values (which may be 1 and 0) to be generated at given time instants (usually equally spaced), e.g. A = 00001, B = 00010, which may be identified by a receiver synchronized so that successive groups of pulses are properly segregated.

An approximation to a continuous waveform can be transmitted in this "telegraphic" fashion provided that it is accepted that only a limited number of discrete values of the variable can be transmitted at regular time intervals, i.e. the signal is sampled, quantized and a code permutation allotted to each quantized value. The coded signal is transmitted as a train of pulses which are decoded on reception, and an approximation of the original waveform is reconstructed from the quantized samples. The approximation may be made as close as desired by the use of a sufficiently fine scale of quantization and a corresponding length of the code permutations. It is this form of transmission which is termed pulse code modulation (P.C.M.). It may be applied to the transmission of any analogue waveforms (telephony, television, etc.).

The elements of a P.C.M. system are indicated in Fig. 1. The analogue signal is restricted in frequency range by a low-pass filter and sampled at regular

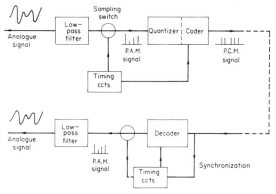

Fig. 1. Elements of a P.C.M. system.

intervals. The samples are quantized to discrete values and the corresponding coded pulse trains are generated by the coder for transmission. On reception, the coded pulse train is decoded into quantized samples, which are passed through a low-pass filter to remove all components above the original frequency range, and amplified to an appropriate level.

Sampling and Quantization

It is known that any band-limited signal can be accurately characterized by instantaneous samples of its amplitude taken at a uniform rate in excess of twice the upper limit of its frequency range, i.e. if the highest frequency in the original signal is f Hz then the sampling rate $>2f$/sec. This is the underlying principle of pulse *amplitude* modulation (P.A.M.) by which a train of pulses is modulated in amplitude by the waveform to be transmitted, and the waveform is recovered after transmission by passing the signal through a low-pass filter with a cut-off frequency of f. For commercial telephony, f may be taken as 3400 Hz and the sampling rate is made 8000 per sec.

Sampling is an essential part of the process of P.C.M., but additionally the samples must be quantized in amplitude so that the appropriate code can be allotted to the sample. In a practical system quantization is combined with the processes of encoding and decoding. The process of sampling and quantization is illustrated in Fig. 2. The range of amplitude of

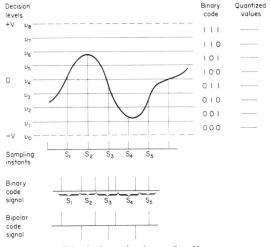

Fig. 2. Quantization and coding.

analogue signal for which the system is designed ($+V$ to $-V$) is divided into a number of sub-ranges bounded by the decision levels v_x. All samples whose amplitudes fall within a sub-range v_j, v_{j+1} cause the same code permutation to be generated, each sub-range giving rise to a different permutation. On reception, the pulse permutations are decoded and pulses with amplitudes corresponding to $(v_j + v_{j+1})/2$ are generated. The reconstituted P.A.M. signal, being a quantized approximation to the original, may be in error up to half the quantum and even after filtering there will be an error in the waveform. The error component has a distributed noise-like spectrum and is termed quantizing noise. In the intervals between words the quantizing noise is in general imperceptible. When the speech is present, if the amplitude sub-ranges are of equal magnitude, the noise volume will be almost constant irrespective of the analogue signal amplitude, and inversely proportional to their number. In the absence of input signal, e.g. between utterances of speech, provided that the background noise is very small, crossing from one sub-range to another should be infrequent and the quantizing noise will become imperceptible. To economize in the length of the code (and hence the bandwidth of the pulse transmission path) for satisfactory telephony, the sub-ranges can be graded, making them smaller for the lower amplitudes of signal. This is equivalent to the use of compandors in analogue transmission for reducing the effect of transmission path noise. If the sizes of the sub-ranges increase exponentially over the half-ranges 0 to $+V$ and 0 to $-V$ the signal to quantizing noise ratio can theoretically be made the same for all levels of signal. This can be achieved either by the action of a non-linear device (compressor) upon the P.A.M. signal or a non-linear coder. Binary codes of length 7 or 8 (corresponding to 128 or 256 quantized values) are sufficient to provide excellent telephone quality even when a considerable number of systems are connected in series.

Coding and Decoding

Of course it is possible to use a code of any radix (e.g. binary, ternary, quaternary, etc.) but in general, as has been assumed earlier, binary coding is the most used and easiest to generate and decode. There are various ways of implementing these processes but only one example will be described in principle, and, for simplicity, a system with only 8 regularly spaced quantizing levels will be assumed. Figure 3 shows the

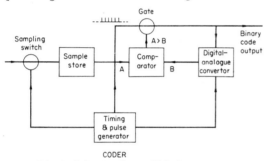

Fig. 3. Coder (analogue–digital convertor).

arrangement. The input waveform sample is stored, and the voltage of the sample is compared with the output of a digital/analogue converter, the input register of which is set initially to 100 which produces an output equal to the discriminating level V_4. If the sample is greater than this, a pulse ($=1$) is fed to the output and also to the d/a converter; if the sample is less, no pulse is generated ($=0$). The corresponding digit is substituted in the first stage of the register and a 1 entered in the next stage. This results in the discriminating voltage being changed to V_6 or V_2 for the determination of the second code pulse, and so on, the code signal being transmitted as that process proceeds. A digital/analogue converter is shown in outline in Fig. 4. The settings of the switches S1–3 (shown as mechanical, but in practice electronic) are controlled by the sequence timing of the system and the code pulses generated as mentioned earlier. The resistors A–D have resistances in the ratios indicated (1, 2, 4, 4), therefore the source current is apportioned 4, 2, 1 in A, B, C, so that the permutations of the switch settings will produce an output potential ranging from 1 to 7 units to provide the discriminating levels to the comparator.

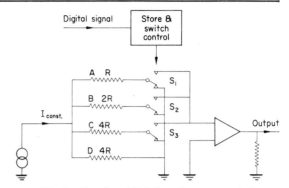

Fig. 4. Decoder (digital–analogue convertor).

The decoder at the receiving end of the circuit is a similar d/a converter with the modification of resistor D being replaced by one of value 8R and an additional resistor of the same value being connected between the source and the input to the current amplifier, so that the output is displaced by a half interval to the middle of the sub-range, which is the correct quantized level.

Line Transmission and Regeneration

For a number of reasons, e.g. isolation, impedance-matching, etc., the pulse signals are required to pass through transformers. Therefore it is desirable that the signal shall have no d.c. component. This can be achieved by the addition of redundant code elements so that an equal number of "1"s and "0"s are produced within a limited period; but to avoid this complication, bipolar transmission is usually employed, i.e. "0"s are transmitted as intervals of no current and "1"s are transmitted alternately as pulses of positive and negative current (see Fig. 2).

In its passage along the transmission circuit, the signal becomes attenuated, the waveshape becomes distorted and noise may be added due to interference. However, the signal may be restored for transmission over the next section of the circuit by the use of a regenerative repeater (Fig. 5). An attenuation equalizer approximately corrects the received waveform for the effects of the attenuation/frequency characteristics of the cable. The signal is then amplified and applied to two discriminating devices which conduct when the input is of the same polarity as a timing pulse and exceeds a critical value. The timing is derived from the signal itself, but since no component of the signal is at the frequency of maximum pulse repetition rate, one is generated by passing the signal through a non-linear device (e.g. a biased full-wave rectifier). This is extracted by a filter, amplified and shaped to produce the timing pulses. The critical value of amplitude for discriminating between 0 and 1 is half the amplitude of the pulse. The pulse discriminators are automatically biased to that value by a circuit which averages the pulse amplitudes over a limited period. The outputs of the pulse discriminators trigger pulse generating circuits to produce the new

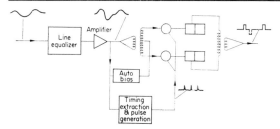

Fig. 5. Regenerative repeater.

signal corresponding to the input. Only if additive noise has changed the amplitude of the signal to such an extent that it cannot be properly identified by the discriminators, will a false signal be produced; otherwise the quality of the regenerated signal is unaffected by the noise.

Time-division Multiplex

A number of telephone channels can be combined in one P.C.M. transmission. The analogue signals from different input circuits can be sampled in turn at a rate which is the appropriate multiple of that required by one channel, the coder dealing with each in sequence. To identify the order of the channels so that they may be properly distributed to the corresponding circuits at the distant end of the transmission path, a distinctive pulse pattern is included in the pulse stream of the aggregated signals for synchronizing the receiver. Additional pulses may also be interpolated for supervisory signalling, the operation of exchange switches, etc. A common order of multiplex transmission for use on multipair junction cables between telephone exchanges is one combining 24 telephone circuits, needing only two pairs of wires, one in each direction. Including the pulses for synchronization and signalling, the pulse rate is approximately 1·5 million/sec.

Advantages of P.C.M.

The main advantage of P.C.M. over analogue transmission is its considerable resistance to noise and interference. In an analogue transmission system the noise accrued in transmission cannot be eliminated from the received signal (although measures can be taken to abate it when no signal is present) but in pulse code transmission it is necessary only to distinguish the presence or absence of the pulse at the appropriate instant and, provided the noise amplitude is insufficient to cause wrong identification, it can have no effect at all on the decoded signal. In transmission along a circuit the quality of the signal can be preserved almost independently of the total distance by the employment of regenerative repeaters spaced so that the noise accumulated in the intervening circuit is insufficient to falsify the signal. This noise immunity makes the employment of P.C.M. very profitable for use in cables where mutual interference between signals on different conductors would prevent the use of other transmission methods.

Other advantages are that multiplexing by time-sharing can be relatively cheap, and because of its digital form it offers the possibility of complete automatic telephone exchanges in which all communication and routing is achieved through electronic digital switches.

Bibliography

CATTERMOLE K. W. (1969) *Principles of Pulse Code Modulation*. London: Iliffe.

HARTLEY G. C., MORNET P., RALPH F. and TARRAN D. J. (1967) *Technique of P.C.M. in Communication Networks*, I.E.E. Monograph, Cambridge: University Press.

Members of the Technical Staff Bell Telephone Laboratories (1964) *Transmission Systems for Communications*. Winston-Salem N.C.: Western Electric Co. Inc., Technical Publications.

L. K. WHEELER

PYROELECTRIC DETECTOR FOR INFRA-RED RADIATION. Of the 32 known crystal classes, 10 out of 21 without a center of symmetry exhibit spontaneous polarization, i.e. alignment of electric dipoles within domains of the crystal. These spontaneously polarized crystals are called pyroelectric since the value of the spontaneous polarization is temperature dependent. A slight change in temperature of a pyroelectric crystal will give rise to electric charges. The pyroelectric detector, based on this effect, is a thermal sensor of infra-red radiation that requires no bias during operation. By analogy with photovoltaic detectors, the pyroelectric detector may be considered a "thermovoltaic" device.

The performance of other radiation detectors is limited by electrical noise which is a function of their resistance. Pyroelectric detectors, however, are in principle pure capacitors and hence theoretically noiseless, being only limited by amplifier noise, and should therefore be able to detect minute amounts of infra-red radiation in thermography, radiation pyrometry and spectroscopy. Operating from d.c. levels to beyond 10 kc/s, the detector in actual practice is noise limited at low frequencies $\omega \ll \omega_e$ by the leakage resistance, by the loss tangent in a middle range for $\omega_e \ll \omega \ll \omega_a$, and the amplifier short circuit noise at high frequencies for $\omega \gg \omega_a$,

where ω_e = frequency where load resistor noise equals loss tangent noise,
ω_a = frequency where loss tangent noise equals amplifier short circuit noise.

The actual sensor is a small flake of a crystal with spontaneous polarization, in which electric charges of opposite sign appear on opposite faces of the crystal. Operating below the Curie point, the spontaneous polarization and dielectric constant of the pyroelectric crystal are temperature dependent. A change in incident power on the detector element causes a change in its temperature, giving rise to a change in polarization and capacitance. Thus, an electric charge appears across the electroded surfaces cut perpendicular to the crystal's ferroelectric axis. Under open circuit condition a voltage is obtained across the capacitor which is ultimately neutralized by current flow through the leakage resistance or load resistor. The detector may also be operated in the current mode by using an amplifier with capacitive feedback.

From pyroelectric materials such as triglycine sulfate (TGS), triglycine fluoberyllate, triglycine selenate, lithium sulfate, Rochelle salt, barium titanate, and various lead zirconate titanates (PZT), TGS has demonstrated the best performance at ambient temperatures of 20° to 40°C. TGS has a Curie point of 49°C, and after a one-time application for about 2 min of 10 kV d.c. per cm of crystal thickness, it exhibits a dielectric constant of 11 at 25°C rising to 35 at 40°C, with a pyroelectric coefficient (i.e. change in spontaneous polarization with temperature) 0.3×10^{-7} C/°C cm² at 25°C and rising to 1.0×10^{-7} C/°C cm² at 40°C.

The evacuated detector package incorporates a blackened thin flake of pyroelectric TGS mounted on a substrate of low thermal and electrical conductivity, a field effect transistor (FET) for impedance matching from $\geq 10^{12}$ to $10^4 \Omega$, a $10^{12} \Omega$ ohm load resistor, and an infra-red transparent window (generally thallium bromide–thallium iodide for 0.5 to 40μ). The electrical hook-up of the detector to an amplifier is shown in Fig. 1 with the actual pyroelectric crystal sensor indicated as a capacitor. Since TGS has virtually zero transmission from 2μ to beyond 35μ and exhibits a high absorptance in this region, a transparent electrode without blackening may be substituted for the usually opaque electrode. This reduces the thermal mass of the detector and facilitates construction of arrays and mosaics on a single-crystal section with negligible thermal and electrical crosstalk. The relative spectral response of a blackened TGS detector is shown in Fig. 2.

Detector flakes ranging in size from 10^{-3} cm² to 1.5 cm² have been built. The detectivity (commonly referred to as D^*), responsivity, and noise as a function of frequency for a 0.4 mm $\times$ 0.4 mm and 5.0 mm $\times$ 5.0 mm detector respectively are shown in Fig. 3.

As long as the detector capacitance is sufficiently high (20 pfd for a 1 mm $\times$ 1 mm detector, poled, at 23°C) and is not shunted materially by the FET capacitance, the A.C. responsivity is essentially invariant with temperature from 20°C to about 40°C, inversely proportional to the area of the detector, and drops 6 db/octave with increasing frequency. A typical d.c. responsivity for a 1 mm by 1 mm detector has approached 10^6 V/W and is a function of the shunt resistor. The actual thermal time constant is of the order of several seconds but this becomes an unimportant parameter since the noise decreases with the frequency (along with signal) out to frequencies much higher than determined by the

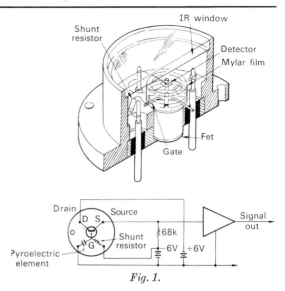

Fig. 1.

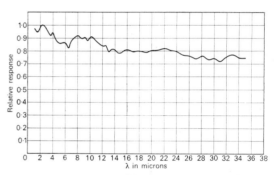

Fig. 2.

thermal time constant. The noise of the detector is inversely proportional to the capacitance and drops at approximately 3 db/octave out to 2 kc/s before being amplifier noise limited. For a constant detector flake thickness at a given frequency, the D^* would increase by the square root of the ratio of two respective areas. However, for practical reasons, the thickness of detectors is increased by the square root of the ratio of two respective areas so that essentially D^* is independent of area. Further improvements in pyroelectric detector sensitivity appear possible since the ultimate noise limit imposed by background radiation falling on the detector has not yet been reached.

Pyroelectric detectors of polycrystalline TGS are feasible. In general, the polycrystalline detector detectivity is lower than that of its monocrystalline counterpart by a factor of 3. However, polycrystalline pyroelectric detectors simplify further the construction of arrays and mosaics, and make it possible to produce detectors with curved surfaces.

Life tests to date indicate that there is no gradual depoling of the detector with time as long as the Curie

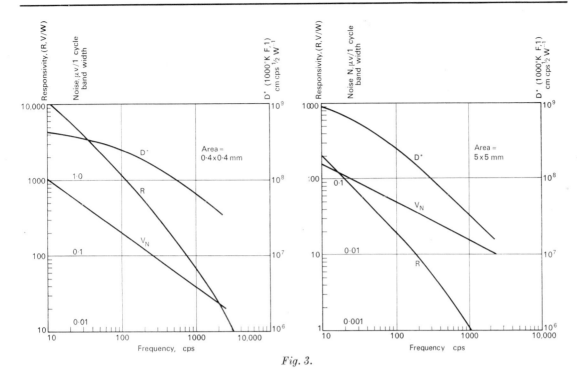

Fig. 3.

point is not exceeded. If it is accidentally depoled, the detector is readily repoled at ambient temperatures by application of a d.c. potential for several minutes. By hooking up the internal resistor terminal of the detector package to the negative 6 volt supply as shown in Fig. 1, the detector will also gradually repole itself if it has been accidentally depoled. Simulation of the electron bombardment to be encountered over a 5-year period in the Van Allen belt has not caused any permanent deterioration.

Typical applications of a pyroelectric detector have been in a reticle scanner using a Nipkow disc image disector, a multi-element line scanner, a mechanical 2-axis thermal imaging camera, spectrometers, and laser calorimeters.

Definition of Additional Term used in Article

D^* (pronounced "dee-star"): an area independent figure of merit used for comparison of infra-red detectors' detectivity under blackbody conditions.

$$D^* = (V_S/V_N)\, \Delta f^{\frac{1}{2}}\, H^{-1} A^{-\frac{1}{2}}$$

where V_S/V_N is the ratio of the r.m.s. signal voltage to the r.m.s. noise voltage in the bandwidth Δf, and H is the r.m.s. value of the irradiance falling on the detector of area A. The measured value of D^* for a detector is written as $D^*(°K, f, 1)$ where °K represents the temperature in degrees Kelvin of a reference blackbody radiation source, f is the chopping frequency in cps, and 1 stands for the reference bandwidth of 1 cps. The units of D^* are cm cps$^{\frac{1}{2}}$W^{-1}.

Bibliography

BEERMAN H. P. (1967) Pyroelectric infrared radiation detector, *Bull. Am. Ceramic Soc.* **46**, 8.

BEERMAN H. P. (1969) The pyroelectric detector of infrared radiation, *IEEE Transactions on Electron Devices*, ED-16, 6, 554–7.

HADNI A. (1967) *Essentials of Modern Physics Applied to the Study of the Infrared*, Oxford: Pergamon Press.

HADNI A., HENNINGER Y., THOMAS R., VERGNAT P., and WYNCKE B. (1965) Sur les propriétés pyroélectriques de quelques matériaux et leur applications à la detection de l'infrarouge, *J. de Physique* **26**.

JONA F. P. and SHIRANE G. (1962) *Ferroelectric Crystals*, New York: Macmillan.

KRUSE P. W., McGLAUCHLIN and McQUISTAN R. B. (1962) *Elements of Infrared Technology: Generation, Transmission, and Detection*, New York: J. Wiley.

PUTLEY E. H. (1966) Solid state devices for infrared detection, *J. Sci. Instrum.* **43**.

WEINER S., JANKOWITZ G., PEARSALL D. and SCHWARZ F. (1968) The pyroelectric detector—recent developments and applications. 16th National IRIS Symposium, May 7–9, 1968. U.S. Army Electronics Command, Fort Monmouth, New Jersey.

H. P. BEERMAN

Q

Q-SWITCHING. In the technique of "Q-spoiling" or "Q-switching" a laser is pumped in a non-resonant mode and then switched into resonance. The technique was originally developed for use with solid state lasers to provide very high pulsed output powers. Two of the earliest methods tried involved using a rotating prism for the switching into resonance or using a Kerr cell as a switch between the reflector and laser rod. The technique is effective where the upper-level lifetime (spontaneous emission lifetime of the upper laser level) of the laser is long in comparison with the photon cavity lifetime. (A photon in the cavity will have some average lifetime before being scattered, emitted or lost in other ways to the optical system. This is called the photon cavity lifetime.) While for gas lasers operating in the visible or near infra-red such conditions do not appear to be satisfied, it is effective for certain molecular lasers such as the carbon dioxide type molecular lasers.

General Principles

In solid state lasers such as the ruby laser, one limitation of the power obtained is that complete population inversion is never achieved. Thus as soon as a certain amount of inversion is obtained, laser action "drains" off the inversion (reduces the population in the upper-level state).

In pumping a laser, spontaneous emission slowly drains off some of the population inversion. If the pumping is too slow, sufficient population inversion may never occur. If pumped rapidly, irregular laser spiking action occurs. (This is an irregular train of pulses randomly spaced. They appear to represent a kind of relaxation oscillation of a complicated nature, not completely understood.) In Q-switching, an optical shutter called a Q-switch or Q-spoiler is placed inside the optical cavity. (This may also be expressed as arranging an amplifying medium in such a way that it may be pumped, but in such a fashion that initially there is no feedback. The atomic inversion can now grow beyond the limit it would have if feedback were initially present.) With the shutter closed, one mirror end of the cavity is not visible from the other mirror of the cavity and the result is a low cavity Q. No oscillations occur even with quite high pump powers and a substantial population inversion. For high pulse power, the Q-switch must open the light path between the two reflectors rapidly.

Typical Q-switches include the Kerr cell, KDP cell, rotating mirrors, and bleachable absorbers. Repeating the pulse usually requires recycling the optical pumping means and having the repetition rate limited by the need to remove heat from the laser cavity and from the optical pumping means.

A simple description of the output from Q-switching is to consider a solid state laser rod of length L and an optical cavity length of l. Let

$$n = N_{upper} - N_{lower}$$

where N_{lower} is the number of atoms per unit volume present in their lower state, and N_{upper} is the number of atoms per unit volume present in their upper state. We may then write the gain per centimeter

$$\alpha = \alpha_0 n / N_0$$

where α_0 is the absorption coefficient for the unpumped laser material, and N_0 is the number of *active* atoms (e.g. chromium in ruby) per unit volume. For laser ruby, α_0, the absorption coefficient is about 0·4/cm. The threshold condition for a laser oscillation to occur is that the loss per pass equal the gain per pass. If the reflectivity of the mirrors are R_1 and R_2 then

$$R_1 R_2 \exp\left[\alpha_0 (N_{upper} - N_{lower})(2L)/N_0\right] > 1$$

and for the threshold condition

$$R_1 R_2 \exp 2\alpha L = 1$$

or
$$2\alpha L + \log (R_1 R_2) = 0$$

or
$$(\alpha L - \gamma) = 0$$

where, for convenience, we define

$$\gamma = -\tfrac{1}{2} \log (R_1 R_2).$$

If the active medium is pumped to an initial gain α_i where noise photons are initially present in the cavity, then the number of these photons grow after the time of a double transit of the cavity, T_1, to

$$\phi_0 \exp 2 (\alpha_i L - \gamma).$$

After a time t, t/T_1 transits take place, increasing the total number of photons to

$$\phi_0 \exp 2(\alpha_i L - \gamma) t/T_1.$$

This illustrates how a giant pulse grows exponentially after a Q-switch is opened and is shown schematically in Fig. 1. The α_i gain continues for a long time since the intensity is too low during the growth period to

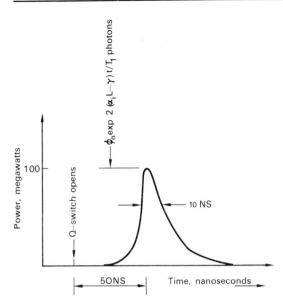

Fig. 1. Typical Q-switched laser pulse (for ruby).

reduce significantly the population inversion, until growth phase ends when the intensity grows to such a level that the inversion is suddenly eliminated. The exponential decay of the energy by loss through the mirrors is then proportional to

$$\exp(-2\gamma t/T_1).$$

Q-switching was first proposed by Hellwarth and was first used with the ruby laser. Q-switching has also been done with gas lasers of long lasing level lifetimes (e.g. carbon dioxide laser where a typical vibrational-rotational transition can be a millisecond or more). In such molecular lasers, the collisional deactivation does not reduce the upper state lifetime by too large a factor or compared with most visible and near infra-red gas lasers. Peak outputs of about 100 kW have been reported with the CO_2 laser with pulse length outputs of the order of 10–20 nsec (1 nsec equals 10^{-9} sec). With ruby lasers, several hundred megawatts peak outputs (approaching 1000 MW) have been reported, while glass lasers, such as neodymium glass lasers have had reported outputs of 50 GW (50,000 MW) using Q-switching.

Bibliography

ELION H. A. (1967) *Laser Systems and Applications*, New York: Pergamon Press, Inc.

GARRETT C. G. B. (1967) *Gas Lasers*, New York: McGraw-Hill Book Co., Inc.

HELLWARTH R. W. (1966) in *Lasers* (A. K. Levine Ed.), p. 253, New York: Marcel Dekker, Inc.

TROUP G. J. (1967) Laser, in *Encyclopaedic Dictionary of Physics* (J. Thewlis Ed.), Suppl. vol. 2, 141, Oxford: Pergamon Press.

H. A. ELION

QUANTUM MECHANICS IN BIOCHEMISTRY.

The recent explosive progress in molecular biology owes much to the application of modern physico-chemical techniques to problems of structure and function of biological macromolecules. Tools such as X-ray crystallography and numerous spectroscopic methods have supplied a large amount of experimental information, while theoretical studies have provided insight into possible modes of biochemical action. Both the experimental and theoretical studies depend centrally upon the electron distributions in the biomolecules involved, as it is the electronic properties which in the first instance govern the chemical behaviour.

Quantum mechanics is the logical tool for the study of the electronic properties of biomolecules, and quantum-mechanical methods have progressed to the stage where meaningful information can be developed for certain biological systems. This article sketches critically the quantum-mechanical methods which have found application to biochemical problems, and surveys the progress which is resulting from their use. The areas of application surveyed include (1) the direct applications of electronic structure calculations to yield energy levels and charge distributions, and to correlate these with chemical reactivity; (2) treatment of intermolecular (and non-bonded intramolecular) forces, to enable discussion of conformational energies of biopolymers, changes in monomer spectra due to polymerization, solvent effects, and the like; and (3) hydrogen bonding, proton tunnelling, and related factors relevant to DNA replication, genetic coding, mutations, and carcinogenesis.

Quantum-mechanical methods. Ideally, we should desire to solve the quantum-mechanical problem of the distribution of all the electrons of a biomolecule in the field of its nuclei. However, the great complexity of biomolecules necessitates the application of quantum mechanics in an approximate form. The approximations may be classified into groups according to their sources. First, it is necessary to introduce approximations which limit the size of the problem to be treated quantum mechanically. This is accomplished by dividing the electrons into two groups, regarding those of the first group, together with the atomic nuclei, as constituting fixed unpolarizable atomic "cores" about which the electrons of the second group are distributed. The best practical approximation of this type results when the inner-shell electrons are assigned to the cores, all valence electrons being retained for explicit consideration. However, a more limited formulation is frequently obtained by including in the cores the relatively immobile electrons defining the skeletal bonding (the sigma electrons), retaining for explicit consideration only the mobile valence electrons (the pi electrons) responsible for the more variable and interesting electronic properties. Unfortunately, this division into sigma and pi electrons, though soundly based on symmetry arguments, is in part artificial when used to limit the scope of calculation. In actuality, both sigma and pi electrons play significant roles in chemical processes.

After restricting explicit consideration to certain electrons, further approximations are introduced to facilitate their description. Invariably each electron is assumed to be distributed in a molecular orbital (MO) formed by linear combination of atomic orbitals (AO's) on the various atoms. In the Huckel type theories, each MO is assigned an energy dependent upon its AO composition, but not upon the detailed distribution of the electrons in other MO's. This assumption means that Huckel theories can describe the inherent variations in energy as an electron moves from one kind of atom to another, but such theories grossly neglect the effect of the repulsion between electrons.

An alternative to the Huckel theories is provided by the antisymmetrized molecular orbital (ASMO) formulation, which includes all energetic interactions in a formally correct fashion. This theory is clearly superior to the Huckel theories, but requires considerably more calculation for its implementation. For many problems, it is found that the Huckel theories yield qualitatively reasonable electron distributions in spite of their obvious deficiencies in energy calculation.

The third class of necessary approximations is in the assignment of values to the energy quantities appearing in the Huckel or ASMO formulations. Some of these assignments can be made empirically, others by approximation formulas. Frequently, simplifications are achieved by the total neglect of certain classes of energy contributions. The most common such approximation is the neglect of contributions arising from the overlap of atomic orbitals on neighbouring atoms. If this overlap is assumed negligible at all points in space, we have the zero differential overlap approximation. In our opinion this approximation, though leading to extremely great computational simplifications, must have serious consequences, as it is just this overlap which is qualitatively essential to an understanding of chemical bonding.

Summarizing, we point out that most of the biochemical applications of quantum mechanics involve one of the following approximation schemes, listed in the order (in our opinion) of their reliability, best first: (i) all valence electrons, ASMO, retention of near-neighbour overlap; (ii) all valence electrons, ASMO, neglect of overlap; (iii) all valence electrons, Huckel; (iv) pi electrons, ASMO; (v) pi electrons, Huckel.

After obtaining a quantum-mechanical description, it must be interpreted in terms of quantities having fundamental or derived significance for chemical structure or reactivity. Such an interpretation can proceed along several lines. First, total molecular energies can be used to compare the stabilities of related compounds; applied to transition states, they can be used to estimate activation energies for chemical reactions. Calculations on suitably chosen parts of molecules can be used to determine electron localization energies, which have been used as measures of chemical reactivity. Individual MO energies can be used to estimate transition energies, and thereby molecular spectra. Moreover, they provide measures of ionization potentials and electron affinities, characterizing electron donating or accepting properties.

From the MO's themselves, it is possible to derive estimates of net atomic charges which, in turn, facilitate discussions of both molecular structure and reactivity. The MO's can also be used to define quantities measuring the character of the bonding between the various atoms (the *bond orders*) and the extent to which individual atoms are unsaturated in their bonding (the *free valence*). (This latter quantity has been correlated with reactivity by mechanisms involving attack by radicals.) The charge distributions can also be processed to yield dipole moments, either for the entire molecule or as a collection of local moments. These quantities are useful in estimating long-range interactions.

Direct applications of electronic structure calculations. The above described methods have by now been applied to a wide variety of biologically important systems. Much of the pioneering work in this area is by Pullman and Pullman, and the reader is referred to their treatise for further details (see the Bibliography at the end of this article). Most of the studies made to date are restricted to pi electrons, and more often by the Huckel method.

Briefly surveying some representative studies, we cite first work on conjugated peptide linkages. ASMO studies by Pullman and Pullman have examined the hypothesis of the existence of energy bands suitable for explaining energy transfer in proteins by a semiconduction mechanism. The calculations contraindicate this conjecture, and point instead toward a resonance mechanism for energy transfer between relatively isolated systems. Turning next to high-energy phosphates, calculations by Grabe and others have indicated that the relative instability of these compounds may be due to the presence of substantial numbers of adjacent electron-deficient atoms. The mutual repulsion of these local concentrations of positive charge would be expected to contribute significantly toward raising the energy above the normal level.

Porphyrins provide further examples of interesting applications. From the highest occupied and lowest unoccupied orbitals of porphyrinmetal complexes, interpretations can be made of the oxidation-reduction properties of these compounds, and of the changes in charge distribution accompanying oxidative or reductive processes. It is also possible to understand the relations between paramagnetism and electronic structure.

Another important problem is the correlation of electronic structure with carcinogenic activity. Pullman and Pullman have investigated a large number of carcinogenic and noncarcinogenic polynuclear hydrocarbons, and have correlated carcinogenic activity with the charge distributions in the highest occupied and lowest unoccupied molecular orbitals. Various aspects of these correlations led to the identification of "K" and "L" regions in these hydrocarbons.

Among the systems receiving the most detailed molecular orbital study are the purine and pyrimidine bases which occur in the nucleic acids. These systems were studied first in the pi electron-Huckel approximation by Pullman and Pullman, and the qualitative results there obtained were confirmed with later pi-electron ASMO studies by Vcillard and Pullman. In later calculations the sigma valence electrons were included both by the Huckel method (Rein and Harris, Pullman) and by ASMO methods (Pullman). The orbital energies and charge densities obtained in these studies are of particular interest because of their potential utility in discussions of the nucleic acids of which these bases are parts.

Intermolecular interactions. Long-range interactions between molecules, or between indirectly connected parts of the same molecule, depend upon induced polarization effects best treated quantum mechanically. These so-called Van der Waals interactions can be discussed with the aid of perturbation theory, which when applied to second order requires a knowledge of matrix elements of charge and dipole moment connecting ground and excited electronic states. Electronic structure calculations can provide the necessary data. In the earlier applications, it was assumed that whole-molecule charges and moments could be used, but these quantities are now known to yield poor results due to the severe violation of the conditions for proper application of multipole expansions. Better results may be obtained by dividing the charges and moments among the atoms, in effect applying multipole expansions locally. By interpreting Huckel or ASMO charge distributions in terms of atomic charges and moments, it is possible to obtain results of practical value. There are several approximations in which local expansions can be applied, and at this writing it is still unclear as to what method will produce satisfactory results with minimum effort.

A typical interaction study is that of Rein, Claverie, and Pollak on the energies involved in the stacking of nucleotide base pairs in the DNA α-helix. These authors obtained stacking energies for each of the possible combinations of adjacent pairs. Data of this sort can be used in discussions of DNA denaturation processes, as these probably proceed with the loss of the regular helical structure as an early step. In particular, differences in stacking energies for different base-pair combinations can explain the dependence of denaturation temperatures upon the base composition of DNA samples.

Hypochromicity effects are further manifestations of intermolecular interactions. Here the interaction influences the strength of the absorption of an electronic transition when the unit undergoing transition is incorporated in a polymer. This phenomenon has been studied by Rhodes and Tinoco, both of whom give formulations of the hypochromic effect in terms of a perturbation theory of inter-unit interactions. Just as for the interaction energies, this effect was first discussed in terms of whole-unit parameters, but more satisfactory treatments result from localized expansion models.

Hydrogen bonding and proton tunnelling. A hydrogen bond occurs when a hydrogen atom forms, in addition to its normal bond to one other atom, a second, usually weak bond to an additional electronegative atom. This situation can be schematically represented by the notation A—H--B. A preliminary description of the bonding can be made in terms of an electron-pair bond between A and H, with an additional pair of electrons on B providing a strong attraction for the proton. A better description treats all four of the electrons involved on a unified basis by including them in a complete molecular orbital calculation. The strength and other characteristics of a hydrogen bond can be studied theoretically by making electronic structure calculations as a function of the A, H and A, B distances. For many hydrogen bonds, it is found that, for a fixed A, B distance, the total energy passes through two minima as the position of H is varied; one, usually the deeper, corresponding to the normal "tautomer" A—H--B, the other corresponding to a rearrangement to a tautomeric form representable as A--H—B. It has been suggested by Lowdin that tautomeric rearrangements in the hydrogen bonds connecting purine and pyrimidine bases in nucleic acids may have genetic consequences, and in particular may play roles in mutagenesis, ageing, and tumour formation. It is thus of great interest to characterize tautomeric interconversion kinetics and equilibria.

The tautomeric rearrangement may be thought of in terms of the motion of a proton in a double-minimum potential well. Because of the relatively small protonic mass, it must be thought of as a quantum-mechanical particle, with the consequence that its motion from one minimum to the other can proceed even when its total energy is insufficient to exceed the intervening maximum. Passage between minima under these conditions is termed "tunnelling", and is the main mechanism for tautomeric conversion in the nucleotide base pairs. We thus see that quantum mechanics is involved in tunnelling problems in two ways: both in the determination of the electronic energy for each proton position, and in the discussion of the tunnelling itself.

ASMO calculations have been made for the energy profiles involved in tautomeric rearrangements of the naturally-occurring nucleotide base pairs; for guanine–cytosine, by Rein and Harris, and for adenine–thymine, by Sperber and Lunell. The tunnelling equilibria deduced from these profiles are qualitatively in excellent agreement with the values required to explain the incidence of spontaneous mutation in terms of the equilibrium concentration of the unusual tautomers in the ground electronic states. The observed radiation-induced mutation rates are also consistent with the altered equilibria calculated for excited and ionic electronic states of the base pairs. Moreover, the calculated proton tunnelling rates are sufficiently high to admit the suggested interpretation.

Bibliography

LOWDIN P. O. (1965) *Advances in Quantum Chemistry* **2**, 216.
PULLMAN B. and PULLMAN A. (1963) *Quantum Biochemistry*, New York: Interscience.
REIN R., CLAVERIE P. and POLLAK M. (1968) *Int. J. Quantum Chem.* **2**, 129.
RHODES W. (1961) *J. Am. Chem. Soc.* **83**, 3609.
TINOCO I. (1960) *J. Am. Chem. Soc.* **82**, 4785.

F. E. HARRIS and R. REIN

QUANTUM METROLOGY.

1. Introduction

Quantum metrology is the phrase which is used to describe the increasing use of phenomena from quantum physics, (a) to define and maintain the units of measurement of the *Système International* (*SI*), and (b) for precise metrology or measurement. One outstanding advantage of atomic quantum phenomena as standards is that it is believed that they are not unique but everywhere similar so that in principle the long chain of comparison between the standards used by industry or research laboratories and those of the national standards laboratories can be eliminated whenever it becomes necessary or convenient to do so.

Most physicists feel intuitively that there must be a natural system of units incorporating certain of the fundamental constants of physics, for example, the electronic charge e, or mass m, the Planck constant of action h, or the velocity of electromagnetic radiation c. At present we do not know enough about atomic physics to define a unique set of units involving these constants, but quantum metrology may be seen to represent the first steps in this direction. The phenomena which are used are chosen on strictly practical rather than philosophical grounds and may bear little relationship to those that will be used ultimately.

It would obviously be inconvenient if it became necessary to change all previous calibrations each time that a material definition was abandoned in favour of one involving quantum metrology. Continuity is therefore ensured by measuring the quantum phenomena to be employed with such precision in terms of the old unit that the whole of the experimental error results from the lack of reproducibility of the latter. This accounts for the apparently irrational choice of the values of the quantum constants described below, which are used to define the metre and the second.

2. Time and Frequency

Probably the most familiar and most precise example of quantum metrology is the "atomic clock". Most of the early work was done at the N.P.L., and the frequency measurement together with the United States Naval Observatory. This makes use of the transition between the hyperfine energy levels $F = 4$, $m_f = 0$ and $F = 3$, $m_f = 0$ of the fundamental state $2S_{1/2}$ of the caesium 133 atom unperturbed by external fields; the frequency assigned to this is 9 192 631 770 hertz (Hz) exactly (1 Hz = 1 c/s), thereby ensuring continuity with measurements made using the ephemeris second. Caesium clocks are believed to be capable of enabling the second to be defined to better than a few parts in 10^{12}. The actual precision is, of course, hard to establish since there is nothing to compare the frequency against with sufficient precision except another caesium standard, or a maser standard possessing a similar stability with poorer accuracy, however, the quoted figure refers to agreement between different methods of manufacture and also when various parameters are changed systematically. The establishment of the precision of an improved method of measurement is a recurrent problem in precise metrology.

The hydrogen maser which uses a quantum transition between the $F = 1$, $m_f = 0$ and $F = 0$, $m_f = 0$ levels of the $1S_{1/2}$ state in atomic hydrogen has not so far proved to be better than the caesium standard, principally because the transition is affected to a greater extent by external fields and there are frequency shifts of about 4 parts in 10^{11} as a result of wall collisions which are not yet very well understood. It has, however, been possible to obtain frequency stabilities of a few parts in 10^{13} as indicated by beating two masers together. The spectral purity resulting from the maser action has already proved useful in some applications. The present value quoted for the hydrogen maser frequency is 1 420 405 751·78 $\pm$ 0·03 Hz.

The atomic second was legally adopted as the SI unit of time by the 1967 General Conference of Weights and Measures. The other definitions of the second, the ephemeris second, with an uncertainty of about 2 in 10^9, or the mean solar second of Universal Time, with an uncertainty of about 5 parts in 10^9 are retained for astronomical purposes. The latter unit, however, has changed by as much as 7 in 10^8 during the last one hundred years as a result of a slow change in the rate of rotation of the Earth.

It is evident from the preceding, possibly as a result of the fundamental role that frequency plays in quantum phenomena, that time or frequency, especially when defined in terms of a quantum transition, is very precisely realizable and measurable. This forms the basis for much of quantum metrology, for measurements which are grounded in atomic and quantum physics can often be reduced to measurements of frequency through the relation

$$E = h\nu.$$

If the difference of energy E between two states of a system can be made to depend on the quantity to be measured, such as a voltage or a magnetic field, the quantity is thereby related to a frequency which can be readily and precisely measured.

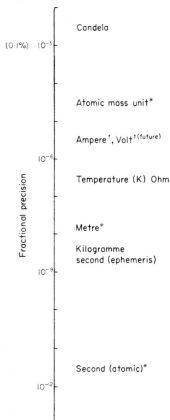

Fig. 1. The present precision with which some of the units of the International System can be realized (quantum effects used to *define and †maintain the unit).

3. Length

The realization of the metre is another example of quantum metrology, for since 1960 the metre has been defined as 1 650 763·73 wavelengths in vacuum of the radiation corresponding to the transition between the $2p_{10}$ and $5d_5$ levels of the ^{86}Kr atom. Prior to this the international prototype of the metre was used which defined the metre as the separation between two transverse lines marked on a bar of platinum–iridium alloy. The precision of intercomparison between the prototype and the standards preserved by certain countries was about 2 parts in 10^7, the precision being limited by the widths of the markings, dispari-ties between edges of the markings, and ultimately by the stability of the metal. By using quantum metrology the metre is now reproducible with a precision of at least 1 in 10^8, which is about 1 per cent of the width of the spectral line. Any improvements are limited by the line width as well as asymmetries in the intensity variation over the line.

At present the metre is realized interferometrically by the method of exact fractions. Two steps are normally required since the width of the line restricts the maximum path difference to about 0·5 m, the visibility of the fringes decreasing rapidly for path differences which are greater than this. It has so far not proved practicable to count these fringes at rates much above 10 to 100 fringes der second, mainly as a result of the low intensity of the krypton-86 lamp. Developments in the laser field have led to sources of high spectral purity, that is only a few Hertz wide, and the high intensity of these already allows fringes to be counted at rates in excess of 10 MHz or 1 m in 0·3 s. Lasers have fascinating technological applications, especially in the field of precise length measurement, but at present although short term stabilities of a few parts in 10^{10} have been achieved, the reproducibility is between two and ten times worse than that of the ^{86}Kr source.

With the advent of infra-red lasers such as the CO_2 laser which use narrower spectral lines, there is now the exciting possibility that the frequency of such a laser line can be measured directly in terms of the caesium frequency. This in turn would lead to a determination of the velocity of electromagnetic radiation c, with the same precision as the metre can be realized and it might then become practicable to define the metre in terms of the caesium frequency and a fixed value of c. Unfortunately our technology has advanced too far for the metre to be redefined such that $c = 3 \times 10^8$ m s^{-1} exactly instead of the presently accepted value of $2 \cdot 997,925 \pm 0 \cdot 000,003 \times 10^8$ m s^{-1}. If the metre were to be defined using a defined, albeit non-integral value for the velocity of electromagnetic radiation, it would, of course, require that any new spectral sources for improving the realization of the metre were such that their frequencies could be measured in terms of the caesium frequency or whatever comes to succeed it.

4. Mass

The unit of mass is the *kilogramme* (kg) and it is defined as the mass of a material standard known as the international prototype of the kilogramme. It is made in the form of a simple cylinder of platinum–iridium alloy whose height is equal to its diameter and other standard kilogrammes can be compared with it with a precision of about 7 parts in 10^9. At present there are no quantum atomic effects involving mass which can be measured within a factor of 10^3 of this precision, for example, the charge to mass ratio of the proton or the nuclidic mass unit are only known to about 2 in 10^5.

5. Electrical Units

The most precisely realizable electrical unit is the ohm, largely as a result of the development of the calculable capacitor at the N.S.L. in Australia and it is now realizable to about 2 in 10^7. It is defined in terms of the metre and the second. The magnetic constant μ_0 is also defined exactly and it is the ampere which is the basic unit of the International System of units and which must be determined in terms of the mechanical force per metre between two conductors which are separated *in vacuo* by 1 metre and carry the same current. In practice the wires are wound in the form of a solenoid and at the N.P.L. the Ayrton-Jones current balance is employed. The electrical forces are balanced by the gravitational force on a precisely known mass and hence the realization of the ampere, as with the unit of pressure, requires a knowledge of the acceleration due to gravity. The latter has been measured recently at the N.P.L. with a precision of 1·3 in 10^7. The electrical force is due to the magnetic field developed by the current and the calculation of the force requires a precise knowledge of the metrology of the coils through which the current is passing and hence heavy coil formers to achieve dimensional stability. The mass of these formers places a severe demand on the precision of the current balance, for they have masses of the order of a kilogramme whereas the masses which are required to balance the electrical forces are of the order of a few grammes. Thus realizing the ampere involves using the balance to a few parts in 10^9 of the total mass suspended from the beam. The ampere is at present realizable to a few parts in 10^6.

Unlike most of the other basic SI units the present definition of the ampere makes its value dependent on the metre and the second and this precludes the use of an independent atomic property for its definition. There are, however, methods using quantum effects which may be used to maintain the ampere. The first and incidentally historically the first example of quantum metrology, used the faraday which is the product of the Avogadro number and the electronic charge. This method of maintaining the ampere has now fallen into disuse; however, the faraday is still of interest in the evaluation of the fundamental constants and has been measured recently at the National Bureau of Standards of the United States. The difficulty of the method arises from the uncertainty in the mass of material which is transferred, as a result of isotopic and other chemical impurities, the formation of a sludge, and falling away of pieces from the electrodes. Despite these and other difficulties the N.B.S. determination has a precision of 7 parts in 10^6. The more favoured quantum effect makes use of the gyromagnetic ratio of the proton which is the constant of proportionality between the spin precession frequency of a proton and the magnetic flux density. If the magnetic flux is produced by passage of a current through a solenoid of precisely measured dimensions, then the ampere may be maintained by measuring the resulting proton spin precession frequency. Good agreement has been obtained between the N.P.L. and N.B.S. determinations of the gyromagnetic ratio of the proton and it has also been established that at the present level of precision the ratio is not dependent on magnetic field.

A more recent possibility has been the use of an effect that was predicted by a young British theoretician, B. D. Josephson, in 1962. If a current is passed between two superconductors which are separated by a small gap $\sim 10^{-9}$ m, which in practice is an oxide layer, then the potential difference between them increases in steps V, which are associated with the emission or absorption of microwave radiation of frequency ν given by

$$2 \, eV = (n/m) \, h\nu \qquad (2e/h = 483 \cdot 6 \text{ MHz}/\mu\text{V}),$$

where n and m are integers and the more favoured steps occur when n/m is also an integer. The energy gap between superconduction and normal conduction electrons is only a few millivolts and this restricts the total potential difference between the two superconductors to the same order. It is still too early to say whether the method will be as precise as the method of obtaining the ampere by the proton spin precession frequency although the present precisions differ only by about a factor of two.

It is only necessary to maintain any two of the three electrical units: current, resistance or voltage, and it is conceivable though not at present practicable, that these may one day be defined in terms of quantum phenomena and some of the other SI units such as mass could then become derived units. Incidentally, the units which were formerly known as the international volt and international ohm have fallen into disuse and the SI electrical units are the *"absolute" Volt*, *"absolute" ohm* and *"absolute" ampere*.

6. Temperature and Luminous Intensity

The unit of temperature is obtained by defining the triple point of water as 273·16°K, with the result that the kelvin is reproducible to about 4 in 10^7. Temperature is not an additive quantity in the same way are the other basic units, for example one cannot perform the equivalent of connecting two batteries in series, and the precision of measurement depends on the temperature which is being measured. Thus at 4°K the precision is 0·01°K and at 400°K it is 0·01°K, which give very different fractional precisions of 0·25 per cent and 25 parts per million respectively. The quantum methods which are being used at present in connection with temperature measurements are γ-ray emission from oriented nuclei for measuring very low temperatures, less than 0·01°K, and Planck's distribution law applied to black-body radiators at higher temperatures. Reliance on the latter also applies to the realization of the candela which is reproducible to $\frac{1}{2}$ per cent, international comparisons agreeing to within $1\frac{1}{2}$ per cent. Work is in hand at the

N.P.L. to calibrate detectors with a greater precision by using the calculable spectral distribution of the radiation emitted by high energy electrons when their trajectory is curved by a magnetic field. This is known as synchrotron radiation and is a form of Bremsstrahlung. It will then be possible to investigate the conformity of sources to black-body radiators in greater detail.

7. Angle

Recent work at the N.B.S. has led to the possibility of using the Moiré fringes formed by the X-ray illumination of two crystals whose atomic planes are at a small angle as an atomic angle standard. The method could lead to an improvement of the precision with which angles can be measured to 0·01 sec of arc.

8. Conclusions

In general, largely as a result of a careful choice of the phenomena which are employed in measurement, quantum metrology is not yet up against the limits set by the quantum theory of measurement which arise essentially from the Heisenberg uncertainty principle. The use of quantum phenomena should enable units of the SI system to be reproduced everywhere with high precision. The chief advantage would be the avoidance, where economic and necessary, of the present chain of comparison of the standards used in science and technology with those maintained by the various national standards laboratories and in turn the international intercomparisons of their standards with some agreed prototype. There is at present no evidence to show that quantum standards vary with time, there is on the other hand plenty of evidence that material standards do; for example metre bars, quartz crystal oscillators, standard cells and resistors all change with time. It is of course important to guard against such a variation and to ensure that a coherent system of units is maintained. One way of achieving this is by measuring as many of the physical and atomic constants as precisely as possible and such measurements are being made increasingly by the various national standards laboratories.

Quantum metrology is also finding increasingly important applications in technology, thus atomic clocks are used for navigational purposes and work at the N.P.L. has led to the development of the Mekometer, an instrument for measuring distances in the region of 10 m to 3 km with a precision of three parts in a million or $\pm 0 \cdot 1$ mm, whichever is the larger, which is of considerable use in surveying. Laser interferometry is being applied to the automatic control of machine tools.

It seems probable that by giving scientists and technologists more precise methods and units of measurement, quantum metrology will lead to discoveries as important as those of the geologists who followed the early continental explorers.

See also: Josephson effect.

Bibliography

COOK A. H. (1968) *J. Sci. Instrum.* (*J. Phys. E.*) Ser. 2, **1**, 73.

ESSEN L. (1961) Microwave frequency standards, in *Encyclopaedic Dictionary of Physics* (J. Thewlis Ed.), **4**, 675, Oxford: Pergamon Press.

DU MOND J. W. M. and COHEN E. R. (1961) Fundamental constants of atomic physics, in *Encyclopaedic Dictionary of Physics* (J. Thewlis Ed.), **3**, 319, Oxford: Pergamon Press.

PROSPERI G. M. (1967) Quantum theory of measurement, in *Encyclopaedic Dictionary of Physics* (J. Thewlis Ed.), Suppl. vol. 2, 275, Oxford: Pergamon Press.

Units and Standards of Measurement Employed at the National Physical Laboratory, Numbers 1 to 4, H.M.S.O.

B. W. PETLEY

QUARKS. Inspection of the spectrum of the hadrons (strongly interacting particles) reveals the existence of various multiplet structures which indicate the validity of certain symmetries. There are, for example, groups of levels of almost equal energy differing only in electric charge; these are the isospin multiplets which may be described by the use of the group SU(2). Considering the proton and the neutron, and their antiparticles, as basic isospin states leads to a means of constructing higher isospin multiplets by forming composite systems of these states. A system composed of a particle and an antiparticle leads to four states which break down into an isosinglet and an isotriplet. Symbolically

$$2 \otimes \bar{2} = 1 \oplus 3.$$

The triplet may be identified with the pion and the singlet with the η° meson. This can mean either that the pion is a bound state of physical particles identified with the basic states, or that the basic states are mathematical objects and the pion merely has the same isospin transformation properties as the combinations of the basic states by which it may be represented.

To incorporate the strange particles in this scheme requires the introduction of one more fundamental object of non-zero strangeness. This leads to a higher symmetry group, SU(3), which is the group of all three by three unitary matrices with determinant equal to one, and which contains SU(2) as a subgroup. The proton, the neutron and the Λ° hyperon were first considered as the basic states of the fundamental triplet, but attempts to form other states from this basis led to disagreements with the observed situation.

Considerable phenomenological success has been achieved by the "Eightfold Way" classification of Gell-Mann and Ne'eman, who placed the simplest known baryons and mesons into the eight-dimensional representation of SU(3) rather than into lower dimensional representations. This classification for the $J^P = \frac{1}{2}^+$ baryons is given in Fig. 1 in the form of a plot of hypercharge Y ($Y =$ baryon number plus strangeness) against the third component of isospin, I_3. Note that the electric charge is given by $Q = I_3 + \frac{1}{2} Y$. There are two isodoublets (p, n and Ξ^-, Ξ°), an isotriplet ($\Sigma^-, \Sigma^\circ, \Sigma^+$) and an isosinglet ($\Lambda^\circ$). The octet for the

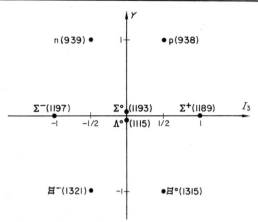

Fig. 1. $J^P = \tfrac{1}{2}^+$ *baryon octet. Masses are indicated in MeV.*

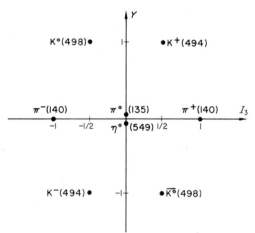

Fig. 2. $J^P = 0^-$ *pseudoscalar meson octet. Masses are indicated in MeV.*

pseudoscalar $J^P = 0^-$ mesons is shown in Fig. 2. Here, the isosinglet is the η°, which was discovered after the proposal of the Eightfold Way. In the $J^P = 1^-$ vector meson octet there are three possible states at $Y = 0$, $I_3 = 0$, instead of the two expected—the ϕ and the ω mesons are both candidates for the isosinglet state. SU(3) allows a pure singlet representation as well as an octet and in the case of the vector mesons these representations have become mixed. The physical ϕ and ω particles are mixtures of the SU(3) octet and singlet states. Justification for this mixing is obtained from considerations of SU(6) symmetry where the intrinsic spin properties of particles are taken into account (see later). There is evidence in the case of the pseudoscalar mesons that the η° is not a pure SU(3) octet state and a small amount of mixing is present. The combination of the meson and baryon octet according to the rules of the SU(3) group gives rise to a set of possible multiplets:

$$8 \otimes 8 = 1 \oplus 8 \oplus 8 \oplus 10 \oplus \overline{10} \oplus 27.$$

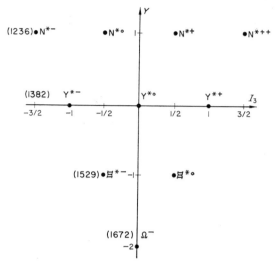

Fig. 3. $J^P = \tfrac{3}{2}^+$ *baryon decuplet. Masses are indicated in MeV.*

The set of $J^P = 3/2^+$ baryon resonances appeared to fit into the decuplet except that one particle was missing. This particle, the Ω^-, has since been observed. The $3/2^+$ baryon decuplet is shown in Fig. 3.

The lowest dimensional representation of SU(3) is the triplet and it is conceivable that particles transforming according to this representation might exist. The existence of these triplets would be the simplest way to understand the SU(3) model. In 1964 Gell-Mann and Zweig independently suggested the existence of an SU(3) triplet of spin $\tfrac{1}{2}$ particles, each with baryon number B equal to $\tfrac{1}{3}$. These three particles are known as quarks. The SU(3) triplet contains an isodoublet ($I = \tfrac{1}{2}$) of strangeness $S = 0$ which is denoted n, p, and an isosinglet ($I = 0$) which is assigned strangeness $S = -1$ and is denoted λ. (These symbols for the quarks must not be confused with the proton, neutron and lambda particles.) The quarks have positive parity by definition. Application of the rule

$$Q = I_3 + \tfrac{1}{2}Y = I_3 + \tfrac{1}{2}B + \tfrac{1}{2}S$$

then gives for the charges e_q of the quarks p, n, λ the fractional values $\tfrac{2}{3}e$, $-\tfrac{1}{3}e$, $-\tfrac{1}{3}e$, respectively, where e is the charge of the proton. The quantum numbers of the quarks are listed in Table 1. For the antiquarks the quantum numbers B, P, I_3, Y, S and e_q are the opposite of those of the corresponding quarks. Quarks can decay weakly into each other; the lightest quark

Table 1. Quantum Numbers of Quarks

	J^P	B	I	I_3	Y	S	e_q/e
p	$\tfrac{1}{2}^+$	$\tfrac{1}{3}$	$\tfrac{1}{2}$	$\tfrac{1}{2}$	$\tfrac{1}{3}$	0	$\tfrac{2}{3}$
n	$\tfrac{1}{2}^+$	$\tfrac{1}{3}$	$\tfrac{1}{2}$	$-\tfrac{1}{2}$	$\tfrac{1}{3}$	0	$-\tfrac{1}{3}$
λ	$\tfrac{1}{2}^+$	$\tfrac{1}{3}$	0	0	$-\tfrac{2}{3}$	-1	$-\tfrac{1}{3}$

state, however, must be stable. The triplet quark and antiquark representations are shown in Fig. 4. Other three-dimensional representations have been proposed which contain integral baryon number and charge; however, these representations always involve more than one triplet and there are no fundamental reasons for choosing them rather than the quark triplet.

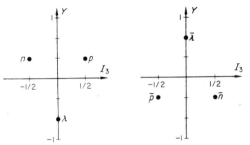

Fig. 4. *Quark and antiquark representations.*

In the quark model the quarks are supposed to be physically realizable objects rather than mere mathematical entities. Hadrons are regarded as being many-quark or quark-antiquark bound states due to some strongly attractive force, the precise nature of which is unknown. Whether quarks can exist as free particles is an independent question; they may, due to some selection principle, exist only in certain bound combinations having integral charge and baryon number, or they may be some kind of quasiparticles, such as phonons which have meaning only inside crystal lattices.

In the quark model the mesons are considered as *quark-antiquark* states (q$\bar{\text{q}}$). These decompose in terms of the dimensions of SU(3) representations according to

$$3 \otimes \bar{3} = 1 \oplus 8.$$

The baryons are composed of three quark states (qqq), which decompose according to

$$3 \otimes 3 \otimes 3 = 1 \oplus 8 \oplus 8 \oplus 10.$$

It is remarkable that all the known hadrons which have been fitted into SU(3) multiplets are in accordance with this scheme. It is assumed that the charge, hypercharge, baryon number and isospin of each hadron is simply the sum of those of the corresponding quarks.

This composite model of hadrons has received further support from the introduction of the higher symmetry group SU(6) which incorporates, in a non-relativistic approximation, the hadron spins with their SU(3) properties. There are now six quark states as each of the previous three states can have spin up or spin down. The hadron spins are supposed to be the vectorial sums of the spins of the constituting quarks, treated in a non-relativistic manner. There is assumed to be no coupling of the spin with the orbital angular momentum of relative motion of the quarks.

In SU(6) there are 36 q$\bar{\text{q}}$ states:

$$6 \otimes \bar{6} = 1 \oplus 35.$$

The content of these states is summarized in Table 2. The group of 24 states and the one of three states are the SU(3) vector meson octet and singlet, respectively; they have total spin one (parallel quark spins). The eight states are the SU(3) pseudoscalar meson octet with spin zero (antiparallel quark spins). The remaining spin zero state is a singlet in both SU(3) and SU(6).

Table 2
Composition of Quark–Antiquark States in SU(6)

SU(6) part	SU(3) part	SU(2) part	Number of states
35 {	8	triplet	$8 \times 3 = 24$
	8	singlet	$8 \times 1 = 8$
	1	triplet	$1 \times 3 = 3$
1	1	singlet	$1 \times 1 = 1$

In the limit of exact SU(6) the SU(3) vector meson octet and singlet are degenerate. Complete $\phi - \omega$ mixing would then be expected—the partial mixing observed is due to the approximate nature of the SU(6) symmetry. For the pseudoscalar mesons the SU(3) singlet is also an SU(6) singlet and so only small mixing with the SU(3) octet is expected.

The p$\bar{\text{q}}$ model allows only octet and singlet SU(3) states for the mesons; only these states are observed. The lowest of these states are those with angular momentum $L = 0$. There are two sets of $L = 0$

Table 3
Quark Content of Pseudoscalar and Vector Mesons

Vector meson	Pseudo-scalar meson	Quark content
ϱ^+	π^+	$p\bar{n}$
ϱ^0	π^0	$\frac{1}{\sqrt{2}}(p\bar{p} - n\bar{n})$
ϱ^-	π^-	$\bar{p}n$
K^{*0}	K^0	$n\bar{\lambda}$
K^{*+}	K^+	$p\bar{\lambda}$
$\bar{K}^{*0}$	$\bar{K}^0$	$\bar{n}\lambda$
K^{*-}	K^-	$\bar{p}\lambda$
ω_8	η_8	$\frac{1}{\sqrt{6}}(\bar{p}p + \bar{n}n - 2\lambda\bar{\lambda})$
ϕ_1	η'_1	$\frac{1}{\sqrt{3}}(\bar{p}p + \bar{n}n + \bar{\lambda}\lambda)$
ω	—	$\frac{1}{\sqrt{2}}(p\bar{p} + n\bar{n})$
ϕ	—	$-\lambda\bar{\lambda}$
—	η	$-\eta_8 \sin 11° + \eta'_1 \cos 11°$
—	X_0	$\eta_8 \cos 11° + \eta'_1 \sin 11°$

states, one with spin zero and the other with spin one, which are associated with the observed nonets of pseudoscalar and vector mesons respectively. The quark content of these mesons is given in Table 3. Higher spin mesons have been observed and have been fitted into nonets which are assumed to be L-excited $q\bar{q}$ systems rather than higher representations of the type $qq\bar{q}\bar{q}$.

The baryon states in the quark model are qqq bound states with total angular momentum L. In SU(6) this gives

$$6 \otimes 6 \otimes 6 = 20 \oplus 56 \oplus 70 \oplus 70.$$

The 56-plet accommodates the low-lying baryon states. It consists of an SU(3) octet with spin $\tfrac{1}{2}$ and an SU(3) decuplet with spin 3/2, again in accordance with observation. The quark content of the particles in these multiplets is given in Table 4; for the $3/2^+$ decuplet all quark spins are parallel whilst the $\tfrac{1}{2}^+$ octet has one pair of quark spins antiparallel.

Table 4

Quark Content of Low-lying Baryon States

Particle		Quark content
$\tfrac{1}{2}^+$ octet	$\tfrac{3}{2}^+$ decuplet	
—	N^{*++}	ppp
p	N^{*+}	ppn
n	N^{*0}	pnn
—	N^{*-}	nnn
Σ^+	Y^{*+}	$pp\lambda$
Σ^0, Λ^0	Y^{*0}	$pn\lambda$
Σ^-	Y^{*-}	$nn\lambda$
Ξ^0	Ξ^{*0}	$p\lambda\lambda$
Ξ^-	Ξ^{*-}	$n\lambda\lambda$
—	Ω^-	$\lambda\lambda\lambda$

There are two puzzling features about the quark model which have so far not been satisfactorily explained. The forces in the qqq system give rise to an $L = 0$ ground state which has an antisymmetric space wave function. For an antisymmetric $L = 0$ state the kinetic energy is high and so it is surprising that this represents the ground state of the system. The second problem is that apparently only $q\bar{q}$ and qqq bound states occur, although there are no fundamental objections to $qqqq$, $qqq\bar{q}$, etc. If the forces between quarks are of two-body type it is difficult to see how one can get attraction for both $q\bar{q}$ and qqq systems. Assuming qqq is a tightly bound system then there must be saturation at $N = 3$ for N-quark systems as no $qqqq$ states are observed. There are indications at present (1968) of the possible existence of positive strangeness resonances in the K$^+$N system; on the quark model these would require combinations such as $qqq\bar{q}q$.

SU(3) symmetry breaking considerations give rise to various mass formulae between particles in SU(3) multiplets. These can all be readily derived on the quark model by assuming that the mass splittings are due to a mass difference Δ between the strange and non-strange quarks:

$$m_n = m_p = m; \quad m_\lambda = m + \Delta; \quad m_q = m_{\bar{q}}.$$

This model predicts a relation between meson and baryon resonance masses

$$m_\phi - m_{K^*} = m_{N^*} - m_{Y^*}$$

which is well obeyed and cannot be obtained from any symmetry considerations.

If the magnetic moment of a quark is assumed to be proportional to its electric charge, and the magnetic moment operators of hadrons are assumed to be additive in the magnetic moment operators of the constituting quarks, then the quark model can be used to predict ratios between the magnetic moments of hadrons. In particular the model predicts $\mu_{proton}/\mu_{neutron} = -3/2$ (experimentally $= -1.47$). This result may also be obtained from SU(6) symmetry, but cannot be achieved by any other theories. Application of the quark model to radiative transition processes involving mesons results in expressions between the transition magnetic moments (and hence the partial decay widths) and the proton magnetic moment. These predictions are in good agreement with experiment and cannot be obtained from any symmetry.

Another type of quark model is that which has been applied to the high energy scattering of hadrons (i.e. for incident laboratory momenta > 4 GeV/c). The basic assumption made is the additivity of scattering amplitudes. The amplitude for the elastic scattering of two hadrons is written as the sum of all possible two-body quark elastic scattering amplitudes multiplied by appropriate form factors for the quarks. For forward elastic processes (zero momentum transfer) the quark form factors reduce to 1. Use of the optical theorem then enables relations between total cross-sections to be obtained, for example,

$$\lim_{\text{energy} \to \infty} \frac{\sigma_T(pp)}{\sigma_T(\pi p)} \to \frac{3}{2}.$$

Extrapolation of known total cross-sections to infinite energy gives a ratio of 36/22. Many other similar results are obtained which are in good agreement with experiment at high energies.

Consideration of non-forward amplitudes shows that the momentum transfer dependence of hadron–hadron diffractive scattering may be controlled by the quark form factors and the asymptotic quark–quark amplitude. From this picture the success of the additivity approximation for scattering amplitudes can be justified in two extreme cases. If the momentum transfer dependence of the diffraction peak comes mainly from the quark–quark amplitude then the quark has a large size for diffraction (of the order of the hadron size) and very weak absorption so that multiple scattering processes would be unlikely. In the opposite case where the dependence is determined mainly by the form factors the quarks are small and separated by distances large compared with their size. Again multiple scattering would be rare.

This "long-range" quark model of bound states much bigger than the quarks is the one usually adopted. The quark–quark force is mediated by a particle of mass much smaller than the mass of the quark. In this model the kinetic energies of the quarks in the bound states are small and the non-relativistic approximation is justified; it is natural on this model to regard hadron states as excited states of the $q\bar{q}$ and qqq systems.

A disadvantage of the quark model is that no quarks have been observed. Present (1968) experimental limits imply that if quarks exist their mass is greater than 4 GeV. Composition of hadrons from quarks then requires massive binding energies. This is unusual as the binding energies of composite states have always been observed to be smaller than the mass of the composite state.

There have been many "quark-hunting" experiments—so far no quarks have been observed. Two separate classes of experiments have been performed. The first class assumes one quark is stable and exists as a dilute constituent of the Earth's crust. The origin of these quarks could be: from production in the initial expansion of the universe (on the "big bang" theory); from acceleration by the same mechanism that produces cosmic rays; or from production in the atmosphere, since the formation of the Earth, from high energy cosmic rays. The quark is assumed to be fractionally charged and bound stably to nuclei in very tight Bohr orbits. This may give rise to peculiar physical or chemical properties of portions of the Earth's crust. Experiments give upper limits for the quark concentration per nucleon (C_q) for various materials. For example,

$$C_q \leq 10^{-17} \text{ for iron meteorites};$$

$$C_q \leq 3 \times 10^{-29} \text{ for sea water}.$$

Theoretical estimates of C_q are unreliable, depending critically on the model used. Cosmological estimates give in general $C_q \approx 10^{-9}$ to 10^{-18}.

The second class of experiments attempts to detect quarks produced from high energy collisions. These experiments lead to upper limits for quark production cross-sections. Estimates of these cross-sections can be made theoretically; they are a strong function of the quark mass. The experimental limits then become lower limits on the quark mass. Accelerator experiments on proton–proton collisions at incident momenta of 30 GeV/c indicate the quark production cross-section is less than 10^{-35} cm^2. This in turn indicates $m_q > 4$ GeV. Cosmic ray experiments have also been performed which set a limit on the quark flux in cosmic rays of $\leq 10^{-10}$ cm^{-2} sterad^{-1} sec^{-1}. These experiments in general rely on the anomalous ionization produced by quarks.

The fact that no quarks have been seen could have several explanations:

(1) They may be very massive.
(2) They may decay strongly into bound quark systems such as qq with mass less than m_q and fractional charge possibly greater than $\frac{2}{3}e$.
(3) They may not exist as free particles.
(4) They may not exist.

Bibliography

DALITZ R. H. (1967) *Thirteenth International Conference on High Energy Physics, Berkeley, 1966*, University of California Press.
GELL-MANN M. and NE'EMAN Y. (1964) *The Eightfold Way*, New York, Amsterdam: W. A. Benjamin.
GOLDBERG H. M. (1967) Particle symmetries, in *Encyclopaedic Dictionary of Physics* (J. Thewlis Ed.), Suppl. vol. 2, 225, Oxford: Pergamon Press.
KERNAN A. (1961) Fundamental particles, in *Encyclopaedic Dictionary of Physics* (J. Thewlis Ed.), **3**, 325, Oxford: Pergamon Press.
LEDERMAN L. (1967) *Comments on Nuclear and Particle Physics* **1**, 155.
LIPKIN H. J. (1967) *Proceedings of the Heidelberg International Conference on Elementary Particles*.
SQUIRES E. J. (1967) *Particle Interactions at High Energies*, p. 380, London, Edinburgh: Oliver & Boyd.
VAN HOVE L. (1967) *Comments on Nuclear and Particle Physics* **1**, 8.
WILLIS W. T. (1967) High energy physics, in *Encyclopaedic Dictionary of Physics* (J. Thewlis Ed.), Suppl. vol. 2, 116, Oxford: Pergamon Press.

S. J. SHARROCK

R

RADAR ASTRONOMY. The word "radar", an acronym for "radio detection and ranging", generally implies the sending and receiving of man-made radio signals. Astronomy by means of radar has included measurements of the Moon, planets and the Sun, made by transmitting powerful radio signals and observing the weak echo that returns. Radar astronomy is distinguished from passive radio astronomy and the more familiar optical astronomy by the fact that the probing electromagnetic radiation is generated by man.

If we exclude radar observations of the Earth's ionosphere and atmospheric ionization caused by meteors or by the aurora, then the Moon is the nearest extra-terrestial radar target. Contact with the Moon by completely Earth-based radar has: (1) told us about general surface roughness and dielectric constant, (2) improved lunar epherides, (3) shown that young craters scatter particularly well, (4) shown differences between radar and optical reflectivity. By preserving radio-frequency Doppler shift and pulse round-trip delay, radar maps of the Moon can be made with the resolution of ground-based optical photographs.

Venus, Mercury and Mars are accessible targets to Earth-based radar in decreasing order of detectability. Study of their orbits, surfaces, atmospheres and rotations has followed the development of powerful transmitters, large antennas and sensitive receivers. Concealed from optical observation by thick clouds, the surface of Venus presents to penetrating radar a non-uniform surface with several strongly scattering regions. Radar has revealed that Venus is in retrograde rotation and, furthermore, the same meridian on Venus faces the Earth during each conjunction (exactly five rotations of Venus with respect to the Sun occurs between conjunctions). This curious "dance" of Venus and the Earth suggests large tidal torque coupling between the two bodies.

Radar was also responsible for the discovery in 1965 that Mercury rotates three times on its axis for every two passages through perihelion. This means that instead of always turning the same face towards the Sun (as the Moon does to the Earth), Mercury rotates so as to present the same face towards the Sun at alternate passages to perihelion (i.e. closest approach). This phenomenon had escaped detection by optical observations, which had always been taken to indicate synchronous rotation and revolution like the Moon around the Earth. Since this discovery, this type of resonance has been shown theoretically to be explained by the large ellipticity of Mercury's orbit.

The large size of the Sun suggests that it would be a detectable target for Earth-based radar. Unlike the Moon, the Sun does not present a solid surface but is surrounded by intense ionization capable of reflecting radio waves well outside the visible photosphere. Short wavelengths penetrate deeply and are absorbed but echoes have been obtained at wavelengths longer than 6 m. The echoes are variable in strength, position and Doppler shift indicating that the plasma reflector is variable and exceedingly complex. Strong echoes are largely correlated with daily solar activity as well as with the 11 year sunspot cycle.

The strength of radar astronomy stems from: (1) ability to measure time periods with high accuracy, (2) selectability of the illuminating radio-frequency transmissions, (3) digital techniques for processing, storing and replay of data.

Bibliography

DYCE R. B. (1966) Radar studies of the solar system, *Discovery* **27**, 22.

EVANS J. V. and HAGFORS T. (1968) *Radar Astronomy*, Massachusetts: M.I.T. Press.

PETTENGILL G. H. (1966) *Radar Astronomy*, International Science and Technology, Oct. 1966, 72.

PETTENGILL G. H. and SHAPIRO I. I. (1965) Radar astronomy, *Annual Review of Astronomy and Astrophysics* **3**, 377.

THOMSON J. H. (1967) Planetary radar, in *Encyclopaedic Dictionary of Physics* (J. Thewlis Ed.), Suppl. vol. 2, 246, Oxford: Pergamon Press.

R. B. DYCE

RADIATION CHEMISTRY, INDUSTRIAL APPLICATIONS OF.

Radiation-initiated Processes

Few chemical systems will react spontaneously; in most cases energy has to be provided before reaction will commence. Industrial reactions are commonly initiated by heat, which is convenient and cheap, but other forms of energy can be used and sometimes bring about specific results. Ultra-violet light, for example, has a selective effect on the chlorination of toluene, producing substitution in the side chain, whereas heat encourages nuclear substitution. The radiations emitted by radioactive substances are highly energetic and can also initiate reactions when absorbed in a chemical system. Table 1 below lists the energy characteristics of various types of radiation.

Despite these high values, the heat-equivalent of radioactive radiations is quite small; 10,000 curies of an isotope with an emission energy of 0·67 MeV will dissipate $2·5 \times 10^{20}$ eV/sec, i.e. less than 10 cal.

Table 1. Quantum Energies (in MeV)

Visible light	10^{-6}
Ultra-violet light	$10^{-6} - 10^{-5}$
X-rays	$5 - 10^{-2}$
Gamma-rays	$10^{-2} - 3·5$
Beta-rays, electrons	$10^{-5} - 250$
Alpha-rays	$2 - 400$

Radioactive radiations transfer their energy to neutral molecules by collision or interaction. The processes involved are complex, but the overall effect of the transfer is to ionize or excite molecules in the path of the incident radiation. The electrons ejected when ionization occurs may also excite or ionize further molecules. Thus the energy transferred at the primary encounter may ultimately be distributed by a cascade of secondary electrons. Successive collisions de-activate these electrons which finally disappear by the combination with a positive ion. The resultant neutral molecule is highly excited since it possesses excess energy at least equivalent to its ionization potential.

These primary processes are almost instantaneous, taking place in a time interval which is short compared with the relaxation time of a chemical bond. The probability of the intermediate charged species entering into chemical reaction is therefore small. There is evidence for the reaction of positive ions with some molecules, but such reactions are of minor interest in applied radiation chemistry. The main chemical effects of radiation are accounted for by the residual excited molecules, which achieve stability by molecular rearrangement or dissociation into free radicals, which in turn initiate chemical reaction.

Radiation chemistry and photochemistry are superficially similar in that free-radical processes are common to both, but there are fundamental differences. Radiation chemistry is usually defined as the study of chemical effects produced by radiations having particle or quantum energies in excess of 50 eV. Photochemical processes are associated with quanta of less than 50 eV which are small enough to be absorbed completely by one molecule or chromophore, and, provided the optical density of the medium is not too great, energy will be deposited relatively uniformly throughout the system. In radiation chemistry, the incident radiation is rarely absorbed in a single encounter and many molecules are affected. These molecules are localized along the track of the incident radiation and the energies involved are such that ionization occurs as well as excitation.

Radiation Sources

The radiations most applicable to industrial processing are gamma-rays and electrons, each of which has special uses. Electrons do not penetrate deeply and are most valuable for the treatment of surfaces and thin films. Since they can be generated in great intensity by machines and their energy is lost in a short distance, very high dose-rates are available. Gamma-rays are highly penetrating. A 1 MeV electron deposits its entire energy on passing through 0·5 cm of water, whereas a gamma-photon of similar energy will only transfer 10 per cent of its energy in traversing 10 cm of water. Gamma radiation is therefore most advantageous for systems requiring treatment in depth or bulk at relatively low dose-rates. The efficiency of many chemical reactions is greatest at low radiation intensities and such reactions are well suited to initiation by gamma radiation. Sources of potential value for radiation processing are: nuclear reactors, fission products, radio-isotopes and electrical accelerators.

Nuclear reactors. Large amounts of energy are released in the fission process and can be made available for chemical synthesis. Proposals have been made for chemonuclear reactors producing industrial quantities of basic chemicals such as nitric acid and hydrazine. Most of the fission energy of a uranium nucleus appears as the kinetic energy of the resulting fission fragments, and this energy could best be used in a reactor with uncanned fuel elements, particularly for gaseous reaction mixtures. A major disadvantage is the contamination of the synthesized material by radioactive fission products. This problem could be overcome in reactors employing canned fuels, but a much smaller fraction of the total fission energy would be available, and the neutron component of the reactor radiation could still induce radioactivity in the product. These considerations limit syntheses to those involving elements of low atomic number which are not readily radioactivated. While the chemonuclear reactor undoubtedly offers the prospect of massive quantities of cheap radiant energy, the problems involved do not encourage an early application.

Fission products. The initial impetus for the industrial application of radiation chemistry was generated by the availability of radioactive fission products from nuclear reactors, but subsequent experience has shown that fission products are not necessarily the most convenient sources of radiation. The utilization of beta-particles from fission products is not easy, since their penetration is limited, and much of their energy is lost by self-absorption, or by absorption in the robust isotope container essential for an industrial process. While such restrictions do not apply to gamma-emitters whose radiation is penetrating, chemical separation of the initial mixed fission products is necessary to obtain a source of uniform and adequate half-life.

The widespread use of fission products has, therefore, not been encouraged except in one particular instance. The periodic purification of nuclear fuels is essential to remove the accumulated fission products which would otherwise inhibit the fission process by neutron absorption. The radioactivity of freshly withdrawn fuel elements is too great for immediate purification and they are usually retained for several months so that the shorter lived fission products may decay. During this period, about half of the total fission product energy is dissipated. Before purification, the fuel elements are stored in water deep enough to provide radiation shielding. With relatively minor modifications, a fuel element storage pond can be converted into an irradiation assembly of considerable strength. Such an assembly requires a regular supply of fresh fuel elements, and is best sited near a nuclear reactor. Means for changing the configuration of the elements under water are also required and the radiation process should be one which can be adapted to underwater operation.

A typical spent fuel assembly of this kind has been in operation in the United Kingdom at the Atomic Energy Research Establishment, Harwell since 1958. As many as 100 fuel elements can be accommodated, each of which may have an initial activity of between 50 and 100 kilocuries. It is in regular commercial use for the radiation treatment of polythene cable.

Radioisotopes. The most widely used source of radiation for industrial processing is the man-made isotope cobalt-60, although caesium-137 is also in use. Cobalt-60 emits two gamma photons per disintegration with convenient energies of 1·17 and 1·33 MeV; its half-life is 5·3 years. Source fabrication presents few problems since cobalt-60 can be used in the metallic form and manufactured by the activation of natural cobalt in a nuclear reactor. The neutron flux in some power reactors is optimized by the insertion of neutron absorbers in the reactor core. The demand for cobalt-60 is now such that natural cobalt is used as the absorber in several reactors, making available a continuous supply of activated material.

Isotopic sources emit energy continuously and are ideally suited for processes requiring radiation for 24 hr each day. Their energy decreases gradually at a rate dependent on the half-life of the radioactive isotope. In the case of cobalt-60, the source activity can be maintained by the addition of 12·5 per cent of the nominal curie strength each year. Considerable experience of cobalt sources has been obtained from plants constructed for the radiation sterilization of medical supplies. It has been estimated that the annual value of material being treated in this way currently amounts to £20,000,000 in the United Kingdom alone, and plants with design loadings of up to 10^6 curies have been in operation for some years. The sources in use and envisaged for direct chemical processing are in general smaller, of the order of several thousand curies.

Electron accelerators. Electrical machines generally provide a means of producing more intense sources of energetic electrons than radioisotopic sources. Most commonly, electrons emitted from a heated filament are accelerated *in vacuo* by an applied potential, before emerging through a thin window. The beam of electrons is then scanned across the working surface in the same way as the image is formed in a television tube. Recently the suitability of electron accelerators for the processing of surface coatings has encouraged the development of simple low-energy machines which provide a continuous curtain of electrons rather than a scanned beam. This approach reduces the instantaneous dose rate to the working surface which is an advantage in many systems. Such machines allow increased working speeds as well as being compact and relatively inexpensive.

Industrial Applications

Yields in radiation chemistry are expressed in terms of the G-value, which is defined as the number of molecules transformed by each 100 eV of absorbed energy. The amount of product formed in any radiation-initiated reaction will thus be directly related to the G-value, the radiation dose and the molecular weight. For a given radiation dose it is apparent that the maximum amount of product will be formed when the G-value or the molecular weight is high. Radiation-initiated reactions in polymeric systems proceed with high efficiency since these have high molecular weights. High G-values are associated with chain reactions where the free radicals formed by radiation react and regenerate further radicals, themselves capable of propagating a sequence of reactions. In such cases radiant energy is most effective, since once the primary dissociation has occurred, the reaction is self-perpetuating. Chain reactions occur in various halogenation and oxidation reactions for which the G-values may exceed 10^6. These two considerations of high molecular weight and G-value have had the greatest influence on the development of industrial radiation chemistry.

Reactions in polymeric systems. Most polymers can be considered as arrangements of long chains of very high molecular weight. The effect of radiation is to break chemical bonds. If bond rupture occurs in the polymer backbone, molecules of lower molecular weight are formed and the polymer is degraded. However, if side chains are attacked it is possible for two adjacent molecules to become united by radical combination, and a bridge or cross-link is formed. In this way an essentially one-dimensional structure is converted into a three-dimensional network. Cross-linking changes the properties of a polymer and it has been shown that for a polymer of molecular weight 50,000, an average of 0·5 cross-links per molecule is sufficient for the new properties to become apparent. At elevated temperatures, uncross-linked polymers exist as viscous liquids, but in the presence of cross-links the three-dimensional network is maintained and the softened polymer has elastic properties.

This effect has been applied to polythene for a variety of uses. Polythene-coated wires have their maximum working temperature increased to 275°F by irradiation. In the same way polythene insulation blocks can be treated to retain or return to their original shape, if deformed during soldering operations. This "memory" effect of irradiated polythene has been used to create novel materials. Irradiated polythene, in the form of either tube or sheet, can be softened by heat and then stretched to a larger, thinner sheet which is cooled under tension. If the film is subsequently heated, it contracts to its original size. Such shrinkable materials find application in food packaging and shrouding electrical conductors. These uses account for the production of several million pounds of these materials each year, in the United States alone.

Polymerization reactions themselves can be initiated by radiation, but although no industrial application of direct polymerization has so far been established, several processes are operational for the modification of existing materials. The irradiation of a pre-formed polymer results in the formation of active radical sites. In the presence of a suitable monomer, these sites initiate further polymerization, and new chains grow from the parent polymer. These added chains are chemically bonded to the original macromolecule, forming graft co-polymers which combine the properties of their constituents. The possibilities of this technique have been studied extensively, particularly for the improvement of natural and synthetic textiles. Early in 1967 it was announced in the United States that a production line had been established for the manufacture of 400,000 yd a week of a grafted cotton-polyester fabric with improved soil release and crease resistance.

A direct polymerization reaction is also employed in the manufacture of wood-plastic composites. These novel materials are prepared by impregnating natural wood with selected vinyl monomers, which are subsequently cured *in situ* by high energy radiation. In this way the natural pore structure of the wood is filled with an inert polymer which is distributed uniformly through the bulk of the timber. Wood-plastic composites combine many of the desirable properties of the natural and synthetic materials used in their manufacture.

They are more dense than untreated wood and their specific gravities usually fall between 0·85 and 1·15. The hardness of the wood is much increased, as is the compressive, bending and shear strength. A disadvantage of natural wood is its tendency to shrink and expand with changing atmospheric conditions, which may result in warping or splitting during fabrication and use. This difficulty is overcome in wood-plastic composites, which possess great dimensional stability and are also resistant to decay, fungi and insect attack.

The materials may be finished to a high gloss, which enhances the natural figuration of the wood grain. The polish is durable and is not affected by cold or hot water. Domestic items fabricated from wood-plastic composites may be repeatedly washed in dishwashing machines and can even withstand boiling water for prolonged periods. The hardness of the surface ensures resistance to abrasion and scuffing. These properties are particularly attractive in pattern-making and for textile shuttles and similar woodware. Cutlery with moisture-resistant handles is now being manufactured in the United Kingdom and the current annual production (1968) amounts to over a quarter of a million handles.

One further application of direct polymerization is the curing of surface coatings without heat. Paints comprising mixtures of vinyl compounds and polyester resins may be hardened to form a three-dimensional network similar to that presented in cross-linked polymers. Accelerated electrons of energies not greater than 300 kV are quite adequate for curing, which is completed in about 0·1 sec. The process is easily applied to heat-sensitive substrates, and can be operated continuously. The need for long curing ovens is eliminated, and the use of solvent- and catalyst-free paints reduces maintenance and makes solvent-recovery equipment unnecessary. Large scale demonstration equipment capable of treating up to 10,000 ft^2 of material per hour is currently being commissioned by the United Kingdom Atomic Energy Authority and the full commercial aqplication of this technique should not be long delayed.

Synthetic uses of radiation. Numerous studies of the use of high energy radiation for industrial synthesis have been reported. Halogenation, sulphochlorination and sulphoxidation reactions appear to be of particular interest. The Dow process for the hydrobromination of ethylene has been operational in the United States since 1963, and has a capacity in excess of 500 tons of ethyl bromide per year. The plant, which contains 3000 curies of cobalt-60, is compact, and adequate radiation shielding is provided by burying the reaction vessel in concrete 4 ft below ground level. The gaseous reactants are introduced into a constantly-cycled stream of ethyl bromide and react on admission to the irradiation vessel. As more ethyl bromide is produced it overflows into a distillation system and is eventually neutralized and dried before shipment in tankers.

One other industrial process has been described. Persulphonic acids may be produced by the reaction of paraffinic hydrocarbons with sulphur dioxide and oxygen under the influence of radiation. Subsequent spontaneous reactions lead to the formation of alkane sulphonic acids by a chain process. The sodium salts of these compounds have attracted considerable interest because they are good detergents and are completely biodegradable. Their susceptibility to bacterial attack has been established by a full-scale demonstration in a sewage system and the manufacturers, Esso, have stated that sodium alkane sulphonates are equivalent in cleansing power to present-day detergents. It is believed that the production of this material is past the pilot plane stage, but full-scale operation will only be achieved when there is a large demand for a completely biodegradable detergent.

The modest requirements of radioactivity for plants such as that for the Dow ethyl bromide process, together with ease of operation, compactness and safety, augur well for the introduction of further processes. Recent comparisons of radiation with other free radical initiators such as chemical catalysts and ultra-violet light, suggest that in many cases radiation is now competitive in cost. These facts, combined with the increasing availability of new synthetic feedstocks from the petrochemical industry, such as n-paraffins and alpha-olefins which are suitable starting materials for many radiation-initiated reactions, may in certain cases offer further economies by process simplification.

Bibliography

The following papers are selected from the proceedings of a symposium on *The Application of Ionising Radiations in the Chemical and Allied Industries* published by the British Nuclear Energy Society, London 1967.

CHARLESBY A. Industrial polymer radiation, p. 7.
DALTON F. L. Coatings and wood-plastics, p. 11.
HARMER D. E. The transition of radiation chemistry to an industrial reality: the Dow ethyl bromide process, p. 26.
MANOWITZ B. Review of industrial radiation chemistry in the United States, p. 16.

P. R. HILLS

RADIATION QUANTITIES, SYSTEMATICS OF. The quantities related to physical aspects of irradiation can be classified as source characteristics, field characteristics, interaction characteristics, medium characteristics and biological characteristics. Within this system one may distinguish: energy aspects, referring to the amount of energy involved, and particle aspects, referring to the number of processes involved. The individual aspects are considered separately below.

Source Characteristics

The *emitted source energy* (E) is the total amount of energy emitted by the source in the time interval under consideration. It is the time integral of the *emissive power of the source* (P) which is defined as the energy emitted per unit time by the source. The *number of emissions* of a source (N) is the number of particles specified by type and/or energy, emitted by the source. This quantity is the time integral of the *emission rate* of the source (R), which gives the number of particles, specified by type and/or energy, emitted per unit time.

When $\bar{E}_i$ is the average energy of particles of type i, and R_i the emission rate for these particles, we have

$$P = \sum_i R_i \bar{E}_i.$$

In cases where the radiation of the source is absorbed partly in the source material itself, a self-absorption correction factor has to be applied.

If a radioactive source consists of only one radionuclide, for which the decay scheme is known, the emission rate R_i of particles i is related to the disintegration rate (or activity) A by

$$R_i = k_i A$$

where k_i denotes the number of specified particles emitted per nuclear disintegration.

The disintegration rate can be determined directly by detectors with 4π geometry. Often one measures not a disintegration rate, but an emission rate. Only when the constituent radionuclides and their decay schemes are known, is it possible to derive the disintegration rate (or activity) from the experimentally determined emission rate. Thus for a transuranium source or for a fission product source for which the composition is unknown one can determine experimentally an alpha-emission rate or a gamma-emission rate for specified particle energies, without being able to determine the total activity of the source. The *specific activity* (symbol α) is equal to the activity per unit mass; the analogous concept emission rate density may refer to the emission rate per unit length, per unit area, per unit volume, or per unit mass, as specified. For a radioactive source containing one gamma emitting nuclide the *specific gamma ray constant* (symbol Γ) is a characteristic quantity: it is the exposure rate $\dot{X}$ for a point source having unit activity at a unit distance in air.

$$\Gamma = \frac{l^2}{A} \dot{X}$$

where A is the activity and l the distance. This quantity is often expressed in units R m^2 h^{-1} Ci^{-1} or in units Cm^2/kg.

Field Characteristics

The fundamental physical quantities describing a radiation field are the energy flux density ψ and the particle flux density ϕ. The relation between these two quantities is

$$\psi = \sum_i \phi_i \bar{E}_i,$$

where $\bar{E}_i$ is the average energy of particles of type i.

The energy fluence Ψ and the particle fluence Φ satisfy the relations

$$\Psi = \int_0^t \psi(t)\,dt$$

$$\Phi = \int_0^t \phi(t)\,dt$$

$$\Psi = \sum_i \Phi_i \overline{E}_i$$

As true field quantities the energy flux density ψ and the particle flux density ϕ (and also the integral concepts Ψ and Φ) can be defined *in vacuo*.

Sometimes a physical quantity describing an interaction characteristic in some reference medium (air, tissue) is used to serve as a field characteristic. In this sense, the concepts kerma rate, dose rate, exposure rate are sometimes used to describe a radiation field in terms of the interaction with air (or tissue). These quantities, however, can in principle not be defined *in vacuo*.

For radioactive sources with uniform composition having a simple geometry in a homogeneous extended medium the emissive power (P) and the emission rate (R) can easily be related to the energy flux density ψ and particle flux density ϕ respectively (e.g. for point, line, disc, slab sources, spherical and cylindrical sources).

For an isotropic point source in a homogeneous absorbing medium:

$$\phi(r) = \frac{R}{4\pi r^2} e^{-\mu_1 r}$$

and

$$\psi(r) = \frac{P}{4\pi r^2} e^{-\mu_2 r}$$

where μ_1 and μ_2 are respectively the attenuation coefficient and the energy absorption coefficient of the medium.

Interaction Characteristics

The *energy imparted to matter* (symbol E_D) in a volume by ionizing radiation is defined as the difference between the sum of energies of all the directly and indirectly ionizing particles which have entered the volume and the sum of the energies of all those which have left it, minus the energy equivalent of any increase in rest mass that has taken place in nuclear or elementary particle reactions within the volume. The energy imparted can also be defined in a simpler way as the energy which appears as ionization, excitation or changes of chemical bond energies. Ultimately the energy imparted will appear mainly as heat and to a small extent as a change in interatomic bond energies.

As, however, during the energy degradation process the energy will diffuse, the spatial distribution of heat may be different from the spatial distribution of imparted energy.

The indirectly ionizing radiation (X- or gamma or neutron radiation) produces in the interaction process with matter directly ionizing (charged particle) radiation, and also to a certain extent indirectly ionizing radiation, which two groups of radiation in general interact with matter at different positions, so that locally the energy transfer and the energy absorption may differ slightly.

The *total ionization* (symbol N_t) is defined as the total number of elementary charges of one sign produced by an ionizing charged particle along its entire path, where the ionizing due to secondary ionizing particles is to be included. When this quantity is taken per unit length of path it is called linear ionization (symbol N_{i1}).

The *exposure* (symbol X) is only defined for photons interacting with air. It is equal to the total electric charge of all the ions of one sign produced per unit mass of air, when all the electrons liberated in small volume element are completely stopped in air.

The *absorbed dose* (symbol D) is the energy imparted by ionizing radiation per unit mass of a specified material at the point of interest.

The *kerma* (symbol K) is the sum of the initial kinetic energies of all charged particles liberated by indirectly ionizing particles per unit mass of specified material. The bremsstrahlung radiated by the charged particles, and the energy of Auger electrons produced in secondary processes are both included in the kerma.

For indirectly ionizing radiation, kerma and exposure are referring to the very first interaction with matter, while absorbed dose and ionization refer to the subsequent interaction processes of charged particles produced by the first interaction of indirectly ionizing radiation.

When there exists *charged particle equilibrium* (C.P.E.) the concept exposure and ionization and also kerma and absorbed dose can be related to each other. This C.P.E. means that the energy imparted per unit mass in a small volume element is equal to the energy liberated per unit mass in this volume element in the form of charged particles.

Charged particle equilibrium would exist at a point at the centre of a small volume under irradiation if

(a) the volume had dimensions which exceed the maximum range of the charged particles,

(b) the energy flux density and the spectrum of the primary radiation were constant throughout this volume,

(c) the energy absorption coefficient for the primary radiation and the stopping power for the secondary charged particles were constant in this volume.

The only situation in which there exists an exact C.P.E. is an infinite homogeneous medium with a uniformly distributed radiation source.

When a source of indirectly ionizing radiation is external to the irradiated material, the flux density of secondary charged particles may differ appreciably from the true equilibrium value. However, at moderate energies the deviation from C.P.E. may not be important for many applications. When the incident indirectly ionizing radiation suffers appreciable attenuation in an absorbing medium over a distance equal to the mean range of the secondary charged particles, a so-called transient equilibrium is established at a depth where the ratio of primary particles to secondary particles reaches a constant value (see Fig. 1).

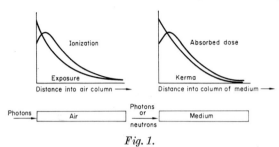

Fig. 1.

Beyond this oint there would be charged particle equilibrium, if the exposure (or kerma respectively) were constant. Since this quantity is not constant and since the emission of the charged particles is anisotropic, there remains a slight difference between exposure and ionization (or kerma and absorbed dose respectively).

In actual applications of the concept kerma, the volume element under consideration should be so small that its introduction does not appreciably disturb the radiation field. This is particularly necessary if the medium for which kerma is determined is different from the ambient medium; if the disturbance is appreciable an appropriate correction must be applied.

Although a fundamental physical description of a radiation field is given by the concepts energy flux density and/or particle flux density, for the purpose of dosimetry, it may be convenient to describe the field of indirectly ionizing particles in terms of the kerma rate for a specified material. A suitable material would be:
  air for electromagnetic radiation of moderate energies;
  tissue for all radiations in medicine or biology;
  any relevant material for studies of radiation effects.

For indirectly ionizing radiation the kerma refers to the *energy transferred* to charged particles. The absorbed dose on the other hand refers to the *energy absorbed*. For indirectly ionizing radiation the kerma is equal to the absorbed dose, provided that
  there exist a charged particle equilibrium at the position and in the material of interest;
  bremsstrahlung losses are negligible;
  both quantities are expressed in the same applicable units.

As exposure is only defined for photons in air, the kerma of photons in air can be related to the exposure, when there is no effect of bremsstrahlung. At a point in air where the exposure is equal to 1 roentgen, the kerma is equal to 86·9 erg/g.

This value is derived by considering the effect of 1R of photons as follows:

$$\left(\frac{1 \text{ esu}}{0 \cdot 001293 \text{ g air}}\right) \times \left(2 \cdot 082 \times 10^9 \frac{\text{electron}}{\text{esu}}\right)$$

$$\times \left(33 \cdot 7 \frac{\text{eV}}{\text{electron or ion pair}}\right)$$

$$\times \left(1 \cdot 602 \times 10^{-12} \frac{\text{erg}}{\text{eV}}\right) = 86 \cdot 9 \text{ erg/g}$$

$$\frac{K_{\text{air}}}{X} = 86 \cdot 9 \frac{\text{erg/g}}{\text{R}}.$$

Or in the appropriate units of the International System:

$$\frac{K_{\text{air}}}{X} = 33 \cdot 7 \frac{\text{J/kg}}{\text{C/kg}}.$$

From this latter relation one sees clearly that the conversion coefficient between the "energy" concept K on one hand, and the "particle" concept X on the other hand, is related to the quantity W, which is defined as the average energy expended in a gas per ion pair formed. For air $W = 33 \cdot 7$ eV and therefore the average energy expended in air per unit charge on all the ions of one sign is 33·7 J/C.

Let K_{med} denote the kerma at the point of interest in a small mass of medium, where when the small mass would be replaced by air an exposure X would be present.

From the relation mentioned above and from

$$\frac{K_{\text{med}}}{K_{\text{air}}} = \frac{(\mu_K/\varrho)_{\text{med}}}{(\mu_K/\varrho)_{\text{air}}}$$

where μ_K/ϱ denotes the mass energy transfer coefficient it follows

$$\frac{K_{\text{med}}}{X} = \frac{(\mu_K/\varrho)_{\text{med}}}{(\mu_K/\varrho)_{\text{air}}} \times 86 \cdot 9 \left(\frac{\text{erg/g}}{\text{R}}\right)$$

$$= \frac{(\mu_K/\varrho)_{\text{med}}}{(\mu_K/\varrho)_{\text{air}}} \times 33 \cdot 7 \left(\frac{\text{J/kg}}{\text{C/kg}}\right)$$

or, in good approximation for tissue irradiated in a photon field:

$$\frac{K_{\text{tissue}}}{X} \approx 94 \left(\frac{\text{erg/g}}{\text{R}}\right)$$

$$\approx 37 \left(\frac{\text{J/kg}}{\text{C/kg}}\right)$$

The energy transfer per unit volume or per unit mass is respectively equal to $\mu_K \Phi E$ or $\frac{\mu_K}{\varrho} \Phi E$, where μ_K is the linear energy transfer coefficient and μ_K/ϱ is the mass energy transfer coefficient.

The relation between particle fluence Φ, energy

fluence Ψ and kerma K may be written as

$$K = \frac{\mu_K}{\varrho} \Psi = \frac{\mu_K}{\varrho} E \Phi$$

When K is expressed in units of 100 erg/g, E in MeV, μ_K/ϱ in cm²/g, and Φ in cm⁻² then

$$K = 16.02 \times 10^{-9} \frac{\mu_K}{\varrho} E \Phi.$$

In general the mass energy-transfer coefficient can be written as

$$(\mu_K/\varrho) = \frac{1}{E} \sum_i N_i \sum_j \varepsilon_{ij}(E) \, \sigma_{ij}(E).$$

The subscript i refers to the target atom and the subscript j refers to the type of interaction (for neutrons: elastic scattering, inelastic scattering, (n, α) etc.; for photons: Compton effect, photoelectric effect, pair production).

N_i denotes the number of atoms of type i per unit mass,

ε_{ij} denotes the amount of energy transferred to kinetic energy of charged particles in interaction of type j,

σ_{ij} denotes the cross-section of the interaction under consideration.

For *roentgen or gamma rays* of energy $h\nu$ one has

$$\frac{\mu_K}{\varrho} = \frac{\tau_a}{\varrho} + \frac{\sigma_a}{\varrho} + \frac{\varkappa_a}{\varrho}$$

where $\dfrac{\tau_a}{\varrho}$ is the mass energy transfer coefficient for the photoelectric process,

$\dfrac{\sigma_a}{\varrho}$ is the mass energy transfer coefficient for the Compton process,

$\dfrac{\varkappa_a}{\varrho}$ is the mass energy transfer coefficient for the pair production process.

The relation between these energy transfer coefficients and the corresponding mass attenuation coefficients is as follows:

$$\frac{\tau_a}{\varrho} = \frac{\tau}{\varrho}\left(1 - \frac{\delta}{h\nu}\right)$$

$$\frac{\sigma_a}{\varrho} = \frac{\sigma}{\varrho} \frac{E_c}{h\nu}$$

$$\frac{\varkappa_a}{\varrho} = \frac{\varkappa}{\varrho}\left(1 - \frac{2m_0 c^2}{h\nu}\right)$$

where τ/ϱ is the mass attenuation coefficient for the photoelectric process,

δ is the average energy emitted as fluorescent radiation per photon absorbed,

σ/ϱ is the mass attenuation coefficient for the Compton process,

E_c is the average energy of the Compton electrons per scattered photon,

$\varkappa/\varrho$ is the mass attenuation coefficient for the pair production process,

$m_0 c^2$ is the rest energy of the electrons.

Values of the mass energy-transfer coefficients for various materials are given in *NBS Handbook 85* (ICRU report 10b, 1962).

In this way one obtains

$$K = 16.02 \times 10^{-9} \Phi_\gamma E_\gamma \frac{\mu_K}{\varrho}$$

where K is the kerma due to interaction with photons (in 100 erg/g),

Φ_γ is the photon fluence (in cm⁻²),

E_γ is the energy of photons (in MeV),

$\dfrac{\mu_K}{\varrho}$ is the mass energy transfer coefficient for photons (in cm²/g).

For the *elastic scattering of fast neutrons of energy* E_n in a material of type i one has

$\dfrac{\mu_K}{\varrho} =$ mass energy-transfer coefficient of material of type i for elastic scattering of neutrons (in cm²/g).

$$\frac{\mu_K}{\varrho} = \frac{1}{\varrho} N_i \sigma_i(E_n) \, \varepsilon_i$$

where N_i is the number of atoms of type i per unit volume (in cm⁻³),

$\sigma_i(E_n)$ is the total cross-section of atoms of type i for elastic scattering of neutrons of energy E_n (in cm²),

ε_i is the fractional neutron energy transfer to the interacting nucleus.

Since $$N = \frac{\varrho N_{Av}}{A_i}$$

where N_{Av} is the Avogadro-constant (6.02×10^{23}/mol),

A_i is atomic mass of atoms of type i,

and $\varepsilon_i = \dfrac{2A_i}{(A_{i+1})^2}$ for isotropic scattering,

one obtains the following relations:

$$K_i = 19.3 \times 10^{-9} \, \Phi_n E_n \frac{\sigma_i(E_n)}{(A_{i+1})^2} \quad \text{(for isotropic scattering)}$$

$$K_i = 19.3 \times 10^{-9} \, \Phi_n E_n \frac{\sigma_i(E_n)}{(A_{i+1})^2} \cdot f_i \quad \text{(for anisotropic scattering)}$$

where K_i is the kerma due to elastic scattering of fast neutrons (in 100 erg/g),

Φ_n is the neutron fluence (in cm⁻²),

E_n is the neutron energy (in MeV),

$\sigma_i(E_n)$ is the total elastic cross-section (in barns),

A_i is the atomic mass of atoms of type i,

$$f_i = \int_0^\pi [\sigma_i(\theta, E_n)/\sigma_i(E_n)] \, (1 - \cos\theta) \, 2\pi \sin\theta \, d\theta$$

$\sigma_i(\theta, E_n)$ is the differential elastic cross-section (in barn/steradian) for neutrons of initial energy E_n scattered at angle θ.

Table 1. System of radiation quantities

	Source characteristics	Field characteristics	Interaction characteristics		Medium characteristics	Biological characteristics
integral concept	emitted source energy (E)	—	energy imparted to matter (E_D)		average energy per ion pair formed (W)	
differential (path)	—	—	linear energy transfer (LET)		linear stopping power (S)	
differential (time)	emissive power (P)	—	—			
differential (space or mass)	—	energy fluence (Ψ)	kerma (K)[a]	absorbed dose (D)		dose equivalent (DE) $1\ \text{rem} = \dfrac{1\ \text{rad}}{(\text{QF})(\text{DF})\ldots}$
differential (space or mass and time)	—	energy flux density (φ)	kerma rate ($\dot{K}$)[a]	absorbed dose rate ($\dot{D}$)		dose equivalent rate ($\dot{\text{DE}}$)
related concepts	—	—	build-up factor		range (R) mass energy transfer coefficient (μ_K/ϱ) mass energy absorption coefficient (μ_{en}/ϱ)	quality factor (QF) distribution factor (DF) time factor (TF)

Energy aspects (referring to amount of energy involved)

Particle aspects (referring to number of processes involved)		number of emissions (N)	number of disintegrations (N)		total ionization $(N_i)^d$	
integral concept		number of emissions (N)	number of disintegrations (N)	—	total ionization $(N_i)^d$	—
differential (path)		emission rate (R)	—	—	linear ionization $(N_{il})^d$	linear attenuation coefficient (μ)
differential (time)		—	activity (A)	—	ionization rate $(\dot{N}_i)^d$	—
differential (space or mass)		—	—	particle fluence (Φ)	exposure $(X)^b$ / ion dose $(X)^c$	—
differential (space or mass and time)		emission rate density (R_s)	specific activity (α)	particle flux density (ϕ)	exposure rate $(\dot{X})^b$ / ion dose rate $(\dot{X})^c$	—
related concepts		specific gamma-ray constant (Γ)^e, self-absorption factor	—	—	build-up factor	mean free path (λ), macroscopic cross-section (Σ), half-thickness (d ½), mass attenuation coefficient (μ/ϱ), self-shielding factor

[a] Kerma is only defined for indirectly ionizing (uncharged) particles; this quantity has sometimes been called "first collision dose".
[b] Exposure is only defined for X- or gamma radiation in air.
[c] Ion dose is only defined in air, but is applicable to any ionizing radiation.
[d] Ionization is only defined for ionizing charged particle radiation.
[e] Specific gamma-ray constant is characteristic for a nuclide emitting gamma radiation.

Medium Characteristics

When the medium is a gas, the average energy W expended per ion pair formed is an important quantity. In general, W is approximately 34 eV.

The *stopping power* S of a medium for charged particles is the average energy lost by a charged particle of specified energy per unit path length.

The mass energy-absorption coefficient (μ_{en}/ϱ) of a medium for indirectly ionizing particles is given by

$$(\mu_{en}/\varrho) = (\mu_K/\varrho)(1-G)$$

where (μ_K/ϱ) is the mass energy-transfer coefficient and G is the proportion of the energy of secondary charged particles that is lost to bremsstrahlung. Other physical quantities which are to a certain extent characteristic for the medium are the attenuation coefficient, the range of charged particles, the mean free path, the cross-section for scattering, absorption, the half thickness, etc.

For an absorbing body the self-shielding factor gives the value by which a radiation quantity (e.g. the flux density) inside this body is reduced by self-shielding.

Biological Characteristics

The dose equivalent (DE) is the product of absorbed dose (D), a quality factor (QF), a dose distribution factor (DF) and other necessary modifying factors.

$$DE = D \times (QF)(DF) \times \cdots$$

The quality factor is dependent on the linear energy transfer and has been introduced to express on a common scale for all ionizing radiations the irradiation incurred by exposed persons.

The distribution factor may be used to express the modifications of biological effect due to non-uniform distribution of internally deposited isotopes.

A time factor may be introduced in some cases to account for the effect of fractionation of the dose delivered. The unit of dose equivalent is the rem:

$$1 \text{ rem} = \frac{1 \text{ rad}}{(QF) \times (DF) \times \cdots}$$

Bibliography

National Bureau of Standards Handbook 84 (1962) *Radiation Quantities and Units*, International Commission on Radiological Units and Measurements, Report 10a (U.S. Government Printing Office, Washington D.C.

National Bureau of Standards Handbook 85 (1962) *Physical Aspects of Irradiation*, Recommendations of the International Commission on Radiological Units and Measurements, Report 10b (U.S. Government Printing Office, Washington D.C.).

DENNIS J. A. (1961) Radiation units and concepts, in *Encyclopaedic Dictionary of Physics* (J. Thewlis Ed.), Suppl. vol. 2, 284, Oxford: Pergamon Press.

REES D. J. (1967) Radiation units, measurement of, in *Encyclopaedic Dictionary of Physics* (J. Thewlis Ed.), Suppl. vol. 2, 289, Oxford: Pergamon Press.

W. L. ZIJP

RADIOEMISSIONS FROM INTERSTELLAR HYDROXYL RADICALS.

Introduction

In October 1963 the microwave lines of interstellar hydroxyl radicals were discovered as absorption lines in the direction of a few radio sources. The lines originate in the interstellar medium but at present (1968) our knowledge of the interstellar medium does not explain all the observations. With the discovery of anomalously excited hydroxyl emission in 1965 the theoretical framework for interpretation of the observations became inadequate and this situation has not yet been resolved.

Microwave Spectrum of OH

The ground electronic state of the OH radical is a $^2\pi$ state, the π indicates that Λ, the quantum number of the angular momentum of the electrons about the internuclear axis, equals one. The superscript 2 equals $(2S+1)$, where S is the net spin angular momentum quantum number, equal to one-half because the molecule has an odd number of electrons which prevents complete pairing. The component of S on the internuclear axis will add to Λ and the quantum number of the resultant electronic angular momentum about the internuclear axis can therefore be 1/2 or 3/2 and results in a split (termed "fine structure") in the $^2\pi$ state into two states $^2\pi_{1/2}$ and $^2\pi_{3/2}$.

In a rotating molecule each rotational level of OH will be further split by lambda doubling. The resulting energy level scheme is shown in Fig. 1.

Further splitting of these levels due to the hyperfine interaction with the unpaired spin of the proton results in four energy levels to each rotational level. In Fig. 2 the energy level diagrams for the $^2\pi_{1/2}$, $J=3/2$ and $^2\pi_{1/2}$, $J=1/2$ states and the allowed electric dipole transitions are shown. The quantum number F represents the total angular momentum, including nuclear spin, for each level. Magnetic dipole transitions between the hyperfine components of each doublet level are allowed but the transition probabilities are about 10^3 times smaller than the electric dipole transitions and are usually ignored.

Under interstellar conditions OH-ion collisions and excitation through 18 cm radiation are the most important processes by which excitation of the OH molecule occurs. Only the $^2\pi_{3/2}$, $J=3/2$ state of OH is therefore expected to be appreciably populated. The frequencies of all four transitions have been measured in the laboratory as well as astronomically and the values are given in Table 1. The laboratory

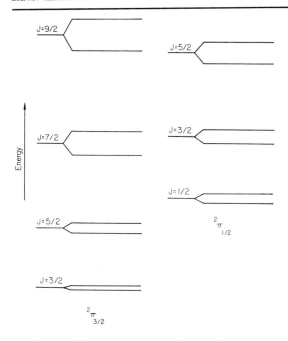

Fig. 1. Schematic representation of the energy levels in the lowest vibrational level of OH. Fine structure and the splitting of each rotational level into Λ-doublets is shown (the splitting is not to scale) (from ref. 1).

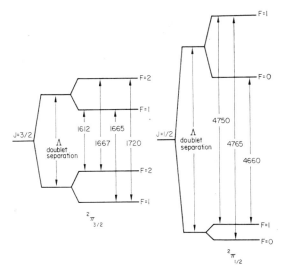

Fig. 2. Energy level diagrams for the $^2\pi_{3/2}$, $J = 3/2$ and $^2\pi_{1/2}$ states of OH showing hyperfine structure. The allowed transitions are shown and for each transition the frequency indicated in MHz.

measurements do not obey the sum rule

1612·231 + 1720·533 (=) 1665·401 + 1667·358

but the astronomically determined frequencies do.

The $^2\pi_{1/2}$, $J = 1/2$ level is metastable and transition frequencies have been measured in the laboratory for this as well as for the $^2\pi_{3/2}$, $J = 5/2$ level. They are quoted in Table 2. Lines from both levels have been observed in interstellar spectra. The lines appear in emission from the OH source in the direction of W 3, one of the radio sources in the Westerhout catalogue (1958).

Absorption Spectra

Line Frequencies

The OH lines observed in the absorption spectrum of the radio source Cassiopeia A confirmed the frequencies of the 1667 MHz and 1665 MHz lines. The satellite lines at 1612 MHz and 1720 MHz in this spectrum do not show the same profiles as the main lines and so do not lead to confirmation of the line frequencies. However, the absorption line profiles in the direction of W 12 are similar and yield the astronomically determined rest frequencies quoted in Table 1. These frequencies obey the sum rule to within the experimental errors and the only correction is in the rest frequency of the 1720 MHz line.

Line Intensities

Under conditions of thermodynamic equilibrium the emission lines of the ground state will have the relative intensities given in Table 1 if the gas producing the lines is of small optical depth. With an increase in optical depth the lines will approach equal intensities if Doppler broadening only is operating. The intensities observed in absorption will depend on the spectrum of the continuum source against which they are observed, the relative sizes of the OH clouds and background source and other factors.

Since no OH has been observed in thermal emission near sources which show OH in absorption, the excitation temperature of the OH cannot be determined. The number of molecules in the line of sight depends on the optical depth and the excitation temperature, and therefore no indication of the abundance of OH relative to hydrogen is available.

From a comparison of the 21 cm hydrogen line absorption profile in the direction of Cassiopeia A with the main OH absorption features, an attempt has been made to separate the turbulent and thermal broadening of the lines. The results are given in Table 3.

In most cases where all four lines have been observed in absorption the populations of the levels show departures from thermal equilibrium, for the intensity ratios of the lines do not appear in the expected ratios. Furthermore, for any population distribution the excitation temperatures T_{ij} corresponding to each

Table 1. Rest Frequencies and Transition Probabilities of the $^2\pi_{3/2}$, $J = 3/2$ State of OH

Transition	Laboratory (MHz)	Astronomical (MHz)	Relative intensity	A (s^{-1})
F 1 ↔ 2	1612·231 ± 0·002	1612·231 ± 0·002	0·90	1·29 × 10^{-11}
1 ↔ 1	1665·401 ± 0·002	1665·401 ± 0·002	4·95	7·11 × 10^{-11}
2 ↔ 2	1667·358 ± 0·002	1667·358 ± 0·002	9·00	7·71 × 10^{-11}
2 ↔ 1	1720·533 ± 0·002	1720·527 ± 0·003	1·10	0·94 × 10^{-11}

Table 2. Transition Frequencies and Probabilities for the $^2\pi_{1/2}$, $J = 1/2$ and $^2\pi_{3/2}$, $J = 5/2$ States of OH

Levels	Transition F ↔ F	Frequency, MHz	Relative intensity	A (s^{-1})
$^2\pi_{1/2}$, $J = 1/2$	1 0	4660·242 ± 0·003	1	2·56 × 10^{-10}
	1 1	4750·656 ± 0·003	2	1·73 × 10^{-10}
	0 1	4765·562 ± 0·003	1	9·10 × 10^{-11}
$^2\pi_{3/2}$, $J = 5/2$	3 2	6016·746 ± 0·008	1	4·17 × 10^{-11}
	2 2	6030·739 ± 0·005	14	5·92 × 10^{-10}
	3 3	6035·085 ± 0·005	20	6·03 × 10^{-10}
	2 3	6049·084 ± 0·008	1	3·03 × 10^{-11}

Table 3. Thermal and Turbulent Broadening of Cassiopeia A Absorption Spectrum

Cloud velocity, km/s	Kinetic temperature, °K	RMS turbulent velocity, km/s
−0·1	120	0·27
−1·5	90	0·24

transition frequency ν must conform to the equation

$$(\nu/T)_{1612} + (\nu/T)_{1720} = (\nu/T)_{1665} + (\nu/T)_{1667}$$

which yields in terms of optical depths τ at velocities V

$$\tau(V)_{1612} + \tau(V)_{1720} = \frac{\tau(V)_{1665}}{5} + \frac{\tau(V)_{1667}}{9}.$$

Observations on W 12, Messier 17 and NGC 6357 do not confirm this. The only way in which the observations can be made to conform to this relation is to postulate concentrations of OH so small that it only partially obscures the radio source. The same hypothesis was found necessary to explain the anomalous absorption in the direction of the galactic centre and in the direction of Cassiopeia A. No polarization has been observed in any of the absorption lines.

Galactic Centre

The OH in the direction of the galactic centre is extended and shows large absorption features that do not correspond to any strong features in the neutral hydrogen profiles. The region has been surveyed at 1667 MHz but when observed at checkpoints the fine structure appeared in all the lines. The ratio τ_{OH}/τ_H also varies by an order of magnitude over the region.

The main contrast with the neutral hydrogen observations is in the fact that the main features of the OH do not show a velocity dependence with galactic longitude. It was just such a dependence in the HI data that led to the model of a rapidly rotating galactic nuclear disc. No dynamical model for the OH observations has been proposed because the distances to the clouds are not known. It is, however, evident that radial motions occur in OH.

Microwave Spectra from Interstellar $O^{18}H^1$

The $F = 2 \leftrightarrow 2$, $^2\pi_{3/2}$, $J = 3/2$ doublet transition of the isotopic species $O^{18}H^1$ has been detected in absorption spectra in the direction of the galactic centre. The line occurs at 1639·460 MHz. The other expected line at 1637·4 MHz has not been observed yet.

Thermal Emission

Normal OH emission has been detected from interstellar dust clouds that show very large optical absorption and lack of excess 21 cm neutral hydrogen emission. Since no satellite lines could be detected above the receiver noise level the line ratios can be assumed to be normal and the optical depth small.

Non-thermal Emission

In the direction of a number of continuum sources non-thermal emission of the OH lines occurs. The emission profiles show very little similarity for any of the four OH lines and absorption may occur at the velocity corresponding to an emission feature for another line. The emission is usually strongest at 1665 MHz but not invariably so, with the features typically 1 kHz to 2 kHz wide.

Most of the OH non-thermal emission regions are close to HII regions and three of the four emission lines are usually seen. Another class of sources shows emission at 1720 MHz at velocities at which absorption occurs at the other lines. We consider this class of sources separately after discussing the strong 1665 MHz and/or 1667 MHz emission sources.

Polarization

Components of the OH non-thermal emission show a high degree of circular polarization, in some cases more than 90 per cent. Although components of elliptical and linear polarization are also found these components do not show the same high degree of polarization. An example of the polarization structure across an OH spectrum is shown in Fig. 3.

source sizes of strong OH emission regions. Individual velocity components in the OH source near W3 arise from regions typically of the order of 0″·04 in diameter separated from each other by a few seconds of arc. The positions of some of the velocity components

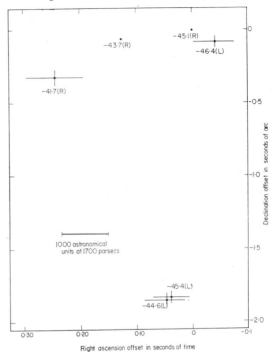

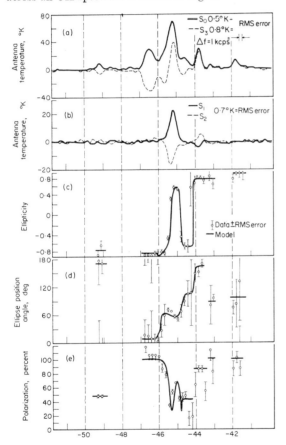

Fig. 3. *Polarization parameters for the 1665 MHz OH emission near W3. (a) and (b) measured values of the Stokes parameters. The points plotted in (c), (d) and (e) were computed from the Stokes parameters. (From Meeks M. L., Ball J. A., Carter J. C., Ingalls R. P., Science, 153, 978–81, 1966.)*

Fig. 4. *The relative positions of some of the features in the spectrum of the W3 OH source at 1665 MHz which are referred to the feature at -45.1 km/s. The velocity (in km/s) of each feature and the sense of circular polarization (right- or left-handed) are indicated. Errors on the left-hand circularly polarized components arise mainly from a comparison of the positions of oppositely polarized features. (From A. J. Cooper, et al., M.N.R.A.S., to be published.)*

in the W3 OH source is shown in Fig. 4. Some of the source sizes are quoted in Table 4.

Identification of Emission Sources

In the position of the OH emission source near W3 a weak continuum radio source has been detected. No familiar optical object can be seen in the position of the source. In general the OH emission comes from regions adjacent to bright nebulosities when these are visible or displaced from the centres of the continuum emission, except in the case of Orion A where the position of the OH emission lies within the extended continuum source. An infra-red source has also been detected at this position in Orion A. Infra-red sources have also been found near the OH sources in W 3 and W 51 and OH emission from the infra-red star NML Cyg.

Sizes of Emission Sources

Interferometers of baselines up to 4×10^7 wavelengths have been used in determinations of the

Table 4. Brightness Temperatures of Some Components in OH Emission Sources

Source	Velocity component, km/s, w.r.t. local standard of rest	Polarization	Source diameter, seconds of arc	Brightness temperature, °K
W3	−43·7	right circular	0·004	6×10^{12}
W3	−45·1	right circular	0·03	3×10^{11}
W49	+16·8	left circular	0·04	7×10^{10}

Brightness Temperatures

The sizes of individual sources of OH emission and intensity measurements yield the brightness temperatures, i.e. the equivalent black body temperatures quoted in Table 4. The very high values of brightness temperature are not consistent with the very narrow line widths if local thermodynamic equilibrium in the source is assumed.

Line Variations

The southern OH source near NGC 6334 appears to show intensity variations against time. The variations are confined to three features of the spectrum but insufficient data exist to confirm any possible periodicity.

1720 MHz Emitters

OH sources showing emission only at 1720 MHz appear in the direction of some radio sources such as Cassiopeia A, W28, W44, W43 and W51. The first three of these are supernova remnants. The absorption lines at the other OH frequencies in these sources are wider than the emission line widths and have anomalous intensity ratios. Characteristic of this class of sources are a low degree of polarization and appreciable linewidths (5 to 20 kHz). The major difference from strong main line emitters appears to be that the 1720 MHz emitters are not necessarily associated with HII regions.

Mechanisms for OH Non-thermal Emission

The high degree of polarization and the high brightness temperatures of the strong main line emission sources make a thermal origin of the emission unlikely. Furthermore if the linewidths arise from thermal broadening the corresponding temperatures should be tens of degrees whereas the observed brightness temperatures are typically of the order 10^{12} °K.

The most plausible explanation suggests a population inversion of the energy levels and subsequent maser amplification of cosmic or galactic background radiation. Spontaneous emission will also be amplified if the excitation temperature is greater than the background temperature. Larger amplification at the line centre than in the wings would result in a narrowing of the emission lines.

The Zeeman effect has been proposed as a possible mechanism by which circular polarization can be generated. The presence of maser action will, however, confuse the simple Zeeman pattern. Observations of some sources in which a single line with a high degree of circular polarization occurs are therefore difficult to interpret on the basis of this mechanism. If the emission is due to maser amplification circularly polarized emission lines may be generated by the saturation of a transition between particular magnetic sublevels and the suppression of others.

If the 1720 MHz emission is maser amplification the continuum source may simply provide a high-level input to the maser and the gain required would be much lower than for the OH sources near HII regions. This would be consistent with greater linewidths and low polarization observed at 1720 MHz.

Possible Pumping Mechanisms

Four possible pumping mechanisms for creating a population inversion have been suggested, viz. pumping by anisotropic microwave radiation, pumping by anisotropic far-infrared radiation, anomalous excitation by ultraviolet radiation and excitation by OH collisions with anisotropic charged particles. Another model proposes that OH could be formed in an excited state and so create a population inversion.

Observations point to maser amplification operating under saturated conditions, i.e. the signal power greater than the pump power. A linear growth in intensity with path length then results if the path length is large. Whether or not the maser is saturated depends on the emission solid angle, for if the emission originates from a directed pulse as the result of avalanche population depletion the possibility of spatially coherent radiation arises. Under these conditions interferometer observations would have to be interpreted with caution. Under heavily saturated conditions line broadening would occur.

At present only the infrared pumping model holds promise of explaining the emission from OH sources.

Bibliography

BARRET A. H. (1964) The detection of the OH and other molecular lines in the radio spectrum of the interstellar medium. *I.E.E.E. Transactions on Antennas and Propagation* **12**, 822–31.

ROBINSON B. J. and McGEE R. X. (1967) OH molecules

in the interstellar medium. *Annual Review of Astronomy and Astrophysics*, vol. 5, Annual Reviews Inc. WESTERHOUT G. (1958) *Bull. Astron. Inst. Netherlands* **14**, 215.

G. DE JAGER

RADIOISOTOPES IN FOOD PROCESSING

Introduction

The present article describes the applications of radioactive tracers and radioisotope instruments in the food processing industries, including their use in connection with the packaging and storage of foods and food materials. The treatment of food and allied products by ionizing radiations from radioactive sources or electrical machines has been described by Longstaff (1968).

Radioactive Tracers

The advantages of radioactive tracers as tools for investigating plant operations in normal and fault conditions, and for laboratory studies, apply to the food industry as much as to any other: the range of possible radioactive tracers, their compatibility with the material to be labelled and the convenience and sensitivity with which they can be detected and measured, are unsurpassed by any other class of tracer. But safety considerations, which are always of overriding importance wherever radioactivity is used, are of very special importance in the food industry: the incorporation of radioactive substances into products which may ultimately be eaten by the public is obviously only acceptable under the most stringently controlled conditions. In addition, consumer reaction against the practice may be expected to be unfavourable, although a properly conducted experiment in fact incurs no public or other hazard. For this reason radioactive tracer studies appear to have been carried out only to a limited extent on actual production plant. Many applications, however, have been made in the laboratory and on experimental or pilot plant where the scale of operations is much smaller, and where the product is not destined for public consumption. The following examples, mostly taken from the published literature, illustrate the scope and potentialities of the techniques. It will be noticed that short-lived tracers have almost always been used, so that storage-time before release or disposal can be kept within reasonable limits.

Plant Kinetics

Radioactive tracers have been used in the sugar refining industry to study a number of problems in plant kinetics, where the scale of production is large enough to make a detailed analysis of performance well worth undertaking. In a 20,000 gallon Bach subsider in an Australian plant bromine-82 (as KBr) was used to study the flow through six outlet pipes; the isotope was introduced in a single injection and the time of its arrival and clearance in each pipe was noted and its concentration measured, using external detectors. Results indicated poor mixing and pointed the way to improvements in operation by modifying the feed system. The countercurrent movements of beet chips and of extracted liquor in a Polish refinery were studied using sodium-24 as soda-glass needles, and lanthanum-140 as an EDTA complex respectively, also with external detectors; it was found that a substantial amount of the clean water entering the diffuser was being carried away with the spent chips, and that movements of liquor and chips varied throughout the diffuser in ways which earlier studies using dyes as tracers had not revealed. In two Dorr thickeners in a Swedish plant the in-going slurry was labelled simultaneously with sodium-24 as bicarbonate for the liquor, and lanthanum-140 co-precipitated as hydroxide during carbonation for the mud; external detectors at the outlet tubes showed that there were significant differences in operating characteristics between the two plants, and in different zones in each plant, as well as giving a quantitative measurement of recycling of the mud. Catalyst recycling has been studied in hydrogenation plants using individually labelled beads and external detectors mounted at suitable points.

Where practicable it is often better to activate a constituent of the process material by neutron bombardment in a nuclear reactor, in preference to introducing a foreign material as a "label". This has been done in establishing the optimum conching time for chocolate (sodium-24), and the efficiency of mixing in tea-blending machines (sodium-24 and potassium-42); in the latter case individual packets filled from the blender were monitored to check uniformity of activity, and a few weeks storage sufficed for the product to have lost all measurable activity. Mineral additives have also been activated to study the effectiveness with which they have been incorporated in a variety of food products.

Filter efficiency can be measured by activating some of the most finely-divided solids that the filter is supposed to retain, or by adding a tracer having the required characteristics, and detecting activity in the filtrate. The movement of air in ventilation systems, or of kiln or oven gases, can often be studied by using radioactive krypton-85 gas as a tracer. A variety of techniques have been developed for locating leaks and blockages in pipes and plant, including concentric, lagged or buried pipe systems. Lubrication and wear in plant and machinery can be studied by activating a bearing or other suitable component and measuring the activity of attrited material in the lubricant. The recently-developed technique of electrically implanting krypton-85 or other radioactive ions in a component adds to the scope of the method, and implanted krypton-85 can also in some cases be used to study problems of localized overheating.

To overcome the drawbacks of introducing radioactivity into food products, while at the same time retaining the sensitivity of radiological measurements, it is sometimes better to add an inactive tracer to the process material and activate the samples afterwards by neutron bombardment in a reactor or neutron generator; however, this loses the essential simplicity and speed of the radioactive tracer method, and it presupposes access to a source of neutrons or to an activation analysis service. Another alternative, which dispenses with a neutron source but involves radiochemical analysis, is to react the sample with a radioactive reagent and assay the reaction-product radiometrically.

Laboratory Applications

Under proper management there need be no question of active laboratory samples finding their way to the public at any time, so the scope for using radioactive tracers for research or for laboratory tests on production materials is the same in the food industry as in any other. A few examples will illustrate the breadth of the field and the variety of applications. The efficacy of different cleansing agents and methods used in the dairying industries have been compared by using radioactively labelled milk or other dairy products, and measuring the amount of activity retained on metallic and other working surfaces; the same method has been used to compare different working materials for ease of cleaning. Sodium-24 has been used to study the diffusion of salt into cheese during manufacture, and calcium-45 to study the physical structure of butter. The incorporation of iodine into medicated table salt has been investigated and improved techniques have been developed, using the isotope iodine-131. Carbon-14 and phosphorus-32 have been used to label insecticides for work on the assay of toxic residues in foodstuffs and on the development of measures to eliminate them. The use of detergents and wetting agents for washing fruit and vegetables prior to canning, and the evaluation of rinsing processes for removing them, have been carried out using labelled detergents. Bacteria labelled *in vivo* with phosphorus-32 have been used as biological tracers for research on the cleaning of food cans before filling, and krypton-85 has been used to measure leaks in the seams of experimental cans.

Safety

As emphasized above, the radiological safety of the consumer must always be a primary consideration when a radioactive material is to be used as a tracer in a food processing plant. Safeguards exist in the form of statutory regulations and codes of practice, with appropriate enforcement and advisory bodies to ensure—and to facilitate—their observance. In the United Kingdom these are the Ministry of Housing and Local Government (and corresponding bodies for Scotland, Wales and Northern Ireland), the Factory Inspectorate of the Ministry of Labour, and the Radiological Protection Service run jointly by the Ministry of Health and the Medical Research Council; the R.P.S. is a valuable source of advice on all aspects of radiological safety and of associated legal questions. A sufficient understanding of radioactivity and of the hazards associated with its use is essential to the successful employment of radioactive tracers, as well as being a statutory obligation on the user. Consultancy and experimental services are available for firms lacking the necessary facilities or experience to undertake the work on their own.

Instruments

The use of radioisotope instruments for process and product control in the food industry is not subject to the same special considerations of radiological safety that limit the use of radioactive tracers; there is no risk of contamination or activation of the product by the radioactive source in an instrument, and the radiation dose to the product is usually well below the exemption limit allowed under the regulations governing radiation treatment of food and food products (ten rads in the United Kingdom). Such limitations as do arise are due mainly to the nature of the process materials involved: in many cases these consist of individual animals or plants, etc., whose natural—and acceptable—variability in size and other properties makes them unsuitable subjects for instrumental control. Moreover, some of the qualities that are of major importance in foods, such as taste, odour and appearance are in any case not susceptible of measurement by radioisotope techniques. The most important classes of instrument in use in the industry are simple on-off radiation switches in their applications as level gauges, package monitors and position-indicators; density gauges are second in importance, while thickness gauges and analytical instruments (including neutron moisture meters) are of much more limited direct application—although isotope thickness-gauges are very widely used in the manufacture of food packaging materials (e.g. paper, plastic sheet and tinplate).

Level Gauges

Installed level gauges are used to monitor or control surface levels in storage or process vessels containing solids, liquids, creams or pastes, particularly where other and more direct methods are less suitable. Examples include bulk storage of grain, sugar-beet or lime in silos or in hoppers, of fats, oils and chemicals in tanks, in intermediate storage containers, or in pressure vessels, and of the contents of processing vessels in which the physical state or chemical nature of the contents favour the method—e.g. where frothing may mask the true level. The gauges may be fitted singly at a critical level or in pairs at maximum and minimum levels; in the latter case they are sometimes used for the automatic metering of ingredients for a mix. Gauges having a horizontal radiation path can operate to precisions of $\pm 1/8$ in. or better. Arrangements having multiple or extended sources or detectors, or a sloping or vertical radiation path, can give continuous readings of level, but usually to a much lower precision. Automatic "hunting" systems combine the precision of a horizontal beam with the advantages of continuous indication, but they are very much more expensive to instal than the simpler systems; they are perhaps most useful for the control of fluid-bed driers, heaters and reaction-vessels, where levels can change significantly in a very short time and where other methods cannot be used satisfactorily. Portable hand-held level gauges are available for checking the contents of cylinders of carbon dioxide in mineral-water plants, etc., and of other pressure-liquified gases such as refrigerants and aerosol propellants; they are also useful for general factory or inventory use on vessels up to about 12 in. across. Isotope level gauges are equally useful for monitoring and controlling interface levels between immiscible liquids of different densities. Some level-gauges operate by backscattered radiation and require access to only one side of the vessel.

Package Monitors

A special example of the horizontal-path level gauge is the package-filling monitor used for automatic production-line checking of the correct filling of cans, cartons, boxes and other containers whose contents are sold by volume rather than weight, or where for other reasons the level of the contents is of major importance. Precisions of $\pm 1/16$ in. are obtainable, and packages can be handled at rates of 200 a minute on a single line. Modifications of the principle, where the amount of radiation passing through the package is measured, can detect faulty filling of collapsible tubes or of packets of biscuits, boxes of confectionery, etc.

Density Gauging

Radioisotope density gauges, operating to precisions of ± 1 per cent or thereabouts, provide a simple method for on-line checking of the composition of many foodstuffs and intermediates, without the need for sampling. Transmission gauges can often be fitted on the outside of existing pipes or plant where they cause no interference with flow, or special flow cells can be used. Such gauges can provide continuous monitoring, for manual or automatic control, of processes involving density changes. Examples include the control of evaporators and other plant used in the manufacture of pastes, syrups and condensed milk, the measurement of over-run (air content) in ice cream, or of the fat content of baby-foods, or the meat content of noodle soup. Concentrations of milk of lime, brine, glycerine, sugar solutions and fruit juices, and the water-to-fat ratios in emulsified products are among other examples reported in the literature.

Thickness Gauges

Although many products and intermediates in the food-processing industry are made in sheet, strip or other form where the thickness is of critical importance, the beta thickness gauge widely used in other industries is generally unsuitable for use here. This is because minor variations in moisture content may give rise to corresponding variations in density, and these might in turn mask changes in physical thickness that are too large to be acceptable. For this reason other methods of thickness measurement, such as mechanical probes or sliding shoes, are commonly used to control the thickness, e.g. of the dough sheet from which biscuits are made. In a biscuit factory in the U.S.S.R. radiation gauges are used to measure and control the gaps between the rolls used in producing the dough sheet, thus indirectly controlling the thickness of the product. Backscatter wall-thickness gauges are useful for plant or pipeline inspection by maintenance staff.

Automatic Weighing

The amount of radiation passing vertically through the material on an automatic conveyor belt can be measured and integrated with the speed of the conveyor, to give the weight of material passing the measuring-point in unit time. This has been applied in the U.S.A. to the continuous bulk weighing of potatoes at up to 30 tons an hour, both for inventory purposes and for process control.

Moisture Content

The moisture content of non-hydrogenous materials, or of materials whose hydrogen content is small and accurately known, can be measured continuously by neutron moisture gauges and these have been suggested for various food products and intermediates. However, because carbohydrates, proteins and fats all contain large amounts of combined hydrogen, and because most foods also contain varying amounts of bound water in addition to the free water which is to be measured, neutron gauges have not generally proved satisfactory and other methods, such as infra-red or microwave-absorption or capacitance have to be used. More usually moisture content is measured directly by loss of weight of a sample on drying.

Analytical Instruments and Methods

The preferential absorption of low-energy X-rays by elements of comparatively high atomic number is widely used in industry to measure the heavier-element content of mixtures (e.g. sulphur in kerosene, lead in motor spirit or chlorine in water). Simple transmission gauges are used, usually in conjunction with flow-cells, and the principle can be applied equally in the food industry: for example, it is being used experimentally for determining the lean to fat ratio in meat samples (based on the higher oxygen content of lean meat), from which it may be developed for process control in sausage manufacture, etc. Another experimental method of determining lean to fat ratio, this time in carcases or even in live animals, is to measure the natural radioactivity emitted by the potassium-40 present in protein but not in fat;

this is currently at an early stage of development. Radioactivity is used indirectly in analytical techniques that are of particular use in the food and associated industries: one type of gas-chromatography detector, which relies for its operation on ionization by a small radioactive source, gives an extremely sensitive means of measuring volatile substances responsible for odours and flavours, or for determining insecticidal or toxic residues in foodstuffs. Radiochemical methods of analysis are as useful in food laboratories as elsewhere, and neutron activation analysis is particularly applicable to toxic or trace element determinations, both in raw materials and in finished products.

Bibliography

BRODA E. and SCHONFELD T. (1966) *Technical Applications of Radioactivity*, Vol. I, p. 257, Oxford: Pergamon Press.

LONGSTAFF R. M. (1968) Ionising radiations in industry, in *Encyclopaedic Dictionary of Physics* (J. Thewlis Ed.), Suppl. vol. 3, 173, Oxford: Pergamon Press.

R. M. LONGSTAFF

RADIO TELESCOPES, THE PRESENT POSITION.

The 1960's have seen the completion of a wide range of powerful and specialized radio telescopes. In addition to many large steerable and fixed reflectors, there are at present in operation several cross- and T-telescopes, and synthesized apertures in which the image formation is carried out in a computer. Interferometers have been operated with the elements separated by a distance of 10^8 wavelengths, the local oscillator being derived from separate rubidium frequency standards at each station. The principal objective of these unconventional instruments is to obtain a high resolving power at radio wavelengths. The situation has now been reached where the radio position of objects has been measured with an accuracy exceeding that normally available optically. Long base line radio interferometers have achieved resolving power far in excess of the resolution of a 200" optical telescope!

In this article, some of the principal types of radio telescope are described with an account of their range of application and limitations.

Large Reflecting Radio Telescopes

The most versatile and familiar type of radio telescope consists of a large metallic reflecting surface in the form of a paraboloid of revolution mounted so that it can be pointed to any part of the sky. Cosmic radio energy is brought to a focus where it is collected by a relatively simple waveguide or aerial system.

The performance of a reflecting telescope is determined by the diameter of the reflector and the accuracy with which it is figured. The angular detail, θ, that can be resolved by a telescope of diameter d at wavelength λ is approximately $\theta = \lambda/d$. Studies of the effects of reflector irregularities show that the efficiency is high for wavelengths longer than about 20ε, where ε is a length characteristic of the scale of the irregularities. At shorter wavelengths the efficiency falls rapidly. Provided there are no large-scale deformities of the reflector, the surface errors produce side lobe responses, but do not much affect the resolution of the main beam.

It is unlikely that fully steerable reflecting radio telescopes with a diameter much greater than 100 m can be economically built on the surface of the Earth.† The 600 ft telescope at Sugar Grove in U.S.A. has been abandoned. It would have needed an elaborate

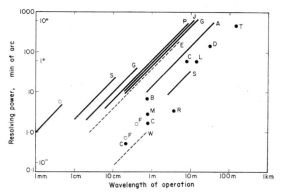

Fig. 1. A comparison of a selection of large pencil beam radio telescopes showing the range of wavelengths over which they operate and the corresponding angular resolution. The 200 in. Mt. Palomar optical telescope is also included. The unaided eye has a resolving power of about 1 min of arc.

KEY.

A Arecibo	L Clark Lake
B Bologna	M Molonglo
C Cambridge	O 200 in. optical tel.
D Penticton	P Parkes
E Effelsberg (being built)	R Radioheliograph
F Fleurs (being built)	S Serpukhov
G Green Bank	T Tasmania
J Jodrell Bank	W Westerbork (being built)

servo system to actively maintain the figure of the reflector. A 100 m telescope would have a resolving power of about 30 min of arc at a wavelength of 1 m, an example which emphasizes the relatively poor resolving power of this type of telescope at metre wavelengths. While higher resolving power can obviously be obtained by using a shorter radio wavelength, a limit is ultimately set by the irregularities of the reflector, the opacity of the atmosphere or the falling flux density of radio sources at high radio frequencies.

The largest fully steerable radio telescopes in use at present (1968) are those at Jodrell Bank (U.K.) and Parkes (N.S.W., Australia). The Jodrell Bank Mark I

---

† Design studies for a fully steerable reflector 120 m in diameter are being carried out at Jodrell Bank.

Table 1. Some Large Reflecting Radio Telescopes

Location	Movement of reflector	Diameter (m)	λ_{min} (cm)	Resolution at λ_{min} (min of arc)	Notes
Algonquin Canada	alt-az	46	3	3	
Arecibo Puerto Rico	fixed	305	50	10	1000 ft sinkhole
Crimea U.S.S.R	alt-az	72	—	—	Under construction
Danville Illinois	fixed	180 × 120	50	10 × 15	Cylindrical paraboloid
Delaware Ohio	alt	111 × 21·3	15	6 × 25	Fixed paraboloid, tiltable plane reflector
Effelsberg W. Germany	alt-az	100	2	1	Under construction
Green Bank W. Virginia	equatorial	43	2	2	140 ft
Green Bank W. Virginia	alt	92	20	10	300 ft
Goldstone California	alt-az	64	12	8	For space tracking
Jodrell Bank England	alt-az	76	20	13	Mk I
Jodrell Bank England	alt-az	38 × 25	6	8 × 12	Mk II (elliptical)
Malvern Enaland	alt-az	two of 25	10	17	Used as interferometer
Nançay France	alt	300 × 35	11	3 × 13	Fixed paraboloid tiltable plane reflector
Owens Valley California	equatorial	two of 27	10	20	Used as interferometer
Parkes N.S.W.	alt-az	64	6	4	210 ft
Scrpukhov U.S.S.R	alt-az	22	1	2	
Westford Massachusetts	alt-az	36	2	2	"Haystack". In radome

Table 2. Some Large Specialized Radio Telescopes

Location	Type	Dimensions (m)	Wavelength	Resolution	Notes
Bologna Italy	T	620 × 36 320 × 50	74 cm	7′	
Bothwell Tasmania	Dipole array	1075 × 1075	144 m	8°	
Cambridge England	Synthesis in declination	420 × 20 and 20 × 20	1·7 m	25′ × 30′	Interferometer 7′·5 lobe spacing
Cambridge England	Synthesis in declination	1000 × 12 and 12 × 12	7·9 m	1°	
Cambridge England	Full synthesis	Three paraboloids diam. 18	74 cm 21 cm	1′·5 0′·5	

[Continued on next page]

Table 2 (continued)

Location	Type	Dimensions (m)	Wavelength	Resolution	Notes
Clark Lake California	T	1560, 780	11·4 m	1°	
Culgoora N.S.W.	Annulus	Diam. 3000	3·7 m	3'·5	96 13-m paraboloids Radioheliograph
Fleurs N.S.W.	Full synthesis	Thirty-two 6 m and two 14 m paraboloids	21 cm 43 cm	0'·7 1'·5	Under construction
Hobart Tasmania	Dipole array	120 × 300	64 m	11° × 3°	
Leningrad U.S.S.R.	Parabolic strip	3 × 120	≥ 3 cm	≥ 1' × 1°	Pulkovo Observatory
Molonglo N.S.W.	Cross	1560 × 12 and 1560 × 12	74 cm	2'·8	
Peking China	Compound interferometer	1 km line of paraboloids	20 cm to 1 m	1' to 5'	Solar work synthesis planned
Penticton Canada	T	1300 × 650	13·5 m	1° × 1°·6	
Penticton Canada	T	1300 × 650	30 m	2°·6 × 2°·4	
Serpukhov U.S.S.R.	Cross	1000 × 40 and 1000 × 40	2·5 m to 10 m	13' at $\lambda = 3\cdot7$ m	
Westerbork Holland	Synthesis	12 paraboloids diam. 25 m	≥ 10 cm	0'·3 at $\lambda = 21$ cm	Under construction

telescope has a diameter of 86 m (250 ft), giving it a resolving power of 45 min of arc at $\lambda = 75$ cm. The shortest wavelength at which it is usable is about 20 cm. The Parkes 64 m (186 ft) telescope has a resolving power very close to that of the Jodrell Bank instrument at the same frequency. It can be used successfully at wavelengths down to about 6 cm. Another notable steerable reflector is the 43 m (120 ft) telescope at Green Bank (U.S.A.). This is the largest equatorially mounted radio telescope in the world. It has an extremely accurate parabolic surface, enabling it to operate with high efficiency at wavelengths as short as 2 cm.

By sacrificing a certain amount of manœuvrability it is possible to build radio telescopes in excess of 100 m. The largest such instrument is the remarkable 1000 ft reflector at Arecibo in Puerto Rico. The reflecting surface is made of wire-netting supported in a valley. The reflector is spherical rather than parabolic in shape, the primary feed is designed to correct for spherical aberration. The beam can be steered to any position within about 20° of the zenith by moving the primary feed, which is suspended 150 m (500 ft) above the mesh. The telescope was initially intended for ionospheric radar. It is used about half the time for radio astronomical work. It has a good efficiency at wavelengths longer than 50 cm.

Steerable reflecting radio telescopes are used in almost every kind of radio astronomical work. In the last few years a number of projects have assumed particular significance and a selection of these is listed below.

(*i*) *Radio spectrum lines.* In addition to the well-known 21 cm line of neutral hydrogen, radio spectrum lines associated with OH and H^+ have been discovered. The observation of these lines represents a major part of the work of most large parabolic reflectors.

(*ii*) *Radio background radiation.* Low resolution surveys have been made at a number of frequencies of the continuum radiation from the sky. At long wavelengths these reveal complicated structure apparently associated with the galaxy but far from the visible Milky Way. At high frequencies an isotropic black body radiation has been discovered which is interpreted by some astronomers as evidence of the primeval fireball.

(*iii*) *Radio polarization.* Linear polarization of the short wavelength radio emission of some radio sources and parts of the background radiation has been discovered, and its study has led to many interesting results. By studying how the polarization changes with wavelength, deductions can be made about the magnetic field and electron density in the galaxy.

(*iv*) *Linear occultations of radio sources.* By observing a radio source as it is occulted by the Moon, very accurate positions and details of the source structure can be obtained. The first quasars were discovered in this way. The very large collecting area of the Arecibo telescope has proved invaluable in these observations.

(v) *Radio source scintillations.* When the radio waves from a radio source pass close to the Sun they are affected by the irregularities in solar wind. This causes sources of small angular diameter to twinkle. Systematic study of such scintillation has provided information about the solar wind and about the diameters of radio sources. Quasars can be recognized by this means, because they have very small diameters.

(vi) *Pulsars.* Extremely large collecting areas are needed for the study of pulsars. This is particularly true for the initial discovery, for which the 20,000 sq. m tiltable E-W arm of the Molonglo Cross has proved outstandingly suitable. For detailed study of pulsars, fully steerable telescopes of the largest possible collecting area are required.

In addition to their usefulness as single radio instruments, steerable paraboloids are used extensively as the basic elements of interferometers and other specialized telescopes. The construction of fully steerable reflectors with high efficiency up to 20 or 25 m in diameter, now presents no serious engineering problems. Several more-or-less standard designs are available. Reflectors of about this size are being used in the Benelux Synthesis Radio Telescope and were proposed for an ambitious "very large array" telescope at Green Bank (U.S.A.). For space tracking, too, it has been argued that high collecting areas are best obtained by means of several 25 m telescopes rather than one large one. It is probable that reflectors of about 20–25 m diameter will be increasingly used in the design of compound radio telescopes.

Arrays, Cross- and T-type Radio Telescopes

The limited resolving power of conventional parabolic reflectors at metre wavelengths makes them unsuitable for some of the major tasks of radio astronomy. Fortunately it is possible to build a radio telescope that has a resolving power equivalent to a very large circular reflector by suitable positioning and interconnection of aerials or small radio telescopes.

One system of this kind is the "dipole array" which consists of an array of dipole aerials, each connected to a "focus" by suitable cabling. Dipole array telescopes are used, notably in Tasmania, to study the very long wavelength ($\lambda > 30$ m) radiation from the sky with resolving powers of a few degrees. At these wavelengths observations depend on favourable ionospheric conditions.

One of the most successful designs for a high resolution radio telescope is the "Mills Cross". A Mills

Fig. 2. The 1 mile Molonglo radio telescope, N.S.W. An aerial view of the Molonglo cross-type radio telescope taken from the north. Each arm is 1560 m long and 12 m wide, in the form of a cylindrical paraboloid. The east–west arm is steered in declination by mechanically tilting it. The north–south arm is steered by introducing suitable gradients of phase into the north–south line feed (not visible in the photograph).

Fig. 3. The 300 m Arecibo radio telescope, Puerto Rico. A catwalk leads from a mountain side to the feed mechanism at the focus 150 m above the reflector of the Arecibo telescope. The feed mechanism is supported by steel cables from three towers arranged symmetrically about the reflector. One of these towers is visible at the left of the photograph.

cross consists of two very long thin radio telescopes arranged at right angles in the form of a $+$. Usually one "arm" of the cross is aligned north–south and the other east–west. Each arm has a fan beam response, i.e. it receives radiation from a narrow strip of sky. If the arms have length d the narrow dimension of these fan beams is about λ/d radians. If the signals from the two arms are suitably combined (correlated) the resulting signal is proportional to the radiation from the overlap area of the two fan beams. This is an area of angular dimensions of approximately $\lambda/d \times \lambda/d$ (radians)^2, i.e. approximately equivalent to the resolving power of a circular aperture of diameter d. The effective collecting area of cross-type telescope is about $\sqrt{2} \times$ the area of one arm. They are usually constructed as meridian transit instruments, steering in declination being obtained by adjusting the phase and delay gradient of the north–south arm. A number of simultaneous beams can be formed by means of sets of suitably phased connections to the elements of one (or both) arms. In principle, any one of the four half-arms of a cross can be omitted, making a T-telescope. In practice the performance of a cross is usually superior.

Cross-type (and other partly filled aperture) telescopes are built to obtain a high resolving power, and they achieve this at a cost of some flexibility. For example they usually operate at one or two fixed wavelengths, and a change to different wavelengths involves changing a substantial part of the aerial system and its associated electronics.

A number of cross- and T-telescopes designed for wavelengths longer than 10 m are in operation in various countries. Arms more than a kilometre long are needed to give resolution better than 1°. One such instrument is a T-telescope which operates at a wavelength of 13 m at the Dominion Radio Astrophysical Observatory (Canada). It consists of 624 full-wave dipoles supported on 1698 wooden poles. The reception pattern has dimensions 1·1° × 1·6°. Instruments of this type provide valuable information on the non-thermal radiations from the Milky Way, and measurements of the strength of radio sources at long wavelengths. Long wavelength measurements are of particular importance in deducing the physical processes responsible for radio source emission.

The cross-type telescope with the highest resolution is the Molonglo Radio Telescope (N.S.W., Australia).

It operates at a wavelength of 74 cm and has a resolving power of 2·8 min of arc. It consists of two mile long (1560 m) cylindrical paraboloid reflectors, each with approximately 4000 dipole aerials at the line focus. This instrument is capable of measuring the position of radio sources to within a few seconds of arc. Eleven simultaneous beams are formed, and these are combined to make an image of the region of sky being surveyed. With this information the all-important identification of the radio source with the corresponding optical object (galaxy, quasar, etc.) can be made with considerable confidence. The main work of the Molonglo cross is to make a detailed survey of southern radio sources and to map the radio emission of the Milky Way in fine detail. The Bologna (Italy) T-telescope also operates at a wavelength of 74 cm. Its resolution is about 7 min of arc. Its work and capabilities in the northern hemisphere are comparable with those of the Molonglo cross in the Southern skies.

As an alternative to the cross- or T-configuration, high resolution may be obtained by building a radio telescope in the form of a ring. The Radioheliograph at Narrabri (N.S.W.) consists of 96 steerable paraboloids 13 m in diameter distributed uniformly round the circumference of a 3 km diameter circle. The instruments operate at 3·75 m wavelength, and the resolving power is 3·5 min of arc. A complete picture of the radio emission associated with sunspots is produced in $\frac{1}{2}$ sec. A map of the normal (quiet) Sun is produced after about 1 min observing.

4. Aperture Synthesis Radio Telescopes

The achievement of very high resolutions by means of cross- (T-)telescopes involves overcoming some formidable problems in mechanical and electronic engineering. An alternative approach has been developed (principally by the Mullard Observatory at the University of Cambridge, U.K.). The method is based on the fact that the radio sky is substantially unchanging, so that the whole telescope does not have to be present at the same time: the incident electromagnetic field in different parts of the telescope aperture may be sampled in sequence.

In practice it is necessary to use two or more small aerials simultaneously. The method can be understood by supposing it is required to synthesize a square telescope aperture of side $D = 8d$ using two small square aerials of side d. Then observations must be made with the two small aerials placed in sequence in every relative position that occurs between any two squares of a chess board (8×8 squares). In this example, the required number of spacings is 120. In general it is $\sim 2D^2/d^2$. It might be thought that this procedure would be extremely slow, but in fact once the set of observations has been made it is possible to combine the data in the computer to produce D^2/d^2 independent beam positions. This is exactly equivalent to "pointing" the complete aperture ($D \times D$) to D^2/d^2 separate points in the sky. The effective signal to noise ratio can be understood by comparing a full aperture using only one beam and a synthesized aperture surveying the same large area of sky in the same time interval. Then the synthesized results have a signal to noise ratio worse by a factor of approximately d/D than that of the full aperture. The principle disadvantage of the aperture synthesis method is that it relies on the constancy of the radio emission from the sky in the beam (field of view) of the small aerials. In practice most radio sources are constant, but the method is not suitable for studying time-dependent emission. A synthesized aperture cannot be used to look at a particular point. Observations must be made at a complete set of points, and then a set of beams synthesized, all lying within the field of view of the small telescopes.

Radio telescopes employing aperture synthesis to achieve a high resolution in declination have been used at Cambridge to map the sky radiation at a long wavelength of 7·9 m and to carry out an accurate survey of radio sources at a wavelength of 1·7 m. A recent and sophisticated aperture synthesis instrument is the Cambridge 1 mile radio telescope which consists of three equatorially mounted paraboloid reflectors 18 m (60 ft) in diameter. Two are fixed a distance of 800 m ($\frac{1}{2}$ mile) apart east–west and the third is mounted on rails extending for a further 800 m. During the course of 12 hr, the baseline is rotated through all possible orientations relative to the sky by the rotation of the Earth. The telescope operates at wavelengths of 74 cm and 21 cm. At 21 cm the synthesized beam has a resolving power of about $\frac{1}{2}$ min of arc. This is the greatest resolving power so far achieved by a pencil beam radio telescope (higher resolution has been obtained by long baseline interferometers). The Cambridge 1 mile telescope has mapped fine detail in the structure of many strong radio sources and surveyed limited regions of the sky for extremely weak (distant) radio sources.

An interesting synthesized aperture telescope is nearing completion at Fleurs (N.S.W.). It is a compound interferometer comprising thirty-two 5·7 m steerable paraboloids.

By recording the signals from every pair of aerials it will be possible to produce a complete synthesized map of the field of the 13·6 m paraboloids after only 12 hr observations. This type of system involves a great deal of complex electronics and computer analysis. However, it might well form the basis of radiotelescopes that eventually extend the pencil beam resolution into the region of a few seconds of arc. The synthesis radio telescope being built at Westerbork in Holland will use twelve 25 m paraboloids with a maximum spacing of 1620 m to provide a pencil beam of 22×22 (sec of arc)^2 at 21 cm wavelength.

5. Long Base-line Interferometers

It is possible to make very high resolution measurements of the angular diameter and structure of radio sources by the use of long base-line interferometry. The information obtained is necessarily less complete than that obtained using a pencil beam. In fact the use of an interferometer at a limited number of base lines corresponds to carrying out an aperture synthesis process but not placing the aerials at every available spacing. An interferometer of fixed base line measures one spatial frequency of the Fourier transform of the intensity distribution of a radio source. If the length of the base line is L, the spatial frequency measured by the interferometer is L/λ cycles per radian, in the direction of base-line projected on to the sky. By making observations at a number of base-lines and hour-angles, a set of points on the two-dimensional transform of the source is obtained. The number of such points used depends on the degree of complexity of the source, but valuable information on the source diameter can be obtained by one measurement at a base line at which the source is partly resolved.

Long base-line interferometer measurements of radio sources have been in progress at Jodrell Bank (U.K.) for many years. The early observations were made at wavelengths of 1·8 m and 74 cm. The 250 ft paraboloid was used as one element of the interferometer and a demountable reflector at a number of remote stations was the other element. The radio frequency connection was via a microwave link. More recently the 130 ft elliptical paraboloid has been used in conjunction with reflectors at Malvern to obtain base lines of one and two million wavelengths at $\lambda = 11$ cm and $\lambda = 6$ cm. A few radio sources were unresolved at these long base lines, showing them to have angular diameters less than 0·025 sec of arc.

An exciting development in radio astronomy has been the successful operation of a long base line interferometer without a radio frequency link between the stations. Some of the first successful observations using this technique were made by a group of Canadian scientists to measure the angular diameter of radio sources at 67 cm wavelength with base lines of 3074 km and 183 km (4·6 million λ and 270,000 λ). The local oscillator at each station was controlled by independent rubidium frequency standards. The signals were recorded on wide-band magnetic video tape recorders. The signals were correlated after the observations were complete by synchronizing the tape recorders during play back. Results showed that the gaussian diameter of one component of the well known quasar 3C273 was less than 0·02 sec of arc, and that the diameter of several other sources was less than 0·3 sec of arc.

Recently (mid 1968) this technique has been used to make interferometer measurements at 6 cm wavelength using radio telescopes in California and Sweden. With this enormous base line, it was shown that the quasar mentioned in the previous paragraph (3C273) has some features with an angular diameter of less than 0·0006 sec of arc.

There is a strong incentive to measure quasars with resolving powers even higher than that obtained at 6 cm wavelength with a baseline equal to the Earth's diameter. Since the atmosphere absorbs radiation of appreciably shorter wavelengths it seems that higher resolutions can be obtained only by increasing the base-line. The next step is a radio telescope on the Moon, or in orbit around the Earth.

Bibliography

Akademiya Nauk S.S.S.R. (1966) *Fizicheski institut imeni P. N. Lebedeva, Proceedings (Trudy)* Vol. 28, *Radiotelescopes*. Authorized translation. Consultants Bureau, New York. (Technical details of a number of Russian radio telescopes.)

Ground Based Astronomy. A Ten Year Program. A report to the American National Academy of Sciences, Washington, D.C. (1964) (Includes details of radio telescopes in the world of diameter greater than 10 m as of 1964.)

Large Steerable Radio Antennas (1964) *Annals of the New York Academy of Sciences* **116,** 1–355, New York. (Technical. Principally design details. Includes information on existing telescopes as of 1964.)

Radio Astronomy (1965) Report of the Fleck Committee, London: H.M.S.O. (Survey of British radio telescopes and proposed future instruments.) *Sky and Telescope* (periodical). Sky Publishing Corporation, Cambridge, Mass., U.S.A. (Appears monthly and includes articles on the majority of new radio telescopes. Principally concerned with amateur optical astronomy.)

SMITH F. G. (1966) *Radio Astronomy*, 3rd ed., London: Penguin Books.

Proc. I.R.E. Australia (1963) Radio Astronomy Issue, **24,** No. 2.

Trans. I.R.E. (1963) Radio Astronomy Issue, **AP 9,** No. 1.

Trans. I.R.E. (1964) Radio and Radar Astronomy Issue, **AP 12,** No. 7.

M. I. LARGE

REFRACTORIES, ULTRA-PURIFICATION OF.

1. Introduction

In many cases traces of certain impurities make materials unsuitable for use in technology, e.g. traces of oxygen reduce the electrical conductivity of copper, traces of hydrogen cause brittle fracture in Ti and 1 in 10^{10} atom fraction of Ni in Ge or of Au in Si reduce the life-time of minority carriers. Ductility and resistance to corrosion of refractory metals and critical temperature and critical field of refractory superconductors are very sensitive to small impurity concentrations. Even if an impurity produces desirable effects, it is important to produce a pure enough material and then dope it with controlled concentration of the desired impurity to obtain reproducible and desirable results. Investigations on materials where impurities obscured true properties have resulted in enormous waste of funds and talent. High purity materials are needed for use as standards in analytical procedure, quality control and other comparative applications. Intensive efforts are therefore being made to produce materials with a very high degree of purity.

The method suitable to purify a given solid depends upon the solid and upon the impurities to be separated. It also depends upon the scale, i.e. whether a few micrograms or several kilograms of the pure material are needed. Zone refining is the most powerful single method of purification. However, in certain cases zone purification is not effective in separating the impurities and alternative methods of purification have to be used.

The methods of purification are discussed briefly in the next two sections.

2. Zone Refining

This method was briefly discussed under the title "Zone melting of Metals" in vol. 7, p. 864 of this encyclopaedia. The discussion given in this section brings up to date the above entry.

(a) Principle of the Method

In zone purification, a small zone at one end of the solid is melted by a heater and is moved from one end to the other by moving one, the solid or the heater, and keeping the other stationary. As the zone travels, the impurity is continuously rejected from the solid into the liquid at the freezing interphase,† and is carried from the starting end to the other end. The process can be repeated any number of times so that the starting end becomes more and more pure and the other end collects more and more impurities. Two or more zones can be passed simultaneously by equal number of ring-shaped heaters traversing the rod. For a detailed description of equipment used, reference may be made to the book *Zone Melting* by Pfann.

The calculation of impurity distribution along the length of the rod after a given number of zone passes is laborious. However, the distribution of impurity after one pass or the ultimate distribution after an infinite number of passes can be represented by simple analytical equations.

(b) Cellular Structure and Dendrites

The effective value of k is defined as the ratio of the concentrations of the impurity in the main body of the solid and in the molten zone away from the interface. If stirring is absent, the steady state value of k becomes unity. As the interface is approached, an exponential gradient of temperature is set up in the diffusion layer which corresponds to a liquidus temperature gradient. If the actual temperature gradient imposed on the diffusion layer is smaller, temperature does not rise fast enough on moving into the liquid from the freezing interface and the diffusion layer is said to be constitutionally super cooled. Constitutional super cooling causes the formation of cellular substructure or dendrites projecting from the interface into the liquid. Constitutional super cooling and growth of cellular substructure can be avoided by low freezing rate, large temperature gradient in the liquid (temperature increasing away from the interface into the liquid), rapid mixing in the liquid and low solute concentrations.

(c) Banding

Fluctuations in solute concentration occur in many cases in zone refined ingots. This phenomenon is known as banding. Spacing between these bands varies from a few tens to a few hundreds of microns. Because of the temperature gradients, density gradients occur which result in thermal convection currents, warmer (and hence lighter) liquid rises and cooler liquid sinks. Even at fairly low temperature gradients, the convection results in a turbulent flow. When these fluctuations reach the freezing interface, they cause fluctuations in the growth rate. Since the concentration of impurity rejected into liquid is a function of growth rate, fluctuations of solute concentrations in the ingot result. The formation of bands can be avoided by imposition of d.c. magnetic fields on the melt. This is essentially eddy current damping and can be used with materials which are good conductors of electricity.

(d) Floating Zone Method

As the melting point rises so does the reactivity. The melts of solids like Si, W, Nb, etc., react with almost all known crucible materials and get contaminated, resulting in the negation of the very purpose of purification. Further, the ingot sticks to the crucible walls and due to differential thermal expansion severe strains are caused in the ingot, the fracture of the ingot or the crucible or both may occur. The floating zone method has been developed to avoid contamination of the melt of refractory solids from the boat material. In this method, crucibles or containers are not used, the melt is supported by surface tension between two vertical rods of the solid to be purified. The maximum height L_m of the zone that can be supported by surface tension against gravity in round rods is given by

$$L_m = 2 \cdot 8 \sqrt{\frac{\gamma}{dg}}$$

where γ is the surface tension, d is the density, g is the acceleration due to gravity. The values of limiting heights for some solids are shown in Table 1. In several cases, different values of melting points have been given by different authors. The values given in Table 1 are taken from *Periodic Chart of Elements* (1965) prepared by H. W. Loverenz of RCA Laboratories, Princeton, N.J., U.S.A.

---

† Usually k (ratio of concentration of impurity in the solid to that in the liquid in equilibrium with each other) is less than unity and the impurities are more soluble in the melt than in the solid. If the impurity is more soluble in the solid, $k > 1$, and the impurity is rejected from the liquid into the solid. If $k = 1$, impurities cannot be separated by zone purification.

Table 1. Maximum Zone Heights and Other Related Data for Several Refractory Solids

Element	Atomic number	M.P. °C	Density at R.T. (g/cm³)	Surface tension at M.P. dynes/cm	Critical stable zone ht (cm)
Co	27	1495	8.85	1880	1.30
Cr	24	1800	7.19	1420	1.25
Fe	26	1539	7.87	1754	1.34
Hf	72	2330	13.09	1460	0.95
Ir	77	2454	22.5	2310	0.90
Mo	42	2625	10.22	2080	1.28
Nb	41	2500	8.57	2030	1.38
Ni	28	1455	8.9	1735	1.25
Os	76	2700	22.57	2450	0.93
Pd	46	1554	12.02	1470	0.99
Pt	78	1773	21.45	1819	0.83
Re	75	3170	21.04	2480	0.79
Rh	45	1966	12.44	1940	1.12
Si	14	1415	2.42	680	1.50
Ta	73	3000	16.6	1910	0.96
Th	90	1845	11.66	1075	0.86
Ti	22	1820	4.51	1390	1.57
V	23	1735	6.1	1710	1.50
W	74	3410	19.30	2300	0.98
Zr	40	1750	6.49	1400	1.32

There is no upper limit on the rod diameter which can be float zoned. However, it is difficult to melt completely large diameter rods without exceeding the limiting zone heights.

The power required to maintain a float zone can be easily calculated if the zone is at a great distance from both ends of the rod. For creating the molten zone of 4 mm length in Nb rod of 4·8 mm diameter, the calculated power is 356 W as compared with the experimental value of 330 W.

Purification of refractories by induction heating has also been done in "cold containers". A boat made of silver or copper pipes is kept cold by circulation of water. The charge is melted by induction heating. A thin layer of the solid material is formed between the melt and the cold boat, and the melt does not react with the boat material.

3. *Purification by Heating in Vacuum*

Some metallic impurities and particularly interstitial gaseous impurities are not efficiently removed by zone purification. In such cases, suitable ambient (vacuum or a desired gas) is used to purify the solid either in the molten state or at very high temperatures near the melting point.

If the vapour pressure of the impurities is much larger than the vapour pressure of the host solid, and if the pumping speed is so large that the equilibrium vapour pressure of the impurities is larger than the actual vapour pressure near the melt, the impurity is removed from the surface of the melt by evaporation. The impurities in the interior now diffuse to the surface and are removed by evaporation. The process can be made faster by mechanical agitation. The vapour pressures of W, Nb and Zr at 2500 °C are 3×10^{-6}, 1×10^{-3}, 2.5×10^{-2} mm Hg respectively. Experiments have confirmed that when a Nb rod melt is heated in vacuum, concentration of tungsten impurities increases (because of preferential evaporation of Nb) and that of zirconium decreases.

Oxygen, nitrogen, hydrogen and the reaction products of oxygen with carbon are best removed from refractory solids by degassing. It is necessary to keep the refractory in the molten state or near the melting temperature for a sufficiently long time in good vacuum (preferably ultra high vacuum) to obtain a low equilibrium concentration of these impurities and then to cool the specimen rapidly to avoid the high equilibrium concentration of the impurities characteristic of low temperatures.

Experimental results on Mo rods have shown that the resistance ratio increases along the entire length of the rod suggesting that purification by distillation of impurities and degassing is significant.

A thin layer of the melt vapour is formed around the molten zone and protects it from the impurity gases present in the vacuum chamber. However, the impurities CH_4, H_2, CO and CO_2, if present in the chamber, can react with the solid some distance away from the molten zone and the carbon content in the solid may increase or decrease depending upon the relative concentration of the carbon containing gases in the chamber.

4. *Heating Methods*

Recently considerable improvements have been made in the techniques of heating the specimen to high temperatures for purification. These methods are discussed briefly in the next section.

The following three heating methods have been used for purification of refractory solids, (1) Induction heating (2) Electron beam heating and (3) Hollow cathode heating. These methods of heating are efficient because the heat is generated directly in the material.

(a) Induction heating is well known and can be used if the resistivity of the material is less than about 100 ohm-cm at least in the molten state. Usually the frequencies used are in the range 450 kc/s to 5 Mc/s. The ambient can be vacuum or any pressure of an inert gas. The gas helps in suppressing the evaporation if the material to be purified is volatile.

(b) In the electron beam heating method, a hot filament, usually a tungsten strip, forms the source of electrons and the anode is the specimen rod containing the molten zone. It works on the principle of a high vacuum diode. Focusing plates are used to concentrate the electrons over a small portion of the specimen rod. The kinetic energy of the bombarding electrons is dissipated as heat energy in the surface layer of the specimen. The interior of the rod is heated by conduction of heat from the surface. It is necessary to have a vacuum better than 10^{-4} mm Hg for an electron gun to operate and therefore the method is not suitable for the purification of volatile materials. The gases evolved in heating or melting the material can interfere with the performance of the electron gun. If the temperature of the filament is higher than that of molten zone, filament material can contaminate the melt. On the other hand, if the filament temperature is lower, the material is deposited from the molten zone on the filament and the emission characteristics of the filament are changed. To avoid this, focussing arrangements have been modified such that the filament does not "see" the molten zone. The main advantages of the electron beam heating method are its high efficiency, and the facts that heating can be confined to a very small area and that insulating rods can also be processed by this method. While purifying insulating solids, a grid is used to capture the secondary electrons from the non-conducting anode.

(c) Very recently a new method of heating known as hollow cathode heating has been described. The method is essentially based on the principle of discharge of electricity through rarefied gases. In the high voltage discharge, a stream of electrons emanates from the cathode which travels large distances without suffering collision. These electrons are known as "run away" electrons and have many characteristics common with the electron beam. It is these electrons which produce heat by bombarding the specimen. The cathode gun consists of an externally shielded water-cooled stainless steel cathode having a spherical hollow geometry. The external shield has a small separation from the cathode and acts as a barrier to regeneration mechanism suppressing the formation of a discharge at all external surfaces of the cathode.

This method is superior to induction heating in that it has a higher efficiency like the electron beam heating. The method can be used both for conductors and insulators. Unlike the electron beam heating, this method can be used with 3 mm Hg pressure of any desired gas. The permitted gas pressure helps in suppressing the evaporation of volatile solids. The ambient gas can also be used to remove certain impurities or to maintain stoichiometry in case of compound solids such as refractory oxides. The degassing does not interfere with the operation of a hollow cathode. The poisoning of the filament by the processing of the material from the melt does not occur and disturbances due to the sag or distortion of the filament are not caused. With the methods of purification discussed above and using one of the three heating methods, many refractory metals (and compounds) have been refined to a very high degree of purity. Concentrations of impurities of a few ppb have been claimed.

5. Effects and Analysis of Impurities in Refractory Solids

The experiments on ultrapure refractory solids have revealed many remarkable properties. An increase in the overall purity of tungsten crystals from 1 to 5 zone melting passes decreases the critical resolved shear stress by about 40,000 psi at -196 °C, increases the ductility and facilitates deformation twinning. Careful experimental work shows that the calculated values of the strengthening of refractory solids by interstitial impurities are at least one order of magnitude lower than those observed experimentally and suggests the need of improvements in current theories. The critical temperature and critical field of superconducting refractories seem to depend upon trace impurity concentration. Small carbon content is found to effect the hardness, microstructure and cold working properties of high purity Nb and other refractory solids. Much work is being done currently on the effect of trace impurities on properties of refractory metals.

The properties of semiconductors Si, Ge and III–V compounds have been studied very extensively and many reviews, books and treatises have been published on the subject. (See the entries "Semiconductor" on page 447, Vol. 6 and "Semiconductors, ionization of" page 454, Vol. 6 of this encyclopaedia.)

With the improvements in techniques to purify materials, the problem of determining the trace impurity concentration has also become important. Proper characterization of material involves the determination of not only the concentration of the impurity but also electronic state and state of aggregation or dispersion of the impurity in the solid. A knowledge of concentration of vacancies, interstitials, dislocations, mosaic boundaries and porosity is also required. Activation analysis, emission spectroscopy, spectrofluorometry, colorimetry, polarography, vacuum fusion and mass spectrometry are the important methods used for determining directly trace concentration of impurities. The concentration of nitrogen impurity is conveniently determined by the well-known Kjeldahl method.

Some impurities introduce large changes in some physical property of the solid. The change in the property can be measured and used to measure the overall concentration of such impurities. The impurities in metals scatter the electrons and increase the resistivity of the metals. The scattering of electrons by thermal oscillations of the lattice becomes negligible at liquid hydrogen or helium temperatures. Hence the ratio of resistivity at room temperature to that at low temperatures (4–20 °K) is a sensitive parameter indicative of the impurity of the metal. The larger the impurity concentration, the larger will be the resistivity ratio. The electrical conductivity of Si and Ge is used to measure the concentration of group III or V impurities. Nuclear magnetic resonance, electron paramagnetic resonance and many other techniques have been used to measure the trace concentration of certain impurities in solids. X-ray diffraction, electron microscopy and electron diffraction, neutron diffraction and optical microscopy techniques have also been used to measure the impurities in the solids.

Bibliography

ALBERT PH. (1960) *Nouvelles properties physiques et chimiques des metaux de très haute pureté*, Paris: Centre National de la Recherche Scientifique.

BROOKS MARVIN S. and KENNEDY JOHN K. (1961). *Ultrapurification of Semiconductor Materials*, New York: Macmillan.

CALI J. P. (Ed.) (1964) *Trace Analysis of Semiconductor Materials*, New York: Macmillan.

CHARLOT G. and BEZIER DENISE (1957) *Quantitative Inorganic Analysis*, London: Methuen.

CLASS W. and NESOR H. R. (1967) *The Growth of Single Crystals of Doped and Undoped Yttria and Yttrium Aluminate in a Hollow Cathode Zone Refiner*, Materials Research Corporation, Technical Report AFML-TR-67-51, Air Force Materials Laboratory, Wright Patterson Air Force Base, Ohio, U.S.A.

GILMAN J. J. (Ed.) (1963) *The Art and Science of Growing Crystals*, New York: John Wiley.

MATERIALS ADVISORY BOARD, NATIONAL RESEARCH COUNCIL (1967) *Characterization of Materials*. Available from Clearing House for Federal Scientific and Technical Information, Springfield, Virginia, U.S.A.

MORRISON G. H. (Ed.) (1965) *Methods of Trace Analysis*, New York: John Wiley.

PARR N. L. (1960) *Zone Refining and Allied Techniques*, London: George Newnes.

PASTERNAK R. A. and EVANS B. (1967) *J. Electrochem. Soc.* **114**, 452–7.

PFANN W. G. (1966) *Zone Melting*, 2nd ed., New York: John Wiley.

RESEARCH MATERIALS INFORMATION CENTER (1965) *Selected References on Groups IV, V and VI Transition Metals*, Oak Ridge National Laboratory, Oak Ridge, Tennessee, U.S.A.

RESEARCH MATERIALS INFORMATION CENTER. *Selected References on Groups IV, V and VI Transition Metals*, Supplement No. 1, Oak Ridge National Laboratory, Oak Ridge, Tennessee, U.S.A.

SCHILDKNECHT H. (1966) *Zone Melting*, New York: Academic Press.

WEINING S. (1967) *Transactions of the First Conference and School on Purification by Zone Refining*, Materials Research Corporation, Orangeburg, New York.

YOE J. H. and KOCH H. J., Jr. (Eds.) (1957) *Trace Analysis*, New York: John Wiley.

ZIEF M. and WILCOX W. R. (1967) *Fractional Solidification*, Vol. 1, New York: Marcel Dekker.

S. C. JAIN and VIJAYA NARAYAN

REVERBERATION UNDER WATER. Any discontinuity in the acoustic properties of a medium intercepts and reradiates a fraction of the acoustic energy incident upon it. This energy reradiation process is known generally as scattering and the acoustic energy obtained by summing the contributions from a large number of scatterers is termed reverberation. The amount of energy reradiated and the directional properties of the reradiated sound depend largely upon the ratio of the dimensions of individual scatterers to the wavelength of the acoustic signal. When the dimensions of the scatterers are small compared with the wavelength the pressure of the scattered sound, which is radiated isotropically, is inversely proportional to the square of the wavelength and directly proportional to the volume of the scatterers, irrespective of their shape. At the other extreme the scattering process is entirely independent of frequency and depends upon the cross-sectional area presented by the scatterer to the incident beam and upon its acoustic properties while its directional characteristics are determined by the shape and size of the scatterer.

Although the term reverberation strictly applies to sound scattered in any direction, it is usual in underwater acoustic engineering to restrict its use to cover that scattered energy which is reradiated back towards the source (and receiver) and it is thus synonymous with the alternative term backscatter. Sound scattered in the direction of propagation of the incident signal is said to be forward scattered and the subsequent process of acoustic interference between the incident and forward scattered signals produces a resultant which exhibits amplitude and phase fluctuations which become progressively more severe as the wave propagates.

In the volume of the sea there exists a wide variety of discontinuities ranging from inanimate objects such as dust, air-bubbles and the sea's own thermal and turbulent microstructures on the one hand to marine life in the form of plankton, fish and the Deep Scattering Layer (D.S.L.) on the other. The D.S.L. is a complex aggregate of marine life in the form of plankton and the higher forms of animal life which feed upon it and each other. It is very easily observed by means of an echo sounder and undergoes a diurnal vertical migration in an attempt to maintain the intensity of light illumination constant at the depth of the layer.

Further randomly disposed discontinuities are to be found at or near the boundaries of the sea. At the surface these are due both to randomly situated and suitably orientated wave facets and to a substantial layer of air-bubbles situated just below the surface originating from the breaking of the surface waves. On the seabed they are due to a random dispersal of its constituents (sand, silt, mud, gravel, shell, rocky outcrops, etc.) upon an uneven surface.

It is therefore possible to identify three distinct types of reverberation, each with its own particular properties and characteristics, namely volume reverberation, surface reverberation, and bottom reverberation.

The Characteristics of Reverberation

On the display of an active sonar reverberation appears as a continuous background signal with the same approximate frequency as that of the transmitted signal against which the target echo has to be identified. It fluctuates rapidly and randomly in amplitude and phase but its mean level is subjected to a slow irregular decay. However, contributions from the sea-surface and the seabed cause marked increases in the overall reverberation level at certain times during this decay. The situation is particularly complex in shallow water when multiple reflections between the two boundaries cause the "tail" of the reverberation to become very drawn out and irregular.

Since reverberation is the sum of the contributions from a large number of individual scatterers its instantaneous amplitude and phase are naturally subject to rapid random fluctuations. Observations show that the instantaneous amplitude is Rayleigh distributed, that is, the probability $P(x)$ that the instantaneous amplitude exceeds a certain value x is

$$P(x) = \exp(-x^2/a^2)$$

where a is the root-mean-square value of x. This is the distribution originally predicted by Rayleigh for the summation of a large number of sinusoids of equal amplitude but random phase, a situation which is more or less applicable to reverberation.

A detailed examination of reverberation shows that it appears to be made up of bursts or blobs of energy of the same duration, approximately, as the transmitted signal. As a consequence of this the autocorrelation coefficient of the reverberation envelope decreases to nearly zero in a time interval equal to the pulse duration. This property is, however, shared by the envelope of the echo from a target and in consequence makes it more difficult to discriminate between them.

Detailed measurements of the frequency spectrum of reverberation show that it is similar to the spectrum of the transmitted signal pulse in that it occupies a band of frequencies with an approximate width of $1/\tau$ Hz, where τ is the transmitted pulse duration, centred about the frequency of the carrier signal. The centre frequencies of the transmitted and reverberation spectra are not coincident, however, because the latter is subject to a number of Doppler shifts. These occur because the scatterers responsible for the reverberation are themselves in continuous motion and when the sonar system is mounted on a moving ship they are necessarily moving relative to the transmitting and receiving transducers. The situation is further complicated by the fact that their horizontal motion relative to the ship is different along different directions within the solid angle of the transmitting and receiving beams. In the case of a fast moving ship this can cause a considerable frequency shift coupled with a marked spreading of the frequency band occupied by the reverberation spectrum.

Volume, Surface and Bottom Reverberation Levels

If multiple scattering (reverberation arising from reverberation) is ignored and it is assumed that the scattering centres are randomly distributed in a statistically homogeneous manner, simple expressions can be obtained for the voltage appearing at the output terminals of the receiving transducer of a pulsed sonar due to each of the three components of reverberation.

(a) *Volume Reverberation*

In the case of volume reverberation the output voltage R_V is of the form:

$$R_V = I_0 m_V s J_V c\tau \exp(-2\alpha R)/2R^2.$$

In this expression I_0 is the source intensity of the sonar transmitter (referred as is customary sonar practice to a unit distance, the so-called index point), s is the receiving transducer sensitivity, c is the velocity of sound in the sea, τ is the transmitted pulse duration, R is the range at which the reverberation is observed and the term $\exp(-2\alpha R)$ represents the absorption loss in the sea characterized by the absorption coefficient α. The reverberant properties of the volume of the sea are characterized by the volume scattering coefficient m_V which is defined as the power scattered and reradiated isotropically by one metre cube of the sea when illuminated by a plane wave (of unit intensity). The composite transmitting and receiving directional response of the sonar system is expressed in terms of the volume reverberation factor

$$J_V = 1/4\pi \int_0^{2\pi} \int_{-\pi/2}^{+\pi/2} X(\theta, \phi) \, Y(\theta, \phi) \, d\theta \, d\phi$$

where $X(\theta, \phi)$ and $Y(\theta, \phi)$ are the individual transmitter and receiver beam patterns.

It is usual to define an equivalent plane wave reverberation level $R_V L$ which is equal to the intensity, on a decibel scale, of an axially incident plane wave which produces the same output voltage at the receiver as the observed reverberation such that

$$R_V L = 10 \log I_0 - 40 \log R + M_V - 2aR + 10 \log V$$

where $M_V = 10 \log m_V$, $a = 10\alpha \log e$ and $V = c\tau/2J_V R^2$ is the instantaneous reverberating volume. (Values of $10 \log J_V$ the volume reverberation index are tabulated by Urick for a number of simple transducer configurations in terms of their dimensions.)

Typical values of the volume scattering index M_V range from a mean of $-58 \cdot 2$ dB in the Deep Scattering Layer to -100 dB for scattering due to the thermal microstructure of the sea alone.

Although the actual sources of volume scattering are still obscure marine animal life at much lower population densities than exist in the D.S.L. are the most probable cause. This would, to some extent, account for the observed variation of M_V with the frequency of the incident acoustic signal which above 10 kHz rises at a rate of between 3 and 5 dB per octave. These figures are substantially lower than the 12 dB per octave predicted by Rayleigh scattering which assumes rigid scatterers, small compared with the acoustic wavelength, and thus indicate that the scatterers are in fact comparable in size with the wavelength.

Below 10 kHz the frequency dependence of M_V becomes more complex and is apparently due to more diffuse scattering layers nearer the surface than the D.S.L.

(b) Surface Reverberation

The equivalent plane wave level of surface reverberation is similar to that for volume reverberation and is

$$R_sL = 10 \log I_0 - 40 \log R + M_s - 2aR + 10 \log A$$

where $A = \frac{1}{2}c\tau J_s R$ is now the reverberating surface area and J_s is the surface reverberation factor. (Values of $10 \log J_s$ are again tabulated by Urick for simple transducer configuration.)

The value of the surface scattering index M_s can lie anywhere in the range -60 to 0 dB since it is dependent simultaneously upon a wide variety of factors the most important of which are (a) wind speed, (b) the angle of incidence of the sound beam upon the surface (and hence elapsed time after transmission or range) and (c) the frequency of the incident acoustic signal.

The variation with angle of incidence may be partially explained in terms of a shadowing of certain reflecting surfaces at or near the wave crests by adjacent, intervening wave troughs which increases as the angle the sound ray makes with the horizontal is decreased. It is also dependent upon the apparent thickness of the sub-surface layer of bubbles which is necessarily a function of the angle of incidence.

The frequency dependence of M_s is in the main due to the variation of the effective scattering cross-sectional area with frequency of individual air-bubbles within the sub-surface layer. The scattering cross-section of a bubble is a maximum when the frequency of the incident acoustic signal is equal to the natural resonant frequency of the bubble. Since the resonant frequency depends, amongst other things, upon the dimensions of the bubble and as a whole range of bubble sizes is to be expected in the sub-surface layer, the frequency dependence of M_s is necessarily complex. (Albers and Urick between them present sufficient tabulated and graphical data for an appropriate value of M_s to be determined for most sonar applications.)

(c) Bottom Reverberation

If M_s and J_s are replaced by the appropriate bottom scattering index M_b and the bottom reverberation index J_b the expression for the surface reverberation level applies equally well to the case of bottom reverberation. It should be pointed out, however, that the R^{-3} range dependence predicted by this expression is not always corroborated by observation and R^{-4} and R^{-5} range dependences have been variously reported. As yet no mathematical formulation has been produced which satisfies these observations.

Values of M_b, and their variation with the angle of incidence of the sound beam on the bottom and frequency, are very much dependent upon the type of bottom encountered. M_b is usually greater over rocky and shell-covered bottoms than it is over sand and mud.

Sand and mud bottoms show a rise in scattering strength of some 3 dB per octave with increase in frequency whereas rock and sand/rock mixtures exhibit little or no frequency dependence. This behaviour is obviously due to a difference in the scale of bottom roughness between the two cases.

Although all types of seabed show a general rise in scattering strength as the grazing angle θ (that is the angle measured with respect to the plane containing the seabed) is increased the actual angular dependence of M_b is very much a function of the bottom composition. There is some evidence that in a limited number of cases Lambert's law is followed. This assumes that the power reradiated by a scattering centre is proportional to the sine of the angle of scattering and leads to a value of M_b given by $M_b = k + 10 \log \sin^2 \theta$ where k is a constant.

Urick and Albers between them summarize the available data relating to the dependence of M_b upon angle, frequency and bottom composition.

A comparison of the volume, surface and bottom reverberation indices shows that the latter usually predominates. The relative values in a typical situation are, very roughly, -80, -40 and -25 dB. Local sound propagation conditions can, however, change this relative order of importance, as for example is the case when downward refraction of the sound beam is experienced. This occurs when the velocity of sound decreases as the depth increases. Under these conditions sound rays radiated from the sonar transmitter are bent away from the surface and as the amount of refraction increases with increase in range the surface scattering index decreases very rapidly with range until it eventually becomes insignificant compared with the volume scattering index.

Examination of the expressions for the volume, surface and bottom reverberation levels shows that the sonar engineer wishing to minimize the reverberation background of his system has only one possible way of achieving this, that is, to affect a reduction in V or A, the reverberating volume or area, whichever is applicable. Reducing the transmitted power does not bring about any improvement as might be expected since the target echo strength is reduced also. A reduction of the reverberating volume or area can be brought about in two ways: (a) by decreasing the beamwidths of both transmitter and receiver (reducing the beamwidth of the receiver is more important in this respect) or (b) by decreasing the duration of the transmitted pulse. Often both of these can only be achieved at the expense of increasing the complexity and cost of the equipment and the final choice of how much reverberation to permit must be a compromise between several conflicting factors.

Bibliography

Albers V. M. (1960) *Underwater Acoustics Handbook*, Pensylvania State Univ. Press.

Horton J. W. (1957) *Fundamentals of Sonar*, Annapolis: United States Naval Institute.

Urick R. J. (1967) *Principles of Underwater Sound for Engineers*, New York: McGraw-Hill.

B. K. Gazey

RUBBER PHYSICS. The primary task of rubber physics is to account for the great extensibility of rubber, its elastic recovery, and their dependence on temperature and rate of strain. An understanding of these properties is basic to the treatment of other important subjects within the scope of rubber physics, notably strength, friction and wear. Great progress has been made in these fields but this article confines itself to the fundamentals of rubber-like elasticity.

All rubbers consist of long-chain polymeric molecules the most commonly used of which are:

Natural rubber	Polyisoprene
SBR	Copolymer of styrene and butadiene
Poly-butadiene	
Nitrile rubber	Copolymer of acrylonitrile and butadiene
Butyl rubber	Copolymer of isobutylene and a small proportion of isoprene
Neoprene	Polychloroprene

The intermolecular forces between the polymer chains are similar in magnitude to those in viscous liquids; a permanent structure is produced by vulcanization, a chemical process by which the chains are cross-linked at rare intervals to form a continuous network. The most frequently employed cross-linking agent is sulphur, in conjunction with accelerators.

Theory of Rubber-Like Elasticity

The chain segments between cross-links are not fully extended but coiled, and fluctuate in Brownian motion between random configurations compatible with the distance between the cross-links. Deformation reduces the number of accessible configurations and hence decreases the entropy of the system. The entropy change ΔS is found by enumerating the possible configurations for a given state of strain, calculating their probability P and then using the Boltzmann equation

$$S = k \ln P \quad (1)$$

The calculation involves replacing the actual molecule by a tractable model, generally a chain of universally jointed rigid links, and assuming that the coordinates of the cross-links change as the principal strain ratios.

Neglecting internal energy changes, the elastically stored energy W per unit volume is

$$W = -T\Delta S \quad (2)$$

where T is the absolute temperature. The principal stresses t_i ($i = 1, 2, 3$) are obtained from equation (2) by equation (3):

$$t_i = \lambda_i(\partial W/\partial \lambda_i) + p \quad (3)$$

where the λ_i are the strain ratios. The addition of a hydrostatic pressure p to the stresses is necessary for the following reason. When deriving equation (3), the experimental finding has been taken account of that no appreciable volume changes accompany straining of rubber under ordinary conditions ($\lambda_1 \lambda_2 \lambda_3 = 1$). Having thus assumed the rubber to be virtually incompressible, an arbitrary hydrostatic pressure will not change the strains, and its presence must be allowed for. Only the difference between the stresses is defined unless the stress is known in at least one of the principal directions.

The chain segments behave like Hookean springs at moderate strains, and the strain energy is in this limiting case

$$W = \tfrac{1}{2} NkT \, (\lambda_1^2 + \lambda_2^2 + \lambda_3^2 - 3) \quad (4)$$

where N is the number of chain segments between cross-links.

The extension force F (per unit original cross-section) in simple elongation to the extension ratio λ is, from equation (4)

$$F = t/\lambda = NkT(\lambda - 1/\lambda^2) \quad (5)$$

and describes the results up to 100 per cent strain reasonably well. Equation (5) is also frequently used for the experimental determination of the cross-link density.

An interesting result is obtained for simple shear. Contrary to textbook statements, a shearing stress is not sufficient to maintain this strain but must be supplemented by a compressive stress normal to the plane of shear. The existence of this effect, which is proportional to the square of the shear and can therefore be demonstrated with a highly extensible material, is seen by twisting a rubber rod: the rod increases in length.

Equations (4) and (5) lose validity at large strains because some of the chain segments approach the limit of their extensibility by becoming fully uncoiled. Theory gives equation (6) for the extension force at large strains

$$F = \tfrac{1}{3} NkT \, n^{\frac{1}{2}} [L^{-1}(\lambda n^{-\frac{1}{2}}) - \lambda^{-3/2} L^{-1}(\lambda^{-\frac{1}{2}} n^{-\frac{1}{2}})] \quad (6)$$

where L^{-1} denotes the inverse Langevin function, and n is the number of links in the chain segment. The continuous lines in Fig. 4 are stress–strain curves calculated from equation (6). Comparison with experimental results is postponed until the effect of time on elasticity has been discussed. The theory of rubber-like elasticity is based on thermodynamic reasoning and applies, strictly speaking, only to equilibrium conditions which are not realized generally.

Dynamic Properties

When rubber is suddenly deformed to a given strain, the stress does not remain constant but decreases with time and eventually reaches a constant value. This phenomenon of stress-relaxation, which governs dynamic behaviour, originates from the finite time necessary for the rubber chains to attain new equilibria on being strained. The retarding forces are viscous in nature, and rubber is referred to as a visco-elastic material.

The measure of stress-relaxation is the relaxation modulus $E(t)$ defined by equation (7):

$$F(t) = f(\varepsilon) E(t) \qquad (7)$$

for the force $F(t)$ at the time t after application of the strain $\varepsilon = \lambda - 1$; $f(\varepsilon)$ depends only on the strain. Experimental stress-relaxation moduli $E(t)$ can be described as the sums of partial moduli E_i, each decreasing exponentially with time at a rate given by their relaxation times τ_i; thus

$$E(t) = \sum_0^\infty E_i \exp(-t/\tau_i) \qquad (8)$$

The spectrum of relaxation times reflects the viscous properties of the rubber, and its temperature dependence determines the temperature dependence of the relaxation modulus. As will be shown presently, it is closely connected with the "glass transition temperature" at which rubber loses all rubber-like properties and becomes glassy. The glass-transition temperature is characterized by a discontinuity in the temperature coefficient of various physical properties, such as the coefficient of thermal expansion.

An empirical finding of great consequence is that all relaxation times in the spectrum have the same temperature dependence. This temperature dependence is, furthermore, identical for all investigated polymers if temperature coefficients are referred to a temperature characteristic of the rubber, the so-called standard reference temperature T_s which lies about 50° above the glass-transition temperature. The temperature dependence of the relaxation spectrum is given quantitatively by equation (9)

$$\log[\tau(T)/\tau(T_s)]$$
$$\equiv \log a_T = -8\cdot 86(T - T_s)/(101\cdot 5 + T - T_s) \qquad (9)$$

The consequence of this result is an intimate connection between time and temperature effects on dynamic properties; for example, the relaxation modulus has the same value for any combination of time and temperature as long as the ratio t/a_T remains constant.

The usefulness of the time (or rate)-temperature equivalence is exemplified by the practically important case of small, sinusoidally varying strains. Rubber obeys Hooke's law at small strains, and its response to a time-dependent strain can be derived from equation (8) by mathematical methods. The stress is found not to be in phase with the strain, and the elastic behaviour of the rubber at the circular frequency ω must be specified by a complex modulus E^*

$$E^* = E' + jE'' \qquad (j = \sqrt{-1}) \qquad (10)$$

where E'' measures the energy loss during a strain cycle. The components of E^* are

$$E' = E_0 + \sum_0^\infty E_i \omega^2 \tau_i^2/(1 + \omega^2 \tau_i^2) \qquad (11)$$

$$E'' = \sum_0^\infty E_i \omega \tau_i/(1 + \omega^2 \tau_i^2) \qquad (12)$$

where E_0 is the static modulus. Time–rate equivalence appears here through ω and τ_i entering the equations only as products $\omega \tau_i = 2\pi \tau_i/t$ (t = period of cycle).

Figures 1 and 2 give the components of the complex shear modulus $G^* (= E^*/3$ for an incompressible

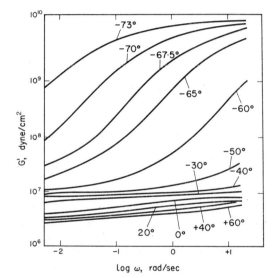

Fig. 1. Frequency dependence of G' for unvulcanized natural rubber at different temperatures. (From A. R. Payne (1958) *Rheology of Elastomers*, London: Pergamon Press.)

material) of unvulcanized natural rubber in a limited frequency range, with temperature as parameter. The families of curves present a bewildering picture of the functional relationships; this is where temperature–rate equivalence helps with the interpretation. According to this principle, a change in temperature from T_1 to T_2 is equivalent to a change in frequency by the factor a_{T_1}/a_{T_2} *at the temperature T_1*. All data in Figs. 1 and 2 can therefore be reduced to one temperature by a change-of-scale of the frequency axis. In a logarithmic plot, the procedure is simply to shift the curves horizontally by appropriate amounts.

The result is shown by the "master curves" in Fig. 3 which give the moduli at constant temperature in a greatly enlarged frequency range, so presenting a coherent picture of their frequency dependence. At a different temperature, the curves are bodily displaced into a different frequency range.

Figure 3, apart from validating rate–temperature equivalence, shows the typical frequency dependence of the complex modulus of rubbers. G' rises from its low static value to a magnitude comparable with that of a glass, marking the transition from the rubbery to

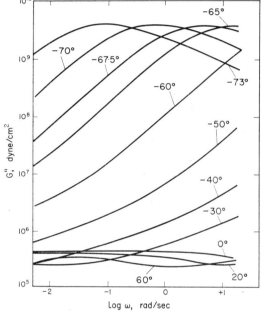

Fig. 2. Frequency dependence of G'' for unvulcanized natural rubber at different temperatures. (From A. R. Payne (1958) *Rheology of Elastomers*, London: Pergamon Press.)

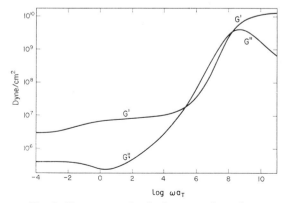

Fig. 3. Master curve for the frequency dependence of G' and G''. (From A. R. Payne (1958) *Rheology of Elastomers*, London: Pergamon Press.)

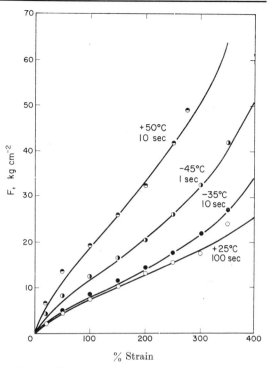

Fig. 4. Continuous curves: stress–strain relationships calculated from equation (6). Experimental points: isochronal stress–strain data for a vulcanizate of natural rubber at different temperatures and extension times. (From Harwood and Schallamach (1967) *J. Appl. Polym. Sci.* **11**, 1835–50.)

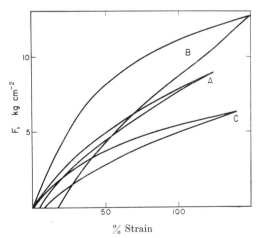

Fig. 5. Stress–strain loops at 20°C. A, natural rubber at 440 per cent/sec; B, Acrylonitrile-butadiene rubber at 440 per cent/sec; C, acrylonitrile-butadiene rubber at 0·044 per cent/sec. (From A. Schallamach and D. B. Sellen, unpublished data.)

the glassy state with increasing frequency. The loss factor G'' has a maximum in the transition region.

It should be mentioned that the temperature and rate dependence of certain other physical properties, for example friction and some kinds of abrasion, are similarly related.

The dynamic behaviour of rubber at large extensions is complicated by the non-linear stress–strain relationship which makes the assessment of time effects difficult. When rubber is stretched at a constant rate of strain, the stress depends at any instant on the stress-relaxation during the elapsed extension time. Because of the wide relaxation spectrum, the various relaxation mechanisms have come into operation to different degrees, and the experimental results are hard to explain directly. It has been found, however, that stresses obtained at constant rates of extension can be factorized as products of a strain function $f(\varepsilon)$ and a time function $\varphi(t)$:

$$F = f(\varepsilon)\,\varphi(t) \qquad (13)$$

where t is now the time taken to reach the strain ε. Equation (13) implies that the rate of stress relaxation is independent of the strain. "Isochronal" stress–strain curves can be constructed for which every strain has been produced during the same extension time t and which then follow the strain function $f(\varepsilon)$ in equation (13). (The necessary data are obtained by carrying out experiments at various rates of strain.)

Figure 4 shows results for a vulcanizate of natural rubber at various temperatures and extension times. The points fall close to the full curves which have been calculated from the statistical theory by means of equation (6), with values of N and n chosen to give the best fit.

The graphs in Fig. 4 establish therefore a bridge between theoretically predicted equilibrium and dynamic stresses, the link between the two being the function $\varphi(t)$ in equation (13) which, it is easily seen, is at the same time the dynamic Young's modulus of the rubber. Similar results have also been obtained for SBR.

Taking rubber through strain cycles produces hysteresis loops, examples of which are seen in Fig. 5.

The area between the ascending and descending branches gives the energy loss during the cycle. The two rubbers referred to in Fig. 5 have different glass-transition, and hence standard reference temperatures, $-20\,°\mathrm{C}$ for natural rubber, and $+20\,°\mathrm{C}$ for acrylonitrile–butadiene rubber. The latter is therefore at room temperature much nearer its glass-transition temperature and exhibits considerably greater losses than natural rubber. The difference between curves B and C shows the polymer effect on both modulus and losses; the curve for natural rubber at the lower extension rate would in this temperature range be little different from that at the high rate.

Expressions for the hysteresis loop can be derived from equation (13) by a modified Boltzmann superposition principle. The results, though closely resembling experimental curves, indicate that not all hysteresis losses can be explained by visco-elastic effects. One additional source of energy loss, crystallization, accounts qualitatively for deviations from the theory in the case of natural rubber at high extensions. Another loss mechanism is a semi-irreversible structural change of the rubber during extension.

Hysteresis is still a field of active research because considerable practical interest attaches to it. Far from being a necessarily objectionable aspect of rubber-like elasticity, it is an essential factor in determining the strength of rubber, and apparently indispensable for good skid resistance on wet roads.

Bibliography

BATEMAN L. (Ed.) (1963) *The Chemistry and Physics of Rubberlike Substances*, London: Maclaren.

FERRY J. D. (1961) *Viscoelastic Properties of Polymers*, New York: John Wiley.

SCHALLAMACH A. (1968) Recent advances in knowledge of rubber friction and tirewear, *Rubber Chem. Tech.* **41**, 209.

TOBOLSKY A. V. (1960) *Properties and Structure of Polymers*, New York: John Wiley.

TRELOAR L. R. G. (1958) *The Physics of Rubber Elasticity*, 2nd edn., Oxford: Clarendon Press.

A. SCHALLAMACH

S

SAILING, PHYSICS OF. The essence of sailing is that the rig produces an aerodynamic force F_a which has a forward component overcoming the resistance to motion, and in general there is an associated lateral component. This is opposed by a keel force at the cost of an addition to the resistance, and the capsizing couple from sail and keel must also be balanced.

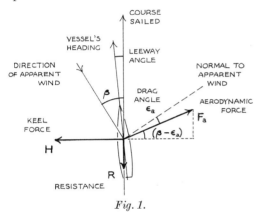

Fig. 1.

Assuming that the sails set are of an area, weight and tension appropriate to the average wind strength on a given occasion, their angle will in general be governed by one sheet for each sail; these could in theory be so coupled that one unified sheet control gave the best combination. In discussing performance quantitatively it is usually convenient to treat the rig as if it were under such a single incidence control. It is justifiable to assume that, for a given apparent wind angle β, hauling in the sheets increases the aerodynamic force F_a, but it continues to act through the same point in the rig, and only the drag angle ε_a changes; alternatively, the sheets being kept at the same trim, an increase of the apparent wind angle β increases F_a and changes ε_a.

On a course at, say, 90° to the true wind (a "reach") the helmsman will only have to control the incidence for the maximum speed. The ratio of the vessel's speed to the true wind speed, V_s/V_T, is substantially independent of wind speed at the lowest wind speeds, and it falls as the wind increases. This decrease of speed relative to the wind speed continues idefinitely for a normal yacht, whereas some craft such as catamarans, dinghies capable of planing, and the few which have ever been fully hydrofoil-borne, benefit from a marked improvement in the resistance-speed relation above a critical speed; hydrofoil craft have sailed at twice the wind speed in winds of some 12 to 15 knots. Ice-yachts may sail at up to four times the wind speed when reaching, and the amplification of the apparent wind by the vessel's own speed is then a dominant factor, but it is by no means negligible for even the slowest of sailing craft on water.

To make good the maximum speed towards the wind ("beating") or away from the wind, the helmsman has to select the best of a range of courses, each with its own best incidence for the rig. The operation of sailing downwind has been subjected to the least analysis, although the design and handling of spinnakers, the sails commonly added for this course, have recently been considerably developed empirically. It is not uncommon for a yacht, even with spinnaker set, to make good a higher speed to leeward by sailing perhaps 25° off the direct line. When an ice-yacht tacks to leeward, however, it sails between 30° and 40° off the direct leeward course. Whereas the direct downwind speed of a yacht using a spinnaker must inevitably be less than the true wind speed—by 45 per cent in even the most favourable case—the wide tacking of an ice-yacht may take it to leeward at two or three times the wind speed.

Tacking to windward, or beating, has received the most attention in yacht research and development, this having been generally regarded as the most important course of sailing. It does afford the most critical test of the aerodynamic and hydrodynamic efficiency of rig and keel, and it also intensifies the conflict between the major factors of design. The most obvious of these conflicts in the conventional yacht is that between, on the one hand, narrowness and lightness, giving an easily driven hull, and, on the other hand, beam and weight, giving power to stand up to the sail forces. Experience is still the best guide on these proportions. The choice of aspect ratio for the rig of any vessel can be more closely guided by mathematical analysis, mainly by analogy from experience with aircraft wings. When sailing to windward the lateral force is at least four times the forward thrust. The taller the rig, the higher its efficiency and hence the greater the forward thrust for a given lateral force. At the same time, the higher the centre of effort of the rig, the greater the angle of heel of a keel yacht for the same lateral force; angles of heel above about 20° pay an increasing penalty in the form of decreasing efficiency in both rig and keel. In the lightest winds there would be virtually no limit to the height of a rig, except the practical difficulties of keeping it up and in the correct shape. For a given capsizing couple, however, the best aspect ratio is not far from 3, so for sailing to windward in moderately fresh winds it does not pay to use a rig much taller than this. The criterion of a given capsizing couple has a precise significance in the case of multi-hulled craft or for ice-yachts and land-yachts, but for a keel yacht it is far from absolute, since the adverse effects of heeling increase only gradually. The effect of heeling on hull performance will be discussed first.

Mere asymmetry of the immersed hull when heeled need not add more than a few percent to the wave-making resistance for heel angles below 30°, and it is common for a hull heeled at 10° or 15° to gain from a reduction of wetted area by a few percent below its value when upright. In the case of a 2–3 ton yacht of waterline length L at speed v_s such that $v_s/\sqrt{gL} \sim 0\cdot 20$, skin friction accounts for 70–80 per cent of the total resistance; in a vessel of 200–300 tons at the same Froude number the fraction is nearer to one-half.

One important factor in hull behaviour which is markedly affected by heel is its efficiency in meeting the unwanted lateral force from the rig. When a yacht of average proportions is being driven hard to windward, the cost of the required keel force (induced drag) may well be 20 per cent of the total resistance; in a broad-beamed boat it may be nearer to 30 per cent of the total. That the keel efficiency falls off by anything between 15 per cent and 30 per cent as heel increases from zero to 30° matters less than that the efficiency at the larger angles of heel should be as high as possible.

Empirical development has given both racing and cruising vessels a keel area which, on average, is close to the ideal for working to windward; a smaller keel area would require a higher angle of leeway, and the consequent increment of wave-making resistance increases rapidly with leeway angle—probably as the square of the angle—whereas an increase of keel area would increase skin friction by more than would be gained from a smaller angle of leeway. The area of keel needed on other points of sailing is appreciably less for the same angle of leeway; reaching, at 90° to the true wind, the required keel area is only 40–50 per cent of that required on a beat. Retractable keels, such as centreboards, permit adjustment of this order at the cost of some complication in construction and operation. Appreciable reduction of wetted area below the classical proportion is being effected in an increasing number of ocean-racers by the combination of a steering rudder, well aft of the keel, with a flap at the rear of the keel; the flap being set to leeward, the effective axis of the keel is skew to that of the hull, enabling a smaller keel to generate the required keel force with little or no leeway on the hull proper.

A typical set of resistance curves for a keel yacht is shown, as derived from tank tests of a model in still water; at each angle of heel the tests cover only a small range of speeds around that to be expected in actual sailing, and this particular presentation applies

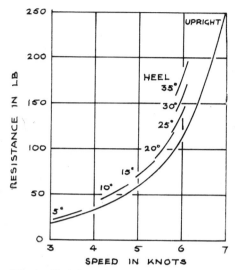

Fig. 2. Heeled resistance of "Gimcrack", the classic case presented by Davidson in 1936.

only to one assumed height for the centre of effort of the rig; the angle of leeway is so adjusted in the test that the yacht generates the keel force which would combine with an equal capsizing force on the assumed rig to provide the heeling moment corresponding to the selected angle of heel. The technique for such tests and their interpretation in terms of performance was devised by K. S. M. Davidson in 1936 and is still followed with relatively little change in commercial testing at the Davidson Laboratory and elsewhere. Research on keel behaviour has now been sufficiently extended (De Saix, 1962; Barkla, 1962; MacLaverty, 1966) to permit initial design studies to be made, however crude.

Comparison of rival designs with similar hulls is believed not to be sensitive to the assumed properties of the rig, and the sail coefficients deduced by Davidson from full-scale trials of a yacht named "Gimcrack" have been so extensively used that each laboratory engaged on such testing has accumulated its own case-histories of development in hull design, in which the final recorded figures are speeds made good to windward, predicted with the aid of these coefficients.

Knowledge of the behaviour of yacht rigs has been much advanced by research in the wind-tunnel of Southampton University (Chapleo, Marchaj, Spens). Ship rigs have been studied in Hamburg, and Thieme's comprehensive survey of sail propulsion spans the whole range from sailing ships to ice-yachts.

Typical aerodynamic characteristics of a yacht with a good rig are shown. The maximum obtainable values of the *lift* coefficient C_L are comparable with those from aircraft wings, but the relative drag is higher. Furthermore the rate of increase of C_L with angle of incidence is 30 or 40 per cent lower than for a rigid wing of the same proportions. Due to the higher parasite drag of the yacht itself, with rigging and crew, the minimum ratio of drag to lift D/L occurs at

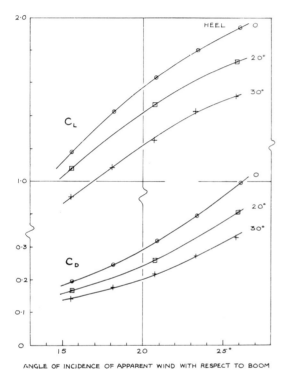

Fig. 3. *Wind-tunnel figures for sloop rig, from Spens (unpublished). "Lift" is defined as the component of aerodynamic force in the horizontal plane and normal to the direction of the apparent wind.*

a higher value of C_L than in the case of an aircraft, but, contrary to a common belief, the condition of minimum D/L is of no significance except in the case of the ice-yacht. For all other craft light winds call for almost the highest obtainable value of C_L: as the wind strength increases, however, progressive wind-spilling, or reduction of C_L, becomes necessary for best performance.

The nearest to a unified theoretical treatment of the complete yacht is that of Crewe, whose aim has been the representation of hull and sail data in an idealized form, which could be fed in to the equations of motion so that the effect of changing any variable might be predicted analytically. Other studies have been made (Barkla, 1961) showing the general pattern of behaviour obtained either (i) by marrying a set of hull characteristics with a set of aerodynamic properties representing the rig more completely than the Gimcrack coefficients, or (ii) by studying the effect of individual variables analytically by simplifying or neglecting the inter-relation of others; this second technique is particularly appropriate in studies assessing the effects of radical changes in design.

The inter-relation of the variables in the case of a heeled yacht is such that the labour of calculation tends to obscure the basic balance of the principal forces. To illustrate this balance and also the effect of the vessel's own speed on the strength and direction of the wind we may consider two simple and only slightly idealized cases, an ice-yacht and a keel yacht in very light winds.

The runners of the ice-yacht will be presumed to experience no resistance whatever, and to take up the lateral force with no equivalent of induced drag. As the sail is sheeted in, speed increases and the apparent

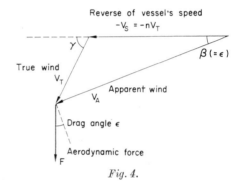

Fig. 4.

wind angle β decreases until it is equal to the drag angle ε_a; the resultant aerodynamic force F_a is then perpendicular to the direction of motion. The ratio n of the vessel's speed to the wind speed is therefore

$$n = \frac{\sin \gamma}{\tan \varepsilon_a} - \cos \gamma.$$

If at first we limit consideration to the case of sailing at 90° to the true wind, i.e. when $\gamma = 90°$,

$$n = \frac{1}{\tan \varepsilon_a}. \qquad (1)$$

This is the simplest case of what is, in theory, the simplest form of sailing. As long as the wind strength is low, so that the limit of stability is not reached, the maximum speed is entirely determined by the minimum possible overall drag/lift ratio for the craft, $n_{\max} = 1/\tan \varepsilon_{\min}$. As soon as the limit of stability is reached, it ceases to be possible to present any such simple formula for the speed.

For a given ice-yacht the drag angle may be expressed as a function of the coefficient of lift C_L of the rig, and in any wind on a given course there is a maximum value of C_L which will give $F = F_{\max}$, the maximum permissible lateral force: this is effectively equal to the total weight multiplied by the ratio of the half-width between runners to the height of the centre of the rig. Expressing force in lb, sail area S in ft², velocity in knots (1 knot = 1·69 ft s⁻¹), and taking the density of air at 0°C, $\varrho_a = 2\cdot 51 \times 10^{-3}$ slug ft⁻³,

$$F = 1\cdot 69^2 \sec \varepsilon_a \frac{\varrho_a}{2} C_L S V_A^2$$
$$= 3\cdot 58 \times 10^{-3} \sec \varepsilon C_L S V_T^2 (n^2 + n \cos \gamma + 1).$$

The range of variation of $\sec \varepsilon_a$ is so low that we may replace it by an average figure. At the condition of capsizing, then, we may write, for the case of $\gamma = 90°$,

$$n = \sqrt{\frac{10^3 \times F_{\max}}{3\cdot 6 C_L S V_T^2} - 1}. \qquad (2)$$

Since the apparent wind will strike the structure from practically the same angle whatever course is sailed, the drag angle may be adequately expressed in the form

$$\tan \varepsilon_a = \frac{C_D}{C_L} = aC_L + b/C_L \qquad (3)$$

where $\dfrac{1}{\pi a}$ is of the order of the aspect ratio of the rig, and b is proportional to the ratio of the projected area of the whole structure and crew to the sail area.

One method of solution is to combine equations (1) and (3) to give one expression for n as a function of C_L, equation (2) providing a second expression, and the two being solved graphically for n.

The procedure is illustrated by application to a Class-E ice-yacht, taking an estimated figure of 620 lb for the capsizing force $F_{\max}$. The 75 ft² rig is assumed to behave as an aerofoil of aspect ratio 3, and

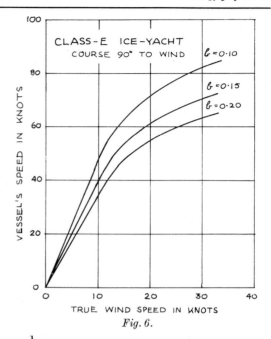

Fig. 6.

$a = \dfrac{1}{3\pi}$. Three different values of b are used, 0·1, 0·15 and 0·20; it seems probable that the middle figure is the best that could be achieved without extreme measures for streamlining.

If the runner resistance were really zero, then for all wind speeds from zero up to about 10 knots the vessel's speed would be directly proportional to wind speed. In moderate to strong winds the sheet has to be eased progressively, the ratio n falling steadily. The figure of 100 mph (87 knots) claimed for these craft would only be attainable with the assumed rig on a course 120° to a 40 knot true wind. The best course to windward in more moderate winds is slightly over 45°, and, for $b = 0\cdot 15$, the speed made good is 28 knots in a 20-knot wind; to leeward 48 knots can be made good in the same wind by sailing just over 30° off the true line of the wind.

While calculation of the full polar diagram of speeds in all stronger winds on all courses is somewhat laborious, the process is exceedingly simple for light

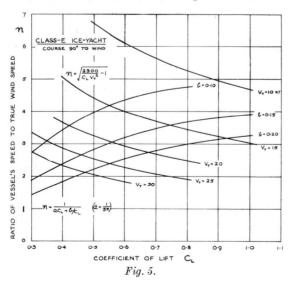

Fig. 5.

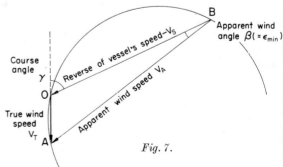

Fig. 7.

winds. Since the apparent wind angle β is constant and equal to the minimum drag angle $\varepsilon_{\min}$ the locus of B is a circle with the true wind vector OA as a chord.

The best course for tacking to windward or leeward can be seen immediately from such a polar diagram, or it may be obtained by differentiating $n \cos \gamma$ with respect to γ, which gives $\tan \gamma = \dfrac{-1}{\tan \varepsilon_n}$. The optimum windward and leeward courses angles are, therefore, $\left(45 + \dfrac{\varepsilon}{2}\right)^\circ$ and $\left(135 + \dfrac{\varepsilon}{2}\right)^\circ$.

The difference between the polar diagrams of absolute speed for the same ice-yacht in winds above and below the limit at which the stability factor becomes operative is illustrated for the case $b = 0.15$ (see Fig. 9).

The assumption of zero resistance which is appropriate to the case of the ice-yacht has no validity for any of the other forms of sailing, but the abrupt limit to stability is similar to that of catamarans. In term of (capsizing force)/weight the ice-yacht has the advantage. Having to overcome water resistance exceeding a quarter of their weight at their top

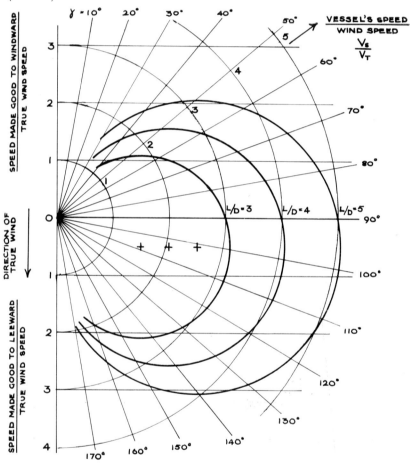

Fig. 8. *Polar diagrams of speeds relative to the true wind speed, for ice-yachts with maximum lift/drag ratios of 3, 4 and 5.*

Applying this to the E-class ice-yacht with $a = \tfrac{1}{3}\pi$ and the same three values of b as before, we find:

	Best course to windward	Best course to leeward
$b = 0.10$	$\gamma = 50.8°$	$\gamma = 140.8°$
0.15	52.1	142.1
0.20	53.1	143.1

speeds, around 25 knots, the fastest catamarans, those of the C-class, need all their permitted sail area of 300 ft²; an efficient rig of this area has a centre of effort whose height above the centre of the keel is considerably greater than the statutory maximum beam, so that, even with the crew weight to windward, the maximum lateral force is under half the total weight (see Fig. 10).

A dinghy in which the crew's weight is disposed to windward on a sliding-seat or a trapeze is limited

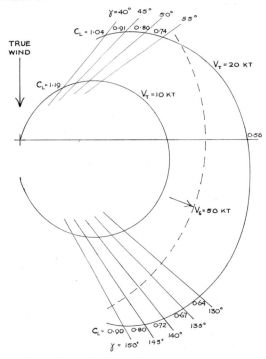

Fig. 9. Polar diagrams of absolute speeds of an ice-yacht in winds of 10 and 20 knots. In the 10-knot wind the rig can be trimmed for minimum drag angle on any course. The 20-knot wind requires a reduction of C_L, which varies with the course.

almost as sharply in tolerable force, and the limit is lower, seldom exceeding 0·3 times the total weight. This naturally reduces the possible forward thrust, so that the very fastest dinghies can only reach about 18 knots.

In winds of strengths which suit them best, generally between 12 and 18 knots, both catamarans and planing dinghies exhibit polar diagrams of speed which resemble those of ice-yachts more nearly than those of conventional keel yachts. The latter approximate more closely to a circle centred on the origin, because of the steep rise in resistance at Froude numbers $v_s/\sqrt{gL} > 0·36$; if the driving force on a reach exceeds that when close-hauled by 50 per cent, the difference in water-speed will be only about 10 per cent. Measurements of the performance of dinghies and catamarans are scarce, but the sketch by Davidson (see Fig. 11), comparing a 6-Metre yacht with a light displacement craft, is fairly representative.

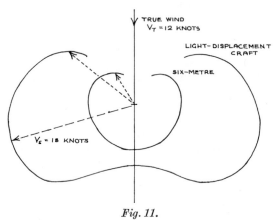

Fig. 11.

Whereas the optimum course angle to windward for a 12-Metre yacht may be as low as 35° or even 33° in a wind of 15 knots, it exceeds 50° for any vessel with a capacity for higher speeds; sailing at a wider angle to the wind, the lighter craft put the increased drive to such good purpose that they still make good a higher

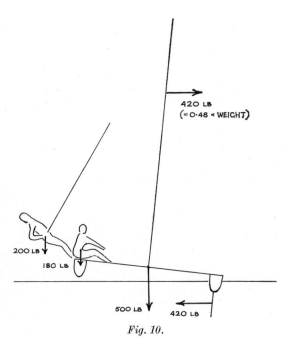

Fig. 10.

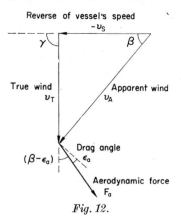

Fig. 12.

speed. This does not mean that they can not sail much closer to the wind if circumstances require it.

The second illustration to be considered quantitatively is the keel yacht in light winds, the condition often called "ghosting".

An idealised yacht, with resistance proportional to the square of the speed, independent of lateral force, and with a rig which is unaffected by heel, is a fair approximation to a real vessel sailing in very light winds. Considering for simplicity the case of such a vessel reaching at $\gamma = 90°$, the driving force from sail area S_a is (see Fig. 12)

$$F_D = \tfrac{1}{2}\varrho_a C_a S_a (v_S^2 + v_T^2) \sin(\beta - \varepsilon_a)$$

where v_S and v_T are the velocities of ship and true wind in consistent units; the lift coefficient C_L used above has here been replaced by the coefficient C_a corresponding to the force F_a; $(C_a = C_L \sec \varepsilon_a)$. The resistance of the yacht may be expressed in terms of the wetted area S_w, water density ϱ_w and a coefficient C_f', which is the normal skin friction coefficient augmented by a small factor to take account of the residual or wave-making resistance. In steady motion

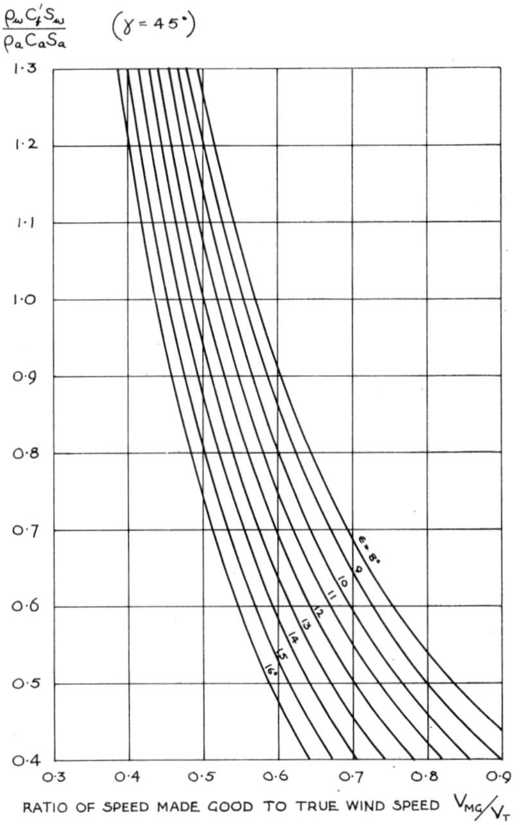

$$\tfrac{1}{2}\varrho_a C_a S_a (v_S^2 + v_T^2) \sin(\beta - \varepsilon_a) = \tfrac{1}{2}\varrho_w C_f' S_w v_S^2$$

and hence

$$\frac{\varrho_w C_f' S_w}{\varrho_a C_a S_a} = [1 + (v_T/v_S)^2] \sin(\beta - \varepsilon_a). \quad (4)$$

In this range of sailing speeds, then, the ratio of vessel's speed to wind speed on any course is independent of wind speed, and it is largely determined by the ratio S_a/S_w between sail area and area of wetted hull; a value of around 2 is common for this ratio, whereas only a few large racing cutters have exceeded 3. At standard atmospheric pressure and 15°C the ratio $\varrho_w/\varrho_a = 837$. For a given yacht, S_a/S_w being fixed, the LHS of equation (4) is a function of C_a only. The RHS is a function of v_S/v_T and the drag angle ε_a, and may be presented in the form of lines of constant ε_a against v_S/v_T, so that, if a relation between ε_a and C_a is known or postulated for the rig, the LHS being calculated for each value of ε_a, the solution in terms of v_S/v_T is read from the graph. Similar plots, such as that shown (see Fig. 13), may be presented in terms of relative speed made good v_{mg}/v_T for any course angle γ. Four such charts were used to predict the performance to windward of a Dragon class yacht, given $S_a/S_w = 2·16$; the coefficient C_f' was taken as 1·23 times the skin friction coefficient derived from the Schoenherr line at $v_s/\sqrt{gL} = 0·15$; four pairs of values of C_a and ε_a were assumed, as indicated (see Fig. 14).

Fig. 13.

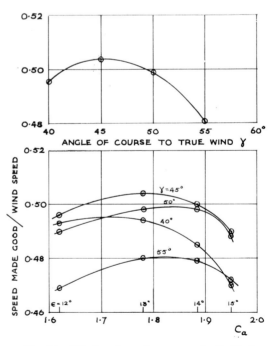

Fig. 14. Windward performance of idealized Dragon in light winds, with varying course angle and sheet trim. The upper plot represents the maximum speed made good, derived from the lower set of curves.

Each course sailed has its own optimum value of C_a, and there is also a best course, which is here almost exactly 45° (see Fig. 14) but, even when the properties of the yacht are as drastically simplified as they are in this treatment, there is no simple relation which gives these optimum quantities. If the absolute values of the speeds shown are in doubt, for lack of more specific data, the general trends are correct, subject to one qualification: there being a limit to the sheeting in of the sails, it is generally practicable, when on optimum course, just to attain the optimum C_a but not appreciably to exceed it.

Although in stronger winds the properties of hull and rig for keel yachts change appreciably with heel and hence with C_a, the corresponding relations differ from those shown only in that the best course angle falls as the wind speed increases and the optimum C_a lies progressively further below that attainable.

Bibliography

BARKLA H. M. (1961) *Trans. R. Inst. Naval Architects* **103**, 1.
BARKLA H. M. (1962) TM-132, Davidson Laboratory, Stevens Institute of Technology.
CREWE P. R. (1964) *Trans. R. Inst. Naval Architects* **106**, 287.
DAVIDSON K. S. M. (1936) *Trans. Soc. Naval Architects and Marine Engineers* 288.
DAVIDSON K. S. M. (1956) Chapter in *Survey of Mechanics* by G. K. Batchelor and R. M. Davies (Eds.), Cambridge University Press.
DE SAIX P. (1962) TM-129, Davidson Laboratory.
MACLAVERTY K. J. (1966) SUYR Report No. 17, University of Southampton.
MARCHAJ C. A. SUYR Reports Nos. 6, 11, 13 and 14, University of Southampton.
THIEME H. (1955) *Jahrbuch der Schiffbautechnischen Gesellschaft.*

H. M. BARKLA

SATELLITES IN SOFT X-RAY SPECTRA. X-ray satellite lines (sometimes called "non-diagram" lines) are X-ray emission lines at wavelengths that do not correspond to differences between pairs of levels in the electron orbital energy level diagram. There may be one or more satellite lines corresponding to each "diagram line". The satellite lines will always occur, in the case of metals at least, on the high energy side of the diagram line.

In 1921 Wentzel correctly explained the positions of many of the satellites on the basis that the atom was ionized in two inner levels rather than only one. Wentzel hypothesized that both of the inner vacancies were created by the initial electron excitation. However, the low probability for double ionization made this hypothesis untenable in the face of the relatively high intensity observed for many of these satellites.

In 1935 Coster and Kronig showed that the double ionization could result from the Auger process. An example of the process is shown in Fig. 1 for the production of the $L\alpha_1$ satellite (Fig. 1b) in rhodium corresponding to the $L\alpha_1$ parent line (Fig. 1a). In the production of the parent line, the original excitation creates a vacancy in the L_3 level. An electron from the M_5 level drops into the L_3 vacancy resulting in the emission of an $L\alpha_1$ photon. In the production of the $L\alpha_1$ satellite, the initial excitation creates a vacancy in the L_1 level. The probability for the

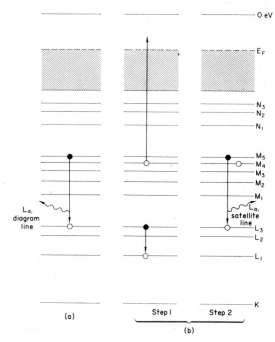

Fig. 1. The Auger process; example: rhodium $(Z = 45)$.

vacancy to be filled by an L_3 electron is high if the energy is transferred to another electron, in this case an M_5 electron, ejecting it from the material rather than the energy being liberated in the form of electromagnetic radiation. Such a filling of the L_1 vacancy by an L_3 electron (written $L_1 \to L_3$) is called a *radiationless* or *Auger transition*, after Prof. P. Auger who, in 1925, first recognized the process. The electron ejected from the M_4 level as a result of the Auger transition is called an Auger electron. An M_5 electron now falls into the L_3 hole exactly like in the production of the $L\alpha_1$ parent line except, in the case of the satellite, the extra vacancy in the M_4 level means that some of the electron screening is removed. This has the effect of increasing the energy difference between the M_5 and L_3 levels causing the $L\alpha_1$ satellite to occur at a higher energy than the parent $L\alpha_1$ line. This Coster–Kronig explanation accounts for both the intensity and position of the satellites of many of the L and M diagram lines.

On the other hand, excitation of any of the five satellites occurring in the K spectral series is exclusively due to double ionization by a single electron impact. By far the most intense of the K series satellites are the $K\alpha^3$ and $K\alpha^4$. The $K\alpha^3$ is the result of a vacancy having been produced in both the K and the L shells by the initial electron excitation. The vacancy in the K shell is then filled by a p-electron from the $L_{2,3}$ levels in the normal manner of producing a $K\alpha$ line except that now with a vacancy already in the L shell, this $K\alpha^3$ satellite line would occur at a higher energy. The production of the $K\alpha^4$ satellite is similar but in this case, the initial electron excitation removes both of the electrons from the K shell.

The satellites discussed thus far correspond to the energies of electron transitions which have been considered to be transitions between single-particle states. As such they differ from diagram lines only in that the initial and final state energies associated with the electron transitions have been perturbed by various mechanisms. However, there is another type of satellite recognized in soft X-ray spectra called the plasmon satellite. To account for this type of satellite we must recognize that the electrons when in either their initial or final state are not operating independently but rather are operating collectively. We now consider a hole in the collective state of lower energy. The entire system may now relax by the emission of a photon together with the generation of a plasmon. A plasmon is a quantized collective oscillation of the conduction electrons, obeying the equation of motion of a simple harmonic oscillator of frequency, ω_p. The plasmon satellite will occur in the spectrum at the energy corresponding to the energy of the single-particle model electron transition, $h\nu$, less the energy of the plasmon, $\hbar\omega_p$. The plasma frequency is obtained from the density per unit volume of conduction electrons, n, their effective mass, m, and the charge on an electron, e, through the relationship,

$$\omega_p = [(4\pi n e^2)/m]^{1/2}.$$

The probability of the system relaxing by the generation of a plasmon together with the emission of a photon is rather small compared with the single emission of a photon. Thus the plasmon satellite intensity is very low and in fact has been observed only by electronic recording methods whereby many scans of the spectrum are summed.

Bibliography

BURHOP E. H. S. (1952) *The Auger Effect*, Cambridge: University Press.

HIRSH F. R., Jr. (1942) *Reviews of Modern Physics* **14**, 45.

J. R. CUTHILL

SEMICONDUCTORS, EFFECTS OF RADIATION ON. Since their electrical properties are markedly sensitive to the degree of perfection or purity of the crystal, most extrinsic semiconductors are strongly affected by lattice imperfections introduced by bombardment with energetic particles. Historically, investigations of radiation effects (radiation damage) in semiconductors go back to 1947 when Lark-Horovitz and co-workers were attempting to introduce acceptor and donor impurities into germanium by nuclear transmutations resulting from bombardment with 10 MeV deuterons: the change in resistivity was much more profound than could be expected from the small amount of chemical impurity introduced and it was found that the original resistivity could be restored by thermal annealing at 450°C. Therefore, it was concluded that interstitial atoms and lattice vacancies created by elastic collisions of the deuterons with lattice atoms were responsible for the changes in electrical behaviour. Subsequently, extensive investigations of many semiconductors have been carried out using a wide variety of energetic radiations capable of displacing lattice atoms. From a basic point of view, this field of research is concerned with the connection between electronic behaviour and the various types of stable lattice imperfections which can be created by the bombarding particles. From a practical viewpoint its aim is to understand those factors which contribute to radiation sensitivity in order that semiconducting devices (diodes, transistors, solar cells, etc.) and integrated circuits may be utilized in radiation fields to best advantage; this is particularly important in space electronics and for applications near nuclear reactors.

Effects on Conductivity and Concentration of Charge Carriers

Lattice defects possess localized electronic states and may behave as acceptors and/or donors depending upon their identity and the semiconductor in question. In germanium the dominant effect of impurities is acceptor action; hence n-type germanium is drastically affected—irradiation decreases the conductivity and sufficient irradiation causes conversion to p-type, whereas effects on p-type germanium are much less pronounced. In silicon both n- and p-type material approach intrinsic behaviour when irradiated; i.e. irradiation tends to decrease the concentration of either electrons or holes. Therefore the radiation acceptors in germanium ionize by cap-

Material	Response to Irradiation
Si	Both n- and p-type approach intrinsic
Ge	n-type converts to p-type
InSb	p-type converts to n-type
GaSb	n-type converts to p-type
AlSb	p-type converts to n-type
InAs	p-type converts to n-type
CdTe	p-type converts to n-type

turing conduction electrons in n-type material and by producing holes in p-type material; in silicon radiation defects "neutralize" electrons and holes in both n- and p-type material, respectively, with the result that the conductivity of both types decreases toward the intrinsic value. Tabulated below are the radiation responses exhibited by several semiconductors.

Types of Damaging Radiation

Any corpuscular radiation capable of imparting an energy in excess of some value E_d called the displacement energy to a lattice atom in an elastic collision can produce lattice defects and alter the electronic properties of semiconductors. For semiconductors E_d lies in the range from 6 to 20 eV. Quite familiar are effects due to fast neutrons encountered in nuclear reactors and energetic ions (1 to 10 MeV protons, deuterons and alphaparticles) available from particle accelerators. Electrons with energies in excess of a few tenths MeV are also capable of displacing atoms, as are the Compton electrons and photo-electrons of this energy produced when gamma-rays are absorbed in a solid. The simplicity of the damage depends upon the energy transferred to the struck atom by the incident particle; e.g. electrons in the MeV range are capable of displacing only one or two atoms whereas a fission neutron can transfer an energy several hundred times E_d, thereby creating a displacement cascade or heavily damaged region whose diameter is 100 Å. The difference in damage can be detected from the effect on physical properties. The simple electron damage usually exhibits well-defined acceptor or donor states (ionization energies), whereas the energy levels associated with the disordered regions created by fast neutrons are quite indistinct and smeared out. In n-type germanium these heavily damaged regions are p-type islands which together with their space charge layer act as an insulating void to block electron current flow, thus reducing the conductivity.

Radiation-sensitive Properties

As mentioned above, the acceptor and donor capacities of radiation defects are exerted through altering the concentrations of electrons and holes. Any property such as conductivity, Hall coefficient, thermoelectric power, etc., is thereby affected. Ionized defects can scatter conduction electrons and holes, thus decreasing carrier mobility.

From the standpoint of device performance one of the most important electronic properties is minority carrier lifetime and this is one of the most sensitive to radiation. The reason for the great sensitivity is that in germanium and silicon, electron-hole pairs in excess of the thermal equilibrium value are forbidden from recombining directly but instead, recombine via recombination centres which have states deep in the forbidden energy gap. Most radiation defects have deep states and are highly efficient as recombination centres. Since device grade material is characterized by a very small recombination centre concentration (a long minority-carrier life time being needed for effective operation of transistors, solar cells, nuclear particle detectors, etc.), a very small concentration of radiation defects, as low as one part in 10^{12}, can be detected through a measurable decrease in minority carrier lifetime in high-quality germanium or silicon crystals.

Radiation defects are also capable of altering the optical absorption spectra of semiconductors and the spectral response of photoconductivity. Of considerable basic interest is the electrons paramagnetic resonance (EPR) spectrum associated with radiation defects because this property gives information about the symmetry and environment of the defects. From a careful analysis of EPR spectra a variety of radiation defects have been identified in silicon (see below). Other radiation sensitive properties include recombination luminescence and low-temperature thermal conductivity. Changes in each of these properties depend upon the nature of the radiation defects in different ways. Therefore in order to characterize as completely as possible the radiation defects, the practice is to examine the radiation-induced changes in as wide a variety of properties as practical, and from the evidence on different aspects of the defect to piece together a complete model.

Radiation Defects

One of the surprises and important contributions of radiation effects investigations is the discovery that simple interstitials and vacancies are unstable at room temperature in germanium and silicon. These simplest defects are highly mobile well below room temperature and either escape from the crystal, recombine with each other, or become trapped at some impurity atom or other imperfection. In electron-bombarded material, it is the composite imperfections such as vacancy-impurity complexes, interstitial impurity atoms, coupled pairs of vacancies and small clusters of defects which induce changes in the electronic behaviour of semiconductors and semiconducting devices. The thermal stability of these more complex forms varies from one to another, but nearly all can be removed (annealed) by thermal treatment. Several hours annealing at 450°C is adequate to remove nearly all traces of radiation damage from the crystal lattices of most semiconductors of interest.

One form of radiation-induced imperfection, namely chemical impurities resulting from nuclear transformations, cannot be removed by annealing. In both germanium and silicon and in many compound semiconductors, thermal neutron capture produces unstable nuclei which decay to stable isotopes with a different chemical nature from that of the original atom. In germanium both donors (^{75}As) and acceptors (^{71}Ga) are produced; in indium antimonide the indium transmutes to ^{114}Sn which is a donor impurity when it replaces indium in the InSb lattice. It is in principle possible to tailor-make semiconductors with desired characteristics by "nuclear doping"; however, careful annealing to restore these atoms to proper lattice sites is required. It is usually less costly to grow crystals doping them either in the melt or by controlled impurity diffusion to produce the desired characteristics rather than to use nuclear methods.

Bibliography

BARUCH P. (Ed.) (1965) *Proceedings of the Royaumont Conference on Radiation Effects in Semiconductors*, Paris: Dunod.

BILLINGTON D. S. and CRAWFORD J. H. (1960) *Radiation Damage in Solids*, Princeton: University Press.

KRUMHANSL J. A. (Ed.) (1959) *Proceedings of the Gatlinburg Conference on Radiation Effects in Semiconductors*, New York: American Institute of Physics.

LARK-HOROVITZ K. (1951) in *Semiconducting Materials* (H. K. Henisch Ed.), Academic Press.

VOOK F. L. (1968) *Proceedings of the Santa Fe Conference, Radiation Effects in Semiconductors*, New York: Plenum Press.

J. H. CRAWFORD, Jr.

SHOCK PRESSURE IN METEORITES. Interest in the shock history of meteorites developed when the hypothesis was proposed that the diamonds found in the Canyon Diablo meteorite were shock produced, and blossomed with the recent reporting of two preterrestrial shock events. Before considering the results of these studies, a brief description of a shock event is desirable.

During impact, a shock wave is generated in the meteoroid and the colliding object, be it another meteoroid or the Earth. The magnitude of the generated shock wave depends on the mass, compressibility, and relative velocities of the colliding objects. The shock wave passes through the meteroid and is reflected from its rear surface. Lightly shocked fragments are spalled from the rear surface and land some distance from the impact site. The compressed material then releases its kinetic energy as an explosion with most of the meteoroid being completely vapourized. The shock intensity felt by the surviving fragments depends on their distance from the exploding region and the attenuation of the shock wave through the material. Therefore, most of any recovered material comes from the rear portion of the meteoroid.

The shock wave both directly and indirectly (through associated high temperatures) causes a number of changes in the microstructure and crystalline character of the constituent minerals of the meteoroid. Many of these expected changes have been confirmed by their observation in controlled artificially shocked iron meteorite fragments. These changes can then be used as shock indicators (barometers). The most striking microstructural change occurs in the kamacite (α–Fe, b.c.c.). When shocked to pressures in the range 130–200 kb, kamacite shows a fine-grained matte structure, which apparently results from the reversion of shock-formed ε-iron (h.c.p.). This transformation structure is more coarse-grained in specimens shocked to 400–600 kb. Kamacite shocked to about 800 kb shows incipient recrystallization and pressures of 1000 kb result in complete recrystallization.

The kamacite of unshocked (<130 kb) meteorites exists in the form of large single crystals. Shock pressures in the range 130–750 kb cause the kamacite to transform to a preferentially oriented polycrystalline aggregate. The preferred orientation most likely results from a shock-induced martensite-like transformation involving ε-Fe as an intermediate. Shock pressures greater than 750 kb cause kamacite to recrystallize. Since this recrystallization results from the shock-associated residual temperature, X-ray analysis does not differentiate it from recrystallization caused by annealing at 1 atm. Both result in X-ray diffraction patterns typical of a randomly oriented polycrystalline aggregate. In general the crystallographic changes closely parallel the microstructural changes. Similar crystallographic alterations occur in cohenite (Fe_3C) and schreibersite (Fe_3P).

Through the use of these and other recently developed shock indicators, two preterrestrial shock events have been discovered. Heymann proposed the origin of most of the hypersthene chondrites. He concluded that at least one-third, and probably two-thirds, of the hypersthene chondrites were definitely part of a single large body which was involved in a collision about 520 million years ago.

From an extensive study of shock effects in iron meteorites, Jaeger and Lipschutz were able to correlate shock effects with both cosmic ray age and chemical (Ga-Ge) composition. Iron meteorites can be classed into four major groups according to their Ga-Ge composition. Surprisingly, over 50 per cent of the iron meteorites were shocked above 130 kb. Nearly all of the shocked meteorites had the same cosmic ray age of 650 $\pm$ 60 million years, and the same gallium–germanium composition (group III).

Due to slight variance in chemical composition, it is not entirely clear whether the iron meteorites involved in the 650 million years collision came from one or two parent bodies. However, the chemical differences are subtle, and are not at all inconsistent with an origin from a single body.

It is possible that the parent body of the Ga-Ge group III shocked iron meteorites collided with the much larger parent body of the hypersthene chondrites. The errors in the dating are large enough so that the 520 and 650 million year dates could possibly refer to the same event. It is also possible that the shocked Ga-Ge group III irons and the hypersthene chondrites were part of a single asteroid which was disrupted by a collision with a smaller asteroid of unspecified composition. It is impossible to eliminate either of the two alternatives, yet, the former seems more widely accepted.

In either case, a very significant percentage of all iron meteorites have been shocked to above 130 kb during preterrestrial collisions. A single collision 650 ± 60 million years ago apparently was responsible for most of the shocked meteorites. The Ga-Ge contents of the iron portion of at least one of the colliding asteroids corresponded to those of group III. The composition of the other asteroid was probably similar to that of the hypersthene chondrites. The collision resulted in the ejection of iron meteoroids into Mars-crossing orbits. Subsequent perturbations by Mars resulted in the continuous ejection of material into Earth-crossing orbits.

Bibliography

HEYMANN D. (1967) On the origin of hypersthene chondrites: Ages and shock-effects of black chondrites, *Icarus* **6**, 189–221.

HEYMANN D., LIPSCHUTZ M. E., NIELSEN B. and ANDERS E. (1966) Canyon Diablo meteorite: Metallographic and mass spectrometric study of 56 fragments, *J. Geophys. Res.* **71**, 617–41.

JAEGER R. R. and LIPSCHUTZ M. E. (1967) Implications of shock effects in iron meteorites, *Geochim. Cosmochim. Acta* **31**, 1811–32.

LIPSCHUTZ M. E. and ANDERS E. (1961) The record in the meteorite. IV. Origin of diamonds in iron meteorites, *Geochim. Cosmochim. Acta* **24**, 83–105.

LIPSCHUTZ M. E. and JAEGER R. R. (1966) X-ray diffraction study of minerals from shocked iron meteorites, *Science* **152**, 1055–7.

WASSON J. T. and KIMBERLIN J. (1967) The chemical classification of iron meteorites. II. Irons and pallasites with germanium concentrations between 8 and 100 ppm, *Geochim. Cosmochim. Acta* **31**, 2065–93.

R. R. JAEGER

SINGING SANDS. Under certain rare conditions dry sand masses undergoing shear emit audible sounds which are limited to narrow ranges of frequency. The frequency is independent of the resonance frequency of any external object, container, etc. The emission persists in a vacuum, being then transmitted via the sand and its container.

The phenomenon is known to occur in two distinct forms.

(i) Certain beach sands emit when dry a brief musical whistle when trodden on or poked with a blunt probe, the pressure being applied normally downwards on a near-horizontal sand surface. The frequency is limited to the neighbourhood of 1000 sec^{-1}.

(ii) In some desert regions sand dunes emit spontaneously a deep booming sound at a frequency in the neighbourhood of 250 sec^{-1}. The sound is emitted from the foot of a slip-face where, and when, successive portions of a sand avalanche down the face are retarded to rest and undergo internal shearing in the process of being telescoped (Fig. 1). The booming sound persists for the duration of the avalanche, i.e. up to 30 sec according to the length of the slipface. It is audible a kilometre away, and the ground tremor is strong enough to initiate an avalanche down another nearby dune. When this happens the frequencies may be so close as to cause a slow beat. In contrast to form (i) the activating pressure is here applied in a direction parallel or nearly so to the sand surface.

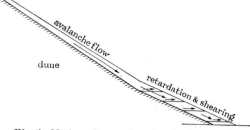

Fig. 1. Motion of a sand avalanche down a dune when the booming sound is emitted.

In both forms the phenomenon can be reproduced by the artificial shearing of a sand mass that has been removed from its natural site; but attempts have failed to produce the sound either from other natural sands or from any other granular material. Nor have any special characteristics been discovered whereby singing sands may be distinguished from others.

The precise mechanism causing the grains to vibrate (a) together in unison, and (b) at a certain frequency only, is not yet definitely established. The following hypothesis appears, however, plausible. It is based on experimental studies of the relation between rate of shear, normal pressure, and degree of dilatation in an array of cohesionless rigid grains. It predicts the observed frequencies to a reasonable approximation.

Condition (a), that the grains tend to move together, is very commonly encountered. It appears to follow from the principle of least effort.

Let the dilatation be defined by

$$1/\lambda = \frac{\text{mean free distances between grains}}{\text{mean grain diameter } D}$$

λ may be called the linear concentration. If C is the corresponding volume concentration

$$\frac{\text{grain-occupied space}}{\text{whole space}}$$

geometry gives

$$1/\lambda = (C_*/C)^{\frac{1}{3}} - 1$$

where C_* is the maximum possible volume concentration of the grain array at rest. When $C = C_*$ the free distance s is zero, the dilatation is zero and $\lambda = \infty$. The array is entirely rigid.

An applied shear stress T creates an internal dilatant pressure P between the grains, normally to the shear direction. Shearing over a single shear surface becomes possible when λ is reduced locally at that surface to a critical value $\lambda_2 = 17$ approx. If the normal component of the applied pressure, due, for example, to the weight of an overburden, is Q, continued steady motion requires that

$$P = Q, \quad \text{and} \quad T = P \tan \varphi \tag{1}$$

where φ is the true friction angle, a constant for the material, measurable as the angle at which a slip-face stands after an avalanche has passed down it and away over a ledge below.

The applied shear stress T_0 needed to initiate motion is, however, greater than that needed to maintain it. For in addition to the stress needed to overcome friction an extra stress is needed in the first instance to dilate the material at the incipient shear surface by raising the whole overburden through a distance $D\left(\dfrac{1}{\lambda_2} - \dfrac{1}{\lambda_1}\right)$ against the applied normal pressure Q. Here λ_1 defines the initial state of compaction of the material, depending on its past history, and λ_2 is the minimum local dilatation at the shear surface necessary to allow movement.

The initial, apparent, friction angle can be expressed as $\tan^{-1} \dfrac{T_0}{Q} = \varphi + \varDelta \varphi$. Thus the greater the value of $\varDelta \varphi$ the greater the tendency for the overburden to move as a rigid body over a single instantaneous shear surface. For simultaneous shearing at a second surface would require more energy in order to create a second dilatation elsewhere.

Condition (b), that the sand mass vibrates at a certain definite frequency can be explained by inference.

It appears from experiment that when, as with dry sand, the effects of the presence of an inter-granular fluid (air) can be neglected,

$$T = P \tan \varphi = a\sigma\lambda^2 U^2 \tag{2}$$

where σ is the density of the grains, U is the velocity of shear of one grain layer over another $\left(= D \dfrac{du}{dy}\right.$ approximately for general shearing$\left.\right)$ and a is a numerical parameter.

At the highest concentrations a is found to decrease very rapidly with decrease in λ to a second critical value $\lambda_3 = 12 \cdot 5$ approx., but to remain constant at all lower concentrations.

Experiments made with uniform concentrations contrived by immersing the grains in a fluid of equal density show that λ_3 marks a distinct change of state, from a substance having the consistency of a paste to a Newtonian liquid having no residual shear resistance at zero shear rate. From this and the above rapid reduction in the shear resistance it is inferred that whereas between λ_2 and λ_3 the sheared grains remain in more or less continuous frictional contact with one another as they are jostled out of each other's way, at λ_3 and beyond they have sufficient freedom to lose contact with one another between successive elastic collisions.

Applying (2) to the slow steady shearing of an overburden at an underlying shear surface where the weight of the overburden exerts a normal pressure Q, it will be seen that since $P = Q = \text{constant}$ and λ is also constant at its maximum mobile value λ_2, the whole overburden should tend to move at a preferred velocity U_c given by

$$U_c = \frac{1}{\lambda_2} \sqrt{\frac{Q \tan \varphi}{a_2 \sigma}} \tag{3}$$

For shearing parallel to the upper free surface of the overburden, $Q = C_1 \sigma g h$, whence

$$U_c = \frac{1}{\lambda_2} \sqrt{\frac{C_1 g h}{a_2}}$$

where the suffix 1 refers to the undisturbed overburden. More generally, in dimensionless terms

$$\frac{U_c}{\sqrt{gh}} = \text{constant}$$

which, like the analogous Froude Number for open liquid flow, is independent of the density.

Recent experiment confirms the existence of a preferred shearing velocity having the predicted value.

Now suppose that owing to a chance shock the velocity U is suddenly increased, to exceed U_c. Inspection of (2) shows that P must momentarily increase to exceed Q. The whole overburden is accelerated upwards. A momentary increase occurs in the dilatation at the shear surface, i.e. a decrease occurs in the local value of λ.

Owing to the very rapid decrease in the value of the parameter a with λ in this limiting region of concentration ($a\lambda^2$ varies as the 11th power of λ, approx.), both the dilatant pressure P and the shear resistance T would virtually disappear at the outset, leaving the overburden free to rise and fall under gravity. Simultaneously it would jump forward till stopped again at the instant of impact.

It is postulated that under certain conditions a vertical oscillation of this kind on the part of the entire overburden could become self-maintaining. If so the preferred frequency is predictable.

The preferred amplitude would be that to a dilatation D/λ_3. This is the minimum dilatation at the underlying shear surface necessary for free frictionless motion between successive impacts. The period of one cycle would be that of rise and fall through this distance under gravity. The frequency f would therefore be

$$f = \sqrt{\frac{\lambda_3 g}{8D}} \tag{4}$$

The frequency would be independent of the thickness and weight of the overburden, and dependent only on the grain size D.

Giving λ_3 its experimental value 12·5, the frequency of the booming sound emitted by a dune of mean sand size $D = 0·02$ cm should be 278 sec^{-1}. The frequency measured with a pitch-pipe in the Kalahari Desert of the boom emitted by a dune of this sand size was reported to be 264 sec^{-1}. The booming is, however, not a pure note, and the sound may well have included undertones corresponding to rather larger amplitudes.

The aurally estimated frequency of the sound from the coarser sands of Libyan Desert and Persian Gulf dunes ($D = 0·034$ cm approx.) is appreciably lower, as (4) would predict.

The foregoing theory has the merit that it explains at once why other sands, though of the same grain size, emit a whistle at a frequency in the neighbourhood of four times higher when subjected to a push delivered normally instead of parallel to the surface.

During a displacement due to a push parallel or nearly so to the free surface (Fig. 1) the normal stress Q on a shear surface is of the same order as the gravity stress corresponding to the thickness of the overburden. So, for any given thickness, the overburden rises and falls freely under the gravity acceleration g. When, however, the push is applied normally to a horizontal sand surface, successive shear surfaces resemble those sketched in Fig. 2. The stress on the face AB of the probe is, for the same thickness of overburden, greater by a large factor K.

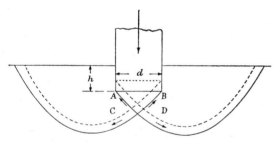

Fig. 2. *Shear surfaces resulting from a vertical push into a horizontal sand surface.*

The material CD which is being sheared at successive surfaces inclined at 45° approx. is subject to normal accelerations $Kg \cos^2 45° = Kg/2$. Consequently the oscillation frequency would be expected to be given by (4) with $Kg/2$ substituted for g. From soil mechanics data on the bearing strength of foundations K appears to vary between 25 and 40 for the conditions sketched in Fig. 2, the variation depending on the degree of compaction of the sand. The oscillation frequency should therefore be between 3·5 and 4·5 times that of the spontaneous booming of the dunes, in satisfactory agreement with observation.

The phenomenon of singing sands cannot, however, be claimed to be understood until it is known what special physical properties a granular material must have in order that a sustained oscillation will occur.

Bibliography

BAGNOLD R. A. (1954) Experiments on a gravity-free dispersion of large solid spheres in a Newtonian fluid under shear. *Proc. Roy. Soc. London* A. **225**, 49–63.

BAGNOLD R. A. (1966) The shearing and dilatation of dry sand and the "singing" mechanism, *Proc. Roy. Soc. London* A. **295**, 219–32.

R. A. BAGNOLD

SNOEK SPECTRA. In body-centred cubic metals interstitial solute atoms such as nitrogen and carbon occupy the octahedral interstices half-way along the unit cell edges and the crystallographically equivalent sites at the centres of the cell faces (see Fig. 1). The presence of an interstitial in the lattice causes a large local tetragonal distortion, but in the absence of an externally applied stress all the sites are equivalent and hence the interstitials are distributed at random throughout the lattice, causing local strain in the

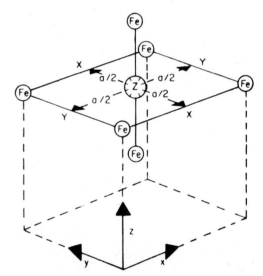

Fig. 1. *A solute atom (shaded) in an interstitial octahedral site of the b.c.c. lattice. The three types of sites are designated X, Y and Z, according to whether the axis of tetragonal distortion is along the x, y or z cube edges. Solvent atoms are denoted by "Fe".*

three principal crystallographic directions (see Fig. 1). If, however, a tensile stress is applied in one of these directions, the sites become inequivalent and interstitials tend to jump into neighbouring sites so that the direction of maximum distortion coincides with the direction of applied stress. This stress-induced ordering of interstitials was first elucidated by Snoek and is referred to as the Snoek effect.

When a single species of interstitial is present in an otherwise pure lattice, stress-induced ordering of interstitials gives rise to a single relaxation in the mechanical relaxation spectrum of the metal. Snoek showed that the resulting internal friction, Q^{-1}, of a cyclicly stressed specimen is related to the angular frequency ω, and the relaxation time τ through the well-known Debye equation:

$$Q^{-1} = \frac{\Delta \omega \tau}{1 + (\omega \tau)^2} \quad (1)$$

where Δ, the dimensionless relaxation strength, is directly proportional to the concentration of interstitial atoms provided the concentration is not too high. Since interstitial migration is a thermally activated process, the relaxation time τ obeys the Arrhenius equation

$$\tau = \tau_0 \exp E/RT \quad (2)$$

where ΔH is the activation energy for diffusion of the interstitials usually expressed in calories per mole, R is the gas constant and T is the absolute temperature.

Thus the Snoek peak can be obtained by measuring internal friction while either continuously varying the frequency at constant temperature or the temperature at constant frequency. Because of the experimental difficulties involved in continuously varying the frequency over the two decades of frequency necessary to trace out the Snoek peak, Snoek phenomena are usually investigated by observing the internal friction versus temperature curve. In this case, the effect of choosing a higher frequency is to shift the peak to a higher temperature. If the relaxation strength, Δ, is considered constant it can be seen that the peak has a maximum when $\omega\tau = 1$. The thermodynamic theory of the relaxation strength requires that a small correction be made to the above condition for a maximum, since it predicts that Δ varies as T^{-1}. Since this variation of Δ with T^{-1} has not been conclusively demonstrated experimentally, it is very often neglected. An experimental Snoek peak for an Fe–N alloy and the corresponding theoretical peak calculated using equation (1) is shown in Fig. 2.

When certain substitutional solute atoms are added to an Fe–N alloy, the internal friction curve becomes complex (see Fig. 3). These complex curves can often be analysed into overlapping single-time-of-relaxation

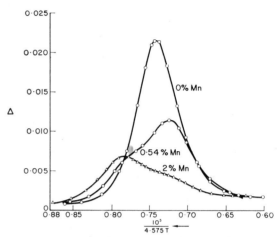

Fig. 3. *Internal friction curves for iron–nitrogen alloys containing 0 per cent Mn, 0·54 per cent Mn and 2 per cent Mn.*

peaks and they are all referred to as Snoek peaks since it is thought that they are all due to stress-induced jumping of interstitials. The relaxations, due only to stress-induced jumping of interstitials, form the Snoek spectrum of the metal.

In the case of Fe–N alloys substitutional additions of Mn, Cr, V and Mo cause the appearance of another relaxation at a higher temperature than the normal Snoek peak found in the binary alloy, and this is believed to be due to the jumping of nitrogen interstitials close to single substitutional atoms. However, this is not the general case because no comparable peak has been found on addition of Ni, Co, Nb, Ti, or Al.

In Fe–Mn–N, Fe–Cr–N and possibly other ternary alloys, there are peaks at lower temperatures than the normal Snoek peak and these have been attributed to nitrogen interstitials associated with clusters of substitutional atoms (see Fig. 3).

With the exception of Fe–Al–C alloys, ternary alloys containing carbon interstitials do not exhibit peaks other than the normal Snoek peak for Fe–C alloys. This difference in the behaviour of nitrogen and carbon interstitials in the presence of substitutional atoms is not well understood, but possibly arises from differences in the state of ionization of the interstitial elements in the crystalline solid solution.

When the concentration of interstitials in binary alloys is large, peaks due to the stress-induced re-

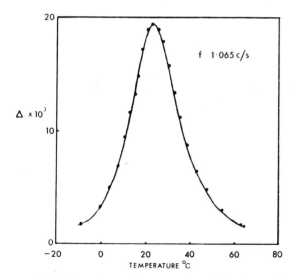

Fig. 2. *Experimental points and calculated line for an iron–nitrogen alloy.*

orientation of interstitial clusters can also occur and overlap with the normal Snoek peak.

Bibliography

BERRY B. S. (1961) Anelasticity, in *Encyclopaedic Dictionary of Physics* (J. Thewlis Ed.), **1**, 173, Oxford: Pergamon Press.

FAST J. D. (1960) Le Frottement Intérieur des Métaux, Colloque de Saint-Germain-en-Laye (I.R.S.I.D. and C.N.R.M.).

MASON W. P. (1962) Relaxation spectra in solids, in *Encyclopaedic Dictionary of Physics* (J. Thewlis Ed.), **4**, 276, Oxford: Pergamon Press.

MASON W. P. (1966) *Physical Acoustics*, Vol. IIIA, New York and London: Academic Press.

NOWICK A. S. (1964) *Resonance and Relaxation in Metals*, American Society for Metals, New York: Plenum Press.

SCHOECK G. (1961) Internal friction, in *Encyclopaedic Dictionary of Physics* (J. Thewlis Ed.), **4**, 5, Oxford: Pergamon Press.

I. G. RITCHIE and R. RAWLINGS

SOFT X-RAY SPECTROSCOPY. It will be assumed that the reader is familiar with the general principles and techniques of X-ray spectroscopy, either from the discussion in Volume 7, pp. 802–40 of this Dictionary, or from other sources.

The soft X-ray range is not precisely defined, but for practical purposes can be said to begin at wavelengths for which an air or helium path and conventional windows cannot be used and extends to wavelengths for which the techniques of optical spectroscopy are applicable (approximately 2·5 to 500 Å). Solid state X-ray spectra in this range involve electron transitions from the valence-conduction band (hereafter referred to simply as the valence band) into one

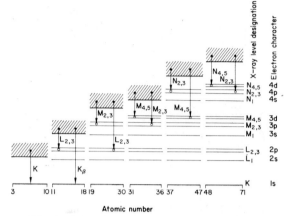

Fig. 1. The electron transitions of principal interest in solid state (valence band) soft X-ray emission spectroscopy.

of the outermost normally full shells. The transitions of particular interest are identified in Fig. 1. The nominal energies (because the spectra are bands rather than lines) corresponding to these transitions, for elements up to $Z = 52$, are plotted in Fig. 2. The corresponding absorption spectra involve transitions from one of the outermost normally closed subshells to available empty states in the valence band. The emission and absorption spectra are thus a measure of the density of occupied and unoccupied states, respectively, in the valence band. The relationship of the absorption and emission spectra to the responsible electron transitions are shown somewhat idealized in Fig. 3.

Soft X-ray spectroscopy is of particular interest in solid state physics because it is one of the few methods

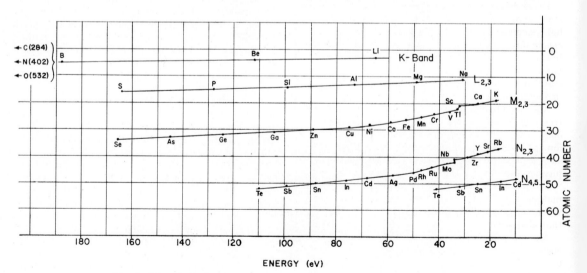

Fig. 2. Plot of the locations of the valence band spectra of principal interest.

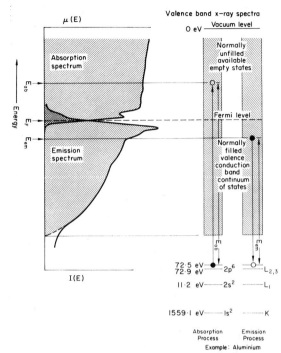

Fig. 3. Idealized representation of the absorption and emission processes and their relation to the observed absorption and emission spectra. (Diagram is patterned after aluminium.)

that give information on the density of states throughout the valence band (rather than at the Fermi energy only). Some of the problems involved in obtaining a quantitative measure of the density of states will be discussed later.

Dependence of the X-ray spectra on the chemical state enters into the spectrochemical analysis of the light elements ($Z < 10$) because, in this range, the K spectrum is the valence band spectrum. The changes in shape and shifts of peaks and edges in the emission band spectra which occur as a result of alloying, valence state, and crystal structure, must be taken into account in spectrochemical analysis.

Another important application of soft X-ray spectroscopy is the study of the radiation in the 1 to 500 Å range resulting from recombination of highly ionized atoms in a plasma. An interesting example of such a plasma is the corona of the Sun. (Some astrophysicists refer to the study of this radiation as "extreme ultra-violet spectroscopy" (XUV), thus making the distinction between transitions that involve only the outermost electrons and transitions that also involve an inner shell.)

An indication of the accomplishments in spectrochemical analysis, solid state physics and astrophysics follows a review of the experimental technique. The experimental problems apply to all three of these research areas.

Experimental Technique

Single-crystal spectrometers and, for better resolution, two-crystal spectrometers are used in the short wavelength end of the soft X-ray range. Grazing incidence grating spectrometers are used for the longer wavelength portion. When equipped with the multi-layer lead stearate soap film type of diffracting element, a crystal spectrometer can be used up to about 100 Å, but the resolution of a grating spectrometer usually surpasses that of even the two-crystal spectrometers above about 50 Å. Both the crystal and grating spectrometers are always high vacuum instruments to minimize absorption and specimen surface contamination. Electron excitation of emission spectra is the usual practice; however, K emission from targets of low atomic number, and bremsstrahlung, are often used in the 2 to 25 Å region. Until recently, line spectra from gas discharge tubes and bremsstrahlung from X-ray tubes were the only light sources available for absorption spectroscopy. However, these sources were never really satisfactory, because of the discrete nature of line spectra and the low intensity of the bremsstrahlung. Now synchrotron light, which is a relatively high intensity continuum, is coming into widespread use for absorption spectroscopy, and improved spectra are being obtained as a result. Sufficiently intense synchrotron light beams should be available in the near future for emission spectroscopy in the 100 to 300 Å region as well. Photon counting has largely replaced film recording to take advantage of the higher sensitivity, better linearity, and adaptability to repeated scanning and summing to improve counting statistics. A schematic diagram of a grazing incidence grating spectrometer together with the associated electronics for recording, is shown in Fig. 4.

High vacuum is essential not only to give a low absorption path, but in solid state spectroscopy, to maintain a clean, oxide-free specimen surface. This is necessary because emission is from a layer of material extending not more than 100 to 1000 Å beneath the surface and any appreciable oxide film could constitute a significant portion of this depth.

Because of absorption, the conventional 1/4 mil Mylar† (polyester film) or beryllium windows so commonly used in conventional X-ray spectrometers cannot be employed. Of course, this would seem to rule out the use of gas-filled detectors requiring windows, such as geiger and proportional counters. However, because these counters are much more efficient than the open structure photomultipliers shown in Figs. 5 and 6, and because the energy resolution capability of the flow proportional counter can be used to eliminate order overlap, flow proportional counters are used up to about 100 Å by equipping them with extremely thin windows. The windows most commonly used are made by floating a film of Formvar (poly-

---

† Trademark of E. I. Dupont de Nemours & Co., U.S.A.

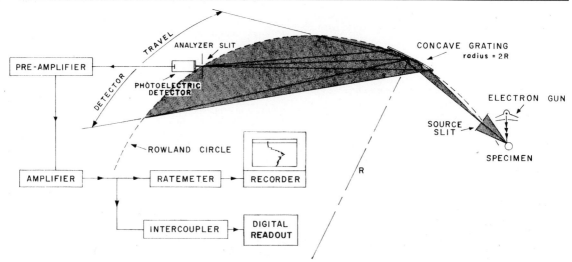

Fig. 4. Schematic diagram of a grazing incidence grating spectrometer with photoelectric detection and with associated electronics for both analogue and digital recording.

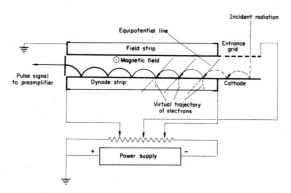

Fig. 5. Diagram of the continuous strip, windowless, magnetic electron multiplier (MEM).

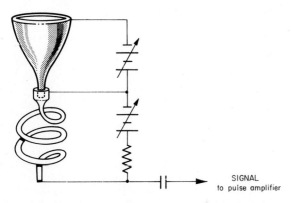

Fig. 6. Schematic diagram of the new channel capillary tube electron multiplier. (Courtesy of Bendix Corporation.)

vinyl formal) or nitrocellulose off the surface of water onto an electron microscopy screen. The thickness of such a film is only about 0·1 μ, and under the normal stray ion bombardment of the gas in the counter, the life of such a window is about a month. Above 100 Å completely windowless detectors must be used. The most common detector, until recently, has been the so-called magnetic electron multiplier (MEM) shown schematically in Fig. 5.

Another type of detector, employing only an electric field gradient, is shown schematically in Fig. 6. This detector is a glass capillary tube with the internal surface treated to become semiconducting. A cone on the front end intercepts the photon beam. The photoelectrons ricochet down the tube under the influence of a voltage gradient and multiply upon each impact with the wall of the tube until the electron avalanche, corresponding to the original photon, strikes the collector at the end of the capillary tube. The principal feature of the channel electron multiplier (CEM) is the uniformity of the output pulse amplitude which is also voltage independent over a broad range of applied voltage. Therefore, if this detector is used in the pulse counting mode the signal degradation with use, observed to varying degrees in any open electron multiplier, is minimized and an improved signal-to-noise ratio is achieved in respect to other detectors.

Like the MEM, the CEM is not energy discriminating. Commercial CEM versions are made by Mullard, Ltd. in England and by the Bendix Corp. in the United States.

A reflection grating at grazing incidence is the only dispersing element useable over the entire soft X-ray range. Aluminium, gold, and platinum coated replica gratings as well as original glass gratings are now being used successfully. Until recently, gratings were all that were available for any portion of the soft X-ray range, but during the past 10 years there has been great progress in preparation of organic crystals such as KAP (potassium acid phthalate, $KHC_8H_4O_4$, $2d = 26.6$ Å) and layered polar molecule lead stearate "crystals" with large effective d spacings which can be used for spectroscopy up to 100 Å. Crystals are used, particularly in analytical spectroscopy, because of the order of magnitude greater intensity obtainable as compared with gratings. In enjoying this greater intensity, however, one sacrifices resolution.

Analytical Spectroscopy

There has been considerable progress in light element analysis during the last 10 years. The wavelength range covered by commercially available equipment now extends to 100 Å or more, and this permits one to obtain emission spectra of elements as low in atomic number as lithium ($Z = 3$). The development of light element analysis has been spearheaded by the electron-probe microanalysts because the design of the instrument lends itself naturally to the extension into the soft X-ray region. Electron excitation, as in the electron-probe microanalyser (*see* Electron-probe microanalysis, *Encyclopaedic Dictionary of Physics*, Suppl. vol. 1, pp. 82–84) is much more effective in the production of soft X-ray spectra than are the hard X-ray sources used in conventional analytical spectrometers. Hard X-radiation, such as say MoK radiation, produces the bulk of the soft X-ray excitation so deep in the specimen that it is virtually all absorbed before escaping the specimen. Also most electron probe microanalysers are designed to function with the spectrometers in vacuum, with the only window in the system being that of the detector. It was relatively easy to extend the range of the micro-analyser into the soft X-ray region by merely substituting suitable crystals as they became available and substituting a suitable detector.

However, the macroanalysts have not stood still. At least one multichannel X-ray spectrometer employing electron excitation for macroanalysis in the soft X-ray region is on the market.† The characteristic radiation from low Z targets such as aluminium, carbon, silicon, etc., in demountable windowless X-ray tubes is also being used as an excitation source in spectrometers for macroanalysis. The ability to change source targets easily permits one to use a target of the next higher atomic number element than the one being analysed, and in so doing, one avoids excitation of any heavier element that may be in the sample. When the optimum target material is a poor thermal conductor, particles of the material can be imbedded in the surface of a water-cooled block of high-thermal-conductivity metal.

Although sufficient intensity can be obtained in the characteristic spectra of light elements down to lithium, considerably more caution must be exercised to assure correct results in light metal analyses than in heavy

† Produced by Telsec Instruments Ltd., Oxford, England.

metal analysis. The effect of chemical state upon the position and shape of the emission peak and gross uncertainties in the absorption coefficients in the soft X-ray range are two of the more important factors. This situation requires the analyst to rely much more on sets of comparison standards that closely resemble the sample under analysis.

However, with careful attention to the many factors plaguing light element analysis, good quantitative results can be obtained. This is evidenced by the example shown in Fig. 7, an analysis for oxygen, nitrogen, and carbon in urea. Also, the valence and crystalline state effects in the light metal spectra are not necessarily a liability. These effects can be an asset. Figure 8 shows the distinct difference between the spectrum of carbon in graphite from that of carbon

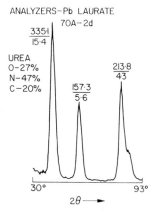

Fig. 7. Quantitative spectrochemical analysis for carbon, oxygen and nitrogen in urea. The numerator of each "fraction" shown at the respective peaks is the peak intensity in counts per second. The denominator is the corresponding background intensity. The 6 per cent unaccounted for is hydrogen. (Courtesy of B. L. Henke, U. of Hawaii.)

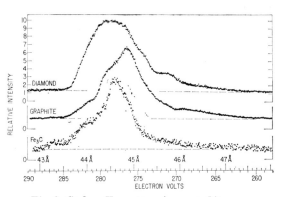

Fig. 8. Carbon–K spectrum from graphite, cementite (Fe_3C) and diamond. (Courtesy of J. E. Holliday, U.S. Steel Corp.)

in cementite (Fe_3C). This difference can be used in determining the ratio of carbon in the form of graphite to that in cementite in steels and cast irons.

Solid State Spectroscopy

In the framework of the independent particle model, we may approximately express the emission intensity, $I(h\nu)$ (in photons/sec/steradian/unit energy) and the absorption coefficient, $\mu(h\nu)$, as follows:

$$I(h\nu) \propto \nu T(E)\, P(E)\, K(E)\, S(E)\, N(E) \quad (1)$$

$$\mu(h\nu) \propto 1/\nu (1 - T(E))\, P(E)\, K(E)\, S(E)\, N(E) \quad (2)$$

Here, E denotes the single particle electron energy in the valence band, and $N(E)$ is the electron density of states in energy. The photon energy, $h\nu = E - E_\alpha$, E_α being the inner level energy. $T(E)$ is the Fermi distribution function, $T(E) = (\exp[(E - E_F)/kT] + 1)^{-1}$. Thus $I(h\nu)$ probes filled states and $\mu(h\nu)$ probes empty states. The factor $P(E)$ is taken to be a mean square of the matrix element $<\psi(E)|p|\psi_\alpha>$ over the initial energy E, where p is the electron momentum operator. (If $P(E)$ is taken as the square of the radial matrix element $\langle\psi(E)|r|\psi_\alpha\rangle$, an extra factor of ν^2 is needed on the right-hand side of (1) and (2).) $S(E)$ and $K(E)$ are instrumental and lifetime broadening functions, respectively. In the latter, the effects of inner state lifetime and, at least approximately, level broadening due to the electron–electron interaction are combined. It should be noted that while we have written S and K as if they were multiplicative factors, they are in fact integral smearing functions, which act separately, in the manner

$$I(h\nu) = \int T(E)\, S(h\nu - E)\, N(E)\, dE.$$

In practice, one hopes to employ equations (1) and (2) to estimate the functions I and μ from band calculations, for comparison with measured spectra. In doing so, the validity of the independent particle approximation for states well off the Fermi energy is as much under test as the particular band calculation. Such work as has been done lends support to this approach.

It is to be noted, however, that the practice of assuming $P(E)$ to be constant or very slowly varying with energy when comparing experimental spectra to calculated densities of states might be considerably in error. As an example, consider a calculation of the quantity $|\langle\psi_d|r|\psi_\alpha\rangle|^2$ for L_3 and M_3 spectra through the d-band of Fe, using $3d$ radial electron wave functions available from augmented plane wave energy band calculations, and atomic $2p$ and $3p$ inner state wave functions. Sample wave functions are shown in Fig. 9. The square of the matrix element increases by 10 per cent in going from the bottom to the top of the d band in the case of the M spectrum, but by 80 per cent in the case of the L. Calculations for Ni and Cu show the same behaviour. This difference in the energy dependence of the matrix elements across the d bands

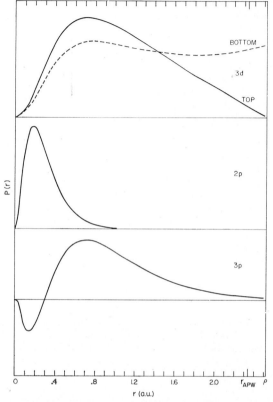

Fig. 9. $P(r)$, r times the radial wave function, for the bottom and top of the Fe–3d band from an APW calculation, and for Fe–2p and 3p atomic orbitals.

appears to account reasonably well for the narrower width of the observed L spectra relative to the M. See Fig. 10 for a comparison of the observed Ni–L_3 and Ni–M_3 as an example.

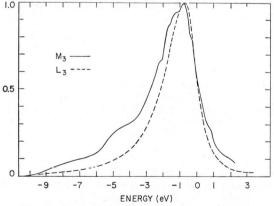

Fig. 10. Comparison of the L_3 and M_3 band widths for nickel before adjustment for transition probability distortion.

Energy band calculations and the electron wave functions and density-of-states distributions derived from them are being refined to a point that meaningful detailed interpretation of the pure metal spectra should now be possible. Unfortunately, energy band calculations for alloys are much less complete.

The smearing function, $K(E)$, has been inserted to represent a variety of lifetime and many body effects that broaden the observed spectrum. The finite lifetime of the inner level vacancy gives a lower limit for the width of this smearing function. Also, the presence of the vacancy causes a perturbation in the other electron energies. Fortunately in the case of metals, screening by the valence electrons permits this readjustment in the levels rapidly enough so that the electron vacancy perturbation is not as serious a problem in metals as in non-metals.

It is no doubt safe to say that the specimen temperature in most soft X-ray spectroscopy thus far has been considerably above room temperature although reported to be room temperature data. The heating is due to the electron beam. Although the specimen is often secured to a water-cooled support, the region of the specimen from which the X-rays are being emitted may often be considerably above room temperature.

If the temperature is known, a measure of its effect on the observed spectrum can be estimated. For instance, the Ni–M spectrum in Fig. 10, from a specimen at 960 °C does not exhibit the sharp peaking at the Fermi energy that is shown in the theoretical density of states plotted in Fig. 11(a). However, modifying the theoretical distribution for the effect of the temperature and lifetime broadening (Fig. 11(b)) experienced in the experimental spectrum results in the disappearance of not only the peak at the Fermi energy but other sharp peaks in the theoretical distribution. It is the temperature broadening that is responsible for washing out completely that intense narrow peak at the Fermi level in the theoretical distribution, but the lifetime broadening becomes the dominant factor causing smearing deeper in the band. To enable detailed direct comparison with theoretical calculations in the vicinity of the Fermi energy, it would be necessary to obtain spectra at low temperature. On the other hand, there are circumstances wherein a spectrum may be obtained at an elevated temperature by choice. Such is the case with the nickel spectrum at 960 °C. At this temperature the nickel is above the decomposition temperature of the bulk oxide on any principal crystallographic plane.

The instrumental broadening, $S(E)$, is negligibly small relative to the transition probability and lifetime broadening corrections in the case of a modern, well-aligned spectrometer. Instrumental broadening is primarily due to the finite slit width, but grating distortions also contribute. The actual smearing function for a particular spectrometer could be determined by recording the indicated shape of an atomic line of known, narrow width. The function will generally be Gaussian in shape and only a few hundredths of an electron volt in half-width.

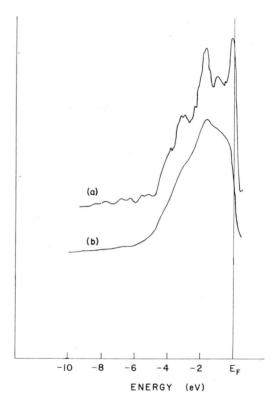

Fig. 11. (a) The theoretical density of states distribution in nickel from an APW calculation. (Courtesy of E. C. Snow, Los Alamos.) (b) The theoretical density of states with Fermi-Dirac temperature correction to 960°C and with smearing function applied to take into account various broadening mechanisms affecting the experimental M_3 spectrum.

The basic relationship of the emission spectrum to the true density of states and the corrections that have been discussed thus far have all been in the framework of the one electron model. Sodium, the most nearly free-electron-like of the elements, would be expected to exhibit a spectrum closely resembling a parabolic distribution up to the sharp cut-off at the Fermi energy. The Na–L spectrum, which involves only s-electron transitions which have energy-independent transition probability (in the free electron approximation) at low energies, is shown in Fig. 12(a). The parabolic character of the spectrum is tested in Fig. 12(b). A true parabolic shape is verified for most of the lower half of the band. Inner level width and many-body effects cause a deviation at the extreme bottom of the band. Recent calculations indicate that the deviation at the top of the band may also be due to many-body interactions.

The K spectrum on the other hand, involves p-electron transitions with a transition probability proportional to energy. We would therefore expect the K spectrum of a free-electron metal to exhibit a spectrum resembling an $E^{3/2}$ distribution. Aluminium

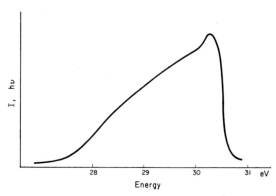

Fig. 12(a). Sodium-$L_{2,3}$ emission spectrum.

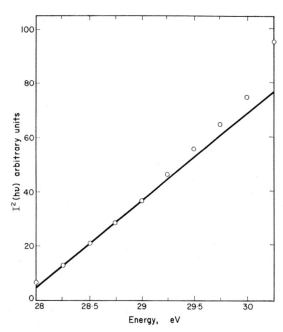

Fig. 12(b). $I^2(h\nu)$ versus E from the sodium-$L_{2,3}$ spectrum (Fig. 12a).

is largely free-electron-like and again the $L_{2,3}$ spectrum of aluminium, Fig. 13(a), shows the $E^{1/2}$-type dependence up to energies at which Brillouin zone boundary effects begin to change the spectrum. The K_β spectrum, Fig. 13(b), shows the expected $E^{3/2}$-

type dependence at the bottom of the band. Aluminium valence band spectra have no doubt been the most studied and best characterized of all the elemental spectra. Some of the structure occurring in the L spectrum has been identified with Van Hove singularities and, therefore, with certain symmetry points in the Brillouin zone.

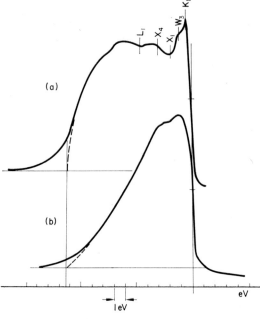

Fig. 13. (a) Aluminium-$L_{2,3}$ emission spectrum. (Courtesy of G. A. Rooke, University of Strathclyde.) (b) Aluminium-K_β emission spectrum. (Courtesy of R. Deslattes, National Bureau of Standards.)

The most universal electronic structure feature distinguishing metals from non-metals is the energy gap that exists in a non-metal between the top of the filled band and the lowest available empty state, and the absence of any such gap in a metal. This characteristic of metals is implied in the somewhat idealized portrayal of the valence band X-ray absorption and emission processes in Fig. 3.

With the availability of the synchrotron light source, very precise absorption spectra can now be obtained. One might expect to be able to locate, in absolute energy, the position of the emission and absorption edges to within ± 0.1 eV and, therefore, band gaps to within ± 0.2 eV, at least in the vicinity of 200 Å. Band gaps in non-conductors have been determined by soft X-ray spectroscopy but determinations thus far have consisted primarily in confirming what band gap data was already known from other methods. However, with the better precision now possible, soft X-ray spectroscopy could become a useful method of measuring band gaps.

Beginning with the start of the first transition series, the Fermi edge becomes increasingly less discernible in the observed emission spectrum because of the increasing favourability of Auger filling and high energy satellites, but a reasonably well-defined edge remains in the absorption spectrum. Satellites, like diagram lines, are due to transitions of electrons from a higher to a lower state, but in the case of satellites, the levels have been shifted because of the presence of an additional vacancy at the time of emission. This additional vacancy can arise by various mechanisms (see Satellites in soft X-ray spectra, this volume). In most cases the various satellites can be identified and subtracted out, although in some instances the satellite intensity may be comparable with the intensity of the parent spectrum simply because of the high probability for the occurrence of the multiple-electron type transition.

In addition to satellites, peaks due to many-body resonances have been predicted in the soft X-ray spectra. In contrast to the satellites, the intensity of these many-body resonance peaks is only 1 per cent or less of the main peak. Nevertheless, by improvement in the counting statistics, some of these predictions have been confirmed with a fair degree of assurance.

In addition to the factors involved in the interpretation of the spectra of the pure elements, other factors arise in the interpretation of spectra from alloys and compounds which are even more difficult to evaluate, such as effects of alloying on transition probabilities, relative excitation efficiencies, and secondary fluorescence. Nevertheless, considerable data have been obtained on alloys and compounds, and some interpretation has been possible. The classes of alloy systems upon which the most data have been reported are the carbides, oxides, and borides of transition metals, aluminium alloys, and intermetallic compounds. The interpretation of alloy data thus far has been limited essentially to empirical correlations of peak shifts and peak intensity ratios with various parameters related to bond strength. For instance, a study of Al–K spectra from aluminium binary systems indicated a correlation of the $K\alpha^4/K\alpha^3$ satellite intensity ratio with difference in the electronegativity of the two components in each case. Correlation of peak shifts and spectrum shape with the difference between the electronegativities of the two components in binary systems, and with change in the valence state of one of the components in various series of compounds, have been established.

Figure 14 is a plot of the titanium–L_3 peak energy shift in TiO_x as a function of x. Other types of measurements show the Ti–O_x bond to be increasingly ionic with increase in x. Ionic character in the bond implies electron transfer, or more precisely, a shift in the electronic centre of gravity of those electrons taking part in the ionic character of the bond. On the basis of this relationship, soft X-ray spectroscopists have related the peak shifts that they observe with the amount of electron transfer.

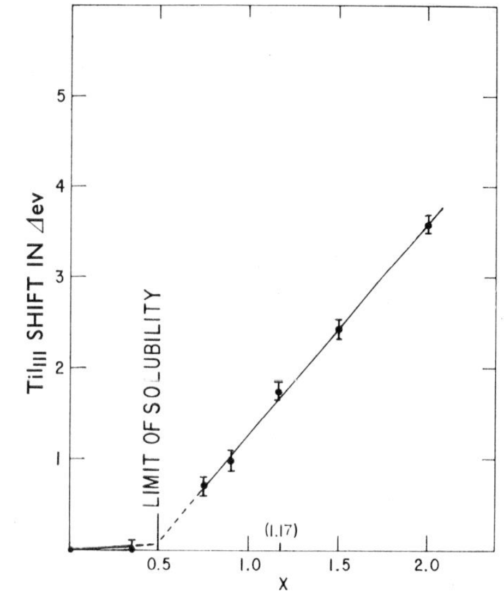

Fig. 14. Plot of the titanium–L_3 peak energy shift in TiO_x as a function of x. (Courtesy of J. E. Holliday, U.S. Steel Corp.)

Figure 15 shows the B–K spectrum from ZrB_2, TiB_2 and VB_2 as well as from pure boron. Note that both the Ti and Zr boride spectra peak at the same energy, but that the peak in the VB_2 is shifted and all are

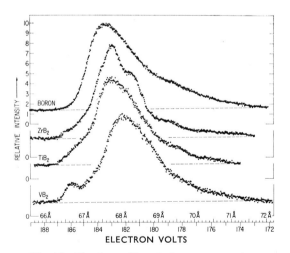

Fig. 15. The boron-K emission spectrum from pure boron and from ZrB_2, TiB_2 and VB_2. (Courtesy of J. E. Holliday, U.S. Steel Corp.)

shifted from the peak in pure boron. Borides of metals from any one subgroup appear to show the same shift in the boron peak from that in pure boron. The differences in the shift between subgroups increases with increase in the separation of the subgroups in the periodic table. Carbides are reported to show similar characteristics. In the case of binary alloys, investigators find that any pair of metals from the same two subgroups will show about the same shift in their respective peaks when alloyed. Furthermore, these shifts will be a minimum when the subgroups are adjacent in the periodic table and will generally increase with increase in the separation of the subgroups. On the basis of these observations, investigators thus far have dwelt on correlating the peak shifts that occur upon alloying with the electronegativity differences between the constituents.

On the other hand, it has just as often been shown that there is no correlation between peak shifts or gross changes in the shape of spectra and the crystal structure *per se*. A typical example is the series of spectra in Fig. 16 traversing the aluminium–nickel system from pure aluminium to pure nickel and

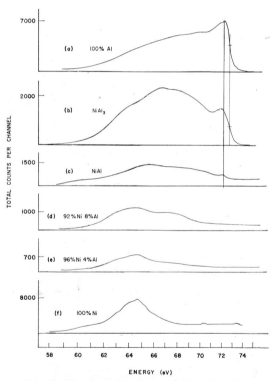

Fig. 16. *The aluminium–$L_{2,3}$ and the nickel–$M_{2,3}$ and the corresponding overlapping emission spectra from four intermediate compositions across the Al–Ni system.*

including four intermetallic compositions across the Al–Ni system. There is in evidence a gradual decrease in prominence of the abrupt aluminium high energy edge and a gradual development of the nickel spectrum shape, but there is no apparent effect of the intermediate changes in crystal structure that have taken place in traversing the system. These crystal structure transformations apparently do not involve gross changes in the electron energy levels, but probably at most involve changes in the lattice vibration modes because of differences in the crystal symmetry. However, when crystal structure differences clearly involve a gross change in the electronic structure, as in the case of carbon (compare the insulator (diamond) structure and the conductor (graphite) structure) one would expect a gross change in the valence band spectrum. Such a gross difference is indeed clearly seen in Fig. 8 between the carbon-K (diamond structure) and the carbon-K (graphite structure) emission spectra.

More fundamental relationships between the electronic structure of alloys and alloy spectra will no doubt come with yet further improvement in both experimental and data reduction techniques, and a parallel improvement in the band theory of alloys. Solid state soft X-ray spectroscopy is about 40 years old. Only now does the detailed correlation of soft X-ray spectra with theoretical density of states in pure metals appear to be nearing fruition. Progress toward this end has been dependent upon developments in experimental techniques to get better data on the one hand, and the development of more accurate band calculations and wave functions on the other, to make possible the interpretation of the spectra. This same parallel development will continue. One of the most intriguing thoughts for the future is the improvement to be realized in the experimental data by using selected portions of a suitable electromagnetic continuum source such as synchrotron light for spectrum excitation. The bremsstrahlung background and most of the heating of the specimen will thus be avoided.

Solar Spectroscopy

Analysis of the soft X-ray radiation from the Sun for the information that this spectral region can give on the ion species present in the solar corona and on the physics of plasmas in general is the newest field of soft X-ray spectroscopy. The problem of analysing solar radiation in the soft X-ray range is absorption by the blanket of air around the Earth. This problem, of course, has been overcome by mounting spectrometers on satellites of the types shown in Fig. 17. Some of the most interesting data have come from spectrometers in the OSO-III satellite (for Orbiting Solar Observatory). Four spectrometers, two using MEM detectors, were on this satellite, as follows:

Type	Wavelength range
LiF crystal	1·3 to 3·1 Å
EDDT crystal	3 to 8 Å
KAP crystal	6 to 25 Å
Grating	20 to 400 Å

Fig. 17. Photograph of the OSO–III satellite showing the spectrometers mounted in the centre and projecting through the panel of solar cells. (Courtesy of W. M. Neupert, NASA.)

The spectrometers were kept pointed at the Sun with a precision of better than 1 min of arc. The satellite was launched in March 1967. Spectral data were received from the KAP and grating spectrometers until their power failed after 6 months, and a year, respectively. Data is still being received (June 1968) from the LiF and EDDT crystal spectrometers. These X-ray spectra are primarily of highly ionized first transition series metals, and originate entirely within the corona which is the outermost layer of highly ionized gas.

Many of the spectra were taken during periods of solar flares. Figures 18 and 19 are examples of the

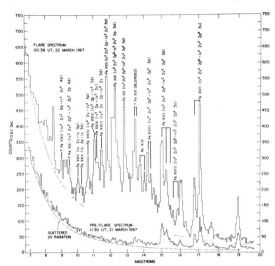

Fig. 18. Spectrum due to the radiation from very highly ionized atoms in the corona of the Sun during a solar flare. (Courtesy of W. M. Neupert, NASA.)

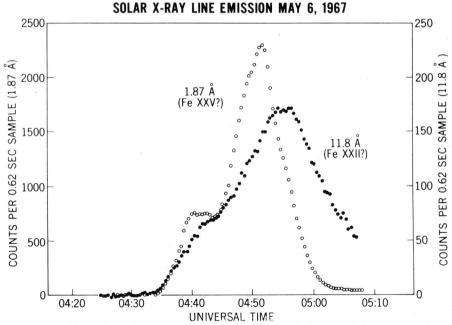

Fig. 19. Change in emission intensity with time, at 1.87 Å and 11.8 Å believed to be due to radiation from $Fe\,XXV$ ions and $Fe\,XXII$ ions, respectively, throughout the duration of a solar flare. (Courtesy of W. M. Neupert, NASA.)

two types of records obtained: Fig. 18 is the spectrum as a function of wavelength during a solar flare; Fig. 19 is the change in the Fe^{+24} and Fe^{+21} peak intensities as a function of time during the period of a flare.

This is the first attempt to either resolve the lines in the X-ray radiation from the corona of the Sun during a solar flare or to record the rate of build-up and decay in the intensities of specific lines throughout the duration of the flare. The identifications of the lines in Fig. 18 are still tentative. The investigators admit that there are many more lines to be expected in the spectrum than are indicated in Fig. 18. Presumably, these are not yet resolved, because of state-of-the-art limitations in satellite instrumentation to date. Without doubt, space research instrumentation will continue to be improved, and as a result, new discoveries will be made that are as spectacular as the advances made thus far.

Bibliography

Advances in X-ray Analysis (1964, 1965, 1966, 1967) **7, 8, 9, 10,** New York: Plenum Press.

BLOKHIN M. A. (1957) *The Physics of X-rays*, Moscow: State Publishing House of Technical-Theoretical Literature (in Russian) [English translation, Document No. AEC-tr-4502, Springfield, Virginia: Clearinghouse for Federal Scientific and Technical Information.]

Bulletin of the Academy of Sciences of the USSR, Physical Series; **21,** no. 10 (1957); **25,** no. 8 (1961); **26,** no. 3 (1962); **27,** no. 3 (1963); **28,** no. 5 (1964); **31,** no. 6 (1967). [English translation, White Plains, New York: Columbia Technical Translations.]

ELWOOD E. C., FABIAN D. J. and WATSON L. M. (1967) *Metals and Materials* **1,** 333.

FABIAN D. J. (1968) *Soft X-ray Band Spectra and the Electronic Structure of Metals and Materials*, London: Academic Press.

FAESSLER A., CAUCHOIS Y. and ZEMANY P. D. (1963) *Proceedings of the 10th Colloquium Spectroscopicum Internationale*, 307, 321, 341, Washington: Spartan Books.

FAESSLER A. (in press) *Compilation of Soft X-ray Spectra and Binding States*, Landolt-Börnstein, New Series, Berlin: Springer-Verlag. (Scheduled for publication as Vol. III/4.)

HOLLIDAY J. E. (1967) *Handbook of X-rays*, Chapter 38, New York: McGraw-Hill Book Company.

HOLLIDAY J. E. (1969) *Techniques in Metals Research*, Vol. III, New York: Wiley-Interscience Inc.

NEUPERT W. M. (1969) *Annual Review of Astronomy and Astrophysics* **7,** 121.

NORELCO REPORTER (1967) **14,** (3–4) Special Issue, Mt. Vernon, New York: Philips Electronics Instruments.

PARRAT L. G. (1959) *Reviews of Modern Physics* **31,** 616.

TOMBOULIAN D. H. (1957) *Handbuch der Physik* **30,** 246, Berlin: Springer (in English).

YAKOWITZ H. and CUTHILL J. R. (1962) *Annotated Bibliography on Soft X-ray Spectroscopy*, NBS Monograph 52, Washington: U.S. Government Printing Office.

J. R. CUTHILL

SOLAR CELL

Introduction

The photovoltaic effect was discovered by Becquerel in 1837 who noticed that a voltage was developed when a liquid electrolyte was illuminated. Subsequently, similar effects were noticed in the rather crude semiconductors of the time, leading to the discovery of the selenium barrier layer cell, which was much used for light-measurement, but was not significant as a power generator. The crucial advance came in 1954 at the Bell laboratories, following their successful work on transistors and similar *p–n* junction semiconductor devices. They were able to demonstrate a silicon cell with 6 per cent conversion efficiency in sunlight. In the years since then the solar cell has been intensively developed particularly as a source of power in spacecraft, though a few terrestrial applications have also been found. Although other materials have been tried—and are discussed later— silicon still appears to lead the field. Today's cell is a very reliable and sophisticated, albeit still expensive, device.

Theory

If a photon of sufficient energy strikes a semiconductor, it may be able to release an electron from the valence band, raising it to the conduction band, leaving a "hole" in the former. The energy of the photon must therefore be at least as great as that of the forbidden energy-gap between the two bands, about 0·7 eV in germanium and 1·1 eV in silicon. Such electrical carriers are responsible for photoconductivity—the rise in conductivity associated with exposure to radiant energy—but a further step is necessary for the creation of a useful external current. This is the formation of a junction, or barrier layer, between two types of semiconductor material. The action of the junction can, however, best be explained by first considering a little further the action of a simple type of material (see also *Thermoelectric generation of electricity*).

In a material which is in equilibrium at any temperature, the product of the electron and hole concentrations is constant,

$$p_0 n_0 = n_i^2, \qquad (1)$$

where n_i is the intrinsic carrier concentration at that temperature. In pure, intrinsic silicon at 300°K (room temperature) $n_i^2 \simeq 10^{21} (n_i \simeq 3 \times 10^{10}$ cm^{-3}). If the material were doped to be strongly *n*-type, n_0 might be (say) 10^{18} and hence p_0 10^3 cm^{-3}. The electrons are called the majority carriers, and the holes are the minority carriers. For a p-type material this situation would be reversed.

If now n and p types are joined, and again are in thermal equilibrium, they must have a common energy level so that there is no net flow of carriers (holes or electrons) across the junction. This is usually expressed by stating that they have a common Fermi level. For an n-type material, there are "donor" levels below the bottom of the conduction band, and in a p-type acceptor levels above the valence band; and in each case the Fermi level is the half-way energy line.

A practical cell consists of a thin layer of p-type material diffused on to an n-type base with contacts to the base and surface layer. The junction between the n- and p-type material might be as shown in Fig. 1(a), and the energy levels as in Fig. 1(b). At the junction there must be a potential barrier, with the p-type material at the higher level. From what has been said, it will be appreciated that the height of the barrier will depend on the level of doping but it is less than the energy gap. If now electron-hole pairs are created in the top layer by the absorption of photons, the equilibrium is upset; although the concentration of the majority carriers is scarcely altered, that of the minority carriers will alter very considerably. The constancy of the product np can be restored either by recombination, or by the diffusion of the minority carriers across the junction where they become majority carriers. Corresponding processes will happen if radiation is absorbed in the lower layer.

all radiation is absorbed. The only radiation which is usefully absorbed, however, is that having a photon energy greater than E_g, and the excess energy of more energetic photons is also wasted. Thus the response of a simple cell to a broad spectrum such as that of the sun is inherently rather inefficient, although the conversion is not subject to the Carnot heat engine limitation. Further, the only minority carriers to be effective will be those which can reach the junction before they can recombine: i.e. this distance must be less than the diffusion length, L say. (Generally therefore $l < L$.)

Suppose a fraction Q of the initially-separated pairs (above or below the junction) is effective. Then the short circuit current j_s (per cm^2 of cell surface) is

$$j_s = \mathbf{e}QN_g \qquad (1)$$

where N_g is the number of photons cm^{-2} sec^{-1} with energy $> E_g$. Any reflection loss R at the surface must also be allowed for: so that (1) must be multiplied by $(1 - R)$. The open-circuit voltage is given by

$$V_0 = \frac{\mathbf{k}T}{\mathbf{e}} \ln\left(\frac{j_s}{j_r}\right) \qquad (2)$$

where j_r is the (saturated) reverse current when a large reverse voltage is applied, and k is Boltzmann's constant.

It is convenient to define

$$j_s/j_r = (r + 1)\exp r \qquad (3)$$

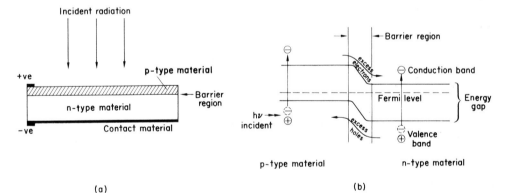

Fig. 1. *The principle of construction (a) and operation (b) of a typical photovoltaic cell.*

The open-circuit voltage is about half the energy gap. If connections are made to an external circuit, the hole and electron movements can create a current. As this current rises, the Fermi levels of both materials change, so that the potential barrier and the output potential fall.

Suppose the radiation has an absorption coefficient α (forgetting for the moment how this varies with wavelength λ) and the top layer has a thickness l: then the fraction absorbed in this layer is $1 - \exp(-\alpha l)$. Attenuation will continue at the same rate through the base material, until (if thick enough)

The relationship between voltage and current is of the general form shown in Fig. 2. The ideal power is $V_0 j_s$, the actual maximum power when matched to an external circuit is $V_m j_m$; and one can define a junction efficiency S as

$$S = V_m j_m / V_0 j_0 = \frac{r - 1}{r + \ln(r + 1)}. \qquad (4)$$

S is always <1 because the cell departs from an ideal diode, and because of internal resistance. The efficiency η is then the ratio of the maximum power output to the total power density falling on the cell.

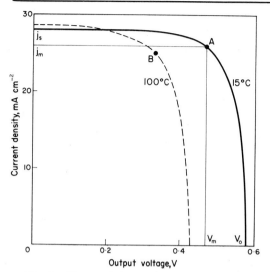

Fig. 2. Characteristic curves of voltage/current for a silicon cell at 15°C and 100°C, exposed to the solar spectrum. The points A and B define the maximum power output in each case.

The latter may be expressed as the product $N_s E_s$, where the two terms are the number of photons cm^{-2} sec^{-1} in the (solar) spectrum and their average energy in eV, respectively. What may be defined as the solar collection efficiency Q_s, the number of carrier pairs produced by each mean solar photon, is given by

$$Q_s = Q N_g / N_s.$$

The final expression arrived at for the maximum efficiency is

$$\eta_{\max} = (r-1)\,\mathbf{k}T(1-R)\,Q_s/E_s. \qquad (5)$$

It can be seen that in a good cell r (and hence the junction efficiency S), N_g and Q must all be high, with R low. Some of the factors involved will now be considered in turn in greater detail.

Junction Efficiency

The junction efficiency S largely depends on the quality of the material, particularly as it affects the internal resistance. The higher the doping level, the higher is the conductivity for the majority carriers, and the smaller is j_r, which is also useful. However, the carrier lifetime, which vitally affects the current-collection efficiency Q, is usually almost inversely proportional to the doping, so this feature sets an upper level.

The circulation of current in an external circuit depends on majority carriers in the surface layer moving *transversely* to the surface to the collector strip. Since the surface layer is thin (typically, as will be seen, a few microns) the effects of its resistance can be minimized by placing on it a surface grid of highly conducting material, to which the external contact is made. The loss of active area exposed to radiation is much more than offset by the lower effective resistance, which in moden 2 × 1 cm cells is about 0·1 Ω, compared with 1 Ω in the early cells.

Cells intended for space use are nowadays almost always "reversed", i.e. they are n- on p-type, and *not* as shown in Fig. 1. Their value was first reported by Mandelkorn in 1960, and it arises from two useful effects. The surface majority carriers in a reversed cell are electrons, with about three times the mobility of holes. Secondly, the minority carriers in the bulk layer are also electrons. Thus, if carriers are generated fairly deeply by radiation absorbed there they have a better chance of reaching the junction before recombining. This is particularly important when the lifetime is degraded, which is a typical effect of particle bombardment, e.g. from the Van Allen belts.

The effects of radiation damage can be minimized by a very thin cover, which absorbs protons, bonded to the cell surface. Reflection losses at the surface can be minimized by quarter-wave blooming, similar to that normally used for optical components, but matched (in the case of silicon) for a refractive index of about 3·5. Since the index for the coating material should be the square root of that of the bloomed material, silicon monoxide ($n = 1 \cdot 95$) is suitable, and is very tough.

Two further refinements are possible, in principle at least. The surface layer must reflect as much of the unwanted radiation as possible, since the performance of a cell deteriorates seriously at a high temperature. Reflection can be increased by half-wave blooming over the unwanted spectral range. The final point is that if the surface coating is fluorescent it can convert short-wave radiation, which might otherwise penetrate too deeply to be useful, into longer-wave radiation. Needless to say a good deal of practical compromise is needed to produce the surface coating since, apart from somewhat conflicting optical requirements, it must not impair the surface electrical properties.

Energy Gap

If the energy gap E_g is low, so is the output voltage, almost in proportion. On the other hand, as it is raised the proportion of photons with greater energy (and hence which are able to release carrier-pairs) falls. It is thus clear that there will be an optimum value, to give the highest efficiency. This can be found, at least approximately, by finding a value for j_r in equation (2) in terms of E_g. A decrease in E_g raises the intrinsic carrier concentration n_i at any temperature, since thermal agitation is more effective in inducing transitions across the forbidden gap;

$$n_i^2 = \text{const.}\, T^3 \exp(-E_g/\mathbf{k}T). \qquad (6)$$

Also, very approximately

$$j_r = \text{const.}\, \mathbf{e} n_i^2, \qquad (7)$$

so that, from equation (2),

$$V_0 \simeq \frac{kT}{e} \ln [j_s/\text{const. e} \exp(-E_g/kT)]. \quad (8)$$

Evaluating the constant, and assuming in comparing one material with another that variations in Q, R and S do not occur,

$$\eta \simeq \frac{SN_g E_g}{N_s E_s}\left[1 + \frac{kT}{E_g}\ln\frac{N_g}{10^{23}}\right]. \quad (9)$$

For values of $E_g > 1$ eV the second term is a very small correction, so that in this approximate expression η is determined mainly by the product $N_g E_g$. The values of N_g can be determined for any value of E_g from the known solar spectrum, on earth, or at any point in space. In the tropics at sea level, the total radiant energy density is close to 1 KW m^{-2} (= 1 "sun"). The way in which the efficiency varies with energy gap according to equation (9), is shown in Fig. 3 based on $R = 0$, $S = 0.8$. The E_g value for some semiconductors is also shown. At sea level, silicon is close to the optimum, but outside the atmosphere, where the spectrum is "bluer", a higher E_g, such as that of gallium arsenide, is more appropriate.

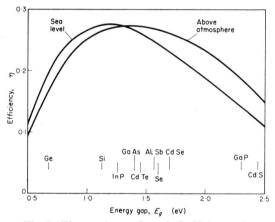

Fig. 3. The maximum theoretical efficiency of a photovoltaic cell as a function of energy gap, in space and sea level conditions. The energy gaps of some potential semiconductor materials are also shown.

Current Collection and Spectral Response

As has already been said, the current collection efficiency Q depends on minority carriers generated on one side of the junction layer, crossing it before recombination can occur. If most of the radiation is to be absorbed in the surface layer the product $l\alpha$ must be around or above unity, and so must the ratio L/l for efficient carrier collection. Any radiation which penetrates (because its value of α at any particular wavelength λ is small) to below the junction can make a useful contribution, provided its photon energy $> E_g$, and it is absorbed within the diffusion length L, around 10 μ for typically doped Si. Figure 4 shows how Q varies with α for some typical values of l/L. The total Q is the sum of the surface and bulk components, Q_f and Q_b, the latter making a useful contribution for $\alpha < 10^3$ cm^{-1}. The best value of l (or rather of l/L) is somewhat dependent on α, but for Si, $l = 3\,\mu$ gives an optimum for typical doping

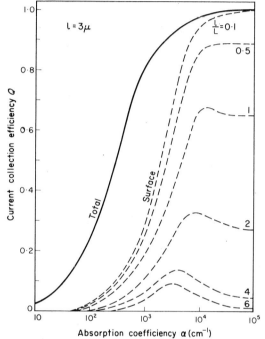

Fig. 4. The variation of current collection efficiency with absorption coefficient, and with the ratio l/L. The dotted lines show the "surface" contribution, the solid line the sum of "surface" and "base" for $l/L = 0.1$.

conditions, and this value has been used in compiling the data of Fig. 4. In GaAs, where the electron and hole diffusion lengths are about 10 and 4 μ, the best value of l is 2 μ.

In order to see how Q varies with wavelength, it is necessary to know the spectral absorption characteristics, which are given in Fig. 5(a) for Si and GaAs. Then as shown in Fig. 5(b), Q can be plotted as a function of λ. To obtain the overall solar collection efficiency Q_s, it is necessary to multiply Q at each wavelength by the relative number of photons at λ, and integrate through the spectrum. The resulting values are about 38 and 42 per cent for Si and GaAs respectively. This is for the solar spectrum above the atmosphere, and neglecting reflection losses. Making

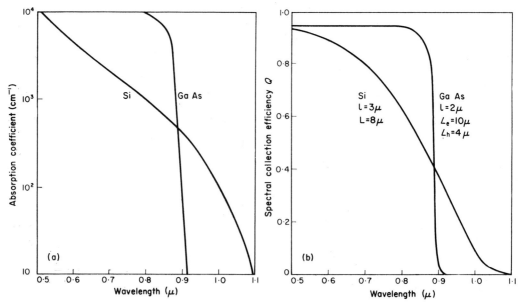

Fig. 5. The absorption coefficient (a), and the collection efficiency (b), as functions of wavelength for Si and GaAs. The data assumed in (b) are currently available only for Si.

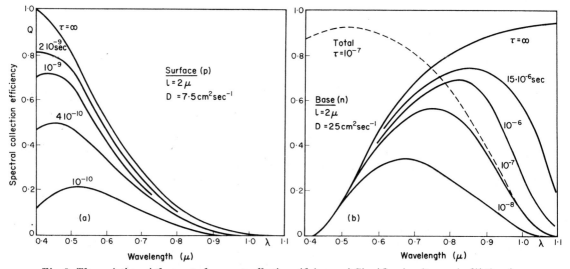

Fig. 6. The variation of the spectral current collection efficiency of Si with minority carrier lifetime for (a) the surface and (b) the base layer. The dotted curve in (b) is the total response, for a minority carrier lifetime of 10^{-7} sec.

this latter allowance, and taking $S = 0.9$ as an upper likely limit, η_{max} values are 15 and 23 per cent respectively. The former is closely approached in modern gridded Si cells, but the technology of GaAs has so far prevented the latter from being obtained. This is because an S value even approaching 0·8 has not been realized; and present carrier lifetimes are well below the 10^{-8} sec assumed in Fig. 5(b). However, a few

experimental cells of about 20 per cent efficiency have been produced. The carrier lifetime τ is related to the diffusion length L by

$$L^2 = D\tau$$

where D is the diffusion constant, determined mainly by the mobility of the electrons (or holes). The effect of a variation of τ is shown in Fig. 6. The effect is negligible on the surface behaviour, unless τ is very much smaller than the 10^{-7} sec usually attained with Si.

For very high values of $\alpha (>10^4 \text{ cm}^{-1})$ the carriers are produced almost on the surface, and surface recombination can be an undesirable complication. It has the effect of reducing the effective Q for wavelengths below 0.5μ in Si. The value of Q does not depend very sensitively on the thickness l of the surface layers, and there is some merit in exceeding the optimum for Q, in order to raise the junction efficiency S. With the reversed GaAs cell, S should be appreciably better because the surface majority carriers are then electrons. The doping of the top layer is, in practice, far from uniform with depth, and the electron field so created does, in fact, help the minority carriers towards the junction.

During use in space, however, the carrier lifetimes may be seriously reduced, as mentioned earlier. Efforts then must be concentrated on developing the short-wave response, since as Fig. 6 shows, this is least affected. The minority carriers have the best chance of reaching the junction if the surface layer thickness is reduced, and in the present so-called super-blue cells it is about 2μ, compared with 3μ in earlier cells. Note that Fig. 6 is based on $l = 2 \mu$, but applies to a p-on-n, and not a reversed cell.

Temperature Dependence

It is important to study the effects of temperature on cell performance, since absorption of radiation, particularly on space missions relatively close to the Sun, will cause a rise. The diffusion constant D varies as $T^{-1/2}$, but τ is practically temperature-invariant. The energy gap E_g falls slightly with temperature (the coefficient for Si is $-4 \times 10^{-4}/°C$) as shown in Fig. 8. The consequent changes in short circuit current j_s are very small. On the other hand the open-circuit voltage V_0 falls seriously with temperature rise. This is because V_0 depends logarithmically on j_r (equation 2) and in turn j_r depends exponentially on T (equations 6 and 7). The net effect is an almost linear decrease of V_0 with temperature as shown in Fig. 7.

Assuming that S is independent of temperature (and also that effective carrier masses, etc., are similarly invariant) the combined effect on the product jV is shown in Fig. 8. This is valid for the spectrum above the atmosphere, and has been normalized so that the maximum efficiency at about 300°K is the same as that in Fig. 3. It can be seen that the major effect is a serious fall in efficiency as the temperature rises, justifying the efforts already mentioned to reflect the unwanted radiation. A secondary effect is that the optimum energy gap rises with temperature, which is a further reason for developing higher gap materials than silicon.

Development of Better and Cheaper Cells

Now that the basic technology and theory of the solar cell are well established, present aims are to increase efficiency and reduce costs per unit of delivered energy; these aims partly tend to conflict.

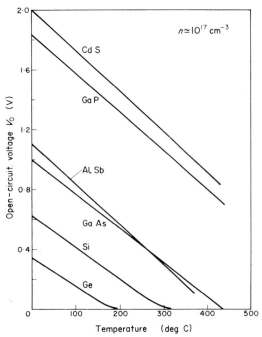

Fig. 7. *The theoretical variation of open-circuit voltage with cell temperature for different materials, exposed to the solar spectrum. Measured values for Si agree closely.*

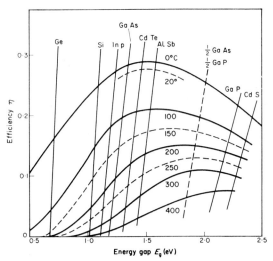

Fig. 8. *The variation of maximum efficiency with energy gap, as a function of temperature. The manner in which the energy gaps of individual semiconductors vary with temperature is also shown.*

The efficiency of a cell made by doping a single material is inherently limited to something under 30 per cent, and to attain a higher value it is necessary to consider using two materials with very different energy gaps, to absorb from two different spectral regions. A simple way, in principle, of achieving this object is to use a dichroic (spectrally-selective) mirror which reflects say the short-wave radiation, and transmits the long. If, for example, the mirror has a "rectangular" cut-off at $0.58\ \mu$, the optimum energy gaps of cells to receive the long- and shortwave radiation respectively are 1.1 and 1.7 eV. Silicon is ideal for the former, and CdSe, if developed, might be suitable for the latter. However, the arrangement is clumsy and not likely to be practicable.

An alternative arrangement would be to use two cells, with the high-gap cell physically on top of the low-gap cell, separated by a transparent but electrically-insulating layer. The upper cell, of course, has to be very thin, so that it is sufficiently transparent to the long-wave radiation. Parasitic losses with this form of construction are such that no net gain over a single-gap cell has been attained.

The most hopeful approach seems to be to develop a single cell with a variable or 2-step energy-gap: for example, the top layer could be of high-gap p-type material, and the base of low-gap n-type. A possible combination is GaAs–GaP (1.35 and 2.3 eV). Attempts have been made to prepare such a cell by diffusing P into GaAs, the resulting cells tending to have a continuously varying gap, decreasing with depth. Although the parasitic radiation losses of such a cell can be very small, the energy gaps of the particular pair just cited are somewhat above the optimum.

Cells of the highest efficiency are produced from pure single crystal material—normally Si, and the resulting cells are usually about 2×1 cm in size. To attain a large electrical output, a "shingled" array is used—like tiles on a roof. Such an arrangement suffers from the disadvantages of cost, and complexity to optimize the electrical output. There has therefore long been an urge to make large uniform areas, perhaps from polycrystalline (i.e. more or less amorphous) material. Further, the single cell is made from a wafer (sliced from a stick of material) which is usually at least $\frac{1}{2}$–1 mm thick. The useful thickness, however, cannot be more than one diffusion-length ($\simeq 10\ \mu$) below the barrier layer, so that there is a weight, as well as a cost penalty, which for space applications is important.

Attempts to produce large-area polycrystalline Si cells have not yet been very successful, partly because the carriers on reaching a grain boundary recombine ineffectually. It is difficult, also, to define a barrier layer, and there is better hope of doing this by depositing Si epitaxially by vapour-phase decomposition of the chloride. Another method which has been suggested is the use of Si spheres about 1 mm in diameter, surface diffused to have a p-layer coating, and embedded in a plastic layer. By lightly grinding the rear surface, contact can be made to the inner n-regions. Efficiencies of around 10 per cent have been attained in this way.

At the time when the first Si cells were developed, efficiencies of about 6 per cent were attained with cadmium sulphide cells, which consist of CdS backed by a layer of copper. The requirements of such a semiconductor-metal junction are rather less stringent than those for a p-n junction, and some cells have been produced by evaporation of CdS on to transparent substrates, followed by the Cu layer. Such cells have a proportion of excess Cd in the CdS, which helps to confer conductivity. Their values of V_0 and j_s are good individually, but because the surface and contact resistance is high, the junction efficiency S is poor. Better doping techniques will probably bring an improvement here.

The energy gap of CdS is 2.4 eV and it is therefore rather surprising that this material is sensitive at an energy well below this—corresponding to a wavelength greater than about $0.5\ \mu$. Part of this photosensitivity is conferred by photoelectrons released from the copper. Another part of the low energy "tail" response can come from transitions which in a real material (as opposed to the theoretical ideal) can take place *within* the conduction or valence bands. (This helps to explain why the spectral cut-off in Si is so much less sharp than in GaAs). Some carriers, too, in a doped material may not be free, but loosely bound to impurity centres or other defects, and these can easily be detached. This tail response can therefore be encouraged by additional doping with a high ionization energy material, provided this can be done without detriment to other properties.

CdS gives promise that effective large-area cells can be developed, although probably of lower efficiency than the shingled Si arrays. However, it is envisaged that sheets of the material could be very light and flexible. For spacecraft use, therefore, a very large area could be tightly rolled up and unfurled in orbit. One of the present difficulties appears to be that these cells are sensitive to water vapour, and it is rather difficult to control their environment on the way to their ultimate destination.

The present Si cells have a specific power of about 20–25 W kg^{-1} (including the supporting structure) compared with about 4 W kg^{-1} for current radio-isotope-thermoelectric generators. The polycrystalline CdS cells have much the same specific weight as the Si cells today (their lower weight is offset by lower efficiency) but it seems likely that outputs up to 100 W kg^{-1} may be possible eventually. The present costs per watt are about £500–£1000, and £200 for Si and CdS respectively, with the hope that the latter may come down to about £120. Even this price, however, is too high for most terrestrial applications.

Despite the superior potential performance of GaAs, the superblue Si cell is in the forefront today, but the CdS cell may catch up before too long.

See also: Space power supplies, Thermionic generation of electricity, Thermoelectric generation of electricity.

Bibliography

COEKIN J. A. (1966) Semiconductor devices, in *Encyclopaedic Dictionary of Physics* (J. Thewlis Ed.), Suppl. vol. 1, 281, Oxford: Pergamon Press.

SPRING K. H. (1965) *Direct Generation of Electricity*, New York: Academic Press.

SPRING K. H. (1969) Direct conversion of heat to electricity, in *Encyclopaedic Dictionary of Physics* (J. Thewlis Ed.), Suppl. vol. 3, 46, Oxford: Pergamon Press.

K. H. SPRING

SOLAR WIND, PLANETARY INTERACTIONS WITH. The solar wind is the name given to the continuous stream of charged particles, mainly protons and electrons, which flows outwards from the Sun. It provides a medium for interaction between the bodies of the solar system as a result of disturbances which they may induce in the stream of particles.

A body possessing no magnetic field will merely form a cavity in the solar wind, with a turbulent wake for some distance downstream, and the whole system will be surrounded by a shock wave whose characteristics are determined by the ratio of the velocity of the body to that of the Alfvén wave in the medium. If the body has a magnetic field, a magnetosphere will be formed which entirely encloses the field and, in turn, is surrounded by a shock wave. Again there is a tail in the downstream direction. The characteristics in the distant part of the tail, and particularly its length, have not yet been determined but by analogy with cometary tails, or with the re-entry of bodies in the Earth's atmosphere where both a long tail and a shock wave are developed, there is reason to believe that the effect of a planet's interference with the solar wind could extend downstream for up to thousands of times the diameter of the planet. Tails of this length in the solar system could lead to interaction between planets, the extent of the effect being a function of the energy of the solar wind which is known to vary with solar activity.

Apart from the Earth, only Jupiter is known to have a significant magnetosphere; from the evidence of its radio emission, this is probably much larger than that of the Earth. In the case of Venus, results obtained by Bridge *et al.* (1967) from *Mariner V* provide clear evidence of a bow shock wave and very low plasma density in the cavity in the solar wind produced by Venus. Space probes have failed to deduce the existence of a magnetosphere around Mars. Dolginov *et al.* (1967) obtained results from *Luna 10*, which implied that the Moon possesses a magnetosphere, but Ness *et al.* (1967) from *Explorer 35*, found no sign of a shock wave from the Moon and only small but detectable disturbances of the magnetic field in the leeward portion of the solar plasma flow around the Moon. A space probe, however, samples only an extremely small volume of space in any region of interest, and deduction by this means of weak time-dependent variations in the solar wind, due to a planet, is necessarily difficult. The Earth itself is likely to be a far more sensitive detector of disturbance produced by the Moon, Mercury or Venus, since it will integrate the effect over a much greater volume of space.

Observational evidence for effects in the case of the Earth has been sought by Bigg (1963) who found a marked decrease in the frequency of the Great Magnetic Storms on the Greenwich catalogue at New Moon (Fig. 1a), and at inferior conjunctions of both Venus and Mercury. (A similar effect had been pointed out earlier by Sucksdorff (1956) and Houtgast (1962) but its physical significance was not appreciated at the time.) Stolov and Cameron (1964) and Bell and Defouw (1964) have shown that there is an even larger effect on the geomagnetic index at Full Moon, which takes the form of a decrease in magnetic activity as the Moon enters the tail, followed by an increase as it emerges (Fig. 1b).

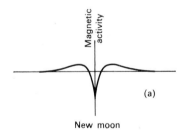

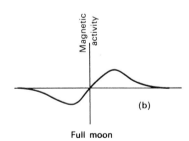

Fig. 1. (a) *General form of change in magnetic activity as the Moon, Venus and Mercury cross the Earth–Sun line.* (b) *General form of change in magnetic activity of the Earth about Full Moon.*

The Earth appears to have an effect on the "Blue Clearings" of Mars. Slipher was aware that blue clearings occur more frequently at time of inferior conjunction of Mars, but Bigg (1963) showed that there were three maxima in the frequency of the blue clearings, one at conjunction and two others, approximately 10° on either side of conjunction (Fig. 2). This might be expected if disturbances from both shock wave and turbulent wake extend to the orbit of Mars. Bigg (1964) has failed to find any significant effects produced on Jupiter by either the Earth or Mars. However, the marked increase in the frequency

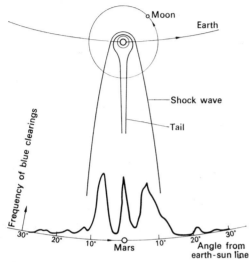

Fig. 2. Variation in frequency of the "Blue Clearings" on Mars in relation to the position of the Earth's tail and shock wave.

of occurrence of Jupiter's decametric radio emission as the Earth approaches the Sun–Jupiter line, may provide evidence of a major downstream effect of the Earth, but could also have an alternative explanation.

Bibliography

BELL B. and DEFOUW R. J. (1964) *J. Geophys. Res.* **69**, 3169.
BIGG E. K. (1963) *J. Geophys. Res.* **68**, 4099.
BIGG E. K. (1964) *Nature* **203**, 1008.
BOWEN E. G. (1967) *Proc. Astron. Soc. Australia* **1**, (1) 5.
BRIDGE H. S., LAZARUS A. J., SNYDER C. W., SMITH E. J., DAVIS L., COLEMAN P. J. and JONES D. E. (1967) *Science* **158**, 1669.
DOLGINOV SH. SH., YEROSHENKO YE. G., ZHUZGOR L. N. and ZHULIN I. A. (1967) *Geomagnetizm i Aerononuya* **7**, 436.
HOUTGAST J. and VAN SLUTTERS (1962) *Nature* **196**, 462.
NESS N. F., BEHANNON K. W., SCEARCE C. S. and CANTARANO S. C. (1967) *J. Geophys. Res.* **72**, 5769.
SLIPHER E. C. (1962) *Mars*, Sky Publishing Corporation, Cambridge, Mass.
STOLOV H. L. and CAMERON A. G. W. (1964) *J. Geophys. Res.* **69**, 4975.
SUCKSDORFF E. (1956) *Geophysica* (Helsinki) **5**, (2), 95.

E. G. BOWEN

SOLID NUCLEUS FORMATION FROM SOLUTION. An initially homogeneous solution may become unstable as the result of a change in temperature or a change in composition following chemical reaction and may then separate into two phases. Two mechanisms may be distinguished, spinodal separation and nucleation.

For a two-component solid solution (mixed crystal) below an upper consolute temperature, the free energy (G) versus mole fraction (x) curve, in the simplest case, is continuous and shows two minima with a maximum between them. Between the minima and the maximum there are inflections at the spinodal compositions s_1 and s_2 for which $\partial^2 G/\partial x^2 = 0$. Between s_1 and s_2, $\partial^2 G/\partial x^2 < 0$ and spinodal phase separation may take place by continuous compositional fluctuations. If the common tangent points to $G(x)$ are at x_1 and x_2, then between x_1 and s_1 and x_2 and s_2 the parent solution is metastable and phase separation then requires nucleation.

When there is no continuity of state between the two phases, the $G(x)$ curve for the solid or liquid solution may have only one minimum but may intersect the $G(x)$ curve for a crystalline phase which also has only one minimum. Solutions with compositions greater than x_1, the common tangent point, are metastable and the precipitation of the crystalline phase can only occur by nucleation.

Crystal nucleation from a liquid solution is qualitatively very similar to *Nucleation phenomena* in liquids (Vol. 5, p. 164) (see also *Freezing process, nucleation in*, Vol. 3, p. 300). The rate of homogeneous nucleation from solution, J, is given by equation 2, p. 300, in Vol. 5, in which ΔG^* the free energy of formation of a critical nucleus assumed to be spherical, is given by equation (3) and σ is now the interfacial tension of the crystalline nucleus–solution interface. Here $\Delta T = T_0 - T$ is the undercooling below the saturation temperature T_0; T_0 is the temperature at which the actual concentration of solute is equal to the solubility and in this approximation $-\Delta S_v$ is the differential entropy of solution per unit volume of the crystal.

ΔG^* may also be written

$$\Delta G^* = 16\pi\sigma^3 v^2/3[kT \ln(\alpha/\alpha_0)]^2$$

where v is the volume per molecule of the crystal and α and α_0 are the activities of the solute in the supersaturated and saturated solutions respectively; thus α/α_0 is the supersaturation activity ratio. For an undissociated solute $\alpha = \gamma_x s = \gamma_m m = \gamma_c c$, etc., where x, m or c is the concentration expressed in mole fraction, molality or molar terms and γ_x, γ_m or γ_c is the corresponding activity coefficient at that concentration. For crystals which dissociate in solution, e.g. alum, the magnitudes α, γ_x, x, etc., may refer to the formula unit $Al_2(SO_4)_3 \, K_2SO_4 \cdot 24 \, H_2O$ and v must then be the volume of this formula unit in the crystal. For sparingly soluble salts, e.g. $BaSO_4$, $\alpha/\alpha_0 = m_+ m_- \gamma_+ \gamma_- / K_{SP}$, where K_{SP} is the thermodynamic solubility product, m_+, m_-, γ_+, γ_- are the molalities and activity coefficients of the positive and negative ions, e.g. Ba^{++} and SO_4^{--}.

In the expression for J the value of the term $n(kT/h) \exp(-\Delta g/kT)$ is often estimated to be of the order 10^{30}, though in extreme cases, e.g. the nucleation of sucrose solutions, Δg, the free energy of activation for the transport of a molecule from the solution into the lattice of the nucleus may be very large. Large values of Δg may result in a time-lag or induction period during which the critical nuclei are built up.

On plotting J versus α/α_0 or m/m_0, J remains negligibly small until, in the neighbourhood of the Ostwald "metastable limit", it increases sharply and then becomes very large. Definitive experiments are rendered difficult by such behaviour; the exclusion of foreign nuclei (see Vol. 5, p. 300) is not easy and nuclei formed in the first moments when the supersaturation is high may redissolve later as the supersaturation falls. Mixing the reacting solutions may be dispensed with by generating one of the ions *in situ* by a chemical reaction, e.g. sulphate ion results from the slow reaction between persulphate and thiosulphate ions. Nucleation experiments indicate that the values of σ range from about 100 erg cm^{-2} for AgCl to about 270 erg cm^{-2} for MgF_2. When the ions are present in non-stoichiometric ratio the changed electrical potential at the nucleus-solution interface may result in a change in the value of σ.

Bibliography

A general discussion on crystal growth, *Disc. Faraday Society* **5**, 1949.

VAN HOOK A. (1961) *Crystallization: Theory and Practice*, New York: Reinhold.

NIELSEN A. E. (1964) *Kinetics of Precipitation*, Oxford: Pergamon Press.

W. J. DUNNING

SONIC BOOM, GENERATION AND SUPPRESSION OF.

The phenomenon known as the sonic boom results from the detection by the ear of the pressure pattern continuously created by aircraft in supersonic flight. The sonic boom is quite different in character from the more common aircraft noise associated with rotating components (propellers, compressors, turbines) and turbulent airstreams. The onset of this sound is quite sudden, often startling, and it has only a short duration. In addition, the boom can persist to extremely large distances from the generating aircraft (in some cases 40 miles or more), distances at which ordinary noise would be inaudible.

The pressure patterns responsible for the sonic boom are caused by the displacement of the air as the aircraft approaches and its replacement as the aircraft passes. At subsonic speeds, these pressure disturbances travelling at the speed of sound are felt well ahead of the slower moving vehicle and are distributed over a wide area. Although the propulsion system noise may be detected, the pressure changes due to the air displacement occur gradually and are not perceived as noise. At supersonic speeds, the aircraft travels faster than the disturbances it creates and the air influenced by the aircraft at a given instant is confined to a region of generally conical shape originating at the aircraft nose. Pressure variations due to air displacement are concentrated within a small region, and the now rapid variations in pressure within that region are detected by the ear and a characteristic noise described as a boom is heard. The noise associated with rotating components and turbulent airstreams may still be present but the noise due to air displacement is predominant.

The sonic boom pressure field if it were visible would resemble the wave pattern created by a high-speed boat in still water. A more detailed examination of the nature of the sonic boom pressure field surrounding an airplane in supersonic flight may be made with the aid of Fig. 1. In this drawing is represented that portion of the pressure field directly below the airplane (a section of the conical region which contains the airplane flight path and its ground track). Close to the supersonic airplane the pressure field as depicted

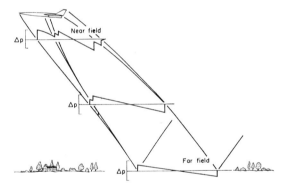

Fig. 1. Airplane pressure field.

by the upper pressure-time history or pressure signature is quite complex, containing separate disturbances from each of the airplane components (fuselage, wing, engine, etc.). This portion of the flow field where the signature shape is dependent on the detailed airplane shape is termed the near field. At greater distances, in the far field, these separate disturbances coalesce to form a pressure signature in the shape of an elongated letter N. This "N" wave is the usual form for ground level signatures of airplanes in supersonic cruise. The ground pressure pattern is characterized by an abrupt pressure rise or bow shock followed by a linear decline in pressure to a value below ambient and a subsequent recompression or tail shock to restore ambient pressure. Normally the observer will hear a percussive noise or boom in response to each of the shocks, but the gradual change between shocks will not be heard. The magnitude of the pressure rise is dependent on the geometry of the airplane, its weight, and its speed and altitude. The shape of the cone-like region of influence depends primarily on the ratio of the airplane's speed to the speed of sound (Mach number). The cone half angle at the airplane nose is only slightly greater than the Mach angle, whose sine is the reciprocal of the Mach number. At speeds greater than the Mach number 1·4 case illustrated, the cone would be more slender.

The ground area swept by the conical pressure field of the supersonic airplane in its passing may be as much as 80 miles wide for high altitude flight. The lateral extent of the boom is limited by the refraction of the shock front in its propagation from high altitudes with low sound speeds to low altitudes with higher sound speeds. Refraction causes the propagation path to bend away from the earth. Beyond the point where the path becomes tangent to the earth, a shadow region is created and no boom is heard.

At ground level the shocks encounter a rigid surface and are reflected upward. When this occurs, the incident and reflected waves are superimposed and an intensification of the pressures take place. For a smooth, and perfectly rigid surface, the free air pressures are increased by a factor of two. For natural terrain the factor is most often somewhat less than two (say 1·9). In cities, where multiple reflective surfaces are encountered, the factor can sometimes be greater than two.

Analytic work conducted by W. D. Hayes in the U.S.A. and G. D. Whitham in Great Britain has provided the foundation for methods of estimating sonic boom intensity. In his doctoral thesis of 1947 Hayes introduced the concept of analysing the drag and distant pressure field characteristics of non-lifting airplane configurations by replacing their complex geometry with equivalent bodies of revolution having the same longitudinal development of cross-sectional area. Hayes' work also covered the treatment of lift effects in terms of equivalent geometry and thus provided a simplified method of treating the net displacement of air as it would influence the pressure field at large distances. It was shown that to be suitably applicable at supersonic speeds, the equivalent body should be determined not by cross-sections normal to the airplane axis but by the frontal projections of areas intercepted by planes inclined at the Mach angle with respect to the flight direction. Thus a given airplane has not one equivalent body, but many. Equivalent body shape depends on the Mach number, the flow field position, and the airplane lift. An airplane and its equivalent body representation for the flow field directly below is illustrated in Fig. 2. The contribution of the actual airplane components to the equivalent body areas is defined by Mach angle cutting planes, one of which is represented in end view by the sloping line. The contribution due to lift at a given station is proportional to the accumulated lifting forces from the nose to that particular station. At the tail of the airplane

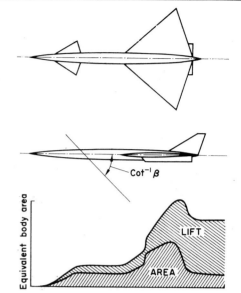

Fig. 2. *Area distribution of a typical airplane equivalent body of revolution.*

where all the lift is accounted for in the equivalent area due to lift, that area is given by $\frac{\beta}{2} C_L S$; where β is the cotangent of the Mach angle or $\sqrt{M^2 - 1}$, C_L is the airplane lift coefficient, and S the area on which the coefficient is based.

Although the foregoing analysis reduced the problem of treating a complete airplane configuration to that of treating a definable equivalent body of revolution, it did not provide for the formation of the shocks known to exist in the real flow. The analytic work of Dr. Whitham provided the missing element. He developed a method for estimating the pressure distribution, including shocks, at any distance in the

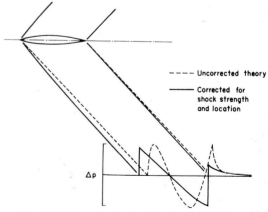

Fig. 3. *Whitham's solution for the flow field of a body of revolution (in this case a simple parabolic body).*

axisymmetrical flow field of slender bodies of revolution. Whitham noted that the basic linearized theory gave a correct approximation for the strength of disturbances, but failed to provide the correct propagation path, an error that accumulated with distance. He introduced a correction to the theory which takes into account local variations in flow field velocity and speed of sound, and provides a good approximation to the true propagation paths and true shock strengths. The Whitham correction applied to the linearized theory flow field of a simple parabolic body of revolution is illustrated in Fig. 3. Whitham's results applied to an arbitrary body of revolution or an equivalent body representing an airplane, give the following formula for the bow pressure rise in the far field where an "N" wave has formed:

$$\Delta p_{max} = \frac{\bar{p} K_r \beta^{0.25}}{(h/l)^{0.75}} \left[K_s \sqrt{\frac{A_{max}}{l^2}} \right] \quad (1)$$

where Δp_{max} is the bow shock pressure rise,
$\bar{p}$ is the ambient pressure, or reference pressure,
K_r is the reflection factor,
β is the cotangent of the Mach angle, $\sqrt{M^2 - 1}$,
h is the airplane altitude,
l is the airplane length,
K_s is the airplane equivalent body shape factor,
A_{max} is the airplane equivalent body maximum area.

The slope of the linear portion of the far field "N" wave may be expressed as

$$\frac{\Delta p}{\Delta X} = 0.583 \frac{K_r \beta}{M^2} \frac{\bar{p}}{h} \quad (2)$$

where ΔX is a horizontal distance, and other terms are as previously defined. Equations (1) and (2) thus permit definition of the complete far-field "N" wave.

It will be noted in equation (1) that the pressure rise is weakly dependent on the Mach number and that only increased altitude h, improved shape factor K_s, increased length l, or decreased equivalent body maximum area can be employed in minimizing the boom. The altitude is the most powerful factor, not only because of the 0·75 power of h in the denominator, but also because of the fact that reference pressure $\bar{p}$, which must represent a mean pressure between the airplane and the ground, decreases rapidly with altitude. However, these two favourable altitude factors are opposed somewhat by the increase of equivalent-body cross-sectional area which, it will be remembered, includes effects of lift.

The term within the bracket in equation (1) is a function of the airplane shape and its flight conditions such as Mach number and lift coefficient. The methods of Hayes and Whitham have now been programed for high-speed digital computers, enabling rapid and accurate evaluation of the shape term. These programs have been exercised for a variety of airplane configurations. The results of these studies for a number of airplane types are summarized in Fig. 4. The shape factors are plotted as a function of a Mach number and lift coefficient factor. In general, the lower portion of

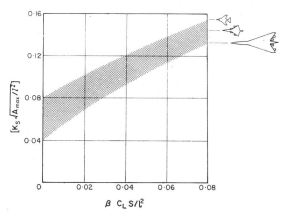

Fig. 4. Representative airplane shape terms applicable to equation (1).

the band is representative of large airplanes such as supersonic transports and bombers which tend to have slender equivalent bodies, and the upper portion of the band is representative of fighters which tend to have short fat equivalent bodies.

A crude estimate of sonic-boom pressure on the ground may be made by extracting a shape term based on the airplane type under consideration from the chart of Fig. 4 and applying it in equation (1) together with the proper values of reflection factor, Mach number, airplane length, altitude, and reference pressure. The average value for a reflection factor is often taken to be 1·9. The reference pressure which accounts for the variation of atmospheric properties from the airplane to the ground has in the past been approximated as the geometric mean pressure. The geometric mean pressure is expressed as $\sqrt{p_a p_g}$, where p_a is the ambient pressure at the airplane altitude and p_g is the atmospheric pressure at ground level.

Estimates made by use of the geometric mean, however, tend to give lower boom pressures than those measured in flight test programs. A better approximation is afforded through use of a computer program devised by Friedman, Kane, and Sigalla which treats the propagation of the shock front through stratified layers of an atmosphere with temperature, pressure, and wind gradients. Program results for a U.S. Standard Atmosphere, 1962, with no wind are depicted in Fig. 5. Reference pressure as determined by the program has been referenced to the geometric mean and this ratio is plotted as a function of altitude and Mach number. Use of a reference pressure obtained by use of this chart normally results in more realistic sonic-boom predictions than does the simple geometric mean approximation.

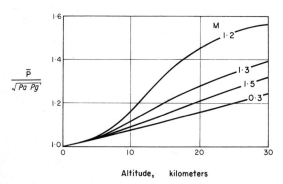

Fig. 5. Ratio of mean atmospheric reference pressure to geometric mean.

Predictions made by employment of the complete method starting with airplane area and lift distributions have generally shown good correlation with flight test results in those cases where a good statistical sampling was available and measured pressures were averaged. The averaging is required because of boom intensity variations introduced by wind, turbulence, and other non-steady, non-uniform atmospheric properties. A sample correlation of theory and flight data is shown in Fig. 6.

The search for methods of suppressing or eliminating the sonic boom has become a popular endeavour. Some of the more imaginative ideas employ radically new and untried airframe and propulsion system design concepts. However, the discussion herein will be restricted to the more conventional approaches, those which have been substantiated in flight or in the wind tunnel.

Suppression of the sonic boom by means of increased altitude was discussed briefly in the examination of the Whitham equation. As can be seen from the curves in Fig. 6, altitude can have a powerful effect in reducing ground pressures. But, as altitude increases, the benefits of additional altitude become less and less. There is, in fact, little benefit to be derived from operation of the airplane at altitudes much above the latitude which gives maximum lift–drag ratio or maximum range. For the same range, an airplane flying at the higher altitudes requires a larger fuel load which compensates for the beneficial effect of the increased distance to the ground. Increased speed, however, raises the altitude for maximum range and the combination of increased speed and altitude can be employed to advantage.

As shown in equation (1), the only other factors which influence the nominal boom are those associated with the airplane design. Clearly, a more slender airplane expressed by the ratio A_{max}/l^2 will tend to have lower sonic booms. The shape factor K_s is also important. A considerable amount of study has been devoted to the search for equivalent body shapes yielding minimum sonic-boom pressures for specified conditions. For the case where a simple far-field "N" wave is known to exist, it has been shown that a lower bound shape has an area that varies as the one half power of the distance from the nose. This blunt shape of course has a high drag, but wind-tunnel experiments support the theoretical contention that in the far field it does indeed have low values of sonic-boom pressure. In Fig. 7 the lower-bound shape is compared with representative supersonic transport

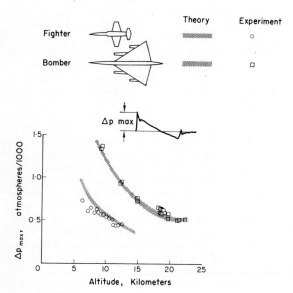

Fig. 6. Correlation of theory with flight test data.

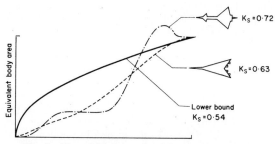

Fig. 7. Far field lower bound shape compared with representative airplane equivalent body shapes.

equivalent body shapes for $M = 3.0$ cruise conditions. A more practical approach to minimization of far-field sonic-boom pressures is represented by the airplane design corresponding to the short dashed line equivalent body area distribution. The other airplane is a rather poor design from a sonic-boom standpoint, not only because of the higher shape factor, but also because of the large maximum equivalent body area.

The discussion of minimization so far has concentrated on reducing the bow shock in the airplane far field, where a well-defined "N" wave has already formed. A practical equivalent body shape for far-field minimization such as that shown in Fig. 7 can under some circumstances have additional benefits. This occurs when the airplane length is great enough and the maximum equivalent area is small enough for a near-field type of signature to extend from the airplane to the ground, where a far-field "N" wave normally would be expected. The near-field signature is particularly sensitive to changes in equivalent body shape and new opportunities for airplane sonic-boom control are presented. It has been found that some supersonic transport designs meet these conditions in the critical transonic acceleration portion of their flight were boom pressures tend to reach a maximum. Whitham's solution for the near-field pressure signature now has also been programed for digital computers. A calculated example of an airplane modification to reduce sonic-boom pressures under near-field conditions is shown in Fig. 8. The left half of the figure represents the original configuration and the right half represents the revised fuselage shape with

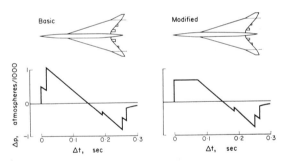

Fig. 8. An example of an airplane modification resulting in an improved near-field signature.

its effect on the signature. Maximum positive pressure is reduced by about 40 per cent. Even further gains can be made if airplanes can be made larger and more slender. In fact, from a purely theoretical standpoint, it is possible that the shock could be eliminated with the formation of a sine wave type of pressure signature. However, this would require an airplane impossibly long and slender by today's standards. Of course, even minimization methods which appear to be of practical use must be considered in relationship to the total airplane design problem, and boom minimization advantages must be weighed against possible penalties in other aspects of the airplane performance. Prospects for boom elimination short of prohibition of supersonic flight appear to be dim and distant.

Bibliography

Proceedings of the Sonic Boom Symposium, sponsored by The Acoustical Society of America, St. Louis, Mo., 3 November, 1965. Available from the Back-Numbers Department, American Institute of Physics, 335 East 45 Street, New York, New York 10017. Price: $ 3.00.

WHITHAM G. B. (1952) The flow pattern of a supersonic projectile, *Commun. Pure Appl. Math.* **5**, No. 3, 301–48.

HAYES W. D. (1947) *Linearized Supersonic Flow*, North American Aviation, Inc., Rept. No. AL-222.

H. W. CARLSON

SPACE POWER SUPPLIES

The Range of Choice

The criteria by which successful space power supplies are judged are generally very different from those applicable to terrestrial generators. In order to minimize the cost of a space mission, the weight (and volume) of the total payload must be minimized. Thus power to weight ratio rather than capital or fuel cost is paramount. High reliability is also crucial, and this tends to put a premium on power generators which have no moving parts.

A further requirement, which in sheer engineering terms may overshadow all others, is the need to provide equipment which can survive the high-g environment of the launch phase, and come successfully into operation in a vacuum in the low or zero-g part of the mission. Temperature extremes may also have to be contended with. The handling of liquid fuels, etc., under these conditions for primary propulsion has provided a fund of knowledge for the design of power systems.

One can infer that, for short missions, the weight of the generator must be kept small, but for prolonged missions the key factor is the amount of fuel to be stored. There are four basic primary "fuels" which can be used. *Solar energy* does not require any fuel to be stored, except perhaps indirectly so that power is available when the space vehicle is in shadow, or when the short term power needs exceed the average supply. Clearly the average energy available is limited to that which can be intercepted by the energy-collecting devices. The other three forms of energy available are *chemical*, *radio-isotope*, and *fission* (and later possibly *fusion*) reactors.

The effectiveness of energy *storage* (as opposed, for the moment, to generation) is shown by Table 1, which gives in round terms the capacity in W hr kg^{-1} for various forms.

The total range is about 14 decades, the advent of nuclear power—which also made the use of radio-isotopes practicable—having added about 6 orders. The obvious tendency is to use methods high in the list, and the only point to be noted in passing is that the table gives no indication of speed or efficiency of conversion. The latter is important, since the total thermal power required varies inversely. This power, whether used effectively or not, has all to be radiated into space. The problem of the heat sink is dealt with briefly by Spring (1969).

Table 1

Form of storage	Example	Energy density W hr kg^{-1}
Fission (fusion)		10^9
Radioisotopes		10^7
Chemical	combustion	10^3
Electrochemical	battery	10^2
Latent heat	steam/water	10
Compressed gas		1
Kinetic	flywheel	1
Mechanical	spring	1
Mechanical	rubber	10^{-1}
Gravitational	pumped-storage	10^{-2}
Magnetic	inductor	10^{-3}
Electrostatic	capacitor	10^{-4}

In principle there is a bewildering array of possible combinations of primary energy source, energy conversion system and storage device if needed. Many of these have been investigated in the past 10 years but the choice has gradually narrowed to about half a dozen developed, or developing arrangements, which are now dealt with serially.

Energy requirements range from 0·05 W with a duration of about a year for the Vanguard satellite of 1960 through about 1 kW for a few days for current manned satellites to say 1000 or more kilowatts for several years for projected manned space stations. At electrical power levels above about 10 kW it is believed that nuclear/direct-conversion systems will be lighter in weight than anything else.

Nuclear Reactor

The nuclear reactor has been considered to power thermoelectric, thermionic and Rankine cycle turbine generators (see also Spring, 1969). The United States is developing a series of SNAP (systems for nuclear auxiliary power) generators, the even-numbered being fuelled by fission reactors, the odd-numbered by radioisotopes (see Craston, 1966).

The thermal neutron reactor being developed for SNAP 2 and SNAP 10 A consists of ceramic coated alloy tubes containing the fuel which is a fully enriched uranium/zirconium hydride, capable of operating at about 750°C, cooled by a 78/22 NaK mixture. The core is about 25 cm diameter, 40 cm long, surrounded by a 5 cm beryllium reflector. SNAP 2 is being designed to use a mercury turbine driving a 3 kW, 120 V a.c. generator, the turbo-alternator (and pump) being designed as a closed, two-bearing unit. The total weight is expected to be about 700 kg. SNAP 10 A utilizes the same reactor with a 500 W, 28·5 V thermoelectric generator, sufficiently shielded from the reactor to give a lifetime of one year. The generator uses SiGe alloy elements, over the temperature range 550–320°C. The design weight, fully shielded, is about 400 kg.

SNAP 8 is a development of SNAP 2/10 A, originally estimated to be ready for flight by the early 1970's, and aimed at attaining an output of 35 kW. SNAP 8 has been beset with many difficulties—mainly in the vaporization and condensation of the Hg, and associated materials and bearing problems. The design weight has increased fourfold from the original value of about 1000 kg. Another development (now cancelled) on a longer time scale was SNAP 50, for distant planet flights. This would have been a lithium-cooled reactor, operating at about 1100°C, with potassium as the turbine working fluid, giving an electrical output of 300–1000 kW. A further concept is SNAP 70 which uses a thermionic converter, with an ultimate goal of about 300 W(e) kg^{-1}. The thermionic converter offers the promise of a higher efficiency than the thermoelectric converter, but with the reliability inherent in a static device.

All these reactors are "thermal". A number of design studies have been initiated for fast-neutron reactors for space use. They pose, of course, formidable problems even for terrestrial use, but offer ultimately the concept of an even more compact core, and the possibility of good fuel burn-up, leading to prolonged life.

A paramount concept with all reactor designs is the ability to withstand the "maximum credible accident", particularly if the reactor may return to earth at the end of its working life full of fission products.

Fission-Fragment Generator

One attractive possibility being examined is the concept of converting directly into electricity the kinetic energy in fission products, thus eliminating the process of degradation into heat, and the need for a thermal-power converter. A further advantage is that, since a heat engine stage is eliminated, the waste-heat radiator can operate at a temperature near that of the reactor core, and hence be smaller in size.

The principle of operation is shown in Fig. 1 and is somewhat analogous to that of the thermionic generator. The fissionable material is thinly coated on the cathode, and the fission fragments emitted under neutron bombardment are collected on the anode. Each fragment carries about 20 positive charges and has an average energy of 80 MeV, so that the open circuit potential is about 4×10^6 V. By suitable matching, any load voltage below this can be obtained. The vacuum insulation must be good enough to ensure that no large neutralizing ionization current is generated. The counter-current of

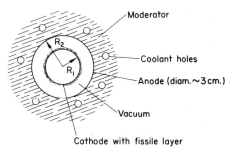

Fig. 1. Schematic arrangement of direct nuclear fission-fragment generator. This figure and Fig. 2 based, by kind permission, on Pederson (1964).

electrons from the fissionable material must also be prevented, possibly by a grid or magnetic field. The reactor core will consist of a number (possibly 1000) of these tubes, embedded in a moderator (possibly BeO_2), with perhaps some additional fissionable material to enable criticality to be obtained within a reasonable size and weight. A core height and diameter of about 1·5 m are indicated.

The electrical efficiency η can be defined as

$$\eta = \frac{Vcn}{tNE}$$

where V is the open-circuit voltage,
c the charge per fragment,
n the number of fragments per unit cathode area/sec,
t the thickness of fissionable layer (in units of fission-fragment range),
N the number of fragments sec^{-1} cm^{-3} of fuel,
E the fission energy per fragment.

The value of n can be calculated for various anode cathode configurations. For the concentric arrangement of Fig. 1, the efficiency is shown in Fig. 2. With very thin layers, and good load matching, efficiencies comparable with those of thermoelectric or thermoionic converters seem possible. The method appears to have potentialities for generating high voltage power, but this is not likely to be required on board the spacecraft in the quantities consistent with the reactor size, and effective (d.c.) voltage conversion will not be easy. The system may, however, come into its own for some of the photon-ejection schemes now being considered for rocket-propulsion for long journeys into space.

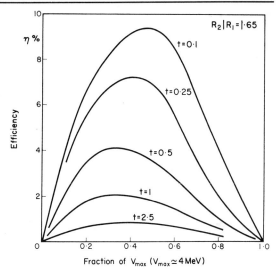

Fig. 2. Theoretical performance of a fission fragment generator. The efficiency is shown as a function of the maximum generated voltage, and the geometrical factors R_1 and R_2 the cathode and anode radii, and t, the thickness of the fissionable layer, in range units.

Radioisotope Generators

Radioisotopes, which provide self-contained high-density energy sources are already in use for many terrestrial applications (such as light-buoys) as well as in the Transit and other satellites. They are rugged and reliable and inherently safer (with proper shielding) than nuclear reactors. The decay rate (half-life) mainly determines the lifetime of the device. Heat is generated in the isotope and its environment by the absorption of α, β and γ radiation. Generally, α-emitters produce more useful heat per disintegration than β-emitters, and the shielding problem is easier, though care is needed to avoid α-neutron reactions. Pu^{238}, Cm^{244}, Cm^{242} and Po^{210} are typical α-emitters, the latter two having thermal outputs of around 1 kW cm^{-3}, and half-lives of around 0·4 year. With current converter efficiencies of about 10 per cent, the fuel cost per W(e) is about £2,000.

Sr^{90}, with its longer half-life (28 years), gives a cost per W hr much less than say Cm^{242}. It is a β-emitter and is very suitable, with heavy shielding, for terrestrial use.

For sizes of about 500 W, for which radioisotopes look most appropriate, the Rankine turbine is not very suitable. At this size its efficiency is low—very low unless an extremely high speed is used, in which case it loses reliability on long missions. SNAP 3b, the first in the odd-numbered series, uses Po^{210} (about 1700 curies initial activity $\sim$0·4 g), with a lead telluride thermoelectric generator, the hot and cold junctions being respectively at 500 °C and

100°C. The electrical output is 3·5 W at an overall efficiency of 5·5 per cent and a weight of 2 kg.

A later generator is the SNAP 11, using Pu^{238} (89 year half-life) to provide about 25 W, used in the Transit satellite with a mission life of about 5 years. Solar cells are considerably damaged in this time by radiation damage from the Van Allen belts. Thermoelectric materials, because of their doping level, are much less affected than solar cell semiconductor material.

SNAP 13 is being designed as a Cm^{242} thermionic converter unit of about 13 W output.

Radioisotopes, like nuclear reactors, require great attention to shielding and containment—especially for manned spacecraft, or radiation-sensitive experiments. However, despite this and the present high cost, they have outstanding advantages of compactness, reliability and continuity—e.g. for experiments during the 14-day lunar night.

Solar Energy

The obvious advantage of solar energy is that no fuel has to be taken on board the spacecraft, provided the average energy that can be collected and converted is adequate for the mission. The solar energy-density in space is $0·13$ W cm^{-2} (at Venus it is twice, and at Mars half, this value). Thus to generate 1 kW at 10 per cent efficiency requires a collector of nearly 10^5 cm^2 (say 3 m square). The Mariner spacecraft has a span of about 7 m. For this reason solar cells are usually mounted on large "paddles" which are arranged to unfold when the spacecraft is above the atmosphere. Nevertheless design studies are being made for 50 kW arrays.

Thermoelectric and thermionic converters have also been considered for conversion of solar energy. With the former it has been found that the maximum temperature differential which can be maintained between the hot and cold junctions is about 120°C, using the best available techniques for energy absorption and rejection, and for thermal insulation. With thermoelectric materials having a ZT value of unity this corresponds to a generator efficiency of only 1–2 per cent, which is an order of magnitude less than that of solar cells. Thus, despite their lower cost, solar thermoelectric generators are not used in space (but they may have terrestrial applications where weight and exposed area are not critical).

Considerable improvements in specific cost can be achieved by using some form of energy-concentrating mirrors. Parabolic cylinders can give a concentration factor of about 10, and paraboloids about 100. The cost per unit area of mirror is low, as accurate figuring is not required. With a thermoelectric generator the gain in performance is almost proportional to the degree of concentration: the cooling losses are increased and the effective ZT is reduced because of the wider temperature range. If cascading has to be used, there is a further loss.

With the photovoltaic solar cell the effects are more complex. The early (1955–9) Si cells had an efficiency of about 10 per cent at 1 sun (1 kW m^{-2}), falling to about 5 per cent at 5 suns. This was because of both internal and surface resistance effects. The use of a surface grid permits the series resistance of a 1×2 cm cell to be reduced tenfold to 0·1 ohm. With no series resistance the voltage developed should rise logarithmically with intensity, leading to about a 5 per cent efficiency improvement at 5 suns. It is now possible to produce cells with no loss of efficiency at these high intensities, and the way would therefore seem clear to use a concentrating mirror. However, the efficiency of a solar cell also depends markedly on its surface temperature (it falls by about a third between 20°C and 100°C for a Si cell). A selective coating on the mirror is of some help, but the cells must also be mounted in good thermal contact with a heat sink, which adds significantly to the mass per unit area. Materials with a high energy-gap more nearly maintain their performance at high temperature and the arrival of the Cds cell (energy gap 2·4 eV) may make the use of concentrators technically worthwhile. The Cds cell may become so much cheaper than the Si cell that the latter will not be economically attractive.

With any solar energy system, the collectors have to face all ways on the spacecraft, or some system of maintaining orientation has to be provided. This is expensive, unless wanted also for some other purpose, and the problem is accentuated with energy concentrators. If it is desired to use solar energy in conjunction with a thermionic converter, for which cathode temperatures in the region of 1500–2000°C are required to attain good efficiencies, orientation must be maintained to better than 0·5°. For this reason, solar/thermionic converters have largely been abandoned for space use, despite their ruggedness and radiation resistance. One model being developed has a design efficiency of 15 per cent, 135 W output and a specific power of 10 W kg^{-1}.

Ferroelectric Generator

A possible way of generating high voltage fairly directly is to use the sudden changes of polarization that recur as a ferroelectric material is cycled about its Curie point (rather analogous to cycling a ferromagnetic material). Barium titanate is available both in single crystal and polycrystalline form, strongly "poled" or oriented to maximize the effects. The material is made the dielectric of a capacitor, which is heated under open circuit conditions and then discharged isothermally. Even with the most effective operating cycle (Spring, 1965) the efficiency is expected to be only about 0·1 per cent).

A method by which a. c. may be generated is shown in Fig. 3 with a multi-unit converter mounted in a spinning satellite. The upper frequency limit is determined by the rate of thermal cycling: with specimens

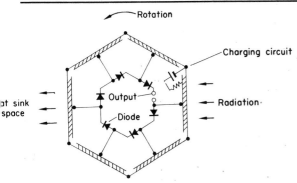

Fig. 3. Generation of low frequency a.c. in a spinning satellite with a ferroelectric converter.

0·1 cm thick it might be about 20 c/s. In space, the equilibrium surface temperature of a black body squarely facing the sun is about 400°K, though the mean temperature of a rotating plate would be about 250°K—well below the Curie point of even the doped titanates, so that some form of optical concentrator would be needed.

Chemical Energy

The primary propulsion energy of all present-generation rockets is supplied by the combustion of solid or liquid fuels and oxidants, and it is perhaps surprising that secondary power needs are not similarly supplied. These latter needs are, however, met far more safely and conveniently by electrochemical conversion, using batteries or fuel cells (Liebhafsky, 1966). All the early spacecraft in fact used batteries or solar cells. Chemical energy systems may be primary or secondary, in which case they depend on re-charging from a primary source such as solar cells.

The lead/acid cell is a convenient standard for comparison, and in its present state of development offers about 20–25 W hr kg^{-1}. The silver–zinc cell offers just over twice this, and although its re-cycling ability is limited, for space use this is not a serious handicap. A number of high-energy, high-temperature battery systems (e.g. sodium/sulphur, lithium/chlorine) are being developed for terrestrial use. These may attain 200–300 W hr kg^{-1}, but are not likely to be suitable for space use.

The Gemini was the first (and until at least 1967 the only) satellite to be equipped with a fuel cell. One of the problems of any H_2/O_2 cell is to remove water, the reaction product. This is normally done by re-circulating the fuel gases at a rate such that at the operating temperature the water formed is just removed as vapour. A neater way of achieving the same object is to use an ion-exchange membrane. In this way the use of pumps may be avoided, and an overall compact arrangement results. In the Gemini cell the membrane consisted of sulphonated cross-linked polythene, which formed the electrolyte. Such sheets are strong and flexible with an electrical resistance similar to that of a weak aqueous electrolyte. As the solid electrolyte acts as a gas separator, the electrodes can be of a simple open-mesh form.

Convenient though the ion exchange cell is, its performance is rather poor, current densities in the range of 50 mA cm^{-2} being obtainable. In contrast the original Bacon cell, with its unique double layer porous Ni electrodes gave densities of around 1000 mA cm^{-2} at the same operating voltage, as shown in Fig. 4. The Bacon cell operated at the relatively high temperature (for an aqueous electrolyte) of 200 °C and was pressurized to about 40 atm and the electrolyte was 40–50 per cent KOH. Polarization losses were thus kept small.

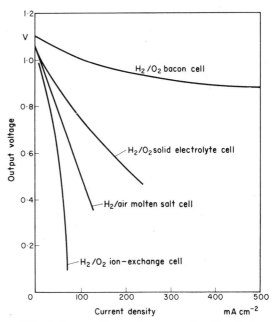

Fig. 4. Comparative performance of various fuel cell systems.

A modified version of the Bacon cell has been developed for the Apollo mooncraft. Second-generation space H_2/O_2 fuel cells are tending to be intermediate in temperature and performance to the Gemini and Apollo series, but as simple in design as possible. One incidental advantage of these cells is that they provide pure, potable water. For obvious reasons, attempts are being made to develop for terrestrial uses fuel cells which will operate on cheap impure fuels (such as kerosone). There is little interest for space use as such fuels tend to require high temperature cells, using molten salt (c. 600°C) or solid oxide (c. 900°C) electrodes, together with complex catalysts. Their reliability and life to date even on earth have not been particularly encouraging, despite the effort put in.

Conclusion

A decade of effort on space power supplies, which has cost well over $100 million in the U.S. alone, has taught many useful, though hard, lessons. In the early days, quite properly almost all combinations of primary energy source and conversion process, direct and indirect, were investigated. Despite persistent efforts, the efficient Rankine and gas-turbine cycles, almost exclusively used for large scale terrestrial power generation, have not been found reliable in zero-pressure zero-g environments. Effort is now tending to concentrate on simple, more reliable, if less efficient schemes—the solar cell, the hydrox fuel cell, and for larger outputs, isotope or reactor-powered thermionic or thermoelectric generators.

See also: Solar cell, Thermionic generation of electricity, Thermoelectric generation of electricity.

Bibliography

CRASTON J. L. (1966) SNAP (Systems for nuclear auxiliary power), in *Encyclopaedic Dictionary of Physics* (J. Thewlis Ed.), Suppl. vol. 1. 294, Oxford: Pergamon Press.

LIEBHASKY H. A. (1966) Fuel cells, in *Encyclopaedic Dictionary of Physics* (J. Thewlis Ed.), Suppl. vol. 1, 110, Oxford: Pergamon Press.

PEDERSON E. S. (1964) *Nuclear Energy in Space*, Prentice-Hall.

SPRING K. H. (1965) (Ed.) *Direct Generation of Electricity*, New York: Academic Press.

SPRING K. H. (1969) Direct conversion of heat to electricity, in *Encyclopaedic Dictionary of Physics* (J. Thewlis Ed.), Suppl. vol. 3, 45.

SWIFT-HOOK D. T. (1962) Magnetohydrodynamic generation, *Encyclopaedic Dictionary of Physics* (J. Thewlis Ed.), Suppl. vol. 1, 162.

K. H. SPRING

SPEAKING MACHINES. A speaking machine is a device that generates speech under the control of a set of instructions or signals which indicate in some manner the sequence of speech units to be actualized. In some kinds of speaking machines these instructions consist of signals that specify how certain parameters within the device are to vary with time. These control signals are analogous to the movements of the tongue, jaw, lips, and other structures in human speech, and vary at roughly the same rates as these movements. For other types of speaking machines, the instructions provide a specification of the sequence of discrete phonetic units or larger units such as syllables or words that are to be generated by the machine. A speaking machine is a *synthesizer* of speech, and must be distinguished from a *reproducing* device, such as a tape recorder, which is capable simply of creating a replica of an utterance that has been generated previously by a human speaker.

The Human Speech-generating Process

Speech sounds are generated by modulation of the flow of air from the lungs. This modulation is accomplished by manipulating the positions of the vocal cords in the larynx and the structures that form the airways above the larynx. The type of modulation that plays the most important role in speech is the quasiperiodic interruption or modulation of the air flow by the vibrating vocal cords in the larynx. The puffs of air that pass through the glottis (the space between the vocal cords) during the interval in the vibrating cycle when the cords are open act as a source of excitation for the vocal tract, i.e. for the acoustic cavities above the larynx. Adjustment of the muscles in the larynx and the pressure in the lungs causes changes in the frequency of vibration of the vocal cords. For an adult male speaker, this frequency of vibration is usually in the range 80–200 cps, while for a female speaker it is about an octave higher.

The cavities above the larynx form acoustic resonators, and the resonances modify the characteristic of the source at the glottis, introducing peaks and valleys in the spectrum of the sound that is radiated from the lips. The spectral peaks are called *formants*, and the frequency positions of the formants can be changed by adjusting the shape of the vocal tract. Different vowels and also some consonants are generated by causing the vocal cords to vibrate and by manipulating the tongue, jaw, lips, and other portions of the anatomy in order to change the spectral characteristics of the sound output.

The first three formant frequencies for an adult male speaker usually lie in the frequency range 250–800, 700–2200, and 1600–3000 cps, respectively. For example, for the vowel in the word *boot*, the lowest three formant frequencies for such a speaker are about 300, 800, and 2400 cps, while the stressed vowel in the word *father* is characterized by formant frequencies of about 750, 1250, and 2500 cps. For a female speaker, these formant frequencies may be 15–20 per cent higher.

Another type of modulation of the air flow that is of importance for some classes of speech sounds is the quasi-random velocity fluctuations that occur when there is turbulence in the air flow in the vicinity of a constriction or obstacle in the vocal tract. This turbulence gives rise to a sound source that is like random noise, and this source acts as acoustic excitation for the cavities of the vocal tract. The sound radiated from the lips has different acoustic properties depending upon the place in the vocal tract at which the constriction is located. Examples of sounds that are generated in this manner are the initial consonants in the words *fill*, *sell*, and *shell*.

Some consonants are produced by using both vocal-cord vibrations and turbulence noise. Examples are the initial consonants in the words *veal* and *zeal*. Still another class of sounds is generated by forming a complete closure at some point along the vocal tract, and then releasing this closure rapidly to produce a transient burst of sound energy, as in the initial consonants in *boat* and *tall*.

When speech sounds are strung together to form a word or a phrase, the acoustic characteristics of the individual sounds are usually greatly influenced by the phonetic context in which they occur. Thus, for example, the actualization of the sound *t* is markedly different in the three words *top*, *butter*, and *cat*. Furthermore, there is a kind of blending of one sound into its neighbour, with the result that there is often no clear boundary in the signal separating one sound from another. Thus a machine that attempts to generate speech simply by concatenating a sequence of standardized sounds cannot provide a natural or intelligible output.

Almost all speaking machines simulate to a greater or lesser degree the mechanism of sound generation in the human vocal tract. These machines include components for generating two kinds of sound sources—a quasiperiodic source, usually controllable in frequency, to simulate the acoustic source at the vocal cords, and a noise source to simulate turbulence noise. The machines also have the capability of shaping the spectral characteristics of the sources. Speaking machines of this type differ primarily in the manner in which the sources are produced and in the procedures that are used for realizing the spectral shaping of the sources. Some speaking machines do not attempt to imitate the operation of the human vocal tract but rather produce a speech output by stringing together a sequence of prerecorded segments drawn from an inventory or library of such segments that is stored within the machine.

Synthesis from Prerecorded Segments

As has been noted, it is not possible to generate intelligible speech by stringing together a sequence of sounds that have been recorded previously when each of these sounds is intended to represent one phonetic segment (e.g. by stringing together three sounds *b*, *i*, and *t* to form the word *bit*). Attempts have been made, however, to use prerecorded units of greater length. One such scheme, proposed by Peterson, Wang and Sivertsen (1953), utilized segments whose length was approximately half a syllable, the syllable being divided in the middle of the vowel and in the middle of the initial and final consonant. A word like *rabbit*, for example, would be generated by concatenating four segments: *ra*, *ab*, *bi*, and *it*.

It has been estimated that the number of these half-syllable elements that must be available in storage in a machine based on this principle is of the order of 8000. One of the main difficulties with such a machine is in providing smooth transitions between the elements. This difficulty imposes a major limitation on this synthesis procedure.

Of more practical interest is a machine in which the prerecorded items are words or short phrases. Highly natural sentences cannot be produced by playing back sequences of prerecorded words since the resulting speech will lack the normal pattern of intonation and of rhythm that characterizes a sentence. Nevertheless, the speech will be intelligible if the sequence of words is not too rapid and if proper care is taken to record the words in the stored library with an appropriate neutral intonation pattern. Such a scheme would require a prohibitively large storage capacity if it were to be capable of producing an arbitrary utterance. In situations requiring a rather limited vocabulary, however, such as the generation of stock market reports or weather information, speech machines utilizing a library of stored words are being employed successfully.

Early Speaking Machines

Most early attempts to synthesize speech were restricted to demonstrations of the production of vowel sounds by means of acoustic resonators excited by vibrating reeds. Several such vowel-generating devices were described by various investigators in the late 18th and 19th centuries. The first reported attempt to produce a machine that could be controlled in a dynamic fashion to produce short utterances consisting of words and phrases was described by von Kempelen in 1791. Von Kempelen's machine, which was controlled by hand, consisted of a bellows for an air supply, a reed source simulating the vocal cords, an adjustable resonator for vowels and vowel-like sounds, and other ancillary valves and noise-making devices. Models based on von Kempelen's description were constructed by Wheatstone, and more sophisticated mechanical vocal tracts that bore at least a partial resemblance to the human vocal tract were described by Alexander Graham Bell in the late 19th century and by Riesz in 1937. Some of these early attempts at speech synthesis are reviewed by Dudley and Tarnoczy (1950), and by Flanagan (1965).

The acoustic resonances of the vocal tract can be simulated by appropriately excited electrical resonant circuits, and this principle was first utilized as early as 1922 to generate various isolated vowels. The first machine to produce continuous speech by means of controllable electrical circuits was the Voder (VOice DEmonstratoR), which was described by Dudley, Riesz, and Watkins in 1939. In the years since this original electrical synthesizer, a large number of speaking machines of various kinds have been constructed.

Source-filter Types of Synthesizers

The voder is an example of a class of electrical speech synthesizers that have the general form shown in Fig. 1. The buzz source simulates the glottal excitation, and its amplitude and frequency are usually controllable, as indicated by the arrows labelled A_b

and F_0, respectively. The noise source generates energy over a broad range of frequencies, and its amplitude A_n can usually be controlled. In different types of synthesizers various procedures are used to shape the spectra of the sources.

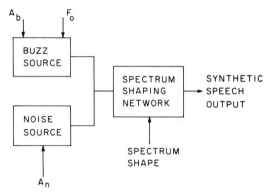

Fig. 1. Block diagram of a speech synthesizer.

In the voder, spectrum shaping was achieved by adjusting the gains of ten band-pass filters covering the important frequency range of speech. To synthesize continuous speech in the original voder, these gains were manipulated by an operator who sat at a keyboard. The synthesizer contained a switch that could be controlled (by the wrist) to select either the buzz source or the noise source, but not both, and the frequency of the buzz source could be controlled from a foot pedal. The voder was demonstrated successfully at the New York World's Fair in 1939.

As was the case with several other electrical speaking machines, the voder was an outgrowth of research on bandwidth compression of speech. A speech synthesizer of the general type shown in Fig. 1 can produce speech under the control of a small set of slowly varying signals, and if these signals can be extracted from natural speech they can be transmitted over a reduced-bandwidth communication link to control a synthesizer located at the receiving end of the link. An analysis-synthesis system of this type is called a *vocoder*. The total bandwidth of the control parameters in the case of a vocoder is of the order of 300 cps, and the channel capacity required to transmit the signals is approximately 2000 bits/sec.

Several synthesizers based on the voder principle have been developed since the original machine was conceived. One of these synthesizers, called a Pattern Playback, controls the amplitudes of the harmonics of a buzz source from patterns painted on a belt that moves under an array of light-sensitive devices. These patterns provide graphical representation of the time variation of the short-time spectrum of an utterance to be synthesized. Different frequencies are displayed across the width of the belt, time is displayed along the length of the belt, and paint is applied to regions in time and frequency in which sound energy is present. This machine has been used extensively in research on the nature of the cues that a listener employs to perceive the sounds of speech. Other more recent versions of the voder principle utilize digital computers to manipulate and to generate the signals that control the synthesizer.

The voder and devices like it accomplish the spectrum shaping of the sources by controlling the levels in each of a set of contiguous bandpass filters. It can be shown, however, that the kind of filtering that the vocal tract imposes on the sources, at least for the glottal source, can be described adequately in terms of the frequencies of the lowest few resonances (usually three) of the vocal tract. Thus if one specifies simply the frequencies of these formants as a function of time, then the spectrum shaping imposed on the glottal source is known at all frequencies (in the range of interest for speech). For sounds generated with a noise source, the spectrum shaping can likewise be approximated by a small number (two or three) of parameters. In the synthesizer scheme shown in Fig. 1, therefore, it is possible to realize the spectrum shaping network by means of a set of about five simple electrical resonant circuits (three for voiced sounds, two for voiceless sounds) whose resonant frequencies are adjustable by means of control voltages. A synthesizer based on these principles is called a formant synthesizer. In order to arrive at a specification of the control signals necessary to synthesize an utterance with such a machine it is necessary to do a careful acoustic analysis of natural utterances to determine how the formant frequencies and source characteristics vary with time.

A number of formant synthesizers have been constructed and demonstrated. One of the first was a machine called PAT (Parametric Artificial Talker), which was designed by W. Lawrence and was demonstrated first in 1953. The high degree of naturalness obtainable with a formant synthesizer was demonstrated with a Swedish machine, which was able to produce appropriately selected samples of synthetic speech (when the parameters were carefully adjusted) that were almost indistinguishable from the original natural speech from which the control parameters were derived. More recently several formant synthesizers have been simulated either completely or in part on a digital computer, and have successfully produced speech of high quality. One difficulty with such machines is that for some consonants it is difficult to produce a highly natural and intelligible output.

Vocal-tract Analogue Synthesizers

While a voder or a formant synthesizer simulates certain aspects of the speech-generating process, there is only an indirect relation between the control parameters of the synthesizer and the movements of the anatomical structures in the vocal tract. A more direct analogue to the human speech process has been achieved by means of a series of electric circuits that simulate the acoustic behaviour of the vocal tract at each point along its length. This simulation is realized by imagining the vocal tract to be a tube of varying cross-sectional area extending from the glottis to the lips, and by approximating this tube in a stepwise fashion by a series of short tubes each of which is fixed in cross-sectional area. Each of these short tubes is simulated by a simple electric circuit (or by a simple difference equation that can be represented by a digital computer) and the effective area of each tube is controlled by adjusting a parameter of the circuit (or of the difference equation). Such an analogue device is excited electrically by a buzz source at the "glottal" end or by a noise source inserted at some point along the tube to represent the noise arising from turbulent air flow. The control signals for a vocal-tract analogue synthesizer specify the frequency and amplitude of the noise source and the cross-sectional area of the vocal tract at a number of points along its length.

Several vocal-tract analogue synthesizers have been constructed or simulated on a digital computer, and have generated a speech output that is of reasonable quality (Rosen, 1958; Kelley and Lochbaum, 1962). Specification of the control signals for such a synthesizer is a difficult problem, however, since it requires a knowledge of the dimensions of the vocal tract and how these dimensions vary with time during speech. Data of this type are not easy to acquire, and consequently vocal-tract analogue synthesizers have not been utilized a great deal in experimental studies of speech production, and have certainly not reached the point where they have practical application.

Synthesis by Rule: Control of Speech Synthesizers from a Sequence of Discrete Linguistic Units

The operation of source-filter types of synthesizer s and of vocal-tract analogue synthesizers requires that appropriate time varying parameters be available to control the elements within the machines. In order to demonstrate the adequacy of these speaking machines, it is usual to derive control parameters by analysing the acoustic or articulatory characteristics of segments of natural speech. A more practical goal, however, is to devise a speaking machine that accepts as instructions a sequence of discrete linguistic units such as phonetic symbols, or even an orthographic representation of the speech, and generates an intelligible and natural speech output. Students of language recognize that a sequence of such units underlies any meaningful utterance, and, in fact, the orthography is simply a representation of these units (although in English the relation between the orthographic and phonetic representations is not always direct and obvious). A speaking machine that accepts a sequence of discrete phonetic units as instructions must include a set of rules to effect a transformation from these units to the quasi-continuous control parameters that provide direct control of the elements within the synthesizer. Synthesis of speech in this manner is called speech synthesis by rule.

Several schemes for speech synthesis by rule have been developed (Holmes, Mattingly and Shearme, 1965; Rabiner, 1968). The general approach to the problem is to specify, for each phonetic segment, a target or idealized set of parameters and some timing information. Movement from one segment to the next is accomplished by manoeuvring these parameters from one set of values to another along appropriate trajectories. When several phonetic units occur in rapid sequence, the parameters may not reach the target values for a given segment before they begin motion toward a set of values appropriate for the next segment. While synthesis schemes of this type can produce speech that is reasonably intelligible, the output is rather unnatural. A major difficulty is that present knowledge of the factors governing rhythm, stress, and intonation in natural speech are not yet understood, and rules for implementing these aspects are thus rather rudimentary.

Some of the difficulties inherent in synthesis-by-rule schemes based on phonetic segments have been circumvented by using as the basic units words or syllables rather than phonetic segments. The synthesizer in this case includes a large library which specifies the waveforms of the control parameters that are required to synthesize each word or syllable. The memory required for such a machine is, of course, much larger than that needed in a synthesis-by-rule procedure based on phonetic segments, but is much less than a synthesizer in which the library consists of prerecorded versions of the speech waveform itself. Furthermore, when the units are stored in the form of control parameters, there is an opportunity for providing smooth and continuous joining of the parameters from one word or syllable to the next one, to obtain smooth and natural transitions from one item to the next.

Machines for synthesis of speech based on stored control parameters for words, and utilizing a voder type of synthesizer are now used in applications that require a limited vocabulary of 1000 words or less (Buron, 1968). Procedures which use syllable-like units and a formant-type synthesizer have also been shown to yield intelligible speech (Estes, Kerby, Maxey and Walker, 1964). Considering the present state of knowledge of the dynamics of the human speech-generating system, it is probable that a synthesis scheme based on stored control signals for word or syllable-length units represents the best compromise between complexity and storage capacity on the one hand and intelligibility and quality of speech output on the other.

Bibliography

Buron R. H. (1968) Generation of a 1000-word vocabulary for a pulse-excited vocoder operating as an audio response unit, *IEEE Transactions on Audio and Electroacoustics,* **AU 16,** 21–25.

Dudley H., Riesz R. R. and Watkins S. A. (1939) A synthetic speaker, *J. Franklin Inst.* **227,** 739–64.

DUDLEY H. and T. H. TARNOCZY (1950) The speaking machine of Wolfgang von Kempelen, *Journal of the Acoustical Society of America* **22**, 151–66.

ESTES S. E., KERBY H. R., MAXEY H. D. and WALKER R. M. (1964) Speech synthesis from stored data, *I.B.M. Journal of Research and Development* **8**, 2–12.

FLANAGAN J. L. (1965) *Speech Analysis, Synthesis and Perception*, Berlin: Springer-Verlag.

HOLMES J., MATTINGLY I. and SHEARME J. (1964) Speech synthesis by rule, *Language and Speech* **7**, 127–43.

KELLY J. L. jr. and LOCHBAUM C. (1962) Speech synthesis, *Proceedings of the Stockholm Speech Communications Seminar*, Royal Institute of Technology, Stockholm, Sweden, September.

KEMPELEN W. v. (1791) *Le Mechanisme de la Parole, suivi de la Description d'une Machine Parlante*, Vienna: J. V. Degen.

LAWRENCE W. (1953) The synthesis of speech from signals which have a low information rate, in *Communication Theory* (W. Jackson Ed.), 460–9, London: Butterworths.

PETERSEN G. E., WANG W. S.-Y. and SIVERTSEN E. (1953) Segmentation techniques in speech synthesis, *Journal of the Acoustical Society of America* **30**, 739–42.

RABINER L. (1968) Speech synthesis by rule: an acoustic domain approach, *Bell System Technical Journal* **47**, 17–37.

ROSEN G. (1958) Dynamic analog speech synthesizer, *Journal of the Acoustical Society of America* **30**, 201–9.

K. N. STEVENS

SPLIT-BEAM MICROWAVE BEACON. With the advent of radio and radar many new types of beacon navigational aids to aircraft and ships have become available.

Sometimes aboard the ship or aircraft, is placed a receiver to receive signals from a ground transmitter which can be coded to indicate the angle of bearing or a divergence from a particular bearing from the transmitter.

In one such system the ground transmitter has its radiated beam, split into two slightly divergent beams overlapping at a particular bearing angle. The receiver on the ship or aircraft can distinguish this bearing by comparison of the amplitude or the modulation of the signals received from the split beams. The advantage of using microwaves is that the beams can be made very sharp to provide greater accuracy at their crossover.

The navigation season has steadily been extended on waterways such as the St. Lawrence Seaway, until it extends well into the winter season. At the start of winter the channel buoys are removed to prevent their being damaged or carried away by ice, and are replaced by wooden spar buoys which do not give a usable radar echo. Floating ice gives a radar return which can conceal the true shoreline, making the radar of limited value for navigation. Under these conditions, fog, rain, or snow will obscure the range lights which are the primary aid to navigation, and shipping comes to a complete halt for hours at a time.

The split-beam microwave beacon can provide accurate guidance of ships in narrow channels, especially during bad weather conditions.

The system includes a fixed shore station located in line with a straight section of channel and a portable battery-operated receiver to be used on board the vessel being guided.

The shore station comprises a low-power x-band pulsed magnetron oscillator operated at 3·2 cm wavelength, with a rotating mechanical chopper to switch the transmitter output between two horn antennae with angular separation. Photoelectric code slots on periphery of the chopper disk alternately gate two low-frequency oscillators which trigger the modulator, producing distinctive modulation patterns on the two beams.

The transmitter is accurately positioned to ensure that the signal strengths from the two beams are equal on the true course line. Figure 1 illustrates the transmitter siting.

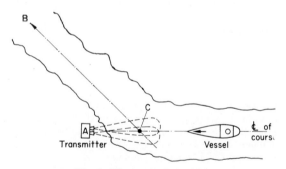

Fig. 1. The transmitter siting.

The small battery-operated receiver is designed to be hand-carried on board the vessel, and uses a small horn receiving antenna with a crystal detector. The antenna can be clamped to the side of the wheel house or other location with an unobstructed view. The incoming trains of pulses are amplified and switched to the appropriate channel depending on the interval between pulses. The amplitudes of the pulses are compared and the result is presented as a visual zero-centre meter indication.

To follow a course consisting of a series of straight-line segments, multiple transmitters are used. When the vessel approaches point C in Fig. 1 the receiver is switched to receive a new pair of repetition rates being used by the transmitter at point B, and the vessel can swing on to the new course in a smooth curve.

A vessel may be held within 0·1° of a true course line at ranges up to 10–15 miles with the standard low-power magnetron. Longer ranges require either a higher-power transmitter or a more directive antenna assembly.

Other types of systems may require range information. The ship or aircraft then also includes a transmitter of pulses to the beacon, which is now designed to receive and retransmit these pulses through its split beams. The ship or aircraft on receiving these pulses can, as before, obtain bearing information, but can also be measuring time delay as a measure of range.

L. G. Cox

SPUTTERING, THEORIES OF. Sputtering is the ejection of atoms from the surface of solids under ion bombardment and was first observed as long ago as the late eighteen hundreds as the result of ion bombardment of cathodes in electrical discharge devices. Originally it was thought that the dissipation of energy from the incident ion was manifest in the formation of a local hot-spot—called a thermal spike—and that sputtering was entirely due to the thermal evaporation of atoms from the surface of such heated regions. As sputtered atoms are generally neutral it was originally impossible to determine experimentally their energy spectra and so check whether or not the ejected atoms had a velocity distribution that was, in fact, appropriate to an evaporation process, i.e. Maxwellian. However, well before such an experiment was performed G. K. Wehner (1955) studied the sputtering of single-crystal targets, and in particular the spatial distribution of the ejected atoms. He found that atoms were not ejected randomly as would be expected from an evaporation process, but were ejected preferentially in certain well defined crystallographic directions (see, for example, Fig. 1). Such a phenomenon was completely inconsistent with any ejection mechanism relying solely on evaporation and the whole question of the mechanism of sputtering was reopened.

At about the same time R. H. Silsbee (1957) first pointed out that the transference of energy and momentum between atoms in a crystalline solid might be strongly influenced by the well-ordered nature of its lattice. He recognized that an ordered atomic array would impose a directional correlation between sucessive collisions and that energy and momentum would be focused into those directions consisting of close-packed rows of atoms. Such a process would give rise to correlated collision sequences—called focused collision sequences—which propagate along certain crystal directions and on the intersection with a free surface could give rise to a preferential ejection in the same directions. By now, it was generally argued that sputtering resulted simply from atomic collisions

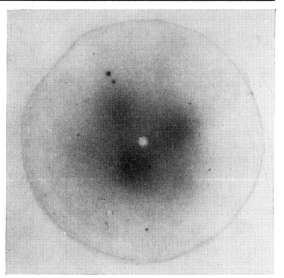

Fig. 1. The deposit pattern produced on a glass plate during the sputtering of a copper single crystal with energetic ions. Preferential ejection from the $\langle 110 \rangle$ close-packed directions of the f.c.c. structure is readily apparent.

which spread through the surface layers of the crystal; but due to the regular atomic arrangement, focused collision sequences gave rise to a strong preferential ejection in certain directions.

As previously mentioned, a measurement of the energies of sputtered atoms should, in principle, provide information on the mechanism of sputtering. Firstly the general form of the energy spectra should immediately confirm that the dominant mechanism responsible for sputtering is one of direct atomic collisions rather than one of evaporation from local hot-spots; and secondly detailed analysis of the ejection from single crystals should provide some insight as to the importance of focusing. However, as sputtered atoms are generally electronically neutral, conventional methods of energy analysis using electrostatic and electromagnetic fields are of no use and sophisticated time of flight techniques using radioactive targets must be used, e.g. Thompson (1968). From theoretical grounds it is to be expected that the flux of recoils having energy in dE at E within the solid (ϕ_i) is proportional to $1/E^2$. It is then simple to show that for ejection normal to the surface the flux of atoms outside the solid (ϕ_0) is given by

$$\phi_0 = \phi_i/(1 + E_b/E), \qquad (1)$$

where E_b is the surface binding energy $\sim$5 eV. Then for $E \gg E_b$, $\phi_0 = \phi_i$, in other words the external flux is identical to the internal flux at high energy. Figure 2 shows the energy spectrum of atoms sputtered from a polycrystalline gold target under 45 keV A$^+$ and Xe$^+$ ion bombardment. It is readily apparent that from energies approaching 10^4 eV down to about 10 eV the probability of ejection $p(E)$, in dE at E follows the $1/E^2$ relationships rather well. Further, below this energy, when the ejection energy is com-

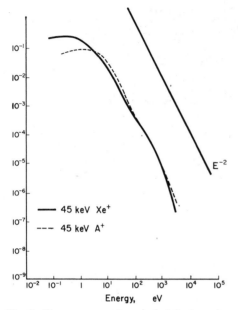

Fig. 2. The energy spectra of ejected atoms from a polycrystalline gold target, showing the E^{-2} dependence.

parable or less than E_b, the spectrum turns over in agreement with that expected from (1). The average energy of sputtered atoms can simply be obtained from the spectrum by integration, i.e.

$$\bar{E} = \int_0^{E_{\max}} E\phi_0 \, dE / \int_0^{E_{\max}} \phi_0 \, dE \qquad (2)$$

then for $E_{\max} \gg E_b$ this leads to

$$\bar{E} = \frac{E_b}{2} \ln (E_{\max}/4E_b) \qquad (3)$$

This expression shows that the average energy is strongly dependent on the binding energy and that as the maximum recoil energy is increased the average energy increases rather slowly. Both features have been substantiated by measurement with low and high energy bombarding ions. We therefore conclude that the main features of sputtering can be explained in terms of the atomic collision cascade intersecting the surface and do not occur as a result of thermal spikes.

The observation that preferential ejection occurs in virtually all materials, in those directions consistent with focusing was originally taken as unequivocal evidence for the existence of focused collision sequences. However, some notable exceptions such as the [20$\bar{2}$3] direction in zinc and the [021] direction in α-uranium, together with the observations at very low energies (<500 eV) where the penetration depth of the incoming particles was only of the order of one lattice spacing, made it clear that purely surface collisions might be quite sufficient to produce preferential ejection. For instance Lehmann and Sigmund (1966) pointed out that because the maximum energy transfer to a surface atom occurred through a near head-on collision by a nearest neighbour atom beneath the surface, preferential ejection was possible even without focusing. Further, a cursary glance at ejection patterns provided by the sputtering of single crystals was therefore unlikely to provide sufficient information to be dogmatic about the importance of focused collision sequences in sputtering.

In an attempt to clarify the relative importance of focusing collisions and simple head-on collisions at the surface, von Jan and Nelson (1968) have developed a model in which the angular distribution of atoms sputtered in close-packed directions is calculated as the cumulative effect of random collisions and collision sequences of different range. Here random ejection means that the surface atom receives its energy by collisions that are random in direction, whereas sputtering by a collision sequence implies that the energy is received in a particular direction as a consequence of the regular nature of the crystal lattice. The main conclusions are that focused collision sequences are important in producing preferential ejection, although in many cases the contribution of sequences with greater than two or three collisions can be very small. Preferential ejection can also occur for just one collision between two near-neighbour atoms. However, this can only occur as the result of a single focusing collision as simple head-on collisions lead to a rather flat distribution. An interesting experimental result which gives direct information on the importance of long range effects has been obtained by studying the preferential ejection from α-uranium. In this structure it is possible to compare directly those directions which constitute ordinary close-packed directions such as the [100], with those which contain only two atoms but with the same interatomic spacing such as the [021]. In the latter case preferential ejection can only occur as a result of single collisions. If the single collision mechanism were dominant even for the case of the [100] direction then these spots would have an intensity ratio 2 : 1 since, due to the nature of the crystal structure one expects twice as many [100] ejections as [021]. However, experiments with 80 keV Xe$^+$ ions specifically designed to study this point have shown that the [100] spots are significantly narrower than the [021] and contain at least four times their intensity. Clearly this difference must be due to collision sequences involving more than two atoms along the [100] direction, and a comparison between the observation and theory suggests that sequences of up to five or more collisions are involved in [100] ejection.

Due to the anisotropic ejection from single crystals, some deviations in the detailed nature of the time of flight spectra of atoms sputtered in different directions from single crystals are to be expected. However, the theoretical treatment suggests that the *total* average energy of atoms sputtered from different crystal faces is rather constant. Figure 3 shows two time of flight spectra for atoms sputtered in different directions from a gold single crystal. Only in one case can ejection occur as the result of collision sequences intersecting the surface, and this is clearly manifest in the rather different nature of the spectra in as much as the curve in Fig. 3(b) can be interpreted as resulting from the sum of both a random component and a rather larger component due to focused collision sequences.

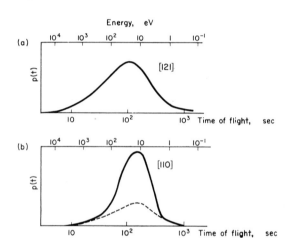

Fig. 3. (a) *Time of flight spectra from the [121] direction of a gold crystal during 45 keV A^+ bombardment.* (b) *Time of flight spectra from the [110] direction of a gold crystal during 45 keV A^+ bombardment, the dashed line indicates the contribution from the two components.*

Bibliography

NELSON R. S. (1968) *The Observation of Atomic Collisions in Crystalline Solids*, North Holland.
SILSBEE R. H. (1957) *J. App. Phys.* **28**, 1246.
THOMPSON M. W. (1967) Focused collision sequence, in *Encyclopaedic Dictionary of Physics* (J. Thewlis Ed.), Suppl. vol. **2**, 94, Oxford: Pergamon Press.
THOMPSON M. W. (1968) *Phil. Mag.* **17**.
VON JAN R. and NELSON R. S. (1968) *Phil. Mag.* **17**, 1017.
WEHNER G. K. (1955) *Phys. Rev.* **102**, 690.

R. S. NELSON

STANDARD EARTH, GEODETIC PARAMETERS FOR. A set of standard geodetic parameters provides a unified framework for many global topics and activities. Such a set should adopt a well defined geocentric coordinate system. The set should include parameters relating survey coordinates on the surface of the Earth to the adopted geocentric system. It should also specify, in the same system, a detailed representation of the gravitational field of the Earth, expressed usually as a series of spherical harmonic functions. A description of the geoid, or the mean sea level surface, follows from this representation (see *Earth, figure of*).

Recently artificial satellites have provided a powerful tool to determine geodetic parameters for a standard earth. The potential for the gravitational field is derived by techniques of celestial mechanics from the observed motions of satellites caused by the irregular mass distribution within the Earth. The same calculations yield accurate geocentric coordinates for the optical or radio instruments used to observe the satellites. Simultaneous observations of a satellite from two or more instruments give independent information of the relative positions of the instruments, even over intercontinental distances. Surface survey coordinates of the same instruments provide the link between the survey datums and the geocentric system.

The Smithsonian Institution published in 1966 a three-volume work, *Geodetic Parameters for a 1966 Smithsonian Institution Standard Earth*. The Astrophysical Observatory of the Institution with headquarters in Cambridge, Massachusetts, produced this work, and plans to prepare revisions every few years. The last issue was dated 1969.

The 1966 Smithsonian Standard Earth adopts a cartesian geocentric coordinate system with its origin at the centre of mass of the Earth. This is possible because satellite orbits provide a means to locate the centre of mass. The coordinate system has its z-axis through the mean pole of 1900–5, as defined by the International Polar Motion Service, and its x-axis implicitly determined by the defined longitude of the United States Naval Observatory to which tabulations of Universal Time correspond. This coordinate system, fixed in the Earth, is related to celestial directions, that is the stars, through the expressions and data given by the international services monitoring the motion of the Earth. The distance scale comes from the adopted value of GM, the multiplicative constant in the geopotential. In future work, accurate ranging to satellites by laser instruments will provide an independent means of finding a distance scale.

The direct geometrical results from satellite observations are the coordinates of the observing instruments in the adopted geocentric system. For the 1966 Standard Earth, the coordinates of the Baker-Nunn satellite tracking cameras were determined globally to an accuracy of about 15 m. This is an order of magnitude improvement over pre-satellite information. The United States Coast and Geodetic Survey is carrying out an extensive program of satellite observations through 1970 designed to locate some forty sites, uniformly distributed around the Earth, to an accuracy better than 10 m.

From accurate instrument coordinates, the relationships of ground survey datums to the geocentric system, and to each other, may be found. The 1966 Standard Earth gives such relationships for several of the major survey datums, for example the North American and European Datums.

The geopotential representation in the 1966 Standard Earth contains coefficients for all the terms through indices 8, 8 in a series of spherical harmonic functions. These functions involve trigonometric functions of latitude and longitude up to $\cos 8\theta$, $\sin 8\theta \cos 8\lambda$, $\sin 8\lambda$ respectively. The representation also contains some terms with higher indices, including several resonant terms, for example 15, 12. The coefficients for the resonant terms can be determined because the effect of these terms on the orbit of a satellite is magnified by a commensurability between the periodic motion of the satellite and the rotation of the Earth. The United States National Geodetic Satellite Program has the future objective of determining the geopotential with all terms through 15, 15.

Bibliography

LUNDQUIST C. A. and VEIS G. (Eds.) (1966) *Geodetic Parameters for a 1966 Smithsonian Institution Standard Earth*, 3 volumes, Special Report 200, Smithsonian Astrophysical Observatory, Cambridge, Massachusetts.

C. A. LUNDQUIST

STREAMER CHAMBERS

1. Principle of the Streamer Chamber

The streamer chamber is a relatively new member of the spark chamber (Roberts, 1966) family. It locates accurately the trajectory of charged particles passing through its sensitive volume (as spark chambers do), but has rather unique properties.

The simplest streamer chamber (Fig. 1) consists of two parallel conducting electrodes with a gas container in between. The gases are a mixture of 90 per cent Ne, 10 per cent He, or pure Ne or pure He.

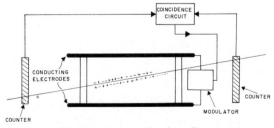

Fig. 1. Simple streamer chamber. Two counters in coincidence trigger the high voltage modulator after the passage of a charged particle in the neon gas.

Very faint streamers have been generated in H_2. Other gases do not seem to have the necessary properties for a streamer chamber operation.

A particle passing through the streamer chamber gas (for instance, a mixture of 90 per cent Ne, 10 per cent He) leaves a trail of ion electron pairs, of the order of 25 per cm at n.t.p., for a minimum ionizing particle. A high pulsed electric field (~ 20 kV/cm) is applied to the electrodes after a short time (2–5×10^{-7} sec) and each individual electron starts to multiply in the direction of the electric field (Townsend avalanche) (Fig. 2). When the dipole field of the primary avalanche reaches a value close to that of the applied electric field, the propagation of the

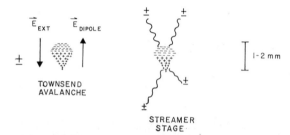

Fig. 2. The initial ion electron pair develops in a Townsend avalanche at first. When $E_{ext} \approx E_{dipole}$ photons start producing new ion electron pairs. Each of this pair will generate other Townsend avalanches in the region close to the tips of the primary avalanche, where the electric field is enhanced.

discharge is primarily by photons in both directions (streamer stage). After a short time in the streamer mode the electric field is removed, and the result is a faint track made of many streamers a few millimetres long in the direction of the field, located close to the original path of the particle. The pulse duration needed is of the order of 10–20×10^{-9} sec; longer pulses at the quoted electric field would bring the chamber into the projection mode, observed by Fukui and Miyamoto for the first time (Fukui and Miyamoto, 1959). During the Townsend avalanche formation, the track shifts 1–2 mm coherently in the direction of the electric field.

2. Property of the Streamer Chamber

The formation of the streamers does not appreciably reduce the applied electric field, except in regions very close (few millimetres) to each track. Therefore many tracks can be detected with very high efficiency.

The probability of forming a track does not depend upon the angle between the track and the electric field, i.e. the streamer chamber is isotropic. The track brightness is not isotropic; tracks parallel to the electric field are roughly two orders of magnitude brighter than tracks perpendicular to the electric field. High multiple track efficiency and detection isotropy are the two properties of the streamer chamber not shared by other spark chambers. The streamer chamber can also yield useful information about the ionizing power, and in particular conditions can be used for measuring the ionizing power with great accuracy (Bulos et al., 1967).

The memory of a streamer chamber is hard to define. After the passage of a particle, if the chamber is not immersed in a magnetic field, the primary electrons start to diffuse isotropically. In the absence of electronegative gases, the high voltage pulse can be applied up to 200–400 μsec later, and still a diffuse track can be seen (Fig. 3). Such a long "memory" is a disadvantage because it limits the number of particles per unit time that can be sent through the

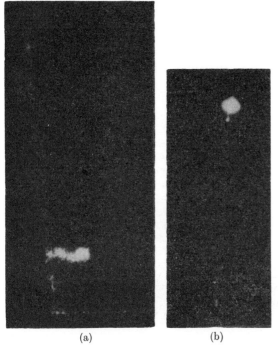

Fig. 3. (a) Photograph of a track where the delay between the passage of the cosmic ray particle and the application of the pulse is 80×10^{-6} sec. The dense short track at the bottom is a highly ionizing delta ray. (b) Photograph of a track with 2×10^{-7} sec delay, in the same chamber as (a).

chamber. Clearing fields cannot be used to reduce the memory because they are too slow in removing primary ionization from large gaps. If one tries to increase the clearing field to speed up the process, secondary multiplication may occur. Careful dosage of electronegative gases in the chamber (like SO_2, few parts per million) reduces the memory to 1–3 μsec. If the pulse is applied with a delay longer than that, only a few streamers per meter of track will be formed.

3. Modulators

The high voltage pulse required to drive a streamer chamber imposes stringent requirements on the modulator. The 10–20 nsec long, several-hundred kilovolt pulse is normally generated by a Marx voltage multiplier circuit (Craggs and Meek, 1956) followed by a pulse-forming network (PFN). This combination proved to be the most reliable and has been used for all operating streamer chambers. The choice of the pulse-forming network depends on the size and configuration of the streamer chamber.

In Fig. 4 are shown three combinations of Marx generators (schematized as a capacity with its own internal inductance and a voltage generator in series) and pulse-forming network. The first pulse-forming network is used for "small" chambers, i.e. for chambers that can be treated as pure capacitors. For large chambers, i.e. for chambers where the transit time of an electrical pulse is comparable with the pulse length, the configurations of Fig. 4(b) and Fig. 4(c) are preferred, since they generate a pulse on a defined characteristic impedance. The chambers are then treated as transmission lines, and terminated in their characteristic impedance. Figure 4(b) consists of a cable with a switch in series. It has the disadvantage that while the charging voltage is V_0, the output is $V_0/2$. Figure 4(c) is a pulse-forming network called a *Blumlein line* (Craggs and Meek, 1956). It gives the full charging voltage as output voltage, and can compensate the coherent shift of the tracks by applying a pulse of opposite polarity before the high voltage pulse is applied.

4. Photography and Data Acquisition

The light emitted by each streamer is very faint. The brilliance of a streamer chamber track is roughly three orders of magnitude smaller than the brilliance of a spark in a spark chamber. Moreover, most of the light emitted is in the orange-red region (6500–5500 Å) where conventional, very fast films are not sensitive. Wide open lenses (f 1 : 2) are required for direct photography, and large demagnifications are necessary to allow sufficient depth of field.

Another approach is to use image intensifiers with large amplification (10^5). Currently available magnetically-focused image intensifiers have sufficient resolution, but suffer from very large distortions. A detailed analysis of the resolution obtainable with direct photography and with image intensifier photography shows that the first is preferable, because of its simplicity, cost and competitive accuracy, unless extremely large depths of field are required. Image intensifiers have been used for the photography of pure helium-filled streamer chambers, where the streamers are three to five times fainter than in Ne. In principle, three stereoscopic views are necessary to reconstruct unambiguously a three dimensional event. The arrangement of the three cameras depends largely upon the configuration of each experiment. Figure 5

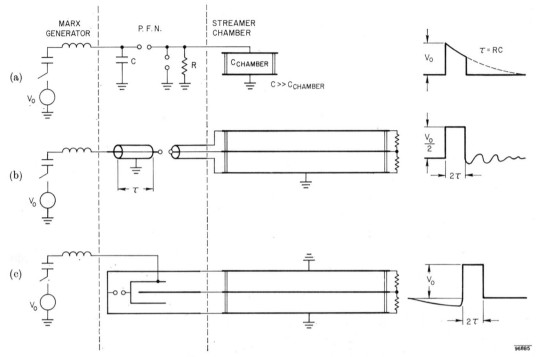

Fig. 4. (a) Marx + capacitor PFN and pulse shape. (b) Marx + coaxial cable PFN and pulse shape. (c) Marx + Blumlein PFN and pulse shape.

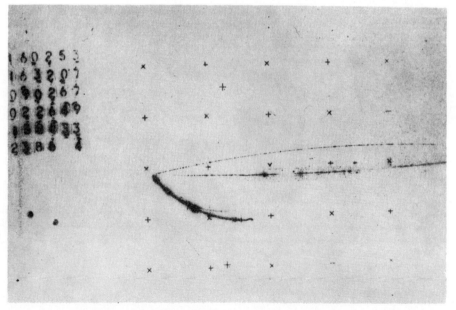

Fig. 5. A photoproduction event in the Stanford Linear Accelerator Center 2-metre streamer chamber. Positive tracks are bending upwards. The distance between fiducial marks is 40 cm in the beam direction (left to right); 30 cm in the perpendicular direction. The heavy track is a proton circa three times minimum ionizing.

shows one of the three stereoscopic views of a high energy photoproduction event ($\gamma + p \rightarrow p + \pi^+ + \pi^-$) obtained in a large streamer chamber. The analysis of the data proceeds from the film in a manner similar to that with ordinary bubble chamber film. Similar digitizing machines and computer programs are used.

The triggering rate of the streamer chamber is ultimately limited by the advancement of the film in the cameras.

5. Streamer Chamber versus Bubble Chamber

Every visual particle detector is normally compared with the bubble chamber. A high data rate and selectivity obtained through triggering on particular events are advangates of all spark chambers. The streamer chamber has the additional property of recording topologically complicated events, usually troublesome in conventional spark chambers. For streamer chamber and bubble chamber of equal size, the spatial accuracy (i.e. the error in measuring one point on one track) is about a factor of two or three worse for the streamer chamber. Furthermore, in the study of high energy interactions with protons (hydrogen), target and detector are a single unit in the bubble chamber. This is not the case with the streamer chamber where the hydrogen (the target) must be separated from the Ne/He mixture (the detector). The vertex of the interaction cannot be seen in the streamer chamber, and this may generate problems in some cases.

At very high energy the momentum measurement accuracy of a bubble chamber is limited by Coulomb scattering and nuclear interactions in liquid hydrogen. This is not the case of the streamer chamber, since for gaseous neon the radiation length and the nuclear interaction length are far greater than for liquid hydrogen.

Bibliography

BULOS F., ODIAN A., VILLA V. and YOUNT D. (1967) *Streamer Chamber Development*, SLAC-Report **74**, Stanford Linear Accelerator Center, Stanford, California.

CRAGGS J. D. and MEEK J. M. (1956) *High Voltage Laboratory Technique*, London: Butterworths.

FUKUI S. and MIYAMOTO S. (1959) *Nuovo Cimento* **11**, 113.

ROBERTS A. (1966) Spark chambers in high energy Physics, in *Encyclopaedic Dictionary of Physics* (J. Thewlis Ed.), Suppl. vol. 1, Oxford: Pergamon Press.

F. VILLA

SUBMILLIMETRE RADIATION, APPLICATIONS OF

Introduction

In general, radiation is used by allowing it to interact with a system in such a way that a comparison of some aspect of the waveform observed in the absence and presence of the system enables us to gain information about it. In the extreme case, of course, the absence of any modification of the waveform can be of use, for example in looking for extremely transparent or highly reflecting materials, but generally a large effect is desirable if the waves are to be put to optimum use. The interactions may be of microscopic, macroscopic or even bulk nature. The microscopic and some of the macroscopic interactions are probed by spectroscopic techniques, whereas the bulk and the remaining macroscopic interactions are revealed by measurements of phase shift, attenuation or some other property of the (generally undispersed) radiation. The spectroscopist can exploit the inherent low energy of *submillimetre radiation* whereas the physicist, engineer or metrologist can make use of the equivalent manifestation of the low energy—the large wavelength. The submillimetre spectral region is generally defined, by (arbitrary) convention, as extending from 0·1 to 1 mm wavelength. In alternative units this corresponds to 10 to 100 cm^{-1}, 0·33 to 3·3 THz, 2×10^{12} to 20×10^{12} rad sec^{-1}, or 10^{-3} to 10^{-2} eV.

Spectroscopic Uses and Applications

Owing to the low frequency of the electromagnetic waves it is clear that the microscopic interactions are confined to those in which the molecular system can respond only relatively slowly to an imposed electromagnetic field (in times of the order 10^{-12} sec). Molecules having an electric dipole moment μ are responsive to an electric field and are capable of rotating, in the gas phase, at frequencies lying in the region near 1 THz. The rotations are governed by the laws of quantum mechanics and have energies corresponding to the differences between rotational levels within a given vibrational state of the molecule, usually its ground state. The exact frequency at which the rotation occurs depends on the shape, size and structure of the molecule, while the magnitude of the total energy withdrawal (that is, the integrated intensity of a rotational line) is dominantly proportional to μ^2. Measurement of the position and intensity of the features in the rotational spectrum of a gas leads directly to quantitative information about the moments of inertia and the dipole moments of the molecules comprising it. In some cases the rotational features are not simple and an analysis of their structure enables further information about the molecular motion to be obtained; for example, the rotational spectrum of ammonia shows splitting and fine structure due, respectively, to an inversion oscillation of the nitrogen atom through the plane containing the three hydrogen atoms and to a consequent non-degeneracy of the rotational levels. A direct application of submillimetre waves to the analysis of molecular constants will generally only be made when none of the spectral features in question lies in the microwave region where extremely precise measurements are possible.

The water molecule in the vapour state has an intense and, due to the asymmetric top structure of the molecule, irregular rotational spectrum which obtrudes into any application of submillimetre waves that involves transmission of radiation through the atmosphere. In many of the uses described here it is necessary to operate the apparatus so that the optical paths are in vacuo.

Absorption spectroscopy may be employed in the submillimetre region to investigate any system responsive to the radiation using a grating spectrometer or, preferably (because of the luminosity, multiplex and ordersorting advantages) a Fourier spectrometer. An advantage of *Fourier transform spectroscopy* employing a Michelson two-beam interferometer is that continuous absorption, refraction and, in some cases, reflection spectra may be determined at any resolution. This is specially significant at the long-wave end of the range where a comparison and overlap with discrete monochromatic microwave measurements can be made with the advantage that the submillimetre investigations give continuous profiles the shapes of which are vital clues to the origin of the processes under study. When such profiles are broad they can be obtained from microwave results only in approximate form by interpolation between the relatively widely spaced data.

Before turning to the condensed phases it is worth noting that submillimetre waves can be applied to the study of the broad profiles resulting from second order or induced effects in compressed non-polar gases. For example, information about quadrupole moments can be gained by recording the submillimetre absorption of a compressed non-dipolar gas, such as carbon dioxide, in which the absorption is due to the induction of dipoles during collision by the large quadrupole fields. A further example, of great theoretical significance, concerns the rare gases, atoms of which are extremely non-polar. Because they are relatively simple they are amenable to detailed quantum mechanical analysis thus quantitative observations of their properties can give considerable insight into quantum processes. Pure rare gases show no submillimetre absorption, mixtures do due to induced dipoles. A quantitative fit, therefore, between the observed and predicted profiles, based on some model, signifies a considerable contribution to our understanding of the behaviour of simple atoms.

Fundamental studies yielding fundamental information about the two condensed phases are no less important. The probing of liquids by submillimetre radiation probably holds the key to our understanding of the real nature of the liquid state. Liquids are frequently thought of as condensed gases or, alternatively, as completely disordered solids. Since both compressed gases and crystalline solids have submillimetre spectra of distinct but different forms any resemblance to each or either of these by the liquid spectrum gives a clue on the nature of the liquid state. Relaxational (Debye) absorption by polar molecules in the liquid state is generally confined to the microwave region but the high-frequency tail extends to submillimetre wavelengths where the recently observed absorption is greater than had been expected and present even in non-polar liquids. Some workers have linked the absorption with the rotational spectrum of the gas, others with the lattice spectrum of the frozen liquid; neither association is, as yet, conclusive. Since the peak of the absorption lies in the region 30 to 70 cm^{-1} for almost every liquid so far studied it is clear that a vitally important application of submillimetre waves is to the study of the structure of liquids and the inter-molecular interactions within them.

Relatively few molecules vibrate internally at frequencies near 1 THz. Those which do contain one or more massive atoms, for example $GeCl_4$ or $CHBr_2:CHBr_2$ exhibit strong broad features observations on which give intramolecular data. Similarly, studies of the spectra of crystal lattices, which typically occur below 200 cm^{-1}, give considerable information on intermolecular forces and the particular modes of vibration—vital for a complete understanding of the crystal structure.

The association of molecules into bound pairs in both the gaseous and liquid states is relatively little investigated. Because of the large overall mass involved the vibrational frequencies of these dimers lie considerably lower than the typical fundamental vibrational frequency of the monomer. The spectral features due to dimeric vibration occur, therefore, at microwave and submillimetre wavelengths. An investigation of the position and intensity of these bands yields valuable information about intermolecular bonding (and in particular, hydrogen bonding) in non-ordered systems.

A spectroscopic application of submillimetre radiation that is of great technological importance is to the study of polymers which can show sharp bands arising in the crystalline regions and broad bands due to the amorphous continuum in which they are immersed; sharp bands due to vibration within the very large molecules may also be seen. The observation of some or all of these absorptions is of importance in making an analysis of the polymeric structure. The absence of such absorption makes the material in question suitable for the construction of submillimetre-wave optical components such as lenses.

Apart from the specific applications already listed there are other spectroscopic aspects which give parameters not readily determined in any other way. These may be summarized by collection under the general heading of resonance phenomena. The more common examples are magnetic resonance, and in particular cyclotron resonance, absorption by superconductors and absorption in antiferromagnetic and ferrimagnetic materials. Because of the magnitude of submillimetre wavelengths high fields are necessary for the study of cyclotron phenomena and quantitative precision depends somewhat on the homogeneity of these, but materials such as impure compound semiconductors (for example gallium arsenide) are inaccessible to alternative methods. Fields of 150 kG and laser radiation of 337 μm wavelength have already been combined to produce valuable information on quantum effects in the degenerate valence bands of germanium.

One class of materials, often described as the fourth state of matter, is the plasma. A plasma is a complicated gaseous assembly of electrons, ions and neutral atoms (or molecules) which is electrically neutral in bulk, but highly ionized. An exact description of the conditions within a plasma and a full understanding of its properties are still wanting and any scientific attack that can be brought to bear upon them is of value. Microwaves, commonly used, can be subject to large attenuation while infra-red waves, from lasers, record only small phase shifts. Submillimetre waves are, therefore, in many respects optimum. In a simple plasma, phase-shift and loss give the electron density and the collisional frequency of electrons and ions while the presence of a magnetic field can be exploited to give measurements of Faraday rotation which also leads to a figure for the electron density.

Atmospheric Applications

In spite of the strong absorption by water vapour, two regions of relative transparency have been found in the atmospheric transmission spectrum. These windows are centred near 22 cm^{-1} and 29 cm^{-1}. Beyond 60 cm^{-1} any atmospheric path more than a few centimetres long is, at normal altitudes, virtually opaque. Of the two atmospheric windows the latter is of immense importance because of an almost exact coincidence with the emission at 29·7 cm^{-1} from the HCN *laser*. Consequently, laboratory experiments with the laser radiation can be done without the provision of evacuated optical paths but, more important, the radiation may be used to probe the atmosphere. In addition, a broad band of radiation centred on 30 cm^{-1} may also be used for atmospheric investigations over limited paths.

The degree of transparency in the atmospheric windows is determined, of course, by the water vapour content of the air thus making it possible to monitor the latter by observing precisely the attenuation of a 337 μm laser beam. More important, probably, than the ability of this submillimetre radiation to penetrate the clear atmosphere is the ability to penetrate fogs because the scattering loss from the water droplets is very low. Beams have been sent over one or two hundred metres but this distance can be considerably increased by the use of homodyne techniques in which the attenuated atmospheric signal is mixed with a strong "local oscillator" laser signal. Application of the submillimetre laser to medium-range communication or direction finding problems would appear to be plausible; strong scattering of the radiation from shorter wavelength lasers is a severe problem while microwaves are not very convenient for short-range applications because of the large transmitting and detecting elements necessary on account of the wavelength.

Submillimetre waves may also be applied to atmospheric studies at higher altitudes using either terrestrial observatories or aircraft as the observation platforms. New atmospheric windows near 44 and 50 cm^{-1} appear, thus extending the range over which submillimetre astronomical observations may be made. The atmospheric absorption features may be observed either by radiation exchange between the Sun and the relatively cool detector or between the sky and the relatively warm detector. The latter has the advantage of requiring no solar guidance system and thus permits stratospheric water vapour to be estimated with much simpler apparatus than has been used in other spectral regions.

There is no reason to suppose that the submillimetre region should be less fruitful to astronomers than any other portion of the spectrum but astronomical applications have, up to now, been small in number due mainly to technical difficulties. Now that compact interferometric spectrometers and efficient detectors are available astronomers can search for new celestial sources other than the Sun, scrutinize the atmosphere in search of new spectral features or even new gaseous constituents and measure angular and amplitude scintillation to gain information on atmospheric turbulence and thus assist the meteorologist.

Metrological Applications

Distance measurements may assume a variety of forms which may be grouped under either of the headings: absolute determination of length or determination of change of length. The precision required for either type of measurement depends on the particular application. In engineering it is common to need to know distances to about 0.5×10^{-3} in (about 12 μm). 12 μm corresponds to about 1/30 of a submillimetre wavelength. Consequently if we can measure interference fringes to this fraction of a wavelength submillimetre waves are inherently suitable for engineering use. Although shorter wavelength radiation is capable, in principle, of yielding higher precision, because of the greater fineness of the scale of the waves, the surfaces used to reflect the radiation need to be proportionately flatter and free from vibration.

Cosinusoidal fringes, produced in a two-beam interference system illuminated with monochromatic radiation, are suitable for the measurement of monotonic change of length. It is essential that the change be continuous, with no finite increments, otherwise the relationship between the observed fringes (which are of identical appearance) cannot be assessed with certainty. Because of this difficulty, laser radiation is only of practical use in those cases where a precision of $\lambda/2$ (equal to about 0·16 mm) is adequate. Such a circumstance obtains in the measurement of a long distance. Remembering that the power available from a 337 μm HCN laser is relatively high and that the atmosphere is not highly absorbing near 340 μm, it is evident that since the bandwidth of the radiation is less than 1 MHz the upper limit of the distance over which interference may theoretically be obtained is given by the coherence length. Hence, path lengths of up to 400 m may be introduced into one arm of a Michelson interferometer and measured to within 0·16 mm. Not only may total length be measured in this way but also change of length and in particular a small increase in a large distance. The precision with which this may be done is increased if the intensity of the detected signal is increased. One way in which this may be achieved is by replacement of the simple Michelson interferometer with a homodyne local oscillator system. In this the audio-modulated c.w. signal is made, by use of a cube-corner, to travel twice over the path whose change of length is to be determined. This beam is then mixed in a square law detector (e.g. a Golay cell) with a relatively intense unmodulated c.w. beam of identical wavelength taken from the same source. The magnitude of the detected intensity is proportional to the square root of the product of the intensities of the d.c. and modulated beams and is sensitive to the phase relation between the two beams. Change of distance may, therefore, be measured in a straightforward way with an accuracy of $\lambda/4 = 0·09$ mm.

Laser radiation is not well-suited to the accurate measurement of small changes in already small distances but polychromatic radiation continuously occupying a finite spectral interval is. If the radiation band stretches from 0 to σ_M (cm^{-1}) and has its intensity maximum at $\sigma_0 = \frac{1}{2}\sigma_M$ the total width at half-intensity is approximately $\frac{1}{2}\sigma_M$. The first dark fringe observed with a two-beam interferometer illuminated by this band is located at a path-difference given approximately by $2/\sigma_M$. The fringe contrast has its maximum value at zero path-difference and, owing to the unique character of polychromatic fringes, path-difference changes of up to $2/\sigma_M$ can be uniquely followed. If, however, instead of the familiar amplitude modulation the detected intensity is modulated by a suitable periodic variation of the phase within the interferometer the detection of fringe shifts is made at least an order of magnitude more sensitive and small shifts up to $1/\sigma_M$ are unambiguously and accurately followed. For example, with $\sigma_0 = 30$ cm^{-1}, changes in path-difference of $1/20\sigma_M = 9$ μm are readily detected. In a workshop application the workpiece is made to constitute one of the mirrors of a two-beam interferometer while the other mirror oscillates with an amplitude $1/8\sigma_0 = 40$ μm about its rest position. Deviations, in the workpiece arm, from the position of zero path may be detected and measured with an accuracy determined by the precision with which the scale attached to the vibrator may be read. The change in size, therefore, of an object undergoing machining can be monitored without resort to any sort of contact gauge and often with no need to arrest the machining process.

Teaching Applications

Demonstrations of the laws and principles of physical optics are difficult if carried out with visible light and not conveniently done on a large scale suitable for the accompaniment of lectures. A common solution to this difficulty is the provision of demonstration kits using microwave radiation. While these solve many of the problems there still remains the snag that such kits do not in many ways resemble optical systems as commonly encountered and there is often a conceptual difficulty for the student at an elementary stage. Submillimetre radiation can be employed in demonstrations of physical optics which depend on components worked to about $\lambda/10$ in optical quality, i.e. to about 30 μm, which is a quality readily attainable on ordinary workshop machinery. There is the further point that a mercury lamp or, more desirably, an HCN *laser* may be used as source. Of course, no direct visual display of the effects is possible (as no submillimetre-video converters currently exist) and indirect display on an oscilloscope or pen recorder is necessary. This disadvantage is outweighed by the convenience of being able to use optical components of manageable size machined from polythene or even a modern polymer transparent to both visible and submillimetre radiation. Lenses, prisms, zone plates, diffraction gratings, etc., are easily and cheaply made and the classic experiments of Young, Fresnel and others can be demonstrated using workshop components once the investment in source and detector is made.

Summary

It is evident that the range of applications of submillimetre waves is no less diverse and rich than any other portion of the electromagnetic spectrum. The reputation gained by submillimetre radiation as being of little use and infrequent application is unjustly due to the misconceived notion that it is hard to produce and detect and will yield little information that cannot be gained with greater ease elsewhere in the spectrum. Clearly there are many unique applications, some of which are already in widespread use while others await exploitation.

J. Chamberlain

SUBMILLIMETRE-WAVE LASERS. The first stimulated emission of submillimetre-wave radiation was observed in 1964 at the National Physical Laboratory by Gebbie and his co-workers. Water vapour gave radiation at 118·8 μm and HCN emitted at 337 μm. In the same year emission from neon at 106·0, 124·4 and 132·8 μm was found by Patel *et al*. Since then nearly fifty laser wavelengths have been observed at wavelengths up to 773·5 μm, but most of these radiations are of low intensity and all are confined to just a few molecular systems, namely compounds of H, C and N (sometimes with I present); H_2O; Ne and deuterated versions of the species named. Most of the lasers were first discovered by using pulsed excitation but a good number have been made to radiate continuous-wave (c.w.) using d.c. excitation. All submillimetre-wave lasers so far discovered (with the exception of neon) employ molecular gases or vapours; no successful solid state generator has been developed.

A common type of arrangement consists of a long tube of length between 1·5 and 10 m and diameter 5 to 10 cm as shown in Fig. 1. Plane mirrors are placed at each end of the tube and electrodes are sealed

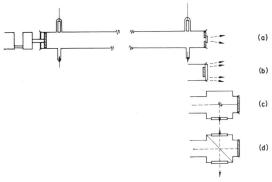

Fig. 1. Typical arrangement for submillimetre-wave laser with (a) hole coupling, (b) circular slit coupling, (c) inclined mirror coupler, (d) dielectric film coupler.

into the side near the ends. A resonator such as this is characterized by the Fresnel number $N = d^2/(4\lambda l)$ where d is the diameter of the mirrors, λ is the wavelength of the observed emission and l is the separation of the mirrors. For submillimetre-wave lasers N is bound to be small, for example with $d = 5$ cm, $l = 10$ m, $N \sim 60/\lambda$ where λ is in μm. When the Fresnel number is low, the loss per pass of the radiation within the cavity and mode behaviour are hard to calculate. When the wavelength is long losses due to diffraction effects become serious. For instance, a cavity 5 cm in diameter and 30 cm in length will emit pulsed 337 μm HCN radiation but, so far, a cavity 2·5 cm in diameter will not emit no matter how long it is. (Greater length increases the gain but reduces, of course, the Fresnel number.) Some suitable gas or vapour is generally passed through the tube at pressures near one torr or less and an electric discharge passed either d.c. or in short pulses. The voltage for c.w. operation is typically in the range 500 to 1000 V with currents of about 0·5 to 1 A; for pulsed operation pulses of a few microseconds duration and peak voltages ranging from 5 to 60 kV may be used. If the mirrors are plane parallel and an exact number of half wavelengths apart, the cavity they form becomes resonant so that stimulated emission may be observed.

One of the mirrors can be moved along the axis of the cavity. A convenient form of mounting is a precise micrometer screw driven by a motor. As the cavity length is increased a succession of resonances occurs; these may correspond to different modes of the same wavelength, or different wavelengths. The two can generally be easily distinguished (Fig. 2): the former have an invariant relative spacing as a function of cavity length while any resonances due to a different wavelength, although themselves a fixed

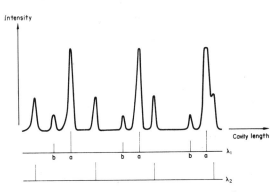

Fig. 2. Example of a cavity interferogram for a laser emitting wavelengths λ_1 and λ_2. There are two modes a and b of the radiation λ_1.

distance apart, have a variable spacing relative to those due to the reference wavelength. There might be ambiguities if the cavity itself is used to analyse the wavelengths of the stimulated emission so it is better to extract the radiation from the cavity and use an independent spectrometer. Ideally, the radiation could be coupled from the resonator by semi-silvering one of the mirrors. But such a procedure is rather difficult for submillimetre radiation which is readily reflected by metallic layers even when they are very thin. A small hole may be pierced in one of the mirrors and covered with a window [Fig. 1(a)], or a gap may be left round one of the mirrors to enable extraction to occur by coupling round the edge (b) or a small mirror may be placed at 45° within the cavity (c). This latter is equivalent to the hole coupling.

In the cases where a small central coupling is used only those modes of the resonator with fields concentrated along the axis are coupled out and such coupling is not, therefore, suitable for mode studies. The circular gap round the periphery of one of the mirrors does not have this disadvantage. A simple additional method which gives a beam of full width and is, in practical effect, closely equivalent to semi-silvering one of the mirrors is to insert a thin dielectric film at 45° within the cavity (d). A small amount of radiation is reflected by this and consequently the film acts as a weak coupler. A suitable material is polyethylene terephthalate (Melinex) as used in far infra-red and submillimetre-wave interferometers for *Fourier transform spectroscopy*. A suitable thickness for the film is 6 μm. The characteristics of such very thin films are independent of wavelength over a wide range but the polarization of the radiation within the resonator and that reflected from the cavity is strongly influenced by the film.

It is not essential that the cavity have a pair of plane mirrors. A plano-concave or a confocal cavity resonator may equally well be used. However, in the case of the hole-coupling of any sort of cavity and in the case of the cavities having spherical mirrors the output beam is very divergent due to diffraction effects occasioned by the dimension of the hole in the former case and due to the focused radiation in the latter case. If it is desirable in a particular application of a laser, to have a parallel beam then the beam-divider coupling is to be recommended as the diffraction effects are determined in that case by a larger dimension—the diameter of the discharge tube window.

The wavelength of the emission from the lasers may be measured with a *grating spectrometer* or a *Fourier spectrometer* as commonly used in the far infra-red or submillimetre-wave spectral regions. The time dependence of pulsed emission and its relation to the imposed current pulse is studied with a fast detector but a Golay cell is adequate for the simpler purpose of monitoring the emissions and estimating the average power in the output. The frequency of the laser radiation is found by comparison with the harmonics of accurately known microwave sources using a diode detector as the mixing element.

Water vapour laser. Emission was first discovered from a pulsed discharge by Mathias and Gebbie at 28 μm but the first submillimetre emission was observed at 118·8 μm by Gebbie. Above 100 μm wavelength there are four lines emitted from H_2O and six from D_2O as shown in the Table 1.

The power of the radiation in the case of H_2O is a few milliwatts while that from D_2O is probably an order of magnitude less. It is found that the intensity of the 118 μm line of H_2O is sufficiently large to enable a beam propagated through atmospheric paths several metres long to be easily detected.

The mechanism of the H_2O laser was, until recently, obscure. It had at one time been suggested that OH was responsible but three independent theoretical contributions coupled with some more recent experimental observations have shown that H_2O is the species responsible. The interpretations are essentially the same and lead to closely similar identifications: agreement is absent only for the 115·42 μm line. The identifications of the remaining three, shown in the table, are based on weak Coriolis interactions of the symmetric stretching vibration v_1 and the assymmetric stretch v_3 with the bending vibration v_2 (see Fig. 3).

HCN laser. Emission was first discovered by Gebbie,

Table 1

	Wavelength (μm)	Pulsed	C.W	Assignment
H_2O	115·42	*		
	a 118·59101	*	*	$(001)6_2 \rightarrow (020)6_5$
	120·08	*		$(001)6_2 \rightarrow (001)6_0$
	a 220·22797		*	$(100)5_{-1} \rightarrow (020)5_5$
D_2O	103·33	*		
	107·71	*	*	
	108·88	*		
	110·46	*		
	111·74	*		
	171·6		*	

a Accurately measured by harmonic frequency mixing with klystron.

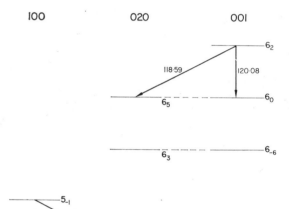

Fig. 3. *Schematic arrangement showing levels responsible for H_2O submillimetre-wave laser lines.* ----- *Coriolis interaction.* (*Wavelengths in μm.*)

Findlay and Stone in 1964 at 337 μm using a pulsed discharge in HCN gas. The discovery of other lines followed, notably by Mathias *et al.* and by Kneubuhl *et al.* There are 21 submillimetric emissions from vapours containing H, C and N and six if D is substituted for H. C.W. operation was first reported in 1966. In addition, stimulated emission has been observed under circumstances in which it is evident that the presence of iodine is essential. All such radiation has been observed by Kneubuhl *et al.* using pulsed excitation. The lines emitted from the various mixtures containing C and N are shown in Table 2.

The mechanism of the water vapour laser remained obscure with only one model proposed for almost four

Table 2

	Wavelength (μm)	Pulsed	C.W.	RI (pulsed)	Assignment
H, C and N	101·257	*			
	110·240	*			
	112·066	*			
	113·311	*			
	116·132	*			
	126·164	*		0·1	$(12^20)27_+ \to (05^10)26_-$
	128·629	*	*	0·1	$(12^20)26_- \to (05^10)25_+$
	130·838	*		0·1	$(12^00)26_+ \to (05^10)25_-$
	134·932	*		0·03	$(12^00)25_- \to (05^10)24_+$
	138·768	*			
	165·150	*			
	201·059			0·001	
	211·001	*	*	0·006	
	222·949	*		0·0004	
	284		*		
	a 309·7140	*	*	0·06	$(11^10)11_- \to (11^10)10_+$
	a 310·8870	*	*	0·2	$(11^10)11_- \to (04^00)10_+$
	335·1831		*		
	336·1	*			
	a 336·5578 (337)	*	*	1	$(11^10)10_+ \to (04^00)9_-$
	a 373·5283	*	*	0·01	$(04^00)9_- \to (04^00)8_+$
D, C and N	181·789	*		0·01	$(22^00)23_- \to (22^00)22_+$
	a 189·949	*		0·01	$(22^00)22_+ \to (09^10)21_-$
	a 190·0080		*		$(22^00)22_+ \to (22^00)21_-$
	a 194·7027	*		0·003	$(22^00)21_- \to (09^10)20_+$
	a 194·7644		*		$(09^10)21_- \to (09^10)20_+$
	a 204·3872	*		0·006	$(09^10)20_+ \to (09^10)19_-$
I, H, C and N	169	*			
	268				
	465	*			
	538·2	*		0·1	
	545·4	*			
	676	*			
	773·5	*		0·03	

RI: Intensity of at peak of output relative to 337 μm HCN radiation.
a Accurately measured by mixing with harmonic frequency of a klystron.

years prior to expositions of what would appear to be correct formulations. This has not been the case with the HCN laser. It has been possible for many workers to account for some of the observed lines using a variety of diverse theories. The most satisfactory and acceptable attributes many of the lines to HCN, including the important ones at 311 and 337 μm. A Coriolis perturbation is responsible for a mixing of rotational levels within the vibrational levels 11^10 and $04°0$ (Fig. 4). This gives rise to population inversions within the two vibrational states and consequently the lines shown in the figure are radiated. Precise frequency measurement of the lines at 310 μm and 337 μm and at 311 μm and 335 μm together with the observation that the 310 μm and 373 μm run simultaneously with the 337 μm while those at 284 μm and 335 μm run simultaneously with 311 μm supports the hypothesis. A similar type of Coriolis coupling can be invoked to explain the lines of HCN emitted near 130 μm and those of DCN (Fig. 5). The identifications are shown in the tables.

The intensities of the emissions vary very considerably with the particular configuration used and also with the gas mixture although it has been shown that for the 337 μm line the laser is always saturated at the peak of a mode—that is, all the molecules that are able to radiate do so. Although the radiations shown in Table 2 are due to the HCN species a variety of compounds, giving rise to HCN after electrical excitation, may be used. The line at 337 μm is generally the strongest while that at 311 μm is about ten times weaker. Peak powers of several watts are common for the 337 μm line while the use of methane and nitrogen as the source of HCN has been shown to lead to a peak power of 100 W. Continuous-wave powers are, of course, less and for the 337 μm line lie between 0·01 and 1 W. The width of the 337 μm HCN radiation is probably less than 700 kHz, a figure based on a long-path interference experiment.

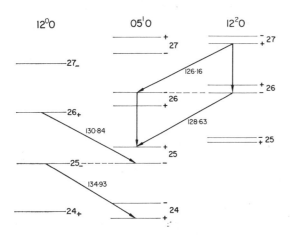

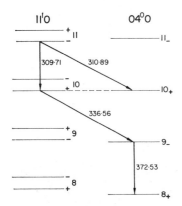

Fig. 4. *Schematic arrangement showing levels responsible for some HCN submillimetre-wave laser lines.*
---- *Coriolis interaction. (Wavelengths in μm.)*

Now that there has been some success not only in the generation of submillimetre stimulated emission but also in many cases in the explanation of its origin it is to be expected that additional sources will be developed. It would be particularly beneficial to have c.w. sources of some power coincident with the "windows" in the submillimetre-wave atmospheric spectrum (the 337 μm line of HCN is the only one falling in a very transparent window).

See also: Submillimetre radiation, applications of. Submillimetre-wave methods.

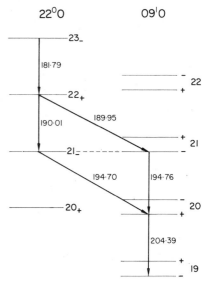

Fig. 5. *Schematic arrangement showing levels responsible for some DCN submillimetre-wave laser lines.*
---- *Coriolis interaction. (Wavelengths in μm.)*

J. CHAMBERLAIN

SUBMILLIMETRE-WAVE METHODS. The techniques that are employed for the generation and detection of submillimetre waves have much in common with both microwave and infra-red methods. This is because submillimetre wavelengths link the millimetre-wave and far infra-red spectral regions and constitute a band in which guided-wave methods become increasingly difficult at the lower wavelength end and free-wave "optical" methods are hard to cope with at the longer wavelengths.

The submillimetre waveband was entered first from the far infra-red region by Rubens and Wood who used a primitive Fourier spectrometer to detect the radiation emitted by a Wellsbach mantle. The longest wavelength detected was 200 μm (50 cm^{-1}). The extension of millimetre-wave methods to shorter wavelengths was first reported by Nichols and Tear, who, in 1925, claimed to have observed overtones of Hertzian oscillators with wavelengths as short as 0·22 mm (45 cm^{-1}). This work was not followed by results of any spectroscopic significance until Gordy and his co-workers achieved, in 1954, the generation and accurate detection at high resolution of wavelengths as short as 0·77 mm (13 cm^{-1}). A harmonic generator driven by a centimetre-wave klystron served as the source. More recently, in 1964, the year in which the first submillimetre-wave laser was invented, Jones and Gordy, using a 5 mm klystron to drive a silicon crystal multiplier, extended high precision microwave spectroscopy to 0·43 mm (23 cm^{-1}). It seems unlikely that waveguide will be made much smaller than the present limit of 0·34 mm which has been applied to a detector of HCN radiation.

It has only recently been possible, using carefully designed spectrometric systems and sensitive liquid helium cooled detectors, to apply infra-red methods up to and beyond 1 mm wavelength and at the same time obtain reliable photometric results.

It is generally necessary in almost every application of submillimetre waves to evacuate, or flush with dry air, the propagation path, as in the infra-red, owing to the strong absorption by water vapour. It is only in the "windows" in the atmospheric spectrum that this is not necessary and it is fortunate that there are windows sufficiently transparent to enable many long-wave experiments to be performed without need for a vacuum provided the air paths are kept short (say 50 to 100 cm). The submillimetre waveband is almost the lowest lying frequency region over which continuous broad-band spectroscopy may be performed. Although broad-band observations are now possible to about 2 cm^{-1} (5 mm), below this frequency only very small ranges in the neighbourhood of a highly defined central frequency may be covered.

A widely used source of submillimetre radiation is the medium pressure mercury lamp in a quartz encapsulation, as employed in the far infra-red. This behaves approximately as a black body at an effective temperature of about 2500° to 3000°K and above 200 μm (below 50 cm^{-1}) follows the Rayleigh-Jeans law according to which the luminance of the radiation of wave-number σ and 1 cm^{-1} bandwidth emitted from 1 cm^2 of source within 1 steradian is approximately $2 \cdot 5\sigma^2$ nW cm^{-1} sr^{-1}. Although the mercury lamp is the best broad-band source so far found it is very poor at long wavelengths. Approximately half the radiation is given from the quartz encapsulation and half from the gas inside. A more convenient source having smaller heat dissipation is the quartz-iodine tungsten lamp in which most of the emission is from the quartz encapsulation.

The broad-band incoherent sources have recently been supplemented by submillimetre-wave gas *lasers*. The most important are the water vapour laser radiating at 119 μm and the HCN laser radiating at 311 and 337 μm. These sources can be operated both pulsed and CW and have line widths of the order 10^{-6} times the resonant frequency. Although the power of the radiation is considerably greater than can be obtained in a comparable bandwidth from any black body the output from these sources is small compared with that obtained from near infra-red and visible lasers: the pulsed output may reach a few watts at peak power but the CW radiation is generally only a few milliwatts. The frequency of the emission cannot, as yet, be varied and this shortcoming renders all but very specialized spectroscopic studies impossible.

Submillimetre radiation produced by harmonic generation does not have this disadvantage and enables very high resolution broad-band spectroscopy to be carried out. One of the more common forms of generator is a microwave rectifier diode driven by some oscillator, such as a carcinotron (backward-wave oscillator), that is itself tunable over a moderate range. These can produce radiation in the millimetre region to something less than a millimetre, lying in the submillimetre region. The frequency of the signal from these harmonic generators is extremely stable and may be accurately referred to a standard radio frequency.

An alternative to the diode has been developed by Froome in the form of the plasma arc in which an arc discharge in argon between platinum electrodes is the non-linear element providing the harmonic generation. This has produced power at 0·3 mm, the twenty ninth harmonic of the fundamental source. One of the primary problems associated with this generator is the maintenance of stability. Also, of course, the power of the higher harmonics is small (a few milliwatts).

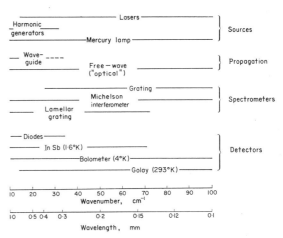

Fig. 1. *Summary of equipment and components for submillimetre waves.*

The most commonly used detector of broad-band submillimetre radiation is the *Golay cell* (also described in *infra-red spectroscopy*). Although the sensitivity of this falls at longer wavelengths owing to the increasing comparability between the magnitude of the wavelength and the size of the detecting element the Golay cell may be used with ease to 750 μm. It is a thermal detector operating at room temperature with a noise equivalent power of about 10^{-10} W for a band of 1 Hz. The response time is 15 msec. There is a tendency to microphony and the detector is limited to fairly slow observations.

Another common thermal detector suitable for submillimetre use is a bolometer which is constructed from a material with a high temperature coefficient of resistance α_r. A bolometer in which the element is pure germanium doped with, for example, gallium or indium can be very sensitive when operated at liquid helium temperatures. Cooling a bolometer in this way improves its performance because of a reduction in thermal noise and a reduction in thermal capacity which gives rise to a large coefficient α_r. Following absorption of the incident radiation the change of resistance is recorded using a bridge. A response time of 1 msec is typical and a noise equivalent power of 10^{-11} W for a 1 Hz bandwidth may be obtained. An alternative bolometer of only slightly lower sensitivity is made from commercial carbon resistors which have a high α_r at liquid helium temperatures.

The most sensitive submillimetre detector is a photoconductive device which also has a relatively fast response time. All the detectors operate near 4°K or below. Elements whose detection ability depends on extrinsic photoconductivity have been constructed from Ge doped with Sb, B or Ga and elements whose detection ability depends on free carrier effects have been made from InSb and Ge. Particularly effective InSb detectors, in which the element resides in a magnetic field of about 7 kG, have been constructed. The magnetic field is not essential but enhances the responsivity. The detector is distinguished by a low response time in the region 0·1 to 1 μsec and operates only for radiation greater in wavelength than about 100 μm. Because of this no complicated near infra-red filtering is necessary—a particular advantage when the unit is used in conjunction with a grating spectrometer in which sophisticated filtering is often necessary for the removal of unwanted orders.

An even faster detector is the point-contact diode in which a metallic whisker is in contact with a semiconductor. Until very recently, however, no diode was made with submillimetre waves specially in mind and consequently all were inefficient detectors of that waveband. This is evident when one considers that the fall-off in sensitivity may lie between 6 and 20 dB per octave. Some success has been achieved with commercially available germanium rectifier crystals (GEC type VX 3352) intended for use at 8 mm. Both pulsed and CW radiation from the HCN laser at 0·34 mm is readily detected but the signal-to-noise ratio is low. This may be improved by piercing a hole in the alumina encapsulation surrounding the whisker and crystal and carefully focusing the radiation on to the junction by means of a lens or mirror. The magnitude of the output signal from the element is critically governed by the size and position of this image.

This difficulty is minimized if a horn-coupling is used and the detecting element is placed in a resonant cavity specially designed for the radiation to be recorded. A GaAs element with a phosphor bronze whisker in 0·34 mm waveguide is available for this purpose. This is relatively insensitive to the exact location of the incident beam but tends to be rather susceptible to mechanical shock. Point contact diodes are essential for use in mixing experiments in which submillimetre laser radiations are compared, in frequency, with precisely controlled microwave emissions. They are also suitable for use in super-heterodyne systems which can be made operational to wavelengths as low as 0·5 mm.

Because of their low sensitivity diodes so far produced are not successful detectors of black-body radiation which is, of necessity, low in power. Only one such observation (in 1957) has been reported in the literature.

A detector, currently under development, having the convenience of the Golay cell and the point contact diode in being capable of operation at room temperature, depends on the pyroelectric effect. It is, however, less sensitive than the Golay cell and slower in response than the diode. A suitable element may be made from polarized triglycine sulphate a few millimetres square. The change in polarization, following absorption of the incident radiation, is recorded.

As in the infra-red, most submillimetre detectors will respond only to a change in the intensity of the incident radiation. Consequently the incident beam is modulated, generally by means of a rotating sectored disc placed before the source or detector. Even when it is not essential a.c. detection is desirable as it means that a sharply tuned amplifier may be used in conjunction with the detector, with a great gain in signal-to-noise. Unmodulated stray radiation is not then detected. A further improvement is to use phase-sensitive amplification in which a reference signal is mixed, in a phase-sensitive amplifier, with the detector signal thus guaranteeing that only radiation from the source under investigation is recorded. The frequency of the modulation varies with the type of detector but in general the higher the frequency the better the performance—frequencies of 400, 800 or 1000 Hz are typical. The Golay cell requires the incident radiation to be modulated at 11 to 13 Hz for optimum performance.

Placed between the generator of the radiation and the detector will be the system under investigation. This must be designed to deal as efficiently as possible with the submillimetre radiation. The system will frequently be a spectrometer of some sort. Whatever its form close attention must be paid to the size of the components if free-wave optical techniques are employed. At the extreme end of the submillimetre range the wavelength is 1 mm and so a component of dimension 2·5 cm is only 25 wavelengths across and consequently diffraction and other effects likely to lead to aberrations cannot be overlooked. A preservation of the standards normally considered desirable for the dimensions of components designed for use with visible light would lead to elements of up to several metres in size. This is not very practical nor, as it turns out, is it generally necessary because images of the quality common in visible optics are rarely required.

Until recently mirrors, having the form of spherical or more esoteric surfaces, were exclusively employed for the controlled propagation of submillimetre waves. These mirrors may be moulded from glass which is subsequently aluminized or constructed from solid metal such as stainless steel or aluminium. The former may be necessary if corrosion is a danger or frequent cleaning is likely to be necessary but the latter is more easily worked. However, aluminium becomes coated with aluminium oxide which shows some submillimetre absorption. The recent discovery of transparent polymeric materials has led to a widespread use of lenses for submillimetre radiation. The commonly used and easily worked material is polyethylene. This has a refractive index of approximately 1·5 giving a reflection loss of 4 per cent at each surface. It is relatively transparent, having an average absorption coefficient of approximately 0·5 neper cm^{-1} which is only slightly greater near 143 μm (70 cm^{-1}) where there is an absorption feature whose strength and width is determined primarily by the crystallinity of the specimen. This band is rarely a problem since the thicknesses commonly used are generally insufficiently large to produce significant absorption. Lenses may readily be constructed from polyethylene, polypropylene and so on, as may windows for absorption cells. However, caution is required in application because there is a danger of absorption of a gaseous or liquid specimen into the softer plastics; moreover, many polymers are soluble in a wide range of organic liquids. Crystalline quartz is transparent to submillimetre radiation having an average absorption coefficient of 0·4 neper cm^{-1}. Its refractive index is about 2·2. Since crystalline quartz is birefringent it should be cut so that the optical axis of the crystal coincides with the direction of propagation of the radiation. This material is considerably more rigid than most polymers.

It we apply standards for optical flatness similar to those commonly expected in the visible region, say one fiftieth of a fringe, we see that even at 100 μm (100 cm^{-1}) this amounts to only 2 μm over a surface. This quality is easily achieved for most materials, even stainless steel. In fact, surfaces intended for use as submillimetre-wave mirrors need not be polished after machining to 2 μm; it is, however, convenient to have them polished as it is then possible to align the apparatus using the visible radiation generally obtained from the source. The easy production of flat surfaces is accompanied by an easy production of plane-parallel layers since the requirements are similar. Unless an experiment specially requires plane-parallel layers these are best avoided since multiple reflection occurs within the layers (unless they are strongly absorbing) and causes considerable difficulties and confusion in the interpretation of the observations. For example, in a spectral investigation multiple interference effects will be manifested as periodic fluctuations in intensity across the spectrum. This modification to the true spectrum is called the channel spectrum (Edser-Butler fringes). The maxima (or minima) are separated by an amount $\Delta\sigma = 1/\{2n(\sigma_0)\,d\}$ near a wave-number σ_0 for a layer of thickness d and refractive index $n(\sigma_0)$. When coherent radiation from a laser or harmonic generator is being used additional similar effects may be present due to multiple interference (alternatively and appropriately termed standing waves) occurring between widely separated parallel surfaces. Care must be taken in order that channel spectra are not interpreted as true absorption effects.

The spectrometric system, for free-wave propagation, may be a Fourier spectrometer of some sort or a grating spectrometer. Prisms are rarely used for submillimetre-wave spectrometers owing to a poor performance in terms of the *luminosity advantage* of Jacquinot. The grating spectrometer will generally be used sequentially, that is the spectrum is scanned element by element. This approach may be advantageous or even necessary under certain circumstances, for example in examining the time behaviour of the emission from a source. However, the *étendue* of a grating spectrometer is low and when high resolution is required (that is, if narrow slits are necessary) the throughput is even lower. Because of this, the use of an interferometer in conjunction with Fourier transformation is essential in a case where strong attenuation is to be studied or when high resolution is required (say, better than 0·2 cm^{-1}). If photometric precision is necessary it will in any case be preferable to exploit the luminosity and multiplex advantages associated with an interferometer in order to offset so far as is possible the limitations imposed by the low intensity of submillimetre sources. At the long-wave end of the submillimetre band the lamellar grating interferometer becomes a serious competitor for the Michelson interferometer with dielectric beam-divider as commonly used in the far infra-red. Taking the transparency of an interferometer as a measure of its efficiency we find that a lamellar grating is nearly 100 per cent efficient below a wavenumber $\sigma_c = f/gw$, where f is the focal length of the collimator, g is the grating constant and w the slit width. σ_c is easily arranged to lie between 10 and 100 cm^{-1}. A Michelson interferometer is less efficient than this, but is much more easily constructed. Moreover, with it it is possible to measure refractive index spectra by placing a plane-parallel specimen within one arm of the interferometer. Simple filters are used to remove most of the unwanted radiation pervading a spectrometric system. This consists mainly of infra-red radiation the bulk of which is removed by a thin film (say 100 μm) of black polyethylene and a few mm of crystalline quartz. Such a combination will transmit radiation above about 60 μm and so extra filtering is necessary to isolate just the submillimetre band or some portion of it.

In the use of grating spectrometers and, to a lesser extent, Fourier spectrometers, the procedures are much as in the infra-red except that the (optical) filters necessary may require different and rather more detailed attention. This is particularly the case with grating spectrometers where filters with a sharp cut-off are required to eliminate unwanted orders and at the same time retain spectral purity. Generally

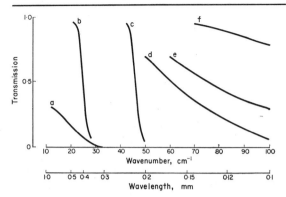

Fig. 2. Transmission of some filters. (a) 5 mm CaF_2. (b), (c) Metal meshes (reflection). (d) Polyethylene + mixed powder, $Al_2O_3 + Cr_2O_3 + K_2CrO_4$. (e) Polyethylene + mixed powder, $Al_2O_3 + Cr_2O_3 + NaCl$. (f) 1 mm crystalline quartz (cuts off near 200 cm^{-1}).

speaking, transmission filters are more convenient than reflection devices. The most effective type of transmission filter is an immersion of powdered inorganic material in polyethylene sheets. These are capable of giving a fairly sharp cut-off but the maximum transmission is rarely more than 80 per cent. A wire mesh may be made to act as a reflection filter. These have sharper cut-offs, the exact wavelength at which these occur being determined by the ratio of the size of the wire of the mesh to the spaces between adjacent strands. If two meshes are placed some distance apart with their planes parallel sharp transmission characteristics are obtained since the configuration behaves much as a Fabry-Perot cavity. The filtering necessary for a Fourier spectrometer is much less critical as the cut-off has to lie below or coincident with some maximum frequency which is determined by how closely the interferogram is sampled (see *Fourier transform spectroscopy*).

When really high resolution spectra are required it is desirable to use a harmonic generator as source, a course of action only possible, at present, at the longer wavelength end of the range.

See also: Submillimetre radiation, applications of.

Bibliography

MARTIN D. H. (Ed.) (1967) *Spectroscopic Techniques*, Amsterdam: North-Holland.

J. CHAMBERLAIN

T

TEMPERATURE-RESISTANT POLYMERS.

Among the many virtues of plastics as useful materials are those of low cost, light weight, high strength: weight ratio and ready fabrication into complex shapes. Thus, plastics are increasingly replacing the more traditional materials, wood and metals, in everyday life. For most applications, the temperature characteristics of these plastics are unimportant although many may be used to around 100°C. Over the past few years there has been a small but persistent demand for synthetic non-metallic materials suitable for use at high temperatures. This demand has been stimulated to a large extent by the requirements of the aerospace programme and the needs of the aircraft, nuclear power and other industries.

Certain of the conventional materials, e.g. glass-filled phenolic or epoxy resins, may be used at very high temperatures (thousands °C) for short periods of time. These reinforced resins have important applications in rocketry and missiles where temperatures as high as 5000°C may be encountered for a few seconds.

However, the main problem is the development of synthetic materials capable of withstanding temperatures up to 400°C for long periods of time without a significant deterioration in their physical properties. A considerable amount of research has been carried out; but despite intensive efforts only a few materials offer any real promise.

The only example of linear polyaddition polymers to find any application at elevated temperatures are the fluorocarbon polymers based on polytetrafluoroethylene ($-CF_2CF_2-)_n$. These materials possess excellent electrical and chemical resistance and may be used continuously up to a maximum of 220–240°C. Several fluorocarbon copolymers, e.g. from vinylidine fluoride and perfluoropropylene, have also been developed for commercial use. These have elastomeric properties and are used as heat stable rubbers up to 220°C. Apart from their temperature limitations, fluorocarbon polymers have low intrinsic strength and are difficult to process. They find use as gaskets, valves, wire and cable insulation, release coatings and the like.

Generally speaking, polycondensation polymers of the thermosetting type are preferred since they possess better heat stability than the polyaddition types. Research has shown that the basic requirements for temperature-resistant plastics are that the material must possess a high softening point, good resistance to pyrolytic attack, and to thermal oxidation. The inclusion of an aromatic or fused heterocyclic ring into a polymer chain (as in aromatic and heteroaromatic polymers) leads to a marked improvement in thermal stability. Several materials have been developed and are now finding a number of commercial applications.

Aromatic Polymers

Several new polymers have been prepared with recurring aromatic units and certain of these withstand temperatures in the region of 500°C. The simplest of these aromatic polymers is *polyphenylene* prepared by linking up benzene units.

Low molecular weight polyphenylenes ($n \leq 6$) are used as high temperature lubricants, hydraulic fluids and heat transfer agents. High molecular weight materials, however, are insoluble and infusible solids and have not found commercial acceptance. Nevertheless, temperature-resistant resins have been developed by chemically bridging the benzene repeating units in the oligomeric polyphenylenes using suitable cross-linking agents under Friedel Crafts conditions. Polyphenylene glass laminates, so prepared, retain strength up to 400°C.

Modifications to the basic structure of polyphenylene, by the introduction of substituents (e.g. methyl groups or fluorine) into the aromatic ring, lead to an improvement in some of the physical properties of the polymer but at the expense of thermal stability.

Improvements in the properties of aromatic polymers may be obtained by the insertion of suitable linking groups between the benzene rings thereby imparting a greater degree of flexibility to the polymer chain. To preserve thermal stability such links must in themselves be resistant both to thermal and oxidative attack. A whole series of aromatic polymers has thus been examined; the most promising of these are polyphenylene oxides, polyphenylene sulphide, polyphenylene sulphone and polyxylylene.

$$\left(-\!\!\left\langle\!\!\bigcirc\!\!\right\rangle\!\!-X-\!\!\left\langle\!\!\bigcirc\!\!\right\rangle\!\!-X-\!\!\left\langle\!\!\bigcirc\!\!\right\rangle\!\!-X-\right)_n$$

X = —O—   polyphenylene oxide
  = —S—   polyphenylene sulphide
  = —SO$_2$— polyphenylene sulphone
  = —CH$_2$CH$_2$— poly xylylene

Polyphenylene oxides

Polyphenylene oxide is the simplest example of the class of polyaromatic ethers developed in recent years. Polymers of high molecular weight are prepared by the controlled polymerization of phenolics. The low molecular weight materials are liquids and find use as high temperature lubricants, nuclear reactor moderator coolants and heat transfer fluids. The best example of the class of polyphenylene oxides so far developed is PPO® and its related plastic Noryl®, developed by the General Electric Company.

$$\left(-\!\!\left\langle\!\!\bigcirc\!\!\right\rangle\!\!-O-\!\!\left\langle\!\!\bigcirc\!\!\right\rangle\!\!-O-\!\!\left\langle\!\!\bigcirc\!\!\right\rangle\!\!-O- \right)_n$$
(with CH$_3$ substituents)

PPO®

PPO has good stiffness, excellent dimensional stability and electrical properties. It can be moulded, shaped and machined with ease. Although heat stable to 400°C in air, PPO is not recommended for commercial use above 190°C due to creep and loss of tensile strength. Several important applications have been

$$\left(-\!\!\left\langle\!\!\bigcirc\!\!\right\rangle\!\!-\underset{CH_3}{\overset{CH_3}{C}}-\!\!\left\langle\!\!\bigcirc\!\!\right\rangle\!\!-O-\!\!\left\langle\!\!\bigcirc\!\!\right\rangle\!\!-SO_2-\!\!\left\langle\!\!\bigcirc\!\!\right\rangle\!\!-O- \right)_n$$

Polysulphone®

suggested including electrical insulators, component parts for electrical and electronic equipment, moulded motor casings and also in the manufacture of surgical equipment.

Closely related to the polyphenylene oxides are the Doryl® resins, first produced by Westinghouse Electric. These materials are prepared by the condensation of diphenyl oxide with formaldehyde in a similar fashion to the classical phenol-formaldehyde condensation. Both straight chain and cross-linked polymers are prepared. Doryl resins possess excellent high temperature properties and are used in the form of varnishes and pre-impregnated sheets for high temperature electrical insulation. Doryl-glass laminates can also be made but very high temperature presses are necessary in order to effect a satisfactory cure.

Polyphenylene sulphide

Polyphenylene sulphide is prepared by the self-condensation of cuprous *p*-bromothiophenoxide in pyridine. The polymer possesses good oxidative stability and is stable in air up to 400°C. It shows an unexpected tenacious adherence to glass, chromium plating, stainless steel and other substrates. For example, on hard stainless steel the polymer exhibits a bond strength in excess of 3000 psi. Polyphenylene sulphide, therefore, shows considerable promise as a high temperature adhesive and laminate. The material also possesses good electrical properties and may be moulded and machined. Several commercial applications have been suggested such as films, fibres, wire insulation for electrical motors, adhesives, laminates and the manufacture of moulded articles.

Polyphenylene sulphone

A number of polyphenylene sulphones containing the

$$\left(-\!\!\left\langle\!\!\bigcirc\!\!\right\rangle\!\!-SO_2- \right)$$

repeating unit have been synthesized and examined. In the laboratory, these materials show good thermal stability withstanding temperatures up to 400°C in air. However, to date the only example of this class of polymer to find any commercial application is Polysulphone® marketed by Union Carbide. This polymer is believed to have a structure (see illustration) consisting of four phenylene groups linked by iso-propylidene, ether and sulphone groups. Although Polysulphone is thermally stable to >300°C, it loses mechanical strength above 150°C. It possesses good electrical properties and is resistant to inorganic acids and alkalis but is readily attacked by polar organic solvents. It has a useful temperature range from —100° to 150°C. The polymer may be injection or blow moulded, extruded or cast into film form.

Polyxylylene

Polyxylylene has been developed by Union Carbide under the trade name Parylene® and is prepared by pyrolysis of *p*-xylene. It is the only example of an aromatic polymer containing an aliphatic linking group ($-CH_2CH_2-$) with any commercial potentiality. The polymer is condensed from the vapour phase onto a cooled substrate and may be used as a protective coat or used in film form. Polyxylylene is resistant to chemical attack and is insoluble in organic solvents up to 150°C. It has outstandingly good electrical properties and is claimed to be stable at 220°C for ten years in an inert atmosphere. However, in air, oxidation takes place fairly readily and the material cannot be used much above 100°C.

A series of co-polymers based on polyxylylene have also been prepared but these have not found commercial acceptance.

Aromatic polyamides

Temperature-resistant aromatic polyamides have been developed and are now offered commercially. The best example of this class of polymer is poly(*m*-phenylene isophthalamide) marketed by Du Pont Co. as HT-1® fibre and Nomex® paper. Polyamides have good thermal stability and may be used to *ca.* 250°C. For example, it is claimed that Nomex retains some 60 per cent of its strength at 200°C. These materials are now finding considerable specialist use for electrical insulation, gas filtration, piping and hose and in the manufacture of high temperature fibres for use in parachutes, flight suits, etc.

Heteroaromatic Polymers

In recent years this general class of polymers has been studied intensively. The main polymer chain contains heterocyclic and aromatic rings, the former usually being formed during polycondensation reactions. Such polymers offer the most promise for the development of temperature-resistant materials.

Polyimides

Of the heteroaromatic polymers developed to date, the most successful are the *polypyromellitimides* or Polyimides. Indeed it is claimed that these materials represent the most thermally stable organic polymer developed and may be used in engineering plastics, films, fibres and adhesives. Their high temperature strength and electrical properties are such that they have numerous potential applications.

Polyimides are prepared in a two-stage condensation reaction between an aromatic dianhydride and a diamine. The intermediate material, a polyamic acid, is produced under mild conditions. This is both soluble and fusible and is used as the precursor in the preparation of films, fibres and laminates. A final bake cure at 200°C is necessary to effect the final intramolecular condensation to the insoluble, infusible polyimide. Moulded articles in polyimide may also be made employing a high pressure, high temperature sintering process. A number of manufacturers are developing these resins. For example, Du Pont Co. market Kapton H® film, Vespel® moulded articles and Pyre-ML® lacquers and adhesives.

Polyimides may be used over a wide temperature range ($-270°$ to 400°C) and are claimed to be flame resistant, charring at $>800°C$. They find numerous applications, e.g. electric motor insulation, magnetic and pressure-sensitive tapes, hoses and pipes, radiation-resistant shields, fuel seals in jet engines, valve components and high temperature resistant lacquers, adhesives and coatings.

Polybenzimidazoles

Polybenzimidazoles are usually prepared by reacting aromatic tetramines with diphenyl esters of aromatic dicarboxylic acids in a melt polymerization. The first stage of the polycondensation gives a soluble intermediate poly(aminoamide) which is heat cured, at *ca.* 280°C, in an inert atmosphere to give the final polybenzimidazole. The cured polymer has good thermal and oxidative stability with useful mechanical and electrical properties up to *ca.* 400°C.

Currently, polybenzimidazoles are offered by the Whittacker Corporation, under the trade name Imidite®, for use as adhesives and glass-filled laminates. The polymer is soluble in highly polar solvents and tough films and fibres have been prepared. A number of applications have been suggested. It is used for thermal and electrical insulation. High temperature radomes, deflectors, turbine blades, radiation shields and the like have also been made.

Polyoxadiazoles

The polyoxadiazoles are of considerable interest since they have excellent fibre-forming properties. They are prepared by condensing aromatic diacid chlorides with aromatic dihydrazides to give polyhydrazides which in themselves are interesting heat stable polymers. On heating, the intermediate polyhydrazides undergo intramolecular cyclization to give the final polybenzoxadiazoles. In some respects, the physical properties of the polybenzoxadiazoles are superior to those of the polybenzimidazoles, since they retain strength to 400–450°C and do not degrade at 450°C in air. However, one of the main disadvantages of the material is that it is often difficult to process since during the polycondensation competing chemical rearrangements occur unless experimental conditions are carefully controlled.

Other systems

A wide variety of other heteroaromatic polymers including polythiazoles, polyquinoxalines, polypyrazoles and polytetrazopyrenes have also been studied. Although many of these show good thermal stability, they generally lack other desirable properties and to date have not found commercial utilization.

Inorganic Polymers

In the belief that the maximum useful life of a wholly organic system is likely to be ca. 400°C, a considerable amount of research effort has been devoted to the development of inorganic polymers for service at higher temperatures. However, to date only the silicones have been developed as a commercially successful inorganic polymer, although these are only stable <250°C.

$$\left(\begin{array}{c} R \\ | \\ -Si-O-Si-O-Si-O- \\ | \\ R \end{array} \begin{array}{c} R \\ | \\ \\ | \\ R \end{array} \begin{array}{c} R \\ | \\ \\ | \\ R \end{array} \right)_n$$

Silicones have been known for quite a number of years and a variety of different types have been examined. In commercial applications, they find numerous uses, e.g. rubbers, sealants, varnishes, electrical insulants, textile coatings, release coatings, water repellants and the like. In themselves, the silicones do not possess any degree of strength and they cannot be used as materials of construction. Their maximum working temperature is about 220°C, failure occurring due to the tendency of the polysiloxane chain to undergo thermal rearrangement to cyclic structures of low molecular weight and thereby causing a diminution in physical properties. Attempts have been made to overcome this cyclization by modification of the polymer either by the insertion of organic groups, e.g. benzene rings as in the polysilphenylenes, or by the incorporation of aluminium, titanium or other suitable metal. Of these only the polysilphenylenes show any real promise. A number of cross-linked laminates based on such polymers are claimed to be stable to 400°C in air and to possess good mechanical properties.

Many other inorganic systems have been studied. The phosphonitrilic polymers $\left(\begin{array}{c} Cl \\ | \\ -P=N- \\ | \\ Cl \end{array} \right)_n$ showed early promise. These are prepared by heating the readily available phosphonitrilic chloride. They have rubber-like properties and are stable in air to 300°C. However, the phosphorus–nitrogen bond is subject to hydrolysis and the phosphonitrilics have not found commercial acceptance. Modification of the chemical structure has produced no marked improvement in properties.

Polyphosphates, polyborazoles, polyboraphanes and analogous systems have been synthesized. Some of these have surprisingly good thermal stability, but they generally lack hydrolytic or oxidative stability and are often brittle and lack toughness.

A number of other inorganic polymer systems have been synthesized, based on beryllium, germanium, titanium, aluminium and other metals. Several show remarkable heat stability and at least one or two may reach the development stage within the next few years.

Ladder Polymers

Most of the examples of heat stable materials possess a linear structure. On degradation, the polymer chain is broken with resultant loss in physical properties. Recently, ladder polymers have been synthesized. As the name implies, the basic structure of such polymers consists of a double chain cross-linked at regular intervals. Both organic and inorganic ladder polymers have been examined. While they possess excellent thermal resistance, so far little commercial application has been found for these materials. The only example to find a strictly limited use is believed to be Black Orlon prepared by the controlled pyrolysis of polyacrylonitrile. This polymer possesses essentially a naphthyridine-type structure.

Black Orlon

A form of Black Orlon is marked by the Minnesota Mining and Manufacturing Co. under the trade name Pluton B® for specialized uses in the manufacture of valve seats, gaskets, seals, brake and clutch linings and the like.

Conclusions

Some of the more promising materials (e.g. polyimides, polybenzimidazoles) used in conjunction with carbon fibres or refractory whiskers will undoubtedly be developed as composites for specialized engineering applications. A few composites are already being used in the manufacture of turbine blades for aero engines, in aircraft structures and in ablative shields where the requirements are for materials with high temperature resistance coupled with low density and high strength.

Although many different organic and inorganic polymer systems are currently under investigation, it is difficult to decide which is likely to be the most successful. What is clear, however, is that no one system combines all the desired properties for continuous high temperature use. Many systems so far studied are insoluble, infusible powders in their final form. New and improved methods of polymer technology must be developed in order to process such materials. Emulsion spinning, high temperature–high pressure sintering processes and the use of highly polar solvents illustrate the new techniques now being used for the preparation of potentially useful temperature-resistant materials from previously intractable substances.

For the future, improved and refined polyaromatics are likely to find a maximum ceiling temperature of 300°C in contrast to the heteroaromatic polymers which will be suitable up to 400–450°C. For temperatures between 400–450°C, ladder polymers and inorganics offer the greatest potentiality but these require a much greater research effort before they can find commercial outlets. Only the ceramics are likely to be of any use for temperatures in excess of 500°C.

Bibliography

BRENNER W., LUM D. and RILEY M. W. (1962) *High Temperature Plastics*, New York: Reinhold Publishing Corp.

FRAZER A. H. (1968) *High Temperature Resistant Polymers*, New York: John Wiley and Sons.

GIMBLETT F. G. R. (1963) *Inorganic Polymer Chemistry*, London: Butterworths.

IDRIS JONES J. (1964) Thermally stable organic polymers, *Reports on the Progress of Applied Chemistry*, **49**, 621–36, London: Society of Chemical Industry; (1968) Synthesis of thermally stable polymers: a progress report, *Reviews in Macromolecular Science*, **C 2**, 303–71, New York: Marcel Dekker.

ROBB J. C. and PEAKER F. W. (1968) *Progress in High Polymers*, **2**, 187–313, London: Iliffe Books Ltd.

SEGAL C. L. (1967) *High Temperature Polymers*, London: Edward Arnold (Publishers).

E. JONES

TEXTILE INDUSTRY, PHYSICS IN. The applications of physics in the textile industry are diverse. Some of them are common to other manufacturing industries: the statics and dynamics of machines; heat, moisture and mass transfer; measurement methods; control techniques; and so on. But the most interesting applications, with which this article will deal, are concerned with the special features of textiles.

The peculiar characteristics which determine the physics of textile material properties and of textile production arise from two main causes. Firstly, textiles are designed to combine strength and flexibility, in contrast to the usual engineering requirement of strength and rigidity. As a result, textiles can suffer large deformations, and the usual approximations of small strain theory cannot be applied. Secondly, in order to achieve the first characteristic, textiles are made of fibres, which are extremely fine, flexible, long units of matter. In the textile fabric, they must be assembled in such a way that the inherent fibre flexibility is maintained and enhanced, while the strength of the fibre along its length is brought into effective use when the fabric is highly stressed.

Many of the physical problems arise from the dimensions of the fibres. A typical textile product, such as a man's suit, contains over 10^7 fibres: it is the job of the textile industry to put them in the right place! Clearly, it is impossible to do this by using conventional engineering design and production processes, in the sense that the position of each unit is specified in a plan and put in the right place. Instead it is necessary to develop both the physics of fibre assemblies in order to give useful general descriptions of fibre arrangements and their relation to properties, and the dynamics of masses of fibres in motion in order to be able to control the formation of particular fibre assemblies. One problem, which has faced physicists, is that the textile industry has acquired vast expertise in thousands of years of development as a craft industry: consequently, the initial attempts at the application of physics tended to do little more than explain scientifically what was already known practically. In the most recent developments, we are now beyond this stage.

Another aspect of the subject is the study of fibre structure and properties. A typical fibre, a few centimetres long, will have a mass of about 10^{-5} g and a diameter of less than 10^{-3} cm, only one order of magnitude greater than the wavelength of light. The study of a piece of matter with these dimensions raises interesting problems. Furthermore, only a limited number of substances, mostly crystalline polymers, are suitable for making fibres. The material properties are unusual (at least to a conventionally educated physicist) and interesting both experimentally and theoretically.

Looking back, we find some early references to interest in the science of textile materials: both Galileo and Leonardo da Vinci make comments on problems of the mechanics of textile structures. But it was not until this century that there was any really serious work on physical properties. There were a few scattered studies made up to 1920, including some good analyses of the mechanics of fibre assemblies, notably by Gégauff and Haas; and then, mainly owing to the establishment of several textile research associations in Britain and later similar developments elsewhere, a concentrated study of the physics of textile materials and textile processes began. This has continued and grown, although now the contribution of industrial and University laboratories is more important than the contribution of the research associations. Among the pioneers of the subject, mention should be made of W. L. Balls and F. T. Peirce.

Two developments in the last twenty years have made textile physics more important. Firstly, whereas at the turn of the century the textile industry used only about six basic types of natural fibre in only a few long-established manufacturing processes, there are now around fifty different fibre types to choose from; and a number of new processes have been introduced or are being developed. The old empiricism is no longer adequate, scientific choices must be made if there is to be rapid progress. Secondly, textiles can be seen as part of materials science generally. Much fibre physics is an important part of the rapidly growing subject of polymer physics. High-strength fibres are being used in reinforced materials with advanced properties. And the study of textiles has led to fundamental advances in subjects such as static electricity, applied mechanics, and transmission properties. Textile physics is no longer an isolated subject. We must now turn to some of its special areas.

Fibre Physics

The physics of fibres can be divided into three main areas: measurement and empirical description of physical properties; investigation and understanding of fibre structure; theories of the relation between structure and properties.

Experimental study of physical properties is largely a discovery of deviations from simple ideal behaviour: Hooke's law is not followed, stress–strain relations are non-linear and show hysteresis and time dependence; Ohm's law is not followed, current is not proportional to voltage; Amonton's laws are not followed, frictional force is not independent of area and not proportional to normal load; fibres are anisotropic, and so doubly refracting; fibres sometimes contract rather than expand on heating; the absorption of moisture and the effects of heat can cause temporary or permanent changes in properties.

First in importance among fibre properties come the dimensions. On an individual basis, these are most conveniently determined by finding the length; the mass, using a specially sensitive balance; and the density usually by flotation in a density gradient column. The transverse vibrations of fibres under a controlled tension can also be used to measure the mass per unit length (linear density) of a fibre: this quantity, which is generally both more convenient to measure and more useful than the diameter or the area of cross-section, is now preferably expressed in units called *tex*, namely g/km, although *denier* (g/9 km) is still widely used.

In order to characterize natural fibres, in which a given sample will contain fibres with a range of dimensions, and to control quality in man-made fibres, there is a great need for routine tests of fibre dimensions and other properties. Traditionally this function has been performed subjectively by skilled graders. Among newer methods, fibre fineness can be estimated from the rate of air-flow through a plug of fibres, and various photo-electric devices have been used for length measurements. However, there is still a great need for improved test methods: the problem has a feature which most physicists do not meet—not only must the test be technically satisfactory, it must also be extremely cheap to run.

Second in importance are the mechanical properties. Commonly, the relation between stress and strain in simple extension is examined: a typical curve, with an indication of properties that are usually calculated, is shown in Fig. 1. Because of the difficulties of direct measurement of area of cross-section,

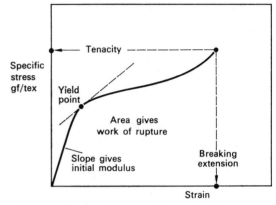

Fig. 1. Stress–strain curve illustrating important fibre properties.

and because it is usually more useful to compare textile materials in terms of comparable masses, the *specific stress*, equal to (force/linear density) in gf/tex or gf/den, is preferred to the use of conventional stress. The relations between these quantities are as follows:

$$1 \text{ gf/tex} \equiv 1/9 \text{ gf/den} \equiv \varrho \text{ kg/mm}^2$$

where ϱ = density in g/cm^3

$$1 \text{ gf/tex} \equiv 9{\cdot}81 \times 10^3 \text{ S.I. units of specific stress} \\ (\text{newton kg}^{-1}\text{ m})$$

Specific stress can be used to make useful comparisons between forces borne by molecular chains, by individual fibres, and by yarns and fibres where much of the volume is occupied by spaces between fibres: it is related to energy per unit mass, in the same way that conventional stress is related to energy per unit volume.

A comparison of the tensile properties of fibres is shown in Fig. 2.

However, there are many other aspects of the study of mechanical properties:

(a) Fibres are anisotropic and more than two elastic constants (or, in general, more than two functional relationships) are needed to characterize their properties. Torsion and bending are fairly easily studied; the former gives a shear modulus; but the latter depends on the tensile modulus, although if the fibre is not homogeneous the averaging will be different, since bending is more affected by the outer layers. Some measurements of bulk moduli, of Poisson's ratios and of transverse moduli have been attempted. Many fibres have, at least approximately, orthotropic symmetry, by this we mean that they are "transversely isotropic" with identical properties in all directions perpendicular to the fibre axis but different properties in other directions, with two principal directions along and perpendicular to the axis. The various elastic constants are indicated in Fig. 3: relations between them reduce the number of independent constants to four.

The electrical properties of fibres include dielectric properties, electrical resistance and electrostatic charging. The dielectric constant depends on the basic chemical constitution, and in particular on the amount of absorbed water present. Electrical conduction is

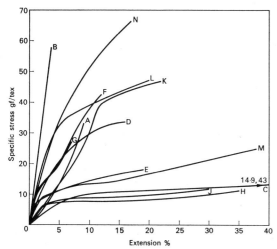

Fig. 2. Stress–strain curves of a range of fibres under standard conditions (65 per cent relative humidity; 20°C).

A American cotton
B flax
C wool
D silk
E continuous filament viscose rayon
F high tenacity viscose rayon
G polynosic rayon
H secondary acetate
J triacetate
K nylon 66
L polyester
M acrylic (Courtelle)
N polypropylene

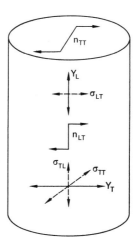

Fig. 3. Elastic constants of a transversely isotropic fibre. Y_L and Y_T are Young's moduli; n_{LT} and n_{TT} are shear moduli; σ_{LT}, σ_{TL} and σ_{TT} are Poisson's ratios.

usually ionic, and, since it depends on the degree of disassociation of the ions present, it is determined mainly by the dielectric constant: the following equation which has a theoretical basis, fits most of the observations:

$$\log R = A/\varepsilon + B$$

where R = resistance, ε = dielectric constant, and A and B are constants.

Studies of static electricity have shown that the inherent tendency to charge separation on surfaces in contact is very great, but the observed charges on a good insulator are reduced by leakage through the air. On a moderate conductor, with a resistance of less than about 10^{10} ohm cm, there is also a rapid discharge through the material itself. However the newer synthetic fibres do have very high resistances, so that static changes are a common nuisance: they can be removed by anti-static dressings or by ionizing the neighbouring air.

(b) Fibres are imperfectly elastic; there is hysteresis in extension and recovery; and permanent deformation occurs where fibres are strained beyond the yield point. The elastic recovery is defined as the ratio of elastic strain to total strain.

(c) The mechanical behaviour is time-dependent. Under a steady load, there will be an immediate elastic extension, followed by primary (recoverable) and secondary (non-recoverable) creep. At constant strain, stress relaxation occurs. The time-dependent behaviour can also be studied by the effect of rate on load–elongation curves, and by dynamic tests with an oscillatory strain.

Friction commonly follows a relation of the form $F = a N^n$, where F is frictional force, N is normal load, and a and n are constants: n is usually between 0·6 and 1.

A collection of typical physical properties of fibres is given in Table 1.

In matter which is capable of giving strength in the form of fine fibres, there must be good continuity of structure along the length of the fibre. Furthermore, in textile fibres, a moderately high degree of exten-

Table 1. Textile Fibres and Their Properties (typical values)

Type	Name	Range of diameters μ	Density g/cm³	Initial modulus g/tex	Tenacity g/tex	Breaking ext. %	Work of rupture g/tex	Elec. res. 65%rh ohm-cm	Moisture regain 65%rh %	Melting point °C	Strength ret. 20 days, 130°C %	Attacked, degraded, dissolved by
Natural vegetable	Cotton	11–22	1·52	500	35	7	1·3	10^7	7	§	38	strong acid, strong alkalis, mildew, light
	Flax	5–40*	1·52	1800	55	3	0·8	10^7	7	§	24	
	Sisal	8–40*	1·52	2500	40	2	0·5	10^7	—	§	—	
	Jute	8–30*	1·52	1800	50	2	0·5	10^7	12	§	—	
Natural animal	Wool	18–44	1·31	250	12	40	3	10^9	14	§	—	strong acids, alkalis, light
	Silk	10–15	1·34	750	40	23	6	10^{10}	10	§	—	
Regenerated fibres	Viscose rayon	12+	1·49	500	20	20	3	10^7	13	§	44	acids, strong alkalis, mildew, light
	High tenacity rayon	12+	1·49	600	45	10	4	10^7	13	§	—	
	Polynosic rayon	12+	1·49	800	30	8	1	10^7	11	§	—	
	Fortisan	5+	1·49	1700	60	6	2	10^7	11	§	28	
	Acetate	15+	1·32	350	13	24	2	10^{12}†	6	230	—	acids, alkalis, acetone, light
	Triacetate	15+	1·32	300	12	30	2	>10^{12}†	4	230	—	acids, alkalis, light
	Casein	17+	1·30	350	10	60	4	.10^9	14	§	—	
Synthetic	Nylon	14+	1·14	250	50	20	6	>10^{12}†	4	250‡	21	strong acids, light
	Terylene	12+	1·39	1000	50	15	5	>10^{12}†	0·4	250	95	strong alkalis
	Acrylic	13+	1·17	650	30	25	5	>10^{12}†	1	250	91	strong alkalis
	Polypropylene	—	0·91	800	60	20	8	>10^{12}†	0·1	165	—	light
Inorganic	Glass	5+	2·54	3000	76	2·5	1	10^9	0	800	100	very resistant
	Asbestos	0·01–0·03*	2·5	1300	—	—	—	—	1	1500	—	very resistant

* Ultimate fibres—usual textile fibres are coarse bundles.
† Usually much lower, due to anti-static agents.
§ Decomposes first.
‡ Nylon 6 melts at 225°C.

sibility is necessary, although not the extremely high values found in rubbers. These requirements are met by a system of long-chain polymer molecules arranged in partly ordered, partly crystalline form. The crystallinity provides the strength and stability; the disorder allows some freedom for extension. Ever since it was established in the 1920's that natural fibres were made of crystallizable polymers, there has been controversy about the detailed fine structure. A very successful view was the fringed micelle model illustrated in Fig. 4: the long chains pass between crystalline

Fig. 4. Typical diagram of fringed micelle theory of fibre structure.

and non-crystalline regions. But more recently this view has been criticized, and structures with fibrillar units, structures with a more uniform distribution of order and disorder, and structures with localized crystal defects have been proposed. The conflicting views can be reconciled when it is realized that described structures are inevitably simplified and idealized while real structures are complicated, and that there will be a great deal of diversity in structure among different fibres. Many fibres also have a larger-scale morphology.

Techniques used to study structure include optical and electron microscopy; optical and infra-red absorption, birefringence and dichroism; X-ray diffraction; and nuclear magnetic resonance. The structures must also be compatible with the general physical properties. In the last 10 years, the study of polymer crystallization has led to the possibility of speculating on what structures would be likely to form, either during the rapid crystallization and subsequent stretching on the extrusion of man-made fibres, or during the slow build-up of natural fibres in living cells.

The complexity of interpretation of properties can be indicated by the different modes of deformation which can occur in fibres: these are listed in Table 2. The strategy of theoretical interpretation is to choose a particular limited aspect of behaviour, and the tactics are to select the most useful model to analyse. The physical form of the structure is important: for instance, by different modes of formation, cellulose can be caused to crystallize in rayon fibres in either a micellar or a fibrillar form with marked differences in the resulting mechanical properties.

Table 2
Modes of Deformation of Polymeric Fibres

Elastic, recoverable	Plastic, permanent
Energy dependent	*Crystallization*
Short-range	recrystallization
crystal lattice	unfolding, refolding
glassy polymer	crystal defects moving
chain	*Molecular*
crosslinked network	glassy polymer chain configuration
Long-range	
crystal transformation	chain slippage
	chain breakage
Entropy dependent	crosslinks breaking,
Long-range	forming
rubber elasticity	and re-forming

As an interesting example of the interaction of thermal and mechanical effects, the behaviour of nylon can be mentioned. The molecule of nylon 66 has a very long repeat:

$$-CO \cdot CH_2 \cdot CH_2 \cdot CH_2 \cdot CH_2 \cdot CO \cdot NH \cdot CH_2 \cdot CH_2$$
$$\cdot CH_2 \cdot CH_2 \cdot CH_2 \cdot CH_2 \cdot NH-$$

In parts of the fibril the chains will be packed in regular crystalline order: elsewhere they will remain disordered, unable to sort themselves out into crystals. At room temperature the $-CH_2-CH_2-$ links will be flexible, and so give some freedom of deformation to the structures. But the freedom will not be too great, because the $-CO \cdot NH-$ groups tend to associate together by forming hydrogen bonds. Above about 50°C, these bonds disassociate: a temporary setting can thus be achieved by deformation, above 50°C, followed by a cooling in which the bonds reform in new places. Permanent heat setting can be obtained at higher temperatures, for at temperatures appreciably below the true melting-point (250°C) small, imperfect crystals will melt and recrystallize or anneal.

We should note that the structure found in a fibre is a metastable state resulting from the particular way in which the tangle of chain molecules has crystallized. There is little chance of the system sorting itself into a state of true stable equilibrium. There are many metastable states with closely related structures, and it is easy by heat effects or mechanical deformation to shift from one metastable state to another. What is unusual in fibres, and other crystalline polymer systems, is that the metastable states correspond to structures which have dimensions approaching the sizes of the molecular repeat units and much less than the whole lengths of chain molecules.

Indeed, one of the interesting features of the relationship between fibre structure and properties is the way in which effects which are commonly regarded as microscopic, such as stress and strain distribution in inhomogeneous materials, must be combined with effects which are molecular. There is much structure at about the 50 Å level, too coarse to consider in terms of individual molecular interactions but too fine to ignore the molecular texture. In wool fibres, for example, microfibrils with a diameter of about 75 Å can undergo a phase change from a crystalline form containing helical chains to another form containing extended chains, with an accompanying extension of 100 per cent. The microfibrils are embedded in a rubber-like matrix. The study of the behaviour of wool in extension thus involves experiments under a variety of procedures and conditions; theoretical discussion of the basic molecular behaviour of the system in terms of moduli of deformation, equilibrium and critical stresses for the phase change, rate effects, and coupling between fibril and matrix; and analyses of the mechanics of the particular composite system.

Fibre Assemblies

At the larger scale of fibre assemblies, whether intermediate forms such as yarns or final forms such as fabrics and made-up products, the physical problems are classical but still complicated. Again both structure and properties are important.

Yarns are usually dealt with on the basis of an idealized model consisting of an assembly of fibres following helical paths around concentric cylinders. Twist is necessary in order to give cohesion to the yarn. The mechanics of this model, including the effects of transverse forces between fibres and of slip at fibre ends has been solved both by equilibrium of force methods and, more recently, by energy methods. In most applications, energy methods are proving more powerful, simpler to use, and easier to express in general terms applicable to many problems.

In real yarns, the fibres follow helices of varying radius. Indeed, this is essential in yarns made of staple (short) fibres, which would otherwise just peel off the surface. The actual fibre paths can be followed by including a few coloured fibres and then immersing the yarn in a liquid of similar refractive index to that of the fibres. The geometry of migrating fibres has been extensively studied.

In fabrics, the problem is to define the equilibrium path taken by the interlacing warp and weft threads in a woven material or the interlooped threads in a knitted structure, and to see how the structure deforms by strain and slip under load.

More recently, non-woven fabrics, made of a web of fibres, either bonded together by adhesive or mechanically entangled by needling, have been developed. Usually these materials do not have an adequate combination of strength, stability and flexibility. When they are strong enough, they are too stiff for many purposes. The understanding of the mechanics of these assemblies is leading to practical advances which are likely to be of great technological importance because of the economic advantage of direct fibre to fabric, or if extrusion is included polymer-to-fabric, production. The problem is to secure adequate freedom for deformation, while developing enough strength under severe conditions.

Besides mechanical properties, the transmission properties of fabrics are of interest, and of technical importance in the comfort of fabrics and for industrial uses such as filtration. One topic of interest is fluid transmission through the inter-fibre spaces. Another is heat and moisture transfer which can occur both in the fibres and in the spaces: the diffusion process is rendered more complicated by the fact that heat and moisture effects are coupled, because a change in temperature causes a change in the amount of water absorbed and because absorption generates heat.

Textile Processes

In much of the traditional design and operation of textile machinery there has been a tendency to regard the textile as an inert passenger to be forced into certain paths. With this approach the physical problems are in the machines. The interesting subjects of study include the control and measurement of tension in running yarns; the effects of eccentricity of rollers; the forces developed in different parts of a loom as it bangs the shuttle across and raises and lowers the threads several times a second; the measurement, control and efficiency of drying; the control of uptake of size and other additives; and the general increase in machine speeds and efficiency. The interaction with the textile is taken into account in finding the optimum settings for machine parameters, and in avoiding faults due to yarn breakage or other causes.

This approach was natural because textiles had developed as a hand craft; and with the industrial revolution, the old hand operations were merely transferred to machines, which have since developed mainly by empirical advances.

But really, the behaviour of the textile should be regarded as the prime feature, and then, once we know how we want the material to behave, the machine design should follow. This point can be well illustrated by weft projection in weaving. The aim is to pass the weft yarn across between alternating sets of warp threads. In the simplest form of hand-weaving, this was quite rightly done by winding the yarn on a bobbin and passing it backwards and forwards. But then, when the process was automated, the bobbin was put in a wooden shuttle, and the development of looms concentrated in this aspect on improving the projection of heavy shuttles. The real aim—just to get the yarn across—had been forgotten; and, what is more, the use of shuttles brought the technical disadvantages that a lot of power is needed to project the shuttle and that the size of bobbin which can be accommodated is limited.

A new fundamental look at the problem, i.e. the consideration of what other physical methods could be used to project yarns across a loom, has led to the introduction of new loom types in recent years.

They all start from the view that it is better to have the yarn on a large package beside the loom, and then to throw across each time only the particular length of yarn needed. This gets rid of the need to throw across the whole package and the shuttle. The problem is how to transmit the yarn: all the methods tried involve interesting problems of mechanics. One commercially successful method is to use a small metal gripper to hold the yarn: the momentum given to the gripper then carries the yarn over. Alternatively yarns can be carried across in jets of water or air. Finally, in a beautiful demonstration, J. J. Vincent has shown that it is possible to accelerate the yarn so that it is carried across by its own momentum: in practice, in order to prevent the yarn piling up in a ball as the front end slows down due to air resistance, the rear end must be released more slowly.

By his training, the physicist is well equipped to take the broad view and look at the fundamental aims of an operation and see what the possibilities are. He is also equipped to deal with the detailed problems of mechanics which occur. In the end, however, he must remember that the choice of method depends on the overall balance of the economics, including capital costs, labour costs, power costs, material costs, output, and interaction with proceding and subsequent processes through the quality of the raw material needed and the quality of the output produced. The process which is physically most satisfying is not necessarily economically the best: shuttles still have their uses!

The example of weft projection has been discussed at some length, because the physical problems of momentum, acceleration, air drag, and so on, while complicated, can nevertheless be comprehended by any physicist. In some other processes, it is more difficult to follow what is happening.

Take the manufacture of cotton yarns as an example: this is significant because cotton still accounts for more than half the weight of textiles produced in the world, and because similar problems arise with other staple fibres.

The cotton is supplied in dense bales. The first series of operations is designed to remove dirt and to separate the fibre into a fine, open web. This is all mechanics, since it involves acting on the fibre masses with aerodynamic forces and with beaters, followed by combing between flat surfaces covered with fine wire points. Most studies in this area have been empirical.

The next stage consists of condensing the web to form a ropelike sliver of loose fibres. This sliver is then drafted by passing between pairs of rollers running at different speeds: there are two purposes here, to pull the fibres into parallel alignment and to attenuate the sliver into a fine yarn. The movement of fibres in the free zone between the rollers involves a complex pattern of acceleration: the fibres enter at the speed of the first rollers, and are then speeded up to leave with the speed of the second rollers. This tends to be an unstable process in which the fibres are accelerated in groups and so give rise to drafting waves. The drafting process has been studied quite extensively, and progress has been made in understanding the physics of what happens. Much of the work concentrates on the study and avoidance of yarn irregularity.

Finally the yarn has to be twisted in order to give it strength. This is usually done by ring spinning, using the principles illustrated in Fig. 5. The yarn

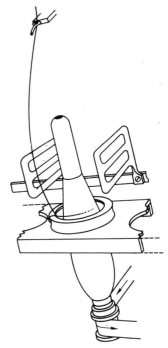

Fig. 5. Ring spindle.

is fed down to a spindle rotating at high speed. The slight drag exerted by a traveller on a ring causes the yarn to wind up on a package. The rate of insertion of twist is thus equal to the difference between the winding speed and the spindle speed. The twist runs back up the yarn, so that it is really inserted in a small V-region close to the rollers through which the strand of fibres is supplied. There are a host of physical problems here: the transmission of twist back along the yarn, with a partial blockage at any yarn guide; the fibre movements at the point of yarn formation, which control the yarn structure; the form and density of the package which is built up; and, most important, the dynamics of the balloon which forms as the yarn whirls round, influenced by inertial forces, aerodynamic forces and frictional forces, and determining yarn tensions, the space needed per spindle, and the limiting and optimum conditions for the whole operation. Aspects of yarn ballooning also occur in any process of high-speed winding of yarns from one package to another. Once again, the problems are not only technical, they are also economic: for example, there are advantages in labour costs in spinning large packages which do not need doffing so often, but this puts up the power needed to rotate the package. The research physicist must bring cost in as a variable in determining the relevant functional relationship.

In traditional methods of twisting such as the one described above yarn is passed between two packages which are rotating relative to one another about the yarn axis. The technical and economic limits (which are not identical) depend on the difficulty and cost of rotating whole packages. This is true of all the usual variants of twisting methods, each of which brings in particular applications of physics.

But now, a new approach is being tried. If twist can be inserted at a break in the yarn, as indicated in Fig. 6, then it is not necessary to rotate the pack-

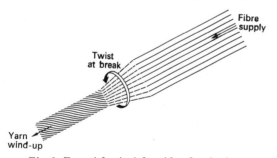

Fig. 6. *Essential principles of break spinning.*

ages, except at the speeds needed to wind the yarn. There seem to be inherent advantages in this: higher speeds and larger packages being possible. A variety of different methods are being tried. The most promising are drum spinning, in which the fibres are fed in to the rim of a drum rotating at high speed and then picked up by the end of the newly formed yarn, and vortex spinning, in which the fibres are fed in by an air-stream and picked up by a free yarn end whirling round in a vortex. The development of these methods involves not only the detailed dynamics of the process, but also the need for basic thought about the mode of yarn formation and the sort of yarn structure which is wanted. The need of twist to give cohesion in a staple fibre yarn is obvious, but traditional methods also control other aspects of yarn structure, in particular the migration which is needed if fibres are not to lie in separate, incoherent layers. The same mechanism will not necessarily operate in the new processes, and so it is necessary to study the fundamentals of fibre assembly in the yarns and the effects on yarn properties.

With continuous filament yarns, there are other problems. Here the yarn itself is made by the fibre producer, though the textile industry (in a narrower sense) may have to perform subsequent winding and twisting processes. In continuous filament yarns, uniformity is of enormous importance: in such a simple regular structure the slightest difference in colour, lustre, size, shape or mechanical properties shows up in a fabric. When we remember how easily yarns can be permanently deformed, we see the importance of maintaining uniform, and preferably low, tensions at all stages of processing. Friction and lubrication are important studies. What is more, yarns will respond differently according to the position on which they are wound on a package: at the centre on a rigid tube, no length changes will be possible, but stress relaxation will occur, while near the surface of a soft package dimensional changes can occur. These differences, which reflect changes in internal structure, can show up as dyeing faults and in other ways.

Creep and relaxation effects also play a part in determining the behaviour of yarn in a loom. During operation a particular sequence of dynamic response of the yarn will be maintained; but if the loom is stopped the response will change, the balance of forces, which affect the structure, will alter, and a fault may appear. This sort of problem is typical of what is found in the application of physics to the whole weaving process.

In knitted fibres, too, the passage of yarn through the machine and the equilibrium fabric structure formed must be studied. In simple hand-knitting it is obvious that the tightness of the material will depend on the tension applied to the yarn. The earliest control methods for machine knitting concentrated on tension control. But there is a better method of control, which a physicist developed. This is a positive feed device, which supplies just the right amount of yarn for each cycle of stitch formation. The fabric structure is thus controlled in terms of directly relevant parameters.

After the fabric has been formed it is subject to a variety of colouring and finishing processes. Much of this processing is predominantly chemical, though all the processes involve some physics problems; and some treatments, such as mechanical stretching, are purely physical.

Heat setting treatments are important in both yarn and fabric processing. In fabrics, setting is designed to achieve smooth, wrinkle free, crease-resistant fabrics or to put in pleats or required creases. In yarns, setting is designed to force the fibres to take up a distorted form, so that an ordinary filament yarn is given bulk, stretch, and texture. All setting involves problems of heat transfer, of fine temperature control, and the carrying out in an industrial process of the annealing processes discussed in the previous section on fibre physics.

The most important of the yarn bulking processes has many interesting physical features, both in the process and in the resulting yarn. The process involves twisting a yarn, setting it, and then untwisting it. The resulting twist liveliness in the filaments is relieved by filament distortion. The theoretical problem in the final yarn is being tackled by finding the equilibrium conformation of lowest energy. For a simple snarled form, there are only two independent variables, and a reasonably good analytical solution has been found, and confirmed by experiment. But, since the filaments are set in a helical form in the yarn, helical conformations are also found to occur: in one common form, consisting of helices twisting in opposite directions, there are six independent variables, the theoretical problem is more complicated, but solutions are being attempted by computer methods.

The original manufacturing process for these yarns was in a series of separate steps. Now it is carried out in a continuous operation: yarn is fed over a heater onto a "false-twist" spindle which causes twist to run back into the yarn on the heater. The inter-related problems of heat and twist transfer and the effects of cooling and of tension all contribute to the physics of the process. The rate of advance in the practical development can be seen by the fact that typical spindle speeds have risen from 50,000 rpm to 500,000 rpm in ten years.

Textiles in Use

The remaining major applications of physics relate to textiles in use. One area is in testing and quality control. Suitable test methods which can be used to ensure the maintenance of quality must be devised. It should be noted that these are not necessarily tests of real, practically relevant properties. Thus tensile strength is often used as a quality control test, though it is rare for a new textile fabric to fail in a direct pull. Failure is due to much more complicated processes of wear. Nevertheless, for a particular type of fabric, strength may be used as a general measure of the reproducibility of a given quality.

The problems are more difficult when new fibres or new constructions are being tried. Then physicists must try to isolate what are the real demands of any application, and to devise relevant tests.

A typical example would be tyre cords, which account for a surprisingly large fibre usage. Here the real criterion is the behaviour of fibres embedded in rubber and subject to intermittent cyclic stresses, and to temperature changes as a result of internal energy losses and the external environment. Because the market is economically very evenly balanced between rayon and nylon, understanding of the physical realities of the situation is important. The fibre producer needs to know what are the important physical properties to aim at.

A more limited example may be more easily comprehended. The rigging-lines of parachutes used as brake parachutes for aircraft suffer a loss in strength with use, and so their useful life is limited. A thorough physical investigation is needed if the causes are to be sorted out. This would include tests in fibres, yarns and cords and the examination of the possible effects of fibre fatigue, inter-fibre wear, chemical deterioration due to heat, specific damage due to snagging or local heating, and other sources of deterioration.

Although the problems in the industrial applications of textiles are complicated, those in the domestic and fashion field are even more so. Here we are concerned not only with objective qualities, such as useful life or efficiency in thermal insulation or in water repulsion, but also with purely subjective, aesthetic qualities like drape, feel, handle, and various aspects of appearance and texture. The physicist tackling this type of subject has a multiple problem: the study of subjective evaluation and ranking; the use of related objective tests; the relation of the complex phenomena to simpler physical properties of the fabric; the relation of these fabric properties to the constituent fibre and yarn properties and fabric structure.

Drape may be taken as an example. The subjective evaluation was examined by asking observers to rank fabrics in skirt form for both magnitude and attractiveness of drape. As would be expected, a plot of attractiveness versus magnitude showed a peak, which varied in position with observer and would change with changing fashion. A more subtle point was some regular deviation between different types of fabric; this suggested that the observer also reacted to the detailed form of the drape. As an objective test, the drape of a circular fabric specimen over a circular stand was used: a parameter derived from the projected area of the draped fabric gives a numerical *drape coefficient*, which correlated well with the subjective assessment. Then it was shown that three-dimensional drape which involved double curvatures of the fabric could be related to a combination of simple bending in one plane and fabric shear. The behaviour of fabrics in bending and shear is part of the basic study of fabric mechanics.

Conclusions

In conclusion, it seems best to go back to the beginning again: "The applications of physics in the textile industry are diverse." They include such fundamental topics as crystallization and the statistical mechanics of polymer chains, the thermodynamics of solids and the behaviour of systems with many accessible metastable states, basic rheology and viscoelasticity, the mechanics of systems with large deformations, the principles of energy methods for the solution of static problems, and, perhaps later, the use of energy methods to find the dynamic path followed by a fibre assembly during processing. They include opportunities for theoretical analysis, for experimental studies of materials and processes with available equipment, for the development of new test methods and instruments, and for the invention and developing of new processes.

Inevitably the above article has been selective, and in its examples biassed by the author's personal experiences, but the aim has been to show the typical ways in which physics has been used, and will be used in the textile industry.

J. W. S. HEARLE

THERMIONIC GENERATION OF ELECTRICITY.

Introduction

The origins of the thermionic converter can be traced back to Edison's experiments in 1883 with vacuum lamps. He found that a small current could be collected from an electrode placed adjacent to a battery-heated filament, provided the electrode was connected to the positive end of the filament. Under these conditions, however, the current was equivalent to that of a thermionic diode rectifier of the type patented by Fleming in 1904. True thermionic conversion caused by the action of heat only, and not an applied field, can only be observed with an indirectly heated cathode, or by measuring the anode current during the off part of the heating-current cycle.

Under these conditions, significant currents are passed only when the anode-cathode separation is less than about 0·5 mm. The mode of operation can probably be best explained by reference to Fig. 1(a). Free electrons in the cathode have a mean energy given by the Fermi level F and in order to release them into the vacuum space they have to be supplied with energy equal to ϕ_c (electron volts); ϕ_c is the cathode *work function*. In the potential situation shown in the figure the electrons can travel unhindered direct to the anode, where they are absorbed, emitting energy ϕ_a electron volts as heat. The Fermi level at the anode is above that at the cathode, so that the electrons still have sufficient potential energy V_e electron volts, to do further work. As we shall see later, by connecting the anode and cathode via a suitable load resistor, external energy can be generated.

In essence, then, the thermionic converter is a device to boil off electrons from a cathode at temperature T_c on to a cooler anode at temperature T'_a; we might expect the maximum efficiency to be that of a heat engine working between these temperatures. However, if an attempt is made to extract any significant power, it is found that the emitted current is severely limited. This is because the electrons in the

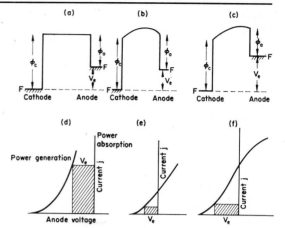

Fig. 1. *The upper figures (a, b and c) show the potential situation between the cathode and anode. The lower figures (d, e and f) show the variation of current with anode voltage; the hatched area represents the maximum power that can be generated in each case. (a) represents the situation with no space charge, (b) with space charge, and (c) as for (b) but with a smaller anode work function.*

inter-electrode space form a space-charge cloud or barrier which repels further electrons trying to emerge from the cathode. The potential situation now looks like that in Fig. 1(b). Apparently more energy must be supplied to the electrons, and it is only the most energetic which can surmount the barrier and reach the anode. The effective elimination of this space charge barrier is therefore essential for a practical generator, and various methods are discussed later.

Richardson, in 1903, gave the first quantitative treatment of emission, based on the classical Drude assumption of a cloud of free electrons in a metal, with a Maxwellian velocity-distribution. This first expression for the current density j was

$$j = AT^{\frac{1}{2}} \exp - \frac{\phi \mathrm{e}}{\mathbf{k}T}, \qquad (1)$$

where A is a constant, e is the electronic charge and $\mathbf{k}$ is Boltzmann's constant. This relationship between j and cathode temperature bore fairly close agreement to measured values.

Later a modified expression was produced by Richardson and others based on thermodynamic reasoning and also on the use of Fermi-Dirac statistics for the velocity distribution of free electrons in a metal. The latter imply a finite electron energy even at absolute zero temperature with a sharply defined maximum, W. The energy distribution at other temperatures is only a slowly varying function, which means that the electron specific heat is small. It is necessary to find the number of electrons with velocity components normal to the surface, and integrate for energies between ϕ_c and ∞. This yields, eventually,

$$j = \frac{4\pi \mathbf{mek}^2}{\mathbf{h}^3} T^2 \exp\left(-\frac{e\phi}{kT}\right), \qquad (2)$$

where $\mathbf{m}$ = electronic mass and $\mathbf{h}$ = Planck's constant.

If this is put in logarithmic form, plotting $\ln(j/T^2)$ against T^{-1} the slope $-e\phi/k$ is a straight line, characteristic of the emitter. Such a graph is known as a Richardson plot.

The discussion above applies strictly only to metals, whereas the most effective emitters are often semiconductors, such as oxides. Although a similar type of characteristic applies, the actual emission is much greater, e.g. for the usual strontium oxide on nickel at 800°C it is about 12 decades above that of pure tungsten at the same temperature (see Fig. 4).

Neutralization of Space Charge

Three principal methods have been investigated for the elimination or reduction of space charge. These are: the use of a very small (typically $\langle 0.01$ mm) electrode spacing, the introduction of positive ions, and the use of "guidance" by some combination of magnetic and electric fields. Clearly, any effective means of neutralization must absorb only a fraction of the total output power.

The close-spaced vacuum diode uses no such power, but the maintenance of the small spacing, uniform over large areas and free from creep at high temperature is an extremely difficult technical problem to solve in a reliable manner. Typical power densities are in the range 0.1 W cm^{-2} for spacings of 0.05 mm, to 1 W cm^{-2} at spacings of 0.005 mm. Current densities are typically in the range 1–4 A cm^{-2}, with oxide cathodes at about 1100°K. However, the vacuum converter has now been practically abandoned—except perhaps for powers of only a few watts. In no case does the current emission exceed about a tenth of the saturated value obtained under conditions of complete space-charge cancellation.

It seems that the auxiliary field devices can also be disposed of fairly quickly. The usual arrangement is shown in Fig. 2, the cathode and anode (collector) being coplanar, and facing an auxiliary electrode, spaced about 1 mm away and charged to positive 200 V, which is enough to draw the saturated emission current. By means of a transverse magnetic field, the

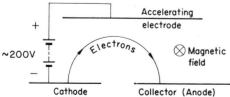

Fig. 2. Elimination of space charge effects by auxiliary fields: the magnetic triode.

electrons are drawn into cycloidal paths on to the collector. The trajectories are symmetrical, so no power is supplied by the accelerator, and the collector receives the full current equivalent to that of a zero-spacing vacuum diode. A further advantage, in principle, is that the accelerator does not collect electrons. Its work function (and temperature) are therefore unimportant, and it can be chosen to reduce thermal radiation losses.

Unfortunately, the attractions of the "magnetic triode" have never been realized in practice. An unacceptably high proportion of the electrons always reaches the accelerator, however stringent the precautions. The other foreseeable problem, of providing a uniform magnetic field cheaply (in energy terms) has never really been tackled.

This leaves, as the only practicable method of neutralization, the introduction of positive ions. From the early days of Langmuir (before 1930) there has been a long series of theoretical investigations of this. The method may itself be subdivided into three groups—volume ionization, ionization by thermal contact, and ionization by the use of auxiliary electrodes. This last sub-group can again be fairly quickly disposed of: it attained some temporary interest because the alkali metals (and most notably Cs), although readily ionized, are chemically very aggressive and alternatives were sought. Gases such as He, Ar, Hg, which are often used in discharge tubes, have higher ionization potentials but can be ionized, for example, by injecting electrons into the inter-electrode space from an auxiliary gun. An alternative method is to allow some of the main thermionic current to reach a positive collector behind the anode, where positive ions are generated by collision and find their way to the main inter-electrode space. Although it has been claimed that these schemes can operate at an energy cost in the region of 0.1–0.2 eV they have not, so far, done so in practice.

The main method of neutralization then is injection of an alkali metal, almost always Cs with the low work function of 1.8 eV. A neutralized diode can operate either in the low or high pressure region. In the former the vapour pressure is typically 10^{-4} mm Hg, and the mean free path 1 cm, i.e. much greater than the electrode spacing. At these pressures, although it is possible to attain complete neutralization it is difficult to obtain efficiencies above a few per cent with plain cathodes.

A different mode of operation becomes possible at pressures around 1 mm Hg—involving rather complex ion-sheath phenomena at the electrodes—but the effect is that complete space-charge cancellation can be obtained, without much voltage drop in the vapour "plasma", provided the anode/cathode spacing is small.

The phenomenon of thermal contact ionization, first observed in 1923, deserves some explanation. The work function of W (4.62 V) exceeds the ionization potential (3.88 V) of Cs, so that caesium on contact with hot tungsten tends to give up electrons and become ionized. For clean W at 1200°K, almost all incident Cs atoms leave as ions. (With thoriated tungsten at this temperature ($\phi = 2\cdot 6$ eV) the contact ionization effect is almost negligible.) However, ionization occurs only when the tungsten surface is less than about 8 per cent covered, as the effect of the Cs is to lower the work function of the W, and thus the emission of positive ions drops. It has been found that the minimum temperature of the surface to be effectively clean in the presence of saturated Cs vapour at a temperature T_s is about $3\cdot 6\, T_s$. Other combinations (e.g. the other alkali metals) with W, Ta or Mo are effective, but none quite so effective as Cs/W because of the combination of work function and ionization potential, and because W can be heated to a very high temperature to expose a clean surface.

In the "high pressure" mode it is possible to obtain a high current density with a relatively low cathode temperature and work function.

Power and Efficiency

Under matched load conditions, and with space charge effects just neutralized, the load voltage V_L is given by

$$V_L = \phi_c - \phi_a. \qquad (3)$$

The power (per unit area of the device) developed in the load is

$$P_L = (\phi_c - \phi_a)\, j, \qquad (4)$$

and the input power is

$$P_c = \phi_c j + R \qquad (5)$$

where R is the parasitic heat loss. The efficiency is thus

$$\eta = \frac{\phi_c - \phi_a}{\phi_c + R/j}. \qquad (6)$$

This immediately shows the importance of a low anode work function, and, of course, low R. Under ideal conditions, if one can set $\phi_c/T_c = \phi_a/T_a$, the efficiency reduces to

$$\eta = \frac{T_c - T_a}{T_c}, \qquad (7)$$

that of an ideal heat engine operating between the cathode and anode temperatures. A further condition necessary is that the heat engine should be *reversible*. This implies that back-emission of electrons from the anode must be possible, though any significant amount must be avoided in practice.

Before deriving the efficiency of a practical generator, we must consider the various forms of energy loss. Probably the most important are the loss of heat from the cathode to the anode, mainly by radiation, and other heat losses, through leads and the outer envelope. The radiation losses can be minimized by using materials with low thermal emissivities, though there is not much control if the primary requirement is good electron emission. There is little

too that can be done about lead losses, since the ratio of thermal to electrical conductivities is determined by the Wiedemann-Franz law. Another important source of loss is known as electron cooling. As already stated the amount of energy ideally needed just to remove the electrons from the cathode is $j\phi_c$. Under these conditions, the electrons would only just emerge, but normally they enter the inter-electrode space with mean energy about $2kT$ each, which has to be supplied from the input heat.

Taking into account electron cooling and radiation losses, it is found that the efficiency becomes

$$\eta = \frac{\phi_c - \phi_a}{\phi_c + \dfrac{2kT_c}{e} + \dfrac{\sigma \varepsilon T_c^4}{j_c}}, \qquad (8)$$

where σ is Stefan's constant, ε is the effective emissivity and j_c is the cathode current density. The latter may be found in terms of ϕ_c from Richardson's equation (equations (1) or (2)). Some results are shown in Fig. 3 for various values of T_c. The anode

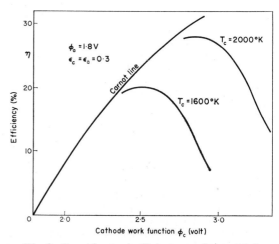

Fig. 3. *Dependence of efficiency, under matched load conditions, on cathode work function.*

work function is assumed to be $1\cdot 8$ eV, and since the matched load condition must be realized at every point ($V_L = \phi_c - \phi_a$), $\eta = 0$ when $\phi_c = \phi_a$.

Efficiencies approaching those of Fig. 3 cannot be obtained for two reasons; first, lead and some other losses have been omitted; secondly no consideration has been given to the way in which the cathode can be heated in a practical generator. In a flame-heated device there are bound to be thermal losses, though with some types of nuclear reactor diodes, in which the cathode is heated directly by nuclear fission, these losses can be small. It is implicit in equation (8) that there is no voltage drop in the "plasma", the ion-containing space between the electrodes. This rarely happens, though the conditions may be fairly closely approached.

Materials

The essential properties of a *cathode* are adequate emission with low evaporation at the intended operating temperature, together with low thermal emissivity (to reduce radiant heat loss), low electrical resistivity and good mechanical properties. Generally (by Richardson's equation) the emission is inversely related to the work function, but considerations such as output voltage also enter.

It has already been suggested that plain metal cathodes are not very suitable. A low-pressure Cs converter with a W filament operating at 2200 °C gives an efficiency of about 1·5 per cent. To obtain an efficiency approaching 15 per cent it would be necessary to go up to about 2700 °C, when the life would be very limited (<1000 hr) (see Fig. 4). With the high pressure Cs diode, in which a monolayer is maintained on the cathode, lower temperatures are satisfactory and there is no evaporation: the effective emitting material is the Cs film which is in equilibrium with its vapour.

A film of Th on W is very effective in reducing the evaporation rate, and the thoriated tungsten cathode is made from a sintered mixture of W with 2 per cent thorium oxide. The atoms diffuse through the matrix to replace those lost from the surface. An extremely useful variant is the Lemmens (L) or dispenser cathode, which consists of porous W containing a core of BaO. At 1100 °C, Ba is dispensed from the core to the surface, lowering its average work function. BaO (and SrO) can also be used to form oxide-coated cathodes, on the surface of a Ni base.

The properties of some cathode materials, including UC and ZrC, which are mentioned later, are given in Table 1.

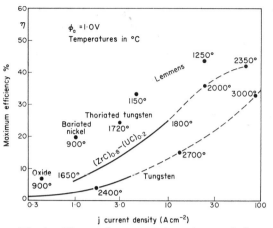

Fig. 4. The performance of various cathode materials, in terms of temperature and current density, for an anode work function of 1·0 V. The dotted portions of the two curves correspond to regions of excessive evaporation.

Table 1

Material	ϕ_c (eV)	j (A cm^{-2})	T (°C)	Evaporation rate (g cm^{-2} sec^{-1})
W	4·59	0·65	2300	10^{-8}
Mo	4·56	0·07	2000	10^{-7}
Ta	4·25	1·0	2100	10^{-8}
W/Cs	2·0	50	1200	10^{-1}(Cs)
W/BaO	2·1	5	1100	10^{-10}
BaO/SrO	1·55	10	800	10^{-11}
UC	3·16	5	1700	10^{-8}
ZrC	3·9	0·1	1700	10^{-11}

Figure 4 shows the converter efficiency attainable with various types of cathode, as a function of current density and temperature. At low current densities (< 1 A cm^{-2}) the efficiency is low, and at high values evaporation is excessive.

Anode materials should have mechanical and electrical properties similar to those of cathode materials, together with a low work-function. This should preferably be below 2 eV, the lower limit being set by back-emission of electrons. Cs at 1·8 eV gives the lowest for any bulk material, but slightly lower values, 1·7 eV, can be obtained by using Cs on W, as already discussed in connection with cathodes. However, at an anode temperature of 300 °C the Cs vapour-pressure required to give this minimum is well below that for space-charge neutralization, though at around 700 °C it becomes a possibility. A monolayer of Cs on Cu or Ni appears, however, to be able to give a low work function at a temperature of about 300 °C. Photo-emissive devices operating at room temperature are well known to use Cs on oxygen on silver anodes, with very low work function. If these are usable at temperatures appropriate to thermionic anodes the work function would be in the region of 1 eV.

Diodes which depend on Cs for space charge neutralization pose difficult envelope and seal problems. Glass, which is attacked to some extent, has been used for most of the experimental devices; and stainless steel has been used for in-pile work. However, fossil-fuel diodes intended for long life will need envelopes of materials such as silicon carbide or alumina. Even these barely possess the necessary combination of corrosion resistance, impermeability, good thermal conductivity, but low thermal expansion and electrical conductivity. However, lives exceeding 1000 hr have now been obtained with flame-heated devices.

Practical Applications

Thermionic generators can be used for "topping" steam plant either coal-fired or nuclear (Spring, 1969) to increase its efficiency, or directly in a nuclear reactor, the first such experiments having taken place at Los Alamos in 1959, and shortly after at Harwell. Thermionic converters are suitable for space power supplies because they are insensitive to radiation damage and reject heat at a high temperature so that a small radiator can be used.

The concept of topping steam plant can be realized by, for example, replacing the conventional water-tube walls by coaxial diodes, the outer tubes being the flame-heated cathode, and the inner tubes the steam or water-cooled anodes. Heat transfer rates in modern boilers are of the order 30 W cm^{-2}, which is appropriate also to a thermionic diode, but no practical designs yet seem to have emerged. This is probably partly because of the difficulty in finding a coating which can prevent a flame-heated cathode from ultimately oxidizing at its operating temperature of 1200–2000°C.

The concept of the nuclear "in-pile" diode is extremely ingenious. Its effectiveness depends on the fact that certain fissile materials, for example the carbides and oxides of uranium, are also good electron emitters. It is thus possible to make an enclosed diode in which the cathode is heated by the fission energy released in it (and also by neutron capture from the general neutron flux). A solid solution of ZrC–UC is also an effective emitter, but with better mechanical properties. A typical design is shown in Fig. 5. Efficiencies of such generators are in the region of 15 per cent.

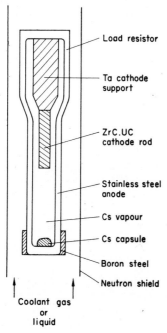

Fig. 5. *Typical design for an in-pile thermionic diode.*

An alternative to the in-pile concept is to heat the diode externally, for instance using a liquid metal heat-exchange loop. A recently developed method of conveying heat into a thermionic generator, which may be useful for nuclear-powered space generators, is the "heat pipe". This carries heat by vaporizing a liquid at one end, and condensing it at the other, whence it returns by capillary action. The effective thermal conductivity is about 3 or 4 orders better than that of Cu, so that the temperature drop in the "pipe" is negligible. The application to radioisotope powered sources must await the development of such fuels able to operate at least at 1200°C and preferably much higher.

Problem Areas

There are a number of problem areas associated with thermionic generators, particularly those intended for in-pile use. A high cathode temperature with adequate emission is required, together with a low evaporation rate. A ratio of heat of vaporization to work function of greater than 2·5 is needed to satisfy these conditions completely—and this is beyond the range of present materials, except for caesium on tungsten, and UC, which with a ratio of 2·3, closely approaches it. The problem of compatibility of materials with caesium vapour has already been mentioned. Tungsten is one of the few metals and alumina one of the few insulators not attacked. Metal-to-ceramic seals remain something of a problem, particulary at the high envelope temperatures associated with fossil fuels.

With an in-pile system a key factor is the development of uniform power throughout the reactor, so that the diodes are under uniform temperature and loading conditions. This may be done in a number of ways, e.g. by local variations in moderator, fuel enrichment, etc. Continued operation causes the build-up of fission products, which may cause swelling and distortion of the fuel elements sufficient to degrade the electrical performance, and even to disintegrate the fuel. This occurs for example with high density 30–70 UC–ZrC operating at around 2000 °C with burn-ups of a few per cent. The solution is to use lower density (∼80 per cent) porous fuel, although this will adversely affect the specific power/weight ratio of the reactor.

Low evaporation rates of the fuel are particularly crucial for in-pile diodes. Even if the fuel is clad, the fission gases have to be vented. Uranium oxide (which when tungsten clad is a good emitter) is now thought to be unsuitable as a fuel, because of its poor electrical (and thermal) conductivity, and because its evaporation rate is about 10^4 that of a UC–ZrC alloy. Even with these alloys (at 80 per cent density) evaporation is rather high and they should preferably be clad. Tungsten is a suitable material, although it has a higher neutron absorption cross-section than Mo. It is difficult to produce a machined can, though it is now possible to deposit W on the fuel by a vapour-phase reaction. Such techniques give the promise of controlled emission and porosity to permit fission-product release. However, considerable progress is also being made with unclad fuel elements. Hot and cold pressing and sintering techniques may be used to produce UC–ZrC single phase structures and Mo–UO_2, W–UO_2 and W–UC cermets. Not enough is yet known, however, of the interaction between fuel and the electron-emitting cladding to be certain of long-term behaviour under the combination of high temperature and irradiation.

There is a possibility that the fission products released into the caesium vapour space may adversely affect the electrode behaviour, e.g. by altering work functions, since, as we have seen, Cs confers special properties on W. Gross fission product contamination will have to be removed, as eventually the partial pressure will rise sufficiently to scatter the electrons.

At the present time although some design studies continue, the "externally-fired" fossil-fuel diode has been largely abandoned. The nuclear thermionic diode still offers considerable promise for space power.

See also: Solar cell. Space power supplies. Thermoelectric generation of electricity.

Bibliography

SPRING K. H. (Ed.) (1965) *Direct Generation of Electricity*, New York: Academic Press.

SPRING K. H. (1969) Direct conversion of heat to electricity, in *Encyclopaedic Dictionary of Physics* (J. Thewlis Ed.), Suppl. vol. 3, 45, Oxford: Pergamon Press.

K. H. SPRING

THERMOELECTRIC COOLING OF INFRA-RED DETECTORS. Seebeck's discovery of thermoelectricity was made in 1821, but it has taken over a century to develop practical thermoelectric devices of acceptable efficiency. Thermoelectrically, miniature components can be cooled silently, with no moving parts.

By passing a d.c. current (less than 10 per cent ripple) through a typical, commercially available thermoelectric couple of doped Bi_2Te_3 with a "figure of merit" Z of $2 \times 10^{-3}°C^{-1}$, one can achieve a ΔT in vacuum of about 65°C (55°C in still air) when the hot side is maintained at an ambient temperature of 25°C. Reversal of the current merely changes the direction of the heat pump. The efficiency (or coefficient of performance) of a Peltier cooler is improved by increasing the number of junctions in a module (Fig. 1(A)). For larger temperature differentials one resorts to cascading modules on top of each other so that they are always in series thermally, and generally (but not necessarily) also in series electrically. A good rule of thumb on cascading modules, employing the same cross-sectional area pellets for each stage, is to decrease the number of couples by a ratio of about 4 in going from the hottest bottom stage to the next cooler stage. A typical 3 stage cascade (Fig. 1(B)) provides a maximum ΔT of 105°C under no thermal load, in vacuum, with the hot side at 25°C. Little is gained in going beyond a three stage design. Typical derating of maximum ΔT under thermal load for a three stage design as in Fig. 1(B) is 0·04°C per milliwatt.

Fig. 1. (A) Thermoelectric module, (B) three-stage cooler, (C) thermoelectrically cooled immersed thermistor bolometer package, and (D) miniature thermoelectrically cooled detector package.

In the field of infra-red (I.R.) radiometry there existed a need for thermoelectrically cooled I.R. detectors. Photoconductive detectors of PbS, PbSe, InSb, and a thermal detector consisting of an "immersed" thermistor bolometer were assembled into thermoelectrically cooled packages (Fig. 1(C) and 1(D)) and examined for improvement in performance on cooling. Thermistor materials for bolometers are chiefly oxides of Mn, Ni, and Co. They were originally developed in several compositions at the Bell Telephone Laboratories. In final sintered form they are semiconductors exhibiting a high negative coefficient of resistance. The composition designated at B.T.L. as material No. 2 was processed into a thermistor flake and coupled optically to a germanium lens through a thin evaporated layer of glass of suitable high index of refraction. Thermistors obey the following relationship:

$$R/R_0 = e^{\beta(1/T - 1/T_0)}$$

where R and R_0 are the resistances at temperature T and T_0 respectively (the resistivity of B.T.L. material No. 2 at 300°K is 250 ohm-cm), and β is a material constant (°K) equal to 3400 for No. 2 material.

For a phenomenological description of photoconductive and thermal detector mechanism see Kruse *et al.* (1962) and DeWaard and Wormser (1958).

The figure of merit generally employed for comparing the detectivity, D^*, of I.R. detectors is expressed by:

$$D^* = (Vs/Vn)(\Delta f^{1/2}) H^{-1} A_D^{-1/2}$$

where D^* is in cm-cps$^{1/2}$ W^{-1},

Vs/Vn = ratio of r.m.s. signal to noise voltage in the bandwidth Δf,

H = r.m.s. value of irradiance (W/cm²) falling on detector area A_D (cm²).

The condition under which D^* was measured is expressed by $D^*(°K, f, 1)$ where °K stands for the temperature of the black-body radiation source in degrees Kelvin, f stands for the chopping frequency, and 1 stands for the standard bandwidth of 1 cps. Figure 2

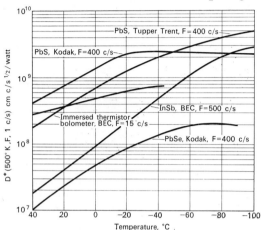

Fig. 2. Detectivity (D^*) of various infra-red detectors vs. temperature.

is a plot for various detectors of D^* vs. temperatures achieved through thermoelectric cooling. Photoconductive I.R. detectors of PbS and PbSe are improved by a factor of 10, photoconductive InSb by a factor of 100, and an immersed thermistor bolometer by a factor of 2. Since I.R. detectors must be maintained at a constant temperature to maintain constant responsivity ($\mathcal{R} = Vs/HA_D$), thermoelectric cooling of these detectors also results in higher responsivities.

More could still be gained in performance for many I.R. detectors if one could achieve liquid nitrogen temperature ($-195°C$) by thermoelectric cooling. No great success has been reported so far at these low temperatures by strictly Peltier cooling although thermoelectromagnetic methods employing the (Nernst) Ettinghausen effect appear promising.

For a complete treatment on the subject of thermoelectricity, readers are advised to consult the works listed below.

Bibliography

CRUMP R. (1963) Thermoelectric cooling, *Prod. Eng.*
DEWAARD R. and WORMSER E. M. (1958) Thermistor infra-red detectors. NAVORD 5495.
EVANS W. G. (1963) Characteristics of thermoelectric materials, *Semicond. Prod.* **6**, 4, 34–39.
HEIKES R. R. and URE R. W. (1961) *Thermoelectricity: Science and Engineering*, New York: Interscience.
IOFFE A. F. (1957) *Semiconductor Thermoelements and Thermoelectric Cooling*. Translated from Russian by A. Gelbtuch. London: Infosearch Ltd.
KRUSE P. W., McGLAUCHLIN L. D. and McQUISTAN R. B. (1962) *Elements of Infrared Technology: Generation, Transmission, and Detection*, New York: John Wiley.
URE R. W. (1963) Theory of materials for thermoelectric and thermomagnetic devices, *Proc. IEEE*, 699–713.
WOLFE RAYMOND (1963) Physics of Thermoelectricity, *Semicond. Prod.* **6**, 4, 23–28.

H. P. BEERMAN

THERMOELECTRIC GENERATION OF ELECTRICITY.

Introduction

Thermoelectric effects were first noticed by Seebeck in 1822 and he was able to arrange the elemental metals into a series. The extreme pair, antimony and bismuth, enabled efficiencies of about 1 per cent or just over to be obtained with appropriate hot and cold junction temperatures. Since these efficiencies were better than those of the rather clumsy steam engines then available, it is perhaps rather surprising, looking back, that no effort was made to develop power generation by this route. However, Faraday's discoveries of electromagnetic induction produced the dynamo, and little progress in thermoelectrics was made until the 1950's, when the semiconductor era arrived. As will be shown later, these permit efficiencies of an order of magnitude greater than metals to be obtained.

The principal advantages of the thermoelectric generator are absence of moving parts, leading to ruggedness and reliability, for example in space or military conditions. For the latter purpose the further feature of silence is sometimes an additional advantage. The upper temperature of operation is well matched to that capable of being readily produced by a radio-isotope source. By contrast, the thermionic generator generally needs rather higher temperatures and tends to lead to more difficult material problems.

Theory of thermoelectric effects. The basic principles of thermoelectricity can be discussed without reference to the nature of the material; but, of course, this must be considered if materials with optimum properties are to be prepared.

The effect named after Seebeck arises because, if there is a temperature difference between the two ends of a semiconductor (or metal) the electrons at the hot end have a higher average kinetic energy than those at the cold end. Crudely, one can see that this leads to two effects—the electrons transfer energy down the material, leading to thermal conduction; and the faster and more energetic electrons try to diffuse towards the cold end leading to a current: under open-circuit conditions of course only for an instant until an equilibrium voltage is set up. This is the Seebeck e.m.f., which can be either positive or negative.

If we now consider two dissimilar materials, joined to form hot and cold junctions at temperatures T_h and T_c respectively, the e.m.f. generated V and the Seebeck coefficient α for the junction are related by

$$V = \alpha(T_h - T_c), \tag{1}$$

provided the temperature difference is small. The value of α for a junction is the algebraic sum of α for each material. As it happens, n-type materials (electron-carriers) have negative values of α, and p-type (hole carriers) are α-positive.

The inverse effect was noticed by Peltier in 1834—namely that when a current is passed through a junction of two dissimilar materials, heat is generated or absorbed (according to the direction of the current). The relationship is

$$Q = \pi j, \qquad (2)$$

where Q is the heat developed, π is the Peltier coefficient and j the current.

Lord Kelvin, from thermodynamic reasoning, deduced a third effect, the generation and absorption of heat in a homogeneous material subject to both a voltage and temperature gradient. All these effects must be distinguished from the normal Joule resistive heating (which depends on the square of the current j and so is independent of direction).

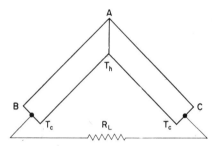

Fig. 1. *A hot junction (A) between two dissimilar thermoelectric materials AB, AC. The load resistance R_L is joined between the two cold ends B, C.*

Consider now the arrangement of Fig. 1, with a hot junction at A and the two cold ends B, C (at the same temperature T_c) joined by a load resistance R_L, with negligible thermoelectric effects between B and C. T_h and T_c are respectively sources and sinks of heat at these two constant temperatures. Altenkirch in 1911 considered the problem of optimizing the power developed. Suppose the materials have lengths l_1, l_2 and areas A_1, A_2, electrical conductivities σ_1, σ_2 and thermal conductivities $\varkappa_1$, $\varkappa_2$; the electrical resistance R of the couple is

$$R = l_1/A_1\sigma_1 + l_2/A_2\sigma_2, \qquad (3)$$

and similarly the thermal "resistance"

$$1/K = l_1/A_1\varkappa_1 + l_2/A_2\varkappa_2. \qquad (4)$$

There is a transfer of heat from the hot to the cold junctions, given by $K(T_h - T_c)$. There is Peltier cooling πj at the hot junction, and Kelvin showed that

$$\pi j = \alpha j T_h. \qquad (5)$$

There is also Joule heating in the couple materials, the net effect of which can be shown to be a transfer to the hot junction of $\tfrac{1}{2}j^2R$. The heat generated H is thus

$$H = \alpha j T_h + K(T_h - T_c) - \tfrac{1}{2}j^2R.$$

The power output W will be highest when the load resistance matches the total couple resistance, so that the load voltage is $\tfrac{1}{2}\alpha(T_h - T_c)$. Thus,

$$W = \frac{\alpha^2}{4R}(T_h - T_c)^2, \qquad (6)$$

and the efficiency η is (on simplifying)

$$\eta = W/H = (T_h - T_c)/\left(\tfrac{3}{2}T_h + \tfrac{1}{2}T_c + \frac{4RK}{\alpha^2}\right). \qquad (7)$$

This is maximized when RK is minimized, for which

$$\left(\frac{l_1 A_2}{l_2 A_1}\right)^2 = \frac{\sigma_1 \varkappa_1}{\sigma_2 \varkappa_2}. \qquad (8)$$

If for this condition α^2/RK is defined as Z, a figure of merit for the junction,

$$\eta = (T_h - T_c)/(\tfrac{3}{2}T_h + \tfrac{1}{2}T_c + 4/Z). \qquad (9)$$

Z depends only on the properties of the junction materials, and is given by

$$Z = \alpha^2 \bigg/ \left[\left(\frac{\varkappa_1}{\sigma_1}\right)^{\frac{1}{2}} + \left(\frac{\varkappa_2}{\sigma_2}\right)^{\frac{1}{2}}\right]. \qquad (10)$$

Equation (9) gives the efficiency at maximum power. Maximum efficiency can be derived by setting $R_L = mR$ and differentiating $(d\eta/dm)$. This gives

$$\frac{(1 + ZT_m)^{\frac{1}{2}} - 1}{(1 + ZT_m)^{\frac{1}{2}} + T_h/T_c} \cdot \frac{T_h - T_c}{T_h} \qquad (11)$$

where T_m is the mean of T_h and T_c. Equation (1) can be recognized as the Carnot efficiency, multiplied by some device factor γ.

It can be seen at once that, for the efficiency to approach that of a Carnot engine, ZT_m must be larger. For the overall efficiency to be high, of course, T_h must also be large. The fraction of Carnot efficiency γ plotted against ZT_m is given in Fig. 2, and the efficiency against Z, and T_h in Fig. 3. It is worth anticipating the subsequent discussion a little to state here that values of ZT_m much above unity have not been realized, so that junction efficiencies over about 20 per cent of Carnot are unattained. Device efficiencies are still lower, and compare unfavourably with say steam or Diesel plant, where very high fractional Carnot efficiencies are obtained—about 0·7 that of the ideal heat engine between the two temperatures T_h and T_c.

Whereas equation (10) defines Z for a junction, that for a *material* of properties, α, σ, $\varkappa$, is given by

$$Z = \frac{\alpha^2 \sigma}{\varkappa},$$

a result originally obtained over 50 years ago by Altenkirch. Z for a junction of two different materials is

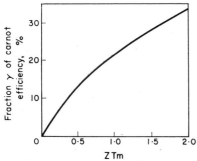

Fig. 2. The fraction of Carnot efficiency attained for an ideal generator (operating at mean temperature T_m) as a function of the dimensionless figure of merit parameter ZT_m.

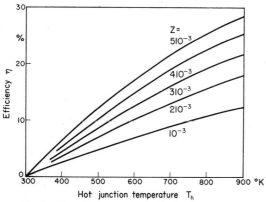

Fig. 3. The efficiency of an ideal generator as a function of hot junction temperature T_h, and of figure of merit Z. The cold junction temperature is assumed to be $300°K$.

obtained by the averaging process of equation (10). The dimensionless parameter ZT is often a more effective and meaningful measure of performance at a temperature T than Z itself.

Properties of Thermoelectric Materials

The electrons in a metal are responsible for both its thermal and electrical conductivity, and these are related by the Wiedemann-Franz law

$$\varkappa = L \frac{\mathbf{k}^2}{e^2} \sigma T \qquad (13)$$

where $\mathbf{k}$ is Boltzmann's constant, e the electron charge and $L(\mathbf{k}^2/e^2)$ is the Lorentz number. For a metal, L is 3·3, and α is in the region of 10 μV/°C so that ZT is around 10^{-2}. Thus even with T_h and T_c separated by several hundred degrees the efficiency is only around 0·1 per cent. The combination antimony/bismuth, already mentioned, yields ZT of about 0·1, and (partly because of their relatively low melting points) efficiencies are limited to about 1 per cent.

With a semiconductor it is relatively easy to produce large values of α, and hence, it might be thought, of Z. However, any change in α must be accomplished by a change of the electrical conductivity σ, and to some extent of the thermal conductivity $\varkappa$. Hence to get the best value of Z requires an optimization of $\alpha^2\sigma$ or $\alpha^2(\sigma/\varkappa)$. To see why this is so, it is necessary to consider a little more deeply the properties of semiconductors.

Suppose one starts with an extremely pure material, in what is called the *intrinsic* state, when its properties are those of the basic material. The electrical conductivity is given by $n_i e \mu$ where n_i is the density of charge carriers (electrons or holes) and μ is their mobility. This carrier density depends on the so-called "energy-gap" E_g, the energy needed to form electron-hole pairs in the material. For silicon, for example, E_g is 1·1 eV. The relationship between n_i and temperature is

$$n_i = \text{const.} \times T^{3/2} \exp(-E_g/2\mathbf{k}T), \qquad (14)$$

and so n_i rises rapidly with T, because of the exponential factor. So too does σ, because any change in μ (which normally falls with temperature) is swamped. The room temperature values of n_i for germanium and silicon are, respectively, about 2×10^{13} and 10^{10} cm^{-3}.

The basic properties can be altered by deliberately adding controlled amounts of impurity—a process known as *doping*. At a sufficiently low temperature, the conductivity and other properties are now determined by the charge carriers released by the impurity atoms. The material is then called extrinsic. If all these carriers are free, σ varies only slowly with temperature but it does, of course, vary with the doping level and is, in fact, almost proportional to it. It is normally found that α is large only when the material is extrinsic. This condition is most likely to be obtained with a large value of E_g, and a high doping level. As the temperature rises, there will be a transition from extrinsic to intrinsic behaviour. It is found that, if a maximum operating temperature T is to be specified,

$$E_g > 4\mathbf{k}T.$$

At 300°K (room temperature) $\mathbf{k}T$ is about $\frac{1}{40}$ eV, and the minimum usable energy gap is thus 0·1 eV.

The general way in which the parameters vary with the doping level is shown in Fig. 4. The highest value of α occurs for a relatively low value of σ, and that of $\alpha^2\sigma$ at a higher value usually in the region of $\sigma = 10^3$ ohm^{-1} cm^{-1}. The precise optimization can be considered by reference to Fig. 5, which illustrates a semiconductor between two metals, at temperatures T_h and T_c. The problem is somewhat analogous to that of thermionic emission. If an electron is transferred from the metal at the left, to enter the conduction band of the semiconductor it must gain both potential energy ($-W$ say) and kinetic energy

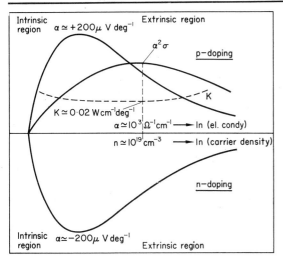

Fig. 4. The way in which the various thermoelectric parameters vary with doping level (carrier density n), which is plotted (logarithmically) increasing from left to right. The electrical conductivity σ is nearly proportional to n. In the upper half of the figure, with p-type doping, α is positive, and in the lower half, with n-type doping, it is negative.

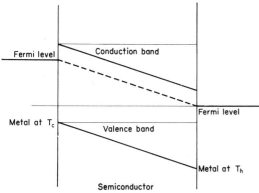

Fig. 5. A semiconductor in contact with two metals, at temperatures T_h and T_c: at each end the Fermi levels must be the same as those in the contacting metals.

(shown in thermionic theory to be $2\,kT_c$). The cooling at this junction, then, due to latent heat of electron evaporation, gives a Peltier coefficient of

$$\pi = \frac{1}{e}(2kT_c - W) \qquad (16)$$

and a rather similar expression for the hot junction. Thus from equation (5)

$$\alpha = \frac{k}{e}(2 - W/kT). \qquad (17)$$

The optimum value of α, α_0 say, can be found by putting $\alpha^2\sigma$ in terms of W, and then differentiating. It is found that $W = 0$ for maximum $\alpha^2\sigma$, and hence $\alpha_0 = 2k/e = 172\ \mu V/°C$, at the intended mean operating temperature.

The relatively simple theory just given applies only to a material in which the electrons behave in "classical" fashion—which of course they do not. In so-called degenerate semiconductors, the kinetic energy of the electrons will be determined by Fermi-Dirac, rather than Maxwellian, statistics. In addition, one has to take into account the way in which the electrons interact with the crystal lattice of the semiconductor. It has been shown by Goldsmid (1960) that the optimum value of $\alpha^2\sigma$ is obtained when the doping is adjusted to make α almost 200 μ/V°C.

To obtain an optimum value of Z, account must now be taken of the further fact that in a semiconductor $\varkappa$ also varies somewhat with the doping level. The thermal conductivity $\varkappa$ arises from two parts—that of the electrons themselves $\varkappa_e$, and that of the crystal lattice $\varkappa_l$. The first part, $\varkappa_e$, is related to the electrical conductivity σ by the Wiedemann-Franz law.

The relationship is like that of equation (13), except that the Lorentz number $L = \dfrac{\pi^2}{3} = 3 \cdot 3$ for metals must be replaced by a constant having a value between 2 and 3·3 depending on the degeneracy.

Now in a semiconductor, unlike a metal, σ is usually fairly small, and so therefore is $\varkappa_e$. However, $\varkappa_l$, which is effectively zero in a metal, is fairly large. The total thermal conductivity of a semiconductor $\varkappa$ is the sum, or more accurately the resultant, of $\varkappa_l$ and $\varkappa_e$ which varies, like σ, with the doping level.

If $\varkappa_l$ is much greater than $\varkappa_e$, $\varkappa$ will vary little with doping level, and the optimum Z will be obtained with $\alpha^2\sigma$ close to its optimum α_0, 200 μV/°C. In other cases Ioffe (1957) has shown that

$$\alpha_k = \alpha_0(1 + \varkappa_e/\varkappa_l).$$

This new, larger, optimum value of α is obtained at a lower doping level, and hence lower σ and $\varkappa_e$. However, since the parameters are all inter-related, the best doping level for any particular material has to be found by experiment to give α in the range from say 150–300 μV/°C.

There are a few empirical rules which are useful in the selection and doping of suitable materials. For example, $\varkappa_l$, and hence $\varkappa$, tends to fall as atomic weight rises. Hence the tellurides of lead and bismuth, which are the "heaviest" semiconductors known, have been extensively studied. A further reduction of $\varkappa_l$ tends to be obtained when a proportion of the atoms in the crystal lattice is replaced. Pb or Bi may, for instance, be replaced by up to 50 per cent Sb, and Te by 50 per cent Se. Similarly, a 50/50 Si–Ge alloy has a low $\varkappa_l$ and good thermoelectric properties (although the individual elements are very poor). Unfortunately as $\varkappa_l$ tends to fall, so too does E_g, and hence the limiting temperature of operation.

In addition to the "broad band" semiconductors

Table 1

Material	σ	α	$\varkappa_e$	$\varkappa_l$	$\alpha^2\sigma$	Z	T	T_{max}
Cu	5.9×10^5	2.5	3.96	—	3.7×10^{-6}	9×10^{-7}		
Bi	8.5×10^3	—75	0.08	—	4.8×10^{-5}	6×10^{-4}		
Ge	10^3	200	0.006	0.63	4×10^{-5}	6×10^{-7}		
Si	5×10^2	200	0.003	1.13	2×10^{-5}	2×10^{-5}		
Ge/Sn (n)				0.03	3×10^{-5}	9×10^{-4}	900	1200
Ge/Si (p)				0.03	2×10^{-5}	6×10^{-4}	900	1200
InAs	2×10^3	200	0.015	0.07	5×10^{-5}	6×10^{-4}	900	1100
Bi$_2$Te$_3$	10^3	220	0.004	0.016	4×10^{-5}	2×10^{-3}	300	450
Bi/SbTe$_3$			0.004	0.010	5×10^{-5}	3×10^{-3}	300	450

σ is the electrical conductivity, ohm^{-1} cm^{-1},

α is the Seebeck coefficient, μV °C^{-1},

$\varkappa_e$ is the electronic component of thermal conductivity, W cm^{-1} °C^{-1},

$\varkappa_l$ is the lattice component of thermal conductivity, W cm^{-1} °C^{-1},

Z is the figure of merit,

T is the temperature at which all properties are measured, °K,

T_{max} is the maximum operating temperature, °K.

considered so far, there are also "narrow band" materials (such as nickel oxide). In these materials, there is little or no overlap of the electron energy levels, and the electrons can move from one ion to another only by jumping. The effective mobility and electrical conductivity is thus low, though rising with temperature. Very high doping levels can be used, so that there is a high density of "hopping" charge carriers. Despite the very different carrier mechanism, α_0 is again about 200 μV/°C: $\varkappa_e$ is effectively zero, but $\varkappa_l$ may be high. Although nickel oxide can be operated up to 1000°C, its value of Z is rather poor.

Table 1 gives the properties of a few elemental metals and semiconductors, followed by those of some good thermoelectric materials. The values quoted for this second group are those with optimum doping to maximize Z at the operating temperature quoted.

The ZT values of some of the best materials are also shown in Figs. 6 and 7.

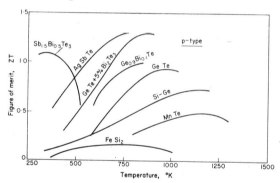

Fig. 7. *Figure of merit ZT as a function of temperature for the best current p-type thermoelectric materials.*

Effects of a Magnetic Field

Before considering the design of practical generators, it is worth considering briefly whether the performance of thermoelectric materials can be improved by the use of a magnetic field. Any semiconductor in a transverse field develops a Hall voltage when a current flows. This is because the field tends to deflect the electrons to one side of the specimen. However, the faster electrons tend to be deflected less than the slow ones, so that there is a temperature gradient in the same sense as the Hall voltage gradient. This is called the Ettingshausen effect (Goldsmid, 1966) which was first used for refrigeration just over 10 years ago. Its inverse, the Nernst effect, implies that if a temperature gradient and a magnetic field are applied to a specimen in say the x and y directions, an e.m.f. will be generated in the z direction. The

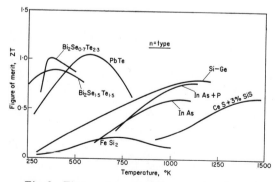

Fig. 6. *Figure of merit ZT as a function of temperature for the best current n-type thermoelectric materials.*

temperature gradient causes electrons to diffuse from the hot end towards the cold, the flow being deflected by the field.

The full theory is very complex, but it appears that for Nernst generation with extrinsic materials, the effective ZT is only about 0·1 at best. With intrinsic materials, good values of ZT ($\sim$1·5) are observed at low temperatures in fields of about 10^4 Oe, but nothing useful at room temperature or above. The method is therefore not very promising, so far, as a source of power, even neglecting the energy needed to produce the field.

Design and Performance of Generators

Since the e.m.f. generated per junction is usually much less than 0·1 V it is desirable to operate junctions thermally in parallel but electrically in series. No material will give its best performance at a single doping level over a wide temperature range, so that either the doping and/or the material itself, should ideally be varied to suit the local temperature. It is difficult to put different materials in series without violating the geometrical requirements (equation 8) so that the idea of a "cascade" (Fig. 8) has been suggested, though for practical reasons it has not been

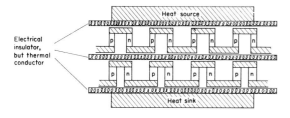

Fig. 8. 'Exploded' diagrammatic arrangement of a 2-stage cascade thermoelectric generator. The thermoelements are in series electrically, but in parallel thermally. Their geometry has to be designed so that there is proper thermal and electrical matching.

widely adopted yet. If a material has a fractional Carnot efficiency γ (see equation 11), the theoretical efficiency of a cascade γ_c is

$$\gamma_c = 1 - (T_c/T_h)\bar{\gamma},$$

where $\bar{\gamma}$ is the geometric mean value of γ over the temperature range. The stages can be joined in series electrically, provided the currents are the same under optimum conditions.

One of the earliest useful generators was the Russian "Ioffe lamp" of the early 1950s, which originally used zinc antimonide (p-type) with a metal, later replaced by n-type bismuth telluride/selenide. Such generators were used to supply power in remote areas, as an alternative to dry batteries.

Much more ambitious were the U.S. SNAP series of generators more fully described in "Space Power Supplies", which use radioisotopes as the heat source. As mentioned lead telluride has been used as the thermoelectric material, and because of its good mechanical properties and relatively high energy gap it is suitable up to about 600°C. With this upper temperature the theoretical efficiency is around 12 per cent; that realized in the practical generators is about 5·5 per cent.

A complementary series of somewhat similar generators has been developed in the last five years in the U.K. by the Atomic Energy Authority. They are intended for terrestrial use, under the code-name RIPPLE (Radio Isotope Powered Prolonged Life Equipment). These have used upper temperatures of about 300°C, with bismuth telluride as the thermoelectric material. The various designs have had outputs from a few milliwatts up to 4 W, with versions up to 50 W still to come. Ripple IV is being installed in a submarine cable repeater, V–VIII are in use (or being constructed) for marine navigational lights and IX in an aircraft ground radio beacon. Ripple generators all use an Sr^{90} source (28 year half-life) in the form of ceramic strontium titanate. It is a β-emitter, requiring considerable shielding, for which depleted uranium (density 18·9 g cm^{-3}) is convenient.

In both the SNAP and RIPPLE generators there is a decrease in power with time, as not only does the specific activity decay, but the upper temperature attained also decreases. In most cases, therefore, deliberate initial over-power has to be designed for. There may also be fluctuations, dependent on the ambient temperature. Batteries (which can also be used to supply a high peak from a low average power) may thus be essential for smoothing. Since the cost of radioisotopes is considerable, much igenuity is needed to get the best efficiency at every stage of design, including that of any voltage-conversion or other "power-conditioning" equipment.

In addition to the radioisotope generators, many others supplied by fossil fuels are now coming on to the market. Iron disilicide is a popular thermoelectric material, since, although its ZT value is only about 0·2 (see Fig. 7) it is cheap and may be used up to about 800°C. It is easily fabricated by hot pressing, and the pressence of foreign metals and oxides does not seem to impair the performance appreciably.

In any flame-heated device there are inevitable losses associated with the heat source. In contrast, the radioisotope device can be so well thermally insulated that practically all the heat generated passes through the thermoelectric element.

Problems and Prospects

Most thermoelectric materials require considerable care in the technology of their preparation and doping to p- or n-type if they are to achieve a Z value near the optimum for the operating temperature T. It is generally considered that they should not be operated at an absolute temperature above 80–90 per cent of the melting point, or there is a danger of the doping agent diffusing along the temperature gradient which (by definition) exists. Pressing, sintering and other techniques must be such that the material is as isotropic and uniformly doped as possible. It is rather difficult to find contact materials which are completely inert, especially in hot conditions. Cu, for example, diffuses rapidly in semiconductors and has to be Ni plated. A further problem is that of finding an electrical insulator at or between the stages which does not present too much of a thermal barrier—a 10°C drop is considered acceptable. Generally the semiconductor modules are encapsulated to confer mechanical strength, and the whole generator may operate in an inert atmosphere to prevent corrosion and oxidation over long-period operation. None of these problems is insuperable, but they tend to create a gap between theoretical and practical efficiencies which is not readily overcome.

It will have been gathered from the earlier discussion that high carrier mobility and low lattice thermal conductivity are among the essential properties of a good thermoelectric material. A further requirement, not specifically mentioned, is that the effective (density-of-states) mass of the carriers should be high. The optimum carrier concentration depends markedly on this effective mass, although the optimum α is always around 200 $\mu V/°C$. As already noted, the energy gap must be such that the material remains extrinsic at the operating temperature: in other words mixed hole and electron conduction must be avoided as their thermoelectric effects will be in opposition. A further reason for avoiding the intrinsic region is that the thermal conductivity is higher than in the extrinsic, as shown in Fig. 4.

If one imagines a hypothetical material with the best values of all these conflicting parameters, a ZT value of about 4 is obtained, and it is perhaps reasonable to hope that a practical value of about 2 may one day be obtained. As Fig. 2 shows, even this will not be particularly exciting by large-scale steam plant standards. It will, however, ensure an increasing role for a versatile generator in which the relatively poor efficiency is offset by other desirable properties.

See also: Solar cell. Space power supplies. Thermionic generation of electricity.

Bibliography

GOLDSMID H. J. (1960) *Applications of Thermoelectricity*, London: Methuen.

HARMAN F. E. and HONIG J. M. (1967) *Thermoelectric and Thermomagnetic Effects and Applications*, McGraw-Hill.

HEIKES R. R. and URE R. W. (1961) *Thermoelectricity: Science and Engineering*, New York: Interscience.

IOFFE A. F. *Semiconductor Elements and Thermoelectric Cooling*, London: Infosearch.

SPRING K. H. (1965) (Ed.) *Direct Generation of Electricity*, New York: Academic Press.

SPRING K. H. (1969) Direct conversion of heat to electricity, in *Encyclopaedic Dictionary of Physics* (J. Thewlis Ed.), Suppl. vol. 3, 46, Oxford: Pergamon Press.

K. H. SPRING

THERMOMAGNETIC GAS TORQUE (THE SCOTT EFFECT). In determining gyromagnetic ratios of solids by the Einstein–de Haas effect (see *Einstein–de Haas Effect* and *Gyromagnetic Ratios*) it is necessary to measure very small torques acting on a suspended sample. In the course of improving measurement techniques, G. G. Scott observed a small anomalous torque when a weak magnetic field was applied along the axis of a cylinder suspended in a rarefied gas at a temperature different from that of the cylinder.

The gas torque is related to other observed effects of magnetic fields on both heat flow and viscous flow in gases. The first observations of the small decrease in both the thermal conductivity and viscosity of paramagnetic gases in a magnetic field were made by Senftleben in 1930. The observation of the same effects in diamagnetic gases and the identification of the role of the rotational gyromagnetic ratio g_J was made by Beenakker in 1962. More recently, transverse effects in both the viscous flow and heat flow of gases in a magnetic field have been observed by Beenakker and coworkers in Leiden and by Gorelik and coworkers in Russia.

The key parameter in the theory of the Senftleben effect is the ratio of the Larmor precession frequency to the collision frequency of the gas molecules. This ratio is proportional to H/p and inversely proportional to g_J, as the gas torque data and the Senftleben effect data demand. Levi and Beenakker have shown that the kinetic theory of aligned, spinning gas molecules (using the second-order Chapman-Enskog approximation) is capable of explaining both the approximate magnitude and shape of the gas torque curves at the higher pressures. L. Waldman has shown that the Scott effect may be explained by the amplification of spin which can occur when oriented, spinning molecules strike the cylinder and bounce sideways. The amplification factor is of the order of the radius of the cylinder divided by the radius of the molecule. A complete explanation of all of the features of the Scott effect is not yet available.

The torque vs. magnetic field data seen in Figs. 1 and 2 illustrate the characteristic features of most of the paramagnetic and diamagnetic gases studied. It should be noted that a magnetic field of only about one Oersted is required to produce the maximum torque in nitric oxide, a paramagnetic gas, while diamagnetic gases such as nitrogen show very similar curves whose maxima are at about one hundred Oersteds.

The direction of the torque depends on the sign of g_J and is a direct indication of this molecular property. When g_J is positive, the torque is in the

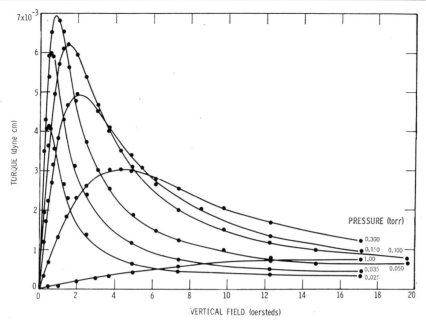

Fig. 1. The torque in NO versus magnetic field at various pressures. The data are for a 1·9 cm diameter, 20 cm long cylinder suspended in a 7·6 cm diameter chamber; the temperature difference was 30°C.

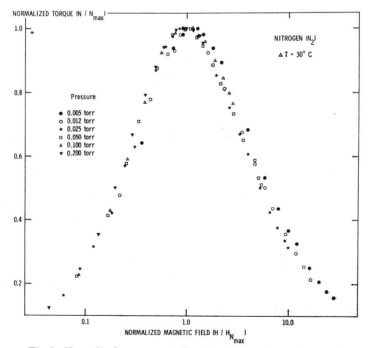

Fig. 2. Normalized torque vs. H/p for N_2 (semilogarithmic plot) at various pressures.

direction of $\vec{H}$ when the radial temperature gradient is negative.

The position of the torque peak occurs at a value of the magnetic field $(H/p) = b[1 + (a/p)]$, where a and b are constants. b is inversely proportional to g_J, and a is inversely proportional to the gas collision cross section and to the cylinder-wall spacing. Figure 2 shows N_2 torque data at various pressures plotted vs. H/p on a semilogarithmic scale. The torque peak heights have been normalized in order to show that the peak is located at a constant value of H/p. It is interesting to note that peak occurs when the Larmor precession frequency is about one eighth the molecular collision frequency.

The magnitude of the torque depends on the non-sphericity of the molecule. O_2 and N_2 give very similar torque maxima, even though the former gas is paramagnetic and the latter diamagnetic. H_2 and D_2 give such smaller torques than does HD, whose torque is comparable to that of N_2 but in the opposite direction. Noble gases give no observable torque. This is the best evidence that the effect is not an artifact.

The torque is a linear function of temperature differential and reverses when the temperature gradient is reversed.

The maximum torque is a peaked function of pressure; and the position of the peak depends on the dimensions of the apparatus. When the radial dimension of the gas is much larger than the mean free path of the molecules, the maximum torque is proportional to $1/p$.

G. W. Smith has studied the effects of a.c. magnetic field modulation. He has found an averaging effect when the modulation frequency is less than the molecular collision frequency and also a resonant absorption at a frequency about twice the Larmor precession frequency.

The torque may be characterized as very large in that any ordinary transfer of angular momentum in the gas directly due to the partial alignment of molecules by the magnetic field is a factor of 10^8 smaller than the observed torque. However, the effect represents a very small perturbation to the velocities of gas molecules in that a deviation of 10^{-6} rad in the average direction of molecular velocities could cause the effect.

In general, the various effects of magnetic fields on the transport properties of gases provide tests for the detailed features of the kinetic theory of gases and may give information about the gyromagnetism and collision cross-sections of molecules. The Scott effect may also give information about the surface collisions of spinning molecules. Numerical model calculations are not yet available for any of the effects of magnetic fields on gases. These will be needed before molecular properties (other than the signs of g_J's) can be deduced. Knowledge of the effects of magnetic fields on dilute and rarefied gases should be valuable in studies of the upper atmosphere.

Bibliography

BEENAKKER J. J. M. (1968) The influence of electric and magnetic fields on the transport properties of polyatomic dilute gases, *Festkörperprobleme* VIII, Freidr. Vieweg and Sohn GmbH, Braunschweig, Germany.

LEVI A. C. and BEENAKKER J. J. M. (1967) Thermomagnetic torques in dilute gases, *Phys. Ltrs.* **25A**, 350.

SCOTT G. G., STURNER H. W. and WILLIAMSON R. M. (1967) Gas torque anomaly in weak magnetic fields, *Phys. Rev.* **158**, 117.

SCOTT G. G., STURNER H. W. and WILLIAMSON R. M. (1967) Gas torque effect for molecular combinations of hydrogen and deuterium, *Phys. Ltrs.* **25A**, 573.

SMITH G. W. and SCOTT G. G. (1968) Measurement of dynamic behavior of the thermomagnetic gas torque effect, *Physical Review Letters*, **20**, 1469.

WALDMANN L. (1967) On a gyro-thermal effect with polyatomic gases in a magnetic field, *Zeit für Naturforschung*, **22a**, 1967.

R. M. WILLIAMSON

THIN FILMS, NUCLEATION AND GROWTH OF.

Introduction

The atomic structure of thin films is of great importance because of the growing demand for cheap and reliable components such as magnetic thin film storage elements, resistors, transistors and many other items whose geometry is an essentially planar configuration.

These components are prepared using a variety of techniques including vapour deposition, electroplating, sputtering and chemical decomposition to name only the most common methods. In all of these techniques, however, one property is common throughout, that is, the resulting film structure after deposition onto a suitable substrate, will depend critically on the parameters involved in the deposition process whichever method is employed. The structure, in turn, dictates the physical properties of the film to a very great extent.

It is not surprising, therefore, that much effort is being put into understanding the interrelationship of the deposition parameters with the resulting film structure. Up to the present time there is a good deal of experimental evidence to suggest that a film grows during deposition by a process in which individual nuclei are formed and that these grow laterally and vertically to produce a coherent film. The kinetics of nucleation and growth have been widely studied particularly in the case of single crystal growth. There has been less study of processes leading to the production of polycrystalline films.

Here the emphasis is placed on applying general nucleation theory to the problem of thin film deposition and the subsequent explanation of the growth of super-critical nuclei in the formation of a continuous film. Films prepared by vacuum evaporation are discussed almost exclusively, but the broad features reviewed here may be applicable to films prepared by other methods, for example, by chemical transport from the vapour phase or from solution.

As a consequence of the mode of growth of a film there is often a large number of crystal defects present and the defect density is usually greater than in a bulk material. Such defects can be studied experimentally using a number of established techniques, the most important being electron microscopy.

Also of great interest, not only to the research worker but to the technologist, whose interests lie in the field of transistors, is the property of a film to establish a definite lattice orientation when grown on a single crystal substrate. This is known as epitaxy and will only occur under favourable film preparation conditions. This will be discussed more fully later, but suffice it to say at this stage that the necessary conditions for epitaxy are not completely clear. The actual individual effects of the preparation conditions are separately understood.

It is worth mentioning at this stage that even though a wealth of information exists on the types of atomic structure to be found in a wide range of films, a great majority of such studies, however, suffer from a lack of correlation of nucleation rates, critical nuclei size, etc., with precise data on the state and structure of the nucleating surface. It is particularly important in planning new studies to evaluate adequately the influence, on such parameters, of contamination layers produced, for example, by the residual gases in the evaporator.

Kinetics of Nucleation and Growth

Here we consider the nucleation and subsequent growth of thin films condensed from the vapour phase on to suitable substrates held at temperatures lower than that of the evaporating source.

Upon impingement at the substrate surface the atoms have essentially three choices—they can adsorb or stick permanently to the substrate, they can adsorb and re-evaporate in a finite time or they can be immediately "reflected" from the surface. The first two cases are the most common. The actual ratio of sublimed to impinging atoms is called the accommodation or "sticking" coefficient, and is a function of the substrate temperature, substrate material and evaporant.

Assuming some atoms have stuck to the substrate surface, they can then migrate over the surface colliding with other atoms and form aggregates or clusters of atoms. These clusters often form at imperfection points of the substrate surface where the local strain is high or at cleavage steps where the free energy of the adsorbed atom can be minimized. If the aggregates are small, the surface-to-volume ratio will be large and the local vapour pressure will be higher than the bulk value. This causes the aggregate to dissociate again. There is a point where the aggregate has reached a critical size of minimum stability and upon adding another atom to an aggregate of critical size makes it more stable.

It is not difficult to calculate this critical size using standard thermodynamic methods but before outlining the technique it is necessary to raise the question as to whether it is valid to apply the macroscopic thermodynamic properties for bulk materials to small clusters of atoms. A completely acceptable solution has not yet been found but Walton et al. (1963) have employed a method which treats the clusters as macromolecules.

The classical analysis examines the difference ΔF between the free energy of the original surface and the free energy of the surface plus adsorbed nucleus in terms of the surface free energies and the latent heat of condensation. It is found (see Fig. 1) that initially

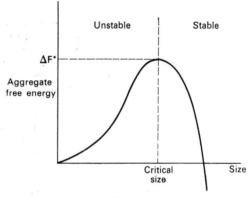

Fig. 1.

this energy increases with aggregate size up to a critical value above which the energy falls. Nuclei with a size greater than the critical size will grow to form a large permanent island. The cases of various shapes of nuclei, for example, hemispheres, cylinders, etc., can readily be calculated.

Critical nuclei usually grow by the addition of atoms diffusing over the substrate surface rather than by direct impingement from the vapour stream. In this case the rate at which critical nuclei grow (the nucleation frequency) is just the product of the number of initial critical nuclei and the rate at which surface diffusing atoms join the cluster. The result is given by:

Nucleation frequency R

$$= CN \exp\left[\frac{-(F_{ad} + F_D + \Delta F^*)}{kT}\right]$$

where C is a constant, N is the deposition rate, F_{ad} is the free energy of adsorption, F_D is the activation energy for surface diffusion of adsorbed atoms, and ΔF^* is the value of ΔF for the nucleus of critical size.

This equation not only shows that the rate of formation of critical nuclei is dependent on the deposition variables—deposition rate, temperature of substrate and surface properties—but that there is a critical concentration of atoms adsorbed on the substrate surface; above this value the nucleation rate is high and below it the rate is very small.

One very important point emerges from the simple analysis above and that is the nucleation frequency is not a function of time. This is unreasonable of course because nucleation sites will gradually be used up as time proceeds since the substrate surface is being covered by adsorbed atoms. Avrami (1939), however, calculated on the basis of random nucleation that the nucleation frequency should decrease exponentially. This is satisfactory as far as it goes but of course assumptions still have to be made concerning the details of the shape of the nuclei.

Epitaxy

When a thin film is deposited upon a single crystal substrate, often a single-crystal film is formed with a narrowly related structure. In addition, the orientation of the newly formed crystal frequently stands in a rigid relationship to that of the substrate. One then speaks of epitaxial growth of the film. The term epitaxy is usually restricted to the growth of one material on a foreign substrate. In the simple case of identical substrate and evaporant materials it is usual to refer to the growth of the film as autoepitaxy.

Epitaxy can extend over macroscopic distances only when at the interface between the two crystals an occasional dislocation of suitable Burgers vector appears, to take up the integrated amount of relative misfit between the two crystal lattices, thereby compensating the internal stresses.

Here the atomic periodicity of the substrate surface is the predominant force influencing the film formation. Earlier evidence due to Royer (1928) suggested that a percentage misfit between the lattice constants of the film and substrate of less than 10–15 per cent was necessary for the occurrence of epitaxy. Subsequent information shows this not to be the case necessarily, particularly for the vacuum evaporation of alkali halides on cleavage surfaces of other alkali halides.

In order to obtain well-defined epitaxy it is necessary to control the deposition variables accurately and these will vary depending on the particular combination of film and substrate materials. Brück (1936) gives the most systematic evidence for the effects of substrate temperature. From his experiments on the growth of metals on sodium chloride substrates, he concludes that an "epitaxial temperature" exists, above which epitaxy is perfect and below which it is imperfect. Again, since his work was published new evidence exists showing that often increasing the substrate temperature can also have a cleaning effect of the substrate surface which can change a good deal depending on the past history of the substrate. It is concluded that it is not in general possible to define an epitaxial temperature but increasing the substrate temperature can improve epitaxy for a wide variety of materials.

The above mentioned ideas are reasonable from a qualitative point of view providing one thinks of epitaxy as arising from a type of heterogeneous nucleation and growth of grains. However, even though a comprehensive study has been made of the experimental conditions of epitaxy on a wide variety of film-substrate combinations, theoretical work has not moved hand in hand and it is not possible as yet to predict the criterion for the initiation of a particular orientation.

Moazed (1966) has treated the problem of epitaxy by extending ideas of heterogeneous nucleation of grains. By assuming the nuclei formed are spherical caps he treats theoretically the heterogeneous equivalent of the free energy theory given above for homogeneous nucleation of grains. He can predict variations of the epitaxial formation of a film as a function of substrate temperature, deposition rate and wetting angle of the critical nucleus (which will be related to the interfacial energy). Again this theory can only be justified when all the parameters involved are known experimentally. Few reliable values exist, for example, of interfacial energies and of course the actual nuclei are unlikely to be simple invariant shapes.

Structural Disorder Introduced During Film Growth

During the growth of an epitaxial film it is possible to artificially divide up the type of structural disorder introduced into three categories—one type of disorder occurring at the interface between film and substrate which is necessary to accommodate the misfit between the lattices, a second type which is produced, not only in an epitaxial film, at the boundary of two grains when they coalesce and thirdly the imperfections and dislocations normally created due to imperfect packing of atomic planes within a grain.

The production of an epitaxial film was originally thought to occur via the formation of a pseudomorphic monolayer which is formed initially upon deposition of the film and upon which the remaining epitaxial film is built. This did not prove to be a completely successful theory providing the criterion for epitaxy. However, it did form the basis of the modern concept of describing the interface between substrate and deposit in terms of a grid of dislocations.

The grid consists of edge dislocations which accommodate the misfit between film and substrate lattices by concentrating the strain produced at the dislocations, leaving good matching in between. Since to a certain extent the dislocation net can be chosen arbitrarily it is important to try to observe the network experimentally. This will be described in the next section.

When two islands grow together whose lattices are slightly rotated relative to one another a grain boundary will usually be formed between them. It will generally be a high angle boundary except where the relative lattice rotations are small as in epitaxial films. The different grains, if their orientations vary, will be clearly seen in an electron microscope by their different contrast.

It is difficult to generalize in any way as to the types of structural disorder expected to be found within a grain of a film because the formation of a particular type of dislocation is related critically to the crystal structure of the film. Very little evidence exists concerning the imperfection structure of polycrystalline films and the majority of evidence has been obtained for single crystal films of face-centred cubic metals, usually formed by epitaxial growth on (100), (110) or (111) faces of alkali halides. Dislocation densities again can vary from film to film but it is usual for a film to contain about 10^{10} dislocations cm^{-2}.

Even though there is now considerable interest in growing various semi-conducting materials on substrates of other semi-conductors, particularly silicon and gallium arsenide, most of the emphasis is being placed on obtaining the experimental conditions for epitaxy rather than studying the imperfections in the deposits. A possible reason for this is that semi-conducting layers are commonly thicker than the metal films mentioned above and are mostly too thick for direct transmission electron microscopy. Silicon layers deposited on single-crystal silicon substrates have been studied using a combination of light optical and etching techniques.

Experimental Studies

A large range of possible experimental techniques has been developed to a fine scale during the last few years for the observation of lattice defects, interface dislocation networks and the early stages of growth of a thin film. Techniques such as electron microscopy, electron diffraction (both high and low energy), X-ray diffraction and field-ion microscopy have contributed largely to our understanding of the mode of formation of a film.

Evidence for the actual formation of discrete islands of material during the initial stages of growth has been obtained by electron microscope studies of films prepared in such a way that their thickness varies continually from zero to approximately 1000 Å from one end of the specimen to the other. The results show quite clearly how as growth proceeds adjacent nuclei coalesce by forming a "bridge" between them. This bridge then acts as a centre of low free energy so that diffusing atoms gradually fill up the remaining areas of substrate until a continuous film is produced. Additional studies on the effects of substrate temperature show that as the temperature is increased, larger grains are produced before coalescing and the thicker the film must be, before a continuous sheet is produced.

It is of interest that upon annealing an incomplete film of small average grain size a recrystallization effect can take place in which the surprisingly liquid-like behaviour of the initial nuclei is clearly seen. The nuclei "run together" to form larger islands of material. This has been directly observed by heating a film in an electron microscope using a special stage.

Further evidence for the island growth concept is given by measuring the surface area of a film when it is very thin. Values as large as one hundred times the geometric areas are not uncommon. The surface area is best measured by a gas adsorption method. The volume of gas adsorbed on a film is proportional to the surface area and its value can be calculated if the area occupied by each gas atom is known. Differences in the surface area for films of the same material can be measured accurately.

Measurements of the rates of nucleation of various film–substrate combinations under carefully controlled conditions can yield values of binding energies, surface diffusion energies and dissociation energies of critical clusters by use of the equations derived from consideration of the kinetics of nucleation and growth. Walton *et al.* (1963) have made such deductions, but further carefully controlled experiments of this kind are needed before full confidence can be given to the results which can be severely affected by contamination of the substrate surface before deposition of the film has commenced.

Of course the granular nature of a film manifests itself in a large number of other physical effects—the electrical resistance of a very thin film can be explained in terms of electron transfer between islands separated by a few tens of angstroms, a process known as tunnelling. The magnetic moment of ferromagnetic thin films whose thickness is less than 100 Å is lower than expected from bulk values and this has been accounted for by regarding the islands as superparamagnetic particles.

Matthews (1961) has made the most outstanding contribution to the study of the interface layer between an epitaxial film and its parent substrate. He studied superposed films of lead sulphide and lead selenide, the former acting as a thin film substrate. Moiré fringe patterns were observed from both the overlap of the two lattice structures and from what appears to be uniquely associated with the dislocation grid at the interface. He has also studied layers of palladium on gold and again observed a dislocation grid. For both of these combinations the lattice misfit was less than 5 per cent.

Conclusions

As a result of many carefully planned experiments particularly during the last five years, thin films can now be produced in forms and with properties very close to specified conditions. Whilst all the factors involved in the growth processes of a film are not yet fully understood, the appearance of the electron microscope has perhaps, more than any other single development, given a good deal of insight into the important phenomenon of epitaxy. This is a particularly important development area, not only from the semiconductor point of view but also because epitaxy is a relatively simple and quick way of producing single crystals of material for fundamental study.

Bibliography

AVRAMI M. (1939) *J. Chem. Phys.* **7,** 1103.
BRÜCK L. (1936) *Ann. Phys. (Leipzig)* **26,** 233.
FRANCOMBE M. H. and SATO H. (1964) *Single Crystal Films*, Oxford: Pergamon Press.
HIRTH J. P. and POUND G. M. (1963) *Condensation and Evaporation*, Oxford: Pergamon Press.
HOLLAND L. (1956) *Vacuum Deposition of Thin Films*, Chapman and Hall.
MATTHEWS J. W. (1961) *Phil. Mag.* **6,** 1347.
MOAZED K. L. (1966) *The Use of Thin Films in Physical Investigations*, p. 203, Academic Press.
ROYER L. (1928) *Bull. Soc. franc. Min.* **51,** 7.
WALTON D., RHODIN T. N. and ROLLINS R. W. (1963) *J. Chem. Phys.* **38,** 2698.

M. J. FOLKES

TUNGSTEN-HALOGEN LAMPS.

Introduction

Recent developments in the field of tungsten filament lamps have resulted in a wide range of lamps of a basically new type. This type has been given various names by different manufacturers, such as Quartz-Iodine, Tungsten-Iodine, Quartz-Iodide, Tungsten-Bromide, etc. The design of this new type is not dependent upon the use of either quartz or iodine and, therefore, in an endeavour to avoid confusion the lamp industry have given the generic term of Tungsten-Halogen to all lamps of this type.

To appreciate fully the principles of tungsten-halogen lamps it is useful to consider the normal incandescent types. In the latter a tungsten coil is held by one or two supports in the centre of a glass bulb. Electrical current flows along the lead-in wires to the coil which is raised to incandescence. A very large proportion of the radiated energy consists of infra-red radiation, whilst only a small amount is emitted in the visible spectrum. The efficiency (expressed in lumens per watt) increases with the coil temperature, therefore making it desirable to attain as high a coil temperature as possible.

When tungsten is heated evaporation takes place and the evaporation rate increases rapidly with temperature so that the actual filament temperature chosen is a compromise between these two factors.

To reduce evaporation, the bulb is filled with an inert gas. The larger the gas molecules, the more the evaporation is suppressed. Normally lamps are filled with argon to which some nitrogen is added to suppress arcing. The life of the lamp is related to the percentage of the tungsten of the coil which is evaporated. This percentage, termed the "dead loss", is generally about 10 per cent for vacuum lamps and 2 per cent for gas filled lamps.

Table 1. The evaporation of a tungsten filament in vacuum

Temperature (°K)	W/cm^2	lm/W	Evaporation rate (g/cm^2/sec)
2000	29	2·8	1·76 × 10^{-13}
2500	70	11·7	2·03 × 10^{-9}
2800	118	20·5	1·10 × 10^{-7}
3000	160	27·2	9·95 × 10^{-7}
3200	214	34·7	6·38 × 10^{-6}
3655 (melting point)	383	53·1	2·28 × 10^{-4}

A normal general lighting service lamp is designed for an average life of 1000 hr from which the efficiency follows automatically. In comparison with a 225 V coil, a 120 V coil has a thicker wire and can be maintained at a higher temperature (higher efficiency) for the same wattage (see Table 2).

Table 2. Comparison of Efficiency for 120 V and 225 V Lamps

	120 V	225 V
100 W	15·6 lm/W	13·8 lm/W
200 W	17·3 lm/W	15·8 lm/W
1000 W	20·2 lm/W	18·8 lm/W

To achieve either a higher efficiency with the same life or a longer life with the same efficiency, the evaporation needs to be suppressed by a gasfilling. A filling pressure of 1 atm of argon reduces the evaporation by a factor of 25, but increased heat losses are introduced by conduction and convection through the gas. If krypton or xenon gases are used the efficiency can be raised, but such gases are expensive to use in large bulbs and similarly, increased filling pressure will reduce evaporation but a high pressure filled bulb of normal general service lamp dimensions is potentially hazardous. Consequently the bulb must be smaller and this requires a "hard" glass with a higher softening point.

The lead-in wires must now be of tungsten because this has a similar coefficient of expansion to that of the normal "hard" glass and this is necessary to obtain a good gas-tight seal. Recent developments have shown that it is possible to make halogen lamps in "hard" glass using only tungsten or molybdenum materials inside the bulb.

The reduction in size obtained with "hard" glass is not sufficient to utilize all the advantages of an halogen lamp. The next step is therefore to use quartz. Quartz has a softening point above 1500°C which permits the use of a much smaller bulb. This, in turn, allows filling pressures to be raised to 5 or 6 atm and, because of the small volume, the more expensive krypton and xenon gases become economically feasible.

One of the difficulties of using quartz concerns the sealing of the lead-in wires. Because of its low coefficient of expansion, there is no metal which will make a matched seal in the normal manner, and so the method adopted involves pinching an etched molybdenum foil into the tube end with two knife-edges.

In most cases the degree of miniaturization is limited by the pinch temperature. Above 350 °C the molybdenum foil oxidizes and this gives rise to small cracks in the quartz which break the gas-tight seal. As quartz is difficult to blow into bulbs most quartz halogen lamps are therefore made out of quartz tubing.

It is clear that quartz lamps, because of their small bulb surface area, would blacken much more quickly than the larger bulb surface area of normal lamps. Even with a filling pressure of 6 atm of krypton a small projection or motorcar lamp will blacken within a few minutes.

The miniaturization of such lamps has been made possible due to a regenerative tungsten cycle, which is maintained in the lamp with the help of halogens. Evaporated tungsten can be returned to the filament by a chemical transport reaction. A halogen or halogen compound is added to the filling gas of the lamp, so forming volatile tungsten halides with the evaporated tungsten. These tungsten halides are chemically stable at the bulb temperatures but decompose in the vicinity of the filament, and in this way the tungsten is returned to the filament and the bulb remains clear.

This regenerative cycle involving halogens is based on the following chemical equilibrium:

$$W + nX \underset{\text{filament temp}}{\overset{\text{bulb temp}}{\rightleftarrows}} WX_n$$

At relatively low temperatures (at the bulb wall) solid tungsten (W) reacts with gaseous atomic halogen (X) to form gaseous tungsten halide WX_n: the equilibrium shifts to the right side.

At relatively high temperatures (at the filament) gaseous tungsten halide dissociates into tungsten and atomic halogen: the equilibrium shifts to the left side.

To realize this regenerative cycle, one can choose any of the four halogens, fluorine, chlorine, bromine and iodine, as they all give volatile tungsten compounds.

Fluorine is theoretically the most suited to a regenerative tungsten cycle because:

(a) Tungsten fluorides only start dissociation at temperatures corresponding to the hottest spots of the filament.
(b) The dissociation of tungsten fluorides, unlike the chlorides, bromides and iodides, is more strongly dependent on temperature than on the vapour pressure of tungsten. This is illustrated by Fig. 1.

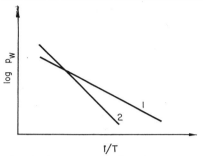

Fig. 1. (1) *Vapour pressure of tungsten.* (2) *Vapour pressure of tungsten by dissociation of tungsten fluoride.*

This means that a thin hot spot on a tungsten filament receives more tungsten from dissociation of the halide than is evaporated from it. Fluorine is able to equalize the diameter of a tungsten wire.

For practical application fluorine gives rise to many difficulties. All parts of the lamp including the bulb and metal parts at relatively low temperatures are strongly attacked by fluorine. This chemical attack at low temperatures becomes less with chlorine and bromide, with iodine having the lowest reactivity towards tungsten.

For this reason the first regenerative cycle tungsten lamps were based on iodine as the "transport" gas. With increasing scale of manufacture of "Quartz Iodine" lamps, the tungsten–iodine cycle proved to be highly sensitive to traces of impurities which, in many cases, caused blackening of the bulb.

In recent years, research has shown that the cycle with iodine can operate well in the presence of a small quantity of oxygen. It was found that the transportation of tungsten from cold to hot is effected by a tungsten–oxygen–iodine cycle, in which an important part is played by a previously unknown compound WO_2I_2.

The correct dosing of the iodine in the lamp creates many technological problems. Iodine is a solid with a low vapour pressure at room temperature and is also corrosive. These considerable disadvantages, as far as trouble-free large scale manufacture is concerned, led to experiments with the other halogens, chlorine and bromine.

It was found that, if the halogen pressure was correctly chosen, it was possible to make a bromine lamp and even a chlorine lamp with normal life. The required pressure is, however, very critical in bromine lamps and even more so when chlorine is used. In practice it was not possible to make the correct dosing reproducible, because a part of the halogen introduced into the lamp is withdrawn from the cycle by impurities.

When using bromine or chlorine as "transport" gases the main problem is therefore: How can the correct dosing be maintained throughout the life of the lamp, considering that only a minute quantity is required for the regenerative cycle? This can be achieved by using (instead of elemental halogens) compounds which maintain the required partial halogen pressure by thermal dissociation. An important factor here is that since the halogen pressure must remain very low, the thermal stability of the compound must be very great and its dissociation reaction must be reversible. Some of the hydrogen halides such as HBr or HCl fulfill these conditions fairly well. Figure 2 shows graphically the degree of dissociation (α) for the reaction $2\,HX \rightleftharpoons H_2 + 2\,X$, and the partial halogen pressure (P_x) as a function of temperature (T); the hydrogen halide pressure is 10^{-2} atm before dissociation.

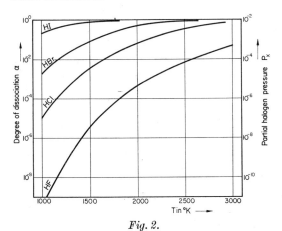

Fig. 2.

For practical application, hydrogen bromide, in particular, yields favourable results. By closing HBr in a pressure range from 4 to 500 Torr, it is possible to make lamps which have both a normal life and do not blacken.

The use of HBr as the "transport" gas offers many advantages over iodine. The dosing is simpler and is not critical as well as the fact that the tungsten-bromine cycle is less affected by impurities. Nevertheless, one cannot say that all technological problems are solved. Hydrogen bromide is hygroscopic and corrosive and if traces of oxygen are present in the lamp, a water-vapour cycle is set up which reduces the life of the lamp.

Both these drawbacks can be obviated in a simple manner by using halogen derivatives of methane, such as CH_3Br, CH_2Br_2, $CHBr_3$, CH_2ClBr and some other substituted hydrocarbons. These compounds are chemically inert and some have a fairly high vapour pressure. Mixed with the filling gas (krypton, argon or nitrogen) they are dosed into the lamp in a suitable proportion. During the initial burning of the lamp they decompose on to the filament to give hydrogen bromide and carbon according to the reaction of the type:

$$CH_2Br_2 \xrightarrow{t\,=\,500°C} C + 2\,BHr.$$

This finely divided carbon acts as a getter absorbing oxygen and water vapour.

The advantages of tungsten-halogen lamps over conventional types are being successfully exploited for lighting, amateur photography, cine and slide projection, floodlighting, film and television studio lighting, infra-red heating, etc.

Bibliography

KOPELMAN B. and VAN WORMER K. A. Jr. (1967) Thermodynamic considerations of tungsten halogen lamps, *Illuminating Engineering*, Sept. 10–14, 1967.

MOORE J. A. and JOLLY C. M. (1962) Quartz-iodine tungsten lamps—mechanism design and performance, *G.E.C. Journal of Science and Technology* **29**, 99.

MOSBY F. A., SCHUPP L. J., STEINER G. G. and ZUBLER E. G. (1966) Incandescent bromine cycle lamps, *Illuminating Engineering*, Aug. 21–26, 1968.

RABENAU A. (1967) Zur Chemie der Glühlampe, *Angew. Chem.* **79**, Jahrgang 1967, No. 1.

SCHILLING W. (1961) Glühlampen mit Jodzusatz, *Elektrotechnische Zeitschrift*, B**13**, 485.

SCHRÖDER (1965) Examples from florine chemistry and possible industrial applications, *Philips Techn. Rev.*, **26**, No. 4/5/6.

STRANGE J. W. and STEWART J. (1963) *Trans. Illum. Engng. Soc. (London)* **28**, 91.

T'JAMPENS G. R. (1967) Regenerative-cycle tungsten lamps, *New Scientist*, Sept. 1967.

T'JAMPENS G. R. and BARDIN J. (1966) Les lampes à incandescence aux halogènes, *Revue Générale de l'Electricité*, **75**, No. 718.

T'JAMPENS G. R. and VAN DE WEIJER M. H. A. (1966) Gas-filled incandescent lamps containing bromine and chlorine, *Philips Techn. Rev.*, **27**.

VAN TIJEN J. W. (1961/2) Iodine incandescent lamps. *Philips Techn. Rev.*, **23**, No. 8/9.

ZUBLER E. G. and MOSBY F. A. (1959) An iodine incandescent lamp with virtually 100% lumen maintenance, *Illuminating Engineering*, **54**, 734.

T. JACOBS and G. T'JAMPENS

ULTRASONIC AMPLIFICATION IN SEMICONDUCTORS.

The processes whereby ultrasonic waves are amplified which are specific to semiconductors all involve the interaction between the waves and free electrons (or holes). Amplification occurs when the electrons are caused, by external fields, to have a mean velocity (drift velocity) greater than the phase velocity of the acoustic wave.

Free electrons interact with the acoustic wave through periodic distortion of the lattice which gives rise to a periodic fluctuation of potential through:
(1) the piezo-electric constant, and
(2) the deformation potential constant.

1. Piezo-electric Coupling

Let the strain, stress, local electric fields, etc., vary as

$$\exp(k_i r) \exp[i(\omega t - k_r r)]$$

where ω is the (real) angular frequency, and k_r and k_i are the real and imaginary parts of the wave vector respectively. In the presence of an electric field which causes electrons to drift with a velocity v, the contribution to k_i due to free carriers is

$$k_i = -\frac{1}{2}\frac{e^2}{c\varepsilon}\frac{\omega_c}{s}\frac{(1-v/s)}{(1-v/s)^2 + \left(\frac{\omega_c}{\omega}\right)^2 (1+k_r^2\lambda^2)^2}. \quad (1)$$

Here s is the velocity of the acoustic wave $\sqrt{c/\varrho}$, ϱ is the density, λ is the Debye screening length and ω_c the reciprocal dielectric relaxation time. (The reciprocal dielectric relaxation time is related to the conductivity, σ, by $\omega_c = \sigma/\varepsilon$.) e, c and ε are elements of the piezoelectric, elastic stiffness and permittivity tensors respectively. The particular elements involved depend upon the polarization and direction of propagation of the wave in the crystal. There is therefore a strong orientation dependence of the effect and distinctive behaviour for longitudinal and shear waves. Equation (1) is valid for weak coupling $\left(\frac{c\varepsilon}{e^2} \ll 1\right)$ and for $k_r l \ll 1$, where l is the mean free path of the electron.

The behaviour of k_i as a function of v/s is shown in Fig. 1. For drift velocities in excess of the velocity of the acoustic wave, k_i is positive and the wave is amplified. For drift velocity less than that of the acoustic wave and also for oppositely directed pro-

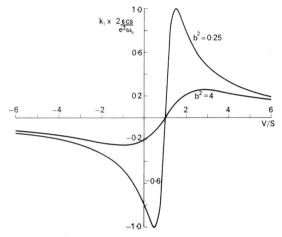

Fig. 1. The normalized amplification constant of acoustic waves due to interaction with free carriers as a function of drift velocity.

$$b = \frac{\omega_c}{\omega}(1 + k_r^2\lambda^2)$$

pagation of the wave and drift of electrons, the wave is attenuated. From equation (1) it follows that the maximum gain occurs at a frequency such that

$$k_r \lambda = 1.$$

Only the contribution to k_i from free carriers has been discussed; to this must be added a contribution from the lattice attenuation.

The attenuation and amplification of the acoustic wave can be understood in terms of momentum (and energy) transfer between the free electrons and the wave. The overall effect of the potential associated with the acoustic wave is to tend to force the electrons to travel at the same velocity as the wave. If, due to an external field, electrons are travelling on average faster than the wave, then momentum is transferred from the electrons to the wave as the electrons are decelerated; consequently the wave is amplified. Conversely, if the electrons are travelling slower than the wave they gain momentum from the wave and the wave is attenuated.

Ultrasonic amplification and related phenomena due to piezo-electric coupling have been observed in

III-V, II-VI compounds and tellurium, notable amongst these being cadmium sulphide.

2. Deformation Potential Coupling

For deformation potential coupling, the expression for k_t is given by equation (1) with

$$\frac{e^2}{\varepsilon\varepsilon} \text{ replaced by } \frac{D^2 k_r^2 \varepsilon}{q^2 c}$$

where D is an element of the deformation potential tensor and q is the charge on the electron. The additional wave vector dependence introduced indicates that this coupling mechanism is only important at high frequencies (greater than 1 GHz, say). Observations of amplification due to this coupling mechanism are usually complicated by the form of the deformation potential which reflects the complexity of the band structure.

Bibliography

McFee J. M. (1968) Transmission and amplification of acoustic waves in piezoelectric semiconductors, in *Physical Acoustics* (W. P. Mason Ed.), vol. 4, part II, New York: Academic Press.

Spector H. N. (1966) Interaction of acoustic waves and conduction electrons, in *Solid State Physics* (F. Seitz and D. Turnbull Eds.), vol. 19, p. 291, New York: Academic Press.

E. G. S. Paige

ULTRASONIC IMAGING. Pictures obtained by ultrasonic radiation are used for a variety of purposes, which include the evaluation of ultrasonic transducers, basic studies of ultrasonic propagation, the diagnostic use of pulse-echo, through-transmission or reflection techniques in medical diagnosis, and other similar methods for the examination of materials. Included in the latter category are nondestructive testing methods for materials and components and ultrasonic imaging methods for use in water, liquid sodium (in nuclear reactor technology), or other opaque media. Most of the reported work involves frequencies in the 0·5 to 10 MHz region, but efforts have been made and reported from sonic frequencies in air (9 KHz) to frequencies well into the megahertz range (at least to 30 MHz).

The through-transmission technique involves a transmitter on one side of the object and a detector on the other side. This technique can be accomplished by a direct-shadow method in which a relatively thin object placed close to the detector is examined, or an ultrasonic lens can be placed between the object and the detector. Pulse-echo images normally make use of a piezoelectric transducer as both transmitter and receiver. The transducer is placed on one side of the object and physically moved to scan an image. Reflection methods usually have a transmitter, a separate receiver and lens, all on the same side of the object, but the method is not as commonly used for imaging. Since ultrasonic lenses for imaging are only in the initial stages of development, the results of recent work in ultrasonic holography are encouraging because the need for the ultrasonic lens in the imaging of thick objects is eliminated.

Much of the interest in ultrasonic imaging is in the detection method. Many of the methods that have been employed to detect ultrasonic images vary not only in technique, but also in sensitivity, required exposure time, complexity, and general image quality. Of the several ways to categorize detection methods, one that offers an advantage in grouping the techniques approximately in terms of threshold sensitivity, or of the minimum ultrasonic intensity needed for image detection, involves four mechanisms: (1) photographic and chemical, (2) thermal, (3) optical and mechanical, and (4) electronic.

The threshold sensitivities required for the various detection methods vary from about 1 W/cm² for the photographic and chemical methods, 0·1 W/cm² for the thermal methods, an average of about 10^{-4} W/cm² for the optical and mechanical methods, to values at least as low as 10^{-11} W/cm² for the electronic techniques.

Ultrasound can be detected because of the direct action of ultrasound energy on a photographic emulsion and because ultrasound accelerates or causes some chemical reactions. The fact that ultrasonic radiation influences a photographic emulsion was reported as early as 1933, but subsequent studies have not clearly revealed the exact mechanism involved. Luminescence and pressure effects do not appear to explain all the existing facts. Although the photographic speed of the film emulsion has little effect on the results, softness of the emulsion, on the other hand, has been a significant factor for the film detection of ultrasound. Film exposed in the dark to ultrasound and developed in the normal manner will yield a useful image with an exposure time of about 4 hr for an ultrasonic intensity of 1 W/cm². If the temperature of the film is raised from 20° to 28 °C, a significant exposure improvement factor can be obtained. If the emulsion is soaked in water at room temperature prior to exposure, about four times less exposure can be used. Both reported improvements were the result of softened emulsion. These film methods, of course, require the use of a darkroom.

Other photographic methods can be used in light. One involves the ultrasound exposure of film in an iodine solution. The effect of the ultrasound exposure on the emulsion is to render the emulsion resistant to fixing in an amount proportional to the exposure. The image becomes visible during the exposure because the emulsion turns a darker yellow. This detection method can be used in light because the exposed film is not developed; light appears to have little influence on emulsion fixing. The speed of this method appears to be independent of photographic film speed at least over the tungsten film-speed range from 6 to 150.

Ultrasonic exposure times in iodine solutions are on the order of 30 min to 1 hr.

A second photographic method that can be performed in light involves the ultrasound exposure of photographic film or paper in a developer solution. The emulsion is exposed to light uniformly, and develops more quickly in areas where the ultrasonic intensity, and therefore agitation of the developer, is highest. One study of this technique with photographic paper indicated that a maximum contrast paper density could be obtained in the exposed areas for a developer-water concentration of 0·2, an exposure time of 90 to 110 sec and an ultrasonic intensity of 0·15 to 0·25 W/cm². The threshold sensitivity was 0·05 W/cm² for a high developer concentration and an exposure time of 40 sec. Exposure times could not be too long or the paper would completely develop. An example of an ultrasonic image produced by the photographic paper technique is shown in Fig. 1.

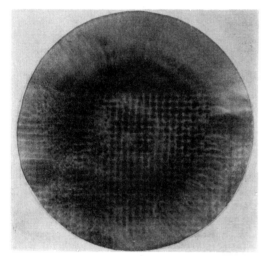

Fig. 1. Ultrasonic image of a 16-mesh wire screen taken by photographic paper in a developer solution. The near-field pattern from the barium titanate transducer is also shown. Exposure time was 30 sec.

One chemical detection method that has been studied by a number of investigators involves potassium iodide–starch solutions. Under ultrasonic irradiation, water saturated with air undergoes an oxidizing reaction to form H_2O_2. This reaction tends to discolour organic dyes. For example, the potassium iodide–starch solution tends to turn blue. This phenomenon has been used to detect ultrasound images by means of an array of boxes that contain this solution. Each liquid-filled box darkens depending upon the ultrasonic intensity. At the threshold ultrasonic intensity of 0·5 W/cm², exposure times are about 2 min. The threshold intensity can be as low as 0·07 W/cm² if small amounts of aliphatic chlorides such as CCl_4 or chloroform are added to the solution.

A similar method involves the exposure of films of starch on glass plates in an iodine solution. Here also, a blue colour is produced in areas of higher ultrasound intensity. Exposure times of about 2 min at 1 W/cm² are common.

A variety of image detection methods described in the literature have been based on the heating effect produced by mechanical vibrations that are generated by ultrasound. By using a thermocouple or thermistor probe to scan the image area, the heat produced by ultrasound can be detected. An array of thermocouple detectors can also be used. Thermocouple junctions as small as 5 μ have been reported. Thermistor studies indicate that ultrasound intensities as low as 0·1 W/cm² can be detected and similar average values are obtainable with thermocouples. The use of a thermopile can, of course, improve sensitivity at the expense of spatial resolution. The detection of ultrasonic intensities as low as 0·01 W/cm² with a thermopile detector has been reported. In terms of temperature sensitivity, the detection of an increase in temperature due to ultrasound as small as 10^{-4}°C should be possible.

Semiconductor materials have shown some response to the heating effects of ultrasound. Sulphides of zinc and cadmium change electrical conductivity with ultrasonic exposure, in an effect said to be caused by heating. Ultrasonic intensities as low as 0·1 W/cm² are detectable by this method.

A similar threshold limitation (0·1 W/cm²) was theorized in a report describing a camera tube that would derive signals from changes in tube target photoemission caused by the heating that results from ultrasound exposure. The sensitivity calculation involved an ultrasonic frequency of 5 MHz.

Another thermal detection method for ultrasound involves compounds that change colour as a function of temperature. Mercury silver iodide is a reversibly chromotropic compound that has been studied for ultrasound image detection. When prepared in a thin screen, this yellow material instantly changes to a bright red when placed at the focus of an ultrasonic transducer. The material was reported to be most sensitive at an ambient temperature of 50°C. Best reported results involve an exposure time on the order of 1 sec at an ultrasonic intensity of 1 W/cm².

The possibility exists that superior results could now be obtained by taking advantage of the recent development of temperature-sensitive liquid crystal material. These materials can be developed to operate within various temperature ranges and for temperature variations of only 1 to 2°C, they display a colour change over the entire spectrum (red to blue) when viewed by reflection. This colour change has been reported for a temperature variation as small as 0·1°C.

A number of luminescent techniques have been described as ultrasound image detectors. The methods include those in which ultrasound-induced heating stimulates luminescence, or extinguishes luminescence stimulated by another radiation such as ultra-violet. The material (ZnS), stimulated by ultra-violet, shows a decrease in luminescence with ultrasound excitation for silver-activated material, or an increase in luminescence under similar conditions if ZnS(Cu) is used.

Heat-induced changes of phosphor persistence have also been investigated for ultrasound image detection. The phosphor (Ca-SrS), which was stimulated by ultra-violet and then exposed to ultrasound, showed increased persistence so that areas were brighter and could be photographed. Ultrasound exposure times as short as 1 min at an intensity of 0·1 W/cm² yield good results. The reported threshold sensitivity is 0·05 W/cm². A spatial resolution of 0·2 mm has been achieved.

Two of the optical and mechanical methods to be described here differ from the remaining techniques in that visual images normally are provided by means of light passing through a sample in the direction perpendicular to the ultrasonic beam direction. Therefore, a side-view photograph of the ultrasonic beam is obtained, as opposed to photographs of the ultrasound area obtained by the other methods. These have been used to evaluate ultrasonic transducers and to study ultrasonic propagation, as in the schlieren photograph shown in Fig. 2. Schlieren methods have been used to display ultrasound and other vibrational phenomena in transparent media. For ultrasound, systems that display beam patterns of continuous-wave and pulsed systems have been described. Colour displays of schlieren techniques have also been reported. Threshold ultrasound intensities as low as 3×10^{-4} W/cm² have been claimed to provide useful results for schlieren ultrasound imaging.

A similar type of ultrasonic photographic technique depends on the fact that some transparent media show temporary birefringence when stressed. Polarized light passing through the medium will be partially depolarized in regions of high ultrasonic intensity. Photographs taken by this method in a medium of fused quartz show good contrast and picture details such as the Fresnel region of a 10 MHz X-cut quartz transducer and beam side lobes. Useful results can be obtained at ultrasonic intensities as low as 0·1 W/cm², which is not necessarily a lower limit.

A relatively sensitive technique for ultrasound detection depends on the fact that a material surface will deform due to the pressure of an acoustic beam. Most work with this technique has involved liquid surface deformation and the observation of this deformation by reflected light from the surface. Some early work was done by transmitting light through the liquid and deformed surface, but this seems to be less sensitive than the surface reflection method. The sensitivity of the technique is inversely proportional to the coefficient of surface tension and to the density of the liquid. At high ultrasonic frequencies (above 10 MHz), the surface tension forces are predominant, while at low frequencies (0·5 MHz) the capillary forces

Fig. 2. *A schlieren photograph of a continuous-wave ultrasound beam from a 5 MHz, 1·87 cm diameter, zirconate titanate transducer (beam enters from the right). The ultrasound beam is shown striking a thin sheet of aluminium. Reflected and transmitted ultrasound beams can be readily observed. Evidence of conversion of the longitudinal beam to shear waves in the aluminium and subsequent re-irradiation as longitudinal waves into the water can also be observed.*

give way to gravitational forces. Between these two extremes, both effects are important. Threshold sensitivity values as low as 10^{-5} W/cm² have been reported. Recent work indicates that the detection of deformation of a metal surface is possible by reflecting a laser beam from the surface. Indications are that, at an ultrasound frequency of 2 MHz, the threshold sensitivity of the technique would approach 10^{-6} W/cm².

The Pohlman cell has been used as a detection method for ultrasound images; the method makes use of a liquid-filled cell in which small aluminium flakes are suspended. In the presence of ultrasound the flakes tend to align in a pattern that reflects light incident on the cell. With no ultrasound present, the flakes assume a random orientation and present a gray background by reflected light. Improved contrast results if a small bias voltage (typically 25 to 30 Vac) is put across the cell. The voltage tends to align the flakes so that no light is reflected in the absence of ultrasound. The threshold sensitivity of the Pohlman cell has been reported to be $2·8 \times 10^{-7}$ W/cm². At such low ultrasound intensities a reaction time on the order of many minutes could be anticipated for this detector. At intensities on the order 10^{-1} to 10^{-3} W/cm², the reported reaction time is 1 sec or less. One major problem with the Pohlman cell is the limited dynamic range (20 dB).

The electronic method of detecting ultrasound images can provide the best sensitivity of all the categories discussed and, for this reason, is the most used detection method. Piezoelectric or electrostrictive transducers have been widely used to detect ultrasound images either on the transducer, by mechanically scanning the transducer(s) or object, or by making use of an array of transducers. The mechanical movement of a transducer or object provides an imaging method that has been applied in medical and industrial nondestructive testing applications. Scanning methods such as B and C scan and complicated motions over curved surfaces have been used to generate images. The image can be presented by recording in sympathy with the mechanical movement of the transducer. Image readout can be on an oscilloscope, film, or paper recording. A through-transmission ultrasonic image is shown in Fig. 3.

The threshold sensitivity or resolution of the electronic systems are difficult to discuss because of the many variations that are possible. Most reported systems are pulse-echo types in the frequency range from 1 to 10 MHz and make use of quartz or ceramic transducers. Resolution capabilities on the order of one to a few millimetres, and threshold sensitivities of 10^{-11} W/cm² or less seem readily achievable with such techniques. Threshold sensitivities of the array system as low as 5×10^{-12} W/cm² have been reported.

Since a piezoelectric transducer actually acts as a mosaic the voltage signal at any location on the piezoelectric detector is proportional to the incident ultrasonic intensity at that point. Images can be prepared without moving the object or detector by suitably recording the voltage variations on a number of locations on a piezoelectric detector of reasonable size. Images have been detected by mechanically moving a small probe across the back of a piezoelectric material.

A faster approach is to scan the piezoelectric target with an electron beam and generate a television-type signal. Television methods for the detection of ultrasound images have been described. Television camera tube systems provide a linear response over a

Fig. 3. An ultrasound, through-transmission image of a human adult hand and fore arm. The two transducers scanned a saw-tooth pattern across the object. The recording shows areas of high ultrasound transmission in the dark lines and low transmission (the bone images) in the white areas. The ultrasound frequency was 1 MHz.

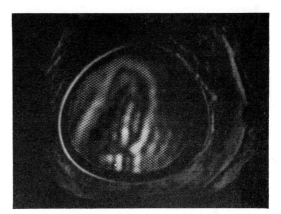

Fig. 4. The ultrasonic image of a finger tip (adult index finger). The side view of the finger was obtained by a direct shadow, through-transmission method at an ultrasonic frequency of 1·9 MHz. Transmission through the tissue (right side of finger) and attenuation (dark areas) in the bone area can be observed. The photograph was taken from the monitor of a continuous-wave ultrasonic television system.

wide range of ultrasonic intensities (3×10^{-9} to 3×10^{-3} W/cm²) and a threshold sensitivity as low as 2×10^{-11} W/cm² has been discussed. Spatial resolution is normally limited by the thickness of the detecting material or by the wavelength of the ultrasound.

Ultrasound television camera tubes have included continuously evacuated and sealed tubes with ultrasound targets of barium titanate or quartz. Tubes of this type were built into the sidewall of an immersion tank to provide television images of incident ultrasound images by direct shadow, through-transmission with lenses, and reflection methods. Continuous-wave and pulsed systems have been described. An example of an ultrasonic television image is shown in Fig. 4.

A summary of ultrasonic imaging detection methods and the ultrasonic intensity required for each method to function is given in Table 1.

Table 1

Summary of Ultrasonic Imaging Detection Methods

	Approximate reported threshold sensitivity, W/cm²
Photographic and chemical methods	
Direct action on film	1
Photographic paper in developer	0·05 to 0·1
Starch plate in iodine solution	1
Film in iodine solution	1
Colour change effects	0·07 to 0·5
Thermal techniques	
Phosphor persistence changes	0·05 to 0·1
Extinction of luminescence	1
Stimulation of luminescence	—
Thermosensitive colour changes	1
Change in photoemission	0·1
Change in electrical conductivity	0·1
Thermocouple and thermistor detectors	0·1
Optical and mechanical methods	
Optical detection of density variations	3×10^{-4} to 10^{-3}
Liquid surface deformation	10^{-5} to 10^{-3}
Solid surface deformation	10^{-6}
Mechanical alignment of flakes in liquid	$2·8 \times 10^{-7}$
Acoustic birefringence	10^{-1}
Electronic methods	
Piezoelectric detector — mechanical movement of transducer or object, or use of an array of transducers to form an image	5×10^{-12} to 10^{-11}
Probe detection of potential on back of piezoelectric receiver	10^{-11}
Electron scan of piezoelectric receiver	2×10^{-11}
Electron scan of piezoresistive receiver	10^{-7}

Acknowledgements

This work was performed under the auspices of the U.S. Atomic Energy Commission.

Bibliography

BERGER H. and DICKENS R. E. (1963) *A Review of Ultrasonic Imaging Methods With A Selected Annotated Bibliography*, Argonne National Laboratory Report ANL–6680, Argonne, Illinois.

BERGMANN L. (1954) *Der Ultraschall*, 6th ed., pp. 248–335, Stuttgart: S. Hinzel.

FREITAG W. (1958) Investigations of an electronic-ultrasonic image converter, *Jenaer Jahrbuch* **1**, 228–74.

GOLDMAN R. (1962) *Ultrasonic Technology*, pp. 211–22, New York: Reinhold Publishing Corp.

JACOBS J. E., BERGER H. and COLLIS W. J. (1963) An investigation of the limitations to the maximum attainable sensitivity in acoustical image converters, *IEEE Transactions on Ultrasonic Engineering* **10**, No. 2, 83–88.

METHERELL A. F. (Ed.) (1969) *Proceedings First International Symposium on Acoustical Holography*, New York: Plenum Press.

OSHCHEPKOV P. K., ROZENBERG L. D. and SEMENNIKOV YU. B. (1965) An electronic-acoustical transducer for the visualization of sound images, *Soviet Physics-Acoustics* **1**, 362–5.

ROZENBERG L. (1955) Survey of methods used for the visualization of ultrasonic fields, *Soviet Physics-Acoustics* **1**, 105–16.

SMYTH C. N., POYNTON F. Y. and SAYERS J. F. (1963) The ultra-sound image camera, *Proceedings IEE* **110**, 16–28.

SPENGLER G. (1953) Problem of acoustic-optical image conversion, *Nachrichtentechnik* **3**, 399–402.

TURNER W. R. (1965) Ultrasonic-imaging, *Ultrasonics* **3**, 182–7.

H. BERGER

ULTRASONIC SPECTROSCOPY. This is the spectral analysis of ultrasonic vibrations induced in a material for the purpose of conducting a non-destructive inspection. The spectroscopic method yields information on

(1) the frequency-dependence of ultrasonic attenuation in a material,

(2) the natural resonances of a specimen,
(3) the frequency-dependence of the ultrasonic energy returning from a reflecting discontinuity,

and can accordingly be used to determine
(1) microstructure,
(2) physical dimension,
(3) defect geometry.

Instrumentation

The spectral analysis of ultrasonic vibrations can be carried out by electronic means if the vibrations are first converted to electrical oscillations. This is achieved by coupling a piezoelectric transducer to the test specimen. A piezoelectric transducer also can be utilized to impart ultrasonic vibrations to the specimen in the first place. In some cases a single transducer can serve both these purposes (see Barron, 1962).

(a) *Choice of transducer.* The transducer is by far the most critical component of the ultrasonic spectroscope because meaningful test results can only be obtained if the transducer has a sufficiently broad frequency response. To broaden the response of a transducer element, it has to be mechanically loaded, for instance, with a suitable backing. Figure 1 shows the broad frequency response characteristic of a heavily damped lithium sulfate transducer suitable for ultrasonic spectroscopy.

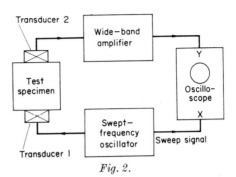

Fig. 2.

is independent of frequency, the energy distribution in the displayed spectrum is a function of the two transducer response characteristics and the ultrasonic attenuation vs. frequency relationship exhibited by the specimen. To determine the latter, the observed spectral amplitude has to be normalized with respect to the spectral curve obtained with the two transducers alone, coupled directly to each other.

Method B employs a pulsed swept-frequency technique and has the advantage to require access to only one surface of the specimen. The experimental arrangement is shown in Fig. 3. The pulsed excitation of the transducer which is produced with an electronic

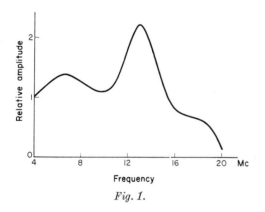

Fig. 1.

(b) *Electronic equipment.* Equipment requirements for ultrasonic spectroscopy vary in accordance with the method that is used.

Method A, which employs a swept-frequency technique, utilizes instrumentation outlined in the schematic block diagram of Fig. 2. Two transducers are coupled to the test specimen. Transducer 1 is energized by a swept-frequency oscillator and introduces ultrasonic vibrations into the specimen. Transducer 2 receives the ultrasonic vibrations that have passed through the specimen and converts them to electrical signals. These signals are amplified by a wide-band amplifier and fed to the Y-deflection plates of a cathode-ray oscilloscope. The X-deflection of the oscilloscope is furnished by the sweep signal of the oscillator. The horizontal axis therefore represents frequency and the oscilloscope display a spectrum. Assuming that the output amplitude of the oscillator

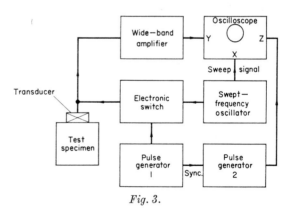

Fig. 3.

switch provides transmission intervals during which the transducer can function as a receiver. Thus, a single transducer suffices for both transmission and reception of ultrasonic signals.

Because pulse generator 1 which actuates the electronic switch operates at a considerably higher repetition rate than the oscillator sweep, the frequency of emitted ultrasonic signals and their echoes varies from pulse to pulse as illustrated in Fig. 4. The first plot of this figure represents the frequency variation of transmitted pulses and received echoes. The second trace shows envelopes of transmitted oscillatory pulses and their echoes (only three pulse trains are shown for

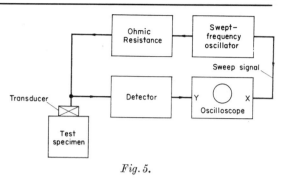

Fig. 5.

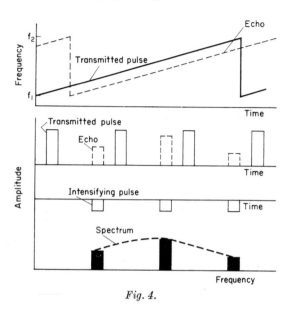

Fig. 4.

the sake of clarity). The third trace depicts negative pulses which are furnished to the Z-axis of the oscilloscope by pulse generator 2 to intensify echo indications, and the last trace shows the intensified echoes and their envelope curve. This curve constitutes the spectrum of the ultrasonic echoes.

In practice the number of switching pulses per oscillator sweep can be made much higher (e.g. 1000) than is shown in Fig. 4 and the spectrum will then consist of numerous lines.

The spectral amplitude function observed with this method depends on three factors, the square of the transducer response function (also called "loop response"), the ultrasonic attenuation in the specimen, and the reflection characteristics of the discontinuity echoing the incident pulse. In contrast to Method A, Method B will therefore yield information on individual discontinuities that are contained in a specimen.

Method C, the third technique employing swept-frequency transducer excitation, is also called ultrasonic resonance testing. It reveals the natural resonances of a specimen which in turn are dependent upon the specimen configuration and the propagation velocity of ultrasound (usually the longitudinal wave mode) in the specimen.

Figure 5 shows the electronic apparatus used for this method. The transducer is connected to a swept-frequency oscillator through an ohmic resistor. As a result, the temporary decrease in transducer current which is encountered at a specimen resonance will produce a momentary rise in signal amplitude. A detector converts the amplitude fluctuation to a d.c. pulse signal which is displayed by the oscilloscope as a vertical indication. Since the horizontal beam deflection of the oscilloscope is furnished by the oscillator sweep, the X-axis of the display represents frequency and a vertical indication therefore constitutes a spectral line.

If more than one specimen resonance occurs within the range covered by the swept-frequency oscillator, several spectral lines can be observed on the screen of the oscilloscope.

Method D. Instead of using swept-frequency excitation, a transducer can also be excited with a rectangular voltage pulse to generate an ultrasonic spectrum. In this case the amplitude A_f of the transducer excitation voltage is no longer independent of frequency but obeys a $\dfrac{\sin x}{x}$ relationship which for a d.c. pulse is

$$A_f = \frac{PW}{\pi} \frac{\sin \pi W f}{\pi W f}.$$

In this equation, P represents pulse amplitude, W pulse width, and f frequency.

Figure 6 illustrates a portion of the theoretically infinite pulse spectrum. Shown are the main lobe, which represents about 90 per cent of the total pulse energy, and two of the side lobes. Since the width of the main spectral lobe is inversely proportional to the pulse width, a reduction in pulse width will effectively broaden the spectrum. But it will also decrease the spectral amplitude which, according to the above equation, is proportional to W. Thus, d.c. pulsing will usually provide less transmission power than the swept-frequency techniques. This is important if the test specimen exhibits high ultrasonic attenuation or the reflecting discontinuity is relatively small.

The equipment used for the d.c. pulse procedure is shown in Fig. 7. Pulse generator 1 supplies a rectangular pulse of short duration to the transducer. Pulse generator 2, which is synchronized with Pulse generator 1, actuates an electronic switch. The switch is employed to gate out the portion of the transducer signal in the time domain which one desires to have processed by the electronic spectrum analyser. The echo reflected from a discontinuity, for example, can

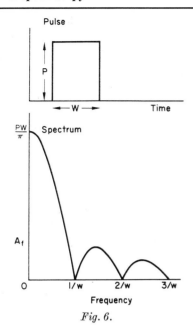

Fig. 6.

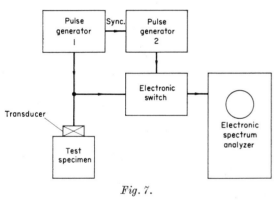

Fig. 7.

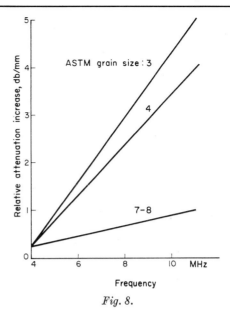

Fig. 8.

be selected for spectral analysis without encountering interference from the much more powerful pulse that is initially transmitted.

The ultrasonic spectrum obtained with the d.c. pulse method is subject to the same influences discussed under Method B and is, in addition, dependent upon the width of the rectangular excitation pulse.

Applications

(a) *Microstructure determination.* Methods A, B, and D can be used to determine the frequency-dependence of ultrasound attenuation in a material which, in some instances, is characteristic of the microstructure.

As an example, Fig. 8 illustrates the attenuation vs. frequency relationships for three steel specimens with average grain sizes of 3, 4, and 7–8 (according to the ASTM scale).

The data plotted in Fig. 8 was derived from back-echo spectra obtained for a frequency range of 4 to 12 Mc/s by Method D. The attenuation observed at the lowest frequency was arbitrarily chosen as 0·25 db/mm for all three specimens. The data, therefore, represent the relative increases in attenuation with frequency, which, as Fig. 8 shows, depend on the average grain size of the material.

(b) *Thickness and velocity measurements.* Method C, also known as the resonance technique, is widely used for thickness measurements. In comparison with mechanical gauges, the ultrasonic method has the significant advantage of requiring access to only one surface of the specimen. Thickness d is determined by noting the frequencies f_n and f_{n+1} of two consecutive spectral lines representing thickness resonances and inserting these in the equation

$$d = \frac{c}{2(f_{n+1} - f_n)}.$$

The longitudinal wave propagation velocity c of the specimen material must be known.

If, on the other hand, the thickness d can be measured by other than ultrasonic means, Method C will yield the longitudinal ultrasonic velocity c according to the equation

$$c = 2d(f_{n+1} - f_n).$$

(c) *Determination of defect geometry.* Methods B and D can be used to determine the effect which a reflecting discontinuity has on the frequency content of the ultrasonic echo return. The configuration of concealed defects can thus be explored. The practical value of this approach can be demonstrated with the two test specimens shown in Fig. 9. The specimens are made

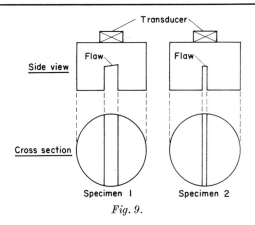

Fig. 9.

of a material exhibiting a low ultrasonic attenuation and are both 75 mm in diameter and 50 mm high. They are provided with machined cuts to simulate internal defects. The defect contained in Specimen 1 is 8 mm wide and oriented at an angle of 10 degrees with respect to the surface to which the transducer is coupled. The defect in Specimen 2 is only 2 mm wide and has a horizontal orientation.

Defect-echo spectra obtained for these specimens by Method D, using a transducer of 10 mm diameter with a frequency response similar to Fig. 1, exhibit significant differences. The curve depicted in Fig. 10 obtained by subtracting the spectrum of the small, horizontal defect from the spectrum of the large,

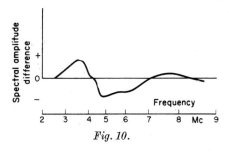

Fig. 10.

angulated defect, demonstrates these differences. Between 2·5 to 4 mc/s the curve assumes positive values indicating that in this frequency range defect width is more important for the reflection of ultrasound than defect orientation. The opposite is true for frequencies of 4 to 7 mc/s where the ultrasonic beam emitted by the transducer is better collimated. At these latter frequencies, the curve shown in Fig. 10 therefore goes negative.

The fact that the dissimilarity of the two examined defects can be revealed by ultrasonic spectroscopy is very significant since the conventional pulse-echo inspection of these defects, conducted with the same ultrasonic transducer, produces indications of exactly equal height. Thus, if similar circumstances were encountered in a practical test situation, an erroneous assessment of relative defect size could easily result unless the spectroscopic method is employed to augment the test data. The presented example was chosen to illustrate the potential importance of the spectroscopic procedure for an improved ultrasonic interrogation of defect geometry.

Bibliography

BARRON K. (1962) Piezoelectric oscillator, in *Encyclopaedic Dictionary of Physics* (J. Thewlis Ed.), **5**, 505, Oxford: Pergamon Press.
GERICKE O. R. (1965) *Materials Research and Standards* **5,** No. 1, 23–30, Oxford: Pergamon Press.
GERICKE O. R. (1966) *Journal of Metals* **18**, No. 8, 932–937.
GERICKE O. R. (1966) *Materials Evaluation* **24**, No. 8, 409–11.
GOLDMAN R. G. (1962) *Ultrasonic Technology*, London: Chapman & Hall.
KRAUTKRAEMER T. and H. (1966) *Ultrasonic Material Inspection*, Heidelberg: Springer.
WEBER E. (1957) *Linear Transient Analysis*, Vol. I, London: Chapman & Hall.

O. R. GERICKE

ULTRA-VIOLET RADIATION, SOME BIOLOGICAL EFFECTS OF. Great progress has been made in recent years in the identification of ultra-violet lethal photoproducts and in the description of cellular recovery phenomena. These important advances are treated in some detail below. However, other ultra-violet radiation effects on biological systems, important but of a different genre, such as fluorescence, phosphorescence, and dye sensitization, have had to be omitted.

A wavelength unit that is becoming increasingly popular is the nanometer (nm). $1 \text{ nm} = 10^{-9} \text{ m} = 1 \text{ m}\mu = 10 \text{ Å}$. This is a logical and unambiguous unit and will be used throughout this article. In biological applications, it is convenient to divide the ultra-violet region into *vacuum ultra-violet* (100–90 nm), *far ultra-violet* (190–300 nm), and *near ultra-violet* (300–80 nm).

1. Far Ultra-violet (190–300 nm) Photoproducts

An important part of the action of far ultra-violet radiation on large cells may be due to absorption in proteins or in lipids. However, in bacteria, which far ultra-violet radiation can easily penetrate, nucleic acid is overwhelmingly the most important chromophore for the biological effects most commonly studied, such as killing and mutation. The genes of a living cell being unique, irreplaceable, and usually essential components, deoxyribonucleic acid (DNA) is considered to be a much more important cellular target than ribonucleic acid (RNA), and this view is supported by many experimental data.

(*a*) *Nucleic acids.* Within DNA, the phosphate and sugar moieties are generally of little importance in ultra-violet effects, primarily because they do not absorb radiation above 200 nm. This leaves the purine and pyrimidine bases, both of which absorb far ultra-violet strongly. The pyrimidines are much more sensitive (quantum yields of the order of 10^{-3}) than the purines (quantum yields of the order of 10^{-4}), and energy absorbed in the latter is generally either dissipated or transferred to the former. Consequently, *the far ultra-violet photobiology of small cells is concerned primarily with effects on the pyrimidines of DNA.*

Ultra-violet irradiation of an aqueous solution of a pyrimidine can cause almost complete loss of the absorption band at 260 nm, an effect discovered in 1949 by Sinsheimer and Hastings. This is often due to the addition of water (*hydration*) at the 5,6 carbons, converting the double bond to a single bond, with consequent loss of the characteristic absorption peak (Fig. 1). This reaction is often easily reversible at extremes of pH or temperature. In some cases the reversal occurs even under normal conditions.

Far ultra-violet irradiation of thymine in frozen solution produces a stable photoproduct, a *thymine dimer*, discovered in 1960 by Beukers and Berends. The effect on the absorption spectrum is similar to that induced by hydration. Subsequent irradiation *even at the same wavelength* in liquid solution will break this dimer, restoring the original thymines and the original absorption peak (Fig. 2). Similar reactions (i.e. frozen induction, liquid reversal) occur with uracil and various derivatives of thymine and uracil, but not with cytosine.

Some of these reactions, as well as others that do

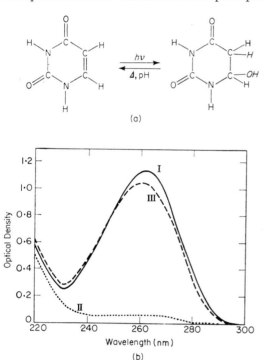

Fig. 1. *Photohydration of pyrimidines.* (a) Addition of a water molecule to the 5,6 carbons of uracil destroys the ring double bond, forming 6-hydroxy-5-hydrouracil. (b) Absorption spectrum of uridylic acid before irradiation (I), after irradiation (II), and after holding at pH 0.8 for 42 hr (III). [Curves in (b) from R. L. Sinsheimer, *Radiation Research* **1**, 505 (1954)].

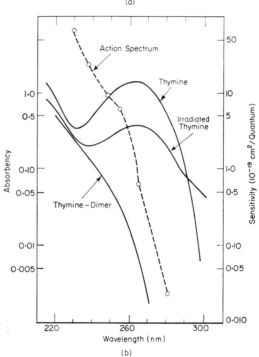

Fig. 2. *Photodimerization of pyrimidines.* (a) Two adjacent thymines are converted to a thymine dimer, with consequent loss of the 5,6 double bonds and formation of a four-carbon cyclobutane ring. In two-stranded polynucleotides, the two thymines are in roughly parallel planes, one beneath the other. (b) Absorption spectra (left ordinate) for thymine, irradiated thymine, and purified thymine dimer. Also shown is the action spectrum (right ordinate) for reversion of the thymine dimer to thymine, which shows that this reversion is caused by absorption in thymine dimer. [(a) is from R. B. Setlow, *Mammalian Cytogenetics and Related Problems in Radiobiology,* Pergamon, p. 291 (1964), and (b) from R. B. Setlow, *Biochim. Biophys. Acta* **49**, 237 (1961)].

not occur in liquid or solid solutions of pyrimidines, take place in polynucleotides. In DNA, thymine-thymine, thymine-cytosine, and cytosine-cytosine dimers are formed. Photochemical splitting of pyrimidine dimers can also occur in these molecules, formation of the dimers being favoured by longer wavelengths (well absorbed by thymine but not by the dimer—Fig. 2) and splitting being favoured by shorter wavelengths (well absorbed by both thymine and thymine dimer—Fig. 2). At sufficiently high doses, an equilibrium is reached between dimer formation and splitting, with the result that the maximum yield of thymine dimers in polydeoxyribose thymidylic acid is only about 15 per cent at 240 nm. This number rises to 35 per cent at 254 nm and to about 70 per cent at 280 nm. Thus one may preferentially produce dimers in polynucleotides around 280 nm and split them around 240 nm (Fig. 2), this sequence being repeatable for several cycles. (Eventually irreversible products accumulate.)

Hydration products have been observed in single-stranded DNA, but not in double-stranded DNA or in polydeoxyribose thymidylic acid. In polyribose uridylic acid, both dimers and hydration products are formed by ultra-violet radiation. The former can be reversed by irradiation at 240 nm (if produced at longer wavelengths) and the latter by heat or extremes of pH. By analogy, one supposes that both dimers and hydration products would also occur in RNA, but only dimers have so far been demonstrated.

It has been shown by R. B. Setlow and J. K. Setlow that biological damage produced *in vitro* in transforming DNA at 280 nm can be reversed by subsequent irradiation *in vitro* at 240 nm. Such reversal probably occurs in the DNA of living cells, but it cannot be observed because biological doses of ultra-violet radiation produce in cells a concentration of dimers far below the equilibrium level, which must be approached for one to observe reversal effects.

The following generalizations regarding ultra-violet photoproducts in polynucleotides may be noted: (1) dimerization occurs within a strand, not between two complementary strands, (2) only pyrimidines adjacent in a chain are likely to dimerize, (3) extent of dimerization varies with wavelength, usually leveling off at doses of the order of 10^4 erg/mm^2, (4) formation and reversal of hydration products is not generally as important as dimerization, but it may complicate the net yield of dimers, (5) small changes in nucleic-acid base ratios can produce large differences in relative photochemical yields of different products, (6) in natural DNAs, thymine dimers are more readily produced (per available pyrimidine pair) in the region 250–80 nm than any of the other dimers, and (7) thymine dimers are very stable, even in DNA. A dose of 1 erg/mm^2 at 260 nm produces about 1 thymine-containing dimer/200 μ DNA. This dose produces about 6 thymine dimers per cell of *Escherichia coli*, whose genome (a single circle of double-stranded DNA) is about 1200 μ long.

DNA molecules may become cross-linked upon ultra-violet irradiation, as evidenced by *in vitro* experiments, as well as by observations on inactivation of bacteria in different humidities or physical states. There is also evidence in bacteria for ultra-violet-induced linkage of DNA to protein, with some indication that this might involve pyrimidine–cysteine bonds. The DNA-protein binding is not photoreactivable, but may account for a major portion of the nonphotoreactivable damage. Ultra-violet-induced binding of RNA to protein has been demonstrated in tobacco mosaic virus.

(b) *Proteins*. Although of minor or even negligible importance for lethal effects in bacteria, the action of far ultra-violet radiation on protein cannot be ignored, since it may play an important role in other biological systems. Per unit weight, proteins absorb only about 1/10 as many photons at 260 nm as nucleic acid, and this is one reason for their secondary importance in ultra-violet effects. (A more important reason is the genetic, and therefore crucial, role of DNA.) Some of the aromatic amino acids are good ultra-violet absorbers, but as a class they do not seem to be important targets in protein inactivation, with the exception of tryptophan. The peptide bond (CONH) is unimportant for protein inactivation at wavelengths above 240 nm; at shorter wavelengths, however, it becomes increasingly important. In general, it appears that ultra-violet inactivation of proteins proceeds by disruption of secondary and tertiary structure, rather than of primary structure. The amino acid cystine has a high quantum yield (0·13) for splitting at 254 nm. It is also the only amino acid that provides covalent bonds for maintenance of the tertiary structure of proteins. Cystine is therefore often the most important ultra-violet target in a protein.

It is now possible to predict, within a factor of 2, the ultra-violet sensitivity of a protein, given knowledge of the amino acid composition and the number of disulphide linkages. The work of McLaren and of R. B. Setlow has led to a rough rule for protein inactivation: *at 254 nm the quantum yields of proteins tend to show an inverse relation to molecular weight and a direct relation to number of cystine residues.*

There is still, however, considerable debate about the details of protein inactivation, and some proteins show an ultra-violet sensitivity that falls wide of predictions. One complication is provided by energy transfer between amino acids and other parts of a protein molecule (such as prosthetic groups or hydrogen bonds), an effect shown to exist, but the extent of which is very difficult to predict.

2. Recovery from Far Ultra-violet Damage

(a) *Photoreactivation*. We now have an elementary understanding of this effect, in which damage induced in biological systems by far ultra-violet radiation may be partly eliminated by subsequent irradiation with near ultra-violet or blue light (300–500 nm). The effective wavelengths generally differ in different systems (Fig. 3). It was shown by Rupert and co-workers in 1958 that extracts of photoreactivable cells can mediate *in vitro* photoreactivation of the biological activity of transforming principles. Later work showed that the active factor is an enzyme. The process takes the following course: (1) the enzyme forms a complex with far ultra-violet-damaged DNA (it is specific for this damage and will not combine with DNAs damaged in other ways), (2) the complex absorbs photoreactivating radiation, and (3) the complex dissociates into enzyme and repaired DNA. The photoreactivating enzyme from baker's yeast has now been purified approximately 6000-fold over the crude extract. It has a molecular weight of about 30,000. It is now known that *this photoreactivating enzyme functions by splitting pyrimidine dimers in DNA*. This appears to be its only function. Details of the reaction are not understood and the chromophore remains unidentified. This enzyme does not operate on ultra-violet-irradiated RNA, although the RNA of plants and of plant viruses is clearly photoreactivable.

Photoreactivation may be of two different types, either or both of which may occur in one biological system. The predominant type is *direct photoenzymatic reactivation*, which utilizes the photoreactivating enzyme, shows a strong dependence upon temperature during photoreactivation, and shows saturation effects at high dose rate of the photoreactivating light. A second type is called *indirect photoreactivation*. This is mediated by a narrower band of wavelengths (300–80 nm, with a peak at 340 nm) and is similar in mechanism to photoprotection (see below), showing little or no dependence upon temperature and no dose-rate-saturation effects.

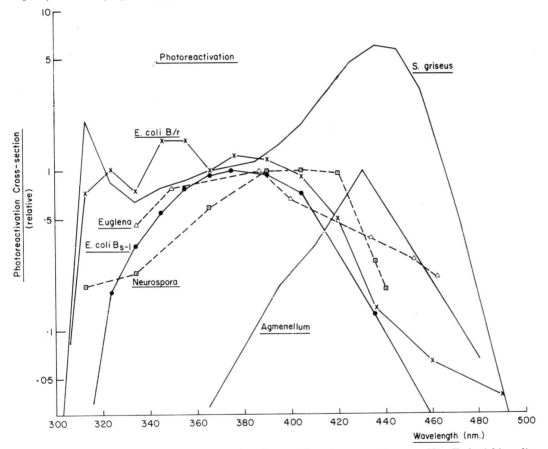

Fig. 3. Action spectra for photoreactivation. Sources: Streptomyces griseus *conidia,* Escherichia coli B_{s-1}, *J. Jagger, R. S. Stafford and J. M. Snow, Photochem. Photobiol.* **10** *(1969);* E. coli *B/r, J. Jagger and R. Latarjet, Ann. Inst. Pasteur* **91**, *858 (1956);* Euglena gracilis, *J. A. Schiff, H. T. Epstein and H. Lyman, Progress in Photobiology, Elsevier, p. 289 (1961);* Neurospora crassa, *C. E. Terry and J. K. Setlow, Photochem. Photobiol.* **6**, *799 (1967);* Agmenellum quadriplaticum, *C. Van Baalen, Plant Physiology* **43**, *1689 (1968). Curves have been normalized to facilitate comparison and thus do not reflect relative efficiencies in different organisms.*

(b) *Excision-resynthesis repair.* There is now much evidence that cells repair, in the absence of photoreactivating light, most of the genetic (DNA) damage inflicted upon them by ultra-violet radiation. It has been shown that pyrimidine dimers are removed from the DNA of ultra-violet-resistant strains but not from the DNA of very sensitive strains (Fig. 4). The process is considered to require (a) production of a single-strand break, or "incision", near the dimer, (b) removal of an oligonucleotide containing the dimer ("excision"), (c) "resynthesis" of the removed segment along the intact complementary strand, which it uses as a template, and (d) "sealing" of the end of the resynthesized strand to form a continuous and "native" double-stranded DNA. The first step of "excision-resynthesis" repair has been demonstrated *in vitro*, using phage DNA and bacterial extracts, and bacterial mutants have been found that are deficient in this step. The excision-resynthesis system appears to be able to repair at least 95 per cent of the far ultra-violet damage in bacteria and it is believed to be the major mechanism operating in *host–cell reactivation* of bacteriophage (a higher survival of some bacteriophage when plated on certain *Escherichia coli* hosts). Thus, a bacterium lacking the ability for host–cell reactivation (hcr^-) is some 30 times more sensitive to ultra-violet killing (compare curves for hcr^-rec^- and hcr^+rec^- in Fig. 4).

(c) *Recombination repair.* Bacteria deficient in the ability to undergo genetic recombination (rec^-) are also some 30 times more sensitive to ultra-violet killing than the wild type (compare curves for hcr^-rec^-

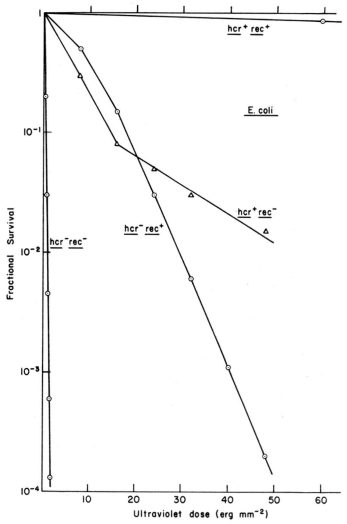

Fig. 4. *Survival at 254 nm for E. coli AB 1157 (hcr^+rec^+), AB 2463 (hcr^+rec^-), AB 2437 (hcr^-rec^+) and AB 2480 (hcr^-rec^-). Data from W. Harm; see also P. Howard-Flanders and R. P. Boyce, Radiation Research Supplement* **6**, *156 (1966).*

and hcr^-rec^+ in Fig. 4). An hcr^-rec^+ strain does not excise pyrimidine dimers. Therefore the recombination-repair system appears to be fundamentally different from the excision-resynthesis system and presumably bears some resemblance to the process of genetic recombination. A cell deficient both in the ability to host-cell-reactivate and in the ability to recombine becomes extremely sensitive to ultra-violet (Fig. 4), showing a 37 per cent survival dose of only about 0·2 ergs/mm², which corresponds to only 1 thymine dimer per cell. This is presumably, therefore, the most ultra-violet-sensitive cell that could exist. It is approximately 1000 times as sensitive as a typical wild-type strain, which implies that bacteria are ordinarily capable of repairing in the dark some 99·9 per cent of their far ultra-violet damage. In addition, photoreactivation can repair 2/3 of the remaining damage, making a total of 99·97 per cent of the cellular damage that can be repaired.

(d) *Repair treatments.* The three repair systems discussed above, namely, direct photoenzymatic reactivation, excision-resynthesis repair, and recombination repair, are repair *mechanisms* that are controlled by specific genes. A variety of repair *treatments* are known that will encourage various of these mechanisms.

Photoprotection is a decreased far ultra-violet sensitivity (slope decrease by factors up to 3) induced by near ultra-violet irradiation *before* the far ultra-violet irradiation. It is similar in mechanism to *indirect photoreactivation* (see above), the initial action being apparently a purely photochemical effect outside of DNA (which does not absorb near ultra-violet). It does not operate by splitting thymine dimers, but is believed to involve inactivation of isoprenoid quinones of the electron-transport system, although this has not been proven. The illumination (effective wavelengths 300–80 nm, with a peak at 340 nm) induces growth delays and division delays, which presumably permit more time for dark-repair processes (repair processes other than photoreactivation) to operate. It is therefore easy to see why the illumination may be either before (photoprotection) or after (indirect photoreactivation) the far ultra-violet irradiation. *Escherichia coli* B_{s-1}, which has very little dark-repair ability, shows only direct photoenzymatic reactivation, and *E. coli* B phr^-, which lacks active photoreactivating enzyme, shows only indirect photoreactivation. Many strains, like *E. coli* B, show both direct and indirect photoreactivation.

Liquid-holding recovery (a higher survival—slope decrease by factors up to 3—of bacteria after holding them in a liquid for a few hours) appears to enhance excision-resynthesis repair. Also, it appears to operate on essentially the same fraction of damage as does photoprotection.

Ultra-violet reactivation is an enhancement of bacteriophage survival induced by ultra-violet irradiation of the host cell. It is believed to result from a change in the excision-resynthesis balance, such that nucleases are less damaging to the phage genomes.

(e) *Filament formation.* Ultra-violet irradiation can induce the formation of long filaments in some bacterial strains, such as *E. coli* B. This effect involves a failure of cytokinesis without major effects on macromolecular synthesis or growth. It apparently contributes to lethality. *Temperature restoration* (higher survivals by holding bacteria for a few hours at elevated temperatures), *neighbour restoration* (higher survivals with higher densities of incubating bacteria), and *pantoyl-lactone restoration* are recovery treatments that appear to act by decreasing filament formation.

3. Biological Spectrum of Far Ultra-violet Molecular Processes

Thymine dimers have been found to be induced by far ultra-violet radiation in all types of living cell that have been tested. Photoreactivation appears to be a quite wide-spread phenomenon, although it has not yet been clearly demonstrated in mammalian systems, in spite of many attempts. In mammalian cells, there is evidence for dark repair, including excision of dimers.

Bacterial studies provide evidence that both photoreactivable and nonphotoreactivable lethal damages can be dark-repaired, and there is some evidence for circumstances in which dark-recovery systems will work on mutational, but not on lethal, damage. This raises the question, still under debate, of the extent to which mutational damage may be similar to lethal damage and the extent to which it may consist of pyrimidine dimers. Mutational damage can generally be photoreactivated, which strongly suggests that much of it is induced by pyrimidine dimers. Some mutational damage, however, may be induced by different photoproducts.

All earlier studies on ultra-violet killing as a function of dose rate (when a constant dose is applied) showed no effect (i.e. the Bunsen-Roscoe Reciprocity Law is obeyed at the cellular level), from which it was correctly concluded that the initial action of a far ultra-violet photon is purely photochemical. Recent studies, however, have shown that ultra-violet-resistant bacteria, which presumably have good dark-repair systems, show a much higher survival for relatively high ultra-violet doses when the dose rate is very low (for example, if the dose is administered over a 24-hr period). This does not reflect a difference in the initial action of the photons, but rather an inefficiency in the dark-repair process at high ultra-violet doses with normal dose rates, under which conditions lesions can accumulate before repair.

4. Near Ultra-violet (300–80 nm) Photoproducts

The killing of small cells by the near ultra-violet radiation (wavelengths above 320 nm) of the solar spectrum,† was discovered late in the last century. The doses required are of the order of 1000 times those needed in the far ultra-violet region. The chromophore(s) has not been identified. Near ultra-violet radiation also induces growth delay and division delay in bacteria (see above). One important chromophore for this process is probably the isoprenoid quinones of the electron-transport system. Such quinones as coenzyme Q and vitamin K_2 are sufficiently sensitive in the near ultra-violet to account for their involvement in these effects. One of the major photoproducts produced *in vitro* is a quinone dimer containing a cyclobutane ring. This product is somewhat analogous in structure to a pyrimidine dimer, but whether it is of biological importance is not yet known.

5. *Vacuum Ultra-violet (100–90 nm) Studies*

Very little work has been done in the region below 190 nm, where absorption by biological molecules quickly becomes nonspecific (around 160 nm). In addition, the absorption by water and oxygen below 190 nm makes experimentation with biological material very difficult. Nevertheless, it has been found that the quantum yield for inactivation of proteins and nucleic acids rises in this region as the wavelength is decreased, approaching unity at the point where the radiation becomes ionizing (around 120 nm). Cells are easily killed by vacuum ultra-violet radiation, but the effects below 190 nm are due to surface absorption in all but the smallest cells. In such very small cells, killing appears to be due primarily to absorption in nucleic acids. In actinomycete conidia, the production of photoreactivable damage falls sharply in this region, reaching zero at about 170 nm. This drop is apparently not due to shielding of DNA by proteins, but represents a decreased efficiency for induction of pyrimidine dimers in DNA.

6. *Microbeams—Effects on Cell Organelles*

Far ultra-violet microbeams as small as 1 micron diameter have been utilized (smaller beams are of little use with cells because of light scattering). Experiments with such microbeams have shown that cytoplasmic damage can be of considerable importance in cells, resulting in effects such as killing and division delay in amoebae and spindle breakdown in newt cells. The cytoplasm of some cells appears to contain important ultra-violet-sensitive recovery factors. Photoreactivation of cytoplasmic damage has also been demonstrated in larger cells, but it is not known if this represents repair of RNA damage, since many cells are now known to have DNA-containing organelles in the cytoplasm. In grasshopper neuroblasts, far ultra-violet inactivation of one of the two nucleoli has shown that, up to late prophase, both nucleoli are required for cell division. In euglenae, chloroplasts can be inactivated by far ultra-violet radiation at doses that are not lethal for the cell. The effect is presumably due to nucleic acid absorption and it is photoreactivable.

Bibliography

GIESE A. C. (Ed.) (1964) *Photophysiology*, Vol. 2, *Action of light on animals and microorganisms; photobiochemical mechanisms; bioluminescence;* Vol. 4, *Current topics*; New York: Academic Press.

JAGGER J. (1967) *Introduction to Research in Ultraviolet Photobiology*, Englewood Cliffs, New Jersey: Prentice-Hall.

McLAREN A. D. and SHUGAR D. (1964) *Photochemistry of Proteins and Nucleic Acids*, New York: Pergamon Press.

ROE E. M. F. (1962) Ultra-violet radiation, biological applications of, in *Encyclopaedic Dictionary of Physics* (J. Thewlis Ed.), **7,** 525, Oxford: Pergamon Press.

J. JAGGER

---

† The solar spectrum also includes a small amount of radiation in the range 290–320 nm, which presumably acts on the DNA of small cells to produce killing.

V

VITREOUS CARBONS, STRUCTURE AND PROPERTIES OF.

Definition

"Vitreous", "glassy", or "impermeable" carbons are non-graphitizing carbons prepared by the thermal degradation of certain cross-linked polymers, including phenolic and furfuryl resins, modified pitches and cellulose. The properties (and probably also the structures) of the carbonized products are, in general, very similar, although dependent to some extent on the start material and the maximum HTT.† These carbons may be distinguished from common industrial carbons and graphites by their conchoidal fracture, high lustre, and ability to ring when struck; their greater strength, hardness and resistance to both erosion and corrosion; and their much lower porosity and permeability.

A material prepared as early as 1923 by the thermal decomposition of hydrocarbon gases was described as "vitreous" carbon: the process, however, resembles that used for the preparation of pyrolytic graphite. The product was hard and shiny, resistant to chemical attack and was thought to bear resemblances both to diamond and graphite.

The following abbreviations will be used for materials currently manufactured:

VC	Vitreous Carbon	Plessey Company Limited and licensees
GC	Glassy Carbon	Tokai Electrode Manufacturing Company
LCV	Le Carbone Vitreux	Société Le Carbone Lorraine
ICC	Impermeable Cellulose Carbon	General Electric Company Ltd. (U.K.)

Relation of the Structure of Vitreous Carbons to Diamond and Graphite

Only two forms of carbon, diamond and graphite, can be completely characterized and these are the true allotropes; both occur naturally. The density of diamond is 3·52 g/cm³ whereas that of ideal (single crystal) graphite is 2·26 g/cm³.

† HTT, heat treatment temperature.

The diamond lattice consists of carbon atoms located at the corners of regular tetrahedra and linked together by covalent bonds with an interatomic distance of 1·54 Å. The carbon atoms in graphite are arranged in hexagonal arrays in flat parallel sheets, each atom within a sheet being linked to three adjacent atoms, at a distance of 1·42 Å, by strong covalent bonds. The sheets are 3·35 Å apart and are linked by weak (van der Waals') forces. Graphite is the stable allotrope and diamond is converted into it by heating. Diamond is the hardest substance

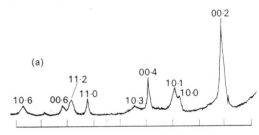

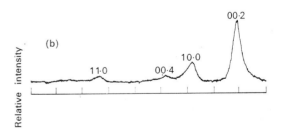

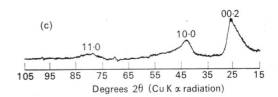

Fig. 1. *X-ray diffraction patterns:* (a) *graphitized petroleum coke,* (b) *VC (HTT 2750°C),* (c) *VC (HTT 1800°C).*

Table 1. Physical Properties of Carbon Products

Property	VC (HTT 1800°C)	GC 20 (2000°C)	LCV (2500°C)	ICC	Typical Electrode Graphite
Apparent density, g/cm^3	1·47	1·47	1·50–1·53	1·55	1·6
Porosity, %	<0·05	1–3	0	0	10–20
Helium permeability, cm^2/sec	<10^{-10}	10^{-10}–10^{-12}	10^{-8}	<10^{-12}	10^{-1}–10^{-2}
Hardness (Shore)	105–110	100–110	120–125	95	60
Thermal expansion, °C^{-1} × 10^{-6}	2 (0–100°C) 3 (100–1000°C)	2	3·2 (50–150°C) 3·5 (600–700°C)	3·4 (0–700°C)	2 (with grain) 3·5 (across grain)
Thermal conductivity, cal/cm sec °C	0·01–0·02	0·02	—	0·01	0·2 (with grain) 0·15 (across grain)
Electrical resistivity, × 10^{-3} Ω cm	3–8	4–4·5	7	4	1·5 (with grain) 2·5 (across grain)
Transverse strength, kg/cm^2	700–2000	750	500–800	1800	70 (with grain) 50 (across grain)
Young's Modulus, kg/mm^2	2000–3000	2700	2200	2800	400–600
Ash, commercial grade, ppm	200	800	200	—	>2000
Ash, pure grade, ppm	10–50	70 (HTT 3000°C)	50	—	10
Oxidation rate in dry air at 600°C, g/m^2/h	1–5 (commercial) 0·15–0·5 (pure)	—	10 (commercial) 0·3–1·0 (pure)	—	20–100 (pure, close-grained)
L_a, Å	50	30	—	45 (HTT 1800°C)	>1000
L_c, Å	15	20	—	25 (HTT 1800°C)	~1000
$C_0/2$, Å	3·5	3·55	—	3·44 (HTT 1800°C)	3·35

known (with the exception of one or two man-made materials) and has a value of 10 on the Mohs' scale, whereas that of graphite is 1.

Industrial carbons and graphites are generally imperfect forms of graphite. They are prepared by heating mixtures of ground coke and pitch and their properties are dependent on those of the original coke grains, the pitch residue and the heat treatment temperature. When taken to about 2500°C graphitization occurs. The X-ray diffraction patterns (Fig. 1) indicate the much greater degree of crystallinity in graphitized petroleum coke than in VC prepared at 2750°C and 1800°C. The diffraction patterns of the latter materials are very diffuse and have been interpreted in terms of a turbostratic (disordered) structure, consisting of graphite-like hexagonal layers in very small packets but with no graphitic orientation between adjacent layers. The packet or crystallite size of VC prepared at 1800°C has been estimated from the broadening of the 10·0 and 00·2 reflections by the Warren and Biscoe technique to be only 50 Å (L_a) and 15 Å (L_c). These figures are somewhat tentative, as higher order lines are too diffuse to measure and so no estimate can be made of strain broadening, but the apparent crystallite size is certainly much less than that of the graphitized coke.

There are reasons for believing, however, that this picture of the structure of VC is, at best, an over-simplification. In view of the low density of vitreous carbons, only 2/3 of the theoretical density of graphite, any structure based even remotely on that of graphite would be expected to be associated with high porosity; yet the density in helium or krypton is the same as in water, showing that any pores present must be inaccessible to these molecules. Moreover, the high strength and hardness of vitreous carbons suggest that the weak interlamellar and inter-crystallite forces in graphite may have been replaced by stronger, chemical bonds. It is possible, for example, that the structure may involve random bridges of tetrahedrally bonded carbon between chains of hexagonal rings, since such bridges (or cross-links) are present in the start material and a proportion of these may persist through the carbonization process. Evidence that this suggestion is at least consistent with the X-ray diffraction pattern of a GC has been obtained by Noda and Inagaki who showed that the shape of the first peak in the radial distribution curve could be reproduced fairly accurately by assuming that the structure consisted of a mixture of trigonal and tetrahedral carbon atoms and that the observed (composite) curve was merely the sum of the component curves due to these two forms of carbon. On the other hand, Kakinoki has postulated an oxygen-bridged structure, apparently supported by analyses indicating that the GC under investigation had an oxygen content of 5–6 per cent; in general, however, VC heat-treated to 1800°C contains less than 0·1 per cent of oxygen, so such a structure is not applicable.

The Properties of Vitreous Carbon

Physical Properties

Unlike graphites, which generally exhibit varying degrees of anisotropy depending on the method of processing, vitreous carbons are (as far as is known) completely isotropic. The properties of various vitreous carbon products are compared with those of a typical graphite in Table 1. The permeabilities are many orders less than that of the graphite and some are similar to those of hard glasses. The open porosity is very low, despite the evolution of large amounts of volatile products during carbonization: this suggests that any pores created by the escaping gases become closed during the later stages of the heat treatment. Comparison of the helium density with the lump density for a wide range of HTT has confirmed that, for at least some of the start materials mentioned, the accessible porosity rises to a maximum below 1000°C and then decreases to a very low level for HTT above about 1500°C.

In the VC process the resistivity drops during carbonization by approximately 17 orders, from 10^{14}–10^{15} ohm cm to about 5×10^{-3} ohm cm; the final resistivity is slightly greater than that of the synthetic graphite. The thermal conductivities are lower than those of graphites, as expected in the light of the structural differences already discussed.

Vitreous carbons are markedly stronger than graphite and are much harder: on the Mohs scale the value for VC is 7, while the Shore figure is 105–110. The Vickers method gives inaccurate results because of elastic recovery. Vitreous carbons are hard enough to scratch window glass and cannot readily be machined except by diamond grinding or ultrasonic techniques.

Chemical Properties

Vitreous carbons can be prepared in a very pure state; the "ash content" of VC is normally about 200 ppm (Table 2) but this figure can be reduced to about 10 ppm by using very pure starting resins.

Table 2. *Typical Impurity Levels in VC (estimated spectrographically)*

Impurity	Quantity (ppm as oxide)
Silicon	80
Calcium	40
Iron	40
Aluminium	20
Barium	5
Sodium	5
Titanium	<5
Strontium	<5
Magnesium	<5
Ash	200

Pure VC is highly resistant to oxidation (Fig. 2), and reaction kinetics studies in various oxidizing gases have shown that this low reactivity (compared with that of commercial graphite) is maintained at temperatures up to at least 3000°K. A provisional figure has been obtained for the heat of combustion (J. B. Lewis, Harwell, unpublished work) which is seen to be significantly greater than those of diamond and graphite, indicating that VC is thermodynamically the least stable.

$$\Delta H_{(VC\ 1800)} = 95 \cdot 25 \text{ kcal/g}$$

$$\Delta H_{(diamond)} = 94 \cdot 48 \text{ kcal/g}$$

$$\Delta H_{(graphite)} = 94 \cdot 04 \text{ kcal/g}$$

Since $\Delta H = \Delta H_v - \Delta H_s$ (where ΔH = heat of combustion of solid carbon, ΔH_v = heat of combustion of carbon vapour, ΔH_s = heat of sublimation) and ΔH_v is constant, the ΔH_s of VC is less than that of either diamond or graphite, and therefore the mean bond energy per atom of carbon is lower.

The rates of attack by fusion agents (Table 3) are low, as expected in view of the low porosity. VC is also

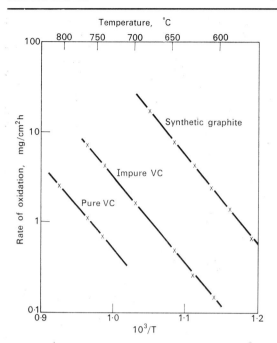

Fig. 2. Variation of oxidation rate with temperature: VC and close-grained synthetic graphite in dry air at 600–800°C.

very inert to many acids including hydrochloric, hydrofluoric, nitric, sulphuric and chromic acids and also to mixtures of nitric acid with other oxidizing agents.

Table 3. Rates of Reaction of VC

Reagent	Temperature (°C)	Weight loss rate (g/cm²h) (apparent surface)
Fused potassium bisulphate	500	2.4×10^{-4}
Fused boric oxide	580	$<5 \times 10^{-6}$
Fused caustic soda	320	$<5 \times 10^{-6}$

Reactions with elements and compounds which give lamellar compounds with graphite are of interest. VC is resistant to attack by bromine and by mixtures of nitric and sulphuric acids, whereas graphites form compounds such as C_8Br and graphite bisulphate $C_{24}^+HSO_4^- \cdot 2\ H_2SO_4$. However, both VC and graphite are attacked by alkali metals.

With non-carbide-forming elements including germanium, silver, gallium, phosphorus, arsenic, tellurium, zinc, tin and lead the rates of attack are small. Like graphite, VC is not wetted by molten aluminium except at temperatures greatly in excess of the melting point.

Applications

Applications have been established which use many of the characteristic properties of vitreous carbons. Some of the more important types of application are listed below, with the relevant properties in parentheses:

1. Apparatus for fusion and leaching operations (inertness, high strength and low porosity). The low reactivity of vitreous carbon makes it superior to such metals as nickel, iron, platinum and tantalum for fusions with both alkalis and acid salts, and also to glass and silica for operations involving the use of alkalis or hydrofluoric acid.

2. Gas-tight carbon–metal and carbon–glass seals (low porosity: coefficient of thermal expansion close to that of borosilicate glasses).

3. Vacuum furniture (ease of outgassing, low permeability and porosity, electrical and thermal properties ideal for use as R.F. susceptors).

4. Electrical applications (electrical properties, erosion resistance, low porosity). Vitreous carbon has performed satisfactorily in the forms of potentiometer brushes, lightning protectors for telephone circuits, and electrodes for spot welding and for the deposition of strippable metal films.

5. Metallurgical equipment for zone refining, casting and crystal growth (inertness, high porosity and strength, non-wettability).

Bibliography

COWLARD F. C. and LEWIS J. C. (1967) *J. Matls. Science* **2**, 507.
DAVIDSON H. W. and LOSTY H. H. W. (1963) *G.E.C. Journal* **30**, 22.
HOFMANN K. A., ROCHLING C. and HOFMANN U. (1923) *Berichte* **56**, 2071.
KAKINOKI J. (1965) *Acta Cryst.* **18**, 578.
LEWIS J. C. (publ. 1966) Industrial carbon and graphite, 2nd S.C.I. Conf., London 1965, p. 258.
LEWIS J. C., REDFERN B., COWLARD F. C. (1963) *Solid State Electronics* **6**, 251.
NODA T. and INAGAKI M. (1964) *Bull. Chem. Soc. Japan*, **37**, 1534.
PARISOT M. J. (1966) Bull. de la Société Française de Céramique No. 3, 13 (1966); *Rev. Hautes Temp. et Refract.* **1**, 171.
YAMADA S. and SATO H. (1962) *Nature* **193**, 261.
YAMADA S., SATO H. and ISHII T. (1964) *Carbon* **2**, 253.

F. C. COWLARD and J. C. A. LEWIS

W

WALSH FUNCTIONS. The system of orthogonal functions introduced by Rademacher (1922) has been the subject of a great deal of study. This system is not a complete one. Its completion was effected by Walsh (1923), who studied some of its Fourier properties, such as convergence and summability. Others, notably Kaczmarz (1929), Kaczmarz and Steinhaus (1930) and Paley (1932), have studied various aspects of the Walsh system before the Second World War. In 1949 Sneider (1949) and Fine (1949) independently proved uniqueness theorems. Since then there has been increasing interest in the system on the part of mathematicians. More recently, significant advances in applications have been made by Harmuth (1968), Pichler (1968), and others.

It is convenient, in defining the functions of the Walsh system, to follow Paley's modification. The Rademacher functions are defined by

$$\varphi_0(x) = \begin{cases} 1 & (0 \leq x < \tfrac{1}{2}), \\ -1 & (\tfrac{1}{2} \leq x < 1), \end{cases}$$
$$\varphi_0(x+1) = \varphi_0(x) \quad (-\infty < x < \infty),$$
$$\varphi_n(x) = \varphi_0(2^n x) \quad (n = 1, 2, \ldots). \tag{1}$$

The Walsh functions are then given by

$$\psi_0(x) = 1,$$
$$\psi_n(x) = \varphi_{n_1}(x) \varphi_{n_2}(x) \cdots \varphi_{n_\nu}(x) \tag{2}$$

for $n = 2^{n_1} + 2^{n_2} + \cdots + 2^{n_\nu}$, where the nonnegative integers n_l are uniquely defined by the condition $n_{l+1} < n_l$. The $\{\psi_n(x)\}_{n=0}^{\infty}$ form a complete orthonormal system. Every periodic function which is integrable (Lebesgue) on [0,1] will have associated with it a Walsh–Fourier series

$$f(x) \sim c_0 + c_1 \psi_1(x) + c_2 \psi_2(x) + \cdots, \tag{3}$$

where the coefficients are given by

$$c_n = \int_0^1 f(t) \psi_n(t) \, dt. \tag{4}$$

The partial sums of the series in (3) are

$$s_n(x) = s_n(x; f) = \sum_{k=0}^{n-1} c_k \psi_k(x)$$
$$= \int_0^1 f(t) D_n(x, t) \, dt, \tag{5}$$

where D_n (the Dirichlet kernel) is given by

$$D_n(x, t) = \sum_{k=0}^{n-1} \psi_k(x) \psi_k(t). \tag{6}$$

For every real number x and for every nonnegative integer n we define $\alpha_n = \alpha_n(x)$ and $\beta_n = \beta_n(x)$ by

$$\alpha_n = m \cdot 2^{-n} \leq x < (m+1) \, 2^{-n} = \beta_n. \tag{7}$$

Then

$$D_{2^n}(x, t) = \prod_{r=0}^{n-1} (1 + \varphi_r(x) \varphi_r(t))$$
$$= \begin{cases} 2^n & (\alpha_n(x) \leq t < \beta_n(x)) \pmod 1, \\ 0 & \text{elsewhere}. \end{cases} \tag{8}$$

It follows that

$$s_{2^n}(x) = 2^n \int_{\alpha_n}^{\beta_n} f(t) \, dt = \frac{F(\beta_n) - F(\alpha_n)}{\beta_n - \alpha_n} \tag{9}$$

where $F(x)$ is an integral of $f(x)$. From (9) it is clear that $s_{2^n}(x)$ converges to $f(x)$ almost everywhere. In particular, we have convergence at every point of continuity of f, and the convergence is uniform in an interval of continuity.

The formula (9) clearly exhibits the partial sum (of order 2^n) as an average of the function over a shrinking interval containing the point x. Thus the ultimate behavior of these special partial sums is determined by the local behavior of f. This localization principle holds also for the complete sequence of partial sums. There are various sufficient conditions for the convergence of this sequence, that is, of the Walsh–Fourier series (3). For example, if f is of bounded variation, then the series converges at every point of continuity and at every dyadic rational $m/2^n$. In striking contrast with trigonometric Fourier series, however, the series diverges at all other points (still under the assumption of bounded variation).

If $(f(t) - c)/(t - x_0)$ is integrable in a neighborhood $|t - x_0| < \delta, \delta > 0$, then the Walsh–Fourier series of f converges to c at the point x_0.

If f is continuous, and if its modulus of continuity satisfies $\omega(\delta; f) = o \left(\log \frac{1}{\delta} \right)^{-1}$ as $\delta \to 0$, then its Walsh–Fourier series converges to $f(x)$ uniformly. In particular, if $f \in \text{Lip } \alpha (0 < \alpha \leq 1)$, we have uniform convergence. If $\alpha > \tfrac{1}{2}$, the convergence is absolute.

The integrals $\int_0^1 |D_n(x, t)|\, dt$ are of obvious importance in the theory. Just as in the trigonometric case, they are independent of x. These *Lebesgue constants*, as in the classical case, are unbounded. It follows from a theorem of Haar that for any given x_0, there exists a continuous function whose Walsh–Fourier series diverges at x_0. However, the series for a continuous function is uniformly $(C, 1)$ summable. For an arbitrary integrable function, the series is $(C, 1)$ summable almost everywhere. Other types of summability have also been studied.

Concerning the order of the Fourier coefficients, the usual theorems can be proved, with one important exception: for every nonconstant absolutely continuous function, $c_n \neq o(1/n)$. If the coefficients converge to zero too rapidly, the jumps of the functions ψ_n cannot be smoothed out in time. Thus the order of the coefficients cannot be improved by smoothness conditions, as in the classical case.

Consider a Walsh series

$$S(x) = \sum_{k=0}^{\infty} a_k \psi_k(x), \qquad (10)$$

not necessarily a Walsh–Fourier series. The problem of finding conditions under which $S(x)$ is actually a Fourier series has been studied. The analogous problem in the theory of trigonometric series has long been a part of classical mathematics. Riemann employed the device of twice integrating a given trigonometric series (formally) and studying the properties of the resulting function, relating it to the original series by means of a generalized second derivative. In the case of Walsh series, the *first* formal integral is the natural tool. It turns out that it converges wherever $S(x)$ does. If $S(x)$ is a Fourier series, then the formal integral coincides with the actual integral $\int_0^x f(t)\, dt$. Using this tool, it has been proved that if $S(x)$ converges except on a countable set to an integrable function f, it is actually the Fourier series of f. In particular, if $f = 0$, the coefficients a_n are all 0. Thus we have the uniqueness theorem: two distinct Walsh series cannot converge to the same values for all but a countable set of points.

In many respects the Walsh system resembles the trigonometric one, $\{e^{2\pi i n x}\}_{-\infty}^{+\infty}$. In addition to being complete orthonormal systems on the interval $[0, 1]$, the most striking property they share is that of being groups under multiplication. It has been shown that any such countable *orthonormal group* is essentially the character group of a compact abelian group. For $\{e^{2\pi i n x}\}$, this compact group is the circle; for $\{\psi_n(x)\}$, it is the countable direct product of the two-element group $\{0, 1\}$, the operation being addition mod 2. Denoting this product by G, we see that its elements are all sequences $(x_1, x_2, x_3, \ldots)$, where each x_i is either 0 or 1, and addition of sequences is componentwise mod 2. The function

$$\lambda: (x_1, x_2, \ldots) \to \sum_{i=1}^{\infty} x_i 2^{-i}$$

maps G continuously onto $[0, 1]$, and is one-to-one except at the dyadic rationals. The Haar measure on G corresponds, under λ, to Lebesgue measure on $[0, 1]$. If χ is a character of G, that is, a continuous homomorphism of G into the unit circle, then with a proper definition of λ^{-1}, $\chi(\lambda^{-1}(x))$ is a well-defined function on $[0, 1]$. When extended periodically to the entire real line, it coincides with one of the Walsh functions, and all of them are uniquely obtainable in this way. This method of constructing an orthonormal group on the unit interval has been widely generalized.

There is another way to generalize the Walsh system. Just as the $\{e^{2\pi i n x}\}$ may be imbedded in the system $\{e^{2\pi i y x}\}$, $-\infty < y < \infty$, so the $\{\psi_n(x)\}$ may be imbedded in a system $\{\psi_y(x)\}$, $0 \le y < \infty$, and there is a corresponding Walsh–Fourier transform theory.

There are other branches of mathematics that are related to the Walsh functions. For example, the integrated Walsh functions have been used to construct a simple example of a continuous, nowhere differentiable function.

When regarded as measurable functions on the unit interval, the Rademacher functions $\varphi_n(x) = \psi_{2^n}(x)$ are independent random variables taking the values ± 1 with equal probability. Thus they serve as concrete models for Bernoulli trials. An example of the type of probabilistic question that may be treated is the following: let (c_n) be a sequence of positive numbers, and let the signs $\pm$ in the series

$$s = \pm c_1 \pm c_2 \pm c_3 \pm \cdots$$

be chosen by a sequence of independent tosses of a coin; what is the probability that s converges? The answer is 1 if $\Sigma c_n^2 < \infty$ and 0 if $\Sigma c_n^2 = \infty$.

The asymptotic distribution of sums of the form

$$\sum_{n=0}^{N} f(2^n x)$$

has been studied by means of the Walsh system. The function f is assumed to satisfy certain reasonable conditions on the order of its Walsh–Fourier coefficients. See Fine (1954).

The concept of frequency has been generalized to that of *sequency*, which leads to a new classification of the Walsh functions in direct analogy with the trigonometric functions. Applications have been made to telephonic multiplexing. Because of the two-valued nature of the Walsh functions, it is reasonable to expect many more applications in modern technology.

Bibliography

AGAER G. N. (1962) A Wiener type theorem for series of Walsh functions (Russian). *Dokl. Akad. Nauk. SSSR* **142**, 751–3 (transl. in *Soviet Math.* **3**, 129–31).

BILLARD P. (1967) Sur la convergence presque partout des séries de Fourier-Walsh des fonctions de l'espace L²(0, 1), *Studia Math.* **28**, 363–88.

CHRESTENSON H. E. (1955) A class of generalized Walsh functions, *Pac. J. Math.* **5**, 17–31.

CIVIN PAUL (1952) Multiplicative closure and the Walsh functions, *Pac. J. Math.* **2**, 291–5.

CIVIN PAUL (1954) Orthonormal cyclic groups, *Pac. J. Math.* **4**, 481–2.

CRITTENDEN RICHARD B. and SHAPIRO VICTOR L. (1956) Sets of uniqueness on the group 2^ω, *Ann. of Math.* (2) **81**, 550–64.

FINE N. J. (1949) On the Walsh functions, *Trans. Amer. Math. Soc.* **65**, 372–414.

FINE N. J. (1950) The generalized Walsh functions, *Trans. Amer. Math. Soc.* **69**, 66–77.

FINE N. J. (1954) On the asymptotic distribution of certain sums, *Proc. Amer. Math. Soc.* **5**, 243–52.

FINE N. J. (1955) Cesàro summability of Walsh Fourier series, *Proc. Nat. Acad. Sci. U.S.A.* **41**, 588–91.

FINE N. J. (1955) On groups of orthonormal functions (I), *Pac. J. Math.* **5**, 51–59.

FINE N. J. (1955) On groups of orthonormal functions (II), *Pac. J. Math.* **5**, 61–65.

FINE N. J. (1957) Fourier-Stieltjes series of Walsh functions, *Trans. Amer. Math. Soc.* **86**, 246–55.

HAMMOND J. L. and JOHNSON R. S. (1962) A review of orthogonal square wave functions and their application to linear networks, *J. Franklin Inst.* **273**, 211–25.

HARMUTH H. F. A generalized concept of frequency and some applications. Defense Dept. of the Federal Republic of Germany, contract T-563-L 203.

HARMUTH H. F. (1964) Die Orthogonalteilung als Verallgemeinerung der Zeit- und Frequenzteilung, *Archiv der elektrischen Übertragung* **18**, 43–50.

HARMUTH H. F. (1964) Verallgemeinerung des Fourier-Integrals und des Begriffes Frequenz, *Archiv der elektrischen Übertragung* **18**, 439–51.

HARMUTH H. F. (1964) Grundzüge einer Filtertheorie für die Maenderfunktionen $A_n(\theta)$. *Archiv der elektrischen Übertragung* **18**, 544–54.

HARMUTH H. F. (1968) Sequenzmultiplexsysteme für Telephonie und Datenübertragung 1. Quadraturmodulation 2. Einseitenbandmodulation, *Archiv der elektrischen Übertragung*, Jan. Feb. 1968.

HARSILADZE F. I. (1962) On Walsh–Fourier series (Russian-Georgian summary). *Gruzin-Politekn Inst. Trudy* **1** (81), 47–53.

HENDERSON K. W. (1964) Some notes on the Walsh functions, *Trans. IEEE* **EC-13**, 50–52.

HIRSCHMAN I. I. (1955) The decomposition of Walsh and Fourier series, *Mem. Amer. Math. Soc.* **15**, 65 pp.

HSIEH P. and HSIAO M. Y. (1964) Several classes of codes generated from orthogonal functions, *Trans. IEEE* **IT-10**, 88–91.

IGARI S. (1964) Sur les facteurs de convergence des séries de Walsh-Fourier, *Proc. Japan Acad.* **40**, 250–2.

KAC M. (1938) Sur les fonctions $2^n t - [2^n t] - 1/2$, *J. London Math. Soc.* **13**, 131–4.

KAC M. (1946) On the distribution of values of sums of the type $\Sigma f(2^n t)$, *Ann. of Math.* **37**, 33–49.

KACMARZ S. (1929) Über ein Orthogonalsystem, *Comptes rendus du premier congres des mathematiciens des pays slaves, Warsaw*, 189–92.

KACMARZ S. and STEINHAUS H. (1930) Le système orthogonal de M. Rademacher, *Studia Math.* **2**, 231–47.

KOKILAŠVILL V. M. (1965) On best approximations by Walsh polynomials and the Walsh–Fourier coefficients (Russian–English summary), *Bull. Acad. Polon. Sci. Ser. Sci. Math. Astron. Phys.* **13**, 231–47.

KOTJAR B. D. (1966) Walsh series and the theorem of Men'šov on "correction" of functions (Russian), *Izv. Akad. Nauk. SSSR Ser. Mat.* **30**, 1193–2000.

LIEDL R. (1964) Über eine spezielle Klasse von stark multiplikativ orthogonalen Funktionensystemen, *Monatshefte der Mathematik* **68**, 130–7.

LIEDL R. (1968) Über gewisse Funktionale im Raum $C^{(v)}$ (0,1) und Walsh-Fourier-koeffizienten, *Monatshefte für Mathematik*.

LÜKE H. D. (1966) Orthogonale Signalalphabete mit speziellen Korrelationseigenschaften, *Archiv der elektrischen Übertragung* **20**, 310–16.

McLAUGHLIN J. R. (1969) Functions represented by integrated Rademacher series, *Colloq. Math.*

MORGENTHALER G. W. (1957) On Walsh–Fourier series, *Trans. Amer. Math. Soc.* **84**, 472–507.

OSIPOV R. I. (1966) Convergence of series with respect to the Walsh system (Russian–English summary), *Izv. Akad. Nauk. Armjan. SSR, Ser. Mat.* **1**, 270–83.

PALEY R. E. A. C. (1932) A remarkable series of orthogonal functions, *Proc. London Math. Soc.* **24**, 241–79.

PALEY R. E. A. C. and WIENER N. (1933) Characters of Abelian groups, *Proc. Nat. Acad. Sci. U.S.A.* **19**, 253–7.

PICHLER F. Walsh-Fourier-Analysis in der Sequenz-Technik. *Archiv der elektrischen Übertragung*, Dissertation, Math. Institut Universität Innsbruck.

PICHLER F. (1967) Das System der sal- und cal-Funktionen als Erweiterung des Systems der Walsh-Funktionen und die Theorie der sal- und cal-Fouriertransformation, Dissertation Universität Innsbruck.

PICHLER F. (1968) Synthese linearer periodisch zeitvariabler Filter mit vorgeschriebenem Sequenzverhalten, *Archiv der elektrischen Übertragung*, March 1968.

PRICE J. J. (1967) A density theorem for Walsh functions, *Proc. Amer. Math. Soc.* **18**, 209–11.

PRICE J. J. Walsh series and adjustment of functions on small sets (preprint).

RADEMACHER H. A. (1922) Einige Sätze über Reihen von allgemeinen Orthogonalenfunktionen, *Math. Ann.* **87**, 112–38.

RUBINSTEIN A. I. (1963) The A-integral and series with respect to a Walsh system (Russian), *Uspekhi Mat. Nauk.* **18**, No. 3 (111), 191–7.

SELFRIDGE R. G. (1955) Generalized Walsh transforms, *Pacific J. Math.* **5**, 451–480.

SHAPIRO VICTOR L. (1965) U(ε)-sets for Walsh series, *Proc. Amer. Math. Soc.* **16**, 867–70.

SLOOK T. H. (1960) A note on Walsh–Fourier series, *Amer. Math. Monthly* **67**, 253–6.

SNEIDER A. A. (1948) On series of Walsh functions with monotonic coefficients, *Izv. Akad. Nauk. SSSR, Ser. Mat.* **12**, 179–92 (Russian).

SNEIDER A. A. (1949) On the uniqueness of expansions in Walsh functions, *Mat. Sbornik* N.S. **24** (66), 279–300 (Russian).

SNEIDER A. A. (1950) On the convergence of subsequences of the partial sums of Fourier series of Walsh functions, *Doklady Akad. Nauk. SSSR* (N.S.) **70**, 969–71 (Russian).

SNEIDER A. A. (1954) On the convergence of Fourier series of Walsh functions, *Mat. Sbornik* N.S. **34** (76), 441–72.

SUNOUCHI GEN-ICHIRÔ (1951) On the Walsh–Kaczmarz series, *Proc. Amer. Math. Soc.* **2**, 5–11.

SUNOUCHI GEN-ICHIRÔ (1964) Strong summability of Walsh–Fourier series, *Tôhoku Math. J.* (2) **16**, 228–37.

TANDORI K. (1966) Über die Divergenz der Walshchen Reihen, *Acta Sci. Math.-Szeged.* **27**, 261–3.

TASTO M. (1967) Analyse von Zeitfunktionen durch Mäandertransformation und durch Fouriertransformation, Diplomarbeit TH Darmstadt.

VILENKIN N. YA. (1947) On a class of complete orthonormal systems, *Izv. Akad. Nauk. SSSR, Ser. Mat.* **11**, 363–400 (English summary).

WALSH J. L. (1923) A closed set of normal orthogonal functions, *Amer. J. Math.* **55**, 5–24.

WATARI CHINAMI (1957) On generalized Walsh Fourier series I, *Proc. Japan Acad.* **33**, 435–8.

WATARI CHINAMI (1963) Best approximation by Walsh polynomials, *Tôhoku Math. J.* (2) **15**, 1–5.

WATARI CHINAMI (1964a) Contraction of Walsh Fourier series, *Proc. Amer. Math. Soc.* **15**, 189–92.

WATARI CHINAMI (1964b) Mean convergence of Walsh–Fourier series, *Tôhoku Math. J.* (2) **16**, 183–8.

WEISS P. (1967a) Die Darstellung der zyklischen Codes mit Hilfe der Walsh-Funktionen, *Archiv der elektrischen Übertragung* **21**.

WEISS P. (1967b) Über die Verwendung von Walsh-Funktionen in der Codierungstheorie, *Archiv der elektrischen Übertragung* **21**, 255–8.

WEISS P. (1967c) Zusammenhang von Walsh-Fourier-Reihen mit Polynomen, *Monatshefte der Mathematik* **71**, 165–79.

YANO SHIGEKI (1951a) On Walsh–Fourier series, *Tôhoku Math. J.* (2), **3**, 223–42.

YANO SHIGEKI (1951b) On approximation by Walsh functions, *Proc. Amer. Math. Soc.* **2**, 962–7.

YANO SHIGEKI (1954) A convergence test for Walsh–Fourier series, *Tôhoku Math. J.* (2) **6**, 226–30.

YANO SHIGEKI (1957) Cesàro summability of Walsh–Fourier series, *Tôhoku Math. J.* (2) **9**, 267–72.

YASTREBOVA M. A. (1966) On approximation of functions satisfying Lipschitz conditions, by arithmetic means of Walsh Fourier series (Russian), *Mat. Sb.* (N.S.) **71** (113), 214–26.

N. J. FINE

WEATHER CONTROL. In 1947 in the United States and also in Australia it was shown for the first time that an ordinary cumulus cloud produced by quite normal natural processes could be interfered with artificially in such a way that rain could be induced. The clouds were isolated and small, and the rainfall produced was utterly trivial and of no economic significance whatsoever. But most technological advances have begun with scientific discoveries of no immediate practical importance (magnetism, electricity, radio, nuclear energy, antibiotics are a few cases in point); meteorologists were not slow to imagine rainmaking of commercial significance, and the whole field of weather control suddenly became a subject of serious scientific study instead of science fiction.

Before going on to outline the position twenty years later, it must first be stated categorically that progress made has been disappointing. Many responsible and eminent meteorologists would say that nothing of commercial importance has yet been established in spite of hundreds of millions of dollars spent on research trials, experiments and much hopeful activity of little scientific value. Other people of standing claim that the evidence is convincing, that in certain circumstances artificial seeding has increased the average rainfall by 10 per cent, with adequate rewards in increased water supplies. The reason for this lack of consensus lies in the nature of weather, which is so complicated, variable and naturally unpredictable that it is extremely difficult to disentangle any effect produced by artificial interference. The middle course, the balanced view, is that weather conditions can be altered to some extent, and that the economic consequences could be tremendous; a considerable expense in continuing research is therefore fully justified. There is also the inescapable fact that the development of methods of altering the weather on the large scale could be employed for good or ill, and that, regarded as a weapon, a "weather capability" could be just as potent as a nuclear capability and possibly much more flexible and sophisticated: many a country could be ruined in a few years if the weather of every year were only made as adverse as it is in a poor year already.

A scientific assessment of the possibilities of weather control, which must be made if effort is to be wisely invested, is beset with difficulties. We may begin with an analogy. The Earth's weather is a mechanical system, a machine if one prefers the word, and to interfere with and control its working scientifically one must understand how it works. Let us then take a simpler man-made machine by way of comparison, say a motor car. To alter its performance we might look first at the basic processes involved: the input of energy through the fuel combustion, the cooling system, the lubrication. Or we may look to the component parts, the pumps, the carburettor, the ignition, the cylinder, the transmission, the wheels, and so on and so forth. Evidently there are almost innumerable points of entry, and, as a matter of fact, automobile engineers are continuously giving their attention to almost every detail and striving for improvements, that is for better control.

Turning now to our weather machine, we may also examine the basic processes, the heat input from the Sun, the way the energy is taken up and lost again through the cooling system, evaporation of moisture from land and sea, processes of condensation and precipitation, radiation and turbulence. Or we may look to the component parts, the individual clouds, showers or thunderstorms, depressions or anticyclones, fogs, frosts or other phenomena. The meteorological engineer of the future might be thought of as examining continuously every aspect of the processes, mechanisms and components of the weather and making beneficial adjustments. The analogy is fair enough, but there are crucial differences. The automobile engineer has at least designed his machine, and should know exactly how it functions. The meteorologist has not designed the weather and still only partially understands its working. Indeed, our ignorance is such that we could hardly be sure what the effect would be of any change in the system, although we may be certain that some effect would occur.

An example or two will help. Suppose we were to increase the Sun's heat by 10 per cent. A rise of temperature would confidently be expected, but also perhaps more evaporation, more rainfall, more winter snow and growing ice caps; perhaps a new ice age, hardly a likely outcome but difficult to disprove. Suppose we were to cover the sea with a film to inhibit evaporation. If this were successful, rainfall also would decrease, for the two must balance in the long run. But what of the temperature? The Sun's heat, not used in evaporation, would raise the sea temperature by many degrees, perhaps ice caps would melt and the whole Earth's climate become less extreme. But in the meantime what of the climates of individual lands. Would the Nile continue to be fed by tropical rains? Would the monsoons still be maintained? At present we have no way of judging with any sort of confidence, but we have reason to fear that interference of this kind may be within our powers. International authorities, such as the World Meteorological Organization, while encouraging research, wisely agree that no large-scale tests of any kind should be carried out until we are much surer of the likely outcome. The effort at present and for the next ten years or so is therefore being strongly directed towards gaining a better knowledge of the working of the weather by two main approaches. First by more full and world-wide observation, the so-called World Weather Watch, which by the middle 1970's will, it is hoped, give us an acceptably complete and continuous survey of the world's atmosphere. The complementary attack is by the development of physical theory and mathematical calculation, especially with the aid of the largest and fastest electronic computers, so that the performance of the machine can be worked out in detail and the consequence of changes can be pre-assessed. There will of course always be some element of trial and error, just as a new aircraft, however carefully designed on the drawing board, cannot be completely predicted without progressive trials. But we look forward to controlled trials based on proved theory, and at present we are far from the goal. Another ten or perhaps twenty years could make a great difference.

Although we do not know what the degree or even nature of the effects would be, a few possibilities are worth mentioning. Volcanic activities in historic time, a recent one in 1963 at Bali, have introduced sufficient dust into the stratosphere to interfere measurably with the strength of the sunshine. There is no doubt that the deliberate use of nuclear explosions to introduce dust and gases into the high stratosphere could, at no prohibitive expense—perhaps some thousands of millions of dollars—have a lasting and profound effect. Then there is the interference with the lower boundary of the atmosphere. The inhibiting of evaporation already mentioned is quite practicable for reservoirs and small lakes, but at present appears unrealistic on the largest scale. More practicable are alterations to the land–sea distribution by carefully conceived engineering works. The closing of the straits of Gibraltar would lead to a large shrinkage of the area of the Mediterranean Sea with consequential benefits and penalties impossible to assess. The closing of the Behring Straits has been advocated as a first step towards freeing the Arctic Ocean of ice.

Amongst other ideas ventilated has been the initiation of new depressions at chosen times and places instead of leaving matters to the free play of nature. Energy considerations make this seem beyond early reach, but the modification of the tracks of destructive tropical cyclones seems more hopeful, and some experiments have already been made. These and many other suggestions are sufficiently plausible to be impossible to put aside in our present state of ignorance, and must be regarded as prizes possibly within our reach.

The account of actual achievement is, however, much less exciting, and is almost entirely confined to some doubtful developments from the first successes with cloud seeding. Very briefly, the position is something as follows. Clouds consist of aggregates of minute water droplets of diameter averaging about 10 microns (1000 microns = 1 millimetre) in concentrations of some hundreds to the cubic centimetre. If all the water in some litres of cloud, say a million droplets, can be combined together, a single raindrop of 1 mm diameter would form and might fall as rain to the ground. Nature achieves this quite effectively, but the process takes time, perhaps about half an hour from the creation of the cloud, and in this time the wind may carry the cloud many miles away where it may mix with dry air and never give rain at all. If one can accelerate the process, more rain might be obtained, and this is what cloud seeding is claimed to do and certainly can accomplish in special conditions. By introducing frozen carbon dioxide crystals or better, crystals of silver iodide, it has been shown beyond doubt that clouds, so long as their temperature is below freezing point, a common state, may be made to precipitate within a few minutes. The early experiments were exciting, but were no scientific justification for the large-scale, almost indiscriminate seeding of clouds of all kinds by commercial operators claiming to increase the rainfall over agricultural lands. This kind of operation, which needy farmers were prevailed upon to support, have probably never been proven, the difficulty being that no-one has yet devised a way, statistical or otherwise, of distinguishing natural from artificial rainfall or of measuring the effect of an operation. It seems at present on the most optimistic view that something like a 10 per cent increment in average annual rainfall can be achieved by regular seeding, but if so it may take another twenty years of statistics to prove the point, and the outcome is still very much in doubt. In these circumstances, it is not of much scientific interest to note that cloud seeding with quite inadequate statistical controls is still going on on a large scale in Australia and the United States especially.

Other applications of seeding have also been studied. Certain types of layer clouds and fog at temperatures below freezing point have been cleared sufficiently to allow aircraft to land in otherwise prohibitive conditions, and certain types of thunderstorm cloud have, according to Russian claims, been seeded by rocket and artillery fire in such a way as to prevent the development of heavy hailstorms. The idea here is that if a large number of freezing nuclei are introduced at a critical stage, the formation of the relatively few but large and damaging stones is prevented. Controlled seeding has even been advocated as a method of guiding the movement of hurricanes, but here the chain of argument is only just plausible.

In brief, enough has been done to encourage further work, but little of practical value so far. More basic knowledge, especially more theoretical understanding, is necessary for systematic progress. At the same time, the risks are too great to allow large scale experiments to be made on the principle of trial and error.

Bibliography

FLETCHER N. H. (1962) *The Physics of Rainclouds*, Cambridge: C.U.P.

MASON B. J. (1962) *Clouds, Rain and Rainmaking*, Cambridge: C.U.P.

U.S.A. National Academy of Sciences and National Research Council (1966) *Weather and Climate Modification: Problems and Prospects*, 2 volumes, Washington.

U.S.A. National Science Foundation (1958–1966) *Weather Modification*, Washington: Series of Annual Reports.

R. C. SUTCLIFFE

WHOLE BODY COUNTING. In recent years increasing attention has been paid to the measurement of the total radioactive content of the human body, originating from natural or man-made sources. The technique of whole body counting seeks to measure the total quantity of radioactivity in the body at any time, rather than that which has accumulated in particular organs, as is the concern of many clinical investigations. Although the methods used are in many cases applicable to other living organisms, the greatest stimulus for research into techniques of whole body counting has come from investigations of radioactivity in humans.

The various methods devised for estimating total body activity may be used for (1) the detection and measurement of naturally occurring radionuclides in the body, (2) the estimation of quantities of radionuclides ingested or inhaled after a nuclear accident or in the presence of radioactive fall-out, and (3) to follow the fate of radioisotopes administered for medical diagnosis or therapy.

Principles

A radioactive substance introduced into the body will be distributed as a result of metabolic processes in a manner dependent on the nature of the substance and its chemical form. This distribution may then remain effectively constant apart from gradual excretion, or it may change rapidly with time. For example, radiocalcium entering the body lodges with non-radioactive calcium, mainly in the bones. It is only removed from there by the slow process of calcium turnover and excretion, so that over any short period of measurement (days or weeks) the distribution remains effectively constant. In contrast, radioiodine administered orally passes rapidly from the stomach via the bloodstream into the thyroid, where it is incorporated into thyroid hormones in which it re-enters the bloodstream and circulation. These hormones are then broken down, some of the iodine returns to the thyroid and is used again, while the rest is excreted via the kidneys. Thus over a similar short period of measurement the distribution of radioiodine through the body does not remain constant.

As far as possible, whatever detector system is used to measure the total activity of any radionuclide within the body, it should be independent of the spatial distribution of the nuclide. If the patient and detector are both to remain stationary, it is important to use a counting geometry where the detector response alters little with the position of the source within the patient. A single detector and a flat couch for the patient do not satisfy this condition very well (Fig. 1a), and various "geometries" have been tried in attempts to improve matters, among them being (a) the arc bed and (b) the tilting chair.

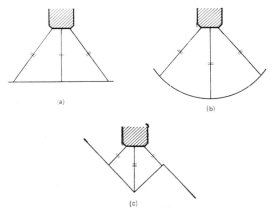

Fig. 1. (a) *Single detector above flat couch: the response of the detector will not be independent of source position.*
(b) *Single detector above arc bed: the response of the detector will be less dependent on source position than in Fig. 1(a).*
(c) *Single detector above tilting chair.*

(a) *Arc bed*

An arrangement whereby all parts of the body of the patient are equidistant from the detector will obviously give the most uniform response obtainable, while the greatest counting rate will be achieved if the patient is as close as possible to the detector (Fig. 1b). However, it has been found that the smallest radius of curvature which can be sustained by the human body for any appreciable length of time is about 50 cm. With this geometry this must be the minimum detector patient distance attainable, which limits the counting rates which can be achieved, and thus sets a limit to the smallest quantity of activity measurable.

(b) *Tilting chair*

Another geometry which has been quite widely used positions the subject in a chair with the detector above the abdomen (Fig. 1c). The chair is then tilted backwards at an angle so that the chest and thighs are equidistant from the detector, but the abdomen is at a greater distance. Hence the response of the detector to a source of activity within the patient will be non-uniform, varying from a maximum for a source in chest or thighs to a minimum in head, abdomen or legs. The variation in response may be as much as ± 20 per cent.

Scanning

The problem of non-uniform response inherent in stationary systems has led to the development of the technique known as scanning, in which the detector and the subject are made to move relative to one another so that the detector travels along a line parallel to the central axis of the patient's body. In most systems the patient lies on a flat couch, which is then made to travel so that the whole body passes through the field of view of the detector, which is collimated to reduce the background count rate. The counts received during the whole scan are totalled. If a single detector is used, the patient may be scanned both prone and supine. This gives a greater total count, and also reduces errors caused by any anterior-posterior movement of the radioactivity, thus increasing the accuracy of measurement.

Multi-detector Systems

Two or more counters may be placed along the length of a stationary subject (Fig. 2a), or one or more pairs of counters may be set in direct opposition above and below a travelling couch (Fig. 2b). In the first case counting should be done with the patient both prone and supine, in the second case this is not necessary.

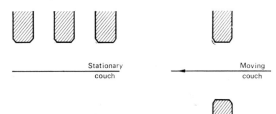

Fig. 2. (a) *Severa detectors above a stationary couch.*
(b) *Two detectors in direct opposition above and below a travelling couch.*

Detector Systems

The normal form of detector used for whole body counting consists of one or more conventional sodium iodide thallium activated crystals with associated photomultiplier tubes, single- or multi-channel pulse height analysers, and some kind of count integrator. There are also some special systems using solid or liquid scintillation compounds which are briefly discussed later.

Whole body counting systems have been designed using crystals ranging in size from 3 in diameter and 3 in thick to about 12 in diameter and 4 in thick. The larger crystals have greater sensitivity, but also give higher background counting rates. Cost also increases quite steeply with crystal size, although the price of even the biggest crystals is only a small part of the overall cost of a whole body counting apparatus.

Spectrometers

The simplest system merely totals the number of photons of all energies received by the detector, displaying them after a given time or at the end of the scan. However, more sophisticated instruments use analysers by which pulses of different energies can be separated. In studies of natural body burdens of radioactivity, or of the results of a nuclear accident, it is useful to be able to examine the relative abundances of the various radionuclides present. For these complex studies multichannel analysers are used, often comprising several hundred channels.

In medical diagnostic tests, it may be required that more than one radionuclide be used so that the metabolism of two or more substances may be studied simultaneously. In this case it is convenient to have a pulse height analyser which has a channel for each radionuclide present.

Background Counting Rates

In many of the studies for which whole body counters are used, the quantity of radioactivity present in the body is very small. This is particularly true when naturally occurring radioisotopes are to be measured. The minimum level of activity detectable will be partly dependent on the background radiation, being approximately that giving twice the background count rate. Therefore reduction of the background count is one of the main problems of this technique if high sensitivity is required. Reduction of the detector angle of view from the wide angle necessary for stationary whole body counting to the slit required for scanning does decrease the background radiation reaching the crystal per unit time. On the other hand, increasing the number of detectors does not necessarily lead to a great increase in sensitivity because the background count will also be raised approximately in proportion.

Two main methods have been used to reduce the background count rates: the construction of special low background areas, and the so-called "shadow-shield" technique.

Low Background Areas

These are usually specially constructed rooms, often underground or in basement areas, with walls made of about 6 to 8 in of steel faced with about $\frac{1}{8}$ in of lead. These "steel rooms", which usually measure about 7 ft each way, weigh about 40 tons.

As an alternative to a steel room, in areas where the subsoil is chalk or some other material with a naturally low background count rate, a "quarry" can be dug out and lined with lead.

Shadow Shielding

This method uses lead shielding 2 to 4 in thick, so arranged that the face of the detector crystal receives no direct radiation from any direction other than that of the subject. These "shadow shields" weigh about 5 tons and are considerably less expensive to build than the special steel rooms, but they do not reduce the background to quite such a low level.

Special Systems

The better and more complete the geometry relating the subject to the detector system, the better the sensitivity that will be achieved. A larger detector will give greater counting rates, and will also reduce any effects due to a redistribution of the source within the subject. Therefore a detector which completely surrounds the patient, giving 4π geometry, will have the best chance of collecting all the radiation emitted.

Large volume whole body counters have been constructed on this principle. A cylindrical shell large enough to contain a human body, and several inches thick, is moulded from plastic organic scintillation material, or a similar shaped tank can be filled with liquid scintillator. Several large photomultiplier tubes are let into the walls of the shell or tank. The subject is then put inside the cylinder, and is thus completely surrounded by the detecting medium. The spatial response characteristics of a system of this kind are very good. A slightly less efficient system uses a semi-cylindrical shell or tank, giving 2π geometry. In this case the subject is turned through 180° after a given length of time.

These counters have a sensitivity about ten times that of the ordinary crystal detector systems. They need to be enclosed in low background rooms because their response to background radiations is also very high. There are very few counters of this type in existence, mainly because the cost of their construction is extremely high, while their great sensitivity is not required in routine applications.

Uses of Whole Body Counters

1. Naturally occurring body burdens of radioactivity

The body contains very small quantities of naturally occurring radionuclides, particularly potassium-40. These can be measured using the more sensitive detectors and low background areas, with multi-channel pulse height analysers. The information obtained is used in estimations of the radiation dose to man from his natural environment.

2. Nuclear accidents

Accidents at nuclear installations, or nuclear incidents leading to radioactive fall-out, can cause situations where it is desirable to measure the quantities of radioisotopes of various elements and different energies which may have been ingested or inhaled, so that the radiation hazard to the individual or the general population may be estimated. Whole body counters with multi-channel analysers can be used to give the required information more easily than it can be obtained by chemical analysis of blood and urine samples, etc.

3. Medical uses

The fate of administered radionuclides may be followed by whole body counting. The two main fields of clinical diagnostic study are (a) cases where there is question of blood loss via the gastro-intestinal tract into the faeces, and (b) cases of suspected malabsorption of vitamin B_{12}. In both of these problems, after administration of suitably labelled compounds, the excreta can be collected and any radioactivity present measured. However, this is unpleasant and subject to errors through incomplete collection. It is considerably easier and more reliable to measure the activity remaining in the body, and thus to deduce how much has been excreted.

Therapeutic doses of radionuclides may also be followed by whole body counting, giving a measure of the length of time the activity remains in the body, and hence of the radiation dose and the rate at which it was delivered, both important factors in radiation therapy.

In the past, the technique of whole body counting has been almost exclusively used by workers interested in establishing natural levels of radioactivity in the body and in studies of radioactivity in the general environment. In recent years there has been increasing interest in whole body counters as clinical tools, and for these applications sensitivity is not such an important criterion as in environmental measurements, as the amount of radioactivity administered to the patient will be at least one order of magnitude greater than that of the natural quantity present in the body. Increasing attention has therefore been given to the development of less expensive systems which have good spatial characteristics although less sensitivity, and it is likely that these instruments will find more and more usefulness in clinical work.

Bibliography

BODDY K. The development and performance of a shadow shield whole body monitor, *Physics in Medicine and Biology* 12, No. 1, 43–51.

I.A.E.A., Vienna (1961) Symposium on whole body counting.

I.A.E.A., Vienna (1965) Clinical uses of whole body counters.

I.A.E.A., Vienna (1964) Directory of whole body radioactivity monitors.

WARNER G. T. and OLIVER R. A whole body counter for clinical measurements utilizing the shadow shield technique, *Physics in Medicine and Biology* 11, No. 1, 83–95.

M. H. GRIEVESON and D. J. REES

INDEX

Aberration, chromatic: *see* High resolution electron microscopy 179
Aberration, spherical: *see* High resolution electron microscopy 179
Abrasion resistance: *see* Glass ceramics 146
Absorbed dose: *see* Radiation quantities, systematics of 367
Absorption and scattering of solar radiation: *see* Daylight, power distribution in 55
Absorption coefficient: *see* Solar cell 426
Absorption cross-section: *see* Interplanetary dust 214
Absorption, infra-red: *see* Infra-red spectroscopy, recent developments in 205
Absorption of liquids, crystals and glasses in the far infra-red: *see* Far infra-red spectroscopy 107
Absorption, OH: *see* Radio emissions from interstellar hydroxyl radicals 373
Absorption, planetary atmospheric: *see* Infra-red spectroscopy, recent developments in 206
Absorption, polar cap: *see* Ionosphere, recent observations of 225
Absorption, relaxational (Debye) by polar molecules: *see* Submillimetre radiation, applications of 452
Absorption, resonance, of acoustic waves: *see* Acousto-electric effects 9
Absorption spectrum: *see* Far infra-red spectroscopy 106; Radio emissions from interstellar hydroxyl radicals 373; Soft X-ray spectroscopy 412
Absorption, spin resonance: *see* Phonon avalanche and bottleneck 318
Abundance, atomic: *see* Geochronology 135
a.c. Josephson effect: *see* Josephson effect 237
Acceleration due to gravity: *see* Quantum metrology 356
Accelerators, electron: *see* Radiation chemistry, industrial application of 364
Acceptor: *see* Semiconductors, effect of radiation on 405; Solar cell 423
Accommodation or "sticking" coefficient: *see* Thin Films, nucleation and growth of 491
Acids, amino-: *see* Proteins, structure of 337
Acids, nucleic: *see* Ultra-violet radiation, some biological effects of 506
Acoustic holography 1; *see also* Lasers, applications of 247
Acoustic resonators: *see* Speaking machines 440
Acoustic wave: *see* Acoustic waveguides 6
Acoustic waveguide: *see* Acoustic waveguides 5

Acoustic waveguide, fluid: *see* Acoustic waveguides 5
Acoustic waveguide, solid: *see* Acoustic waveguides 7
Acoustical phonons, resonant: *see* Phonon avalanche and bottleneck 316
Acoustics, underwater: *see* Ambient noise in the sea 14
Acousto-electric current: *see* Acousto-electric effects 8
Acousto-electric effects 8
Acousto-electric field: *see* Acousto-electric effects 8
Acousto-electric voltage: *see* Acousto-electric effects 8
Activity: *see* Radiation quantities, systematics of 366
Activity, carcinogenic: *see* Quantum mechanics in biochemistry 352
Activity, optical: *see* Pockels effect 329
Activity, specific: *see* Radiation quantities, systematics of 366
Additively coloured crystals: *see* Metallic colloids in non-metallic crystals 282
Additives, concentration of: *see* Metal film preparation by electroless (chemical) reduction 279
Adiabatic fast passage: *see* Fast passage in paramagnetic resonance 109
Adler–Weisberger sum rule: *see* Current algebras and their applications 53
Aerofoil: *see* High-lift devices 170
"Aerofoil amplifier": *see* Fluidics 121
Aerofoil theory: *see* Jet flap 234
Aeroplane: *see* High lift devices 173
Aeroplane equivalent body of revolution: *see* Sonic boom, generation and suppression of 432
Affine deformation: *see* Polymer physics 337
Affinity, electron: *see* Metallic colloids in non-metallic crystals 283
AGA infra-red camera: *see* Infra-red cameras and infra-red TV 198
Age, cosmic ray: *see* Shock pressure in meteorites 407
Age equation: *see* Geochronology 136
Age of the earth: *see* Geochronology 133
Agglomeritic ice: *see* Hydrometeorology 192
Agyroscopic term: *see* Hamilton's principle 162
Airborne collisions prevention devices 9
Aircraft, STOL: *see* Jet flap 233
Aircraft, VTOL: *see* Jet flap 233
Algebra, current: *see* Current algebras and their applications 51

Aliasing: *see* Interferometric spectrometers 209
Alkali halides: *see* Far infra-red spectroscopy 107
Alkyl chain: *see* Liquid crystals and molecular stucture 259
"ALL" or "MULTIPLE AND" element: *see* Fluidics 118
Alloy magnets, properties of: *see* Permanent magnet technology 311
Alloy spectra: *see* Soft X-ray spectroscopy 420
Alloys and metals, microstructure of: *see* Electron metallography, applications of 89
Alloys and metals, surface topography of: *see* Electron metallography, applications of 89
Alloys, electronic structure of: *see* Soft X-ray spectroscopy 420
Alpha-Chapman layer: *see* Ionospheres of Mars and Venus 230
Alpha-helix: *see* Proteins, structure of 338
Alpha-keratin: *see* Proteins, structure of 339
Alpha-polypeptides: *see* Proteins, structure of 338
Alpha-proteins: *see* Proteins, structure of 339
Alpha-structure of proteins: *see* Proteins, structure of 338
Alpha-uranium: *see* Sputtering, theories of 446
Ambient noise in the sea 13
Amino-acids: *see* Proteins, structure of 337
Amino-boranes and borohydride as reducing agents: *see* Metal film preparation by electroless (chemical) reduction 281
Amorphous polymers: *see* High polymers, mechanical properties of 175
Ampere: *see* Quantum metrology 356
Amplification, maser: *see* Radioemissions from interstellar hydroxyl radicals 376
Amplification, phase-sensitive: *see* Submillimetre-wave methods 460
Amplification, ultrasonic: *see* Acousto-electric effects 8
Amplification (ultrasonic) in semiconductors: *see* Ultrasonic amplification in semiconductors 497
"Amplifier, aerofoil": *see* Fluidics 121
Amplifier, bistable: *see* Fluidics 119
Amplifier, controlled separation (elbow): *see* Fluidics 121
Amplifier, edgetone: *see* Fluidics 121
Amplifier, induction: *see* Fluidics 120
Amplifier, proportional fluid: *see* Fluidics 119
Amplifier, turbulence: *see* Fluidics 120
Amplifier, vented momentum: *see* Fluidics 118
Amplitude grating: *see* Holography 184
Amplitude modulation, pulse: *see* Pulse code modulation 345
Analogue computer: *see* Hall effect, applications of 157
Analogue-digital converter (coder): *see* Pulse code modulation 346
Analogue transmission: *see* Pulse code modulation 344
Analyser, pulse height: *see* Whole body counting 524

Analysers for process control: *see* Infra-red spectroscopy, recent developments in 206
Anchor-ice: *see* Hydrometeorology 191
"AND" element: *see* Fluidics 118
"AND MULTIPLE" or "ALL" element: *see* Fluidics 118
Anelasticity: *see* Microplasticity 287
Angle: *see* Quantum metrology 357
Angle, Cabibbo: *see* Current algebras and their applications 52
Angular resolution: *see* Radio telescopes, the present position 380
Anharmonic forces: *see* Melting, modern theories of 276
Animal fibres, natural: *see* Textile industry, physics in 470
Anisotropic effects in polymers: *see* High polymers, mechanical properties of 178
Anisotropic molecules: *see* Liquid crystals and molecular structure 257
Anisotropy constant, stress: *see* Magnetic thin films 267
Anisotropy constants: *see* Magnetic thin films 266
Anisotropy dispersion: *see* Magnetic thin films 267
Anisotropy energy: *see* Magnetic thin films 266
Anisotropy field: *see* Magnetic thin films 266
Anisotropy, field-induced (or M-induced): *see* Magnetic thin films 266
Anisotropy, magnetocrystalline: *see* Magnetic thin films 265; Piezomagnetic effect 326
Anisotropy, magnetostrictive: *see* Magnetic thin films 266
Anisotropy, uniaxial: *see* Magnetic thin films 266
Anode: *see* Thermionic generation of electricity 479
Anomalous dispersion: *see* Infra-red spectroscopy, recent developments in 204
Anomalous mode: *see* IMPATT diode microwave generators 196
Anomaly, mean: *see* Astrodynamics 23
Anomaly, true: *see* Astrodynamics 23
Antiferromagnetic materials, electron diffraction in: *see* Electron diffraction from periodic magnetic fields 87
Antiferromagnetism: *see* Far infra-red spectroscopy 108
Antineutrino: *see* Helicity 163
Antiparticle: *see* Charge conjugation 46; Elementary particles, nomenclature of 94
Antiquark: *see* Quarks 358
Anti-reciprocity relation: *see* Gyroscopic coupler 151
Anti-Stokes lines: *see* Lasers, tunable 251
Antisymmetrized molecular orbital: *see* Quantum mechanics in biochemistry 352
Aperture synthesis radio telescopes: *see* Radio telescopes, the present position 385
Apodization: *see* Interferometric spectrometers 210
Apodizing or weighting function: *see* Fourier transform spectroscopy 123
Apofocus: *see* Astrodynamics 22

Apparatus function: *see* Fourier transform spectroscopy 123
Appleton, or geomagnetic anomaly: *see* Ionosphere, recent observations of 225
Arc bed: *see* Whole body counting 523
Arc, mercury: *see* Far infra-red spectroscopy 104
Arc, plasma: *see* Submillimetre-wave methods 459
Archaeology, astro-: *see* Astro-archaeology 16
Archimedes principle: *see* Ferrohydrodynamics 116
Areal precipitation: *see* Hydrometeorology 188
Argon: *see* Laser-induced sparks 243
Argon laser: *see* Lasers, applications of 246
Argon potassium dating: *see* Geochronology 135
Aromatic polyamides: *see* Temperature-resistant polymers 465
Aromatic polymers: *see* Temperature-resistant polymers 463
Arrhenius equation: *see* Snoek spectra 411
Artificial rainfall: *see* Weather control 522
Ascending node, longitude of the: *see* Astrodynamics 23
Astigmatic image: *see* Mirage and looming, optics of 294
Astigmatism: *see* High resolution electron microscopy 179
Astro-archaeology 16
Astrodynamics 17
Astronomy, radar: *see* Radar astronomy 362
Astronomy, radio: *see* Radio telescopes, the present position 386
Atmospheric absorption, planetary: *see* Infra-red spectroscopy, recent developments in 206
Atmospheric applications of submillimetre radiation: *see* Submillimetre radiation, applications of 453
Atmospheric drag: *see* Astrodynamics 21
Atmospheric refractivity profiles: *see* Ionospheres of Mars and Venus 230
Atmospheric transmission spectrum: *see* Submillimetre radiation, applications of 453
Atomic abundance: *see* Geochronology 135
Atomic clock: *see* Quantum metrology 354
Atomic collision cascade: *see* Sputtering, theories of 446
Atomic cores: *see* Quantum mechanics in biochemistry 351
Atomic energy level schemes: *see* Laser-induced sparks 244
Atomic energy levels (shells): *see* Auger emission spectroscopy 25
Atomic mass unit: *see* Quantum metrology 355
Atomic number: *see* Chart of the nuclides, trilinear 47
Atomic orbital: *see* Quantum mechanics in biochemistry 352
Atomic scattering factor: *see* Liquid metals 261
Atomic second: *see* Quantum metrology 354
Atoms, ejection of, from the surface of solids: *see* Sputtering, theories of 445
Atoms, interstitial: *see* Snoek spectra 410

Atoms, thermal evaporation of: *see* Sputtering, theories of 445
ATR (or internal reflection) techniques: *see* Infra-red spectroscopy, recent developments in 204
Attachment cross-sections: *see* Ionization and attachment cross sections of atoms and molecules by electron impact 219
Attenuation, sound wave: *see* Melting, modern theories of 276
Attenuation, ultrasonic, in a material, frequency-dependence of: *see* Ultrasonic spectroscopy 502
Attractions, intermolecular: *see* Liquid crystals and molecular structure 257
Attractions, van der Waals: *see* Proteins, structure of 341
Auger effect: *see* Auger emission spectroscopy 23
Auger electron: *see* Satellites in soft X-ray spectra 404
Auger emission spectroscopy 23
Auger ionization: *see* Ionization and attachment cross sections of atoms and molecules by electron impact 219
Auger process: *see* Satellites in soft X-ray spectra 404
Auger spectra: *see* Auger emission spectroscopy 25
Auger transition: *see* Satellites in soft X-ray spectra 404
Auger-type process: *see* Ion-neutralization spectroscopy 220
Aurorae: *see* Auroral electrons 27
Auroral electrons 27
Auroral luminosity: *see* Auroral electrons 27
Auroral precipitation: *see* Auroral electrons 28
Auroral zone: *see* Auroral electrons 28
Auto- and cross-correlation: *see* Optical information processing 303
Autoepitaxy: *see* Epitaxy, recent work on 98; Thin films, nucleation and growth of 492
Auto-ionization: *see* Ionization and attachment cross sections of atoms and molecules by electron impact 219
Avalanche current multiplication: *see* Microwave solid state oscillators, recent 289
Avalanche, ionizing (Townsend cascade): *see* Laser-induced sparks 243
Avalanche microwave generators: *see* IMPATT diode microwave generators 195
Avalanche, phonon: *see* Phonon avalanche and bottleneck 315
Avalanche process: *see* Impatt diode microwave generators 194
Avalanche, sand: *see* Singing sands 408
Avalanche threshold: *see* Microwave solid state oscillators, recent 289
Avalanche, Townsend: *see* Streamer chambers 448
Avalanching pn diode: *see* IMPATT diode microwave generators 195
Avogadro number: *see* Quantum metrology 356
Avoidance manoeuvres: *see* Airborne collision prevention devices 12

Avoidance systems, collision: *see* Airborne collision prevention devices 9
Axes, rotor: *see* Dynamical *n*-beam theory of electron diffraction 72
Axes, screw: *see* Dynamical *n*-beam theory of electron diffraction 72
Axis, semi-major: *see* Astrodynamics 23

Background counting rates: *see* Whole body counting 524
Background radiation, radio: *see* Radio telescopes, the present position 382
Backscatter, incoherent: *see* Ionosphere, recent observations of 228
Bacon cell: *see* Space power supplies 439
Bacteria: *see* Ultra-violet radiation, some biological effects of 506
Bacteriophage, host-cell reactivation of: *see* Ultra-violet radiation, some biological effects of 510
Band, energy, calculations: *see* Soft X-ray spectroscopy 417
Band gaps: *see* Soft X-ray spectroscopy 418
Band, slip: *see* Fracture of brittle materials 128
Band structure, photoemission and: *see* Photoemission and band structure 319
Band, valence, spectra: *see* Soft X-ray spectroscopy 412
Banding: *see* Refractories, ultra-purification of 387
Barium ferrite permanent magnets: *see* Permanent magnet technology 311
Barium ferrite, temperature coefficient of magnetization of: *see* Permanent magnet technology 313
Barnes infra-red camera: *see* Infra-red cameras and infra-red TV 197
Barrier crack: *see* Fracture of brittle materials 127
Baryon: *see* Elementary particles, nomenclature of 94; Quarks 357
Baryon number: *see* Elementary particles, nomenclature of 95
Baryon resonances: *see* Quarks 358
Basal-plane pseudomorphism: *see* Expitaxy, recent work on 99
Batch data processing: *see* Astrodynamics 19
Beacon, split-beam microwave: *see* Split-beam microwave beacon 444
Beam condensers: *see* Infra-red spectroscopy, recent developments in 204
Beam splitter: *see* Interferometric spectrometers 210
Beating: *see* Sailing, physics of 397
Bed, arc: *see* Whole body counting 523
Belts, trapped radiation (Van Allen): *see* Auroral electrons 29
Bending and stretching vibrations: *see* Far infra-red spectroscopy 106
Bernoulli theorem, generalized: *see* Ferrohydrodynamics 112
Beta-Chapman layer: *see* Ionospheres of Mars and Venus 230

Beta-decay: *see* Bosons, intermediate 35; Current algebras and their applications 53
Beta-decay coupling constant: *see* Bosons, intermediate 35
Beta-keratin: *see* Proteins, structure of 339
Beta-polypeptides: *see* Proteins, structure of 339
Beta-proteins: *see* Proteins, structure of 339
Beta-structure of proteins: *see* Proteins, structure of 338
Binary coding: *see* Pulse code modulation 346
Biochemistry, quantum mechanics in: *see* Quantum mechanics in biochemistry 351
Biological effects of ultra-violet radiation: *see* Ultra-violet radiation, some biological effects of 506
Biological electron microscopy: *see* High resolution electron microscopy 183
Biological noise in the sea: *see* Ambient noise in the sea 16
Biological pressure transensors 31
Biomolecules: *see* Quantum mechanics in biochemistry 351
Bipolar transmission: *see* Pulse code modulation 346
Birefringence: *see* Ultrasonic imaging 500
Bismuth probes: *see* Magnetoresistivity 272
Bistable amplifier: *see* Fluidics 119
Black body: *see* Far infra-red spectroscopy 105
Black Orlon: *see* Temperature resistant polymers 466
Black-body emission: *see* Infra-red cameras and infra-red TV 196
Black-body radiation: *see* Emissivity at high temperatures 95
Blast wave: *see* Laser-induced sparks 243
Bleeds: *see* Fluidics 119
Bloch wave functions: *see* Photoemission and band structure 320
Blocking oscillator: *see* Cardiac pacemaker, external 40
Blocking, proton: *see* Proton scattering microscopy 342
Blood pressure: *see* Biological pressure transensors 34
Blooming: *see* Solar cell 424
Blow stroke: *see* Glandless reciprocating-jet pump 140
Blowing: *see* High lift devices 172
Blown flap: *see* High lift devices 171
"Blue Clearings": *see* Solar wind, planetary interactions with 429
Blumlein circuit: *see* High intensity pulsed radiation sources 165
Blumlein line: *see* Streamer chambers 449
Body, radioactivity in the: *see* Whole body counting 522
Body, whole, counting: *see* Whole body counting 522
Boeing 727: *see* High lift devices 171
Bolometer: *see* Far infra-red spectroscopy 105; Infra-red cameras and infra-red TV 197; Submillimetre-wave methods 459
Bolometer, germanium: *see* Far infra-red spectroscopy 105

Bolometer, thermistor: *see* Thermoelectric cooling of infra-red detectors 481
Boltzmann equation: *see* Rubber physics 393
Boltzmann superposition principle: *see* High polymers, mechanical properties of 177
Boltzmann transformation: *see* Diffusion in non-uniform media 59
Bombardment, ion: *see* Sputtering, theories of 445
Bombyx mori silk: *see* Proteins, structure of 339
Bond, covalent: *see* Proteins, structure of 341
Bond, hydrogen: *see* Proteins, structure of 337
Bond, ionic: *see* Proteins, structure of 341
Bond lengths: *see* Polymer physics 332
Bond order: *see* Quantum mechanics in biochemistry 352
Bond, peptide: *see* Proteins, structure of 337
Bond rupture: *see* Radiation chemistry, industrial applications of 364
Bonding, hydrogen: *see* Quantum mechanics in biochemistry 353
Bonding, inter-molecular: *see* Submillimetre radiation, applications of 452
Bordoni peak 34; *see also* Peierls stress 306
Born series: *see* Dynamical n-beam theory of electron diffraction 70
Borohydride and amino-boranes as reducing agents: *see* Metal film preparation by electroless (chemical) reduction 281
Boson: *see* Elementary particles, nomenclature of 95
Boson field: *see* Bosons, intermediate 35
Boson, intermediate: *see* Bosons, intermediate 34
Bottleneck and avalanche, phonon: *see* Phonon avalanche and bottleneck 315
Bottlenecked decay: *see* Phonon avalanche and bottleneck 316
Bottomside ionosphere: *see* Ionosphere, recent developments of 224
Boundary conditions, magnetic field: *see* Ferrohydrodynamics 113
Boundary, grain: *see* Fracture of brittle materials 126
Boundary, grain, fracture: *see* Fracture of brittle materials 128
Boundary, phase: *see* Diffusion in non-uniform media 60
Bow shock pressure rise: *see* Sonic boom, generation and suppression of 433
Bragg-effect holograms: *see* Holography 184
Branches: *see* Kirchhoff's laws (electrical circuits) 240
Brayton cycle: *see* Electrogasdynamics 75
Break spinning: *see* Textile industry, physics in 474
Breakdown of gases, laser-induced: *see* Laser-induced sparks 242
Breakdown of water gaps, electrical: *see* Electrohydraulic crushing 81
Breakdown strength, dielectric: *see* Glass ceramics 147
Breakdown strength, electric: *see* Electrogasdynamics 76

Breathalyser: *see* Law enforcement science and technology 254
Bremsstrahlung: *see* Quantum metrology 357
Bright arc aurorae: *see* Auroral electrons 27
Brightness temperature: *see* Radioemissions from interstellar hydroxyl radicals 376
Brillouin scattering: *see* Phonon avalanche and bottleneck 318
Brillouin scattering, stimulated: *see* Lasers, applications of 248
Brillouin zone: *see* Soft X-ray spectroscopy 418
Brittle crack: *see* Fracture of brittle materials 124
Brittle fracture: *see* Fracture of brittle materials 124
Brittle materials, fracture of: *see* Fracture of brittle materials 124
Brittleness in ceramics 37
"Broad band" semiconductors: *see* Thermoelectric generation of electricity 485
Broad beam ion gun: *see* Glow discharge electron and ion guns for material processing 150
Broadening, Doppler: *see* Radioemissions from interstellar hydroxyl radicals 373
Broadening, instrumental: *see* Soft X-ray spectroscopy 417
Broadening, lifetime: *see* Soft X-ray spectroscopy 417
Broadening, temperature: *see* Soft X-ray spectroscopy 417
Broadening, thermal: *see* Radioemissions from interstellar hydroxyl radicals 373
Broadening, turbulent: *see* Radioemissions from interstellar hydroxyl radicals 373
Bromine, tungsten-, cycle: *see* Tungsten halogen lamps 496
Bubble chamber: *see* Streamer chambers 451
Bubble transensor, plastic: *see* Biological pressure transensors 32
Bulb-venturi pump: *see* Glandless reciprocating-jet pump 141
Bulk negative differential resistivity in semiconductors: *see* Microwave solid state oscillators, recent 289

C invariance, violation of: *see* Charge conjugation 46
C region: *see* Ionosphere, recent observations of 224
C, velocity of electromagnetic radiation: *see* Quantum metrology 355
Cabibbo angle: *see* Current algebras and their applications 52
Caesium: *see* Laser-induced sparks 244; Thermionic generation of electricity 477
Caesium 133 atom: *see* Quantum metrology 354
Caesium clocks: *see* Quantum metrology 354
Caesium time standard: *see* Airborne collision prevention devices 11
Camera, infra-red: *see* Infra-red cameras and infra-red TV 196
Camera tubes, ultrasound television: *see* Ultrasonic imaging 502
Cancer research: *see* Lasers, applications of 246

Candela: *see* Quantum metrology 356
Capacitor, coaxial gas: *see* High intensity pulsed radiation sources 165
Capillary cell: *see* Infra-red spectroscopy, recent developments in 205
Capture, resonance: *see* Ionization and attachment cross-sections of atoms and molecules by electron impact 219
Carbon chain: *see* Liquid crystals and molecular structure 259
Carbon dating: *see* Geochronology 139
Carbon dioxide, frozen: *see* Weather control 522
Carbon extraction replica: *see* Electron metallography, applications of 89
Carbon, glassy: *see* Vitreous carbons, structure and properties of 513
Carbon, impermeable: *see* Vitreous carbons, structure and properties of 513
Carbon, vitreous, structure and properties of: *see* Vitreous carbons, structure and properties of 513
Carboxypeptidase: *see* Proteins, structure of 340
Carcinogenic activity: *see* Quantum mechanics in biochemistry 352
Carcinotron: *see* Infra-red spectroscopy, recent developments in 204; Submillimetre-wave methods 459
Cardiac catheterization: *see* Cardiac pacemaker, external 39
Cardiac pacemaker, external 39
Carnot efficiency: *see* Thermoelectric generation of electricity 483
Carrier, charge, concentration in semiconductors, radiation effects on: *see* Semiconductors, effect of radiation on 405
Carrier concentration, intrinsic: *see* Solar cell 422
Carrier, drift of: *see* Impatt diode microwave generators 194
Carrier generation: *see* Impatt diode microwave generators 194
Carrier lifetime: *see* Solar cell 424
Carrier lifetime, minority: *see* Semiconductors, effect of radiation on 406
Carrier, majority: *see* Solar cell 422
Carrier, minority: *see* Solar cell 422
Carrier mobility: *see* Hall effect, applications of 156; Thermoelectric generation of electricity 484
Cascade, atomic collision: *see* Sputtering, theories of 446
Cascade thermoelectric generator: *see* Thermoelectric generation of electricity 487
Cascade, Townsend (ionizing avalanche): *see* Laser-induced sparks 243
Cassiopeia A: *see* Radioemissions from interstellar hydroxyl radicals 373
Casting, continuous: *see* Freeze-coating 130
Casting, slip: *see* Ceramics production, physics in 41
Catalytic surface: *see* Metal film preparation by electroless (chemical) reduction 278
Catheterization, cardiac: *see* Cardiac pacemaker, external 39
Cathode: *see* Thermionic generation of electricity 479

Cathode discharge, cold: *see* Glow discharge electron and ion guns for material processing 149
Cathode fall: *see* Glow discharge electron and ion guns for material processing 148
Cathode glow discharge, warm: *see* Glow discharge electron and ion guns for material processing 149
Cathode gun: *see* Refractories, ultra-purification of 389
Cathode, hollow, heating: *see* Refractories, ultra-purification of 388
Cavitation: *see* Fracture of brittle materials 129
Cavitation effect: Electrohydraulic crushing 82
Cavity interferogram: *see* Submillimetre-wave lasers 455
Cavity lifetime, photon: *see* Q-switching 350
Celestial mechanics: *see* Astrodynamics 17
Cell, Bacon: *see* Space power supplies 439
Cell, capillary: *see* Infra-red spectroscopy, recent developments in 205
Cell, fuel: *see* Space power supplies 439
Cell, Golay: *see* Submillimetre-wave lasers 456; Submillimetre-wave methods 459
Cell, gridded: *see* Solar cell 426
Cell, ion exchange: *see* Space power supplies 439
Cell, multi-reflection gas: *see* Infra-red spectroscopy, recent developments in 206
Cell organelles, effect of microbeams on: *see* Ultraviolet radiation, some biological effects of 512
Cell, photoconductive: *see* Infra-red cameras and infra-red TV 197
Cell, photovoltaic: *see* Solar cell 423
Cell, photovoltaic solar: *see* Space power supplies 438
Cell, Pohlman: *see* Ultrasonic imaging 500
Cell, polycrystalline: *see* Solar cell 428
Cell, reversed: *see* Solar cell 424
Cell, silver chloride: *see* Infra-red spectroscopy, recent developments in 205
Cell, solar: *see* Solar cell 422; Space power supplies 438
Cell, super-blue: *see* Solar cell 427
Cell, variable gap: *see* Solar cell 428
Cellular structure: *see* Refractories, ultra-purification of 387
Ceramic magnets: *see* Permanent magnet technology 313
Ceramic slips: *see* Ceramics production, physics in 41
Ceramics brittleness in: *see* Brittleness in ceramics 37
Ceramics, firing of: *see* Ceramics production, physics in 43
Ceramics, glass: *see* Glass ceramics 143
Ceramics, microcrystalline glass: *see* Glass ceramics 145
Ceramics production, physics in 41
Chain, alkyl: *see* Liquid crystals and molecular structure 259
Chain, carbon: *see* Liquid crystals and molecular structure 259
Chain deformation: *see* Polymer physics 332
Chain molecular: *see* High polymers, mechanical properties of 175

Chain, polymer: *see* High polymers, mechanical properties of 175; Polymer physics 331
Chain, polypeptide: *see* Proteins, structure of 337
Chain tension: *see* Polymer physics 332
Chair, tilting: *see* Whole body counting 523
Chamber, bubble: *see* Streamer chambers 451
Chamber, spark: *see* Streamer chambers 448
Chamber, streamer: *see* Streamer chambers 448
Channel electron multiplier: *see* Soft X-ray spectroscopy 414
Channel spectrum: *see* Submillimetre-wave methods 461
Charge carrier concentration in semiconductors, radiation effects on: *see* Semiconductors, effect of radiation on 405
Charge conjugation 45
Charge conjugation in variance: *see* Charge conjugation 45
Charge injector, external: *see* Electrogasdynamics 78
Charge of W meson: *see* Bosons, intermediate 35
Charge pacemaker, constant: *see* Cardiac pacemaker, external 40
Charge, space: *see* High intensity pulsed radiation sources 164
Charged particle equilibrium (C.P.E.): *see* Radiation quantities, systematics of 367
Charged particle, trajectory of: *see* Streamer chambers 448
Charging, electrostatic: *see* Textile industry, physics in 469
Chart of the nuclides, trilinear 46
Chemical (electroless) reduction, metal film preparation by: *see* Metal film preparation by electroless (chemical) reduction 277
Chemical reactivity: *see* Quantum mechanics in biochemistry 351
Chemistry, radiation, industrial applications of: *see* Radiation chemistry, industrial applications of 362
Chemonuclear reactors: *see* Radiation chemistry, industrial applications of 363
Cholesteric liquid crystals: *see* Liquid crystals and molecular structure 257
Chondritic meteorites: *see* Interplanetary dust 214
Chords: *see* Kirchhoff's laws 240
Chromatic aberration: *see* High resolution electron microscopy 179
Chromosphere: *see* Daylight, power distribution in 55
Classical drude theory: *see* Thermionic generation of electricity 476
Clay: *see* Ceramics production, physics in 41
Cleavage crack: *see* Fracture of brittle materials 125
Cleavage fracture: *see* Fracture of brittle materials 125
Clock, atomic: *see* Quantum metrology 354
Clock, caesium: *see* Quantum metrology 354
Closed-cycle plant: *see* Electrogasdynamics 76
Close-spaced vacuum diode: *see* Thermionic generation of electricity 477
Cloud seeding: *see* Weather control 521

"Coanda effect": *see* Fluidics 117; High lift devices 172; Jet flap 233
Coarsening, particle: *see* Dispersion and fibre strengthening of materials 63
Coating, freeze-: *see* Freeze-coating 130
Coating, surface, curing of: *see* Radiation chemistry, industrial applications of 365
Coaxial gas capacitor: *see* High intensity pulsed radiation sources 165
Coaxial transmission line: *see* High intensity pulsed radiation sources 165
Cobalt: *see* Magnetic thin films 266
Cobalt-60: *see* Radiation chemistry, industrial applications of 364
Cobalt-phosphorus alloys: *see* Metal film preparation by electroless (chemical) reduction 282
Code modulation, pulse: *see* Pulse code modulation 344
Coded system of communications: *see* Pulse code modulation 344
Coder (analogue–digital convertor): *see* Pulse code modulation 346
Coefficient of performance for refrigeration: *see* Peltier refrigerators 307
Coefficient of thermal expansion: *see* Glass ceramics 145
Coercive field: *see* Magnetic thin films 268
Coercivity: *see* Permanent magnet technology 311
Coherence length: *see* Submillimetre radiation, applications of 454
Coherent length: *see* Holography 184
Coherent light: *see* Optical information processing 303
Cold cathode discharge: *see* Glow discharge electron and ion guns for material processing 49
Cold drawing: *see* High polymers, mechanical properties of 179
Collagen structure of proteins: *see* Proteins, structure of 338
Collision cascade, atomic: *see* Sputtering, theories of 446
Collision cross-section, gas: *see* Thermomagnetic gas torque (The Scott effect) 490
Collision drag effects: *see* Acousto-electric effects 8
Collision frequency of gas molecules: *see* Thermomagnetic gas torque (The Scott effect) 488
Collision prevention devices, airborne: *see* Airborne collision prevention devices 9
Collision sequences, focused: *see* Sputtering, theories of 445
Colloidal ions: *see* Electrogasdynamics 77
Colloidal particles: *see* Metallic colloids in non-metallic crystals 282
Colloids, metallic, in non-metallic crystals: *see* Metallic colloids in non-metallic crystals 282
Colour temperature, correlated: *see* Daylight, power distribution in 56
Colour temperature range of daylight: *see* Daylight, power distribution in 57

Coloured crystals, additively: see Metallic colloids in non-metallic crystals 282
Coloured glasses: see Glass ceramics 144
Cometary debris theory: see Interplanetary dust 216
Communications: see Lasers, applications of 245; Law enforcement science and technology 255
Communications, coded system of: see Pulse code modulation 344
Communications density: see Airborne collision prevention devices 10
Communications, inter-space: see Lasers, applications of 249
Commutators, current: see Current algebras and their applications 52
Complex power: see Hamilton's principle 163
Complex shear modulus: see Rubber physics 394
Complex transformation: see Interferometric spectrometers 209
Compressed nonpolar gases: see Submillimetre radiation, applications of 452
Compression, efficiency of gas: see Glandless reciprocating jet pump 142
Compton process: see Radiation quantities, systematics of 369
Computer, analogue: see Hall effect, applications of 157
Computer, memories: see Magnetic thin films 265
Computer, optical: see Lasers, applications of 249
Computer or data processing: see Lasers, applications of 245
Computer simulation of the melting process: see Melting, modern theories of 275
Computerized criminal data file: see Law enforcement science and technology 255
Concentration of additives: see Metal film preparation by electroless (chemical) reduction 279
Concentration of reagents: see Metal film preparation by electroless (chemical) reduction 279
Concentrator, flux: see Hall effect, applications of 156
Condensers, beam: see Infra-red spectroscopy, recent developments in 204
Conduction electron spin resonance, CESR: see Metallic colloids in non-metallic crystals 283
Conduction, heat: see Conductivity at high temperatures 48; Freeze-coating 130
Conduction in liquid metals: see Liquid metals 262
Conductivity at high temperatures 48
Conductivity, electrical: see Conductivity at high temperatures 48; Thermoelectric generation of electricity 484
Conductivity, electronic: see Metallic colloids in non-metallic crystals 284
Conductivity, electronic thermal: see Conductivity at high temperatures 48
Conductivity in semiconductors, radiation effects on: see Semiconductors, effect of radiation on 405
Conductivity of ferromagnetic metals: see Conductivity at high temperatures 50
Conductivity, thermal: see Conductivity at high temperatures 48; Glass ceramics 147; Thermoelectric generation of electricity 484

Conductivity, thermal, electronic: see Thermoelectric generation of electricity 485
Conductivity, thermal lattice: see Conductivity at high temperatures 48; Thermoelectric generation of electricity 485
Conductivity, thermal, of semiconductors: see Peltier refrigerators 308
Cones, refractory: see Ceramics production, physics in 43
Configurations: see Proteins, structure of 338
Conformations: see Proteins, structure of 338
Conjugation, charge (C): see Charge conjugation 45
Conjugation, charge, invariance: see Charge conjugation 45
Conservation of parity: see Electric dipole moment of the neutron 74
Conservative system: see Hamilton's principle 161
Conserved-vector-current theory (C.V.C. theory): see Current algebras and their applications 52
Constant charge pacemaker: see Cardiac pacemaker, external 40
Constant current pacemaker: see Cardiac pacemaker, external 40
Constant voltage pacemaker: see Cardiac pacemaker, external 40
Constants, fundamental, of physics: see Quantum metrology 354
Contamination, surface: see Auger emission spectroscopy 26
Continuous casting: see Freeze-coating 130
Continuum theory, elastic: see Fracture of brittle materials 124
Contrast, diffraction: see High resolution electron microscopy 181
Contrast, image: see High resolution electron microscopy 182
Contrast, phase: see High resolution electron microscopy 182
Control and transport of mechanical power: see Electroviscous fluids, applications of 93
Controlled separation (elbow) amplifier: see Fluidics 121
Convergence: see Walsh functions 517
Convergent beam electron diffraction: see Dynamic electron diffraction, application of 65
Conversion efficiency, generator: see Electrogasdynamics 77
Converter, thermionic: see Space power supplies 438
Converter, thermoelectric: see Space power supplies 438
Convolution: see Optical information processing 303
Convolution theorem: see Interferometric spectrometers 208
Cooler, Peltier: see Thermoelectric cooling of infra-red detectors 481
Cooling effect, Peltier: see Peltier refrigerators 307
Cooling, electron: see Thermionic generation of electricity 478
Cooling, thermoelectric, of infra-red detectors: see Thermoelectric cooling of infra-red detectors 481

Cooper pairs: *see* Josephson effect 235
Coordinate system, geocentric: *see* Standard earth, geodetic parameters for 447
Copper: *see* Ion-neutralization spectroscopy 224
Copper, band structure of: *see* Photoemission and band structure 319
Copper, electroless: *see* Metal film preparation by electroless (chemical) reduction 280
Copper-formaldehyde system: *see* Metal film preparation by electroless (chemical) reduction 280
Corbino disc: *see* Magnetoresistivity 270
Cores, atomic: *see* Quantum mechanics in biochemistry 351
Coriolis force: *see* Gyroscopic coupler 153; Physical limnology 322
Coriolis interactions: *see* Submillimetre-wave lasers 456
Coriolis perturbation: *see* Submillimetre-wave lasers 457
Corona discharge: *see* Electrogasdynamics 75
Corona, solar: *see* Soft X-ray spectroscopy 420
Correlated colour temperature: *see* Daylight, power distribution in 56
Corrosion cracking, stress: *see* Fracture of brittle materials 129
Cosine transform: *see* Fourier transform spectroscopy 124; Interferometric spectrometers 207
Cosmic ray age: *see* Shock pressure in meteorites 407
Cosmic rays, galactic: *see* Ionosphere, recent observations of 225
Cotton yarns: *see* Textile industry, physics in 473
Coulomb forces: *see* Melting, modern theories of 275
Coulomb potential, screened: *see* Proton scattering microscopy 343
Counter, flow proportional: *see* Soft X-ray spectroscopy 413
Counter, phonon-phonon quantum: *see* Phonon avalanche and bottleneck 319
Counterflow diode: *see* Fluidics 122
Counting rates, background: *see* Whole body counting 524
Counting, whole body: *see* Whole body counting 522
Coupler, gyroscopic (gyrator): *see* Gyroscopic coupler (gyrator) 151
Coupling constant, beta-decay: *see* Bosons, intermediate 35
Coupling constant, quadrupole: *see* Nuclear quadrupole resonance 301
Coupling, deformation potential: *see* Acousto-electric effects 9
Coupling, electromagnetic: *see* Acousto-electric effects 9
Coupling, piezo-electric: *see* Acousto-electric effects 9
Covalent bonds: *see* Proteins, structure of 341
Cowell method: *see* Astrodynamics 21
CP invariance, violation of: *see* Charge conjugation 46
CPT invariance: *see* Charge conjugation 46
Crack: *see* Fracture of brittle materials 124
Crack, Griffith-: *see* Electrohydraulic crushing 82
Crack growth: *see* Fracture of brittle materials 129
Crack nucleation: *see* Fracture of brittle materials 127
Crack propagation: *see* Fracture of brittle materials 125
Crack tip: *see* Fracture of brittle materials 124
Crazing process: *see* High polymers, mechanical properties of 178
Creep: *see* High polymers, mechanical properties of 177; Textile industry, physics in 469
Criminal data file, computerized: *see* Law enforcement science and technology 255
Criminalistics: *see* Law enforcement science and technology 253
Criminology, science of: *see* Law enforcement science and technology 253
Critical point: *see* Melting, modern theories of 275
Critical snowmelt rate: *see* Hydrometeorology 190
Cross- and auto-correlation: *see* Optical information processing 303
Cross beta: *see* Proteins, structure of 339
Cross-section: *see* Ionization and attachment cross-section of atoms and molecules by electron impact 218
Cross-section, absorption: *see* Interplanetary dust 214
Cross-section, momentum transfer for low-energy electrons: *see* Momentum transfer cross-section for low-energy electrons: determination by swarm techniques 296
Cross-section, quark production: *see* Quarks 361
Cross-section, scattering: *see* Interplanetary dust 214
Cross-section, W meson production: *see* Bosons, intermediate 37
Cross-type telescopes: *see* Radio telescopes, the present position 384
Crushing by shock waves: *see* Electrohydraulic crushing 82
Crushing, electrohydraulic: *see* Electrohydraulic crushing 80
Crushing zone: *see* Electrohydraulic crushing 82
Crystal, absorption of a, in the far infra-red: *see* Far infra-red spectroscopy 107
Crystal, additively coloured: *see* Metallic colloids in non-metallic crystals 282
Crystal defects: *see* Thin films, nucleation and growth of 491
Crystal, germanium rectifier: *see* Submillimetre-wave methods 460
Crystal growing: *see* Glow discharge electron and ion guns for material processing 149
Crystal growth: *see* Glass ceramics 143
Crystal lattice periodicities, observation of: *see* High resolution electron microscopy 181
Crystal, liquid: *see* Liquid crystals and molecular structure 257
Crystal, multislice model of: *see* Dynamical n-beam theory of electron diffraction 69
Crystal non-metallic, metallic colloids in: *see* Metallic colloids in non-metallic crystals 282
Crystal orientation: *see* Proton scattering microscopy 344

Crystal phases: *see* Glass ceramics 143
Crystal, scattering of protons from a: *see* Proton scattering microscopy 342
Crystal, sodium iodide thallium activated: *see* Whole body counting 523
Crystal spectrometer: *see* Soft X-ray spectroscopy 413
Crystal, uniaxial: *see* Magnetic thin films 266
Crystalline form of polymers: *see* Polymer physics 332
Crystalline quartz: *see* Submillimetre-wave methods 461
Crystallizable polymers: *see* Textile industry, physics in 471
Crystallization: *see* Polymer physics 335
Crystallization of glasses: *see* Glass ceramics 143
Crystallization, selective: *see* Glass ceramics 144
Crystallization, spherulitic: *see* Polymer physics 335
Curie point: *see* Conductivity at high temperatures 51
Curing of surface coatings: *see* Radiation chemistry, industrial applications of 365
Current algebras: *see* Current algebras and their applications 51
Current commutators: *see* Current algebras and their applications 52
Current law: *see* Kirchhoff's laws (electrical circuits) 239
Current multiplication, avalanche: *see* Microwave solid state oscillators, recent 289
Current theory, conserved-vector- (C.V.C. theory): *see* Current algebras and their applications 52
Currents: *see* Hydrometeorology 192
Curtain-type aurorae: *see* Auroral electrons 27
Cut-off frequency of acoustic waveguide: *see* Acoustic waveguides 6
Cycle, Brayton: *see* Electrogasdynamics 75
Cyclotron, ion, whistlers: *see* Ionosphere, recent observations of 228
Cyclotron, proton, whistler: *see* Ionosphere, recent observations of 228
Cyclotron resonance, using acoustic waves: *see* Acousto-electric effects 9
Cylindrical lenses: *see* Optical information processing 303
Cystine: *see* Ultra-violet radiation, some biological effects of 508
Cytherian (Venusian) ionosphere: *see* Ionosphere, recent observations of 229
Cytoplasmic damage: *see* Ultra-violet radiation, some biological effects of 512

D region: *see* Ionosphere, recent observations of 225
Dam spillway: *see* Hydrometeorology 189
Damage, cytoplasmic: *see* Ultra-violet radiation, some biological effects of 512
Damage, radiation: *see* High resolution electron microscopy 183

Damper, gyroscopic vibration: *see* Gyroscopic coupler 152
Data or computer processing: *see* Lasers, applications of 245
Data processing: *see* Astrodynamics 19
Dating, radiometric (geochronology): *see* Geochronology (radiometric dating) 133
Daylight, colour temperature range of: *see* Daylight, power distribution in 57
Daylight, spectral power distribution in: *see* Daylight, power distribution in 55
Day-shift: *see* Metallic colloids in non-metallic crystals 285
d.c. pulse procedure: *see* Ultrasonic spectroscopy 504
DCN submillimetre-wave laser lines: *see* Submillimetre-wave lasers 458
de Broglie wavelength: *see* Liquid metals 262
"Dead loss": *see* Tungsten halogen lamps 494
Debris, cometary, theory: *see* Interplanetary dust 216
Debye equation: *see* Snoek spectra 411
Debye (relaxational) absorption by polar molecules: *see* Submillimetre radiation, applications of 452
Decay, beta: *see* Bosons, intermediate 35; Current algebras and their applications 53
Decay, bottlenecked: *see* Phonon avalanche and bottleneck 316
Decay constants: *see* Geochronology 133
Decay coupling constant, beta: *see* Bosons, intermediate 35
Decay, muon: *see* Neutretto 300
Decay, non-leptonic: *see* Current algebras and their applications 54
Decay of a radionuclide: *see* Geochronology 133
Decay of a W meson: *see* Bosons, intermediate 35
Decay rate, population: *see* Phonon avalanche and bottleneck 316
Decoder (digital-analogue convertor): *see* Pulse code modulation 346
Deep scattering layer (D.S.L.): *see* Reverberation under water 390
Defect-echo spectra: *see* Ultrasonic spectroscopy 506
Defect geometry, determination of: *see* Ultrasonic spectroscopy 505
Defects, crystal: *see* Thin films, nucleation and growth of 491
Defects, radiation: *see* Semiconductors, effect of radiation on 406
Deflected jet: *see* High lift devices 174
Deflected thrust: *see* High lift devices 174
Deformable particles: *see* Dispersion and fibre strengthening of materials 63
Deformation: *see* Microplasticity 286
Deformation, affine: *see* Polymer physics 337
Deformation bands: *see* High polymers, mechanical properties of 179
Deformation, chain: *see* Polymer physics 332
Deformation, liquid surface: *see* Ultrasonic imaging 500

Deformation modes of polymeric fibres: see Textile industry, physics in 471
Deformation of polymers, modes of: see Polymer physics 336
Deformation, piezo-electric: see Pockels effect 327
Deformation, plastic: see Brittleness in ceramics 37; Dispersion and fibre strengthening of materials 62; Fracture of brittle materials 125
Deformation potential constant: see Ultrasonic amplification in semiconductors 497
Deformation potential coupling: see Acousto-electric effects 9
Degassing: see Refractories, ultra-purification of 388
Delay line, ultrasonic: see Acoustic waveguides 5
Delay time for underwater discharge: see Electrohydraulic crushing 81
Demagnetization field: see Magnetic thin films 268
Demagnetizing field: see Permanent magnet technology 311
Demand pacing: see Cardiac pacemaker, external 39
Dendrites: see Refractories, ultra-purification of 387
Density gauging: see Radioisotopes in food processing 379
Density-of-states distributions: see Soft X-ray spectroscopy 417
Density of states, optical: see Photoemission and band structure 320
Deoxyribonucleic acid (DNA): see Quantum mechanics in biochemistry 353; Ultra-violet radiation, some biological effects of 506
Design flood: see Hydrometeorology 189
Detection and generation of submillimetre waves: see Submillimetre-wave methods 458
Detection systems, infra-red: see Airborne collision prevention devices 12
Detectivity, infra-red detectors': see Pyroelectric detector for infra-red radiation 349
Detector, far infra-red: see Far infra-red spectroscopy 105
Detector, Golay: see Far infra-red spectroscopy 105; Infra-red spectroscopy, recent developments in 202
Detector, infra-red, thermoelectric cooling of: see Thermoelectric cooling of infra-red detectors 481
Detector, lie: see Law enforcement science and technology 257
Detector, particle: see Streamer chambers 451
Detector, photoconductive: see Thermoelectric cooling of infra-red detectors 481
Detector, pneumatic: see Far infra-red spectroscopy 105
Detector, pyroelectric: see Infra-red cameras and infra-red TV 197
Detector, pyroelectric, of infra-red radiation: see Pyroelectric detector of infra-red radiation 347
Detector, quantum: see Infra-red cameras and infra-red TV 197
Detector, radiation: see Infra-red spectroscopy, recent developments in 204
Detector systems: see Whole body counting 523

Detector, thermal: see Infra-red cameras and infra-red TV 197; Submillimetre-wave methods 459
Diamond: see Polymer physics 331; Vitreous carbons, structure and properties of 513
Diamond pyramid hardness: see Glass ceramics 146
Dichroic mirror: see Solar cell 428
Dielectric breakdown strength: see Glass ceramics 147
Dielectric constant: see Far infra-red spectroscopy 108; Pockels effect 327
Dielectric loss: see Glass ceramics 145
Dielectric properties: see Textile industry, physics in 469
Dielectric, pulsed liquid: see High intensity pulsed radiation sources 166
Dielectric relaxation time, reciprocal: see Ultrasonic amplification in semiconductors 497
Differential correction, weighted least squares: see Astrodynamics 19
Differential resistivity, bulk negative, in semiconductors: see Microwave solid state oscillators, recent 289
Diffraction: see Holography 184
Diffraction contrast: see High resolution electron microscopy 181
Diffraction, electron: see Dynamic electron diffraction, application of 65; Dynamical n-beam theory of electron diffraction 69; Electron diffraction from periodic magnetic fields 86; Electron metallography, applications of 90
Diffraction error: see High resolution electron microscopy 179
Diffraction, low energy electron: see Auger emission spectroscopy 24; Ion neutralization spectroscopy 221
Diffraction, X-ray: see Dynamic electron diffraction, application of 67; Proteins, structure of 338
Diffractometer: see Optical information processing 303
Diffuser/nozzle loss coefficient: see Glandless reciprocating jet pump 142
Diffusion coefficient: see Diffusion in non-uniform media 58
Diffusion constant: see Solar cell 426
Diffusion energies, surface: see Thin films, nucleation and growth of 493
Diffusion in non-uniform media 58
Diffusion length: see Solar cell 423
Diffusion, surface: see Epitaxy, recent work on 98
Diffusivity: see Conductivity at high temperatures 50
Digital-analogue converter (decoder): see Pulse code modulation 346
Digital elements: see Fluidics 118
Digital information, optical: see Lasers, applications of 249
Dilatant pressure: see Singing sands 408
Dilation: see Singing sands 408
Dilution analysis, isotope: see Geochronology 135
Dilution refrigerators, He^3–He^4: see Polarized proton targets 329

Dimer: *see* Submillimetre radiation, applications of 452

Dimer, thymine: *see* Ultra-violet radiation, some biological effects of 507

Dimpled fracture: *see* Fracture of brittle materials 128

Diode, avalanching *pn*: *see* Impatt diode microwave generators 195

Diode, close-spaced vacuum: *see* Thermionic generation of electricity 477

Diode, counterflow: *see* Fluidics 122

Diode, field emission: *see* High intensity pulsed radiation sources 163

Diode, fluid: *see* Fluidics 122

Diode, IMPATT (Impact avalanche transit-time): *see* Microwave solid state oscillators, recent 288

Diode, jet: *see* Fluidics 122

Diode microwave generators, IMPATT: *see* IMPATT diode microwave generators 194

Diode, microwave rectifier: *see* Submillimetre-wave methods 459

Diode, nuclear "in-pile": *see* Thermionic generation of electricity 480

Diode, point-contact: *see* Submillimetre-wave methods 460

Diode, Read: *see* IMPATT diode microwave generators 194

Diode, semiconductor: *see* IMPATT diode microwave generators 194

Diode, turbulence: *see* Fluidics 122

Diode, vortex: *see* Fluidics 122

Diode, wall attachment: *see* Fluidics 122

Dip-forming: *see* Freeze-coating 130

Dipole array telescopes: *see* Radio telescopes, the present position 383

Dipole, electric, transitions: *see* Radioemissions from interstellar hydroxyl radicals 372

Dipole, induced: *see* Submillimetre radiation, applications of 452

Dipole moment, electric, of the neutron: *see* Electric dipole moment of the neutron 74

Dipole moments of molecules: *see* Submillimetre radiation, applications of 451

Dirac equation: *see* Charge conjugation 45

Direct (or vertical) transitions: *see* Photoemission and band structure 320

Direct photoenzymatic reactivation: *see* Ultra-violet radiation, some biological effects of 509

Dirichlet kernel: *see* Walsh functions 517

Disc, Corbino: *see* Magnetoresistivity 270

Disc of confusion: *see* High resolution electron microscopy 179

Discharge, cold cathode: *see* Glow discharge electron and ion guns for material processing 149

Discharge, corona: *see* Electrogasdynamics 75

Discharge, delay time for underwater: *see* Electro-hydraulic crushing 81

Discharge electron, glow, and ion guns for material processing: *see* Glow discharge electron and ion guns for material processing 148

Discharge, spark: *see* Laser-induced sparks 243

Discharge supplies, pulsed: *see* High intensity pulsed radiation sources 164

Discharge, warm cathode glow: *see* Glow discharge electron and ion guns for material processing 149

Discriminators, pulse: *see* Pulse code modulation 346

Disintegration rate: *see* Radiation quantities, systematics of 366

Dislocation: *see* Dispersion and fibre strengthening of materials 62; High resolution electron microscopy 181

Dislocation, edge: *see* Peierls stress 305; Thin films, nucleation and growth of 492

Dislocation, energy of a: *see* Peierls stress 306

Dislocation etch pits: *see* Microplasticity 287

Dislocation jog: *see* Dislocation kinks 62

Dislocation kinks 62

Dislocation, misfit: *see* Epitaxy, recent work on 99

Dislocation motion, slip by: *see* Brittleness in ceramics 37

Dislocation, screw: *see* Fracture of brittle materials 126; Peierls stress 307

Dislocations, grid of: *see* Thin films, nucleation and growth of 492

Disorder, thermal: *see* Liquid metals 262

Dispersion and fibre strengthening of materials 62

Dispersion, anisotropy: *see* Magnetic thin films 267

Dispersion, anomalous: *see* Infra-red spectroscopy, recent developments in 204

Dispersion, easy axis: *see* Magnetic thin films 267

Dispersion, electroviscous: *see* Electroviscous fluid 92

Dispersion, magnitude: *see* Magnetic thin films 267

Dispersion of acoustic wave: *see* Acoustic waveguides 6

Displacement energy: *see* Semiconductors, effect of radiation on 406

Dissipation function, Rayleigh's: *see* Hamilton's principle 162

Dissociation energies: *see* Thin films, nucleation and growth of 493

Dissociation, thermal: *see* Tungsten halogen lamps 496

Dissociative attachment: *see* Ionization and attachment cross-sections of atoms and molecules by electron impact 219

Distance measurements: *see* Lasers, applications of 248; Submillimetre radiation, application of 453

Distillation: *see* Refractories, ultra-purification of 388

Distortion, modulating envelope: *see* Hall effect, applications of 158

Distribution function: *see* Liquid metals 261

DNA (deoxyribonucleic acid): *see* Quantum mechanics in biochemistry 353; Ultra-violet radiation, some biological effects of 506

Domain, high field: *see* Microwave solid state oscillators, recent 289

Domain, magnetic: *see* Electron diffraction from periodic magnetic fields 88

Domain walls: *see* Magnetic thin films 268

Donor: *see* Semiconductors, effect of radiation on 405; Solar cell 423
Doping: *see* Solar cell 423; Thermoelectric generation of electricity 484
Doping, nuclear: *see* Semiconductors, effect of radiation on 407
Doppler broadening: *see* Radioemissions from interstellar hydroxyl radicals 373
Doppler frequency shift: *see* Airborne collision prevention devices 11
Doppler shift: *see* Lasers, applications of 248; Reverberation under water 391
Doryl resins: *see* Temperature resistant polymers 464
Dose, absorbed: *see* Radiation quantities, systematics of 367
Dose distribution factor: *see* Radiation quantities, systematics of 372
Dose equivalent: *see* Radiation quantities, systematics of 372
Dose, X-radiation: *see* High intensity pulsed radiation sources 167
Doublet, Kramers: *see* Far infra-red lasers 103
Downwash: *see* Jet flap 233
Drag: *see* Sailing, physics of 399
Drag, atmospheric: *see* Astrodynamics 21
Drag coefficient: *see* Astrodynamics 23
Drag induced, coefficient of: *see* Jet flap 234
Drape coefficient: *see* Textile industry, physics in 475
Drapery-type aurorae: *see* Auroral electrons 27
Draw ratio, natural: *see* High polymers, mechanical properties of 179
Drawing: *see* High polymers, mechanical properties of 178
Drift, littoral: *see* Hydrometeorology 192
Drift of carriers: *see* IMPATT diode microwave generators 194
Drift velocity: *see* Momentum transfer cross section for low energy electrons: determination by swarm techniques 297; Ultrasonic amplification in semiconductors 497
Drilling, electron beam: *see* Electron beam machining 84
Drude theory, classical: *see* Thermionic generation of electricity 476
Drum spinning: *see* Textile industry, physics in 474
Dust-cloud theory, primaeval: *see* Interplanetary dust 216
Dust-free zone: *see* Interplanetary dust 217
Dust, interplanetary: *see* Interplanetary dust 213
Dye-lasers: *see* Lasers, tunable 252
Dynamic electron diffraction, application of 65
Dynamical n-beam theory of electron diffraction 69
Dynamical Peierls stress: *see* Peierls stress 307
Dynamical two-beam approximation of electron diffraction: *see* Dynamic electron diffraction, application of 66

E models: *see* Ionospheres of Mars and Venus 230
E region: *see* Ionosphere, recent observations of 225

Earth, age of the: *see* Geochronology 133
Earth, standard, geodetic parameters for: *see* Standard earth, geodetic parameters for 447
Easy axis dispersion: *see* Magnetic thin films 267
Easy direction of magnetization: *see* Magnetic thin films 265
Eccentricity: *see* Astrodynamics 22
Echelette grating: *see* Far infra-red spectroscopy 105
Echelette grating spectrometer: *see* Infra-red spectroscopy, recent developments in 203
Echo, pulse-, images: *see* Ultrasonic imaging 498
Echo, ultrasonic: *see* Ultrasonic spectroscopy 505
Edge dislocation: *see* Peierls stress 305; Thin films, nucleation and growth of 492
Edgegraphy: *see* Infra-red cameras and infra-red TV 199
Edgetone amplifier: *see* Fluidics 121
Edser-Butler fringes: *see* Submillimetre-wave methods 461
Efficiency: *see* Solar cell 423; Thermionic generation of electricity 478; Thermoelectric generation of electricity 483
Efficiency, Carnot: *see* Thermoelectric generation of electricity 483
Efficiency, generator: *see* Electrogasdynamics 77
Efficiency, hydrodynamic or pumping: *see* Glandless reciprocating jet pump 142
Efficiency, junction: *see* Solar cell 423
Efficiency of gas compression: *see* Glandless reciprocating jet pump 142
Efficiency, pneumatic: *see* Glandless reciprocating jet pump 142
Efficiency, solar collection: *see* Solar cell 424
Efficiency, thermal: *see* Electrogasdynamics 78
Eightfold way: *see* Quarks 357
Ejection of atoms from the surface of solids: *see* Sputtering, theories of 445
Ekman motion: *see* Physical limnology 322
Elastic constants: *see* Microplasticity 286
Elastic constants of a fibre: *see* Textile industry, physics in 469
Elastic continuum theory: *see* Fracture of brittle materials 124
Elastic energy: *see* Microplasticity 286
Elastic fracture: *see* Fracture of brittle materials 125
Elastic moduli: *see* Glass ceramics 146
Elastic recovery: *see* Textile industry, physics in 469
Elastic scattering of electrons: *see* Momentum transfer cross-section for low energy electrons: determination by swarm techniques 296
Elastic scattering of fast neutrons: *see* Radiation quantities, systematics of 369
Elasticity: *see* Microplasticity 285
Elasticity, rubber: *see* Polymer physics 333
Elasticity, rubber, molecular theory of: *see* High polymers, mechanical properties of 176
Elasticity, rubber-like: *see* Rubber physics 393
Electric breakdown strength: *see* Electrogasdynamics 76

Electric dipole moment of the neutron 74
Electric dipole transitions: *see* Radioemissions from interstellar hydroxyl radicals 372
Electric field gradients in solids: *see* Nuclear quadrupole resonance 301
Electric network theory: *see* Kirchhoff's laws (Electrical circuits) 239
Electric quadrupole interaction: *see* Nuclear quadrupole resonance 301
Electric quadrupole moment, nuclear: *see* Nuclear quadrupole resonance 300
Electrical breakdown of water gaps: *see* Electrohydraulic crushing 81
Electrical conductivity: *see* Conductivity at high temperatures 48; Thermoelectric generation of electricity 484
Electrical noise: *see* Pyroelectric detector for infra-red radiation 347
Electrical porcelain: *see* Ceramics production, physics in 44
Electrical resistance: *see* Textile industry, physics in 469
Electrical resistance of electroviscous fluids: *see* Electroviscous fluids, applications of 92
Electrical resistance of liquid metals: *see* Liquid metals 263
Electrical resistivity: *see* Conductivity at high temperatures 50; Glass ceramics 145
Electrical units: *see* Quantum metrology 356
Electricity, thermionic generation of: *see* Thermionic generation of electricity 476
Electricity, thermoelectric generation of: *see* Thermoelectric generation of electricity 482
Electrodynamic transducer: *see* Gyroscopic coupler 154
Electrogasdynamic generator: *see* Electrogasdynamics 75
Electrogasdynamics (EGD): *see* Electrogasdynamics 75
Electrohydraulic crushing 80
Electrohydraulic forming of metals 83
Electrojet, equatorial: *see* Ionosphere, recent observations of 225
Electrokinetic effect: *see* Electroviscous fluids, applications of 93
Electroless (chemical) reduction, metal film preparation by: *see* Metal film preparation by electroless (chemical) reduction 277
Electroless copper: *see* Metal film preparation by electroless (chemical) reduction 280
Electroless plating: *see* Metal film preparation by electroless (chemical) reduction 277
Electroluminescent screens: *see* Infra-red cameras and infra-red TV 200
Electroluminescent semiconductor lasers: *see* Far infra-red lasers 103
Electrolytes, diffusion in: *see* Diffusion in non-uniform media 59
Electromagnetic coupling: *see* Acousto-electric effects 9

Electromagnetic interactions: *see* Charge conjugation 46; Current algebras and their applications 51
Electromagnetic radiation c, velocity of: *see* Quantum metrology 355
Electron: *see* Radiation chemistry, industrial applications of 363
Electron accelerators: *see* Radiation chemistry, industrial applications of 364
Electron affinity: *see* Metallic colloids in non-metallic crystals 283
Electron auger: *see* Satellites in soft X-ray spectra 404
Electron, auroral: *see* Auroral electrons 26
Electron beam: *see* High intensity pulsed radiation sources 166
Electron beam analysis: *see* Infra-red cameras and infra-red TV 199
Electron beam drilling: *see* Electron beam machining 85
Electron beam generating column: *see* Electron beam machining 84
Electron beam heating: *see* Refractories, ultra-purification of 388
Electron beam machining 83
Electron beam milling: *see* Electron beam machining 85
Electron cooling: *see* Thermionic generation of electricity 478
Electron density, ionospheric: *see* Ionosphere, recent observations of 226
Electron diffraction analysis, selected area: *see* Electron metallography, applications of 90
Electron diffraction, convergent beam: *see* Dynamic electron diffraction, application of 65
Electron diffraction, dynamic, application of: *see* Dynamic electron diffraction, application of 65
Electron diffraction, dynamical n-beam theory of: *see* Dynamical n-beam theory of electron diffraction 69
Electron diffraction, dynamical two-beam approximation of: *see* Dynamic electron diffraction, application of 66
Electron diffraction from periodic magnetic fields 85
Electron diffraction in antiferromagnetic materials: *see* Electron diffraction from periodic magnetic fields 88
Electron diffraction in Ferromagnetic materials: *see* Electron diffraction from periodic magnetic fields 88
Electron diffraction in type II superconductors: *see* Electron diffraction from periodic magnetic fields 89
Electron diffraction, low energy: *see* Auger emission spectroscopy 24; Ion-neutralization spectroscopy 221
Electron, elastic scattering of: *see* Momentum transfer cross-section for low-energy electrons: determination by swarm techniques 296
Electron-electron scattering: *see* Photoemission and band structure 319

Electron emission spectroscopy, secondary: *see* Auger emission spectroscopy 24
Electron energy analysis: *see* Auger emission spectroscopy 24
Electron escape probability: *see* Ion-neutralization spectroscopy 223
Electron field, second-quantized: *see* Charge conjugation 45
Electron, glow discharge, and ion guns for material processing: *see* Glow discharge electron and ion guns for material processing 148
Electron gun: *see* Electron beam machining 84
Electron-hole pair: *see* IMPATT diode microwave generator 194; Microwave solid state oscillators, recent 289
Electron impact, ionization and attachment cross-sections of atoms and molecules by: *see* Ionization and attachment cross-sections of atoms and molecules by electron impact 217
Electron, inelastic scattering of: *see* Momentum transfer cross-section for low energy electrons: determination by swarm techniques 297
Electron-ion interaction: *see* Liquid metals 261
Electron, ion, pairs: *see* Streamer chambers 448
Electron lenses: *see* High-resolution electron microscopy 179
Electron, low-energy, momentum transfer cross-section for: *see* Momentum transfer cross-section for low energy electrons: determination by swarm techniques 296
Electron metallography, applications of 89
Electron microscope: *see* Dynamic electron diffraction, application of 65; High resolution electron microscopy 179
Electron microscopy: *see* Epitaxy, recent work on 98
Electron microscopy, biological: *see* High resolution electron microscopy 183
Electron microscopy, reflection: *see* Electron metallography, applications of 91
Electron microscopy, scanning: *see* Electron metallography, applications of 91
Electron microscopy, transmission: *see* Electron metallography, applications of 89
Electron multiplier: *see* Soft X-ray spectroscopy 414
Electron, one, model: *see* Soft X-ray spectroscopy 417
Electron orbital energy level diagram: *see* Satellites in soft X-ray spectra 404
Electron-phonon scattering: *see* Photoemission and band structure 319
Electron, pi: *see* Quantum mechanics in biochemistry 351
Electron-probe microanalyser: *see* Soft X-ray spectroscopy 415
Electron, sigma: *see* Quantum mechanics in biochemistry 351
Electron spectrometers: *see* Auger emission spectroscopy 24
Electron spin resonance, conduction (CESR): *see* Metallic colloids in non-metallic crystals 283

Electron states in liquid metals: *see* Liquid metals 262
Electron, valence: *see* Quantum mechanics in biochemistry 351
Electron wave functions: *see* Soft X-ray spectroscopy 417
Electronic conductivity: *see* Metallic colloids in non-metallic crystals 284
Electronic resonance frequency: *see* Polarized proton targets 330
Electronic structure calculations: *see* Quantum mechanics in biochemistry 351
Electronic structure of alloys: *see* Soft X-ray spectroscopy 420
Electronic thermal conductivity: *see* Conductivity at high temperature 48; Thermoelectric generation of electricity 485
Electronic transition: *see* Ion-neutralization spectroscopy 220
Electronics, opto-: *see* Optical information processing 302
Electro-optic coefficients: *see* Pockels effect 327
Electro-optic effect: *see* Lasers, tunable 253
Electro-optic effect, linear: *see* Pockels effect 327
Electrostatic charging: *see* Textile industry, physics in 469
Electrostatic thruster: *see* Electrogasdynamics 79
Electrostatic transducer: *see* Gyroscopic coupler 154
Electrostrictive transducers: *see* Ultrasonic imaging 501
Electroviscosity: *see* Electroviscous fluid 91
Electroviscous dispersion: *see* Electroviscous fluid 92
Electroviscous fluid 91; *see also* Electroviscous fluids, applications of 92
Electroviscous fluids, applications of 92
Elementary particles: *see* Current algebras and their applications 51; Quarks 357
Elementary particles, nomenclature of 94
Elements, fluidic: *see* Fluidics 118
Ellipsoid, error: *see* Astrodynamics 19
Elongated molecules: *see* Liquid crystals and molecular structure 257
Emf, Seebeck: *see* Thermoelectric generation of electricity 482
Emission, photoelectric: *see* Metallic colloids in non-metallic crystals 282
Emission, radio, from interstellar hydroxyl radicals: *see* Radioemissions from interstellar hydroxyl radicals 372
Emission rate: *see* Radiation quantities, systematics of 366
Emission, secondary: *see* Infra-red cameras and infra-red TV 201
Emission spectra: *see* Soft X-ray spectroscopy 412
Emission spectroscopy, Auger: *see* Auger emission spectroscopy 23
Emission spectroscopy, secondary electron: *see* Auger emission spectroscopy 24
Emission, spontaneous: *see* Lasers, tunable 250
Emission, stimulated: *see* Lasers, tunable 250

Emission, submillimetre stimulated: *see* Submillimetre-wave lasers 458

Emission, thermal, OH: *see* Radioemissions from interstellar hydroxyl radicals 374

Emission, thermionic: *see* Metallic colloids in non-metallic crystals 284

Emission tube, field: *see* High intensity pulsed radiation sources 163

Emissive power: *see* Radiation quantities, systematics of 366

Emissivity at high temperatures 95

Emissivity, spectral: *see* Emissivity at high temperatures 95

Emissivity, temperature coefficient of: *see* Emissivity at high temperatures 96

Emittance, spectral: *see* Emissivity at high temperatures 95

Emitted Josephson radiation: *see* Josephson effect 237

Emitted source energy: *see* Radiation quantities, systematics of 366

Emulsion, photographic: *see* Infra-red cameras and infra-red TV 199; Ultrasonic imaging 498

Energy, anisotropy: *see* Magnetic thin films 266

Energy band calculations: *see* Soft X-ray spectroscopy 417

Energy-concentrating mirrors: *see* Space power supplies 438

Energy concentration: *see* Lasers, applications of 245

Energy, dissociation: *see* Thin films, nucleation and growth of 493

Energy, elastic: *see* Microplasticity 286

Energy, Fermi: *see* Josephson effect 235

Energy flux density: *see* Radiation quantities, systematics of 366

Energy-gap: *see* Solar cell 422; Thermoelectric generation of electricity 484

Energy imparted to matter: *see* Radiation quantities, systematics of 367

Energy level diagram: *see* Satellites in soft X-ray spectra 404

Energy level schemes, atomic: *see* Laser-induced **sparks** 244

Energy levels of OH: *see* Radioemissions from interstellar hydroxyl radicals 373

Energy levels (shells), atomic: *see* Auger emission spectroscopy 25

Energy levels, vibration-rotation: *see* Far infra-red lasers 102

Energy, mass, transfer coefficient: *see* Radiation quantities, systematics of 368

Energy, neutralization: *see* Ion-neutralization spectroscopy 220

Energy of a dislocation: *see* Peierls stress 306

Energy product: *see* Permanent magnet technology 309

Energy reradiation: *see* Reverberation under water 390

Energy, solar: *see* Space power supplies 435

Energy spectra of ejected atoms: *see* Sputtering, theories of 446

Energy storage: *see* Space power supplies 435

Energy, strain: *see* Rubber physics 393

Energy, strain, release rate: *see* Fracture of brittle materials 125

Energy, surface: *see* Brittleness in ceramics 37; Fracture of brittle materials 125

Energy, surface diffusion: *see* Thin films, nucleation and growth of 493

Energy, surface free: *see* Thin films, nucleation and growth of 491

Enthalpy: *see* Electrogasdynamics 75

Entropy: *see* Polymer physics 331; Rubber physics 393

Envelope, modulating, distortion: *see* Hall effect, applications of 158

Enzyme action: *see* Proteins, structure of 341

Enzyme, photoreactivating: *see* Ultra-violet radiation, some biological effects of 509

Ephemeris: *see* Astrodynamics 22

Ephemeris second: *see* Quantum metrology 354

Epilimnion: *see* Physical limnology 322

Epitaxial deposits, internal structure of: *see* Epitaxy, recent work on 99

Epitaxial growth: *see* Hall effect, applications of 159; Thin films, nucleation and growth of 492

Epitaxial planar devices: *see* Infra-red spectroscopy, recent developments in 206

Epitaxial temperature: *see* Epitaxy, recent work on 99; Thin films, nucleation and growth of 492

Epitaxy, *see* Thin films, nucleation and growth of 491

Epitaxy, recent work on 98

Epoch: *see* Airborne collision prevention devices 11; Astrodynamics 22

Epoch start pulse: *see* Airborne collision prevention devices 11

Equations of motion: *see* Hamilton's principle 161

Equator: *see* Proteins, structure of 341

Equatorial electrojet: *see* Ionosphere, recent observations of 225

Equilibrium, heterogeneous: *see* Metallic colloids in non-metallic crystals 283

Equinox, vernal: *see* Astrodynamics 23

Equivalent body of revolution, aeroplane: *see* Sonic boom, generation and suppression of 432

Equivalent plane wave reverberation level: *see* Reverberation under water 391

Error ellipsoid: *see* Astrodynamics 19

Errors, excitation: *see* Dynamical n-beam theory of electron diffraction 70

Escape probability, electron: *see* Ion-neutralization spectroscopy 223

Etalon, Fabry-Perot: *see* Infra-red spectroscopy, recent developments in 202; Luminosity of a spectrometer 264

Etch pits, dislocation: *see* Microplasticity 287

Ettingshausen effect: *see* Thermoelectric generation of electricity 486

Evanescent modes of acoustic waveguide: *see* Acoustic waveguides 6

Evaporated film junction: *see* Josephson effect 238

Evaporated Hall plates: *see* Hall effect, applications of 159
Evaporation: *see* Hydrometeorology 192
Evaporation from lakes: *see* Physical limnology 323
Evaporation, thermal, of atoms: *see* Sputtering, theories of 445
Evaporation, vacuum: *see* Thin films, nucleation and growth of 491
Evaporation, vacuum, technique: *see* Epitaxy, recent work on 98
Evaporography: *see* Infra-red cameras and infra-red TV 198
Evolution, organic: *see* Geochronology 133
Ewald sphere: *see* Dynamical n-beam theory of electron diffraction 70
Exchange field: *see* Far infra-red spectroscopy 108
Excision-resynthesis repair: *see* Ultra-violet radiation, some biological effects of 510
Excitation errors: *see* Dynamical n-beam theory of electron diffraction 70
Excitation, optical: *see* Lasers, tunable 252
Excitation, phonon: *see* Phonon avalanche and bottleneck 317
Exclusion principle, Pauli: *see* Charge conjugation 45; Josephson effect 235; Liquid metals 262; Melting, modern theories of 275
Expansion, coefficient of thermal: *see* Glass ceramics 145
Exploding wires: *see* Electrohydraulic forming of metals 81; Electrohydraulic crushing 83
Exposure: *see* Radiation quantities, systematics of 367
External cardiac pacemaker: *see* Cardiac pacemaker, external 39
External charge injector: *see* Electrogasdynamics 78
Extraction replica, carbon: *see* Electron metallography, applications of 90
Extrinsic semiconductor: *see* Thermoelectric generation of electricity 484

f and g series: *see* Astrodynamics 23
F centres: *see* Metallic colloids in non-metallic crystals 282
F region: *see* Ionosphere, recent observations of 225
F_1-layer: *see* Ionospheres of Mars and Venus 230
F_1 models: *see* Ionospheres of Mars and Venus 230
F_2 models: *see* Ionospheres of Mars and Venus 230
Fabrics, non-woven: *see* Textile industry, physics in 472
Fabry-Perot etalon: *see* Infra-red spectroscopy, recent developments in 202; Luminosity of a spectrometer 264
Fabry-Perot interferometer: *see* Far infra-red lasers 103
Fabry-Perot interferometric spectrometer: *see* Interferometric spectrometers 212
False-twist spindle: *see* Textile industry, physics in 475
Fan in capability: *see* Fluidics 121

Fan out capability: *see* Fluidics 118
Far field: *see* Sonic boom, generation and suppression of 431
Far infra-red, absorption of liquids, crystals and glasses in: *see* Far infra-red spectroscopy 107
Far infra-red detectors: *see* Far infra-red spectroscopy 105
Far infra-red laser spectrometer: *see* Far infra-red lasers 104
Far infra-red lasers 101
Far infra-red lasers, solid state: *see* Far infra-red lasers 103
Far infra-red molecular spectroscopy: *see* Far infra-red spectroscopy 106
Far infra-red, refractive index measurement in the: *see* Interferometric spectrometers 211
Far infra-red sources: *see* Far infra-red spectroscopy 105
Far infra-red spectroscopy 104
Far ultra-violet: *see* Ultra-violet radiation, some biological effects of 506
Far ultra-violet molecular processes, biological spectrum of: *see* Ultra-violet radiation, some biological effects of 511
Far ultra-violet photoproducts: *see* Ultra-violet radiation, some biological effects of 506
Faraday: *see* Quantum metrology 356
Faraday rotation: *see* Ionosphere, recent observations of 228
Fast-neutron reactors: *see* Space power supplies 436
Fast neutrons, elastic scattering of: *see* Radiation quantities, systematics of 369
Fast passage: *see* Phonon avalanche and bottleneck 318
Fast passage, adiabatic: *see* Fast passage in paramagnetic resonance 109
Fast passage in paramagnetic resonance 109
Fatigue: *see* High resolution electron microscopy 183
Fatigue fracture: *see* Fracture of brittle materials 129
Fatigue, metal: *see* Microplasticity 286
Feather keratin: *see* Proteins, structure of 339
Feedback, negative: *see* Cardiac pacemaker, external 40
Fellgett's advantage: *see* Far infra-red spectroscopy 105; Multiplex principle (of Fellgett) 299
Fellgett's multiplex principle: *see* Multiplex principle (of Fellgett) 298
Fermi-Dirac statistics: *see* Josephson effect 235; Thermionic generation of electricity 476
Fermi distribution function: *see* Soft X-ray spectroscopy 416
Fermi edge: *see* Soft X-ray spectroscopy 419
Fermi energy: *see* Josephson effect 235
Fermi gas: *see* Josephson effect 235
Fermi level: *see* Photoemission and band structure 320; Solar cell 423; Thermionic generation of electricity 476; Thermoelectric generation of electricity 485
Fermi wave number: *see* Liquid metals 262

Fermion: *see* Elementary particles, nomenclature of 95
Feroba magnets, properties of: *see* Permanent magnet technology 311
Ferrimagnetism: *see* Far infra-red spectroscopy 108
Ferrite thin films: *see* Magnetic thin films 268
Ferroelectric generators: *see* Space power supplies 438
Ferroelectric retina: *see* Infra-red cameras and infra-red TV 200
Ferroelectricity: *see* Far infra-red spectroscopy 109
Ferrohydrodynamic levitation: *see* Ferrohydrodynamics 116
Ferrohydrodynamics 111
Ferromagnetic films: *see* Magnetic thin films 268
Ferromagnetic fluid: *see* Ferrohydrodynamics 112
Ferromagnetic materials: *see* Piezomagnetic effect 325
Ferromagnetic metals, conductivities of: *see* Conductivity at high temperatures 50
Ferromagnetic materials, electron diffraction in: *see* Electron diffraction from periodic magnetic fields 88
Ferromagnetism: *see* Far infra-red spectroscopy 108
Feynman diagram: *see* Current algebras and their applications 53
Fibre assemblies: *see* Textile industry, physics in 472
Fibre composites, high-strength: *see* Dispersion and fibre strengthening of materials 64
Fibre physics: *see* Textile industry, physics in 468
Fibre reinforced metal: *see* Freeze-coating 132
Fibre strengthening of materials, dispersion and: *see* Dispersion and fibre strengthening of materials 62
Fibre structure and properties: *see* Textile industry, physics in 467
Fibrous proteins: *see* Proteins, structure of 338
Fick's Law: *see* Diffusion in non-uniform media 58
Field, acousto-electric: *see* Acousto-electric effects 8
Field, anisotropy: *see* Magnetic thin films 266
Field, boson: *see* Bosons, intermediate 35
Field, coercive: *see* Magnetic thin films 268
Field, demagnetization: *see* Magnetic thin films 268
Field, demagnetizing: *see* Permanent magnet technology 311
Field emission diode: *see* High intensity pulsed radiation sources 163
Field emission tube: *see* High intensity pulsed radiation sources 163
Field, exchange: *see* Far infra-red spectroscopy 108
Field, far: *see* Sonic boom, generation and suppression of 431
Field gradients, electric, in solids: *see* Nuclear quadrupole resonance 301
Field-induced anisotropy (or M-induced anisotropy): *see* Magnetic thin films 266
Field-induced polarization: *see* Pockels effect 327
Field-induced rheological properties of electroviscous fluids: *see* Electroviscous fluids, applications of 93
Field, magnetic, boundary conditions: *see* Ferrohydrodynamics 113

Field, near: *see* Sonic boom, generation and suppression of 431
Field, periodic magnetic, electron diffraction from a: *see* Electron diffraction from periodic magnetic fields 86
Field plates: *see* Magnetoresistivity 271
Field, second-quantized electron: *see* Charge conjugation 45
Field, $W^{\pm}$: *see* Bosons, intermediate 35
Figure of merit: *see* Peltier refrigerators 307; Thermoelectric cooling of infra-red detectors 481
Filament formation: *see* Ultra-violet radiation, some biological effects of 511
Filament lamps, tungsten: *see* Tungsten halogen lamps 494
Film growth, structural disorder introduced during: *see* Thin films, nucleation and growth of 492
Film, metal, preparation: *see* Metal film preparation by electroless (chemical) reduction 277
Films, ferrite thin: *see* Magnetic thin films 268
Films, ferromagnetic: *see* Magnetic thin films 268
Films, magnetic, poly crystalline: *see* Magnetic thin films 265
Films, magnetic thin: *see* Magnetic thin films 265
Films, thin, nucleation and growth of: *see* Nucleation and growth of thin films 490
Films, thin, proton scattering from: *see* Proton scattering microscopy 344
Filter, germanium: *see* Infra-red cameras and infra-red TV 197
Filter, interference: *see* Infra-red spectroscopy, recent developments in 202
Filter, low-pass: *see* Infra-red spectroscopy, recent developments in 202; Pulse code modulation 345
Filter, reflection: *see* Far infra-red spectroscopy 106; Submillimetre-wave methods 462
Filter, Reststrahlen: *see* Infra-red spectroscopy, recent developments in 202
Filter, transmission: *see* Submillimetre-wave methods 462
Filtering, Kalman: *see* Astrodynamics 20
Filtering, spatial: *see* Lasers, applications of 247; Optical information processing 303
Fine structure: *see* Radioemissions from interstellar hydroxyl radicals 372
Fine structure of ionization cross-section: *see* Ionization and attachment cross-sections of atoms and molecules by electron impact 219
Firing of ceramics: *see* Ceramics production, physics in 43
Fission: *see* Radiation chemistry, industrial applications of 363
Fission-fragment generator: *see* Space power supplies 436
Fission products: *see* Thermionic generation of electricity 480
Fission reactors: *see* Space power supplies 436
Fission track dating, nuclear: *see* Geochronology 137
Flaps: *see* High lift devices 170; Jet flap 232
Flare, solar: *see* Soft X-ray spectroscopy 421

Flash radiation, supervoltage: *see* High intensity pulsed radiation sources 167
Flash radiography: *see* High intensity pulsed radiation sources 167
Flash X-ray equipments: *see* High intensity pulsed radiation sources 163
Flexural waves: *see* Acoustic waveguides 7
Flight, supersonic: *see* High lift devices 174
Floating zone method: *see* Refractories, ultra-purification of 387
Flood, design: *see* Hydrometeorology 189
Flood, probable maximum: *see* Hydrometeorology 189
Flow, fluid: *see* Fluidics 117
Flow gain: *see* Fluidics 118
Flow, plastic: *see* Dispersion and fibre strengthening of materials 62; Fracture of brittle materials 124; Microplasticity 286
Flow proportional counter: *see* Soft X-ray spectroscopy 413
Flow resistance, linear: *see* Fluidics 122
Flow stress: *see* Peierls stress 306
Fluid acoustic waveguide: *see* Acoustic waveguides 5
Fluid amplifier, proportional: *see* Fluidics 119
Fluid diode: *see* Fluidics 122
Fluid, electroviscous: *see* Electroviscous fluid 92; Electroviscous fluids, applications of 93
Fluid energy mill: *see* Ceramics production, physics in 42
Fluid, ferromagnetic: *see* Ferrohydrodynamics 112
Fluid flow: *see* Fluidics 117
Fluid jet, magnetic: *see* Ferrohydrodynamics 114
Fluid magnetic pressure: *see* Ferrohydrodynamics 113
Fluid media, magnetizable: *see* Ferrohydrodynamics 111
Fluid seals, magnetic: *see* Ferrohydrodynamics 115
Fluid surface stability, magnetic: *see* Ferrohydrodynamics 114
Fluidic elements: *see* Fluidics 118
Fluidic logic devices: *see* Fluidics 118
Fluidics 117
Fluorescence: *see* Infra-red cameras and infra-red TV 199
Fluorides, tungsten: *see* Tungsten halogen lamps 495
Fluorocarbon polymers: *see* Temperature-resistant polymers 463
Flux concentrator: *see* Hall effect, applications of 156
Flux density, energy: *see* Radiation quantities, systematics of 366
Flux density, particle: *see* Radiation quantities, systematics of 366
Flux density, recoil: *see* Permanent magnet technology 312
Flux, luminous: *see* Luminosity of a spectrometer 264
Flux, magnetic: *see* Magnetoresistivity 272
Flux quantization, magnetic: *see* Josephson effect 235
Flux quantum: *see* Josephson effect 236
Fluxmeter: *see* Josephson effect 238

"Fly's eye lens": *see* Lasers, applications of 247
Focal-conic texture of smectic liquid crystals: *see* Liquid crystals and molecular structure 257
Focused collision sequences: *see* Sputtering, theories of 445
Folded band aurorae: *see* Auroral electrons 27
Food processing, radioisotopes in: *see* Radioisotopes in food processing 377
Force, anharmonic: *see* Melting, modern theories of 276
Force, Coriolis: *see* Gyroscopic coupler 153; Physical Limnology 322
Force, interionic: *see* Liquid metals 261
Force, intermolecular: *see* Polymer physics 333; Quantum mechanics in biochemistry 351
Force, Lorentz: *see* Electron diffraction from periodic magnetic fields 86
Force, Peierls: *see* Peierls stress 305
Force, quark-quark: *see* Quarks 361
Force, repulsive: *see* Melting, modern theories of 275
Force, van der Waals: *see* Liquid metals 262
Forensic science: *see* Law enforcement, science and technology 253
Form factors, quark: *see* Quarks 360
Formaldehyde-copper system: *see* Metal film preparation by electroless (chemical) reduction 280
Formant synthesizer: *see* Speaking machines 442
Formants: *see* Speaking machines 440
Fourier spectrometer: *see* Submillimetre-wave lasers 456; Submillimetre-wave methods 461
Fourier transform: *see* Optical information processing 303
Fourier transform holograms: *see* Holography 185
Fourier transform spectroscopy 122; *see also* Far infra-red spectroscopy 105; Submillimetre radiation, applications of 452
Fourier transformation: *see* Dynamical *n*-beam theory of electron diffraction 69; Interferometric spectrometers 207
Fourier, Walsh-, series: *see* Walsh functions 517
Fourier's inversion theorem: *see* Fourier transform spectroscopy 123
Fowler flap: *see* High lift devices 171
Fracture: *see* Dispersion and fibre strengthening of materials 62; Fracture of brittle materials 124
Fracture and yield of polymers: *see* High polymers, mechanical properties of 178
Fracture, Hopkinson: *see* Electrohydraulic crushing 82
Fracture propagation: *see* Glass ceramics 146
Fracture stress: *see* Brittleness in ceramics 37
"Fracture, work of": *see* Fracture of brittle materials 125
Fraunhofer hologram: *see* Holography 185
Fraunhofer lines: *see* Daylight, power distribution in 55
Frazil ice: *see* Hydrometeorology 191
Free energy, surface: *see* Thin films, nucleation and growth of 491

Free path, momentum transfer: *see* Momentum transfer cross-section for low energy electrons: determination by swarm techniques 298
Free valence: *see* Quantum mechanics in biochemistry 352
Freeze-coating 130
Freezing: *see* Physical limnology 322
Frequency: *see* Quantum metrology 354
Frequency analyses of rainfall: *see* Hydrometeorology 188
Frequency-dependence of ultrasonic attenuation in a material: *see* Ultrasonic spectroscopy 502
Frequency distribution, Gaussian: *see* Interferometric spectrometers 208
Frequency, Larmor precession: *see* Thermomagnetic gas torque (The Scott effect) 488
Frequency, nucleation: *see* Thin films, nucleation and growth of 491
Frequency, plasma: *see* Satellites in soft X-ray spectra 405
Frequency shift, Doppler: *see* Airborne collision prevention devices 11
Fresnel hologram: *see* Holography 185
Fresnel number: *see* Submillimetre-wave lasers 455
Friction angle: *see* Singing sands 409
Friction, internal: *see* Snoek spectra 411
Friedel's law: *see* Dynamical n-beam theory of electron diffraction 72
Fringed micelle model: *see* Polymer physics 335; Textile industry, physics in 471
Fringes, Edser–Butler: *see* Submillimetre-wave methods 461
Fringes, interference: *see* Interferometric spectrometers 207
Fringes, Moiré: *see* Interferometric spectrometers 207; Quantum metrology 357
Fringes, polychromatic: *see* Submillimetre radiation, applications of 454
Frozen carbon dioxide: *see* Weather control 522
Fuel: *see* Thermionic generation of electricity 480
Fuel cells: *see* Space power supplies 439
Fuels, nuclear, purification of: *see* Radiation chemistry, industrial application of 364
Fundamental constants of physics: *see* Quantum metrology 354

G-value: *see* Radiation chemistry, industrial applications of 364
(Ga-Ge) composition: *see* Shock pressure in meteorites 407
Gain, flow: *see* Fluidics 118
Gain, power: *see* Fluidics 118
Gain, pressure: *see* Fluidics 118
Galactic centre: *see* Radioemissions from interstellar hydroxyl radicals 374
Galactic cosmic rays: *see* Ionosphere, recent observations of 225
Galena dating: *see* Geochronology 137
Galvanizing: *see* Freeze-coating 130

Galvanomagnetic effects: *see* Magnetoresistivity 270
Gamma-ray constant, specific: *see* Radiation quantities, systematics of 366
Gamma-rays: *see* Radiation chemistry, industrial applications of 363
Gap, band: *see* Soft X-ray spectroscopy 418
Gap, energy-: *see* Solar cell 422
Garbuny's image convertor: *see* Infra-red cameras and infra-red TV 201
Gas capacitor, coaxial: *see* High intensity pulsed radiation sources 165
Gas cell, multi-reflection: *see* Infra-red spectroscopy, recent developments in 206
Gas collision cross section: *see* Thermomagnetic gas torque (The Scott effect) 490
Gas, compressed (non-polar): *see* Submillimetre radiation, applications of 452
Gas compression, efficiency of: *see* Glandless reciprocating jet pump 142
Gas, Fermi: *see* Josephson effect 235
Gas in a magnetic field: *see* Thermomagnetic gas torque (The Scott effect) 488
Gas, laser-induced breakdown of a: *see* Laser-induced sparks 242
Gas lasers: *see* Infra-red spectroscopy, recent developments in 203
Gas lasers, rare: *see* Far infra-red lasers 101
Gas lasers, submillimetre-wave: *see* Submillimetre-wave methods 459
Gas molecules, collision frequency of: *see* Thermomagnetic gas torque (The Scott effect) 488
Gas, rotational spectrum of a: *see* Submillimetre radiation, applications of 451
Gas torque, thermomagnetic: *see* Thermomagnetic gas torque (The Scott effect) 488
Gas velocity, measurement of: *see* Laser-induced sparks 244
Gaussian frequency distribution: *see* Interferometric spectrometers 208
Generalized Bernoulli theorem: *see* Ferrohydrodynamics 112
Generating column, electron beam: *see* Electron beam machining 84
Generation and detection of submillimetre waves: *see* Submillimetre-wave methods 458
Generation and suppression of sonic boom: *see* Sonic boom, generation and suppression of 431
Generation of electricity, thermionic: *see* Thermionic generation of electricity 476
Generation of electricity, thermoelectric: *see* Thermoelectric generation of electricity 482
Generator, cascade thermoelectric: *see* Thermoelectric generation of electricity 487
Generator conversion efficiency: *see* Electrogasdynamics 77
Generator, electrogasdynamic: *see* Electrogasdynamics 75
Generator, ferroelectric: *see* Space power supplies 438
Generator, fission-fragment: *see* Space power supplies 436

Generator, harmonic: *see* Submillimetre-wave methods 459
Generator isentropic efficiency: *see* Electrogasdynamics 77
Generator, Marx: *see* Streamer chambers 449
Generator, Marx-Surge: *see* High intensity pulsed radiation sources 165
Generator, microwave, avalanche: *see* IMPATT diode microwave generators 195
Generator, microwave, IMPATT diode: *see* IMPATT diode microwave generators 194
Generator, radioisotope: *see* Space power supplies 437; Thermoelectric generation of electricity 487
Generator, single sideband: *see* Hall effect, applications of 158
Generators, RIPPLE: *see* Thermoelectric generation of electricity 487
Generators, SNAP: *see* Thermoelectric generation of electricity 487
Geocentric coordinate system: *see* Standard earth, goedetic parameters for 447
Geochronology: *see* Geochronology (radiometric dating) 133
Geodetic parameters for the standard earth: *see* Standard earth, geodetic parameters for 447
Geoid: *see* Standard earth, geodetic parameters for 447
Geologic time scale: *see* Geochronology 138
Geomagnetic or Appleton anomaly: *see* Ionosphere, recent observations of 225
Geometric resonance, using acoustic waves: *see* Acousto-electric effects 9
Geometrical kinks: *see* Peierls stress 307
Geopotential: *see* Standard earth, geodetic parameters for 448
Geostrophic motion: *see* Physical limnology 322
Germanium: *see* Magnetoresistivity 269
Germanium bolometer: *see* Far infra-red spectroscopy 105
Germanium filter: *see* Infra-red cameras and infra-red TV 197
Germanium rectifier crystals: *see* Submillimetre-wave methods 460
Ghosting: *see* Sailing, physics of 403
Gibbsian method for preliminary orbit determination: *see* Astrodynamics 18
Glandless reciprocating jet pump 140
Glass ceramics 143
Glass composition: *see* Glass ceramics 144
Glass region: *see* Melting, modern theories of 277
Glass, rubber-, transition: *see* Polymer physics 334
Glass transition temperature: *see* High polymers, mechanical properties of 176; Rubber physics 394
Glasses, absorption of, in the far infra-red: *see* Far infra-red spectroscopy 107
Glasses, coloured: *see* Glass ceramics 144
Glasses, crystallization of: *see* Glass ceramics 143
Glasses, immiscibility in: *see* Glass ceramics 143

Glassy carbons: *see* Vitreous carbons, structure and properties of 513
Glassy form of polymers: *see* Polymer physics 332
Glassy state: *see* High polymers, mechanical properties of 176
Glide plane: *see* Dynamical n-beam theory of electron diffraction 72; Peierls stress 305
Globar: *see* Infra-red spectroscopy, recent developments in 202
Globular proteins: *see* Proteins, structure of 338
Glottis: *see* Speaking machines 440
Glow discharge electron and ion guns for material processing 148
Glow-to-arc transition: *see* Glow discharge electron and ion guns for material processing 149
Golay cell: *see* Submillimetre-wave lasers 456; Submillimetre-wave methods 459
Golay detector: *see* Far infra-red spectroscopy 105; Infra-red spectroscopy, recent developments in 202
Gold: *see* Proton scattering microscopy 342; Sputtering, theories of 445
Gradient motion: *see* Physical limnology 322
Grain boundary: *see* Fracture of brittle materials 126; Thin films, nucleation and growth of 493
Grain boundary fracture: *see* Fracture of brittle materials 128
Grains, heterogeneous nucleation of: *see* Thin films, nucleation and growth of 492
Graphite: *see* Polymer physics 332; Vitreous carbons, structure and properties of 513
Graphitization: *see* Vitreous carbons, structure and properties of 514
Graphitized coke: *see* Vitreous carbons, structure and properties of 514
Grating, amplitude: *see* Holography 184
Grating, echelette: *see* Far infra-red spectroscopy 105
Grating/interferometer comparison: *see* Interferometric spectrometers 211
Grating interferometer, lamellar: *see* Interferometric spectrometers 211
Grating, lamellar: *see* Far infra-red spectroscopy 105
Grating, lamellar, interferometer: *see* Submillimetre-wave methods 461
Grating, phase: *see* Electron diffraction from periodic magnetic fields 86; Holography 184
Grating, phase, series: *see* Dynamical n-beam theory of electron diffraction 70
Grating, reflection: *see* Soft X-ray spectroscopy 414
Grating reflectors: *see* Infra-red spectroscopy, recent developments in 202
Grating spectrometer: *see* Interferometric spectrometers 211; Infra-red spectroscopy, recent developments in 202; Luminosity of a spectrometer 264; Submillimetre-wave lasers 456; Submillimetre-wave methods 461
Grating spectrometer, echelette: *see* Infra-red spectroscopy, recent developments in 203
Grating spectrometer, grazing incidence: *see* Soft X-ray spectroscopy 413

35*

Grating spectroscopy: *see* Far infra-red spectroscopy 105

Gravity, acceleration due to: *see* Quantum metrology 356

Grazing incidence grating spectrometers: *see* Soft X-ray spectroscopy 413

Grid of dislocations: *see* Thin films, nucleation and growth of 492

Gridded cells: *see* Solar cell 426

Griffith-crack: *see* Electrohydraulic crushing 82

Ground bounce technique: *see* Airborne collision prevention devices 10

Ground effect on jet flap: *see* Jet flap 233

Group, orthonormal: *see* Walsh functions 518

Group velocity in acoustic waveguides: *see* Acoustic waveguides 6

Growth and nucleation of surface layers: *see* Epitaxy, recent work on 98

Growth and nucleation of thin films: *see* Thin films, nucleation and growth of 490

Growth, crack: *see* Fracture of brittle materials 129

Growth, crystal: *see* Glass ceramics 143

Growth, epitaxial: *see* Hall effect, applications of 159; Thin films, nucleation and growth of 492

Growth, recrystallization during: *see* Epitaxy, recent work on 99

Gun, broad beam ion: *see* Glow discharge electron and ion guns for material processing 150

Gun, cathode: *see* Refractories, ultra-purification of 389

Gun, electron: *see* Electron beam machining 85

Gun, glow discharge electron, and ion, for material processing: *see* Glow discharge electron and ion guns for material processing 148

Gunn oscillation: *see* Microwave solid state oscillators, recent 290

Gunn oscillator: *see* Microwave solid state oscillators, recent 288

Gutzeit mechanism: *see* Metal film preparation by electroless (chemical) reduction 280

Gyrator (gyroscopic coupler): *see* Gyroscopic coupler (gyrator) 151

Gyromagnetic ratio: *see* Thermomagnetic gas torque (The Scott effect) 488

Gyromagnetic ratio of the proton: *see* Quantum metrology 356

Gyroscope, laser (ring laser): *see* Lasers, applications of 248

Gyroscopic coupler (gyrator) 151

Gyroscopic pendulum: *see* Gyroscopic coupler 152

Gyroscopic vibration damper: *see* Gyroscopic coupler 152

Gyrostat: *see* Gyroscopic coupler 151

Hadron: *see* Current algebras and their applications 52; Elementary particles, nomenclature of 95; Quarks 357

Haemoglobin: *see* Proteins, structure of 341

Hagen–Rubens relation: *see* Emissivity at high temperatures 95

Half-life: *see* Geochronology 133

Half-wave retardation voltage: *see* Pockels effect 327

Halides, alkali: *see* Far infra-red spectroscopy 107

Hall angle: *see* Magnetoresistivity 270

Hall coefficient: *see* Hall effect, applications of 156

Hall constant: *see* Hall effect, applications of 156

Hall effect: *see* Magnetoresistivity 269

Hall effect, applications of 156

Hall plates, evaporated: *see* Hall effect, applications of 159

Hall voltage: *see* Hall effect, applications of 157; Magnetoresistivity 270; Thermoelectric generation of electricity 486

Halogen, tungsten, lamps: *see* Tungsten halogen lamps 494

Hamilton's principle 161

Hard direction of magnetization: *see* Magnetic thin films 265

Hardening, precipitation: *see* Permanent magnet technology 311

Harding's image converter: *see* Infra-red cameras and infra-red TV 199

Hardness, diamond pyramid: *see* Glass ceramics 146

Harmonic generators: *see* Submillimetre-wave methods 459

HCN laser: *see* Submillimetre radiation, applications of 453; Submillimetre-wave lasers 456; Submillimetre-wave methods 459

HCN laser lines: *see* Far infra-red lasers 102

He^3–He^4 dilution refrigerators: *see* Polarized proton targets 329

Heat capacity: *see* Melting, modern theories of 276

Heat conduction: *see* Conductivity at high temperatures 48; Freeze-coating 130

Heat pipe: *see* Thermionic generation of electricity 480

Heating, electron beam: *see* Refractories, ultra-purification of 388

Heating, hollow cathode: *see* Refractories, ultra-purification of 388

Heating, induction: *see* Refractories, ultra-purification of 388

Heating, Joule: *see* Thermoelectric generation of electricity 483

Heavy ion forming: *see* Electrogasdynamics 78

Heeling: *see* Sailing, physics of 398

Helicity 163

Helium: *see* Far infra-red lasers 101; Geochronology 136; Laser-induced sparks 243

Helium 4: *see* Melting, modern theories of 276

Helium, liquid: *see* Josephson effect 238

Helium, superfluid: *see* Josephson effect 235

Helix, alpha-: *see* Proteins, structure of 338

Hereditary materials: *see* High polymers, mechanical properties of 178

Hertzian oscillator: *see* Far infra-red spectroscopy 104

Heteroaromatic polymers: see Temperature resistant polymers 465
Heterodyne detection: see Far infra-red spectroscopy 105
Heterogeneous equilibrium: see Metallic colloids in non-metallic crystals 283
Heterogeneous nucleation: see Glass ceramics 143
Heterogeneous nucleation of grains: see Thin films, nucleation and growth of 492
Hexamine: see Pockels effect 329
High energy scattering of hadrons: see Quarks 360
High field domain: see Microwave solid state oscillators, recent 289
High intensity pulsed radiation sources 163
High lift devices 170
High polymers, mechanical properties of 175
High pressure mercury lamp: see Infra-red spectroscopy, recent developments in 202
High resolution electron microscopy 179
High-strength fibre composites: see Dispersion and fibre strengthening of materials 64
High temperatures, conductivity at: see Conductivity at high temperatures 48
High temperatures, emissivity at: see Emissivity at high temperatures 95
Hole-electron pair: see IMPATT diode microwave generators 194; Microwave solid state oscillators, recent 289
Hole punching: see Lasers, applications of 246
Hole theory: see Charge conjugation 45
Hollow cathode heating: see Refractories, ultra-purification of 388
Hologram: see Acoustic holography 2; Holography 184; Lasers, applications of 246
Holography 184
Holography, acoustic: see Acoustic holography 1; Lasers, applications of 247
Holography, laser: see Lasers, applications of 246
Holography, ultrasonic: see Holography 187
Homodyne local oscillator system: see Submillimetre radiation, applications of 454
Homogeneous nucleation: see Glass ceramics 143
Hoop strain: see Electrohydraulic crushing 82
Hopkinson fracture: see Electrohydraulic crushing 82
Host-cell reactivation of bacteriophage: see Ultraviolet radiation, some biological effects of 510
Huckel theories: see Quantum mechanics in biochemistry 352
Hull performance: see Sailing, physics of 398
Human speech-generating process: see Speaking machines 440
Hydraulic vibrators: see Electroviscous fluids, applications of 93
Hydrazine as a reducing agent: see Metal film preparation by electroless (chemical) reduction 281
Hydrodynamic or pumping efficiency: see Glandless reciprocating jet pump 142
Hydrofoil: see Sailing, physics of 397

Hydrogen bonding: see Quantum mechanics in biochemistry 353
Hydrogen bonds: see Proteins, structure of 337
Hydrogen bromide: see Tungsten halogen lamps 496
Hydrogen maser: see Quantum metrology 354
Hydrograph, unit: see Hydrometeorology 191
Hydrologic cycle 187; see also Hydrometeorology 188
Hydrological data, time series of: see Hydrometeorology 189
Hydrology: see Hydrologic cycle 188
Hydrometeorology 188
Hydrophone: see Ambient noise in the sea 14
Hydroxyl radicals, interstellar, radioemissions from: see Radioemissions from interstellar hydroxyl radicals 372
Hypercharge: see Current algebras and their applications 52; Elementary particles, nomenclature of 95; Quarks 357
Hyperfine interaction: see Radioemissions from interstellar hydroxyl radicals 372
Hyperon: see Elementary particles, nomenclature of 95
Hypochromicity: see Quantum mechanics in biochemistry 353
Hypolimnion: see Physical limnology 322
Hypophosphite/nickel system: see Metal film preparation by electroless (chemical) reduction 277
Hysteresis: see Textile industry, physics in 468
Hysteresis loops: see Rubber physics 396

Ice: see Hydrometeorology 191
Ice, lake: see Physical limnology 323
Ice-yachts: see Sailing, physics of 397
Idler waves: see Lasers, tunable 251
"IF" element: see Fluidics 118
Illuminants, standard: see Daylight, power distribution in 56
Image: see Gyroscopic coupler 152
Image, astigmatic: see Mirage and looming, optics of 294
Image contrast: see High resolution electron microscopy 182
Image converter: see Infra-red cameras and infra-red TV 199
Image, pulse-echo: see Ultrasonic imaging 498
Image reconstruction procedure: see Holography 185
Imaging, schlieren ultrasound: see Ultrasonic imaging 500
Imaging, three-dimensional: see Acoustic holography 1; Holography 184
Imaging, ultrasonic: see Ultrasonic imaging 498
Imidite: see Temperature resistant polymers 465
Immiscibility in glasses: see Glass ceramics 143
Impact ionization: see IMPATT diode microwave generators 194
Impact modulator: see Fluidics 120
Impact parameter: see Proton scattering microscopy 343

IMPATT diode (Impact Avalanche Transit-time): *see* Microwave solid state oscillators, recent 288

IMPATT diode microwave generators 194

Imperfections, lattice: *see* Epitaxy, recent work on 100

Imperfections, lattice, in semiconductors: *see* Semiconductors, effect of radiation on 405

Impermeable carbons: *see* Vitreous carbons, structure and properties of 513

Impurity concentration, measurement of: *see* Refractories, ultra-purification of 389

Impurity levels in vitreous carbon: *see* Vitreous carbons, structure and properties of 515

Incandescent lamps: *see* Tungsten halogen lamps 494

Inclination: *see* Astrodynamics 22

Incoherent backscatter: *see* Ionosphere, recent observations of 228

Independent particle model: *see* Soft X-ray spectroscopy 416

Index point: *see* Reverberation under water 391

Indirect photoreactivation: *see* Ultra-violet radiation, some biological effects of 509

Indium antimonide: *see* Magnetoresistivity 270

Induced dipoles: *see* Submillimetre radiation, applications of 452

Induced drag, coefficient of: *see* Jet flap 234

Induction amplifier: *see* Fluidics 120

Induction heating: *see* Refractories, ultra-purification of 388

Induction, magnetic: *see* Magnetoresistivity 273

Inelastic scattering of electrons: *see* Momentum transfer cross section for low energy electrons: determination by swarm techniques 297

Inertia, moments of: *see* Submillimetre radiation, applications of 451

Inertial motion: *see* Physical limnology 322

Information processing, optical: *see* Optical information processing 302

Information storage, three-dimensional optical: *see* Lasers, applications of 249

Infra-red absorption: *see* Infra-red spectroscopy, recent developments in 205

Infra-red cameras: *see* Infra-red cameras and infra-red TV 196

Infra-red detection system: *see* Airborne collision prevention devices 12

Infra-red detector's detectivity: *see* Pyroelectric detector for infra-red radiation 349

Infra-red detectors, thermoelectric cooling of: *see* Thermoelectric cooling of infra-red detectors 481

Infra-red, far, absorption of liquids, crystals and glasses in the: *see* Far infra-red spectroscopy 107

Infra-red, far, detectors: *see* Far infra-red spectroscopy 105

Infra-red, far, laser: *see* Far infra-red lasers 101

Infra-red, far, laser spectrometer: *see* Far infra-red lasers 104

Infra-red, far molecular spectroscopy *see* Far Infra-red spectroscopy 106

Infra-red, far, refractive index measurement in the: *see* Interferometric spectrometers 211

Infra-red, far, solid state lasers: *see* Far infra-red lasers 103

Infra-red, far, sources: *see* Far infra-red spectroscopy 105

Infra-red, far, spectroscopy: *see* Far infra-red spectroscopy 104

Infra-red lasers: *see* Quantum metrology 355

Infra-red radiation, pyroelectric detector of: *see* Pyroelectric detector of infra-red radiation 347

Infra-red retina: *see* Infra-red cameras and infra-red TV 197

Infra-red spectrometers: *see* Infra-red spectroscopy, recent developments in 204

Infra-red spectroscopy, recent developments in 202

Infra-red television: *see* Infra-red cameras and infra-red TV 196

Injector, external charge: *see* Electrogasdynamics 78

Inorganic fibres: *see* Textile industry, physics in 470

Inorganic polymers: *see* Temperature resistant polymers 466

Instability: *see* Melting, modern theories of 276

Instrumental broadening: *see* Soft X-ray spectroscopy 417

Integrating by optical processing: *see* Optical information processing 302

Intensity, luminous: *see* Quantum metrology 356

Interaction, crack: *see* Fracture of brittle materials 127

Interaction, electromagnetic: *see* Charge conjugation 46; Current algebras and their applications 51

Interaction, electron-ion: *see* Liquid metals 261

Interaction, hyperfine: *see* Radioemissions from interstellar hydroxyl radicals 372

Interaction, inter-molecular: *see* Submillimetre radiation, applications of 452

Interaction, spin-spin: *see* Fast passage in paramagnetic resonance 110

Interaction, strong: *see* Charge conjugation 46

Interaction, van der Waals: *see* Quantum mechanics in biochemistry 353

Interaction, weak: *see* Bosons, intermediate 34; Charge conjugation 46; Current algebras and their applications 51

Interference: *see* Fourier transform spectroscopy 122; Holography 184

Interference filters: *see* Infra-red spectroscopy, recent developments in 202

Interference fringes: *see* Interferometric spectrometers 207

Interferogram: *see* Fourier transform spectroscopy 123; Interferometric spectrometer 207

Interferogram, cavity: *see* Submillimetre-wave lasers 455

Interferometer: *see* Fourier transform spectroscopy 122

Interferometer, Fabry–Perot: *see* Far infra-red lasers 103

Interferometer/grating comparison: *see* Interferometric spectrometers 211
Interferometer, lamellar grating: *see* Interferometric spectrometers 211; Submillimetre-wave methods 461
Interferometer, long base-line: *see* Radio telescopes, the present position 385
Interferometer, Michelson: *see* Fourier transform spectroscopy 122; Luminosity of a spectrometer 264; Submillimetre radiation, applications of 454; Submillimetre-wave methods 461
Interferometer, Michelson's stellar: *see* Optical information processing 304
Interferometer, mock: *see* Interferometric spectrometers 213
Interferometric spectrometer 206; *see also* Multiplex principle (of Fellgett) 299
Interferometric spectrometer, Fabry–Perot: *see* Interferometric spectrometers 212
Interferometric spectrometer, Michelson: *see* Infra-red spectroscopy, recent developments in 203; Interferometric spectrometer 207
Interionic forces: *see* Liquid metals 261
Intermediate bosons: *see* Bosons, intermediate 34
Intermolecular attractions: *see* Liquid crystals and molecular structure 257
Intermolecular bonding: *see* Submillimetre radiation, applications of 452
Intermolecular forces: *see* Polymer physics 333; Quantum mechanics in biochemistry 351
Intermolecular interactions: *see* Submillimetre radiation, applications of 452
Internal cardiac pacemaker: *see* Cardiac pacemaker, external 39
Internal friction: *see* Snoek spectra 411
Internal reflection (ATR) techniques: *see* Infra-red spectroscopy, recent developments in 204
Internal waves: *see* Physical limnology 323
International system of units (SI): *see* Quantum metrology 356
Interplanetary dust 213
Interrogator-transponder technique: *see* Airborne collision prevention devices 10
Inter-space communications: *see* Lasers, applications of 249
Interstellar hydroxyl radicals, radioemissions from: *see* Radioemissions from interstellar hydroxyl radicals 372
Interstitial atoms: *see* Snoek spectra 410
Intracranial pressures: *see* Biological pressure transensors 31
Intra-ocular pressure: *see* Biological pressure transensors 31
Intrinsic carrier concentration: *see* Solar cell 422
Intrinsic semiconductor: *see* Thermoelectric generation of electricity 484
Invariance, charge conjugation: *see* Charge conjugation 45
Invariance, CPT: *see* Charge conjugation 46

Invariance, time-reversal: *see* Electric dipole moment of the neutron 74
Invariance, violation of C: *see* Charge conjugation 46
Invariance, violation of CP: *see* Charge conjugation 46
Inversion: *see* Dynamical n-beam theory of electron defraction 72
Inversion, population: *see* Q-switching 350; Submillimetre-wave lasers 457
Inversion theorem, Fourier's: *see* Fourier transform spectroscopy 123
Inverted population: *see* Far infra-red lasers 101; Lasers, tunable 251
Iodine, quartz-, tungsten lamp: *see* Submillimetre-wave methods 459
Iodine, tungsten-, cycle: *see* Tungsten halogen lamps 495
Ioffe lamp: *see* Thermoelectric generation of electricity 487
Ion bombardment: *see* Sputtering, theories of 445
Ion, colloidal: *see* Electrogasdynamics 77
Ion cyclotron whistlers: *see* Ionosphere, recent observations of 228
Ion-electron interaction: *see* Liquid metals 261
Ion-electron pairs: *see* Streamer chambers 448
Ion exchange cell: *see* Space power supplies 439
Ion guns: *see* Glow discharge electron and ion guns for material processing 148
Ion, heavy, forming: *see* Electrogasdynamics 78
Ion, mobility of an: *see* Electrogasdynamics 76
Ion, molecular: *see* Electrogasdynamics 77
Ion-neutralization spectroscopy (INS): *see* Ion-neutralization spectroscopy 220
Ion-phonon transition: *see* Phonon avalanche and bottleneck 316
Ion-sheath phenomena: *see* Thermionic generation of electricity 477
Ion slip: *see* Electrogasdynamics 75
Ion source, corona-discharge: *see* Electrogasdynamics 78
Ion temperatures: *see* Ionosphere, recent observations of 226
Ionic bonds: *see* Proteins, structure of 341
Ionic temperature: *see* Phonon avalanche and bottleneck 316
Ionization: *see* Laser-induced sparks 244; Radiation chemistry, industrial applications of 363; Thermionic generation of electricity 477
Ionization and attachment cross-sections of atoms and molecules by electron impact 217
Ionization, Auger: *see* Ionization and attachment cross-sections of atoms and molecules by electron impact 219
Ionization energy: *see* Laser-induced sparks 243
Ionization, impact: *see* IMPATT diode microwave generators 194
Ionization, thermal contact: *see* Thermionic generation of electricity 477
Ionization threshold: *see* Ionization and attachment cross-sections of atoms and molecules by electron impact 219

Ionization, total: see Radiation quantities, systematics of 367
Ionizing avalanche (Townsend cascade): see Laser-induced sparks 243
Ionizing power: see Streamer chambers 449
Ionogram: see Ionosphere, recent observations of 227
Ionosphere: see Ionosphere, recent observations of 224; Ionospheres of Mars and Venus 229
Ionospheric electron density: see Ionosphere, recent observations of 226
Iricon: see Infra-red cameras and infra-red TV 200
Iron: see Magnetic thin films 265
Iron meteorite: see Shock pressure in meteorites 407
Irradiation, laser: see Lasers, applications of 246
Irradiation of polythene: see Radiation chemistry, industrial applications of 365
Irtran: see Infra-red spectroscopy, recent developments in 203
Isentropic efficiency, generator: see Electrogasdynamics 77
Ishibashi mechanism: see Metal film preparation by electroless (chemical) reduction 280
Isobaric spin: see Elementary particles, nomenclature of 95
Isobars: see Chart of the nuclides, trilinear 47
Isochron: see Geochronology 135
Isohyetal analysis: see Hydrometeorology 188
Isospin: see Current algebras and their applications 52; Elementary particles, nomenclature of 95
Isospin multiplets: see Quarks 357
Isotope dilution analysis: see Geochronology 135
Isotope level gauges: see Radioisotopes in food processing 378
Isotopic geochronology: see Geochronology 133
Isotopic spin: see Elementary particles, nomenclature of 95

Jet coefficient: see Jet flap 232
Jet, deflected: see High lift devices 174
Jet diode: see Fluidics 122
Jet flap 232; see also High lift devices 171
Jet, magnetic fluid: see Ferrohydrodynamics 114
Jet pump, glandless reciprocating: see Glandless reciprocating jet pump 140
Jet stream: see Jet flap 233
Jet, vertical: see High lift devices 174
Jog, dislocation: see Dislocation kinks 62
Josephson current: see Josephson effect 236
Josephson effect 235; see also Quantum metrology 356
Josephson effect, a.c.: see Josephson effect 237
Josephson equations: see Josephson effect 235
Josephson radiation, emitted: see Josephson effect 237
Joukowski formula: see High lift devices 170
Joule heating: see Thermoelectric generation of electricity 483
Junction efficiency: see Solar cell 423
Junction, evaporated film: see Josephson effect 238

Junction, p-n: see Microwave solid-state oscillators, recent 288; Solar cell 423
Junction, point contact: see Josephson effect 238
Junction, tunnel: see Josephson effect 236

Kalman filtering: see Astrodynamics 20
Kamacite: see Shock pressure in meteorites 407
Kanigen nickel alloy: see Metal film preparation by electroless (chemical) reduction 281
KBr disc technique: see Infra-red spectroscopy, recent developments in 204
Keel yacht: see Sailing, physics of 398
Kelvin: see Quantum metrology 356
Kelvin effect: see Thermoelectric generation of electricity 483
Keratin: see Proteins, structure of 339
Kerma: see Radiation quantities, systematics of 367
Kerr effect: see Pockels effect 327
Kiln: see Ceramics production, physics in 41
Kilogramme: see Quantum metrology 355
Kink: see Peierls stress 306
Kink, dislocation: see Dislocation kinks 62
Kink, geometrical: see Peierls stress 307
Kink pairs, nucleating: see Peierls stress 306
Kirchhoff's laws (electrical circuits) 239
Knee: see Ionosphere, recent observations of 229
Knitted fibres: see Textile industry, physics in 474
Korteweg—Helmholtz expression: see Ferrohydrodynamics 112
Kramers doublet: see Far infra-red lasers 103
Krypton 86 lamp: see Quantum metrology 355

Ladder polymers: see Temperature resistant polymers 466
Lagrangian integrand function: see Hamilton's principle 161
Lake studies: see Physical limnology 321
Lakes: see Hydrometeorology 191
Lambert's law: see Reverberation under water 392
Lamellar grating: see Far infra-red spectroscopy 105
Lamellar grating interferometer: see Interferometric spectrometers 211; Submillimetre-wave methods 461
Laminar restrictor: see Fluidics 122
Lamp, incandescent: see Tungsten halogen lamps 494
Lamp, Ioffe: see Thermoelectric generation of electricity 487
Lamp, krypton-86: see Quantum metrology 355
Lamp, mercury, medium pressure: see Submillimetre-wave methods 459
Lamp, quartz-iodine tungsten: see Submillimetre-wave methods 459
Lamp, tungsten filament: see Tungsten halogen lamps 494
Lamps, tungsten halogen: see Tungsten halogen lamps 494
Langevin function: see Polymer physics 332
Lanthanum magnesium nitrate (LMN): see Polarized proton targets 330

Laplace transformation: *see* Optical information processing 303
Laplacian method for preliminary orbit determination: *see* Astrodynamics 17
Larmor precession: *see* Electric dipole moment of the neutron 74
Larmor precession frequency: *see* Thermomagnetic gas torque (The Scott effect) 488
Laser: *see* Holography 184; Optical information processing 303; Q-switching 350; Quantum metrology 355
Laser, argon: *see* Lasers, application of 246
Laser, dye-: *see* Lasers, tunable 252
Laser, electroluminescent semiconductor: *see* Far infra-red lasers 103
Laser, far infra-red: *see* Far infra-red lasers 101
Laser, gas: *see* Infra-red spectroscopy, recent developments in 203
Laser, gas, submillimetre-wave: *see* Submillimetre-wave methods 459
Laser gyroscope (ring laser): *see* Lasers, applications of 248
Laser, HCN: *see* Submillimetre radiation, applications of 453; Submillimetre-wave lasers 456; Submillimetre-wave methods 459
Laser holography: *see* Lasers, applications of 246
Laser-induced breakdown of gases: *see* Laser-induced sparks 242
Laser-induced sparks 242
Laser, infra-red: *see* Quantum metrology 355
Laser irradiation: *see* Lasers, applications of 246
Laser lines, HCN: *see* Far infra-red lasers 102
Laser microprobe: *see* Laser-induced sparks 245
Laser, molecular: *see* Far infra-red lasers 101; Far infra-red spectroscopy 105
Laser, neodymium: *see* Lasers, tunable 251
Laser oscillation, threshold condition for a: *see* Q-switching 350
Laser photocoagulation technique: *see* Lasers, applications of 246
Laser, pulsed: *see* Lasers, applications of 246
Laser, Q-switched: *see* Lasers, applications of 246; Laser-induced sparks 245
Laser, rare gas: *see* Far infra-red lasers 101
Laser, ring (laser gyroscope): *see* Lasers, applications of 248
Laser, ruby: *see* Q-switching 350; Laser-induced sparks 243; Lasers, applications of 246; Lasers, tunable 251
Laser seismometers: *see* Lasers, applications of 248
Laser, semiconductor: *see* Lasers, applications of 249
Laser, solid: *see* Infra-red spectroscopy, recent developments in 203
Laser, solid state: *see* Q-switching 350
Laser, solid state far infra-red: *see* Far infra-red lasers 103
Laser sparks, triggering of: *see* Laser-induced sparks 243
Laser spectrometer, far infra-red: *see* Far infra-red lasers 104

Laser submillimetre: *see* Submillimetre radiation, applications of 453
Laser, submillimetre-wave: *see* Submillimetre-wave lasers 454
Laser systems, strain measuring: *see* Lasers, applications of 247
Laser, tunable: *see* Lasers, tunable 250; Lasers, applications of 250
Laser, water vapour: *see* Submillimetre-wave lasers 456; Submillimetre-wave methods 459
Lasers, applications of 245
Lateral intermolecular attractions: *see* Liquid crystals and molecular structure 259
Lattice, crystal, periodicities, observation of: *see* High resolution electron microscopy 181
Lattice imperfections: *see* Epitaxy, recent work on 100
Lattice imperfections in semiconductors: *see* Semiconductors, effect of radiation on 405
Lattice, reciprocal: *see* Dynamical n-beam theory of electron diffraction 70
Lattice thermal conductivity: *see* Conductivity at high temperatures 48; Thermoelectric generation of electricity 485
Law enforcement science and technology 253
Layer, alpha-Chapman: *see* Ionospheres of Mars and Venus 230
Layer, beta-Chapman: *see* Ionospheres of Mars and Venus 230
Layer F_1: *see* Ionospheres of Mars and Venus 230
Layer-lines: *see* Proteins, structure of 341
Lead dating: *see* Geochronology 137
Leading edge slat: *see* High lift devices 171
Least-squares differential correction, weighted: *see* Astrodynamics 19
Least-squares method: *see* Astrodynamics 18
Lebesque constants: *see* Walsh functions 518
Leith—Upatnieks holograms: *see* Holography 185
Length: *see* Quantum metrology 355
Lens, cylindrical: *see* Optical information processing 303
Lens defects: *see* High resolution electron microscopy 179
Lens, electron: *see* High resolution electron microscopy 179
Lens, fly's eye: *see* Lasers, applications of 247
Lens, ultrasonic: *see* Ultrasonic imaging 498
Lepton: *see* Current algebras and their applications 52; Elementary particles, nomenclature of 95
Lepton number: *see* Elementary particles, nomenclature of 95
Level, isotope, gauges: *see* Radioisotopes in food processing 378
Levitation, ferrohydrodynamic: *see* Ferrohydrodynamics 116
Lie detector: *see* Law enforcement science and technology 257
Lifetime broadening: *see* Soft X-ray spectroscopy 417
Lifetime, carrier: *see* Solar cell 424

Lifetime, minority carrier: *see* Semiconductors, effect of radiation on 406
Lift: *see* Jet flap 232; Sailing, physics of 399
Lift coefficient: *see* High lift devices 170; Jet flap 232
Lift devices, high: *see* High lift devices 170
Lift on rotating cylinder: *see* High lift devices 173
Light modulation: *see* Pockels effect 327
Light sources, tunable: *see* Infra-red spectroscopy, recent developments in 204
Light, velocity of: *see* Far infra-red lasers 102; Quantum metrology 355
Light, zodiacal: *see* Interplanetary dust 213
Limnology, physical: *see* Physical limnology 321
Lindemann law: *see* Melting, modern theories of 275
Line of nodes: *see* Astrodynamics 23
Line of stability: *see* Chart of the nuclides, trilinear 47
Linear electro-optic effect: *see* Pockels effect 327
Linear flow resistance: *see* Fluidics 122
Linear occultations of radio sources: *see* Radio telescopes, the present position 383
Linear polymers: *see* Polymer physics 332
Linear system, nonconservative: *see* Hamilton's principle 162
Lines, Fraunhofer: *see* Daylight, power distribution in 55
Lippmann–Bragg holograms: *see* Holography 184
Liquid, absorption of a, in the far infra-red: *see* Far infra-red spectroscopy 107
Liquid crystals and molecular structure 257
Liquid dielectrics, pulsed: *see* High intensity pulsed radiation sources 166
Liquid helium: *see* Josephson effect 238
Liquid-holding recovery: *see* Ultra-violet radiation, some biological effects of 511
Liquid metals 261
Liquid structure: *see* Liquid metals 261
Liquid structure factor: *see* Liquid metals 261
Liquid surface deformation: *see* Ultrasonic imaging 500
Liquids, non-polar: *see* Submillimetre radiation, applications of 452
Liquidus temperature: *see* Glass ceramics 143
Littoral drift: *see* Hydrometeorology 192
Logic devices, fluidic: *see* Fluidics 118
Longitude of the ascending node: *see* Astrodynamics 23
Longitudinal mode of acoustic waveguides: *see* Acoustic waveguides 7
Loom: *see* Textile industry, physics in 472
Looming and mirage, optics of: *see* Mirage and looming, optics of 290
"Loop response": *see* Ultrasonic spectroscopy 504
Loop, van der Waals-like: *see* Melting, modern theories of 275
Lorentz deflection: *see* Electron diffraction from periodic magnetic fields 87
Lorentz force: *see* Electron diffraction from periodic magnetic fields 85

Lorentz number: *see* Thermoelectric generation of electricity 484
Lorentz peaks: *see* Electron diffraction from periodic magnetic fields 87
Loss coefficient, nozzle/diffuser: *see* Glandless reciprocating jet pump 142
Loss modulus: *see* Polymer physics 337
Loudspeaker magnet assemblies: *see* Permanent magnet technology 313
Low energy electron diffraction (LEED): *see* Auger emission spectroscopy 24; Ion-neutralization spectroscopy 221
Low-pass filters: *see* Infra-red spectroscopy, recent developments in 202; Pulse code modulation 345
LSA (Limited space-charge accumulation) oscillator: *see* Microwave solid-state oscillators, recent 288
Lukes mechanism: *see* Metal film preparation by electroless (chemical) reduction 280
Luminescence: *see* Ultrasonic imaging 499
Luminosity: *see* Fourier transform spectroscopy 122
Luminosity advantage: *see* Luminosity of a spectrometer 264; Submillimetre-wave methods 461
Luminosity, auroral: *see* Auroral electrons 27
Luminosity of a spectrometer 264; *see also* Interferometric spectrometers 213
Luminous flux: *see* Luminosity of a spectrometer 264
Luminous intensity: *see* Quantum metrology 356
Lysozyme: *see* Proteins, structure of 337

M-induced (or field-induced) anisotropy: *see* Magnetic thin films 266
Mach angle: *see* Sonic boom, generation and suppression of 432
Mach number: *see* Sonic boom, generation and suppression of 432
Machining, electron beam: *see* Electron beam machining 84
Macroscopic element: *see* Microplasticity 286
Magnet, permanent, technology: *see* Permanent magnet technology 309
Magnet, superconducting: *see* Polarized proton targets 329
Magnetic activity: *see* Solar wind, planetary interactions with 429
Magnetic domains: *see* Electron diffraction from periodic magnetic fields 87
Magnetic electron multiplier: *see* Soft X-ray spectroscopy 414
Magnetic field boundary conditions: *see* Ferrohydrodynamics 113
Magnetic field, gases in a: *see* Thermomagnetic gas torque (The Scott effect) 488
Magnetic fields, periodic, electron diffraction from: *see* Electron diffraction from periodic magnetic fields 86
Magnetic films, polycrystalline: *see* Magnetic thin films 265
Magnetic fluid jet: *see* Ferrohydrodynamics 114
Magnetic fluid seals: *see* Ferrohydrodynamics 115

Magnetic fluid surface stability: see Ferrohydrodynamics 114
Magnetic flux: see Magnetoresistivity 272
Magnetic flux quantization: see Josephson effect 235
Magnetic induction: see Magnetoresistivity 273
Magnetic moment of a quark: see Quarks 360
Magnetic pressure, fluid: see Ferrohydrodynamics 113
Magnetic properties, stress-dependent: see Piezomagnetic effect 326
Magnetic resonance: see Far infra-red spectroscopy 108; Submillimetre radiation, applications of 452
Magnetic resonance, nuclear (NMR): see Metallic colloids in non-metallic crystals 283; Polarized proton targets 330
Magnetic spectrometer: see High intensity pulsed radiation sources 167
Magnetic stress: see Ferrohydrodynamics 111
Magnetic susceptibility: see Piezomagnetic effect 325
Magnetic thin films 265
Magnetic triode: see Thermionic generation of electricity 477
Magnetite: see Piezomagnetic effect 326
Magnetizable fluid media: see Ferrohydrodynamics 111
Magnetization: see Ferrohydrodynamics 111; Magnetic thin films 265
Magnetization, easy direction of: see Magnetic thin films 265
Magnetization, hard direction of: see Magnetic thin films 265
Magnetization, remanent (permanent): see Piezomagnetic effect 326
Magnetization, saturation: see Ferrohydrodynamics 112
Magnetization, temperature coefficient of, for barium ferrite: see Permanent magnet technology 313
Magnetocrystalline anisotropy: see Magnetic thin films 265; Piezomagnetic effect 326
Magneto-elastic effect: see Piezomagnetic effect 326
Magnetoresistance of semiconductors: see Magnetoresistivity 274
Magnetoresistance wattmeter: see Magnetoresistivity 273
Magnetoresistivity 269
Magnetoresistors: see Magnetoresistivity 272
Magneto-seismic effect: see Piezomagnetic effect 326
Magnetosphere: see Solar wind, planetary interactions with 429
Magnetostriction coefficient: see Magnetic thin films 266
Magnetostrictive anisotropy: see Magnetic thin films 266
Magnetron oscillator, pulsed: see Split-beam microwave beacon 444
Magnitude dispersion: see Magnetic thin films 267
Magnus effect: see High lift devices 172
Majority carriers: see Solar cell 422
Manganese bismuth: see Lasers, applications of 249
Manoeuvres, avoidance: see Airborne collision prevention devices 12
Manometry: see Biological pressure transensors 31
Mariner 4: see Ionospheres of Mars and Venus 229
Mariner 5: see Ionospheres of Mars and Venus 230
Mars, ionosphere of: see Ionospheres of Mars and Venus 229
Martian ionosphere: see Ionosphere, recent observations of 229
Marx generators: see Streamer chambers 449
Marx–Surge generator: see High intensity pulsed radiation sources 165
Maser: see Far infra-red lasers 101
Maser amplification: see Radioemissions from interstellar hydroxyl radicals 376
Maser, hydrogen: see Quantum metrology 354
Maser, solid state: see Far infra-red lasers 103
Mass: see Quantum metrology 355
Mass attenuation coefficients: see Radiation quantities, systematics of 369
Mass curves of rainfall: see Hydrometeorology 189
Mass energy transfer coefficient: see Radiation quantities, systematics of 368
Mass number (A): see Chart of the nuclides, trilinear 47
Mass, quark: see Quarks 361
Mass spectrometry: see Geochronology 135
Mass transfer method: see Hydrometeorology 193
Mass unit, atomic: see Quantum metrology 355
Mass unit, nuclidic: see Quantum metrology 356
Matano–Boltzmann method for solution of diffusion equations: see Diffusion in non-uniform media 59
Materials, brittle, fracture of: see Fracture of brittle materials 124
Materials, refractory: see Ceramics production, physics in 43
Matrix, weighting: see Astrodynamics 19
Maxwell stress tensor: see Ferrohydrodynamics 112
Mean anomaly: see Astrodynamics 23
Mean solar second: see Quantum metrology 354
Mechanical power, transport and control of: see Electroviscous fluids, applications of 93
Mechanics, celestial: see Astrodynamics 17
Mechanics, fracture: see Fracture of brittle materials 124
Medical uses of whole body counting: see Whole body counting 525
Mekometer: see Quantum metrology 357
Melting, modern theories of 275
Melting point: see Liquid crystals and molecular structure 258
Melting process, computer simulation of the: see Melting, modern theories of 275
Memories, computer: see Magnetic thin films 265
Mercury: see Radar astronomy 362
Mercury arc: see Far infra-red spectroscopy 104
Mercury lamp, high pressure: see Infra-red spectroscopy, recent developments in 202
Mercury lamp, medium pressure: see Submillimetre-wave methods 459
Mercury silver iodide: see Ultrasonic imaging 499
Meridian: see Proteins, structure of 341

Meshes: *see* Kirchhoff's law (electrical circuits) 240
Mesomorphic states: *see* Liquid crystals and molecular structure 257
Meson: *see* Elementary particles, nomenclature of 95; Quarks 357
Meson, W: *see* Bosons, intermediate 35
Mesophase, smectic: *see* Liquid crystals and molecular structure 257
Mesorphic thermal stability: *see* Liquid crystals and molecular structure 259
Metal fatigue: *see* Microplasticity 286
Metal, fibre reinforced: *see* Freeze-coating 132
Metal film preparation by electroless (chemical) reduction 277
Metallic colloids in non-metallic crystals 282
Metallic dispersions: *see* Glass ceramics 144
Metallic nucleating agent: *see* Glass ceramics 143
Metallography, electron, applications of: *see* Electron metallography, applications of 89
Metals and alloys: *see* Electron metallography, applications of 89
Metals and non-metals, whiskers of: *see* Dispersion and fibre strengthening of materials 65
Metals, electrohydraulic forming of: *see* Electrohydraulic forming of metals 83
Metals, ferromagnetic, conductivities of: *see* Conductivity at high temperatures 50
Metals, liquid: *see* Liquid metals 261
Metals, refractory: *see* Refractories, ultra-purification of 386
Metamorphisms: *see* Geochronology 136
Metastable limit, Ostwald: *see* Solid nucleus formation from solution 431
Meteorites, chrondritic: *see* Interplanetary dust 214
Meteorites, iron: *see* Shock pressure in meteorites 407
Meteorites, shock pressure in: *see* Shock pressure in meteorites 407
Meteorology: *see* Hydrologic cycle; Hydrometeorology 188
Metre: *see* Quantum metrology 355
Metrological applications of submillimetre radiation: *see* Submillimetre radiation, applications of 453
Metrology, quantum: *see* Quantum metrology 354
Micelle, fringed, model: *see* Polymer physics 335; Textile industry, physics in 471
Michelson interferometer: *see* Fourier transform spectroscopy 122; Luminosity of a spectrometer 264; Submillimetre radiation, applications of 454; Submillimetre-wave methods 461
Michelson interferometric spectrometer: *see* Infra-red spectroscopy, recent developments in 203; Interferometric spectrometers 207
Michelson stellar interferometer: *see* Optical information processing 304
Microanalyser, electron-probe: *see* Soft X-ray spectroscopy 415
Microbeams, effect of, on cell organelles: *see* Ultraviolet radiation, some biological effects of 512
Microcell: *see* Infra-red spectroscopy, recent developments in 204

Microcrystalline glass ceramic: *see* Glass ceramics 145
Microplasticity 285
Microprobe, laser: *see* Laser-induced sparks 245
Microscope, electron: *see* Dynamic electron diffraction, application of 65; High resolution electron microscopy 179
Microscope, reflecting: *see* Infra-red spectroscopy, recent developments in 204
Microscopic elements: *see* Microplasticity 286
Microscopy, biological electron: *see* High resolution electron microscopy 183
Microscopy, electron: *see* Epitaxy, recent works on 98
Microscopy, electron transmission: *see* Electron metallography, applications of 89
Microscopy, high resolution electron: *see* High resolution electron microscopy 179
Microscopy, proton scattering: *see* Proton scattering microscopy 342
Microscopy, reflection electron: *see* Electron metallography, applications of 91
Microscopy, scanning electron: *see* Electron metallography, applications of 91
Microstructure determination: *see* Ultrasonic spectroscopy 505
Microstructure of metals and alloys: *see* Electron metallography, applications of 89
Microvoids: *see* Fracture of brittle materials 124
Microwave beacon, split-beam: *see* Split-beam microwave beacon 444
Microwave generators, avalanche: *see* IMPATT diode microwave generators 195
Microwave generators, IMPATT diode: *see* IMPATT diode microwave generators 194
Microwave rectifier diode: *see* Submillimetre-wave methods 459
Microwave solid state oscillators, recent 288
Microwave spectrum of OH: *see* Radioemissions from interstellar hydroxyl radicals 372
Microwaves: *see* Infra-red spectroscopy, recent developments in 204
Mie scattering: *see* Interplanetary dust 214
Migration, interstitial: *see* Snoek spectra 411
Milk analyser: *see* Infra-red spectroscopy, recent developments in 206
Mill, fluid energy: *see* Ceramics production, physics in 42
Miller's rule: *see* Pockels effect 328
Milling, electron beam: *see* Electron beam machining 84
Mills Cross radio telescope: *see* Radio telescopes, the present position 383
Minority carrier lifetime: *see* Semiconductors, effect of radiation on 406
Minority carriers: *see* Solar cell 423
Mirage and looming, optics of 290
Mirror, dichroic: *see* Solar cell 428
Mirror, energy-concentrating: *see* Space power supplies 438
Misfit dislocations: *see* Epitaxy, recent work on 99

Mistors: *see* Magnetoresistivity 272
Mobility, carrier: *see* Hall effect, applications of 156; Thermoelectric generation of electricity 484
Mobility of ions: *see* Electrogasdynamics 76
Mock interferometer: *see* Interferometric spectrometers 213
Mode, anomalous: *see* IMPATT diode microwave generators 196
Model ships, photoelectric tracking of 294
Modes of deformation of polymers: *see* Polymer physics 336
Modes of propagation of acoustic waves in waveguides: *see* Acoustic waveguides 6
Modes, resonator: *see* Far infra-red lasers 102
Modulating envelope distortion: *see* Hall effect, applications of 158
Modulation, light: *see* Pockels effect 327
Modulation, pulse amplitude: *see* Pulse code modulation 345
Modulation, pulse code: *see* Pulse code modulation 344
Modulator, Hall effect radio-frequency: *see* Hall effect, applications of 157
Modulator, impact: *see* Fluidics 120
Modulator, vortex (or valve): *see* Fluidics 120
Modulators: *see* Streamer chambers 449
Module, thermoelectric: *see* Thermoelectric cooling of infra-red detectors 481
Moduli, elastic: *see* Glass ceramics 146
Modulus, loss: *see* Polymer physics 337
Modulus of rupture: *see* Glass ceramics 146
Modulus, real: *see* Polymer physics 337
Modulus, relaxation: *see* Polymer physics 335
Modulus, Young's: *see* Brittleness in ceramics 37
Moiré fringes: *see* Interferometric spectrometers 207; Quantum metrology 357
Moiré pattern: *see* High resolution electron microscopy 181
Molecular breadth: *see* Liquid crystals and molecular structure 259
Molecular chains: *see* High polymers, mechanical properties of 175
Molecular ions: *see* Electrogasdynamics 77
Molecular lasers: *see* Far infra-red lasers 101; Far infra-red spectroscopy 105
Molecular orbital: *see* Quantum mechanics in biochemistry 352
Molecular spectroscopy in the far infra-red: *see* Far infra-red spectroscopy 106
Molecular structure: *see* Liquid crystals and molecular structure 257
Molecular theory of rubber elasticity: *see* High polymers, mechanical properties of 176
Molecular vibrations: *see* Far infra-red spectroscopy 106
Molecules, anisotropic: *see* Liquid crystals and molecular structure 257
Molecules, dipole moments of: *see* Submillimetre radiation, applications of 451
Molecules, elongated: *see* Liquid crystals and molecular structure 257
Molecules, gas, collision frequency of: *see* Thermomagnetic gas torque (The Scott effect) 488
Molecules, polar, relaxational (Debye) absorption by: *see* Submillimetre radiation, applications of 452
Molecules, polymer: *see* Polymer physics 331; Rubber physics 393
Molecules tropocollagen: *see* Proteins, structure of 340
Moment, dipole, of molecules: *see* Submillimetre radiation, applications of 451
Moment, magnetic, of a quark: *see* Quarks 360
Moment, nuclear electric quadrupole: *see* Nuclear quadrupole moment 300
Moment of the neutron, electric dipole: *see* Electric dipole moment of the neutron 74
Moment, quadrupole: *see* Submillimetre radiation, applications of 452
Moments of inertia: *see* Submillimetre radiation, applications of 451
Momentum coefficient: *see* High lift devices 171
Momentum, orbital, quenching of: *see* Magnetic thin films 266
Momentum transfer cross-section for low energy electrons: *see* Momentum transfer cross-section for low energy electrons: determination by swarm techniques 296
Momentum transfer free path: *see* Momentum transfer cross-section for low energy electrons: determination by swarm techniques 298
Monochromatic radiation sources: *see* Infra-red spectroscopy, recent developments in 203
Moon: *see* Radar astronomy 362
Motion, equations of: *see* Hamilton's principle 161
Moving coil instruments, magnets for: *see* Permanent magnet technology 314
Multichannel analysers: *see* Whole body counting 524
Multicomponent diffusion: *see* Diffusion in non-uniform media 59
Multiphase systems, diffusion in: *see* Diffusion in non-uniform media 60
Multi-photon absorption: *see* Laser-induced sparks 243
"MULTIPLE AND" or "ALL" element: *see* Fluidics 118
Multiple reflection technique: *see* Infra-red spectroscopy, recent developments in 205
Multiple scattering diagrams: *see* Dynamical n-beam theory of electron diffraction 71
Multiplets, isospin: *see* Quarks 357
Multiplex advantage: *see* Submillimetre-wave methods 461
Multiplex (or simultaneous) spectrometer: *see* Multiplex principle (of Fellgett) 298
Multiplex principle: *see* Fourier transform spectroscopy 122
Multiplex principle (of Fellgett) 298
Multiplication by optical processing: *see* Optical information processing 302

Multiplier, electron, channel: see Soft X-ray spectroscopy 414
Multiplier, electron, magnetic: see Soft X-ray spectroscopy 414
Multiplier, Hall effect: see Hall effect, applications of 156
Multiplier, three-dimensional: see Magnetoresistivity 274
Multi-reflection gas cell: see Infra-red spectroscopy, recent developments in 206
Multislice model of crystal: see Dynamical n-beam theory of electron diffraction 69
Muon: see Bosons, intermediate 36; Neutretto 300
Muon decay: see Neutretto 300
Mutation, spontaneous: see Quantum mechanics in biochemistry 354
Myoglobin: see Proteins, structure of 340

N-beam theory, dynamical, of electron diffraction: see Dynamical n-beam theory of electron diffraction 69
"N" wave: see Sonic boom, generation and suppression of 431
Nanometer: see Ultra-violet radiation, some biological effects of 506
"Narrow band" semiconductors: see Thermoelectric generation of electricity 486
Natural animal fibres: see Textile industry, physics in 470
Natural draw ratio: see High polymers, mechanical properties of 179
Natural vegetable fibres: see Textile industry, physics in 470
Near field: see Sonic boom, generation and suppression of 431
Near ultra-violet: see Ultra-violet radiation, some biological effects of 506
Near ultra-violet photoproducts: see Ultra-violet radiation, some biological effects of 511
Néel temperature: see Far infra-red spectroscopy 108
Negative binomial (Polya) distribution: see Laser-induced sparks 244
Negative differential resistivity, bulk, in semiconductors: see Microwave solid state oscillators, recent 289
Negative feedback: see Cardiac pacemaker, external 40
Negative resistance: see IMPATT diode microwave generators 194
Negative temperature: see Far infra-red lasers 101
Neighbour restoration: see Ultra-violet radiation, some biological effects of 511
Nematic liquid crystals: see Liquid crystals and molecular structure 257
Neodymium laser: see Lasers, tunable 251
Neon: see Far infra-red lasers 101; Laser-induced sparks 243
Nernst effect: see Thermoelectric generation of electricity 486

Network theory, electrical: see Kirchhoff's law (electrical circuits) 239
Neutralization energy: see Ion-neutralization spectroscopy 220
Neutralization of space charge: see Thermionic generation of electricity 477
Neutretto 300
Neutrino: see Neutretto 300
Neutrino, two-component theory of the: see Helicity 163
Neutrinos, production of W meson by: see Bosons, intermediate 35
Neutron, electric dipole moment of the: see Electric dipole moment of the neutron 74
Neutron, fast-, elastic scattering of: see Radiation quantities, systematics of 369
Neutron, fast-, reactors: see Space power supplies 436
Neutron number (N): see Chart of the nuclides, trilinear 47
Nickel: see Ion-neutralization spectroscopy 224
Nickel alloy, Kanigen: see Metal film preparation by electroless (chemical) reduction 281
Nickel/hypophosphite system: see Metal film preparation by electroless (chemical) reduction 277
Nickel-phosphorus alloy: see Metal film preparation by electroless (chemical) reduction 279
Node, ascending, longitude of the: see Astrodynamics 23
Nodes: see Kirchhoff's law (electrical circuits) 240
Nodes, line of: see Astrodynamics 23
Noise: see Fluidics 122; Jet flap 234
Noise analysis of particle size: see Ceramics production, physics in 42
Noise, electrical: see Pyroelectric detector for infra-red radiation 347
Noise equivalent power: see Far infra-red spectroscopy 105
Noise in the sea, ambient: see Ambient noise in the sea 13
Noise performance: see IMPATT diode microwave generators 195
Noise, quantizing: see Pulse code modulation 345
Noise spectrum level: see Ambient noise in the sea 13
Noise, transmission path: see Pulse code modulation 346
Noise, turbulence: see Speaking machines 440
Nonconservative linear system: see Hamilton's principle 162
Non-deformable particles: see Dispersion and fibre strengthening of materials 63
Non-direct transitions: see Photoemission and band structure 320
Non-leptonic decay: see Current algebras and their applications 54
Non-linear optical effects: see Lasers, tunable 250
Non-metallic crystals, metallic colloids in: see Metallic colloids in non-metallic crystals 282
Non-metals and metals, whiskers of: see Dispersion and fibre strengthening of materials 65

Non-polar gases, compressed: *see* Submillimetre radiation, applications of 452
Non-polar liquids: *see* Submillimetre radiation, applications of 452
Non-reciprocal systems: *see* Gyroscopic coupler 154
Non-uniform media, diffusion in: *see* Diffusion in non-uniform media 58
Non-woven fabrics: *see* Textile industry, physics in 472
"NOR" element: *see* Fluidics 118
Noryl: *see* Temperature resistant polymers 464
Norton's theorem: *see* Gyroscopic coupler 153
Nose flaps: *see* High lift devices 171
"NOT" element: *see* Fluidics 118
"NOT" function: *see* Fluidics 117
Notches: *see* Fracture of brittle materials 130
Nozzle diffuser loss coefficient: *see* Glandless reciprocating jet pump 142
Nuclear accidents: *see* Whole body counting 524
Nuclear doping: *see* Semiconductors, effect of radiation on 407
Nuclear electric quadrupole moment: *see* Nuclear quadrupole resonance 300
Nuclear fission track dating: *see* Geochronology 137
Nuclear fuels, purification of: *see* Radiation chemistry, industrial application of 364
Nuclear "in pile" diode: *see* Thermionic generation of electricity 480
Nuclear magnetic resonance (NMR): *see* Metallic colloids in non-metallic crystals 283; Polarized proton targets 330
Nuclear polarization: *see* Polarized proton targets 329
Nuclear power: *see* Space power supplies 435
Nuclear quadrupole resonance (NQR) 300
Nuclear reactor: *see* Liquid metals 261; Radiation chemistry, industrial applications of 363; Space power supplies 436
Nuclear scattering experiments: *see* Polarized proton targets 329
Nucleating agent: *see* Glass ceramics 143
Nucleating kink pairs: *see* Peierls stress 306
Nucleation: *see* Glass ceramics 143; Solid nucleus formation from solution 430
Nucleation and growth of surface layers: *see* Epitaxy, recent work on 98
Nucleation and growth of thin films: *see* Thin films, nucleation and growth of 490
Nucleation, crack: *see* Fracture of brittle materials 127
Nucleation frequency: *see* Thin films, nucleation and growth of 491
Nucleic acids: *see* Ultra-violet radiation, some biological effects of 506
Nucleon: *see* Elementary particles, nomenclature of 95
Nucleon, quark concentration per: *see* Quarks 361
Nucleus, crack: *see* Fracture of brittle materials 128
Nucleus, solid, formation from solution: *see* Solid nucleus formation from solution 430
Nuclides, trilinear chart of the 47

Nuclidic mass unit: *see* Quantum metrology 356
Nylon: *see* Polymer physics 335; Textile industry, physics in 471

Observation range for a mirage: *see* Mirage and looming, optics of 292
Observation range for looming: *see* Mirage and looming, optics of 292
Occultation: *see* Ionospheres of Mars and Venus 230
Occultations, linear, of radio sources: *see* Radio telescopes, the present position 383
Oceanic traffic noise in the sea: *see* Ambient noise in the sea 15
OH absorption: *see* Radioemissions from interstellar hydroxyl radicals 373
OH, energy levels of: *see* Radioemissions from interstellar hydroxyl radicals 373
OH, microwave spectrum of: *see* Radioemissions from interstellar hydroxyl radicals 372
OH source, W3: *see* Radioemissions from interstellar hydroxyl radicals 375
OH thermal emission: *see* Radioemissions from interstellar hydroxyl radicals 374
ohm: *see* Quantum metrology 356
One electron model: *see* Soft X-ray spectroscopy 417
One-sided interferogram: *see* Interferometric spectrometers 209
Onsager relationships: *see* Diffusion in non-uniform media 59
Open-cycle plant: *see* Electrogasdynamics 76
Optic, electro-, effect, linear: *see* Pockels effect 326
Optical activity: *see* Pockels effect 329
Optical computers: *see* Lasers, applications of 249
Optical density of states: *see* Photoemission and band structure 320
Optical digital information: *see* Lasers, applications of 249
Optical effects, non-linear: *see* Lasers, tunable 250
Optical excitation: *see* Lasers, tunable 252
Optical information processing 302
Optical information storage, three-dimensional: *see* Lasers, applications of 249
Optical materials: *see* Infra-red spectroscopy, recent developments in 203
Optical medium, refractive properties of an: *see* Pockels effect 327
Optical parametric oscillator: *see* Lasers, tunable 251
Optical pumping: *see* Q-switching 350; Radioemissions from interstellar hydroxyl radicals 376
Optical pyrometer: *see* Conductivity at high temperatures 50; Emissivity at high temperatures 96
Optical shutter: *see* Q-switching 350
Optical studies of colloidal particles: *see* Metallic colloids in non-metallic crystals 282
Optical triggering devices: *see* Laser-induced sparks 244
Optics of mirage and looming: *see* Mirage and looming, optics of 290

Opto-electronics: *see* Optical information processing 302
"OR" element: *see* Fluidics 118
Orbit determination: *see* Astrodynamics 17
Orbit prediction: *see* Astrodynamics 20
Orbit, reference: *see* Astrodynamics 23
Orbit, two-body: *see* Astrodynamics 23
Orbit, variant: *see* Astrodynamics 23
Orbital, antisymmetrized molecular: *see* Quantum mechanics in biochemistry 352
Orbital, atomic: *see* Quantum mechanics in biochemistry 352
Orbital, electron, energy level diagram: *see* Satellites in soft X-ray spectra 404
Orbital, molecular: *see* Quantum mechanics in biochemistry 352
Orbital momentum, quenching of: *see* Magnetic thin films 266
Organelles, cell, effect of microbeams on: *see* Ultra-violet radiation, some biological effects of 512
Organic evolution: *see* Geochronology 133
Orientation angles: *see* Astrodynamics 23
Orientation, crystal: *see* Proton scattering microscopy 344
Origin of the solar system: *see* Interplanetary dust 216
Orion A: *see* Radioemissions from interstellar hydroxyl radicals 375
Orlon, Black: *see* Temperature resistant polymers 466
Orthonormal group: *see* Walsh functions 518
Orthotropic symmetry: *see* Textile industry, physics in 469
Oscillation, laser, threshold condition for a: *see* Q-switching 350
Oscillation, transit-time: *see* Microwave solid-state oscillators, recent 289
Oscillator: *see* IMPATT diode microwave generators 195
Oscillator, blocking: *see* Cardiac pacemaker, external 40
Oscillator, Gunn: *see* Microwave solid-state oscillators, recent 288
Oscillator, Hertzian: *see* Far infra-red spectroscopy 104
Oscillator, LSA (limited space-charge accummulation): *see* Microwave solid-state oscillators, recent 288
Oscillator, microwave solid-state: *see* Microwave solid-state oscillators, recent 288
Oscillator, optical parametric: *see* Lasers, tunable 251
Oscillator, pulsed magnetron: *see* Split-beam microwave beacon 444
Oscillator, Raman: *see* Lasers, tunable 251
Oscillator, swept frequency: *see* Ultrasonic spectroscopy 503
Oscillator system, homodyne local: *see* Submillimetre radiation, applications of 454
Oscillator, vortex: *see* Fluidics 121
Osculating element: *see* Astrodynamics 23
Osmium/Rhenium dating: *see* Geochronology 139

Ostwald metastable limit: *see* Solid nucleus formation from solution 431
Overhauser effect: *see* Metallic colloids in non-metallic crystals 285
Oxide nucleating agents: *see* Glass ceramics 144
Oxides, polyphenylene: *see* Temperature resistant polymers 464

Pacemaker, cardiac, external: *see* Cardiac pacemaker, external 39
Pacemaker, cardiac, internal: *see* Cardiac pacemaker, external 39
Pacing, demand: *see* Cardiac pacemaker, external 39
Pair, Cooper: *see* Josephson effect 235
Pair, hole-electron: *see* Microwave solid state oscillators, recent 289
Pair, nucleating kink: *see* Peierls stress 306
Pair production: *see* Radiation quantities, systematics of 369
Pantoyl-lactone restoration: *see* Ultra-violet radiation, some biological effects of 511
Paramagnetic crystals: *see* Far infra-red spectroscopy 108
Paramagnetic resonance, fast passage in: *see* Fast passage in paramagnetic resonance 109
Parametric oscillator, optical: *see* Lasers, tunable 251
Parity: *see* Current algebras and their applications 52
Parity conservations: *see* Electric dipole moment of the neutron 74
Partial ionization cross-sections: *see* Ionization and attachment cross-sections of atoms and molecules by electron impact 218
Particle, charged, equilibrium (CPE): *see* Radiation quantities, systematics of 367
Particle coarsening: *see* Dispersion and fibre strengthening of materials 63
Particle, colloidal: *see* Metallic colloids in non-metallic crystals 282
Particle, deformable and non-deformable: *see* Dispersion and fibre strengthening of materials 63
Particle detector: *see* Streamer chambers 451
Particle, elementary: *see* Current algebras and their applications 51; Quarks 357
Particle, elementary, nomenclature of: *see* Elementary particles, nomenclature of 94
Particle flux density: *see* Radiation quantities, systematics of 366
Particle, independent, model: *see* Soft X-ray spectroscopy 416
Particle size, analysis of: *see* Ceramics production, physics in 42
Particle size distribution: *see* Interplanetary dust 216
Particle, strange: *see* Quarks 357
Parylene: *see* Temperature resistant polymers 464
Passive transensor: *see* Biological pressure transensors 31
Patrol activities: *see* Law enforcement science and technology 254
Pattern playback: *see* Speaking machines 442

Pauli exclusion principle: *see* Charge conjugation 45; Josephson effect 235; Liquid metals 262; Melting, modern theories of 275
Peierls forces: *see* Peierls stress 305
Peierls stress 305
Peierls trough: *see* Peierls stress 306
Peltier coefficient: *see* Peltier refrigerators 307; Thermoelectric generation of electricity 483
Peltier cooler: *see* Thermoelectric cooling of infra-red detectors 481
Peltier cooling effect: *see* Peltier refrigerators 307
Peltier effect: *see* Thermoelectric generation of electricity 483
Peltier refrigerators 307
Pendulum, gyroscopic: *see* Gyroscopic coupler 152
Peptide bond: *see* Proteins, structure of 337; Ultra-violet radiation, some biological effects of 508
Peptide linkages: *see* Quantum mechanics in biochemistry 352
Perifocus: *see* Astrodynamics 23
Perihelion: *see* Radar astronomy 362
Periodic magnetic fields, electron diffraction from: *see* Electron diffraction from periodic magnetic fields 86
Periodicities, crystal lattice, observation of: *see* High resolution electron microscopy 181
Permanent magnet technology 309
Permanent (remanent) magnetization: *see* Piezomagnetic effect 326
Permeance: *see* Permanent magnet technology 312
Permittivity: *see* Glass ceramics 146
Perturbation, Coriolis: *see* Submillimetre-wave lasers 457
Perturbations: *see* Astrodynamics 20
pH: *see* Metal film preparation by electroless (chemical) reduction 279
Phase boundary: *see* Diffusion in non-uniform media 60
Phase contrast: *see* High resolution electron microscopy 182
Phase grating: *see* Electron diffraction from periodic magnetic fields 85; Holography 184
Phase grating series: *see* Dynamical *n*-beam theory of electron diffraction 70
Phase hologram: *see* Holography 184
Phase matching: *see* Lasers, tunable 251
Phase-sensitive amplification: *see* Submillimetre-wave methods 460
Phase separation: *see* Glass ceramics 144
Phase velocity in acoustic waveguides: *see* Acoustic waveguides 6
Phases, crystal: *see* Glass ceramics 143
Phasors: *see* Kirchhoff's law (electrical circuits) 239
Phonon: *see* Conductivity at high temperatures 48
Phonon avalanche and bottleneck 315
Phonon-electron scattering: *see* Photoemission and band structure 319
Phonon excitation: *see* Phonon avalanche and bottleneck 317
Phonon interaction: *see* Josephson effect 235

Phonon, ion-, transition: *see* Phonon avalanche and bottleneck 316
Phonon-phonon quantum counter: *see* Phonon avalanche and bottleneck 319
Phonon, resonant acoustical: *see* Phonon avalanche and bottleneck 316
Phosphates: *see* Quantum mechanics in biochemistry 352
Phosphor persistence: *see* Ultrasonic imaging 500
Phosphorusc–obalt alloys: *see* Metal film preparation by electroless (chemical) reduction 282
Phosphorus–nickel alloy: *see* Metal film preparation by electroless (chemical) reduction 279
Photochemistry: *see* Radiation chemistry, industrial applications of 363
Photocoagulation technique, laser: *see* Lasers, applications of 246
Photoconductive cells: *see* Infra-red cameras and infra-red TV 197
Photoconductive detectors: *see* Thermoelectric cooling of infra-red detectors 481
Photoconductive device: *see* Submillimetre-wave methods 460
Photoconductive retinae: *see* Infra-red cameras and infra-red TV 199
Photoconductivity: *see* Solar cell 422
Photodimerization of pyrimidines: *see* Ultra-violet radiation, some biological effects of 507
Photo-elastic constants: *see* Pockels effect 328
Photoelectric effect: *see* Laser-induced sparks 243
Photoelectric emission: *see* Metallic colloids in non-metallic crystals 282
Photoelectric process: *see* Radiation quantities, systematics of 369
Photoelectric tracking of model ships: *see* Model ships, photoelectric tracking of 294
Photoemission and band structure 319
Photoemissive retinae: *see* Infra-red cameras and infra-red TV 201
Photoenzymatic reactivation, direct: *see* Ultra-violet radiation, some biological effects of 509
Photoexcitation: *see* Photoemission and band structure 319
Photograph paper technique: *see* Ultrasonic imaging 499
Photograph, thermal: *see* Infra-red cameras and infra-red TV 197
Photographic emulsions: *see* Infra-red cameras and infra-red TV 199; Ultrasonic imaging 498
Photohydration of pyrimidines: *see* Ultra-violet radiation, some biological effects of 507
Photoionization: *see* Ionosphere, recent observations of 225
Photomultiplier tubes: *see* Whole body counting 523
Photon absorption, multi-: *see* Laser-induced sparks 243
Photon cavity lifetime: *see* Q-switching 350
Photoproducts, far ultra-violet: *see* Ultra-violet radiation, some biological effects of 506

Photoproducts, near ultra-violet: *see* Ultra-violet radiation, some biological effects of 511
Photoprotection: *see* Ultra-violet radiation, some biological effects of 511
Photoreactivation: *see* Ultra-violet radiation, some biological effects of 508
Photosphere: *see* Daylight, power distribution in 55
Photovoltaic cell: *see* Solar cell 423
Photovoltaic effect: *see* Solar cell 422
Photovoltaic solar cell: *see* Space power supplies 438
Physical limnology 321
Physics, fundamental constants of: *see* Quantum metrology 354
Pi electrons: *see* Quantum mechanics in biochemistry 351
Piezoelectric constant: *see* Pockels effect 328; Ultrasonic amplification in semiconductors 497
Piezoelectric coupling: *see* Acousto-electric effects 9; Ultrasonic amplification in semiconductors 497
Piezoelectric deformation: *see* Pockels effect 327
Piezoelectric transducer: *see* Ultrasonic imaging 501; Ultrasonic spectroscopy 503
Piezomagnetic effect 325
Pilot warning instruments: *see* Airborne collision prevention devices 9
"Pinch" effect: *see* High intensity pulsed radiation sources 166
Pion: *see* Bosons, intermediate 35; Neutretto 300; Quarks 357
Pion, soft-, theorems: *see* Current algebras and their applications 53
Pitch angle distribution of auroral electrons: *see* Auroral electrons 28
Pits, dislocation etch: *see* Microplasticity 287
Plain flap: *see* High lift devices 170
Planar (thin) holograms: *see* Holography 184
Planckian radiator: *see* Daylight, power distribution in 55
Planck's law: *see* Emissivity at high temperatures 95
Plane wave reverberation level, equivalent: *see* Reverberation under water 391
Planetary atmospheric absorption: *see* Infra-red spectroscopy, recent developments in 206
Planetary interactions with the solar wind: *see* Solar wind, planetary interactions with 429
Plant, closed-cycle: *see* Electrogasdynamics 76
Plant, open-cycle: *see* Electrogasdynamics 76
Plasma: *see* Ionosphere, recent observations of 226; Submillimetre radiation, applications of 453; Thermionic generation of electricity 477
Plasma arc: *see* Submillimetre-wave methods 459
Plasma frequency: *see* Satellites in soft X-ray spectra 405
Plasmapause: *see* Ionosphere, recent observations of 229
Plasmon satellite: *see* Satellites in soft X-ray spectra 405
Plastic bubble transensor: *see* Biological pressure transensors 32

Plastic deformation: *see* Brittleness in ceramics 37; Dispersion and fibre strengthening of materials 62; Fracture of brittle materials 125
Plastic flow: *see* Dispersion and fibre strengthening of materials 62; Fracture of brittle materials 124; Microplasticity 286
Plastic shear: *see* Fracture of brittle materials 125
Plastic, wood-, composites: *see* Radiation chemistry, industrial applications of 365
Plastic yield: *see* Fracture of brittle materials 124
Plasticity: *see* Microplasticity 285
Plasticizers: *see* High polymers, mechanical properties of 176
Plastics: *see* High polymers, mechanical properties of 175; Temperature-resistant polymers 463
Plastics, thermoset: *see* Polymer physics 331
Plating, electroless: *see* Metal film preparation by electroless (chemical) reduction 277
Playback, pattern: *see* Speaking machines 442
***pn* diode, avalanching:** *see* IMPATT diode microwave generators 195
***p-n* junction:** *see* Microwave solid state oscillators, recent 288; Solar cell 423
Pneumatic detector: *see* Far infra-red spectroscopy 105
Pneumatic efficiency: *see* Glandless reciprocating jet pump 142
Pneumatic-tube-activated speed timer: *see* Law enforcement science and technology 255
Pockels coefficient: *see* Pockels effect 328
Pockels effect 327
Pohlman cell: *see* Ultrasonic imaging 500
Point-contact diode: *see* Submillimetre-wave methods 460
Point-contact junction: *see* Josephson effect 238
Poisson's equation: *see* IMPATT diode microwave generator 194
Polar cap absorption (PCA): *see* Ionosphere, recent observations of 225
Polar molecules, relaxational (Debye) absorption by: *see* Submillimetre radiation, applications of 452
Polarization: *see* Ferrohydrodynamics 112; Radio-emissions from interstellar hydroxyl radicals 375
Polarization, field-induced: *see* Pockels effect 327
Polarization, nuclear: *see* Polarized proton targets 329
Polarization, radio: *see* Radio telescopes, the present position 382
Polarization, spin: *see* Polarized proton targets 329
Polarization, spontaneous: *see* Pyroelectric detector for infra-red radiation 347
Polarized proton targets 329
Poly xylylene: *see* Temperature resistant polymers 464
Polya (negative binomial) distribution: *see* Laser-induced sparks 244
Polyaddition polymers: *see* Temperature-resistant polymers 463
Polyamides, aromatic: *see* Temperature-resistant polymers 465

Polybenzimidazoles: *see* Temperature-resistant polymers 465
Polychromatic fringes: *see* Submillimetre radiation, applications of 454
Polycondensation polymers: *see* Temperature-resistant polymers 463
Polycrystalline cells: *see* Solar cell 428
Polycrystalline magnetic films: *see* Magnetic thin films 265
Polyethylene: *see* Infra-red spectroscopy, recent developments in 203; Submillimetre-wave methods 461
Polygon, Thiessen, technique: *see* Hydrometeorology 188
Polygraph: *see* Law enforcement science and technology 257
Polyimides: *see* Temperature-resistant polymers 465
Polymer: *see* Proteins, structure of 337; Submillimetre radiation, applications of 452
Polymer, amorphous: *see* High polymers, mechanical properties of 175
Polymer, anisotropic effects in a: *see* High polymers, mechanical properties of 178
Polymer, aromatic: *see* Temperature-resistant polymers 463
Polymer chain: *see* High polymers, mechanical properties of 175; Polymer physics 331
Polymer, crystallizable: *see* Textile industry, physics in 471
Polymer, fluorocarbon: *see* Temperature-resistant polymers 463
Polymer, heteroaromatic: *see* Temperature-resistant polymers 465
Polymer, high mechanical properties of: *see* High polymers, mechanical properties of 175
Polymer in solution and melt: *see* Polymer physics 333
Polymer, inorganic: *see* Temperature-resistant polymers 466
Polymer, ladder: *see* Temperature-resistant polymers 466
Polymer, linear: *see* Polymer physics 332
Polymer, modes of deformation of a: *see* Polymer physics 336
Polymer molecule: *see* Polymer physics 331
Polymer physics 331
Polymer, polyaddition: *see* Temperature-resistant polymers 463
Polymer, polycondensation: *see* Temperature-resistant polymers 463
Polymer, temperature-resistant: *see* Temperature-resistant polymers 463
Polymers, yield and fracture of a: *see* High polymers, mechanical properties of 178
Polymeric fibres, deformation modes of: *see* Textile industry, physics in 471
Polymeric molecules: *see* Rubber physics 393
Polymeric systems, reactions in: *see* Radiation chemistry, industrial applications of 364
Polymerization, degree of: *see* Polymer physics 337

Polyoxadiazoles: *see* Temperature-resistant polymers 465
Polypeptide: *see* Proteins, structure of 337
Polyphenylene: *see* Temperature-resistant polymers 463
Polypyromellitimides: *see* Temperature-resistant polymers 465
Polystyrene: *see* Infra-red spectroscopy, recent developments in 203; Polymer physics 335
Polysulphone: *see* Temperature-resistant polymers 464
Polythene, irradiation of: *see* Radiation chemistry, industrial applications of 365
Population decay rate: *see* Phonon avalanche and bottleneck 316
Population inversion: *see* Q-switching 350; Submillimetre-wave lasers 457
Population, inverted: *see* Far infra-red lasers 101; Lasers, tunable 252
Porcelain, electrical: *see* Ceramics production, physics in 44
Porosity of vitreous carbon: *see* Vitreous carbons, structure and properties of 515
Porphyrins: *see* Quantum mechanics in biochemistry 352
Positron: *see* Charge conjugation 45
Positronium: *see* Charge conjugation 45
Potassium 40: *see* Geochronology 133
Potassium/argon dating: *see* Geochronology 135
Potassium iodide-starch solutions: *see* Ultrasonic imaging 499
Potential, deformation, constant: *see* Ultrasonic amplification in semiconductors 497
Potential, screened Coulomb: *see* Proton scattering microscopy 343
Potential, Thomas-Fermi: *see* Proton scattering microscopy 343
Power: *see* Thermionic generation of electricity 478; Thermoelectric generation of electricity 483
Power, complex: *see* Hamilton's principle 163
Power, emissive: *see* Radiation quantities, systematics of 366
Power gain: *see* Fluidics 118
Power generation: *see* Electrogasdynamics 77
Power, ionizing: *see* Streamer chambers 449
Power, mechanical, transport and control of: *see* Electroviscous fluids, applications of 93
Power, nuclear: *see* Space power supplies 435
Power, on-board rocket: *see* Electrogasdynamics 79
Power, radiative: *see* Emissivity at high temperatures 95
Power, resolving: *see* High resolution electron microscopy 179; Infra-red spectroscopy, recent developments in 202; Radio telescopes, the present position 380
Power, space, supplies: *see* Space power supplies 435
Power, spectral, distribution in daylight: *see* Daylight, power distribution in 55
Power, thermoelectric: *see* Liquid metals 263; Peltier refrigerators 307

36*

Power transducer: *see* Hall effect, applications of 160
Poynting—Robertson effect: *see* Interplanetary dust 215
PPO: *see* Temperature-resistant polymers 464
Precession frequency, Larmor: *see* Thermomagnetic gas torque (The Scott effect) 488
Precession frequency, proton spin: *see* Quantum metrology 356
Precession, Larmor: *see* Electric dipole moment of the neutron 74
Precipitation: *see* Hydrometeorology 192; Physical limnology 323
Precipitation, areal: *see* Hydrometeorology 188
Precipitation, auroral: *see* Auroral electrons 28
Precipitation due to diffusion: *see* Diffusion in non-uniform media 61
Precipitation hardening: *see* Permanent magnet technology 131
Precipitation strengthening: *see* Dispersion and fibre strengthening of materials 62
Pressure, blood: *see* Biological pressure transensors 34
Pressure, dilatant: *see* Singing sands 408
Pressure field, sonic boom: *see* Sonic boom, generation and suppression of 431
Pressure gain: *see* Fluidics 118
Pressure, intracranial: *see* Biological pressure transensors 31
Pressure, intra-ocular: *see* Biological pressure transensors 31
Pressure, magnetic fluid: *see* Ferrohydrodynamics 113
Pressure, radiation: *see* Interplanetary dust 214
Pressure rise, bow shock: *see* Sonic boom, generation and suppression of 433
Pressure, shock: *see* Electrohydraulic crushing 81
Pressure, shock, in meteorites: *see* Shock pressure in meteorites 407
Pressure signature: *see* Sonic boom, generation and suppression of 431
Pressure transensors, biological: *see* Biological pressure transensors 31
Primaeval dust-cloud theory: *see* Interplanetary dust 216
Primary structure: *see* Proteins, structure of 337
Principal function: *see* Hamilton's principle 161
Prism spectrometer: *see* Luminosity of a spectrometer 264
Probe, bismuth: *see* Magnetoresistivity 272
Probe, Hall effect: *see* Hall effect, applications of 156
Probe, thermistor: *see* Ultrasonic imaging 499
Propagation, crack: *see* Fracture of brittle materials 125
Propagation vector: *see* Photoemission and band structure 320
Proportional elements: *see* Fluidics 118
Proportional fluid amplifier: *see* Fluidics 119
Proportional limit: *see* Microplasticity 286
Propulsive efficiency: *see* High lift devices 174

Protein: *see* Proteins, structure of 337; Ultra-violet radiation, some biological effects of 508
Proton blocking: *see* Proton scattering microscopy 342
Proton cyclotron whistler: *see* Ionosphere, recent observations of 228
Proton, gyromagnetic ratio of the: *see* Quantum metrology 356
Proton–nucleon collisions, production of W meson in: *see* Bosons, intermediate 36
Proton scattering from thin films: *see* Proton scattering microscopy 344
Proton scattering microscopy 342
Proton, scattering of, from crystals: *see* Proton scattering microscopy 342
Proton spin precession frequency: *see* Quantum metrology 356
Proton targets, polarized: *see* Polarized proton targets 329
Proton tunnelling: *see* Quantum mechanics in biochemistry 353
Pseudomorphism, basal-plane: *see* Epitaxy, recent work on 99
Pseudoscalar mesons: *see* Quarks 359
Pseudo-potentials, weak scattering: *see* Liquid metals 262
Puckering vibrations: *see* Far infra-red spectroscopy 106
Pulsating aurorae: *see* Auroral electrons 27
Pulse amplitude modulation (P.A.M.): *see* Pulse code modulation 345
Pulse code modulation 344
Pulse, d.c., procedure: *see* Ultrasonic spectroscopy 504
Pulse discriminators: *see* Pulse code modulation 346
Pulse-echo images: *see* Ultrasonic imaging 498
Pulse height analyser: *see* Whole body counting 524
Pulse radiolysis: *see* High intensity pulsed radiation sources 169
Pulse, shock: *see* Electrohydraulic crushing 81
Pulse transformer: *see* High intensity pulsed radiation sources 164
Pulsed discharge supplies: *see* High intensity pulsed radiation sources 164
Pulsed laser: *see* Lasers, applications of 246
Pulsed liquid dielectrics: *see* High intensity pulsed radiation sources 166
Pulsed magnetron oscillator: *see* Split-beam microwave beacon 444
Pulsed radiation sources, high intensity: *see* High intensity pulsed radiation sources 163
Pulsed swept-frequency technique: *see* Ultrasonic spectroscopy 503
Pump, glandless reciprocating jet: *see* Glandless reciprocating jet pump 140
Pump stagnation head: *see* Glandless reciprocating jet pump 142
Pump, "Tree": *see* Glandless reciprocating jet pump 140

Pump, venturi-bulb: *see* Glandless reciprocating jet pump 141
Pump wave: *see* Lasers, tunable 251
Pumping, optical: *see* Q-switching 350; Radio-emissions from interstellar hydroxyl radicals 376
Pumping or hydrodynamic efficiency: *see* Glandless reciprocating jet pump 142
Purification of nuclear fuels: *see* Radiation chemistry, industrial applications of 364
Purification, zone: *see* Refractories, ultra-purification of 387
Purines: *see* Ultra-violet radiation, some biological effects of 507
Pyrimidines: *see* Ultra-violet radiation, some biological effects of 507
Pyroelectric coefficient: *see* Pyroelectric detector for infra-red radiation 348
Pyroelectric detector: *see* Infra-red cameras and infra-red TV 197
Pyroelectric detector for infra-red radiation 347
Pyroelectric effect: *see* Submillimetre-wave methods 460
Pyroelectric retinae: *see* Infra-red cameras and infra-red TV 200
Pyroelectricity: *see* Far infra-red spectroscopy 105
Pyrolysis unit: *see* Infra-red spectroscopy, recent developments in 205
Pyrometer, optical: *see* Conductivity at high temperatures 50; Emissivity at high temperatures 96

Q-spoiling: *see* Q-switching 350
Q-switched laser: *see* Laser-induced sparks 245; Lasers, applications of 246
Q-switching 350
Quadrupole coupling constant: *see* Nuclear quadrupole resonance 301
Quadrupole, electric, interaction: *see* Nuclear quadrupole resonance 301
Quadrupole moment, nuclear electric: *see* Nuclear quadrupole moment 300
Quadrupole moments: *see* Submillimetre radiation, applications of 452
Quadrupole, nuclear, resonance (NQR): *see* Nuclear quadrupole resonance 300
Quality factor: *see* Radiation quantities, systematics of 372
Quantizing noise: *see* Pulse code modulation 345
Quantum constants: *see* Quantum metrology 354
Quantum counter, phonon-phonon: *see* Phonon avalanche and bottleneck 319
Quantum detectors: *see* Infra-red cameras and infra-red TV 197
Quantum, flux: *see* Josephson effect 236
Quantum mechanics in biochemistry 351
Quantum metrology 354
Quantum numbers of quarks: *see* Quarks 358
Quantum resonances, using acoustic waves: *see* Acousto-electric effects 9
Quarks 357

Quartz: *see* Pockels effect 328
Quartz, crystalline: *see* Submillimetre-wave methods 461
Quartz-iodine tungsten lamp: *see* Submillimetre-wave methods 459
Quasers: *see* Radio telescopes, the present position 383
Quaternary structure: *see* Proteins, structure of 338
Quenching of orbital momentum: *see* Magnetic thin films 266

Radar astronomy 362
Radar speed control devices: *see* Law enforcement science and technology 253
Radar tracking: *see* Lasers, applications of 248
Rademacher functions: *see* Walsh functions 517
Radial distribution function: *see* Liquid metals 261
Radiation belts, trapped (van Allen): *see* Auroral electrons 29
Radiation belts, van Allen: *see* Ionosphere, recent observations of 225
Radiation, black body: *see* Emissivity at high temperatures 95
Radiation chemistry, industrial applications of 362
Radiation chopper, selective: *see* Infra-red spectroscopy, recent developments in 202
Radiation damage: *see* High resolution electron microscopy 183; Semiconductors, effect of radiation on 405
Radiation defects: *see* Semiconductors, effect of radiation on 406
Radiation detectors: *see* Infra-red spectroscopy, recent developments in 204
Radiation, effects of, on semiconductors: *see* Semiconductors, effects of radiation on 405
Radiation, emitted Josephson: *see* Josephson effect 237
Radiation hazard: *see* Whole body counting 524
Radiation pressure: *see* Interplanetary dust 214
Radiation quantities, systematics of 366
Radiation, radio background: *see* Radio telescopes, the present position 382
Radiation, radioactive: *see* Radiation chemistry, industrial applications of 363
Radiation, solar, absorption and scattering of: *see* Daylight, power distribution in 55
Radiation sources: *see* Radiation chemistry, industrial applications of 363
Radiation sources, high intensity pulsed: *see* High intensity pulsed radiation sources 163
Radiation sources, monochromatic: *see* Infra-red spectroscopy, recent developments in 203
Radiation, submillimetre, applications of: *see* Submillimetre radiation, applications of 451
Radiation, synchrotron: *see* Quantum metrology 357
Radiation, ultra-violet, some biological effects of: *see* Ultra-violet radiation, some biological effects of 506

Radiationless or Auger transition: *see* Satellites in soft X-ray spectra 404

Radiative power: *see* Emissivity at high temperatures 95

Radiator, Planckian: *see* Daylight, power distribution in 55

Radicals, interstellar hydroxyl, radioemissions from: *see* Radioemissions from interstellar hydroxyl radicals 372

Radio astronomy: *see* Radio telescopes, the present position 386

Radio background radiation: *see* Radio telescopes, the present position 382

Radio frequency modulator, Hall effect: *see* Hall effect, applications of 157

Radio pill technique: *see* Biological pressure transensors 31

Radio polarization: *see* Radio telescopes, the present position 382

Radio source: *see* Radio telescopes, the present position 383

Radio spectrum lines: *see* Radio telescopes, the present position 382

Radio telescope: *see* Radio telescopes, the present position 380

Radio, two-way: *see* Law enforcement science and technology 254

Radioactive radiations: *see* Radiation chemistry, industrial applications of 363

Radioactive tracers: *see* Radioisotopes in food processing 377

Radioactivity: *see* Geochronology 133; Radioisotopes in food processing 377

Radioactivity in the body: *see* Whole body counting 522

Radiocalcium: *see* Whole body counting 522

Radioemissions from interstellar hydroxyl radicals 372

Radiogenic: *see* Geochronology 135

Radiography, flash: *see* High intensity pulsed radiation sources 167

Radioheliograph: *see* Radio telescopes, the present position 385

Radioiodine: *see* Whole body counting 522

Radioionosonde: *see* Auroral electrons 28

Radioisotope generators: *see* Space power supplies 437; Thermoelectric generation of electricity 487

Radioisotopes: *see* Radiation chemistry, industrial applications of 364; Space power supplies 436; Whole body counting 522

Radioisotopes in food processing 377

Radiolysis, pulse: *see* High intensity pulsed radiation sources 169

Radiometer: *see* Infra-red spectroscopy, recent developments in 206

Radiometric dating: *see* Geochronology (radiometric dating) 133

Radionuclide, decay of: *see* Geochronology 133

Radionuclides: *see* Whole body counting 522

Radomes: *see* Glass ceramics 147

Rainfall: *see* Hydrometeorology 188

Rainfall, artificial: *see* Weather control 522

Raman active materials: *see* Lasers, tunable 251

Raman oscillators: *see* Lasers, tunable 251

Raman scattering: *see* Lasers, applications of 247

Raman scattering, stimulated: *see* Lasers, tunable 250

Raman spectroscopy: *see* Lasers, applications of 247

Ramsauer–Townsend effect: *see* Momentum transfer cross-section for low energy electrons determination by warm techniques 296

Range of observation of a mirage: *see* Mirage and looming, optics of 292

Range of observation of looming: *see* Mirage and looming, optics of 292

Ranging: *see* Lasers, applications of 247

Rankine turbine: *see* Space power supplies 437

Rare gas lasers: *see* Far infra-red lasers 101

Rasterplatte: *see* Magnetoresistivity 271

Ratio recording with interferometer: *see* Interferometric spectrometers 210

Ray tracing in a stratified medium: *see* Mirage and looming, optics of 291

Rayleigh distribution: *see* Reverberation under water 391

Rayleigh–Jeans law: *see* Submillimetre-wave methods 459

Rayleigh scattering: *see* Daylight, power distribution in 55; Interplanetary dust 214; Reverberation under water 391

Rayleigh's dissipation function: *see* Hamilton's principle 162

Reach: *see* Sailing, physics of 397

Reactions in polymeric systems: *see* Radiation chemistry, industrial applications of 364

Reactions, thermonuclear: *see* Lasers, applications of 246

Reactions, thermonuclear, controlled: *see* Laser-induced sparks 244

Reactivation, direct photoenzymatic: *see* Ultra-violet radiation, some biological effects of 509

Reactivation, host-cell, of bacteriophage: *see* Ultra-violet radiation, some biological effects of 510

Reactivation, ultra-violet: *see* Ultra-violet radiation, some biological effects of 511

Reactivity, chemical: *see* Quantum mechanics in biochemistry 351

Reactors, chemonuclear: *see* Radiation chemistry, industrial applications of 363

Reactors, fast-neutron: *see* Space power supplies 436

Reactors, fission: *see* Space power supplies 436

Reactors, nuclear: *see* Liquid metals 261; Radiation chemistry, industrial applications of 363; Space power supplies 436

Read diode: *see* IMPATT diode microwave generators 194

Reagents, concentration of: *see* Metal film preparation by electroless (chemical) reduction 279

Real modulus: *see* Polymer physics 337

Reciprocal dielectric relaxation time: *see* Ultrasonic amplification in semiconductors 497
Reciprocal lattice: *see* Dynamical n-beam theory of electron diffraction 70
Reciprocating jet pump, glandless: *see* Glandless reciprocating jet pump 140
Reciprocity: *see* Dynamical n-beam theory of electron diffraction 72
Reciprocity relation: *see* Gyroscopic coupler 151
Recoil energy of a magnet: *see* Permanent magnet technology 312
Recoil flux density: *see* Permanent magnet technology 312
Recombination centres: *see* Semiconductors, effect of radiation on 406
Recombination repair: *see* Ultra-violet radiation, some biological effects of 510
Recombination, surface: *see* Solar cell 427
Recovery, liquid-holding: *see* Ultra-violet radiation, some biological effects of 511
Recrystallization during growth: *see* Epitaxy, recent work on 99
Rectifier crystals, germanium: *see* Submillimetre-wave methods 460
Red aurorae: *see* Auroral electrons 27
Reducing agent: *see* Metal film preparation by electroless (chemical) reduction 277
Reduction, electroless (chemical), metal film preparation by: *see* Metal film preparation by electroless (chemical) reduction 277
Reference orbit: *see* Astrodynamics 23
Refining, sugar: *see* Radioisotopes in food processing 377
Refining, zone: *see* Refractories, ultra-purification of 387
Reflecting microscope: *see* Infra-red spectroscopy, recent developments in 204
Reflecting radio telescopes: *see* Radio telescopes, the present position 380
Reflection electron microscopy: *see* Electron metallography, applications of 90
Reflection filter: *see* Far infra-red spectroscopy 106; Submillimetre-wave methods 462
Reflection grating: *see* Soft X-ray spectroscopy 414
Reflection methods: *see* Ultrasonic imaging 498
Reflection, multiple, technique: *see* Infra-red spectroscopy, recent developments in 205
Reflectivity: *see* Emissivity at high temperatures 96
Reflector, grating: *see* Infra-red spectroscopy, recent developments in 202
Reflector, wire gauge: *see* Infra-red spectroscopy, recent developments in 202
Refraction of light: *see* Mirage and looming, optics of 290
Refractive index measurement in far infra-red: *see* Interferometric spectrometers 211
Refractive index, variations of, with height: *see* Mirage and looming, optics of 291
Refractive properties of an optical medium: *see* Pockels effect 327

Refractivity profiles, atmospheric: *see* Ionospheres of Mars and Venus 230
Refractories, superconducting: *see* Refractories, ultra-purification of 389
Refractories, ultra-purification of 386
Refractoriness: *see* Ceramics production, physics in 43
Refractory cones: *see* Ceramics production, physics in 43
Refractory materials: *see* Ceramics production, physics in 43
Refractory metals: *see* Refractories, ultra-purification of 386
Refrigeration, coefficient of performance for: *see* Peltier refrigerators 307
Refrigerator, He^3–He^4: *see* Polarized proton targets 329
Refrigerator, Peltier: *see* Peltier refrigerators 307
Refrigerator, thermoelectric: *see* Peltier refrigerators 307
Regenerated fibres: *see* Textile industry, physics in 470
Regenerative repeater: *see* Pulse code modulation 346
Reinforced metal, fibre: *see* Freeze-coating 130
Reinforcing fillers: *see* High polymers, mechanical properties of 178
Relaxation modulus: *see* Polymer physics 335
Relaxation, spin-lattice: *see* Phonon avalanche and bottleneck 316; Polarized proton targets 329
Relaxation strength: *see* Snoek spectra 411
Relaxation, stress: *see* Rubber physics 394; Textile industry, physics in 469
Relaxation time: *see* Magnetoresistivity 269
Relaxation time, reciprocal dielectric: *see* Ultrasonic amplification in semiconductors 497
Relaxational (Debye) absorption by polar molecules: *see* Submillimetre radiation, applications of 452
Remanence: *see* Permanent magnet technology 311
Remanent (permanent) magnetization: *see* Piezomagnetic effect 326
Repair, excision – resynthesis: *see* Ultra-violet radiation, some biological effects of 510
Repair, recombination: *see* Ultra-violet radiation, some biological effects of 510
Repair treatments: *see* Ultra-violet radiation, some biological effects of 511
Replica, carbon extraction: *see* Electron metallography, applications of 90
Replica technique: *see* Electron metallography, applications of 89
Repulsive forces: *see* Melting, modern theories of 275
Reservoir: *see* Hydrometeorology 189
Residuals: *see* Astrodynamics 23
Resins, Doryl: *see* Temperature-resistant polymers 464
Resistance, abrasion: *see* Glass ceramics 146
Resistance, electrical: *see* Textile industry, physics in 469

Resistance, electrical, of electroviscous fluids: *see* Electroviscous fluids, applications of 93
Resistance, electrical, of liquid metals: *see* Liquid metals 263
Resistance, linear flow: *see* Fluidics 122
Resistance, negative: *see* IMPATT diode microwave generators 194
Resistance, shear: *see* Singing sands 409
Resistance, thermal shock: *see* Glass ceramics 147
Resistivity, bulk negative differential, in semiconductors: *see* Microwave solid-state oscillators, recent 289
Resistivity, electrical: *see* Conductivity at high temperatures 50; Glass ceramics 145
Resistors, variable: *see* Magnetoresistivity 272
Resolution, angular: *see* Radio telescopes, the present position 380
Resolution limit: *see* High resolution electron microscopy 181
Resolving power: *see* High resolution electron microscopy 179; Infra-red spectroscopy, recent developments in 202; Radio telescopes, the present position 380
Resonance absorption of acoustic waves: *see* Acousto-electric effects 9
Resonance absorption, spin: *see* Phonon avalanche and bottleneck 318
Resonance, baryon: *see* Quarks 358
Resonance capture: *see* Ionization and attachment cross-sections of atoms and molecules by electron impact 219
Resonance, conduction electron spin (CESR): *see* Metallic colloids in nonmetallic crystals 283
Resonance, cyclotron, using acoustic waves: *see* Acousto-electric effects 9
Resonance frequency, electronic: *see* Polarized proton targets 330
Resonance, geometric, using acoustic waves: *see* Acousto-electric effects 9
Resonance, magnetic: *see* Far infra-red spectroscopy 108; Submillimetre radiation, applications of 452
Resonance, nuclear magnetic (NMR): *see* Metallic colloids in non-metallic crystals 283; Polarized proton targets 330
Resonance, nuclear quadrupole (NQR): *see* Nuclear quadrupole resonance 300
Resonance, paramagnetic, fast passage in: *see* Fast passage in paramagnetic resonance 109
Resonance, quantum, using acoustic waves: *see* Acousto-electric effects 9
Resonance, spin: *see* Fast passage in paramagnetic resonance 109
Resonance, ultrasonic, testing: *see* Ultrasonic spectroscopy 504
Resonant acoustical phonons: *see* Phonon avalanche and bottleneck 316
Resonant frequency detection of a passive transensor: *see* Biological pressure transensors 32
Resonator: *see* Acoustic waveguides 5; Submillimetre-wave lasers 455
Resonator, acoustic: *see* Speaking machines 440
Resonator modes: *see* Far infra-red lasers 102
Restrictor, laminar: *see* Fluidics 122
Reststrahlen filters: *see* Infra-red spectroscopy, recent developments in 202
Resynthesis, excision-, repair: *see* Ultra-violet radiation, some biological effects of 510
Retardation voltage, half-wave: *see* Pockels effect 327
Retinae: *see* Infra-red cameras and infra-red TV 197
Reverberation under water 390
Reversed cell: *see* Solar cell 424
Revolution, aeroplane equivalent body of: *see* Sonic boom, generation and suppression of 432
Rhenium/osmium dating: *see* Geochronology 139
Rheological properties of electroviscous fluids, field-induced: *see* Electroviscous fluids, applications of 94
Ribonucleic acid (RNA): *see* Ultra-violet radiation, some biological effects of 506
Richardson equation: *see* Thermionic generation of electricity 476
Richardson plot: *see* Thermionic generation of electricity 477
Rig: *see* Sailing, physics of 397
Ring laser (laser gyroscope): *see* Lasers, applications of 248
Ring pattern: *see* Electron metallography, applications of 90
Ring spindle: *see* Textile industry, physics in 473
Ring spinning: *see* Textile industry, physics in 473
Riometer: *see* Auroral electrons 28
Riot control: *see* Law enforcement science and technology 256
RIPPLE (Radio Isotope Powered Prolonged Life) generators: *see* Thermoelectric generation of electricity 487
Ripple structure: *see* Magnetic thin films 268
River forecasting: *see* Hydrometeorology 191
River lines: *see* Fracture of brittle materials 127
Rocket measurements: *see* Ionosphere, recent observations of 225
Rocket power, on-board: *see* Electrogasdynamics 79
Rotary interference filter: *see* Infra-red spectroscopy, recent developments in 202
Rotating cylinder, lift on: *see* High lift devices 173
Rotating flap: *see* High lift devices 173
Rotation, Faraday: *see* Ionosphere, recent observations of 228
Rotation spectrum: *see* Far infra-red spectroscopy 106
Rotation-vibration energy levels: *see* Far infra-red lasers 102
Rotational spectra: *see* Infra-red spectroscopy, recent developments in 205
Rotational spectrum of a gas: *see* Submillimetre radiation, applications of 451
Rotor axes: *see* Dynamical n-beam theory of electron diffraction 72
Row lines: *see* Proteins, structure of 341
Rubber elasticity: *see* Polymer physics 333

Rubber elasticity, molecular theory of: *see* High polymers, mechanical properties of 176
Rubber-glass transition: *see* Polymer physics 334
Rubber-like elasticity: *see* Rubber physics 393
Rubber-like state: *see* High polymers, mechanical properties of 175
Rubber network, stress-strain curves for a: *see* High polymers, mechanical properties of 175
Rubber physics 393
Rubber stress–strain curve: *see* Polymer physics 334; Rubber physics 393
Rubber, vulcanization process in: *see* High polymers, mechanical properties of 175
Rubber, Young's modulus of: *see* Rubber physics 396
Rubbery form of polymers: *see* Polymer physics 332
Rubidium/strontium dating: *see* Geochronology 136
Ruby laser: *see* Q-switching 350; Laser-induced sparks 243; Lasers, applications of 246; Lasers, tunable 251
Rupture, modulus of: *see* Glass ceramics 146

Sailing, physics of 397
Sampling: *see* Proteins, structure of 341
Sampling, signal: *see* Pulse code modulation 345
Sand avalanche: *see* Singing sands 408
Sands, singing: *see* Singing sands 408
Satellite experiments: *see* Ionosphere, recent observations of 225
Satellite, plasmon: *see* Satellites in soft X-ray spectra 405
Satellites: *see* Standard earth, geodetic parameters for 447
Satellites in soft X-ray spectra 404; *see also* Soft X-ray spectroscopy 419
Saturation magnetization: *see* Ferrohydrodynamics 112
Scabbing: *see* Electrohydraulic crushing 82
Scanning: *see* Whole body counting 523
Scanning electron microscopy: *see* Electron metallography, applications of 91
Scanning function: *see* Interferometric spectrometers 211
Scanning systems: *see* Infra-red cameras and infra-red TV 197
Scatter plates: *see* Infra-red spectroscopy, recent developments in 202
Scattering: *see* Melting, modern theories of 276
Scattering and absorption of solar radiation: *see* Daylight, power distribution in 55
Scattering, Brillouin: *see* Phonon avalanche and bottleneck 318
Scattering cross-section: *see* Interplanetary dust 214
Scattering, elastic, of fast neutrons: *see* Radiation quantities, systematics of 369
Scattering, electron–electron: *see* Photoemission, and band structure 319
Scattering, electron–phonon: *see* Photoemission and band structure 319
Scattering factor, atomic: *see* Liquid metals 261

Scattering, geometrical: *see* Interplanetary dust 214
Scattering layer, deep: *see* Reverberation under water 390
Scattering, Mie: *see* Interplanetary dust 214
Scattering, multiple, diagrams: *see* Dynamical n-beam theory of electron diffraction 71
Scattering, nuclear, experiments: *see* Polarized proton targets 329
Scattering of electrons: *see* Momentum transfer cross-section for low-energy electrons: determination by swarm techniques 296
Scattering of hadrons, high energy: *see* Quarks 360
Scattering of protons from crystals: *see* Proton scattering microscopy 342
Scattering of sound: *see* Reverberation under water 390
Scattering, proton, from thin films: *see* Proton scattering microscopy 344
Scattering, proton, microscopy: *see* Proton scattering microscopy 342
Scattering, Raman: *see* Lasers, applications of 247
Scattering, Rayleigh: *see* Daylight, power distribution in 55; Interplanetary dust 214; Reverberation under water 391
Scattering, stimulated Brillouin: *see* Lasers, applications of 248
Scattering, stimulated Raman: *see* Lasers, tunable 250
Scattering, Tyndall: *see* Interplanetary dust 214; Metallic colloids in non-metallic crystals 282
Scattering, weak, (pseudo)-potentials: *see* Liquid metals 262
Scattering, X-ray: *see* Liquid metals 261
Schlieren ultrasound imaging: *see* Ultrasonic imaging 500
Scott effect: *see* Thermomagnetic gas torque (The Scott effect) 488
Scott R' centres: *see* Metallic colloids in non-metallic crystals 284
Screened Coulomb potential: *see* Proton scattering microscopy 343
Screening length: *see* Proton scattering microscopy 343
Screw axes: *see* Dynamical n-beam theory of electron diffraction 72
Screw dislocation: *see* Fracture of brittle materials 126; Peierls stress 307
Sea, noise in the: *see* Ambient noise in the sea 13
Seals, magnetic fluid: *see* Ferrohydrodynamics 115
Second: *see* Quantum metrology 354
Secondary electron emission spectroscopy: *see* Auger emission spectroscopy 24
Secondary emission: Infra-red cameras and infra-red TV 201
Secondary structure: *see* Proteins, structure of 337
Second-quantized electron field: *see* Charge conjugation 45
Sedimentation analysis of particle size: *see* Ceramics production, physics in 42

Seebeck coefficient (thermoelectric power): *see* Peltier refrigerators 307; Thermoelectric generation of electricity 482
Seebeck effect: *see* Thermoelectric generation of electricity 482
Seebeck emf: *see* Thermoelectric generation of electricity 482
Seiche: *see* Hydrometeorology 192; Physical limnology 322
Seismic noise in the sea: *see* Ambient noise in the sea 15
Seismomagnetic effect: *see* Piezomagnetic effect 326
Seismometers, laser: *see* Lasers, applications of 248
Selected area electron diffraction analysis: *see* Electron metallography, applications of 89
Selective crystallization: *see* Glass ceramics 144
Selenium: *see* Infra-red cameras and infra-red TV 199
Semiconducting layers: *see* Thin films, nucleation and growth of 493
Semiconductor: *see* Peltier refrigerators 308; Refractories, ultra-purification of 389; Solar cell 422; Thermoelectric generation of electricity 484
Semiconductor, "broad band": *see* Thermoelectric generation of electricity 485
Semiconductor, bulk negative differential resistivity in a: *see* Microwave solid-state oscillators, recent 289
Semiconductor diodes: *see* IMPATT diode microwave generators 194
Semiconductor, effect of radiation on a: *see* Semiconductors, effect of radiation on 405
Semiconductor, extrinsic: *see* Thermoelectric generation of electricity 484
Semiconductor, intrinsic: *see* Thermoelectric generation of electricity 484
Semiconductor lasers: *see* Lasers, applications of 249
Semiconductor lasers, electroluminescent: *see* Far infra-red lasers 103
Semiconductor, lattice imperfections in a: *see* Semiconductors, effect of radiation on 405
Semiconductor, magnetoresistance of a: *see* Magneto-resistivity 274
Semiconductor, "narrow band": *see* Thermoelectric generation of electricity 486
Semiconductor, thermal conductivity of a: *see* Peltier refrigerators 308
Semiconductor, ultrasonic amplification in a: *see* Ultrasonic amplification in semiconductors 497
Semi-major axis: *see* Astrodynamics 23
Senftleben effect: *see* Thermomagnetic gas torque (The Scott effect) 488
Sensor, vortex: *see* Fluidics 121
Separation, phase: *see* Glass ceramics 144
Separation, spinodal: *see* Solid nucleus formation from solution 430
Sequency: *see* Walsh functions 518
Sequential data processing: *see* Astrodynamics 20
Sequential spectrometer: *see* Multiplex principle (of Fellgett) 298

Shadow shielding: *see* Whole body counting 524
Shape function: *see* Fast passage in paramagnetic resonance 111
Shear mode, transverse: *see* Melting, modern theories of 276
Shear modulus, complex: *see* Rubber physics 394
Shear, plastic: *see* Fracture of brittle materials 125
Shear resistance: *see* Singing sands 409
Shear stress: *see* Peierls stress 305
Shear surface: *see* Singing sands 409
Shear waves: *see* Acoustic waveguides 7
Sheet ice: *see* Hydrometeorology 191
Shielding, shadow: *see* Whole body counting 524
Shift, Doppler frequency: *see* Airborne collision prevention devices 11
Ships, model, photoelectric tracking of: *see* Model ships, photoelectric tracking of 294
Shock, bow, pressure rise: *see* Sonic boom, generation and suppression of 433
Shock front: *see* Electrohydraulic crushing 82
Shock indicators: *see* Shock pressure in meteorites 407
Shock pressure: *see* Electrohydraulic crushing 81
Shock pressure in meteorites 407
Shock pulse: *see* Electrohydraulic crushing 81
Shock resistance, thermal: *see* Glass ceramics 147
Shock wave: *see* Electrohydraulic crushing 80; Laser-induced sparks 244; Shock pressure in meteorites 407
Shock wave method: *see* Melting, modern theories of 275
Shrouded-jet flap: *see* Jet flap 232
Shutter, optical: *see* Q-switching 350
SI units: *see* Quantum metrology 356
Sideband generator, single: *see* Hall effect, applications of 158
Sideband holograms: *see* Holography 185
Side-chains: *see* Proteins, structure of 337
Sieve analysis of particle size: *see* Ceramics production, physics in 42
Sigma electrons: *see* Quantum mechanics in biochemistry 351
Signal sampling: *see* Pulse code modulation 345
Signature, pressure: *see* Sonic boom, generation and suppression of 431
Silicon: *see* Magnetoresistivity 269; Solar cell 422
Silicones: *see* Temperature resistant polymers 466
Silks: *see* Proteins, structure of 339
Silver chloride cell: *see* Infra-red spectroscopy, recent developments in 205
Silver iodide: *see* Weather control 522
Simultaneous (or multiplex) spectrometer: *see* Multiplex principle (of Fellgett) 299
Sine transform: *see* Fourier transform spectroscopy 124
Singing sands 408
Single phase system, diffusion in: *see* Diffusion in non-uniform media 59
Single sideband generator: *see* Hall effect, applications of 158

Size, analysis of particle: *see* Ceramics production, physics in 42
Slat, leading edge: *see* High lift devices 171
Slip band: *see* Fracture of brittle materials 128
Slip by dislocation motion: *see* Brittleness in ceramics 37
Slip casting: *see* Ceramics production, physics in 41
Slip, ceramic: *see* Ceramics production, physics in 41
Slip, ion: *see* Electrogasdynamics 75
Slot: *see* High lift devices 171
Slotted flap: *see* High lift devices 171
Smectic liquid crystals: *see* Liquid crystals and molecular structure 257
Smectic mesophase: *see* Liquid crystals and molecular structure 257
Smectic thermal stability: *see* Liquid crystals and molecular structure 259
SNAP generators: *see* Thermoelectric generation of electricity 487
Snell's law: *see* Mirage and looming, optics of 291
Snoek effect: *see* Snoek spectra 410
Snoek spectra 410
Snow storm maximization method: *see* Hydrometeorology 190
Snowmelt: *see* Hydrometeorology 190
Sodium iodide thallium activated crystals: *see* Whole body counting 523
Soft-pion theorems: *see* Current algebras and their applications 53
Soft X-ray spectra, satellites in: *see* Satellites in soft X-ray spectra 404; Soft X-ray spectroscopy 419
Soft X-ray spectroscopy 412
Solar cell, 422; *see also* Space power supplies 438
Solar cell, photovoltaic: *see* Space power supplies 438
Solar collection efficiency: *see* Solar cell 424
Solar corona: *see* Soft X-ray spectroscopy 420
Solar energy: *see* Space power supplies 435
Solar flares: *see* Soft X-ray spectroscopy 421
Solar radiation, scattering and absorption of: *see* Daylight, power distribution in 55
Solar second, mean: *see* Quantum metrology 354
Solar spectroscopy: *see* Soft X-ray spectroscopy 420
Solar system, origin of: *see* Interplanetary dust 216
Solar wind: *see* Auroral electrons 30; Interplanetary dust 213; Ionosphere, recent observations of 229
Solar wind, planetary interactions with 429
Solid acoustic waveguides: *see* Acoustic waveguides 7
Solid effect: *see* Polarized proton targets 330
Solid lasers: *see* Infra-red spectroscopy, recent developments in 203
Solid nucleus formation from solution 430
Solid state far infra-red lasers: *see* Far infra-red lasers 103
Solid state lasers: *see* Q-switching 350
Solid state masers: *see* Far infra-red lasers 103
Solid state oscillators, microwave: *see* Microwave solid state oscillators, recent 288
Solid state spectroscopy: *see* Soft X-ray spectroscopy 416

Solid state X-ray spectra: *see* Soft X-ray spectroscopy 412
Sonic boom, generation and suppression of 431
Sonic boom pressure field: *see* Sonic boom, generation and suppression of 431
Sound, scattering of: *see* Reverberation under water 390
Sound wave: *see* Acoustic holography 1
Sound wave attenuation: *see* Melting, modern theories of 276
Source energy, emitted: *see* Radiation quantities, systematics of 366
Sources, high intensity pulsed radiation: *see* High intensity pulsed radiation sources 163
Sources in the far infra-red: *see* Far infra-red spectroscopy 105
Sources, monochromatic radiation: *see* Infra-red spectroscopy, recent developments in 203
Sources, radiation: *see* Radiation chemistry, industrial applications of 363
Sources, radio: *see* Radio telescopes, the present position 383
Sources, tunable light: *see* Infra-red spectroscopy, recent developments in 204
Space charge: *see* High intensity pulsed radiation sources 164; Thermionic generation of electricity 476
Space power supplies 435
Spark chamber: *see* Streamer chambers 448
Spark channel: *see* Electrohydraulic crushing 81
Spark circuits, underwater: *see* Electrohydraulic crushing 81
Spark discharges: *see* Laser-induced sparks 243
Spark-gap: *see* Laser-induced sparks 243
Spark, laser-induced: *see* Laser-induced sparks 242
Spark, underwater: *see* Electrohydraulic forming of metals 80; Electrohydraulic crushing 83
Spatial filtering: *see* Lasers, applications of 247; Optical information processing 303
Speaking machines 440
Specific activity: *see* Radiation quantities, systematics of 366
Specific gamma ray constant: *see* Radiation quantities, systematics of 366
Specific gravity: *see* Glass ceramics 145
Specific heat: *see* Emissivity at high temperatures 96; Glass ceramics 147
Specific stress: *see* Textile industry, physics in 468
Spectra, absorption: *see* Radioemissions from interstellar hydroxyl radicals 373; Soft X-ray spectroscopy 412
Spectra, alloy: *see* Soft X-ray spectroscopy 420
Spectra, Auger: *see* Auger emission spectroscopy 25
Spectra, defect-echo: *see* Ultrasonic spectroscopy 506
Spectra, emission: *see* Soft X-ray spectroscopy 412
Spectra, energy, of ejected atoms: *see* Sputtering, theories of 446
Spectra, rotational: *see* Infra-red spectroscopy, recent developments in 205
Spectra, Snoek: *see* Snoek spectra 410

Spectra, soft X-ray, satellites in: *see* Satellites in soft X-ray spectra 404; Soft X-ray spectroscopy 419
Spectra, valence band: *see* Soft X-ray spectroscopy 412
Spectra, vibrational: *see* Infra-red spectroscopy, recent developments in 205
Spectra, X-ray: *see* Auger emission spectroscopy 25
Spectra, X-ray, solid state: *see* Soft X-ray spectroscopy 412
Spectral emissivity: *see* Emissivity at high temperatures 95
Spectral emittance: *see* Emissivity at high temperatures 95
Spectral power distribution in daylight: *see* Daylight, power distribution in 55
Spectral response of the solar cell: *see* Solar cell 425
Spectral window: *see* Fourier transform spectroscopy 123
Spectran: *see* Infra-red spectroscopy, recent developments in 205
Spectrograph: *see* Multiplex principle (of Fellgett) 299
Spectrometer: *see* Whole body counting 524
Spectrometer, crystal: *see* Soft X-ray spectroscopy 413
Spectrometer, echelette grating: *see* Infra-red spectroscopy, recent developments in 203
Spectrometer, electron: *see* Auger emission spectroscopy 24
Spectrometer, Fabry–Perot interferometric: *see* Interferometric spectrometers 212
Spectrometer, far infra-red laser: *see* Far infra-red lasers 104
Spectrometer, Fourier: *see* Submillimetre-wave lasers 456; Submillimetre-wave methods 461
Spectrometer, grating: *see* Infra-red spectroscopy, recent developments in 202; Interferometric spectrometers 211; Luminosity of a spectrometer 264; Submillimetre-wave lasers 456; Submillimetre-wave methods 461
Spectrometer, grazing incidence grating: *see* Soft X-ray spectroscopy 413
Spectrometer, infra-red: *see* Infra-red spectroscopy, recent developments in 204
Spectrometer, interferometric: *see* Interferometric spectrometers 206; Multiplex principle (of Fellgett) 298
Spectrometer, luminosity of a: *see* Interferometric spectrometers 213; Luminosity of a spectrometer 264
Spectrometer, magnetic: *see* High intensity pulsed radiation sources 167
Spectrometer, Michelson interferometric: *see* Infra-red spectroscopy, recent developments in 203; Interferometric spectrometers 207
Spectrometer, prism: *see* Luminosity of a spectrometer 264
Spectrometer, sequential: *see* Multiplex principle (of Fellgett) 298

Spectrometer, simultaneous (or multiplex): *see* Multiplex principle (of Fellgett) 298
Spectrometry, mass: *see* Geochronology 135
Spectroscope, ultrasonic: *see* Ultrasonic spectroscopy 503
Spectroscopy, Auger emission: *see* Auger emission spectroscopy 23
Spectroscopy, far infra-red: *see* Far infra-red spectroscopy 104
Spectroscopy, Fourier transform: *see* Far infra-red spectroscopy 105; Fourier transform spectroscopy 122; Submillimetre radiation, applications of 452
Spectroscopy, grating: *see* Far infra-red spectroscopy 105
Spectroscopy, infra-red, recent developments in: *see* Infra-red spectroscopy, recent developments in 202
Spectroscopy, ion-neutralization: *see* Ion-neutralization spectroscopy 220
Spectroscopy, molecular, in the far infra-red: *see* Far infra-red spectroscopy 106
Spectroscopy, Raman: *see* Lasers, applications of 247
Spectroscopy, secondary electron emission: *see* Auger emission spectroscopy 24
Spectroscopy, soft X-ray: *see* Soft X-ray spectroscopy 412
Spectroscopy, solar: *see* Soft X-ray spectroscopy 420
Spectroscopy, solid state: *see* Soft X-ray spectroscopy 416
Spectroscopy, ultrasonic: *see* Ultrasonic spectroscopy 502
Spectrum, absorption: *see* Far infra-red spectroscopy 106
Spectrum, atmospheric transmission: *see* Submillimetre radiation, applications of 453
Spectrum, biological, of far ultra-violet molecular processes: *see* Ultra-violet radiation, some biological effects of 511
Spectrum, channel: *see* Submillimetre-wave methods 461
Spectrum, radio, lines: *see* Radio telescopes, the present position 382
Spectrum, rotation: *see* Far infra-red spectroscopy 106
Spectrum, rotational, of a gas: *see* Submillimetre radiation, applications of 451
Spectrum, transmittance: *see* Interferometric spectrometers 209
Speech-generating process, human: *see* Speaking machines 440
Speech synthesizer: *see* Speaking machines 440
Speed control devices, radar: *see* Law enforcement science and technology 253
Speed timer, pneumatic-tube-activated: *see* Law enforcement science and technology 255
Sphere, Ewald: *see* Dynamical n-beam theory of electron diffraction 70
Spherical aberration: *see* High resolution electron microscopy 179
Spherulitic crystallization: *see* Polymer physics 335

Spike, thermal: *see* Sputtering, theories of 445
Spillway, dam: *see* Hydrometeorology 189
Spin, isobaric: *see* Elementary particles, nomenclature of 95
Spin, isotopic: *see* Elementary particles, nomenclature of 95
Spin-lattice relaxation: *see* Phonon avalanche and bottleneck 316; Polarized proton targets 329
Spin of W meson: *see* Bosons, intermediate 35
Spin polarization: *see* Polarized proton targets 329
Spin, proton, precession frequency: *see* Quantum metrology 356
Spin, quark: *see* Quarks 360
Spin resonance: *see* Fast passage in paramagnetic resonance 109
Spin resonance absorption: *see* Phonon avalanche and bottleneck 318
Spin resonance, conduction electron (CESR): *see* Metallic colloids in non-metallic crystals 283
Spin-spin interaction: *see* Fast passage in paramagnetic resonance 110
Spin temperature: *see* Fast passage in paramagnetic resonance 110
Spindle, false-twist: *see* Textile industry, physics in 475
Spindle, ring: *see* Textile industry, physics in 473
Spinnakers: *see* Sailing, physics of 397
Spinning: *see* Textile industry, physics in 473
Spinodal separation: *see* Solid nucleus formation from solution 430
Split-beam microwave beacon 444
Split flap: *see* High lift devices 171
Splitter: *see* Fluidics 119
Spoiler: *see* High lift devices 171
Spoiling, Q-: *see* Q-switching 350
Spontaneous emission: *see* Lasers, tunable 250
Spontaneous mutation: *see* Quantum mechanics in biochemistry 354
Spontaneous polarization: *see* Pyroelectric detector for infra-red radiation 347
Sporadic E: *see* Ionosphere, recent observations of 225
Spot pattern: *see* Electron metallography, applications of 90
Sputtering, theories of 445
Stability, line of: *see* Chart of the nuclides, trilinear 47
Stability, mesomorphic thermal: *see* Liquid crystals and molecular structure 259
Stability, smectic thermal: *see* Liquid crystals and molecular structure 259
Stability, surface, magnetic fluid: *see* Ferrohydrodynamics 114
Stabilization of protein structures: *see* Proteins, structure of 341
Stagnation head, pump: *see* Glandless reciprocating jet pump 142
Stalling stability: *see* Jet flap 233
Standard earth, geodetic parameters for 447
Standard illuminants: *see* Daylight, power distribution in 56

Standard time, cesium: *see* Airborne collision prevention devices 11
Starch, potassium iodide-, solutions: *see* Ultrasonic imaging 499
Statistics, Fermi–Dirac: *see* Josephson effect 235; Thermionic generation of electricity 476
Stefan–Boltzmann law: *see* Emissivity at high temperatures 95
Stellar interferometer, Michelson's: *see* Optical information processing 304
"Sticking" or accommodation coefficient: *see* Thin films, nucleation and growth of 491
Stigmator: *see* High resolution electron microscopy 179
Stimulated Brillouin scattering: *see* Lasers, applications of 247
Stimulated emission: *see* Lasers, tunable 250
Stimulated emission, submillimetre: *see* Submillimetre-wave lasers 458
Stimulated Raman scattering: *see* Lasers, tunable 250
Stokes lines: *see* Lasers, tunable 251
STOL Aircraft (short take off and landing): *see* High lift devices 174; Jet flap 233
Stopping power: *see* Radiation quantities, systematics of 372
Storm rainfall: *see* Hydrometeorology 189
Strain energy: *see* Rubber physics 393
Strain energy release rate: *see* Fracture of brittle materials 125
Strain, hoop: *see* Electrohydraulic crushing 82
Strain measuring laser systems: *see* Lasers, applications of 247
Strain–stress curve, fibre: *see* Textile industry, physics in 468
Strain–stress curves for rubber: *see* Polymer physics 334; Rubber physics 393
Strain–stress properties of materials: *see* Microplasticity 286
Strange particles: *see* Quarks 357
Strangeness: *see* Current algebras and their applications 52; Elementary particles, nomenclature of 95; Quarks 357
Stratified medium, ray tracing in a: *see* Mirage and looming, optics of 291
Streamer chambers 448
Strength, electric breakdown: *see* Electrogasdynamics 76
Strength, yield: *see* Dispersion and fibre strengthening of materials 62
Strengthening of materials, dispersion and fibre: *see* Dispersion and fibre strengthening of materials 62
Stress anisotropy constant: *see* Magnetic thin films 267
Stress corrosion cracking: *see* Fracture of brittle materials 129
Stress-dependent magnetic properties: *see* Piezomagnetic effect 326
Stress, flow: *see* Peierls stress 306
Stress, fracture: *see* Brittleness in ceramics 37

Stress-induced ordering of interstitials: *see* Snoek spectra 410

Stress intensity factor: *see* Fracture of brittle materials 125

Stress, magnetic: *see* Ferrohydrodynamics 111

Stress, Peierls: *see* Peierls stress 305

Stress relaxation: *see* Rubber physics 394; Textile industry, physics in 469

Stress sensitivity: *see* Piezomagnetic effect 326

Stress, shear: *see* Peierls stress 305

Stress, specific: *see* Textile industry, physics in 468

Stress—strain curves, fibre: *see* Textile industry, physics in 468

Stress—strain curves for a rubber network: *see* High polymers, mechanical properties of 175

Stress—strain curves, rubber: *see* Polymer physics 334; Rubber physics 393

Stress—strain properties of materials: *see* Microplasticity 286

Stress tensor, Maxwell: *see* Ferrohydrodynamics 112

Stress, yield: *see* Dispersion and fibre strengthening of materials 63; High polymers, mechanical properties of 178

Stretching and bending vibrations: *see* Far infra-red spectroscopy 106

Strong interactions: *see* Charge conjugation 46

Strontium 87: *see* Geochronology 135

Strontium/rubidium dating: *see* Geochronology 136

Structural disorder introduced during film growth: *see* Thin films, nucleation and growth of 492

Structure factor, liquid: *see* Liquid metals 261

Structure factors, refinement of: *see* Dynamic electron diffraction, application of 67

SU(2) symmetry: *see* Quarks 357

SU(3) symmetry: *see* Current algebras and their applications 52; Quarks 357

SU(6) symmetry: *see* Quarks 358

Submillimetre laser: *see* Submillimetre radiation, applications of 453

Submillimetre radiation, applications of 451

Submillimetre stimulated emission: *see* Submillimetre-wave lasers 458

Submillimetre-wave gas lasers: *see* Submillimetre-wave methods 459

Submillimetre-wave lasers 454

Submillimetre-wave methods 458

Suck stroke: *see* Glandless reciprocating-jet pump 140

Suction: *see* High lift devices 171

Sugar refining: *see* Radioisotopes in food processing 377

Sum rule, Adler—Weisberger: *see* Current algebras and their applications 53

Summability: *see* Walsh functions 517

Sun: *see* Radar astronomy 362

Super-blue cells: *see* Solar cell 427

Supercirculation: *see* High lift devices 170

Superconducting magnets: *see* Polarized proton targets 329

Superconducting refractories: *see* Refractories, ultrapurification of 389

Superconductivity: *see* Far infra-red spectroscopy 108; Josephson effect 235

Superconductors, electron diffraction in type II: *see* Electron diffraction from periodic magnetic fields 89

Supercurrent: *see* Josephson effect 236

Superfluid helium: *see* Josephson effect 235

Superfluidity: *see* Josephson effect 235

Superposition principle: *see* Dynamical n-beam theory of electron diffraction 69

Superposition principle, Boltzmann: *see* High polymers, mechanical properties of 177

Supersaturation activity ratio: *see* Solid nucleus formation from solution 430

Supersonic flight: *see* High lift devices 174; Sonic boom, generation and suppression of 431

Supervoltage flash radiation: *see* High intensity pulsed radiation sources 167

Suppression and generation of sonic boom: *see* Sonic boom, generation and suppression of 431

Surface, catalytic: *see* Metal film preparation by electroless (chemical) reduction 278

Surface coatings, curing of: *see* Radiation chemistry, industrial applications of 365

Surface contamination: *see* Auger emission spectroscopy 26

Surface diffusion: *see* Epitaxy, recent work on 98

Surface diffusion energies: *see* Thin films, nucleation and growth of 493

Surface energy: *see* Brittleness in ceramics 37; Fracture of brittle materials 125

Surface free energy: *see* Thin films, nucleation and growth of 491

Surface layers, nucleation and growth of: *see* Epitaxy, recent work on 98

Surface, liquid, deformation: *see* Ultrasonic imaging 500

Surface of solids, ejection of atoms from the: *see* Sputtering, theories of 445

Surface recombination: *see* Solar cell 427

Surface, stability, magnetic fluid: *see* Ferrohydrodynamics 114

Surface topography of metals and alloys: *see* Electron metallography, applications of 89

Surveying: *see* Lasers, applications of 247

Susceptibility: *see* Pockels effect 328

Susceptibility, magnetic: *see* Piezomagnetic effect 325

Swarm experiments: *see* Momentum transfer cross section for low energy electrons: determination by swarm techniques 297

Sweptfrequency oscillator: *see* Ultrasonic spectroscopy 503

Sweptfrequency, pulsed, technique: *see* Ultrasonic spectroscopy 503

Sweptfrequency transducer excitation: *see* Ultrasonic spectroscopy 504

Switching: *see* Fluidics 117; High intensity pulsed radiation sources 166

Switching, Q-: *see* Q-switching 350

Symmetry: *see* Dynamical n-beam theory of electron diffraction 71
Symmetry determination: *see* Dynamic electron diffraction, application of 66
Symmetry, orthotropic: *see* Textile industry, physics in 469
Symmetry SU(2): *see* Quarks 357
Symmetry, SU(3): *see* Current algebras and their applications 52; Quarks 357
Symmetry, SU(6): *see* Quarks 358
Synchronized time and frequency technique: *see* Airborne collision prevention devices 10
Synchro-transmitter: *see* Model ships, photoelectric tracking of 294
Synchrotron light: *see* Soft X-ray spectroscopy 413
Synchrotron radiation: *see* Quantum metrology 357
Synthetic fibres: *see* Textile industry, physics in 470

T-telescope: *see* Radio telescopes, the present position 384
Tacking: *see* Sailing, physics of 397
Tail response: *see* Solar cell 428
Targets, polarized proton: *see* Polarized proton targets 329
Tau: *see* Airborne collision prevention devices 10
Tautomer: *see* Quantum mechanics in biochemistry 353
Telecommunications: *see* Hall effect, applications of 158
Telegraphy: *see* Pulse code modulation 344
Telephony: *see* Pulse code modulation 344
Telescope: *see* Radio telescopes, the present position 380
Television camera tubes, ultrasound: *see* Ultrasonic imaging 502
Television, infra-red: *see* Infra-red cameras and infra-red TV 196
Temperature: *see* Quantum metrology 356
Temperature, brightness: *see* Radioemissions from interstellar hydroxyl radicals 376
Temperature broadening: *see* Soft X-ray spectroscopy 417
Temperature coefficient of emissivity: *see* Emissivity at high temperatures 95
Temperature coefficient of magnetization of barium ferrite: *see* Permanent magnet technology 313
Temperature, colour, range of daylight: *see* Daylight, power distribution in 57
Temperature, correlated colour: *see* Daylight, power distribution in 56
Temperature dependence of solar cells: *see* Solar cell 427
Temperature, epitaxial: *see* Epitaxy, recent work on 99; Thin films, nucleation and growth of 492
Temperature, glass transition: *see* High polymers, mechanical properties of 176; Rubber physics 394
Temperature, high, conductivities at: *see* Conductivity at high temperatures 48

Temperature, high, emissivity at: *see* Emissivity at high temperatures 95
Temperature, ionic: *see* Ionosphere, recent observations of 226; Phonon avalanche and bottleneck 316
Temperature, liquidus: *see* Glass ceramics 143
Temperature, Néel: *see* Far infra-red spectroscopy 108
Temperature, negative: *see* Far infra-red lasers 101
Temperature resistant polymers 463
Temperature restoration: *see* Ultra-violet radiation, some biological effects of 511
Temperature, spin: *see* Fast passage in paramagnetic resonance 110
Tension, chain: *see* Polymer physics 332
Terminal intermolecular attractions: *see* Liquid crystals and molecular structure 259
Tertiary structure: *see* Proteins, structure of 337
Textile industry, physics in 467
Thallium activated crystals, sodium iodide: *see* Whole body counting 523
Thermal broadening: *see* Radioemissions from interstellar hydroxyl radicals 373
Thermal conductivity: *see* Conductivity at high temperature 48; Glass ceramics 147; Thermoelectric generation of electricity 484
Thermal conductivity of semiconductors: *see* Peltier refrigerators 308
Thermal contact ionization: *see* Thermionic generation of electricity 477
Thermal detector: *see* Infra-red cameras and infra-red TV 197; Submillimetre-wave methods 459
Thermal disorder: *see* Liquid metals 262
Thermal dissociation: *see* Tungsten halogen lamps 496
Thermal efficiency: *see* Electrogasdynamics 78
Thermal emission, OH: *see* Radioemissions from interstellar hydroxyl radicals 374
Thermal evaporation of atoms: *see* Sputtering, theories of 445
Thermal expansion, coefficient of: *see* Glass ceramics 145
Thermal image converter: *see* Infra-red cameras and infra-red TV 201
Thermal noise in the sea: *see* Ambient noise in the sea 14
Thermal photographs: *see* Infra-red cameras and infra-red TV 197
Thermal shock resistance: *see* Glass ceramics 147
Thermal spike: *see* Sputtering, theories of 445
Thermal stability, mesomorphic: *see* Liquid crystals and molecular structure 259
Thermal stability, smectic: *see* Liquid crystals and molecular structure 259
Thermal vision: *see* Infra-red cameras and infra-red TV 197
Thermicon: *see* Infra-red cameras and infra-red TV 200
Thermionic converter: *see* Space power supplies 438; Thermionic generation of electricity 476

Thermionic emission: see Metallic colloids in non-metallic crystals 284
Thermionic emitter: see Electrogasdynamics 78
Thermionic generation of electricity 476
Thermionic work function: see Metallic colloids in non-metallic crystals 285
Thermistor: see Infra-red cameras and infra-red TV 197
Thermistor bolometer: see Thermoelectric cooling of infra-red detectors 481
Thermistor probe: see Ultrasonic imaging 499
Thermochromism: see Infra-red cameras and infra-red TV 199
Thermocline: see Physical limnology 322
Thermocouple: see Conductivity at high temperatures 50; Emissivity at high temperatures 96; Infra-red spectroscopy, recent developments in 202; Peltier refrigerators 307; Thermoelectric cooling of infra-red detectors 481; Ultrasonic imaging 499
Thermoelastic inversion point: see Polymer physics 334
Thermoelectric converters: see Space power supplies 438
Thermoelectric cooling of infra-red detectors 481
Thermoelectric effects: see Thermoelectric generation of electricity 482
Thermoelectric generation of electricity 482
Thermoelectric generator, cascade: see Thermoelectric generation of electricity 487
Thermoelectric materials: see Peltier refrigerators 308
Thermoelectric module: see Thermoelectric cooling of infra-red detectors 481
Thermoelectric power: see Liquid metals 263; Peltier refrigerators 307
Thermoelectric refrigeration: see Peltier refrigerators 307
Thermoelectricity: see Thermoelectric generation of electricity 482
Thermomagnetic gas torque (The Scott effect) 488
Thermonuclear reactions: see Lasers, applications of 246
Thermonuclear reactions, controlled: see Laser-induced sparks 244
Thermopile: see Infra-red cameras and infra-red TV 197; Ultrasonic imaging 499
Thermoradiography: see Infra-red cameras and infra-red TV 199
Thermoscope: see Ceramics production, physics in 43
Thermoset plastics: see Polymer physics 331
Thickness gauges: see Radioisotopes in food processing 379
Thickness measurements: see Ultrasonic spectroscopy 505
Thiessen polygon technique: see Hydrometeorology 188
Thin films, ferrite: see Magnetic thin films 268
Thin films, magnetic: see Magnetic thin films 265
Thin films, nucleation and growth of 490
Thin films, proton scattering from: see Proton scattering microscopy 344

Thin foil technique: see Electron metallography, applications of 89
Thixotropy: see Ceramics production, physics in 43
Thomas–Fermi potential: see Proton scattering microscopy 343
Thorium/uranium dating: see Geochronology 136
Threads, warp: see Textile industry, physics in 472
Threads, weft: see Textile industry, physics in 472
Threat evaluation technique: see Airborne collision prevention devices 11
Three-atom distribution function: see Liquid metals 261
Three-dimensional imaging: see Acoustic holography 1; Holography 184
Three-dimensional multiplier: see Magnetoresistivity 274
Three-dimensional optical information storage: see Lasers, applications of 249
Threshold, avalanche: see Microwave solid-state oscillators, recent 289
Threshold condition for a laser oscillation: see Q-switching 350
Threshold intensity: see Lasers, tunable 250
Through-transmission technique: see Ultrasonic imaging 498
Thrust, deflected: see High lift devices 174
Thruster, electrostatic: see Electrogasdynamics 79
Thwaites flap: see High lift devices 173
Thymine: see Ultra-violet radiation, some biological effects of 507
Thymine dimer: see Ultra-violet radiation, some biological effects of 507
Tilting chair: see Whole body counting 523
Time: see Quantum metrology 354
Time and frequency technique, synchronized: see Airborne collision prevention devices 10
Time delay, transit: see Microwave solid-state oscillators, recent 289
Time-reversal-invariance: see Electric dipole moment of the neutron 74
Time scale, geologic: see Geochronology 138
Time series of hydrological data: see Hydrometeorology 189
Time standard, cesium: see Airborne collision prevention devices 11
Time, universal: see Quantum metrology 354
Tin-plating: see Freeze-coating 130
Tip, crack: see Fracture of brittle materials 124
Tonometry: see Biological pressure transensors 31
Top hat kiln: see Ceramics production, physics in 41
Topocentric: see Astrodynamics 23
Topography, surface, of metals and alloys: see Electron metallography, applications of 89
Topping: see Thermionic generation of electricity 479
Topside ionosphere: see Ionosphere, recent observations of 225
Topside ionospheric trough: see Ionosphere, recent observations of 229

Torque, thermomagnetic gas: *see* Thermomagnetic gas torque (The Scott effect) 488
Torsion viscometer: *see* Ceramics production, physics in 43
Torsional waves: *see* Acoustic waveguides 7
Torsions: *see* Far infra-red spectroscopy 106
Total ionization: *see* Radiation quantities, systematics of 367
Total ionization cross-sections: *see* Ionization and attachment cross-sections of atoms and molecules by electron impact 217
Toughness, fracture: *see* Dispersion and fibre strengthening of materials 65
Toughness parameters, fracture: *see* Fracture of brittle materials 125
Townsend avalanche: *see* Streamer chambers 448
Townsend cascade (ionizing avalanche): *see* Laser-induced sparks 243
Tracers, radioactive: *see* Radioisotopes in food processing 377
Track dating, nuclear fission: *see* Geochronology 137
Tracking: *see* Lasers, applications of 247
Tracking, photoelectric, of model ships: *see* Model ships, photoelectric tracking of 294
Traffic control: *see* Law enforcement science and technology 254
Traffic lights: *see* Law enforcement science and technology 254
Trailing-edge flap: *see* High lift devices 172; Jet flap 232
Trajectory: *see* Astrodynamics 23
Trajectory of charged particles: *see* Streamer chambers 448
Transducer, electrodynamic: *see* Gyroscopic coupler 154
Transducer, electrostatic: *see* Gyroscopic coupler 154
Transducer, electrostrictive: *see* Ultrasonic imaging 501
Transducer, piezoelectric: *see* Ultrasonic imaging 501; Ultrasonic spectroscopy 503
Transducer, swept-frequency, excitation: *see* Ultrasonic spectroscopy 504
Transducers, watt: *see* Hall effect, applications of 160
Transensor, biological pressure: *see* Biological pressure transensors 31
Transform, Boltzmann: *see* Diffusion in non-uniform media 59
Transform, Fourier: *see* Optical information processing 303
Transform, Fourier, spectroscopy: *see* Fourier transform spectroscopy 122
Transform, Laplace: *see* Optical information processing 303
Transformer, pulse: *see* High intensity pulsed radiation sources 164
Transit time delay: *see* Microwave solid state oscillators, recent 289
Transit time oscillation: *see* Microwave solid state oscillators, recent 289

Transition density: *see* Ion-neutralization spectroscopy 220
Transition, direct (or vertical): *see* Photoemission and band structure 320
Transition, electric dipole: *see* Radioemissions from interstellar hydroxyl radicals 372
Transition, electronic: *see* Ion-neutralization spectroscopy 220
Transition, non-direct: *see* Photoemission and band structure 320
Transition, radiation-less or Auger: *see* Satellites in soft X-ray spectra 404
Transmission, analogue: *see* Pulse code modulation 344
Transmission, bipolar: *see* Pulse code modulation 346
Transmission electron microscopy: *see* Electron metallography, applications of 89
Transmission filters: *see* Submillimetre-wave methods 462
Transmission line, coaxial: *see* High intensity pulsed radiation sources 165
Transmission path noise: *see* Pulse code modulation 346
Transmittance spectrum: *see* Interferometric spectrometers 209
Transmitter, synchro: *see* Model ships, photoelectric tracking of 294
Transparency: *see* Glass ceramics 145
Transponder technique, interrogator: *see* Airborne collision prevention devices 10
Transport and control of mechanical power: *see* Electroviscous fluids, applications of 92
Transport coefficients: *see* Momentum transfer cross-section for low energy electrons: determination by swarm techniques 297
Transverse shear mode: *see* Melting, modern theories of 276
Trapped radiation belts (Van Allen): *see* Auroral electrons 29
Tree: *see* Kirchhoff's laws 240
"Tree" pump: *see* Glandless reciprocating jet pump 140
Trident: *see* High lift devices 171
Trigatron: *see* High intensity pulsed radiation sources 166
Triggering of laser sparks: *see* Laser-induced sparks 243
Triglycine sulfate (TGS): *see* Pyroelectric detector for infra-red radiation 348
Trilinear chart of the nuclides: *see* Chart of the nuclides, trilinear 47
Triode, magnetic: *see* Thermionic generation of electricity 477
Triple point of water: *see* Quantum metrology 356
Tropocollagen molecule: *see* Proteins, structure of 340
Trough, Peierls: *see* Peierls stress 306
Trough, topside ionospheric: *see* Ionosphere, recent observations of 229
True anomaly: *see* Astrodynamics 23
True elastic limit: *see* Microplasticity 288

Tube, field emission: *see* High intensity pulsed radiation sources 163
Tunable lasers: *see* Lasers, applications of 250; Lasers, tunable 250
Tunable light sources: *see* Infra-red spectroscopy, recent developments in 204
Tungsten: *see* Thermionic generation of electricity 478
Tungsten-bromine cycle: *see* Tungsten halogen lamps 496
Tungsten filament lamps: *see* Tungsten halogen lamps 494
Tungsten fluorides: *see* Tungsten halogen lamps 495
Tungsten halogen lamps 494
Tungsten-iodine cycle: *see* Tungsten halogen lamps 495
Tungsten lamp, quartz-iodine: *see* Submillimetre-wave methods 459
Tunnel junction: *see* Josephson effect 236
Tunnel kilns: *see* Ceramics production, physics in 41
Tunnelling: *see* Josephson effect 235; Thin films, nucleation and growth of 493
Tunnelling, proton: *see* Quantum mechanics in biochemistry 353
Turbine, Rankine: *see* Space power supplies 437
Turbostratic structure: *see* Vitreous carbons, structure and properties of 514
Turbulence amplifier: *see* Fluidics 120
Turbulence diode: *see* Fluidics 122
Turbulence noise: *see* Speaking machines 440
Turbulence noise in the sea: *see* Ambient noise in the sea 15
Turbulent broadening: *see* Radioemissions from interstellar hydroxyl radicals 373
Turbulent wake: *see* Solar wind, planetary interactions with 430
Tussah silk: *see* Proteins, structure of 339
Twisting of yarn: *see* Textile industry, physics in 472
Two-beam approximation of dynamic electron diffraction: *see* Dynamic electron diffraction, applications of 66; Dynamical *n*-beam theory of electron diffraction 71
Two-body orbit: *see* Astrodynamics 23
Two-component theory of the neutrino: *see* Helicity 163
Two-way radio: *see* Law enforcement science and technology 254
Tyndall scattering: *see* Interplanetary dust 214; Metallic colloids in non-metallic crystals 282
Type II superconductors, electron diffraction in: *see* Electron diffraction from periodic magnetic fields 87
Tyre cords: *see* Textile industry, physics in 475

Ultrasonic amplification: *see* Acousto-electric effects 8
Ultrasonic amplification in semiconductors 497
Ultrasonic attenuation in a material, frequency-dependence of: *see* Ultrasonic spectroscopy 502

Ultrasonic delay line: *see* Acoustic waveguides 5
Ultrasonic echo: *see* Ultrasonic spectroscopy 505
Ultrasonic holography: *see* Holography 187
Ultrasonic image conversion: *see* Acoustic holography 4
Ultrasonic imaging 498
Ultrasonic lens: *see* Ultrasonic imaging 498
Ultrasonic resonance testing: *see* Ultrasonic spectroscopy 504
Ultrasonic spectroscope: *see* Ultrasonic spectroscopy 503
Ultrasonic spectroscopy 502
Ultrasound imaging, schlieren: *see* Ultrasonic imaging 500
Ultrasound television camera tubes: *see* Ultrasonic imaging 502
Ultra-violet radiation, some biological effects of 506
Ultra-violet reactivation: *see* Ultra-violet radiation, some biological effects of 511
Uncertainty principle: *see* Laser-induced sparks 243
Underwater acoustics: *see* Ambient noise in the sea 14
Underwater discharge, delay time for: *see* Electrohydraulic crushing 81
Underwater reverberation: *see* Reverberation under water 390
Underwater sparks: *see* Electrohydraulic forming of metals 80; Electrohydraulic crushing 83
Unfolding: *see* Ion-neutralization spectroscopy 223
Uniaxial anisotropy: *see* Magnetic thin films 266
Uniaxial crystal: *see* Magnetic thin films 266
Unit hydrograph: *see* Hydrometeorology 191
Units, electrical: *see* Quantum metrology 356
Units, International System of (SI): *see* Quantum metrology 356
Universal time: *see* Quantum metrology 354
Uranium: *see* Radiation chemistry, industrial applications of 363
Uranium, alpha-: *see* Sputtering, theories of 446
Uranium/thorium dating: *see* Geochronology 136

Vacancies: *see* Melting, modern theories of 276
Vacuum diode, close-spaced: *see* Thermionic generation of electricity 477
Vacuum evaporation: *see* Thin films, nucleation and growth of 491
Vacuum evaporation technique: *see* Epitaxy, recent work on 98
Vacuum ultra-violet: *see* Ultra-violet radiation, some biological effects of 506
Valence band spectra: *see* Soft X-ray spectroscopy 412
Valence electrons: *see* Quantum mechanics in biochemistry 351
Valence, free: *see* Quantum mechanics in biochemistry 352
Valve, vortex modulator or: *see* Fluidics 120
Van Allen radiation belts: *see* Ionosphere, recent observations of 225

Van der Waals attractions: *see* Proteins, structure of 341
Van der Waals forces: *see* Liquid metals 262
Van der Waals interactions: *see* Quantum mechanics in biochemistry 353
Van der Waals-like loop: *see* Melting, modern theories of 275
Van Hove singularities: *see* Soft X-ray spectroscopy 418
Variable gap cell: *see* Solar cell 428
Variable resistors: *see* Magnetoresistivity 272
Variant orbits: *see* Astrodynamics 23
Vector-current theory, conserved- (C.V.C. theory): *see* Current algebras and their applications 52
Vector mesons: *see* Quarks 359
Vectorial hologram: *see* Lasers, applications of 247
Vegetable fibres, natural: *see* Textile industry, physics in 470
Velocity, drift: *see* Momentum transfer cross-section for low energy electrons: determination by swarm techniques 297; Ultrasonic amplification in semiconductors 497
Velocity, gas, measurement of: *see* Laser-induced sparks 244
Velocity measurements: *see* Ultrasonic spectroscopy 505
Velocity of electromagnetic radiation *c*: *see* Quantum metrology 355
Velocity of light: *see* Far infra-red lasers 102; Quantum metrology 355
Vented momentum amplifier: *see* Fluidics 118
Venturi pump, bulb-: *see* Glandless reciprocating jet pump 141
Venus: *see* Radar astronomy 362
Venus, ionosphere of: *see* Ionospheres of Mars and Venus 229
Venusian (Cytherian) ionosphere: *see* Ionosphere, recent observations of 229
Vernal equinox: *see* Astrodynamics 23
Verneuil process: *see* Glow discharge electron and ion guns for material processing 149
Vertical jet: *see* High lift devices 174
Vertical (or direct) transitions: *see* Photoemission and band structure 320
Vibration damper, gyroscopic: *see* Gyroscopic coupler 152
Vibration-rotation energy levels: *see* Far infra-red lasers 102
Vibrational spectra: *see* Infra-red spectroscopy, recent developments in 205
Vibrations, bending and stretching: *see* Far infra-red spectroscopy 106
Vibrations, molecular: *see* Far infra-red spectroscopy 106
Vibrations, puckering: *see* Far infra-red spectroscopy 106
Vibrators, hydraulic: *see* Electroviscous fluids, applications of 94

Vidicon: *see* Infra-red cameras and infra-red TV 200
Violation of C invariance: *see* Charge conjugation 46
Violation of CP invariance: *see* Charge conjugation 46
Viscoelastic material: *see* Rubber physics 394
Viscoelasticity: *see* High polymers, mechanical properties of 176; Polymer physics 334
Viscometer, torsion: *see* Ceramics production, physics in 43
Vision in the dark: *see* Infra-red cameras and infra-red TV 197
Vision, thermal: *see* Infra-red cameras and infra-red TV 197
Vitreous carbons, structure and properties of 513
Vocal cords: *see* Speaking machines 440
Vocal tract: *see* Speaking machines 440
Vocal tract analogue synthesizers: *see* Speaking machines 442
Vocoder: *see* Speaking machines 442
Voder (VOice DEmonstratoR): *see* Speaking machines 441
Void formation: *see* Fracture of brittle materials 128
Volt: *see* Josephson effect 237; Quantum metrology 356
Voltage, acousto-electric: *see* Acousto-electric effects 8
Voltage, Hall: *see* Hall effect, applications of 157; Magnetoresistivity 270; Thermoelectric generation of electricity 486
Voltage law: *see* Kirchhoff's law (electrical circuits) 239
Voltage pacemaker, constant: *see* Cardiac pacemaker, external 40
Voltage standardization: *see* Josephson effect 238
Volume (thick) holograms: *see* Holography 184
Von Mises criterion: *see* Brittleness in ceramics 38
Vortex diode: *see* Fluidics 122
Vortex modulator or valve: *see* Fluidics 120
Vortex oscillator: *see* Fluidics 121
Vortex sensor: *see* Fluidics 121
Vortex spinning: *see* Textile industry, physics in 474
Vowel-generating devices: *see* Speaking machines 441
VTOL Aircraft (Vertical Take Off and Landing): *see* High lift devices 174; Jet flap 233
Vulcanization: *see* Rubber physics 393
Vulcanization process in rubber: *see* High polymers, mechanical properties of 175

W$^\pm$ field: *see* Bosons, intermediate 35
W meson: *see* Bosons, intermediate 35
W3 OH source: *see* Radioemissions from interstellar hydroxyl radicals 375
Wake, turbulent: *see* Solar wind, planetary interactions with 430
Wall attachment diode: *see* Fluidics 122
Wall effect: *see* Fluidics 121
Walls, domain: *see* Magnetic thin films 268
Walsh-Fourier series: *see* Walsh functions 517
Walsh functions 517

Walsh series: see Walsh functions 518
Warm cathode glow discharge: see Glow discharge electron and ion guns for material processing 149
Warning instruments, pilot: see Airborne collision prevention devices 9
Warp threads: see Textile industry, physics in 472
Water: see Far infra-red lasers 101
Water budget method: see Hydrometeorology 193
Water control structures: see Hydrometeorology 189
Water gaps, electrical breakdown of: see Electrohydraulic crushing 81
Water, lake, density: see Physical limnology 322
Water levels: see Hydrometeorology 192
Water, triple point of: see Quantum metrology 356
Water vapour laser: see Submillimetre-wave lasers 456; Submillimetre-wave methods 459
Watt transducers: see Hall effect, applications of 160
Wattmeter, Hall effect: see Hall effect, applications of 160
Wattmeter, magnetoresistance: see Magnetoresistivity 273
Wave analyser: see Interferometric spectrometers 209
Wave analysis: see Hydrometeorology 192
Wave, blast: see Laser-induced sparks 243
Wave, flexural: see Acoustic waveguides 7
Wave functions, Bloch: see Photoemission and band structure 320
Wave functions, electron: see Soft X-ray spectroscopy 417
Wave, idler: see Lasers, tunable 251
Wave, "N": see Sonic boom, generation and suppression of 431
Wave number, Fermi: see Liquid metals 262
Wave, pump: see Lasers, tunable 251
Wave, shear: see Acoustic waveguides 7
Wave, shock: see Laser-induced sparks 244; Shock pressure in meteorites 407
Wave, shock, method: see Melting, modern theories of 275
Wave, sound: see Acoustic holography 1
Wave, sound, attenuation: see Melting, modern theories of 276
Wave, torsional: see Acoustic waveguides 7
Waveguide, acoustic: see Acoustic waveguides 5
Wavelength, de Broglie: see Liquid metals 262
Weak interactions: see Bosons, intermediate 34; Charge conjugation 46; Current algebras and their applications 51
Weak scattering (pseudo)-potentials: see Liquid metals 262
Weather control 520
Weaving: see Textile industry, physics in 472
Wedge-type cracking: see Fracture of brittle materials 129
Weft projection: see Textile industry, physics in 473
Weft threads: see Textile industry, physics in 472
Weighted least-squares differential correction: see Astrodynamics 19

Weighting or apodizing function: see Fourier transform spectroscopy 123
Weighting matrix: see Astrodynamics 19
Weinreich relation: see Acousto-electric effects 8
Welding, pulsed laser: see Lasers, applications of 246
Whiskers of metals and non-metals: see Dispersion and fibre strengthening of materials 65
Whistlers: see Ionosphere, recent observations of 225
Whole body counting 522
Wiedemann–Franz law: see Peltier refrigerators 308; Thermoelectric generation of electricity 484; Thermionic generation of electricity 478
Wien's law: see Emissivity at high temperatures 95
Wind dependent noise in the sea: see Ambient noise in the sea 14
Wind set-up: see Hydrometeorology 192
Wind, solar: see Auroral electrons 30; Interplanetary dust 213; Ionosphere, recent observations of 229
Wind, solar, planetary interactions with: see Solar wind, planetary interactions with 429
Window, spectral: see Fourier transform spectroscopy 123
Wing section: see High lift devices 170
Wing-tip slots: see High lift devices 171
Winter anomaly: see Ionosphere, recent observations of 225
Wire gauge reflector: see Infra-red spectroscopy, recent developments in 202
Wires, exploding: see Electrohydraulic forming of metals 81; Electrohydraulic crushing 83
W–L–F (Williams, Landel and Ferry) equation: see Polymer physics 334
Wood-plastic composites: see Radiation chemistry, industrial applications of 365
Work function: see Metallic colloids in non-metallic crystals 283; Thermionic generation of electricity 476
Work function, thermionic: see Metallic colloids in non-metallic crystals 285
"Work of Fracture": see Fracture of brittle materials 125

X-radiation dose: see High intensity pulsed radiation sources 167
X-ray diffraction: see Dynamic electron diffraction, application of 67; Proteins, structure of 338
X-ray equipments, flash: see High intensity pulsed radiation sources 163
X-ray scattering: see Liquid metals 261
X-ray spectra: see Auger emission spectroscopy 25
X-ray spectra, soft, satellites in: see Soft X-ray spectroscopy 419
X-ray spectra, solid state: see Soft X-ray spectroscopy 412
X-ray spectroscopy, soft: see Soft X-ray spectroscopy 412
X-rays: see Auroral electrons 29

Yacht, keel: see Sailing, physics of 398
Yarns: see Textile industry, physics in 472
Yield and fracture of polymers: see High polymers, mechanical properties of 178
Yield, plastic: see Fracture of brittle materials 124
Yield point: see Microplasticity 286
Yield strength: see Dispersion and fibre strengthening of materials 62
Yield stress: see Dispersion and fibre strengthening of materials 63; High polymers, mechanical properties of 178
Young's modulus: see Brittleness in ceramics 37
Young's modulus of rubber: see Rubber physics 396

Zeeman effect: see Radioemissions from interstellar hydroxyl radicals 376
Zero differential overlap approximation: see Quantum mechanics in biochemistry 352
Zodiacal light: see Interplanetary dust 213
Zone, auroral: see Auroral electrons 28
Zone, Brillouin: see Soft X-ray spectroscopy 418
Zone, crushing: see Electrohydraulic crushing 82
Zone, floating, method: see Refractories, ultra-purification of 387
Zone purification: see Refractories, ultra-purification of 387
Zone refining: see Refractories, ultra-purification of 387

ENCYCLOPAEDIC DICTIONARY OF PHYSICS

SCOPE OF THE DICTIONARY

For convenience in planning, and to provide a framework on which the Dictionary could be erected, physics and its related subjects have been divided into upwards of sixty sections. The sections are listed below, but, as the Dictionary is arranged alphabetically, they do not appear as sections in the completed work.

Acoustics
Astronomy
Astrophysics
Atomic and molecular beams
Atomic and nuclear structure
Biophysics
Cathode rays
Chemical analysis
Chemical reactions, phenomena and processes
Chemical substances
Colloids
Cosmic rays
Counters and discharge tubes
Crystallography
Dielectrics
Elasticity and strength of materials
Electrical conduction and currents
Electrical discharges
Electrical measurements
Electrochemistry

Electromagnetism and electrodynamics
Electrostatics
Engineering metrology
General mechanics
Geodesy
Geomagnetism
Geophysics
Heat
Hospital and medical physics
Industrial processes
Ionization
Isotopes
Laboratory apparatus
Low-temperature physics
Magnetic effects
Magnetism
Mathematics
Mechanics of fluids
Mechanics of gases
Mechanics of solids
Mesons
Meteorology
Molecular structure

Molecular theory of gases
Molecular theory of liquids
Neutron physics
Nuclear reactions
Optics
Particle accelerators
Phase equilibria
Photochemistry and radiation chemistry
Photography
Physical metallurgy
Physical metrology
Positive rays
Radar
Radiation
Radioactivity
Reactor physics
Rheology
Solid-state theory
Spectra
Structure of solids
Thermionics
Thermodynamics
Vacuum Physics
X rays